Multinational Corporations

This book extends current macroeconomic theories of international production by examining the emergence and evolution of multinational corporations (MNCs) from a broad range of developed and developing countries. *Multinational Corporations* presents case studies of the emergence and evolution of MNCs based in 11 developed and developing countries of widely divergent patterns of national development.

Paz Estrella Tolentino's research agenda was to determine whether variations exist in the pattern of the early stages of outward foreign direct investment (FDI) across different countries, and in their developmental paths over time that are determined by distinctive patterns of national economic development in each country. The countries covered include Brazil, Germany, Hong Kong, Japan, Singapore, South Korea, Sweden, Switzerland, Taiwan, United Kingdom and the United States. This book provides a comprehensive theoretical framework on the emergence and evolution of MNCs from a macroeconomic perspective based on the categorization of MNCs in three groups in accordance with distinctive patterns of national economic development of their home countries: resource-abundant countries, resource-scarce large countries and resource-scarce small countries.

Multinational Corporations presents a radical conceptual framework in which to analyse MNCs from a wide range of countries and also promotes the advancement of current knowledge of an important aspect of international business history.

Paz Estrella Tolentino is Lecturer in International Business at the School of Management and Organizational Psychology, Birkbeck College, University of London. Her previous book, *Technological Innovation and Third World Multinationals* – based on her PhD dissertation which received the 1989 Academy of International Business Richard N. Farmer prize for the best PhD thesis on international business – laid the foundations for the elaboration of a concept of stages of development in explaining international production.

Routledge Studies in International Business and the World Economy

Multinational Corporations
Emergence and evolution

Paz Estrella Tolentino

London and New York

First published 2000
by Routledge
11 New Fetter Lane, London EC4P 4EE

Simultaneously published in the USA and Canada
by Routledge
29 West 35th Street, New York, NY 10001

Routledge is an imprint of the Taylor & Francis Group

© 2000 Paz Estrella Tolentino

Typeset in Baskerville by Keyword Publishing Services Ltd
Printed and bound in Great Britain by MPG Books Ltd, Bodmin

British Library Cataloguing in Publication Data
A catalogue record for this book is available from the British Library

Library of Congress Cataloging in Publication Data
Tolentino, Paz Estrella E.
    Multinational corporations : emergence and evolution / Paz Estrella
Tolentino.
        p. cm. – (Routledge studies in international business and the world
        economy ; 15)
    Includes bibliographical references and index.
    1. International business enterprises–Case studies. 2. Investments,
Foreign–Case studies. 3. Technological innovations–Economic aspects–
Case studies. I. Title. II. Series.

    HD2755.5 .T6 2000
    338.8′8–dc21   00-030429

ISBN 0-415-14575-9

To Alexandra Maissa Farrah T. Soumati
with love

Contents

PART III
Multinational corporations from the resource-scarce large countries

PART IV
Multinational corporations from the resource-scarce small countries

PART V
Conclusion

Tables

Foreword

Paz Tolentino has written a worthy follow-up book to her excellent first manuscript, *Technological Innovation and Third World Multinationals* which was published by Routledge in 1993. In one sense this volume has its origins in her continuation of a critique of John Dunning's notion of an investment–development cycle of countries. This is a project that she and I had discussed and initially worked on together in Reading, but which she has now taken very much further, to the point at which a new range of ideas has emerged to enable her to generalize more accurately and meaningfully about the international direct investment position of countries. She is to be congratulated on the sterling effort with which she has collated such a wide breadth of evidence and organized it coherently, so as to succeed in deepening our understanding of the variety of evolutionary paths that foreign direct investment (FDI) may follow at the national macroeconomic level.

The essential idea of the investment–development cycle is that the stylized or 'typical' path of a country is that inward FDI takes off and enhances the earlier stages of development (so the net outward investment [NOI] position of countries becomes steadily more negative as GDP per capita rises), while past some point outward FDI takes off too and grows in accompaniment to the later stages of economic development (so the NOI position becomes gradually more balanced again, and ultimately becomes positive in countries that have attained the highest levels of GDP per capita). Thus, as shown by Dunning, if we plot the NOI position of countries against their GDP per capita in a cross-section from 1970s data, we observe a 'J-shaped' schedule. Paz Tolentino showed in her previous book that the shape of this function changed in the 1980s, since outward FDI tends to come now at an earlier stage of national development. Third World firms engage in outward FDI at an earlier stage of national development and experience a faster transformation in the industrial structure of this outward FDI, although it has still become easier to distinguish countries at different levels of development by the industrial composition of their outward FDI than by a simple calculation of their NOI position. As development rises, the industrial structure of outward FDI is widened and evolves towards more complex types of activity.

The explanation of the greater speed of corporate internationalization depends on two elements – an account of a change in the international economic environment, and an analysis of the accumulation of technological competence in firms at varying ages or degrees of maturity. The first of these can be taken from Ray Vernon's discussion of the growth of FDI with the fall in transport and communications costs in the post-war period, together with the more recent emergence (from the 1970s onwards) of a new techno-economic paradigm based on new organizational forms linked to new information and communication technologies as suggested by Chris Freeman. The second part can be derived from the modern evolutionary theory of technological change as a localized and firm-specific process, as pioneered by Richard Nelson and Sidney Winter. Firms build up their technological competences or what are termed in the international business field (following John Dunning's eclectic paradigm) 'ownership advantages', through locally differentiated learning activities.

Thus, in the 1980s those of us that continued to insist on the concept of ownership advantages as a condition for international corporate expansion reformulated the concept more precisely. Critics of the concept had focused their attack on an interpretation of Hymer's earliest discussion of ownership advantages, as a net cost advantage of foreign-owned over indigenous firms in the relevant local market (and this is still often the version that critics prefer to disparage today, seemingly unaware that the discussion, like the real world, has moved on!) The first and obvious revision is that ownership advantages must be thought of in relation to the international competition mainly from other multinational corporations (MNCs) rather than relative to domestic companies in a particular host country. Today MNCs are generally competing with one another in international markets, they are usually not in the earliest stages of internationalization and their investments are not all of a local market-oriented kind (unlike in Hymer's case, in which he had addressed the more specific question of why firms initially go abroad and begin to engage in FDI). Second, and more importantly in the present context, innovation and hence innovative advantages are differentiated and relative concepts, not indicative of some notional technology frontier. All surviving MNCs have some distinctive competitive edges, and it is these differentiated firm-specific strengths that constitute each firm's ownership advantages rather than some overall absolute cost advantage. Hence, Third World MNCs may have as well ownership advantages especially in operating in certain kinds of less developed conditions, and with a change in techno-economic paradigm some of them have been able to upgrade these advantages more rapidly than they otherwise would, encouraging and facilitating a faster internationalization.

Paz Tolentino has now returned to a further issue that she had previously to leave to one side. This issue is that the investment-development cycle was an interpretation of cross-sectional evidence in search of a time series framework of analysis. The argument that countries' investment positions evolve faster than before is a special case of the more general contention that each country follows some nationally specific path in its FDI, and there is no reason to believe that each country moves along a similar J-shaped path in its NOI position. Indeed, we know that countries like Japan and Korea have witnessed major take-offs in their outward FDI without much prior inward investment, whereas the take-offs in outward FDI in many European countries such as France or Germany had been partially a response to the earlier penetration of US inward FDI, and their consequent interaction with US MNCs. At one level it is fashionable to attribute such variations to policy differences, but the formation of the policies themselves and the way in which they have worked depend very much on the national institutional environment of countries, which varies greatly.

Differences in institutional systems and in the related industrial structures of countries may be dependent on a range of variables, but two that seem to matter most are the national degrees of resource availability or scarcity, and country size. Paz Tolentino has used these variables to construct a taxonomy of country types, and to organize her voluminous material in a fruitful way. She demonstrates here quite compellingly that the variation in national paths of FDI that we observe can be explained in good measure by whether a country is resource-scarce or resource-abundant, and whether it is large or small. Her book has taken us a good way forward in understanding how the industrial structure of FDI evolves in these different types of country, and on how the form of interaction between inward and outward FDI may (or may not) influence the accumulation of ownership advantages on the part of domestically owned firms, and hence the nature of their own internationalization. I am happy to commend this book to anyone interested in deepening their appreciation of the variety of historical determinants that have shaped and are shaping the growth of international direct investment and the modern multinational firm.

John Cantwell

Preface

Intention

In 1991, John Cantwell argued that there are four frameworks in which to analyse the major theories of international production: the microeconomic approach which examines the international growth of individual firms or multinational corporations (MNCs), the mesoeconomic approach which considers the interaction between firms at an industry level, the macroeconomic approach which studies broad national and international trends, and the eclectic paradigm which is not an alternative analytical framework but an overall organizing paradigm which incorporates elements from all the other three types of approaches (Cantwell, 1991). The various theoretical strands comprising the macroeconomic development approach to international production are the earliest version of the product cycle model (PCM Mark I) of Vernon (1966); the integrated theory of trade and direct investment of the Japanese economists, Kojima (1973, 1978) and Ozawa (1982); the concept of an investment-development cycle and path of Dunning (1982); and the stages-of-development approach associated with Tolentino (1993) and Cantwell and Tolentino (1990). However, although the origins of the more modern macroeconomic theories of international production could be traced to the mid-1960s with the formulation of the PCM Mark I, the current stage of development of this set of theories is regarded to be more rudimentary by comparison to the microeconomic or mesoeconomic theories whose origins can be traced similarly to the 1960s. This is owing partly to the demise of the PCM as an analytical tool to explain the more balanced process of technological competition between the United States, Europe and Japan over the last 35 years away from the previous technological hegemonic role of the United States during Pax Americana (Giddy, 1978; Vernon, 1979; Tolentino, 1993) and the more industry-based as opposed to country-based theory of trade and direct investment developed by Kojima couched in the neoclassical Hecksher-Ohlin-Samuelson theory of trade. In addition, although the advancement of more general macroeconomic theories of international production began with the elaboration of the concept of an investment-development cycle of international production by Dunning (1982)

which relates the *level* of inward and outward foreign direct investment (FDI)
and the national stage of development of home and host countries, empirical
evidence that I had gathered in my PhD thesis on the general trend of
internationalization of firms since the mid-1970s showed that this formulation
required some qualification (Tolentino, 1987). The refinement of the concept
by Dunning to denote the presence of an investment-development path which
considers that both the *level* as well as the *character and composition* of the outward FDI of a country's *firms* vary with the national stage of development has
been extended since by John Cantwell and myself to consider the developmental course of the outward FDI of *countries* over time (see Cantwell and
Tolentino, 1990; Tolentino, 1993). Indeed, the growth of MNCs from a
historical perspective shows a developmental course in the pattern of outward
FDI from an initial emphasis on trading, resource-based or simple market-
seeking investments where the competence of leading national firms is
embedded in basic engineering skills, complementary organizational routines
and structures. Over time as industrial development of home counties proceed and as their firms gain maturity as MNCs, there is the gradual progression of their outward FDI towards more sophisticated forms of manufacturing
and services activities that embody greater technological and organizational
complexities. Indeed, the dominant sectoral pattern of outward FDI at the
global level had been associated with primary commodity production before
1939, and only since the Second World War did outward FDI in manufacturing and services come into its own in a major way, bringing forth the
growth in international technological competition in the industrialized countries and the global integration of FDI.

Such an evolution describes the growth of the mature MNCs of Europe and
the United States over the last two centuries. The evolution of Japanese
MNCs since the Second World War, on the other hand, has been compressed
into a much shorter time span. Investments in resource-based activities and
import substituting manufacturing in South East Asia led the way in the
1950s until the early 1970s, but since the late 1960s the interest of Japanese
firms shifted to more sophisticated manufacturing production in Europe and
the United States (Ozawa, 1991). Although MNCs from the Asian newly
industrialized countries are nowhere near as technologically advanced as the
more established MNCs from Europe and the United States as well as the
modern Japanese MNCs owing to the less advanced stage of the industrial
development of their home countries, the lower forms of technological competence of their leading national firms and their lack of maturity as MNCs,
the sectoral complexity and geographical scope of their international production activities are evolving much more rapidly than that of the more historical
home countries of FDI. The main reason for this is the general trend towards
the internationalization of business since the mid-1970s which has been common to firms of all countries. Domestic enterprises from the developing countries have, in general, embarked on international production and became

MNCs at an earlier stage of their development than industrialized country firms (see Tolentino, 1993 for further support of this argument).

Despite the general trend towards the internationalization of business which has been common to firms of all countries and the more rapid pace in the emergence of MNCs from the newer home countries and in the evolution of the pattern of their international production activities, it has become increasingly apparent that the pattern of the emergence and evolution of MNCs are also determined by factors other than the *stages* of development or maturity of countries and firms. In particular, there seem to be variations in the pattern of the early stages of outward FDI across different types of home countries as well as their developmental paths over time that are determined by distinctive *patterns* of national economic development. Since each country or group of countries has a unique pattern of national economic development owing in part to different endowments of natural resources, different sizes of the domestic market and different types of development path pursued in achieving industrial development, the developmental course of international production in each country or group of countries is likely to be unique and differentiated. Thus, any effort to advance the modern macro-economic theories on the basis of general principles must be comprehensive in its attempt to explain the emergence and evolution of MNCs from a broad range of home countries. This is the major challenge to research that I have chosen to face at the dawn of the twenty-first century and the third millennium. In advancing a conceptual framework to analyse the emergence and evolution of MNCs from a broad range of countries over the course of more than two centuries, the research also hopes to advance current knowledge of an important aspect of international business history – the growth and evolution of MNCs.

Background

The initial idea behind the major research that eventually led to the production of this book was nurtured at the start of my academic career as Lecturer in International Business at Birkbeck College, Univeristy of London, at the beginning of 1995. My research plan at that time was far less ambitious by comparison to the major research that I ended up pursuing. At that early stage, my objective had been simply to advance further the concept of stages of development of international production that I had elaborated in my first book through an analysis of the dynamic and developmental process of international production or the way in which stages of development or maturity of countries and firms determine the pattern of international production of MNCs based in developing countries – a body of research that I had already worked on previously to a considerable extent. On that basis, I completed a paper in the spring of 1995 with the title *Third World Multinationals: Emergence and Evolution* which was accepted for a competitive session at the Annual Meeting of the Academy of International Business. The conference paper also formed the basis to solicit a contract to publish a book with the same title from Routledge – a contract that I obtained successfully in the spring of 1995.

There have been two major factors that determined the direction of research since. The first factor was the realization fairly early on that a confinement of the research to the emergence and evolution of MNCs based in developing countries would have limited use if this was not analysed in relation to that of the more mature and far more significant MNCs based in the developed countries. Not only do MNCs based in developing countries account for less than 10 per cent of the global stock of outward FDI in 1998 (UNCTAD, 1999), but the delimitation of the empirical evidence to this narrow group of countries would preclude the elaboration of a comprehensive macroeconomic theory of international production on the basis of general principles. The second factor was brought about by the opportune time in 1995 when John Cantwell was preparing a research paper with the title *Globalization and Development in Africa*, in which he argued that variations in the early stages of *inward FDI* across different types of countries and their developmental paths over time are determined by distinctive patterns of national development. This idea was applied by Cantwell in the context of mapping the potential development paths of inward FDI in Africa (see Cantwell, 1997).

The combination of these two forces dictated the direction of the research since 1996. The research began to build on Cantwell's idea from the perspective of *outward FDI*. In particular, the research sought to explore the possibility of the more general application of the proposition that patterns of national development (or industrial development) tend to be associated with equivalent phases of MNC expansion from a broad range of countries. The pursuit of this research agenda required that the empirical evidence be gathered on the growth patterns of MNCs from a significant number of countries that collectively would cover the three distinctive patterns of national economic development envisaged of home countries: the resource-abundant countries; the resource-scarce large countries; and the resource-scarce small countries. After a process of careful selection, I decided to focus on 11 countries to include Brazil, Germany, Hong Kong, Japan, Singapore, South Korea, Sweden, Switzerland, Taiwan, United Kingdom and the United States. During the course of the research, I was always tempted to add more country case studies and, in particular, to include the study of the emergence and evolution of MNCs from the Netherlands or France; but on the other hand I thought that I may already be driving home the point far enough with the existing 11 country case studies and thus to further belabour the point would be totally unnecessary. Besides, I would like to leave some aspect of the research open for the future.

With the pursuit of a more ambitious research agenda, it had become rapidly obvious that the conceptual basis of the research would encompass the macroeconomic theories of international production as well as international business history, the latter in both its more traditional microeconomic perspective and more modern macroeconomic perspective. The use of both perspectives has been quite useful in the conduct of the research. The growth and evolution of MNCs from a particular country has often been more clearly illustrated by the analysis of the histories of particularly prominent MNCs based in that country where this was warranted.

In 1996 I completed another paper with the title *Patterns of Growth of Multinational Enterprises: The Case of the Resource Scarce Large Countries* which was accepted for a competitive session at the 22nd Annual Conference of the European International Business Academy.[1] With some aspects of the growth of MNCs based in resource-scarce small countries and resource-scarce large countries set in place in these two seminal papers, the research in 1997 and 1998 focused on analysing the growth of MNCs based in resource-abundant countries.

As the new year dawned in 1999, I made a firm resolution to shift the major research into high gear and to realize the production of the book within a year. First, I informed Routledge of an important revision of the title of the book to *Multinational Corporations: Emergence and Evolution* to reflect more accurately the more comprehensive research that I had embarked on. Secondly, I began to pursue the research with some vigour and haste country after country, repeating the almost never endless and exhausting process of research, drafting, revision, editing and polishing chapter after chapter in the period between January and November 1999. Although the two seminal papers that I prepared in 1995 and 1996 enabled me to have a first draft of seven chapters, the drafts were far from being acceptable as proper book chapters and had to be improved considerably by further research and various rounds of drafting. Those chapters in their final state represent a significant progression far beyond recognition from the seminal papers from which they emerged.

This major research project did not have research funding simply because I chose to use the limited time to pursue actual research than to devote it to the speculative and often frustrating exercise of applying for a research grant. In any event, the research consumed so much of my interests that I was determined to pursue it to the end, even if I had to be jack-of-all-trades and be responsible for all aspects of the execution of the research and the production of the book. The success of such determination as seen in this final product is a testament to what can be achieved solely on the basis of my own effort. If there is a debt of gratitude that I have, it is to John Cantwell for two reasons: first, for showing me a possible theoretical framework with which I could explore and eventually found quite useful; and second, for agreeing so graciously to write the Foreword long before he ever saw the book in its final state. Beyond that, I have to thank some Divine Providence for keeping me safe and healthy to conduct this research and to see it through to its end, and to derive genuine pleasure in the final product. However, the greatest joys had been derived from the learning experience as a result of conducting the research.

The period of research behind this book also encompassed my journey into parenthood for the first time. Although it has not always been the easiest of tasks to fulfil my multiple roles as a mother and a jack-of-all-trades in the production of this book while also attending to a demanding teaching and administrative position at my current employment, I derived immense inspiration from my daughter who has grown up rapidly over the course of the research. While at the young and tender age of three years she may not have a clear idea of what her mother had pursued so relentlessly in 1999, she is fully aware that the book is dedicated entirely to her. The completion of this book at the dawn of the twenty-first century and the third millennium is symbolic of both the passage of time and a time well spent.

Paz Estrella Tolentino
November 1999

Note

1 The revised versions of both seminal papers that formed the basis of more major research for this book appear as working papers. These are *Third World Multinationals: Emergence and Evolution*, Birkbeck College, Department of Management and Business Studies, Working Paper 96/02, 1996; and *Patterns of Growth of Multinational Enterprises: The Case of the Resource Scarce Large Countries*, Birkbeck College, Department of Management, Working Paper 97/01, 1997. ISSN 1461 4669.

Part I

Introduction

1 The theoretical foundations

Introduction

This book examines the emergence and evolution of multinational corporations (MNCs). In general, historical observations of the growth of MNCs suggest a developmental course in the pattern of outward foreign direct investment (FDI) from an initial emphasis on trading, resource-based or simple market-seeking investments where the competence of leading national firms is embedded in basic engineering skills, complementary organizational routines and structures. Over time, as industrial development of home countries proceed and as their firms gain maturity as MNCs, there is the gradual progression of their outward FDI towards more sophisticated forms of manufacturing and services activities that embody greater technological and organizational complexities. Indeed, the historical analysis of the developments in the sectoral pattern of outward FDI at the global level shows that the bulk of this investment was associated with primary commodity production before 1939, and only since the Second World War did outward FDI in manufacturing and services come into its own in a major way, bringing forth the growth in international technological competition in the industrialized countries and the global integration of FDI.

Such an evolution describes the growth of the mature MNCs of Europe and the United States over the last two centuries. The evolution of Japanese MNCs since the Second World War, on the other hand, has been compressed into a much shorter time span. Investments in resource-based activities and import substituting manufacturing in South East Asia led the way in the 1950s until the early 1970s, but since the late 1960s the interest of Japanese firms shifted to more sophisticated manufacturing production in Europe and the United States (Ozawa, 1991). Although MNCs from the Asian newly industrialized countries (NICs) are nowhere near as technologically advanced as the more established multinationals from Europe and the United States as well as the modern Japanese MNCs owing to the less advanced stage of the industrial development of their home countries, the lower forms of technological competence of their leading national firms and their lack of maturity as MNCs, the sectoral complexity and geographical scope of their international production activities are evolving much more rapidly than that of the more historical home countries of FDI. The main reason for this is the general trend towards the internationalization of business since the mid-1970s which has been common to firms of all countries. Domestic enterprises from the developing countries have, in general, embarked on international production and became MNCs at an earlier stage of their development than industrialized country firms (see Tolentino, 1993 for further support of this argument).

Despite the general trend towards the internationalization of business which has been common to the firms of all countries and the more rapid pace in the emergence of MNCs from the newer home countries and in the evolution of the pattern of their international production activities, there seem to be variations in the pattern of the early stages of outward FDI across different types of countries as well as their developmental paths over time that are determined by distinctive patterns of national economic development. Since each country or group of countries has a unique pattern of national economic development owing in part to different endowments of natural resources, different sizes of the domestic market and different types of development path pursued in achieving industrial development, the developmental course of international production in each country or group of countries is likely to be unique and differentiated. This is the central theme that permeates the analysis of the emergence and evolution of MNCs in this book, and the hypothesis that the research seeks to validate *vis-à-vis* the empirical evidence.

Since the conceptual framework relating outward FDI and patterns of national development belongs to the set of theories comprising the macroeconomic developmental approaches to international production, the theoretical foundations underpinning the analysis of the emergence and evolution of MNCs proceed from the various theoretical strands of this approach. The development of these various theoretical strands from their origins in the 1960s is therefore provided in the proceeding section of this chapter. The various theories developed have sought to relate the level, character and composition of inward and outward FDI of a country's firms to the stage of the product life cycle (Vernon, 1966), the comparative advantage of countries (Kojima 1973, 1975, 1978, 1982), the national stage of development (Dunning, 1982, 1986a, 1986c) or the process of domestic industrial development. The latter has been useful in elaborating the concept of stages of development in international production of countries generally (Tolentino, 1993; Cantwell and Tolentino, 1990) as well as the developmental course of outward FDI of particular countries such as Japan (Ozawa, 1979a, 1979b, 1982) and the developing countries (Tolentino, 1993; Cantwell and Tolentino, 1990). By relating the emergence and evolution of international production to distinctive patterns of national economic development in different types of countries, the present research aims to refine the stages of development concept in international production, and therefore contribute to the advancement of more modern macroeconomic theories of international production.

The macroeconomic development approaches to explaining international production

The macroeconomic development theories of international production describe the dynamic and developmental process or the way in which stages of development or maturity of countries and firms affect their international production activities. The major theoretical strands comprising this framework are the product cycle model (PCM) Mark I advanced by Vernon (1966) to explain the patterns of American FDI in Europe in the 1960s, the integrated theory of trade and FDI of Kojima and Ozawa which explains the patterns of outward FDI of Japan, and the two more general concepts associated with the investment-development cycle and path identified with Dunning, and the stages of development in international production promoted by Tolentino (1993) and Cantwell and Tolentino (1990).

The product cycle model

The earliest version of the PCM Mark I, advanced by Vernon (1966) was the first theoretical strand to emerge among the modern macroeconomic theories of international production. The model was developed owing to the limitations of the conventional neoclassical Hecksher-Ohlin-Samuelson theory of international trade in explaining the growth since the Second World War of trade and international production between the United States and Europe with similar proportional factor endowments.

In developing the PCM Mark I, Vernon drew on the newer trade theories promoted by Leontief (1954), Johnson (1958), Linder (1961) and Posner (1961). Leontief, Johnson and Posner emphasized the important role of technological factors in explaining American trade patterns. In particular, Leontief alluded to the embodiment of higher skills in American export products, while Johnson referred to the presence of a slower rate of innovation in Europe compared to the United States in explaining the existence of a persistent dollar shortage in Europe. The presence of a technological gap as an important element in explaining trade patterns became more entrenched in the early 1960s with the pioneering work of Posner in developing a theory of trade based on technology gaps, and in particular the different rates of innovation and learning among different firms and countries. At the same time, Linder also proposed that similarity of income levels, factor endowments and demand patterns were the important determinants of the pattern of trade flows.

The fundamental principle behind the PCM Mark I is that the extent and form of innovation and product development are determined by demand and relative factor prices which exist in the market particular to the home country of the innovating firm. The presence of a large market, for example, favours entrepreneurial opportunities in the research and development, production and marketing of new products and processes. Furthermore, the presence of high income levels in the United States in the 1950s and 1960s that engender new wants encourages the generation of ownership advantages of American firms in the production of high-value consumer goods and industrial products. The high unit labour costs in the United States relative to production also creates a specific kind of entrepreneurial innovation, i.e. labour-saving innovation.

Drawing on the assumptions that products are capable of standardization at various income levels and encounter foreseeable changes in production technology and marketing methods, and that production processes proceed through phases over time and inevitably achieve scale economies, the PCM Mark I delineates three principal stages in the life cycle of a product. The first stage is that of the *innovative new product*, resulting from the awareness of unique entrepreneurial opportunities and the identification of a novel demand, or the adoption of new methods of production. The PCM further postulates that American entrepreneurs are first aware of opportunities to fulfil new wants by new products concomitant with high average income levels or high unit labour costs. This stems from the model's other assumption that the entrepreneurs' consciousness of, and responsiveness to, entrepreneurial opportunities are a function of ease of communication with the market place which in turn is a function of geographical proximity. As a result, American entrepreneurs are expected to have a consistently higher rate of expenditure on product development than entrepreneurs from other countries, at least in product lines that fulfil high income wants and that substitute capital for labour.

The unstandardized nature of the new product, the high degree of product differentiation or the existence of monopoly in the early stages and the need for expeditious communication between producers, customers, suppliers and competitors underscore the importance of a location of production that favours external economies and the minimization of communication costs. Since the model assumed that these costs increase directly with geographical distance, a location of production which is close to the market is favoured in the first stage.

The second stage in the life cycle of a product is that of the *maturing product*, the result of a certain degree of standardization. The importance of flexibility – arising from the integration of research, production and marketing activities at the site of innovation in the first stage – decreases. Instead, the possibility of economies of scale through mass production increases with the specification of product and process technology. Thus, in contrast to the first stage where product specifications were fundamental, production costs now become far more important. Moreover, demand for the product increases correspondingly and becomes more price elastic with increasing buyer knowledge. In time, the demand for the product increases in other relatively advanced countries as Western Europe with similar demand patterns, especially since the product has a high income elasticity of demand and is labour saving. These foreign markets are first served through exports until the marginal production and transport costs of the goods exported from the home market are below the average cost of establishing a production facility in the export market, where factor costs, appropriate technology and scale economies are divergent from those in the home market.

Apart from cost considerations, any threat to the large-scale export business in manufactured products in the form of tariff protection imposed by export markets to promote growth or balance trade as well as the emergence of local competition within the export market becomes a powerful 'galvanizing force' propelling the initial import substituting international production of an established firm. The subsequent growth of production of rival firms may result in a threat manifested in the form of declining global share of the market with respect to the initial investor. Besides, the relocation of production abroad increases the possibility of exports to third country markets and even the home market should differences in factor costs surpass transport costs.

The third and final stage in the product life cycle is that of the *standardized product*. The nature of the product at this stage means that accessibility to market information is greater and competition is largely, if not solely, on the basis of price. The search for the lowest cost source of supply therefore becomes the priority of investor firms. At this stage, the ownership advantages of the firm are based mainly on marketing and distribution, unlike the earlier stages where ownership advantages were based on the abilities of the firm to engage in technological innovation.

A major feature of the PCM is its implicit reply to the *Leontief paradox* that American firms export more labour intensive goods instead of capital intensive goods with which the United States has a comparative advantage. The PCM describes the research intensive innovative stage of a product and the establishment of a pilot plant as particularly labour intensive because of the demand for research staff and marketing personnel. However, as the product reaches the standardized stage, scale economies become far more important. The mass production of the standardized product necessitates greater capital intensity compared to the greater labour intensity of the innovative stage.

In addition to the substitution of capital intensive means of production for labour intensive means of production in the standardized product stage, there is also a substitution or displacement of higher skilled labour by less skilled or unskilled labour. A cost determined equilibrium regulates the shift of these lower skilled and unskilled labour stages of production of standardized products to developing countries where labour costs are lowest and where incomes begin to 'catch up'. A fourth stage in the product life cycle can therefore be envisaged in which there is a shift in the location of production to developing countries and specifically to the NICs. However, there is a major difference in the nature of American FDI in the developing countries at this stage and that of American FDI in developed countries at an earlier stage. International production by American MNCs in the developing countries is more likely to be of an export-oriented kind that is not driven by local market demand. By comparison, international production by American MNCs in Europe is more likely to be import substituting that is determined largely by demand factors in the host country.

The PCM is re-stated clearly in Hufbauer (1965, 1970), particularly its interrelationship with technological gap theories, while the further applications of the PCM in explaining American trade and international production patterns are found most notably in Knickerbocker (1973), Stobaugh (1968) and Wells (1972). Knickerbocker (1973), in particular, used the framework of the PCM to identify the capabilities of American manufacturing companies which enabled their expansion abroad as MNCs, and as a critical determinant of the industrial structure in the process of international expansion. The empirical support for the PCM has extended beyond the analysis of the patterns of American trade and international production. Graham (1975, 1978) and Franko (1976) have examined the significance of the PCM and other theories relevant to an understanding of the development of foreign manufacturing operations of American firms in Europe to the growth in the United States of some of the largest industrial firms based in the western part of Continental Europe. Graham (1978), Flowers (1976) and Hymer and Rowthorn (1970) have also extended the PCM to consider the concept of rivalry between firms from different countries.

The theory of Japanese FDI: the contributions of Kojima and Ozawa

Like the PCM, Kojima's integrated theory of international trade and international production also analyses the interaction between ownership advantages and the changing location of production. However, Kojima's theory differs in the theory of trade upon which it is based. The theory is based on the Hecksher-Ohlin principle of comparative advantage (or costs) (Kojima, 1973, 1975, 1978, 1982, 1990). Kojima's basic theorem is that FDI should originate from the comparatively disadvantaged (or marginal) industry of the home country which leads to lower cost and expanded volume of exports from the host country. This type of FDI is referred to as *pro-trade, Japanese-type FDI*. The non-equity forms of Japanese resource-based investment in resource-rich countries are regarded as trade-oriented investments because of the assurance of a supply quota or production sharing arrangements with indigenous enterprises in the host countries.

Such investments contrast with the wholly owned, vertically integrated resource-based production of firms from the United States and major European countries which originate from the comparatively advantaged industries of the home country and lead to misallocation of resources and a decreased volume of exports from the host country. This type of FDI is

referred to as *anti-trade, American-type FDI*. The oligopolistic and technologically advanced American firms that engage in FDI are seen to be motivated
by the defence of their oligopolistic positions, the exploitation of factor markets and the presence of tariff barriers in the developed countries.

An important criticism of Kojima's theory is the way in which import
substituting international production is regarded as anti-trade oriented.
While import substituting international production may be considered anti-
trade oriented at the microeconomic level, this may not necessarily be so at
the macroeconomic level unless very restrictive assumptions are introduced in
the model. The growth of outward FDI from the United States, Germany
and Japan is often accompanied by an increasing level of exports. There is
also evidence to suggest that owing to their potentially enclave nature export-
oriented international production may play a less significant role in industrial
adjustment or in increasing welfare of the host country (Dunning and
Cantwell, 1990).

Kojima's theory is essentially an industry-based theory as opposed to a
country-based theory as he claims, in as much as Japanese and American
MNCs were concentrated in different industries which helps to explain why
their international production was aimed principally at serving either export
or local markets. The extent to which the theory is country based derives
essentially from the general way in which the stage of national development
helps to explain the industrial structure of indigenous firms. Although further
support for Kojima's macroeconomic approach to international production
was provided by Ozawa (1971, 1977, 1979a, 1979b, 1985), he acknowledges
in his later work that the distinctive characteristics of Japanese FDI from
1950 to the early 1970s in resource-based and labour intensive or technologically standardized industries in developing countries belonged to the first
two phases of Japanese MNC growth. Since the late 1960s, Japanese FDI has
evolved to the third phase involving import substitution and the recycling of
trade surplus resulting from the rapid process of industrial restructuring in
Japan. Japanese FDI in this phase is focused on the mass production of
assembly based consumer durables in high-income countries such as the
United States and Europe, supported by a network of subcontractors. A
fourth phase of Japanese FDI was discerned to pertain since the early
1980s. International production by Japanese firms in this phase has predominated in the flexible manufacturing of highly differentiated products, involving the application of computer-aided designing (CAD), computer-aided
engineering (CAE) and computer-aided manufacturing (CAM) (Ozawa,
1991).

As with American MNCs, the growth of Japanese MNCs has therefore
followed an evolutionary course from resource based and simple manufacturing towards more technologically sophisticated forms of international production. The essential difference between the evolution of American and
Japanese MNCs lies in the swiftness with which Japanese MNCs have
made the transition through the evolutionary path. Thus, although

American and Japanese firms and industries are at a different stage of evolution, Japanese firms have demonstrated their ability to catch up rapidly, keep pace and, in some cases, even surpass technological advancements in the West.

In his more recent work, Kojima (1990) acknowledges that more recent American FDI in Japan is largely akin to a Japanese-type FDI as earlier described, while Japanese FDI in the United States has taken on the characteristics of American-type FDI. Such a rapid pace in the evolution of Japanese FDI may be attributed in part to the efforts of the Japanese government to change systematically the composition of the country's comparative advantage, aided in part by licensed foreign technology (Ozawa, 1974). The view of pro-trade, Japanese FDI thus appears to be a reflection of the early stages of development of Japanese MNCs. The history of Japan's trade development since the Second World War does not lend support to this form of application of the static theory of comparative advantage which prescribes that Japan's trade patterns should conform to its comparative advantage then, i.e. the production and export of labour intensive goods and the import of capital and technology intensive goods. Instead, the increasing technological competitiveness and trade surplus of Japan in technologically intensive products provide support for the development of future-oriented technologies, so that the industries and sectors in which a country enjoys the greatest potential for innovation and in which investment may be most beneficial are not necessarily those in which it currently has a comparative advantage (Blumenthal and Teubal, 1975; Pasinetti, 1981). Major new investments by Japanese companies in the electronics components industry and the diversification into new product lines of those in the consumer electronics sector, for example, provides evidence of Japan's continuing efforts to build an industrial structure based on industries and sectors in which the country has no current comparative advantage (Dunning, 1986b). As a necessary extension of the argument, Kojima and Ozawa (1985) argue that global welfare is increased where international production helps to restructure industries in line with dynamic comparative advantage.

Although most MNCs based in developing countries are unlikely to develop as rapidly as their Japanese counterparts who have drawn on frontier technologies, the trend is towards a persistent upgrading of activities for which these firms are responsible (Tolentino, 1993). The phases of development suggested by Ozawa for Japanese MNCs were thus rendered relevant to the study of MNCs based in developing countries. The first two phases of Japanese outward FDI from the 1950s to the early 1970s in resource-based and labour intensive international production in developing countries are pertinent to explaining the early stages of the growth of newer sources of outward FDI from the developing countries. In the early stages, as domestic industrial development proceeds with domestic firms in the developing countries upgrading their domestic production activities, comparatively disadvantaged (or marginal) industries are relocated abroad as locational advantages accrue in foreign countries at an earlier stage of development. In later stages, more sophisticated manufacturing activities are transferred abroad, even to the developed countries. The overseas activities of the NICs in the developed countries since the 1970s, although based on a less research intensive form of technological innovation than the Japanese MNCs, can nevertheless be described as being in the early phases of the import substituting or surplus recycling stage of international production.

The underlying development process of international production that permeates the Kojima–Ozawa theory echoes Vernon's PCM and Dunning's concept of an investment-development path. Such similarity exists between Vernon's PCM and Kojima's integrated theory of international trade and production despite the different theories of trade upon which the two models are based (Mason, 1980). The comparative advantage based model of Kojima and Ozawa presents a useful explanatory framework within which to view the emergence and evolution of international production from countries undergoing rapid growth such as Japan, Germany and the NICs in more recent years. Both models explain the process of relocation of production of mature or technologically standardized industries as locational advantages favour foreign countries at an earlier stage of development. Such relocation of production is undertaken while the domestic firms still have the technological and organizational advantages associated with lower technology and more labour intensive production activities which can be exploited more profitably in foreign countries with lower levels of technological capacities and production costs. The common theme of a developmental process of international production in the two models is obscured by the misleading theoretical framework adopted by Kojima in his analysis.

The concept of an investment-development cycle and path advanced by Dunning shows in more general terms the impact of the national stage of development on both the level and character and composition of international production.

The concept of an investment-development cycle and path

The idea of an investment-development cycle has been advanced by Dunning (1981a, 1981b, 1982, 1986a, 1986c, 1988). The early versions of the concept proposed that the level of inward and outward FDI of different countries, and the balance between the two, is a function of their stage of development as measured by gross national product (GNP) per capita. It was further suggested that the plotted data of the net outward investment (NOI) and GNP of different countries, both variables normalized by the size of the population, reveal the presence of a J-shaped investment-development curve with countries classified as belonging into four main groups that correspond to four stages of development. However, a fifth stage corresponding to the fifth stage of development was later added (Dunning, 1988).

Countries at the lowest or first stage of development have little or no inward or outward FDI, and consequently a level of net outward investment (NOI) that is close to zero. Countries at somewhat higher levels of development attract significant amounts of inward FDI, but as the outward FDI of their own firms is still limited, NOI is negative. The argument then is that, past some threshold stage of development, outward FDI increases for countries at yet higher level of development. The continued growth of their outward FDI at a fourth stage results in a positive NOI, and the balance between inward and outward FDI in major industrialized countries consistent with the growth of cross-investments and intra-industry production results in the return of their NOI to zero at the fifth stage of development.

Although the concept of an investment-development cycle proposed a relationship between NOI and a country's relative stage of development with a balanced investment position in the early and late stages of the cycle, the term 'cycle' seems inappropriate when referring to the presence of a J-shaped investment-development curve. The term 'stages' or 'path' seems more appropriate as countries at higher levels of development have not returned to the NOI position of countries at lower levels of development.

Empirical evidence provided in Tolentino (1993) for the period since the mid-1970s showed the existence of a structural change in the relationship between NOI and the country's relative stage of development as result of the general rise in the internationalization of firms from countries at intermediate stages of development. The growth of newer MNCs from Japan, Germany and smaller developed countries, as well as some of the higher income developing countries, reflects their firms' capacity to follow the earlier expansion of MNCs from the traditional home countries, the United States and the United Kingdom, at a much earlier stage of their national development. The increased significance of outward FDI from these newer home countries provides first-hand evidence of the general trend towards internationalization so that the national stage of development no longer becomes a good predictor of a country's overall NOI.

More recent versions of the concept of an investment-development path have shown that apart from the level of NOI, the character and composition of outward FDI of a country's *firms* varies with the national stage of development (Dunning, 1986a, 1986c). The early forms of foreign investment are frequently resource based or sometimes import substituting, and in each case a specific location is associated with a particular type of activity. However, as firms mature their outward FDI evolves from a single activity or product in a particular location, and adopt a more international perspective on the location of their different types of production activities. Cross-investments between countries become more common at this stage, and the visible hand of the direct organization of an international division of labour by firms replaces increasingly the invisible hand of coordination by international market transactions. The character of international production and the ownership advantages of firms become less determined by conditions peculiar to their home countries, but increasingly by firm-specific factors.

The concept of stages of development in international production

My earlier works have extended the perception of the later versions of the concept to the investment-development path by proposing that the outward FDI of *countries* themselves follows a developmental or evolutionary course over time (see Tolentino, 1993; Cantwell and Tolentino, 1990). Thus, outward FDI of countries has tended to predominate initially in trading, resource based or simple forms of manufacturing that embody limited technological and organizational requirements in the earlier stages of development. Factors specific to the home country such as the abundance or scarcity of natural resources or the size of the domestic market have tended to engender particular ownership advantages of MNCs of different national origins. For example, the early forms of ownership advantages of American and British firms in wood processing and metal and coal processing respectively were acquired owing to the abundant availability of timber and coal in the United States and the United Kingdom (Rosenberg, 1976).

As home countries advance through progressively higher stages of industrial development and as their firms gain maturity as MNCs, the technological and organizational embodiment of their outward FDI activities becomes more significant and more complex. This is evident partly in the growth of their research intensive investments in centres of innovation in the developed countries through which firms gain access to more advanced but complementary forms of foreign technology which can be adapted and integrated within their indigenously created programme of technology generation. As a result, the developmental course of MNCs from the developing countries that are of more recent vintage has also been evolving rapidly, but has a distinctive technological tradition compared to the more mature MNCs from Europe, the United States and Japan owing in part to the earlier stage of their national development at which domestic firms in developing countries have acquired the capacity and the incentive to become MNCs.

The changing complexity of outward FDI as development proceeds is associated closely with a changing geographical scope. The simpler forms of investment in resource-based activity or simple manufacturing are frequently undertaken in host countries that are rich in natural resources or have an abundant supply of low-cost labour, or in countries with a close psychic distance, while more complex research intensive investments are undertaken in more industrialized countries. The sectoral and geographical development of international production of countries associated with the innovative capacities of domestic firms is therefore a gradual process over time, and one that is predictable to some extent.

The emergence and evolution of international production in relation to distinctive patterns of national development

More recent refinements of the concept of stages of development in international production are predicated on the premise that variations in the early stages of *inward FDI* across different types of countries and their developmental paths over time are determined by distinctive patterns of national development (see Cantwell, 1997). As mentioned, the developmental course of inward FDI in each country or group of countries is likely to be unique and differentiated since each country or group of countries has a unique pattern of national economic development owing in part to different endowments of natural resources, different sizes of the domestic market and different types of development path pursued in achieving industrial development. This idea was applied by Cantwell (1997) in the context of mapping the potential development paths of inward FDI in Africa.

The present research builds on this idea from the perspective of *outward FDI*. In particular, the research seeks to explore the possibility of the more general application of the proposition that patterns of national development (or industrial development) tend to be associated with equivalent phases of MNC expansion from a broad range of countries.[1] This dictum has antecedents in the eclectic paradigm of Dunning (1981a, 1988), and in the work of Swedenborg (1979), Clegg (1987), Wilkins (1988b), Porter (1990) and Lane (1998) among others. In the analysis of ownership-specific advantages in the eclectic paradigm of international production, Dunning distinguishes between those advantages that are determined by country-specific factors (i.e. those that accrue to all firms of one nationality over those of other nationalities), in addition to those that are determined by industry-specific factors (i.e. those that accrue to all firms within a given industry) and firm-specific factors (i.e. those that enable a firm to compete successfully with other firms within their own industry both in domestic and international markets). The country-specific factors may be generated by the size of a country's market, level of income, resource endowments, educational system, government policy toward R&D, patent and trade mark legislation, etc. Swedenborg (1979) considered such home country characteristics as firm-specific knowledge, size of the home market, resource endowment, distance to major markets, etc. as relevant determinants of both the industrial and geographical distribution of outward FDI. Similarly, Clegg (1987) cited country-specific variations in licensing, exports and outward FDI which are explained by the indigenous environment and resource advantages of home countries and also its institutions, government policy and the maturity of firms in international economic involvement. Furthermore, Wilkins (1988b) argued that all companies with outward FDI have been shaped by economic and other conditions in their home country, and only subsequently by economic and other conditions in their host countries. Unique national characteristics in each home country have thus an impact on the nature and extent of the outward FDI of firms. This includes factor costs, level and pace of industrialization, areas of technological expertise, size and nature of the domestic market, relationships between banks and industries, national endowments of and requirements for natural resources, the availability of professional education, the country's position as an exporter or importer of capital, government policies, geographical position, trade patterns, emigration, and culture and taste. Porter (1990) also developed the proposition that the national environment plays a central role in the competitive success of firms and industries. In that view, the home nation influences the ability of its firms to succeed in particular industries. This relates closely to the theory of Lane (1998) that the extent of embeddedness of MNCs to their home countries and in particular their degree of implantation into national economic and policy networks and national business systems influences their internationalization strategy – the degree of outward FDI undertaken, the kind of competitive advantages the MNCs possess and the kind of competitive advantages MNCs derive from FDI and the way in which nationally based and globally based activities are combined.

In the analysis of the relationship between national economic development and the development paths of outward FDI in this research, countries are classified into groups in the way that Cantwell (1997) had described when analysing developmental paths of *inward FDI*, the justification being that the sectoral or industrial patterns of outward FDI bear a close analogy to that of inward FDI in an earlier period. The analogy emerges partly from the role of inward FDI as a major modality of technology transfer to the host country (Findlay, 1978; Lall, 1983a), as a result of which there is developmental upgrading of innovative domestic industries and the enhancement of indigenous technological competence of host country firms that are able to respond competitively to the presence of foreign based MNCs by undertaking outward FDI at a later stage (see Dunning and Cantwell, 1982 and Tolentino, 1993 for further elaboration and support of this argument).

The country groups vary according to the distinctive patterns of national economic development of *home countries* which take into account endowment of natural resources, the size of the domestic market and the type of development path pursued in achieving industrial development. Three country groups are described: resource-abundant countries, resource-scarce large countries (with resource-intensive production) and resource-scarce small countries (with non-resource-intensive production) (Table 1.1).[2] The organization of countries in these groups facilitates the identification of the dominant form of earliest outward FDI, as well as the type of locally based firm that initiated outward FDI. This assumption is predicated on the premise that since patterns of national economic development are distinctive between country groups, the distribution of firm-specific knowledge across industries in different countries would also tend to differ – at least in the early stages. This means that there may be a peculiar home country-specific element to the analysis of outward FDI since ownership advantages – although these may

Table 1.1 Variations in the early stages of outward direct investment across different types of country

Categorization of national development	Examples of countries	Dominant form of earliest outward FDI	Type of locally based MNC
Resource-abundant countries	Brazil Bolivia (tin) Chile (copper) Indonesia Malaysia Philippines Thailand Canada Sweden Russia(oil) United States	Resource oriented, and local market oriented in large countries	Resource-based firms Manufacturing firms, backwardly integrating
Resource-scarce large countries (with resource intensive production)	Argentina South Korea Taiwan Germany Japan United Kingdom	Local market oriented, trade related	Trading companies Manufacturing firms, serving local markets
Resource-scarce small countries (with non-resource intensive production)	Hong Kong Singapore Belgium Netherlands Switzerland	Export oriented, service based	Offshore producers Service firms (trade, shipping, finance)

Source: Author's adaptation based on Cantwell (1997).

remain the property of firms – vary in a systematic manner between countries (Clegg, 1987) owing to differences in national economic structures, values, cultures, institutions and histories. The generation and maintenance of competitive advantages thus tends to be a localized process (Porter, 1990).

The evolution of outward FDI, i.e. its development path, is closely associated with domestic industrial development regardless of country groupings according to patterns of national development. However, the precise form of the relationship varies among country groups (Table 1.2). Thus, while locally based MNCs from *resource-abundant* countries can be expected to diversify from an initial concentration in resource extraction towards downstream processing of natural resources in resource-rich host countries, locally based MNCs from *resource-scarce large* countries are expected to upgrade their international production steadily from resource processing towards more capital and technology intensive industries. The rapid evolution of outward FDI is associated with industrial upgrading and the generation of strong production and export position in the home country. Finally, the evolution of outward FDI by

Table 1.2 Potential development paths for outward direct investment, and their association with local industrialization across different types of country

Categorization of national development	Link between domestic development and the growth of outward FDI	Type of locally based MNC
Resource-abundant countries	Related diversification (for example from mining), downstream processing (for example, metal processing, wood products, petrochemicals, agribusiness), with some other upgrading of industry in large countries	Resource-based firms Manufacturing firms
Resource-scarce large countries	Industrial upgrading and export growth (as wages rise following productivity growth), growth in importance of services	Manufacturing firms Services firms
Resource-scarce small countries	Shift away from simpler manufacturing activity to some industrial upgrading, but more towards a greater service orientation	Manufacturing firms, relocating activity International service companies

Source: Author's adaptation based on Cantwell (1997).

locally based MNCs from *resource-scarce* small countries are expected to be far more limited and although there may be some industrial upgrading, there is a marked tendency towards greater service orientation. Locally based MNCs from these resource-scarce smaller countries are thus expected to have significant FDI in the service industries owing to the development of their home countries as service economies.

Thus, the pattern of domestic industrial development influences greatly the emergence and evolution of outward FDI, and the main avenue through which this occurs is through the development of local technological competence of leading national firms. The emergence and evolution of local technological capacities are expected to vary according to patterns of national development and influence the type of outward FDI and its industrial course over time (Table 1.3).

Structure of the book

The conduct of the research of the emergence and evolution of MNCs on the basis of three main country groups that accord with distinctive patterns of national economic development dictated the structure of the book. Part II analyses the MNCs from the resource-abundant countries and, in particular, those based in the United States (Chapters 2 and 3), Sweden (Chapter 4) and Brazil (Chapter 5). Part III investigates the MNCs from the resource-scarce large countries and, in particular, those based in the United Kingdom

Table 1.3 Technological accumulation and the national course of outward direct investment

	Stages of national development		
	(1)	*(2)*	*(3)*
Form of technological competence of leading indigenous firms	Basic engineering skills, complementary organizational routines and structures	More sophisticated engineering practices, basic scientific knowledge, more complex organizational methods	More science-based advanced engineering, organizational structures reflect needs of coordination
Type of outward direct investment	Early resource- or market-seeking investment	More advanced resource-oriented or market-targeted investment	Research-related investment and integration into international networks
Industrial course of outward direct investment	Resource-based (extractive MNCs or backward vertical integration) and simple manufacturing	More forward processing of resources, wider local market-oriented and export-platform manufacturing	More sophisticated manufacturing systems, international integration of investment

Source: Author's adaptation based on Cantwell (1997).

(Chapter 7), Germany (Chapter 8), Japan (Chapter 9), Taiwan (Chapter 10) and South Korea (Chapter 11). Part IV examines MNCs from the resource-scarce small countries and, in particular, those based in Switzerland (Chapter 13), Hong Kong (Chapter 14) and Singapore (Chapter 15). These 11 home countries, comprising six developed countries and five developing countries, accounted collectively for two-thirds of the global stock of outward FDI in 1998. The six developed countries accounted for an equivalent share of the stock of outward FDI of the developed countries, while the five developing countries accounted for 69.5 per cent of the stock of outward FDI of the developing countries (based on data in UNCTAD, 1999). The selection of these 11 home countries of widely divergent national characteristics was necessary to demonstrate the relevance of the inter-relationships between the developmental process of outward FDI and distinctive patterns of national development and to advance the development of a comprehensive theory of the emergence and evolution of MNCs. Current theories in the macroeconomic development approach which are based on research of MNCs based in only one or two countries such as the United States (the PCM) and Japan (Kojima and Ozawa's theory) can hardly form the basis of general principles.

Each country chapter analyses the developmental course of outward FDI of MNCs based in that country over time. It examines over time the fundamental determinants of outward FDI, its dominant industrial and geographical patterns, market orientation and the kind of locally based firms undertaking the outward FDI. The main conclusions drawn from the evolutionary analysis of MNCs from each of the country groups are presented in Chapters 6, 12 and 16. The research seeks to determine whether distinctive patterns in the emergence and evolution of MNCs indeed exist between MNCs in the group of resource-abundant countries, the group of resource-scarce large countries and the group of resource-scarce small countries owing to unique patterns of their national economic development. Finally, the implications of the study of the history of MNCs for theory development are explored in Chapter 17 in Part V.

Notes

1 In fact, there is increasing evidence to suggest that a more appropriate unit of analysis is the sub-nation or regions and not the country. See, for example, the works of Cantwell and Iammarino (1998a, 1998b, 1999), Dunning (1998) and O'Farrell *et al.* (1996). This is because there are regional differences within countries in terms of technological innovation and in their ability to attract inward FDI and to generate outward FDI. For further discussion of the analysis of the location of innovative activities of MNCs in regions within Italy, the United Kingdom and Europe, see the various papers by Cantwell and Iammarino. Their papers suggest that investments in innovation at the national level are determined by those at the regional level. Since investments in innovation at the regional level are the 'building blocks' of innovation at the national level, then it follows that the regional scope of innovation is much more important than the national one. The effects of regional differences on the internationalization behaviour of local firms in the business services sector are discussed in O'Farrell *et al.* (1996).

 While recognizing the importance of regional specificities in determining the emergence and evolution of MNCs based in a particular country, this has not been the main object of the research which focuses on country-specific patterns that differentiate MNCs based in different countries. In the pursuit of this research question, the country is, therefore, the appropriate unit of analysis.

2 Thus, countries rich in natural resources are included in the group of resource-abundant countries. Resource-scarce countries, on the contrary, are those that are less abundantly endowed with natural resources. In analysing the development path in inward FDI of the resource-scarce countries, it became apparent that there was the need to differentiate countries belonging to this group according to size (Cantwell, 1997). This is because resource-scarce countries with large domestic markets tended to pursue resource intensive domestic and international production, while resource-scarce countries with smaller domestic markets pursued non-resource intensive domestic and international production.

Part II

Multinational corporations from the resource-abundant countries

2 The emergence of multinational corporations from the United States

The period until 1914

Introduction

The United States accounted for no more than 18 per cent of the global stock of outward FDI in 1914 – a share much lower than that of the largest home country, the United Kingdom, with more than 45 per cent (Dunning, 1983). Mexico, Cuba, Panama (and perhaps some other Central American countries) were probably the only host countries in the world where American foreign direct and portfolio investments in 1914 exceeded the British stake (Wilkins, 1970). Nevertheless, the share of the United States in the global stock of outward FDI grew rapidly to reach almost 28 per cent by 1938 and more than 47 per cent in 1960 (Dunning, 1983). Although the relative importance of the United States as a home country for FDI has declined to 24–25 per cent of the global stock since 1990 owing to the rise in the absolute importance of outward FDI from newer source countries such as Germany and Japan, the United States has, nevertheless, remained the single largest source country of outward FDI since the Second World War.[1] Thus, no historical account of the emergence and evolution of international business would ever be complete without the analysis of the growth of American MNCs.

Given the long history of American MNCs and the presence of a rich and abundant historical literature on the subject, the analysis of their emergence and evolution is conducted in two chapters. This chapter covers the period of their emergence until 1914, while the following chapter covers the period of their evolution in the period since 1914. The account of the growth of American MNCs in these two chapters, although based mainly on the detailed and comprehensive data and information contained in Wilkins (1970) and Wilkins (1974), has aimed to distill the essence of the dynamic changes in the determinants and the sectoral and geographical patterns of American FDI abroad from its origins, with the help in some cases by references to other sources.

The emergence of American MNCs can be traced to the pioneering foreign investments of American merchants in the late seventeenth and especially the eighteenth century. By the late nineteenth century, railroads accounted for the largest share of American FDI at almost 23 per cent. These foreign investments in railroads paved the way for the emergence and growth of American FDI in mining which at that time accounted for the second largest share. The manufacturing sector was the third-largest recipient of American FDI in the late nineteenth century, followed by the petroleum and agricultural industries. Although their relative importance has changed in the period leading to 1914, these remained the five most important industries of American FDI in 1914 (see Table 2.1).

The analysis of the emergence of American MNCs in this chapter begins with the foreign investments by American traders in the colonial times and the early years of American independence. The discussion then proceeds with the analysis of American participation in foreign railroads and other transportation. The general factors behind the emergence and growth of American FDI in mining, manufacturing, petroleum and agricultural industries between 1893 and 1914 is then considered, followed by a more detailed examination of American FDI in each of those major industries. Finally, some main conclusions are drawn about the changing industrial and geographical patterns of American FDI in the period until 1914.

The evolution of the early American traders to American foreign direct investors

The installation of American mercantile houses overseas as well as the foreign branches of American merchants, and the transformation of traders into investors in other foreign businesses constituted the initial phase in the history of American business abroad (Wilkins, 1970). The colonial merchants of the late seventeenth and especially of the eighteenth century pioneered in planting America's first investment stakes in foreign lands.

By the eighteenth century the typical colonial trader that engaged in foreign commerce used either independent agents abroad – generally British but on occasion transplanted Americans – or appointed a member of his own family as an overseas agent. American merchants would use British independent agents in foreign countries where the volume of business was not large enough to warrant an outpost.[2] In other cases, transplanted Americans were used: these formed one aspect of American business abroad. These were individual Americans that migrated to Europe or to the West Indies and established foreign enterprises to handle the overseas trade of American merchants and firms. The enterprises established abroad had no parent company in the United States and were not, therefore, branches of American companies. Such transplanted Americans started the first American firms overseas in the 1730s and 1740s.[3]

Some colonial merchants, preferring not to rely on independent agents abroad, sent members of their families to handle their overseas trade. This was the earliest evidence of the establishment of a foreign branch house, as in most cases the overseas representative was financed by an American

Table 2.1 Estimates of the stock of outward foreign direct investment of the United States, 1897, 1908 and 1914 (book value, US$ million)

Country or Region	1 Total[a]			2 Railroads			3 Utilities			4 Petroleum[b]			5 Mining[c]			6 Agriculture			7 Manufacturing			8 Sales Organizations[d]		
	1897	1908	1914	1897	1908	1914	1897	1908	1914	1897	1908	1914	1897	1908	1914	1897	1908	1914	1897	1908	1914	1897	1908	1914
Mexico	200	416	587	111	57	110	6	22	33	1	50	85	68	234	302	12	40	37	–	10	10	2	2	4
Canada & Newfoundland	160	405	618	13	51	69	2	5	8	6	15	25	55	136	159	18	25	101	55	155	221	10	15	27
Cuba & other W. Indies	49	196	281	2	43	24	–	24	58	2	5	6	3	6	15	34	92	144	3	18	20	4	5	9
Central America	21	38	90	16	9	38	–	1	3	–	–	–	2	10	11	4	18	37	–	–	–	–	1	1
South America	38	104	323	2	1	4	4	5	4	5	15	42	6	53	221	9	11	25	–	2	7	10	16	20
Europe	131	369	573	–	–	–	10	13	11	55	99	138	–	3	5	–	–	–	35	100	200	25	30	85
Asia	23	75	120	–	–	10	–	15	16	14	36	40	–	1	3	–	–	12	–	5	10	6	12	15
Africa	1	5	13	–	–	–	–	–	–	1	2	5	–	2	4	–	–	–	–	–	–	–	1	4
Oceania	2	10	17	–	–	–	–	–	–	1	2	2	–	–	–	–	–	–	1	6	10	–	2	5
Banking	10	20	30	–	–	–	–	–	–	–	–	–	–	–	–	–	–	–	–	–	–	–	–	–
Total	635	1638	2652	144	161	255	22	85	133	85	224	343	134	445	720	77	186	356	94	296	478	57	84	170

Source: Wilkins (1970).

Notes

a Total includes sum of colums 2 through 8 plus miscellaneous investments.
b Petroleum includes exploration, production, refining, and distribution; the bulk of this is in distribution.
c Mining and smelting.
d Excluding petroleum distribution; includes trading companies and sales branches and subsidiaries of large corporations.

enterprise. As with the independent agent, the overseas representative often acted for a number of American traders, not just his own family firm.

The choice of using independent agents or establishing their own trade representation abroad or 'markets and hierarchies' (Williamson, 1975, 1986) as alternative modes for conducting overseas trade transactions was based on the adequacy of existing independent agents, their potential to exhibit opportunistic behaviour, the frequency and volume of trade and the profit opportunities in a particular foreign market. If there were an adequate number of independent agents present in a foreign country that were trustworthy and if the trade volume and profit opportunities were small, colonial merchants preferred generally to use existing independent agents (either European or transplanted American) who would buy and sell on order or consignment. In such circumstances no investment was made by the American enterprise, apart from the goods involved in each trading transaction. The market transaction between colonial merchants and independent agents took the form of 'relational contracting' (Williamson, 1986) and informal enforcement, where long-term relationships between the colonial merchants and independent agents substituted for common ownership.

On the other hand, should the volume of business in a foreign country be large and the profit opportunities tempting, there were increased risks of market failure or high transaction costs associated with information impactedness and bounded rationality (i.e. the unavailability of perfect and full information at zero costs leading to limited rationality of individuals and the high costs of negotiating and concluding complete contracts) and the greater tendency for opportunistic behaviour on the part of independent agents (Coase, 1937). This was particularly the case in the profitable trade with England and the West Indies. In these important export markets, the seller often felt that uncertainty could be overcome and greater control over foreign business could be had if the transaction was brought under 'unified governance' (Williamson, 1986) through the establishment of the firm's own overseas representative.[4] This is an historical application of the modern theory of 'internalization' of markets as applied in the case of international production by Buckley and Casson (1976). The close supervision afforded by the overseas representative would enable the foreign business to expand more rapidly *inter alia* because trading risks were minimized. The American merchant or trading firm could expect more personal attention to its goods, more satisfactory storage facilities, more information on markets, more beneficial credit arrangements and, more importantly, an opportunity to retain profits within the firm. However, there were costs involved with the establishment of an overseas representative: the salary of the representative abroad and the expenses of an office and warehouse, the costs of acquiring knowledge (at least initially) of a particular market and the loss of trade financing if the American merchant obtained this from an independent foreign agent (Wilkins, 1970).

With the withdrawal of preferential trade treatment by the British empire after American Independence, new products for export as well as new and more distant markets were sought by American merchants. Cotton surpassed tobacco in importance and became the largest export of the United States by 1803. Closely associated with the diversification of American exports destined to wider markets was the establishment of more transplant American trade agents overseas. At the end of the 1830s or earlier, transplant Americans who acted for American traders were resident in Madagascar, Zanzibar, Smyrna and Bombay. There was also a collection of independent American trading agencies and branches of American commercial concerns in Cuba, the West Indies, Canada and Newfoundland in the 1820s and 1830s and in many Pacific islands and Canton in the late eighteenth century. Meanwhile, new trading commitments were made in the United Kingdom. In Liverpool, there were both merchants collecting goods for sale and some representatives of American mercantile houses handling cotton. By the 1830s, a large number of American merchants resided in England. Nonetheless, in spite of the increase in American trading outposts in foreign countries, American merchants continued to use independent European – mainly British – houses as their foreign agents in certain foreign ports (Wilkins, 1970).

Although outward FDI by individual traders and trading firms did not account for the bulk of capital outflows of the United States in the colonial period and in the early years of the republic, Wilkins (1970) argues that American mercantile activity affected the later growth of American international business in a number of significant ways.

First, the accumulated surplus capital from the trading ventures was directed in a large part towards American domestic enterprises in real estate, transportation, industry and mining. The development of an American national market, made possible by the establishment of railroads that linked different parts of the United States, led to the emergence of national firms that later evolved to became international firms. Foreign ventures by American firms in railroads, mining, manufacturing and natural resource extraction later started their own wave of American FDI which became the most important sectors of American FDI by the late nineteenth century.

Second, while *none* of the American FDI made either by independent agents or by members of a trader's family in England and islands in the West Indies during the colonial period lasted, a small number of the pioneering American businesses abroad in the post-colonial era continued for years as permanent trading posts (Wilkins, 1970). By the mid-nineteenth century, the continuity of such trading companies became more common. These American trading firms became agents of large American industrial companies in their early forays in foreign sales. It was only when foreign business in American manufactured products expanded in a major way that American industrial firms begin to handle their overseas marketing themselves.

Third, there was the metamorphosis or merger of particular mercantile businesses into more diversified and sometimes entirely different enterprises

engaged in lines of businesses besides commerce. Some broadened their functions by integrating into mining, agriculture, industry, transportation and
banking. Sometimes this transformation took place abroad and sometimes
in the United States. Of most interest was the change that occurred overseas.
The first case of an an American trader turned foreign direct investor was
David Beekman, the son of the New York merchant Gerardus Beekman, who
went to St Croix to represent his family's sugar trading firm in the 1750s.
Profits earned from sugar trading provided the capital for the purchase of
sugar estates on the island. Similarly, the Boston Fruit Company – the predecessor of United Fruit Company – invested its earnings from trading and
shipping concerns in the Caribbean in banana plantations in the region. This
pattern of American traders becoming foreign investors through vertical integration in agriculture, mining, and to some extent in railroads and industry
became typical in years following. The evolution of the nineteenth century
trader into an investor was a result of either accident (i.e. through extending
credit and acquiring property as a result of the debtor's default on debt) or
made on purpose through the intentional integration of the business.[5] In
neither case was much, if any, capital exported from the United States.
Foreign capital investment was made either through the credit line or the
reinvestment of profits accrued abroad (Wilkins, 1970).

The earliest American private banks to venture overseas also had their
genesis in trading companies. The House of Morgan, with its London connection, evolved in 1837 from the mercantile business of George Peabody.
Samuel B. Hale's Argentine trading firm also moved into banking whose
main business was the flotation of loans for the Argentine national government and several provincial governments.[6] From the late 1840s onwards, the
financial dealings of the American firms Lazard Frères, Seligman, and
Morton, Bliss & Co. that set up branch banks in Europe also emerged out
of mercantile endeavours. These firms found trading in money to be more
lucrative than trading in goods.

By number of ventures, the foreign trader turned foreign investor was the
most significant form of American FDI during the colonial times and in the
first 75 years of the republic (Wilkins, 1970).

American participation in foreign railroads

By size of investment, railroads accounted for the largest share ($144 million
or 23 per cent) of American FDI in 1897. Although the level of American
FDI in railroads grew steadily, reaching some $161 million in 1908 and $255
million in 1914, its share in total American FDI declined to less than 10 per
cent in those years (see Table 2.1).

American capital investment in transportation was most significant in
Panama and Nicaragua in the period before the American Civil War
(1861–65) and in Mexico, Canada and the Caribbean in the period between
1880 and 1914. Mexico, Canada and the Caribbean together accounted for

99 per cent of all American FDI in railroads in 1897 and 1908 and even by 1914 these neighbouring areas continued to account for more than 94 per cent of American FDI in railroads. American participation in foreign railroads came in the form of *horizontal integration* in the western hemisphere by American railroad companies or by the *forward vertical integration* of agricultural and mineral companies in the ownership of railroads to facilitate the transport of agricultural produce, mineral ores or mineral products. The United Fruit Company, the Cerro de Pasco Mining Company and the mineral investments of the Guggenheims were some of the excellent cases in point.[7] In some cases and, perhaps more significant in number, was the American participation in railroads in the Caribbean, Canada and South America where their main contribution came not from the provision of American capital but in the ingenuity, talent and expertise provided by American entrepreneurs, syndicates and small companies in the construction of railroads.[8]

The railroads investments in the western hemisphere were considered the international extension of the domestic business plans of American railroad companies (Wilkins, 1970). Such was the case with the construction of the Panama railroad and the Nicaraguan carriage-steamboat line by the Panama Railroad Company and the Accessory Transit Company, respectively. The stake in the Panama railroad by the Panama Railroad Company (chartered in the New York legislature on 7 April 1849) with authorized capital of $1.5 million (later increased to $5 million) was considered an exceptional and the first truly large American capital investment in a foreign land. The enterprise had the unique feature of combining the export of American skills, techniques and capital on what was for the time a mammoth scale (Wilkins, 1970). After 1856, with no other route across the Central American isthmus, and until the completion of a United States transcontinental railroad in 1869, the Panama railroad provided the undisputed best means of transport from the American East to the American West. Similarly, the carriage-steamboat Nicaraguan route established for $2 million was the best approach to the American West in the period from 1851 to 1855, carrying some 20,000 people annually across the isthmus.[9]

The establishment of foreign railroad networks in Mexico, Canada and the Caribbean by American firms and individuals enabled the further expansion of American business overseas. The Panama railroad stimulated significant business opportunities as well as trade between the American East and the American West, to China, Australia, Europe and the East Indies. Similarly, the railroads constructed by American capital in Mexico supported the emergence and development in that country of business opportunities for American companies in mining and agriculture. A similar stimulus to business and trade opportunities brought about by the railroads was evident in South America and the Caribbean where, despite the lower level of American capital investment in railroads, the development of railroads in those regions paved the way for the growth of further American investments in mining and agriculture, respectively. At $221 million, the level of American capital investment in mining accounted for more than two-thirds of American capital investment in South America in 1914. On the other hand, American capital investment in agriculture, amounting to some $181 million, accounted for almost half of American capital investment in the Caribbean in 1914.

American FDI in mining, manufacturing, petroleum and agriculture in the post-Civil War period until 1914

The period since the American Civil War and in particular from 1893 to 1914 described the large-scale expansion of American FDI. The phenomenal growth of American MNC activity abroad made the United States a net exporter of capital from 1898 to 1905 (Wilkins, 1970).[10] Factors relating to the home country, host countries and the firm had a pronounced influence on the growth of American MNCs during this time. The push factors associated with the growth of the firm and of conditions in the home country (i.e. the economic environment of the United States, including actions by the United States government) combined with the pull of profitable business prospects abroad (the growth of foreign markets, the availability of foreign sources of supplies and actions by host country governments) contributed to the rising levels of American FDI in the decades before 1914. The following sections analyse these factors more closely.

Home country-specific factors and firm-specific factors

There were several home country-specific factors as well as firm-specific factors that were fundamental to the expansion of American FDI in the period before 1914.

First, the emergence and expansion of American FDI in mining, petroleum and agriculture can be linked to the presence of rich and abundant natural resources in the United States. The United States was the world's largest producer of copper, lead and petroleum in 1900, and was the second largest producer of bauxite, gold and zinc (Jones, 1996). Combined with a high agricultural productivity, this enabled the United States to shape its economic destiny on the basis of natural resources. American firms thus developed management and organizational skills and technologies in natural resource extraction and processing which were exploited profitably abroad.[11] The accumulated skills and technology of American firms in natural resource extraction and processing were particularly important when the United States demanded increasing amounts of minerals, petroleum and foodstuffs which could not be met fully by domestic sources of supply.

Similarly, the emergence and expansion of American FDI in manufacturing can be linked to the rapid development in the 1850s of American technologies in certain metallurgical industries (i.e. machine tools, guns, reapers, and sewing machines) linked with mass production in which the United

States had already acquired world leadership (Wilkins, 1970). The American companies with novel products and whose entrepreneurs exhibited farsighted leadership grew rapidly and made the most far-ranging investments in foreign countries. These features describe consistently the growth of American MNCs over time, particularly in the manufacturing sector.

The development of American skills and American technology were key to the growth of their FDI – whether in search of supplies or in search of markets.

Second, the slow growth of domestic demand brought about by the depression in the United States between 1893 and 1897 led American companies with surplus production to turn to foreign markets as an outlet for goods that could not be sold at home. As their exports to foreign markets started to rise, companies established sales branches abroad initially and then factories at a later stage to maintain and increase their foreign sales.

Third, there was the development of large enterprises resulting from the trend towards domestic mergers at the end of the nineteenth century. The trend was facilitated by the development of the American capital market and encouraged by the Sherman Antitrust Act of 1890 that forbade agreements among independent companies to restrain trade. Indeed, the domestic firms that survived the depression were the very large enterprises that resulted from the merger movement during this period. The modern industrial enterprise became large not merely by expanding its size – which the trend towards domestic mergers at the end of the nineteenth century facilitated – but also by broadening its scope through horizontal and vertical integration (Chandler, 1980). The consequence was that not only did productive assets in many industries become concentrated in a few large business firms, those few large business firms that inherited the foreign businesses of their predecessor companies became international companies from their existence. Many of these were resource-based companies.[12] Thus, the same entrepreneurs involved in the creation of large domestic businesses led the expansion of American businesses abroad, seeking markets and raw materials in foreign lands. Outward FDI in turn afforded another route by which these large American companies further increased their size and scope.

Fourth, actions by the United States government also favoured the growth of international production by American firms, particularly in resource-based industries. For example, the duty free status accorded to Canadian newsprint in 1911 as a result of strong pressures from American newspaper publishers, combined with the industrialization policy of Canada which encouraged local processing of Canadian natural resources made lumber and paper and pulp mills account for at least one-third of American FDI in manufacturing in Canada in the period between 1897 and 1914.[13] Similarly, the establishment of smelters across the Rio Grande was a profitable venture for many American investors as a result of the passage in the United States of the McKinley tariff in 1890 which imposed new duties on the exports of lead ores to the United States. The duty free status accorded to Cuban sugar under the McKinley tariff and other actions by both the Cuban and American governments also led to the growth of American FDI in the Cuban sugar industry – the most important industry for American FDI in that country.[14] In addition, the worldwide surge in demand for rubber from the early 1900s led President William McKinley to encourage American firms to cultivate rubber overseas to meet the domestic requirements of the United States. This stimulated the growth of American investments in Mexican rubber and by 1912 Americans controlled 68 per cent of Mexico's rubber business (Schell, 1990; Wilkins, 1970).

Host country-specific factors

Locational advantages of host countries favoured the international expansion of American firms in the mining, manufacturing, petroleum and agricultural industries in the post-Civil War period until 1914. In the case of those firms in the resource-based industries, the presence in foreign lands, particularly in neighbouring countries, of rich and abundant natural resources, including those that were not available in the United States or were available at lower costs, low labour costs and the provision of favourable incentives by host country governments to foreign investment, posed profitable FDI opportunities for the largest American mining, agricultural and petroleum companies. The lower costs of minerals prospecting and development in Mexico, for example, was associated with lower capital investments by American companies when compared to the size of their capital investments for similar activities in the United States.[15] In these industries, American companies dominated domestic production in their host countries and, in some cases, became the world's largest producers. For example, large copper companies based in the United States – the world's leading producer and consumer of copper – secured world leadership in that industry in part by their control of rich foreign ore deposits in Mexico, Canada and South America. In 1911, the three largest American copper groups combined controlled 48 per cent of world copper refining output (Schmitz, 1986). In a similar fashion, the Canadian Copper Company established by the American entrepreneur Samuel J. Ritchie eventually became the world's largest nickel producer (Wilkins, 1970). In other cases, the dominant participation of American companies enabled the host countries to become the world's largest producers. For example, American investments enabled Mexico to surpass the United States as the world's largest silver producer in 1902 and the second-largest copper producer after the United States in 1904 (Jones, 1996). American investments also enabled Mexico to become the world's third-largest oil producing country in 1911, following the United States and Russia (Wilkins, 1970).

In the case of market-oriented American FDI particularly significant in Canada, Europe and, at a later period, Japan, the high incomes of those countries, their growing demand for American products and the emergence of a competitive fringe of local firms producing similar products which posed a threat to the continued growth of the American firm through exports were the fundamental locational determinants favouring international production. The product cycle model considered these factors to be important in explaining the growth of international production of American manufacturing firms in Europe in the immediate period after the Second World War (Vernon, 1966), but evidence presented in this chapter and the next trace the relevance of the same set of factors further back to the emergence of American manufacturing MNCs prior to 1914.

Apart from the traditional location-specific variables related to the presence of abundant natural resources, low labour costs and large and rapidly growing foreign markets, actions by foreign governments in important export markets in the form of incentives to FDI or obstacles to exports or particular value-added activities by MNCs favoured the decision to either switch from exports to international production or to engage in higher value added activities in their foreign operations. These were of different kinds.

First, there were nationalistic policies of foreign governments involving the imposition of tariffs, the demand for local production of imports, or local processing of extracted raw materials that provoked American businesses to engage in international production in Canada, Europe and Japan. Foreign customs duties which made imports uncompetitive with locally produced goods prompted international production behind tariff walls. American manufacturers of sewing machines, cars, car accessories, rubber goods, harvesters, radiators and enamelware established factories to sell within the Dominion in order to avoid full customs duties. Similarly, differential customs duties encouraged foreign petroleum refining by American firms. Through international production, such companies were able not only to overcome tariff barriers but also to meet existing or, more often, preclude potential competition.

Furthermore, insistence by foreign governments in Europe for local purchases of equipment stimulated American firms in the electrical industry to produce in Europe, for example. The industrialization policy of Canada which encouraged local processing of Canadian natural resources favouring the expansion of American FDI in the Canadian paper and pulp industry was already mentioned. Similarly, nationalist sentiments and special product needs abroad prompted American firms to manufacture in a foreign country.

Indeed, the international production of American companies in the oil, tobacco, electrical equipment, car, and fertilizer industries was influenced by host government tariffs, preferences for national businesses, support of local cartels, etc. In these industries American firms were affected profoundly both by Canadian and European business and political conditions. The sensitivity of some American businesses abroad to local traditions and their respect for local jurisdictions and policies was manifested in part in the hiring of local personnel, the purchase of local supplies and the adoption of 'national' titles. This was true of many American manufacturing firms that invested in Canada (Wilkins, 1970).

Second, foreign government action in the form of rigid patent requirements particularly in Canada and Germany forced compliance by American firms in order to maintain their businesses. Patent laws require that local manufacture be initiated and rendered adequate to meet the demands of the local market within a set period of time (three years in Germany and two years in Canada) from the date of the patent grant. This prompted the American controlled Bell Telephone Company of Canada, for example, to begin to manufacture telephone equipment in Montreal in 1882.

Third, the introduction of unilateral concessions to goods made in the British empire affected the growth of American FDI in Canada. The Canadian expansion of the Ford Motor Company was planned on the basis of preferential trading arrangements within the empire.

Fourth, host governments in Mexico, Canada and the Caribbean offered incentives in the hope that their economies would benefit from the influx of American capital. To encourage economic growth, Mexico not only extended subsidies to American railroad companies, but also implemented favourable mining laws, offered tax incentives to new foreign enterprises, and cancelled import duties on materials for mills and smelters and provided police protection to mining projects and plantations. In addition, many of the export levies on silver and gold, imposts on certain business transactions, and property taxes could be reduced or waived through negotiated concessions. The Canadian government likewise extended subsidies, loans and bounties to American investors. The Chilean government also encouraged American capital and expertise in line with the country's objective to regain its position as the world's largest copper producer. Indeed, with the large American capital stakes eventually invested in Chilean copper, that country obtained more American capital than any other South American country before the First World War. The favourable tax incentives provided in the Peruvian Mining Law of 8 November 1890 also led to the influx of large American capital to exploit and develop Peruvian mines, particularly copper, vanadium and gold.[16] Many Caribbean governments also gave concessions to new American businesses, including low taxes, land, and freedom of workers from conscription (Wilkins, 1970).

The subsequent sections examine more closely the emergence of American MNC activity in each of the mining, manufacturing, petroleum and agriculture industries in the period until 1914.

American participation in foreign mining

As mentioned, the development of foreign railroads paved the way for the growth and development of new business ventures including mining, particularly in Mexico and South America. The railroad enabled travel to remote mining districts, lowered freight charges on the transport of ores from the mines, and made feasible the introduction to the mining regions of bulky and heavy modern machinery (Wilkins, 1970).

In the period before the end of the nineteenth century, mining and smelting was the second largest sector of American FDI after railroads accounting for some 21 per cent or $134 million of American FDI in 1897. Since that time and until 1914, the large capital requirements of mining prospecting and development made the mining and smelting sector surpass railroads and become the most important sector of American FDI, with a share of some 27 per cent of American FDI in 1908 and 1914. By 1914, American FDI in mining and smelting reached $720 million.[17]

As with American investments in foreign railroads, American businesses that ventured abroad in mining and smelting tended to concentrate in the neighbouring areas, particularly Mexico, Canada and South America. These areas accounted for some 95 per cent of all American FDI in this sector between 1897 and 1914. Mexico received the dominant share of American FDI in the mining and smelting sector in the period until 1914. The establishment of the Mexican railways system combined with the attractive business opportunities posed by the rich and undeveloped Mexican mineral resources particularly of silver, gold, lead, zinc and copper as well as coal led to the influx of hundreds of American mining companies in Mexico. These location advantages combined with the lower costs of minerals prospecting and development and various forms of host government incentives and the nationalization of the key railroad lines between 1903 and 1909 made the mining and smelting sector surpass the previously dominant share of railroads and become the most important sector of American FDI in Mexico by 1908. Thus, although in 1897 railroads accounted for the largest share (56 per cent) of American FDI in that country, by 1908 mining and smelting accounted for that share. The early American mining ventures in Mexico often took the form of acquisition of old abandoned mines from Mexican owners.

Canada was the second largest recipient of American FDI in mining and smelting in the period before 1914, at least until the South American mines were discovered. The richness of Canada's mineral resources encouraged many American entrepreneurs to acquire interests in mining copper, nickel, iron ore, asbestos, lead, zinc, coal, gypsum, manganese, antimony and phosphate and precious metals such as gold and silver. In a majority of cases the mining investments made by American entrepreneurs took the form of the establishment or acquisition of small companies, some of which were incorporated in the United States and some in Canada.[18]

The development of South American mines by American firms made that region an important recipient of American mining FDI by 1914, with a share of more than 30 per cent. American interests in South American mining and smelting before the end of the nineteenth century were small and many were short-lived.[19] Indeed, the region accounted for less than 5 per cent of American FDI in mining and smelting worldwide in 1897. Some of the mining stakes evolved from the activities of traders. The presence of rich mineral resources in the region combined with the development of the American capital market by the early 1900s made the mining and processing of various ores account for the largest share of American capital in South America in the period from the early 1900s to 1914. The vast copper resources of Chile and Peru made these countries the most important recipients of American FDI in mining and smelting in the region at the turn of the century. In these two countries American companies developed the richest copper properties and American capital came to dominate the copper mining industry.[20] It also enabled South America to become the second largest recipient of American FDI in mining and smelting by 1914, after Mexico. Indeed, the region became associated with mining for American firms. As mentioned, this sector accounted for $221 million or 68 per cent of American FDI in that region in 1914 – a considerable increase from the $6 million invested in the industry in 1897 (Table 2.1).

For the most part, the output of foreign mines served to supplement the domestic mineral resources of the United States. Thus, the activities of American mining MNCs involved primarily the extraction of mineral ores to be exported in an unprocessed state to the United States for smelting and refining.[21] The smelting of mining ores alongside the extraction of minerals occurred only in a major way in Mexico and Chile. The smelting operations in Mexico were a response to the passage in the United States of the McKinley tariff in 1890 which imposed new duties on the exports of lead ores into the United States as mentioned, while that in Chile was encouraged by the introduction of the flotation process in 1912 as a new technique for concentrating copper ores which facilitated mechanization and the simplification of tasks that could be matched with semi-skilled, low-cost labour (O'Brien, 1989).[22] The processing of Mexican and Chilean ores usually went only so far as smelting, then the product would be exported.[23] With the ascendance of mining and smelting as the most important American FDI in Mexico in the early 1900s, the United States increasingly looked at this country as the most important and convenient 'across-the-border' source of supply for minerals.

Investments by American manufacturing firms or their stockholders in overseas sales organizations and branch factories

The manufacturing sector accounted for between 15 and 18 per cent of the stock of American FDI worldwide between 1897 and 1914 (Table 2.1). It was the third-largest sector after railroads and mining in 1897, but since the early 1900s it became the second-largest sector after mining by size of investment. However, the largest *number* of foreign stakes of American-controlled international corporations in the period prior to 1914 was in the direction of sales and manufacturing. Indeed, the much larger capital investments required in multinational extractive and agricultural holdings meant that there were far fewer American-controlled international corporations in these activities compared with those that had sales or manufacturing stakes (Wilkins, 1970).

At the turn of the century, the new wave of American corporations that engaged in foreign expansion through export marketing and international production invested mainly in their most important foreign markets – Canada, Europe and, after 1899, Japan where per capita income was high and the ability to consume was great.[24] Some three-fifths of American FDI in sales worldwide and 96 per cent of American FDI in manufacturing worldwide in industries other than petroleum were directed to Canada and Europe in 1897. Largely owing to the consideration of the Canadian market as an extension of the American market, Canada was the single most important host country for international production by American firms – accounting for between 46 and 59 per cent of American FDI in the manufacturing sector worldwide between 1897 and 1914. Europe, however, was a far more significant host country than Canada for American FDI in sales. The influx of American manufactured goods in Europe between 1897 and 1902 raised concerns about an American invasion of Europe of which the works of McKenzie (1901), Stead (1902) and Thwaite (1902) provide an indication.[25] The geographical concentration of foreign sales and foreign manufacturing activities by American firms in the market-rich economies of Europe, Canada and Japan was different from that of American businesses that ventured abroad to build railroads or search for natural resources. As seen in the previous sections, the latter investments generally gravitated towards the resource-rich neighbouring countries of Mexico, Canada, South America and the Caribbean.

The analysis of American FDI in manufacturing which follows begins with the earliest investments in the pre-Civil War period, followed by those in the period since 1865 with the emergence of sustained American FDI in foreign branch factories.

Investments in overseas manufacturing in the pre-Civil War period

In the period between 1800 and 1860, there were investments by American manufacturing firms or their stockholders in overseas branch factories to supply a foreign market. As mentioned, the rapid development of American technology in certain metallurgical industries (machine tools, guns, reapers, and sewing machines) in the 1850s, linked with mass production in which the United States had already acquired world leadership, explain the pioneering role of American metallurgical firms in international production. The emergence of 'classic' or modern American MNCs – firms which had grown in the domestic market owing to their managerial and technological competences prior to embarking on international expansion – can be traced to the pre-Civil War period. The first foreign manufactory of an American company was that of Samuel Colt in London, England in 1852. This foreign manufactory of firearms is regarded to have been the first foreign branch plant of any American company (Wilkins, 1970). The motive behind its establishment was to protect the firm from unfair competition as a result of the introduction of counterfeit firearms in England where the firm had no patent. The second foreign manufactory of an American company was that of J. Ford and Company that built a factory for vulcanized rubber in Edinburgh, Scotland in 1856. The motive behind its establishment was the opportunity to earn higher profits from foreign expansion and to prevent the exploitation of a patent in Scotland where English patents were not protected.

In the product cycle model, the emergence of a competitive fringe of local firms in relatively advanced follower countries that are able to catch up through the imitation of innovation of the technologically leading American firms is one of the factors associated with the growth in demand for a 'maturing' product in these countries. The threat to the large-scale export business in the form of local competition becomes a powerful 'galvanizing force' for the initiation of import substituting international production by American firms (Vernon, 1966).

The establishment of these two foreign manufactories – although unprofitable and eventually sold to British interests – were exceptional American FDI in foreign branch factories before the Civil War. There appear to have been no others (Wilkins, 1970).[26]

The period since 1865

The level of American FDI before the Civil War was generally low because rapid means of transportation and communication both within the United States as well as between the United States and foreign countries had not been developed fully to enable domestic firms to explore new and wider markets. The period since the American Civil War brought sweeping changes in the development of domestic and foreign transportation and communication facilities.[27] As mentioned, the development of railroads in the United States enabled domestic companies to become national firms as distinct from local, state or regional firms. It was American companies with national sales plans and unique products that discovered the attractions of business abroad and were the first to be successful in undertaking such activities (Wilkins, 1970). This was associated closely with the beginning of the era of Pax Americana with the pioneering role of American firms in the second generation of industrial discoveries around the 1870s which encouraged the development of fabricating industries such as motor vehicles, office machinery, electrical goods, synthetic chemicals and others. This was unlike the technical and organizational advances of the First Industrial Revolution which led to the development of processing industries in which the United Kingdom had the technological hegemony (see Chapter 7). The Second Industrial Revolution exerted major repercussions on the organization of production as the development of fabricating industries required more hierarchical organizations to operate successfully and the modern industrial company grew both by horizontal and vertical integration (Chandler, 1980). As the process of firm growth encompassed integration across national boundaries, the implications of the later technical and organizational advances became truly transcontinental.

Plans for the further growth of the American industrial firm through international expansion sometimes coincided with, initiated shortly after, or long delayed after national plans. Inventors (such as Thomas Edison and Alexander Graham Bell) and manufacturing companies sought not only to cater to but also create foreign demand on the basis of their strengths in making new and innovative products, manufacturing methods, sales and advertising techniques. Unlike foreign investments to obtain sources of supply or raw materials, foreign marketing investments involved typically only small amounts of capital investment, if any; and could be initiated and maintained with virtually no outflow of capital from the United States. In fact, should additional capital to support export growth be required, as in the case of some American firms in the electrical equipment and film industries, such funds were raised typically abroad.

Generally, the foreign activity of American manufacturing firms in Europe or Canada tended to be either *forward integration into sales* or *horizontal integration in manufacturing*, but over time the former kind often led to the latter kind. The initial foreign investments established by American manufacturing (and also petroleum) companies in the period after the American Civil War were mainly marketing or sales oriented – made primarily to broaden markets, obtain better control over distribution, meet competition as well as to avoid national business cycles (Wilkins, 1970). This conforms with the product cycle model that suggests that the growing demand in other relatively advanced countries for American products are first served typically through exports on account of the lower costs of domestic production and transportation compared to the costs of establishing a production facility in the export market.

Although most international extensions by American firms were in marketing, some companies began to manufacture and to refine outside the United States during this period. As mentioned, the initial forays in foreign marketing activity often led to the later expansion of finishing, assembly manufacturing or refining abroad, as there were instances when merely to have a stake in a foreign sales establishment was insufficient to meet the objectives of the firm and the foreign market. The precise reasons for this varied across firms. Sometimes international production was made to save on costs associated with transportation, warehousing and domestic production, to obtain superior customer service and avoid damage in shipping. However, the decision to engage in international production was often spurred by the emergence of a competitive fringe of local firms producing similar products and the nationalistic policies of foreign governments in important export markets which made the costs of exporting more expensive. Similar reasons have explained the growth of American FDI in Europe in the 1960s in the product cycle model (Vernon, 1966).

Generally, each factory served the particular foreign country in which it was located, except in the case of resource-based manufacturing in Canada, Mexico and South America that was geared towards exports to the United States and sometimes also Europe. Other American firms in Canada had broader goals. Some started in the Dominion in order to cater to British empire markets and hence take advantage of expected preferential duties (as in the case of Ford Motor Company already mentioned), to use as a base for further foreign investment (the case of Sherwin-Williams whose Canadian affiliate owned the company's English subsidiary), or as an export platform (the Canadian investments of International Harvester was prompted by the desire to sell to France after the Canadian–French commercial treaty of 1907). There were also investments prompted by the need to escape American antitrust legislation (the case of Alcoa),[28] or to avail of tax incentives (in 1913 there were no corporate income taxes in Canada). These incidental uses of the Canadian subsidiary or affiliate rarely seemed to have been the fundamental motives, but nevertheless enhanced Canada's advantages as a location for international production (Wilkins, 1970). There were also investments to sell services – mainly utilities and insurance – to cater to the needs of foreign consumers.

Metallurgical companies continued to lead the way in international expansion in the period since the Civil War. Given that Singer was the first American-based corporation with investments in foreign branches, subsidiaries and/or affiliates in the period after the American Civil War, a brief account of the international business history of the company is of interest as a basis for comparison with other American MNCs in the manufacturing sector.

Singer's forays in international business developed gradually in a way that does not conform exactly to the product cycle model. The firm initiated its international business in 1855 by selling its French patent for the single-thread machine to a French merchant, Charles Callebaut, in exchange for royalties. Between 1860 and 1861, the company engaged in exports, mainly to independent franchised agents in Mexico, Canada, Cuba, Curaçao, Germany, Venezuela, Uruguay, Peru and Puerto Rico that sold and advertised Singer sewing machines. By this time Singer had also sent salaried representatives to Glasgow – the site of the company's initial British sales headquarters for Great Britain and Ireland – and London. The London branch office soon became Singer's British sales centre as well as the seat of extensive marketing programmes in England, Spain, Portugal, Italy, Belgium, and in the 1870s in France. In 1863, the company established branch offices in Hamburg and Sweden which enabled it to extend businesses throughout Germany, Scandinavia, Russia and Austria-Hungary. The firm built its first foreign factory in Glasgow in the spring of 1867 largely owing to the restoration of the American dollar to its normal specie value in the period after the Civil War and increasing labour costs in the United States, both of which rendered American exports uncompetitive. Furthermore, Singer expected major economies in domestic production costs, freight bills, storage and other incidental expenses with the establishment of its first foreign plant (Wilkins, 1970).

The first factory in Glasgow was initially an assembly operation, receiving parts from America in a partly finished state. At first the orders were for 100 machines every other week. By 1869 the company required a larger factory, with tools sent to Glasgow to manufacture all the parts required to produce 600 machines a week. In 1872, an additional factory was established in Glasgow, and six years later the firm acquired additional space. All this expansion was the result of a continued growth in foreign demand for Singer sewing machines. Indeed, by 1879 – 12 years after the first factory opened – the foreign sales of Singer's sewing machines produced in Glasgow outnumbered domestic sales (Carstensen, 1984). By the end of the 1870s Singer's worldwide sales surpassed those of the former industry leader, Wheeler & Wilson. Singer had become pre-eminent in the industry both in the United States and worldwide, particularly by the 1880s as its vast foreign sales network and foreign manufacturing plants had been established.[29]

Singer's commitment to expand abroad was not unique. In the 1870s and 1880s many American companies in the metallurgical industries sought export markets to dispose of surplus output and obtain economies of scale. This was true of companies producing screws, harvesters, cash registers, elevators, steam pumps, locomotives, locks and guns whose quality was superior to those produced and sold by European firms. Thus, many American metal products companies after having established national sales organizations created international sales networks in Canada and across the Atlantic which provided the most customers. Wilkins (1970) then describes an evolutionary pattern in the growth of the foreign business of most of these metal-working companies (although companies might skip a step or several steps in the process).

In the *first stage*, the American concern sold abroad through independent agents (through an export person or export or commission houses in New York City) or on occasion filled orders directly from abroad. However, companies frequently started to export using the facilities of international trading firms. In the *second stage*, the company appointed a salaried export manager, an existing export agency and its contacts, or independent agencies in foreign countries to represent the company. In the *third stage*, the company either installed one or more salaried representatives, or a sales branch, or a distribution subsidiary abroad, or it purchased a formerly independent agent located in a foreign country. At this point, for the *first* time, the company made a foreign investment. In the *fourth stage*, a finishing, assembly or manufacturing plant might be established to meet the needs of a foreign market. By the mid-1880s all these stages had emerged. In Canada, 47 verified American branch, subsidiary, or controlled affiliated manufacturers were established between

1876 and 1887, many of which were in the metal-working fields (Wilkins, 1970).

While the process of growth of the foreign business of most metal-working companies conforms to the pattern described by Wilkins (1970), such a pattern does not describe the growth process of the foreign business of all metal-working companies, let alone the American manufacturing companies in other industries that also became international firms in the period before 1914. Indeed, the period between 1890 and 1914 described the international expansion of American firms in such diverse manufacturing industries as bicycles, cars, food and drink, electrical equipment, tobacco, films and pharmaceuticals, among other industries. An analysis of the history of international growth of some of these American companies suggests the following trends.

First, the process of growth of international business varies across firms, and even within the same firm in different host countries. The case of the Ford Motor Company is an excellent case in point. Faced with high Canadian tariff barriers, the company – as with other American car companies or their affiliates that followed – started at once to establish its first foreign factory in Canada. However, the expansion of Ford in England conforms more to the model described by Wilkins above. There, the firm began with agents, before developing a sales branch, and an assembly plant in 1911–12, with a small amount of manufacturing. By 1914 the Model T had become the best seller in England. By then Ford also had a small assembly unit in France and, with a sales branch established in Argentina in that same year, it became the first American car company to establish a direct sales outlet in Latin America.

Second, the sequence described by Wilkins (1970) may be reversed by some firms. For example, the Pittsburgh Reduction Company – the predecessor to the Aluminum Company of America (Alcoa) – started its initial foray in international business by building a plant in France in 1891–92 to enhance its competitiveness in Europe, however unsuccessfully. The firm then initiated aggressive marketing activity in Europe in the period after 1896, and in 1899 established a plant at Shawinigan Falls, Quebec, motivated in part by the Canadian tariff. To circumvent the Sherman Antitrust Act, the company also formed a new subsidiary in Canada – the Northern Aluminum Company – in 1900 to act on behalf of the American parent, including the handling of all the company's export business and entering into accords with European aluminium producers to regulate foreign business. By 1912, Alcoa established a foreign subsidiary, Bauxite du Midi, a French bauxite company to procure raw materials, and by 1913 Alcoa was looking to South American sources of supply. The investment in foreign bauxite came after the company had been involved in international business for more than two decades. Increasingly, by 1914 certain American corporations were beginning to make *both* market-oriented investments (sales outposts, manufacturing plants and oil refineries) and supply-oriented investments (mines, oil wells and refineries, farms, packing plants, plus purchasing outposts, etc.) (Wilkins, 1970).

Third, the growth process of international business described by Wilkins (1970) does not take into account the foreign licensing agreements that many American firms engaged in as part of their international business activities particularly in the early stages. In fact, the process of international growth of I. M. Singer & Company or its successor, the Singer Manufacturing Company, described above, showed that the firm initiated its international business in 1855 by selling its French patent to a French merchant, Charles Callebaut in exchange for royalties. Similarly, C. H. McCormick, discovering that exports of American reapers were not competitive in Europe, entered into a licensing agreement with the British firm, Burgess & Key, in 1851 to manufacture and sell reapers. The Edison Electric Light Company Ltd, formed in England in 1881, also had as its fundamental objective to sell, install and license sub-companies. There are other numerous examples of American firms that engaged in international licensing at the early stages of their international business.

American participation in foreign petroleum

The petroleum industry accounted for some 13 to 14 per cent of American FDI in the period between 1897 and 1914 (Table 2.1).[30] Europe and Asia were the major recipients at the turn of the nineteenth century, but Mexico and South America also became important host countries in the 1900s. These four areas accounted for some 89 per cent of American FDI in the petroleum industry worldwide in 1914.

The position of the United States as the world's largest oil producing country through to the First World War (Jones, 1996) explains the high export propensity of the American petroleum industry.[31] However, despite the increasingly dominant proportion of domestically refined oil that was exported, most of the exports continued to be sold in domestic transactions to independent export merchants or representatives of foreign importers in the 1860s and 1870s (Wilkins, 1970). It was not until the early 1880s that a more advanced stage in the growth of foreign business of American petroleum companies became evident. This was manifested in the integration of Standard Oil – the best known of all American-based international companies – in petroleum – in export marketing and, in a limited number of cases, in foreign refining in response to the imposition by export markets of tariff barriers to refined oil or when there was keen local competition. A brief account of the expansion of the international business activities of this company is instructive. As will become evident, the evolutionary pattern in the growth of the foreign business of American petroleum companies conforms to that described by Wilkins (1970) for metal-working companies.

In 1881, Standard Oil of Ohio – the first Standard Oil Company formed in 1870 – acquired an interest in Meissner, Ackerman & Company of New York and Hamburg, a partnership of export merchants with a long history in the oil business. New York Standard, organized in 1882, acquired Ohio

Standard's earlier investment in Meissner, Ackerman & Company and through this company consigned the company's refined products to Henry Funck & Co. in the United Kingdom. Waters-Pierce, a Standard Oil affiliate, also established an extensive marketing network in Mexico in the mid-1880s.

In the mid-1880s with the increased competitiveness of Russian oil in Europe associated with the completion of the Baku-Tiflis line, Standard Oil responded initially by cutting prices, but when this proved inadequate, the firm strengthened its marketing organization in countries where the rivalry was strongest. Thus, foreign subsidiaries were established in the United Kingdom (Anglo-American Oil Company Ltd) and Germany in 1888 and 1890 to enable the firm to observe petroleum trade closely. In both countries, the affiliates marketed American refined oil and placed local business under the company's direction and control. In addition, since the British affiliate found that it could handle the East of Suez oil trade more efficiently than the American companies, it began to consign in its name Standard Oil products to traders in the key ports East of Suez.

The period 1893 to 1914 saw the Standard Oil companies handling the bulk of American oil exports. Unlike in the period until the mid-1880s, most of its foreign business was increasingly undertaken by its own companies abroad. The firm expanded its own foreign distribution networks through the establishment or acquisition of marketing firms that the firm owned in part or in whole. The result was effective control and management of its own distribution.

There were fewer instances of the establishment or acquisition of foreign refineries by Standard Oil in the period before 1914. The establishment of the first foreign refinery by Standard Oil of Ohio in Galicia was recorded in 1879, but this was closed down in 1886 owing to antagonism against Americans by local producers. Waters-Pierce also began to refine imported Pennsylvania crude oil in Mexico City and Vera Cruz because of the high import duties on refined oil. Similar reasons prompted Archbold and Conill to establish two refineries in Cuba. These refineries in Mexico and Cuba were built with marketing considerations of the host country in mind. The establishment of foreign refining facilities to overcome local competition occurred in 1898 when Standard Oil bought a controlling interest (75 per cent of the stock) of Imperial Oil Company, a Canadian competitor company that had the largest Canadian refinery and a national marketing network in the 1890s. In turn, the acquisition by Imperial Oil Company of most of the companies and the plants affiliated with Standard Oil in the Dominion and the facilities of small competing refineries assured Standard Oil's pre-eminent position in Canada.

In sum, although Standard Oil had been pushed by competitive pressures in the 1880s and 1890s to extend its foreign investments, this consisted mainly of marketing and, to a lesser extent, refining. The amount of capital investment remained relatively small as the company had not yet made investments in the highly capital intensive foreign oil production. By the turn of the century, Standard Oil was a large MNC, and beginning for the first time to consider purchasing foreign oil-producing properties as well as buying foreign oil to supplement American supplies of oil. The company had established or acquired control of 55 foreign enterprises with an approximate capitalization of $37 million by the end of 1907.[32] In most of these enterprises, Standard Oil held over 50 per cent of the stock, directly or indirectly. Although most foreign enterprises were for marketing, some were transportation facilities, some were refineries and two were fully integrated foreign producers. When in 1911 the United States Supreme Court ordered the dissolution of the Standard Oil monopoly into 34 separate companies, three of the nine companies that retained foreign facilities developed into the world's largest oil companies, namely: Standard Oil of New Jersey (later known as Jersey Standard, Esso and Exxon), Standard Oil of New York (later known as Mobil) and Standard Oil of California. Jersey Standard obtained the largest foreign assets including the oil fields and refineries in Rumania and Canada, refineries in Germany and Cuba and the marketing network in Canada, Latin America (except Mexico) and Western Europe (except Britain). New York Standard, on the other hand, took over the Far Eastern oil distribution companies. Practically all the nine companies that retained foreign facilities continued to expand abroad after 1911 (Wilkins, 1970).[33]

Other American oil companies initiated foreign oil exploration activities beginning from the 1890s. Edward L. Doheny, an independent oil man from California, incorporated the Mexican Petroleum Company in California in 1890, and pioneered the exploration and discovery of Mexican oil in 1891.[34] The success of Doheny's Mexican Petroleum Company and the Mexican Eagle Oil Company owned by Sir Weetname Pearson – the largest oil company in Mexico – led to the further expansion of American and British capital investment in Mexican oil. No doubt stimulated by Doheny's efforts, the Waters-Pierce Oil Company – one of the nine companies resulting from the dissolution of Standard Oil that retained foreign facilities – whose primary interests was in refining and marketing, acquired some oil lands in 1902. By 1908, with the ascendance of Mexico as an important site of oil extraction and processing, that country became the second largest recipient of American FDI in petroleum worldwide after Europe. By 1911 Mexican oil production reached 34,000 barrels per day, more than half of which was American owned. As mentioned, this enabled Mexico to assume third place in the world's oil industry in that year, following the United States and Russia.

Between 1911 and 1914, many more American oil companies including the Texas company, Gulf Oil, and the Magnolia Oil Company began to invest in the prolific oil fields of Mexico. Neither civil strife, disorder, forced loans nor taxes associated with the Mexican Revolution starting in 1913 affected adversely the activities of American oil firms in Mexico. Challenged by wells that were then the most prolific producers in the world's history, their production and exports mounted.

South America also became an important region of American FDI in petroleum, particularly after 1900. Among the most important American petroleum investment in the region was that of Standard Oil of New Jersey in Peruvian oil production with the objective to sell to the west coast of South America. Other American petroleum investments were made in the nascent Colombian and Venezuelan oil industries. By 1914, American petroleum investments in the region surpassed that in Asia and reached $42 million, or more than 12 per cent of American FDI in petroleum worldwide. Although the share remained small, the investments paved the way for larger petroleum investments by American firms in the region in the decades to follow.

American participation in foreign agriculture

Agriculture was the fifth most important sector of American FDI before the end of the nineteenth century, accounting for some 12 per cent of American FDI worldwide in 1897. The relative importance of the sector was maintained until 1914. As with American FDI in railroads and mining and smelting, the bulk of American FDI in agriculture in the period until 1914 was directed to nearby regions, namely, Cuba and other West Indies, Canada and Mexico. These areas accounted for between 79 per cent and 84 per cent of American FDI in agriculture worldwide between 1897 and 1914. By 1914 Central America also became a significant recipient of American FDI in agriculture accounting for a share of more than 10 per cent, making the region as important to American agricultural FDI as Mexico. Competition in local agricultural businesses in these nearby regions was often among several American firms or among American and European firms rather than with local firms.

Agriculture was the most important sector of American capital in the Caribbean even before 1898. With an average share of some 50 per cent of American FDI in agriculture worldwide between 1897 and 1914, the region received the largest share of American capital investments in agriculture because of the richness of its natural resources, its proximity to the United States and the provision of American tariff preferences particularly with respect to Cuban sugar. Indeed, in the less developed Caribbean regions, Americans went mainly in search of tropical agricultural produce. Sugar and tobacco were the most important recipients of American FDI in the region and particularly in Cuba at the conclusion of the Spanish–American War. Tropical fruit was also another important area of American FDI in agriculture in the region, with the United Fruit Company being a major investor. Much of the early American FDI in the sector was made by American commodity traders that integrated into agriculture, particularly bananas and sugar. Only the rare agricultural stakes in the region were made by American industrial companies integrating backward into agriculture production (Wilkins, 1970).

The business history of the United Fruit Company provides an excellent case study of the international expansion of an American firm through backward and forward integration. While its predecessor company – the Boston Fruit Company – acquired the bulk of its fruits through purchases from independent farmers in Jamaica which were then shipped and sold in Boston, the United Fruit Company as it evolved in 1887 engaged in backward vertical integration by acquiring four banana plantations spanning 1,300 acres in Jamaica, and 40,000 acres in Santo Domingo in order to secure a reliable source of fruit. With the formation of the United Fruit Company in 1899 the business owned or leased over 320,000 acres of land in the Caribbean, including Jamaica, Cuba, Costa Rica, Colombia and Nicaragua – properties previously acquired by the Boston Fruit Company and Minor C. Keith, First Vice-President of the company. By 1913, the company owned or leased 852,560 acres, of which 221,837 acres were under cultivation. Although the land was used primarily for the cultivation of bananas, it was also used for the cultivation of orange groves, coconut trees, rubber trees, sugar, cacao as well as the raising of cattle. The dispersion of its properties throughout the Caribbean enabled the firm to spread the risks associated with natural calamities, banana disasters as well as revolution, riot and political unrest. To support these investments, the company invested heavily in infrastructure by installing drainage and water systems and radio communications, and establishing company towns and hospitals in former jungle areas. The company also integrated further into transportation not only by steamship but also through ownership of railroads as previously mentioned or contracts with existing railroads to facilitate the efficient transport of its perishable products (Wilkins, 1970). The fruits cultivated by the firm were shipped primarily to the United States, but the firm also began to ship fruits to Europe facilitated by its acquisition in 1904 of a large stake in Elders and Fyffes – the leading firm in the European banana trade – and its acquisition of full control in 1913 (Davies, 1990).

The integration of the company into fruits cultivation and transportation, with the former enhanced by the introduction of mass production cultivation techniques, enabled the company to become increasingly independent in the supply of its own fruits rather than continue to become dependent on relational contracting with independent companies engaged in fruit production or transportation. Using internalization theory as applied in the case of international production, unified governance provided the most efficient means with which transactions could be organized, involving lower costs compared with those of a market exchange.[35] As firms increased in size and scope, the banana industry became organized increasingly as an oligopoly with stiff barriers to entry (Read, 1983, 1986).

Investments in foreign agriculture assumed further importance for American firms and the American economy around the early 1900s. This was associated with the worldwide surge in demand for rubber from the early 1900s, and the encouragement provided by President William

McKinley to American firms to cultivate rubber overseas to meet the domestic requirements of the United States. Although this led to the emergence of American FDI in rubber, particularly in Mexico and Brazil, the investments then were regarded as speculative (Wilkins, 1974).

Canada was the second most important host region, accounting for 23 per cent of American FDI in agriculture in 1897, and although that share declined to 13 per cent in 1908 owing to the surge of American agricultural investments in Cuba and the West Indies, Canada's share of American agricultural FDI grew to more than 28 per cent in 1914 (Table 2.1). Unlike in the Caribbean, where American agricultural investments were directed to sugar, tobacco and a variety of tropical fruits (primarily bananas), and in Mexico where American agricultural investments was directed to rubber, chicle, cattle and some tropical fruits, American capital investments in the primary sector of Canada were mainly in timber and timberlands.

Conclusion

The period until 1914 was associated with the emergence of American MNCs. With the exception of Standard Oil of New Jersey, Singer Manufacturing Company, International Harvester, New York Life and perhaps a handful of other large companies, many American firms had obtained a foreign stake in only one foreign country before 1914, and international business did not make a substantial contribution to the profits of American enterprises (Wilkins, 1970).

Before the end of the nineteenth century in 1897, the major host countries for American FDI were Mexico (with a share of 31 per cent of the stock of American FDI worldwide), Canada (25 per cent), Europe (21 per cent), the Caribbean (11 per cent) and South America (6 per cent). These countries continued in the same order to be the five largest recipients of American FDI by 1908. The primacy of Mexico as a host country until 1908 was associated with the importance of railroads as one of the pioneering sectors of American FDI in 1897 followed by, and associated with, mining and smelting in 1908. In both of these most important sectors of early American FDI, Mexico was the most important host country. By 1914, American stakes in Mexican oil production was also growing rapidly. The other nearby developing regions in the western hemisphere – the Caribbean islands, Central America and South America – were also important recipients of supply-oriented investments by American MNCs in the period until 1914. While American FDI in the Caribbean and Central America was concentrated in agriculture (sugar, fruits, rubber, tobacco, etc.), that in South America was concentrated in mining.

By contrast, the substantial American FDI in Europe until 1914 was concentrated in refining and distribution of petroleum and in the manufacturing sector. Petroleum was the leading American industry in Europe in 1897, accounting for some 42 per cent of the stock of American FDI in the region, followed far behind by manufacturing with a share of 27 per cent. By 1908, the relative importance of both these sectors was about equal at 27 per cent, but by 1914, manufacturing surpassed petroleum in relative importance with shares of 35 per cent and 24 per cent of the stock of American FDI in the region, respectively. The high per capita income of the European consumers, their growing demand for American products, the imposition of trade barriers, and the emergence of a competitive fringe of local firms favoured the expansion of American MNCs in European manufacturing. This provides evidence of the relevance of the product cycle model – formulated primarily to explain the growth of American MNCs in Europe in the immediate period after the Second World War – to explain the emergence of American MNCs prior to 1914.

Changes in the relative importance of host countries became evident in 1914, when Canada surpassed Mexico as the largest host country of American FDI, but the relative position of the other host countries remained unchanged. The geographical proximity of the Dominion combined with its high per capita income and imposition of trade barriers led American companies to expand and secure markets through the establishment of a large number of American manufactories in Canada. In addition, the proximity and the presence of rich natural resources there meant that the country also became a convenient and important source of mineral and agricultural supplies (particularly timber) for American firms. These reasons, combined with the difficulties encountered by American businesses in the civil war in Mexico between 1911 and 1914, led to the curtailment of many American businesses in that country, particularly in mining and agriculture. Only petroleum investments in Mexico continued to flourish.

American FDI in more distant areas – Asia, Oceania and Africa – in the period until 1914 was more limited, owing largely to the large risks and costs associated with investments in areas of farther geographical distance, the inaccessibility to American firms of the European colonies in Asia and Africa, low incomes (Asia other than Japan, Oceania other than Australia, Africa), and small markets (Oceania) (Wilkins, 1970).

The principal industries of American FDI in 1897 were railroads (23 per cent of the stock of American FDI worldwide), mining (21 per cent), manufacturing (15 per cent), petroleum exploration, production, refining and distribution (13 per cent) and agriculture (12 per cent). By 1908, railroads descended to become the fifth most significant industry of American FDI, and the other industries moved up one notch in importance in the same order. Thus, mining ascended to become the most important activity of American MNCs abroad with a share of more than 27 per cent, followed far behind by manufacturing (18 per cent), petroleum (14 per cent) and agriculture (11 per cent). The same order of importance was maintained until 1914, except that the significant expansion of American FDI in agriculture at the time led to that sector surpassing petroleum in importance and becoming the third largest industry of American FDI abroad in 1914. In addition, the continuing position of the United States as a significant oil exporter made American FDI in petroleum until 1914 remain primarily in distribution and, to a lesser extent, some refining in consumer countries, although some oil companies had begun to make some investments in foreign oil extraction from the end of the nineteenth century.

The ascendance of mining as the most important sector of American FDI at the beginning of the twentieth century until 1914 is associated with the large capital requirements involved in the initiation of mining projects. This is not a feature that is associated with the emergence of American FDI in the manufacturing sector where investments have grown gradually, in many cases on the basis of reinvested profits. Nevertheless, despite the smaller size of American FDI in the manufacturing sector compared to that in mining in the period until 1914, the foundations had been laid for the expansion of American FDI in the manufacturing sector in subsequent periods.

Notes

1 Based on data obtained from UNCTAD (1999).
2 Such was the case of the New York merchant John Van Cortlandt, who, when planning to send some 3,000 bushels of wheat from Virginia to Madeira, notified Newton & Gordon, a British concern in Madeira, to sell the wheat and ship the proceeds in good Madeira wines (Wilkins, 1970).
3 For example, the American-born Francis Wilks served the Boston merchant Thomas Hancock in England in the 1730s, followed by Christopher Kilby in 1740. These merchants also represented the Massachusetts Colony in London, combining business and politics (Wilkins, 1970).
4 Sons, cousins, nephews and brothers of the head of the firm went to Jamaica, Curaçao, Antigua, St Eustatius, St Croix, and other islands in the Caribbean to serve the New York houses of Lloyd, Ludlow, Cruger, Livingston, Van Ranst, Cuyler, Beekman and Gouverneur (Wilkins, 1970).
5 For example, defaults on loans extended by American merchants to local (usually Spanish) sugar dealers and sugar estate owners after the end of the Cuban war in 1868–78 led to the emergence of American stakes in Cuban sugar plantations. A case in point is the Atkins family involved in Cuban sugar trade since 1838. By 1882 E. Atkins & Company obtained land holdings owing to a default on a debt to the company by a sugar estate owner (Wilkins, 1970).
6 The flotation of loans was made in association with Baring Brothers, J. P. Morgan and Morton Rose & Company (Wilkins, 1970).
7 The ownership of railroads wherever the United Fruit Company planted or purchased bananas was made to overcome the risks of market failure in the transportation of the rapidly perishable fruit to their final markets. Thus, whereas in its early years the United Fruit Company owned only 71 miles of railroad in the Caribbean, the company had a total of 669 miles of railroad in Costa Rica, Guatemala, Honduras, Panama, Colombia, Cuba and Jamaica by 1914. Including the 163 miles of track of the Costa Rica Railway Company operated by Northern Railway, the United Fruit Company controlled some 833 miles of railroads in tropical America. The Cerro de Pasco Mining Company also had to build 83 miles of railroad to connect with an existing line in the process of mining the ore bodies in the Peruvian Andes which was rich in copper, silver, lead, gold and other minerals (Wilkins, 1970). The Guggenheims similarly added railroads

to their Mexican and Chilean holdings to support their extraction and smelting operations in those countries (O'Brien, 1989).

8 American capital was the predominant source of capital in the construction of railroads in Mexico. The other main sources of finance in the construction of foreign railroads was Europe (United Kingdom, Spain and Germany) in the case of the Caribbean countries and South America, and the United Kingdom in the case of Canada (Wilkins, 1970).

9 Other examples of American capital investments in railroads were those of the Sonora Railway Company (formed by a group of Boston capitalists) in Mexico to break the monopoly of Collis P. Huntington by providing an alternative route to the American West through the extension of the Atchison, Topeka and Santa Fé to the Pacific Coast. In 1881 Huntington and Southern Pacific interests obtained a concession. Over time, Jay Gould, Russell Sage and E. H. Harriman among others made investments in Mexican railways. Similarly, the American railroad entrepreneurs (Erastus Corning, John Murray Forbes and John W. Brooks) invested in the Great Western Railway in Canada in 1849–50 with the objective of establishing a connection between New York Central and the Michigan Central Railroads. The same can be said of the investments by the Northern Pacific Railroad which sought to obtain a Canadian charter to extend its railroads into the Canadian prairies as well as to construct a railway in eastern Canada to connect with the Vermont Central, providing the company with a route to Boston via Montreal (Wilkins, 1970).

10 The largest American capital exports were made in the years between 1898 and 1901 but the United States continued to be a net capital exporter through 1905 even though it remained a debtor nation on international accounts (Wilkins, 1970).

11 For example, technological innovations by American firms in the copper industry since 1879 eventually made possible for the first time the profitable mining and smelting of low-grade ores in Chile. This, combined with the high price of copper in 1904 and the risk-taking nature of American entrepreneurs in their roles as industrialists and financial capitalists, stimulated the interest of American companies in copper mining in Chile (Wilkins, 1970; O'Brien, 1989).

12 Some of these firms were the American Smelting and Refining Company, International Paper Company, International Nickel Company, United Fruit Company, and the General Asphalt Company (Wilkins, 1970).

13 In 1886, the Dominion government raised the export duty on timber to encourage domestic sawmill construction. In 1897, the Ontario government insisted that all timber logged on Crown lands in that province be manufactured in Canada; this prompted the building of American mills in Ontario. Next came the prohibition by the Ontario government on the export of pulpwoods from the Crown lands in 1900, followed by the Dominion Parliament which forbade pulp wood export from Dominion Crown lands in the Prairie provinces in 1907. Other provinces followed suit: Quebec (in 1910), New Brunswick (in 1911) and British Columbia (in 1913) (Wilkins, 1970).

14 The Platt Amendment to the Army Appropriation Bill of 1901, the Cuban constitution drafted in late 1901, and the treaty between Cuba and the United States in 1903 allowed for the intervention of the United States in the preservation of Cuban independence and in the maintenance of a government adequate for the protection of life, property and individual liberty (Wilkins, 1970).

15 A ton of ore could be mined in Mexico for 40 per cent of the cost of mining similar ore in the United States, assuming that the American miner did twice as much manual work as a Mexican miner (Wilkins, 1970).

16 The Law specified that for 25 years (or until 1915) there would be no new taxes or tax increases in the mining industry (Wilkins, 1970).

17 Since the development of mines required large amounts of capital, typically a few Americans would form a syndicate and raise money from a number of investors or seek funds from Boston or New York financial houses (Wilkins, 1970).

18 Among the most important American investors in Canadian mining between 1870 and 1914 was the Orford Nickel and Copper Company organized in March 1878 after several Americans acquired a mining property in Orford, Quebec in 1877. This small company developed the *first* Canadian nickel deposits to be used commercially. In 1881, Samuel J. Ritchie established the Canadian Copper Company as well as the Anglo-American Iron Company in Ohio to handle the Canadian business. With the recognition that its mines possessed both nickel as copper, the Canadian Copper Company became eventually the world's largest nickel producer. In the merger movement occurring in the United States at the turn of the nineteenth century, the $24 million International Nickel Company was formed in New Jersey in 1902 from the merger of the Canadian Copper Company, the Orford Nickel and Copper Company and some other companies. By 1910, American capital also came to dominate the asbestos industry in Canada when the Johns-Manville Company of New York acquired control of a leading Dominion asbestos company. Other significant participants in Canadian mining were the Guggenheims whose interests were in precious metals (Wilkins, 1970).

19 Evidence for this is found, for example, in Alsop & Co. that lost substantial holdings in Bolivian mining when Chileans occupied the Bolivian silver mines at Caracoles in 1879. A handful of Americans began to mine gold in Ecuador in the 1880s, but these were short-lived activities (Wilkins, 1970).

20 The first large-scale American capital investment in mining in South America was that of William C. Braden who acquired the Rancagua low-grade copper mine in north central Chile in 1904 for $100,000. Braden acquired the property on the basis of $625,000 raised on the preference shares of his company, which was also used to build a 35-mile road, establish a community and construct a concentrating plant and a hydroelectric plant. The Chilean investment of W. C. Braden was especially important because of the large size of its initial investment for the period, its long history and the fact that it was the first major entry of American capital into the Chilean copper mining industry – an industry in which Americans played eventually a pre-eminent role. While Braden succeeded initially in developing the mine on his own, he eventually turned to the Guggenheims for financial support. As a result, the Braden Copper Company was established by the Guggenheims in 1908 with a capitalization of $23 million. With the introduction of new techniques, El Teniente – the principal Braden mine – was the first copper company in the world to utilize the flotation process in concentrating low-grade copper ores. The success of this process enticed the Guggenheims to seek more copper properties in Chile. This resulted in the organization by the Guggenheims of the Chile Exploration Company (Chilex) in 1912 with $95 million in equity, of which $25 million was used to acquire a large porphyry copper ore deposit at Chuquicamata in northern Chile controlled by British interests. A further infusion of an additional $12 million in development funds raised through a bond issue floated in New York was necessary for the construction of a railroad, power plant and port. By 1914 development was well under way at both El Teniente and Chuquicamata – two of the largest Chilean copper properties of the Guggenheims – with the latter developed into the largest open pit mine in the world and which introduced a new copper concentration process utilizing sulfuric acid and electrolytic precipitation. In that year, American investment in Chilean copper stood at $169 million (practically all of it in these two Guggenheim companies) (Wilkins, 1970; O'Brien, 1989).

21 It was perhaps only the Guggenheims that looked to copper markets in both the United States and Europe. No American extractive enterprise envisaged the host country or region as its primary market (Wilkins, 1970).

22 For example, the Guggenheim brothers obtained a Mexican government concession in October 1890 to build smelters, the first one of which was for lead solely and a second one for copper and lead. Similarly, Robert S. Towne, of the Consolidated Kansas City Smelting and Refining Company, built a smelter in the 1890s. For a time the largest American company involved in Mexican mining and smelting was the American Smelting and Refining Company (ASARCO), a trust formed in 1899 by all the major lead and silver smelters in the United States with a capitalization of $65 million. The firm acquired Towne's Consolidated Kansas City Smelting and Refining Company and in April 1901 the Mexican smelters of the Guggenheims. In turn, the acquisition of control of ASARCO by the Guggenheim brothers in that year enabled them to become the largest American investors in Mexico with 64 or so mining properties owned or operated by them (Wilkins, 1970; O'Brien, 1989).

23 The investment in Mexico's first steel plant by the New Yorker Eugene Kelly was exceptional. In a rare joint venture with the Mexicans, he organized in May 1900 the Compañia Fundidora de Fierro y Acero de Monterey with a capital of $10 million (Wilkins, 1970).

24 The final abandonment of unequal treaties between Japan and the United States gave Japan the opportunity to introduce tariff protection and forced American businessmen to either invest in Japanese manufacturing or lose the Japanese market (Wilkins, 1970).

25 These publications predate that of J. J. Servan-Schreiber (1967), *The American Challenge*, London: Hamish Hamilton.

26 The first branch manufactory located across the border from the United States in Canada was not started until 1870 (Wilkins, 1970).

27 In 1866 the first transatlantic cable was completed and other cables followed in the period between 1865 and 1892, making possible speedy communication in different parts of the world. By the 1880s, passenger-carrying steamships made the trip from United States to Europe in five to six days, compared to the steamships in the 1850s which took nine to ten days (Wilkins, 1970).

28 A separate entity organized in a foreign country was useful for discussions with European competitors to control production and prices – discussions from which the American company was barred (Wilkins, 1970).

29 By 1881, as Singer's three Glasgow factories became congested and inadequate the company decided to erect in Kilbowie, near Glasgow, a modern plant equipped with the latest American machine tools and with a capacity equal to the company's largest American factory. The company soon realized that it could meet the demands of Europe and many other markets more cheaply by producing in Scotland than in the United States. In 1883, Canadian customs duties led to the establishment of a small manufacturing plant in Montreal to supply the Dominion market. A similar reason compelled the manufacture of stands in Austria in the same year. The Russian factory was established in 1901, and employed over 3,000 workers by 1914. The operations in Canada and Austria were small compared to those in Kilbowie and in Russia. However, Singer's sales in Russia alone accounted for over 30 per cent of its total worldwide sales in 1913 (Carstensen, 1984).

30 This includes the exploration, production, refining and distribution of petroleum. In the period until 1914, the bulk of this was in distribution (see Table 2.1).

31 In the years from 1862 to 1865, between 28 and 59 per cent of domestically refined oil was exported. This increased to 69 per cent in 1866, and never dropped below

64 per cent in the years to 1885. In fact, it soared as high as 77 per cent in 1871 (Wilkins, 1970).

32 The expansion of the company in the Far East proved problematic with the obstacles posed by Royal Dutch Shell (which prevented Standard Oil from buying into its shares in 1898), the Dutch government (which prevented concessions in the Dutch East Indies in 1899), the British (which prevented exploration in Burma in 1902) and large expenses (investments in Japan) (Wilkins, 1970).

33 This was especially true of Jersey Standard and New York Standard. The former enlarged its foreign marketing, refining and producing facilities because it feared antitrust action with sustained domestic expansion. Its foreign deliveries exceeded its domestic deliveries of refined products. A majority of its foreign business consisted of exports of American refined products sold through wholly owned or affiliated sales companies or branches, but between 1912 and 1914 the output of its foreign refineries rose rapidly. Through its Dutch marketing affiliate, Jersey Standard was able to make its first investment in oil concessions in the Dutch East Indies in 1912. In 1913 the firm engaged in oil production in Peru. New York Standard expanded its sales network throughout the Orient and sought to acquire oil-producing properties in Palestine, Syria and Asian Minor. At the time of the dissolution of Standard Oil in 1911, the separate companies were not competitive with one another; each under the old structure had its specific region, function or products which it retained. Standard of California, which would become one of the major international oil companies, had no foreign business in 1911 (Wilkins, 1970).

34 In 1904 the company drilled a well that flowed for nine years and which yielded about 3.5 million barrels of heavy fuel oil. This was in the so-called 'Golden Lane' – a strip of land about one mile long and 25 miles wide on the Gulf of Mexico – which became the principal source of Mexican oil. The company's major bonanza, however, was discovered in September 1910 when the well, Juan Casiano no. 7, yielded 80 million barrels (Wilkins, 1970).

35 Independent of United Fruit Company (of which he was First Vice President), Minor C. Keith – a key American entrepreneur in railroads – raised revenues for the Costa Rican railways by cultivating bananas along its route in the late 1870s. In the 1880s and 1890s the entrepreneur, financed by British capital, acquired additional lands to cultivate bananas, obtaining 10,000 acres of jungle land near Bluefield, Nicaragua, 10,000 acres on the Caribbean side of Panama, 15,000 acres near Santa Marta, Colombia and some holdings in Honduras. Indeed, Keith – more than any other American entrepreneur – was responsible for the expansion of American farms throughout Central America. Before the end of the nineteenth century, Keith was also involved in cattle raising, mining and railroads (Wilkins, 1970).

3 The evolution of multinational corporations from the United States

The period since 1914

Introduction

In the period until 1914 – the period of *emergence* of American MNCs – the principal industries of American FDI were mining and processing of mineral ores, manufacturing, oil production and refining, agriculture and processing of agricultural produce, and railroads. The importance of these American business activities abroad – which flourished under conditions of war and the post-First World War recession – remained broadly unchanged until 1929. Hence, the period since 1914 is associated with the *growth and evolution* of American MNCs (Wilkins, 1974).

Despite the stability in the industrial structure of American FDI in the period until 1929, there were significant shifts in the relative importance of the major host countries and in the position of the United States from a debtor nation in 1914 (receiving more portfolio and FDI from abroad compared to American portfolio and FDI in foreign countries) to a creditor nation in 1919 on account largely of the growth of American FDI abroad. Such growth enabled American MNCs to pose increasingly a challenge to European hegemony, although European investors still controlled essential raw materials such as rubber, tin and nitrates at the end of the First World War. The American challenge was particularly felt in Canada and Latin America where the gap was narrowing between American and European interests. With the breakdown of European monopolies by 1929, the American business challenge to European investors was profound in the western hemisphere where American FDI ranked supreme and in the United Kingdom.

The analysis of the evolving industrial patterns of American MNCs shows the *dominant* role of the pioneering investments abroad by American merchants in the late seventeenth and especially the eighteenth century, railroads and mining in the late nineteenth century (which accounted for 23 per cent and 21 per cent of the stock of American FDI abroad, respectively), mining in 1908 (which accounted for 27 per cent), manufacturing in 1929 (which accounted for 24 per cent) and services in 1989 (which accounted for 47 per cent).

This chapter on the evolution of American MNCs analyses these dynamic changes in the industrial patterns and their impact on the geographical destination of American FDI in the periods from 1914 to 1929, from 1929 to the Second World War, and since the Second World War.

The expansion of American MNCs from 1914 to 1929

The industrial pattern of American FDI

As mentioned, there had been relative stability in the principal industries of American FDI from around 1908 to around 1929. Mining and processing of mineral ores, manufacturing, oil production and refining, and agriculture and processing of agricultural produce continued to be the predominant activities abroad of American MNCs. However, there were important changes in the pattern of American MNC activities within these industries between 1914 and 1929. The following sections examine some of those important changes.

The mining and agriculture industries

The previous chapter showed that early American FDI in agriculture prior to 1914 was determined largely by profitable investment opportunities abroad. The presence of abundant and relatively cheap natural resources in foreign countries in the western hemisphere enticed the emergence of American FDI in this industry. Evidence for this lies in the fact that much of the early American FDI in the industry was made by American commodity traders such as David Beekman and the Boston Fruit Company that integrated backwardly from trading into agriculture production, particularly of bananas and sugar. As mentioned, only the rare agricultural stakes in the region prior to 1914 were by American industrial companies integrating backward into agriculture production (Wilkins, 1970).

Investments in foreign agriculture by American MNCs after 1914 was determined by the need to overcome anticipated potential shortages of raw materials in the United States or to gain access to raw materials that were absent completely in the United States (such as rubber, nickel and nitrates), in addition to the exploitation of profitable investment opportunities in host countries. The size of American FDI in agriculture as well as mining grew both through the expansion of existing industries and the emergence of new ones.

The demands of the First World War provided greater emphasis to the importance of certain agricultural and mining investments to procure nitrates, copper, iron ore, aluminium, nickel, sugar and meat.[1] The direct involvement of the United States in the war between 1917 and 1918 increased the country's requirements for all these commodities as well as rubber, tin, tungsten and petroleum and led to the further expansion of American FDI in these industries. The need to overcome the total reliance upon London for the purchase of crude rubber and tin forced some American companies to make direct purchases or to extract/cultivate these commodities in foreign countries to fulfil the needs of American domestic enterprises. American firms purchased crude tin directly from Bolivia for smelting and refining in the United States.[2] On the other hand, the three major American rubber tyre companies – U.S. Rubber Company, Goodyear and Firestone – engaged in backward vertical integration into rubber production or acquired rubber abroad directly to protect their rubber supplies. Their foreign investments were directed strategically to the countries of the Far East that were important sources of supply to the British and Dutch.[3] The growth of American FDI in rubber in the inter-war period was, therefore, different in terms of its determinants and host countries from that in the pre-1914 period when American enterprises speculated in rubber in Mexico and Brazil. The civil strife in Mexico between 1911 and 1914 and the adverse investment climate in that country after 1914 combined with the unsatisfactory experience of the U.S. Rubber Company with respect to their investments in Brazil discouraged the further expansion of American FDI in rubber in these countries.

Other American firms enlarged their foreign stakes seeking fruit (primarily bananas), tobacco, sugar, timber, hemp and other agricultural and forestry products in their quest to become more independent of foreign producers as well as to make greater profits. At the end of the First World War and through much of the 1920s, American investors engaged in agriculture and mining abroad extended their wartime investments as commodity prices remained high.

Apart from the increasing complexity in the determinants and the greater number of host countries of American FDI in agriculture and mining from 1914 to 1929, the investments were also different in another respect. Unlike in the early history of the emergence of American FDI in the primary sector when the investments represented the vertical integration by American commodity traders and the horizontal integration abroad of American mining and agricultural companies, much of the expansion of American FDI in the sector since 1914 was in the form of vertical integration of *industrial* companies in agriculture production or mineral extraction.

While Cuba and other West Indies, Central America and Canada were the major host countries for American stakes in foreign agriculture until 1914, the same set of host regions with the exception of Canada continued to be important for American FDI in agriculture until 1929. The importance of the Caribbean region as a location became more emphasized with a share of almost 80 per cent of American FDI worldwide in this industry in 1929, particularly in light of the expansion of investments by fruit growers and buyers throughout Central America, the Caribbean Islands and Colombia. The largest among these continued to be the United Fruit Company whose major expansion from the Caribbean to the Pacific coast of Central America in the 1920s enabled the company to be much larger in comparison to its host countries (Wilkins, 1974). The declined importance of Canada as a host country for American FDI in agriculture has stemmed largely from the forward integration of American firms from the extraction of timber to the processing of paper and pulp in that country, as previously mentioned. The relative importance of Mexico as a host country for American FDI in this sector also declined with the nationalization of American landholdings in the 1920s. On the other hand, Asia and the Far East and, to a lesser extent, Africa (Liberia particularly) grew in importance with the expansion of American tyre companies in the extraction of rubber in the region.

American FDI in mining ore processing continued to be directed to South America, Mexico and Canada from 1914 to 1929, although the relative importance of Mexico – the site of large-scale American investments in mining in the period until 1914 – declined since owing to the unstable political situation in that country. Thus, only 14 of the 110 American mining companies with properties in Mexico were able to continue in operation from 1914 to 1919 (Wilkins, 1974). Meanwhile, American mining interests in Chile grew dramatically, accounting for 87 per cent of Chile's copper output by 1918 (O'Brien, 1989). Indeed, Chile became the single most important host country of American FDI in mining in the 1920s. The foreign mining activities of American firms by the 1920s included asbestos, bauxite, chrome, coal, copper, diamonds, iron ore, lead, manganese, nickel, nitrates, platinum, potash, tin, tungsten, vanadium, and zinc, as well as gold and silver. Investors such as the Guggenheims (directly and through their control of the American Smelting and Refining Company and Kennecott), American Metal Company, Anaconda, and Newmont Mining were important MNCs in the 1920s, all of which with the exception of Kennecott were increasingly becoming more multinational in their operations through their mining or exploration of ores in three or more foreign countries. During the 1920s, the growth of American mining in Latin America was greatest in nitrates and copper in Chile, copper, lead and zinc in Peru, and tin in Bolivia. In addition, American businesses invested in iron ore (Chile and Brazil), manganese (Chile and Brazil), vanadium (Peru), tungsten (Bolivia), gold (Colombia and Ecuador), platinum (Colombia) and bauxite (Dutch Guiana and British Guiana). By 1929, the bulk of the productive mineral resources of South America was owned by American interests (Bain and Read, 1934).

The manufacturing sector

The previous chapter showed that the manufacturing sector has always been an important sector of American FDI, accounting for some 15 per cent of the stock of American FDI abroad in 1897, and some 18 per cent in both 1908 and 1914 (see Table 2.1). Thus, while representing the third-largest sector of American FDI in 1897, the sector became the second single largest sector of

American FDI from 1908 to 1928. In 1929, the manufacturing sector became the most important sector of American FDI.

The rapid importance of market-oriented FDI by American MNCs since 1914 was associated with relatively new industries. Hence, the previous importance of metallurgical firms in American FDI in manufacturing gave way in the 1920s to the expansion of industries and firms with trademarked or branded merchandise advertised widely in the United States (such as food and drink, textiles and clothing), as well as firms with distinctive products. American industries with worldwide technological leadership gained abroad from the transfer of techniques in product design and engineering and organization of production (electrical industry, motor vehicle industry, certain metal products, petroleum), as have companies using advanced marketing methods (motor vehicle industry, metal products, petroleum). As in the period before 1914, large companies – typically exporting companies in such industries as electrical equipment, telephone and telegraph, motor vehicles and petroleum – led the expansion of American manufacturing activities abroad in the 1920s, with the exception of suppliers to the motor vehicle industry that followed their American customers, the motor vehicle companies, in their investment forays overseas. More typically, the reason for market-oriented FDI lay in the business opportunities to be met in the host country and in the fact that the best way to meet those business opportunities was through outward FDI. As in the past, the largest market-oriented stakes in the 1920s tended to be directed to the most affluent nations of the world, those with industrialized and technologically advanced economies and where the markets were largest – Canada and Europe.

The predominance of Canada as a host country for American FDI in manufacturing and its relative importance remained largely unchanged from 1897 to 1929. As in the mining and agriculture industries, the period from 1914 to 1929 was associated with the growth of already existing American factories as well as the emergence of new manufacturing industries in Canadian manufacturing. For example, the emergence of the American abrasives industry in Canada from 1914 to 1916 was prompted by the disruption in 1914 of American imports from Europe of aluminium oxide, emery and corundum – essential raw materials for grinding wheels and abrasives. War-associated factors also prompted the emergence and expansion of FDI in Canada by American companies in the explosives, chemicals and petroleum refining industries, including that of Procter & Gamble whose first foreign plant – a $1 million Canadian plant completed in 1915 – was made to counter the virtual domination of the Lever Company of the Canadian economy at a time when the British were preoccupied with the war. Participation in the Dominion war effort also led to two new entries by American firms in the Canadian car industry – that of Chalmers Motor Corp (predecessor of Chrysler) and Willys-Overland – and the expansion of the Ford Motor Company. The inflation in the United States brought on by the war boom also led to the further expansion of American investments in Canadian pulp and paper mills to service the needs of American domestic enterprises that were major users of paper. Four new, large American pulp and paper mills were constructed across the northern border in Canada from 1914 to 1916, attracted by the presence of an abundant supply of inexpensive timber and hydroelectric power, relatively cheap labour and the absence of American duties on Canadian newsprint since 1911. American manufacturing firms continued to expand until the end of that decade, opening more than 200 branch factories in 1919 alone and a large number in 1920 (Wilkins, 1974).

With the search for markets becoming an important facet of the American expansion abroad in the 1920s, there was a sharp rise in the market-oriented endeavours of American MNCs in Canada in that decade. As in times past, the imposition of tariff barriers to trade and the maintenance of a liberal policy on the movements of labour, capital and technology enabled Canada to encourage the growth of American FDI and secure further the position of American business in the Canadian economy. American stakes in Canada in both manufacturing and utilities became overwhelmingly local market oriented.

The position of Europe as the second-largest host region for American FDI in manufacturing remained stable from 1897 to 1929. While some American manufacturing businesses in Europe ceased functioning during the First World War, and in particular in the period between the summer of 1914 and the spring of 1917, many affiliates of American metallurgical companies in Europe were redirected to war production to fulfil the host nation's or occupying power's needs.[4] The affiliates were used either as temporary barracks, or were made to produce airplane parts, shells, munitions and other weapons for the military on both sides of the conflict. The operations of the production subsidiaries on both sides of the conflict during the First World War proved to be profitable. Although the American parent companies retained direct contact with their European subsidiaries and affiliates on both sides of the divide during this period, the various European units usually had no contacts with one another (Wilkins, 1974).

The involvement of the United States in the war between 6 April 1917 and 11 November 1918 had dramatic implications for American business in the territory of the Central Powers. Some 159 American businesses and properties in Germany had been sequestrated by 21 October 1918, and although many of these were not substantial, often comprising a sales branch, a warehouse and a small inventory, a few were large investments such as the German plants of the American Radiator Company, Steinway & Sons and Singer (Wilkins, 1974). On the other hand, American enterprises in the Allied and most neutral nations, including Latin America, Japan and China flourished during this period to meet wartime demands.

American manufacturers took steps in the 1920s both to regain control over their European operations and to initiate new investments in England and in the continent in response to both opportunities for expansion of investments and threats to exports. The promise of European economic growth in the mid-1920s as normality returned to much of Europe (excluding the Soviet Union) was followed toward that decade's end by growing nationalism and rising tariffs in Germany, France, Italy, Spain, Austria and England. Thus, much of the European stakes by American firms by 1929 reflected the need of American companies to overcome tariffs and to operate plants close to their customers.

By comparison to Canada and Europe, South America was a far less important site for American FDI in manufacturing with a share of between 6 and 9 per cent of American FDI in this sector worldwide in 1919 and 1929. Nevertheless, the region became an important site for major new stakes in manufacturing by American companies since 1914. Unlike in Canada and Europe, American FDI in the manufacturing sector of South America between 1914 and 1917 was less influenced by the war, and became the basis for the further expansion of American businesses in the region particularly in the motor vehicles industry and supplier industries such as rubber tyres. For example, the Ford Motor Company that established a sales branch in Argentina in 1914 also established in that country the first car assembly plant in Latin America in 1916. Studebaker similarly established a sales room, a repair shop, a stock room and an assembly plant in Argentina in 1915. Associated with these investments by American car companies in Argentina were those of the Goodyear Tire and Rubber Company and the U.S. Rubber Company that opened sales branches in Argentina in 1915 and 1916, respectively (Wilkins, 1974).

Apart from Latin America (primarily Argentina), many of the same large American industrial enterprises that sold and then manufactured (or refined oil) in Europe and Canada during the 1920s also invested in Asia, Oceania (primarily Australia) and Africa, but made far smaller capital commitments to search for markets. In these continents, neither opportunities for sale nor threats to exports attracted large investments by American companies seeking foreign markets. Instead, in many instances, Americans established sales outlets and built service, assembly, packaging and mixing plants rather than full-fledged manufactories.[5] There were no market-oriented oil refineries owned by American companies in all of Asia, Oceania and Africa in the 1920s.

The petroleum industry

By comparison to the period before the First World War when marketing was the predominant motive behind American stakes in petroleum abroad, the period since described the increasing motivations of American petroleum companies to expand abroad in light of both marketing and supply considerations. As had been the case with companies in the agricultural, mineral and manufacturing sectors, the international expansion of petroleum companies in the immediate period after 1914 was influenced initially by the demands of the war in Europe and by rising petroleum prices. This was particularly the case during the direct involvement of the United States in the First World War between 1917 and 1918, but even at the end of that war fears of an oil shortage in the United States influenced the expansion abroad of American petroleum companies in search of foreign oil. Thus, foreign investments in petroleum in the period since 1914 was geared increasingly to supply the petroleum requirements of the United States as well as the established marketing outlets of American petroleum companies worldwide.

These reasons propelled the expansion of American petroleum companies in Mexico and South America – a country and region that had always been important sites for petroleum exploration by American companies since the 1890s with the pioneering investments of Jersey Standard. Although Mexico persisted as the single most dominant host country for American companies in foreign oil production with a share of one-third of American FDI in petroleum worldwide in 1919, a number of investment obstacles in Mexico in the early 1920s – involving the imposition of government decrees on oil nationalization and new Mexican export taxes on petroleum combined with substantial capital expenditures in exploration – encouraged American oil enterprises at the start of the 1920s to seek foreign crude oil in every continent (Wilkins, 1974).[6] As a result, American corporations sought oil in Central America, Colombia, Peru, Argentina, Brazil, Ecuador, Bolivia, Canada, in the vast Soviet oil resources in Europe, the Far East and the Middle East, although American stakes in oil production remained most significant in Mexico and Venezuela by 1929. Indeed, American petroleum companies became involved in oil production in each of the ten largest oil producing countries by 1929, with the exception of the USSR and Iran.[7]

As the assumed petroleum shortage abated in the late 1920s, the enthusiasm for foreign oil development subsided by the end of that decade. At that time, negotiations with governments of host countries began to characterize the entry and expansion of American petroleum companies in many foreign countries.[8] Despite these investment obstacles and the declining prices of crude rubber, sugar, nitrates, manganese and oil in 1929, American supply-oriented ventures in these industries continued to rise in number and scale in order to diversify and control the sources of supply.

The geographical destination of American FDI

Despite the relative stability in the importance of mining, manufacturing, petroleum and agricultural sectors in American FDI from 1897 to 1929, there were significant shifts in the relative importance of host countries. The problems with Mexico as mentioned above led to the slow growth of American FDI in that country. Thus, from being the most important host country of American FDI in the period until the start of the Mexican Civil War in 1911 and the second-largest host country in 1914 after Canada, Mexico had become only the fifth-largest host country for American FDI in 1919. As a result, the former major host regions ascended in importance in the same order with Europe becoming the second-largest host region after

Canada, followed by the Caribbean and South America. The growth of new American FDI between 1914 and 1917 outside Mexico, Russia and Continental Europe flourished with the opportunities and the temporary cessation of European competition owing to the war. The majority of American FDI in these regions continued to expand even after the end of the First World War.

With the breakdown of European monopolies by 1929, South America became the largest host country for American FDI, followed by Canada, Europe, the Caribbean and Mexico. The ascendance of South America was largely owing to the expansion in the region of American FDI in petroleum and utilities primarily, and also mining and manufacturing.

The growth of American FDI from 1929 to the Second World War

As noted in the introduction to this chapter, the period around 1929 marked a structural change in the industrial pattern of American FDI. It was the period in which manufacturing became the most important sector of American FDI, accounting for 24 per cent of the stock of American FDI. The other four most important industries in 1929 were petroleum (18 per cent), mining (16 per cent), utilities (14 per cent) and agriculture (13 per cent).[9] Railroads, which was the most important sector of American FDI in 1897, became only the fifth important sector in 1908 and declined further to seventh place by the end of the 1920s. The importance of these industries remained stable until around the end of the Second World War when another structural change in the industrial pattern of American FDI became evident, although less dramatic than that observed in 1929. Some of the salient determinants of American FDI between 1929 and the Second World War are discussed below.

The expansion of American businesses abroad between 1929 and the end of the Second World War was influenced by a number of factors. While the depression of the late 1920s and the stock market crash of 1929 either curtailed or caused many American enterprises to retreat from business abroad, some manufacturing companies felt compelled to initiate international production that would not have been made otherwise had there been no actions by host country governments to achieve nationalism, autarky and industrialization. This included trade restraints of various forms (tariffs, exchange controls, quotas and sometimes embargoes), as well as policies specifically directed to affect inward FDI either in terms of conditions of entry or operating requirements.[10] This was the case in the United Kingdom and throughout the European continent, Canada and Australia where higher tariffs were imposed in the early 1930s, as well as in Japan where trade restrictions were imposed and emphasis was placed on rapid domestic industrialization by the Japanese government. Indeed, host country government regulations increasingly became the *sine qua non* of international production by American firms (Wilkins, 1974).

However, the surge in the establishment of American-owned factories abroad in this period proved temporary as many ventures failed owing to the depressed economic conditions. American firms in agriculture and mining also reduced their FDI in the 1930s as prices of both mineral and agricultural products correlated directly with the fluctuations in the business cycle. The decade of the 1930s represented the beginning of *retrenchment and retreat* of American FDI (Wilkins, 1974), a trend that was to continue until the end of the Second World War. To the extent that international production took place, this was determined on the basis of negotiations with governments in each host country and by international agreements concluded by American companies.

The unfavourable investment climate in the various host countries in the 1930s also affected the international expansion of American petroleum companies.[11] Nevertheless, the sustained growth of American FDI in this sector was driven by an increasing demand for petroleum as an energy source for modern industry, the need to diversify risks, the availability of capital, the potential for profit and, above all, the need to overcome or stay ahead of competing investments by other oil companies, whether American or foreign.

Several features describe the evolution of American FDI in the petroleum industry in the 1930s. First, there was the reorganization of trade and FDI by American firms as a result of the oil glut, the possibility of sourcing cheap crude oil from Venezuela, the low price of petroleum associated with the economic depression, and the imposition of American tariffs on the entry of foreign oil in 1932. This meant that supplies of refined oil previously exported from the United States to the marketing and distribution outlets of American petroleum companies in Europe were replaced increasingly by cheaper supplies of oil exported from Venezuela.[12] Venezuela remained the predominant source of petroleum in Europe until 1948, when crude oil from the Middle East began to displace oil from Venezuela in importance. Second, the form of international expansion by American petroleum companies in the 1930s was favoured, in some cases, by joint ventures concluded mainly with other western oil companies in light of the prevailing economic depression, the increased political and commercial risks and the high capital requirements for investment.[13] Foreign oil companies in Europe have been expected to participate in national cartel arrangements by many nationalist European governments.

By 1939, American oil companies had invested in 58 refineries abroad, of which 39 were in consumer countries outside of Africa, Asia and Oceania (with the exception of Japan) that imported crude oil, and 19 were in or near oil-producing countries.[14] Ten American oil companies represented well over 90 per cent of American petroleum investments abroad, of which five stood out as large integrated MNCs: Jersey Standard, Socony-Vacuum, Gulf, Standard of California and Texaco. Those five American petroleum companies together with the British-owned Anglo-Iranian Oil Company and the British and Dutch-owned Royal Dutch Shell comprised the seven largest oil companies in the world at the end of the 1930s (Wilkins, 1974).[15]

The Second World War exerted a variable impact on the expansion of American MNCs. American businesses abroad encountered new interventions by host country governments or occupying power, as well as increasing government interventions both in the United States and in foreign countries. On the one hand, there was the expansion of American businesses in Canada and Europe where subsidiaries and affiliates of American companies served to fulfil the military and defence requirements of the Allied countries – a theme resonant of the events around the First World War.[16] This made the position of American MNCs in the western hemisphere to grow significantly in relation to European MNCs. On the other hand, there was the fragmentation of American businesses in the Axis countries as well as in the occupied countries of Asia and Europe where American plants that had not been sold or destroyed by bombings were geared to serve the Axis war effort. At the same time, the restrictive international agreements concluded by some American MNCs abroad regarded to have an effect on the foreign commerce of the United States came under increased scrutiny by the United States government as a violation of the Sherman Antitrust Act. All these factors meant that the level of American FDI in the immediate aftermath of the Second World War – at \$7.2 billion in 1946 – remained below its 1929 level of \$7.6 billion.

The dominance of American MNCs since the Second World War

Explaining the expansion of American FDI since the Second World War: the role of financial factors

Despite the political, economic and military uncertainties abroad in the immediate aftermath of the Second World War, the United States government sought to promote the expansion of all American private investment abroad in 'politically friendly' countries starting in Europe and in less developed countries in the 1950s and 1960s.[17] The promotion of American private capital abroad served both economic and political objectives: first, it was an important means to recycle the balance-of-payments surplus of the United States in the immediate period after the Second World War; and second, it was an important means to implement American foreign policy to contain communism and to promote democracy in host countries by its impact on economic growth. The prevailing ideology was that economic growth fostered by the export of capital, technology, skills and management by American MNCs was the foundation of a democratic world.

Even as the balance-of-payments position of the United States turned from a surplus to a deficit in the 1950s and more so in the 1960s, American stakes abroad nevertheless continued to rise as indicated by the near tripling of the outward stock of American FDI between 1950 and 1960 from $11.8 billion to $31.9 billion, with the growth particularly high in developed countries.[18] One explanation for this is a financial one, and stems from the United States dollar remaining a strong currency not only on account of the overall balance-of-payments surplus of the United States in the early post-war period, but also on account of the unique role of the dollar as a key currency and a medium of international monetary reserve in the Bretton Woods adjustable-peg system of exchange rates that prevailed from 1944 until 1971. Under this system of exchange rates the dollar came to be overvalued substantially in relation to the currencies of its major trading partners as its fixed exchange rate overstated the worth of the dollar when compared to its market or equilibrium rate. The obligations of the United States to keep the dollar convertible to gold at $35 per ounce meant that it could not devalue the dollar in terms of gold without undermining confidence in the existing system of exchange rates. The overvaluation of the dollar made foreign currencies and, by extension, foreign assets, goods and services cheap in terms of the dollar and contributed to yet higher levels of outward FDI by American firms (Aliber, 1970). Indeed, the growth of American FDI was so rapid that by 1960 the stock of American FDI at $31.9 billion not only exceeded the stock of FDI abroad of the United Kingdom – the largest source country of FDI until then – whose value reached $12.4 billion in 1960, but also the total private investment overseas of the United Kingdom at $26.4 billion (Wilkins, 1974). This enabled the United States to become the largest home country of FDI since the Second World War, a position it has maintained consistently for more than half a century to the present time.

In response to the growing balance-of-payments deficits in the 1960s, mandatory controls on the outflow of American FDI as well as controls on the reinvestment of profits of American businesses earned abroad was imposed for the first time on 1 January 1968.[19] The aim was not to limit American FDI as such, but to limit its negative balance-of-payments effect (Wilkins, 1974). This had important implications on the financing of American FDI during this period. The growth of foreign stakes in market-seeking manufacturing activities associated with the growth of foreign demand was financed almost entirely from reinvested earnings of established foreign subsidiaries and affiliates.[20] Borrowing abroad also became a popular mode to finance FDI not only to relieve the balance-of-payments deficit, but also as a hedge against devaluation of currencies, a safeguard against blocked payments and, often, as a convenience.[21]

The industrial pattern of American FDI since the Second World War

The United States emerged from the Second World War as the leading industrial nation and American firms persisted as the dominant source of new innovations and thus had a decisive lead in the production of a whole range of technologically intensive producer goods and high-income consumer goods until around the late 1960s and 1970s with the decline of Pax Americana (Dunning, 1985). Manufacturing thus remained the most important sector of American FDI for sixty years from 1929 through to 1989, although there were other more subtle changes in the industrial pattern of American FDI that defined the era since the Second World War.

The most important industries of American FDI in 1950 were manufacturing which accounted for one-third of the stock of American FDI, petroleum (with a share of 29 per cent), public utilities (12 per cent), mining (10 per cent), and trade (6 per cent). Thus, wholesale and retail trade was defined as a significant sector of American FDI in 1950 associated with the expansion abroad of American mass merchandisers, while American FDI in public utilities and especially agriculture began to become less important. Consequently, although the expansion of American MNCs in public utilities (in particular those in communications and power and light) starting in the 1920s was favoured initially by the strong support of the United States government and the ownership advantages of American firms in the form of capital, technology, management and marketing expertise, their rate of growth slowed down after 1950.[22] American public utility companies in Latin America, for example, sold their assets owing to low profits and national pressures in host countries (Wilkins, 1974).

More significantly, American FDI in agriculture which featured consistently among the five most important industries of American FDI, at least until 1929, declined considerably in importance and never since regained influence as a major industry of American FDI (US Department of Commerce, Office of Business Economics, 1960). This owed much to the depression of the late 1920s and early 1930s which made investments in raw materials a liability as prices of agricultural produce plummeted and agricultural companies recorded losses. The nationalization and expropriation in Latin America from 1933 to 1939 served to exacerbate the adverse investment conditions for American investors holding foreign agricultural properties. By 1950, FDI in agriculture amounted to only $589 million, of which 60 per cent was in sugar production in Cuba and the West Indies mainly, and another 26 per cent was in fruit (principally in bananas) in Central America (US Department of Commerce, Office of Business Economics, 1950).

The bulk of American FDI in manufacturing since the Second World War continued to be directed to western Europe and Canada. The end of the Second World War and the economic recovery in Europe was considered a watershed in the history of American manufacturing in that region (Wilkins, 1974).[23] Indeed, manufacturing became the most important sector of American FDI in Europe by size of investment starting from the 1950s and exceeded the level of American FDI in the same sector in Canada by 1964. This represented a notable change from the previously prevailing pattern when Canada was a more important host country for American manufacturing FDI by comparison to Europe which was a more important host country for American FDI in sales.

The kinds of American manufacturing enterprises that invested in plants abroad in the late 1950s and 1960s resembled closely the investors of earlier years: these were leading firms in their industries in the United States that had advantages in technology, unique products, and a long history of international economic orientation. The fact that an industry was technologically advanced did not *ipso facto* guarantee large FDI, but it generally meant that leading companies in that industry would, in time, and after finding exports could not continue to fill foreign demand, show an interest in extending their business through investments abroad (Wilkins, 1974).[24] This was the era of American FDI which inspired the development of the product cycle model of Vernon (1966). In the product cycle model, a technologically innovative firm would serve a growing foreign demand for a maturing product through exports initially, and may switch to international production at a later stage when threats to the export business emerge. Similarly, the rank of a firm in American industry was in itself an inadequate explanation of the growth of FDI of American companies in manufacturing in the late 1950s and 1960s.[25] At least two of these attributes seem to have been a prerequisite for large-scale American FDI (Wilkins, 1974).

Most formal analyses of the determinants of the outward FDI intensity or the propensity of the United States to invest abroad find statistically significant positive roles for R&D intensity and advertising intensity, reflecting the importance of technology and marketing capabilities as key elements of firm-specific intangible assets of American MNCs (see among others, Caves, 1982).[26] Other determinants of outward FDI intensity found to be also important were managerial capabilities and capital cost advantages (Pugel, 1981). Similar findings were concluded by Clegg (1987) who showed that the degree of innovation and the creation of technological ownership advantages as well as capital intensity had highly significant positive influences on American FDI in manufacturing, while the skill level of production workers had a significant negative effect. A related study of the industry-specific determinants of the export competitiveness of American MNCs showed the statistically significant positive roles of R&D intensity and human capital intensity and a statistically significant negative role of labour intensity – an indication of the importance of both research and skilled labour, but not unskilled labour, as driving forces behind the high shares of American MNCs in world exports of manufactures. High advertising intensity, although contributing to the export competitiveness of American MNCs as a whole, fulfilled a relatively more important role in boosting the export competitiveness of foreign affiliates than of their parent companies (Kravis and Lipsey, 1992).[27]

The most important manufacturing industries of American FDI during most of the 1960s when measured in terms of the size of sales of American foreign manufacturing affiliates were in declining order: transportation equipment (primarily motor vehicles and equipment), chemicals, machinery (excluding electrical), food products, electrical machinery, and primary and fabricated metals. Thus, those firms in the vanguard of technology in the late 1950s and 1960s were both in the same (albeit transformed) industries of past decades as well as in new manufacturing industries which emerged rapidly in importance since the Second World War. The acquisition of undisputed world leadership in technology by American chemical companies as well as the severance of restrictive ties between these companies and their British and German counterparts owing to antitrust action and wartime conflict enabled American MNCs in the chemical industry to come into their own after the Second World War. Similarly, the large investments in international production by American car manufacturers in the late 1950s and 1960s reflected not only the strengths of the leading firms in this industry, but also the prominent position of the car industry in all of American industry.[28] Their significant presence in foreign markets made the sales of foreign manufacturing affiliates of American firms in this industry exceed that of all other foreign manufacturing affiliates of American firms in the 1960s (Wilkins, 1974).

Transportation equipment, chemicals and allied products, machinery (excluding electrical), food and related products and electrical and electronic equipment remained the five most important manufacturing industries of American FDI through the 1980s and 1990s, even though the relative importance of the individual industries in American FDI altered since the 1960s.[29] As in decades past, the expansion of American manufacturing firms through international production enabled the firms to maintain their long-established position in foreign countries, fulfil a growing foreign demand and meet rising foreign competition – explanations that are in line with the product cycle model. The pursuit of profits was not seen as a crucial motivation (Wilkins, 1974). Some market-oriented investments were made to defend existing markets, but many of the investments since the late 1950s and 1960s were aggressive new stakes designed to penetrate new overseas markets. To acquire and hold these markets, international production as opposed to exports from the United States seemed almost a necessity.

Petroleum remained the second-largest sector of American FDI from 1929 until the mid-1960s, when its relative importance began to decline from 34 per cent of the stock of American FDI in 1960 to 26 per cent in 1970, 13 per cent in 1989, and 10 per cent in 1997.[30] Despite the disadvantageous foreign investment climate, the aggressive nature of foreign investments of American oil companies in every facet of the petroleum business in the immediate years after the Second World War owed largely to the growing demand for oil as the most important single source of energy, the perceived shortage of oil resources of the United States as well as profits.[31] Starting in the late 1950s and 1960s, the American oil companies exported not so much new products as improved methods associated with significant process innovations in refining, transportation, production and exploration. While the main stakes in petroleum marketing and market-related activities continued to be concentrated in developed nations where markets were largest, investments in petroleum exploration and production were concentrated in oil-rich developing countries especially in Venezuela and elsewhere in Latin America, the Middle East, north of Africa, west Africa and Indonesia. In the Middle East, in particular, American oil companies triumphed over British–Dutch oil interests starting from the early 1950s (Wilkins, 1974).

Similar demands placed on America's mineral resources by the Second World War and the frequent forecasts of domestic shortages of minerals (particularly iron ore) ensured that mining remained an important sector of American FDI in the period since the Second World War. Investments in mining iron ore to supply American steel producers were the most significant new stakes by Americans in foreign mining in the immediate postwar years, followed by uranium owing to developments in atomic energy. Over time, the mining of copper, lead, zinc, gold and silver ores accounted for the bulk of investments in the sector by American MNCs. The investments in foreign mining were made by mining, manufacturing and, occasionally, oil companies.[32] A marked change in the number of host countries of American FDI in mining occurred starting in the late 1950s and 1960s owing much to the unfavourable investment environment in Latin America which led American MNCs to seek minerals in Canada and Australia. Latin America, Canada, Australia and Africa continued to be the most important host countries for mineral investments by American MNCs by 1997.

Apart from the five major sectors of American FDI of manufacturing, petroleum, trade, mining and public utilities, American FDI since the late 1950s and 1960s emerged and expanded in finance, insurance and real estate, a wide range of services (business services, car rental and leasing, hotels and lodging, motion pictures, etc.), communication, and other services-related industries. Their emergence abroad demonstrated a cluster complex in which American companies and American travellers abroad enticed their traditional suppliers into new investment stakes. By around 1989, another structural change in the industrial pattern of American FDI became evident with the ascendance of services – with a share of 47 per cent of the stock of American FDI abroad – as the single most important economic sector for American MNCs. This is associated with the increasing service orientation of the American economy – a feature of the post-industrial status of most developed market economies.

The geographical destination of American FDI since the Second World War

Despite American policies in the late 1950s and 1960s to foster the growth of American FDI in less developed countries, an increasing amount of American FDI since the Second World War was directed to developed countries.[33] Thus, although by 1950 the book value of American FDI stock in both developed and developing countries was roughly the same at $5.7 billion, the faster growth of American FDI in developed countries since made American FDI in developed countries reach $19.3 billion in 1960 and $51.8 billion in 1970 – levels that were 1.7 times larger and 2.7 times larger than their corresponding investments in developing countries in those years, respectively. The growing importance of market-oriented FDI over supply-oriented FDI which started in the 1920s and continued in a major way since the Second World War largely explain this trend.

In particular, the wealthiest nations of the world – those of western Europe and Canada – where there were high levels of income, rapid economic growth, large markets, resources and a favourable investment climate attracted American FDI. The attractiveness of Europe in general, and the large countries of the European Economic Community in particular, since the late 1950s was favoured by the emerging regional prosperity and the movements towards regional economic integration.[34] Notwithstanding the persistent dollar shortage in Europe after the Second World War linked to the lag in which innovation in that region followed that in the United States (Johnson, 1958), financial factors in Europe also played a role: the convertibility of the currencies of many European countries in 1958 and the formation and development of a Eurodollar market and Eurodollar bond market facilitated borrowing and lending in dollars and helped to change the dollar shortage in Europe to one of a dollar surplus. By 1960, the level of American FDI in Europe surpassed that in Canada for the first time. The growing presence and control of American MNCs in Europe since the Second World War as forcefully argued in Vernon (1971, 1974) and Servan-Schreiber (1967) among others was reminiscent of a similar pattern of events exhibited around the turn of the century when the influx of American manufactured goods in Europe precipitated the works of McKenzie (1901), Thwaite (1902) and Stead (1902).

Nevertheless, Canada remained an important host country for American FDI in the period since the Second World War. American FDI both fulfilled the needs of growing Canadian markets and helped to develop its newly discovered mineral and oil resources. By contrast, although American companies desired to expand their investments in Japan, controls on inward FDI associated with policies of indigenous technological development would not permit American corporations to play a leading role in the Japanese economy.

In general, developing countries did not attract high levels of American FDI since the Second World War owing to their economic and political instabilities which made the investment climate in these countries unsatisfactory. In the developing countries of Asia, Africa and Latin America, the dramatic rise in the economic power of national governments had a profound impact on the course of American businesses in petroleum, mining, agriculture and manufacturing in the 1960s and 1970s. It became increasingly evident that the growth of American FDI in developing countries would be under terms set by the host governments. In Latin America, government measures served to stimulate the growth of American FDI in some sectors such as manufacturing and discourage new stakes in other sectors. American FDI in the developing countries of Asia and Africa increased, particularly those that took advantage of abundant labour at lower costs or did not benefit from economies of scale.[35] However, with the exception of the Philippines, the size and degree of integration of American stakes in manufacturing in these countries were far smaller than in the developed countries or Latin America – host regions of long interest to American investors. This is largely due to unfamiliarity of American firms with investment in these territories, government restrictions and bureaucracy, and limited markets.[36] By 1987, the book value of American FDI stock in developed countries reached $237.5 billion – a level some 3.3 times higher than American FDI in developing countries of $73 billion. By 1997, the book value of American FDI stock in developed countries further increased to $590 billion – a level some 2.2 times higher than American FDI in developing countries of $267.5 billion.[37] Thus, although there seems to be some evidence to suggest that the gaps between the stock of American FDI in developed and developing countries narrowed between 1987 and 1997 owing to the greater locational advantages of these countries for FDI explained partly in the case of Mexico in the context of the North America Free Trade Agreement concluded in 1994, the developed countries remain the overwhelmingly dominant recipient of American FDI with a share of 69 per cent of the stock of American FDI in 1997.

Conclusion

The period since 1914 described the period of the growth and evolution of American MNCs. The principal industries of American FDI in 1908 consisting of mining and processing of mineral ores, manufacturing, oil production and refining, agriculture and processing of agricultural produce, and railroads remained essentially stable until 1929. Thus, in essence, the primary sector was the dominant sector of American FDI until 1929 associated with the importance of supply-oriented FDI for American MNCs in resource-rich developing countries. In 1929, a significant structural change in the sectoral pattern of American FDI became evident with the ascendance of the manufacturing sector as the most important sector of American FDI associated with the growth in importance of market-oriented FDI in market-rich developed countries of Western Europe and Canada. This pattern persisted for 60 years until 1989 when another structural change in the sectoral pattern of American FDI became evident with the dominance of the services sector in American FDI. Thus, the dominant sectoral pattern of American FDI traversed the primary, secondary and tertiary sectors of economic activity over the course of more than a century.

Such significant structural shifts in the sectoral pattern of American FDI correlated directly with shifts in the geographical pattern of American

FDI. Thus, in the period before 1929 when the primary sector was the dominant sector of American FDI, the resource-rich countries of the Caribbean, South America, Mexico and Canada received the majority of American FDI. The ascendance of the manufacturing sector as the dominant sector of American FDI starting from 1929 led to an increase in the share of American FDI directed to Canada, Western Europe and other market-rich developed countries. These countries have received the majority share of the stock of American FDI abroad since the Second World War and will continue to do so in the future.

Notes

1 The destruction brought by war of American sugar-producing properties in north-ern France, Belgium and south-western Russia, and the end of sugar exports from Germany and Austria-Hungary led to the further expansion of American invest-ments in Cuban cane sugar to meet the demands of the Allied powers as well as the United States. The war demands for nitrates used for explosives and fertilizers led to the acquisition in 1916 by W. R. Grace & Co. – an American company pro-minent in the nitrate trade in Chile – of the Tarapaca & Tocopilla Nitrate Company in Chile from its British owners for $3 million. This represented the first backward integration of the company into the ownership of nitrate refineries or *oficinas*. By selling their Chuquicamata copper interests in Chile, the Guggenheim brothers also acquired an interest in Chilean nitrates, developing in the process a new nitrate production system in 1923 based on refrigeration which permitted treatment in large lots and which out-performed the prevailing Shanks system in its ability to treat ores with a lower assay of 8 per cent as opposed to a minimum of 15 per cent. Similar war demands led to the further expansion of American copper, steel, nickel and aluminium companies as well as the emergence of new ones. Also, the vast meat requirements of European armies led to the expansion of the established meat packing plants of Swift, Armour, Morris and Wilson in South America, as well as the establishment of new ones in the southern part of the western hemisphere and in Australia (Wilkins, 1974; O'Brien, 1989).
2 To support the objective of obtaining tin from Bolivia, W. R. Grace & Co., which was for years established in trade and investment on the west coast of South America, organized the International Mining Company – a large producer of Bolivian tungsten and, to a lesser extent, tin (Wilkins, 1974).
3 With the continued expansion of rubber plantations in the Far East by American tyre manufacturers, the companies obtained some protection against the antici-pated higher price of rubber as a result of the Stevenson Rubber Scheme advanced by Great Britain in 1922 to assist British planters in Ceylon and Malaya by restricting output and raising prices. Firestone also sought to develop rubber plantations in Liberia, tied to a private loan to that country and the improvement of the Liberian harbour to aid rubber exports. This direct investment was excep-tional in the involvement of the American government in the loan arrangements. Henry Ford also resolved to grow his own rubber and made substantial invest-ments in planting rubber trees in Brazil in 1927–1928. However, these companies that integrated backward into rubber cultivation were not protected against the fall in rubber prices that occurred in 1929 (Wilkins, 1974).
4 Nowhere was this more evident than in the operations of the various European affiliates of the American Radiator Company. Its factory in Dôle was requisitioned by the French state to provide temporary barracks and to produce shells for their

nation's military needs. The Italian unit of the company pledged to sell semicast steel shells to the Italian government, while their two German plants made cast iron shells for the imperial regime. The plant in Austria manufactured munitions, while that in England – the National Radiator Company Ltd – agreed to supply the Belgians with hand grenades and concluded other contracts with British authorities. The British government took over the large sewing machine plant of Singer in Scotland, where sewing machine production ceased and the plant manufactured airplane parts and munitions instead (Wilkins, 1974).

5 For example, there existed a variety of small American stakes in miscellaneous manufacturing for export such as carpets, hand-woven tapestries and textile mills in China, embroidery in the Philippines, carpet manufacture in Turkey, and jute in India. The various foreign interests of American firms in textiles were exceptional in their motive to gain access to cheap labour abroad with a high degree of skill and dexterity. Most of the investments were made by small American businesses resident abroad, rather than by MNCs (Wilkins, 1974).

6 American petroleum companies, however, were undeterred and continued to expand in the rich and profitable Mexican oil resources. The oil companies escaped major property damage associated with the Mexican Revolution owing to their location along the coastal periphery of the country and by paying for protection (Wright, 1971). Mexico became the second-largest oil producer after the United States in 1918, mainly owing to the large-scale American FDI in the industry. In 1919, practically every sizable American oil company had oil lands in Mexico with total FDI of at least $200 million. Although American FDI in mining in Mexico was higher at $222 million, the petroleum revenues of American companies exceeded the mining revenues of American companies (Wilkins, 1974).

7 The ten largest oil producing countries were the United States, Venezuela, USSR, Mexico, Iran, Dutch East Indies, Rumania, Colombia, Peru and Argentina (Wilkins, 1974).

8 By 1929, efforts by American investors to develop petroleum resources in many countries had been thwarted by host government policies or policies of colonial or imperial governing powers. This had been the case in Asia and Africa where the British and, to a lesser extent, other European governments tried to limit American stakes. The Japanese had also interfered with American FDI in Soviet-held Northern Sakhalin. The United States government often, but not always, tried to assist American companies in their negotiations with foreign governments (Wilkins, 1974).

9 Based on data contained in Wilkins (1974).

10 This included *inter alia* restrictions on profit remittances, controls on capital repatriation, extensive state intervention, higher taxes, state companies that competed with private enterprises, and nationalization (Wilkins, 1974).

11 For example, American companies encountered an entirely new investment environment in the infantile oil industry of the Middle East. Nationalist fervours in Latin America also worked against large, foreign-owned petroleum enterprises. Foreign oil companies had also become enmeshed in a Mexican government-backed labour dispute. In Japan, the Petroleum Control Bill passed in 1934 introduced price controls, new conditions on government sales and a requirement that six months of oil stocks be held in Japan by all importers and refineries in order to boost Japanese oil reserves. Government policies in Europe also confronted foreign oil companies. Import quotas, exchange restrictions, price-fixing regulations, enforced use of alcohol and other petroleum substitutes, new taxation, blocking of profit remittances, export subsidies and reciprocal trade agreements between foreign countries affected directly the petroleum operations of foreign companies (Wilkins, 1974).

12 Jersey Standard had become the largest American oil producer in Venezuela through its purchase of the foreign properties of the Pan American Petroleum & Transport Company. Although much of this oil was refined in Aruba and then shipped to Europe, some was shipped in crude form at a later stage with the establishment of refineries in Europe (Wilkins, 1974).

12 The principal new joint ventures in petroleum concluded abroad in the 1930s involved the following activities: (1) refining as in the cases of the Société Franco-Américaine de Raffinage and Mitsubishi Oil Company; (2) marketing and seeking oil supplies as in the cases of Stanvac and Caltex; and (3) petroleum exploration and production as in the cases of cooperative efforts in the Middle East, the Dutch East Indies and Latin America (Wilkins, 1974).

14 Of the 58 foreign refineries, 28 refineries belonged to the Standard Oil of New Jersey, 15 refineries belonged to Socony-Vacuum, 7 refineries belonged to Texaco and the rest were refineries of other American petroleum companies. At the end of the 1930s, there were still no American refineries in consumer countries in Africa, Asia and Oceania where, with the exception of Japan, American petroleum companies continued to sell imported refined oil (Wilkins, 1974).

15 By contrast, there had been only three large companies in the world oil industry in 1929: Standard Oil of New Jersey, Royal Dutch Shell and the Anglo-Iranian Oil Company (Wilkins, 1974). The increase in the number of large companies by 1939 had thus been brought about by the emergence and expansion abroad of other American petroleum companies.

16 American subsidiaries and affiliates in India, Australia, Northern Rhodesia, the Gold Coast, Latin America and elsewhere fulfilled a similar purpose (Wilkins, 1974).

17 Wartime destruction had left much of Europe and Japan in ruins and dollar shortages delayed economic recovery. As a result, government restraints in commerce and exchange transactions became the norm, as with currency depreciation in relation to the dollar. American investors feared the spread of communism as well as the general rise of public sector activities abroad. American businesses in foreign countries experienced further sizable losses as a consequence of expropriation in both communist and non-communist countries. These uncertainties abroad more than the threat of antitrust actions in the United States rendered American MNCs to be cautious about FDI. To the extent that antitrust cases affected American FDI, it made American firms realize that the crutch of private international agreements as a means of survival in foreign markets was unsustainable. Thus, as private international agreements regulating trade was past, the way was open for more complex multinational organizations. The development of these more complex organizational structures of American industrial MNCs administered through managerial hierarchies was described by Chandler (1980).

18 Figures were derived from the most recently revised data contained in the US Department of Commerce, *Survey of Current Business*, February 1981.

19 The imposition of mandatory controls was owing to the ineffectiveness of the voluntary balance-of-payments (BOP) programme announced in February 1965 to cope with BOP deficits. This voluntary programme urged some 500 large corporations to improve their individual BOP by raising exports, bringing in more income from abroad, repatriating short-term assets and borrowing in developed countries instead of exporting funds from the United States or reinvesting monies earned overseas. The guidelines applied initially to investments in developed countries, excluding Canada but the guidelines were extended to Canada and the less developed oil-producing countries in 1966. The imposition of mandatory controls on FDI in 1968, however, did not apply to Canada (where American companies were allowed unlimited FDI but were required to report their Canadian transactions) and to less developed countries. In certain other

countries (United Kingdom, Australia, Japan and the oil-producing countries of the Middle East), investments would be curtailed; and in industrial Europe no new capital would be allowed to flow from the United States, but a specified percentage share of the profits earned abroad could be reinvested (Wilkins, 1974).

20 Apart from convenience, the use of reinvested earnings to finance FDI also served to circumvent controls on profit remittances by the host country as well as taxes imposed by the United States on dividends, and fulfilled the need of foreign subsidiaries to reinvest before the funds depreciated in value. Thus, despite the paradox of dollar shortages, import licensing, restrictions of remittances, and similar obstacles to foreign trade and FDI in the postwar years, the level of American FDI continued to increase.

21 For a further discussion of the financing of American FDI in the period since the Second World War, see Sestáková (1989).

22 Among the most significant American firms in public utilities were International Telephone and Telegraph and the American & Foreign Power Company. American FDI in this sector tended to be concentrated in Spain, Italy and the developing countries where the demand for public utilities could not be met by indigenous firms (Wilkins, 1974).

23 In England, where a Labour government had been elected, American companies reconverted to peacetime production while facing state intervention unprecedented in peacetime Britain. On the European continent, war had disrupted operations and it was not until 1946–1948 did the prospects for recovery seem at all promising. Nonetheless, Cold War anxieties, restrictions on foreign remittances and trade as well as government attempts at planning persisted and hindered American FDI in European manufacturing. Nevertheless, some American companies with sales subsidiaries and the desire to keep, renew, or establish markets in soft currency areas did establish new production facilities in Europe (Wilkins, 1974).

24 A case in point was the technologically advanced aviation industry of the United States which was exceptionally slow in embarking in FDI until the end of the 1960s (Wilkins, 1974).

25 For example, the US Steel Corporation which ranked among the top 12 companies in sales in American industry for several decades was not a prominent player in international production. In the period of the company's technological leadership in the early 1900s, the firm met foreign demand through exports or the establishment of sales outlets abroad but did not engage in international production to a great extent. By the late 1950s and 1960s the company lacked the technological leadership in world industry as well as a network of foreign plants to provide a basis for major expansion, unlike the leading American car companies. The growth of its outward FDI since has been more significant in mining than in manufacturing (Wilkins, 1974).

26 The FDI intensity is measured typically in one of two ways: the first is in terms of the ratio of some measure of the size of FDI activity in the industry to the total size of the industry. Size can be measured by value added, assets, employment or profits (Pugel, 1981). Alternatively, FDI intensity can be measured in terms of degree of penetration in particular host countries (Pugel, 1985).

27 Export competitiveness was measured as the share of American MNCs in exports of manufactures of the market economies (Kravis and Lipsey, 1992).

28 In a similar fashion, the pioneering American firms in the meat packing industry that had gone abroad before the First World War was a reflection of the fact that the industry once belonged to the five largest American industries in revenues. The industry declined in importance by the 1960s, and its leading firms no longer made significant foreign investments, thus mirroring their loss of position in American industry (Wilkins, 1974).

29 For example, the most important manufacturing industries for American MNCs by the size of outward FDI stock in 1987 were in declining order: chemicals and allied products, machinery except electrical, transportation equipment, food and related products, and electrical and electronic equipment. By 1997, the most important manufacturing industries of American MNCs were in declining order: chemicals and allied products, food and related products, transportation equipment, electronic and other electrical equipment, and industrial machinery and equipment. The above statements were based on data contained in US Department of Commerce, *Survey of Current Business*, August 1992 and October 1998 for the 1987 data and 1997 data, respectively.

30 The information for 1960 and 1970 was calculated from data contained in US Department of Commerce, *Survey of Current Business*, February 1981. The information for 1989 was calculated from data contained in *United States Direct Investment Abroad: 1989 Benchmark Survey, Final Results*, Washington, DC: US Government Printing Office, October 1992. The data for 1997 was calculated from data contained in US Department of Commerce, *Survey of Current Business*, October 1998.

31 The investment obstacles faced by American petroleum companies ranged from antitrust action at home to expropriation risks (in Eastern Europe and China), creeping exports (in many countries), political instability, labour difficulties, dollar shortages, restraints on trade and payments, investment control laws, legislation excluding their participation, foreign government intervention, etc. Despite this, the stock of American FDI in petroleum grew the fastest – by 142 per cent – between 1946 and 1950, while that in manufacturing and mining grew more modestly at 60 per cent and 41 per cent, respectively (based on data in Wilkins, 1974).

32 As in the past, mining companies engaged in foreign mining when faced with shortages at home and the need to supplement domestic or world resources with cheaper sources of supply. Manufacturing companies invested in mining abroad to obtain raw materials. For example, the aluminium companies sought bauxite, the steel enterprises sought iron ore and various types of manufacturers sought specialty metals such as manganese and chrome. Oil companies moved into foreign mining as part of their diversification strategies (Wilkins, 1974).

33 The United States government guaranty programme for new investments that was initiated with the European recovery plans in the immediate aftermath of the Second World War was applied exclusively to underdeveloped nations in the late 1950s and 1960s. The Agency for International Development acquired other functions to encourage American FDI in less developed countries: making dollar loans to private American investors abroad, conducting foreign investment surveys and giving information on investment projects. This was complemented by government measures to protect investments such as, for example, the Hickenlooper Amendment to the Foreign Assistance Act in which foreign aid to a particular nation would be withdrawn should any expropriation of American properties after 1 January 1962 not be followed by prompt and adequate compensation (Wilkins, 1974).

34 The Treaty of Rome creating the European Economic Community (EC) was signed by the governments of Benelux, West Germany, France and Italy on 25 March 1957. When negotiations for the United Kingdom's entry into the EC collapsed in January 1963, American businesses that expected to export from Britain to the continent began to make new investments in the EC (Wilkins, 1974).

35 For example, American investors made some supply-oriented investments in electronic plants in South Korea, Taiwan and Hong Kong in the 1960s to take advantage of cheap labour with a high degree of dexterity. The output was sent to the United States or was incorporated in final products in a third country

(generally Japan) for sale in that country, the United States or even in a fourth country (Henderson, 1989).

36 American companies also gained entry in European colonial territories, although the level of their stakes did not exceed those of the imperial power. Generally, American investors acted through companies incorporated in the imperial nation and used managers of the nationality of the imperial power. With independence, Americans invested directly and dealt with national rather than imperial sovereignties (Wilkins, 1974).

37 Based on data contained in US Department of Commerce, *Survey of Current Business*, October 1998.

4 The emergence and evolution of multinational corporations from Sweden

Introduction

Although a far smaller country than the United States, the resource abundant country of Sweden has always been engaged actively in outward FDI and engendered one of the oldest MNCs in the world economy. By 1960, Sweden accounted for 0.6 per cent share of the global stock of outward FDI which increased significantly to 1.7 per cent in 1975 and 2.9 per cent in 1990, but its share declined slightly to 2.3 per cent in 1998 despite the near doubling of its stock of outward FDI between 1990 and 1998.[1] Notwithstanding the low relative importance of Swedish FDI, the study of the emergence and evolution of Swedish MNCs is of interest not the least because although Sweden has a far smaller domestic market, the growth pattern of their MNCs can be compared to that of American MNCs whose home country shares a similar natural resource abundance.

Swedish MNCs feature among the most important companies in Sweden – accounting for more than half of the domestic industrial workforce and more than half of the country's exports in the 1990s (Olsson, 1993). Although an overwhelming proportion of its outward FDI has always been concentrated in a few firms, the population of Swedish MNCs consist of many small firms.[2] Thus, the analysis of Swedish MNCs shows that size of firms is not a useful indicator of the propensity to engage in outward FDI, and that the determinants of international production does not always derive from oligopolistic market conditions. Rather, it is the size of the firm relative to its home market that determines the propensity to engage in international production. Firm size is more a reflection of a firm-specific advantage than a major source of such an advantage (Swedenborg, 1979).

There is a rapid process of internationalization through exports and international production by Swedish firms owing to the need to overcome the limited size of the domestic market. Internationalization of sales and manufacturing operations were motivated mainly by the desire to expand markets and attain economies of scale. Indeed, Sweden – as with small countries in general – has had earlier forays into international markets and a relatively larger international economic involvement by comparison to the United

States (Swedenborg, 1979).[3] Hence, Swedish firms, as with firms based in other small countries such as Switzerland and Singapore, have always had an offensive and aggressive market expansion strategy (see Chapters 13 and 15). Newly established Swedish firms with a significant competitive advantage (often a superior technology) became international companies typically after only a few years, and many technology intensive smaller firms felt the need to increase sales and sustain growth outside the confines of a rapidly saturated small domestic market. The choice between exports and international production tended to be determined by the balance of locational advantages in the home and host countries. International production in host countries was favoured by trade barriers and/or high transportation costs, the need to establish closer ties with local customers to support sales and the need for a local presence. Foreign plants served to maintain or increase foreign sales through the final assembly of products and/or modifications or adaptations of products designed essentially for the Swedish market. The local adaptations associated with international production have led, in some cases, to a broadening of the product range, but rarely had an impact on basic product design or new product development for the Swedish firm. Neither did international production involve the most skilled labour or relationships with local research institutions in foreign countries (Sölvell *et al.*, 1991).

In the interest of achieving consistency with Olsson (1993), the analysis of the history of Swedish MNCs in this chapter covering the major determinants of Swedish FDI and their industrial and geographical patterns is conducted in three time frames: the late nineteenth century to the First World War, from the First World War to 1960, and from 1960 to the present.

The emergence of Swedish MNCs from the late nineteenth century to the First World War

During the period of the emergence of Swedish MNCs in the late 1800s and until the beginning of the twentieth century, there were two main groups of Swedish export industries (Olsson, 1993). The dominant group of Swedish export industries were those based on raw materials (wood, pulp and metal ores) whose principal markets for centuries were the more industrialized countries of Europe. Between the beginning of the twentieth century and the First World War, half of Sweden's exports comprised forestry products sold mainly to Great Britain and iron and steel sold to Germany, and a further third by agricultural products such as oats and butter (see also Hörnell and Vahlne, 1986). Indeed, it was the industrialization of Great Britain and the subsequent expansion of building and construction activities in that country that led to the substantial increase in demand for timber products from Sweden between 1850 and 1880. Similarly, industrialization in other parts of Europe (particularly Germany) led to increased demand for iron and steel where Sweden had a comparative advantage in production owing to its favoured access to energy (charcoal). Since Swedish legislation prohibited foreigners from owning Swedish land, these staple industries remained largely in the domain of domestic firms. Raw materials continued to describe the export pattern of Sweden even until 1953 when it accounted for some 43 per cent share of merchandise exports; such share, however, has declined rapidly since (based on data in Söderström, 1980). Firms in the industries of mining, metal production and forestry developed their domestic production based on numerous small units in close proximity to the sources of raw materials and energy, and their exports were handled typically by merchant houses based in the major ports of Stockholm and Gothenburg that also provided working capital (Carlson, 1977). These firms became MNCs only to a very limited extent.

The second group of export industries comprised those of modern engineering generally consisting of metal manufactures, machinery and transport equipment whose growth was fostered by the presence of abundant natural resources in Sweden. In combination with export earnings from wood and iron products which financed the establishment of new industrial companies, the technical and metallurgical know-how developed from iron and steel production proved useful to the development of the modern engineering industry in the industrialization phase between 1870 and the First World War (Lundström, 1986). The long tradition of metal manufacturing based originally on iron and steel production from the local presence of abundant high-quality iron ore was enhanced by other country-specific advantages to include major investments in technological education and training, the large degree of trade orientation and the close personal contacts within the Swedish industrial establishment which favoured the rapid absorption and adaptation of technological innovations, and their application in modern industrial production in the engineering industry (Nabseth, 1974). The growth of specialized engineering firms was a manifestation of an emerging structural change in the domestic production and export patterns of Sweden. Since these firms were geared towards exports at an early stage in their development, the country developed rapidly a more advanced export industry based on engineering products in addition to the more traditional, staple oriented ones (Olsson, 1993). However, unlike the firms in the staple industries, the high-quality steel mills and the new industrial companies based on mechanical engineering established direct contacts with foreign markets by employing travelling salesmen (Hörnell and Vahlne, 1986) owing to increased product differentiation (e.g. production of special steel versus ordinary steel), and the necessity to have intimate knowledge of market developments and to adapt products to customers' particular needs (Carlson, 1977). Their growth enabled Swedish exports to have a wider geographical scope to include not only the leading industrial countries of Europe, but also developing countries. Russia was, in fact, the largest foreign market for the Swedish engineering industry at the turn of the century (Lundström, 1986). The rapid change in the industrial and geographical pattern of its exports enabled Sweden to occupy an important position in international trade.

Unlike the firms based in the raw materials industry, most of the early FDI of the early exporting firms in the engineering industry stemmed from the need to establish a firm market presence abroad. Indeed, Swedish MNCs have come to emerge and evolve in the engineering industry and *not* in the raw materials industry. Thus, during the period of the emergence of Swedish MNCs in the late 1800s there seems to have been a disparity in the industrial distribution of exports and outward FDI of Sweden and, hence, greater scope for substitution between these two alternative modes of internationalization.[4] Indeed, one striking difference between Sweden and the larger sized but also resource-rich United States lies in the relative importance of outward FDI to exploit foreign raw materials during the period of emergence of their MNCs. This is explained in the context of the development of competitive advantages by Swedish industry in the extraction and processing of domestically available raw materials in a small country in relation to the large scale in which most investment in raw materials is undertaken typically (Swedenborg, 1979). The lesser significance of outward FDI in raw materials and the increased propensity of engineering firms to engage in outward FDI has made the manufacturing sector the most important sector for Swedish MNCs during the period of their emergence.

The growth of the engineering industry as an outcome of the Swedish industrialization phase between 1870 and the First World War can thus be regarded as the period of emergence of Swedish MNCs. The first Swedish MNCs consisting of technologically innovative firms in the engineering industry came into their own starting around the late 1800s, and their emergence can be explained as much by the need of firms from small countries to pursue an offensive and aggressive strategy to search for markets as by the need to overcome the threat posed to exports growth by the trend towards international industrial protectionism and high transportation costs.[5] Through international production supported by large, internationally oriented investment banks and the government, the process of internationalization of Swedish firms proceeded in this period of dynamic industrial change.

The traditional pattern of a gradual process of growth and development of innovative Swedish engineering firms can be regarded as consistent with that of firms from larger domestic markets: after mechanizing and specializing in a technically advanced product, firms secured the home base and engaged in exports to achieve scale economies; this was associated with the use of a network of agents or representatives (or travelling salesmen) and then by the establishment of a sales subsidiary abroad; and, finally, by the initiation of international production. This incremental process of establishment which has been described as an 'establishment chain' by Johanson and Wiedersheim-Paul (1975) can also be analysed within the framework of the product cycle model of Vernon (1966). There were, of course, exceptions to this growth pattern in some Swedish firms. The substantial growth of Svenska

Kullagerfabriken (SKF) – the ball and roller bearing company formed in 1907 – was initiated and established in the most important large-sized foreign markets of Britain, France, United States and Russia. It was only with the standardization of production and the industrialization of Sweden that the home market assumed significance to the firm. This was owing to the preference of Swedish consumers and industries for the old type of ball bearings made in Germany, unlike the consumers in Britain, France and the United States that preferred ball bearings made in Sweden (Lundström, 1986). Some other firms often had to wait for the domestic market to be large enough or modern enough to become the market base for a more specialized product (Olsson, 1993). Other exceptions emerged in cases where there was a necessity to engage in international production at once and bypass exports in order to gain access to an important foreign market. This was the case when the foreign customer was a public authority and therefore it became an outright requirement to have a local presence by the initiation of international production. This factor influenced largely the rapid development of some Swedish firms to become MNCs in this period. Among these were firms engaged in the production of generators and turbines for hydroelectric plants (ASEA), beacons and lighthouses equipment (AGA), and telephone systems (Ericsson) (Olsson, 1993; Lundström, 1986).

To the limited extent that Swedish firms in industries based on raw materials became MNCs in this period, this took the form of backward vertical integration in foreign countries. This was evident in the case of some firms in the forestry industry which tried to enlarge their own timber supplies by establishing subsidiaries in Finland and Russia, and to manufacturing firms that aimed to secure the supply of aspen splint for match fabrication (Swedish Match), cork (Wicander), or that attempted to produce chromium for steel production (Sandvik) (Olsson, 1993; Lundström, 1986). Forward integration by firms in staple industries to establish production closer to final markets abroad as well as horizontal integration to exploit accumulated know-how in resource extraction and processing in foreign markets belonged to a much later stage in the history of the growth and development of Swedish MNCs (Olsson, 1993).

The growth of Swedish MNCs from the First World War to 1960

The First World War reinforced the importance of international production both for political reasons and because of the physical barriers to trade that resulted from the hostilities. The political neutrality of Sweden during the war enabled the continued expansion of their MNCs in a way that was less possible or more difficult for firms from aligned countries.[6] The Russian Revolution of 1917 had perhaps a more profound negative impact on Swedish MNCs than the First World War since this led to the disappearance of an important export market (Russia), and the nationalization of several subsidiaries in that country without compensation (see Lundström, 1986).[7]

Despite the slow growth in the 1920s, Sweden continued its industrial transformation through technological knowledge accumulation and, as a consequence, developed into a modern industrial nation. Several new Swedish MNCs were established in the export-oriented era of the 1920s, and in particular by firms within groups of industrial companies created in the inter-war period by the leading commercial banks.[8] The rationale for international production was to circumvent trade barriers and to build commitment in important foreign markets. As a result, Sweden started to became a net exporter of capital, much of it in the form of FDI. By the end of the 1920s, there were at least 50 companies with production subsidiaries abroad (Lundström, 1986). The most notable of these companies was the Swedish Match Company which had no less than 144 producing units abroad in 33 different countries. In many ways, the match empire established by the Swedish entrepreneur Ivar Kreuger was an exceptional Swedish MNC, not only in terms of its large size but also in terms of its method of internationalization. The latter ran the gamut of large-scale acquisitions to funding large state loans in return for obtaining national match monopolies in no fewer than 15 countries, and other means of ruthless oligopolistic power play including negotiated price and trade agreements with the other large match companies in the world (Hassbring, 1979; Lindgren, 1979; Wikander, 1980).

The international depression of the 1930s led to the reduction in significance of international trade owing to increasing tariff barriers and trade wars. The share of Swedish exports to GNP declined substantially in this decade by comparison to the two preceding decades, and not until the 1960s was the share back to its level in the 1920s. There was also a smaller number of new Swedish MNCs established in this decade than during the previous decades (Sölvell *et al.*, 1991).[9] For those few firms that initiated international production in this decade and for already established MNCs, the impetus to international production was provided by host country-related locational advantages owing to trade barriers, high transportation costs and political factors. Indeed, most Swedish MNCs reported relatively high financial gains between the mid-1930s and 1944 comparable to that of the period up to and including the First World War (Wohlert, 1989). The tendency for fewer Swedish manufacturing MNCs to be established grew even stronger in the 1940s and through the Second World War. Recovery was not forthcoming until the 1950s when the number of new manufacturing MNCs established reached those before the crisis of the 1930s.

Unlike in the First World War, the Second World War led to the destruction and loss of property of Swedish MNCs in Eastern Europe, and involved risks to Swedish MNCs generally despite the continuing political neutrality of Sweden.[10] Despite this and the reduced foreign trade during this period, the volume of domestic production and profits in the engineering industry was

maintained owing to the industry's strategic importance to the national rearmament process initiated in the mid-1930s. In particular, the production of
defence material provided an important technological impetus to the industry
and many firms were established with the support of the government.[11]
Indeed, some of the most research intensive firms of the engineering industry
in the period after the Second World War profited from large state orders
associated with the heavy investment in a domestically based technologically
advanced armaments production involving leading engineering firms
(Olsson, 1977). The period of rearmament thus formed an important basis
for the further growth and development of Swedish industry and MNCs in
future years (Olsson, 1993).

The development of Swedish MNCs from 1960 to the present

The development of Sweden into a technologically sophisticated industrial
country since the Second World War (Wohlert, 1989) in combination with
international economic growth and the liberalization of world trade had
profound and positive effects on its exports and international production.
Although the share of exports to GNP reached 35 per cent during the
1970s, foreign trade was forming a smaller part of the international economic
activities of Swedish firms compared to international production that was
playing an increasingly important role. Indeed, Swedish outward FDI
increased strongly from the 1950s onwards (Hörnell and Vahlne, 1986).

As in previous eras, international production in this period was carried out
mainly by industrial firms reflecting the stability in the industrial structure of
Swedish MNCs, although firms from other sectors have increased their outward FDI in more recent decades. In the mid-1980s, some three-quarters of
the total outward stock of Swedish FDI continued to be accounted for by
industrial firms, while the remaining was owned by companies in the financial and service sectors. This owed partly to the liberalization of foreign
exchange controls in Sweden as well as investments in financial assets and
property abroad which resulted in much of the annual growth in Swedish
outward FDI since 1987 accounted for by large property investments abroad
(Olsson, 1993).

The period since the 1960s described a metamorphosis in the process of
internationalization by some Swedish manufacturing companies away from
the traditional and gradual pattern of an establishment chain described by
Johanson and Wiedersheim-Paul (1975) towards a more rapid and direct
process described by Nordström (1991) owing to a combination of factors
to include firm characteristics (other than knowledge and experience), host
country characteristics and home, host and international industry structures.
As a result, there has been an increasing tendency for Swedish manufacturing
companies that initiated the internationalization process since the 1960s to
leapfrog some of the stages of establishment by having sales as well as production subsidiaries abroad at an early stage, to invest in relatively more distant

foreign countries that have market potential (such as the United States) at an early stage with the use in many cases of the modes of acquisitions and cooperative ventures rather than greenfield investments and the establishment of wholly owned foreign subsidiaries.

Thus, while acquisitions accounted for less than 40 per cent of Swedish outward FDI in the 1960s, this proportion increased rapidly to over 60 per cent in the 1970s (Hörnell and Vahlne, 1986). Some 75 per cent of all foreign subsidiaries established by Swedish firms between 1979 and 1986 were acquired rather than started as greenfield projects, including those developed out of sales or service companies. This is associated with the need of Swedish firms to accelerate rapidly the internationalization process by obtaining new technologies, a product range or brand name, a net of subcontractors, a sales organization, or to gain rapid market share in both very fast and very slow growing industries (Olsson, 1993; Nordström, 1991). Sometimes, foreign acquisitions were made as part of structural rationalizations of companies in which firms focus their efforts narrowly on products and applications that emerge from their original technological specializations. Acquisitions of other enterprises within the same lines of production served to enhance the specialization of inventive Swedish enterprises within the engineering industry. However, even less R&D intensive industries or firms (e.g. Esselte) have chosen to move away from unprofitable business activities towards more profitable ones by the acquisition of other enterprises in more profitable lines of activities (Söderström, 1980).

The 1960s and 1970s marked the rapid growth in international production of the forestry firms in the form of vertical integration achieved through the full or partial acquisition of their major foreign customers – a move in response to the entry of the North American pulp and paper firms in Western Europe at the end of the 1950s (Carlson, 1977). Billerud, presently owned by Stora and SCA, featured among the most internationally active Swedish firms in the pulp and paper industry in the 1960s. Other firms with foreign plants included ASSI, Fiskeby, Korsnäs-Marma, NCB and MoDo (Sölvell *et al.*, 1991).[12] Apart from the few examples of firms in the forestry industries engaging in outward FDI to gain access to raw materials in Canada and the United States, and the production of pulp based on domestic raw materials in Portugal, Canada and Brazil (Söderström, 1980; Olsson, 1993) to safeguard their future expansion, the more recent pattern of outward FDI by Swedish firms in resource-based industries was quite unlike their more limited FDI in the form of backward vertical integration in resource extraction between the late 1800s and 1914. Forward integration by firms in staple industries to establish production closer to the final markets abroad as well as horizontal integration to exploit accumulated know-how in resource extraction and processing in foreign markets was motivated as much by the profitable exploitation of accumulated know-how and skills in extraction and processing of natural resources (Swedenborg, 1979) as by the need to gain market knowledge and to influence product development in the pulp- and paper-consuming industries (Carlson, 1977). In the early 1970s, the Swedish pulp and paper manufacturers had established more than 40 manufacturing plants abroad, mostly in the United Kingdom and Germany, and the further growth of their European investments in the 1970s and 1980s enabled the Swedish pulp and paper companies to build dominant market positions (Sölvell *et al.*, 1991).

After the domestic industrial crisis of the 1970s when the traditional industries producing pulp and paper, minerals, metal and steel, shipbuilding, etc. experienced heavy setbacks and underwent painful restructuring, the technologically more advanced and more internationalized engineering firms displayed better ability to adjust. The emphasis thereafter laid on upgrading domestic production as reflected in the significant reduction in the share of merchandise exports derived from raw materials (wood, pulp and metal ores) from 43 per cent in 1953 to 12 per cent in 1978, and a dramatic increase in the share derived from specialized and rather advanced engineering products from 24 per cent to 47 per cent over the same period (based on data in Söderström, 1980). As a result, there had been a rapidly increasing congruence in the industrial pattern of exports and outward FDI of Sweden since 1953.

Developed countries continued to receive a dominant share of Swedish FDI in the period since 1960, with some 83 per cent of employment in Swedish manufacturing subsidiaries abroad (Olsson, 1993). This was because Swedish industry regarded it as necessary to invest within the tariff walls of the emerging regional market of the European Community (EC) during the 1950s and 1960s, while Sweden as a member of the European Free Trade Area (EFTA) could service the markets of the other EFTA member states more efficiently through exports (Hörnell and Vahlne, 1986). The free trade agreement concluded between EFTA and the EC in 1972 combined with the expectation of slower growth in Europe and the high market shares already held by Swedish firms led to a slowdown in the growth of Swedish FDI in the EC, and to a corresponding growth of FDI in the United States during the 1970s and 1980s.[13] Regardless of the physical and psychic distance, the large size, technical sophistication and growth potential of the United States market attracted not only small and inexperienced Swedish manufacturing firms that began their internationalization process after 1977 but also large, already well-established Swedish MNCs particularly in industries with a high degree of international competition (Nordström, 1991).

The renewed importance of Europe for Swedish FDI since 1986 took place in the context of the formation of the Single European Market and the expectation of rapid growth and dynamic industrial development as well as the continuing uncertainty about Sweden's future relations with the EC. Pulp and paper firms have been inclined to invest in the EC, as mentioned, particularly in the last stage of the production process.[14] Thus, through vertical integration, Swedish pulp and paper firms secured important segments of the final markets in Europe, and some firms, such as Stora and SCA, consolidated their leading positions in Europe through larger acquisitions (Sölvell *et al.*, 1991). Similarly, the more established skills-intensive engineering firms carried on investing in Europe to gain larger shares of a growing market, while the younger, highly specialized and R&D intensive firms became MNCs to gain a foothold in the European market. As a result, there was a doubling of outward FDI flows in the EC in every year from 1985 to 1988 (Olsson, 1993).

Since the 1980s, the importance of production and sales in Sweden for Swedish manufacturing companies – including those in the resource-based industries – continued to decline by comparison to their production and sales abroad.[15] This confirms a general trend that domestic and foreign production do not grow at the same rate over time, with domestic growth preceding growth in foreign markets typically in the early stages, but growth through foreign production once initiated exceeds domestic growth at later stages (Swedenborg, 1979). Although this trend applies rather more to MNCs in the engineering industry that have a longer history of international production, it is also applicable to the resource-based industrial firms whose international production had become increasingly important since the 1960s. Indeed, by the late 1980s, the leading pulp and paper firms – Stora, MoDo, SCA and ASSI – had 50 per cent of their total employment abroad (Sölvell *et al.*, 1991). Apart from these firms, other firms in the mining and steel industries have also invested in major foreign operations abroad in the 1980s (Sölvell *et al.*, 1991).[16]

Not only has there been a more rapid pace in the internationalization process of Swedish firms in more recent decades with the tendency to leapfrog some of the stages of establishment and the approach of relatively more distant markets early on (Nördstrom, 1991) but also a change in the determinants of internationalization. The rapid internationalization of young technology-based firms in recent decades reflect efforts not only to increase sales but to establish closer contacts with markets where important technological developments in the industry take place (Lindqvist, 1991). In a limited number of Swedish industries, sophisticated foreign demand has substituted for having advanced customers in the home market in providing the crucial pressures and challenges for new product development; but this was selective and could not have worked without the support from a dynamic home base (Sölvell *et al.*, 1991).[17]

More importantly, the high international orientation of Swedish firms and MNCs have fostered learning and upgrading of their core skills and technologies assisted by the establishment of specialized product centres in different countries, the establishment of core R&D operations tied to major foreign plants, the relocation of the corporate headquarters or the divisional and business units to another country, and the use of new modes of international investment, including mergers and acquisitions as well as the formation of strategic alliances with foreign firms. By establishing specialized product centres or several home bases for each product or technology area, leading Swedish MNCs such as SKF, Atlas Copco, Electrolux, Ericsson and ABB and Fläkt have attained scale economies and gained from the concentration of research and/or production in the most advantageous location. The 20 largest Swedish MNCs perform some one-quarter of all R&D abroad and, in addition, have been involved heavily in international research programmes to have an influence in the standards set by industry organizations, to enhance technological skills and to establish contacts with competitors and customers (Håkanson and Nobel, 1989; Håkanson, 1990). The establishment of foreign-based research and/or production in the electronics and pharmaceuticals industries in particular enabled Swedish firms to have listening posts in centres of innovation around the world, to tap into special research competencies in certain countries, to establish networks with foreign research institutions, and to recruit foreign specialists (Sölvell *et al.*, 1991).

With the increasing importance of international production, the firm-specific advantages of Swedish firms and MNCs have drawn increasingly from their ability to develop new technologies in a narrow range of products that formed the basis of their specialization and world leadership owing to quality, service and market knowledge. Their growth strategies are evolving rapidly as seen both in the faster pace of the process of their internationalization and in the rationalization of the structure of their international production activities that facilitate cross-border product or process specialization and learning from producing in different environments. Since some foreign affiliates have become increasingly instrumental in the strategy formulation of the MNC of which they are a part, the organizational pattern of some Swedish MNCs have changed from a hierarchical system of decision making to a heterarchical system associated with globally integrated MNCs (Hedlund, 1984).

Conclusion

This chapter analysed the emergence and evolution of Swedish MNCs since the late nineteenth century. It showed the more rapid process of internationalization through exports and outward FDI of firms based in small countries compared to firms based in larger countries as an important means to expand markets and attain economies of scale. Although there has been a closer similarity in the industrial distribution of exports and outward FDI of Sweden since 1953, there was a much greater industrial disparity and hence greater scope for substitution between trade and outward FDI of Sweden during the period of the emergence of Swedish MNCs in the late 1800s. Indeed, firms in the dominant group of Swedish export industries based on raw materials (wood, pulp and metal ores) became MNCs only to a very limited extent during the period of emergence of Swedish MNCs, while those engaged in the domestic production and exports in the engineering industry consisting of metal manufactures, machinery and transport equipment, engendered the emergence of the first Swedish MNCs in a more significant way in the late 1800s. The increasing importance of international production by the more resource-based industrial firms became evident in a major way only since the 1960s. Forward integration by firms in staple industries to establish production closer to the final markets abroad as well as horizontal integration to exploit accumulated know-how in resource extraction and processing in foreign markets was motivated as much by the profitable exploitation of accumulated know-how and skills in extraction and processing of natural resources (Swedenborg, 1979) as by the need to gain market knowledge and to influence market development in the pulp- and paper-consuming industries (Carlson, 1977).

The period since the 1960s also describes the metamorphosis in the process of internationalization by some Swedish manufacturing companies away from the traditional and gradual pattern of an 'establishment chain' described by Johanson and Wiedersheim-Paul (1975) towards a more rapid and direct process described by Nördstrom (1991) with the marked tendency of Swedish MNCs that initiated international production since that decade to leapfrog some of the stages of establishment by having sales as well as production subsidiaries abroad at a much earlier stage of the internationalization process, to invest in relatively more distant foreign countries that have market potential (such as the United States) at an early stage with the use in many cases of the modes of acquisitions and cooperative ventures rather than greenfield investments and the establishment of wholly-owned foreign subsidiaries. This is associated with the changed determinants of outward FDI as a means not only to expand markets and attain scale economies but to establish closer contracts with markets where important technological developments in the industry take place (Lindqvist, 1991). As a result, the firm-specific advantages of Swedish MNCs have derived increasingly from their ability to develop new technologies in a narrow range of products that formed the basis of their specialization and world leadership.

Notes

1 Based on data contained in Tolentino (1993) for 1960 and 1975, and UNCTAD (1999) for 1990 and 1998.
2 In terms of number, there were some 82 Swedish firms with foreign production facilities in 1965, and 118 such firms in 1978 employing some 300,000 people abroad. Some 64 of the 118 Swedish MNCs were small companies with foreign employment of less than 2,000. There is a high degree of concentration within Swedish MNCs. In 1978, the 20 largest Swedish MNCs operating abroad were responsible for more than 80 per cent of all Swedish enterprises' employment abroad (Hörnell and Vahlne, 1986). By 1982, 47 Swedish MNCs with more than 500 employees abroad accounted for 92 per cent of all foreign employment. The 15 largest accounted for 81 per cent (Sölvell *et al.*, 1991).
3 In a cross-sectional analysis of practically all Swedish firms in the manufacturing sector in 1974, Swedenborg (1979) found that Swedish industry had a higher export intensity but a lower international production intensity than American industry. The higher export intensity of Swedish industry is explained by the need to exploit economies of scale in production in the face of a relatively small domestic market. The relatively small distance to major export markets favoured

the exploitation of such scale economies. The corollary is that the lower export intensity and higher international production intensity of American industry is explained by the larger size of the United States in both an economic and geographic sense.

4 This is a finding that seems to be an exception to the general principle of a greater correlation in the sectoral distribution of the growth of exports and outward FDI for relatively newer investor countries such as Germany and Japan (see Cantwell, 1989a). By comparison, there is a much greater disparity in the sectoral distribution of growth of exports and outward FDI and, hence, greater scope for substitution between trade and outward FDI for the more mature and established international investors such as the United Kingdom and the United States.

5 Among the pioneering Swedish MNCs were Alfred Nobel's corporation, Nitroglycerin AB, which had already initiated production in ten different countries in the 1870s, as well as Alfa-Laval (through its forerunner company, AB Separator) which established its first foreign plant (milk separators) in the United States in 1883. Nobel (explosives) established a plant in Germany in 1886. Wikander (cork and linoleum floors) established plants in Finland and Russia in the late 1800s. Perstop (Skånska Ättiksfabriken) built a plant for acetic acid production in Norway in 1898. ASEA installed a factory in the United Kingdom in 1898 to adapt electrical machinery to local needs, and had equity ties with licensees in Finland (1897), Norway (1898) and Denmark (1900). Ericsson set up major manufacturing plants in Russia around the turn of the century. SKF, the most internationalized Swedish firm in manufacturing, expanded production around the First World War. Plants were established in the United Kingdom (1911), Germany (1914), United States (1916) and France (1917). AGA established a plant for lighthouse production in England in 1914 (Sölvell *et al.*, 1991; Hörnell and Vahlne, 1986).

6 For example, the investments of SKF in the United States continued to flourish, while German producers of ball bearings in that country did not (Olsson, 1993).

7 As mentioned, Russia appeared to be the most promising market for Swedish industry during the period of Swedish industrialization, especially for the young engineering industry. Swedish exports to Russia soared and many Swedish companies founded Russian subsidiaries encouraged by the great success of the Nobels, Wicander, Ericsson and SAT (Lundström, 1986).

8 Leading commercial banks became owners of industrial shares used as collateral for loans. Thus Enskilda Bank, owned by the Wallenberg family, gradually became the nucleus of the largest industrial group (Olsson, 1993).

9 Among the Swedish firms that initiated international production in the 1920s and 1930s were Esab which opened up plants for welding consumables (electrical welding machinery) across Europe in the 1920s, with the first manufacturing unit initiated in Germany where superior electrical welding know-how existed. Svenska Flätfabriken (now ABB Fläkt) opened its foreign plants in the 1930s in Finland, Norway and France, and expansion continued in the 1950s in Europe, United States and Mexico. Atlas Copco established plants in the United Kingdom (1939), South Africa (1947), Brazil (1960), India (1962) and Germany (1970) (Sölvell *et al.*, 1991).

10 During this time, SKF with production subsidiaries in Germany and the United States could not escape the conflict (Olsson, 1993).

11 An excellent case in point is SAAB (later SAAB-Scania) that was given a privileged start as a producer of aircraft by the government (Olsson, 1993).

12 Almost all foreign paper plants of Swedish firms were specialized in either sack paper or paper board (Sölvell *et al.*, 1991).

13 The value of international production by Swedish firms in the United States increased eightfold between 1979 and 1986 (the peak) compared to the value of Swedish exports to that country which increased only fourfold (Olsson, 1993).

14 It has not been possible to move the large scale basic processes out of Sweden since these had to remain close to the source of raw materials (Olsson, 1993).

15 Production in Sweden declined from 73 per cent to 53 per cent between 1965 and 1986 while production abroad increased from 27 per cent to 47 per cent over the same period. Sales in Sweden declined from 47 per cent in 1965 to 23 per cent in 1986, while sales abroad increased correspondingly from 53 per cent to 77 per cent (Olsson, 1993). The ten most internationalized Swedish MNCs reduced their domestic investments further by 14 per cent between 1985 and 1988 (Sölvell *et al.*, 1991).

16 Gränges has operated mines abroad since the 1950s, notably the LAMCO mine in Liberia. Boliden acquired Greenes in Denmark in 1986 and Falconbridge in Canada in 1989, and operates mines in several countries (Sölvell *et al.*, 1991).

17 Influences derived from international customers described the development of Inter-Innovation (the Swedish manufacturer of teller-operated cash dispensers) and Wallenius (the Swedish manufacturer of carrier ships for cars and trucks). Although Inter-Innovation developed within a strong domestic cluster of industries based on precision electro-mechanical know-how, domestic customers did not play a decisive role in the growth of the company. The growth impetus came from a large order from Citibank in 1979, followed by contracts from British and Spanish banks for teller-operated cash dispensers. Stringent product demands by foreign users in terms of quality and technological sophistication pressured the firm to upgrade its technology. By 1987, the firm derived only 5 per cent of its total sales in Sweden. Similarly, in the Swedish car shipping industry, contacts with foreign customers influenced the development of shipping design by Wallenius, helped by the presence of a strong domestic cluster of shipbuilding industries. Continuous technical development of carrier ships by Wallenius that led to the development of combined car-bulk carriers in 1974 and the pure car and truck carriers in 1977 was driven primarily by foreign customer needs (Sölvell *et al.*, 1991).

5 The emergence and evolution of multinational corporations from Brazil

Introduction

By comparison to the long and rich history of MNCs based in other resource-abundant countries such as the United States and Sweden, the history of Brazilian MNCs whose origins can be traced to the late 1960s is much more recent and contemporary. Besides, the stock of Brazilian FDI which reached an estimated $9.8 billion in 1998 is far more modest and represented some 0.2 per cent of the global stock of outward FDI. Nevertheless, in relation to the stock of outward FDI from developing countries in that year, Brazil assumed greater relative importance with a share of 2.5 per cent (based on data in UNCTAD, 1999).

Despite the low relative importance of Brazilian FDI, the study of the emergence and evolution of Brazilian MNCs is of interest not the least because Brazil, like the United States and Sweden, is a resource-abundant country. Thus, although the history of Brazilian MNCs may be of more recent vintage and is in many respects still in the stage of emergence and their outward FDI remains low in both absolute and relative terms, the growth pattern of their MNCs as it has been evolving over the last 30 years or so can be compared to that of MNCs from the United States and Sweden whose home countries share similar patterns of national economic development. The analysis of the history of Brazilian MNCs in this chapter is divided in two distinct phases: the period of emergence from the late 1960s to 1975, and the period of development in the period since 1975.

The emergence of Brazilian MNCs in the late 1960s to 1975

The pioneering Brazilian MNCs spanned the three sectors of economic activity. In the primary sector, there was the oil exploration and drilling activities by the state-owned firm, Petrobrás, while in the secondary sector there was the initiation of international production by local firms in bicycles (Calói), electrical equipment (Gradiente Electronics) and motor vehicle parts (Eluma and Marcopolo). Thus, although Brazil had one of the largest and most developed manufacturing sectors among developing countries, its outward

FDI did not feature prominently in the manufacturing sector in the emergent phase of Brazilian MNCs. Indeed, apart from Petrobrás, the main thrust of the earliest outward FDI by Brazilian firms had been in services. Of these, the most significant were those firms involved in heavy civil construction in Nigeria, the Middle East and Latin America and the engineering consulting companies that set up foreign subsidiaries to build refineries, to design projects, etc.

The emergence of Brazilian MNCs in the primary sector

Concerns about chronic balance-of-payments deficits during the end of the era of vertical import substitution industrialization (ISI) (1955–70) prompted a political decision by the Brazilian government to search for new sources of foreign oil in light of Brazil's position as a major importer of petroleum. Towards that end, Braspetro – a subsidiary of the state-owned oil monopolist Petrobrás – was created in 1972 to fulfil a strategic national interest of securing oil supplies abroad through risk contracts and close relations with OPEC countries. It became responsible for the foreign exploration, production, commerce, transportation and storage of oil and its products, as well as for the execution of technical and administrative services related to those activities (Guimaraes, 1986). Unlike American MNCs that emerged in a major way in outward FDI in the primary sector to exploit firm-specific assets in natural resource extraction and processing nurtured in the natural resource abundancy of the United States, the security of the supply of raw materials or natural resources as a motive for outward FDI in the emergent phase of Brazilian MNCs was rather unique to the Brazilian state-owned company Petrobrás, and specific to one natural resource – oil – given Brazil's rich agricultural and mineral resources. The only other exceptional case of Brazilian resource-based FDI was the joint venture established with the Colombian government by the Brazilian state-owned steel enterprise, Siderbras, to exploit coal in Colombia with the objective of gaining independence from the United States for such raw material (White, 1981). In this respect, the Brazilian MNCs share more common features with Swedish MNCs where firms in the industries of mining, metal production and forestry became MNCs only to a very limited extent (see Chapter 4).

The geographical scope of the foreign activities of Braspetro covered at least 14 countries. Exploration contracts were undertaken frequently in association with other companies, and primarily with state-owned firms in Iran, Egypt, Colombia, Algeria, Iraq, Angola and Congo of which in some cases it had some equity participation, as well as with large multinational oil companies such as Texaco, Mobil Oil, Cities Services, BP, Elf Aquitaine and Total (Guimaraes, 1986; White, 1981). In its various foreign activities, Braspetro drew upon the accumulated experience of its parent company in oil and gas exploration, geophysics and drilling – an advantage that had been enhanced by a training centre for engineers and middle-range professionals in the company's Centre for Research and Development in Rio de Janeiro. Those assets in addition to learning by doing on its own outside Brazil enabled Braspetro to build up even more advanced engineering and managerial skills related to petroleum exploration and to progress over time to become its own profit centre and not simply an instrument of national policy. Evidence for this was found in its increasing ability to enter into risk investments and to render technical services in the construction and installation of oil rigs, refineries, storage systems and pipelines, and to participate in the oil trading market. It is also reflected in its use of deep-sea oil production know-how in the North Sea and Gulf of Mexico – former turfs of major oil companies (Villela, 1983; Wells, 1988).

The emergence of Brazilian manufacturing MNCs

As mentioned, outward FDI by Brazilian manufacturing companies in the emergent phase was rather limited. International production by the earliest Brazilian manufacturing MNCs in bicycles (Calói), electrical products (Gradiente Electronics) and motor vehicle parts (Eluma and Marcopolo) which consisted mainly of assembly operations was a means to overcome the high degree of protection in export markets. In the mould of the product cycle model of Vernon (1966), both Gradiente and Calói developed some export experience (although never more than 20 per cent of total sales according to Wells, 1988) with a network of dealers and service centres when the firms established assembly operations in Mexico and in the Andean Pact countries, respectively. The assembly operations in Bolivia and Colombia by Calói (the leading Brazilian firm in bicycles for leisure) and Eluma (the leading Brazilian firm in copper products and a major car parts manufacturer) circumvented the prohibition on imports stipulated in the Andean Pact regulation by local production with a high domestic value added (no less than 64 per cent of the value of a product).[1] Although similar protectionist tendencies in Mexico prompted the assembly of audio/stereo equipment in that country by Gradiente Electronics (Brazil's largest manufacturer of sound equipment), the assembly plant imported parts and components from the parent company's factory at the Manaus Free Trade Zone in Brazil. An attempt to encourage greater local value added in Mexico met with limited success as the assembly plant closed down in 1986 (Peres Núñez, 1993).

The emergence of these pioneering Brazilian manufacturing MNCs can be viewed in the context of the diversified export promotion industrialization phase of Brazil in the period since the mid-1970s. The elaborate system of export promotion established in this phase served to enhance the growth and diversification of exports particularly of manufactures and technical services and to reduce dependency on primary products.[2] More importantly, the system also favoured the decision of some domestic firms to become MNCs.[3] While the elaborate export promotion system did not provide the basic incentive to export capital to establish foreign production capacity largely because of Brazil's large foreign debt and the need to conserve foreign exchange, the financial and credit incentives offered under the system added to the broad range of incentives already enjoyed by many manufacturing companies (mainly assembly industries interested in foreign markets) with the establishment of the Free Trade Zone in Manaus in the mid-1960s. Export incentives in the Free Trade Zone, including the widespread use of drawbacks, exemptions from customs duties on imports of capital goods for industries under installation and capital financing at subsidized rates, stimulated Brazilian manufacturing companies to establish facilities in the Zone, and in the case of Gradiente and Calói to also establish plants in border countries such as Bolivia, Colombia and even Mexico. Typically, Brazilian manufacturing firms that later became MNCs had facilities already located in the Free Trade Zone (Villela, 1983).

The emergence of Brazilian services MNCs

Brazilian companies in the construction and consulting engineering industries established foreign subsidiaries to execute projects in heavy civil construction (building expressways, railways, irrigation dams, hydroelectric plants, hotels and university cities, etc.) in Africa, the Middle East and Latin America. Among the most prominent of these companies were Mendes Junior (which ranks among the world's largest construction companies in terms of foreign sales), Camargo Correa, Construtora Rabello, Ecisa, Esusa and Sisal. Construction and consulting engineering firms operated frequently together in foreign markets, with the consulting engineering firms often spearheading construction companies in developing countries with imports of machinery and equipment from Brazil. This occurred in many countries of Latin America and Africa, mainly in the construction of highways and dams. There were some 26 Brazilian consulting engineering firms that had overseas activities between 1975 and 1979 with an average annual sales per company of $18.5 million (Villela, 1983).

The international forays of almost all Brazilian construction and consulting engineering companies have been initiated in the form of exports of services to Latin America – a move that was partly in response to the country's export promotion policy and partly by the need to overcome the firms' vulnerabilities to the vagaries of the domestic market. At the same time, the tax breaks and subsidized credits associated with the elaborate exports promotion system geared to promote exports of capital goods prompted consulting engineering and construction companies to establish foreign subsidiaries to compete for contracts in overseas projects involving telecommunications networks, ports, hydroelectric power stations, dams, highways, etc. Exports and international production became a dual means to expand markets as many of these companies had achieved a size that required a minimum level of demand to use their installed capacity effectively.[4]

The expansion abroad of Brazilian construction and consulting engineering companies can be attributed to their ability to adapt technologies to the special needs of developing countries. Most of the Brazilian engineering companies active abroad acquired their expertise from previous associations with foreign companies that had come to Brazil to carry out specific jobs such as the design and assembly of hydroelectric plants, nuclear plants, petrochemicals, steel mills, telecommunications systems and so on (Villela, 1983).[5] However, the requisite technologies to execute more sophisticated projects involving petrochemical complexes, offshore drilling rigs and ferroalloy plants were acquired through former licensing arrangements (Sercovich, 1984).[6]

Apart from their technological skills, Brazilian construction and consulting engineering companies have the ability to raise capital (particularly important in winning contracts in non-oil producing developing countries where bids for the execution of jobs were sometimes linked to the concession of credits at competitive interest rates), and have sophisticated management structures to mobilize a large number of workers, to procure from many suppliers across the world, and to complete projects rapidly and efficiently (Villela, 1983).

The major expansion of Brazilian MNCs since 1975

While Brazilian MNCs emerged in all three major sectors of economic activity, the main thrust of outward FDI undertaken by Brazilian MNCs during the period of their emergence from the late 1960s until 1975 was accounted for by Petrobrás and the construction and consulting engineering companies, as mentioned. The limited amounts of outward FDI in the manufacturing sector were undertaken by the pioneering Brazilian manufacturing MNCs such as Calói (bicycles), Gradiente Electronics (electrical equipment) and Eluma (copper products and motor vehicle parts). By contrast, the second phase of Brazilian MNC growth which pertain to the period since 1975 can be distinguished clearly as the phase of the major emergence of Brazilian MNCs in the manufacturing and banking sectors and, to a lesser extent, the emergence of overseas mineral prospecting activities by Brazilian state and privately owned resource-based companies. The period also described the continued international expansion of Brazilian MNCs in construction and consulting engineering and the emergence of Globo, a Brazilian MNC in TV broadcasting.[7] Some examples of Brazilian MNCs that have come into existence are provided in Table 5.1.

The emergence of Brazilian MNCs in banking

A major trend towards the internationalization of both public and private Brazilian commercial banks had been observed particularly since 1974, and the pace was so rapid that by June 1981 14 public and private Brazilian commercial banks had operations abroad consisting of 190 branches in 48

Table 5.1 Some examples of Brazilian multinational corporations

Companies (by main industry)	Foreign activity	Country	Year	Notes
A. Primary				
Oil				
Petrobrás (through Braspetro and Interbras)	Oil exploration	Iraq	1972	First contract abroad. It discovered the giant oil fields of Majnoor and Nahr Umr
	Oil exploration	Algeria	1974	
	Oil exploration	South Yemen	1982	
	Oil exploration	China, India	1983	
	Oil services	Norway	1984	Joint venture
	Oil services	Angola	1985	$8 million contract
	Oil exploration	United States	1987	Joint venture with Texaco to work in the Gulf of Mexico
Minerals				
Siderbras	Coal exploration	Colombia	1970s	Joint venture
Companhia de Pesquisa de Recursos Minerais (CPRM)	Mineral prospection	Africa	1979	Contract. No subsidiaries
	Gold prospection	Angola	1984	Contract. No subsidiaries
Paranapanema	Gold prospection	Guyana	1984	Joint venture
B. Manufacturing				
Food products				
Copersucar (a cooperative formed by about 70 of the largest sugar and coffee producers)	Soluble coffee	United States	1976	Acquired Hills' Brothers Coffee, the fourth-largest coffee processing plant in the USA, for $150 million. The purpose was to vertically integrate into coffee processing and marketing of coffee as well as cocoa and sugar using the Hills' Brothers brand name. Sold to Nestlé in 1986 after failing to make adequate returns

Cotia	Cattle ranches and meat packing, cold storage facilities	Nigeria	late 1970s	Joint venture. Prompted by highly protected host country market
	Soft drinks factory	Nigeria	late 1970s	Joint venture. Prompted by highly protected host country market. Plant for bottling guaraná (a traditional Brazilian soft drink), of which the syrup is imported from Brazil
Cica	Canned foods	Argentina	1979	
Cacique	Instant coffee bottling	China	1985	The leading exporter of Brazilian instant coffee invested $0.5 million in a joint venture to move away from exporting unpackaged instant coffee, and to sell most of the bottled coffee to Japan, Australia and other Asian markets
Brahma	Malt production	Argentina	1987	Joint venture
Cerval Alimentos	Soya oil and meal plant	Portugal	1993	Joint investments with other Brazilian groups Investment: $40 million
Textiles, clothing and footwear				
Grendene	Plastic shoes	Argentina, Colombia, Mexico	1986	Joint ventures with Argentine partner
Hering	Cotton knitwear licensing	Argentina	1986	Joint venture
	Shirt factory	Spain	1988	To serve a market segment demanding rapid deliveries. 780 workers in 1991
Vacchi	Tannery	United Kingdom	1986	Acquired British tannery
SP Alpargatas	Jeans factory	Spain	1989	Penetration of market segments with bigger profit margins. Flexibility in meeting clients' needs. Investment: less than $14 million

(continued)

Table 5.1 (continued)

Companies (by main industry)	Foreign activity	Country	Year	Notes
Textiles, clothing and footwear (continued)				
Staroup	Jeans factory	Portugal	1989	Add value to product. With Portuguese partner
Packaging				
ITAP	Food packaging factory	United States (Buffalo, New York)	1983	Production and finishing of products exported from Brazil Project for a plant in Portugal
Toga	Paper and cardboard packaging factory	United States	1988	Bryce Corporation provided 40% of capital
Wood and furniture				
Securit	Assembly and showroom	United States	1979	Investment of $0.5 million in a joint venture to bypass furniture dealers and sell directly to customers, taking advantage of Brazilian wood and labour as well as the oil boom and large property investments in Texas. Local competitive pressures forced the firm to close down the operation
Bergamo	Furniture assembly	Colombia	before 1980	Invested abroad to offset the contraction of the domestic market which led to firm losses in 1977 and 1979 Investment in response to Andean Pact legislation. Export of components from Brazil
Labra	Pencil factory	Portugal	1985	Acquired pencil factory for $130 million to increase exports to the European Economic Community
Duratex (Itau Group)	Factory	Germany (Hanover)	1990	Increase value added. Joint investment. 1990 sales: $14 million

Bicycles				
Calói	Bicycle production	Bolivia	1974	Joint venture to overcome protected market. Local production with a high domestic value added in response to Andean Pact regulation
	Bicycle production	Colombia	1978	Joint venture to overcome protected market. Local production with a high domestic value added in response to Andean Pact regulation
	Licensing for bicycles	Guyana	1980	Royalties agreement
Lifts				
Villares	Lifts subsidiary	Chile	1977	
	Lifts subsidiary	Colombia, Uruguay	1978	
	Lifts subsidiary	Mexico	1980	
Electrical products				
Gradiente Electronics	Assembly of audio/ stereo equipment	Mexico	1973	Investment prompted by highly protected market. Imports parts from the parent company's factory at the Manaus Free Trade Zone. Closed down in 1986
	Laboratory and trademark	United Kingdom	1979	Acquired the Plessey subsidiary, Garrard, for $2.5 million. The factory in Swindon, UK, was closed down in 1982, and production transferred to Gradiente's factory at the Manaus Free Trade Zone. The main objective was to become the owner of a world-known trademark and a distribution network in various countries for sound equipment made in Brazil
Inepar	Electrical control equipment	Chile	1977	Investment of $0.4 million in a joint venture

(continued)

Table 5.1 (continued)

Companies (by main industry)	Foreign activity	Country	Year	Notes
Electrical products (continued)				
Inbrac	Cables	Paraguay	1978	Joint venture
	Cables	Ecuador	1985	Joint venture. Total investment in this venture and that in Paraguay: $1.364 million
Nansen	Electricity meters	Colombia	1985	Investment of $0.3 million in a joint venture plant to serve as a springboard for Third World country sales
Brastemp	Refrigerator production	Argentina	1990	Acquired 40% of former Philips affiliate. Joint investment with Whirlpool. Investment: $10 million
Steel products and capital goods				
Gerdau	Steelworks	Uruguay	1981	Produces 39,000 tons per year. To sell in protected market
	Steelworks	Canada (Ontario)	1989	Produces 250,000 tons per year
Companhia Vale do Rio Doce (CVRD)	Steel production	United States (Fontana, California)	1984	The world's top iron ore producer acquired 25% of the Fontana steel mill from Kaiser Corporation for $20 million. Together with Michael Wilkinson who acquired 50% and Kawasaki Steel Corp. of Japan which acquired the remaining 25%, the California Steel Company was formed. The purpose of CVRD in this investment was to cash in on the value added to iron ore at the processing stage

(continued)

Bardella	Various capital goods	United States	1985	Acquired 50% of Schuler Inc. – an American offshoot of Germany's Schuler GmbH – for $4 million in order to have a significant local presence in the USA to win contracts to supply capital goods and to provide a service network. Since the acquisition, Schuler Inc. has become the principal supplier of stamping presses to Chrysler Corporation
Motor vehicle parts				
Eluma	Auto parts	Argentina, South Africa	1968	Established foreign joint ventures with Bundy – an American firm that has been its long-term partner in Brazil since the birth of the motor vehicle industry in the 1950s
	Auto parts	Colombia Venezuela	1971	As above
Marcopolo	Assembly of bus bodies	Venezuela	1971	
	Assembly of bus bodies	Ghana	1974	
Cotia	Auto parts	Nigeria	1979	Joint venture. Prompted by highly protected host country market
Metal Leve	Piston factory	United States (South Carolina)	1989	Clients demand technological solutions and 'just-in-time' inventory management system rather than merely supplying parts or components. Investment: $15 million. Sales: $10 million. Plans to establish a piston factory in Europe and a gasket factory in the USA
	R&D centre	United States (Ann Arbor, Michigan)	1988	

(continued)

Table 5.1 (continued)

Companies (by main industry)	Foreign activity	Country	Year	Notes
Motor vehicle parts (continued)				
SIFCO	Shaft machining plant	United States	1989	To respond to 'just-in-time' operation. Planned second machining plant in 1991
COFAP	Engine part factory	Portugal	1991	Investment in response to Single Europe. Investment: $150 million
	Engine parts	Argentina	1991	Acquired 50% of INDUFREN to eliminate potential competition in MERCOSUR
Aircraft				
Embraer	Aircraft assembly	Egypt	1983	$181 million contract to build Tucano military training aircrafts for the Egyptian Air Force. Local production necessary
	Aircraft assembly	United Kingdom	1985	$150 million contract to Tucano military training aircrafts in Northern Ireland for the Royal Air Force. Local production necessary
Other manufacturing				
Cotia	Nail making factory	Nigeria	1970s	Joint venture. Prompted by highly protected host country market
Banking				
14 public and private Brazilian commercial banks including Banco do Brasil, Banco de Estado de São Paulo and Banco Real	190 bank branches	48 countries	between 1974 and 1981	Foreign expansion of banks associated with the growth of Brazil's industrial exports, and to gain access to international capital markets

Construction services

Mendes Junior	Construction	Mauritania	1974	Construction of a $120 million highway (606 km)
	Construction	Uruguay	1980s	Completed a $157.8 million hydroelectric plant
	Construction	Iraq	1970s	Sole contractor for the construction of a $1.2 billion, 1,040 kilometre railway, between Baghdad and Akashat; and the construction of a 128.5 kilometre expressway at the cost of $380 million. As of 1981, the company made investments worth $340 million in machinery and equipment
Constutora Rabello	Construction	Algeria	early 1970s	A medium-sized company that designed, built and expanded the university cities of Constantin and Algiers. It also won a $200 million contract for the construction of an irrigation dam
Camargo Correa	Construction	Venezuela	1980s	Involved with two other international firms in the construction of the Gury hydroelectric plant
Odebrecht	Construction	Peru, Ecuador	1970s	
	Construction	Portugal	1988	Acquired a construction firm to enhance its grip of the local market and to break into EEC work
Ecisa	Construction	Tanzania	1970s	A medium-sized company that built a $250 million highway, and a $150 million highway
Esusa	Construction	Iraq	1980s	Built hotel networks worth $150 million
Sisal	Construction	Angola	1980s	Built hotel networks worth $100 million
Veplan	Construction	Chile	1980s	Invested $25 million in a joint venture to build a shopping centre – Parque Arauco – in Santiago

(continued)

Table 5.1 (continued)

Companies (by main industry)	Foreign activity	Country	Year	Notes
Engineering services				
Tenenge	Subsidiary	Paraguay	1975	
Some 26 companies	93 projects	22 countries in Latin America, Africa and the Middle East.	1975–79	Average annual sales per company of $18.5 million
Retail trade				
Pão de Açúcar	Supermarket	Portugal	1970	Joint venture. It is the leading supermarket chain in Portugal. It was expropriated by the Portuguese government
	Supermarket	Angola	1973	
	Supermarket	Spain	1975	Franchise
Broadcasting				
Globo	TV broadcasting	Italy	1985	Acquired Telemontecarlo TV station to apply its programming skills to developed countries. After searching the world for an acquisition, Telemontecarlo was chosen based on price and expansion prospects

Sources: Compiled from information gathered from Peres Núñez (1993) from primary data obtained from Viviane Ventura, *A internacionalização das empresas brasileiras*, ECLAC/UNDP Regional Project RLA/88/039, Santiago, Chile: ECLAC, 1991; Wells (1988); Villela (1983); and White (1981).

countries (Villela, 1983). Two public banks stand out: Banco do Brasil with 78 foreign branches, and the Banco de Estado de São Paulo, while the Banco Real with 55 foreign branches was by far the largest and most forceful Brazilian private bank abroad. The establishment of an overseas network by Brazilian state-owned and privately owned banks in the developed countries enabled the banks to attract resources from international capital markets, and their expansion in Latin America was allied closely with the rapid growth of Brazilian exports to the region (Guimaraes, 1986).

The major expansion of Brazilian manufacturing MNCs

Some 121 locally owned manufacturing firms engaged in outward FDI between 1977 and 1982 (calculated from data in Guimaraes, 1986). The investments were, however, highly concentrated: some 58 firms accounted for 98 per cent of Brazilian outward FDI in the manufacturing sector based on authorization certificates issued by the Central Bank in that period, and many of those firms were leaders in their respective industries. Thus, the process of multinationalization of Brazilian manufacturing firms can be analysed in the context of oligopolistic market structures in which the benefits of economies of scale could be further exploited through international production. A sign of the emergent stage of their multinationalization is seen in the fact that over 80 per cent of those 58 locally owned firms invested in only one foreign country between 1977 and 1982 (Guimaraes, 1986), and sales by foreign subsidiaries and licensed production abroad made up less than a tenth of total sales of the companies in the late 1980s (Wells, 1988). Exports thus continues to be the main modality of serving foreign markets by Brazilian manufacturing companies.

The quarter-century since 1975 described the emergence of an increasing number of Brazilian manufacturing MNCs that span a broad range of industries to include food products, textiles, clothing and footwear, paper packaging, wood and furniture, bicycles, lifts, electrical products, steel products and capital goods, motor vehicle parts, and aircrafts (see Table 5.1). To overcome the high degree of protection in export markets continued to be an important determinant behind the emergence of Brazilian manufacturing MNCs – at least until 1980. In the period since 1980, however, international production became a tool of firms as active agents to fulfil several objectives: to penetrate the markets or market segments of developed countries as in the cases of Cacique (instant coffee), Copersucar (coffee), Gradiente Electronics (stereo equipment), Hering and Alpargatas (clothing), Securit and Duratex (furniture) and Labra (pencils); to have a significant local presence in order to obtain contracts in host country markets as in the cases of Bardella (capital goods), Metal Leve (pistons), SIFCO (shaft machining), COFAP (engine parts), and Embraer (aircrafts); to broaden the geographical scope of their exports as in the case of Nansen (electricity meters); to add value to their product as in the cases of ITAP and Toga (packaging), as well as Cacique (instant coffee); and to integrate in more profitable higher value added activities in foreign markets as in the case of CVRD (steel). The rate of expansion overseas accelerated rapidly in the 1980s, along with the accumulation of large trade surpluses associated with the export drive spurred by government policy after 1984 and the domestic market slowdown after 1982 (Villela, 1983).

The emergence and growth of Brazilian MNCs in both resource-based industries and more capital intensive industries is a reflection of Brazil's large and developed manufacturing sector among developing countries that builds upon its rich agricultural, forestry and mineral resources as well as efforts to expand industrial capacity in capital goods industries. Although there was no specific sequential pattern in the emergence of the different industries that engendered Brazilian manufacturing MNCs, the firms from the more capital intensive industries such as bicycles, electrical equipment and motor vehicle parts, were the pioneering Brazilian manufacturing MNCs in the period before 1975 as mentioned, while firms in the resource-based industries of food, wood products and furniture, paper packaging and textiles, leather and clothing emerged later (see Table 5.2). The next sections explain further the emergence of Brazilian MNCs in the resource-based and capital intensive manufacturing industries.

The expansion of Brazilian manufacturing MNCs in resource-based industries

Brazilian manufacturing MNCs came to emerge in such natural resource-based industries as food products, furniture and wood products, paper packaging, and textiles, leather and clothing. Firms in the food products industry showed a consistent pattern of emergence as MNCs since the pioneering steps in international production by Copersucar (the industry leader in sugar and coffee) in 1976 as well as Cica (the leading firm in canned food products) and Cotia (cattle ranches, meat packing, soft drinks) in the late 1970s.[8] This was followed by Cacique (the leading firm in instant coffee) in 1985, Brahma (one of two industry leaders in beer brewing) in 1987 and other firms, including Cerval Alimentos in 1993. Firms in the wood products and furniture industry also displayed a consistent pattern of growth starting with Securit (the leading Brazilian firm in office furniture) and Bergamo that became MNCs before 1980. Other firms in the industry followed suit, including Labra (one of two leading firms in pencils production) in 1985 and Duratex (Itau Group) in 1990.

On the other hand, firms in the packaging and textiles, leather and clothing industries emerged later than those in the food products and wood products industries. Multinational corporations in the packaging industry such as ITAP and Toga emerged in 1983 and 1988, respectively and firms in the textiles, leather and clothing industries that have always been significant exporters were perhaps the latest group of Brazilian firms to internationalize their production activities. Firms in these sectors became MNCs only after the

Table 5.2 The emergence of Brazilian multinational corporations by period and industry

Period to 1975	1976–1980	1981–1985	1986–present
A. Primary			
Oil exploration Petrobrás (through Braspetro and Interbras			
	Coal exploration Siderbras		
		Minerals exploration Parapanema	
B. Manufacturing			
	Food products Copersucar Cica Cotia	*Food products* Cacique	*Food products* Brahma Cerval Alimentos
			Textiles, clothing and footwear Grendene Hering Vacchi SP Alpargatas Staroup
		Packaging ITAP	*Packaging* Toga
	Wood and furniture Securit Bergamo	*Wood and furniture* Labra	*Wood and furniture* Duratex (Itau group)
Bicycles Calói			
	Lifts Villares		
Electrical products Gradiente Electronics	*Electrical products* Inepar Inbrac	*Electrical products* Nansen	*Electrical products* Brastemp
		Steel products and capital goods Gerdau Companhia Vale do Rio Doce (CVRD)	
Motor vehicle parts Eluma Marcopolo	*Motor vehicle parts* Cotia		*Motor vehicle parts* Metal Leve SIFCO COFAP

(continued)

Table 5.2 (continued)

Period to 1975	1976–1980	1981–1985	1986–present
		Aircraft Embraer	
	Other manufacturing Cotia		
C. Services	*Banking* 14 public and private Brazilian commercial banks including Banco do Brasil, Banco de Estado de São Paulo and Banco Real		
Construction services Mendes Junior Constutora Rabello Odebrecht Ecisa		*Construction services* Camargo Correa Esusa Sisal Veplan	
Engineering services Tenenge	*Engineering services* some 26 companies		
Retail Pão de Açucar			
		Broadcasting Globo	

Source: Derived from Table 5.1.

mid-1980s starting with Grendene (the leading firm in plastic footwear), Hering (the leading firm in cotton knitwear and T-shirts) and Vacchi (a company with a long experience in the exports of leather products to Europe) in 1986, followed by SP Alpargatas and Staroup (both manufacturers of jeans) in 1989 (see Table 5.2). Through international production, Brazilian firms in the clothing and leather industries have been able to exploit profitable higher value added activities and market niches in the developed countries (Peres Núñez, 1993).

The competitive advantages of Brazilian MNCs in these resource-based industries had been nurtured in the richness of Brazil's natural resources. Indeed, primary products formed the basis of Brazilian industrial development between 1800 and 1930 (Gereffi and Evans, 1981). The sustained strengths of Brazil in primary product sectors is a legacy of its primary product export economy industrialization phase in which exports of agricultural produce (particularly coffee) – that provided over 90 per cent of Brazil's foreign exchange before 1945 – was the principal driver of economic growth (Newfarmer, 1980). Indeed, of all primary commodities that Brazil produced and exported to include sugar, cocoa, wood, iron ore and others, coffee became its principal export product and Brazil reigned supreme in the production and export of coffee worldwide.

Unlike in other developing countries where foreign capital controlled primary sector production, coffee production in Brazil was dominated by Brazilians (Gereffi and Evans, 1981). The comparative advantage of Brazil in coffee production thus led to the development of competitive advantages of local producers and major exporters in the coffee industry, some of which like Copersucar (a consortium of the leading firms in sugar and coffee production) and Cacique (the leader in instant coffee) became MNCs.[9] The key strength brought to bear by these coffee companies in their vertically integrated outward FDI activities is the assurance of a steady and abundant supply of coffee and other primary commodities for which the firms in particular, and Brazil in general, is responsible for a considerable proportion of world output.[10] Similarly, Companhia Vale do Rio Doce – the Brazilian leading company in iron ore mining and regarded as the world's top iron ore producer – secured a foothold in the United States by partly acquiring a steel mill in Fontana, California, in order to profit from the value added to iron ore at the processing stage. The company's steady supply of low-cost iron ore from Brazil brought to the United States through its own shipping fleets have been instrumental in bringing down the costs of steel produced by the Fontana mill (Wells, 1988).

Forward vertical integration in foreign markets, particularly in developed countries, thus became a strong driving force behind the outward FDI of these companies to assure themselves of steady buyers. Their strong advantages based on the control of abundant agricultural or forestry based raw materials or mineral resources enabled their integration into the more advanced stage in the value added chain as an important means to earn higher profits and expand markets (Diaz-Alejandro, 1977).

Similar strengths obtained on the basis of abundant raw materials in Brazil are evident among the Brazilian manufacturers in such industries as furniture, pencils, paper and cardboard packaging and textiles, clothing and leather whose origins can be traced to the export economy phase that fostered industrialization in a broad range of industries that had low capital and technological barriers to entry. Indeed, among the industries that thrived during the primary product export phase were textiles and foodstuffs (which supplied nearly three-quarters of factory production), paper, glass, cigarettes, soap, matches as well as several foundries that manufactured hardware, agricultural machinery and railway wagons (Newfarmer, 1980). Many of these industries spawned Brazilian MNCs. For example, Securit – the largest office furniture maker in Brazil – opened a showroom and assembly plant in Houston, Texas, in 1980 with the objective of selling directly to the customer while taking advantage of cheap Brazilian wood and labour and capitalizing on the oil boom in Texas and the large Arab real estate investments in the state (Wells, 1988).[11] Similarly, the pencils produced at the

Portuguese factory of Labra (one of two leading Brazilian companies in pencil production) were sold to European markets at a competitive price on the basis of the company's access to low-cost wood from Brazilian forests, its large-scale production, licensed technology from Europe and cheap labour in both Portugal and in Brazil (Wells, 1988).

The growth of Brazilian firms and MNCs in the textiles industry can be traced to the origins of the domestic textiles industry as one of the oldest and most successful locally controlled Brazilian industry spawned during the primary product export industrialization phase. At the turn of the twentieth century Brazilian textile companies already enjoyed some degree of success despite competition from European and North American imports, and by 1920 Brazilian factories had expanded to a scale comparable to their foreign competitors and met virtually all the local demand for textiles (Stein, 1957). The rapid growth of Brazilian textile companies in this period was attributed partly to their access to relatively inexpensive British textile technology that had become available increasingly with the decline of the comparative advantage of Great Britain in textiles production *vis-à-vis* the United States and other Europe (Hobsbawn, 1968).

The abundant natural resources of Brazil not only helps to explain the emergence of Brazilian manufacturing MNCs in the resource-based industries but also the involvement of Brazilian resource-based companies in foreign mineral prospecting activities. While these activities did not always involve the establishment of subsidiaries (as in the case of the state-owned Companhia de Pesquisa de Recursos Minerais (CPRM) whose mineral prospecting activities in Africa was undertaken through contracts), the accumulated experience of these companies nurtured in the vastness and natural resource abundance of Brazil was applicable directly to their overseas mineral prospecting activities. Gold mining is a particular forte especially for CPRM and the privately owned company, Parapanema. Some of the firms were so advanced in their field as shown, for example, in the case of Metalur that supplied Brazilian equipment to its Norwegian plant producing high-grade magnesium (Wells, 1988).

The expansion of Brazilian manufacturing MNCs in capital intensive industries

Apart from the resource-based industries, Brazilian manufacturing MNCs came to emerge in more capital intensive industries such as bicycles, lifts, electrical products, steel and capital goods, motor vehicle parts and aircrafts. The electrical products industry was one of the first industries to engender Brazilian MNCs starting with the international production by the industry leader in stereo equipment, Gradiente Electronics in 1973 as mentioned, and has since proven to be one of the more important industries of Brazilian FDI. The outward FDI by Gradiente was followed by the other firms in the industry, including smaller- and medium-sized firms such as Inepar (producer of electrical control equipment) and Inbrac (producer of cables and wires) that initiated international production in 1977 and 1978, respectively, and Nansen (producer of electricity meters) and Brastemp (producer of electrical appliances) that initiated international production in 1985 and 1990, respectively.

Villares (the leading firm in the production of mechanical lifts) initiated the establishment of foreign subsidiaries in 1977, while firms in the steel products and capital goods industries initiated international production in the first half of the 1980s starting with Gerdau in 1981, Companhia Vale do Rio Doce (the leader in iron ore mining) in 1984, and Bardella in 1985. The Brazilian state firm, Embraer (the leading firm in aircrafts) initiated aircrafts assembly abroad in 1983. Firms in the motor vehicles parts industry, on the other hand, displayed a rather more sporadic pattern of growth. Thus, the pioneering forays into international production by Eluma in 1968 and Marcopolo in 1971 was followed by Cotia in 1979, and then much later by Metal Leve and SIFCO in 1989 and COFAP in 1991 (see Table 5.2).

The emergence of strengths by Brazilian firms in these industries not directly related to the presence of abundant natural resources can be explained in the context of the switch in industrialization development policy at the time of the Great Depression. The growth in public debt which became an overwhelming burden in the era of declining coffee prices around the Great Depression led to the realization that a total reliance on export-oriented growth based on primary products was unsustainable.[12] The decline in export earnings accompanied by the sharp devaluations of the local currency precipitated a shift in industrial development policy away from growth based on exports towards one based on import substitution industrialization.[13] The objective of the new industrial development policy in effect between 1930 and 1955 starting with the horizontal ISI phase was to foster the local production of consumer non-durables (light consumer goods) and the local assembly of consumer durables. This new industrialization strategy was implemented largely through state entrepreneurship in basic industries like steel to ensure domestic manufacturers a steady supply of locally available inputs while at the same time serving to limit foreign control in these strategic industries; the provision of artificial support to coffee prices to ensure the presence of an adequate local demand for manufactured goods; the imposition of pressures on foreign subsidiaries to increase local content; and, perhaps most important of all, the raising of tariff barriers against manufactured goods (Gereffi and Evans, 1981).[14] The latter led to the adoption of a five-tiered import licensing system with machinery and raw materials given preference (Bergsman, 1970).

Although the horizontal ISI phase did not become the dominant industrial development strategy until 1930, the seeds that enabled this strategy to flourish eventually had been planted in the primary product export industrialization phase in two important ways: first, as previously mentioned, the export economy fostered some industrialization in industries that had low capital and technological barriers to entry. Although most of the industries that thrived in the primary product export economy phase produced non-durable consumer goods such as textiles and foodstuffs, paper, glass, cigarettes, soap and matches, there were also several noteworthy foundries that manufactured hardware, agricultural machinery and railway wagons (Newfarmer, 1980).

Second, local capital in the primary export sector, particularly from coffee, played a strong role in the emergence of consumer goods industries that were the focus of horizontal ISI. Indeed, most industrial capital originated with planters or merchants who viewed industry as complementary to their activities (Newfarmer, 1980). The planters either provided the liquid capital from their coffee sales, which in combination with the government subsidies to the coffee industry financed the development of new import-competing industrial enterprises, or the planters themselves moved into complementary industries such as sugar mills and textile factories that processed the crops they cultivated. The important roles of planters as well as merchants that also moved into industry in sectors in which protection could be derived from tariffs or had high international transport costs meant that industrial production in Brazil fell less than 10 per cent in the early depression years and by 1933 had regained its 1929 levels (Baer, 1965). In addition, there was even the renationalization of some local manufacturing industries as shown by the acquisition during the Depression of the Votorantim rayon mill from the British by a local group of companies – the Ermirio de Moraes group (Gereffi and Evans, 1981). This manifested the growing accumulation of strengths by local textile manufacturers which was already in evidence during the turn of the twentieth century when these manufacturers became capable of replacing British imports, as mentioned.

In order of economic importance, the state, private Brazilian firms and foreign firms participated in the horizontal ISI process (Cardoso, 1973). With the strong role of local capital and the development of locally controlled firms and industries, the period of horizontal ISI was consistent with a period of diminished dependency for Brazil (Gereffi and Evans, 1981).[15]

By the mid-1950s, horizontal ISI was superseded by a phase of vertical ISI in which the emphasis was on internalizing all phases in the manufacture of consumer goods (both non-durables and durables) and integrating backwards into the domestic production of intermediate products and capital goods that were causing a significant drain on the balance of payments. The investments required greater technological sophistication and capital intensity which made foreign based MNCs rather than local firms the main instrument of industrial development. The role of foreign investment in providing capital and technologies in the industrialization process in both the basic industries and the leading industrial sectors meant that Brazil had taken on an associated-dependent pattern of development (Cardoso, 1973).[16] The inability of local capital to penetrate the technologically sophisticated and capital intensive industries combined with a military regime that supported the interests and modes of organization of international capitalism led Evans (1976) and Cardoso and Falleto (1979) to describe the vertical ISI phase of

Brazil as the internalization of imperialism and the internationalization of the internal market, respectively. Thus, unlike in the horizontal ISI phase, dependent development became re-established as the dominant mode of economic growth in the vertical ISI phase (Gereffi and Evans, 1981).[17]

In the midst of the dependent pattern of development, the industrial strengths of the local bourgeoisie in a number of consumer goods industries and in some state-controlled basic industries producing steel as well as machinery, machine tools and other mechanical engineering sectors was increasing rapidly. To explain the growth of Brazilian manufacturing MNCs in capital intensive industries is essentially to explain the growth of Brazilian manufacturing firms in two major groups of industries: the metalworking and mechanical engineering (or non-electrical machinery) industries and the electrical equipment industry.

THE GROWTH OF BRAZILIAN FIRMS AND MNCs IN THE METALWORKING AND
MECHANICAL ENGINEERING (OR NON-ELECTRICAL MACHINERY) INDUSTRIES

As mentioned, the origins of Brazilian firms in this sector can be traced to the several foundries that manufactured hardware, agricultural machinery and railway wagons during the primary product economy phase. In particular, the entry of Brazilian industrialists in mechanical engineering or non-electrical machinery industries where technological barriers to entry were relatively low and profits were relatively high can be traced to the period after the First World War. The 'brilliant mechanics' had captured an important share of the domestic market by 1920 and accounted for approximately 61 per cent of all Brazilian capital goods by 1947–49 (Leff, 1968).

The protection accorded to the metalworking and mechanical engineering industries in the large domestic market of Brazil during the period of vertical ISI (1955–1970) induced industrial development by enabling some firms to attain the scale economies of international plants, to approach rapidly the fairly stable technological frontier in their industrial fields, and to increase their exports of goods and industrial technology significantly.[18] This was the case particularly of such firms as Romi and Metal Leve, the producers of conventional lathes and pistons, respectively (Katz, 1984) and also of other Brazilian firms in steel production (Gerdau) and in the mechanical engineering industries such as bicycles (Calói), lifts (Villares), motor vehicle parts (Eluma, Marcopolo, SIFCO, COFAP), equipment for paper production (Pilao), capital goods (Bardella) and aircrafts (Embraer) that eventually became MNCs.

The growth of locally controlled firms in the motor vehicle parts industry that later became MNCs is of particular interest. Indeed, it was the existence of some 250 locally owned firms producing 8,000 motor vehicle parts in the early 1950s that gave government policy makers the impression that Brazil could benefit from the high linkage potential of the foreign-controlled motor vehicle industry of Brazil (Shapiro, 1994). In principle, the motor vehicle industry could serve as the hub of an integrated industrial structure by triggering the domestic production and technological advancement of suppliers of parts and components, as well as those firms in the more complementary basic industries of steel, glass, rubber and plastics in part through the provision of finance, training and technical assistance by the foreign car producers (Dahlman, 1984). However, the extent to which this happened was limited as the motor vehicle parts industry in time became dominated by foreign firms as a result of the superior market power of the foreign MNCs which enabled them to control their supply channels either through the vertical integration of car manufacturers or through inward FDI by foreign supplier companies. A bifurcated parts sector emerged in which foreign firms predominated in large, capital and technology intensive operations producing parts in a mainly concentrated market and sold most of their output to the terminal sector (car manufacturers). Brazilian firms, on the other hand, were relegated mainly to small- and medium-sized operations that produced standardized parts in a more competitive market and which sold a greater portion of their output as replacement parts (Shapiro, 1994).

Despite the market control exercised by the foreign based MNCs in both the motor vehicle and motor vehicle parts industries, some locally owned firms in motor vehicle parts continued to flourish as reflected in the growth of their exports and outward FDI. The recent impetus for their international production has been provided by the rapid changes over the last ten years in the organizational and technical structure of their user industries particularly in the United States and Europe. Thus, in response to the demand of their clients for suppliers that do not merely provide parts or components but also assist in the development of designs, offer technological solutions and fit in with their just-in-time inventory management systems, Brazilian motor vehicle part makers such as Metal Leve and COFAP as well as the capital goods maker Bardella established foreign plants to be closer to their user industries in the developed countries (ECLAC, 1992).

The emergence of Embraer – a Brazilian mixed state–private firm and an MNC in the aircrafts industry – also stirs particular interest. Indeed, to describe the development of Embraer is to describe the development of indigenous technological capacity in the Brazilian aircrafts industry and the desire to achieve local control over this strategic industry (Dahlman, 1984). The creation of Embraer stemmed from the need to have a monopolist in the domestic aircraft industry in which scale economies are only achievable at very large volumes of output and which required intense technological effort. The eventual success of Embraer was reflected in its ability to produce a broad range of different airplane models for military and commercial uses with a high percentage of domestic content, and to export these to both developed and developing countries.[19] Its development as an MNC was prompted by the need for a local presence in foreign markets in order to obtain contracts to build military training aircrafts for the national Air

Force. This explains the firm's establishment of aircraft assembly operations in Cairo and Belfast (Wells, 1988).

THE GROWTH OF BRAZILIAN MNCs IN THE ELECTRICAL EQUIPMENT INDUSTRY

Although MNCs in the electrical equipment industry were among the pioneering MNCs in manufacturing along with those in the bicycles and motor vehicle parts industries, the origins of Brazilian firms in the electrical machinery industry have a more recent vintage compared to those of the metalworking and mechanical engineering industries. Indeed, the origins of Brazilian producers of electrical products can be traced only after the Depression (coinciding with the onset of the horizontal ISI phase), despite the rapid growth of electrical generating capacity and the final demand for electrical goods in the period prior to the Depression (Newfarmer, 1980).[20] The effective barriers to imports of consumer products imposed as a result of the scarcity of foreign exchange during the Depression created a fortuitous environment for the entry of domestic firms into production relatively free of competition (at least until 1955 when foreign based MNCs began to manufacture appreciable quantities of electrical machinery in Brazil). The entry of domestic firms was, however, not in heavy electrical machinery, where high technological barriers impeded their entry into production, but was rather confined to a segment of the industry that had relatively low technological barriers to entry. Thus, most of the domestic entrants had undertaken the assembly production of electrical wire, small parts and light electrical appliances such as radios, television sets and other small appliances, and eventually their products were rapidly replacing imports in these product lines. Technologies for production of these simple consumer electrical products were acquired either through purchases from firms in the developed countries which could not reach the Brazilian market with the onset of horizontal ISI after the Depression, or through reverse engineering by which Brazilian firms copied the design of products previously imported. The significant growth of local participation in the electrical equipment industry (or a segment of it) between 1930 and 1955 led to above average growth rates for the industry as a whole.

The developments since the onset of the period of vertical ISI had negative repercussions on the growth of Brazilian electrical producers. First, the implementation of SUMOC Instruction 113 in January 1955 which granted foreign MNCs cheap access to foreign machinery favoured the entry of foreign MNCs that sustained industrial growth at the expense of denationalization.[21] Thus, some two-thirds of the Brazilian electrical industry was in the hands of foreign MNCs by 1961 when SUMOC Instruction 113 expired, and this share grew further to over 75 per cent by 1974 (Newfarmer, 1980). The denationalization owed partly to the takeovers or mergers of many Brazilian firms with foreign MNCs. Those firms that were not acquired or merged with foreign companies were either compelled to associate with the foreign MNCs to secure favourable import terms or to seek foreign licences from the United States to keep abreast of latest technologies. Thus, Brazilian firms came to produce such electrical appliances as Emerson, Kelvinator, Westinghouse and Whirlpool under licences.[22] Second, the dominance of foreign MNCs in the electrical equipment industry, particularly in the heavy electrical equipment sector, and the behaviour of these firms to maintain market control, both weakened price competition and erected a further barrier to entry to domestically owned firms quite apart from the high technological requirements of this segment of the industry.[23] It meant that Brazilian firms remained confined to positions of high technological dependency or marginal suppliers in their own domestic market.

This historical review of the origins of Brazilian firms in the electrical equipment industry helps to shed light on the emergence of Brazilian MNCs in this industry to include Gradiente Electronics (the leader in stereo equipment), Inepar (electrical control equipment), Inbrac (cables and wires), Nansen (electricity meters) and Brastemp (refrigerators). It also helps to explain why Brazil had not spawned firms or MNCs in the more dynamic segments of the electrical equipment industry.

Conclusion

This chapter analysed the emergence and evolution of Brazilian MNCs over the last 30 years. It examined the main determinants of outward FDI and their industrial and geographical patterns in two distinct time phases: the period of emergence from the late 1960s to 1975, and the period of development in the period since 1975. Although the pioneering Brazilian MNCs spanned the three sectors of economic activity, the main thrust of outward FDI by Brazilian MNCs had been in services by construction and consulting engineering companies. Unlike American MNCs that emerged in a major way in outward FDI in the primary sector to exploit firm-specific assets in resource extraction and processing nurtured in the natural resource abundancy of the United States, the security of the supply of raw materials or natural resources as a motive for outward FDI in the emergent phase of Brazilian MNCs was rather unique to the Brazilian state-owned company, Petrobrás, and specific to one natural resource – oil – given Brazil's rich agricultural and mineral resources. The only other exception of Brazilian resource-based FDI was the joint venture established with the Colombian government by the Brazilian state steel enterprise, Siderbras, to exploit coal in Colombia with the objective of gaining independence from the United States for such raw material (White, 1981). In this respect, the Brazilian MNCs share more common features with Swedish MNCs where firms in the industries of mining, metal production and forestry became MNCs only to a very limited extent (see Chapter 4).

Outward FDI by Brazilian manufacturing companies in the emergent phase was also rather limited. International production by the earliest

Brazilian manufacturing MNCs in bicycles (Calói), electrical products (Gradiente Electronics) and motor vehicle parts (Eluma and Marcopolo) which consisted mainly of assembly operations was a means to overcome the high degree of protection in export markets. The period since 1975 described the major expansion of Brazilian MNCs particularly in manufacturing and banking. The establishment of an overseas network by Brazilian state-owned and privately owned banks in the developed countries enabled the banks to attract resources from international capital markets, and their expansion in Latin America was allied closely with the rapid growth of Brazilian exports to the region (Guimaraes, 1986). The period also witnessed the emergence of Brazilian manufacturing MNCs in a major way, many of which were accounted for by leading firms in their respective industries. In many respects, the extent of the multinationality of these firms remains emergent as reflected in the small share of sales by foreign subsidiaries and licensed production abroad in the total sales of the companies and the continuing importance of exports as the main modality of serving foreign markets.

The quarter-century since 1975 described the emergence of an increasing number of Brazilian manufacturing MNCs that span a broad range of industries to include food products, textiles, clothing and footwear, paper packaging, wood and furniture, bicycles, lifts, electrical products, steel products and capital goods, motor vehicle parts, and aircrafts. To overcome the high degree of protection in export markets continued to be an important determinant behind the emergence of Brazilian manufacturing MNCs – at least until 1980. In the period since 1980, however, international production became a tool of firms as active agents to fulfil several objectives: to penetrate the markets or market segments of developed countries as in the cases of Cacique (instant coffee), Copersucar (coffee), Gradiente Electronics (stereo equipment), Hering and Alpargatas (clothing), Securit and Duratex (furniture) and Labra (pencils); to have a significant local presence in order to obtain contracts in host country markets as in the cases of Bardella (capital goods), Metal Leve (pistons), SIFCO (shaft machining), COFAP (engine parts), and Embraer (aircrafts); to broaden the geographical scope of their exports as in the case of Nansen (electricity meters); to add value to their product as in the cases of ITAP and Toga (packaging), as well as Cacique (instant coffee); and to integrate in more profitable higher value added activities in foreign markets as in the case of CVRD (steel). The emergence and growth of Brazilian MNCs in both resource-based industries and more capital intensive industries is a reflection of Brazil's large and developed manufacturing sector among developing countries that builds upon its rich agricultural, forestry and mineral resources as well as efforts to expand industrial capacity in capital goods industries during the primary product economy industrialization phase and import substitution industrialization phase of the Brazilian economy.

Notes

1 The investments by Calói in Bolivia were particularly favoured when a Bolivian firm supported by the government began to look for foreign partners for the domestic production of bicycles in response to the prospects of high profits from protection (White, 1981).

2 As a result of the aggressive system of export promotion, Brazilian industrial exports increased in real terms by 2.7 times between 1975 and 1980, a rate of growth much faster than that of total exports which grew in real terms by 1.7 times over the same period (Villela, 1983).

3 The Brazilian system of export promotion contributed greatly to the decision of Brazilian firms to become MNCs. The system consisted of a large number of financial and credit incentives, of which the most important in stimulating outward FDI were the financing of exports of capital and durable goods, the financing of sales of services abroad (including the purchase in Brazil of machinery, equipment, vehicles, instruments, etc. for the execution of construction jobs overseas), the financing of the establishment of enterprises abroad by manufacturing and exporting companies for a period of up to three years, and lines of credit (some 53 of them) to facilitate the purchase of goods and services in Brazil. Interest rates charged on all the above described credits were highly subsidized (Villela, 1983).

4 An indication of the high degree of reliance on foreign markets is seen in Mendes Junior (one of the largest privately owned Brazilian construction firms) that earned more than a third of its revenues abroad in the mid-1980s (Wells, 1988).

5 For example, Engevix – a leading Brazilian consulting engineering firm – designed part of the underground transportation system of Baghdad using expertise gained when it helped plan the São Paulo Metro handled by the Soferail company (France) in the late 1970s. The firm absorbed the technology and adapted it to Iraq (Wells, 1988).

6 Such is the case, for example, with Tenenge – an industrial engineering firm with subsidiaries in Chile and Paraguay – that can deliver state-of-the-art refineries through composite technologies developed by the firm from the fusion of imported technologies obtained through former licensing agreements and its own technology (Wells, 1988).

7 An indication of the continuing international expansion of construction and consulting engineering firms since the period of their emergence was seen in the growing number of firms engaged in outward FDI. For example, some 31 of these companies were involved in outward FDI between 1977 to 1982. The investments, however, were highly concentrated with the three largest firms accounting for 68 per cent of the total investment, and 19 firms for 99 per cent of the total investment (Guimaraes, 1986).

8 Cotia Comércio Exportação e Importação is a trading company founded by the Paulo Brito group – a family company owning cattle ranches and cold storage facilities for meat in São Paulo (Villela, 1983).

9 In 1976, Copersucar acquired Hills' Brothers Coffee, the fourth-largest processor of coffee in the United States in order to integrate vertically into coffee processing as well as the marketing of coffee, cocoa and sugar in the United States through an established brand name. However, the need for a marketing outlet for sugar disappeared unexpectedly as Brazil's energy programme generated a rapid growth in demand for sugar to be made into alcohol for fuel (Wells, 1983). On the other hand, Cacique – the leading exporter of Brazilian instant coffee – entered into a joint venture with a Chinese state farm to bottle coffee starting in 1987. The company's aim was to progress beyond exporting unpackaged instant coffee and to sell most of the bottled coffee to Japan, Australia and other Asian markets (Wells, 1988).

10 This strength does not necessarily guarantee success, as shown by the outward FDI of Copersucar in the processing of coffee and the marketing of coffee, cocoa and sugar in the United States which failed to make adequate returns and led the company to sell out to Nestlé in 1986 (see Wells, 1988).

11 However, pressure from local competition forced Securit to close down this foreign operation (Wells, 1988).

12 The massive public debt incurred in the process of improving urban infrastructure became an overwhelming burden as public debt service soared to 43 per cent of export earnings in 1932-33 (Baklanoff, 1971).

13 It was the conviction of Getúlio Vargas to abandon Brazil's agricultural vocation and thus to pursue policies supporting ISI (Gereffi and Evans, 1981).

14 Symbolic of the state investment in basic industry was the formation of the huge Volta Redonda steel complex (Cardoso, 1973).

15 However, although local capital may have played a leading role in the shift from primary product export phase to the horizontal ISI phase during the Great Depression, inward FDI also assisted in the transition process. This is shown in particular in the growth of FDI in the manufacturing sector in Brazil during the period. Indeed, almost one quarter of American FDI in Brazil in 1929 was in manufacturing (Gereffi and Evans, 1981).

16 Evidence of the role of foreign capital and technologies in the development of even basic industry is seen in the steel industry. Usiminas, one of the three large state-owned integrated steel mills in Brazil, initially obtained the technology necessary for steel production through a Japanese joint venture in the early 1960s (Dahlman, 1984).

17 The dependent development arose largely from the absence of a strong industrial class whose interests were distinct from agricultural and commerce, and the political supremacy of an export-oriented oligarchy that precluded a more nationalist government policy that allowed for the indigenous development of national industry in the more sophisticated products, but instead left the door open to foreign goods (or foreign firms) (Newfarmer, 1980). Thus, instead of changing their comparative advantage systematically in more dynamic industries, Brazilian producers chose to maintain comparative advantages in coffee and consumer non-durables of low technology and capital intensity, while allowing imports or foreign MNCs to dominate the rapidly growing industrial sectors. The cycle of cumulative decline became self-perpetuating as the market power exercised by foreign MNCs over the supply of products, production technology, international finance, etc. further precluded the emergence of strong indigenous firms that could effectively compete.

18 Up to 70 per cent of the output of the capital goods industry was exported in 1984 (Wells, 1988). In addition, some 27 of the 58 Brazilian manufacturing firms that accounted for 69 of the 112 overseas operations involving exports of industrial technology between 1976 and 1981 were capital goods manufacturers. Capital goods manufacturers provided the dominant source of industrial technology exports in such industries as machinery, equipment and components, sugar and alcohol and paper and pulp (Sercovich, 1983).

19 Owing to its policy to increase the participation of local firms in the manufacture of airplanes, Embraer encouraged the development of the local aircraft parts industry that became important exporters in their own right (Dahlman, 1984).

20 This is not to say that Brazilian industrialists in the electrical equipment sector never came to existence in the period before the Depression. The firm, Guinle and Gaffree, that owned one of the largest Brazilian enterprises of the early twentieth century – the docks at Santos – established a substantial system of public utility companies dominating such major cities as Salvador, Bahia and Port Alegre and Rio Grande do Sul but these were eventually sold in the late 1920s to The Electric Bond and Share Company (a finance subsidiary of the United States-based

General Electric Company). The sale was owing largely to the inability of Guinle and Gaffree to raise sufficient capital from the North American capital markets to modernize and expand their systems. Thus by 1920, foreign-owned companies either built or acquired most of the electrical generating capacity in Brazil and provided the greatest demand for heavy electrical equipment. Their demands were largely met through imports because until 1920 there were virtually no electrical products produced locally despite the creation of a partial and temporary protection to foster the growth of local companies and to encourage the entry of foreign MNCs (Newfarmer, 1980).

21 More than 80 per cent of all foreign direct equity flows entered as machinery imports under Instruction 113 between 1956 and 1960. About 80 per cent of these were invested in plants to produce heavy electrical equipment, refrigerators and compressors, telecommunications equipment, cable and conductors and electronic apparatus (Newfarmer, 1980).

22 It is therefore not surprising to discover that the refrigerator production facility established in 1990 in Argentina by the Brazilian firm, Brastemp, through a part acquisition of a former Philips affiliate represented a joint investment with the American electrical equipment company, Whirlpool, with which it had a long association.

23 The expansion of foreign MNCs in the Brazilian electrical industry was not predicated solely on their superior technical efficiency, but to their market tactics and strategies made possible by their superior market power: interlocking directorships, mutual forbearance, control of supply channels, cross-subsidization and predation, formal and informal collusion, formal political ties, product differentiation and acquisition (Newfarmer, 1980).

6 Conclusion

The emergence and evolution of multinational corporations from the resource-abundant countries

This part of the book analysed the emergence and evolution of MNCs based in resource-abundant countries by focusing on three countries of different sizes. In particular, it compared the growth of MNCs from resource-abundant large countries such as the United States and Brazil with that of MNCs from an resource-abundant smaller country of Sweden. The main rationale was to determine whether there exists a common pattern in the emergence and evolution of MNCs from resource-abundant countries. The general conclusion that can be drawn is that to some extent there is a common pattern in the emergence and evolution of MNCs from the three resource-abundant countries. The similarities are closest when considering the pattern and growth of their outward FDI in the manufacturing sector and the mining sector (see Table 6.1).

Regardless of the size of the countries concerned, the emergence of manufacturing firms and MNCs based in the three resource-abundant countries derive commonly from cumulative strengths in the engineering industry generally comprising metal products, machinery and transportation equipment. The basis of their affinity is the abundance of mineral resources in the three countries which fostered a well-entrenched tradition of industrialization based on metals processing, the technical and metallurgical know-how of which spilled over into related sectors of the engineering industry such as machinery and transport equipment.

As has been learned in the individual country case studies, the emergence and expansion of American FDI in manufacturing was linked to the rapid development in the 1850s of American technologies in certain metallurgical industries (i.e. machine tools, guns, reapers, and sewing machines) linked with mass production in which the United States had already acquired world leadership (Wilkins, 1970). Many American companies in the metallurgical industries sought export markets to dispose of surplus output and obtain economies of scale, and fulfilled a pioneering role in international production prior to the American Civil War. Similarly, the growth of the Swedish engineering industry was an outcome of the Swedish industrialization phase between 1870 and the First World War that built upon cumulative strengths in iron and steel production based on indigenously abundant

Table 6.1 Resource-abundant countries: variations in the early stages of outward direct investment across different countries

Examples of countries	Dominant form of earliest outward FDI	Type of locally based MNC
United States	The installation of mercantile houses overseas as well as the foreign branches of American merchants in the late seventeenth century and eighteenth century	Colonial merchants
	The ventures overseas of American private banks which had their genesis in trading companies	Banks and financial institutions
	Railroads construction in the late nineteenth century in Mexico, Canada and the Caribbean. The investments were regarded as the international extension of the domestic business plans of the railroad companies and a means to assure the efficient transport of minerals and agricultural produce.	Railroad companies horizontally integrating Agricultural and mineral companies, forwardly integrating
	Mining investments in the late nineteenth century attracted by the business opportunities posed by the presence of rich mineral resources and the lower costs of minerals prospecting and development in Mexico, Canada and South America, as well as the need to supplement domestic mineral resources	Mining and manufacturing firms
	Sales and manufacturing investments in the late nineteenth century in Canada, Europe and Japan. International production motivated by the need to protect the firm from unfair local competition, the opportunity to earn higher profits, to save on the costs of transportation, warehousing and domestic production and to overcome trade barriers	Manufacturing firms, forwardly integrating or horizontally integrating
	Sales subsidiaries in petroleum and, in a few cases, in petroleum refining and exploration in the late nineteenth century in Europe, Asia, Mexico and South America. Foreign refining motivated by high import duties on refined oil in export markets and to overcome local competition. Foreign exploration motivated by the need to supplement American oil supplies	Petroleum firms
	Agricultural investments in the late nineteenth century in the Caribbean (Cuba and the West Indies), Canada and Mexico. American FDI motivated by the richness of foreign natural resources located in close proximity to the United States. The establishment of agricultural estates was supported by investments in transportation by steamship or railroads	Commodity traders, backwardly integrating Agricultural firms Manufacturing firms, backwardly integrating (rarely)

Table 6.1 (continued)

Examples of countries	Dominant form of earliest outward FDI	Type of locally based MNC
Sweden	Establishment of sales subsidiaries to sell the products of the engineering industry consisting of metal manufactures, machinery and transport equipment	Manufacturing firms, forwardly integrating
	Establishment of production subsidiaries to produce for the host country market the products of the engineering industry consisting of metal manufactures, machinery and transport equipment	Manufacturing firms, backwardly or horizontally integrating
	To a limited extent, backward vertical integration in resource extraction in resource-rich countries	Resource-based firms (mainly forestry firms), backwardly integrating
		Manufacturing firms, backwardly integrating.
Brazil	Oil exploration and drilling activities. International production in bicycles, electrical equipment and motor vehicle parts	State-owned firm. Manufacturing firms, typically exporting firms.
	Investments in civil construction and consulting engineering	Construction and consulting engineering firms

Source: Author's compilation based on the analysis contained in the country chapters.

sources of high quality iron ore. In combination with other country-specific advantages to include major investments in technological education and training, the large degree of trade orientation and the close personal contacts within the Swedish industrial establishment, there was the rapid absorption and adaptation of technological innovations and their application in modern industrial production in the engineering industry. Technologically innovative engineering firms exported at an early stage of their firms' history and became MNCs rapidly to enable a celeritous establishment in foreign markets. Finally, the origins of Brazilian firms and MNCs in the metalworking and mechanical engineering industries can be traced to the primary product export economy industrialization phase of the Brazilian economy between 1800 and 1930 that fostered some industrialization in sectors that had low capital and technological barriers to entry, among which were several noteworthy foundries that manufactured hardware, agricultural machinery and railway wagons (Newfarmer, 1980). The 'brilliant mechanics' of Brazil had captured an important share of the domestic market by 1920, accounted for approximately 61 per cent of all Brazilian capital goods by 1947–49 (Leff,

1968) and eventually became MNCs in such diverse industries as bicycles, lifts, steel products and capital goods, motor vehicle parts and aircrafts.

Apart from the common emergence of manufacturing MNCs based in the engineering industry broadly defined, the process of international expansion of firms in this industry has been fairly common, despite the seemingly disparate attempts to model such process in the case of American metallurgical companies and Swedish engineering companies by Wilkins (1970) and Johanson and Wiedersheim-Paul (1975), respectively. Four stages were described by Wilkins (1970). In the *first stage*, the American concern sold abroad through independent agents (through an export person or export or commission houses in New York City) or on occasion filled orders directly from abroad. In general, companies started to export using the facilities of international trading firms. In the *second stage*, the company appointed a salaried export manager, an existing export agency and its contacts, or independent agencies in foreign countries to represent the company. In the *third stage*, the company either installed one or more salaried representatives, or a sales branch, or a distribution subsidiary abroad, or it purchased a formerly independent agent located in a foreign country. At this point, for the *first* time, the company made a foreign investment. In the *fourth stage*, a finishing, assembly or manufacturing plant might be established to meet the needs of a foreign market.

Such incremental process of establishment in foreign markets by American metallurgical firms based in a large domestic market has many common elements with the concept of an 'establishment chain' proposed by Johanson and Wiedersheim-Paul (1975) in describing the gradual process of penetration in foreign markets of Swedish engineering firms based in a much smaller domestic market. Thus, after mechanizing and specializing in a technically advanced product, firms secured the home base and engaged in exports to achieve scale economies; this was associated with the use of a network of agents or representatives (or travelling salesmen) employed by the firms and then by the establishment of a sales subsidiary abroad; and, finally, by the initiation of international production.

Despite the broad analogy in the process of international expansion of American and Swedish firms in the manufacturing sector, two main factors differentiate the American and Swedish models. Firstly, the first stage described in the American model of exports being handled by independent agents or international trading firms was often bypassed in the Swedish case. This was because, unlike their counterparts in the staple raw materials-based industries, the high-quality Swedish steel mills and the new industrial Swedish companies based on mechanical engineering had to establish direct contacts with foreign markets by employing company-appointed representatives or agents or travelling salesmen (Hörnell and Vahlne, 1986) owing to increased product differentiation (e.g. production of special steel versus ordinary steel), and the necessity to have intimate knowledge of market developments and to adapt products to customers' particular needs (Carlson, 1977).

Secondly, by comparison to the United States that is another resource-rich country of a larger size, Sweden – as with all small countries in general – has had earlier forays into international markets through exports (Swedenborg, 1979) associated with their offensive and aggressive strategy to overcome the limited size of its domestic market.

Another common pattern in the growth of MNCs from the three resource-abundant countries lies in the importance of the mining sector, once again a feature of the presence of rich and abundant mineral resources in the three countries which enabled American, Swedish and Brazilian firms to develop management and organizational skills and technologies in mineral resource extraction and processing which were exploited profitably abroad. Indeed, mineral extraction and processing featured at some stage in the history of the growth of MNCs based in the three resource-abundant countries; however, this type of investment assumed the highest prominence in the case of the United States and one that emerged at an early stage of development of American MNCs. Although the mining sector accounted for the second largest share of American FDI until the late nineteenth century after railroads, the large capital requirements of mining prospecting and development enabled the mining and smelting sector to become the most important sector of American FDI between the late nineteenth century and 1929. The main purpose of these mining investments by American firms was to profit from the attractive business opportunities arising from the rich and undeveloped mineral resources of Mexico, Canada and South America and to supplement the domestic mineral resources of the United States.

By contrast, Swedish firms in the mining sector became MNCs only to a very limited extent. Thus, a striking difference between Sweden and the larger sized but also resource-rich United States was in the relative importance of outward FDI to exploit foreign raw materials. This is explained in the context of the development of competitive advantages by Swedish industry in the extraction and processing of domestically available raw materials in a small country in relation to the large scale in which most investment in raw materials is typically undertaken (Swedenborg, 1979). Apart from the Swedish mining firm, Gränges, that operated foreign mines since the 1950s, and notably the LAMCO mine in Liberia, Boliden is one of the few other major firms in the mining and steel industries that invested in major foreign operations in the 1980s (Sölvell *et al.*, 1991).

In broad similarity to Sweden, Brazilian firms in the mining sector became MNCs only to a very limited extent and although their emergence can be traced to the 1970s, their major growth took place in the 1980s. While their overseas activities in mineral prospecting did not always involve the establishment of subsidiaries (as in the case of the state-owned Companhia Pesquisa de Recursos Minerais (CPRM) whose mineral prospecting activities in Africa was undertaken through contracts), the accumulated experience of Brazilian companies nurtured within the vast and resource-abundant environment of Brazil was directly applicable to their overseas mineral prospecting activities.

Analysing the pattern of MNC growth of the three resource-abundant countries in sectors other than manufacturing and mining shows some further resemblance even though the resemblance may not be universal to all three countries but to two of them. This is the case when examining outward FDI in the petroleum and forestry industries in particular. While the petroleum industry has assumed greater importance for MNCs from the United States and Brazil, the forestry industry has assumed greater importance for MNCs from Sweden and Brazil.

The growth of MNCs in the petroleum industry seems to have featured prominently in the resource-abundant large countries of the United States and Brazil but not in the resource-abundant small country of Sweden. However, despite the common importance of the petroleum industry for American and Brazilian MNCs, the determinants of outward FDI in the industry differed in the two large countries. The emergence and expansion of American FDI in petroleum was linked to the presence of rich and abundant petroleum resources in the United States which enabled American petroleum companies to develop management and organizational skills and technologies in petroleum extraction and processing which were exploited profitably abroad. In sharp contrast, the emergence and expansion of Brazilian FDI in petroleum arose from the necessity to search for new sources of foreign oil in light of Brazil's position as a major importer of petroleum. Towards that end, Braspetro – a subsidiary of the state-owned oil monopolist Petrobrás – was created in 1972 to fulfil a strategic national interest of securing oil supplies abroad through risk contracts and close relations with OPEC countries. It became responsible for the foreign exploration, production, commerce, transportation and storage of oil and its products, as well as for the execution of technical and administrative services related to these activities (Guimaraes, 1986). Thus, although Braspetro was based in a petroleum scarce country, it drew upon the accumulated experience of its parent company in oil and gas exploration, geophysics and drilling in addition to learning by doing on its own in its various foreign activities. This enabled the firm to build up even more advanced engineering and managerial skills related to petroleum exploration and to progress over time to become its own profit centre and not simply an instrument of national policy. Evidence for this was found in its increasing ability to enter into risk investments, to render technical services in the construction and installation of oil rigs, refineries, storage systems and pipelines, to participate in the oil trading market and to use deep-sea oil production know-how in the North Sea and Gulf of Mexico – former turfs of major oil companies (Villela, 1983; Wells, 1988).

The growth of MNCs based in the forestry industry seems to be closely similar in the case of both Sweden and Brazil but not the United States. In both Sweden and Brazil that have rich forestry resources, backward vertical integration in resource extraction in resource-rich foreign countries was not a prominent activity of their firms, particularly so in the case of Brazil. In both these countries, the extraction and processing of rich forestry resources in the home country was largely an activity dominated by local firms and under-taken in close proximity to local sources of raw materials and energy. Firms from both these countries experienced growth of international production but these took different forms in each country. In the case of Sweden, the rapid growth in international production of firms based in the forestry or paper and pulp industries took mainly the form of *forward* vertical integration achieved through the full or partial acquisition of their major foreign customers – a move in response to the entry of the North American pulp and paper firms in Western Europe at the end of the 1950s (Carlson, 1977). The forward integration in foreign markets of largely domestic production oriented Swedish forestry and paper and pulp firms was motivated as much by the profitable exploitation of accumulated know-how and skills in extraction and processing of natural resources (Swedenborg, 1979) as by the need to gain market knowledge and to influence product development in the pulp- and paper-consuming industries (Carlson, 1977). In the early 1970s, the Swedish pulp and paper manufacturers had established more than 40 manufacturing plants abroad, mostly in the United Kingdom and Germany (Sölvell *et al.*, 1991), but there have also been outward FDI to gain access to raw materials in Canada and the United States and the production of pulp based on domestic raw materials in Portugal, Canada and Brazil (Söderström, 1980; Olsson, 1993) to safeguard their future expansion. The further growth of their European investments in the 1970s and 1980s enabled the Swedish pulp and paper companies to build dominant market positions (Sölvell *et al.*, 1991).

By contrast, the rapid growth in international production of Brazilian firms based in or with close links to the forestry industry took mainly the form of *horizontal* integration by manufacturers in foreign markets in such industries as wood products and furniture, pencils, paper and cardboard packaging – a move to support their objective to increase their foreign sales through direct sales and, in some cases, international production, while still taking advantage of an abundant and low-cost supply of wood from Brazilian forests.

On the other hand, the foreign stakes of American firms in the forestry industry was geared to seek timber and other forestry-based products in their quest to become more independent of foreign producers as well as to make greater profits. Canada was an important host country for these types of investments by American firms owing to the presence of an abundant supply of inexpensive timber, good water power, relatively cheap labour and the absence of American duties on Canadian newsprint since 1911. The forward integration of these firms towards the secondary processing of extracted natural resources in Canada in response to the requirements of host country industrialization policy made lumber and paper and pulp mills account for at least one-third of American FDI in manufacturing in Canada between 1897 and 1914. The inflation in the United States brought on by the war boom led to the further expansion of American FDI in Canadian pulp and paper mills to service the needs of American domestic enterprises that were major users of paper.

Beyond the similarity in the pattern of MNC growth in particular industries must be mentioned some of the unique sectors in which any one of the
three resource-abundant countries have spawned significant MNC activity
that was distinctive and incomparable to the growth of MNCs from the other
resource-abundant countries. To take just a few illustrative examples, outward FDI in civil construction and consulting engineering seems to have been
a peculiar element of Brazilian MNCs that was unmatched in American and
Swedish MNCs. In addition, the early outward FDI in the railroads sector
and the metamorphosis or merger of the early mercantile businesses of firms
into more diversified and sometimes entirely different enterprises engaged in
lines of businesses besides commerce (to include mining, agriculture, industry,
transportation and banking) were some of the notable features distinguishing
American MNCs from MNCs from Sweden or Brazil. Outward FDI in agriculture was also associated solely with American MNCs particularly during
the period of their emergence. Thus, despite the similar importance of the
agricultural industry in the history of the domestic industrial development
and export pattern of Sweden and Brazil, this sector did not form a significant
basis of the outward FDI of Swedish and Brazilian firms. Not only did the
industry remain largely in the domain of domestic firms in both these countries, the firms became important international companies through exports
but not through outward foreign direct investment. Unlike American MNCs,
the search for agricultural produce in foreign countries with high agricultural
productivity did not provide a driving force for Swedish and Brazilian MNCs.

Going beyond the examination of specific industries of importance to particular resource-abundant countries into a broader examination of the
breadth of industries that MNCs from resource-abundant countries have
emerged and evolved shows that MNCs from the resource-abundant large
countries of the United States and Brazil were involved in a wider breadth of
industries by comparison to the MNCs from the resource-abundant small
country of Sweden. Considering the manufacturing sector alone, American
MNCs have emerged first in metallurgical industries as mentioned, and then
evolved in the 1920s in industries that competed on the basis of product
differentiation (as in the case of food and drink, textiles and clothing), as
well as in industries with distinctive products. American industries with
worldwide technological leadership gained from the transfer abroad of techniques in product design and engineering and organization of production
(electrical industry, motor vehicle industry, certain metal products, petroleum), as have companies with advanced marketing methods (motor vehicle
industry, metal products, petroleum). The kinds of American manufacturing
enterprises that invested abroad in later decades resembled closely the investors of earlier years: these were leading firms in their industries in the United
States that had advantages in technology, unique products and a long history
of international economic orientation (see Table 6.2). The fact that an industry was technologically advanced did not *ipso facto* guarantee large FDI, but it
generally meant that leading companies in that industry would in time, after

Table 6.2 Resource-abundant countries: actual development paths for outward direct investment, and their association with local industrialization across different countries

Examples of countries	Link between domestic development and the growth of outward FDI	Type of locally based MNC
United States	Slow growth of domestic demand brought about by the depression between 1893 and 1897 led firms with surplus production to turn to foreign markets as an outlet for surplus goods that could not be sold at home. Exports led to the establishment of foreign sales branches and then foreign factories at a later stage.	Manufacturing firms
	The trend towards domestic mergers at the end of the nineteenth century associated with the development of domestic capital markets and the Sherman Antitrust Act that forebade agreements among independent firms to restrain trade.	Large manufacturing firms, typically resource based Agricultural firms
	Actions by the United States government such as the duty free status accorded to exports of natural resources in their unprocessed or processed state.	Manufacturing firms, typically resource based
	The rapid development in the 1850s of American technologies in certain metallurgical industries (i.e. machine tools, guns, reapers, and sewing machines) linked with mass production in which the United States had already acquired world leadership.	Large manufacturing firms, typically exporting firms
	The rapid importance of market-oriented FDI by American firms in the 1920s was associated with relatively new industries. The previous importance of metallurgical firms gave way in the 1920s to the expansion of firms and industries with trademarked or branded merchandise (such as food and drink, textile and clothing), or firms with distinctive products and techniques in product design, engineering and organization of production (electrical equipment, motor vehicles, certain metal products, petroleum) or advanced marketing methods (motor vehicle industry, metal products, petroleum). International production was prompted by prospects of profitable business opportunities in the host country which could not be fulfilled by exports owing to trade restraints.	Large manufacturing firms, typically exporting firms Small- and medium sized suppliers to the motor vehicle industry

(continued)

Table 6.2 (Continued)

Examples of countries	Link between domestic development and the growth of outward FDI	Type of locally based MNC
	The parallel expansion of American firms in public utilities in both domestic and foreign markets between 1914 and 1929 was favoured by the strong support of the United States government and the growth of ownership advantages of American firms in the form of capital, technology, management and marketing expertise.	Utility companies
	The kinds of American manufacturing enterprises that invested in plants abroad since the Second World War resembled closely the investors of earlier years: these were leading firms in their industries in the United States that had advantages in technology, unique products, and a long history of international economic orientation. Among these were firms in transportation equipment, chemicals, machinery, food products, electrical machinery and primary and fabricated metals. The fact that an industry was technologically advanced did not *ipso facto* guarantee large FDI, but it generally meant that leading companies in that industry would in time, after finding exports could not continue to fill foreign demand, show an interest in extending their business abroad through outward FDI. This enabled the firms to penetrate new overseas markets as well as to maintain their long-established position in foreign countries and to meet rising foreign competition.	Large manufacturing firms, typically exporting firms
	Further industrial upgrading towards the services sector in the domestic economy and associated growth of exports and outward FDI in services	Services firms (primarily banks and brokerage houses, professional services, hotels and tourism-related services, car rentals, etc.)

(*continued*)

Table 6.2 (Continued)

Examples of countries	Link between domestic development and the growth of outward FDI	Type of locally based MNC
Sweden	Export earnings from wood and iron products financed the establishment of new industrial companies of the late eighteenth century. The growth of the engineering industry generally consisting of metal manufactures, machinery and transport equipment was an outcome of the industrialization phase between 1870 and the First World War. In this phase, the technical and metallurgical know-how in metals production based originally on iron and steel production from local sources of high quality iron ore spilled over into the engineering industry. Technologically innovative engineering firms that exported at an early stage of their firms' history became MNCs rapidly to pursue markets, defend export markets, overcome high transportation costs and establish a local presence	Manufacturing firms in the engineering industry
	Vertical integration in the 1960s and 1970s to establish production closer to final markets abroad and exploit accumulated know-how in resource extraction and processing in foreign markets, to gain market knowledge and influence product development in the pulp- and paper-consuming industries, and to respond effectively to the entry of North American firms in the pulp and paper industry in Western Europe at the end of the 1950s.	Forestry firms, forwardly integrating
	The relaxation of foreign exchange controls in Sweden since 1987 and controls on investment in financial assets and property abroad	Services firms and individuals (largely in property and finance)
Brazil	The rapid growth of Brazilian exports to the Latin American region and the need to have a presence in the developed countries to attract resources from international capital markets led to the overseas expansion of Brazilian banks particularly between 1974 and the early 1980s	Banks

(continued)

Table 6.2 (Continued)

Examples of countries	Link between domestic development and the growth of outward FDI	Type of locally based MNC
	Growth of international production in a wide range of industries in the 1980s to include resource-based and capital intensive industries associated with the large size and breadth of the Brazilian economy that built upon both its rich agricultural and mineral resources as well as efforts to expand industrial capacity in capital goods sectors. The sustained strengths in primary product-based industries is a legacy of its primary product export economy industrialization phase, while that in capital goods-based industries was a product of the various phases of import substitution industrialization.	Manufacturing firms
	Overseas mineral prospecting activities in the 1980s based on accumulated experience nurtured in the vastness and resource abundancy of Brazil	Resource-based (mainly minerals) firms

Source: Author's compilation based on the analysis contained in the country chapters.

finding exports could not continue to fulfil foreign demand, show an interest in extending their investments abroad. The American companies whose entrepreneurs exhibited farsighted leadership grew rapidly and made the most far-ranging investments in foreign countries. These features describe consistently the growth of American MNCs over time, particularly in the manufacturing sector.

By comparison, Swedish manufacturing MNCs have had a narrower industrial focus in the engineering industry generally and, to a lesser extent, in the pulp and paper industry. Despite their narrower industrial focus, Swedish firms in these industries were technologically intensive, had distinctive products and a longer history of international economic orientation compared to the United States owing to their small country status. Many Swedish firms have become world leaders in their product niches on the basis of product design, engineering and organization of production and advanced marketing methods.

Although Brazilian manufacturing MNCs – like those from the United States and other large countries generally – were also involved in a wide breadth of industries to include food products, textiles, clothing and footwear, paper packaging, wood and furniture, bicycles, lifts, electrical products, steel products and capital goods, motor vehicle parts, and aircrafts (see Table 6.2), these firms did not compete on the same basis as American or Swedish

Table 6.3 Resource-abundant countries: technological accumulation and the national course of outward direct investment

	Stages of national development		
	(1)	*(2)*	*(3)*
Form of technological competence of leading indigenous firms	Basic engineering skills, complementary organizational routines and structures	More sophisticated engineering practices, basic scientific knowledge, more complex organizational methods	More science-based advanced engineering, organizational structures reflect needs of coordination
Type of outward direct investment	Trading, railroads, early resource- and market-seeking FDI, services	More advanced resource-oriented or market-targeted investments in manufacturing and services	Research-related investment and integration into international networks
Industrial course of outward direct investment	Trading, railroads, resource-based (extractive MNCs or backward vertical integration), manufacturing, construction and consulting engineering	More forward processing of resources, or growth of fabricating industries for local markets or exports; growth of services	More sophisticated manufacturing and services systems, international integration of investment
Stage of development United States	starting in the late seventeenth century	starting in 1929	starting in the period since the Second World War
Sweden	starting in the late 1800s	between 1870 and 1914	starting in the 1960s
Brazil	period from the late 1960s to 1975	period since 1975	currently still unreached

Source: Author's compilation based on the analysis contained in the country chapters.

manufacturing MNCs. Although Brazilian manufacturing firms in the metal-working, mechanical engineering and electrical equipment sectors had fairly advanced foundry skills and skills in the organization of production, these firms did not feature prominently in trademarked or branded merchandise widely advertised in Brazil. In fact, as the experiences of some Brazilian MNCs such as Copersucar and Gradiente Electronics showed, some of the outward FDI was geared to penetrate the markets of developed countries by acquiring an established trade name abroad. Neither has any Brazilian industry or firm attained significant worldwide technological leadership (the closest it had achieved was the approach to the world technological frontier where this was fairly stable in some metalworking and mechanical engineering industries) nor developed sophisticated technological advantages and advanced marketing methods. Thus, in relating stages of national development to the form of technological competence of leading indigenous firms, the type of outward FDI and its industrial course over time, MNCs from Brazil are at an intermediate stage of development compared to the more advanced American and Swedish MNCs, despite the similarity in some patterns of their MNC growth (see Table 6.3).

Part III

Multinational corporations from the resource-scarce large countries

7 The emergence and evolution of multinational corporations from the United Kingdom

Introduction

During the era of Pax Britannica that lasted until the 1860s and 1870s, the United Kingdom was the world's premier industrial nation and the head of an imperial empire that spanned vast territories in developing countries or newly settled primary producing areas (Dunning, 1985). Britain was the seat of the First Industrial Revolution and had technological hegemony. Of the 327 important inventions discovered between 1750 and 1850, Britain was responsible for 38 per cent, France for 24 per cent, Germany for 12 per cent, and the United States for 16 per cent (Streit, 1949).

The status of Britain as a world imperial power and the first industrial nation provided the country with the opportunity and the capacity to engage in various forms of international business at an early stage. Exports of portfolio, migratory, merchant and financial capital from the United Kingdom date back to the late sixteenth century (Houston and Dunning, 1976). These early overseas investments which entailed some degree of managerial control were directed mainly towards the primary sector (i.e. mines, plantations, etc.), and services such as railways, utilities, banking, trade and commerce.

One of the important consequences of the First Industrial Revolution that started with England was the development of processing industries. As a result, domestic industry demand for new sources of energy, industrial raw materials and minerals was created, and the country became a major consumer of particular commodities and foodstuffs for its increasingly urbanized population with high real incomes.[1] As the United Kingdom is a country scarce in natural resources, British factories needed to import raw materials and intermediate products financed by the export of finished manufactured products. Trade became the handmaiden of domestic economic growth, but this was often complemented by an outflow of tangible and intangible assets in the form of capital, technology and management to extract the minerals from abroad and provide the raw materials and foodstuffs for the needs of its domestic industries and consumers. With the development of a highly sophisticated international capital market, some of these assets were transferred abroad through the market in the form of foreign portfolio investments and some were transferred through firms in the form of FDI. Both forms of foreign investment were financed from a balance-of-payments surplus accumulated from the growth of Britain's industrial exports (Dunning, 1985).

Although FDI was often the preferred route over spot-market transactions or long-term contracts to obtain primary products as a means to minimize the risks of supply disruptions and price increases and to ensure product quality, the amount of FDI actually undertaken in absolute terms remained very small for most of the nineteenth century and also in relation to other kinds of international capital movements. As the economic activities abroad financed by FDI were complementary with, rather than competitive to, the domestic activities of British MNCs and were directed in large volumes to the Commonwealth countries (Chandler, 1986), the pattern of British FDI that developed was not dissimilar to that of its trade and could be explained by the neoclassical Heckscher–Ohlin theory of comparative advantage.

The United Kingdom was one of the pioneering home countries of MNCs, and dominated the process of MNC expansion prior to the Second World War. The country accounted for 46 per cent and 40 per cent of the global stock of direct investment abroad in 1914 and 1938, respectively (Dunning, 1983), and was responsible for the highest number of overseas FDI operations (Wilkins, 1988b). This chapter is devoted to exploring the emergence and evolution of British MNCs, and the changing determinants of British FDI as it affected the sectoral and geographical patterns of British FDI over time. The historical excursion is conducted in three time frames: the period prior to 1914, the inter-war period and the period since the Second World War.

The emergence of British MNCs in the period prior to 1914

There were four types of British enterprises that operated overseas in the period prior to 1914. First and foremost were the *free-standing companies* (Wilkins, 1986a, 1988a). There were literally thousands of free-standing companies, many more than any other type of British enterprise in the period prior to 1914. These companies which did not undertake any prior production in the United Kingdom before investing abroad were registered in the United Kingdom and floated on the London capital market primarily for the purpose of undertaking business exclusively or mainly abroad (Jones, 1996). The second type consisted of companies that developed and established their business pre-eminently in Britain and then expanded overseas through FDI, thus extending their existing domestic operations (Jones, 1986a). This type of firm, which was the precursor of the 'classic' or modern MNC, started to evolve in the 1860s and 1870s, strongly influenced by the rapid growth of international trade and the technological, organizational and institutional developments of the second half of the nineteenth century. The third type was the *migrating multinational* (Jones, 1986a). This was a company whose headquarters was based originally in one foreign country, invested in Britain and then evolved to become a British-headquartered MNC over time. Such was the case of Borax Consolidated Ltd (Travis and Cocks, 1984) and British-American Tobacco (BAT) (Jones, 1986a). The fourth variant was that of a firm established abroad with no registration in the United Kingdom but nevertheless attracted both British capital (sometimes in the form of large overseas deposits) and British management, the latter often provided by immigrants or expatriates. This type of firm can be distinguished from a free-standing company by the absence of legal headquarters in the United Kingdom (Wilkins, 1988b). Some overseas-registered companies seem to have been established abroad by British trading companies which acted as the centre of an investment portfolio and provided management and finance. These types of firms, which were labelled as '*investment groups*' by Chapman (1985), typically displayed opportunistic patterns of portfolio diversification in their overseas operations.

The foundation of the free-standing companies was based on the profitable exploitation of abundant and relatively cheap capital in Great Britain in lucrative investment opportunities abroad.[2] Free-standing firms were the institutional device to maintain control over the capital transferred (Wilkins, 1988b). Although each company had a board of directors in Britain charged with the management of the business overseas, the main business was conducted overseas. The free-standing companies established either invested abroad directly or served as a holding company through its ownership of the securities of a locally incorporated company. Each free-standing company that was established primarily to conduct foreign business in a single foreign country usually also operated in a single economic activity. Collectively, the activities abroad of these companies spanned the whole economic spectrum from the primary sector, manufacturing and services (including public utilities, transportation and banking services) (Wilkins, 1988b).[3] The enterprises were both local market oriented and supply oriented in providing for the needs of Britain or, less often, third countries.

Apart from the free-standing companies, the emergence of the modern British MNC starting in the 1860s became responsible increasingly for British FDI. These companies which similarly invested abroad in pursuit of more markets and/or sources of supply comprised manufacturing companies as well as producers of services (trade, shipping, insurance, accounting, engineering and so forth) (Wilkins, 1988b).

Resource-based FDI prior to 1914

As in the case of the United States, the majority of British FDI in the period prior to 1914 were comprised largely of supply-oriented resource-based investments and associated investments in services (trading, distribution and transportation) geared primarily to support both the expansion of the domestic processing industries whose growth was spurred by the First Industrial Revolution, and the needs of domestic consumers. These investments in basic inputs and other agricultural products were apparently conducted more commonly by free-standing companies, and being that the United Kingdom was resource scarce, there were often no comparable domestic activities (Wilkins, 1988b). Apart from the free-standing companies, these primary sector investments were also conducted by British companies in manufacturing, petroleum and trading that extended their investment interests outside the territorial borders of the United Kingdom through backward vertical integration into the establishment or acquisition of plantations, farms and mines in foreign countries. Some of the best known examples are discussed below.

The investments were often vertically integrated to cover extraction/cultivation, processing and supporting service sector investments in finance, insurance, transport and distribution that facilitated international trade in raw materials and foodstuffs (Jones, 1996). Such early evidence of the presence of an international hierarchical organization served to protect companies from vulnerabilities to sudden price increases and/or supply interruptions (Vernon, 1983). This can be seen as the rationale behind the many resource-seeking overseas investments of resource-based companies such as British Petroleum, Shell, Burmah Oil, Rio Tinto Zinc and the Charter Consolidated Company over more than a century (Dunning and Archer, 1987). Such imperfections in the primary product markets in the nineteenth and early twentieth centuries also explain the motivations of several resource-based British manufacturing companies to internalize the markets for their required natural resources.

The location of resource-based investments by British companies was determined by the presence of the required natural resources in abundance, but other variables such as exploration and extraction costs, land rents, transport costs and host government attitudes towards foreign ownership of natural resources were also important considerations (Dunning and Archer, 1987). In general, although the investments were spread widely in resource-rich developing countries, the United States and Latin America, there was a marked concentration in the British colonies and Commonwealth countries as a result of Imperial policies and the position of Great Britain as head of British empire (Wilkins, 1988b).

The demand for minerals by domestic industry and the position of the United Kingdom as the centre of the international mining industry explain both the early emergence of the free-standing British mining MNCs, and the massive growth of their FDI in mining. Although the hundreds of British free-standing mining companies were usually small, the St John d'el Rey Mining Company and Rio Tinto Zinc Company were exceptions. Six British-owned companies were exploiting Brazilian goldfields by the 1830s, a process that began in the 1820s. By the middle of the nineteenth century there were a significant number of direct investments in minerals in various European countries (Jones, 1996). For example, there were at least 174 British mining companies that owned or controlled copper-pyrites, iron, lead and silver mines in Spain between 1851 and 1913. French, German and British interests dominated the Spanish minerals industry (Harvey and Taylor, 1987). There were another 659 British-registered companies in mining ventures in the United States between 1880 and 1904 (Wilkins, 1989). Foreign-owned (mainly British) companies accounted for more than 25 per cent of the copper mining output of the United States by 1889. South Africa was also an important host country after 1913, with the discovery of goldfields in that country. Over 20 per cent of the British firms in mining, and 40 per cent of the capital invested was in the southern African gold industry (Harvey and Press, 1990).

The position of the United Kingdom as the world's largest producer of tin, copper and lead until around 1850 provided its firms with mining skills and technology which were exploited abroad. Thus, although by 1900 the country was no longer an important producer of any mineral except coal, the success of its overseas mining activities in the late nineteenth century and afterwards was enhanced by the accumulated mining skills and technology from the old mining region of Cornwall (Jones, 1996).

Apart from minerals, another important area of supply-oriented investments by the United Kingdom was petroleum. Investments in this industry dwarfed all other resource-based investments (Dunning, 1992). The Royal Dutch Shell Group first engaged in FDI in 1890 and had invested in oilfields, refineries and distribution in the United States, Venezuela, Dutch East Indies, Russia and Europe by 1914 (Jones, 1996). British Petroleum began as one of dozens of British free-standing companies established in 1901 to search for oil in various parts of the world. Unlike Shell, the company held a much smaller share of the world oil market, and its competitive position rested on the control of the rich Iranian oil reserves (Bamberg, 1994).

The establishment of rubber plantations overseas by British companies was stimulated by the worldwide surge in demand for rubber from the early 1900s, the high prices fetched by that commodity before the First World War, and the supply constraints subsequent to the various restrictive cartel agreements that came into force in the inter-war years. European-owned plantation companies were established throughout the colonial possessions in South East Asia, which by the First World War accounted for two-thirds of the total world output of rubber (Jones, 1996). The investments in plantation rubber overseas was conducted by both free-standing companies and classic MNCs. Among the latter was the British rubber and tyre manufacturing company, Dunlop, which was engaged in backward vertical integration in rubber plantation through the acquisition of rubber estates in Ceylon and Malaya before the First World War (Jones, 1984a). By 1917, Dunlop owned about 60,000 acres of land devoted to rubber plantation in the Malay peninsula and 2,000 acres in Ceylon (Jones, 1986b).

There were other instances of backward vertical integration by British manufacturing companies in the other resource-based industries. Among the most notable were the investments in the 1900s by Lever Brothers – Unilever's British predecessor – which began to acquire palm oil and copra plantations overseas in the Solomon Isles, the Belgian Congo and Nigeria in order to guarantee supplies for their soap manufacturing business in the face of a predicted shortage of vegetable oils (Fieldhouse, 1978; Wilson, 1954). There is also Cadbury's investments in cocoa plantations in the Gold Coast and Trinidad; Imperial Tobacco's activities in tobacco leaf plantations in Nyasaland; Tate & Lyle's investments in sugar estates in Jamaica, Trinidad, Belize, Zambia and Zimbabwe; and Turner and Newall's investments in asbestos mines in Rhodesia and South Africa, among other numerous examples (Dunning, 1992; Dunning and Archer, 1987).[4] In some cases the cross-border vertical integration was undertaken by a trading firm: this was the case for example with Booker McConnell, a trading company which bought sugar plantations in British Guyana in the 1830s, and accounted for 70 per cent of Guyana's sugar output by the early 1950s (Chalmin, 1990).

There were other primary sector-based activities that engaged British MNCs in foreign countries in the late nineteenth century. However, unlike their investments in minerals, oil and industrial raw materials to serve the needs of rapidly growing domestic industries, some of the agriculturally based overseas investments often arose from entrepreneurial perceptions of profitable opportunities. This was the case with cattle raising which attracted considerable FDI by many British free-standing companies. Attracted by the very high profits earned by indigenous cattle firms in the United States, these companies acquired large acreages of cattle ranges in Texas, Wyoming, Colorado and New Mexico (Lewis, 1938). There were some 41 acquisitions by British companies of cattle ranges in the American West between 1879 and 1889 representing over £10 million of investment, but most of the ranches reverted to American ownership by 1914 (Wilkins, 1989). No less important were the large land companies established by British companies in Latin America, especially in Argentina from the 1880s, for the purpose of raising livestock. While some of these were investments by free-standing companies, other investments were undertaken by large integrated enterprises, e.g. Liebig's Extract of Meat Company which owned vast cattle estates in Argentina and Uruguay by 1913 (Stopford, 1974).

Services sector FDI prior to 1914

The most prominent of British MNCs in services prior to 1914 were the trading companies (Yonekawa and Yoshihara, 1987). As an island economy with scarce natural resources, Britain has always been dependent on international trade and hence the emergence of well-developed and numerous British trading companies with important roles in the development of British business abroad. As mentioned, this could take the form of trading companies acting as the core of British-based investment groups established before 1914 (Chapman, 1985). There were also British individuals involved in shipping as part of free-standing companies (Porter, 1986), and consulting and managing engineers such as John Taylor & Sons that managed free-standing mining companies around the world (Harvey and Taylor, 1987). There was also the expansion abroad of British banks which formed a significant aspect of British business overseas. The lack of British 'universal' banks (Cottrell, 1991) precluded the possibility of the British banking sector to assist directly in the expansion of British business abroad. Instead, it was often the investment group with a trading company at its core that fulfilled a more significant and major role (Chapman, 1985). Many of the British international and imperial banks began as free-standing companies, of which the Imperial Bank of Persia is an example (Jones, 1986c).

Manufacturing sector FDI prior to 1914

Between 1870 and 1900, resource-based extractive investments grew more slowly as a new type of British industry more closely related to new consumer needs began to emerge. International production was confined to a limited number of companies before 1880 (Jones, 1996), and some of these early forays by British MNCs in the manufacturing sector were undertaken by free-standing firms. There were thousands of British free-standing firms active in the United States in the late nineteenth century, and there were more than 100 British parent companies that built or acquired some 255 manufacturing plants in the United States before 1914. Most of the British free-standing firms and approximately one-third of the factories established by the classic MNCs proved unsuccessful, short-lived or were no longer British owned by 1914 (Wilkins, 1989).

Over time, the direct capital exports of the United Kingdom consisted more of the establishment of foreign subsidiaries and branches by enterprises already operating in their home countries – essentially the kind of foreign activity of modern classic MNCs which mainly predominates in modern time. The managerial and technological competences developed in their home countries gave firms the ability to initiate and sustain their international production activities. Unlike the earlier resource-based investments which were directed primarily to resource-rich developing countries in the colonies and the Commonwealth, the emergent import substituting manufacturing investments prior to the First World War displayed a preference for high-income markets, but with some bias towards countries belonging to the British empire owing to political and other psychic ties. This shift in the pattern of MNC activity developed rapidly around 1875, and became established firmly in 1914 (Dunning, 1983).

At least 15 major British MNCs emerged by 1914, of which 14 were manufacturing companies. These were Babcock & Wilcox (industrial machinery), British American Tobacco (tobacco), Bryant & May (matches), J. & P. Coats (cotton thread), Courtaulds (rayon), Dunlop (rubber tyres), English Sewing Cotton (cotton thread), Gramophone (records), Lever Brothers (soap), Nobel Explosives (chemicals), Pilkington Brothers (glass), Reckitt & Sons (household products), Royal Dutch Shell (oil), Vickers (armaments) and Burroughs Wellcome & Co. (pharmaceuticals) (Dunning and Archer, 1987). All of these major pioneering British MNCs that emerged pre-1914 held strong oligopolistic positions in their domestic markets, and several were members of international cartels or market-sharing agreements (examples included Babcock & Wilcox, British American Tobacco, Bryant & May, Gramophone and Nobel Explosives). These agreements both allowed the participants favoured access to certain markets (usually the markets of the British empire and/or Europe) and provided protection from competition from other firms. A further advantage of some of these pioneering British MNCs was their privileged access to essential inputs or raw materials as mentioned earlier in this chapter (the case with Dunlop, Lever Brothers and Royal Dutch Shell, for example).

Perhaps the most outstanding feature of the ownership advantages of the major pioneering British MNCs was the role of individual entrepreneurs or a small group of entrepreneurs that were often also the owner-managers in determining the course of both domestic and overseas expansion of their respective companies. This included Sir James Kemnal (Babcock & Wilcox), J. B. Duke (British American Tobacco), Gilbert Bartholomew and George W. Paton (Bryant & May), D. E. Philippi (J. P. Coats), Henry Tetley (Courtaulds), du Cros family (Dunlop), A. Dewhurst (English Sewing Cotton), Fred Gaisberg (Gramophone), William Lever (Lever Brothers), Thomas Johnson, Lord Melchett and Sir H. D. McGowan (Nobel Explosives), Pilkington family (Pilkington Brothers), T. R. Ferens (Reckitt & Sons), Marcus Samuel and Henri Deterding (Shell), Basil Zaharoff (Vickers) and Henry Wellcome (Wellcome) (Archer, 1990).[5]

The concentration of domestic and international production of a significant number of British manufacturing MNCs in branded consumer goods (Dicken, 1992; Chandler, 1986) was a reflection to a large extent of the comparative advantages of the United Kingdom in labour intensive, capital neutral and human capital-scarce products (Crafts and Thomas, 1986) and, in some cases, the technological hegemony of the United Kingdom in the industries associated with the First Industrial Revolution. Many of these companies were in the textiles or textiles-related industries that had grown first and foremost in the United Kingdom and had begun to invest in a major way in the United States, Canada, France, Germany and Russia (Wilkins, 1989).[6] The ownership advantages of British MNCs in consumer goods industries derived generally from their ability to supply differentiated and high-quality consumer products and their control of selling outlets (Dunning and Archer, 1987).

Only to a limited extent did British manufacturing MNCs emerge from the more technology intensive and knowledge intensive industries spurred by the Second Industrial Revolution in the second half of the nineteenth century. This was the case with Babcock & Wilcox (industrial machinery), Nobel Explosives (chemicals), Royal Dutch Shell (oil), Vickers (armaments) and Burroughs Wellcome (pharmaceuticals). The Second Industrial Revolution witnessed further organizational and technical innovations (such as electricity and the internal combustion engine, the inter-changeability of parts and the introduction of new continuous processing machinery) that not only enhanced the capacity of firms to create or acquire proprietary rights and to produce and distribute at a much larger scale of output through mass production and mass distribution, but also provided firms with opportunities to become multi-product, multinational and multi-regional entities (Cantwell, 1989b). The implications of the later advances were truly trans-continental. The modern industrial enterprise grew both by horizontal and vertical integration (Chandler, 1980). Unlike the technical and organiza-tional advances of the First Industrial Revolution which had a greater effect on the development of processing industries, those of the Second Industrial Revolution of the mid- and late nineteenth centuries encouraged the devel-opment of fabricating industries such as motor vehicles, office machinery, electrical goods, synthetic chemicals and others.

As the United Kingdom had already surrendered much of its earlier lead by the last quarter of that century, it missed out on many of the more recent developments brought by the Second Industrial Revolution and hence the development of the more modern growth industries was influenced heavily by American and, to a lesser extent, German innovations and practices (Dunning and Archer, 1987). Indeed, by comparison to the period between 1750 and 1850 the share of the United States in the important inventions discovered in the second half of the nineteenth century doubled to 32 per cent and the share of Germany increased significantly from 12 to 21 per cent, while the shares of Britain and France declined substantially from 38 per cent to 16 per cent and from 24 per cent to 19 per cent, respectively (Streit, 1949).

In addition, an examination of the industrial distribution of the 200 largest manufacturing firms in selected countries at the time of the First World War showed that some 50 per cent of American firms were in the newer or mainly *producer goods industries* compared with only 28 per cent of British firms. The respective shares for American and British firms in the older or mainly *consumer goods industries* were 50 per cent and 72 per cent, respectively (Chandler, 1980). The failure of the United Kingdom to catch up in the more modern fabricating industries had important implications for both their domestic and international production. Several of the British MNCs whose competitiveness was based on the more recently developed products and/or processes of the Second Industrial Revolution, although displaying some innovatory strength based on long experience, cannot be considered to have advanced technol-ogy.[7] The crucible steel makers are an excellent case in point (Tweedale, 1987). In fact, many were reliant on technology and knowledge acquired from overseas, mainly but not exclusively from the United States. Among these were Babcock & Wilcox (industrial machinery), Gramophone (records), Nobel Explosives (chemicals) and Wellcome (pharmaceuticals) (Dunning and Archer, 1987), and also Courtaulds (rayon), Burroughs Wellcome (pharmaceuticals), Brunner, Mond (alkalies), and Marconi (radio installations).[8] Thus, British business innovation in the modern industries has been promoted by foreigners rather than by those in mainstream Britain (Wilkins, 1988b). Furthermore, for some of the companies that made producer goods (Bradford Dyers, United Alkali, H. & G. Bullock) the home industries were often related to textiles.

The lacuna that developed in the domestic economy as a result of the inability of domestic firms to emerge in the more modern growth industries could only be filled through imports and inward FDI by foreign based MNCs at a later stage. The balance of evidence suggests that the new product and process innovations introduced through inward FDI has enabled foreign MNCs to steer the domestic economic structure of the United Kingdom consistently towards the technologically more advanced and internationally oriented sectors (Dunning, 1958, 1985).

The determinants of international production by British manufacturing MNCs in both the traditional and modern industries was the inability of important foreign markets to be supplied, or supplied as cheaply, through exports (Dunning, 1985). In some cases, the initial decision to go abroad before 1914 rested on more favourable production costs in the foreign location, the provision of host government incentives or patent legislation but more often than not import restrictions imposed by the host governments was the key element that rendered exports uncompetitive in foreign markets (Coram, 1967; Buckley and Roberts, 1982).[9] After 1880, imported manufactured products faced higher tariffs in the United States, Canada and most European countries (Jones, 1996).[10] Indeed, high foreign tariffs were the single most important factor leading to international production by British market-seeking FDI before 1914 (Archer, 1986; Jones, 1986a).[11] In addition, high transport costs encouraged foreign production by Babcock & Wilcox, Gramophone and Nobel Explosives whose products were high volume/low value or dangerous to export over long distances (Nobel). Tariffs and transport costs often combined to prompt British manufacturers to establish overseas subsidiaries, particularly in countries where there was strong or emerging indigenous competition (Dunning and Archer, 1987).

In light of these high transfer costs, three options were considered by exporting firms: the first was the complete abandonment of the export market; the second was the licensing of intangible assets to foreign firms; and the third was international production in the export market (Dunning, 1985). All of these routes were chosen by different companies, but it was the last route which enabled the emergence of some of the more enduring MNCs (Stopford, 1974). Where the exporting firms decided to maintain their business in foreign markets, FDI was often chosen over licensing as a mode of entry in foreign markets because of the expectation of capturing a fuller economic rent by exerting greater control over their proprietary rights (Dunning, 1985). Indeed, few British manufacturing companies seem to have chosen the licensing option before the Second World War due to the lack of enforceable patent legislation, the difficulties of monitoring the licensee's business or the early stages of development of foreign markets which necessitated the total control over operations from the start. Other factors that contributed to the internalization of firms producing high-quality consumer goods were the lack of licensees with the necessary capabilities and trust, while for those firms in high technology industries it was the inappropriate use of the market for the transfer of knowledge that was non-codifiable, idiosyncratic or tacit (Dunning and Archer, 1987).

All these factors compelled several British manufacturing firms, many of whose trade had been with these countries, to move the whole or part of their plants to these countries in order to maintain and expand their business (Mason, 1920). In some cases, the manufacturing firms had already established foreign sales branches in the export markets to either promote their exports of differentiated consumer products in foreign countries with high per capita incomes based on marketing expertise developed in the home market or to ensure and stabilize the demand for producer goods where firms typically incur large sunk costs (Dunning and Archer, 1987). These overseas sales branches have been rendered ineffective in the face of trade barriers and/or where there was strong or emerging indigenous competition. The product cycle model is a useful framework to explain the shift of location of production from the home country in the innovative new stage (or innovation-based oligopoly) to other relatively advanced countries and previous export markets in the maturing product stage (or mature oligopoly) in the presence of high transfer costs in the form of tariffs or threat of new competition. While these served as powerful 'galvanizing forces' to international production (Vernon, 1966), international production also provided the opportunity for firms to have a direct presence in the market and to cater to the specific and special needs of local customers (Dunning, 1992).

Another contributory factor that prompted the shift from exports to international production by British manufacturing firms particularly those in the more modern fabricating industries was the need to sustain a process of competition between firms in oligopolistic industries. In such cases, the motive for FDI was not determined by profits but by the need to protect the firm's overall competitive position through precluding rivals from gaining a foothold in a foreign market or through engaging in cross investments. Examples abound showing the influence of the behaviour, or anticipated behaviour, of competitor firms as a determinant of international production, including the establishment of Dunlop's plant in Japan in 1899 as a pre-emptive move against American tyre companies as well as its loss-making plant in France to combat the firm's rival firm, Michelin; Royal Dutch Shell's investment in the United States in 1912; and Pilkington's investments in Canada in 1913, among other examples (Jones, 1986b; Dunning and Archer, 1987).

The growth of British MNCs in the inter-war years

The inter-war years saw the collapse of international capital markets in the late 1920s and early 1930s, and was an era marked by political instability, economic depression and market fragmentation. At least 17 new major firms emerged to become MNCs between 1915 and 1939 (Dunning and Archer, 1987) in an atmosphere where industrial concentration, rationalization and cartelization became the norm. The competitive advantages of British MNCs during this period were very similar to those of the previous era which derived from asset ownership advantages, despite the stark difference in the under-lying economic and political environment in which FDI was undertaken in the two eras. In addition, the newer MNCs also tended to hold a similar prominent position in the domestic market, to have been established for a long time, and to be led by owner/entrepreneurs with similar drives and visions so important in the determination of the growth path of their companies. Thus, unlike in the case of the American companies (Chandler, 1980), managerial hierarchies in British companies during the inter-war period failed to develop and only a small proportion of British companies had established a multi-divisional structure even by 1950 (Channon, 1973).

While British FDI in the manufacturing sector was becoming increasingly significant with a share of some 25 per cent of outward FDI stock in 1938 compared to 15 per cent in 1914 (Dunning, 1983), international production by British manufacturers remained oriented towards the mature, relatively low technology sectors whose competitiveness emanated from the high income and large size of the British market. These were firms of large size and established technological advantages that derive strengths from product differentiation, quality, and marketing and managerial skills and experience. These characteristics favoured the continuing pre-eminent role of consumer goods firms as the new British MNCs during the inter-war years. Among these were the confectionery firms Cadbury, Rowntree, Pascall and Fry, as well as Peek Frean (biscuits), Cantrell & Cochrane (soft drinks), Distillers (alcoholic drink) and Yardley (soap) (Corley, 1989).

Several of the new British MNCs that emerged in the higher technology industries, including Coates Brothers and Metal Box (metal products) continued to depend greatly on new technological developments from the United States. The slow adjustment of British firms to the more rapid growth opportunities offered by the modern industries was regarded to be a function of the peculiar nature of the British economy which posed a host of institutional barriers that prevented industrial restructuring towards the growth-oriented sectors, including investments in innovation that would sustain those industries. Such barriers included, among others, the lack of provision for commercial studies and for any kind of technical education for managers and industrial staff (Ashworth, 1960; Chandler, 1980).[12] Other contributory factors were the risk averse strategy of British firms which continued to emphasize industries and sectors in which past successes had been based, the presence of a protected home market and the continuing preferential market access in the empire and Commonwealth markets during the 1940s and 1950s (Dunning and Archer, 1987). As the technological leading position of the United States strengthened in the inter-war years, the ratio of its outward FDI stake to its inward FDI stake rose from 1:8 in 1914 to 4:1 in 1938, while that of the United Kingdom declined from 33:1 to 15:1 over the same period (Dunning, 1983).

The prevailing economic and political climate of the inter-war period served to strengthen the protectionist stance of host country governments and, as a result, import restrictions continued to be the main locational factor determining MNC expansion during this period (Corley, 1989; Jones, 1984b; Nicholas, 1982; Chandler, 1980; Stopford, 1974).[13] However, the transportation difficulties that British exporters faced in the years between 1914 and 1918 was also a contributory factor (Dunning and Archer, 1987).

With the growth of industrial concentration, rationalization and cartelization in the inter-war years, the operating decisions of most large British MNCs, including their decisions to engage in international production, were taken in light of intense international oligopolistic rivalry. The actions or anticipated actions of their major international competitors was therefore key for both new and existing MNCs such as the resource-based companies of Shell and Imperial Tobacco, the high technology manufacturers such as APOC, Pilkington Brothers, Courtaulds, Dunlop, and George Kent and even for some manufacturers of consumer products such as the confectionery companies, Cadbury and Rowntree, that previously relied on exports to serve foreign markets prior to 1914.[14] These companies began to establish factories in Australia, Canada, New Zealand, South Africa and Ireland in the inter-war years (Corley, 1989; Jones, 1986a). Such atmosphere of intense rivalry somewhat abated during the 1940s and 1950s when British companies with neither the resources nor the competitive strengths became less generally influenced in their FDI decisions by the behaviour of their competitors, with the exception perhaps of Pilkington Brothers, Unilever, British Petroleum and Shell (Dunning and Archer, 1987).

The preference of British FDI in the empire and Commonwealth was reinforced during the inter-war years. Apart from having political stability, large and/or growing markets, and a transportation and communications infrastructure, these markets were favoured by psychic proximity, traditional ties and indirect enforcement (Svedberg, 1981). This led many British companies to both develop important trading links with these countries and to regard these markets, particularly the White Dominions and India, as a natural extension of their domestic markets in much the same way that Canada had been for American MNCs.

The evolution of British MNCs since the Second World War

As countries recovered from the effects of the Second World War, the international and economic and political environment became increasingly favourable to the expansion of FDI. The early post-war period was associated with the surge in technological developments, the further improvement in international communications and stable exchange rates. The stability, growth and full employment of the first 20 years of the post-war period was consistent with the 'Golden Age of Capitalism' (Kitson and Michie, 1995). It was also the period most closely associated with the peak of United States hegemony. As a result, the ratio of the outward FDI stake to the inward FDI stake of the United States rose further from 4:1 in 1938 to 6:1 in 1971, while that of the United Kingdom declined further from 15:1 to 8:1 over the same period. By 1980, the respective ratios both fell to 3.2:1 and 1.7:1 (Dunning, 1983). Just as the fall of the ratio in the case of the United Kingdom was a reflection of the long-term decline in its technological leadership since the First Industrial Revolution, the fall in the ratio of the United States beginning in the 1970s was a reflection of the gradual decline in that country's industrial hegemony in favour of several European countries (Germany and Sweden, in particular) and Japan.

At least another 26 new major firms emerged to become British MNCs between 1940 and 1959 (Dunning and Archer, 1987). Although their emergence has contributed to the continuing increase in the absolute value of British FDI, the share of the United Kingdom in world FDI stake has declined considerably. Indeed, the baton of leadership in world FDI passed from the United Kingdom to the United States in the period since the Second World War. The new British MNCs that emerged in the period since the Second World War displayed characteristics that were essentially more of the same as that of the pioneering British MNCs that emerged prior to 1914 and the inter-war period. British MNCs still did not feature prominently in the more modern and growth-oriented fabricating industries, a reflection in part of the country's adherence to imperial ties established in the past. Indeed, 61 per cent of the net foreign assets owned by British MNCs in 1960 in industries outside oil, banking and insurance was directed towards the Commonwealth countries (Dunning, 1985). Although the imperial legacy of the United Kingdom still bears some influence in the more recent geographical pattern of British FDI, there has been a re-direction in the geographical destination of British FDI since 1960 with a greater focus placed on the United States and Western Europe. An increasing proportion of British FDI was directed towards the United States even in light of intense competitive pressures and in the original six member states of the European Economic Community (EC) as recovery and expansion proceeded and as closer regional economic integration progressed, culminating in the accession of the country to the EC in 1973.[15] Thus, there has been an increasing propensity for British MNCs since the 1960s to invest in countries with higher per capita incomes and larger markets than itself (Stopford, 1976). It is also a reflection of the objective of several British companies to establish a presence in the major centres of technological excellence in light of the growth of global competition (Cantwell, 1989a).

Continuing structural shifts in the pattern of outward FDI resulted in the dominance of the manufacturing sector in the period since the Second World War. The decline of supply-oriented FDI by British MNCs since 1960 can be seen as arising from the increased efficiency of commodity and futures markets and, more significantly, the increasingly hostile stance by host country governments in the resource-rich developing countries to the foreign ownership and exploitation of strategic resources. This was evident in the spate of nationalization and expropriation of foreign property in resource-rich developing countries starting from the 1960s which reached a peak in the mid-1970s (UNCTC, 1988). Nevertheless, British mining investments abroad continued at an average rate, and some mining investments by British MNCs in non-ferrous metals and petroleum even grew rapidly. The share of the manufacturing sector in the total FDI stock of the United Kingdom (and the United States) increased further to about 35 per cent in 1960 from a share of 25 per cent in 1938 (Dunning, 1983). Such 35 per cent share of manufacturing in total outward FDI stock of the United Kingdom was maintained until the early 1990s.

The determinants of international production by British MNCs in the more modern industries in the period since Second World War appear to be categorically distinct from those in the more traditional industries. For those in the more modern industries, the establishment of local subsidiaries in certain key markets during the 1970s and 1980s was an integral part of the pursuit of oligopolistic strategies on a global scale. The concepts of 'exchange of threats' and 'follow the leader' associated with intra-industry production in oligopolistic competition described by Graham (1975, 1978, 1985) and Knickerbocker (1973) respectively appear to be highly relevant in explaining the international production patterns of Beecham, British Petroleum, Imperial Chemical Industries, Pilkington Brothers, Plessey, Unilever, Redland, Allied Colloids and Tace in the period since the Second World War. The cross-penetration of national markets in the advanced industrialized nations by MNCs is linked to the convergence in the structure of production and taste patterns in these countries and facilitated by the economies of large-scale production (Dunning and Norman, 1985).

It is also these MNCs that have been pursuing more rationalized or efficiency-seeking investments since the 1970s associated with an increasing amount of intra-firm trade and product and process specialization. These investments were motivated by the need to capture the gains from the economies of specialization and integration through taking advantage of differences in factor endowments and costs across different countries as well as to maintain international competitiveness by securing a presence in the major growth markets of the world. For these MNCs that have a higher degree and extent of multinationality, their asset ownership advantages (Oa) are complemented by ownership advantages that arise from multinationality (Ot) leading to transaction cost reduction or the attainment of transaction benefits in international production. It is the pursuit of the latter form of ownership advantages that has provided the incentive for MNCs to both widen and deepen their networks of international production.

On the other hand, for the British MNCs in traditional consumer goods industries that neither engaged in rationalized production and investment nor sought to benefit from transaction cost advantages, their incentives to internationalize in the period since the Second World War were not dissimilar to that in the earlier periods. These firms remained keen to exploit their preferences to produce in the markets of the empire and Commonwealth until the early 1960s when the exporting route became difficult or no longer practical.

Such dualistic pattern of the international production of British firms is a reflection of the domestic economic structure of the United Kingdom which although geared towards the industries of high and medium technology is propelled by the investments of foreign-based MNCs. The more traditional low technology and consumer goods manufacturing sectors on the other hand are where the indigenous strengths of British firms lie which explain the dominant role of these sectors in the foreign activities of British MNCs. Inward and outward FDI in and from the United Kingdom has, therefore, been directed towards different industries. Nowhere is this more evident than in the report prepared by Reddaway *et al.* (1968), which showed that, following a historical trend, 71 per cent of the net foreign assets owned by the leading British manufacturing MNCs in 1964 were in the less technology intensive sectors of food, drink and tobacco, household products, paper, metal products, building materials and textiles, while 29 per cent were in the more technology intensive sectors of chemicals, engineering, electronics and motor vehicles. By contrast, some 67 per cent of the net assets of foreign (mainly American) firms in the United Kingdom in 1965 were in the more technology intensive sectors and only 33 per cent were in less technology intensive sectors (Dunning, 1985).

This also helps to explain the different patterns of exports of the United Kingdom and that of the international production of British MNCs. The industrial structure of British manufactured exports in the years between 1965, 1970 and 1975 continued to differ sharply from that of its manufacturing FDI but for an entirely different reason than that pertaining to the end of the nineteenth century when international production displayed a higher technological intensity compared to that of exports. A study conducted by Clegg (1987) indicated that capital intensity and the skill level of managerial manpower exerted significant positive influences on British FDI in manufacturing between 1965 and 1975, while the skill level of production workers exerted a highly significant negative influence. This contrasts sharply in the case of British exports of manufacturing in which technological intensity exerted a highly significant positive influence.

Britain's present day comparative advantage which rests on the production of labour intensive, capital neutral and human capital-scarce products has remained essentially stable since 1870 (Crafts and Thomas, 1986). Firms in a broad range of industries continue to be more concerned with the production of low-cost standardized goods than in high-quality, technology intensive niche products (Porter, 1990). The employment pattern developed on this basis may have even become further entrenched during the 1990s (Nolan and Harvie, 1995).

Until the 1970s foreign affiliates established by British MNCs were largely truncated replicas of their parent companies and MNCs considered their network of foreign affiliates as a federated group of firms, each of which was designed to produce and sell products for the particular national markets in which it operated (Dunning and Archer, 1987). Although several British MNCs since the 1970s and 1980s have become increasingly aware of the attainment of transaction cost reducing advantages (Ot) arising from their geographical diversification, the number of British companies that have attained this in a significant way through the rationalization of their production and markets across national frontiers has remained essentially confined to British MNCs in internationally oriented high technology industries. Among the more globalized British firms are Imperial Chemical Industries, Glaxo, Unilever and Shell. These companies have adopted a transnational strategy in at least some of their value-chain activities, including R&D (Lane, 1998).

Some evidence of industrial upgrading in the pattern of British FDI in manufacturing may have become evident over the last decade. Products of coal, petroleum, plastics and chemicals; food, drink and tobacco; and electrical machinery accounted for 60 per cent of the total outward FDI stock of the United Kingdom in the manufacturing sector or 22 per cent of their total outward FDI stock in all industries in 1988. By 1997, the most important manufacturing industries for British MNCs were food, drink and tobacco; coal, petroleum, plastics and chemicals; transport equipment; and metals and mechanical engineering. These industries accounted for almost 72 per cent of the total outward FDI stock of the United Kingdom in the manufacturing sector or 27 per cent of their total outward FDI stock in all industries in that year (based on data in UNCTAD, 1999).

From a concentration in the primary sector in the era before 1914, and the growth of manufacturing sector particularly since the Second World War, the relative importance of the primary, secondary and tertiary sectors in British FDI continues to evolve. The attainment of almost equal shares of the three sectors in the total outward FDI stock of the United Kingdom in the mid-1980s has given way since to another major sectoral shift in the pattern of British FDI. With the increasingly larger share of services in the outward FDI stock of the United Kingdom at the expense of the primary sector, this sector became the dominant sector of activity of British MNCs accounting for 46 per cent of total British outward FDI stock in 1991. Thus, over the course of some 200 years, the dominant sectoral pattern of British FDI has spanned the primary, secondary and tertiary sectors.

Conclusion

This chapter examined the emergence and evolution of British MNCs. The United Kingdom was one of the pioneering home countries of MNCs, and dominated the process of MNC expansion prior to the Second World War. The origins of British MNCs in the period prior to 1914 were seen in four types of enterprises: the free-standing companies, the classic or modern MNCs, the migrating multinationals and the investment groups. These institutions provided the means to profitably exploit abundant and relatively cheap capital in Great Britain in lucrative investment opportunities abroad as well as to maintain control over the capital transferred.

The majority of British FDI in the period prior to 1914 comprised largely supply-oriented resource-based investments and associated investments in services (trading, distribution, transportation, finance and insurance) geared primarily to support both the expansion of the domestic processing industries whose growth was spurred by the First Industrial Revolution, and the needs of domestic consumers. These investments were spread widely in resource-rich developing countries, the United States and Latin America, although there was a marked concentration of British FDI in the British colonies and Commonwealth countries as a result of Imperial policies and the position of the Great Britain as head of the British empire.

Between 1870 and 1900, resource-based extractive investments grew more slowly as a new type of British industry more closely related to new consumer needs began to emerge. Import substituting manufacturing investments abroad by British firms displayed a preference for high-income markets, but with some bias towards countries belonging to the British empire owing to political and other psychic ties. The concentration of both the domestic and international production of a significant number of British manufacturing MNCs in branded consumer goods was a reflection to a large extent of the comparative advantages of the United Kingdom in labour intensive, capital neutral and human capital-scarce products and, in some cases, the technological hegemony of the United Kingdom in the industries associated with the First Industrial Revolution. Such comparative advantages of Great Britain has remained essentially stable to the present time.

The importance of British FDI in manufacturing increased in the period since the Second World War. The share of the sector in the outward stock of British FDI has remained around 35 per cent since 1960. The services sector became the dominant sector of British FDI since 1991. Thus, over the course of some 200 years, the dominant sectoral pattern of British FDI has shifted from the primary sector in the period prior to 1914, towards the secondary sector for much of the period since the Second World War, and then the tertiary sector since the 1990s.

Notes

1 The extensive British activity abroad in sugar and tea plantations reflected that country's very high level of sugar and tea consumption (Chalmin, 1990).
2 The free-standing companies attracted not only British capital, but also French, German, Belgian and other foreign investors (Wilkins, 1988b).
3 The different activities of British free-standing companies included, among others, rubber cultivation in Malaya, copper mining in Russia, cattle ranches in the United States, meat packing in Argentina, nitrate mines in Chile, railroads in Brazil, hotels in Egypt and mortgage companies in Australia (Wilkins, 1988b).
4 Although the plantation estates of Cadbury in Trinidad fulfilled a quality-control function, the vast majority of the requirements of the firm for cocoa beans was fulfilled by direct purchases from independent producers in West Africa (Williams, 1931).
5 It has been argued that Wellcome's failure to continue MNC growth after the First World War can be attributed to the loss of much of Henry Wellcome's former commercial drive (Archer, 1990).
6 The largest British textile companies were engaged in overseas FDI by 1914. Among numerous other companies were J. & P. Coats which had mills in the United States, Canada and Russia. English Sewing Cotton produced through affiliated companies in the United States, France and Russia. Linen Thread Company had factories in the United States and France. The Fine Cotton Spinners' & Doublers' Association acquired around the turn of the century a dominant interest in its most prominent French competitor, La Société Anonyme des Filatures Delebart Mallet Fils. Bradford Dyers had a major plant in the United States and one in Germany. Nairn Linoleum (linoleum is made of jute or burlap and thus can be classified as a textile) had factories in the United States, France and Germany (Wilkins, 1989).
7 Evidence of some technological strength is seen in the study of Nicholas (1982) of 119 British manufacturing firms that undertook international production in the period between 1870 and 1939. His study showed that one half of those firms established foreign plants to exploit a perceived technological advantage.
8 The technology of Courtaulds in synthetic textiles was developed in part outside Britain or with the aid of foreign ideas. Burroughs Wellcome was a company founded by Americans which sold patent medicines and ethical drugs. Brunner, Mond was a firm that was part of a Belgian MNC. It was the Belgian MNC rather than the British counterpart that was the driving force behind the company's MNC expansion. Marconi was founded by a man with an Italian father and a British mother and although the firm once held technological leadership, the company suffered rapidly from technological conservatism which threatened its leadership by 1914 (Aitken, 1985).
9 There is little reason to suppose that British companies were influenced by lower production costs abroad in undertaking FDI before 1914. International production seldom occurred in very labour intensive industries, and hence low labour costs abroad were rarely an enticement, except perhaps in a limited number of cases as in the international production activities of Lever Brothers and Courtaulds in the 1900s (Dunning and Archer, 1987). British American Tobacco's investments in cigarette production in China to take advantage of low-cost labour is another example. With the use of labour intensive rather than capital intensive production

techniques, the firm employed thousands of unskilled labourers in China to perform tasks that were already mechanized in the United States (Cochran, 1980).

10 Trade protectionism was a major influence on the rapid growth of British manufacturing FDI after 1880 as the worldwide trend towards free trade went into
reverse. The need to raise government revenue to finance the American Civil
War (1861–1865) led to a substantial rise in American tariffs. The wartime tariffs
were retained after 1865, increased in the 1880s, and then raised by the McKinley
Act of 1890 to an average level of 50 per cent on protected commodities. A brief
lowering of American tariffs in 1894 was followed three years later by an increase
to 57 per cent. Similarly, there was a return to protectionism in Europe after 1880,
stimulated by a severe recession in the previous decade, the growth of European
nationalism and the emergence of new European nation states such as Germany
and Italy. By 1914, the only remaining countries pursuing free trade policies were
the United Kingdom, the Netherlands and Denmark (Jones, 1996).

11 This was the case, for example, with the British cotton thread manufacturer, J. &
P. Coats. Before the American Civil War, three-quarters of the trade of this firm
had been with the United States. A 50 per cent tariff imposed by the United States
in 1864 forced the company and other cotton thread producers to engage in
production in the United States. By the outbreak of the First World War,
British subsidiaries accounted for 80 per cent of cotton thread production in the
United States (Wilkins, 1989).

12 Indeed, at the more advanced educational levels, technical and scientific instruction and inquiry remained 'poor cousins in the family of higher learning' in the
United Kingdom in the 1900s (Murphy, 1973). This was so unlike in the United
States where the need for trained managers, production and marketing specialists
in the technologically advanced machinery, electrical and chemical industries had
been recognized more rapidly and catered for by universities and business schools
(Chandler, 1980). It was not until 1947 that the British Institute of Management
was formed, and only in the 1960s were the London and Manchester Business
Schools founded (Dunning and Archer, 1987).

13 After the First World War protectionism spread. By the early 1920s tariffs in the
United States had been raised to their highest ever levels by the Fordney-
McCumber tariff. Australia, India and some Latin American countries were
among those countries that adopted import substitution industrialization strategies, and tariffs, import quotas and other trade barriers were imposed on imports
to foster the development of infant industries. The Smoot-Hawly Act of June 1930
increased the American tariff level substantially, and other countries followed suit.
By the end of the 1930s almost half of the the world's trade was restricted by tariffs
(Jones, 1996). Tariffs served to encourage foreign firms to establish local manufacturing in host countries, but also strengthened the bargaining position of
national firms in international cartel negotiations (Wurm, 1993).

14 For example, the decision of Courtaulds to manufacture in France was owing to
fears that its French counterparts may produce in the United Kingdom to jump
the tariff imposed on imported artificial silk in 1925 (Jones, 1986b). The investments of Dunlop in Eire, India and South Africa in 1930 were similarly prompted
by oligopolistic considerations (Dunning and Archer, 1987).

15 Tariff and non-tariff trade barriers caused British companies to establish subsidiaries in the EC well before the accession of the United Kingdom to the EC
(Dunning and Archer, 1987).

8 The emergence and evolution of multinational corporations from Germany

Introduction

The origins of German MNCs can be traced to the large firms in the chemical, electrical engineering and metal-producing industries that engaged in FDI during the nineteenth century. Even at that time, the basis of their competitiveness stemmed from a concentration in low volume but high value added segments and niches (Porter, 1990), where cost control traditionally ceded place to technological perfection, and where production of standardized goods had been far less dominant than in Great Britain. Indeed, a comparatively high level of R&D intensity describe German firms and MNCs from the beginning (Chandler, 1986).

Although its outward FDI was focused narrowly on Western and Eastern Europe, Germany accounted for around 10 per cent of the global stock of outward FDI by 1914 which made it the fourth-largest home country of FDI after the United Kingdom, the United States and France (Dunning, 1983). The defeat of Germany in the two world wars led to a dramatic decline in German FDI and a broad-based and dynamic growth of German MNCs was not resumed until the 1970s (Bostock and Jones, 1994). This helps to explain the high degree of embeddedness of German firms and MNCs to their home country and the predominant role of exports in their internationalization strategies (Lane, 1998). Although Germany has been the third-largest home country of FDI since 1980, outward FDI has remained a secondary strategy for German firms despite their rapid growth over the last 30 years (Heiduk and Hodges, 1992).

This chapter tells the long and discontinuous history of German MNCs. The historical excursion into the determinants of German FDI and their industrial and geographical patterns is conducted in three time frames: the period prior to 1914, the inter-war period and the period since the Second World War.

The emergence of German MNCs in the period prior to 1914

Unlike the variety of historical forms of the earliest British MNCs in the period prior to 1914, the prototype of the earliest German MNCs was predominantly the 'classic' type. These were German companies that began to do business in Germany, and then extended their business abroad through exports and international production to reach foreign markets and to obtain raw materials. Thus, the international business activities of German firms has, in general, always tended to be complementary to their domestic activities in Germany (Juhl, 1985).

Indeed, there seem to have been few German free-standing companies that operated in foreign countries. Even migrating MNCs that featured significantly in the history of British MNCs did not appear to have been a particularly important feature in the origins of German MNCs, and neither has there been much evidence to support the presence of German investment groups abroad, i.e. independent German companies registered in the host country with German capital and management (Wilkins, 1988b).[1] What was perhaps more common owing to the importance of cartels in the economic and international business history of Germany was FDI by German cartel representatives to encourage exports. For example, the potash cartel had a sales company in the United States prior to 1914 (Wilkins, 1989).

German FDI prior to 1914 was significant in a wide range of industries spanning the primary, secondary and tertiary sectors of economic activity.

Resource-based FDI prior to 1914

In broad similarity to MNCs from the United Kingdom, the emergence of German MNC activity can be traced to natural resource-based activities, particularly in mining and petroleum. Owing to the natural resource scarcity of Germany, many of the primary products required by the German economy but not produced there were obtained primarily through imports. Nevertheless, there were also extensive FDI to obtain raw materials by German manufacturing companies, metal trading companies and the Deutsche Bank in oil in the period before the First World War.

Among the most significant German manufacturing companies that integrated backward to obtain raw materials abroad were the iron and steel manufacturers (Franko, 1976). However, the large-scale corporate players in world metals before 1914 were the three German metal trading companies: Aron Hirsch and Sohn, Beer, Sondheimer and Co., and Metallgesellschaft. Metallgesellschaft – the largest of the three trading companies founded in 1881 – diversified rapidly into the mining, processing and distribution of copper, lead and zinc. The firm undertook its first FDI in 1887 which included the establishment of its American subsidiary, the American Metal Company, whose principal activities were in coal mining, smelting, refining and distribution. The firm also had mining operations and processing companies in other European countries and in 1912 it built a refinery in Belgium – the largest one in Europe – to process ore from the Belgian Congo (Jones, 1996). A large trading operation in London was also established. The German metal traders succeeded individually and jointly in vertically integrating on an international scale the mining, smelting, refining and sales of all the most important non-ferrous metals. In certain metals, such as lead, zinc, copper and nickel, the firms exerted a very significant influence on world prices (Wilkins 1989; Chandler, 1990). There were also German investments in nitrate mines in Chile (Wilkins, 1988b), as well as a short-lived participation in a borax mine in Chile by the small German pharmaceutical company, Schering of Berlin, in 1897 (Hertner, 1986a).

The oil industry was another important area of early resource-based FDI by Germany led by the Deutsche Bank and its holding companies. For example, the bank gained control of a leading oil producer in Rumania in 1903 which it placed under the control of a holding company in which it had a 50 per cent shareholding and overall management control. A vertically integrated oil business was created which included distribution companies in a number of European countries. The defeat of Germany in the First World War led to the sequestration of German foreign assets and the nationalization of Russian oil. This eliminated the Deutsche Bank from the industry (Pohl, 1989), and a significant German presence in the international oil industry never re-emerged. Even Metallgesellschaft did not resume substantial FDI in mining production until the 1970s.

Services-based FDI prior to 1914

Early German FDI in services were significant in trading and banking. Apart from the foreign trading activities of the metal traders, there were the trading companies or mercantile houses that established trading outlets throughout Europe in the period prior to 1914. The German trading house, Schuchardt and Schutte, which was regarded as the most prestigious distributor of machine tools in Europe had outlets in Germany, Austria, Belgium and Russia to promote German exports, and corresponding outlets in Guatemala and Turkey that played a more important role in facilitating German imports than in selling German exports (Feldenkirchen, 1987).

And then there were the ubiquitous German banks active in Europe, North and South America, Asia and Africa prior to 1914, encouraging and representing German business abroad (Tilly, 1991). The role of Deutsche Bank in relation to the German search abroad for oil in Rumania and the Middle East was already mentioned. The role of the large German banks that established foreign branches helped to support the expansion of German business abroad. This was unlike the roles of banks in the United States and the United Kingdom. In the case of the United States, national banks could not establish foreign branches until after the passage of the Federal Reserve Act of 1913 while private banks established outlets in London and Paris principally to encourage the flow of European monies to America (Wilkins, 1988b). The previous chapter has shown that the lack of British 'universal' banks (Cottrell, 1991) precluded the possibility of the British banking sector to assist directly in the expansion of British business abroad.

Manufacturing-based FDI prior to 1914

Manufacturing was perhaps the more important sector of early German FDI. The early emergence of German manufacturing MNCs can be closely associated with the position of Germany as the birthplace of modern science in the late nineteenth century (Porter, 1990). This helped the country to develop a deep scientific and technical knowledge base drawing on an abundance of skilled workers and professionals which proved instrumental in its efforts to upgrade domestic industry in the new, skills intensive, technically advanced and fast growing industries of the Second Industrial Revolution in the late nineteenth century. German firms developed strengths in the chemicals, pharmaceuticals, machinery, electrotechnical and motor vehicles industries and became significant world producers and exporters of their new products on the basis of accumulated expertise and proprietary technology.[2] Such strengths combined with access to international capital markets at least until the end of the First World War enabled German firms to invest abroad on a substantial scale, and to exert a significant economic impact within their host economies.

The predominance of typically large German firms and MNCs in industries producing the newest and most technologically advanced products can also be attributed to the rapid emergence of administrative hierarchies in German business organizations (Wilkins, 1988b). Although these technologically innovative large-sized firms led the process of German MNC expansion in high technology industries introducing new products and processes through FDI, German FDI prior to 1914 was also present in capital goods industries such as iron and steel manufacture, and relatively smaller and more moderate-sized German MNCs were involved in consumer goods industries such as textiles particularly in woollens and silk and, similar to the British MNCs, in a number of trademarked consumer products. Among the most notable of these was the pharmaceutical company, E. Merck, that established a sales affiliate in New York in 1887 and a production subsidiary in Rahway, New York in 1899; Schering of Berlin that established a partnership in the New York firm of Schering & Glatz established in the late 1870s; the A. W. Faber Pencil Company which was an MNC by the 1870s with its main factory in Stein (near Nürnberg), a large slate facility at Geroldsgrun, Bavaria, branches in Paris and London, an agency in Vienna and a pencil factory in Brooklyn, New York; and a number of other German companies that were involved in the international production of trademarked food products such as chocolate and coffee, among which were Kathreiner's Malzkaffee-Fabriken, a company founded in 1892 to produce malt coffee

(Hertner, 1986a; Wilkins, 1988b). The experiences of E. Merck, Schering of Berlin, the A. W. Faber Pencil Company and Kathreiner's Malzkaffee-Fabriken, among other companies, show that it was not necessarily the large oligopolistic German firm that engaged in FDI.[3] Ownership advantages have also resided in smaller companies particularly in the early phases of development of a specific market, and these advantages could be maintained with the increasing differentiation of the market (Hertner, 1986a).

The following sections analyse the growth of the mainstream German MNCs in their most important industries – chemicals, electrotechnical industry, and machinery and motor vehicles industries.

The chemicals industry

As mentioned, Germany dominated world production and exports in the chemicals industry other than petroleum refining prior to 1914, particularly in such sectors as pharmaceuticals and artificial dyestuffs. The growth of the German chemical industry which started in the 1860s and organized in an oligopolistic market structure owed much to home country-specific advantages such as the German educational system which created a pool of highly trained chemists and chemical engineers. The scientific orientation of German professional education was especially relevant to the dyestuffs sector of the German chemicals industry where thousands of products were produced in small batches by large firms (Wilkins, 1988b).

ARTIFICIAL DYESTUFFS

The German artificial dyestuffs industry prior to 1914 comprised three large firms (BASF, Bayer and Hoechst), and three small firms (Cassella, Kalle and AGFA) (Hertner, 1986a). The three largest firms initiated international production. As Russia was the most important export market for German chemicals accounting for 21 per cent of exports prior to 1914, BASF, Bayer and Hoechst founded or participated in production affiliates in that country starting in 1877, 1883 and 1885 respectively in response to growing protectionism after the imposition of tariffs in 1877. The smaller chemical firms, Cassella, Kalle and AGFA, followed suit shortly after. Their foreign affiliate production covered only the final stages with parent firms providing intermediate products that attracted lower duties in host countries compared to final products. Indeed, parent firms contributed up to 80 per cent of the value added of the foreign affiliates (Haber, 1971). Despite the low domestic value added of the foreign affiliates, their production accounted for around 80 per cent of the German dyes sold on the Russian market in 1913 (Jones, 1996).[4]

Apart from tariffs, stringent patent legislation in foreign countries that required the initiation of local production was also an important determinant. For example, the French patent law which required that patent holders initiate local production immediately led to the foundation by Hoechst in

1881 of the Compagnie Parisienne de Couleurs d'Aniline to produce aniline dyes and pharmaceutical products at Creil (Oise) starting in 1884. BASF similarly established a plant at Neuville-sur-Saône near Lyon in 1878. A change in British patent legislation in 1907 which required that a foreign patent registered in Britain be exploited in that country lest it would be revoked prompted Hoechst, acting on behalf of Casella and Kalle, to establish a plant in Ellesmore Port near Liverpool and BASF, acting for the *Dreibund*, to found the Mersey Chemical Works not far from there and to transfer nearly 20 per cent of its new patents to this new affiliate.[5] Over the next few years, the Hoechst affiliate at Ellesmore Port was practically responsible for the domestic production of indigo, while the BASF subsidiary focused on the production of other aniline dyes and eventually became less reliant on patent transfers from Germany owing to the lack of enforcement of the 1907 British Patent and Designs Act (Hertner, 1986a).

By comparison, the more lax patent legislation in the United States in which there was no legal obligation to use registered patents in the country itself served not to coerce German chemical firms to establish production facilities if this was not otherwise warranted. For example, while having registered about 1,000 patents in the period until 1914, BASF did not establish a factory in the United States. By contrast, Bayer, which registered roughly the same number of patents over the same period, owned a production facility in Albany, New York and eventually came to own three of the seven dyestuffs plants in that country (Jones, 1996). The Synthetic Patent Company Inc. which held all the Bayer patents registered in the United States provided licences to the American Bayer Company. This enabled the Synthetic Patent Company to reduce its tax burden owing to the lower taxes levied on gains from licences compared to those levied on gains from production (Hertner, 1986a). All other German dyestuff companies while not having production facilities in the United States had quite extensive commercial organizations.

PHARMACEUTICALS

Like the German artificial dyestuffs industry, the German pharmaceutical industry also comprised smaller firms such as E. Merck and Schering of Berlin and larger firms such as Bayer and Hoechst, all of which became MNCs prior to 1914. In terms of size of employment, E. Merck was perhaps the largest of the specialized pharmaceutical firms existing in Germany prior to 1914, followed by Schering of Berlin. E. Merck had always been actively engaged in exports which accounted for 77 per cent and 67 per cent of its total sales in 1900/01 and 1912, respectively (Hertner, 1986a). The firm's most significant foreign export markets in 1900/01 were Russia (with a share of 18 per cent of total sales), followed by Latin America (11 per cent), United States (11 per cent), Great Britain (6 per cent), Austria-Hungary (5 per cent), Spain and Portugal (5 per cent), and Italy (4 per cent). The growth in importance of the German market in 1912 was accompanied by a diminished share of exports directed to Russia (12 per cent) and the United States and Canada (8 per cent).

The lesser significance of Russia and the United States as important export markets for E. Merck can be largely explained by the substitution of exports by FDI. Thus, owing to high tariffs in Russia, its principal export market, E. Merck regarded FDI as inevitable even though the extent of its FDI went only as far as establishing a bottling and packing department at its Moscow agency in 1906, with no manufacturing initiated at the time of the First World War. In a similar fashion, the diminishing role of the United States as an export market can be considered a consequence of the migration of George Merck, a family member, to the United States which led to the foundation of a sales affiliate in New York in 1887, and the establishment of the firm of Merck & Co in 1899 which coincided with the initiation of production of pharmaceuticals in Rahway, New York. The major determinants of international production in the United States was the relatively high American import duties on pharmaceuticals, the rapid growth as well as the large size of the American market and the peculiar nature of marketing pharmaceutical products in the United States in which chemists' shops were not supervised by scientifically trained pharmacists with the consequence that preference was often accorded to the sales of ready-made and packed drugs (Hertner, 1986a).

Although Europe was a less important export market for the firm, it nevertheless also established a number of small sales agencies in Genoa, Florence, Rome, Livorno, Naples, Palermo and Catania between 1898 and 1900, all of which was centrally orchestrated from Milan. The addition of a depot in the Milan agency around 1900 led to the substantial decline of the company's exports to Italy from its factory in Darmstadt from 87 per cent in 1890/91 to 34 per cent in 1913. A corresponding initiative was made in London in 1901 and in Paris in 1902 where a limited joint venture with a local firm, Bousquet, was concluded enabling the local firm to use E. Merck's brand names for some pharmaceutical products bottled and packed in France. This arrangement was provoked by the French prohibition against the import of pharmaceuticals in tablet or capsule form as well as the French patent law of 1844 which did not allow the patenting of any type of medicine. In 1910 the firm decided to initiate production itself by acquiring a small chemical factory at Montereau near Paris owing partly to the frequent increases in French tariffs to protect the infant domestic pharmaceutical industry. Shortly before the outbreak of the First World War, some 90 per cent of the fine chemicals required for French pharmaceutical production were either exported from Germany or passed through French affiliates of German industry. The outbreak of the First World War led to the seizure of the Montereau factory by the French authorities in 1914, and when it was taken into custody under the Trading with the Enemy Act in 1917, the factory's capital was valued at $1 million and was a fully integrated production facility which no longer depended exclusively on the imports of German intermediate products (Hertner, 1986a).

The extent of international production of Schering of Berlin, E. Merck's major competitor, included a partnership in the New York firm of Schering & Glatz established in the late 1870s; the establishment of a pharmaceutical factory in Moscow in 1905 using imports of chemical raw materials and intermediates from Germany; as well as a charcoal plant using the immense resources of Russian timber built at Wydriza in the Mogilev province (Hertner, 1986a).

In addition to the international production of the relatively small German pharmaceutical firms were those of the large chemical companies, among which were Bayer and Hoechst which devoted part of their activities to the manufacture of pharmaceuticals in both domestic and foreign markets. Apart from dyestuffs, Bayer developed aspirin which it also produced and sold as a trademarked product in the United States until 1921. Bayer thus became a substantial chemicals manufacturer in the United States (Jones, 1996).

The electrical equipment industry

The growth of the German MNCs in the electrical equipment industry is strongly associated with the virtual duopoly of the German electrotechnical industry comprising Siemens and Allgemeine Elektricitäts-Gesellschaft (AEG) after 1903, and some specialist firms in the industry such as Bosch.

Germany is the home country of Siemens, one of the pioneering German MNCs in manufacturing. The firm, founded as Siemens & Halske in 1847, pioneered the development of telegraph and cable equipment. Indeed, the firm owed much of its success in the first three decades of its existence to the telegraph business, and only gradually to the telephone business. The firm can be considered to be a 'classic' MNC as it began by undertaking value-added activities in its home market before initiating international production. The firm established a factory at St Petersburg in 1855 as a consequence of considerable orders from the Russian government to construct and maintain a telegraph network. The factory which undertook the assembly of parts sourced from Berlin was directed by Carl Siemens, brother of Werner Siemens, the firm's founder. Another brother, William, represented the company's business interests in Britain from 1850 and directed the British Siemens subsidiary founded in 1858. This subsidiary initiated the production of sea cables in 1863 (Scott, 1958). Expansion abroad continued in 1903 when there were 30 so-called technical *bureaux* installed, eight of which were located in European countries. As with the first foreign production affiliate established in Russia in 1855, some of the foreign production affiliates started with the assembly of parts imported from the parent firm. By 1914, Siemens had built ten foreign factories spread over five countries. Almost one-fifth of the firm's total workforce of 80,000 was employed outside Germany (Von Weiher and Goetzeler, 1977).

The origins of Siemens' main competitor, Allgemeine Elektricitäts-Gesellschaft (AEG), can be traced to 1883 in its position as the German licence holder of the American Edison Company. During the first decade of its existence, the firm not only depended on the technical ties with the American Edison Company but also on the equity participation of Deutsche Bank and Siemens. The company always derived its growth from the production of high-voltage electrical equipment particularly since it gained complete independence from Edison and Siemens in 1894. The growth of the company's FDI was determined not as much by the increasing tariff duties under Tsarist Russia as by the growing pressures from the Russian administration which insisted on domestic production for continuing state orders of public utilities. Despite these pressures, AEG continued to look after its Russian business through local agents but founded a sales company based in Berlin in 1898 for its Russian business interests. This was short-lived as the company had to transfer its legal seat to St Petersburg in 1902 to avoid the loss of the Russian market. Only after its merger with the Union Elektricitäts-Gesellschaft (previously controlled by General Electric) in 1905 did the AEG take hold of the Union's electrotechnical factory at Riga (Hertner, 1986a). The firm also acquired factories in Austria-Hungary and established joint venture production subsidiaries with General Electric in Italy and with Siemens in Russia (Chandler, 1990).

The German firm of Bosch which had grown out of a small mechanical and electrotechnical workshop founded by Robert Bosch in Stuttgart in 1886 gained prominence in the production since 1898 of magneto ignitions for cars and, to a lesser extent, for airplanes. Shortly before the First World War, the firm diversified into the production of other electrical equipment for the car industry such as starters and headlights. The growth of the company as a MNC was initiated with the foundation in France in 1899 of the Automatic Magneto Electro Ignition Company Ltd. which was responsible for the company's sales in France and Belgium.[6] Such affiliate initiated the production of magneto ignitions for trucks in 1907 in light of the subsidies extended by the French state to producers of those vehicles whose parts were solely made in France. The French car industry, the most important car industry in pre-1914 Europe, became a highly important customer for Bosch, and hence the company had a quasi-monopoly in the French market with annual sales of 10 million francs before the outbreak of the First World War (Hertner, 1986a). A similar trend towards domestic production of components used for racing cars participating in certain racing events in Great Britain led to the initiation of assembly production by its affiliate in London, the Bosch Magneto Company Ltd, founded in 1907.[7] In the first half of 1914, the local affiliate satisfied approximately 85 to 95 per cent of British demand for magnetos and spark plugs (Hertner, 1986a).

Bosch also founded sales affiliates in New York in 1906, in Chicago in 1908, in San Francisco in 1909, and in Detroit in 1910. However, the high cost of freight and the 45 per cent tariff on the value of magnetos led to the construction of a magneto factory in Springfield, Massachusetts in 1910 where production started in 1912. This was accompanied by a spate of acquisitions in 1912 including the 45 per cent share of the equity capital of the Eisemann Magneto Company in 1912, the American affiliate of its principal German competitor, and the full acquisition of the the Boonton Rubber Manufacturing Company of New York in the same year. The latter was the largest producer of moulded insulation, an essential material to the magneto industry.[8]

By the first half of 1914, the Bosch company exported some 88 per cent of its domestic production and had established agencies and factories in 25 countries (Hertner, 1986a). Bosch and Eisemann combined accounted for at least half of all the magnetos sold in the United States before the First World War. At the seizure of the company's properties in the United States in 1918, the combined capital and surplus of Bosch's assets in the United States exceeded $6.5 million which included branches in Detroit, Chicago, and San Francisco as well as agencies and supply depots in over 100 American cities.

There are other examples of German firms in electrotechnical industry whose FDI was motivated by a common set of factors. While state intervention in the form of tariffs or non-tariff barriers was a major explanatory factor, particularly if the state was an important customer as in the case of Russia, there were other contributory factors of which the most significant was the access to international capital markets of the *Unternehmergeschäft*, the large German trusts that founded local and regional power, tramway and lighting companies in Russia, Italy, Spain and Latin America. Such capital access was provided in the form of acquisition of financial holdings in each of the electrotechnical producer companies by large banks in Germany and international capital markets, some of which were based in Switzerland or Belgium owing to the very liberal company laws and stock exchange regulations in those countries. Such establishment of intermediate financial holding companies enhanced the capacity of the newly created German public utility companies abroad to engage in FDI as it provided a means to overcome the chronic lack of capital of their major customers – the local public authorities in foreign countries – and to counteract the liquidity problems of the electrotechnical firms themselves associated with the accumulation of a growing volume of equity capital and bonds in their portfolio. Thus the equity capital and bonds of the newly created public utility companies were held typically in the portfolio of the financial holding companies during the periods of establishment and initial development, and sold to the general public at a later stage when profits could be earned (Hertner, 1986b). This enabled about 50 per cent and 40 per cent of the local and regional public utility companies in Italy and Russia respectively to be controlled by German capital in 1913. The loss of this access to international capital markets at the end of the First World War led to a similar loss of a major part of German FDI, particularly in the technologically mature high-voltage sector of the business where German firms faced growing competition in world markets. By contrast, the growth of the more rapidly technologically advancing low-voltage sector of the business (telephone installation) continued, and exports of products in this sector displayed a better performance in the inter-war period (Hertner, 1986a).

The machinery and transport equipment industries

The machinery and transport equipment industries were also significant industries that spawned German MNCs. Among these were Mannesmann, Accumulatoren-Fabrik AG and the Daimler Company.

Based on the highly original invention patented in 1886 for producing seamless rolled tubes, the company formed by Reinhard and Max Mannesmann was practically an MNC from its inception (Teichova, 1983). The history of the company shows the establishment of three formally independent subsidiary companies between 1887 and 1889, in addition to the company's original plant. Each of the three subsidiary companies was located in Saarbrücken, Germany; Austro-Hungarian Bohemia; and in Landore in Britain. While the Landore Mannesmann Tube Company was taken over by its parent company in 1899 to gain control over the British market for seamless tubes and to profit from trade with the British empire and dominions, the two other independent companies located in Germany and Austria were amalgamated in 1890 in the Deutsch-Österreichische Mannesmannröhren-Werke with a capital of 35 million marks and was brought under the financial guidance of the Deutsche Bank.[9] The two inventor brothers received half of the new shares of the new amalgamated company in return for their patents. Their equity participation was sold in 1900 allowing Deutsche Bank to become the uncontested entrepreneur and financier of the company. The Austro-Hungarian Bohemia affiliate which eventually represented the most important foreign asset of the firm gained from high Austro-Hungarian tariffs and succeeded in attaining a share of 35 per cent in the cartel of rolled tubes in that country. The affiliate accounted for one-third of the Mannesmann combine's total turnover in the last pre-war year when all non-German subsidiaries accounted for 45 per cent of the overall volume (Teichova, 1983).[10]

A similar trend is evident in the other machinery and motor vehicle industries which spawned the generation of early German MNCs. Among these were Accumulatoren-Fabrik AG and the Daimler Company. From the 1890s, Accumulatoren-Fabrik AG active in standardized light machinery built a network of storage battery factories across Europe from Britain to Russia, and an international marketing network that extended from Buenos Aires to Cairo and Tokyo. The Daimler Company, which began manufacturing in Austria in 1902, claims to be one of the world's first multinational motor companies (Jones, 1996).

The interrupted growth of German FDI in the inter-war period

The emergence of German FDI prior to 1914 thus makes German MNCs share a common origin in time to MNCs based in other developed countries, even though the exact patterns of the emergence or the pace of the evolution of MNCs based in the different countries may not necessarily coincide. The First World War marked a period that was to differentiate the history of German MNCs from that of MNCs based in other developed countries.

German FDI slowed down dramatically in the period after the end of the First World War. The sequestration of German tangible and intangible assets abroad after the Versailles Treaty discouraged the further international expansion of their most dynamic set of firms – the manufacturing companies – that nevertheless possessed a substantial amount of surplus capacity. The confiscation of assets crippled firms in the chemicals industry the most whose intangible assets such as brand names and patents were invaluable in relation to physical assets.[11] As a result, German manufacturing firms became more risk averse in the inter-war years and to the extent that any FDI was undertaken, the emphasis by the larger German enterprises was often on re-building extensive international distribution networks after the First World War, while their foreign production remained modest. Such was the strategy, for example, of IG Farben, a large chemicals company formed in the mid-1920s by the merger of eight firms including Bayer, BASF and Hoechst.[12] In addition, the choice of a host country was based not merely on promising economic expectations, but rather on the stability of their political and legal systems that assure the safety of German assets. Owing to these criteria, German FDI was generally confined to Eastern Europe and Scandinavia. The geographical and psychological proximity of Scandinavia was especially favoured owing to its modern networks of communication and transportation, its rapidly growing markets with a high per capita income and as a key source of wartime supplies, particularly of food and raw materials, during the Nazi period. The Scandinavian countries, therefore, formed a major part of Germany's *Großraumwirtschaft* (expanded economic sphere) in the event of war (Schröter, 1988).

The loss of overseas holdings at the end of the First World War and the severe constraints posed by the shortage of capital in the inter-war period forced German firms to replace FDI with other modes of international economic expansion that conserved the use of capital and entailed less risks, primarily political risks. In the period between 1918 and 1939 these other modes were principally cartels and long-term contracts (including licensing agreements) in which German industry had experienced considerable success before the First World War. These tools came to be used more widely by German firms than by firms of another nationality (Schröter, 1988).

To ensure the success of the new tools of international economic expansion, the strengths of German industries was consolidated on a nationwide basis.

This process was carried out in the mid-1920s when large national cartels were formed and when major mergers created very large companies such as IG Farben already mentioned and the Vereinigte Stahlwerke. Such a process of domestic consolidation provided Germany with clout to mount a major export drive or to conclude international cartels and long-term contracts as a means to recapture their pre-war share of world markets.

For much of the inter-war period, the foreign economic policy of Germany aimed at four important objectives: the long-term security in the supply of raw materials; the maintenance of access to markets or the search for markets; the security against competition through a strategic market presence; and influence in political decision making in either the home or host host country or both focusing mainly on customs duties, tariffs, and legal matters relating to ownership, but sometimes also including attempts to influence national economic and even defence policy (Schröter, 1988). The importance of FDI, international cartels and long-term contracts as instruments of foreign economic policy varied according to the objective.

In the pursuit of the objective to attain long-term security in the supply of raw materials, FDI represented the most important tool. Although long-term contracts were the traditional means employed by Germany to obtain security in raw material supplies, this represented only a second-best solution. The high capital intensity as well as high political risks associated with FDI was outweighed by the desire of large resource-based German industries and firms (e.g. iron and steel and paper) to control the long-term supply of vital raw materials from nearby sources. However, Scandinavian laws that limited foreign ownership of natural resources prevented German firms from attaining the high degree of control desired. Thus, their FDI was often complemented by comprehensive long-term delivery contracts to bind their local partners to supply vital raw materials on favourable terms. The success of these long-term contracts was assured by the large size of German national cartels that participated in these contracts that often intervened to prevent any significant attempts of their Scandinavian partners to diversify their customer base in order to remain their biggest and most influential buyer (Schröter, 1988).

In the pursuit of the objective to secure markets, German companies used a combination of FDI, international cartels and long-term contracts in a complementary manner. International production was not the dominant means to achieve this objective except perhaps for Osram, the largest light bulb supplier formed from the merger of the incandescent lamp sectors of the three major German electrical firms: Siemens & Halske, AEG and Auer. The light bulb industry was an exceptional German industry that was not affected by the events of the First World War. Osram established several new foreign subsidiaries between 1914 and 1922 in Denmark, Norway and Sweden to serve three main purposes: to supply the indigenous market; to prevent the entry of other firms in the industry; and, most importantly, to strengthen the position of Germany in international negotiations for the revival of the worldwide bulb cartel (Schröter, 1988). Thus the Phoebus agreement was signed in 1924 (Schröter, 1986). The Osram investments in Scandinavia while geared primarily to secure foreign markets was also key to the attainment of the other objectives of the foreign economic policy of Germany.

Cartels and long-term contracts were perhaps the more significant means employed by Germany to secure markets. Two of the cartels concluded by German firms operating in Scandinavia were geared for this purpose: the International Coke Convention signed on 11 June 1937 and valid from 31 March 1940 to 1 April 1947 and the cartel for certain electrical commodities covering the Swedish market from 1925 onwards concluded between the dominant suppliers of electrical equipment and public utilities in Scandinavia: the Swedish firm, ASEA, and the German companies, AEG and Siemens. On the other hand, long-term contracts were preferred in instances where a foreign enterprise enjoyed a degree of influence in a particular market. An example of such contract geared to generate growth of sales in a related line of business in foreign markets was that concluded between IG Farben and the Finnish 'Valio' dairy cooperative. The basis of the reciprocity was the patents obtained by the Finnish dairy cooperative for the AIV process used to conserve livestock fodder and the required chemicals for such process provided by IG Farben (Schröter, 1988).

In the pursuit of the objective of security against competition through a strategic market presence, German companies used a combination of FDI, international cartels and long-term contracts, with perhaps the first two exerting greater importance relative to long-term contracts. A mixture of defensive factors (to defend loss of foreign markets) and offensive factors (to prevent the entry of indigenous and other foreign firms in a market to avoid cut-throat price competition and retaliation) underlined the motive to overcome competition in foreign markets that went beyond concerns about supplies, sales and profits. The mode of FDI was used by IG Farben, then the biggest nitrogen producer in the world, to maintain its dominance and control over the world nitrogen industry. This was implemented by an exchange of shares between the Norwegian firm, Norsk Hydro, and IG Farben and the centralization of the sales organization of the two firms in the German *Stickstoff-Syndikat* (nitrogen syndicate).[13] The investment enabled IG Farben to overcome the problems of overcapacity through its control of Norsk Hydro's exports and their channelling through the German sales network. This enabled IG Farben's market power in foreign markets to rise along with its clout in international negotiations. Such was the case in the export cartel concluded with Imperial Chemical Industries in 1929 which formed the nucleus of all international nitrogen agreements (Schröter, 1988).

IG Farben also pursued a defensive and offensive stance in international markets through cartels. The most notable example was the ferrosilicon cartel it formed in 1927 with other firms based in Sweden, Norway, Yugoslavia and Switzerland which endured until 1939. Its success led to the cartelization of firms in other iron alloys. Another important cartel in which IG Farben was a participant along with the American company Du Pont was that involving the world market in synthetic fibres in effect in 1938. The cartel arrangements involved the construction of a plant in Germany by IG Farben with a designated minimum output to ensure the ability of the cartel to meet fully the anticipated world demand. This was reinforced by cross-licensing agreements that served to prevent the entry of new firms in the industry. The cartel for certain electrical commodities comprising ASEA, AEG and Siemens mentioned earlier, the chlorine cartel formed between the Chlorstelle (Germany), Imperial Chemical Industries (Great Britain) and Solvay (Belgium) in November 1938 and the participation of IG Farben in the worldwide nitrogen cartel – the Convention de l'Industrie de l'Azote – all succeeded in preventing the entry of new competing firms or the development of new industries in non-member countries.[14] To the extent that long-term contracts were used to stem the tide of competition, German firms licensed valuable patents to only a few reliable and well-known Swedish chemical firms before the Second World War (Schröter, 1988).

In the pursuit of the objective of influencing political decisions in both domestic and foreign markets, FDI and long-term contracts were perhaps the principal instruments used by German companies. Through its Swedish subsidiary, the SB Anilinkompani in Götenborg, IG Farben played an influential role in resisting the increase or at least preventing the substantial increase in Swedish import duties on a variety of products of interest to IG Farben. And by way of contracts, a secret organization – the Deutsch-Schwedische Ausschuß – was founded and financed by major German businesses in Sweden including AEG, Siemens, IG Farben, and Vereinigte Stahlwerke. The organization was a form of propaganda in Scandinavia working with the Swedish media to stem the tide of anti-German sentiments (Schröter, 1988).

In sum, German companies in the inter-war period used a variety of means to pursue their foreign economic policy objectives. Although FDI was ideal in terms of enabling a high degree of control, it presented disadvantages by way of requiring long-term capital from a country suffering from economic disruption and severe shortage of capital, and entailed an unacceptably high level of political risks for many German inter-war enterprises. This helps to explain why FDI was really only used in a major way to fulfil the foreign economic policy objective to secure the long-term supply of raw materials where the advantages of FDI more than outweighed the disadvantages. In the pursuit of the other objectives of foreign economic policy, international cartels was the most important tool used by German firms in the inter-war period and long-term contracts and FDI were relatively less important. Thus, from being the fourth-largest home country of FDI responsible for more than 10 per cent share of the global stock of outward FDI in 1914, the share of Germany declined dramatically to a mere 1.3 per cent share in 1938 (Dunning, 1983).

The evolution of German MNCs since the Second World War

As Germany experienced a second round of defeat in the Second World War, German firms increasingly displayed a stronger degree of reliance on their home base. The loss of some of the most modern parts of its industrial base as well as natural resources and the confiscation of German patents and foreign assets after the two world wars created strong pressures to upgrade German industry in advanced, knowledge-based industries and to foster technological innovation in indigenous firms (Porter, 1990). Such domestic embeddedness of German firms is manifested in their continued preference for exporting over FDI as an internationalization strategy even to the present time, and in the continuing concentration of their outward FDI in Western Europe. Even in highly internationalized industries such as chemicals, German companies remain more deeply embedded in their home market than their British counterparts in the same industry by whatever measure, thus underscoring the importance of national factors over industry influences in explaining differences in behaviour of firms of different nationalities in the same industry (Lane, 1998).

Indeed, German manufacturing firms undertook comparatively little FDI until two decades after the end of the Second World War, preferring to use export strategies to take advantage of the fast growth in world trade and the opportunities offered by European economic integration (Jones, 1996). Thus, the share of Germany in the global stock of outward FDI in 1960 at 1.2 per cent share was as insignificant as its share of 1.3 per cent share in 1938 (based on data in Dunning, 1983), and a considerable share of this FDI consisted of sales and service foreign affiliates in support of export activity (Heiduk and Hodges, 1992).

The recovery of Germany as an important source of outward FDI became evident only in a major way since the 1970s. The comparatively high level of R&D intensity characteristic of German firms and MNCs from the beginning remained their most consistent distinguishing feature that sets them apart from MNCs of other nationalities. Indeed, a regression analysis of the industrial advantages of Germany on pooled cross-sectional sets of data for the years 1965, 1970 and 1975 showed that the degree of innovation and the creation of technological ownership advantages as well as the skill level of managerial manpower exerted highly significant *positive* influences on German FDI in manufacturing, while the skill level of production workers had a significant *negative* effect. This roughly mirrors the findings for German manufactured exports which showed that the technological intensity and the skill level of production workers exerted highly significant *positive* influences, while capital intensity and the complexity of management exerted significant *negative* influences (Clegg, 1987). The findings show that the ownership advantages of German MNCs, unlike MNCs based in the United States or the United Kingdom, were based more on technology and skilled labour than on capital intensity. During this period, there was no evidence to show that access to capital was an ownership advantage of German MNCs in the way that it was during the period prior to 1914. In the concentration of German FDI in Western Europe, German MNCs also have a more narrow geographical focus compared to American or British MNCs, and have been significant in a narrow spectrum of medium- to high-technology industries in which domestic activity is also strong.

By 1980, Germany became the third-largest home country of FDI after the United States and the United Kingdom with a share of more than 8 per cent of the global stock of outward FDI. While maintaining the third-largest position as a source country of FDI in 1998, the level of German outward FDI stock grew almost ninefold since 1980 and as a result the share of Germany in the global stock of outward FDI increased to almost 9.5 per cent. Manufacturing continues to be an important sector of German FDI, accounting for 44 per cent of the stock of German outward FDI in 1997 and predominated in the same set of industries that German firms and MNCs have demonstrated industrial strengths since their history: transport equipment, chemicals, electrical machinery and mechanical engineering. In fact, until the mid-1970s the manufacturing and services sectors each accounted for almost equal shares of between 47 per cent and 48 per cent of total German outward FDI stock. However, manufacturing surpassed services in importance by 1988 with a share of 61 per cent of the stock of German FDI. The dominant role of the services sector in the stock of German outward FDI evident in the case of the United States since 1989 and in the United Kingdom in 1991 has also been evident in the case of Germany in the 1990s. By 1997, services accounted for almost 56 per cent of the stock of German FDI.[15]

In terms of the geographical destination of German FDI, although the United States and South East Asia grew in importance as host countries in more recent years, German FDI continued to predominate in neighbouring European countries such as Belgium, the Netherlands, France, Switzerland, Austria, and the United Kingdom. In the chemical industry, the extent of German implantation into a European network is so high that Europe will continue to be a highly important host region for German FDI in the next century (Lane, 1998). The traditional multi-domestic strategy continues to be adopted by both large- and medium-sized German MNCs that regard FDI as a means to be in or near important markets and customers, to adapt products to fit the exact needs of their buyers as well as to provide effective systems solutions (Heiduk and Hodges, 1992).

Despite the recovery in the importance of FDI as mode of international economic involvement for Germany in the last 30 years, it has remained a secondary strategy for internationalization as a majority of large German companies continue to serve foreign markets primarily through exports. Indeed, since Germany generates the second-largest export volume in the world economy it is a relatively more important source of exports than of FDI (Heiduk and Hodges, 1992). The greater relative importance of exports as well as the limited geographical scope and the concentration of German MNCs in a narrower spectrum of industries suggest that German companies have remained essentially as nationally embedded firms pursuing a well-entrenched localization strategy and proceeding rather slowly along a continuum to become globally oriented MNCs (Lane, 1998). But there are already emerging signs that the progression along the continuum is proceeding very rapidly. First, German firms are utilizing other modes to internationalize their business activities other than through exports and FDI. This includes strategic alliances and joint ventures which have become particularly dense in high technology industry segments where German firms are heavily involved. Second, international competition is forcing German firms to combine their traditional competitive strengths based on technological excellence and quality with price competitiveness. This has made foreign sourcing and FDI in neighbouring countries of Western and Eastern Europe more popular in recent years, particularly in the less skill intensive parts of the production chain that profit from access to abundant unskilled and semi-skilled labour of low cost. Third, and perhaps more importantly, there has been an increasing need to extend the R&D networks of German firms in foreign countries particularly in the chemicals and information technology industries where innovation is best utilized and transformed in proximity to foreign customers, competitors and research centres (Dörrenbächer and Wortmann, 1991). The need to gain access to complementary innovations in the same industry segment combined with the difficult process of obtaining German government permission for sensitive research in biotechnology has also tended to pull MNCs away from a sole focus on the home market as a prime determinant of innovation (Cantwell, 1989a).

Conclusion

This chapter narrated the long and discontinuous history of German MNCs. Despite the discontinuous pattern in their evolution owing to the political defeat of Germany in the two world wars, certain features of German MNCs persist throughout their history. Although FDI has historically served the purposes of obtaining vital raw materials for domestic industry in light of Germany's scarcity of natural resources and maintaining or securing access to important foreign markets, the overwhelming objective of German FDI has always been to maintain or secure access to foreign markets. Manufacturing has always been an important sector of German FDI and has concentrated the most in chemicals, motor vehicles, electrical engineering and mechanical engineering for at least 145 years. German firms have also consistently derived their competitiveness from their high degree of technological and skills intensity, and their predominant host countries have constantly been neighbouring European countries. Some evidence of a structural change in the sectoral pattern of German FDI became evident in the 1990s when the services sector became the dominant sector at the expense of the manufacturing sector.

Notes

1 The limited evidence of migrating MNCs as an originating form of German MNCs is seen in Rheinishche Stahlwerke AG established in 1869 by Frenchmen and Belgians whose headquarters was once in Paris, France. In addition, the predecessor of the German firm, Allgemeine Elektricitäts Gesellschaft (AEG) was once part of an American business abroad (Wilkins, 1988b).

2 In world exports in the pharmaceutical industry in 1913, Germany had the dominant share of 30 per cent followed by United Kingdom (21 per cent), the United States (13 per cent) and France (12 per cent). Excluding petroleum refining which was particularly important in the American chemical sector, the German chemical industry retained a clear lead in world chemical production and its exports accounted for an estimated 28 per cent of world exports in chemicals (Hertner, 1986a). In the artificial dyestuffs industry, the total value produced by eight German firms and their foreign subsidiaries amounted to 75–80 per cent of world production in 1913, and some 85 per cent of German production was exported (Haber, 1971). In the electrical industry, Germany had a share of between 31 and 35 per cent of world production, while the United States had a share of between 29 and 35 per cent. German industry accounted for a dominant share of 46 per cent of world exports of electrotechnical material, followed by Great Britain (22 per cent), and the United States (16 per cent) (Hertner, 1986a).

3 The view that size of firms is not a good indicator of the propensity of firms to become MNCs is shared by Swedenborg (1979) in analysing the growth of Swedish MNCs. See Chapter 4 of this book.

4 By contrast, there was no German artificial dyestuffs production in Italy until the First World War because by providing concessions to Italy's exports of agricultural products to Germany, German trade negotiators succeeded after 1878 to keep Germany's exports of dyestuffs and related chemicals to Italy totally free of duties (Hertner, 1986a).

5 The diminution of firm-specific advantages after 1900 in the dyestuffs sector resulted in the formation of two *Interressengenmeinschaften*, the so-called *Dreibund* (BASF, Bayer and AGFA), and Hoechst which controlled Casella and Kalle (Hertner, 1986a).

6 This affiliate in France later changed its name to Compagnie des Magnétos Simms-Bosch to reflect its partnership with Fredric R. Simms. The dissolution of the partnership in 1906 led to a further change of name of the French affiliate in 1908 to Société des Magnétos Bosch to reflect the new ownership situation (Hertner, 1986a).

7 The international business activities of Bosch in London predate 1907 when its British partner Fredric R. Simms represented the Stuttgart firm as an agent in London, an arrangement that ceased to exist in 1907 (Hertner, 1986a).

8 Another significant acquisition by Bosch in the United States in 1914 was the plant, business and goodwill of the Rusmore Dynamo Works at Plainfield, New Jersey, for $750,000. This factory was shut down and dismantled shortly thereafter (Hertner, 1986a).

9 As a national firm, the Landore Mannesmann Tube Company became one of the main suppliers of boiler pipes for the British Navy. In 1913 the construction of a second and larger rolling mill at Newport near Cardiff was approved for the intended purpose of producing large-diameter tubes as a replacement for imports

from the German parent firm. Production had not begun when the First World War broke out (Hertner, 1986a).

10 Mannesmann also established a rolling mill in Italy in 1906 not so much on account of the Italian tariff, but in the hope that the mill could profit from state orders for the railways, the navy or the new programme of municipal aqueducts in southern Italy. The rolling mill was established as part of a company founded under Italian law and with an Italian metallurgical firm as a minority partner. Production started in 1909, but improvements in output and profitability came only in 1912. The company was seized when Italy entered the war in May 1915, and was eventually sold to a group of Italian banks in 1916 – a transaction made with the agreement of the Italian government via neutral Switzerland (Hertner, 1986a).

11 Bayer, for example, recovered the right to use its own brand name in the North American market only in 1994 at a cost of $1 billion (Jones, 1996).

12 Two-thirds of the 726 subsidiaries or partly-owned affiliates of IG Farben in foreign countries by the end of the 1930s were sales agencies, and most of the remainder were engaged in the finishing and packaging of pharmaceuticals and dyes (Schröter, 1990, as cited in Jones, 1996). The dismantling of IG Farben after the end of the Second World War led to the resumption of foreign production by BASF, Bayer and Hoechst as independent companies. Their initial overseas investments were directed to Latin America, often re-purchasing plants lost in the war. Afterwards, investments grew elsewhere in Western Europe and North America. By 1965, German chemical companies had 150 foreign production plants (Jones, 1996).

13 Norsk Hydro obtained a minor holding in IG Farben (3.6 per cent share in IG Basle, a subsidiary of IG Farben in Switzerland) and IG Farben received 25 per cent of Norsk Hydro's stock in return. The German influence in Norsk Hydro was much greater than the 25 per cent share indicated (Schröter, 1988).

14 The cartel in electrical equipment succeeded in hampering the development of the Finnish electrotechnical enterprise F. Strömberg OY which emerged after the First World War; the chlorine cartel served to delay the the construction by the Finnish government of a chlorine factory in Finland; and the nitrogen cartel averted the construction of a nitrogen factory in Denmark, and caused the delay in the construction of a nitrogen factory in Finland (Schröter, 1988).

15 The analysis in this paragraph is based on data contained in UNCTAD (1999).

9 The emergence and evolution of multinational corporations from Japan

Introduction

This chapter is devoted to the history of Japanese MNCs. The distinctive nature of Japan lies in its rapid development as a significant source of outward FDI in the world economy in the period since the Second World War accompanied by an equally rapid process of industrial transformation of the Japanese economy and of Japanese MNCs. In 1914, Japan accounted for an almost insignificant share of 0.1 per cent of the stock of outward FDI worldwide, and even by 1960 their share remained low at 0.7 per cent. By 1998, Japan stood as the fourth-largest home country of FDI after the United States, United Kingdom and Germany, and accounted for some 7 per cent of the stock of outward FDI worldwide.[1] Notwithstanding the more recent growth in significance of Japanese MNCs, their earliest emergence can be traced to the late nineteenth century in much the same way as German MNCs. As in the previous two chapters on the growth of MNCs from the United Kingdom and Germany, the historical excursion into the determinants of Japanese FDI and their industrial and geographical patterns is conducted in three time frames: the period prior to 1914, the inter-war period and the period since the Second World War. As will become evident, the growth of Japanese MNCs is in many respects *sui generis* in the growth of modern MNCs.

The emergence of Japanese MNCs in the period prior to 1914

The existence of Japanese MNCs since the late nineteenth century is distinctive for many reasons. Perhaps the most fundamental feature distinguishing the emergence of Japanese MNCs from MNCs from the other more historical home countries is that the period in which it occurred – the late nineteenth century – was one in which the Japanese economy was still relatively undeveloped. Indeed, even by 1913 Japan accounted for a mere 1.2 per cent of world manufacturing output by comparison to the United States which accounted for some 35.8 per cent (Wilkins, 1986b). The emergence of Japanese FDI at that time cannot be said to be related to strong technological, organizational or capital intensities that describe the more modern Japanese MNCs of the present time. Rather, the early emergence of Japanese MNCs can be explained in terms of their important role in sustaining the economic development of Japan in which trade had always been regarded as the engine of growth.

In the late nineteenth and early twentieth century when Japanese MNCs first took form, Japan overcame the problems of its resource scarcity through imports of basic raw materials such as raw cotton and iron financed by the exports of other raw materials, primarily coal and raw silk.[2] The important role of trade in the Japanese economy thus required the presence of an infrastructure in support of trade – banks, shipping companies, marine insurance companies and, above all, trading companies. Without such companies, Japanese economic development could not occur (Hirschmeier and Yui, 1975). Between 1858 and the late nineteenth century, such international trade infrastructure was provided entirely by foreign merchants encouraged by the opening of Japanese ports to foreign trade. Even by 1887 some nine-tenths of Japan's external trade was handled by foreigners, mainly British houses (Maddison, 1969) and foreign companies also provided auxiliary services in shipping, banking and marine insurance facilities (see for example, Wray, 1984). The realization of the central role of trade in the Japanese economy and that the dominant role of foreign companies posed a threat to Japan's economic security and prosperity not the least because of their role in exacerbating balance-of-payments difficulties led the Japanese government to persuade the most prominent local businessmen to initiate the creation of large specialized trading firms that was to become the general trading companies (Yonekawa, 1985). This had historical antecedents in the growth of trading companies, managing agencies, consulting engineers and shipping companies in Great Britain in the late nineteenth and early twentieth century (Wilkins, 1986b).[3] The advent of the Japanese general trading companies – *the sogo shosha* – engaged in the import and export of a variety of commodities on a global scale (Yonekawa, 1985) supported by the development of their auxiliary businesses in shipping and insurance and close ties with banks implied that the responsibility for the bulk of the expanding volume of Japan's international trade fell increasingly in their hands. Their large size was in contrast to the numerous domestic firms of small size whose international trade it handled. Thus, the development of an indigenous international business infrastructure in Japan implied that production and trade was conducted at least initially by two different companies, the domestic-based manufacturing companies and the general trading companies, whose business relationships was of mutual interdependence. The large Japanese trading companies with an expertise in international trade did not have specialized knowledge of the production of the various goods that constituted their trading business, while on the other hand each of the numerous small-sized manufacturing companies had extensive production experience but no international business experience. In later years, trading companies not only promoted the trade of manufacturing companies but also often acted on their behalf abroad by establishing small plants or joint ventures with Japanese producers and with nationals in host countries. The extreme example is provided by Naigaiwata Company that originated as a trading company in its company's history but integrated into domestic production in cotton spinning and then eventually established international production in that industry, thus developing to become a 'classic' MNC. In time, Japanese manufacturers bypassed the trading company and acted on their own in international production. Thus, two major types of Japanese firms expanded overseas (Wilkins, 1986b).

The above historical account of Japanese economic development explains the dominance of emergent Japanese FDI in the services sector, particularly in trading and in complementary services in support of trade such as banking, marine insurance and transportation equipment such as shipping and railroads (see Sekiguchi, 1979; Tsurumi, 1976). Nevertheless, the foundations of Japanese FDI in manufacturing and resource extraction had already been laid in the period prior to 1914. The earliest Japanese FDI were in geographically close and relatively familiar regions. Of the estimated Japanese foreign investments (in the form of both FDI and portfolio) in 1914 of between $227 and $296 million and of which an overwhelming proportion was FDI, some 70 per cent was directed to China (including Manchuria).[4] This was encouraged by the Shimoneseki Treaty concluded between Japan and China in 1895 which permitted Japan to manufacture in Chinese treaty ports for the first time (Wilkins, 1986b). The second-largest host country prior to 1914 was the United States which accounted for some 10 per cent of Japanese FDI. Japanese FDI was also present elsewhere in Asia, though not as large as in China. For example, there were sizable Japanese investments in banking, trade and shipping facilities, as well as in railroads and certain agricultural ventures in Korea well before it became a Japanese colony in 1910, and investments in that country grew after colonialization (Duus, 1984). The Yokohama Specie Bank, the Nippon Yusen Kaisha Shipping Company and the trading companies – Mitsui & Company, the Naigaiwata Company and Japan Cotton Trading Company – also established branches in India (Wilkins, 1986b).

The following sections which look at these important industries in the history of Japanese MNCs show the close link of trade and outward FDI in the international business of Japan made responsible by the cooperative effort between Japanese manufacturers, the trading companies, the shipping, banking and insurance enterprises facilitated by the strong role of the Japanese government.[5] The four most significant Japanese companies involved in international business before 1914 were Mitsui Bussan (trading), Yokohama Specie Bank, Nippon Yusen Kaisha (shipping) and Tokio Marine Insurance (Wray, 1984).

Trading FDI prior to 1914

The Mitsui Company was one of the traditional Japanese merchant companies whose origins can be traced to cloth trading in the late seventeenth century (Yonekawa, 1985). Its foreign trading arm, Mitsui Bussan established in 1876, was one of the pioneering Japanese companies that developed a sizable trading business in China prior to 1914. Mitsui Bussan opened its first overseas branch in Shanghai in 1877 – a year after its foundation – for the initial purpose of facilitating the sales of Japanese exports of coal in China, particularly to Jardine, Matheson & Company and Butterfield & Swire, the British trading companies that were its principal customers. By 1886 the trading functions of the Shanghai branch of Mitsui Bussan was expanded to include the importation of Chinese raw cotton for the Osaka Spinning Mills closely associated with the company, as well as re-oriented to the sales of Japanese cotton yarn and fabrics in China.[6] It also established offices with trading interests in Hong Kong, Paris, Milan and New York.[7] The New York office handled Japanese exports of raw silk to the United States as well as Japanese imports of raw cotton, railroad equipment and machinery from the United States. On the basis of its New York office and the wholly owned subsidiary it established in Houston in 1911 – the Southern Products Company – to facilitate the flow of Japanese imports of American raw cotton, Mitsui Bussan came to handle more than 30 per cent of the raw cotton imported into Japan from the United States by 1914. It was also responsible for some 33.6 per cent of Japanese exports of silk to the United States (Wilkins, 1986b). Before 1914 Mitsui Bussan had more than 30 branches in Asia, Europe, Australia and the United States, in addition to their manufacturing affiliates in China (Jones, 1996). The trading interests of Mitsui Bussan across the world was facilitated by its ownership of a shipping fleet, supported by chartered ships from the government-controlled Nippon Yusen Kaisha Shipping Company (Mitsui & Co., 1977).

In a similar objective to facilitate Japanese imports of American raw cotton, the Japan Cotton Trading Company also established a subsidiary in Fort Worth in 1910 and the Gosho Company opened an office in San Antonio in 1913. Thus, all the three major Japanese trading firms participating in the raw cotton trade between Japan and the United States had Texas-based offices before the First World War. Most Japanese trading companies were also involved in facilitating the growth of exports from Japan and China to the United States such as, for example, medicinal and aromatic products (Wilkins, 1986b).

Transportation (shipping and railroads) prior to 1914

Closely related to the emergence of Japanese general trading companies prior to 1914 was FDI in transportation. The investments in shipping constituted not only the acquisition of ships which enabled Japanese companies to assume importance in certain international shipping routes at the expense of foreign shipping companies. More important from the viewpoint of FDI was the establishment of branch offices, wharves and warehouse facilities in important trading ports in China, the United States and Europe. The investments in shipping were undertaken by both specialized shipping companies as well as by trading companies that had auxiliary investments in shipping.

Japanese companies began to assume importance in the shipping route between Japan and China when Mitsubishi initiated a weekly shipping service between Yokohama and Shanghai in February 1875. This rapidly developed when in October 1875 Mitsubishi acquired four ships from the Pacific Mail Steamship company which abandoned the Yokohama to Shanghai shipping service in favour of its new rival (Wray, 1984). The first shipments of coal of Mitsui Bussan to China was on board Mitsubishi ships, but even after the company had established its own sailing ship company to handle freight between domestic ports and China and Korea, it still continued to use Mitsubishi shipping vessels (Wilkins, 1986b).

The establishment in 1885 of Nippon Yusen Kaisha (NYK), the shipping company in which the government had a large interest, led to the diminution in importance of Mitsubishi as a shipping company. Although NYK was organized at origin as a domestic coastal shipping company, it had established three major shipping services to nearby areas: Shanghai, Vladivostok and Inch'on to advance the trading interests of Japan. It fulfilled an important role in the China trade as it developed close ties with the Japanese cotton spinners in the late nineteenth and early twentieth century.

Other important Japanese shipping companies with a significant role in Chinese trade was Osaka Shōsen Kaisha (OSK) organized in 1884 which initiated the Shanghai-Hankow shipping service in January 1898 with a subsidy from the Japanese government and thus became the first major Japanese shipping company to operate on the Yangtze River. With the subsequent participation of NYK in this shipping service, the Nisshin Kisen Kaisha was formed in 1906 to organize Yangtze river shipping under the dominant influence of the two Japanese shipping companies operating the route, the OSK and NYK. The growth in importance of Japanese shipping companies was such that although 52 per cent of the tonnage of the foreign ships calling at Chinese ports were accounted for by British shipping companies by 1913, Japanese companies accounted for almost 32 per cent (Wray, 1984).

The importance of Japanese shipping companies in trans-Pacific trade also became well established by 1914. The passenger ships of NYK which made 24 trips between Japan and Hawaii in the years between 1885 and 1894 was crucial in the delivery of 27,000 Japanese immigrants. The cargo ships of NYK, which was already playing a key role in the provision of shipping services between Yokohama and Shanghai, broadened its shipping services to the United States indirectly by through-freight agreements it concluded with the Pacific Mail Steamship Company and the Occidental & Oriental

Steamship Company in May 1886 which provided connecting shipping services between Yokohama and San Francisco. Such arrangements were maintained until 1896 when NYK made arrangements with Great Northern Railroad which enabled it to quote through-freight rates for shipping between Asia and the American midwest and onwards to the American east coast (Wilkins, 1986b). By 1914, Japanese shipping companies were providing regular shipping services to Seattle (operated by NYK), San Francisco (operated by Toyo Kishen Kaisha) and Tacoma (operated by OSK), and the opening of the Panama Canal on August 1914 served to expand the growth opportunities for Japanese shipping companies.

The development of shipping services to Europe was regarded as more important than that of the trans-Pacific route as seen by the much larger annual government subsidies received by NYK to build its shipping routes to Europe. Thus, by 1903 NYK had 12 ships providing shipping services to Europe and only six ships providing shipping services to the United States (Wray, 1984). The growth and development of Japanese shipping companies enabled these companies to be responsible for 51 per cent of the ships entering Japanese ports by 1913, and to account for 52 per cent of Japanese exports and 47 per cent of Japanese imports (Wilkins, 1986b).

Apart from shipping as a form of transportation to support trade, the Japanese government also made substantial investments in the development of the South Manchurian Railway in 1906 (Remer, 1933; Hou, 1965). Indeed, some 36 per cent of the total Japanese FDI in China in 1914 was in transportation, chiefly in the South Manchurian Railway. This railway played an important role in the development of the richest coal mines in Manchuria located in Fushin, and in the transport of coal to Japan.

Banking and insurance FDI prior to 1914

Unlike in the case of the United States or the United Kingdom, domestic banks in Japan played a strategic role in the expansion of Japanese business abroad. In no way is this more evident than in the case of the Yokohama Specie Bank whose history was associated intimately with the growth of Japanese business worldwide. The development by the bank of an extensive network of offices, branches and sub-branches worldwide and its role in the intermediation of the flow of foreign capital for use in Japan or in the various business activities of Japan abroad enabled it to play a key role in the growth of Japanese trade and FDI. The bank provided general information, assisted in foreign exchange activities, provided trade financing and even long-term financing of the raw material procurement needs of Japanese industry. An important example of the latter was the loans extended to the Hanyehping Coal and Iron Company to extract coal in Manchuria. In its establishment of a foreign office in New York in 1880, the very same year of its foundation, this government-supported bank became an international bank at origin. The bank then set up a representative office in Shanghai in 1893 at the request of Japanese businessmen to provide banking services to support the triangular cotton trade between Japan, India and China. It also established a branch in Hawaii and San Francisco in 1899, a sub-branch in Los Angeles in 1913 and an important office in London (Wilkins, 1986b). It was ubiquitous in Japanese international business and played a key role in supporting its expansion in a manner similar to German banks but dissimilar to any American or British bank in their respective roles in supporting the growth of German, American and British business abroad (see Chapters 2, 3, 7 and 8 of this book). The role of the Yokohama Specie Bank in aiding the expansion of Japanese international business was further facilitated by its close business relationships with Mitsui & Co. and with NYK, Japan's largest shipping enterprise (Mitsui & Co., 1977; Wray, 1984).

The Industrial Bank of Japan organized in 1900 similarly helped to finance government-sponsored investment ventures, extended loans to the Tayeh mines that became a part of the Hanyehping Company complex and raised finance for the South Manchurian Railway and other Japanese investments in China. Indeed, of the 294.4 million yen in foreign portfolio capital underwritten by the bank between 1902 and 1913, 46 per cent was exported as semi-governmental direct investment to China and Korea, mainly. By 1902 the Yasuda Bank, a private Japanese bank, also had Chinese interests (Patrick, 1967).

Other banks that established foreign branches in countries other than China was the Dai Ichi Ginko which by opening a branch in Korea in 1878 became the first Japanese bank to branch abroad. An Osaka-based bank also established a branch in Formosa (now Taiwan) before it became a Japanese possession in 1895, and the Bank of Japan followed suit in 1896. The Bank of Taiwan established three years later by the Osaka-based bank in Taiwan had in turn established by 1914 branches in San Francisco, Manila, Singapore, Calcutta, Bombay, 7 points in China and 14 points in Japan and its dependencies (Wilkins, 1986b).

In the insurance industry, at least one Japanese company, Tokio Marine Insurance Company Ltd, invested in the United States prior to 1914. The company formed part of a closely integrated network of banks, shipping and trading companies associated with Japanese–American trade (Wilkins, 1986b).

Manufacturing FDI prior to 1914

The development of the domestic textiles industry with the aid of British technicians, technology and machinery played an important role in the emerging pattern of Japanese foreign investment in manufacturing. As the Japanese cotton textile industry developed, interest in export markets grew and Japanese trade in cotton manufactures came to be very important and China became an integral trade partner. Between 1888 and the time of the First World War Japan was an importer of raw cotton from China and an exporter of cotton yarn and cotton cloth to China. In 1913, exports of cotton manufactures and yarn totalled 33.9 per cent of Japanese exports to China and 11.7 per cent of total Japanese exports (Remer, 1933).

The impetus to switch from exports to international production stemmed from the Shimoneseki Treaty of 1895 which allowed foreigners to manufacture in Chinese treaty ports for the first time, as previously mentioned. This prompted Jardine Matheson & Company, the British trading company, to establish at once the Ewo Cotton Spinning and Weaving Company in Shanghai and three foreign textile firms from Great Britain, the United States and Germany followed suit (Wilkins, 1986b). The establishment of these foreign-owned manufacturing facilities in China combined with the incipient growth of Chinese investments in spinning mills, and the sale in China of cheap Indian yarn posed a threat to the continued growth of the Chinese export market for both Japanese spinners and trading companies. In addition, much of the Japanese FDI in China in the early twentieth century may have been made initially to ascertain local costs of production in China and to keep close watch over the Chinese textile market. The determinants of the initial international production activities of Japanese companies was peculiar to Japanese MNCs and not observed in the case of the early manufacturing MNCs based in the United States, the United Kingdom and Germany in which international production was spurred by the inability of important foreign markets to be supplied, or supplied as cheaply, through exports owing more often than not to tariff barriers imposed by foreign governments (see Chapters 2, 7 and 8).

Not only were the determinants of the initial international production activities of Japanese companies peculiar to Japanese MNCs but more fundamentally the fact that Japanese MNCs emerged in simple, labour intensive and technologically standardized manufacturing activities directed towards developing countries is a feature that has not been observed in the emergence of American, Swedish, British or German MNCs. The divergent growth pattern of Japanese FDI in manufacturing can be explained by the earlier stage of development in which Japan entered international production. In the late nineteenth and early twentieth century when Japanese manufacturing MNCs emerged, textiles was the dominant domestic industry in Japan and consequently was the industry in which Japanese manufacturing MNCs initiated international production. The theoretical implication of these facts is that MNCs do not always emerge in technologically advanced industries organized in an oligopolistic market structure with firms that have highly developed firm-specific advantages.

International production by Japanese firms in textiles which served to substitute for exports from Japan to China was pioneered by two Japanese spinners that established plants in Shanghai. These initial ventures failed on account of inadequate evaluation of the market, ineffective management and inland taxes on raw cotton (Kuwahara, 1982; Wilkins, 1986b). The earliest sign of a more successful Japanese FDI in the industry came in 1902 when

J. Yamamoto, the Shanghai branch manager of the merchant company, Mitsui & Co., purchased a Chinese cotton mill and re-named it the Shanghai Cotton Spinning Company Ltd. The company acquired another Chinese cotton mill, the Santai Cotton Spinning Company Ltd., four years later. On 5 December 1908 Mitsui & Co. organized a new local subsidiary, the Shanghai Cotton Manufacturing Company, which owned the two spinning mills and placed the new subsidiary under the charge of J. Yamamoto. The separate management of the subsidiary enabled Mitsui & Co. to combine the benefits of earning high commissions from the trade intermediation between buyers and sellers while not being involved directly in cotton spinning. The Shanghai Cotton Manufacturing Company proved eventually to be a very profitable venture. By 1914, Mitsui & Co. had also established 886 looms, and thus participated in weaving as well as spinning in China (Yasumuro, 1984).

Another example of a trading company whose initial interest in China was in the export of raw cotton from China to Japan but integrated into international production through acquisition was the Japan Cotton Trading Company that had a cotton spinning mill in Shanghai by 1914 (Wilkins, 1986b). The process of expansion of the international production activities of the Naigaiwata Company – which was at origin a trading company with a similar interest in the raw cotton trade between China and Japan – was different. The company acquired two spinning companies in Japan in 1909, and established itself as a domestic spinner in Japan before it began to manufacture in China (Yasumuro, 1984). While it was not one of the largest spinning establishments in Japan, it built a new and powerful spinning mill in Shanghai using its own designs, equipped it with almost advanced machinery and managed it by a Japanese staff sent from the parent company. The investments was thus distinctive in its role as the first Japanese spinner that constructed its own mills abroad as opposed to acquiring existing foreign mills typical of the form of investment of the trading companies, and also as the first Japanese spinner to be a successful manufacturer in China. By 1913, the Shanghai mill of the Naigaiwata Company was far more efficient than any Chinese-owned mill owing to the company's management knowledge, experience in buying raw cotton and practice of blending raw cotton from different sources in order to lower the costs of raw materials (Kuwahara, 1982). By 1914, the Naigaiwata Company ranked nineteenth among the 100 largest Japanese mining and manufacturing companies based on the size of its Japanese and Chinese assets. Indeed, the company was the most successful of the early Japanese investors in China (Wilkins, 1986b).

The success of the Japanese textiles investments in China can be attributed to several factors. Perhaps the most important was the composite technologies developed by Japanese companies from the fusion of imported and domestic R&D. In this respect, Great Britain played a crucial role not only in the development of the Japanese textile industry, but also in the success of transplanted Japanese spinning mills and looms in China as Japanese companies brought as part of their investments British technology and equipment. The looms installed in Chinese mills were made by Platt Brothers in Britain (Kuwahara, 1982). Combined with this composite technologies of Japanese companies was their organizational skills in production harnessed in the domestic textiles industry which were readily applied to their foreign plants in China. The Japanese staff adapted rapidly to living and working in China, probably owing to the high degree of cultural and linguistic affinities between the two countries (Wilkins, 1986b).

The proliferation of Japanese spinning mills in China enabled Japan to account for an increasing share of spindles and power looms in that country. Thus, whereas 63 per cent of the spindles in China in 1897 were Chinese owned and 37 per cent Western owned, the ratios were 60 per cent Chinese owned, 27 per cent Western owned and 13 per cent Japanese owned by 1913. For power looms, the ratios were 70 per cent Chinese owned and 30 per cent Western owned in 1897, but by 1913 the ratios were 56 per cent Chinese owned, 25 per cent Western owned and 19 per cent Japanese owned (Hou, 1965).[8]

Although Japanese companies had not yet dominated the Chinese textiles industry, Japanese mills were already displaying some level of superiority over both Chinese and British mills (Chao, 1977). In the early twentieth century, Japanese firms in the domestic cotton textile industry of China exhibited excellent management skills compared to other foreign companies, demonstrated marketing strengths owing to the direct involvement of trading companies in FDI or the earlier trading company experiences of many Japanese manufacturing companies, and displayed superiority in raw material procurement. As mentioned, many Japanese companies have established the practice of direct purchase of raw cotton from growers in India, China and the United States; this plus the technologies in blending raw cotton from their different sources served to improve the quality while lowering the cost of raw material supplies. Indeed, the Japanese investments in Chinese spinning and weaving prior to the First World War were harbingers of far larger and more important ones to follow (Wilkins, 1986b).

Although textiles was the most important manufacturing industry of Japanese firms in China, there were other manufacturing industries that preoccupied Japanese companies in China even though the size of their investments in these other industries were far less significant than in textiles. This included, for example, the manufacture of bean oil by Nisshin which founded a mill in Dairen, Manchuria in 1907 for 3.5 million yen and the manufacture of flour by the Mitsui & Co. in Shanghai and by the South Manchuria Milling Company in Tiehlin, Manchuria founded in 1906.[9] There were also five match manufacturing plants financed by Japanese investments in Manchuria and in north, central and south China in 1914 (Remer, 1933).

The cumulative growth of Japanese investments in China enabled Japan to account for the largest number of foreign and Sino-foreign manufacturing and mining firms established in China between 1895 and 1913. However, although Japan was responsible for 49 of 136 such firms, firms from the United Kingdom accounted for a significantly larger share of 48 per cent of the value of initial capitalization of the foreign and Sino-foreign manufacturing and mining firms established in China in that period compared to that of Japan of 25 per cent (Wilkins, 1986b).

Last but not the least, mention should also be made in the history of Japanese manufacturing FDI of the one-of-a-kind investment in 1892 of Kikkoman in a factory in Denver, Colorado to produce soy sauce for Japanese emigrants (Kinugasa, 1984).

Resource-based investments prior to 1914

Although far less important than Japanese FDI in trade-related services and manufacturing, resource-based FDI as a means of obtaining scarce natural resources needed by domestic industries of Japan was of some importance in the period prior to 1914. The most significant of these Japanese resource-based FDI was that of Hanyehping Coal and Iron Company which mined the rich coal mines of Fushin, Manchuria on the basis of supporting investments by the Yokohama Specie Bank that extended long-term loans to finance the investments in resource extraction and railroads (the South Manchurian Railway earlier mentioned) to facilitate the development of the mines and the transport of coal and iron ore required by the Japanese government-owned Yawata Iron and Steelworks in Kyushu. Indeed, half of the 2.5 million long tons of coal produced by the Fushin mines in 1913–14 was exported to Japan (Wilkins, 1986b).

The growth of Japanese FDI in the inter-war period

The inter-war period marked the continuing growth in importance of Japanese FDI in the trading and manufacturing sectors.

Trading FDI in the inter-war period

The 1920s was a decade of great difficulties for Japanese trading companies. Among the large trading firms that became insolvent during this period were Mogi Shoten, Furukawa Shoji, Kuhara Shoji, Takada Shokai and Suzuki Shoten (Yamazaki, 1989). Those trading firms which escaped bankruptcy often suffered from wide fluctuations in their incomes or from stagnation at low income levels. Nevertheless, other large trading companies – particularly the general trading companies – continued to prosper in the inter-war period.

The continued growth in the success of general trading companies can be explained in the context of the *zaibatsu* business group of which these companies are often but not always a part (Yonekawa, 1985). If the general trading companies developed from each *zaibatsu* business group's selling department, as in the case for example of Mitsubishi and Sumitomo trading companies, the industrial companies in the *zaibatsu* relied entirely on their member trading companies for the foreign procurement of vital raw materials at the lowest cost and for the foreign sales of a diversified product line in the most lucrative export markets. The general trading companies were thus a principal means to overcome the scarcity of raw materials in Japan and the need to sustain the growth of exports.

Such general trading companies were able to fulfil such functions effectively and profitably by internalizing some functions of international trade which decreased the risks of international trade transactions. Apart from internalizing vital functions of international trade, the general companies also developed new organizational structures and management systems that combined decentralized divisional structures along product lines and geographical areas with strong centralized control. The products that the general trading companies handled were carefully selected to include those in which an effective control could be attained from suppliers to customers either through the supply of funds, sole agency contracts, and/or investments in suppliers. The most important trading services these firms facilitated were the import of raw materials from large foreign suppliers to large local customers; the import of highly mechanized high value products, particularly machinery and technology, for the member companies of *zaibatsu;* and the export of undifferentiated non-perishable miscellaneous goods to foreign markets that were eventually sold through various channels. But apart from being part of a large *zaibatsu* group, the general trading company also played a vital role in the consolidation of a large number of small manufacturers in both Japanese and foreign markets into groups of large transaction units, and sold their products either to American brokers or Japanese importers who in turn sold through various marketing channels. Through an effective system of internalized trade transactions supported by an organizational structure with centralized control, Japanese general trading companies have succeeded in dominating trade flows between the United States and Japan before the Second World War, in spite of the restrictions imposed by the American government and strong competition from established American and European business interests (Kawabe, 1989).

Among the most successful general trading company in the inter-war period was Mitsui Bussan that continuously maintained high profit rates in an era of insolvency of other trading companies. Unlike other general trading companies that developed from each *zaibatsu* business group's selling department, the emergence of the Mitsui Bussan as an offshoot of the Mitsui Company and its eventual growth helped in the formation of the Mitsui group of companies rather than vice-versa. The formation of Mitsui Bussan as a managerial enterprise separate from the Mitsui Company enabled it to develop free from most restraints that plague more traditional mercantile business (Yonekawa, 1985). As mentioned in the previous section, Mitsui Bussan was not only a general trading company but was also involved in auxiliary investments in support of their main trading businesses which included not only marine transportation and shipbuilding but also agencies for ocean transport and insurance, and a holding company. The superb business performance of Mitsui Bussan during the 1920s was attributed to a number of closely inter-related factors. First, the company based a dominant proportion of their profits in the 1920s on trade in a wide range of staple commodities to include coal, machinery, raw silk, sugar, metals, timber and cotton that were closely linked to key resource-based Japanese industries that were experiencing rapid growth during the 1920s. The diversification of Mitsui Bussan in all these main commodities linked to their monopolistic transactions with the manufacturing companies belonging to the Mitsui *zaibatsu* enabled it to attain economies of scale in commodity trade. The second element of its success was their characteristic strategy of developing strong business linkages through sole agency contracts with the leading manufacturing companies both in Japan and abroad which helped to sustain its dominant share in each of the commodity markets. Their monopolistic transactions in Japan with the manufacturing companies of the Mitsui *zaibatsu* which gave Mitsui Bussan sole rights to sell the products manufactured by these companies and to buy the raw materials these companies required was further strengthened by its position in most cases as the exclusive agent for leading foreign manufacturers. Mitsui Bussan thus enjoyed economies of scale in trade in a wide range of products which enabled it to maintain a position of strength *vis-à-vis* its competitors. The third element of its success was its large number of talented and highly educated employees and abundant capital resources. All these factors combined with an organizational structure compartmentalized along lines of commodities it handled enabled Mitsui Bussan to occupy an advantageous position in each commodity trade which contributed to its stable and high level of profits (Yamazaki, 1989; Yonekawa, 1985).

Manufacturing FDI in the inter-war period

The inter-war period described the continuing expansion of MNCs based in the Japanese cotton industry. Indeed, the amount of Japanese FDI in manufacturing in the period prior to 1914 was relatively small in relation to that in the inter-war period. The growth was such that for the first time ever in 1919 the number of Japanese-owned cotton spindles in China exceeded those of Western-owned mills and by 1936 approximately 44 per cent of the cotton spindles in China were Japanese owned (Chao, 1977; Remer, 1933).

In 1914, about 40 per cent of the total sales of the major Japanese cotton spinners was generated from exports which was directed almost completely towards China. Japan had attained a dominant position in the cotton imports of China, accounting for 55 per cent of total imports of cotton yarn in 1914 and 57 per cent of total imports of coarse cotton cloth (Kuwahara, 1989). At the same time, the Chinese modern cotton industry began to expand its production capacity on a large scale during the First World War enabling the country to decrease their imports from Lancashire and to increase the price of cotton goods considerably (Kuwahara, 1982). The rapid growth in the self-sufficiency of China in the production of coarse cotton yarns and coarse cotton cloth had a deleterious impact on the Japanese cotton spinning industry which had established a dominant position in the Chinese market in 1914. Thus, unlike in the period before the 1914 when the growth of foreign manufacturers in Chinese treaty ports posed the threat to the export growth of Japanese cotton spinners in China, in the period during and after the First World War the threat to export growth of Japanese cotton spinners was presented by the emergence and rapid expansion of the indigenous cotton spinning industry. To halt the decline in their market shares acquired through rapid export growth, most of the major Japanese cotton spinners established local production bases in China for the first time in the immediate period after the First World War. The four largest Japanese cotton spinning companies that initiated FDI in China in this period were the Toyo Cotton Spinning Company, the Kanegafuchi Cotton Spinning Company, the Dainippon Cotton Spinning Company and the Fuji Gasu Cotton Spinning Company. These four firms accounted for 58 per cent of all spindles installed in Japan in 1918 (Kuwahara, 1989). Since coarse cotton yarns could only be sold in the Chinese market, most of the local mills established by the large Japanese cotton spinning companies produced coarse cotton yarns. Conversely, since alternative markets for coarse cotton cloth could be found outside China, there were far fewer local mills of Japanese companies that produced coarse cotton cloth.

International production financed by FDI was the common strategy adopted by the four major Japanese cotton spinning companies to defend their market shares in China. Once a major Japanese cotton spinner embarked on the construction of local cotton spinning mills, a 'bandwagon' effect described the subsequent growth of investments of other major Japanese cotton spinning companies that displayed a follow-the-leader behaviour typical of firms in industries organized in an oligopolistic market structure (Knickerbocker, 1973). Nevertheless, there were two other equally important strategies employed by the major Japanese cotton spinning companies in response to the problem of excess capacity posed by the new threat to the sales of coarse cotton yarn and coarse cotton cloth in their major export market. The first was the diversification of products in the home market. This came in the form either of increasing the amount of weaving looms as a means of increasing consumption of coarse cotton yarn within the company or increasing the production of higher value added cotton goods such as middle-count yarns at the expense of coarse cotton yarns. This was the other strategy employed by the Kanegafuchi Cotton Spinning Company apart from FDI. The second other strategy was the diversification of export markets. This was the strategy employed by the Toyo Cotton Spinning Company in light of the increase of Chinese output of coarse cotton yarn and coarse cotton cloth which had a devastating effect on the company's exports to China. The company diversified their export market rapidly after the First World War away from China towards countries of South East Asia, the Middle East and Africa that had large markets for coarse cotton cloth. By 1921, 72 per cent of company exports of cotton sheeting was destined to export markets outside China, while in China itself the company increased exports of fine cotton cloth (Kuwahara, 1989).

The growth of Japanese FDI since the Second World War

The period since the Second World War was associated with the rapid domestic industrial restructuring of the Japanese economy and the expansion of Japanese FDI in manufacturing. In this period more than ever before, Japan's outward FDI was a crucial instrument or catalyst for the rapid process of domestic industrial upgrading (Ozawa, 1985; Kojima and Ozawa, 1985). The rapid industrial transformation from primary to secondary sectors, and within the secondary sector from labour intensive light manufacturing to heavy and chemical manufacturing to knowledge intensive, assembly-based fabricating industries have led to a swift rise in labour productivity and wages at the end of each phase. Shifting patterns of trade competitiveness and sectoral resource allocation have been the dominant features of this process, and the advance of Japanese manufacturers overseas has been influenced to a significant degree by these changes (Ozawa, 1985).

Ozawa (1991) analysed the industrialization process of the Japanese economy since the Second World War in four sequential stages that correspond with equivalent phases or waves in the growth of Japanese cross-border production.

Phase I Expansion of labour intensive manufacturing in textiles, sundries and other low-wage goods in 1950 to the mid-1960s. This industrialization phase corresponded with the 'elementary' stage of Japanese offshore production in Heckscher-Ohlin industries.

Phase II Scale economies-based modernization of heavy and chemical industries such as steel, aluminium, shipbuilding, petrochemicals and synthetic fibres in the late 1950s to the early 1970s. This industrialization phase corresponded with the Ricardo-Hicksian trap stage of Japanese multinationalism associated with the non-differentiated Smithian industries.

Phase III Assembly-based, sub-contracting dependent, mass production of consumer durables such as cars and electrical/electronics goods in the late 1960s to the present. This industrialization phase corresponded with the export substituting-cum-surplus recycling stage of Japanese multinationalism associated with the differentiated Smithian industries.

Phase IV Mechatronics-based, flexible manufacturing of highly differentiated goods involving the application of computer-aided designing (CAD), computer-aided engineering (CAE) and computer-aided manufacturing

(CAM), along with technological breakthroughs such as high definition TV, new materials, fine chemicals and more advanced micro-chips in the early 1980s onwards. While Ozawa (1991) did not identify the corresponding phase of Japanese multinationalism associated with this industrialization phase as he regarded this type of Japanese FDI to be still speculative, this fourth phase of Japanese multinationalism which has yet to evolve fully in the future can be labelled as the robotics and new materials stage of Japanese multinationalism associated with the Schumpeterian industries.

The following sections examine more closely each of the four major phases or waves in the growth of Japanese crossborder production in relation to the sequential phases of Japanese industrial development in the era since the Second World War.

Phase I The 'elementary' stage of Japanese offshore production in Heckscher-Ohlin industries

The first wave of Japanese FDI was associated with the continuing strengths of Japanese industry in traditional labour intensive light manufactures, notably textiles and clothing and other low-wage goods whose trade competitiveness derived from the presence in Japan in the early post-Second World War period of both an abundant labour force at relatively low cost as well as an undervalued national currency. Indeed, by 1950, Japan has sustained a comparative advantage in the textiles industry for more than 70 years.[10]

The emergence of the 'elementary' stage of Japanese offshore production in labour intensive industries for which Japan in the early post-Second World War period continued to have a comparative advantage was influenced largely by the new Japanese industrial and trade strategy adopted in the 1950s to develop modern and capital intensive heavy and chemical industries as domestic infant industries protected by import substitution industrialization. Such process of domestic industrial upgrading has been made possible by the continual acquisition and successful absorption of new and advanced foreign technology. Such advanced foreign technologies from the West which Japan was eager to absorb was obtained mainly in the form of licensing agreements as the Japanese government was always mindful of the possible domination of Japanese industry by foreign MNCs through inward FDI which it thus restricted.[11] As a result, foreign licensing agreements became a far more important vehicle relative to inward FDI in acquiring new and advanced industrial knowledge that revolutionized Japanese industry.

The acquisition of advanced foreign technologies mainly through licensing agreements in the period between the late 1950s and the early 1970s was geared to support the growth and development of the domestic heavy and chemical industries in the producer goods sector. Four industries accounted for the 70.6 per cent of the 1,029 technology import contracts that the government approved during the 1950s: non-electric machinery, electric machinery, chemicals, and metals (including steel) (Ozawa, 1985). Technology imports were complemented by indigenous efforts at adaptive R&D which enabled the commercialization of foreign industrial technologies and their significant improvement, particularly in production processes.[12] The growth of new capital investments in new production facilities in the modern heavy and chemical industries by manufacturers and closely related firms through business linkages engendered by the importation of advanced technologies and complementary R&D efforts of Japanese firms contributed to the success of efforts to structurally upgrade Japanese industry. Domestic production capacities increased which was initially geared to meet the rising domestic demand for such products away from imported manufactured goods, and the further growth of domestic production and increased competitiveness led to the growth of exports by Japanese firms in more modern manufactures.

The success of domestic industrial restructuring towards modern capital intensive heavy and chemical industries at the end of the 1960s enhanced labour productivity owing to high capital to labour ratios and this perpetuated a continuous cycle of rising wages (Cohen, 1975; Kojima, 1978; Ozawa, 1979b). In combination with the sharp upward revaluation of the yen in 1971, it led to the weakening of the traditional labour intensive sectors. This adversely affected the textiles industry that was the mainstay of the Japanese economy over more than 70 years, but it also affected those small- and medium-sized enterprises producing a whole range of other labour intensive products.

In light of the rapid loss of comparative advantage of Japan as a location of traditional labour intensive light manufacturing production, Japanese manufacturers in those industries were faced with three options: upgrade their products qualitatively, diversify into new industries or shift production to labour abundant developing countries suitable for labour intensive manufacturing (Ozawa, 1985). The three options were not mutually exclusive so that while many firms chose the option of international production, many also sought to pursue the option of upgrading the quality of their labour intensive products and some diversified into new industries helped by the government emphasis on domestic industrial upgrading in more modern manufacturing industries.

The growth of Japanese FDI in this phase was critically determined by the liberalization of Japanese government policy on outward FDI away from a ban in place for much of the 1940s and the tight regulation in place for some two decades since the implementation of the Foreign Exchange and Trade Control Act of 1949. The latter was implemented in light of the persistent trade deficits of Japan and the emphasis placed on achieving rapid industrial growth of the domestic economy. The substantial increases in domestic labour costs in the late 1960s, the appreciation of the yen, the emerging balance-of-payments surpluses and the growing natural resource requirements of a rapidly expanding resource scarce industrial economy contributed to the gradual liberalization of controls over outward FDI between 1969 and 1978 (Randerson and Dent, 1996).

Thus, the 'elementary' stage of Japanese offshore production took place in a major way between 1971 and 1973, with the investments in 1972 and 1973 alone more than doubling the existing stock of Japanese FDI. This stage of Japanese overseas production was associated with the transfer of standardized, low technology, labour intensive production from Japan to developing countries such as Taiwan, South Korea, Hong Kong, Thailand and other Asian economies in close geographical proximity and had an abundant labour supply (Ozawa, 1991). China did not seem to be as important a host country in this period in the way that it was in the period prior to 1914 and in the inter-war years. This is confirmed by Vaupel and Curhan (1974) who indicated that of the 105 Japanese manufacturing plants established abroad in the period between the end of the Second World War and the early 1970s, 98 plants were located in developing countries, mainly in the Asia and Pacific region. Some one-third of the overseas FDI in manufacturing were made by small- and medium-sized enterprises, with 88 per cent of these small-scale investment ventures concentrated in other Asian countries (Ozawa, 1985). In many instances, the general trading companies took the initiative to shift the production activities of small- and medium-sized enterprises by investing jointly and offering organizational, financial and managerial assistance to implement overseas production and marketing. Small firms that invested abroad also availed of the overseas scanning services provided by the government's Overseas Trade Development Association (Roemer, 1976; Ozawa, 1979a). The four most labour intensive industries transferred abroad by a large number of small- and medium-sized enterprises were textiles, metal products, electrical machinery and other miscellaneous manufactured products. These industries accounted for 65.5 per cent of the total number and 55.4 per cent of the total value of new Japanese overseas investments made in 1972 and 1973 (Ozawa, 1985).

A formal analysis of the industrial advantages of Japan between 1965 and 1975 showed that the degree of innovation and the creation of technological ownership advantages as well as capital intensity exerted highly significant *negative* influences on Japanese FDI in manufacturing, while the skill level of managerial power had a significant *positive* effect. This contrasts sharply with the findings for Japanese exports of manufacturing which showed that the technological intensity and the skill level of production workers exerted highly significant *positive* influences, while capital intensity, skill level of managerial manpower and complexity of management exerted highly significant *negative* influences (Clegg, 1987). These findings of Japan's comparative advantage in exports of high-technology manufactured goods and international production in goods of low technology and low capital intensity that require high levels of managerial skills confirm at least during this period the normative conclusions of Kojima's theory.[13] During this period, the marked contrast in the type of production transferred abroad by typically small

Japanese firms in standardized product and labour intensive industries and those with production located in Japan for export is a reflection of the reorganization of the location for production for different industries in line with Japan's changing comparative advantage.

Phase II   The Ricardo-Hicksian trap stage of multinationalism in non-differentiated Smithian industries

The second stage of outward FDI was associated with the substantial domestic expansion of heavy and chemical industries at the end of the 1960s which demanded natural resources that Japan lacked. The expansion of such domestic industries turned Japan, one of the world's most resource-scarce countries, into one of the highest natural resource consuming nations. In this respect, trade continued to be assigned the primary role by Japanese industrial policy with outward FDI fulfilling a complementary or supportive role in respect of trade (Kojima 1973, 1978; Ozawa, 1979a, 1979b; Kojima and Ozawa, 1984). This explains why Japan became one of the world's leading importer of many resources, accounting for about 40 per cent of the world's imports of iron ore, 25 per cent of the world's imports of coal and copper ores, 20 per cent of the world's imports of bauxite and 10 per cent of the world's imports of timber based on data contained in MITI (1982). These imports were financed by the growing exports of the heavy and chemical industries. Exports were also regarded as an important means to achieve scale economies and to reduce production costs – the important elements of competitiveness in these industries.

However, despite the emphasis placed on trade as an instrument of Japanese industrial policy, there were factors that threatened its growth. The first factor was that market failure of various forms implied that trade based upon arms-length transactions cannot always be relied upon to secure the vital natural resources and raw materials from abroad to support the expansion of domestic resource-consuming industries. The second factor was that the very growth of these heavy and chemical industries led to the shortage of industrial space and brought forth problems of pollution, congestion and ecological destruction. The oil crisis of the 1970s served to exacerbate the situation. Drawing on the analytical insights of Ricardo and Hicks in explaining the irremovable scarcity of land or labour as constraints in modern economic growth, Ozawa (1991) regarded this phase of Japanese FDI as a response to the Ricardo-Hicksian trap of industrialism. The resource scarcity of Japan, the inadequacy of imports to secure the natural resources from abroad to sustain domestic industrialization, the shortage of industrial space and the oil crisis of the 1970s which aggravated the comparative disadvantage of Japan in resource intensive and pollution-prone industries also constrained the economic growth of Japan.

Typical of the behaviour of Japanese firms throughout the course of its history, the response of Japanese firms to the economic difficulties of the time consisted of outward FDI as a short-term solution combined with a continuing process of domestic industrial upgrading over the long term. Thus, while Japanese outward FDI continued to emphasize on trading to support the comparative advantage of Japan in the production of heavy and chemical products, outward FDI was also used to finance resource extraction to secure stable supplies of industrial raw materials (minerals, oil and other natural resources) in foreign countries (Ozawa, 1977, 1982). Thus, as much as 41.6 per cent of the total value of Japanese outward FDI at the end of March 1980 was in the tertiary (mainly trade-related commerce) sector and 23.1 per cent was in the primary sector (of which 2.6 per cent was in agriculture, fishery and forestry and 20.5 per cent in mining) (Ozawa, 1985). In addition, Japanese FDI in manufacturing continued to grow with the transfer overseas of some of the resource intensive and often pollution-prone industries. The demand for intermediate industrial goods associated with the rapid industrialization in the newly industrialized countries especially South Korea, Hong Kong, Singapore, Taiwan and Brazil provided an additional incentive to FDI by Japanese capital goods producers. In sum, Japanese outward FDI in this phase was an important means to overcome the Ricardo-Hicksian trap of industrialization and economic growth.

At the same time, the process of domestic industrial restructuring in Japan proceeded with new emphasis placed on the development of domestic industries that consumed less natural resources, were more environment friendly and had higher knowledge intensities. Assembly based and R&D intensive industries that produced higher value added goods as electronic goods, cars and machine tools began to emerge and rapidly became the 'sunrise' industries, while the conventional heavy and chemical industries began to recede as the 'sunset' industries (Ozawa, 1991).

Phase III   *The export substituting-cum-surplus recycling stage of Japanese multinationalism in differentiated Smithian industries*

The growing competitiveness of Japanese firms in assembly-based, fabricating industries of high R&D intensities since the late 1960s derive from a pyramidal and multi-layered system of subcontracting that typically consisted of around a dozen major assemblers at the top that are served by a primary (first tier) subcontractors which are in turn supported by their own cohorts of secondary (second tier) subcontractors which in turn subcontract work to the next rung of subcontractors and so on. In this manner, a diverse range of input items are produced under the most cost-effective conditions using different factor intensities and wage structures. This subcontracting system became a important source of static and dynamic advantages to Japanese firms in the modern industries along with the widespread adoption of organizational technology (the just-in-time inventory system and total quality

control); the fusion of imported and domestic R&D in new composite technologies; the feedback from domestic consumers that have exacting standards
of performance, design and quality; and vigorous inter-firm competition
(Ozawa, 1991).

Although the period since the late 1960s describes the significant increase of
Japan's share of world exports of manufactures, Japan improved her revealed
comparative advantage in international export markets the most in electrical
equipment and motor vehicles between 1974 and 1982, while her export
position in the older industries of chemicals, shipbuilding and textiles declined
(Dunning and Cantwell, 1991). The substantial growth of Japanese exports
of new assembly-based consumer durables had two important consequences.
The first consequence was the rising tide of protectionism in export markets
with the implementation of trade barriers on final goods combined with
policies enforcing local content requirements. There was the imposition by
the United States of trade barriers in the form of tariffs such as countervailing
duties and anti-dumping duties as well as non-tariff barriers in the form of
voluntary export restraints to restrain the growth of American imports from
Japan and the Asian newly industrialized countries. In addition to imposing
these trade barriers, the United States insisted on a 'levelling of the playing
field' by placing pressures on Japan to liberalize its domestic markets and
practice fair trade.[14] In a wide-ranging move to attempt to correct its trade
deficits with its major trading partners, the United States initiated action
against alleged dumping practices by Japanese firms in order to curb the
growth of its exports and the United States Congress passed the Omnibus
Trade and Competitiveness Act in 1988. Section 301 of this legislation
required the United States Trade Representative to identify those foreign
countries that have erected systematic barriers against exports of the
United States and to launch mandatory investigations against every identified trade barrier and unfair trade practice. Although this law expired in
1990, the United States persisted with unilateral trade restrictions (Strange,
1993). Thus in the midst of rising foreign demand for sub-compact cars
associated with the oil crises and oil price hikes was also the rising clamour
against the rapid and growing export penetration by Japanese firms of
Western countries' markets. Indeed, for the first time in the history of
Japanese MNCs, tariffs and other trade barriers became the major factor
precipitating the shift from exports to international production. In common
with the protectionist trend starting from the 1880s which spurred the growth
of American, British and German manufacturing MNCs, the growth of 'new
protectionism' from the 1970s stimulated the growth of Japanese manufacturing FDI in new industries (Jones, 1996).

The second consequence was financial in nature and stemmed from the
accumulation of large surpluses created out of domestic savings and net
exports as well as the sharp appreciation of the yen.[15] In turn, counteracting
measures to deal with the recessionary impact of the rapid appreciation of the
yen through the injection of liquidity into the economy by the Bank of Japan

and the large fiscal expenditures of the government resulted in the emergence of a 'bubble' economy in which asset and land prices increased enormously. This exacerbated the wealth effect created by the yen appreciation and provided a further financial incentive for the growth of outward FDI.[16] These two consequences which were mutually reinforcing made international production a more important means to service major foreign markets financed by large cash reserves of Japanese banks and financial institutions accumulated from high levels of domestic savings and net exports.

Thus, unlike international production in the first two phases which were determined primarily by the loss of comparative advantage of Japan as a location of production for traditional labour intensive light manufacturing industries as well as a site for heavy and chemical industries, international production in the third phase was not determined by a loss of comparative advantage of Japan in assembly-based, R&D intensive industries. Indeed in the third industrialization phase, the first best option for Japanese firms was to retain production in Japan and export to overseas markets. However, the consequences of rapid export expansion outlined above served to coerce Japanese companies to engage in international production that were otherwise reluctant owing to the continuing comparative advantages of Japan as a location of production. Besides the threat of protectionism and the need to recycle capital surpluses, international production enabled Japanese manufacturing companies to fulfil their objective of participating in oligopolistic competition on the basis of well-differentiated products and good marketing techniques (Baba, 1987).

The expansion of Japanese manufacturing MNCs abroad had been led by the electronics companies in the 1970s, followed by the car companies in the 1980s and semiconductor manufacturers from the end of the 1980s (Yoffie, 1993). Their international production were mainly in the form of knock-down assembly of cars and electrical/electronic goods in mass production and the partial production of components. This explains the second boom of Japanese FDI in the era since the Second World War which occurred between the end of the 1970s and the end of the 1980s.

Associated with the rapid industrial transformation of Japanese FDI in the third phase was the geographical shift in the late 1970s and through the 1980s away from a concentration in developing Asian countries towards the advanced, industrialized countries, particularly in the United States and Europe (Ozawa, 1985; Franko, 1983). Indeed, while Asia accounted for a share of 30 per cent of Japanese cumulative net outward FDI as of 1981 reflecting the hollowing out of labour intensive industries as well as some heavy and chemical industries of Japan and the need to secure vital natural resources and raw materials, such share almost halved by 1994 as the United and Europe became the the principal host regions for Japanese FDI. These developed regions became critical locations for recycling and safeguarding excess Japanese funds in the form of both portfolio and FDI. The latter was manifested in the rapid expansion of Japanese overseas manufacturing investments in electrical equipment and motor vehicles industries that were at the core of the technological strengths of Japanese firms, and it had been in these industries that their involvement and market shares in the United States and Europe have been expanding fastest (see Dunning and Cantwell, 1991; Lamoriello, 1992). The imposition by the United States of voluntary export constraints and orderly marketing arrangements for Japanese exports of textiles, cars and electronics in the late 1970s led to the sudden upsurge of Japanese FDI in these industries (Encarnation, 1992). In addition, the initial surge of outward FDI in Europe particularly in the period between the mid-1980s and the formation of the Single European Market in 1993 can be considered a response to the threat of protectionism to non-EU-based firms associated with the formation of the Single European Market in 1993 (Heitger and Stehn, 1990). A further stimulus to the growth of Japanese FDI in Europe has been the attraction of the unified large and prosperous market which provided an economically justifiable basis for local assembly since local content requirements though strongly voiced were less stringently applied, at least initially (Ozawa, 1991; Strange, 1993). In addition, the expansion of Japanese FDI in Europe while enabling the exploitation of ownership-specific advantages of Japanese MNCs in a region of the Triad and supporting the industrial restructuring and technological upgrading of the Japanese economy is also enhancing the globalization strategies of Japanese MNCs (Nicolaides and Thomsen, 1991; Thomsen, 1993).

The electrical equipment and transport equipment industries accounted for no less than 64 per cent of the total sales of Japanese companies in the European Community in 1986 deriving from both exports and international production (Dunning and Cantwell, 1991). A unique feature of the Japanese participation in Europe is their orientation towards the integrated European market as a whole and not to a particular national market in which local production is undertaken thus reflecting the pursuit of regional investment strategies by Japanese MNCs as part of their globalization strategy (see also Mason, 1992).

Although the establishment of assembly-based, subcontracting-dependent Japanese industries in the United States and Europe have concentrated at least initially on final stage assembly, this has been accompanied over time by the upstream production of intermediate goods. Thus, slowly but surely, the well-entrenched industrial system of subcontracting in research-based Japanese industries had been transplanted abroad increasingly owing to the high local content requirements in the United States and Europe. This process has been and is being accomplished either through the FDI of their parts suppliers or by sourcing from Spain and Portugal where a vertical division of labour thorough subcontracting can be implemented. In addition, the rapid introduction and adoption of Japanese just-in-time inventory and quality assurance practices in Europe, especially in the United Kingdom, are helping to create an encouraging investment environment for Japanese parts makers (Ozawa, 1991).[17] In so doing, Japanese MNCs in Europe have progressed rapidly from the pursuit of simple globalization strategies towards global localization strategies (Morris, 1991).

Thus, in contrast to Kojima's theory, the pattern of more recent Japanese FDI resemble more closely the American type of FDI in industry and geographical composition and firm behaviour (Roemer, 1976; Sekiguchi and Krause, 1980). The analysis of the trade development of Japan since the Second World War does not lend support to the static theory of comparative advantage which half a century ago would have suggested that Japan should concentrate in the production of goods for which the country is comparatively advantaged (such as labour intensive goods) and import those for which Japan has a comparative disadvantage (such as capital intensive goods). Indeed, the increasing technological competitiveness and trade surplus of Japan in technologically intensive products provides support for the view that the industries in which a country enjoys the greatest potential for innovation and in which investment can be most beneficial are not necessarily those in which a country has a current comparative advantage (Pasinetti, 1981). Major new investments by Japanese companies in the electronics components industry and the diversification into new product lines as well as efforts to sustain the development of new technologies associated with the growth industries show that Japan is seeking to continually build an industrial structure based on industries in which the country has currently no comparative advantage (see also Dunning, 1986b).

Notwithstanding the dominance of North America and Europe in the current phase of Japanese MNC expansion, the developing countries of Asia remained important to Japanese MNCs and received the third-largest share of Japanese FDI flows since 1984. While the recession in Europe and Japan were contributory factors, the major determinants of Japanese FDI in Asia have been the region's robust economic growth, low unit labour costs and trade and FDI liberalization and pro-FDI policy. Indeed, Japanese FDI in Asia has undergone geographical and industrial shifts since the mid-1980s away from the Asian newly industrialized economies that have experienced rapid wage increases and local currency appreciation towards member states of the Association of South East Asian Nations (ASEAN), and to China and other Asian countries that have become more cost competitive locations for Japanese FDI more recently. In addition, the previous importance of heavy and chemical industries such as chemical products, general machinery and transport machinery in the newly industrialized countries as well as transport machinery, iron and steel and textiles in ASEAN as the significant industries of Japanese FDI spawned during the second phase of Japanese MNC expansion has given way since the mid-1980s to the growth in importance of electrical machinery (including electronics) and motor vehicle production geared as much to local sales as the need to export to Japan and third countries (Urata, 1998). As in their investments in North America and Europe, Japanese MNCs have successfully implemented a regionally integrated production network in the electrical and electronics industry as well as motor vehicle industry in Asia with a fairly sophisticated regional division of labour and decentralized control (Kim, 1996).[18]

Despite the recurrently important role of outward FDI as a short-term solution to the threat of export growth, the long-term process of industrial restructuring in new growth industries proceeds towards the fourth phase. This is based on the development of new growth industries based on the domestic R&D of 'next generation' technologies that combine mechanics and electronics (mechatronics). It is envisaged that in this phase industrial goods would be more customer-tailored with a high content of soft technologies, while consumer goods would be increasingly designed to appeal to ever changing tastes of high-income consumers. These goods will be produced increasingly in factories run by highly skilled labour using computer-aided design, computer-aided manufacturing and computer-aided engineering and with small-scale and flexible manufacturing methods using numerically controlled machine tools and robots (artificial intelligence) (Ozawa, 1991). This may later give rise to a fourth phase of Japanese based multinationalism based on robotics and new materials.

Conclusion

This chapter devoted to the history of Japanese MNCs showed that although Japan only became a significant source of outward FDI in the world economy since 1960 and was the fourth-largest home country of FDI after the United States, United Kingdom and Germany in 1998, their earliest emergence can be traced to the late nineteenth century in much the same way as German MNCs. The early emergence of Japanese MNCs can be explained in the context of their important role in sustaining the economic development of Japan in which trade had always been regarded as the engine of growth. This explains the predominance of emergent Japanese FDI in the services sector, particularly in trading and in complementary services in support of trade such as banking, marine insurance and transportation equipment such as shipping and railroads prior to 1914 and through the inter-war period. Nevertheless, the foundations of Japanese FDI in manufacturing and resource extraction had already been laid prior to 1914.

The period since the Second World War was associated with the rapid domestic industrial restructuring of the Japanese economy and the expansion of Japanese FDI in manufacturing. Four phases or waves in the growth of Japanese cross-border production corresponded closely with the industrialization process of the Japanese economy in the period since the Second World War, and shifting patterns of trade competitiveness and sectoral resource allocation have been the dominant features of such process. These phases were the 'elementary' stage of Japanese offshore production in Heckscher-Ohlin industries; the Ricardo-Hicksian trap stage of Japanese multinationalism associated with the non-differentiated Smithian industries; the export substituting-cum-surplus recycling stage of Japanese multinationalism associated with the differentiated Smithian industries; and the robotics and new materials stage of Japanese multinationalism associated with the Schumpeterian industries.

The early emergence of Japanese MNCs at a period in which the Japanese economy was still relatively undeveloped and the predominant role of simple, labour intensive and technologically standardized industries in their early FDI in manufacturing directed towards host developing countries is a feature that has not been observed in the emergence of American, Swedish, British or German MNCs. Nevertheless, the rapid process of catching up of Japanese MNCs since the Second World War have made the pattern of more recent Japanese FDI resemble more closely the American type of FDI in industry and geographical composition and firm behaviour. The growth of Japanese MNCs is thus in many respects *sui generis* in the growth of modern MNCs.

Notes

1 Based on data provided in UNCTAD (1999).
2 Coal eventually became inadequate for Japan's own needs, and hence this commodity had to be imported in later years (Wilkins, 1986b).
3 Yet, an important distinction exists between the East India companies of the seventeenth century and the Japanese trading firms: the resilience of the Japanese trading firms. These firms have continued to survive in modern times as viable and important economic entities, taking advantage of the modern innovations in transportation and communication (Wilkins, 1986b).
4 Based on data contained in Wilkins (1986b).
5 The government was involved both directly and indirectly in the growth of international business of Japan. It was directly involved in Nippon Yusen Kaisha (shipping), the Yokohama Specie Bank and the South Manchurian Railway (Wilkins, 1986b).
6 Prior to 1911, China was a more important source of raw cotton imports to Japan compared to the United States. This changed after 1912 when raw cotton imports from the United States grew rapidly. But the most important source of raw cotton to Japan was India (Wilkins, 1986b).
7 There were 14 Japanese trading companies that had branches in New York as early as 1881 (Mitsui & Co., 1977).
8 By 1913, the British were still the more significant foreign investor in the industry in China even in spinning where the overwhelming proportion of Japanese FDI lay. British-owned spinning mills accounted for 138,000 spindles in China compared to 111,900 spindles owned by Japanese spinning mills (Wilkins, 1986b).
9 The flour mill of the Mitsui & Co. in Shanghai is regarded as the oldest flour mill in China (Remer, 1933).
10 In the period prior to 1914 and in the inter-war period international production by firms in the Japanese cotton textile industry – the main industry of Japanese FDI in manufacturing – was determined by the emergence of foreign and domestic competition in China, the major export market. While this led to declining advantages of Japan as a location of production in the cotton textiles industry, Japan continued to maintain a comparative advantage in labour intensive production until the end of the 1960s because of an abundant domestic labour force of relatively low cost.

11 Although inward FDI had a smaller role to play in Japanese industrial development owing largely to government restrictions, the absolute low levels of inward FDI since the Second World War has been biased towards technology intensive industries such as chemicals and allied products, mechanical and instrument engineering and electrical engineering. Foreign firms were channelled towards the chemicals industry in the 1950s and 1960s to harness Japanese competence in this heavy supplier industry (Clegg, 1987). Through the employment of advanced methods, the provision of valuable knowledge about western technologies and management practices, the stimulus to the growth of intermediate supplier industries, the provision of employment and training and skill development, the association with the *zaibatsu* and in raising export levels of Japan, foreign firms engaged in inward FDI in Japan influenced the development of major Japanese business enterprises and Japanese MNCs (Mason, 1987).

12 The share of technology import payments to R&D expenditures rose throughout the 1950s, reaching a peak of 18.5 per cent in 1960 but has been on the decline since owing to the growing importance of indigenous R&D efforts in relation to imported technologies (Ozawa, 1985).

13 See Chapter 1 for the elaboration of Kojima's theory.

14 This is owing to the allegation by the United States of adverse trading practices exercised by Japan. Thus, although Japan has enjoyed relatively free access to the United States market, it was accused (and to a large extent is still being accused) of restricting the imports of American goods and services through import tariffs and non-tariff barriers such as licensing regulations, technical standards, export subsidies and restricted access to foreign participation in the banking, insurance and securities industries (Chia, 1989).

15 Indeed, by comparison to the annual average net exports of goods and services of 791 billion yen in 1968 and 1969, the annual average net exports of goods and services reached 1,029.6 billion yen between 1970 and 1979, 6,425.9 billion yen between 1980 and 1989 and 7,231.9 billion yen between 1990 and 1997. Thus, the annual average foreign exchange reserves increased from $2.4 billion between 1968 and 1969, $14.5 billion between 1970 and 1979, $41.2 billion between 1980 and 1989 and $123.1 billion between 1990 and 1997. In addition, from an annual average exchange rate of 360 yen per United States dollar in 1968 and 1969, the yen appreciated considerably *vis-à-vis* the United States dollar since to reach an annual average exchange rate of 286.91 yen per dollar between 1970 and 1979, 198.92 yen per dollar between 1980 and 1989 and 117.92 yen per dollar between 1990 and 1997. Based on data contained in the International Monetary Fund, *International Financial Statistics Yearbook, 1998*, Washington, DC: IMF.

16 This also led Redies (1990) to argue that the expansion of Japanese FDI in recent decades was an exercise in financial power.

17 In particular, Unipart, the auto component manufacturer in the United Kingdom, has been at the forefront in embracing Japanese-style production and management methods in Europe (Randerson and Dent, 1996).

18 For further analysis of the regional networks of TNCs established by Japanese MNCs in East Asia, see UNCTC (1991).

10 The emergence and evolution of multinational corporations from Taiwan

Introduction

By comparison to the long and rich history of MNCs based in the developed countries, the history of Taiwanese MNCs whose origins can be traced to the late 1950s and early 1960s is much more recent and contemporary. Taiwan accounted for some 0.7 per cent of the stock of outward FDI worldwide in 1990, a share equivalent to that of Japan in 1960. The stock of Taiwanese FDI which reached an estimated $38 billion in 1998 represented an annual rate of growth of more than 39 per cent since 1980, thus making Taiwan one of the most rapidly growing home countries of FDI in developing countries, if not the whole world. Despite the rapid growth, however, the share of Taiwan in the global stock of outward FDI in 1998 remained low at 0.9 per cent. Nevertheless in relation to the stock of outward FDI from developing countries in the same year, Taiwan assumed greater relative importance with a share of 9.7 per cent.[1]

Despite the low relative importance of Taiwanese FDI, the study of the emergence and evolution of Taiwanese MNCs is of interest not the least because Taiwan, like the United Kingdom, Germany, Japan and South Korea, is a resource-scarce country with a large domestic market. Thus, although the history of Taiwanese MNCs may be of more recent vintage and is in many respects still in the stage of emergence, the growth pattern of their MNCs as it has been evolving over the last 40 years can be compared to the growth pattern of MNCs from other countries that share similar patterns of national economic development. The analysis of the history of Taiwanese MNCs in this chapter is divided into four time frames: the 1950s and 1960s, the 1970s, the 1980s and the 1990s.

The emergence of Taiwanese MNCs in the 1950s and 1960s

The origins of Taiwanese FDI can be traced to the late 1950s and early 1960s. Such period of emergence of Taiwanese MNCs which took place in an era of significant economic growth of Taiwan also coincided with the adoption of an export-led development strategy distinct from the import substituting industrialization strategy of the 1950s.[2] Government guidelines on outward FDI in the 1960s and 1970s was designed to promote the sales of domestic products, to make available raw materials required by domestic industries, to help facilitate the export of technical know-how that may increase foreign exchange earnings, and to promote international economic cooperation (Schive and Hsueh, 1985).

The earliest official outward FDI on record was that of a local cement company that invested in a cement plant in Malaysia in 1959 with $100,000 worth of machinery (Ting and Schive, 1981; Chen, 1986).[3] In 1962, an investment of $492,000 was made by a jute-bag manufacturer which established a plant in Thailand (Schive and Hsueh, 1985; Ting and Schive, 1981). The size of approved FDI grew rapidly since. In 1963, some $1.4 million of Taiwanese FDI was approved and the investments continued to grow throughout the rest of the 1960s. The major recipients of Taiwanese FDI from 1959 to 1969 were Thailand, Malaysia and Singapore which accounted for 53 per cent of approved Taiwanese FDI over the period, with countries other than Thailand, Malaysia, Singapore, Philippines, Indonesia and the United States accounting for the bulk of the remaining proportion.[4] The annual average approved FDI flows from 1959 to 1969 was around $690,000.

At the time of the emergence of Taiwanese MNCs, Taiwan had a comparative advantage in low-cost, labour intensive production – an advantage that it continued to sustain for 30 years until the late 1980s. This helps to explain why unlike in the case of Japan the pioneering local manufacturing industries which ventured overseas during the 1960s and 1970s was not mainly by firms in the small-scale manufacture of toys and textiles in which Taiwan had a comparative advantage, but rather by larger firms concentrated in heavy, natural resource intensive manufacturing industries such as pulp and paper, cement, rubber, food and drink as well as chemicals and cable and wire (Bamford, 1993; Schive and Hsueh, 1985).[5] The latter set of firms had the capital to fulfil their investment objective to gain access to raw materials and cheap land in Thailand, Singapore, Malaysia, Philippines and other countries (Bamford, 1993). Manufacturing was thus the most important sector of the early Taiwanese FDI, while outward FDI in the primary and tertiary sectors were far less important. Thus, in comparison to the history of MNCs from the United Kingdom, Germany and Japan, resource extractive investments abroad to support domestic industrial expansion of the home country was not a prominent feature in the history of MNCs from Taiwan where instead emergent FDI consisted of the relocation of natural resource manufacturing industries to neighbouring resource-rich host countries.

Apart from the impetus to outward FDI that government guidelines provided, the early emergence of Taiwanese FDI at a time when the Taiwanese economy was still largely undeveloped and a recipient of substantial amounts of economic aid from the United States can be explained in part by the presence of numbers of overseas Chinese businessmen in South East Asian countries that shared an ethnic heritage with those of Taiwan and which often took the initiative in approaching businessmen in Taiwan or Hong Kong to undertake investments in the region (Wells, 1983). In fact, some 89 per cent of approved Taiwanese FDI before 1976 was aimed at South East Asian countries where the largest number of overseas Chinese reside (Schive and Hsueh, 1985). The powerful alliances formed by the Chinese business community in different countries of Asia which provided business information as well as financial and marketing assistance enabled the Chinese to play an economic hegemonic role in the business and commerce of South East Asia. The emergence and growth of Taiwanese FDI in South East Asia thus served to reinforce that hegemonic role.

The growth of Taiwanese MNCs in the 1970s

While the annual average approved FDI flows from 1959 to 1969 was around $690,000, the annual average approved FDI flows from 1970 to 1979 was around $5.2 million.[6] As a result, the stock of approved outward FDI in 1969 of $7.6 million grew at an annual average rate of 23 per cent to reach some $59.3 million in 1979. Indeed, major investments abroad by Taiwanese firms began in the 1970s (UN, TCMD, 1993a). By 1979, the member states of the Association of South East Asian Nations (ASEAN) consisting of Thailand, Malaysia, Singapore, Philippines and Indonesia accounted for 52 per cent of the stock of approved outward FDI of Taiwan, while the United States accounted for 15 per cent and other countries for 33 per cent. The geographical distribution of the stock of approved Taiwanese FDI in 1979 thus differed considerably from that in 1969 in two main respects. First, although the member states of the ASEAN accounted for some 56 per cent of the stock of approved Taiwanese FDI in 1969 and some 52 per cent in 1979, the relative importance of host countries within ASEAN differed markedly between 1969 and 1979. Thus, while the most important host countries for Taiwanese FDI in ASEAN in 1969 was Thailand (which accounted for 28 per cent of the total stock of approved Taiwanese FDI), followed far behind by Singapore (13 per cent), Malaysia (12 per cent) and Philippines (3 per cent), the most important host countries in 1979 was the Philippines (17 per cent) and Indonesia (15 per cent), followed far behind by Thailand (8 per cent), Singapore (7 per cent) and Malaysia (5 per cent). Second, the 1970s marked the entry of Taiwanese firms in developed countries, particularly the United States. The investment in developed countries in the 1970s by Tatung, one of Taiwan's largest electrical equipment manufacturers, serves to mark Taiwan's first true investment venture in the developed countries (Chen, 1986).

In order of importance, the five most important industries of Taiwanese FDI in 1979 were plastic and plastic materials (24 per cent), textiles (13 per cent), food and beverages (13 per cent), non-metallic minerals (11 per cent) and electronics and electrical appliances (8 per cent). Thus manufacturing continued to be the most important sector of economic activity of Taiwanese MNCs in the 1970s with a share of 78 per cent of the stock of approved Taiwanese FDI in 1979, followed far behind by the services sector (14 per cent) and the primary sector (8 per cent).

Unlike the early forms of Taiwanese FDI in South East Asia in the 1960s which were motivated by the need to gain access to raw materials and cheap land, their outward FDI in other parts of the world since the 1970s was motivated more by the need to overcome limited opportunities for growth in the home country by either seeking new markets abroad or establishing foreign market positions *inter alia* through the promotion of brand products or maintaining access to host country as well as third country markets in the midst of rising international protectionism (Schive and Hsueh, 1985; Ting and Schive, 1981; Chen, 1986). Protectionism in foreign markets has come in several guises, the most severe being the import quota. By avoiding import quota restrictions through off-shore production in either the foreign country imposing the quota or in third countries where the import quota restrictions did not apply or were not utilized fully, many indigenous Taiwanese firms became MNCs. One of the most important of these import quotas was that imposed under the Multi-Fibre Agreement which became more restrictive in the 1970s and curbed the export expansion of the Asian NICs, including Taiwan. Investments offshore in response to this were undertaken by Taiwanese firms in the member states of the ASEAN, South Asia, Central America and the Caribbean (Chia, 1989).

Other important objectives of Taiwanese FDI were the desire to export indigenous technology abroad (of which the most important was technology embodied in machinery), the enhancement of the firm's image in the home market and to keep pace with other firms based in the home country that have already invested abroad through a similar pursuit of an imitative, follow-the-leader type of investment behaviour (Schive and Hsueh, 1985). Image enhancement can be considered a means of product differentiation (Caves, 1971), while the follow-the-leader investments are indicative of the behaviour of firms in oligopolistic industries (see Knickerbocker, 1973). Indeed, Taiwanese MNCs were large firms relative to the other firms in their respective industries. The largest companies in all industries – with the exception of those in timber and bamboo products, metals and construction – became MNCs at one stage or another. Moreover, all Taiwanese MNCs but five were among the top 500 firms in Taiwan in 1980 (Schive and Hsueh, 1985). The main competitive advantages of Taiwanese MNCs *vis-à-vis* MNCs from developed countries in host countries in order of declining importance were the provision of products better suited to host market, lower price of product, better marketing, greater adaptability to local environment, use of easier-to-operate equipment and lower operating costs (Schive, 1982 as quoted in Schive and Hsueh, 1985).

As in the 1960s, the least important motivations for Taiwanese FDI in the 1970s was access to cheap raw materials, access to cheap labour and diversification of business risks (Schive and Hsueh, 1985). Despite uncertainties over access to resources in the 1970s which had spurred investments abroad in resource-rich countries of South East Asia, Australia, Canada and Latin America to develop and process the products of agriculture, forestry and mining, the investments were very small, particularly in comparison to Japan and other resource-scarce large countries such as the United Kingdom, Germany and even South Korea. This is due in part to the resource nationalism of host countries in the 1970s which placed restrictive conditions on foreign extractive investments, and in part to the large financial requirements of forestry and mining projects (Chia, 1989).

Thus, there were five main types of investments undertaken by Taiwanese MNCs in the decade of the 1970s in order of declining importance: first, to supply host country markets in developing countries; second, to facilitate growth of Taiwanese exports through export platform investments in developing countries; third, to promote growth of Taiwanese exports through outward FDI in trading, sales, distribution and marketing in major foreign markets; and fourth, to secure supplies of essential raw materials and fifth, to supply host country markets in developed countries and to gain access to advanced technologies. Each of these types of Taiwanese FDI is discussed below.

Import substituting FDI in developing countries

Considering Taiwanese FDI in such manufacturing industries as food and drink, synthetic fibre processing, paper, plastic and plastic materials, non-metallic minerals, primary metals and machinery as well as construction, finance, and other services oriented to supply the domestic markets of host developing countries, import substituting FDI by Taiwan in developing countries accounted for an overwhelmingly large share of at least 56 per cent of the stock of approved Taiwanese FDI in 1979.[7]

Since the emergence of Taiwanese FDI in developing countries and throughout the 1970s, the predominant motivations of Taiwanese MNCs in South East Asia was the search for markets, security of export markets and raw material supplies and access to cheaper energy and land costs. This was confirmed in the formal analysis of Joe (1990) of the various determinants of the level of Taiwanese FDI in manufacturing in Thailand, Malaysia and Singapore in the period between 1970 and 1985 which showed that other location factors in the host country such as wage rates and skills, the level of exports of Taiwan to the host country and Taiwan's balance of trade were not significant explanatory factors of Taiwanese FDI.

Taiwanese firms involved in import substituting FDI in developing countries have the common feature of having a long production experience and the requisite technology in their respective industries to compete with local firms and other MNC subsidiaries in host developing countries. For example, the MNCs in the monosodium glutamate (food seasoning) and cement industries were already established by the 1950s and the MNCs in the polyvinyl chloride (PVC) and polyethylene (PE) plastics industries were well developed in the late 1950s and late 1960s, respectively.[8] The synthetic fibre industry also experienced rapid growth in the late 1960s and grew to an extent that enabled the industry to supply up to 90 per cent of the domestic market demand of Taiwan (Ting and Schive, 1981).

As for geographical destination, import substituting FDI in food and drink in developing countries was directed to countries other than the member states of ASEAN and the United States (57 per cent), Indonesia (20 per cent) and Thailand (19 per cent), while those of synthetic fibre processing and, in particular, that of the Tuntex Fiber Company was directed mainly towards Indonesia to avail of its large oil reserves from which to derive the petrochemical-based polyester chips for the production of synthetic fibres. Outward FDI in paper processing was similarly directed solely to Indonesia to avail of its rich timber resources. Outward FDI in plastic and plastic materials was directed overwhelmingly to the Philippines (68 per cent), as was FDI in non-metallic minerals which were directed towards countries other than the member states of ASEAN and the United States (76 per cent). Taiwanese FDI in the primary metal and machinery industries was more evenly distributed among Thailand (34 per cent), countries other than the member states of ASEAN and the United States (26 per cent), and Malaysia (14 per cent), while those of construction, finance, and other services were directed towards countries other than the member states of ASEAN and the United States (45 per cent) and the United States (37 per cent).

Taiwanese FDI in the member states of ASEAN countries shared several peculiar features. First, a considerable amount of Taiwanese FDI was made without the transfer of capital funds. Instead, the form of investment of Taiwanese MNCs typically took the form of the provision of either second-hand or locally manufactured machinery to their overseas subsidiaries. In certain cases, capitalized patents and certain technological know-how were also used as capital for the investments. These forms of investments were most prevalent in the metal products, chemicals and food and drink processing industries (Schive and Hsueh, 1985).

Export platform investments in developing countries

Considering Taiwanese FDI in developing countries in such manufacturing industries as textiles (other than synthetic fibre processing), clothing and footwear and electronics and electrical equipment geared to supply export markets, export platform FDI in developing countries accounted for as much as 19 per cent of the stock of approved Taiwanese FDI in 1979.

As for geographical destination, export platform FDI in textiles in developing countries was directed largely to Indonesia (51 per cent) and to countries other than the member states of ASEAN and the United States (25 per cent), while FDI in garments and footwear was directed overwhelmingly to Singapore (75 per cent). Finally, export-oriented FDI in electronics and electrical appliances was directed to Singapore (38 per cent) and Thailand (33 per cent).

Export-oriented FDI in trading, sales, distribution and marketing and service

This type of investment, which was aimed primarily to facilitate the growth of manufactured exports of Taiwan in major export markets, accounted for as much as 11 per cent of the stock of approved Taiwanese FDI in 1979. Thus, since the United States was the major export market of Taiwan absorbing some 40 per cent of Taiwan's total exports in the 1970s, export-promoting FDI in trade, sales, distribution and marketing and service facilities was most considerable in that country which received close to 80 per cent of approved Taiwanese FDI of this type in 1979. It was often the large exporting manufacturers which were most keen to set up trading offices in their major export markets. For example, Tatung, the largest Taiwanese electronics and electrical appliance producer already mentioned, established marketing affiliates in the United States and Singapore in 1972 before initiating international production of electric fans in the United States in 1975 and of colour television sets in 1977. The firm also established another colour television plant in the United Kingdom in 1981 (UN, TCMD, 1993a). Similarly, FDI in distribution was the main priority behind the initial FDI of Sampo, Taiwan's third-largest home appliance manufacturer behind Tatung and Matshushita (Taiwan). Despite facing escalating cost structures and tremendous pressure from domestic and foreign competition, Sampo did not consider international production as a primary response but instead chose to exploit key foreign markets through the establishment of sales and distribution offices. Thus, in 1976 the firm set up Sampo Corporation of America based in Chicago as a distribution centre for its television sets. It was only in 1979 when the American government accused Sampo of illegal dumping and imposed severe penalties and quotas on the firm's exports of television sets did Sampo respond by establishing its own manufacturing base in the United States enhanced by its long-term relationship and technical assistance provided by the Japanese company, Sharp. This explains why Sampo's production facility in the United States was established in the same Atlanta location as an already existing Sharp facility (Bamford, 1993). As was the case with other Taiwanese producers in the developed countries, the production facilities of Sampo was initially in the assembly of Taiwan-made chassis shipped to Atlanta where these were combined with United States-made picture tubes and cabinets to make the final product – the television sets (Joe, 1990).

As in the case of Japan, Taiwanese FDI in trading, sales, distribution and marketing were often enhanced by complementary investments in shipping. Thus, the Taiwanese shipping company, the Evergreen Marine Corporation, founded on 1 September 1968 commenced its first liner service in August 1969 to the Middle East Gulf. In 1972, the company commenced a second liner service to the Caribbean, and a third conventional liner service linking the Red Sea and Mediterranean began in 1977 after the re-opening of the Suez Canal. The company implemented a full container service starting with shipments to the East Coast of the United States in 1975 and with shipments to the West Coast of the United States in 1976, and the conventional shipping services to other destinations followed suit between 1977 and 1979. The initiation of the Far East-North Europe service in 1979 enabled the Evergreen Marine Corporation to launch a 'round-the-world' shipping service to streamline its operations. And in 1989, the company was granted permission by the Taiwanese government to establish an international airline offering passenger and cargo services to United States, Europe and Asia starting in 1991. Tightly managed and well financed, Evergreen is one of the world's largest and most successful surface transportation companies. With an extensive network of overseas services and agents, it provides regional and around-the-world container shipping services.[9]

FDI to secure supplies of essential raw materials

Investments of this type which have been most significant for Taiwanese firms in the agricultural, forestry (including lumber and bamboo products), and fishery industries explain 8 per cent of the stock of approved Taiwanese FDI by 1979. Examples of Taiwanese MNCs of this type were the producers of tinned pineapples engaged in backward vertical integration into the cultivation of pineapples, the fishing companies, the plywood producers, the paper manufacturers, the producers of basic metal products such as Taiwan Metal Mining and Taiwan Aluminum to gain access not only to primary metals and bauxite but also to cheaper energy resources abroad, the participation of Taiwan Power Corporation in a urea plant in Saudi Arabia and its investments in coal mining in Canada, and the investments of the Formosa Plastics Group in the United States to gain access to lower-cost crude oil and its derivatives for plastics production. The natural resource-rich countries of Costa Rica, Malaysia, Indonesia and Thailand were the major host countries to these extractive ventures and, in some cases, also processing ventures (Ting and Schive, 1981), although as seen from the examples above the Middle East, Canada and the United States were also significant recipients of Taiwanese FDI of this type.

Import substituting FDI in developed countries

This type of Taiwanese FDI, which was aimed at protecting or defending access to existing markets of Taiwanese goods by supplying host country markets in developed countries through international production in the case of manufacturing or the provision of services aimed at meeting local market demand, accounted for some 6 per cent of the stock of approved Taiwanese FDI in 1979. The most substantial of this type of FDI were those of electronics and electrical appliances by firms such as Tatung and the two other largest Taiwanese electronics and electrical appliance producers that had penetrated rapidly the markets of developed countries markets through exports (Ting and Schive, 1981). Indeed, some 47 per cent of the stock of approved Taiwanese FDI in 1979 in the electronics and electrical appliances industry was directed to the United States. The establishment of production subsidiaries in the United States by these firms was a means to overcome trade barriers imposed by the United States against goods produced by Taiwanese firms in this industry owing to their rapid export expansion and the consequent growth of trade deficits of the United States *vis-à-vis* Taiwan. Indeed, Taiwanese FDI in manufacturing in the United States in the period 1970–85 was deemed to have been determined solely and significantly by the trade imbalances between Taiwan and the United States. Taiwanese FDI in the United States thus fulfilled the objective of circumventing trade protectionism and ensure continued access to the United States market (Joe, 1990). Note, however, that the manufacturing subsidiaries were geared initially as assembly points for intermediate products imported into the United States from Taiwan or third countries which were not as tightly restricted as final products. Another subsidiary objective of Taiwanese FDI in electronics in the United States was to upgrade technology by gaining access to more advanced forms of technologies being developed in the Silicon Valley, for example (World Bank, 1989).

It is important to emphasize that although in principle five different types of Taiwanese FDI can be identified in the 1970s, it was often the case that one type of FDI typically led to another type over time as in the cases of Tatung and Sampo analysed above. More commonly, Taiwanese MNCs were engaged in more than one type of FDI. This was the case, for example, of an unidentified leading Taiwanese electrical and electronics manufacturer that is a multi-divisional company with a broad range of product lines ranging from consumer household appliances, telecommunications equipment, electronics products and computers, to heavy electrical equipment and instruments, steel, machinery and chemicals. Beginning in the early 1970s, the firm established a network of manufacturing subsidiaries in Japan, Singapore, Hong Kong and the United States in addition to a global network of sales and purchasing offices in Europe, the Middle East and Africa to facilitate export sales (Ting and Schive, 1981).[10] The success it found in the domestic market appeared to have conferred in the firm the ability to invest abroad as well as engendered a need for the firm to internationalize to maintain the image of leadership and prestige among its competitors. A closely parallel reason was the reinforcement of its brand name in the international markets to which it had already been exporting. The immediate motive of circumventing tariffs and quota restrictions in the United States, in countries of the former European Community and in high-tariff developing countries were also contributory factors. This led to import substituting international production in the countries imposing the trade barriers or in export platform investment in a third country, typically a developing country that had no quotas or had excess quotas. For instance, the television plant established by the firm in Singapore was made to circumvent the quota imposed by the former European Community as well as to supply completely knocked-down units to high-tariff countries in the surrounding region (Ting and Schive, 1981).

Similar multiple objectives for outward FDI applied in the case of TECO Machinery which as an industry leader established affiliates in developed countries in 1976 to advance the marketing, sales and service of its motor vehicle products and also established affiliates in developing countries in the 1970s including in Singapore in 1976 for the purposes of manufacturing, marketing and after-sales service (Bamford, 1993).

The Taiwanese MNCs that emerged in the period until 1979 were firms with long production experience and their technologies though not the most advanced by world standards in their industries enabled these firms to initiate and sustain their investments abroad. Their ownership advantages were enhanced by the rapid industrialization of Taiwan, and its investments in formal R&D, training as well as the presence of a supportive science-and-technology infrastructure (UN, TCMD, 1993a). Thus, although 52 per cent of the stock of approved Taiwanese FDI was directed to countries of South East Asia in 1979, this started changing in 1980 when Taiwan FDI grew even more rapidly, particularly in the United States. In that year, the stock of approved Taiwanese FDI in the United States at $44 billion surpassed for the first time that in countries of South East Asia at $34 billion. The United States became the single largest host country, accounting for 43 per cent of the stock of approved Taiwanese FDI in 1980. Thus 1980 is considered to have been a turning point in Taiwan's FDI. The only ASEAN countries where Taiwanese FDI increased considerably since 1980 were Singapore and Indonesia (Esho, 1985).

The further expansion of Taiwanese FDI in the 1980s

The decade of the 1980s represented an era of phenomenal growth in Taiwanese FDI. On an approved basis, the annual average outward FDI flows of Taiwan reached $146 million between 1980 and 1989 which was 28 times as large as the annual average FDI outflows of $5.2 million in the period from 1970 to 1979 and more than 211 times as large as the annual average FDI outflows of $690,000 in the period from 1959 to 1969. In 1989, the stock of approved Taiwanese FDI at $1.5 billion represented an annual average rate of growth of more than 38 per cent since 1979 – a rate considerably faster than the annual average rate of growth of 23 per cent between 1969 and 1979. However, in terms of actual FDI outflows based on balance-of-payments data, the annual average FDI outflows of Taiwan was about $1.2 billion in the period between 1980 and 1989 compared to $2.4 million in the period between 1970 and 1979, and the estimated stock of Taiwanese FDI stood at $7.6 billion in 1989 compared to $97 million in 1980. In fact, 1988 marked the year in which Taiwan started to become a net outward investor in terms of FDI flows.[11]

A combination of political and economic factors explain the dramatic expansion of Taiwanese FDI in the 1980s. There were two political aspects, one of which involved changes in Taiwan's external relations with the rest of the world and the other involved the wide-ranging political reforms implemented by the Taiwanese government, including the liberalization of outward FDI policy and foreign exchange controls.

The externally generated political factor that encouraged the expansion of Taiwanese FDI was the severance of official diplomatic ties between Taiwan and 55 countries (including the United States and Japan) in the period between 1971 and 1988 with the official diplomatic recognition by those countries and the United Nations of the People's Republic of China. Since only a few countries maintained formal diplomatic ties with Taiwan, the government of Taiwan sought to cultivate and expand its cultural and economic links with the rest of the world as a means to ensure the survival of the Taiwan state. In this respect, external trade (including the strengthening of bilateral trade relations with foreign countries) and outward FDI were regarded as the key mediums to fulfil that objective while also helping to increase national wealth and fulfil the doctrines of San Min Chu Yi and the Kuomintang regime (Kapellas and Liu, 1990).

The other important aspect of political change which perhaps had a greater impact on increasing Taiwanese FDI outflows particularly since 1987 was the wide-ranging political reforms undertaken by the formerly authoritarian Taiwanese regime between 1986 and 1988.[12] Among the political reforms were the lifting of martial law, the easing of restrictions on foreign travel and, in an attempt to to curb the inflationary pressures presented by the large and growing foreign exchange reserves, there was the liberalization of outward FDI policy between 1986 and 1988, including the gradual liberalization of foreign exchange regulations in 1987, as well as other efforts by the government to promote outward FDI that met certain criteria.[13] These include: projects acquiring natural resources and raw materials, projects that will improve regional trade imbalances, FDI promoting the inflow of technical know-how and cooperation, and projects instrumental to the structural adjustment and upgrading of domestic industries. However, despite these wide-ranging political reforms to promote outward FDI, the absence of diplomatic ties between Taiwan and a large number of countries impeded the conclusion of reciprocal trade and investment agreements crucial to the growth and stability of trade and FDI, particularly with their most trade and investment partner countries (Chia, 1989).

The economic factors that contributed to the dramatic growth of Taiwanese FDI in the 1980s was attributable mainly to its rapid export expansion in the 1970s and 1980s which made Taiwan the world's 13th largest trading nation in 1987 (*Central Daily News*, 31 December 1987), and enabled Taiwan to accumulate a balance-of-trade surplus since 1975 to reach some $11 billion in 1988 as well as foreign exchange reserves of around $74 billion – the world's second largest after Japan.[14] This led Taiwan's major trading partners, particularly the United States, to impose protectionist trade pressures in the mid-1980s as a result of the rapid export expansion particularly of electronics and electrical appliances from Taiwan leading to the worsening of the deficits of the United States in its trade with Taiwan.

For Taiwan and South Korea, the problem was compounded by the sharp appreciation of their currencies since 1985 resulting from current account surpluses, excess liquidity and inflationary pressures. However, exacerbating the impact of economic fundamentals on local currency appreciation in Taiwan were demands by the United States that Taiwan manipulate deliberately the exchange rate to allow the New Taiwan dollar to rise.[15] As a result, the New Taiwan dollar appreciated by the 35 per cent between 1986 and 1988.

The strength of the domestic currency then provided an additional motive for outward FDI by local firms. It provides support for Aliber's theory of FDI emanating from strong currency areas as a determinant of the timing of FDI and its growth (and particularly that of foreign takeovers) as well as fluctuations of outward FDI flows around a long-term trend (Aliber, 1970). The financial explanations of the growth of FDI owing to strong domestic currencies has historical antecedents in Britain in the nineteenth century, the United States in the early post-war period and Japan and Germany in the 1970s and 1980s (Cantwell, 1989a). However, in the case of Taiwan, the financial explanations behind FDI expansion over the 1980s extended beyond the accumulation of net exports and the upward revaluation of the New Taiwan dollar. As in the case of Japan, particularly during the third phase of export-substituting-cum-surplus recycling stage of its FDI that started in the late 1960s, the accumulation of large financial surpluses of Taiwan in the 1980s was created not only from net exports and the sharp appreciation of the New Taiwan dollar but also from the escalation in acquired wealth in Taiwan during that decade brought about by the more than 30 per cent savings rates and the late 1980s boom in the Taiwan stock market. Outward FDI flows were also facilitated by the role of domestic financial services industry formed during the era of financial market liberalization in the late 1980s as a financial recycling mechanism of surplus domestic funds.

Apart from currency appreciation which affected adversely the cost competitiveness of Taiwanese exports, there were other exacerbating factors that threatened the export sales and domestic employment of Taiwan. This was the falling overseas demand and declining oil prices which served to increase further the balance-of-trade surplus of Taiwan and contribute to the rapid accumulation of its foreign exchange reserves.

In the 1980s, Taiwan responded to these mutually reinforcing political and economic forces in two ways. The first response was an attempt by the government of Taiwan ongoing since the 1970s to diversify the country's export markets away from an excessive reliance on the United States owing to both trade pressures and the severance of official diplomatic relations between the two countries. In November 1988, although the Board of Foreign Trade regarded Japan and Western Europe as the important markets in which to promote Taiwan's exports, attempts to penetrate those markets largely failed. Instead, trade with and outward FDI in the countries of ASEAN grew far more rapidly in accordance with the dictats of free market forces and without explicit impetus and guidance from the government of Taiwan (Kapellas and Liu, 1990).

The second response of Taiwan to the threat to export growth was to accelerate the growth of outward FDI in its major export market, the United States, and to a far lesser extent in Western Europe to fulfil the objectives of substituting international production for exports to meet local market demand and to gain access to more advanced forms of foreign technologies. There was also the growth of export platform FDI by Taiwan in developing countries to target the markets of the United States and Europe. Indeed, outward FDI of Taiwan over the 1980s fulfilled an important role in maintaining foreign relations and economic prosperity. This means was enabled by Taiwan's corporate wealth which allowed it to pursue outward FDI in what was in some cases regarded as 'reckless abandon' (Bamford, 1993).[16] Quite apart from the economic benefits involved, outward FDI furthers Taipei's foreign policy aims to counteract the island's diplomatic isolation.

The above factors explain why in a marked contrast to the 1960s and 1970s, the developed countries constituting the Organization for Economic Cooperation and Development (OECD) accounted for more than two-thirds of the approved stock of Taiwanese FDI by the end of the 1980s, and the developing countries of Asia accounted for only 30 per cent. The 1980s thus marked a sharp reversal in the geographical destination of Taiwanese FDI away from the previous dominance of developing countries of Asia towards the developed countries which became the far more important recipient of Taiwanese FDI.

The manufacturing sector continued to be the most important sector of Taiwanese FDI at the end of the 1980s with a share of 63 per cent of the stock of approved Taiwanese FDI, followed by services (36 per cent) and the primary sector (1 per cent). This also reflects some considerable change from that in 1979 when manufacturing accounted for 78 per cent of the stock of approved Taiwanese FDI, followed far behind by the services sector (14 per cent) and the primary sector (8 per cent). Thus, although the manufacturing sector continued to be the dominant sector of Taiwanese FDI by the end of the 1980s, the lesser importance of the sector in Taiwanese FDI in relative terms as well as that of the primary sector contrasted with the higher relative importance of the services sector in Taiwanese FDI. This reflected the growth of Taiwanese FDI in sales, distribution, marketing and after-sales services as well as in trade and banking and finance which were particularly concentrated in the United States. In manufacturing, the dominant industry of Taiwanese FDI was electronics and electrical appliances which accounted for 46 per cent of the total stock of approved Taiwanese FDI in manufacturing. The other significant manufacturing industries of Taiwanese FDI at the end of the 1980s in declining order of importance were chemicals (21 per cent), plastic and rubber products (7 per cent), paper products (6 per cent), non-metallic minerals (5 per cent), food and drink (4 per cent) and textiles (4 per cent). By comparison, as mentioned in the previous section, the five most important industries of Taiwanese FDI in 1979 in order of declining importance were plastic and plastic materials, textiles, food and drink, non-metallic minerals and electronic and electrical appliances. Thus the 1980s was also associated with significant shifts in the importance of manufacturing industries in Taiwanese FDI with the electronics and electrical appliances industry coming to the fore as the single most dominant industry of manufacturing FDI away from the more traditional manufacturing industries of plastics, textiles, food and drink, and non-metallic minerals.

The five main types of investments undertaken by Taiwanese MNCs in the decade of the 1970s continued to be relevant in the 1980s. However, the relative importance of the five main types of Taiwanese FDI differed significantly at the end of two decades. As analysed in the previous section, the five main types of Taiwanese FDI at the end of the 1970s in order of declining importance were: first, to supply host country markets in developing countries; second, to facilitate growth of Taiwanese exports through export platform investments in developing countries; third, to promote growth of Taiwanese exports through outward FDI in trading, sales, distribution and marketing in major foreign markets; and fourth, to secure supplies of essential raw materials and fifth, to supply host country markets in developed countries and to gain access to advanced technologies.

By contrast, by the end of the 1980s – a short span of ten years – the order of importance of the five main types had been sharply reversed. In order of declining importance, the five main types of Taiwanese FDI at the end of the 1980s were: first, investments in developed countries to supply local markets and to gain access to advanced technologies; second, to promote growth of Taiwanese exports through outward FDI in trading, sales, distribution and marketing in major foreign markets; third, to supply host country markets in developing countries; fourth, to facilitate growth of Taiwanese exports through export platform investments in developing countries; and fifth, to secure supplies of essential raw materials. Each of these types of Taiwanese FDI pertaining to the decade of the 1980s is discussed below.

FDI in developed countries to supply local markets and to gain access to advanced technologies

Taiwanese FDI in the developed countries in the 1980s of the type described above had three important determinants. The first determinant was to protect or retain existing export markets of Taiwanese goods by supplying host country markets in developed countries through international production in the face of protectionist trade barriers, primarily in the case of electrical and electronics products targeted to the export market of the United States.[17] Indeed, Taiwanese FDI in the developed countries over the 1980s continued to be associated primarily with the maintenance of access to markets, the avoidance of trade barriers, and the relief of trade surpluses (Joe, 1990; Chen, 1986). The second determinant of FDI of this type derive from Taiwan's interest to gain access to more sophisticated and advanced forms of manufacturing technology to support rapid industrial development (World Bank, 1989; Brody, 1986). The third determinant was to establish new markets in the developed countries by the provision of services (such as banking and finance, construction, etc.) aimed at meeting local market demand.

This type of FDI, which was the least important at the end of the 1970s with a share of some 6 per cent of the stock of approved Taiwanese FDI in 1979, became the most important type of Taiwanese FDI at the end of 1980s with a share of 40 per cent of the stock of approved Taiwanese FDI in 1989. Indeed, some 90 per cent of Taiwanese FDI in developed countries involved the manufacturing sector to fulfil the above two determinants and only 10 per cent were in services.

Import substituting FDI in the developed countries

As in the 1970s when Taiwanese FDI first emerged in the developed countries, the single most important industry of Taiwanese FDI in the developed countries in the 1980s was electronics and electrical appliances which accounted for some 62 per cent of the stock of approved Taiwanese FDI in the developed countries in 1989. Indeed, while some 47 per cent of the stock of approved Taiwanese FDI in 1979 in the electronics and electrical appliances industry was directed to the United States, that proportion had increased dramatically to 83 per cent in 1989. This owed largely to the trade barriers imposed by the United States starting in the mid-1980s as mentioned above owing to Taiwan's export boom in the 1980s coupled with the direct competition posed by Taiwanese companies with American manufacturers in certain industry segments of high technology and telecommunications (Kapellas and Liu, 1990).[18] These barriers came in the form of tariffs such as countervailing duties and anti-dumping duties as well as non-tariff barriers in the form of voluntary export restraints to restrain the growth of American imports from Japan and the Asian newly industrialized countries. In addition to imposing these trade barriers, the United States insisted on a 'levelling of the playing field' by placing pressures on these countries to liberalize their domestic markets and practise fair trade.[19] In a wide-ranging move to attempt to correct its trade deficits with its major trading partners, the United States initiated action against alleged dumping practices by Taiwanese firms in order to curb the growth of Taiwanese exports (Kapellas and Liu, 1990) and the United States Congress passed the Omnibus Trade and Competitiveness Act in 1988. Section 301 of this legislation required the United States Trade Representative to identify those foreign countries that have erected systematic barriers against exports of the United States and to launch mandatory investigations against every identified trade barrier and unfair trade practice. Although this law expired in 1990, the United States persisted with unilateral trade restrictions (Strange, 1993). In addition, with effect from January 1989 the four Asian newly industrialized countries were graduated from the Generalized System of Preferences of the United States on the grounds of having achieved a certain level of economic development and competitiveness (Chia, 1989).

This explains why Taiwanese FDI in the United States grew at an annual rate of 52 per cent between 1979 and 1989, a rate almost three times faster by comparison to the 19 per cent growth of Taiwanese FDI in other parts of the world. As a result, the United States became an even more important host country, accounting for a dominant share of 57 per cent of the stock of approved Taiwanese FDI worldwide in 1989 compared to its 15 per cent share in 1979.

Three important features describe the import substituting FDI by Taiwanese firms in developed countries. First, although the United States was the dominant recipient of Taiwanese FDI worldwide in 1989, other developed countries such as Europe and Japan, though far less important, were also recipients of Taiwanese FDI in 1989. Europe and Japan had shares of 5 per cent and 1 per cent of the stock of Taiwanese FDI worldwide in 1989, respectively. The most important sources of competitiveness of Taiwanese MNCs in Europe were low prices, good sales or marketing capability, punctual delivery, and good after-sales service (Wang and Hsu, 1992). Taiwanese MNCs in Europe have ready access to capital and, in a manner broadly similar to Japanese firms, capitalize on their close relationships with their low-cost and relatively high-quality component manufacturers in Taiwan that supply inputs to their overseas manufacturing facilities (Bamford, 1993). Secondly, the manufacturing facilities established by Taiwanese firms in the developed countries were often rather primitive. In most cases, the value added in the host country was often only a step more advanced than simple screwdriver operations. However, local content requirements imposed by the developed countries have meant that mere final assembly in the host country was unsustainable. Thirdly, the establishment of production facilities by Taiwanese firms in the developed countries facilitated a two-way technology transfer between Taiwan and the foreign affiliate with the parent company often providing the foreign affiliate with more efficient cost cutting and streamlining techniques, and the foreign affiliate providing the parent firm with more advanced forms of complementary manufacturing technologies unavailable in Taiwan.

Unlike in the 1960s and 1970s when Europe was not a very significant host region for Taiwanese FDI owing to the long geographical distance and the presence of language and cultural barriers, this changed rapidly since 1986. Some 200 Taiwanese companies have set up branches, subsidiaries or distribution centres in Europe, mainly in Germany or the Netherlands, and mostly in the late 1980s in preparation for the unification of European markets by the end of the 1992. The most important host country for Taiwanese MNCs in Europe had been Germany, Taiwan's largest trading partner in Europe. Among the manufacturing companies, Taiwan's computer industry had been most active in the investment expansion in Europe.[20] As with other foreign firms and, most notably, the Japanese and Korean firms that did not have a significant outward FDI in Europe in the 1980s, the determinants of the outward FDI by Taiwanese MNCs in Europe since the mid- to late 1980s proceed from the necessity to be closer to their customers, and to hedge against fears that the European Union could turn to a protectionist trade fortress. For Taiwanese MNCs, FDI in Europe fulfils an additional motive to narrow the trade surplus of Taiwan with the United States.[21] As a result, Taiwanese FDI in Europe soared to $227.3 million during the first 11 months of 1990, more than three times the level in 1989 of $73.3 million and more than seven times the level of cumulative FDI from 1980 to 1988 of $31.5 million. This included the Taiwanese FDI of $34 million in the 200-acre Far East Industrial Park in Cork, Ireland, expected to accommodate 50 to 60 small- and medium-sized manufacturers.[22]

FDI in developed countries to gain access to advanced technologies

As in the case of the industrial development of Japan, access to advanced foreign technologies is crucial in enhancing the indigenous innovation of Taiwan and the industrial upgrading of the Taiwanese economy (Ranis, 1992). Such access to advanced foreign technologies through outward FDI and technical cooperation agreements with more advanced foreign firms are complementary to larger investments in R&D in Taiwan and the support provided by favourable government technology policies such as the creation of the Hsin-chu Science Park as a cornerstone of the Taiwanese government's drive to upgrade manufacturing technology and encourage high-technology industries to take root in Taiwan. Outward FDI offers a more aggressive and assured means to gain access to foreign technologies, particularly through the mode of foreign acquisitions. Such mode has allowed Taiwanese firms to leapfrog the process of technological development and, in some cases, to gain rapid entry into completely new although complementary lines of business. Some notable examples of this was the acquisition in 1980 by Formosa Plastics of a plant belonging to Imperial Chemical Industries (UK) located in Baton Rouge, Louisiana. Perhaps more sizable in scale was the acquisition in 1987 by the Taiwanese computer company, Acer, of Counterpoint Computers Inc., a Silicon Valley minicomputer maker; the acquisition by the Continental Engineering Company of the American Bridge Company in 1989 for $200 million (UN, TCMD, 1993a); and that by Channel International (a Taiwanese consortium funded in part by the government of Taiwan) of Wyse Technology Inc., an ailing American computer company, in 1989 for $156.7 million.[23] Indeed, the late 1980s witnessed a rush of large-scale acquisitions abroad by Taiwanese firms which led western observers to wonder whether Taiwan was repeating the behaviour of Japan in acquiring some of the West's most prominent and sacred corporate properties.[24]

Research-based FDI have typically the following features: first, the amount of investment in this type of FDI has most often been significantly larger than other types of FDI; second, this type of FDI has started to occur with some regularity for Taiwan since the late 1980s; third, it usually involved full or majority ownership to ensure an effective means of technology transfer from the foreign affiliate to the Taiwan parent; fourth, the investments were often in a business sector which exists outside the country's main operations; and fifth, the immediate objective of Taiwanese firms in embarking on this type of FDI was to gain access to the technology of the acquired firm and not in the acquired firm's profitability or market share. Indeed, the aim in most cases was to use the newly acquired technology to exploit new Asian markets where Taiwanese firms may have some advantages in marketing, distribution, local knowledge, etc. (Bamford, 1993).

Apart from the major foreign acquisitions, some of these research-oriented investments of Taiwanese firms in the developed countries were manifested in the finance that Taiwanese firms provided to a substantial number of companies in the Silicon Valley since the late 1980s as well as the joint ventures concluded by Taiwanese firms with American companies.[25] For detailed examples, see Chapter 9 of Tolentino (1993).

FDI to establish new markets in the developed countries by the provision of services to meet local market demand

Although services constituted only 10 per cent of this type of Taiwanese FDI at the end of the 1980s, the activities of Taiwanese MNCs in this sector are worthy of mention. Most important of all were the Taiwanese banks, the most notable group of Taiwanese service firms that invested in the member states of the OECD and particularly in Europe in the 1980s. The Bank of Communications, the International Commercial Bank of China and Chang

Hwa Commercial Bank planned to begin operations in Amsterdam in the 1980s, while China Trust Co. considered providing banking and financial services in London. Three objectives governed the FDI of Taiwanese banks over the 1980s. The first objective was to service the financial needs of individual Chinese resident overseas as well as home country firms, particularly Taiwanese exporters, by way of providing credit evaluations, export documentation and other associated export financing services. The second motivation was to service the needs of foreign firms wishing to invest in Taiwan. In this respect, the competitiveness of Taiwanese banks derive less from being a source of capital than from being a wellspring of local knowledge and inside market connections unavailable to other banks, including foreign bank subsidiaries in Taiwan. A third motivation was to service the needs of the banks themselves. Thus in the management of foreign exchange, the placement of funds in foreign financial markets, the collection of information and in the enhancement of the banks' reputation, Taiwanese banks have found it convenient and profitable to have on-site presence and control achieved through FDI which were otherwise unobtainable in a correspondent relationship with foreign financial institutions (Bamford, 1993).

Export-oriented FDI in trading, sales, distribution and marketing and service

Outward FDI in response to the need to facilitate exports in major export markets and to meet customer needs in sales, service and distribution grew in importance in the 1980s and accounted for 31 per cent of the stock of approved Taiwanese FDI worldwide at that decade's end. It represented the second most important type of Taiwanese FDI at the end of the 1980s. Unlike in the case of import substituting FDI in developed countries which is defensive in nature, i.e. to circumvent trade protectionism in major export markets, this type of FDI is offensive in its approach to promote exports of Taiwan. This type of FDI is in response to the recognition by Taiwanese firms that increasing sales and market share in world markets demands much more than maintaining price competitiveness but also product promotion through advertising, brand-name recognition, the presence of a knowledgeable sales force, personal relationships, after-sales service, and the maintenance of the firm's reliable reputation. Many Taiwanese firms, particularly the computer and electrical equipment manufacturers and car producers, have recognized the importance of this objective since the late 1980s in order to alter the reputation of Taiwan-made goods that suffer from poor brand-name recognition and poor quality. Apart from promoting sales, marketing and distribution and providing customer service, these sales and distribution centres established in developed countries also carry out other functions such as information gathering (i.e. to determine what competitors are doing or to ascertain local market tastes). At other times, this type of FDI represent an intermediary step to the establishment of a manufacturing base by affording an opportunity to assess market potential and increase customers, establish personal relationships, and identify potential joint venture partners before undertaking larger investments including in production (Bamford, 1993).

Among the most notable example of a Taiwanese MNC engaged heavily in this type of FDI is Sampo, Taiwan's third-largest home appliance manufacturer behind Tatung and Matshushita (Taiwan) already previously mentioned. Since the 1980s, the outward FDI undertaken by Sampo has been associated solely with the establishment of sales and distribution centres worldwide. In 1988 and 1989, it set up offices in Germany and France to handle and monitor the personal computer business of the company in those countries. A similar objective was pursued in 1989 in South Korea, Singapore and Thailand. In a similar fashion, the early thrusts in outward FDI by Aquarius Systems – a small and relatively young and inexperienced Taiwanese MNC engaged in computer production – was in sales and distribution. The firm established such outlets in West Germany, Netherlands and Paris in 1988 and 1989, with the West German outlet serving both as the company's overseas headquarters and also responsible for the distribution of the company's products throughout Europe (Bamford, 1993).

Import substituting FDI in developing countries

Considering Taiwanese FDI in such manufacturing industries such as food and drink, synthetic fibre processing, wood products, paper products, plastic and rubber products, chemicals, non-metallic minerals, basic metal products, machinery as well as construction, finance, and other services oriented to supply the domestic markets of host developing countries, import substituting FDI by Taiwan in developing countries accounted for 20 per cent of the stock of approved Taiwanese FDI in 1989. This type of FDI represented the third most important type of Taiwanese FDI at the end of the 1980s.

Owing to the growth of Taiwanese FDI in developed countries over the 1980s, developing countries took second place in importance at the end of this decade. Investments in the developing countries of South East Asia still predominated, accounting for 30 per cent of the stock of approved Taiwanese FDI in 1989, a share much lower than it had in 1978 of 53 per cent. These investments predominated in such industries as extraction and processing of natural resources such as pulp and paper, cement and rubber, metal mining and metal production, plastics production, and chemicals production. As in the earlier periods and until 1988, the major motivations of Taiwanese FDI in developing countries of South East Asia was the search for markets, security of export markets and raw material supplies and access to cheaper energy and land costs. These findings are corroborated by Chen (1986) who found that cheap labour in developing countries did not seem to be an important determinant of the international production of Taiwanese MNCs in developing countries in the mid-1980s.

Export platform investments in developing countries

Considering Taiwanese FDI in developing countries in such manufacturing industries as textiles (other than synthetic fibre processing), clothing and footwear and electronics and electrical equipment geared to supply export markets, export platform FDI in developing countries accounted for a much lower share of less than 8 per cent of the stock of approved Taiwanese FDI in 1989 compared to its much higher share of 19 per cent in 1979. It was only the fourth most important type of Taiwanese FDI at the end of the 1980s.

As in the 1960s and 1970s, textiles have been the most important industries of export platform investments of Taiwanese FDI in developing countries. Some 90 per cent of these textile investments by Taiwanese MNCs have been directed towards the developing countries of South East Asia. The geographical distribution of these investments in the 1980s were essentially unchanged from that in the 1970s. Indonesia still accounted for the largest share at 47 per cent in 1989 (compared to a 51 per cent in 1979), Singapore (24 per cent) and Thailand (11 per cent). By contrast, although Taiwanese export platform FDI in clothing and footwear were far less important than its FDI in textiles, the investments have been far more significant in developing countries outside of South East Asia in the 1980s. Unlike in the 1970s in which Singapore received some three-quarters of Taiwanese FDI of this type, some 79 per cent of the outward FDI in clothing and footwear as of 1989 have been directed to countries of the Caribbean Basin since some 12 of the 25 countries that maintained formal diplomatic relations with Taiwan belonged to this region. It is with these few countries that Taiwan concluded reciprocal trade and investment agreements (Chen, 1986; Korea Institute for Foreign Investment, 1989). Investments in the Caribbean Basin also offered savings in transportation costs and transit time in shipping owing to their geographical proximity to the United States. Finally, export platform FDI in electronics and electrical appliances was directed overwhelmingly to the developing countries of Asia that accounted for 92 per cent of FDI of this type. Indeed, evidence has been found of rationalized production and product segmentation associated with the transfer of electronics and electrical production abroad by Taiwanese firms. For example, one Taiwanese firm simultaneously assembled electronic filters and magnetic devices in Thailand, technology- and skill-intensive power supply units in Mexico, and final personal computer units in Taiwan (World Bank, 1989).

FDI to secure supplies of essential raw materials

The importance of this type of FDI declined further from its share of 8 per cent of the stock of approved Taiwanese FDI in 1979 to 1 per cent in 1989. Resource-oriented investments by Taiwan had become the least important type of Taiwanese FDI in the decade of 1980s as concerns over resource supply and security declined with the slump in prices in world market for commodities and oil and as investments due to cost-push and market-pull factors became more dominant (Chia, 1989).

Taiwanese FDI in the 1990s

The decade of the 1990s seem to represent a period of even faster growth of Taiwanese FDI. Based on balance-of-payments data, actual FDI outflows from Taiwan reached an annual average level of $3.3 billion in the period between 1990 and 1998 compared to $1.2 billion in the period between 1980 and 1989 and $2.4 million in the period between 1970 and 1979. The estimated stock of Taiwanese FDI stood at $38 billion in 1998 compared to $7.6 billion in 1989 and $97 million in 1980.[26]

Economic factors explain the even more rapid rate of expansion of Taiwanese FDI in the 1990s. The most fundamental perhaps was the accumulation by Taiwan of the world's largest foreign exchange reserves in the early 1990s which enabled the country to rapidly become one of the principal sources of investment capital in the world.[27] The determinants of such rapid growth was the declining competitiveness of domestic labour intensive industries as well as heavy and chemical industries as a result of increasing production costs in Taiwan and the increasingly adverse international trading environment. This involved major structural changes in Taiwan which altered the relevance of cheap labour and an undervalued currency as the basis of its economic success until the late 1980s. Economic pressures were exerted from various angles which increased domestic costs of production. Rapid economic growth and the political liberalization of Taiwan led to low unemployment rates and demands for higher wages and better work benefits thus undermining the low wage levels that formed an essential part of the Taiwanese economic miracle of the past.[28] Indeed, labour costs in Taiwan in the late 1980s had become five times as high as those of Malaysia, Thailand and the People's Republic of China (*Free China Journal*, 15 February 1988).

Although the problem of rising labour costs is broadly parallel to economic developments in Japan at the end of the 1960s which in combination with the sharp upward revaluation of the local currency led to the weakening of the traditional labour intensive industries (see Chapter 9 of this book), the basis of the rise in labour costs differed in the case of Taiwan and Japan. In the case of Japan, the success of domestic industrial restructuring towards modern capital intensive heavy and chemical industries at the end of the 1960s enhanced labour productivity owing to high capital to labour ratios and this perpetuated a continuous cycle of rising wages (Cohen, 1975; Kojima, 1978; Ozawa, 1979b). In Taiwan, on the other hand, not only did the shortage of low-skilled labour create pressures that drove up industrial wages but political democratization led to increased labour organization and activism and wage demands by labour unions.[29] Unlike in Japan, the wage increases in Taiwan were not matched by productivity growth (see van Hoesel, 1997).

Of course, the rising labour costs in both Taiwan and Japan have been spurred by local currency appreciation. In the case of Taiwan, the further upward revaluation of the New Taiwan dollar was not only allied to strong export growth but was a result of the deliberate imposition of a restrictive monetary policy to combat inflation by raising interest rates and tightening domestic credit as well as pressures from the United States to redress the persistent trade imbalances between the United States and Taiwan. Thus, in the period between 1985 and 1992, the New Taiwan dollar appreciated 60 per cent against the United States dollar.[30]

Economic pressures not only came in the form of rising wages and an appreciating local currency. As in the case of Japan during its Ricardo-Hicksian trap stage of multinationalism between the late 1950s and early 1970s, the very growth of Taiwan's heavy and chemical industries led to the shortage of industrial space and brought forth problems of pollution, congestion and ecological destruction. In response to a public outcry over the decrepit state of Taiwan's environment, the Kuomintang regime established a Bureau of Environment Protection in 1985 to monitor and enforce compliance to environmental standards. High property costs propelled by a shortage of land, strict zoning regulations, public opposition to the expansion of 'smokestack factories' and a high degree of land speculation contributed to the difficulties faced by many local firms in the expansion of their domestic operations (Kapellas and Liu, 1990).

These economic forces affected the competitiveness of two sets of domestic-based industries of Taiwan: the traditional labour intensive industries composed of many small- and medium-sized firms and the heavy and chemical industries composed of fewer, larger-sized firms. High labour costs and local currency appreciation have undermined export-oriented manufacturing, especially manufacturers in labour intensive industries such as textiles, footwear and other industries as well as original equipment manufacturers (OEM) in the electronics and electrical appliances industry.[31] Profit margins in the computer industry have also been affected adversely by the high labour costs and competition from China and South East Asia.[32] Similarly, the rapid domestic expansion of firms in the heavy and chemical industries that led to the shortage of industrial space, high property costs and pollution, congestion and ecological destruction in turn posed a threat to their further domestic expansion in a manner reminiscent of Japanese industrial history in the late 1960s.

In considering the problem of a growing comparative disadvantage in labour intensive industries, Taiwan had two alternatives. The first alternative was the importation of cheap foreign labour to perform the low paid, menial and assembly-line type tasks that domestic labour have increasingly been unwilling to perform. Support for this course of action had been provided by small- and medium-sized companies and particularly those in textiles-related industries (*Free China Journal*, 11 April 1988; 15 August 1988). However, the feasibility of pursuing this alternative had been hampered by persistent attempts by organized labour to prevent any threat to undermine their members' position and the already large numbers of illegal immigrants in Taiwan for which the government had continually been reluctant to provide education, medical and insurance coverage. The second alternative was the preferred option by the government and one that had the greatest impact on the growth of Taiwanese FDI in this period. It entailed the restructuring of the Taiwanese economy towards more capital intensive and technology intensive industries.[33] This would require the shedding of those industries which have low technology, low productivity or a low value added component in the production process or their relocation abroad to more cost competitive host countries. The movement offshore of these industries would create opportunities for the growth of more dynamic industries in Taiwan, thus enhancing the growth of the industrial sector while enabling the maintenance of international competitiveness. This alternative, which treats outward FDI by labour intensive industries and domestic industrial upgrading in Taiwan as a mutually reinforcing process, is in line with the views of Kojima and Ozawa (1985).[34] To support this objective, 'economically inefficient' companies relocating abroad to less developed countries with cheap labour can avail of tax rebates from the Taiwanese government on the export of manufacturing facilities and semi-finished products (*Free China Journal*, 11 January 1988). These companies are typically small- and medium-sized firms in the labour intensive industries of textiles, clothing and footwear as well as consumer electronic and electrical production which – unlike the cash-rich, large sized Taiwanese investors of the 1970s that concentrated their FDI in the developed countries – had relatively small amounts of available capital for FDI.[35] This type of Taiwanese FDI is consistent with the first phase of Japanese offshore production in Hecksher-Ohlin industries prevalent in the case of Japan from 1950 to the mid-1960s.

As for the heavy and chemical industries, there had been no other recourse but to embark in outward FDI also in less developed countries to counteract the comparative disadvantage of Taiwan in resource intensive and pollution-prone industries. This type of Taiwanese FDI is reminiscent of the second phase of Japanese multinationalization in response to the Ricardo-Hicksian trap of industrialization that pertained between the late 1950s and early 1970s. Thus, with the simultaneous international expansion of firms in both labour intensive industries and heavy and chemical industries, Taiwanese FDI in the 1990s share features of the first two phases of Japanese overseas production in the period since the Second World War.

Thus, the types of Taiwanese FDI prevalent in the 1990s in declining order of importance were: first, export platform investments in developing countries; second, to supply host country markets in developing countries; third, FDI in developed countries to supply local markets and to gain access to advanced technologies; fourth, to promote growth of Taiwanese exports through outward FDI in trading, sales, distribution and marketing in major foreign markets; and fifth, to secure supplies of essential raw materials.[36]

Thus, although manufacturing remained the dominant sector of Taiwanese FDI in the 1990s, the declining importance of the sector in Taiwanese FDI continued as in decades past and in the 1990s accounted for less than 60 per cent of the stock of approved FDI, while services displayed further growth in Taiwanese FDI with a share of over 40 per cent over the decade. With the predominance once again of host developing countries, Taiwanese FDI in the 1990s displayed characteristics more consistent with the first two decades of the history of Taiwanese MNCs (the 1960s and 1970s) and a reversal of the trend at the end of the 1980s. Indeed, in 1995 some 78 per cent of the stock of approved Taiwanese FDI was in developing countries (of which 61 per cent was in Asia and 17 per cent in Latin America), while only 21 per cent was in the developed countries (of which 17 per cent was in North America and 4 per cent in Europe). The most important feature to note in the geographical destination of Taiwanese FDI in the 1990s was the recovery in the importance of the developing countries of Asia. Thus while these countries accounted for only 30 per cent of the approved stock of Taiwanese FDI at the end of the 1980s, their share doubled to 61 per cent in 1995. As a result of the rapid growth of Taiwanese FDI in the developing countries of Asia starting in the late 1980s, the trade of Taiwan with these countries increased phenomenally and the region became one of Taiwan's major trade partners which allowed for the reduction in importance of the United States as a trading partner.

Each of these types of Taiwanese FDI pertaining to the decade of the 1990s is discussed below in order of importance.

Export platform FDI in developing countries

This type of FDI has become the most important type of FDI for Taiwanese MNCs in the 1990s. The main firms and industries bringing this type of FDI to the fore are small- and medium-sized companies in labour intensive industries involved in the production of textiles, clothing, footwear, handbags, toys, and consumer electronics whose objective in the pursuit of outward FDI is the search for low-cost bases for manufacturing products destined for the United States and European markets. Although the small- and medium-sized companies had been affected most adversely by the comparative disadvantage of Taiwan, the absence of large, cumbersome bureaucracies had also given these firms the nimbleness and flexibility to relocate production in foreign countries when their survival was at stake. This type of FDI had been directed to the countries of South East Asia and, more recently, in China and Vietnam.

Export platform FDI in South East Asia

As shown in the previous sections, Taiwanese FDI of this type had always been largely drawn towards the countries of South East Asia. This had not changed in the 1990s on account of several factors acting in combination. The geographical proximity of these countries, the cultural affinity with Chinese businessmen in various countries of South East Asia whose common Hokkien dialect and business culture facilitated business transactions and negotiations across the region, the presence of an abundant labour force at low cost, the retention of trade privileges of countries of South East Asia under the Generalized System of Preferences, continuing political stability as well as the provision of various investment incentives by governments of countries in the region acted in unison to make the region a continuing favoured location of production by Taiwanese firms in labour intensive industries.[37] In addition, Taiwanese FDI in South East Asia and Hong Kong assumes particular economic and political importance in light of the illegality of direct trade and FDI between Taiwan and mainland China. Thus not only has South East Asia and Hong Kong acted as alternative locations to China, Taiwanese FDI in those host regions is an important means of circumventing direct trade and FDI with China. The favoured locations in South East Asia were Malaysia, Indonesia and Thailand where cost factors are reinforced by political stability and availability of infrastructure. The Philippines also became a favoured location as political stability was gradually restored (Chia, 1989). While most of Taiwan's textile producers had expanded in the Philippines, Indonesia and Thailand, many electronics companies such as Chung Hwa Picture Tube, Acer and Teco Electric had established factories in Malaysia.[38]

Taiwanese FDI in South East Asia particularly in electrical and electronics production share features not dissimilar to Japanese FDI in the same region and in the same industries. The first shared feature is that Taiwanese FDI in South East Asia has also been in the form of joint FDI not only by producers but also by their satellite companies or components suppliers, although perhaps at a smaller scale than in the case of Japanese MNCs. Indeed, small- and medium-sized firms have always worked closely together with other members of their industry when considering FDI, including in the selection of offshore locations. Apart from relieving supply constraints in host countries through the extension of their established and long-standing supplier-purchaser relationships in the home country to their business ventures abroad and the assurance of quality control, these joint FDI provide some measure of insurance against the risks of FDI for small-scale companies inexperienced in operating in foreign territories (Kapellas and Liu, 1990).[39] The second feature making Taiwanese FDI in electrical and electronics production in the region somewhat similar to Japanese FDI is the presence of regionally integrated international production networks. Though perhaps not as developed and complex as the international production networks of Japanese MNCs, foreign affiliates of Taiwanese MNCs in various host countries in Asia along with affiliated and non-affiliated companies acting as suppliers and subcontractors have led to the formation of regionalized international production networks by Taiwanese MNCs (see Kim, 1996).

In the period between 1987 and March 1993, more than 4,000 Taiwanese companies had set up operations in South East Asia with investments of around $12 billion or almost half of the $25 billion capital outflows from Taiwan during this period. Indeed, Taiwan had become in the 1990s the second-largest source of FDI in Malaysia and Indonesia and the fourth-largest in Thailand.[40] However, there were several economic and political factors that threatened the locational advantages of countries of South East Asia to labour intensive production by Taiwan. The economic factors have to do with the inevitable rise of wages as a byproduct of economic development of countries of the region and the loss of their trade privileges under the Generalized System of Preferences with the attainment of a certain level of economic development and the rapid growth of their own exports. The political threat is posed by the growing diplomatic tensions between countries of South East Asia and the People's Republic of China over the growth of Taiwanese trade and investment (Kapellas and Liu, 1990).

Export platform FDI in China and Vietnam

Although an estimated 2,500 Taiwanese-backed companies had already invested between $2 billion and $3 billion in the People's Republic of China in the period between the mid-1980s and 1990 despite the fact that indirect FDI in China was not legalized until early 1991 and the prohibition on direct investment continued,[41] there were a number of factors that enabled China to become a more attractive location for export platform FDI by Taiwanese MNCs in labour intensive industries from around 1990. The first set of factors concerned the growing locational advantages of mainland China which with its abundant supplies of inexpensive labour and land was also eager to encourage FDI in the development of light manufacturing industries. In combination with strong family, cultural and linguistic ties between mainland China and Taiwan plus their very close geographical proximity notwithstanding their political differences, it made economic sense for Taiwan to consider mainland China seriously as a site for its labour intensive production. The second set of factors contributing to the increasing locational advantages of China for Taiwan's labour intensive production from around 1990 was the rising wages in their traditional host countries for this type of FDI. When the initial wave of Taiwanese FDI in labour intensive production began in the late 1980s, wages in countries such as Thailand and Malaysia were one-tenth of those in Taiwan. By around 1992 or 1993, wages in those countries were only one-fifth to one-third cheaper, while China offered labour cost differentials close to what was available in South East Asia in the late 1980s. The two sets of factors acted in combination to cause Taiwanese FDI to fall sharply in Malaysia and

Thailand in the early 1990s, and to rise considerably in China.[42] By investing in China, Taiwanese companies could re-create in essence the conditions of 1960s Taiwan with the presence of a skilled, motivated and inexpensive labour force to produce labour intensive products for export to the world's most advanced markets in the United States and Europe (Bamford, 1993).

Owing to the prohibition of direct investment by Taiwan in mainland China, most Taiwanese firms have channelled their investments in China through Hong Kong which is in close geographical proximity to both mainland China and Taiwan, and also has the advantage of providing investments made from that territory a certain degree of legal protection.[43] By 1994, around $19 billion had been invested in China channelled via Hong Kong, of which more than 80 per cent had been made since 1991.[44] The growth of Taiwanese FDI in China had two important economic implications. The first was that two-way trade between the two countries conducted across the Taiwan Strait increased considerably over the 1990s with China becoming Taiwan's fourth-largest export market. Secondly, Taiwan's large trade surplus with the United States had been shifted to China. Thus as Taiwan's surplus with the United States narrowed from $19 billion in 1987 to $9.8 billion in 1991, China's surplus with the United States grew from $3.4 billion to $12.7 billion over the same period.[45]

However, despite the continuing popularity of China as a site for labour intensive production by Taiwan in the 1990s, economic and political forces had tended to somewhat dampen the attraction of the country as an investment location for Taiwan. The economic factor had to do with the high import propensity of the Chinese affiliates of Taiwanese MNCs for required inputs particularly in the major industries of footwear, toys and clothing. In combination with unproductive workers, uncertain transport links and power shortages, it served to partially negate the advantage of lower wage and land costs in China.[46] The political factor had to do with the persistent political differences between the two countries which made a high degree of trade and FDI by Taiwan in China risky and undesirable.[47]

To address the concerns of Taiwan with respect to the growth of trade and particularly FDI in China, Taiwan sought other low labour cost locations in South East Asia and found Vietnam to be a suitable location. Indeed, Vietnam as a new low-wage location became the fastest growing host country of Taiwanese FDI since the early 1990s (Borrmann and Jungnickel, 1992). The locational advantages of Vietnam apart from the presence of an abundant and cheap hard-working labour force included the presence of a large ethnic Chinese community, geographical proximity to Taiwan and the tax breaks extended to foreign investors.[48] The investments have flowed through both official channels as well as informally through ethnic Chinese relatives or friends in Ho Chi Minh City which are financing the growing number of small workshops in the city. Two Taiwanese firms – the Central Trading & Development Company and and Pan Viet Group – which are controlled by Taiwan's ruling Kuomintang regime have constructed Ho Chi Minh City's first export processing zone. The 89-hectare site accommodating some 200 light manufacturing companies accommodates many firms from Taiwan. Indeed, Taiwanese FDI are almost exclusively in former South Vietnam – Taiwan's third-largest Asian trade partner after Hong Kong and Singapore during the 1960s. The sweatshops established by Taiwan in Vietnam are typically similar to that in China with their concentration in the production of clothing, textiles, shoes, foodstuffs and other consumer goods. By 1991, Taiwan had become the largest source of FDI in Vietnam.[49]

Import substituting FDI in developing countries

This type of FDI is the second largest type of FDI for Taiwanese MNCs in the 1990s. As mentioned, unlike the export platform FDI in developing countries conducted by small- and medium-sized Taiwanese firms, this type of FDI has been conducted by larger-sized firms concerned about the growing comparative disadvantage of Taiwan for natural resource-intensive and pollution-prone industries.

As in decades past, this type of Taiwanese FDI has been directed principally to countries of South East Asia. The region offers stable governments, natural resources, relatively low cost of land and wages, more lax environmental standards and continuing political stability which made the region a haven for Taiwanese businesses seeking to escape the disadvantages of the home country in heavy and chemical industries that are resource intensive and/or pollution prone. Finally, the pace of economic development and the presence within the region of increasingly affluent consumers also provided Taiwanese investors with direct access to an expanding and prosperous market at a time of increasing trade tensions between their country and the United States.

Thus, apart from being a major base for export platform FDI by Taiwan, the importance of South East Asia for FDI by large firms needing to circumvent the Ricardo-Hicksian trap of industrialism in Taiwan posed by the inadequacy of its natural resources to sustain domestic industrialization, the shortage of industrial space and increasing environment concerns had been accentuated in the 1990s. This was owing not only to the growing comparative disadvantage of Taiwan for production in heavy and chemical industries but also the need to gain access to raw materials and to search for new markets.[50]

Although import substituting FDI in China has not been significant owing to the political risks attached to Taiwanese FDI in that country, Taiwanese FDI of this type has been growing rapidly in Vietnam. The extension of tax incentives by the Vietnamese government combined with the provision of aid from Taiwan to Vietnam and the training to Vietnamese government officials and businessmen has enabled Taiwan to transplant not only labour intensive industries but also pollution-prone industries for which Taiwan has a comparative disadvantage.[51] Vietnam has also become important to other import substituting manufacturing industries of Taiwan wishing to cater to the demands of its domestic market.[52]

Apart from manufacturing, there has also been in the 1990s the growth of services FDI by Taiwanese MNCs in developing countries to cater to the market demands of the host country. In a parallel trend to the growth of Taiwanese banks in United States and Europe in the 1980s, there has also been the establishment of representative offices and foreign branches by Taiwanese banks in Hong Kong in the 1990s. The major rationale for the growth of banking sector FDI by Taiwan in Hong Kong derives from the central position of Hong Kong as a conduit for trade and FDI between Taiwan, China and South East Asia. Thus, after having already set up operations in Europe and the United States, the three largest state-owned banks – Hua Nan Commercial Bank, Chang Hua Commercial Bank and First Commercial Bank – established representative offices in Hong Kong in 1991, while private banks such as Overseas Chinese Commercial Banking Corporation and the World Chinese Union Commercial Bank are also planning similar moves.[53]

FDI in developed countries to supply local markets and to gain access to advanced technologies

Although Taiwanese FDI in the United States did not grow as rapidly as that in Asia since 1989, import substituting and research intensive manufacturing FDI as well as services sector FDI geared to the local market demand in host developed countries represented the third most important type of Taiwanese FDI in the 1990s. The motivations behind developed country investments were technology acquisition, tariff circumvention, marketing and distribution needs, after-sales services, exploitation of firm's technical know-how, need for diversification and risk avoidance and a means of managing corporate assets. Thus, unlike their investments in developing countries where Taiwanese FDI was propelled by the declining comparative disadvantage of Taiwan for production in labour intensive industries and heavy and chemical industries, Taiwanese FDI in developed countries have been determined by the needs of Taiwanese firms to overcome trade barriers and to gain access to investment opportunities in those countries.

As mentioned, North America and particularly the United States continued to be the most important host country for Taiwanese FDI in the developed regions of the world with a share of 17 per cent of the approved stock of Taiwanese FDI worldwide in 1995. Investments in Europe in manufacturing and services have also grown significantly in the late 1980s prior to the unification of the European Community in 1992.[54] Europe accounted for 4 per cent of the approved stock of Taiwanese FDI worldwide in 1995.

Two notable features describe Taiwanese FDI in the developed countries in the 1990s. The first is that as in decades past, those firms that have displayed a high propensity to engage in FDI in developed countries had almost always been the largest firms in their respective industries. However, a few smaller firms have also been active investors in the developed countries in the 1990s. Aquarius Systems, Supertron Computers and Yong Hsin Pharmaceuticals are some examples of Taiwanese companies with less than 1,000 employees that have embarked on sizable manufacturing FDI in the developed countries (Bamford, 1993).

Import substituting FDI in developed countries

Import substituting FDI in the developed countries in the 1990s have been concentrated in the United States as always and in Europe. Although the levels of Taiwanese FDI of this type have remained far larger in the United States than in Europe with the motivation of Taiwanese firms to recycle their trade surplus in their most important trading partner country, FDI in both these locations have been determined by the protectionism. In the United States, the continuing protectionist tendencies have been influenced by the persistent trade deficits of that country with Taiwan and the further growth of foreign exchange reserves of Taiwan, while in Europe fears about protectionism arose in association with the formation of the Single European Market in 1992. The import substituting FDI by Taiwanese manufacturing firms in the United States and Europe has also been concentrated in different industries. Thus, while the largest scale FDI in the United States to serve its domestic market had been made by large-sized firms in the plastics products and food products industries, the size of Taiwanese FDI in European manufacturing have been far more modest and, except for Acer, one of the more prominent firms in the Taiwanese computer industry, Taiwanese FDI in Europe have been led by smaller-sized firms and the investments have been directed most notably to the electrical and electronics products industry and, in particular, computers in much the same way as Taiwanese manufacturing FDI in the United States in the 1970s and 1980s.

IMPORT SUBSTITUTING FDI IN THE UNITED STATES

As mentioned above, among the most prominent Taiwanese FDI in the United States in the 1990s were made by Taiwanese manufacturing firms in plastics products and food products. In both these industries, the FDI by Taiwanese firms have been geared in general to penetrate the United States in terms of sales with the majority of the FDI undertaken through acquisitions. As indicated in the previous sections, the firms in the plastics industry had already existing investments in the United States before their large-scale expansion in the 1990s of close to $2 billion.[55] This is demonstrated in the case of the Formosa Plastics Group that had already been producing plastic resins, pipes and films in about a dozen plants in the United States with an annual revenue of around $630 million. Facilitated by the strength of the New Taiwan dollar, the further growth of FDI of this company as well as that of the China General Plastics Corporation in the late 1980s and early 1990s had been prompted by the allure of the rich and large market of the United States as well as the need to integrate backwardly into the production of ethylene – a petrochemical building block of plastics production. Thus, apart from aiming to increase sales of plastics products in the United States through the establishment of downstream plants for plastics production, the backward vertical integration of Taiwanese plastic products companies is also a resource extractive investment since a considerable proportion of the ethylene production in their American plants feeds the input requirements of plastics production in Taiwan (at least initially) or their foreign plants in the United States.[56]

Other large-scale investments by Taiwanese manufacturing firms have been established in the food products sector propelled by the uncertainty of basing further growth of sales on exports based in Taiwan owing to continuing trade pressures from the United States, the increasing costs of land and labour in Taiwan since the late 1980s and problems in raw material procurements sourced from Taiwan's state-dominated procurement agencies. Thus from a previous concentration of their import substituting FDI in the large countries of South East Asia and, in particular, Indonesia, Thailand and the Philippines, President Enterprises – Taiwan's leading food-processing company – undertook its first major overseas FDI in June 1990 with an investment of $335 million to acquire the United States' third largest producer of cookies, Wyndham Biscuits, based in Atlanta, Georgia. The acquisition enabled President Enterprises to gain a marketing network for the distribution and sales of Taiwanese food products in the United States and a production network consisting of eight manufacturing plants based in the low wage southeastern part of the United States as well as to profit from the presence of cheaper supplies of sugar and lard in the United States. It also enabled the company to increase the share of foreign sales in its total sales from 2 per cent to nearly a quarter.[57] Apart from Wyndham Biscuits, President Enterprises also acquired 7-Eleven convenience store chain which also served as an important sales and distribution channel for the Taiwanese company's food products.[58]

IMPORT SUBSTITUTING FDI IN EUROPE

As in the 1980s, Taiwanese computer companies lead the growth of Taiwanese FDI in Europe in the 1990s. However, in comparison to the size and scale of investments by Korean computer companies in Europe such as Samsung and Lucky Goldstar, those of Taiwanese companies are relatively more small scale. The 1990s marked the establishment of production facilities in Europe by two Taiwanese computer companies, Acer and First International Computer. In particular, Acer which gained prominence as one of the most successful producers of clones of IBM personal computers initiated assembly line production of personal computer (PC) systems and notebook computers in the Netherlands in 1990/2 and in Germany between 1985 and 1993 as well as assembly line production of personal computer systems in the United Kingdom between 1988 and 1994. This is in addition to the establishment of sales, warehousing, service facilities in these countries and a number of other countries in Europe in the first half of the 1990s. Similarly, First International Computer initiated the assembly line production of PCs and notebook computers in the Netherlands in 1990, along with sales, warehousing, service facilities in that country and in Spain, Czech Republic and France between 1990 and 1994. The company is one of the largest motherboard manufacturers in the world, and unlike most other Taiwanese computer companies that were primarily focused on the United States as their major foreign market, Europe had been by far the most important market of the company accounting for half of its total revenues in the 1990s. However, owing to its shorter international experience and emphasis on the large sales volume of components (motherboards and add-on cards) than on final products (PC systems and notebooks), including as original equipment manufacturer of PC systems and notebook computers, its market presence in Europe is less extensive than that of Acer (van Hoesel, 1997).

International production combined with facilities in sales, marketing, distribution and after-sales service has enabled the Taiwanese computer companies not only to overcome fears of protectionism associated with Europe 1992 but also to have a physical presence near customers in an important market that has rapidly changing demands for computer configurations. Although production is presently limited to the local assembly of computer components sourced from Taiwan and efforts at marketing networks including the promotion and establishment of brand names have yet to evolve fully, the presence of production and sales, marketing and after-sales service infrastructure in Europe has enhanced the capacity of Taiwanese computer companies to cater more effectively to changing consumer demand and to adapt products as well as offer products at lower cost. However, technological and marketing weaknesses *vis-à-vis* leading companies from more established companies in the consumer electronics and computer industry are major obstacles that have yet to be overcome. While Acer invested substantial resources to establish in a slow and gradual manner a sound market position in Europe including the promotion of its brand name, this has not yet been the strategic priority of First International Computer and other Taiwanese computer companies whose main business has remained essentially in original equipment manufacturing, subcontracting relationships with other leading companies and in the supply of computer components. This has served to hamper their development as MNCs in their own right (van Hoesel, 1997).

Apart from being at the early stage of growth in their FDI in Europe, Taiwanese FDI differs from Japanese FDI in Europe in the same industry even though the growth of FDI in Europe by both these countries have occurred around the same time – since the mid- to late 1980s. This has to do with the market orientation associated with the extent of regional integration strategies adopted by Taiwanese MNCs and Japanese MNCs. As mentioned in Chapter 9, a unique feature of Japanese participation in Europe is their orientation towards the integrated European market as a whole and not to a particular national market in which local production is undertaken. By contrast, the early stages of the development of the outward FDI of Taiwan explain their consideration of Europe as consisting of separate national markets, notwithstanding Europe 1992.

FDI in developed countries to gain access to advanced technologies

Taiwanese FDI to gain access to more advanced forms of technology have accelerated throughout the 1990s, and have occurred mainly as in the past decades through large-scale acquisitions of American firms, although there have also been waves of smaller scale acquisitions and joint ventures.[59] As in the 1980s, this type of FDI has been most most popular in the electrical and electronics industries and in particular computers, as well as in the chemicals and motor vehicle industries.

FDI of this type by Taiwanese firms in the electrical and electronics industries and, in particular, computers have been concentrated in the Silicon Valley of the United States. In addition to its acquisition of Counterpoint Computers in 1987 and its semiconductor joint venture with Texas Instruments Inc. in the mass production of four-megabit dynamic random access memory chips, Acer acquired in 1990 for $94 million Altos Computer Systems, a maker of multi-user computer networks, based in San Jose, California. The acquisition has been motivated as much by the need of Acer to establish a manufacturing facility in the United States and acquire complementary technologies as by the need to acquire an existing distribution network.[60] Indeed, a considerable number of American companies based in the Silicon Valley, and particularly those involved in personal computers and semiconductors, received crucial early stage funding from Taiwan in the 1990s encouraged by Taiwan's venture capital industry which provide tax breaks and other incentives for investments in certain technologies. For the Taiwanese companies, the finance provided to Silicon Valley companies enables their direct access and rapid adoption of more advanced forms of technologies in their industries. Apart from accelerating their investments in R&D, research-linked FDI with small American design houses that specialize in new chip technologies has enabled Taiwanese semiconductor companies, for example, to broaden their focus away from a narrow segment of a very large market – in application-specific integrated circuits or Asics – which are not directly competitive with established foreign companies as well as to overcome overcapacity and overreliance of sales on the local market.[61] However, Acer's microcomputer manufacturing in the United States and some of the other high technology overseas ventures such as Mospec

Semiconductor, Tainic Technology and Sigma Computer have yet to earn a profit (Hu, 1995).[62]

Other large-scale FDI of this type in the 1990s have been made by Yue Tyan Motors, Taiwan's fourth-largest car manufacturer and maker of motor-cycles and industrial machinery, that acquired in 1991 majority control in a California-based aerospace firm, Advanced Aerospace Systems Inc., for $50 million. The main purpose of the acquisition was to gain access to comple-mentary technologies and know-how in aerospace as a means of upgrading the technological development of the company's production of cars, motor-cycles and industrial machinery in Taiwan as well as support the objective of company to penetrate new but complementary business lines such as the production of cargo aircraft and rockets.[63]

In the chemicals industry, mention must be made of the acquisition in the early 1990s by China Synthetic Rubber of a penicillin factory in Northumberland, England owned by the Glaxo Group which supplied some 10 per cent of the world's requirements for raw penicillin. The acquisi-tion enabled China Synthetic Rubber to diversify into drug manufacture (penicillin production) – a investment opportunity complementary to the company's core business of chemical production and, in particular, in the company's capacity as Taiwan's sole producer of carbon black, the primary material used in the production of rubber tyres (Bamford, 1993).

Export-oriented FDI in trading, sales, distribution and marketing and service

Outward FDI in response to the need to facilitate exports in major export markets and to meet customer needs in sales, service and distribution grew in importance and was the fourth most important type of Taiwanese FDI in the 1990s. This type of FDI has played an important supporting role in the international expansion in the 1990s of small companies such as Aquarius Computer Systems and First International Computer as mentioned in the previous section of this chapter, as well as larger companies such as Sampo and Acer, among others. While the export-promoting FDI of Taiwanese firms had been directed largely to the major markets of the United States and Europe, some of the FDI had also been directed to the larger countries of South East Asia (Indonesia and Malaysia) and South Africa where Taiwanese companies have major trade and FDI interests.

FDI to secure supplies of essential raw materials

As in the 1980s, resource extractive investments by Taiwan has remained the least important type of Taiwanese FDI in the decade of 1990s in both abso-lute and relative terms. Nevertheless, this type of FDI continued to be impor-tant for firms associated with industries in the primary sector as well as firms in those manufacturing industries involved in the processing of natural resources. The case of USI Far East that invested in a $370 million petro-chemical complex in the Philippines is an illustrative case in point. The rationale for the company's FDI had less to do with the need to relocate a pollution intensive industrial process (although that was part of it) as by the need of the company to secure steady supplies of petrochemical products.[64]

Conclusion

This chapter analysed the emergence and evolution of Taiwanese MNCs over the last 40 years. The major types of FDI undertaken by these firms and associated features such as the determinants of outward FDI, their industrial pattern and geographical destination vacillated in the 1970s, 1980s and 1990s. Thus, while import substituting FDI in developing countries and export platform FDI in developing countries were the predominant types of FDI from the period of their emergence in the late 1950s to the end of the 1970s, import substituting FDI in developed countries and export-oriented FDI in trading, sales, distribution and marketing and service also mainly in developed countries were the predominant types of Taiwanese FDI at the end of the 1980s. At the end of the 1990s, export platform FDI in the developing countries and import substituting FDI in the developing countries came to the fore in a manner more consistent with the first two decades of the history of Taiwanese MNCs (the 1960s and 1970s) and reversing the trend at the end of the 1980s.

To the extent that Taiwanese FDI could be compared with the pattern of Japanese FDI since the Second World War, the first and second phases of Japanese FDI in labour intensive manufacturing in textiles, sundries and other low-wage goods (the first phase) and in heavy and chemical industries during the Ricardo-Hicksian trap stage of Japanese FDI (the second phase) had been transposed in the case of the history of Taiwanese MNCs. This is associated with a prolonged dependency or sustained comparative advantage of the Taiwanese economy on labour intensive production until the late 1980s. The basis of the prolonged competitiveness of these industries derived from the Taiwan state's policy of wage suppression and the political subordi-nation of popular sectors (i.e. middle stratum of professionals, skilled workers, businessmen and farmers) to the state which was brought to an end with the political democratization in Taiwan from around 1987. The emphasis on the domestic industrial development of heavy and chemical industries in Taiwan in the 1960s and 1970s led to the dominant role of outward FDI to relocate abroad some of the more resource intensive and often pollution-prone indus-tries in the 1970s and 1980s. Taiwanese FDI in labour intensive industries emerged in the late 1980s and grew in a major way in the 1990s.

Two unique features of the history of Taiwanese FDI stand out. The first is the lesser importance of outward FDI to extract natural resources compared to other countries that share features of its national economic development – the United Kingdom, Germany, Japan and even South Korea – in which resource extractive FDI played a more prominent role in the pattern of outward FDI particularly in the early phases. In the case of Taiwan, outward FDI to secure supplies of essential natural resources and raw materials even at its peak in the 1960s and 1970s did not account for more than 10 per cent of the stock of approved FDI. Two factors may have contributed to such small share: the rising tide of resource nationalism in developing countries since the late 1960s and the high capital intensity of FDI of this type which Taiwanese MNCs cannot provide. Perhaps the importance of Taiwanese FDI in resource extraction has been masked by outward FDI in resource processing manufacturing industries that have either been relocated to resource-rich host developing countries in South East Asia to serve host country markets and/ or engage in the export of semi-processed raw materials from the resource-rich host countries to Taiwan.

The second peculiar feature is that the early penetration of Taiwanese MNCs in developed countries was premature and unsustainable. With the lack of technological capabilities (including the ability to produce products geared to higher-income markets) and the inability to establish linkages with local component suppliers in the developed countries, the initial thrusts in the growth of Taiwanese FDI in the developed countries in the 1970s and 1980s proved unprofitable and unsuccessful. This helps to explain why international production in developing countries that were the predominant host countries of Taiwanese FDI in the 1960s and 1970s once again received the bulk of Taiwanese FDI in the 1990s.

Notes

1 The data in this paragraph were sourced from Tolentino (1993) for 1960, and UNCTAD (1999) for 1980, 1990 and 1998.
2 During Taiwan's import substitution stage of the 1950s, the Taiwanese government was actively involved in determining the country's industrial structure, owning most of the largest businesses and managing much of the trade flows through protectionism, exchange rate manipulation, and price setting. The development of locally controlled infant industries was enhanced by government subsidies and protection from foreign competition through the imposition of high tariff barriers. Although this led to the creation of a burgeoning class of entrepreneurs, the gains from an artificially created comparative advantage in heavy industry had become exhausted rapidly and hence around 1960, Taiwan's policy of import substitution had given rise to one of export promotion (Aggarwal and Agmon, 1990).
3 Unless otherwise indicated, the data on Taiwanese FDI refer to the approved value of Taiwanese FDI as provided by the Taiwanese Ministry of Economic Affairs, Investment Commission. All data in this chapter have been derived from this source and from various issues of the *Statistics on Outward Investment and Outward Technical Cooperation, R. O. C.* The stock of approved Taiwanese outward FDI represent cumulative flows of approved outward FDI since 1959. This should not affect the time-series analysis of the broad trends in the industrial and geographical patterns of Taiwanese FDI. However, it is important to appreciate that the amount of actual FDI is more often than not several orders of magnitude larger in comparison to FDI based on registration or approvals. This is because

Taiwanese firms, at least until the political and economic reforms between 1986 and 1988, were not allowed to invest abroad more than 40 per cent of their paid-up capital; have contempt for transparent accounting; and both large and small companies have a preference to raise finance through the network of overseas Chinese to avoid official scrutiny.

4 Thailand was by far the most important host country of Taiwanese FDI in the period from 1959 to 1969. It accounted for a share of 27.9 per cent of the stock of approved Taiwanese FDI in 1969, followed by Singapore with a share of 12.6 per cent and Malaysia with 12.2 per cent. The official data does not allow the dis-aggregation of the category of other countries. All that can be deduced is that this category includes all countries other than Thailand, Malaysia, Singapore, Philippines, Indonesia and the United States.

5 To the extent that there was outward FDI by Taiwanese textile manufacturers in the 1960s, it represented defensive investments in developing countries in response to highly restrictive national quotas placed on their exports from the home base. The investments were directed largely to Singapore and Malaysia where such quotas were yet to be imposed (Chia, 1989).

6 In terms of actual FDI outflows based on balance-of-payments data, the annual average FDI outflows of Taiwan was $2.37 million in the period between 1970 and 1979.

7 Within the food and drinks industry, the major investors were the monosodium glutamate (MSG) producers, while in the plastics industry the major investors were the producers of polyvinyl chloride (PVC) and polyethylene. Most of the firms in the non-metallic industries were cement manufactures (Ting and Schive, 1981).

8 The plastics plant established by a Taiwanese PVC producer in 1975 with an annual capacity of 240,000 tons was considered the fifth-largest PVC plant in the world at that time (Ting and Schive, 1981).

9 Based on information contained in *Far Eastern Economic Review*, Special Advertising Section on The Republic of China in 1990.

10 Its production subsidiaries in Japan, Singapore and the United States primarily assemble household appliances and electronic goods in the production of which the firm had attained success in its home market (Ting and Schive, 1981).

11 Based on UN, *World Investment Report*, various issues.

12 The political reforms were undertaken in light of the increasing affluence of the population, the general rise in national education levels, the influence of Western political ideas, the expansion of economic and cultural contacts abroad as well as public indignation over a series of government scandals, all of which contributed to increase popular pressure for change in the country's political environment (Kapellas and Liu, 1990). The political reforms included the formation of a legal political opposition party and the corresponding growth in importance of public opinion rendered the government increasingly accountable for its actions, thus restricting its ability to impose stringent controls over the outward flow of tourists and capital. The set of rules governing outward FDI from Taiwan until 1986 was contained in a set of regulations – Regulations Governing the Screening and Handling of Outward Investment and Outward Technical Cooperating Projects – and some relevant articles in several statutes. Exchange controls and strict regulation of capital exports were in place to avoid capital flight (Chen, 1986) which served to curtail the transfer of capital overseas (at least officially). As the Taiwanese government regulated outward investments strictly to assess their benefit to the home economy, substantial investments were undertaken through unregulated and unapproved channels (Chia, 1989).

13 As of March 1989, individuals were permitted to transfer a maximum of $5 million per year offshore, an increase from $500,000 earlier that year and $200,000 in

1988. In addition, outward FDI that meets certain criteria become eligible for Export-Import Bank credit and insurance (Chia, 1989). The Central Bank also started to make some foreign reserves available to firms that undertake FDI ('Foreign acquisitions get the nod', *Euromoney*, October 1990). In addition, Chen Li-an, the Minister of Economic Affairs of Taiwan, announced a plan in October 1988 to promote the establishment of 10 to 20 MNCs to accelerate the process of internationalization of local enterprises. More importantly, the Executive Yuan established in April 1988 a $1.04 billion Overseas Development Assistance Fund designed both to aid developing nations and to assist Taiwanese enterprises in overseas investment and the expansion of foreign markets. In addition, faced with a financial infrastructure inadequate for the task of promoting capital intensive and high technology development projects, the government reformed financial regulations to ease access to credit and finance capital (Kapellas and Liu, 1990). Financial market liberalization led to the formation of a domestic financial services industry.

14 Data obtained from 'Investment by Taiwan: The embarrassment of riches', *The Economist*, 25 March 1989. In fact, except in 1974–75 as a result of the oil crises, Taiwan enjoyed a trade surplus since 1971 (Chen, 1986).

15 On attempts by Taiwan's trading partners to demand that Taiwan allow its currency to rise further, see 'Investment by Taiwan: The embarrassment of riches', *The Economist*, 25 March 1989.

16 For example, in some cases outward FDI has been contemplated simply because the head of the company regarded that outward FDI was a 'good thing.' In other cases, the outward FDI of Taiwanese companies had been in sectors outside of their experience or expertise. This was the case in the acquisition by Pacific Wire & Cable of eight failing savings and loan associations in the United States in 1990 for $53 million and the acquisition of Omni Bank by China Rebar which had no prior experience nor expertise in finance (Bamford, 1993).

17 It is sometimes argued that unlike South Korea, Taiwan was able to avoid much of the trade retaliation brought about by the increasing trade pressures owing to the small size of its firms which provided a degree of anonymity, and the important role of many Taiwanese companies as original equipment manufacturers (OEM) whose fate was linked inextricably to that of domestic producers in the developed countries. This made trade retaliation by developed countries a process of self-inflicting pain (Wang and Hsu, 1992).

18 Between 1979 and 1986, Taiwan's exports to the United States rose 336 per cent from $5.65 billion to $18.99 billion, resulting in the growth of the United States trade deficit with Taiwan from $5.33 billion to $13.58 billion. The deficit grew further to $17.44 billion in 1987 (Kapellas and Liu, 1990).

19 This is owing to the allegation by the United States of adverse trading practices exercised by Taiwan and South Korea. Thus, although Taiwan and South Korea have relatively free access to the United States market, these countries were accused of restricting the imports of American goods and services in their countries through import tariffs and non-tariff barriers such as licensing regulations, technical standards, export subsidies and restricted access to foreign participation in the banking, insurance and securities industries. American criticism of Taiwan's trade practices also stem from the latter's large and growing foreign currency reserves second only to Japan which although regarded as an instrument by Taiwan to maintain political and economic security was enough to cover almost three years of import needs. Furthermore, the United States was concerned that Taiwan manipulated their exchange rate to maintain the cost competitiveness of its exports. Many of Taiwan's efforts to correct the problem were seen to be not effective enough (Chia, 1989).

20 For example, Acer, regarded as Taiwan's IBM, had established 11 branches in Europe in the 1980s, including that for production in Helmond, the Netherlands. Delta Electronic Industrial Co. also unveiled its first European expansion project – a $28 million investment to set up a manufacturing facility near Glasgow over a period of five years. Mitac International Corp. also planned to build a plant in the United Kingdom and Plus & Plus Co. and Microtek International Inc. had been studying plans for establishing production lines in Europe. See 'Taiwan companies race to Europe to establish niche ahead of 1992', *The Asian Wall Street Journal Weekly*, 31 December 1990.

21 As a result in part of the growth of Taiwanese FDI in Europe in the 1980s, Taiwan's bilateral trade with Europe accounted for 16.3 per cent of its total foreign trade in 1989, a significant increase from 10.7 per cent in 1985. Correspondingly, the share of the United States in Taiwan's total foreign trade declined to 30.4 per cent in 1989 from 38.4 per cent in 1985. See 'Taiwan companies race to Europe to establish niche ahead of 1992', *The Asian Wall Street Journal Weekly*, 31 December 1990.

22 See 'Taiwan companies race to Europe to establish niche ahead of 1992', *The Asian Wall Street Journal Weekly*, 31 December 1990.

23 The principals in Channel International are the privately owned China Trust Group, Pacific Petrochemical, USI Far East, the Mitac Group (a leading manufacturer of personal computers) and a new development fund set up by the Taiwan government. See 'Taiwanese buy Wyse Technology', *Financial Times*, 12 December 1989; 'Other Asian investors', *The Asian Wall Street Journal Weekly*, 14 January 1991. See also 'Despite setbacks, Taiwan's Acer group intensifies its efforts to expand globally', *The Asian Wall Street Journal Weekly*, 27 May 1991.

24 See 'Why Taiwan is not another Japan', *The Financial Times*, 12 September 1990.

25 Based on a statement of Lip-bu Tan, general partner of the Walden Group of Venture Capital Funds, San Francisco, USA, as contained in 'Other Asian investors follow Japanese to US pouring cash into high-technology companies', *The Asian Wall Street Journal Weekly*, 14 January 1991.

26 Based on United Nations, *World Investment Report*, various issues.

27 See 'A Leading Source of Investment Capital', *International Herald Tribune*, 10 October 1994.

28 In 1987, Taiwan claimed to have the lowest unemployment rate of 1.86 per cent compared to that of South Korea of 2.3 per cent and Japan of 2.7 per cent (*Free China Journal*, 11 January 1988).

29 Political liberalism tolerated a rash of strikes – almost illegal – by workers demanding higher bonuses. See 'Investment by Taiwan: The embarrassment of riches', *The Economist*, 25 March 1989. As a result, wages in the industrial sector increased by 10.9 per cent in in 1988, 14.6 per cent in 1989 and 13.5 per cent in 1990 (van Hoesel, 1997).

30 In fact, it rose by 19 per cent between 1991 and 1992. The increasing pressures exerted by the United States on Taiwan to appreciate the New Taiwan dollar against the American dollar has been made in light of the allegation made by the United States Treasury in May 1992 that Taiwan had been manipulating its exchange rates to reduce upward pressures on the currency and to maintain international competitiveness as reflected in its rising foreign exchange reserves and trade surpluses. See 'Too strong too long: Taiwan exporters rail at rising currency', *Far Eastern Economic Review*, 30 July 1992.

31 The threat to exports was substantial given that labour intensive manufactures comprised roughly 45 per cent of Taiwan's total exports in 1985 (Kapellas and Liu, 1990).

32 See 'Too strong too long: Taiwan exporters rail at rising currency', *Far Eastern Economic Review*, 30 July 1992.

33 The Taiwanese government had identified ten industries of the future and offered many incentives to investors in these areas, including through inward FDI. This included telecommunications, information, consumer electronics, semiconductors, precision machinery and automation, aerospace, advanced materials, fine chemicals and pharmaceuticals, health care and pollution control. See 'A leading source of investment capital', *International Herald Tribune*, 10 October 1994.

34 For example, an electronics company that sets up a low-end assembly line in Thailand may refocus its Taiwan factories on making higher-technology components which can be exported to the Thailand assembly plant. See 'The upstart taipans', *Far Eastern Economic Review*, 19 April 1990.

35 International production by Taiwanese firms in textiles grew in a significant way starting in 1989. Between 1988 and 1992, the stock of approved Taiwanese FDI in textiles increased at an annual average rate of 97 per cent. Taiwanese FDI in electronic and electrical appliances which in 1992 accounted for one-third of the stock of approved Taiwanese FDI in manufacturing (or 19 per cent of the stock of Taiwanese FDI in all industries) grew by 57 per cent.

36 Such ranking of the major types of Taiwanese FDI in the 1990s is largely consistent with the findings of a 1992 survey by the Ministry of Economic Affairs which analysed the motivations of Taiwanese FDI: access to an abundant supply of cheap labour, developing overseas markets, promotion of sales to the host or neighbouring markets, more effective utilization of existing company equipment, more effective utilization of company personnel, access to raw materials, reduction of foreign exchange rate fluctuation, and securing of orders (Bamford, 1993).

37 Taiwanese manufacturers, particularly those who qualify as pioneer industries, may receive income credits or exemptions on income and investment taxes for periods up to eight years after the start of operations. The ASEAN governments also offer import and export duty exemptions and simplified customs procedures on imported raw materials and manufacturing equipment for firms engaged in export manufacturing. Firms which invest in remote or targeted locations may also receive subsidies or discounts on utilities and local licence fees (Korea Institute for Foreign Investment, 1989; Chia, 1989).

38 See 'Taiwan's offshore empire', *Far Eastern Economic Review*, 18 March 1993; 'No more Mr Nice Guy', *Far Eastern Economic Review*, 18 March 1993; and 'Taiwan's offshore empire', *Far Eastern Economic Review*, 18 March 1993. About 400 Taiwan-based manufacturers of all types were present in Malaysia in 1992. Of these, 110 firms were in electronics engaged in the production of both parts and final products such as electronic calculators, facsimile machines, telephones, audio equipment and television screens. Most of these were clustered in industrial parks in the northern state of Penang. Others are involved in steel, textile or paper-based products. However, Taiwan's popularity in Malaysia has waned because its aggressive business presence led to rising rents and their investments were regarded as footloose and either concentrated in labour intensive industries or in pollution intensive industries. From Taiwan's point of view, Malaysia is also losing its locational advantages in production with high labour costs, industrial thefts, social problems, difficulty in obtaining visas, etc. Thus Taiwan has fallen from its pinnacle as one of Malaysia's largest investor in the manufacturing sector to the one most likely to quit in the face of a beckoning Chinese market that instituted economic reforms, and the increase in investment incentives in Vietnam, Indonesia and Thailand. See 'No more Mr Nice Guy', *Far Eastern Economic Review*, 18 March 1993.

39 See also *Business Asia*, 30 September 1991.

40 See 'A leading source of investment capital', *International Herald Tribune*, 10 October 1994.

41 See *Business Asia*, 30 September 1991.

42 In 1991, while there were more than 2,000 Taiwanese firms in China, there were only more than 1,000 factories established in Southeast Asia. See 'Taiwan's trade pattern turns upside down', *Financial Times*, 4 June 1991; 'Taiwan firms' zeal for investing in Southeast Asia may be ebbing', *The Wall Street Journal*, 18 January 1991; 'Taiwan's offshore empire', *Far Eastern Economic Review*, 18 March 1993.

43 See *Business Asia*, 30 September 1991.

44 See 'A leading source of investment capital', *International Herald Tribune*, 10 October 1994.

45 See 'Strait ahead', *Far Eastern Economic Review*, 5 March 1992.

46 See 'Strait ahead', *Far Eastern Economic Review*, 5 March 1992.

47 See 'Investment by Taiwan: The embarrassment of riches', *The Economist*, 25 March 1989.

48 See 'The Vietnam option', *Far Eastern Economic Review*, 31 October 1991; 'Open for business: Vietnam's economy gets big lift from Taiwanese', *Far Eastern Economic Review*, 18 March 1993.

49 See 'A leading source of investment capital', *International Herald Tribune*, 10 October 1994. By comparison, Taiwan is far behind Japan and the United States in terms of the size of its FDI in China. See 'Open for business: Vietnam's economy gets big lift from Taiwanese', *Far Eastern Economic Review*, 18 March 1993.

50 One example of this type of FDI by Taiwanese MNCs was that of China General Plastics Corp which participated in a $480 million petrochemical project in Malaysia. The three largest pulp and paper companies of Taiwan have also set up mills in Indonesia, Thailand and Vietnam. The case of Yuen Foong Yu Paper Manufacturing Co. which invested $400 million to expand its existing paper and pulp mills in Indonesia in light of that country's rich natural resources, cheap labour, relatively weak pollution laws and large domestic market is an excellent case in point. See 'The upstart Taipans', *Far Eastern Economic Review*, 19 April 1990; 'Taiwan's offshore empire', *Far Eastern Economic Review*, 18 March 1993.

51 See 'The Vietnam option', *Far Eastern Economic Review*, 31 October 1991.

52 Some large-sized Taiwanese firms have penetrated the Vietnamese market. For example, Ching Fong Investment received approval for two motor cycle factories and a $288 million joint venture for a cement plant in the northern port city of Haiphong. This made Ching Fong the largest foreign investor in Taiwan. See 'Open for business: Vietnam's economy gets big lift from Taiwanese', *Far Eastern Economic Review*, 18 March 1993.

53 See 'Eye on China', *Far Eastern Economic Review*, 12 September 1991.

54 See 'Taiwan companies race to Europe to establish niche ahead of 1992', *The Asian Wall Street Journal Weekly*, 31 December 1990.

55 This consisted of the $1.7 billion investment of the Formosa Plastics Group to include a plant that produces ethylene as well as at least six downstream plants that processes ethylene into polyethylene and other derivative products. There were further intentions of establishing more petrochemical processing plants in the United States over the 1990s, including in Louisiana. The size of the FDI of the China General Plastics Corporation in its American affiliate, Westlake Polymers Corporation, was far more modest at more than $200 million. See 'Taiwan's US strategy', *Forbes*, 29 May 1989. The investment of the Formosa Plastics Group was considered the largest outward FDI by a Taiwanese company approved by the Ministry of Economic Affairs at the time. See 'Taiwan plastics group to set up US units', *Far Eastern Economic Review*, 24 August 1989.

56 In plastics production, ethylene is converted into polyethylene and other derivative products. The intention of the Formosa Plastics Group at the time of its investment expansion was to export to Taiwan some 70 per cent of its ethylene production in Texas. However, over time the company expects such exports to

decline as more plastic-based consumer goods are produced in the United States. On the other hand, the China General Plastics Corporation through its United States affiliate, Westlake Polymers Corporation, intends to use a similar percentage of its ethylene output to feed the input requirements of the two polyethylene plants the company acquired from Cities Service in 1986. See 'Taiwan's US strategy', *Forbes*, 29 May 1989.

57 Of the various gains from the acquisition of Wyndham Biscuits, the access to the marketing networks was regarded to be the most important gain to President Enterprises. The marketing network of Wyndham Biscuits was considered impeccably grassroots with more than a third of the company's products distributed and sold through delivery vans (with on-board computers for inventory control) that travel along operator-owned franchise routes. Another 25 per cent of the company's products were sold through annual door-to-door fund-raising by legions of American Girl Scouts. This marketing approach serves the interests of President Enterprises best compared to competing head-on with established name brands in food products at the national level. See 'Recipe for success', *Far Eastern Economic Review*, 21 March 1991.

58 See 'Recipe for success', *Far Eastern Economic Review*, 21 March 1991.

59 This included among others the $1 million investment of Tatung Co., a Taiwanese consumer electronics company, in GraphOn Corporation, an eight-year-old maker of computer window displays based in San Jose. The purpose of Tatung was to use GraphOn's products in the production of its computer systems. Hualon Microelectronics Corporation has only purchased 10 per cent of Seeq Technology Inc., a computer chip maker also based in San Jose, for $5.3 million. As part of the deal, Seeq chips will be made in Taiwan, thus assisting Hualon Microelectronics Corporation to use more of its capacity as a foundry for semiconductors. There is also the joint venture between a research institute funded by the Taiwanese government and IBM in International Integrated Systems Inc., a software company. See 'Despite setbacks, Taiwan's Acer group intensifies its efforts to expand globally', *The Asian Wall Street Journal Weekly*, 27 May 1991; 'Other Asian investors follow Japanese to U.S. pouring cash into high-technology companies', *The Asian Wall Street Journal Weekly*, 14 January 1991.

60 See 'Despite setbacks, Taiwan's Acer group intensifies its efforts to expand globally', *The Asian Wall Street Journal Weekly*, 27 May 1991; 'Other Asian investors follow Japanese to U.S. pouring cash into high-technology companies', *The Asian Wall Street Journal Weekly*, 14 January 1991.

61 This is with the exception of Macronix which has developed a different product mix including chips for facsimile machine modems, chipsets to power graphic displays in PCs, sophisticated programmable memory devices called Eproms and data communications semiconductors for use in computer networks. See 'Macronix hopes marketing skill will overcome local pitfalls', *The Asian Wall Street Journal Weekly*, 25 November 1991.

62 See 'Recipe for success', *Far Eastern Economic Review*, 21 March 1991.

63 See *Business Taiwan*, 28 October 1991.

64 See 'The upstart Taipans', *Far Eastern Economic Review*, 19 April 1990.

11 The emergence and evolution of multinational corporations from South Korea

Introduction

As with Taiwanese MNCs, the origins of South Korean MNCs can be traced to the late 1950s and early 1960s. However, the stock of outward FDI from South Korea in 1998 at $21.5 billion is much more modest compared to that of Taiwan of $38 billion. Although the level of South Korean FDI stock in 1998 was roughly comparable to that of Japan in 1980, its share in the global stock of outward FDI was low at 0.5 per cent and even lower than the share of Japan in 1960 at 0.7 per cent. Nevertheless, in relation to the stock of outward FDI from developing countries in the same year, South Korea assumed greater relative importance with a share of 5.5 per cent. Based on the size of outward FDI stock in 1998, South Korea was only the fifth-largest home country based in developing countries after Hong Kong, Singapore, Taiwan and China.[1]

Despite the low relative importance of South Korean FDI, the study of the emergence and evolution of South Korean MNCs is of interest as another case study of MNCs from a resource-scarce developing country with a large domestic market. The growth pattern of South Korean MNCs as it has been evolving over the last 40 years can be compared most directly to those of Taiwan, another newly industrialized country in Asia, or more broadly to those of MNCs from the developed countries such as the United Kingdom, Germany and Japan whose home countries share similar patterns of national economic development and whose histories have been analysed in the previous chapters in this part of the book. The analysis of the development of South Korean MNCs in this chapter is divided in three time frames: from the 1950s to the 1970s, the 1980s and the 1990s.

The emergence of South Korean MNCs from the 1950s to the 1970s

The emergence of South Korean MNCs has to be understood in the context of the industrialization of its domestic economy. At the conclusion of the Korean War (1950–53), South Korea had an open dualistic economy with an unfavourable natural resource endowment but a large supply of surplus labour with a relatively high level of education and skills. Typical of the industrial development of most developing countries, the Korean economy comprised a large traditional subsistence agricultural sector relative to its non-agricultural sectors in spite of its inherent natural resource scarcity. With foreign trade as an important agent of economic growth in post war South Korea, the pattern of domestic production and exports developed rapidly from traditional agricultural- or land-based products towards more non-traditional labour-based industrial products in which the country had been building a comparative advantage (Jo, 1981).

Such process of domestic industrial transformation was supported by an import substitution development strategy implemented in various stages with the initial phase pertinent from the end of the Korean War to 1965. In this early phase of South Korea's industrial development, the policy of import substitution was aimed at the development of domestically controlled industries producing consumer goods that had previously been imported in relatively large quantities but which could be produced domestically with relatively simple technologies. Some of the pioneering industries that developed with imports of capital goods and raw materials were industries processing natural resources to include flour milling, sugar refining, cotton and wool spinning and weaving, food processing, plywood manufacture and so forth. Thus, as in the history of the industrial development of the United Kingdom, South Korea cultivated a comparative advantage in labour intensive, capital neutral and human capital-scarce products. Such process of import substitution was 'deepened' in a small number of large-scale activities in earlier years, as well as expanded to a large number of small-scale activities in later years.[2] For example, the industrial development policy of the early 1970s emphasized the development through import substitution of consumer durables, intermediate inputs, and capital goods (Jo, 1981).

With the rapid growth and development of natural resource-based consumer goods industries in the mid-1960s, the traditional pattern of South Korean exports based on primary products was replaced rapidly by exports of consumer goods produced with the use of relatively unskilled labour and modest amounts of capital. Although a number of new firms were established specifically for the purpose of generating these exports, the majority of the exports were made possible by the large number of firms whose competitiveness and growth of production was achieved from the policy of import substitution but whose growth of production could not be absorbed fully by domestic demand. The year 1965 is thus regarded as the turning point of South Korean industrialization and trade policies away from import substitution towards export promotion starting initially with the domestic production and exports of labour intensive consumer goods. This era coincided with the complete shift in government policy away from direct controls towards economic liberalization consistent with a more market-oriented and export-oriented economy.[3]

It was during the tail end of the import substitution industrialization phase that the first South Korean MNCs emerged. In broad parallelism with the early foreign activities of the Japanese *sogo shosha*, the pioneering outward FDI on record involved the establishment in 1959 of a branch office in the United States by the Korean Traders Association through the acquisition of a commercial building in New York City. Such acquisition was financed from foreign exchange earnings derived from South Korean tungsten exports to the United States (Jo, 1981). This investment foreshadowed the beginning of the export-oriented industrialization phase of the South Korean economy.

The second case of South Korean FDI involved a timber operation in Malaysia initiated by a South Korean resident in 1963 (Jo, 1981). The investment was motivated by the necessity of securing a supply of timber to sustain the growth and development of the South Korean-based plywood industry, one of the consumer goods industries fostered by the import substitution industrialization phase whose products had rapidly become one of the country's principal exports. Indeed, resource extractive outward FDI was a central feature in the emergence of South Korean MNCs, unlike in the case of Taiwan. Outward FDI of this type accounted for as much as one-quarter of the total stock of approved South Korean FDI at the end of the 1970s.

The insignificant amounts of South Korean FDI in the import substitution industrialization phase was in part a reflection of the authorizations for outward FDI which only began to be provided starting in 1968 with the realization by the South Korean government of the important role of outward FDI in securing access to raw materials, expanding exports and promoting international economic cooperation with developed and developing countries (Jo, 1981). In fact, there has been a broad parallelism in the development of outward FDI policy in Japan and South Korea (Randerson and Dent, 1996). The Foreign Exchange and Trade Control Act of 1949 which tightly regulated Japanese overseas investment for two decades had its equivalent in the Foreign Exchange Control Regulations of 1968 which fulfilled a similar function in South Korea until the gradual liberalization of South Korean outward investment policy since 1986. Both foreign exchange acts were implemented at an early phase of industrialization in both countries with the presence of persistent balance-of-payments deficits and the emphasis placed on directing financial resources to achieving rapid domestic industrial growth, both of which served to restrict the availability of investment capital.[4] The strict controls on outward FDI to emphasize domestic industrial development was reinforced in both countries by the restrictions on inward FDI to insulate domestic firms from foreign competition. Despite the tight regulation, however, the overseas investment projects that were approved were extended a number of investment incentives, including low-cost loans from the Korea Eximbank or the Overseas Resource Development Fund (in the case of projects that develop mineral resources overseas), payment guarantees by South Korean commercial banks, protection from investment risks, and an information service (Koo, 1984).

By 1979, the stock of approved South Korean FDI reached some \$126.4 million.[5] Although relatively small in amount, it was more than double the stock of approved Taiwanese FDI in 1979 of \$59.3 million. Similarly, a comparison of the actual FDI outflows of the two countries based on balance-of-payments data shows that the annual average FDI outflows of South Korea of \$9.8 million between 1970 and 1979 was more than four times that of Taiwan of \$2.4 million in the same period, and the estimated stock of South Korean FDI at \$142 million in 1980 was almost one and one-half times larger than that of Taiwan at \$97 million.[6] This provides evidence of the greater relative importance of South Korean FDI *vis-à-vis* Taiwanese FDI at the end of the 1970s.

The services sector was the most important sector of South Korean FDI at the end of the 1970s with a share of 57 per cent, followed by the primary sector (26 per cent) and the manufacturing sector (17 per cent). The importance of services in South Korean FDI reflected the prominence of trading (20 per cent), construction (13 per cent) and property (9 per cent). In terms of geographical destination, developing countries accounted for 72 per cent of South Korean FDI and developed countries for 28 per cent. Within developing countries, the most important regions were South East Asia (43 per cent), Africa (21 per cent) and the Middle East (7 per cent). By contrast, FDI in developed countries had been far more concentrated in North America (23 per cent) than in Europe (3 per cent).

Four main types of outward FDI undertaken by South Korean MNCs was evident by the end of the 1970s in order of declining importance: first, outward FDI to secure supplies of essential raw materials to serve the South Korea-based industrial production complex; second, outward FDI in trading, warehousing and transportation in major foreign markets to promote growth of South Korean exports; third, manufacturing FDI to supply host country markets or as export platform; and fourth, outward FDI in civil construction and engineering-related works to facilitate growth of South Korean exports of services of skilled and semi-skilled labour. Each of these types of South Korean FDI in the 1970s is discussed below.

FDI to secure supplies of essential raw materials

The inherent poverty of South Korea in natural resources, the relatively poor performance of the agricultural sector combined with the rapid growth of consumer goods industries intensive in the use of natural resources as raw materials served to underscore one of the ironies in South Korean industrial development in its high degree of dependence on imports of natural resources. Indeed, the share of natural resources-related imports in the country's total commodity imports reached nearly 50 per cent in the 1970s and 1980s (Jo, 1981). As in the United Kingdom, Germany and Japan, the threat to the role of arms-length trade in securing the vital natural resources and raw materials abroad to support domestic industrial expansion precipitated the growth of resource extractive FDI by local firms owing to market failure of various forms. In addition, there were other factors that determined the growth of this type of FDI in the case of South Korea to do with the worldwide energy crisis in the 1970s and the growing trend towards resource nationalism on the part of resource-rich countries. Such changes in the international economic environment prompted South Korean firms to intensify their efforts to secure access to overseas natural resources, in most cases through the establishment of joint ventures with local partners.

This explains why the primary sector, which accounted for some 26 per cent of the stock of approved South Korean FDI at the end of the 1970s, was the most important type of South Korean FDI. Of this share, the forestry sector accounted for 19 per cent, the fisheries sector for 7 per cent, while mining accounted for an almost insignificant share of 0.35 per cent. The extraction of timber in countries of South East Asia that started in 1963 as previously mentioned became the predominant form of FDI of this type with the intention of gaining assured access to timber supplies to support the growth of production of the South Korean-based plywood industry which was also a major export industry until the late 1970s. In this industry, joint investment ventures with majority South Korean ownership prevailed since the investments which were concentrated in Indonesia were implemented prior to the strengthening of resource nationalism in that country in the late 1970s. South Korean FDI in the fisheries industry, though far less important by comparison to FDI in timber extraction, was spread more evenly across the world except the Middle East and Oceania. About 80 per cent of South Korean FDI in this industry at the end of the 1970s was concentrated in Africa, 12 per cent in Latin America and about 6.5 per cent in North America. In their resource extractive FDI, South Korean firms provided capital and some technicians, although in some cases heavy equipment was also exported to their foreign subsidiaries (Koo, 1984).

Export-oriented FDI in trading, banking, warehousing, transportation and distribution

South Korean FDI in trading as well as FDI in transportation and warehousing which began in the late 1970s with the growing volume of South Korea's international cargo (Koo, 1984) accounted for some 20 per cent of the stock of approved South Korean FDI at the end of the 1970s and thus was the second most important type of FDI during this period. The main geographical destination for South Korean FDI of this type was Africa (43 per cent), North America (27 per cent), Europe (12 per cent), and South East Asia (12 per cent).

This type of FDI had been a natural outcome as well as a cause of the rapid growth of industrial exports associated with the export-oriented industrialization phase of the South Korean economy.[7] There were two factors that fostered the growth of South Korea's industrial exports as well as export-oriented

FDI in trading, banking, warehousing, transportation, distribution and marketing in support of such export expansion. The first factor was the important role of industrial exports in the finance of imports of natural resources, raw materials and intermediate inputs as well as food. The second related factor was the growth after 1974 in the aftermath of the first oil shock of protectionism in developed countries against industrial exports from developing countries. This served to undermine the basic premise of South Korea's export-oriented growth that world markets would continue to purchase a constantly growing volume of South Korean industrial exports and provided the basis for the defensive motivation by many South Korean manufacturing companies and general trading companies as trading arms of the large South Korean conglomerates (*chaebol*) to set up overseas branch offices in sales or trading, warehousing facilities, and distribution channels in major export markets as a means to promote South Korean exports of consumer and producer goods.

This type of FDI can sometimes be regarded as a first step in the internationalization of export-oriented South Korean manufacturing firms. To ensure the continued growth of industrial exports in the face of the growth of protectionism and competition in major export markets, a number of South Korean exporting firms have attempted to integrate forward through the establishment of overseas branch offices in sales or trading, warehousing facilities, and distribution channels in their foreign markets. While the great majority of trading subsidiaries abroad were wholly owned or majority owned by South Korean firms, in a few cases joint investments were made with local partners to combine their production capabilities with their local partners' marketing skills or to overcome unknown marketing systems in remote countries.[8] Outward FDI of this type helped to increase South Korean exports substantially and in many cases also served to facilitate its imports (Koo, 1984).[9]

Mention should also be made in the context of export-oriented FDI of South Korean banks that have similarly embarked in outward FDI in support of the growth of industrial exports of South Korea. A number of South Korean banks have set up overseas branch offices and in some cases through joint ventures in world financial and trading centres to tap foreign capital resources to support South Korean industrial growth and to accommodate the financial needs of South Korean exporters and investors (Jo, 1981).

FDI in manufacturing

This type of FDI which accounted for some 17 per cent of the stock of approved FDI in 1979 was the third most important type of South Korean FDI at the end of the 1970s. The first FDI of this type occurred in 1973 to manufacture food chemicals in Indonesia as well as cement in Singapore with investment stakes of $1.2 million and $1.3 million, respectively.[10] By the end of that decade, the major geographical destination of South Korean manufacturing FDI had been South East Asia (49 per cent), Africa (38 per cent) and Oceania (7 per cent). Although the investments followed a product cycle trend in its concentration in lesser industrialized countries than South Korea in the same way as in the emergence of Japanese and Taiwanese MNCs, there were at least two exceptions in the South Korean case: the printing plant established in Japan in 1975 in which a South Korean firm had a 50 per cent investment share and a pulp project in New Zealand initiated in 1977 in which a South Korean firm had a 10 per cent investment share (Kumar and Kim, 1984).

The most important factors determining the location of international production by Korean manufacturing firms had been the availability in relative abundance of the necessary raw materials for industrial production such as *inter alia* lumber, limestone, pulp, and molasses. The majority of manufacturing FDI in the 1970s had been directed to developing countries in which South Korean firms had served previously through exports (Jo, 1981).[11] Outward FDI in manufacturing had been predominantly domestic market oriented rather than a sourcing type of investment as seen by the far larger size of exports of South Korean parent companies to their foreign subsidiaries of raw materials or intermediate inputs by comparison to imports to South Korea from their foreign subsidiaries (Koo, 1984). Unlike in the case of Taiwanese manufacturing MNCs, only in a few cases have South Korean manufacturing MNCs initiated overseas manufacturing activities for the export of semi-processed raw materials to South Korea. This was the case with Sun Kyong Company that established an Indonesian subsidiary to manufacture plywood, part of which was exported to South Korea in anticipation of an impending prohibition by South Korea on the export of lumber (Kumar and Kim, 1984). What was far more common was export platform type of FDI in host countries as a base to export to third countries. Indeed, nearly one-third of South Korean overseas manufacturing projects as of 1980 have been primarily involved in export platform type of FDI facilitated by the ability of some South Korean manufacturing firms and trading companies to market merchandise exports in international markets (Kumar and Kim, 1984).

The most important manufacturing industries of South Korean FDI at the end of the 1970s were non-metallic mineral products, food, beverages and tobacco, paper and paper products, and basic metal (steel) products. These industries which accounted for 78 per cent of cumulative FDI in manufacturing as of 1977 were the industries whose growth and development was spawned by the later phases of import substitution industrialization of South Korea in the early 1970s and included not only consumer goods industries produced using labour intensive means of production with modest amounts of capital but also heavy and chemical producer and capital goods industries. These were industries in which South Korea's comparative advantage was eroding rapidly because of its natural resource scarcity and the tide of resource nationalism in resource-rich foreign countries (Jo, 1981). An analysis of the type of firms engaged in outward FDI of this type showed that some 30 of the 37 manufacturing projects as of 1982 represented the horizontal integration of manufacturing firms, while only 7 projects were projects undertaken by general trading companies to secure semi-processed raw materials such as pulp and wool (Koo, 1984).

In particular, medium-sized manufacturing firms engaged in the semi-skilled labour intensive domestic production of resource intensive products in South Korea predominated in generating outward FDI of this type in the 1970s. Owing in part to their new status as MNCs and unfamiliarity with the conduct of international business in foreign countries as well as the restrictions on foreign ownership particularly in developing countries, some 65 per cent of the overseas manufacturing projects of South Korean firms in this period (or 24 of the 37 manufacturing projects in 1982) and some 71 per cent of those in developing countries were those in which South Korean firms had minority equity participation. Nevertheless, although South Korean firms owned less than 50 per cent of the equity, in most cases the South Korean wielded control over the foreign operation as the major shareholder and the provider of superior skills in production, management, and marketing. In other cases, economic cooperation with host country government resulted in joint ventures of equal equity participation between the public corporations in the country and the South Korean firm. Only 8 of the 28 manufacturing projects in developing countries were either majority owned or wholly owned by South Korean firms. Majority-owned FDI occurred in the international production of food seasoning, steel rods, fountain pens, apparel, shoes, sanitary rubber products, cement and consumer electronic products which used mature technologies in their production and in which South Korean firms enjoyed a competitive advantage *vis-à-vis* firms in both developed and less developed countries (Koo, 1984).

The self-perceived ownership-specific assets of South Korean manufacturing MNCs were analysed by Kumar and Kim (1984) to lie in order of importance: the ability to initiate and operate overseas projects at relatively low costs, the suitability of their operating technology, the lower costs of their expatriate staff, the suitability of their products, and their skills in marketing. In fact, these firm-specific assets are not mutually exclusive as the price competitiveness of the firms is closely related to their lower costs of production (including lower costs of expatriate staff), the suitability of their operating technologies and products and their international marketing networks including those established by the South Korean general trading companies affiliated to several leading business groups which have also been instrumental in establishing necessary contacts between South Korean firms and their collaborators in host countries. Indeed, one of the principal sources of competitiveness of South Korean manufacturing MNCs derived from the replication in foreign countries of production methods based on the adaptation of foreign technology and/or standardized process to relatively small scale of operations and some small or minor adaptation of product designs to the requirements of developing countries. Such technological advantages had emanated from the machine shops and assembly lines of South Korean plants during import substitution industrialization and had been cultivated through the process of learning by doing. Most of the technical modifications consist of the greater use of labour in the core process, including handling, packaging and storage together with greater use of manual quality control, more intensive machine maintenance, and the upgrading of lower quality raw materials into quality inputs via manual sorting (for example, wool and cotton yarn). South Korean manufacturing MNCs had also developed a greater cost competitiveness owing to the lower wages of technicians and semi-skilled and unskilled workers in host developing countries, and their more flexible business attitudes associated with their relatively small size and informal organization (Jo, 1981).

FDI in civil construction and engineering

Outward FDI by South Korea in civil construction and engineering which accounted for 13 per cent of the stock of approved FDI at the end of the 1970s was the fourth most important type of South Korean FDI in this period. The United States (43 per cent), the Middle East (35 per cent) and South East Asia (21 per cent) were the major host countries of South Korean FDI of this type.

As in the cases of the second and third types of South Korean FDI above, South Korean FDI in civil construction and engineering had been regarded by the South Korean government as another instrument to promote the growth of exports of South Korean goods and services generally as well as to compensate for the high import dependence of the South Korean economy for natural resources and raw materials as well as for agricultural produce. The direct exchange of crude oil for exports of construction services as part of outward FDI in civil construction and engineering in the nations comprising the Organization of Petroleum Exporting Countries (OPEC) in the 1970s is an excellent case in point. In addition, some South Korean consulting engineering companies also embarked on international operations by the end of the 1970s, particularly in the Middle East to meet the growing technical needs of South Korean and local firms. These firms have played a catalytic role in the promotion of joint ventures on the basis of their broad international contacts and accumulated technical expertise (Jo, 1981).

The competitive assets of South Korean civil construction contractors and workers derived from accumulated technical and organizational capabilities as well as long experience in large-scale overseas civil construction projects developed during the Vietnam conflict. South Korean contractors had competitive advantages over their foreign competitors in the supply of skilled and semi-skilled manpower and sophisticated and specialized modern construction equipment and construction materials through exports from their parent companies, the use of modern labour intensive technology for construction and engineering-related operations, and flexible attitudes in dealing with local authorities (Jo, 1981). The provision of capital goods, however, was not a strong feature of a majority of South Korean construction companies; it was evident in only 4 of the 33 overseas construction projects of South Korean firms in 1982 (Koo, 1984).

As in the case of overseas manufacturing projects, some 64 per cent of overseas construction projects of South Korean firms in 1982 (or 21 of the 33 projects) were those in which the South Korean civil construction companies and engineering companies had minority equity participation. In particular, less than majority owned joint ventures had been particularly concentrated in Saudi Arabia where the largest investment stakes had been made (16 of the 21 projects) (Koo, 1984). Their main reason for establishing joint ventures with local Arab partners was that local participation was a legal requirement to be eligible for competitive bidding in many classes of contracts. In other cases, where local participation was not formally required but in which there was the practice of treating local firms more favourably to foreign firms in awarding construction contracts, South Korean contractors also concluded joint venture arrangements with local partners. Nevertheless, despite the minority equity participation of South Korean construction companies it was most often the case that these companies had effective control since their local joint venture partners lacked necessary skills, capital and experience to carry out large-scale development projects. Thus, the formation of joint ventures can be regarded in most cases as mere formalities to bypass host governments' restrictions and facilitate the conduct of business in foreign countries (Koo, 1984). In other cases, joint ventures fulfil a more important role in overcoming the weaknesses of South Korean firms in heavy engineering and chemical processing technology. This explains the joint venture between one South Korean firm and an American contractor in order to combine South Korean manpower with advanced foreign technology and engineering know-how (Jo, 1981).

The expansion of South Korean FDI in the 1980s

The decade of the 1980s represented an era of rapid growth of South Korean FDI. Based on balance-of-payments data, the annual average FDI outflows of South Korea was about $114.6 million in the period between 1980 and 1989 compared to $9.8 million in the period between 1970 and 1979, and the estimated stock of actual South Korean FDI reached $1.4 billion in 1989 compared to $141.9 million in 1980. Despite the rapid growth in South Korean FDI over the 1980s, such growth was not as rapid as that of Taiwanese FDI which made Taiwan assume greater relative importance as a home country for FDI at the end of the decade. On an approved basis, the stock of South Korean FDI at also $1.4 billion in 1989 compared to that of Taiwanese FDI at $1.5 billion. The greater relative importance of Taiwanese FDI compared to South Korean FDI at the end of the 1980s is even more evident in the analysis of actual FDI flows in terms of balance-of-payments data and the size of the outward FDI stock. The annual average FDI outflows of Taiwan of \$1.2 billion in the period between 1980 and 1989 as indicated in the previous chapter was more than ten times as large as that of South Korea at \$114.6 million and thus the estimated stock of outward FDI of Taiwan which reached \$7.6 billion in 1989 was more than five times as large as that of South Korea at \$1.3 billion. Indeed, the 1980s marked the decade in which the Taiwanese MNCs had overtaken the South Korean MNCs in terms of the size of their outward FDI.

Despite the less dramatic rate of growth of South Korean MNCs compared to Taiwanese MNCs over the decade of the 1980s, there were important economic and political forces that explain the growth as well as major structural changes in the pattern of South Korean FDI in that decade. The economic pressures that contributed to this phenomenon were similar to those faced by Taiwan in the same decade but perhaps on a much lesser scale in the case of South Korea. South Korea similarly experienced rapid export expansion in the 1970s and 1980s which made South Korea one of the world's most significant trading nations, and enabled the country to accumulate a balance-of-trade surplus since 1986.[12] As in the case of Taiwan, this led South Korea's major trading partners, particularly the United States, to impose protectionist trade pressures in the mid-1980s to circumvent the growing deficits of the United States in its trade with South Korea. For both Taiwan and South Korea, the problem was compounded by the sharp appreciation of their currencies since 1985 resulting from strong export growth and current account surpluses, the deliberate imposition of a restrictive monetary policy to combat inflation by raising interest rates and tightening domestic credit as well as pressures from the United States to redress the persistent trade imbalances between their countries by allowing both the New Taiwan dollar and the Korean won to rise.[13] Between the end of 1985 and the end of 1990, the Korean won appreciated 19.5 per cent against the United States dollar.[14]

The strength of the Korean won thus provided the financial motivation for outward FDI by South Korean firms. This once again lends support to Aliber's theory of FDI emanating from strong currency areas as a determinant of the timing of FDI and its growth (and particularly that of foreign takeovers) as well as fluctuations of outward FDI flows around a long-term trend (Aliber, 1970). The financial explanations of the growth of FDI owing to strong domestic currencies has historical antecedents in Britain in the nineteenth century, the United States in the early post-war period and Japan and Germany in the 1970s and 1980s (Cantwell, 1989a) as well as Taiwan as shown in the previous chapter. However, in the case of South Korea, the growth of outward FDI in the 1980s was less of an 'exercise in financial power' (Redies, 1990) compared to Japan in the 1970s and 1980s and Taiwan in the 1980s and 1990s where the financial factors in explaining FDI expansion extended beyond large financial surpluses created from net exports, the upward revaluation of the Japanese yen and New Taiwan dollar and the accumulation of the world's largest foreign exchange reserves. In the case of Japan and Taiwan, the accumulation of large financial surpluses in those years was also created from the high domestic savings rates and the escalation in acquired wealth owing in part to the emergence particularly in Japan of a 'bubble' economy in which asset and land prices increased enormously. This was not the case in South Korea where there was a severe shortage of domestic savings and a large demand for foreign exchange until the late 1980s (see Kim, 1990).

In turn, local currency appreciation spurred rising labour costs in South Korea as in Japan and Taiwan which in addition to the increasingly adverse international trading environment undermined the competitiveness of domestic labour intensive industries of South Korea. The basis of competitiveness of these industries derived from the state's policy of wage suppression and the political subordination of popular sectors (i.e. middle stratum of professionals, skilled workers, businessmen and farmers) to the state (Shin and Lee, 1995). Although the problem of rising labour costs is broadly parallel to economic developments in Japan at the end of the 1960s and in Taiwan in the late 1980s which in combination with the sharp upward revaluation of the local currency led to the weakening of the traditional labour intensive industries (see Chapters 9 and 10 of this book), the basis of the rise in labour costs differed in the case of Japan on the one hand, and Taiwan and South Korea on the other. In the case of Japan, the success of domestic industrial restructuring towards modern capital intensive heavy and chemical industries at the end of the 1960s enhanced labour productivity owing to high capital to labour ratios and this perpetuated a continuous cycle of rising wages (Cohen, 1975; Kojima, 1978; Ozawa, 1979b). In Taiwan and South Korea, on the other hand, not only did the shortage of low skilled labour create pressures that drove up industrial wages but also democratization and political liberalization led to increased labour organization and activism and wage demands by labour unions.[15] The presence of severe labour unrest in South Korea (more than 7,000 labour strikes between 1987 and 1989) following democratization measures implemented by President Roh Tae-Woo in the Declaration of 29 June 1987 combined with South Korea's rapid economic growth of more than 9 per cent between 1982 and 1991 caused marked wage increases in South Korea since 1987 (Shin and Lee, 1995). Unlike in Japan, the wage increases in Taiwan and South Korea in the late 1980s were not matched by productivity growth (see van Hoesel, 1997; Kim, 1990).

South Korea responded to these compelling economic and political forces by encouraging outward investment, particularly by manufacturing firms and sustaining export growth. A significant shift in government policy favouring outward FDI in South Korea in light of the new requirements of international competition and changing comparative advantage was regarded as consistent with parallel efforts on domestic industrial restructuring towards more modern manufacturing industries in the long term. This entailed not only the gradual liberalization of foreign investment policy starting from 1986 but also the adoption of a more pro-active role by the South Korean government to promote outward FDI. Just as there had been broad parallelisms in the implementation of foreign exchange controls in Japan in 1949 and in South Korea in 1968, there had also been parallel trends in the liberalization of outward FDI policy in the two countries. As shown in Chapter 9, the substantial increases in domestic labour costs in Japan in the late 1960s, the appreciation of the yen, the emerging balance-of-payments surpluses and the growing natural resource requirements of a rapidly expanding resource-scarce industrial economy contributed to the gradual liberalization of controls over outward FDI in Japan between 1969 and 1978. Such gradual liberalization of Japanese FDI policy was replicated in content and style to that of South Korean FDI policy since 1986 and culminated in the seventh five-year plan of the *segyehwa* (globalization in Korean) programme over the period 1993 and 1997.[16] Having created the necessary policy conditions with the South Korean government playing a less *dirigiste* role, South Korean FDI successfully flourished starting in the late 1980s. Indeed, the stock of approved South Korean FDI tripled between 1985 and 1989 from $476 million to $1.4 billion.[17]

Not only was there rapid growth in South Korean FDI in the 1980s but major structural changes were manifested in the pattern of such FDI by the end of that decade. By comparison to the end of the 1970s when services was the most important sector of South Korean FDI with a share of 57 per cent, followed by the primary sector (26 per cent) and manufacturing (17 per cent), the emergence of mineral extraction as a strategic priority for South Korean resource development companies made the primary sector become the pre-eminent sector of South Korean FDI at the end of the 1980s with a share of 43 per cent, followed by manufacturing (33 per cent) and services (24 per cent). The supremacy of the primary sector at the end of the 1980s was on account of the necessity to acquire stable raw material supplies (particularly minerals) for the sustained development of many domestic industries in the face of strengthening resource nationalism in many resource-rich developing countries (Koo, 1984).

In addition, the ascendancy of the manufacturing sector has been due to the erosion of South Korea's comparative advantages in labour intensive industries as a result of economic and political forces of the decade: rising balance-of-payments surplus, trade barriers, appreciating Korean won, rising wage costs and increasing labour conflicts. While these economic pressures affected adversely the small- and medium-sized enterprises in labour intensive industries which determined their major surge as MNCs in this decade as opposed to their Taiwanese counterparts which emerged in a more major way as MNCs in the 1990s (see Chapter 10), the trade barriers and appreciating Korean won also served to affect adversely the *chaebols*, the large South Korean conglomerates whose growth posed a direct competition to producers in South Korea's major trading partner countries. Thus, two new types of manufacturing firms were responsible for the growth of South Korean manufacturing FDI in the 1980s: the small- and medium-sized firms that transplanted their labour intensive production to developing countries in South East Asia with cheaper labour costs, and the *chaebols* that invested in tariff factories in the United States and Europe. The different determinants of South Korean manufacturing FDI in developed and developing countries is supported by a formal analysis of the motivations of South Korean FDI in the manufacturing sector of both developed and developing countries as of 31 December 1989. Based on firm-level data, Jeon (1992) showed that firm size, the real growth rate of the South Korean economy and non-tariff barriers in the developed countries exerted highly significant *positive* influences on South Korean FDI in manufacturing in developed countries. This implied that large South Korean manufacturing firms at a time of rapid growth of the South Korean economy established production facilities in the developed countries to evade the rising tide of trade protectionism.[18] On the other hand, the exploitation of the cheap labour exerted a highly significant *positive* influence on South Korean FDI in manufacturing in developing countries since 1987. But it was the initiation and rapid growth of major investments in manufacturing by South Korea in the 1980s by the *chaebols* whose investments *en masse* in the United States to secure their largest foreign market that explains the greater geographical concentration of South Korean FDI in the developed countries at the end of the 1980s.

Thus, the shifts in the sectoral pattern of South Korean FDI had also been reflected in the geographical destination of South Korean FDI. Accordingly, while in 1979 developing countries accounted for 72 per cent of South Korean FDI and developed countries accounted for 28 per cent, developed countries became the dominant recipient of South Korean FDI in 1989 with a share of 54 per cent, while developing countries had a share of 46 per cent. The dominance of the developed countries reflected not only the large-scale growth of manufacturing FDI by the *chaebols* but also the large-scale mining development projects by South Korean firms in the resource-rich developed countries. As a whole, the relative importance of North America as a host region increased significantly from 23 per cent of the stock of approved South Korean FDI in 1979 to 41 per cent in 1989. Other important developed host regions were Oceania (9 per cent) and Europe (4 per cent). In developing countries, the major recipients continued to be South East Asia (28 per cent) and the Middle East (12 per cent), the latter owing to the construction boom and the large petrochemical projects by South Korean firms in the OPEC countries (Kim, 1990).

Five main types of outward FDI undertaken by South Korean MNCs could be distinguished at the end of the 1980s in order of declining importance: first, outward FDI to secure supplies of essential raw materials to serve the South Korea-based industrial production complex; second, outward FDI to supply host country markets in developed countries and to gain access to advanced technologies; third, outward FDI in trading, warehousing and transportation in major foreign markets to promote growth of South Korean exports; fourth, manufacturing FDI in developing countries to supply host country markets or as export platform; and fifth, outward FDI in civil construction and engineering-related works to facilitate growth of South Korean exports of services of skilled and semi-skilled labour. Each of these types of South Korean FDI in that decade is discussed below.

FDI to secure supplies of essential raw materials

The primary sector which accounted for some 26 per cent of the stock of approved South Korean FDI at the end of the 1970s became an even more important sector by the end of the 1980s with a dominant share of 42 per cent of the stock of approved South Korean FDI. As mentioned, the predominant role of FDI of this type owed much to the emergence of mining (including oil) as an important sector starting in the late 1970s and through the 1980s in the midst of strengthening resource nationalism in many resource-rich developing countries to secure stable mineral supplies for the domestic heavy and chemical industries whose growth had been given priority in South Korea in the 1980s. This reflected the situation of Japan in the late 1950s to the early 1970s which prompted the second phase or the Ricardo-Hicksian trap stage of Japanese internationalization. Thus, while in 1979 the forestry sector accounted for 19 per cent of the stock of approved South Korean FDI, the fisheries sector for 7 per cent, and mining accounted for an almost insignificant share of 0.35 per cent, by 1989 the mining sector accounted for one-third of the stock of approved South Korean FDI, while forestry accounted for 6 per cent and fishing, 4 per cent. Mining was the single most important sector of South Korean FDI. The magnitude of the overseas mining investments by South Korean public and private industrial companies as well as in some cases by trading companies dwarfed investments in all other industries and resulted in the sizable increase in the scale of South Korean FDI over that decade.

As of 1989, there were some 30 resource extractive FDI projects by South Korean firms conducted in different regions of the world: 7 projects for bitumenous (soft) coal; 2 projects for anthracite (hard coal); 5 projects for uranium mining; 9 projects for oil; 2 projects for timber development; 1 project each for the mining of manganese, sulfur, iron and steel, chrome and talc (Korea Institute for Foreign Investment, 1989). These were the primary products most crucial to South Korea's industrial development. Whereas the South Korean state companies were responsible for the resource extractive investments in oil and uranium, the private companies were more active investors in other mining projects.[19] The mining projects were located all over the world in developed and developing countries where the required mineral resources and petroleum were present in abundance. Among the major host countries were Indonesia, Sri Lanka, the Philippines, the United States, Australia, Canada, North Yemen, Malaysia, Thailand, China and the Soviet Union (Korea Institute for Foreign Investment, 1989).

The majority of the mining projects undertaken by South Korean firms in the 1970s and 1980s were in the form of wholly owned subsidiaries particularly where the parent companies were the direct buyers or users of the extracted resources (as was the case with manufacturing companies). The 100 per cent ownership enabled the security of obtaining stable supplies of raw materials at the most favourable prices. On the other hand, the minority of overseas mining investment projects undertaken in the form of joint ventures were those initiated by a trading company or, more often, several trading companies undertaking joint investment projects which were not direct users of the natural resources, but nevertheless provided finance in the form of loans for mineral exploration (Koo, 1984).[20]

Investments in the forestry sector was the second most important South Korean FDI of this type. This involved the continuing importance of assuring timber and wood supplies to meet the requirements of the furniture industry whose domestic production and exports had been rising faster than that of the plywood industry. Although as previously mentioned the plywood industry was a major South Korean export industry until the late 1970s, it suffered a loss of comparative advantage since owing to the rising competition from timber-rich countries (Koo, 1984).

Finally, mention must be made of South Korean FDI in agriculture in the 1980s of which the most significant were for the purpose of raising cattle and livestock in Australia, Canada and the United States. Haitai Dairy Corporation, one of the largest food companies in South Korea, has long been engaged in livestock farming in Australia; this venture has provided a stable source of cattle and livestock products to its parent company in South Korea (Korea Institute for Foreign Investment, 1989).

FDI in developed countries to supply local markets and to gain access to advanced technologies

South Korean FDI in the developed countries in the 1980s of the type described above had three important determinants. The first determinant was to protect or retain existing export markets of South Korean goods by supplying host country markets in developed countries through international production in the face of protectionist trade barriers, primarily in the case of electrical and electronics products targeted to the export market of the United States.[21] Indeed, as indicated above in the work of Jeon (1992), South Korean FDI in the developed countries over the 1980s was associated primarily with the maintenance of access to markets in the face of trade barriers. The second determinant of FDI of this type stems from South Korea's interest to gain access to more sophisticated and advanced forms of manufacturing technology to support rapid industrial development (World Bank, 1989; Brody, 1986). The third determinant was to establish new markets in the developed countries by the provision of services (such as banking and finance, construction, hotels, etc.) aimed at meeting local market demand.

This type of FDI which was not a significant aspect of South Korean FDI at the end of the 1970s was a new type of South Korean FDI that developed in the 1980s. In fact, it became the second most important type of South Korean FDI at the end of that decade with a share of at least 25 per cent of the stock of approved South Korean FDI in 1989. Some 81 per cent of South Korean FDI in developed countries involved the manufacturing sector to fulfil the above two determinants and only 19 per cent were in services.[22]

Import substituting FDI in the developed countries

Two types of import substituting FDI in developed countries can be distinguished. The first type is manufacturing FDI to overcome trade restrictions in major export markets. This involved FDI by major South Korean firms producing television sets, video cassette recorders (VCRs), computers, microwave ovens, steel and compact cars. The second type is FDI by horizontally integrated South Korean manufacturing firms that produce for segmented markets or niche markets in the developed countries of North America and Europe specific idiosyncratic products such as high fashion goods (women's clothes or men's suits), textiles, furniture, noodles and other food products, etc. The first type, which is by far the more important type of South Korean import substituting FDI in the developed countries and one that grew considerably over the 1980s and the 1990s, is discussed below.

The major recipients of both types of import substituting FDI over the 1980s were the United States, Canada and Europe. North America was the dominant recipient as shown by their 55 per cent share of the stock of approved South Korean FDI in the manufacturing sector in 1989. The growth of South Korean manufacturing FDI in the United States, in particular in the 1980s, owed much to the growth of the domestic consumer electronics industry in South Korea whose size is second largest in the world after Japan (van Hoesel, 1997). The growth of domestic production and export boom in the industry led to the suspension by the United States of South Korean export privileges in electronics products under the Generalized System of Preferences in January 1988 (Kim, 1996) and the imposition by the United States of trade barriers in the form of tariffs such as countervailing duties and anti-dumping duties as well as non-tariff barriers such as voluntary export restraints to restrain the growth of American imports from Japan and the Asian newly industrialized countries. In addition to imposing these trade barriers, the United States insisted on a 'levelling of the playing field' by placing pressures on these countries to liberalize their domestic markets and practice fair trade.[23] In a wide-ranging move to attempt to correct its trade deficits with its major trading partners, the United States initiated action against alleged dumping practices by South Korean firms in order to curb the growth of South Korean exports, and the United States Congress passed the Omnibus Trade and Competitiveness Act in 1988. Section 301 of this legislation required the United States Trade Representative to identify those foreign countries that have erected systematic barriers against exports of the United States and to launch mandatory investigations against every identified trade barrier and unfair trade practice. Although this law expired in 1990, the United States persisted with unilateral trade restrictions (Strange, 1993). In addition, with effect from January 1989 the four Asian newly industrialized countries were graduated from the Generalized System of Preferences of the United States on grounds of having achieved a certain level of economic development and competitiveness (Chia, 1989).

The South Korean firms responsible for the large-scale import substituting FDI geared to overcome trade restrictions in the developed countries were the *chaebols* whose emergence and growth was fostered by the South Korean government in its effort to accelerate domestic industrial development in large-scale, complex and technologically advanced industries through high industrial concentration. This has historical antecedents in Japan where the Japanese government similarly supported the emergence and development of the *keiretsu* to perform a similar function.[24]

The 'tariff factories' in the developed countries were established by firms producing consumer electrical and electronic goods, cars and steel products. In consumer electrical and electronics production, FDI in the United States and Europe was led by the three largest companies in the highly oligopolistic South Korean consumer electronics industry: Samsung Electronics (SEC), the Goldstar Company (now known as Lucky Goldstar (LG) Electronics) and Daewoo Electronics.[25] In the car industry, Hyundai Motor Company's FDI in Canada was also initiated in the 1980s (Kim, 1990).[26] However, one of the largest manufacturing projects undertaken by South Korean firms during the 1980s was that of the state-owned Pohang Iron and Steel Company in a joint venture with US Steel Corporation to escape trade restrictions on Pohang's export of iron and steel products to the United States (Korea Institute for Foreign Investment, 1989). The large scale of the manufacturing FDI by the *chaebols* combined with the even larger scale of mining projects overseas by South Korean industrial companies and general trading companies ensured that the overall scale of South Korean FDI in the 1980s was significantly larger than that in the 1970s.

In the first instance, the *chaebols* conducted their final assembly production in the developed countries as a means of overcoming trade restrictions. Although the cost of labour in developed countries was higher relative to that of South Korea, the final products remained competitive in those markets since the parent companies in South Korea were the main source of raw materials, intermediate products or components at relatively reasonable costs. The initiation of South Korean FDI of this type in defence of export markets threatened by trade barriers was premature, as seen in the higher cost of local production in a host country by comparison to the cost of exporting from the home country prior to the imposition of trade barriers (Jun, 1990).[27]

Although Europe has also been a major export market of South Korea, it was a far less important destination for South Korean manufacturing FDI at the end of the 1980s, with a share of less than 6 per cent of the stock of approved South Korean FDI in 1989. The growth of South Korean FDI in

Europe during the late 1980s was induced significantly by protectionism in the former European Community and anti-dumping duties rather than by the impending formation of the Single European Market (Young *et al.*, 1991). This view is supported by the fact that peak period of South Korean FDI flows in Europe – 1988 and 1989 – coincided with the highest levels of actual or threatened imposition of anti-dumping duties by the European Commission on South Korean exports to Europe. SEC, LG Electronics and Daewoo Electronics were once again the prominent *chaebols* to invest in Europe in electrical and electronics production (van Hoesel, 1997; Korea Institute for Foreign Investment, 1989).[28]

The emergence in a major way of South Korean manufacturing FDI in the developed countries seems to suggest that the pattern of South Korean FDI in the 1980s reached the third stage of Japanese FDI in assembly-based, sub-contracting dependent, mass production of consumer durables such as cars and electrical and electronic goods. The principal ownership-specific advantages of South Korean *chaebols* stem from a cost advantage, commitment and dedication of their managers, appropriate technologies for small-scale production, and support and willingness of parent firms to absorb initial operating losses of foreign subsidiaries. However, unlike the strong technological innovatory capacities and marketing skills of American and Japanese MNCs, the firm-specific advantages of South Korean manufacturing firms have proven to be far from sustainable. This helps to explain the low and declining profitability of South Korean electrical and electronics firms in the United States and Europe over the 1980s, and the virtual closure of the American production plants of SEC and LG Electronics in 1989 and their relocation to Mexico (Choi and Kenny, 1995).

The imposition of stricter local content requirements by the United States and Europe for consumer electrical products and cars, including the levy in the late 1980s of anti-dumping duties by the United States government on the intra-firm trade of intermediate products such as picture tubes for television sets between the South Korean parent companies and their assembly subsidiaries in the United States had served to exacerbate the effects of the continued appreciation of the won and labour cost increases in South Korea in the second half of the 1980s in threatening the cost advantages of products produced by final assembly plants of South Korean firms in these countries. Similar pressures for greater local value added by foreign manufacturers in Europe have been driven by the Screw Drive Regulation in Europe, a trade agreement that binds Japanese, South Korean and other foreign manufacturing plants to increase their local procurement of parts or intermediate products from 20 per cent to 40 per cent in 1989 based on added value (van Hoesel, 1997).

At the end of the 1980s, South Korean manufacturing firms responded to demands for higher localization in the United States and Europe in various ways. The four electrical and electronics goods producers – SEC, Daewoo Electronics, LG Electronics and Saehan Media Co. – have either completed or planned to construct six factories in the European nations to produce parts of final goods at the end of the 1980s. The cost competitiveness of South Korean products in Europe will also be enhanced by their FDI in Central and Eastern Europe as a low-cost production base from which to supply the prosperous markets of Western Europe owing to the low labour costs, preferential access to European Economic Area and anticipated future membership of the European Union (Randerson and Dent, 1996). In North America, LG Electronics set up a plant in Mexico to produce colour TV chassis for its final assembly plant in Huntsville, Alabama, thus availing of Mexico's duty-free access to the United States. These companies and other companies have also responded through the establishment of plants in South East Asia as a means to build their market shares in the region or to use as an export platform to counter their weakening competitiveness and avoid the increasing protectionism in the advanced countries (Korea Institute for Foreign Investment, 1989).

FDI in developed countries to gain access to advanced technologies

An important aspect of South Korean FDI in the developed countries which started to emerge in the 1970s but grew significantly in the 1980s was investments in R&D firms in the United States to develop sophisticated technology and know-how to support South Korea's objective to upgrade its domestic production and exports of products embodying higher skills and advanced technologies (Nishimizu and Robinson, 1984; Euh and Min, 1986). In the 1970s the form of the investment involved the takeover of an American research intensive company by a medium-sized South Korean company which is then used as a base for the development of appropriate technical knowledge, new processes and new product designs as well as the assembly of sophisticated technical components for export to South Korea.

In the 1980s, South Korean FDI of this type grew in importance and manifested mainly in the establishment of R&D laboratory facilities in the Silicon Valley of the United States through acquisition. This included Zymos Corporation, Cordata Technology Inc. of Daewoo Group (computer products and design), Lucky-Biotech Corporation, Maxon System (computer) and Goldstar Technology, among other numerous examples (Korea Institute for Foreign Investment, 1989). The *chaebols* have been at the forefront of FDI of this type which enabled their entry into more technology intensive products such as semiconductors and telecommunications products and led to the domestic industrial upgrading of the South Korean economy (see also Westphal *et al.*, 1984).

Nowhere is the role of research-based FDI key to the creation of capabilities of South Korean firms in more advanced products such as semiconductors than in the case of SEC. Samsung established an R&D institute, Samsung Semiconductor Inc. (SSI), at Silicon Valley in May 1983 to serve as a platform to develop 64K and 256K dynamic random access memory (DRAM) chips in its role as a training post for South Korean experts and a collector of information about the latest developments in technology and markets for semiconductors. Through SSI, SEC recruited several United States-trained South Korean experts which eventually played a key role in developing and commercializing DRAM chips with the help of designs obtained from Micron Technology, a medium-sized DRAM chip producer in the United States, as well as from Sharp – the only source of process technology for 16K SRAM and 256K ROM technology – for the process development of 64K DRAM chips whose mass production was started in mid-1984.

For the development of 256K DRAM, the designs of Micron Technology were subjected to extensive reverse engineering (Ernst, 1994), a means of technological learning adopted by South Korean firms in the development of VCRs and microwave ovens in the late 1970s (Kim, 1996). This process of extensive reverse engineering in the development of 256K DRAM chips involved the adoption of a dual strategy in which two teams based in Silicon Valley and in a South Korea-based laboratory started the same work simultaneously. The 256K DRAM chip sample created by the South Korea-based team in October 1984 was developed by the Silicon Valley team in early 1985, and mass production was started in April 1986 (Ernst, 1994). Subsequently, Samsung developed more advanced DRAM products in a similar way: 1M DRAM (July 1986), 4M DRAM (February 1988), 16M DRAM (September 1990) (Kim, 1996).

FDI to establish new markets in the developed countries by the provision of services to meet local market demand

This type of FDI had become significant in the 1980s particularly in the banking and hotels industries. More South Korean banks have tried to establish foreign affiliates worldwide in the 1980s in accordance with the expansion of their business area, as have large *chaebols* which began in the 1980s to participate in the hotel business overseas. This is reflected in the acquisition by the Hanil Development Corporation of the Inter-Alaska Hotel Inc. in Alaska in the second half of 1988, and the development of a 200-room hotel in California by the South Korean construction company, Ssangyoung (Korea Institute for Foreign Investment, 1989).

Export-oriented FDI in trading, warehousing and transportation

South Korean FDI of this type continued to be important as a means to continually promote industrial exports in the face of protectionism and competition in major export markets as well as to facilitate imports of vital raw materials and intermediate products for South Korean industrial development. Thus, in the 1980s South Korean industrial enterprises established more overseas trading companies or increased their equity investment in existing overseas trading subsidiaries as a means also of supporting the requirements of higher local content in the developed countries (Korea Institute for Foreign Investment, 1989).

Manufacturing FDI in developing countries to supply host country markets or as export platform

This is the fourth most important type of South Korean FDI with a share of some 12 per cent of the stock of South Korean FDI at the end of the 1980s. Two major types of manufacturing FDI in developing countries can be distinguished. The first type is manufacturing FDI by firms in major export industries in the form of export platform to overcome trade restrictions in export markets and to circumvent the declining comparative advantages of South Korea for labour intensive production. The second type is manufacturing FDI to produce standardized products such as fountain pen, adhesives, construction materials, cement, plastic products, photo albums, batteries, gas ranges, food chemicals and other products for developing countries in South East Asia, South Asian countries, Middle East and China (Korea Institute for Foreign Investment, 1989). While this second type of South Korean manufacturing FDI had been the important type of South Korean FDI when South Korean manufacturing MNCs first emerged in the 1970s and continued to grow in the 1980s, it is the first type that has grown most rapidly in the second half of the 1980s and became characteristic of the major type of South Korean manufacturing FDI in developing countries.

Two kinds of South Korean firms invested in the export platform type of manufacturing FDI in developing countries.[29] First, the smaller- and medium-sized firms in labour intensive industries relocating their production to more cost competitive labour abundant developing countries. This included firms in such industries as textiles, clothing and footwear, toys, etc. Second, the larger firms and the *chaebols* in more capital intensive industries that relocate the labour intensive stages of their production processes in labour abundant developing countries. This included the producers of TV sets, refrigerators, washing machines, computers and other industrial electronics components. In the 1980s, the major host countries for South Korean export platform FDI had been Indonesia, Thailand, Malaysia, Sri Lanka, Philippines that are in close geographical and cultural proximity to South Korea, and also countries of Central America and the Caribbean (Korea Institute for Foreign Investment, 1989).[30] The export platform international production in these countries enabled the reduction of labour costs, the circumvention of trade barriers and access to any preferential trading arrangements such as Generalized System of Preferences of host countries with South Korean export markets. In addition, the investments in the Caribbean countries have the added advantage of being in close geographical proximity to North America, thus saving on transportation costs and transit time in shipping the export products to that market (Tolentino, 1993).

South Korean FDI by small- and medium-sized firms in labour intensive industries had increased steadily since the second half of the 1980s. While accounting for less than 2 per cent of the stock of approved FDI before 1985, their share increased to 4.5 per cent in 1988 and 31.5 per cent in the first quarter of 1989 (Jun, 1990). A multinomial logistic regression model estimated by the maximum likelihood method indicated that outward FDI of a sample of 151 South Korean textile and clothing companies in 1988 (taken from a population of 673 firms) was determined significantly by tariffs, labour costs and export experience but not by the appreciating Korean won (Jong, 1990). A similar set of determinants have propelled the expansion of outward FDI by major South Korean companies in the electrical and elec-tronics industries (Korea Institute for Foreign Investment, 1989), including SEC. SEC established two production affiliates for final goods in South East Asia in the late 1980s: the first one was TSE in Thailand founded in 1988 for the production of colour TVs, VCRs and washing machines, and the second one was SMI in Indonesia founded in 1989 for the production of refrigerators (Kim, 1996).

FDI in civil construction and engineering

This was the fifth most important type of South Korean FDI with a share of at least 3 per cent of the stock of approved South Korean FDI at the end of the 1980s. As in the previous decades, the investments were concentrated in the Middle East to participate in the construction projects of the OPEC countries (Kim, 1990), but investments in other parts of the world have also been notable as seen in the $56.5 million investment of Daewoo in the construction of a tourist hotel in Algeria (Korea Institute for Foreign Investment, 1989).

The growth of South Korean FDI in the 1990s

The 1990s represented the decade of the most rapid expansion of South Korean FDI. Based on balance-of-payments data, the annual average FDI outflows of South Korea was $2.7 billion in the period between 1990 and 1998 compared to $114.6 million in the period between 1980 and 1989 and $9.8 million in the period between 1970 and 1979. Thus, the estimated stock of South Korean FDI reached $21.5 billion in 1998, some 57 per cent of the stock of Taiwanese FDI of $38 billion.

Unlike in the 1980s when financial factors influenced the expansion of South Korean FDI owing to financial surpluses created from net exports and the upward revaluation of the Korean won between 1986 and 1990, this situation was overturned over the 1990s with the slower growth of exports and the depreciation of the Korean won. In fact, the growth of South Korean FDI in the 1990s was sustained despite an inherent financial weakness of the Korean *chaebols*. Thus, unlike the 16.8 per cent appreciation of the Korean won against the United States dollar between the end of 1986 and the end of

1990, there was a 136.6 per cent depreciation of the Korean won against the United States dollar between the end of 1990 and the end of 1997.[31] To the extent that financial factors played a role in explaining the most rapid expansion of Korean MNCs in the 1990s, it owed much to the role of the *chaebols* as institutions created by the South Korean government to accelerate domestic industrial development in large-scale, complex and technologically advanced industries through high industrial concentration. The close relationships between the government and the *chaebols* was manifested in the traditional roles of the former in setting policies and in the control of the latter's access to capital to finance their growth and diversification, including through outward FDI.[32] Notwithstanding the envisaged restructuring of the *chaebols* under the *segyehwa* movement between 1993 and 1997, the *chaebols* became more powerful and larger with their continued expansion into new industries and markets financed on the basis of low-interest bearing loans from their government. However, the capacity of the over-leveraged *chaebols* to service their amassed debts have been constrained by the onset of foreign competition associated with domestic market deregulation in the *segyehwa* movement and the weaker markets for computer chips in 1996 which led to domestic economic difficulties in South Korea as well as a 27 per cent deterioration in the South Korean terms of trade in the three years prior to September 1997 and associated current account deficits (World Bank, 1998). These negative factors have been exacerbated by the Asian financial crisis that started in 1997 which affected South Korea greatly and limited the capacity of the government to fulfil its traditional role as a provider of low-cost capital to the *chaebols* – the basis of the country's past economic miracles and a contributory factor to its economic débâcles in the late 1990s (Tolentino, 1997). While the full repercussions of all these compelling forces on South Korean FDI have yet to be observed and analysed fully in the future, Korean FDI outflows in the period since the financial crisis would be determined by the need to overcome the inherent financial weakness of the Korean *chaebols* through more aggressive and extensive penetration of foreign markets through outward FDI as a means to service large amounts of accumulated debt as well as the need to seek alternative sources of finance in international capital markets.

Not only was there rapid growth in South Korean FDI in the 1990s but major structural changes was manifested in the pattern of South Korean FDI in that decade. By 1995, manufacturing became not only the largest economic sector of South Korean FDI but also the dominant sector with a share of 59 per cent of the stock of approved South Korean FDI, a significant growth from its one-third share at the end of the 1980s, and 17 per cent share at the end of the 1970s. Thus, in contrast to the growth of Taiwanese MNCs in which the relative importance of manufacturing has been declining throughout the course of their history (see Chapter 10), the opposite trend is observed in the case of South Korean MNCs. Services was the second most important sector of South Korean FDI with a share of one-third of the stock of approved South Korean FDI in 1995, compared to its share of 24 per cent at the end of the 1980s and 57 per cent at the end of the 1970s. Finally, the primary sector was the least important sector of South Korean FDI with a share of 8.5 per cent of the stock of approved FDI in 1995, compared to its share of 43 per cent at the end of the 1980s and 26 per cent at the end of the 1970s. Thus, the 1990s marked the decade in which resource extractive FDI fell significantly in importance from being the most important type of South Korean FDI at the end of the 1970s and 1980s.

There had also been changes in the geographical destination of South Korean FDI as of 1995 with investments in developing countries taking the slightly larger share of the stock of approved South Korean FDI (51 per cent) compared to developed countries (48 per cent). This is in comparison to 1989 when developed countries had the clear lead as the dominant recipient of South Korean FDI in 1989 with a share of 54 per cent, while developing countries had a share of 46 per cent. However, although developing countries may have received slightly more of the stock of approved South Korean FDI by 1995 compared to developed countries, the share of developing countries was not as high as that in 1979 at 72 per cent. Thus, there has been no strong evidence to suggest that the geographical pattern of South Korean FDI in the 1990s has shifted back to the pattern that prevailed at the end of 1970s, unlike in the case of Taiwan.[33] Thus, although the share of developed countries in the stock of approved Taiwanese FDI at at the end of the 1980s at two-thirds was much higher compared to the share of developed countries in the stock of approved South Korean FDI in the same period at 55 per cent, such high share of developed countries in Taiwanese FDI had dropped significantly to 21 per cent in 1995, and so did the share of developed countries in South Korean FDI but not as dramatically to 48 per cent in 1995. This reflects the greater ability of South Korean MNCs compared to Taiwanese MNCs to sustain existing FDI in the developed countries, as well as to implement new FDI projects in those countries, despite the fact that South Korean MNCs in developed countries emerged and grew in a major way in the 1980s, a full decade behind the major assault of Taiwanese MNCs in the developed countries in the 1970s. The fact that South Korean FDI in developed countries have been led by the *chaebols* may explain the greater resilience of South Korean MNCs to sustain FDI in developed countries compared to the Taiwanese MNCs that invested in developed countries which, although featuring as the largest companies in Taiwan, are not as large and diversified as the South Korean *chaebols*.

South East Asia was the dominant host region of South Korean FDI in 1995, with a share of 45 per cent of the stock of approved South Korean FDI, followed by North America (31 per cent) and Europe (15 per cent). Thus, South East Asia increased considerably in relative importance in comparison to its 1989 share of 28 per cent, while the share of North America declined from its 1989 share of 41 per cent. Europe, on the other hand, emerged over the 1990s as an increasingly significant recipient of South Korean FDI whose share in the stock of approved FDI increased from 4 per cent in 1989 to 15 per cent in 1995.

Five main types of outward FDI undertaken by South Korean MNCs could be distinguished as of 1995 in order of declining importance: first, manufacturing FDI in developing countries to supply host country markets or as export platform; second, FDI to supply host country markets in developed countries and to gain access to advanced technologies; third, outward FDI in trading, warehousing and transportation in major foreign markets to promote growth of South Korean exports; fourth, FDI to secure supplies of essential raw materials to serve the South Korea-based industrial production complex; and fifth, outward FDI in civil construction and engineering-related works to facilitate growth of South Korean exports of services of skilled and semi-skilled labour. Thus, in comparison to the 1980s, resource extractive FDI declined considerably in importance as mentioned, and manufacturing FDI in developing countries superseded the previously greater importance of FDI in developed countries and FDI in trading, warehousing and transportation.

Since the first two main types have been the most important types of South Korean FDI in the 1990s and the pattern of such investments have changed considerably since the 1980s, the discussion below focuses on these two most important types of FDI.

Manufacturing FDI in developing countries to supply host country markets or as export platform

This type of FDI accounted for more than 37 per cent of the stock of approved South Korean FDI as of 1995, and was the most important FDI type in the 1990s. The important change in the pattern of South Korean manufacturing FDI in developing countries over the 1990s was in their market orientation. Thus, while in the 1980s the growth of South Korean manufacturing FDI in these countries was largely of the export platform type driven by the need to gain access to low-cost labour and overcome trade barriers in major export markets through the relocation of labour intensive industries or labour intensive processes of more capital intensive industries, in the 1990s the growth of manufacturing FDI in developing countries and particularly in South East Asia had been spurred by the objectives of Korean MNCs to cater to local, regional and world markets. The growth of import substituting FDI in South East Asia has been influenced by the high economic growth rates of the region as well as the emergence of trade protectionism in the region. As a case in point, the rapid growth of FDI by the South Korean car producers (Daewoo Motor Co., Hyundai Motor Co. and Kia Motor Co.) in motor vehicle assembly in South East Asia in the 1990s had been a response to the policy of most countries in that region to restrain imports of foreign cars under administrative guidance.[34] Asia is not only the dominant recipient of South Korean manufacturing FDI in developing countries, but also of South Korean manufacturing FDI worldwide with a share of 59 per cent of the stock of approved South Korean FDI in manufacturing in 1995.

The most prominent feature of South Korean manufacturing FDI in developing countries was the formation of integrated regional production networks by South Korean consumer electronics companies, of which the shining example is SEC. Throughout the 1990s, SEC has sought to build regional production networks in South East Asia and China associated not only with the final goods assembly but also the upstream production of parts, components and intermediate goods made possible by the growth of corresponding FDI of affiliated supplier companies in the 1990s and the establishment of rapid linkages between final goods assemblers and their affiliated parts and component manufacturing subsidiaries. Indeed, during the 1990s international production by SEC in Asia was associated with an expanding production of consumer electrical products (colour TVs, VCRs, microwave ovens, refrigerators) and more technologically sophisticated products such as telecommunications products and semiconductors as well as components such as colour picture tubes, computer display tubes, tuners, VCR motors, heads and drums (Kim, 1997). This is reminiscent of the growth of Japanese FDI in the electrical and electronics industry as well as the motor vehicle industry in North America, Europe and Asia in which regionally integrated production networks with a fairly sophisticated regional division of labour and decentralized control have been developed by Japanese MNCs (Kim, 1996; UNCTC, 1991).

In South East Asia, for example, the final goods assembly plants established by SEC in Thailand (TSE) in 1988 for the production of colour TVs, VCRs and washing machines and in Indonesia (SMI) in 1989 for the production of refrigerators have both been strengthened by the establishment by SEC of an intermediate products plant in Thailand (SEM-Thailand) in 1990 for the production of colour TV and VCR components. Over the 1990s, SEM-Thailand have not only provided intermediate products for TSE and SMI but also other affiliated final goods assembly plants established by SEC in Indonesia (SME) in 1991 for the production of VCRs and audio products, and in Malaysia (SEMA) in 1991 for the production of microwave ovens. Evidence of further backward vertical integration in South East Asia is found in the establishment by SEC of plant in Malaysia (SED-Malaysia) in 1991 for the production of parts such as colour picture tubes used by its intermediate goods supplier SEM-Thailand in the production of colour TV components. This plant also supplies directly the final goods assembly plant established by SEC in China (TTSEC) in 1995 for the production of colour TVs. Still further, another plant was established by SEC in Malaysia (SC-Malaysia) in 1992 for the production of more sophisticated components such as cathode ray tube glass bulbs used in the production of colour picture tubes by SED-Malaysia (Kim, 1996). A similar regionally integrated production network had been evolving in Tianjin, China for VCR production and in Jiangsu and Kwangdong, China for audio products over the course of the 1990s. SEC envisages the development of such fairly sophisticated regional division of labour to emerge in its other regional networks in Japan, Europe and America in the future (Kim, 1997).

However, despite the similarities in the development of regional production networks established by Japanese and South Korean MNCs in South East Asia and China, there are fundamental differences between the kind of regional production networks established by Japanese and South Korean MNCs apart from the greater sophistication and complexity of regional production networks of the former by comparison to the latter. Perhaps the most important of all is the fact that unlike the foreign affiliates of Japanese firms in both regions that have continually upgraded their product change capability through cross-functional linkages between design and development, production, marketing and service in China and South East Asia, the network of foreign affiliates of South Korea's foreign affiliates (as exemplified by that of SEC) display much weaker linkages between production, marketing and design and development functions (Kim, 1997). This is because the latter two functions are still centralized functions controlled by the South Korean parent company.

The second major difference is that unlike in the case of Japanese parts and components producers whose businesses are very much dependent on Japanese final goods assemblers owing to their quasi-integrated subcontracting networks and long-term inter-firm trading arrangements, this has not been the case of the business relationship between parts and components producers and final goods assemblers in South Korea, except perhaps in the case of SC Malaysia.[35] As a result, even the parts, components and intermediate products produced by affiliated suppliers of South Korean *chaebols* in South East Asia and China have not been for the exclusive use of South Korean final goods assemblers. In fact, a significant proportion is sold to rival Japanese final goods assemblers; and, in addition, South Korean firms producing components and intermediate products rely on Japanese parts suppliers such as NEG and Asahi to provide glass bulbs for the colour picture tube production of SED-Malaysia (Kim, 1996).

Thirdly, while South Korean *chaebols* such as SEC in their regionally integrated production networks in South East Asia and China have somewhat progressed sufficiently beyond the simple globalization stage in which FDI comprises final assembly operations controlled closely by the parent companies in Korea with limited domestic value added in host countries, Japanese *keiretsus* have progressed much farther to global localization in their regionally integrated production networks in North America, Europe, South East Asia and China with much higher local content ratios achieved through local sourcing or local production of both final products and a fuller range of sophisticated parts and components (Morris, 1991).

FDI in developed countries to supply local markets and to gain access to advanced technologies

This type of FDI accounted for some 21 per cent of the stock of approved South Korean FDI as of 1995, and was the second most important FDI type. As mentioned, the most significant change in the 1990s was the decline in relative importance of North America whose share in the stock of approved South Korean FDI fell from 41 per cent in 1989 to 31 per cent in 1995. Europe, on the other hand, emerged over the 1990s as an increasingly significant recipient of South Korean FDI whose share increased from 4 per cent to 15 per cent over the same period. Based on the stock of approved FDI in 1995, South Korea had a larger FDI stake in Europe at more than $1.5 billion compared to Taiwan at less than $556 million.

As in the 1980s, South Korean FDI in the developed countries in the 1980s of the type described above had three important determinants. The first determinant was defensive manufacturing FDI to protect or retain existing export markets of South Korean goods by supplying host country markets in developed countries through international production in the face of protectionist trade barriers, primarily in the case of electrical and electronics products targeted to the United States and Europe. However, as will be noted below, the further expansion of South Korean FDI in Europe in particular also fulfils the pursuit of offensive strategies of South Korean MNCs to build a significant market presence in a Triad location and to benefit from being an insider in a large integrated market.

The second determinant of FDI of this type stems from South Korea's interest to gain access to more sophisticated and advanced forms of manufacturing technology and marketing expertise to support rapid industrial development in South Korea, to build technological capacities of South Korean firms in new products, to increase their capacity to engage in backward vertical integration through the establishment of local supply chain networks in the developed countries as a means to enhance the competitiveness of their affiliate production in developed countries and thus meet higher local content requirements in those countries, as well as to engage in forward vertical integration through the establishment of marketing networks to sell South Korean products in these countries.

The third determinant of this type emerged from the need to establish new markets in the developed countries by the provision of services (such as banking and finance, construction, hotels, etc.) aimed at meeting local market demand. The discussion below focuses on the first two determinants as described above since these were the principal influences in the growth and changing pattern of South Korean FDI in developed countries over the 1990s.

Import substituting FDI in the developed countries

As mentioned above, the importance of Europe as a host region of South Korean FDI grew significantly over the 1990s in both absolute and relative terms. The growth of South Korean FDI in that region in that decade was spurred by further incentives for defensive investments owing to protectionism as shown in part in the European Commission's decision to suspend South Korea's trade privileges under the Generalized System of Preferences between 1989 and 1992, and their graduation from the scheme in 1998. This resulted in the proliferation of quotas on South Korean exports to Europe of 'sensitive' products to include VCRs, microwave ovens, small screen TVs, videotapes, polyester film and yarns, oxamic and glutamic acids. It was also reflected in the anti-dumping duties that had been imposed against South Korean exports and which had continued unabated over the 1990s. Anti-dumping duties were imposed on car radios in 1992, synthetic polyester fibres, DRAM chips, monosodium glutamate and electronic weighting scales in 1993, electronic capacitors and microdisks in 1994, and large screen TVs and microwave ovens in 1995 (Randerson and Dent, 1996).

However, although the initial phase of the expansion of South Korean FDI in Europe starting in the late 1980s had been motivated mainly by defensive considerations in response to trade protectionism, the investments also enabled the South Korean *chaebol* to fulfil their offensive market penetration strategies over the long term in several ways: first, it enables the South Korean *chaebol* to develop a strategic position in another Triad location apart from the United States and thus compete effectively in globalized industries. Second, a presence in the Single European Market through FDI enables South Korean firms to benefit from the large integrated market of Europe. Third, FDI in Europe enables access to high levels of technological expertise and the possibility of gaining from the wider diffusion of technological exchanges associated with the Single European Market. A market presence is also facilitating the collection of marketing information to support the growth of South Korean exports to the continent. These objectives of South Korean MNCs to upgrade their technological and marketing capacities will be discussed in the next section.

Over the 1990s, SEC expanded the number of its foreign affiliates in Europe. This included the establishment of warehousing and sales facilities in Spain in 1990, Italy in 1991, Sweden in 1992, and in Portugal in 1993. In addition, the firm established production and service facilities for refrigerators in Slovakia in a joint venture with Caltex established in 1991, and in Portugal for memory chips in a joint venture with Texas Instruments in 1994. Thus, SEC had established six production facilities in Europe as of 1995: colour TVs have been produced in the United Kingdom since 1987 and in Hungary and Turkey since 1989, VCRs have been produced in Spain since 1989, refrigerators have been produced in Slovakia since 1991, and memory chips in Portugal since 1994. In addition to its six sales and warehousing facilities in

France, Spain, Italy, Sweden and Portugal and headquarters in Germany, the firm also has a small R&D centre in the United Kingdom which monitors local market developments in telecommunications systems, audio-visual systems and home appliances, thus serving to support more fundamental research taking place in South Korea. Furthermore, the firm acquired in 1994, through its sister company Samsung Corning, the TV glass bulb manufacturer FGT located in former East Germany as a means of transforming its European production operations from screwdriver plants into more highly localized manufacturing plants. As a result, SEC had built up a considerable market presence in Europe where 21 per cent of its total sales by 1994 was realized (van Hoesel, 1997).

LG Electronics similarly expanded its European presence in the 1990s though not as extensively as SEC. The firm established a production facility in Italy in 1990, warehousing, sales and service facilities in France in 1990 and in Hungary in 1992, as well as a design facility (LG Design Technology) in Ireland in 1992 where products are designed to tailor to European lifestyles and local consumers. Thus, LG Electronics had established three manufacturing facilities in Europe as of 1995 located in Germany (established in 1986), the United Kingdom (1988) and Italy (1990), four sales subsidiaries and a research centre. In contrast to SEC, the firm had not yet taken concrete measures to expedite backward vertical integration to increase the local value added of its assembly production subsidiaries. Nevertheless, the firm realized some 19 per cent of its total sales in Europe in 1994 compared to SEC's 21 per cent (van Hoesel, 1997).

Typical of firms from countries that are late industrializers, the South Korean *chaebols* have strong production capacities whose competitiveness derive from low production costs and a product specialization in the low end of the consumer electronics market where there is no direct competition with established MNCs (Bloom, 1992; Hobday, 1995). Such product specialization can be explained by their weak technological capacities in product design and development as well as weak marketing capabilities. South Korean products suffer from poor brand recognition in Europe, a manifestation of the large dependence of South Korean consumer electrical and electronics companies on original equipment manufacturing (OEM) contracts in which products are sold under the labels and brands of their OEM buyers. These factors help to explain the closure of the car manufacturing plant of Hyundai Motor Co. in Canada in 1993 (Kim, 1996), among other examples of failed or low-profitability FDI projects of South Korean firms in the developed countries.

FDI in developed countries to gain access to advanced technologies and marketing expertise

Owing to their weak technological capacities in product design and development, FDI of this type by the *chaebols* became even more important over the 1990s. Acquiring a more independent technological capability was an explicit element of the *segyehwa* policy between 1993 and 1997 in which overseas FDI was regarded as a bridge over which domestic industries can access the necessary advanced technologies generated abroad through selected mergers, acquisitions and joint ventures with incumbent firms in the Triad. Indeed, the large burden of technological dependence on foreign companies has placed *chaebols* as agents of domestic industrial transformation at a disadvantage given that certain foreign technologies have been increasingly difficult to obtain, particularly from Japan since South Korean firms have become manifest rivals in the same industries in major regional markets (Ursacki and Vertinsky, 1994).

As in the 1980s, the main route in which South Korean MNCs have fulfilled this objective is through acquisitions of foreign firms in the Triad. SEC has been at the forefront of acquiring foreign firms mainly in the United States that have the most advanced technologies in research, production and/or marketing of semiconductors, consumer electronics, telecommunications and computers.[36] In addition, SEC's interest in acquiring Fokker (the Dutch aerospace group) and Rollei (the German camera maker), and Daewoo's interest in Lotus (sports car manufacturers), their 65 per cent stake in Steyr-Daimler-Puch (producers of advanced engines) and their acquisition of FSO and Rodae (cars) also illustrate the objective of South Korean companies to augment their technological expertise and achieve rapid market presence through the acquisition of foreign firms in Europe (Randerson and Dent, 1996).[37] But apart from outright acquisitions, outward FDI in developed countries are enabling South Korean MNCs to gain access to high levels of technological expertise and widens the possibility of gaining from the wider diffusion of technological exchanges. For example, the European Commission's role in the development of new technologies and the existence of long-term development programmes like ESPRIT (European Strategic Programme for Information Technology) have provided an incentive for the South Korean companies to either conclude joint ventures with European firms or acquire European firms. In addition, the market presence of South Korean MNCs in Europe is facilitating access to the technological and logistical capabilities of European small- and medium-sized firms, particularly those in the car components industry which comprises some of the world's most advanced suppliers. The access and exposure to developments at the forefronts or close to the forefronts of technological developments through various modes is serving to enhance the establishment of the *chaebol's* local supply chains in Europe, thus enabling these firms to respond favourably to the higher local content requirements of the EU while helping to improve the competitiveness of their foreign plants in Europe (Randerson and Dent, 1996).

In addition, a related objective of their FDI in Europe has been to overcome the weak marketing capacities of South Korean MNCs through the collection of marketing information to support the growth of South Korean exports to the continent. Examples of these include, as mentioned above, the design centres in the United Kingdom and Ireland established by SEC and

LG Electronics to monitor local market developments in consumer electronics and telecommunications in Europe. In the food industry, the 'Euro-lab' was established in Frankfurt by South Korean firms where, among other objectives, food habits in Europe are being investigated as a means to adapt existing food products or develop new ones for the European market (Randerson and Dent, 1996).

Conclusion

This chapter analysed the determinants of the emergence of South Korean FDI over the last 40 years. The major types of FDI undertaken by South Korean MNCs, the determinants of outward FDI, and their industrial and geographical patterns vacillated in the period until the 1970s, 1980s and 1990s. Thus, while outward FDI to secure supplies of essential raw materials to serve the South Korea-based industrial production complex, outward FDI in trading, banking, warehousing, transportation and distribution in major foreign markets, and manufacturing FDI in developing countries to supply host country markets or as export platforms were the predominant types of South Korean FDI from the period of their emergence in the late 1950s to the end of the 1970s, outward FDI to secure supplies of essential raw materials, outward FDI to supply host country markets in developed countries and to gain access to advanced technologies, as well as outward FDI in trading, warehousing and transportation in major foreign markets were the predominant types of South Korean FDI at the end of the 1980s. At the end of the 1990s, manufacturing FDI in developing countries to supply host country markets or as export platform, outward FDI to supply host country markets in developed countries and to gain access to advanced technologies, and outward FDI in trading, warehousing and transportation in major foreign markets predominated, thus representing a partial return to a concentration of outward FDI in developing countries consistent with the pattern of Korean FDI pertinent until the end of the 1970s.

To the extent that South Korean FDI could be compared with the pattern of Japanese FDI since the Second World War, the first and second phases of Japanese FDI in labour intensive manufacturing in textiles, sundries and other low wage goods (the first phase) and in heavy and chemical industries during the Ricardo-Hicksian trap stage of Japanese FDI (the second phase) had been transposed in the case of the history of South Korean MNCs and, as shown in the last chapter, Taiwanese MNCs. This is associated with a prolonged dependency or sustained comparative advantage of the South Korean economy on labour intensive production until the late 1980s. The basis of the prolonged competitiveness of these industries derived from the South Korean state's policy of wage suppression and the political subordination of popular sectors (i.e. middle stratum of professionals, skilled workers, businessmen and farmers) to the state (Shin and Lee, 1995) which was brought to an end with political democratization in both South Korea and Taiwan from around

1987. The emphasis on the domestic industrial development of heavy and chemical industries in South Korea in the 1970s and 1980s led to the dominant role of outward FDI in natural resource extraction in South Korean FDI in those decades.

South Korea has reached the third phase of Japanese FDI of assembly-based, subcontracting-dependent, mass production in foreign markets in the same set of consumer durables industries – cars, consumer electronics and semiconductors – that have been at the core of the *chaebol's* global strategies in the 1980s and 1990s. However, the wider range of more technologically advanced consumer durable goods produced by Japanese *keiretsus* in their international production networks worldwide from around the mid-1970s onwards contrasted with the much narrower product range, less technologically advanced and lower brand recognition of South Korean products produced in foreign markets by the *chaebols* in the 1980s and 1990s. Indeed, the technological strengths combined with advanced marketing capabilities achieved through product differentiation, well-known brand names and extensive distribution channels by Japanese MNCs in the electrical equipment and motor vehicles industries meant that it had been in those industries in which their involvement and market shares in United States and Europe have been expanding fastest both through exports and international production (Dunning and Cantwell, 1991). By contrast, the far weaker product change and product design capabilities as well as relatively weak marketing capabilities of South Korean MNCs – a feature inherent in firms from countries that are late industrializers as well as the long history of South Korean firms as OEM suppliers in the electrical equipment industry – have constrained their ability to capture substantial market shares for their brand products in the developed countries, particularly in the larger European countries as Germany and France (van Hoesel, 1997).

Moreover, the regional production networks of South Korean *chaebols* which has progressed the fastest in South East Asia and China in the 1990s differ from the regional production networks of Japanese *keiretsus* in North America, Europe, South East Asia and China not only in the greater sophistication and complexity of regional production networks of Japanese MNCs by comparison to those of South Korean MNCs which has enabled Japanese MNCs to progress towards global localization to a much farther extent than South Korean MNCs. The high product change capabilities of Japanese firms attained through cross-functional linkages between design and development, production, marketing and service in the regionally integrated production networks of Japanese MNCs worldwide with fairly sophisticated regional division of labour and decentralized control contrast with the much weaker linkages between production, marketing and design and development functions in the regional production networks of the Korean *chaebols* such as SEC owing to continuing centralization of control from parent companies in Korea for the higher value added functions. Another major difference arises from the unique pyramidal and multi-layered system of subcontracting relationships between Japanese final goods assemblers and their suppliers which has essentially been closed to external economic actors unlike those of the relationships between affiliated suppliers and final goods assemblers of the Korean *chaebols* which has sold components to Japanese final goods assemblers and bought parts from Japanese parts suppliers.

At the end of the 1990s Korean MNCs have not reached the fourth phase of Japanese multinationalization – the mechatronics-based, flexible manufacturing of highly differentiated goods involving the application of the most advanced computer technologies in design, manufacturing and engineering along with the use of more advanced technological breakthroughs associated with high-definition TV, new materials, fine chemicals and more advanced micro-chips, etc. – and their capabilities of reaching this stage will be defined by significant progress yet to be made with the development of more advanced product innovation as well as marketing capabilities.

Thus, the dynamic analysis of the growth of Korean FDI suggests some similarities in the evolutionary course of Japanese FDI and Taiwanese FDI. However, the evolution of South Korean FDI over time is slower compared to that of Japanese FDI. Japanese MNCs seem to have graduated from the first two phases to the third phase within a span of about 15 years or so years (1950 to the late 1960s), while Korean MNEs have taken around 30 years (1960s to the late 1980s). Therefore, despite the similarity in the patterns of their outward FDI, the pace of evolution in the pattern of outward FDI and the character of that evolution has differed considerably between the two countries.

Notes

1 The data in this paragraph were sourced from Tolentino (1993) for 1960, and UNCTAD (1999) for 1980 and 1998.
2 In order to induce private entrepreneurs to initiate investments in manufacturing, the South Korean government implemented import-restrictive and industry-protective policy measures including import prohibitions, quotas, differential tariff rates, overvalued domestic currency, foreign exchange controls, and a multiple exchange rate system. It also extended low interest loans and low grain prices and introduced other forms of relative price distortions. Such a variety of policy measures for industrial protection guaranteed high profit margins by raising the domestic prices of industrial consumer goods while minimizing the domestic costs of production through the low costs of equipment and raw materials imports and the low real wage rates of unskilled workers (Jo, 1981).
3 The liberalization measures imposed between late 1964 and 1965 including the devaluation of the domestic currency, interest-rate reforms, partial lessening of import restrictions and dismantling of other controls (Jo, 1981) were completely at odds with those measures imposed under the import substitution policy.
4 The Foreign Exchange Control Regulations of South Korea in its original version defined outward foreign investment to include exports on deferred payment base (over one year), loans to non-residents or foreign nationals, provision of technical services the royalty period of which exceeds one year, purchase of stocks issued in foreign countries and the purchase of real estate in foreign countries. The eligible countries were limited to those having diplomatic relations with South Korea and

the projects were limited to those generating foreign exchange for South Korea either through import substitution or exports, and to those contributing to long-term stable supplies of essential raw materials. (In 1978, such list of encouraged overseas investment projects was expanded to include those to secure offshore fishing bases; and those investments facilitating relocation abroad due to eroding domestic competitiveness.) Individual projects had to be approved by the Ministry of Finance which had jurisdiction for foreign exchange management in South Korea (delegated to the Bank of Korea in 1975) (Koo, 1984).

5 Unless otherwise indicated, the data on South Korean FDI refer to the approved value of outward FDI as provided by the Bank of Korea. The stock of approved South Korean FDI represent cumulative flows of approved outward FDI since 1968. Outward FDI as defined by the Bank of Korea includes indirect investment and comprises the following: (1) the purchase of 20 per cent or more of a firm's securities issued abroad; (2) long-term loans; (3) real estate investment; (4) investment in foreign natural resources or in high-technology development projects; and (5) investment in private enterprises in foreign countries. These caveats should not affect the time-series analysis of the broad trends in the industrial and geographical patterns of South Korean FDI.

6 Based on data in United Nations, *World Investment Report*, various issues.

7 South Korean industrial exports grew at an annual average rate of 40 per cent during 1970–1978 (Jo, 1981).

8 There were only 18 less than majority owned trading subsidiaries of a total number of 194 subsidiaries of this type in 1982 (Koo, 1984).

9 Establishment of export-facilitating subsidiaries abroad allowed parent companies to export to the host country even without a clear commitment by host country buyers as inventories could be kept by the subsidiaries until actual sales were made (Koo, 1984).

10 Based on data contained in Kumar and Kim (1984).

11 The extent to which the investment decisions by South Korean firms was influenced by actual and potential tariffs in foreign markets had not been investigated by Jo (1981). However, in the late 1960s, tariff barriers were imposed by many South East Asian countries that pursued import substitution industrialization (UN, ESCAP, 1988). Since South East Asia was an important export market for South Korea, then it is likely that trade barriers also determined international production by South Korea in South East Asia.

12 Based on various issues of the IMF, *Direction of Trade Statistics Yearbook*, merchandise exports of South Korea grew by an annual average rate of 36.5 per cent in the 1970s, and although this slowed down to 6.3 per cent in the period between 1981 and 1985, it grew by more than 30 per cent in the second half of the 1980s. The balance of trade of South Korea was in surplus by $4.6 billion in 1986, and grew further to $9.8 billion in 1987 and $14.3 billion in 1988 (Jun, 1990). Thus, by comparison to the annual average net exports of goods and services of −231 billion won in 1968 and 1969 which further worsened to −589.6 billion won between 1970 and 1979, the annual average net exports of goods and services increased significantly to 2,322.1 billion won between 1980 and 1989 before declining dramatically to −3,987.9 billion won between 1990 and 1997. The annual average foreign exchange reserves nevertheless increased from $468.6 million between 1968 and 1969, $1.4 billion between 1970 and 1979, $5 billion between 1980 and 1989 and $21.7 billion between 1990 and 1997. Based on data contained in the International Monetary Fund, *International Financial Statistics Yearbook 1998*, Washington, DC: IMF.

13 On attempts by the United States to demand that Taiwan and South Korea allow their currencies to rise, see 'Investment by Taiwan: The embarrassment of riches', *The Economist*, 25 March 1989; and Kwag (1987).

14 However, the annual average exchange rate of the won to the United States dollar had been on a long-term decline from an annual average exchange rate of 282.4 won per United States dollar between 1968 and 1969 to 427.3 won per United States dollar between 1970 and 1979, 757.8 won per United States dollar between 1980 and 1989, and 794.3 won per United States dollar between 1990 and 1997. Based on data contained in the International Monetary Fund, *International Financial Statistics Yearbook 1998*, Washington, DC: IMF.

15 Political liberalism tolerated a rash of strikes – almost illegal – by workers demanding higher bonuses. See 'Investment by Taiwan: The embarrassment of riches', *The Economist*, 25 March 1989. As a result, wages in the industrial sector in Taiwan in the three years from 1988 to 1990 increased by 10.9 per cent in 1988, 14.6 per cent in 1989 and 13.5 per cent in 1990. In South Korea, the corresponding figures were 19.6 per cent, 25.1 per cent and 20.2 per cent, respectively (van Hoesel, 1997). This in turn led to declining rates of South Korean export growth from 24.8 per cent in 1988, 2.8 per cent in 1989, 4.2 per cent in 1990, 10.5 per cent in 1991 and 6.8 per cent in 1992 (Shin and Lee, 1995).

16 The first step in the liberalization of outward FDI policy in South Korea was implemented in December 1986 and was manifested in the following forms. While previously all overseas investment projects had been subject to case-by-case approval, investment projects amounting to less than $200,000 did not require approval as of December 1986. Such amount was increased to $500,000 in September 1987, $1 million in December 1987 and to $2 million in November 1988. Apart from a continuing trend towards liberalization of foreign exchange regulations, the South Korean government also sought to eliminate cumbersome restrictive measures in its many facets by liberalizing investor's qualification requirements, allowing investments to be made in foreign countries that did not have diplomatic relations with South Korea, eliminating the restricted or recommended categories of eligible overseas investment projects, excluding technical services or deferred export as outward investment, and simplifying approval procedures. The prevailing system of investment incentives for approved overseas investment projects including access to overseas investment funds in the Korean Eximbank and other financial incentives, tax exemptions and provision of information on overseas investments to South Korean investors was also strengthened since 1986 (Kim, 1990; Koo, 1984). To support small- and medium-sized firms in labour intensive industries to restructure their industries towards high value added, technology intensive industries and to encourage the transfer of labour intensive industries or production lines to developing countries with cheap labour, the South Korean government provided both restructuring and investment funds, as well as access to its Economic Development Cooperation Fund for equity investment ventures in developing countries. The government also planned at the end of the 1980s to create an industrial complex in South East Asia to facilitate the transfer of labour intensive production of South Korean manufacturing firms and thus avail of cheap labour and circumvent trade restrictions. Furthermore, apart from the roles of the Bank of Korea, the Korean Trade Promotion Corporation (KOTRA), and the Small and Medium Industries Promotion Corporation in providing foreign investment information services, the Overseas Investment Service Centre was inaugurated in the first half of 1988 to provide a package of services in consulting, investment approval, investment financing and investment insurance for domestic businesses and entrepreneurs who plan outward FDI (Korea Institute for Foreign Investment, 1989). The *segyehwa* movement between 1993 and 1997 embodied the government recognition of the need for greater integration in the world economy through the encouragement of outward FDI and the introduction of economic liberalization, including the opening up of the Korean economy to

greater foreign competition. The movement involved the implementation of reforms in three major areas: first, political and social reforms to conform to a freer and more mature democratic society; second, economic renewal and the strengthening of economic competitiveness; and third, cultural development. The main tools of the movement were increased deregulation of enterprises, enhanced market liberalization, reduced reliance by the government as an economic partner, greater support for small- and medium-sized firms, and the pursuit of a more equitable partnership between management and labour (Ungson *et al.*, 1997).

17 Based on UNCTAD (1999), the estimated stock of actual FDI of South Korea increased by almost five times between 1985 and 1990 from $461 million to $2.3 billion.

18 Jeon (1992) found no evidence to support the notion that South Korean FDI in manufacturing in the developed countries reflected a defensive motive by oligopolistic firms in South Korea to maintain foreign market shares.

19 The state firms Korea Petroleum Development Corporation and KODECO formed a consortium to develop oilfields in North Yemen and East Madura, Indonesia. In addition, Korea Electric Co., another public sector company, also participated in the resource development of uranium in the United States and Canada (Korea Institute for Foreign Investment, 1989).

20 This was the case, for example, in the mining investment project organized as a joint venture in Alaska to explore oil and coal initiated by several South Korean trading companies. While the South Korean firms provided the financing for exploration in the form of loans, their American counterparts provided exploration rights and sites for the development of the resource at a later stage, if necessary (Koo, 1984).

21 It is sometimes argued that, unlike South Korea, Taiwan was able to avoid much of the trade retaliation brought about by the increasing trade pressures owing to the small size of its firms which provided a degree of anonymity, and the important role of many Taiwanese companies as original equipment manufacturers (OEM) whose fate was linked inextricably to that of domestic producers in the developed countries. This made trade retaliation by developed countries a process of self-inflicting pain (Wang and Hsu, 1992).

22 It is important to take note that although the great majority of South Korean manufacturing FDI in developed countries was geared to cater to local market demand in those countries, in exceptional cases South Korean manufacturing FDI in those countries served as an export platform type of FDI. This was the case, for example, with a South Korean manufacturing project in Portugal in which the subsidiary was used as an export base in Europe (Koo, 1984).

23 This is owing to the allegation by the United States of adverse trading practices exercised by Taiwan and South Korea. Thus, although Taiwan and South Korea have relatively free access to the United States market, these countries were accused of restricting the imports of American goods and services in their countries through import tariffs and non-tariff barriers such as licensing regulations, technical standards, export subsidies and restricted access to foreign participation in the banking, insurance and securities industries. The United States particularly feared that South Korea with the largest industrial capacity of the four Asian newly industrialized countries had the capability to become another economic power like Japan (Chia, 1989).

24 However, unlike in the case of Korea, the Japanese government was not a provider of finance to the *keiretsus* who have their own banks as part of their large organizations. By comparison, the relationship between government and business in Korea is such that the government sets the policies and traditionally controlled the *chaebols'* access to capital. The *chaebols* were not allowed to own banks until the 1990s (Ungson *et al.*, 1997).

25 LG Electronics established a colour TV plant in a suburb of Huntsville, Alabama, in 1982. Threatened by LG Electronics' strategic move, SEC similarly established a few months later its first FDI in a pilot plant in Portugal organized in the form of a joint venture with Portuguese and British partners as a means of gaining international production experience. In 1984, SEC then established a colour TV manufacturing plant in Rousbery, New Jersey, in 1984 with an investment of $25 million (Korea Institute for Foreign Investment, 1989; Kim, 1996). This illustrates that South Korean MNCs emerging from domestic oligopolistic industries also exhibit similar patterns of oligopolistic behaviour such as 'follow my leader' so clearly observed in MNCs of the developed countries (see Knickerbocker, 1973).

26 Supported by initial generous tax relief and low selling price, Hyundai Motor Company's subcompact car, Pony, became the best-selling foreign car in Canada within two years of its 1984 introduction. However, a 36 per cent average anti-dumping duty was imposed on Hyundai car exports to Canada in 1987 in response to a ruling by Canada's federal tax department of dumping practices by the Hyundai Motor Company. See 'Hyundai reined in: South Korean carmaker penalised for dumping in Canada', *Far Eastern Economic Review*, 17 December 1987.

27 In the case of colour TVs, production cost in the United States was calculated to be about 7 per cent higher than the pre-trade barrier export cost. In the case of microwave ovens, local production in Europe costs 4 to 6 per cent more than the pre-trade barrier export cost. Such cost differentials are significant in view of the low profit margin of the export business of South Korean firms (normally less than 3 per cent of sales) (Jun, 1990).

28 The establishment by SEC of a production plant for colour TVs in Portugal in 1982 although no longer in operation was the first production facility established by the company outside South Korea. The company also set up headquarters and a warehousing facility in Germany in the same year, a sales office in the United Kingdom in 1984, production and service facilities in the United Kingdom in 1987 to produce colour TVs, a warehousing and sales facility in France in 1988, and production and service facilities in Spain (1990) and Turkey (1989) to produce VCRs and colour TVs, respectively. As with SEC, LG Electronics similarly established headquarters in Germany in 1980 where it also has warehousing, sales and service facilities. It also established a production facility in that country in 1986 to produce colour TVs and VCRs, warehousing, sales and service facilities in the United Kingdom in 1987 as well as a production facility since 1988 to produce microwave ovens and colour TVs (van Hoesel, 1997). Daewoo Electronics had established a production plant for microwave ovens in Longwy, France, with an annual production capacity of 300,000. Since 1985, Daewoo had been supplying microwave ovens to its joint venture partner JCB which accounted for some 12 per cent of the French market for microwave ovens in the late 1980s (Korea Institute for Foreign Investment, 1989).

29 It is important to take note that some of the export platform manufacturing FDI by South Korean firms in developing countries are also serving the needs of domestic markets of their host countries. The data, however, does not allow the disaggregation of the market orientation of South Korean manufacturing FDI according to export markets and local markets in host countries.

30 Indeed, Indonesia was a highly favoured location for export platform FDI by South Korean firms owing not only to its low wage levels but also due to the state's authoritarian control of the labour force. Indeed, Indonesia was the second-largest host country of Korean FDI in the late 1980s and early 1990s after the United States (Shin and Lee, 1995).

31 Based on the IMF, *International Financial Statistics Yearbook 1998*, Washington, DC: IMF.

32 The relationships between the government and business sectors in Japan is different. In particular, the Japanese *keiretsu* have their own banks as part of their organizational structure which provides the member companies with a reliable source of credit and finance, including in the facilitation of outward FDI. By contrast, the South Korean *chaebols* have not been allowed to own banks until the mid-1990s (Ungson *et al.*, 1997).

33 As shown in Chapter 10, some 79 per cent of the stock of approved Taiwanese FDI in 1995 was in developing countries and 21 per cent was in the developed countries.

34 'Auto makers in South Korea set plant plans', *The Wall Street Journal*, 11 November 1991; 'Korean car makers proceed with plans for overseas plants', *The Asian Wall Street Journal*, 18 November 1991.

35 Indeed, it is sometimes the case that the *chaebol* sources parts, components and intermediate products from independent suppliers because the cost of supplies from their affiliated suppliers was much higher (Kim, 1996).

36 SEC acquired the following foreign companies between 1993 and 1995. The first was the 20 per cent acquisition of Array Company in the United States in 1993 to establish cooperative arrangements in digital process chip technology used in multimedia products. The second acquisition in the same year and in the same country was Harris Microwave Semiconductor, one of the world's leading makers of optical semiconductors with a specialization in gallium arsenide chips. The third acquisition was 51 per cent of of the Japanese company, LUX, a producer of hi-fi audio equipment, in 1994. The objective of the acquisition was to concentrate development and sales in LUX and manufacturing and sales in SEC. The fourth acquisition was 51 per cent of the American company, Control Automation Inc., a CAD/CAM software technology company, in 1994. The fifth acquisition was 15.1 per cent of the Chilean company, ENTEL, the largest operator of telecommunications systems, in 1994. The sixth acquisition was the American company, Integrated Telecom Technology, that specializes in ATN technology, in 1994. The seventh acquisition in 1995 was 4 per cent of the shares of the American company, Integral, that specializes in HDD technology, for the purpose of joint development of HDD products. The eighth acquisition was 40.25 per cent share of another American company, AST Research, a computer company. The purpose of the acquisition is to establish a broad range of commercial relationships including supply and pricing of critical components, joint product development, cross-OEM arrangements and cross-licensing of patent (Kim, 1996).

37 The acquisition of FSO and Rodae has enabled Daewoo to have a rapid means of establishing car production in Europe to advance the company's ambition to be one of the top car producers by 2005 (Randerson and Dent, 1996).

12 Conclusion

The emergence and evolution of multinational corporations from resource-scarce large countries

The general conclusion that can be drawn from the analyses in this part of the book is that there is a common pattern in the emergence of MNCs from the five resource-scarce large countries (United Kingdom, Germany, Japan, Taiwan and South Korea) that despite their inherent natural resource scarcity based their domestic industrial development on resource intensive production. Multinational corporations based in these countries share a common origin in the following types of outward FDI: resource extractive FDI to secure natural resources to support domestic industrial expansion and meet domestic consumer needs, manufacturing FDI and, given the common importance of trade as an engine of growth in the national economies of the five countries analysed, FDI in trading, distribution, marketing and after-sales services (see Table 12.1). Despite the close similarity in the types of FDI in which MNCs from the five resource-scarce large countries emerged, there were important differences with respect to the types of firms that undertook the initial FDI across the different FDI types, the actual development paths of outward FDI as it related to local industrialization in each country (Table 12.2) and the form of technological accumulation of leading national firms in relation to the natural course of outward FDI in each country (Table 12.3). This chapter seeks to reflect on the similarities and differences in the emergence and evolution of MNCs from the resource-scarce large countries.

The emergence of multinational corporations

Resource extractive FDI

Outward FDI in resource-based activities in the agriculture, forestry, fishery and oil sectors were stimulated by the demand of domestic processing industries and consumers for primary products. Because of their resource scarcity, these vital inputs to domestic industrial production or consumer requirements either do not exist in the home country or are available in inadequate amounts to support industrialization or meet consumer requirements. The shortage of natural resources in the large economies of the United Kingdom, Germany, Japan, South Korea and Taiwan provided the need

Table 12.1 Resource-scarce large countries: variations in the early stages of outward direct investment across different countries

Examples of countries	Dominant form of earliest outward FDI	Type of locally based MNC
United Kingdom	Resource-based extraction and processing in agriculture, minerals and petroleum to support domestic industrial expansion and meet domestic consumer needs	Free-standing firms Resource-based firms Firms in manufacturing, petroleum and trading, engaging in backward vertical integration
	Complementary service investments to resource-based investments (finance, insurance, trading, transportation and distribution)	Banks Insurance companies Trading companies Transportation companies (shipping and railroads)
	Agriculturally based investments based on entrepreneurial perceptions of profitable investment opportunities	Free-standing firms Large land companies
	Investments in trading	Trading companies, including those acting as core for British-based investment groups
	Investments in transportation (railroads, shipping, etc.)	Railroad companies Shipping companies Free-standing companies
	Investments in banking and insurance	International and imperial banks, some of which began as free-standing companies Insurance companies
	FDI in maufacturing in branded consumer goods industries mainly and some producer goods industries (industrial machinery, chemicals, oil, armaments and pharmaceuticals)	Free-standing companies Manufacturing companies Trading companies
Germany	Resource-based extraction in mining and petroleum and complementary service investments (trading and distribution) to support domestic industrial expansion	Metal trading companies. Manufacturing companies. Deutsche Bank and its holding companies
	Services FDI in trading	Metal trading companies Trading companies or Mercantile houses

Table 12.1 (continued)

Examples of countries	Dominant form of earliest outward FDI	Type of locally based MNC
	FDI in banking to support the expansion of German business abroad	Banks
	FDI in local market-oriented manufacturing industries of the Second Industrial Revolution (chemicals, pharmaceuticals, machinery, electro-technical products and motor vehicles)	Manufacturing companies, large sized with high technological intensity
	FDI in local market-oriented manufacturing in processing industries (iron and steel, textiles, branded consumer goods)	Manufacturing companies, relatively smaller and more moderate sized
Japan	Trading investments in major export markets to promote the growth of Japanese exports. This was supported by the development of auxiliary businesses in banking, insurance and transportation (shipping and railroads)	*Sogo shosha* (Japanese trading firms) Banks Marine insurance firms Shipping companies Railroad companies
	Banking investments to gain access to international capital markets and to promote the growth of Japanese exports and FDI	Banks
	Investments in small- to medium-scale labour intensive manufacturing (e.g. cotton textiles and food processing) in the face of rising domestic production costs	Manufacturing companies Trading companies
	Resource-related extraction and processing investments in agriculture, mining and petroleum and complementary services (banks, railroads) to support domestic industrial expansion and meet consumer needs	Manufacturing companies Banks Railroad companies
Taiwan	FDI in natural resource intensive manufacturing (pulp and paper, cement, rubber, food and drink) and in heavy (cable and wire, etc.) and chemical industries	Manufacturing companies

(continued)

Table 12.1 (continued)

Examples of countries	Dominant form of earliest outward FDI	Type of locally based MNC
	Export-oriented FDI in trading, sales, distribution, marketing and services to facilitate the growth of manufactured exports of Taiwan	Large exporting manufacturing companies, forwardly integrating
	FDI in shipping to facilitate the exports and imports of Taiwan	Shipping companies
	Banking investments to gain access to international capital markets and to promote the growth of Taiwanese exports and FDI	Banks
	Resource-based extraction in agriculture, forestry and fishery industries	Manufacturing companies, engaged in backward vertical integration
South Korea	Resource-based extractive investments in agriculture, forestry, fisheries and mining to support domestic industrial expansion and meet domestic consumer needs	Large manufacturing companies, backwardly integrating Resource-based firms
	FDI in trading, warehousing and transportation in major foreign markets to promote the growth of South Korean exports	Manufacturing companies, forwardly integrating. General trading companies as trading arms of the large conglomerates (*chaebols*)
	Banking investments to gain access to international capital markets and to promote the growth of South Korean exports	Banks
	Manufacturing FDI in non-metallic mineral products, food, drink and tobacco, paper products and basic metal products to supply host country markets or as export platform	Manufacturing companies General trading companies
	FDI in civil construction and engineering to promote export of skilled and semi-skilled labour	Construction companies Engineering companies

Source: Author's compilation based on the analysis contained in the country chapters.

Table 12.2 Resource-scarce large countries: actual development paths for outward direct investment, and their association with local industrialization across different countries

Examples of countries	Link between domestic development and the growth of outward FDI	Type of locally based MNC
United Kingdom	International production by British manufacturers remained oriented towards the mature, relatively low-technology sectors whose competitiveness emanated from the high income and large British market: large size, established technological strengths, product differentiation, quality, and marketing and managerial skills and experience. These characteristics favoured the continuing pre-eminent role of consumer goods firms as the new British MNCs during the inter-war years and post-Second World War period	Manufacturing firms in consumer goods industries, serving local markets
	Host of institutional barriers that prevented industrial restructuring towards the growth-oriented, higher technology intensive industries, including investments in innovation that would sustain those industries	Slower growth of manufacturing firms in higher technology industries with high foreign technological dependence
	The development of the services sector in the domestic economy and associated growth of exports and outward FDI in services in the 1990s	Services firms (primarily finance, trade and other services)
Germany	The defeat of Germany in the two world wars and the sequestration of foreign assets led to comparatively little international production until two decades after the end of the Second World War. The loss of some of the most modern parts of its industrial base as well as natural resources and the confiscation of German patents and foreign assets after the two world wars created strong pressures to upgrade German industry in advanced, knowledge-based industries and to foster technological innovation in indigenous firms. However, domestic embeddedness of German manufacturing firms, remain, as manifested in their	Manufacturing firms, serving local markets

(continued)

Table 12.2 (continued)

Examples of countries	Link between domestic development and the growth of outward FDI	Type of locally based MNC
Germany (*continued*)	continued preference for exporting over FDI as an internationalization strategy, and in the continuing concentration of their outward FDI in Western Europe	
	The development of the services sector in the domestic economy and associated growth of exports and outward FDI in services in the 1990s	Services firms (finance, business services, etc.)
Japan	Weakening of the traditional labour intensive manufacturing industries owing to a continuous cycle of enhanced labour productivity and rising wages and strengthening of the Japanese yen	Small and medium-sized manufacturing companies
	Scale economies-based modernization of heavy and chemical industries in the late 1950s to the early 1970s which prompted outward FDI to overcome the Ricardo-Hickson trap of industrialization due to natural resource scarcity, the shortage of industrial space and environmental problems	Manufacturing companies
	Industrial upgrading in more complex fabricating industries (motor vehicles and electronics) since the late 1960s. International production based on the mass production of assembly-based consumer durables, supported by a network of subcontractors. International production prompted by protectionism and recycling of trade surplus in the United States in the 1970s and by the threat of a fortress Europe in the 1980s	Large conglomerate companies (*keiretsus*) and their suppliers of parts and components
	Industrial upgrading in even more complex manufacturing of highly differentiated goods, involving the application of computer-aided designing, computer-aided engineering and computer-aided manufacturing in the 1980s, and associated international production. International production prompted both by protectionism and oligopolistic competition between firms in international industries	Large conglomerate companies (*keiretsus*)

(continued)

Table 12.2 (continued)

Examples of countries	Link between domestic development and the growth of outward FDI	Type of locally based MNC
	The development of the services sector in the domestic economy and associated growth of exports and outward FDI in services in the 1990s	Services firms (finance, property, trading, general services)
Taiwan	Domestic industrial upgrading in heavy and chemical-based industries (plastic and plastic materials, synthetic fibres, etc.) International production in the 1970s motivated by the need to seek new markets abroad, secure access to export markets in the midst of rising international protectionism, secure raw materials supplies and maintain access to cheaper energy and land costs, and overcome domestic environmental problems	Large manufacturing companies
	Further advancement in electrical and electronics production and chemicals production in the domestic economy fostered by industrialization policy. Led to further growth of exports and international production in these industries primarily in the United States in the 1970s and 1980s and in Europe in 1980s and 1990s prompted by the strengthening of the New Taiwan dollar, the avoidance of trade barriers and the recycling of trade surpluses and the need to gain access to advanced foreign technologies and marketing expertise	Large manufacturing companies
	Rapid export growth accompanied by the rise of wages faster than productivity growth, appreciating New Taiwan dollar combined with the avoidance of trade barriers spurred the emergence in the late 1980s of export-oriented international production in labour intensive production of textiles, clothing and other labour intensive goods, including electronic and electrical products in the developing countries of Asia and the Caribbean	Small- and medium-sized manufacturing companies

(continued)

Table 12.2 (continued)

Examples of countries	Link between domestic development and the growth of outward FDI	Type of locally based MNC
South Korea	Industrial upgrading in domestic metals production precipitated the growth of outward FDI in minerals extraction between 1978 and 1982 prompted by the strengthening resource nationalism in many mineral-rich countries and concerns about the stability and security of mineral supplies	Manufacturing firms Mining companies
	Rapid export growth accompanied by the rise of wages faster than productivity growth, appreciating South Korean won combined with the avoidance of trade barriers spurred the emergence in the late 1980s of export-oriented international production in labour intensive production of textiles, clothing and other labour intensive goods, including electronic and electrical products in the developing countries of Asia	Small- and medium-sized manufacturing companies
	Domestic industrial upgrading, rapid economic growth, trade barriers in export markets and appreciating Korean won threatened the exports of consumer durables (electrical and electronics products and cars) as well as steel products. This led to international production in assembly of final goods in the developed countries initially in the 1980s and in developing countries of Asia in the late 1980s and 1990s	Large conglomerate companies (*chaebols*)
	Domestic industrial upgrading in more advanced manufacturing industries precipitating FDI in developed countries to gain access to advanced foreign technologies and marketing expertise	Large conglomerate companies (*chaebols*)

Source: Author's compilation based on the analysis contained in the country chapters.

Table 12.3 Resource-scarce large countries: technological accumulation and the national course of outward direct investment

	Stages of national development		
	(1)	*(2)*	*(3)*
Form of technological competence of leading indigenous firms	Basic engineering skills, complementary organizational routines and structures	More sophisticated engineering practices, basic scientific knowledge, more complex organizational methods	More science-based advanced engineering, organizational structures reflect needs of coordination
Type of outward direct investment	Early resource-seeking market-seeking or export-oriented investments in manufacturing and services	More advanced resource-oriented, market-targeted or export-oriented investments in manufacturing and services	Research-related investment and integration into international networks
Industrial course of outward direct investment	Resources based (extractive MNCs or backward vertical integration), simple manufacturing, trading, banking, insurance, transportation and construction	More forward processing of resources, or growth of fabricating industries for local markets or exports; growth of services	More sophisticated manufacturing and services systems, international integration of investment
Stage of development			
United Kingdom	starting in the 1820s	between 1870 and 1900	since the Second World War
Germany	starting in the 1880s	starting in 1855	starting in the 1970s
Japan	starting in 1877	starting in the 1970s	starting in the 1980s
Taiwan	starting in the late 1950s	starting in the 1970s and 1980s	currently still unreached
South Korea	starting in the 1960s	starting in the 1980s	currently still unreached

Source: Author's compilation based on the analysis contained in the country chapters.

to search for, and gain control over, sources of natural resources in resource-rich foreign countries. This objective could not be effectively pursued through trade owing to market failure of various forms. Indeed, much of the FDI by firms from the United Kingdom and Germany in the nineteenth century, as well as those of Japan in the 1950s and 1960s and those of South Korea and Taiwan in the 1960s, 1970s and 1980s was of this type.

However, outward FDI to extract natural resources for MNCs based in Taiwan has been less important compared to MNCs from other countries that share features of its national economic development – the United Kingdom, Germany, Japan and even South Korea – and in which resource extractive FDI played a more prominent role in the pattern of outward FDI particularly in the early phases. In the case of Taiwan, outward FDI to secure supplies of essential natural resources and raw materials even at its peak in the 1960s and 1970s did not account for more than 10 per cent of the stock of approved FDI. Two factors may have contributed to such small share: the rising tide of resource nationalism in developing countries since the late 1960s and the high capital intensity of FDI of this type which Taiwanese MNCs cannot provide. Perhaps the importance of Taiwanese FDI in resource extraction has been masked by outward FDI in resource processing manufacturing industries that have either been relocated to resource-rich host developing countries in South East Asia to serve host country markets and/or engage in the export of semi-processed raw materials from the resource-rich host countries to Taiwan.

So important was resource extractive FDI for the United Kingdom that among the oldest British MNCs were 'free-standing' firms, i.e. firms which did not undertake any prior production in the United Kingdom before investing abroad but were created primarily for the purpose of undertaking resource-related business exclusively or mainly abroad. This was a prominent feature of the emergence of British MNCs (Wilkins, 1988a) which was not evident in the emergence of MNCs from Germany, Japan, South Korea and Taiwan where for the most part resource-based FDI were undertaken by manufacturing firms, trading companies and even banks (as seen in the historical oil-related activities of Deutsche Bank and its holding companies). However, apart from free-standing firms, resource-based firms as well as manufacturing companies, petroleum companies and trading companies of the United Kingdom similarly engaged in resource extractive FDI.

Manufacturing FDI

Although manufacturing was an important sector in the origins of MNCs from the five resource-scarce large countries analysed in this part of the book, the manufacturing FDI that emerged differed across the five countries in the kinds of industries that spawned the earliest MNCs, the kinds of firms that initiated international production, the determinants of such international production and their principal host countries.

Manufacturing industries in which outward FDI emerged

The manufacturing industries that spawned the earliest MNCs differed across the five countries. The international production of a significant number of British manufacturing MNCs emerged in branded consumer goods (Dicken, 1992; Chandler, 1986) which was a reflection to a large extent of the comparative advantages of the United Kingdom in labour intensive, capital neutral and human capital-scarce products (Crafts and Thomas, 1986) and the technological hegemony of the United Kingdom in the industries associated with the First Industrial Revolution. Many of these companies were in the textiles or textiles-related industries that had grown first and foremost in the United Kingdom and had began to invest in a major way in the United States, Canada, France, Germany, Russia (Wilkins, 1989). Some of the other early British manufacturing MNCs in consumer goods industries were British American Tobacco (tobacco), Bryant & May (matches), J. & P. Coats (cotton thread), Courtaulds (rayon), Dunlop (rubber tyres), English Sewing Cotton (cotton thread), Gramophone (records), Lever Brothers (soap), Pilkington Brothers (glass) and Reckitt & Sons (household products). Only to a limited extent did British manufacturing MNCs emerge from the more technology intensive and knowledge intensive industries spurred by the Second Industrial Revolution in the second half of the nineteenth century. This was the case with Babcock & Wilcox (industrial machinery), Nobel Explosives (chemicals), Royal Dutch Shell (oil), Vickers (armaments) and Burroughs Wellcome (pharmaceuticals) (Dunning and Archer, 1987).

By contrast, the industries that spawned the earliest German FDI in manufacturing were the new, skills intensive, technically advanced and fast growing industries of the Second Industrial Revolution in the late nineteenth century: chemicals, pharmaceuticals, machinery, electro-technical and motor vehicles. German firms rapidly became significant world producers and exporters of their new products on the basis of accumulated expertise and proprietary technology. Such strengths, combined with access to international capital markets at least until the end of the First World War, enabled German firms to invest abroad on a substantial scale, and to exert a significant economic impact within their host economies. However, German FDI prior to 1914 was also present in capital goods industries such as iron and steel manufacture, and in consumer goods industries such as textiles, particularly in woollens and silk and, similar to the British MNCs, in a number of trademarked consumer products (consumer pharmaceuticals, pencils and food products).

The emergence of Japanese MNCs, Taiwanese MNCs and South Korean MNCs in manufacturing bear a closer resemblance to the pioneering British MNCs in their concentration on industries reflecting a comparative advantage in labour intensive, capital neutral and human capital-scarce products rather than in the higher technological intensity of the early German MNCs. The emergence of manufacturing FDI in Japan perhaps bears the closest similarity to that of the United Kingdom in their common basis in textiles or textiles-related industries. Indeed, just as the textile industry had its historical origins in the United Kingdom and brought forth some of the earliest British MNCs in manufacturing, textiles was the dominant domestic industry of Japan in the late nineteenth and early twentieth century and it was the first industry in which Japanese manufacturing MNCs and some trading companies initiated international production.

By contrast, the pioneering industries of Taiwanese and South Korean manufacturing FDI were closely similar. These were food and drink, synthetic fibre processing, paper products, plastic and plastic materials, non-metallic minerals, primary metals and machinery in the case of Taiwan, and non-metallic mineral products, food, beverages and tobacco, paper and paper products, and basic metal (steel) products in the case of South Korea. These were generally processing industries of the First Industrial Revolution in which the United Kingdom had a technological hegemony but at the time of the emergence of the Taiwanese and South Korean MNCs were already fairly technologically standardized industries. The different industrial structure of the earliest manufacturing FDI of Japan, Taiwan and South Korea by comparison to those of the United Kingdom and Germany can be explained by the earlier stage of national development in which Japan, Taiwan and South Korea engendered MNCs of their own and thus the more basic industries in which domestic firms developed ownership-specific advantages. Thus, MNCs have not always emerged in technologically advanced industries with firms that have highly developed firm-specific advantages.

Types of firms that initiated the pioneering manufacturing FDI

Although some of the historical British manufacturing FDI were initiated by free-standing companies and trading companies such as Jardine Matheson & Company that established the Ewo Cotton Spinning and Weaving Company in Shanghai before 1914 (Wilkins, 1986b), direct capital exports of the United Kingdom consisted more of the establishment of foreign subsidiaries and branches by manufacturing companies mentioned above that were already operating in their home countries – essentially the kind of foreign activity of the modern classic MNCs which mainly predominates in modern time (Dunning and Archer, 1987). All of the major pioneering British MNCs that emerged pre-1914 held strong oligopolistic positions in their domestic markets, and several were members of international cartels or market-sharing agreements (examples included Babcock & Wilcox, British American Tobacco, Bryant & May, Gramophone and Nobel Explosives). The managerial and technological competences developed in their home countries gave firms the ability to initiate and sustain their international production activities.

The German firms that initiated manufacturing FDI were also predominantly large manufacturing firms in industries producing the newest and most technologically advanced products whose emergence can also be attributed to the rapid development of administrative hierarchies in German business organizations (Wilkins, 1988b). These technologically innovative large-sized manufacturing firms led the process of German MNC expansion in high-technology industries introducing new products and processes through FDI. However, there were also large-sized but less technologically intensive manufacturing firms that were responsible for German FDI in capital goods industries such as iron and steel manufacture, and there were also relatively smaller and more moderate-sized German MNCs in trademarked consumer goods industries prior to 1914.

As in the German case, manufacturing firms were also solely responsible for the earliest Taiwanese manufacturing FDI. These firms were often the largest firms in their respective industries (Schive and Hsueh, 1985) but were perhaps not as large as the pioneering German or British manufacturing companies and were not in the high-technology industries that spawned German manufacturing MNCs nor branded consumer goods industries that bred most of the pioneering British manufacturing MNCs.

By contrast, two types of firms initiated Japanese and South Korean manufacturing FDI: manufacturing companies and trading companies. The emergence of Japanese FDI in manufacturing were initiated by the cotton spinning companies and trading companies, thus bearing a closer similarity to the types of firms that generated the earliest British manufacturing FDI. Indeed, one of the earliest successful Japanese FDI in the cotton textiles industry were those initiated by the Japanese general trading company (*sogo shosha*), Mitsui & Co., which acquired two Chinese cotton spinning mills between 1902 and 1906 through a local subsidiary – the Shanghai Cotton Manufacturing Company – organized by the trading company in 1908. Other prominent Japanese trading companies that initiated manufacturing FDI in the industry were the Japan Cotton Trading Company and the Naigaiwata Company (Wilkins, 1986b; Yasumuro, 1984).

In the case of South Korea, manufacturing FDI was pioneered to a far larger extent by manufacturing companies than by trading companies. An analysis of the type of firms engaged in Korean FDI in manufacturing showed that some 30 of the 37 manufacturing projects as of 1982 represented the horizontal integration of manufacturing firms, while only 7 projects were undertaken by general trading companies to secure semi-processed raw materials such as pulp and wool (Koo, 1984). These were predominantly medium-sized manufacturing firms engaged in semi-skilled labour intensive domestic production of resource intensive products in South Korea. Owing in part to their new status as MNCs and unfamiliarity with the conduct of international business in foreign countries as well as the restrictions on foreign ownership, particularly in developing countries, some 65 per cent of the overseas manufacturing projects of South Korean firms in 1982 (or 24 of the 37 manufacturing projects) and some 71 per cent of those in developing countries were those in which South Korean firms had minority equity participation (Koo, 1984).

Determinants of international production

The determinants of international production by firms from the five countries in both the traditional and modern industries have a common basis in the inability of important foreign markets to be supplied, or supplied as cheaply, through exports. The precise reason for the inadequacy of arms-length trade to service foreign markets differed across the five countries. In the case of the United Kingdom, import restrictions imposed by the host governments more often than not were the key element that rendered exports uncompetitive in foreign markets. Indeed, imported manufactured products faced higher tariffs in the United States, Canada and most European countries after 1880 (Jones, 1996). Only in limited cases was the initial decision by British firms to engage in international production before 1914 based on more favourable production costs in the foreign location, the provision of host government incentives or patent legislation (Coram, 1967; Buckley and Roberts, 1982). In addition, high transport costs encouraged foreign production by Babcock & Wilcox, Gramophone and Nobel Explosives whose products were high volume/low value or dangerous to export over long distances (as in the case of Nobel Explosives). Tariffs and transport costs often combined to prompt British manufacturers to establish overseas subsidiaries, particularly in countries where there was strong or emerging indigenous competition. Another contributory factor that prompted the shift from exports to international production by British manufacturing firms particularly those in the more modern fabricating industries was the need to sustain a process of competition between firms in oligopolistic industries (Dunning and Archer, 1987).

By contrast, the determinants of the earliest international production by German manufacturing MNCs was spurred by a more complex set of factors. In the chemicals industry, the major determinants were the imposition of tariffs after 1877 in Russia – the most important export market for German chemicals – as well as the relatively high import duties in the United States on pharmaceutical products or the total prohibition against the import of pharmaceuticals in tablet or capsule form in France. In some cases, non-tariff barriers such as the peculiar nature of marketing pharmaceutical products in the United States in which chemists' shops were not supervised by scientifically trained pharmacists with the consequence that preference was often accorded to the sales of ready-made and packed drugs prompted the international production of German pharmaceutical companies. In contrast to British manufacturing FDI, other major determinants that affected the emergence of the highly technologically intensive German manufacturing MNCs was stringent patent legislation in foreign countries that required the initiation of local production after a period of time and in an extreme case the French patent law of 1844 which did not allow the patenting of any type of medicine. This coerced the German pharmaceutical company E. Merck to conclude a limited joint venture with a local firm to market E. Merck's brand names for some pharmaceutical products bottled and packed in France (Hertner, 1986a).

In the electro-technical industry, the establishment of foreign factories was often initiated as a consequence of considerable demand from the Russian government to construct and maintain a telegraph network (the case of Siemens), the growing pressures from the Russian administration which insisted on domestic production for continuing state orders of public utilities [the case of Allgemeine Elektricitäts-Gesellschaft (AEG)], the subsidies extended by the French state to domestic producers of trucks whose parts were made solely in France, the need to service the needs of the French car industry (the most important car industry in pre-1914 Europe and a highly important customer), and the need to respond to the requirement for domestic production of components used for racing cars participating in certain racing events in Great Britain (the case of Bosch that initiated international production in France and the United Kingdom of magneto ignitions for cars and trucks). In addition, the high cost of freight and the 45 per cent tariff on the value of magnetos led to the construction of a magneto factory in Springfield, Massachusetts in 1910 where production started in 1912.

While state intervention in the form of tariffs or non-tariff barriers as well as patent legislation in host countries were the major explanatory factors, particularly if the state was an important customer as in the case of Russia, there were other contributory factors of which the most significant was the access to international capital markets of the *Unternehmergeschäft*, the large German trusts that founded local and regional power, tramway and lighting companies in Russia, Italy, Spain and Latin America. Such establishment of intermediate financial holding companies enhanced the capacity of the newly created German public utility companies abroad to engage in FDI as it provided a means to overcome the chronic lack of capital of their major customers – the local public authorities in foreign countries – and to counteract the liquidity problems of the electro-technical firms themselves associated with the accumulation of a growing volume of equity capital and bonds in their portfolio.

The determinants of the switch from exports to international production in the case of the pioneering Japanese MNCs in manufacturing in the cotton textiles industry stemmed from the Shimoneseki Treaty of 1895 which allowed foreigners to manufacture in Chinese treaty ports for the first time. The consequent establishment of foreign-owned cotton spinning factories in China by British trading companies such as Jardine Matheson & Company and three foreign textile firms from Great Britain, the United States and Germany (Wilkins, 1986b) combined with the incipient growth of Chinese investments in spinning mills, and the sale in China of cheap Indian yarn posed a threat to the continued growth of the Chinese export market for both Japanese cotton spinners and trading companies. In addition, much of the Japanese FDI in

China in the early twentieth century may have been made initially to ascertain local costs of production in China and to keep close watch over the Chinese textile market. This contrasts with the case of the pioneering British and German manufacturing MNCs in which there was little reason to suppose that lower production costs abroad influenced the initiation of outward FDI before 1914 since international production seldom occurred in very labour intensive industries, and hence low labour costs abroad were rarely an enticement.[1]

The predominant motivations behind international production of the earliest Taiwanese MNCs and South Korean MNCs were in a class of their own. Given their concentration on processing industries and heavy and chemical industries, the earliest manufacturing projects initiated by Taiwanese and South Korean firms was determined by the search for markets, the security of export markets and access to abundant raw material supplies, cheaper energy and land costs.

Host countries of manufacturing FDI

In general, the host countries of the earliest FDI in manufacturing in the five countries were their major export markets, a finding that provides broader empirical support for the product cycle model of Vernon (1966) in explaining the initiation of international production by American MNCs in their major export markets. The strong ownership-specific advantages of British and German MNCs and the high income elasticity of demand for their export products formed the basis for the establishment of international production in developed host countries. The emergent import substituting manufacturing investments by British firms prior to the First World War displayed a preference for high-income markets, but with some bias towards countries belonging to the British empire owing to political and other psychic ties. Similarly, German manufacturing FDI were largely directed apart from Russia towards the more industrialized countries of Europe (France, the United Kingdom, etc.) and the United States. By contrast, the geographical destination of the earliest international production activities by Japan, Taiwan and South Korea was directed towards developing countries of South East Asia and China predominantly, and can be explained by the early stage of national development at which Japan, Taiwan and South Korea engendered MNCs of their own and thus the more simple ownership-specific advantages possessed by the leading domestic firms. The pioneering manufacturing FDI by Japan was directed to China, the major export market for Japanese exports of cotton manufactures and yarn. Similarly, there was a marked concentration of the earliest manufacturing FDI by Taiwan and South Korea in their major export markets in South East Asia primarily, particularly in those countries with abundant natural resources. Africa and Oceania were also significant host countries for South Korean manufacturing FDI.

FDI in trading, distribution, marketing and after-sales services

The scarcity of natural resources of the five countries and the consequent large dependence on international trade as both an engine and handmaiden of economic growth led to a common emergence of a domestically based infra-structure in support of trade – banks, shipping companies, marine insurance companies and, above all, trading companies. The United Kingdom bred the oldest trading companies serving world markets directly established or sup-ported by the state, of which the shining example was the British East India Company chartered in 1602 (Dunning, 1993). The proliferation of British trading companies had important consequences on the development of British business abroad not only in initiating FDI in trading, transportation and distribution to execute their roles in trade intermediation but also in initiating some international production. This could take the form of trading companies acting as the core of British-based investment groups established before 1914 (Chapman, 1985). Trading companies or mercantile houses that established trading outlets throughout Europe in the period prior to 1914 also formed a significant aspect of the early German MNCs. In fact, the German trading house – Schuchardt and Schutte – was regarded as the most presti-gious distributor of machine tools in Europe (Feldenkirchen, 1987).

The role of general trading companies in the history of Japanese and South Korean MNCs was similar. The Japanese general trading company Mitsui Bussan – the foreign trading arm of The Mitsui Company – was one of the pioneering Japanese companies that developed a sizable trading business in China prior to 1914. Mitsui Bussan opened its first overseas branch in Shanghai in 1877 – a year after its foundation – for the initial purpose of facilitating the sales of Japanese exports of coal in China and later expanded to include the importation of Chinese raw cotton for the Osaka Spinning Mills closely associated with the company, as well as re-oriented to the sales of Japanese cotton yarn and fabrics in China. It also established offices with trading interests in Hong Kong, Paris, Milan and New York. Before 1914 Mitsui Bussan had more than 30 branches in Asia, Europe, Australia and the United States, in addition to their manufacturing affiliates in China (Jones, 1996). The other two major Japanese general trading firms participating in the raw cotton trade between Japan and the United States – Japan Cotton Trading Company and the Gosho Company – similarly established trading offices in Texas before the First World War. Most Japanese trading companies were also involved in facilitating the growth of exports from Japan and China to the United States in other products. Such was the case, for example, in medicinal and aromatic products (Wilkins, 1986b).

In a similar fashion, the Korean general trading companies as trading arms of the large conglomerates (*chaebols*) invested in FDI in trading, warehousing and transportation in major foreign markets to promote the growth of South Korean exports and facilitate its imports. However, unlike in the case of the United Kingdom, Germany and Japan where the trading companies dominated trade-promoting FDI, this was not the case with South Korea where apart from the general trading companies, a number of export-oriented South Korean manufacturing firms have integrated forward through the establishment of overseas branch offices in sales or trading, warehousing facilities, and distribution channels in their foreign markets. This was considered a first step in the internationalization process of South Korean manufacturing firms to ensure the continued growth of industrial exports in the face of the growth of protectionism and competition in major export markets.

By comparison to South Korea in which two types of firms were responsible for outward FDI in trading, marketing and distribution, in the Taiwan case outward FDI of this type was solely generated by the large export-oriented manufacturing companies which were most keen to set up trading offices in their major export markets. This was similarly considered a first step in the internationalization process of Tatung, the largest Taiwanese electronics and electrical appliance producer, which established marketing affiliates in the United States and Singapore in 1972 before initiating international production of electric fans in the United States in 1975 and of colour television sets in 1977. The firm also established another colour television plant in the United Kingdom in 1981 (UN, TCMD, 1993a). However, for other Taiwanese manufacturers such as Sampo – Taiwan's third largest home appliance manufacturer behind Tatung and Matshushita (Taiwan) – FDI in sales and distribution was the sole means employed by the company to exploit foreign markets (see Bamford, 1993).

Related to export-oriented FDI was the expansion abroad of German, Japanese, Taiwanese and to a lesser extent South Korean banks which formed a significant aspect of their country's business overseas. The role of the large German banks that established foreign branches in Europe, North and South America, Asia and Africa prior to 1914 helped to support the expansion of German business abroad. Domestic banks in Japan similarly played a strategic role in the expansion of Japanese business abroad. In no way is this more evident than in the case of the Yokohama Specie Bank whose history was intimately associated with the growth of Japanese business worldwide. The development by the bank of an extensive network of offices, branches and sub-branches worldwide and its role in the intermediation of the flow of foreign capital for use in Japan or in the various business activities of Japan abroad enabled it to play a key role in the growth of Japanese trade and FDI. The Industrial Bank of Japan organized in 1900 similarly helped to finance government-sponsored investment ventures, extended loans to the Tayeh mines that became a part of the Hanyehping Company complex and raised finance for the South Manchurian Railway and other Japanese investments in China. The Yasuda Bank, a private Japanese bank, also had Chinese interests (Patrick, 1967). Other Japanese banks that established foreign branches in countries other than China was the Dai Ichi Ginko which by opening a branch in Korea in 1878 became the first Japanese bank to branch abroad. An Osaka-based bank also established a branch in Taiwan before it became a

Japanese possession in 1895, and the Bank of Japan followed suit in 1896. The Bank of Taiwan, established three years later by the Osaka-based bank in Taiwan, had in turn established by 1914 branches in San Francisco, Manila, Singapore, Calcutta, Bombay, 7 points in China and 14 points in Japan and its dependencies (Wilkins, 1986b). Similarly, one of the main objectives that governed the FDI of Taiwanese banks such as Bank of Communications, the International Commercial Bank of China and Chang Hwa Commercial Bank in the 1980s was the need to service the financial needs of individual Chinese residents overseas as well as home country firms, particularly Taiwanese exporters, by way of providing credit evaluations, export documentation and other associated export financing services (Bamford, 1993). Indeed, banks and other financial institutions helped to support the growth of the international business of the resource-scarce large countries, particularly of Germany, Japan and Taiwan.

Such strategic role of domestic banks in the expansion of the business abroad of Germany, Japan, and Taiwan differed in the case of the United States and the United Kingdom. In the case of the United States, national banks could not establish foreign branches until after the passage of the Federal Reserve Act of 1913 while private banks established outlets in London and Paris principally to encourage the flow of European monies to America (Wilkins, 1988b). The lack of British 'universal' banks (Cottrell, 1991) precluded the possibility of the British banking sector to assist directly in the expansion of British business abroad. Instead, it was often the investment group with a trading company at its core that fulfilled a more significant and major role (Chapman, 1985). In the case of South Korea, the government provided finance to eligible FDI projects as part of a package of investment incentives to include low-cost loans from the Korea Eximbank or the Overseas Resource Development Fund (in the case of projects that seek to develop mineral resources overseas) and payment guarantees by South Korean commercial banks (Koo, 1984).

The evolution of MNCs

Despite the common origins of the MNCs based in the five resource-scarce large countries in certain types of outward FDI, the evolution of their outward FDI while sharing some similarities particularly with respect to MNCs based in Japan, Taiwan and South Korea have also some notable differences. At the broad sectoral level, the rapidly growing importance in recent decades of the services sector in the domestic economies of the United Kingdom, Germany and Japan has been reflected in the sectoral pattern of their outward FDI. As shown in the individual country chapters, while the share of the manufacturing sector in the outward stock of British FDI has remained around 35 per cent between 1960 and the early 1990s, the services sector became the dominant sector of British FDI since 1991. Thus over the course of some 200 years, the dominant sectoral pattern of British FDI has shifted from the primary sector in the period prior to 1914, towards the secondary sector for much of the period since the Second World War, and then the tertiary sector since the 1990s. Similarly, although the manufacturing sector continues to be an important area of investment by German MNCs accounting for around 44 per cent of total German outward FDI in 1997, outward FDI in the services sector since the 1990s has become increasingly dominant at the expense of the secondary sector. The same trend is evident in the case of Japanese FDI when the services sector accounted for 62.2 per cent of the stock of approved FDI as of 1997. Finance-related FDI has been at the forefront of Japanese FDI in recent years and its motivations have expanded progressively beyond its traditional role in supporting the growth of Japanese trade towards supporting the growth of the largest banks of Japan as one of the world's largest banks and financial institutions (Randerson and Dent, 1996).

By contrast, the manufacturing sector remains the dominant sector of economic activity in the outward FDI of both Taiwan and South Korea, a reflection once again of the earlier stage of their national economic development. In 1995, the sector accounted for some 59 per cent of the stock of approved Korean FDI, a significant growth from its one-third share at the end of the 1980s, and 17 per cent share at the end of the 1970s. In Taiwan, the manufacturing sector accounted for less than 60 per cent of the stock of approved FDI in the 1990s but the share of the sector had been declining from 63 per cent at the end of the 1980s, and 78 per cent at the end of the 1970s, in a trend opposite to South Korea.

The analysis below compares and contrasts the evolution of MNCs based in the five countries, and their association with domestic industrialization across different countries with a focus on the manufacturing sector where some of the most important changes have taken place throughout the course of their history (Table 12.2).

International production by British manufacturers has essentially remained oriented towards the mature, relatively low technology sectors whose competitiveness emanated from the high income and large size of the British market. These were firms of large size and established technological advantages which derive strengths from product differentiation, quality, and marketing and managerial skills and experience. These characteristics favoured the continuing pre-eminent role of consumer goods firms as British MNCs during the inter-war years and for much of the period since the Second World War.

Several of the British MNCs that emerged in the higher technology industries continued to depend greatly on new technological developments from the United States. The slow adjustment of British firms to the more rapid growth opportunities offered by the modern industries of the Second Industrial Revolution was regarded to be a function of the peculiar nature of the British economy which posed a host of institutional barriers that prevented industrial restructuring towards the growth-oriented sectors, including investments in innovation that would sustain those industries. Such barriers included, among others, the lack of provision for commercial studies and for any kind of technical education for managers and industrial staff (Ashworth, 1960; Chandler, 1980).[2] Other contributory factors were the risk averse strategy of British firms which continued to emphasize industries and sectors in which past successes had been based, the presence of a protected home market and the continuing preferential market access in the empire and Commonwealth markets during the 1940s and 1950s (Dunning and Archer, 1987). As a result, British MNCs still have not featured prominently in the more modern and growth-oriented fabricating industries, but the few that emerged have been the most active in pursuing more rationalized or efficiency seeking FDI since the 1970s associated with an increasing amount of intra-firm trade and product and process specialization. Among the more globalized British firms are Imperial Chemical Industries, Glaxo, Unilever and Shell. These companies have adopted a transnational strategy in at least some of their value-chain activities, including R&D (Lane, 1998).

On the other hand, for the British MNCs in traditional consumer goods industries that neither engaged in rationalized production and investment nor sought to benefit from transaction cost advantages, their incentives to internationalize in the period since the Second World War were not dissimilar to that in the earlier periods. These firms remained keen to exploit their preferences to produce in the markets of the empire and Commonwealth until the early 1960s when the exporting route became difficult or no longer practical. As the imperial legacy of the United Kingdom bore less of an influence on the geographical pattern of British FDI, there has been a re-direction of outward FDI towards the United States and Western Europe since 1960.

Such dualistic pattern of the British international production is a reflection of the domestic economic structure of the United Kingdom which although geared towards the industries of high and medium technology is propelled by the investments of foreign-based MNCs. The more traditional low technology and consumer goods manufacturing sectors on the other hand are where the indigenous strengths of British firms lie which explain the dominant role of these sectors in the foreign activities of British MNCs. Inward and outward FDI in and from the United Kingdom has, therefore, been directed towards different industries. This also helps to explain the different patterns of exports of the United Kingdom and that of the international production of British MNCs. The industrial structure of British manufactured exports in the years between 1965, 1970 and 1975 continued to differ sharply from that of its manufacturing FDI but for an entirely different reason than that pertaining to the end of the nineteenth century when international production displayed a higher technological intensity compared to that of exports. A study conducted by Clegg (1987) indicated that capital intensity and the skill level of managerial manpower exerted significant positive influences on British FDI in manufacturing between 1965 and 1975, while the skill level of production workers exerted a highly significant negative influence. This contrasts sharply in the case of British exports of manufacturing in which technological intensity exerted a highly significant positive influence.

Britain's present day comparative advantage which rests on the production of labour intensive, capital neutral and human capital-scarce products has remained essentially stable since 1870 (Crafts and Thomas, 1986). Firms in a broad range of industries continue to be more concerned with the production of low-cost standardized goods than with high-quality, technology intensive niche products (Porter, 1990). The employment pattern developed on this basis may have even become further entrenched during the 1990s (Nolan and Harvie, 1995).

By contrast, the high technological intensity that has always consistently described German manufacturing MNCs throughout the course of their history stems from the position of Germany as the birthplace of modern science in the late nineteenth century (Porter, 1990). This helped the country to develop a deep scientific and technical knowledge base drawing on an abundance of skilled workers and professionals which proved instrumental in their efforts to upgrade domestic industry in the new, skills intensive, technically advanced and fast growing industries of the Second Industrial Revolution in the late nineteenth century. German strengths in the chemicals, pharmaceuticals, machinery, electro-technical and motor vehicles industries continued to grow and German firms became significant world producers and exporters of their new products on the basis of accumulated expertise and proprietary technology. Such strengths combined with access to international capital markets at least until the end of the First World War enabled German firms to invest abroad on a substantial scale, and to exert a significant economic impact within their host economies.

The defeat of Germany in the two world wars and the sequestration of tangible and intangible assets abroad after the Versailles Treaty discouraged the further international expansion of German FDI until two decades after the Second World War. This affected adversely their most dynamic set of firms – the manufacturing companies – that nevertheless possessed a substantial amount of surplus capacity. The confiscation of assets crippled firms in the chemicals industry the most whose intangible assets such as brand names and patents were invaluable in relation to physical assets. The loss of overseas holdings at the end of the First World War and the severe constraints posed by the shortage of capital in the inter-war period forced German firms to replace FDI with other modes of international economic expansion that conserved the use of capital and entailed less risks, primarily political risks. In the period between 1918 and 1939 these other modes were principally cartels and long-term contracts (including licensing agreements) in which German industry had experienced considerable success before the First World War. These tools came to be used more widely by German firms than by firms of another nationality (Schröter, 1988).

Notwithstanding the second round of defeat of Germany in the Second World War and the second round of loss of some of the most modern parts of its industrial base and natural resources along with the confiscation of German patents and foreign assets, strong pressures were created to upgrade

German industry in advanced, knowledge-based industries and to foster technological innovation in indigenous firms (Porter, 1990). However, German firms remained essentially domestically embedded in their country as manifested in their continued preference for exporting over FDI as an internationalization strategy even to the present time, and in the continuing concentration of their outward FDI in Western Europe (Lane, 1998). Some recovery of Germany as an important home country of FDI became evident only in a major way since the 1970s. The comparatively high level of R&D intensity characteristic of German firms and MNCs from the beginning remained their most consistent distinguishing feature that sets them apart from MNCs of other nationalities. Indeed, a regression analysis of the industrial advantages of Germany on pooled cross-sectional sets of data for the years 1965, 1970 and 1975 showed that the degree of innovation and the creation of technological ownership advantages as well as the skill level of managerial manpower exerted highly significant *positive* influences on German FDI in manufacturing, while the skill level of production workers had a significant *negative* effect. This roughly mirrors the findings for German manufactured exports which showed that the technological intensity and the skill level of production workers exerted highly significant *positive* influences, while capital intensity and the complexity of management exerted significant *negative* influences (Clegg, 1987). The findings show that the ownership advantages of German MNCs unlike those of MNCs based in the United States or the United Kingdom were based more on technology and skilled labour than on capital intensity. During this period, there was no evidence to show that access to capital was an ownership advantage of German MNCs in the way that it was during the period prior to 1914. In the concentration of German FDI in Western Europe, German MNCs also have a more narrow geographical focus compared to American or British MNCs, and have taken place in a narrow spectrum of medium- to high-technology industries in which domestic industry activity is also strong.

The evolution of Japanese MNCs in manufacturing is in many respects *sui generis* in the growth of modern MNCs. Although the origins of Japanese MNCs can be traced to the late nineteenth century, the early pattern of Japanese FDI remained essentially the same from the period prior to 1914 through to the inter-war period. The early stage of domestic industrial development in which Japan generated MNCs of their own explains why Japanese MNCs from the period of their emergence until the Second World War did not derive their ownership-specific advantages from technological strengths and organizational competence, the possession of brand and trademarks and the ability to supply high-quality, differentiated goods which described the manufacturing MNCs from the United Kingdom and Germany since their emergence in the nineteenth century. However, the period since the Second World War was associated with the rapid domestic industrial transformation of the Japanese economy accompanied by the rapid evolution of Japanese FDI in manufacturing. In this period more than ever before, Japan's outward

FDI was a crucial instrument or catalyst for the rapid process of domestic industrial upgrading (Ozawa, 1985; Kojima and Ozawa, 1985). The rapid industrial transformation from a concentration on the primary sector towards the secondary and tertiary economic sectors, and within the secondary sector from labour intensive light manufacturing and heavy and chemical manufacturing to knowledge intensive, fabricating assembly-based industries and mechatronics-based, flexible manufacturing of highly differentiated goods accompanied by a continual rapid rise in labour productivity and wages at the end of each phase have led to shifting patterns of production and trade competitiveness for Japanese manufacturing companies as well as evolving patterns of Japanese MNC activities.

Ozawa (1991) analysed the industrialization process of the Japanese economy since the Second World War in four sequential stages that corresponded with equivalent phases or waves in the growth of Japanese cross-border production.

Phase I   Expansion of labour intensive manufacturing in textiles, sundries and other low-wage goods in 1950 to the mid-1960s. This industrialization phase corresponded with the 'elementary' stage of Japanese offshore production in Heckscher-Ohlin industries. This stage of Japanese overseas production took place in a major way between 1971 and 1973, with the investments in 1972 and 1973 alone more than doubling the existing stock of Japanese FDI. The stage was associated with the transfer of standardized, low-technology labour intensive production from Japan to developing countries such as Taiwan, South Korea, Hong Kong, Thailand and other Asian economies in close geographical proximity and had an abundant labour supply (Ozawa, 1991). China did not seem to be as important a host country in this period in the way that it was in the period prior to 1914 and in the inter-war years.

Phase II   Scale economies-based modernization of heavy and chemical industries such as steel, aluminium, shipbuilding, petrochemicals and synthetic fibres in the late 1950s to the early 1970s. This industrialization phase corresponded with the Ricardo-Hicksian trap stage of Japanese multinationalization associated with the non-differentiated Smithian industries. Thus, while Japanese FDI in both trading and in resource extraction was both geared to secure stable supplies of industrial raw materials (minerals, oil and other natural resources) in foreign countries (Ozawa, 1977, 1982), FDI in manufacturing also grew with the transfer overseas of some of the resource intensive and often pollution-prone industries. The demand for intermediate industrial goods associated with the rapid industrialization in the newly industrialized countries, especially South Korea, Hong Kong, Singapore, Taiwan and Brazil provided an additional incentive to FDI by Japanese capital goods producers. In sum, Japanese outward FDI in this phase was an important means to overcome the Ricardo-Hicksian trap of industrialization and economic growth.

Phase III Assembly-based, sub-contracting dependent, mass production of consumer durables such as cars and electrical/electronics goods from the late 1960s to the present. This industrialization phase corresponded with the export substituting-cum-surplus recycling stage of Japanese multinationalization associated with the differentiated Smithian industries. Indeed, in this third phase of Japanese multinationalization and for the first time in the history of Japanese MNCs, tariffs and other trade barriers became the major factor precipitating the shift from exports to international production. In common with the protectionist trend starting from the 1880s which spurred the growth of American, British and German manufacturing MNCs, the growth of 'new protectionism' from the 1970s stimulated the growth of Japanese manufacturing FDI in new industries (Jones, 1996).

Phase IV Mechatronics-based, flexible manufacturing of highly differentiated goods involving the application of computer-aided designing (CAD), computer-aided engineering (CAE) and computer-aided manufacturing (CAM), along with technological breakthroughs such as high definition TV, new materials, fine chemicals and more advanced micro-chips in the early 1980s onwards. While Ozawa (1991) did not identify the corresponding phase of Japanese multinationalization associated with this industrialization phase as he regarded this type of Japanese FDI to be still speculative, this fourth phase of Japanese multinationalization which has yet to evolve fully in the future can be labelled as the robotics and new materials stage of Japanese multinationalization associated with the Schumpeterian industries.

Thus, over time Japanese MNCs have evolved towards an increasingly advanced phase of international production owing largely to efforts at rapid domestic industrial transformation and technological advancement in Japan. As a result, Japanese MNCs over the course of their evolution have also drawn strength from the same set of strong ownership advantages that propelled the emergence and growth of the more mature MNCs from the United Kingdom and Germany.

To the extent that Taiwan manufacturing FDI and South Korean manufacturing FDI could be compared with that of Japan since the Second World War, the first and second phases of Japanese FDI in labour intensive manufacturing in textiles, sundries and other low wage goods (the first phase) and in heavy and chemical industries during the Ricardo-Hicksian trap stage of Japanese FDI (the second phase) had been transposed in the case of the history of both Taiwanese and South Korean manufacturing MNCs. This is associated with a prolonged dependency or sustained comparative advantage of both the Taiwanese and South Korean economy on labour intensive production until the late 1980s. The basis of the prolonged competitiveness of these industries derived from a policy in both states of wage suppression and the political subordination of popular sectors (i.e. middle stratum of professionals, skilled workers, businessmen and farmers) to the state (Shin and Lee, 1995) which was brought to an end with political democratization in both countries from around 1987. Thus, the pioneering manufacturing industries in the outward FDI of Taiwan and South Korea were the heavy and chemical industries whose comparative advantage had been declining most rapidly in the 1960s and 1970s owing to their countries' natural resource scarcity, the shortage of industrial space and the ensuing problems of pollution, congestion and ecological destruction. In both countries, outward FDI in labour intensive manufacturing only came into its own in a major way in the late 1980s and 1990s.

Both Taiwan and South Korea had reached the third phase of Japanese FDI in assembly-based, subcontracting dependent, mass production in foreign markets in the same set of consumer durables industries – cars, consumer electronics and semiconductors – that have been at the core of the global strategies of the large Taiwanese manufacturing firms and the South Korean large conglomerate companies (*chaebols*) in the 1980s and 1990s. The South Korean *chaebols* much like their Japanese counterparts – the *keiretsus* – spearheaded the large-scale import substituting FDI geared to overcome trade restrictions in the developed countries. The emergence and growth of the *chaebols* and the *keiretsus* had been fostered by their respective governments in their effort to accelerate domestic industrial development in large-scale, complex and technologically advanced industries through high industrial concentration.[3] Indeed, a common theme in the three resource-scarce large countries of East Asia is the leading role of their governments in directing shifts in their countries' dynamic comparative advantage (see also Aggarwal and Agmon, 1990). Towards this end, the development of indigenous skills and technological capacities was emphasized, and growth was based on knowledge-based industries (Crawford, 1987). Where South Korea and Taiwan differ is in the pace in which domestic industrial development evolved towards more advanced industries. The smaller size of its domestic market and less interventionist role of the government of Taiwan in indigenous technological development at an early stage rendered the process of industrial upgrading to proceed at a slower pace in Taiwan compared to that in South Korea. Indeed, the technological depth and competitive prowess in which South Korea's industry has developed is not observed in any other developing country (UN, TCMD, 1993a).

Despite the arrival of two Asian newly industrialized countries in the third evolutionary phase of Japanese FDI in manufacturing, there are important differences between Japanese MNCs on the one hand and the South Korean and Taiwanese MNCs on the other. The broader range of more technologically advanced consumer durable goods produced by Japanese *keiretsus* in their international production networks worldwide from around the mid-1970s onwards combined with their advanced marketing capabilities (high brand name recognition, product differentiation, extensive distribution channels, high advertising and promotion expenditures, etc.) contrasted with the much narrower product range, less technologically advanced and far less advanced marketing capabilities of South Korean and Taiwanese MNCs in major foreign markets in the 1980s and 1990s. Indeed, the technological strengths combined with advanced marketing capabilities of Japanese MNCs in the electrical equipment and motor vehicles industries meant that it had been in those industries in which their involvement and market shares in United States and Europe have been expanding fastest both through exports and international production (Dunning and Cantwell, 1991). By contrast, the far weaker product change and product design capabilities as well as relatively weak marketing capabilities of South Korean MNCs and Taiwanese MNCs even more so – a feature inherent in firms from countries that are late industrializers as well as the long history of their firms as OEM suppliers in the electrical equipment industry – have constrained their ability to capture substantial market shares for their brand products in the developed countries, particularly in the larger European countries as Germany and France (van Hoesel, 1997). Thus in relating stages of national development to the form of technological competence of leading indigenous firms, the type of outward FDI and its industrial course over time, MNCs from South Korea and Taiwan are at an intermediate stage of development compared to the more advanced British, German or Japanese MNCs, despite some similarity in the pattern of their MNC growth (see Table 12.3).

While Taiwanese MNCs reached the third stage in the 1970s with the major surge of FDI by large-sized Taiwanese industrial companies in the United States, a full decade earlier than the major expansion of manufacturing FDI by the Korean *chaebols* in the United States, the early penetration of Taiwanese MNCs in developed countries was premature and unsustainable. With the lack of technological capabilities (including the ability to produce products geared to higher-income markets), the inability to establish linkages rapidly with local component suppliers in the developed countries or the ability to increase local value added through the upstream production of parts, components and intermediate products in the face of increasing localization requirements in the United States and Europe, the cost competitiveness of Taiwanese MNCs in the developed countries was lost rapidly which rendered their early FDI thrusts in the developed countries in the 1970s and 1980s unprofitable and unsuccessful. This helps to explain why international production in developing countries that were the predominant host countries of Taiwanese FDI in the 1960s and 1970s once again received the bulk of Taiwanese FDI in the 1990s. By contrast, South Korean MNCs have displayed a greater ability compared to Taiwanese MNCs to sustain existing FDI in the developed countries, as well as to implement new FDI projects in those countries, despite the fact that South Korean MNCs in developed countries emerged and grew in a major way a decade later than Taiwanese MNCs. The more rapid pace of domestic industrial upgrading in South Korea earlier mentioned, the greater success at increasing local value added in their developed host countries through backward vertical integration or local sourcing and the fact that South Korean FDI in developed countries have been led by the *chaebols* may explain the greater resilience of South Korean MNCs to sustain FDI in developed countries. By comparison, Taiwanese MNCs that invested in developed countries had experienced greater difficulties in achieving the objectives of higher localization in their host countries, and although Taiwanese MNCs featured among the largest companies in Taiwan these firms were not as large and diversified as the South Korean *chaebols*.

The success with which the *chaebols* and *keiretsus* have transplanted a great part of their production system in foreign markets through backward and forward integration has enabled both Japan and Korea through their MNCs to establish rapidly regionally integrated production networks across the world: North America, Europe, South East Asia and China. However, the regional production networks of South Korean *chaebols* which has developed the fastest in South East Asia and China in the 1990s differ from the regional production networks of Japanese *keiretsus* in North America, Europe, South East Asia and China not only in the greater sophistication and complexity of regional production networks of Japanese MNCs by comparison to those of South Korean MNCs which has enabled Japanese MNCs to progress towards global localization to a much farther extent than South Korean MNCs. The high product change capabilities of Japanese firms attained through cross-functional linkages between design and development, production, marketing and service in their regionally integrated production networks worldwide with fairly sophisticated regional division of labour and decentralized control contrast with the much weaker linkages between production, marketing and design and development functions in the regional production networks of the Korean *chaebols* such as SEC owing to continuing centralization of control from parent companies in Korea for the higher value added functions. Another major difference arises from the unique pyramidal and multi-layered system of subcontracting relationships between Japanese final goods assemblers and their suppliers which has essentially been closed to external economic actors, unlike those of the relationships between affiliated suppliers and final goods assemblers of the Korean *chaebols* which has sold components to Japanese final goods assemblers and bought parts from Japanese suppliers.

At the end of the 1990s Korean MNCs have not reached the fourth phase of Japanese multinationalization – the mechatronics-based, flexible manufacturing of highly differentiated goods involving the application of the most advanced computer technologies in design, manufacturing and engineering along with the use of more advanced technological breakthroughs associated with high-definition TV, new materials, fine chemicals and more advanced micro-chips, etc. – and their capabilities of reaching this stage will be defined by significant progress yet to be made with the development of more advanced product innovation as well as marketing capabilities.

Thus, the dynamic analysis suggests some similarities in the evolutionary course of outward FDI from Japan, South Korea and Taiwan, particularly in manufacturing. However, the evolution of South Korean FDI over time is slower compared to that of Japanese FDI. Japanese MNCs seem to have graduated from the first two phases to the third phase within a span of about 15 years or so years (1950 to the late 1960s), while Korean MNCs have taken around 30 years (1960s to the late 1980s). Although Taiwanese MNCs may have made the same transformation within a similar span of 15 years (1960s to the mid-1970s) as Japan, these firms have yet to improve the sustainability of their FDI in consumer durables production in the developed countries. Therefore, despite the similarity in the patterns of their outward FDI, the pace of evolution in the pattern of outward FDI and the character of that evolution has differed considerably between the three countries.

By way of concluding the analysis of the emergence and evolution of MNCs from resource-scarce large countries, the last part of this chapter analyses the role of financial factors as a determinant of the growth of FDI in the five country case studies. The growth of FDI from the five resource-scarce large countries bears support to Aliber's theory of FDI emanating from strong currency areas as a determinant of the timing of FDI and its growth (and particularly that of foreign takeovers) as well as fluctuations of outward FDI flows around a long-term trend (Aliber, 1970). The financial explanations to the growth of FDI owing to strong domestic currencies has historical antecedents in Britain in the nineteenth century, the United States in the early post-war period and Japan and Germany in the 1970s and 1980s (Cantwell, 1989a) as well as South Korea in the mid-1980s and Taiwan in the mid-1980s and 1990s. However, in the case of South Korea, the growth of outward FDI in the 1980s was less of an 'exercise in financial power' (Redies, 1990) compared to Japan in the 1970s and 1980s and Taiwan in the 1980s and 1990s where the financial factors in explaining FDI expansion extended beyond large financial surpluses created from net exports, the upward revaluation in the Japanese yen and New Taiwan dollar and the accumulation of the world's largest foreign exchange reserves. In the case of Japan and Taiwan, the accumulation of large financial surpluses in those years was also created from the high domestic savings rates and the escalation in acquired wealth owing in part to the emergence particularly in Japan of a 'bubble' economy in which asset and land prices increased enormously. This was not the case in South Korea where there was a severe shortage of domestic savings and a large demand for foreign exchange until the late 1980s (see Kim, 1990).

The growth of Korean FDI in the 1990s has been sustained despite an inherent financial weakness in the Korean *chaebols*. Unlike in the 1980s when financial factors influenced the expansion of South Korean FDI owing to financial surpluses created from net exports and the upward revaluation in the Korean won between 1986 and 1990, this situation was overturned over the 1990s with the slower growth of exports and the depreciation of the Korean won. Thus, unlike the 16.8 per cent appreciation of the Korean won against the United States dollar between the end of 1986 and the end of 1990, there was a 136.6 per cent depreciation of the Korean won against the United States dollar between the end of 1990 and the end of 1997. To the extent that financial factors played a role in explaining the most rapid expansion of Korean FDI in the 1990s (the most rapid decade of South Korean

MNC growth), it owed much to the role of the *chaebols* as institutions created by the South Korean government to accelerate domestic industrial development in large-scale, complex and technologically advanced industries through high industrial concentration. The close relationships between the government and the *chaebols* was manifested in the traditional roles of the former in setting policies and in the control of the latter's access to capital to finance their growth and diversification, including through outward FDI.[4] Notwithstanding the envisaged restructuring of the *chaebols* under the *segyehwa* movement between 1993 and 1997, the *chaebols* became more powerful and larger with their continued expansion into new industries and markets financed on the basis of low-interest bearing loans from their government. However, the capacity of the over-leveraged *chaebols* to service their amassed debts have been constrained by the onset of foreign competition associated with domestic market deregulation in the *segyehwa* movement and the weaker markets for computer chips in 1996 which led to domestic economic difficulties in South Korea as well as a 27 per cent deterioration in the South Korean terms of trade in the three years prior to September 1997 and associated current account deficits (World Bank, 1998). These negative factors have been exacerbated by the Asian financial crisis that started in 1997 which affected South Korea greatly and limited the capacity of the government to fulfil its traditional role as a provider of low-cost capital to the *chaebols* – the basis of the country's past economic miracles and a contributory factor to its economic débâcles in the late 1990s (Tolentino, 1997). While the full repercussions of all these compelling forces on South Korean FDI have yet to be observed and analysed fully in the future, Korean FDI outflows in the period since the financial crisis would be determined by the need to overcome the inherent financial weakness of the Korean *chaebols* through more aggressive and extensive penetration of foreign markets through outward FDI as a means to service large amounts of accumulated debt as well as the need to seek alternative sources of finance in international capital markets.

Notes

1 Among the exceptions were the international production activities of Lever Brothers and Courtaulds in the 1900s (Dunning and Archer, 1987). British American Tobacco's investments in cigarette production in China to take advantage of low-cost labour is another example. With the use of labour intensive rather than capital intensive production techniques, the firm employed thousands of unskilled labourers in China to perform tasks that were already mechanized in the United States (Cochran, 1980).

2 Indeed, at the more advanced educational levels, technical and scientific instruction and inquiry remained 'poor cousins in the family of higher learning' in the United Kingdom in the 1900s (Murphy, 1973). This was so unlike in the United States where the need for trained managers, production and marketing specialists in the technologically advanced machinery, electrical and chemical industries had been more rapidly recognized and catered for by universities and business schools (Chandler, 1980). It was not until 1947 that the British Institute of Management

was formed, and only in the 1960s were the London and Manchester Business Schools founded (Dunning and Archer, 1987).

3 However, unlike in the case of Korea, the Japanese government was not a provider of finance to the *keiretsu* who have their own banks as part of their large organizations. By comparison, the relationship between government and business in Korea is such that the government sets the policies and traditionally controlled the *chaebols'* access to capital (Ungson *et al.*, 1997).

4 The relationships between the government and business sectors in Japan is different. In particular, the Japanese *keiretsu* have their own banks as part of their organizational structure as mentioned which provides the member companies with a reliable source of credit and finance, including in the facilitation of outward FDI. By contrast, the South Korean *chaebols* have not been allowed to own banks until the mid-1990s (Ungson *et al.*, 1997).

Multinational corporations from the resource-scarce small countries

13 The emergence and evolution of multinational corporations from Switzerland

Introduction

With the origins of Swiss MNCs tracing back to around 1750, Switzerland is one of the pioneering sources of MNCs in the world economy. Switzerland has always been a home country of FDI of some consequence with a share of around 3.4 per cent of the global stock of outward FDI in 1960 and 4.3 per cent in 1998. Indeed, with an outward FDI stock of $176.7 billion in 1998 the country was the world's seventh largest home country of FDI in that year after the United States (with a share of 24.1 per cent of the global stock of outward FDI), United Kingdom (12.1 per cent), Germany (9.5 per cent), Japan (7.2 per cent), the Netherlands (6.4 per cent), and France (5.9 per cent).[1]

The early and rapid process of internationalization of Swiss firms both through exports and international production is consistent with their offensive and aggressive market expansion strategies to overcome the limited size of their domestic market – a feature shared by firms from Sweden (see Chapter 4), Singapore (see Chapter 15) and by firms from small countries more generally.[2] Some Swiss MNCs – the Anglo-Swiss Condensed Milk Company, Brown, Boveri & Cie. and Alusuisse for examples – typically became MNCs almost since their companies' inception.

The choice between exports and international production tended to determined by the balance of locational advantages in the home and host countries. International production in host countries was favoured in most cases by tariffs (particularly significant in the emergence of Swiss MNCs in textiles and food products) and non-tariff trade barriers (particularly important in explaining the growth of Swiss MNCs in electrical equipment and chemicals), high transportation costs, low production costs, the necessity to overcome restrictive legislation of the home country or to establish closer ties with local customers to support sales and the need for a local presence in foreign markets.

The largest firms in Switzerland are typically MNCs, with the degree of multinationality (as measured by share of foreign employment in total employment) becoming less pronounced with diminishing firm size.

Switzerland, as with Sweden and the Netherlands, is thus a prototype small country to large MNCs (Niehans, 1977). In a close resemblance to Swedish MNCs, the largest 50 firms in Switzerland account for over half of domestic and international production by Swiss-owned firms in the 1990s (Sally, 1993), although the population of Swiss MNCs also consist of many small- and medium-sized firms, particularly in the more historical periods.

The six largest Swiss MNCs – comprising the three major Basel-based chemical companies (Ciba-Geigy, Sandoz and Hoffman-La Roche), Nestlé, Asea Brown Boveri (the result of a merger between the Swiss company, Brown Boveri & Cie. and the Swedish company, Asea) and Alusuisse – are among the world's firms with a high degree of multinationality. Although the Basel-based chemical MNCs belong to the ten largest European pharmaceutical companies in terms of sales, such firms are considerably smaller than the more diversified German chemical companies such as Hoechst, Bayer and BASF. Despite their relatively smaller size, however, the Swiss chemical companies have maintained consistently a competitive edge in their specialization in high value-added differentiated product segments of the chemicals and pharmaceuticals industries that have high profit margins and are unaffected by cyclical price changes (Sally, 1993).

However, unlike MNCs from Sweden that account for more than half of Sweden's exports in the 1990s (Olsson, 1993), the 35 largest Swiss manufacturing firms – including most of the larger MNCs – are not more export-oriented than the rest of the Swiss economy. This is because Swiss MNCs, apart from those in the chemicals and pharmaceuticals industries, typically regard international production and international trade as substitutes (e.g. Nestlé has no significant exports from Switzerland at all) (Niehans, 1977).

The historical excursion into the determinants of Swiss FDI and their industrial and geographical patterns in this Chapter is conducted in two time frames: the period of emergence until 1914 and the period of expansion covering the inter-war years and since the Second World War.

The emergence of Swiss MNCs and the period until 1914

The origins of Swiss FDI can be traced to the Napoleonic Wars (Masnata, 1924) and the early process of industrialization of the Swiss economy. Wavre (1988) describes two general periods in the historical development of Swiss investments in Italy: 1750 to 1850–60 and 1860–70 to 1920–30. These phases apply to the historical development of Swiss FDI more generally.

Swiss FDI in the period between 1750 and 1850–60

The earliest Swiss FDI on record were made by prominent families engaged in trade, commerce and banking as well as by small insurance companies. The Jenny family, who owned the Wienerhandlung, opened agencies in Ancona and Bologna and established an important banking and commercial business at Trieste that dealt in corn and manufactured goods. The Blumers were another leading commercial family with enterprises throughout Europe, including a trading house in Ancona. Swiss capital was also invested in a commercial company in Sicily and in a Genoese importer named Ricardi (Wavre, 1988). There was also the investment abroad in marine insurance by the Swiss Assurance Company of Livorno between 1820 and 1830 (von Niephans, 1976). The pioneering role of trading, commerce and insurance in Swiss FDI reflected Switzerland's historical position as a trading, commercial and financial power (Bergier, 1968). Such position was favoured by the country's central geographical location on major European trade routes and political neutrality that allowed its firms to benefit from the maintenance of commercial contacts with each of the major European power centres even during times of conflict (France, Germany and Great Britain) (Porter, 1990).

Apart from these activities abroad in trade, commerce, banking and insurance, Swiss FDI around the turn of the nineteenth century grew rapidly in the manufacturing sector, starting with the textiles industry which was developing into an important industry of the Swiss economy and a significant source of its export earnings in the nineteenth century.[3] Swiss investments abroad in that industry, and cotton manufacturing particularly, had begun in Italy around the late eighteenth century and became the most important industry of Swiss FDI in that country between 1750 and 1850–60 (Wavre, 1988).[4] The cotton industry assumed greater prominence as an industry of Swiss FDI with the formation of the German customs union in 1834 which threatened the continued growth of Swiss firms in the cotton industry that exported most of its products. In response, new foreign-based factories were built just across the Swiss border in Southern Germany in a region of inhabitants linguistically and in other ways identical to the Swiss.

At the turn of the nineteenth century, FDI also emerged from the silk ribbon industry of Basel (Masnata, 1924). The investments abroad by otherwise reluctant Swiss silk companies was brought on mainly by foreign tariffs and high wages at home (Schwarzenbach, 1934). By 1904, Robert Schwarzenbach, a Swiss pioneer in silk power looms, had established silk manufacturing plants not only in Switzerland, but also in the United States, Italy, France and Germany (Wilkins, 1988b).

The concentration of the early Swiss investments in Italy in the cotton industry was promoted by the gradual disengagement of Switzerland from the French market during the early nineteenth century. The locational advantages of each of the Italian states varied greatly on account of the different customs and tariff systems, but in general there was a large concentration of Swiss enterprises in Lombardy, around Bergamo, Como and Luino and Turin. Swiss capital in this period was thus attracted primarily to the traditional heartlands of Italian textile production (Wavre, 1988).

The investments in Italy in this period were quite random and made by individual adventurers or families with established interests in the field who were looking for new outlets and avail of lower land and wage costs. Of these factors, the lower cost of the factors of production was probably the most critical locational advantage of Italy between 1750 and 1850-60 (Wavre, 1988). At the time of the discussions of a new commercial treaty between Italy and Switzerland in 1868, Swiss textile investments in Italy comprised seven cotton-spinning establishments with 108,000 spindles and another ten weaving enterprises that employed 2,050 looms. By 1870, some 23 Swiss MNCs and 25 foreign manufacturing affiliates had been traced, most of which were from the textiles industry (Schröter, 1993).

Swiss FDI in the period between 1860–70 and 1914

The growth of these investments since 1850–60 owed much to the new Italian tariffs of 1 March 1888 which led to the significant decline in the value of Swiss exports to Italy, including cotton fabric.[5] The imposition of more rigourous labour legislation in Switzerland in 1898, with the prohibition of employment of school children also forced the relocation to Italy of Swiss companies in silk spinning.[6] This was because the costs of spinning silk were largely determined by labour costs (even though the work was performed mainly by women and children), and by the proximity to raw materials. Although there had been a weaker tendency to relocate abroad by Swiss silk weaving companies, these manufacturers also sought to overcome the higher protective tariffs in force in Italy through international production, but the capital involved was far less than in the case of silk spinning. Neither were there major Swiss investments in silk ribbon weaving or embroidery abroad, and those that did exist were located in Austria, France and Germany. There were also no significant foreign investments in other branches of the textile and clothing industry, although a Swiss straw plaiting manufacturer was established in Italy in 1854 to take advantage of cheap labour, and there was a broad range of specialist products made by Swiss firms in the region surrounding Florence (Wavre, 1988).

By 1900, there were 46 Swiss cotton manufacturers in Italy operating mills of spinning, reeling, weaving and cotton printing. Nonetheless, contact with Switzerland declined rapidly, and the Swiss concerns were acquired in haste by Italians either as a result of the formation of public companies or increased capitalization. By 1922, only 24 of the 46 Swiss cotton manufacturers that existed in 1900 survived, and of these only ten remained in Swiss hands (Masnata, 1924). The value of Swiss investment in the Italian cotton industry amounted to 150 million lire in 1922 (Beck, 1922).

In sum, several features describe the earliest Swiss investments abroad. First, the important industries of Swiss FDI were trading and commerce, banking and insurance, and textiles. There is little indication of Swiss investments in other industries, apart from that by Terra-Film AG in film distribution and Villiger in cigar making (Schröter, 1993). Swiss FDI in cotton manufacturing in particular was driven by the attainment of lower production costs abroad and to mitigate the threat to exports presented by Italian import duties and the German customs union. While Swiss FDI in silk spinning was similarly driven by lower labour costs abroad, other equally important determinants arose from the need to overcome rigourous labour legislation in Switzerland concerning the employment of school children and the proximity of raw materials. The historical Swiss FDI in the textiles sector was, therefore, spurred fundamentally by location-specific advantages that favoured foreign countries. Secondly, Swiss FDI in this period was focused towards countries that were geographically close and culturally similar to Switzerland. Although Britain was an important recipient of Swiss investments (Wavre, 1988), substantial investments by Swiss MNCs had been directed to Germany, France and Italy (Schröter, 1993) that constituted the first export markets of Switzerland. Third, the sums of Swiss capital invested abroad in the period was modest, with a majority of the investment emerging from individual adventurers or families with established interests in the textile industry or small- and medium-sized firms.[7] Fourth, the emigration of industry associated with Swiss FDI until the 1860s and 1870s made Swiss FDI in this period consistent essentially with migratory capital. After the initial connections to Switzerland receded, the investment became independent and detached from managerial control from the home country. Capital connections usually ceased at a later stage, together with family links (Bonnant, 1976). However, only a few examples of this phenomenon were found after the 1860s and 1870s, and therefore the years around 1870 can be traced as the period of the emergence of the modern Swiss MNC (Schröter, 1993).

Although the number of Swiss foreign manufacturing affiliates in the textiles industry doubled between 1870 and 1914 from 25 to 50 (Schröter, 1993), the importance of the industry in both domestic production in Switzerland and international production by Swiss MNCs started to decline even before the turn of the twentieth century owing to the process of domestic industrial upgrading in Switzerland and changes in fashion trends in foreign markets. As a result, the producers of the cotton industry, and later those of the silk industry, embroideries and products of straw lost core competitive advantages. From 1890 onwards, substantial FDI by such firms as Alusuisse-Lonza, von Roll, Georg Fischer and other firms belonging to the *processed metal products industries* began to emerge and grow in importance. Some of these were in the iron and steel industries, the bulk of which went to Germany, France and, to a lesser extent, Italy.[8]

Alusuisse – one of the world's major firms in primary metals and producer of aluminium foil and one of Switzerland's largest manufacturing firms in terms of sales – is a prototype of a resource-oriented MNC. That the company is an MNC partly arises from comparative advantages which favour different countries for different factors and partly from the small and imperfect markets for the intermediate inputs at the various stages of production, both of which would render a non-integrated firm vulnerable to pressures from monopolistic suppliers or customers and to the risk of disruption in the flow of raw materials and products (Niehans, 1977). Thus the firm by necessity became an MNC from its inception, and therefore the pattern of firm growth described in the product cycle model does not apply.

The location of the international production facilities of Alusuisse was determined fundamentally by four major factors: first, abundant sources of bauxite as the primary raw material; second, low-cost electric power for the reduction of alumina to aluminum; third, the presence of markets for products; and fourth, management expertise for the planning and scheduling of complex international production and transportation operations. Given that the first three locational factors favoured a location in the countries of Europe, Alusuisse became a fully internationally integrated company with operations throughout Europe before the Second World War, all of which were under the centralized control of its headquarters in Switzerland.[9]

The loss of the company's bauxite base in Eastern Europe in the aftermath of the Second World War and the search for new sources of primary raw material in Sierra Leone, Guinea, Australia and Central America led to the transformation of the company from a regional to a worldwide enterprise. In addition to diversifying its bauxite base, the firm accomplished its motive of product diversification by acquiring the Swiss firm, Lonza Ltd, in the 1970s. This enabled Alusuisse to become an important producer of organic and inorganic chemicals, plastics and agro-chemicals.

While the firm continued to grow through product and geographical diversification including in other mining ventures, the firm also became an important provider of commercial advisory services which derived from its assembly of a large and varied engineering staff that served its extensive prospective, planning and engineering operations worldwide. The firm's growing role as a provider of services has, however, somewhat lessened the importance of FDI as a modality for international expansion (Niehans, 1977).

Apart from MNCs in the processed metal industries, there was also the emergence and development of firms and industries and eventually MNCs in *machine building and electro-technics* allied closely with the strengths of Swiss firms in processed metal products and textiles. The simultaneous development of both the textiles and textile machine-building industries was attributed to the pioneering roles of Honegger in the first half of the nineteenth century, and Rieter 50 years later. Although Rieter became an MNC only after the Second World War, other firms, such as Sastig with its FDI in the marketing of both embroidery and embroidery machines, grew to become the biggest Swiss MNCs in terms of share capital in 1914 (Schröter, 1993).[10] Similar to the case in Sweden, the Swiss engineering and machine-building industries – apart perhaps from the textile machinery manufacturers – had always been directed abroad both in terms of sales and the location of its main centres of production owing to their strategy to penetrate foreign markets in durable goods.[11] Virtually every leading Swiss durable goods producer engaged in international production in the late nineteenth and early twentieth century. This included the Dubied Company which opened a foreign subsidiary in

Milan, the Sulzer Company which similarly opened a new factory at Milan in 1916, Esche-Wyss which entered into partnership with an existing company at Schio to form the Pretto-Eschwyss Company, and the Gardy Company which established a business in Turin. This list is far from exhaustive, but there is no comprehensive record of these investments (Wavre, 1988).

In the electrical and hydroelectrical industry, Swiss FDI and foreign portfolio investments were closely inter-related, and largely accounted for by holding companies developed in the 1890s. The foreign subsidiaries of these holding companies were particularly prominent in the hydroelectric industry, with no comparable concentration in other industries (Wavre, 1988).

The history of Brown, Boveri & Cie. (BBC) of Baden – which eventually developed into one of the world's largest producers of high-performance electrical machines for generation and transformation – involved creating a network of foreign subsidiaries in Mannheim in 1898, in Paris in 1902, and in Milan in 1904 where it acquired the Tecnomasio Italiano Brown-Boveri (Wavre, 1988). Having established its German subsidiary within two years of its foundation, the BBC became a MNC almost from its inception (Niehans, 1977). Its strategy of expanding abroad through international production was adopted by Motor A. G. für angewandte Elektricität, a combined engineering and financial consortium that depended entirely on the BBC and later became known as the Motor-Columbus AG (Wavre, 1988).[12] As with the Swedish firms in mechanical engineering and transportation equipment, the principal motives for international production by Swiss firms in mechanical and electrical equipment proceeded from trade barriers associated with the buying practices of major customer groups abroad which included public authorities, utility companies and large private firms for which large pieces of equipment were typically purpose built. These buyers often gave preference to domestic suppliers, sometimes being compelled to do so by national regulations. Compared to these non-tariff trade barriers, tariff barriers had played a secondary role.[13] Other relevant motives for MNC expansion were the high transportation costs for heavy equipment, the regional or national differences in technical standards in certain fields, and the need to maintain a sales and service organization in each market area. These facilities were complemented in many instances by local production, particularly of relatively standardized items (Niehans, 1977).

Italy became favoured increasingly as a country in which to establish new Swiss-owned factories abroad in the engineering and electrical industries owing to the situation in Germany which limited the possibility of further expansion of Swiss investments in that country (Masnata, 1924). However, no longer were Swiss investments attracted to the traditional heartlands of Italian textile production but to the three geographical poles consisting of Naples, Upper Italy (Anza, Màira, Cairasca, Orobia) and Milan, and the Genoese Riviera and Liguria. These regions became the strongest foreign footholds for Swiss investors and entrepreneurs, with much of the capital raised from Switzerland (Himmel, 1922). In this relatively circumscribed geographical area within Italy, Swiss FDI predominated in the mechanical engineering and electrical equipment industries in their objective to penetrate and exploit the rapidly growing Italian market.

Despite the importance of Swiss FDI in the mechanical engineering and electrical equipment industries between 1860–70 and 1920–30, these were not the only industries in which Swiss FDI emerged historically. The industrial clusters of *processed food and chemicals* were also of considerable importance for Swiss firms which began to expand abroad at the turn of the twentieth century. The investments abroad in both these industries were dominated by only a handful of MNCs who sought to internationalize rapidly their sales, production and R&D activities (Schröter, 1993). One of the first Swiss industries to internationalize production in the food products sector was milk products, accounted largely for by the Anglo-Swiss Condensed Milk Company which became an MNC only a few years after the company's inception with its establishment of four plants abroad in 1880 (three in England and one in Germany) (Heer, 1966). The rapid expansion into international production was owing to the high sugar content of condensed milk which rendered the product vulnerable to high import duties in an increasing number of countries (Niehans, 1977).

The international production of baby food products by the Henri Nestlé Company also had a long vintage. The growth of this company followed the product cycle pattern in its initial emphasis on the home market, the development of the export business at a later stage, and then finally growth through international production. In the first few decades of the firm's history, the rationale behind the expansion of the MNC activities arose from three main factors: to overcome rising average costs with increasing plant size,[14] to overcome tariff protection in export markets, and to guarantee a steady supply at the retail level of essential consumer commodities by having a local presence. By 1900, factories had already been established in Christiania in Norway, Edlitz in Austria, Tutbury in England and Fulton in New York (Wilkins, 1988b).[15]

Thus, at the merger of the Henri Nestlé Company and the Anglo-Swiss Condensed Milk Company in 1905 to form Nestlé and Anglo-Swiss, both firms had long been MNCs. The post-merger expansion of international production arose from the need to adapt products to local tastes, and to overcome the risk of exchange rate fluctuations and exchange controls which assumed increasing importance during the inter-war period (Niehans, 1977).[16]

The corporate growth of Nestlé had two distinct but closely related elements (Niehans, 1977). The first component involved geographical diversification. In a theme resonant of the product cycle model, the international businesses of the Henri Nestlé Company spanned an increasing number of countries, progressing from importing agents to the establishment of branch offices and production subsidiaries in more than 50 countries as local markets in foreign countries expanded. As this process advanced, the share of aggregate exports in the firm's total sales declined, providing evidence of the strategy of Swiss MNCs to regard exports and international production as substitutes. The second component of corporate growth involved horizontal diversification as the economies of scope warranted their extension to a wider range of food products. This dictated the 1905 merger of the two companies which enabled milk products to become the core of the merged company's business until around the First World War. The company branched out to the chocolate industry in 1911 by taking a direct influence in the Swiss choco-late-producing MNCs – Peter Callier, Kohler and Chocolats Suisses SA (Heer, 1966). With the addition of instant coffee in the 1930s, the main business of the company became the processing of three agricultural com-modities comprising milk, cocoa and coffee into manufactured food products protected by trademarks and patents, and the marketing of these products to the retail trade. An aggressive strategy in the postwar years led to a spate of mergers and acquisitions which extended the breadth of the company's product range sold through retail food stores to include soups, frozen foods, canned goods, bakery products, mineral water and cosmetics. By contrast, the company neither engaged in agricultural production nor retailing, hence the corporate strategy did not encompass vertical integration. The horizon-tal and geographical diversification strategies of the company in combina-tion enabled Nestlé to become the largest Swiss MNC since the First World War (Schröter, 1993) and one of the world's largest food processing companies (Niehans, 1977).

The chocolate industry began to expand rapidly in Switzerland at the very end of the nineteenth century, and the development of foreign markets involved substantial Swiss FDI. Some of the notable foreign investments in this industry apart from that of Nestlé was that of the Tobler Company which purchased the Michele Talmone Chocolate Company of Turin in 1905, and the establishment of the headquarters of the Italo-Swiss Chocolate Manufacturers' Union in Varese in 1921, with a capital of 5 million lire and a management board that included three Swiss citizens (Wavre, 1988).

Swiss MNCs in the jam and preserves industries of Switzerland also began to take off at the start of the nineteenth century, propelled by the narrowness of the Swiss market and the need by Swiss producers to search for foreign markets, including the establishment of factories abroad. The Société Générale de Conserves Alimentaires at Saxon, for example, was a holding company that owned until 1918 the Cirio Società Generale di Conserve Alimentari of Naples, with factories in Mondragone, Paestum and Taranto. This was a short-lived investment as its majority shareholdings were sold to indigenous firms in a decision by the company's board in that year (Masnata, 1924).

Swiss FDI in the *chemicals industry* was also gaining significance rapidly around the early twentieth century. As an example, the Hoffman-La Roche pharmaceuticals company opened a subsidiary in Milan around 1910 (Wavre, 1988). The bulk of Swiss FDI in this industry was, however, attracted mainly to France, Germany, Russia and the United States in an attempt to overcome the small size of the domestic market, reduce risks and to overcome the tariff and non-tariff protection accorded to pharmaceuticals and agro-chemicals in foreign countries.[17] The non-tariff barriers inherent in the widely divergent national drug laws in different countries had always been the main obstacle to the export expansion of chemicals and pharmaceuticals companies generally, and the Swiss companies were no exception. However, despite their high degree of multinationality, there remains considerable concentration in Switzerland (and in Basel in particular) of high-value research and production of specialty chemicals and pharmaceuticals by the major Swiss chemical MNCs such as Ciba-Geigy, Sandoz and Hoffman-la Roche as part of a corporate strategy reinforced since the 1970s.[18] Thus, unlike the other manufacturing industries in Switzerland, the chemicals industry remains in the 1990s a major contributor of Swiss industrial GDP with a share of more than 12 per cent and is responsible for more than 20 per cent of national industrial exports. Indeed, with no less than 85 per cent of domestic production in the industry that is exported, the industry continues to have a sizeable trade surplus (Sally, 1993).

The corporate strategies of the major Basel chemical firms are marked by a high degree of forward integration from basic research to the final product and, in some cases, also backward integration to eliminate the inherent risks associated with obtaining raw materials and intermediate products through arms-length transactions.[19] This would typically involve the concentration in the home country of parts of the value added chain from research to the production of semi-finished products which are, in turn, exported to a network of foreign subsidiaries through intra-firm trade for the final stages of production, packaging and marketing. The marketing part of the value added chain is typically localized in each host country since these need to be differentiated according to local needs and languages, and to respond to host government policy or consumer preferences for drugs with a high domestic value added (Niehans, 1977). Indeed, some two-thirds of exports of the three Basel chemical MNCs had been intra-firm to foreign subsidiaries. However, the Basel-focused strategy of the Swiss chemical MNCs has had to be reconsidered since the late 1980s owing to the limited possibilities for further expansion in the increasingly crowded Basel region, the unfavourable domestic political and regulatory climates, and the perceived need to decentralize high-value production closer to its major markets in the European Union, North America and Japan. This has resulted in the faster pace of growth of their capital and research investments in foreign markets in recent years by comparison to that in Switzerland, including the trimming down of some previously domestically based activities (Sally, 1993).

Other less significant industries of Swiss FDI were *paper making and graphics*, mostly concentrated in Italy, and in gas, transportation and the timber industry.[20] Swiss FDI in the hotels industry was perhaps more considerable, reaching some 16.5 million Swiss francs in 1905 and 19.9 million Swiss francs in 1913 (Himmel, 1922). Nearly all the Swiss hotel companies that expanded abroad were based in Lucerne, among which was the Italienisch-schweiz Hotelgesellschaft (Wavre, 1988). By 1914, there were five Swiss hotel companies that operated hotels in Italy, France and North Africa (Wilkins, 1988b).

By 1913, the significant recipients of Swiss FDI were France, North America (principally the USA), Germany and Italy (Wilkins, 1988b). Thus, apart from the United States, Swiss investors continued to focus on their closest neighbours for investment opportunities. In particular, the considerable Swiss investments in Italy in the 1870s and 1880s was facilitated by improved communications and transport (the San Gothard line opened in 1882), and encouraged by the expectations of an expanding market, the lower costs of production and the introduction of tariff protection (Wavre, 1988). Indeed, Swiss FDI was driven by a diverse range of factors since 1860–70. Thus, although location factors such as tariffs and lower production costs in foreign countries as well as restrictive employment legislation in Switzerland provided the major thrust to the emergence of Swiss FDI between 1750 and 1850–60, the opportunities offered by expanding foreign markets since 1850–60 gradually drew Swiss financiers and industrialists to become MNCs. Nevertheless, factors related to the factor market (29 per cent) were still relatively more important in explaining Swiss FDI before 1914, by comparison to factors related to the sales market (27 per cent) and government intervention such as tariffs mainly and patent legislation to a lesser degree, preferences for local production for state orders, etc. (25 per cent) or strategic considerations, i.e. diversification of location of production and presence in certain national markets (18 per cent) (Schröter, 1993). Regardless of the factors involved in the emergence and growth of Swiss FDI, the size of investments abroad by Swiss businesses between 1898 and 1919 was greater than that in Switzerland (Himmel, 1922).

The expansion of Swiss FDI since 1914

The First World War gave a great boost to Swiss FDI even though Switzerland faced considerable problems in obtaining enough food, fuel and raw materials. With the politically neutral status of Switzerland, considerable export earnings were derived from trading with both countries on both sides of the political divide. As in the Netherlands but unlike in Sweden, many Swiss firms increased both their share capital and FDI, mainly in neutral and (anti-German) entente states (Schröter, 1993).

Although Swiss FDI did not grow as rapidly during the inter-war period as before the First World War owing to the adoption of an overall cautious attitude of Swiss firms, the period described the sharp decline of FDI in the textile industry, while existing MNCs in the chemical, the electro-technical and food industries further increased the levels of their FDI and new Swiss firms ventured abroad and became MNCs. The performance of Swiss FDI in the period around the Second World War was significantly different from that in the First World War. As Switzerland became detached from the world market, there were limited opportunities for the expansion of trade and FDI with the continuing overall cautious attitude of Swiss firms which carried on well into the 1960s. Significant levels of Swiss FDI emerged only since the 1960s and in both the same set of industries as before the Second World War (with the exception of the textiles industry), as well as from new ones. Indeed, the period since the Second World War was an era of emergence of many new Swiss MNCs as shown by the empirical survey conducted by Bürgenmeier (1986). In that survey, some 17 per cent of the Swiss firms investing abroad were organized internationally from their foundations in the nineteenth century, while some 59 per cent of firms diversified internationally after the Second World War in a period of declining tariff barriers in general and the formation of the Single European Market. The increase in Swiss FDI associated with the emergence of more Swiss MNCs since the Second World War may thus be regarded as a strategy by Swiss firms to transcend non-tariff barriers through FDI (Bürgenmeier, 1986).

Swiss FDI after the Second World War was also regarded as a means of survival by Swiss firms in world markets (Borner and Wehrle, 1984). In this context, considerations regarding exports or international production as alternative modes of serving foreign markets had become increasingly rare, with greater emphasis laid instead on the choice between international production or the loss of business opportunities in foreign markets. The increased importance of outward FDI had emerged not only from a defensive viewpoint, i.e. the economic necessity to invest abroad brought about by the small size of the Swiss market, rising production costs in Switzerland and the need to engage in FDI to capture world markets, but also from more aggressive reasons, i.e. the growing ability of Swiss firms to apply their core competencies and become successful MNCs in foreign markets, and to gain access to complementary foreign-based technologies.

However, other modalities of internationalization apart from FDI – to include licensing, sub-contracting, consulting, industrial cooperation (co-production), joint ventures and group investment – have also been used either as the main means of internationalization by small- and medium-sized firms or as a direct trade-off to international production with 100 per cent or majority ownership by larger-sized firms. These alternative forms of international investment have been shown to be particularly suitable to some Swiss machinery firms as well as firms in certain segments of the chemical and pharmaceutical industries, but not to firms in the electronics industry (Borner, 1986).[21]

The decades since the Second World War also described the increasing importance of the services sector to both the Swiss economy and to Swiss MNCs. By shifting an increasing part of their manufacturing operations to foreign subsidiaries and the progressive concentration of their domestic operations on services such as marketing, technology, management and finance, Swiss MNCs have contributed significantly to the general shift of the Swiss economy from manufacturing to services (Niehans, 1977).[22] The importance of the services sector to Swiss MNCs has also been manifested increasingly in more recent decades in the expansion of firms in the banking, finance and insurance industries – an embodiment of Switzerland's historical position as a financial power (Bergier, 1968), as previously stated. However, despite the substantial foreign businesses of firms in these industries, FDI had not been the common method of internationalization in these industries before the 1960s, with the exception of those in the insurance industry. Swiss insurance and reinsurance companies have always had significant investments abroad as seen in the cases of the Swiss Assurance Company of Livorno that became a MNC between 1820 and 1830 as previously mentioned and Swiss Reinsurance.

Swiss Reinsurance – one of the world's largest professional reinsurers – has always been a highly international company to implement efficient risk and portfolio management essential to the reinsurance business.[23] The company established its first branch office in New York in 1910 to achieve marketing advantages in light of government regulation and accounting benefits, and later in the United Kingdom (in recognition of its traditional position as an insurance centre), Canada (for marketing and fiscal considerations), Australia (fiscal considerations) and South Africa (originally planned as a potential emergency headquarters in case of war) (Niehans, 1977).

By contrast, Swiss banks had less than one dozen branches abroad in the period until the Second World War. This was because much of the international banking before the developments of the Euromarkets in the 1960s could be conducted without extensive multinational branching (Cassis, 1990). As a result of the need since the 1960s for a presence by commercial banks in at least the major financial centres of the world, Swiss banks have increased the number of their foreign branches, including a series of large acquisitions in the financial sectors of the United Kingdom, the United States and Germany in the 1980s. This is a reflection of the development of the Euromarkets since the 1960s, the moves of competitors towards global banking and the creation of the Single European Market (Schröter, 1993). As a result of the expansion of Swiss firms in the banking, finance, insurance and other services industries, the share of the services sector in the stock of recent Swiss FDI had increased steadily from a share of more than 30 per cent in 1986, some 32 per cent in 1987 (Bürgenmeier, 1991), 38 per cent in 1988 and 56 per cent in 1997 (UNCTAD, 1999).

The 1980s and 1990s were thus decades of continued growth and expansion of Swiss FDI of which at least half continues to be directed to countries of the European Union that have always played a traditional key role in Swiss international economic relations. In a continuing historical trend, the largest 50 MNCs produced more than twice as much abroad than in Switzerland in 1980, employing a workforce of 535,270 abroad and some 233,120 in Switzerland (Brauchlin, 1986). By 1989, however, the 50 largest MNCs doubled their workforce abroad to 1,167,845 while the corresponding number of workforce in Switzerland remained essentially at the same level (Schröter,

1993). Indeed, Switzerland continues to rank as one of countries with a high degree of multinationality when also measured by the level of its outward FDI per capita.

Swiss FDI in more recent decades have been determined essentially by the same set of factors that encouraged Swiss FDI historically. The three motives related to state intervention abroad, foreign and national conditions of the labour market, and economic policy in Switzerland are recurrent themes (Bürgenmeier, 1986). The increase in the levels of Swiss FDI – whether through the continuing expansion of existing MNCs or the emergence of new firms as MNCs – derive from the need to search for new markets (weighted frequency of 47 per cent), to benefit from from lower production costs abroad (32 per cent) and fiscal advantages (21 per cent) (Nankobogo, 1989).

Conclusion

This chapter analysed the history of Swiss MNCs since around 1750 until the present time. The earliest Swiss investments abroad had several distinctive features. First, the important industries of Swiss FDI were trading and commerce, banking and insurance, and textiles. Swiss FDI in cotton manufacturing in particular was driven by the attainment of lower production costs abroad and to mitigate the threat to exports presented by Italian import duties and the German customs union. While Swiss FDI in silk spinning was similarly driven by lower labour costs abroad, other equally important determinants arose from the need to overcome rigorous labour legislation in Switzerland concerning the employment of school children and the proximity of raw materials. The historical Swiss FDI in the textiles sector was, therefore, spurred fundamentally by location-specific advantages that favoured foreign countries. Secondly, Swiss FDI in this period was focused towards countries that were geographically close and culturally similar to Switzerland. Although Britain was an important recipient of Swiss investments (Wavre, 1988), substantial investments by Swiss MNCs had been directed to Germany, France and Italy (Schröter, 1993) that constituted the first export markets of Switzerland. Third, the sums of Swiss capital invested abroad in the period was modest, with a majority of the investment emerging from individual adventurers or families with established interests in the textile industry or small- and medium-sized firms. Fourth, the emigration of industry associated with Swiss FDI until the 1860s and 1870s made Swiss FDI in this period consistent essentially with migratory capital. After the initial connections to Switzerland receded, the investment became independent and detached from managerial control from the home country. Capital connections usually ceased at a later stage, together with family links (Bonnant, 1976). However, only a few examples of FDI in the form of migratory capital was found after the 1860s and 1870s, and therefore the years around 1870 can be traced as the period of the emergence of the modern Swiss MNC (Schröter,

1993). It was also the period in which Swiss FDI emerged in a number of other industries to include processed metal products, machine building and electro-technics, processed foods and chemicals as well as hotels.

The importance of the textiles industry in both domestic production in Switzerland and international production by Swiss MNCs started to decline even before the turn of the twentieth century owing to the process of domestic industrial upgrading in Switzerland and changes in fashion trends in foreign markets. From 1890 onwards, substantial FDI by such firms as Alusuisse-Lonza, von Roll, Georg Fischer and other firms belonging to the processed metal products industries began to emerge and grow in importance. Some of these were in the iron and steel industries, the bulk of which went to Germany, France and, to a lesser extent, Italy. Apart from MNCs in the processed metal industries, there was also the emergence and development of firms and industries and eventually MNCs in machine building and electro-technics allied closely with the strengths of Swiss firms in processed metal products and textiles. Similar to the case in Sweden, the Swiss engineering and machine-building industries – apart perhaps from the textile machinery manufacturers – had always been directed abroad both in terms of sales and the location of its main centres of production owing to their strategy to penetrate foreign markets in durable goods. Virtually every leading Swiss durable goods producer engaged in international production in the late nineteenth and early twentieth century. In the electrical and hydroelectrical industry, on the other hand, Swiss FDI and foreign portfolio investments were closely inter-related, and largely accounted for by holding companies developed in the 1890s.

As with the Swedish firms in mechanical engineering and transportation equipment, the principal motives for international production by Swiss firms in mechanical and electrical equipment proceeded from trade barriers associated with the buying practices of major customer groups abroad to include public authorities, utility companies and large private firms for which large pieces of equipment were typically purpose built. These buyers often gave preference to domestic suppliers, sometimes being compelled to do so by national regulations. Compared to these non-tariff trade barriers, tariff barriers had played a secondary role. Other relevant motives for MNC expansion were the high transportation costs for heavy equipment, the regional or national differences in technical standards in certain fields, and the need to maintain a sales and service organization in each market area. These facilities were complemented in many instances by local production, particularly of relatively standardized items (Niehans, 1977). Italy became favoured increasingly as a country in which to establish new Swiss-owned factories abroad in the engineering and electrical industries owing to the situation in Germany which limited the possibility of further expansion of Swiss investments in that country (Masnata, 1924).

Other industries in which Swiss FDI emerged historically were processed foods and chemicals, industries which were of considerable importance for Swiss firms that began to expand abroad at the turn of the twentieth century.

The investments abroad in both these industries were dominated by only a handful of MNCs who sought to internationalize rapidly their sales, production and R&D activities (Schröter, 1993). While the need to search for foreign markets as a means of overcoming the limited size of the domestic market was a common objective behind the international production of Swiss food companies, the high import duties in an increasing number of countries was also an important contributory factor (the case with the Anglo-Swiss Condensed Milk Company and the Henri Nestlé Company), in addition to the need to overcome rising average costs with increasing plant size and to guarantee a steady supply at the retail level of essential consumer commodities by having a local presence (the Henri Nestlé Company).

Swiss FDI in the chemicals industry was also gaining significance rapidly around the early twentieth century. The bulk of Swiss FDI in this industry was attracted mainly to France, Germany, Russia and the United States in an attempt to overcome the small size of the domestic market, reduce risks and to overcome the tariffs and, perhaps more importantly, non-tariff protection accorded to pharmaceuticals and agro-chemicals in foreign countries. However, despite their high degree of multinationality, there remains considerable concentration in Switzerland (and in Basel in particular) of high-value research and production of specialty chemicals and pharmaceuticals by the major Swiss chemical MNCs such as Ciba-Geigy, Sandoz and Hoffman-La Roche as part of a corporate strategy reinforced since the 1970s. Other less significant industries of Swiss FDI were paper making and graphics, mostly concentrated in Italy, and the gas, transportation and the timber industry. Swiss FDI in the hotels industry was perhaps more considerable.

By 1913, the significant recipients of Swiss FDI were France, North America (principally the USA), Germany and Italy (Wilkins, 1988b). Thus, apart from the United States, Swiss investors continued to focus on their closest neighbours for investment opportunities. In particular, the considerable Swiss investments in Italy in the 1870s and 1880s were facilitated by improved communications and transport, and encouraged by the expectations of an expanding market, the lower costs of production and the introduction of tariff protection (Wavre, 1988). Although location factors such as tariffs and lower production costs in foreign countries as well as restrictive employment legislation in Switzerland provided the major thrust to the emergence of Swiss FDI in the textiles industry between 1750 and 1850–60, the opportunities offered by expanding foreign markets in addition to tariff and non-tariff trade barriers gradually drew Swiss financiers and industrialists to become MNCs since 1850–60. Nevertheless, factors related to the factor market were still relatively more important in explaining Swiss FDI before 1914, by comparison to factors related to the sales market, government intervention (such as tariffs mainly and to a minor degree patent legislation, preferences for local production for state orders, etc.) or strategic considerations, i.e. diversification of location of production and presence in certain national markets (Schröter, 1993).

Although Swiss FDI did not grow as rapidly during the inter-war period as before the First World War owing to the adoption of an overall cautious attitude of Swiss firms, the period described the sharp decline of FDI in the textile industry, while existing MNCs in the chemical, the electro-technical and food industries further increased the levels of their FDI and new Swiss firms ventured abroad and became MNCs. The period since the Second World War was an era of emergence of many new Swiss MNCs in manufacturing as well as in services in light of the increasing dominant role of the services sector to both the Swiss economy and to Swiss MNCs. Indeed, Swiss MNCs have contributed significantly to the general shift of the Swiss economy from manufacturing to services (Niehans, 1977). The importance of the services sector to Swiss MNCs has been manifested increasingly in more recent decades in the expansion of firms in the banking, finance and insurance industries – an embodiment of Switzerland's historical position as a financial power (Bergier, 1968).

Swiss FDI in more recent decades have been determined essentially by the same set of factors that encouraged Swiss FDI historically. The three motives related to state intervention abroad, foreign and national conditions of the labour market, and economic policy in Switzerland are recurrent themes (Bürgenmeier, 1986). The increase in the levels of Swiss FDI – whether through the continuing expansion of existing MNCs or the emergence of new firms as MNCs – derive from the need to search for new markets, to benefit from lower production costs abroad and fiscal advantages (Nankobogo, 1989).

By way of comparison, the basis of the sustained competitiveness of Swiss firms and MNCs are very similar to those of Germany (see Chapter 8) and Sweden (see Chapter 4). Apart from the silk industry which at the time of its peak development before the First World War concentrated on products of medium quality and therefore not in direct competition with the famous French industry of Lyon, Swiss products like those of Germany and Sweden have generally been focused on quality or based on extensive R&D and technical expertise (Schröter, 1993). A major part of the explanation lies in the low demand in the Swiss economy for lower-quality, mass-produced consumer goods (Porter, 1990). This has driven Switzerland to have the highest intensity of industrial R&D in the OECD, with Swiss MNCs dominating R&D expenditures: the total R&D budgets of the three major Basel chemical companies and Asea Brown Boveri represent more than 80 per cent of gross R&D expenditures in Switzerland (Sally, 1993).[24]

Indeed, both Switzerland and Germany have developed analogous strengths in chemicals, machinery, machine tools, precision mechanical goods, optical products and textiles, with Germany tending to have a broader product range compared to the limited specialization of

Switzerland in the most sophisticated segments of industry (Porter, 1990). However, while the standards for general and technical education and the well-developed apprenticeship system as a pool for human resource development are similar in both Switzerland and Germany, the relations between banks and industry are closer, industrial relations are more stable and investment policies are geared more towards the longer term in the former country than in the latter. In this respect, Sweden seems a closer comparison. In contrast to the German case, the reason for the large amounts and the early forays into international production by Swiss and Swedish MNCs was evidently related to the small size of their domestic markets, particularly so in the case of Switzerland. The land-locked geography of Switzerland and its multi-cultural and multi-lingual population have also enhanced both its international orientation and its successful operations in foreign markets.

However, unlike the Swedish firms generally, Swiss firms in general and the major Basel chemical MNCs in particular display a greater degree of embeddedness to their home country. The sustained concentration of high value added research and production in Switzerland indicated the continuing reliance by Swiss firms on their home base to maintain and upgrade their competitive advantages. Such firms depended on the educational and research institutes of the country as mentioned for vocational training and qualified personnel and in the provision of the R&D infrastructure for the advancement of technological innovation. There was also the tight links between the major Basel chemical companies and their constellation of small suppliers dependent overwhelmingly on the home market. The favourable conditions enjoyed in the home country by these firms explain their strong embeddedness to their country and region – a feature which will take some degree of difficulty and time to replicate abroad even in the bordering areas surrounding Basel in the territories of France and Germany (Sally, 1993).

The analysis of the history of Swiss MNCs leads one to support the view advanced by Schröter (1993) that Swiss FDI possesses three distinctive features: continuity, scale and scope. The continuity and scale features stem from its long history of FDI since 1750, and to an extent that has enabled Switzerland to consistently be one of countries of the world with a high degree of multinational business. The scope of Swiss FDI, on the other hand, is seen in the extraordinarily wide breadth of advanced manufacturing and services industries in which Switzerland spawned MNC activities – a remarkable feat for a nation of a small size and population that had not been observed in other small countries apart perhaps from Singapore (see Chapter 15). This is a reflection of the diverse clusters of industries in which Switzerland had become highly competitive: health care-related industries (including pharmaceuticals), textiles, speciality chemicals, processed food products, processed metal products (including machine tools) and general business services (including trading, banking, finance, insurance, management consultancy, etc.) (Porter, 1990).

Notes

1 Based on data contained in UNCTAD (1999).
2 Indeed, apart from Swiss MNCs having early origins, Switzerland also ranked first and second on the world's list of national exports per capita in 1860 and 1913, respectively (Bairoch, 1973).
3 The textiles industry accounted for 70 per cent of the country's total exports around 1840 and 50 per cent of industrial value added as late as 1880 (Wavre, 1988).
4 Examples of Swiss investment in the textiles industry in Italy in this period included the printed textile factory that J. Speich founded at Cornigliano in 1789 which survived until about 1850 (Masnata, 1924). There was also the cotton industry founded in the Kingdom of Naples at Piedimonete d'Alife by J. J. Egg of Zürich which led to the import ban imposed by the Kingdom of Naples on 17 September 1816 on certain types of cotton fabrics (*balazores*) with the aim of protecting the industry established by J. J. Egg. This company had 500 looms in operation in 1834 and employed 1,300 workers (Clough, 1964). J. R. Glarner founded a calico factory at Sarno in the 1830s, while K. Blumer also opened a factory at Messina. Indeed, cotton plants owned by Swiss and directed by Swiss technicians proliferated in Italy during this period and grew rapidly (Wavre, 1988).
5 The Swiss Federal Customs Department reported in 1888 that as a result of the new tariff regime in Italy, the value of Swiss exports declined from 65 to 51 million Swiss francs, that is by 21 per cent. Three-quarters of the decline in export value was explained by the new increased duties in force in Italy. In particular, it led to a loss of 3.4 million Swiss francs of export revenue on cotton fabric and a further loss of 3.7 million Swiss francs of export revenue on cheese (Wavre, 1988).
6 This factor alone forced one Ticinese spinning works to transfer production to Italy in 1898. A number of Zürich silk companies opened Italian subsidiaries in 1907 (Masnata, 1924).
7 This pattern of Swiss FDI being driven by medium-sized firms continued throughout the inter-war period, and even after the Second World War (Schröter, 1993).
8 A Genevan company opened an ironworks plant in Turin in 1920, and there is reference to two other Swiss iron and steel foundries in Turin (Wavre, 1988).
9 On the basis of central direction from its Swiss headquarters, bauxite extracted in France, Hungary, Rumania and Yugoslavia was shipped to the sites of cheap coal in Germany for the extraction of alumina which was then reduced to aluminium by hydroelectric power in Switzerland, Germany and Italy. The final products were sold mainly in the European market (Niehans, 1977).
10 It is sometimes a contestable point whether or not the machinery for embroideries used by Sastig was entirely Swiss in origin. This is because although the invention of the machinery was mainly Swiss, the machinery was built in Saxony, Germany, on the basis of orders received from Sastig's plants in Switzerland and the United States. As in the case of the chemicals industry, the home market for the Swiss machine-building industry included Germany. It is partly for these reasons that Switzerland is regarded as a special case of Porter's home nation-based theory (Borner *et al.*, 1991).
11 After the decline in importance of the textiles industry in Swiss FDI, the international orientation of firms engaged in textile machinery production was confined mainly to exports until the 1960s. It was only in the decades since the 1960s that firms producing special machinery for the textiles industry emerged as MNCs (Schröter, 1993).
12 Although the Motor company operated exclusively in Switzerland until the early 1900s where it played a leading role in the electrical equipment industry and

accounted for most of the 100 million Swiss francs invested in the industry (Himmel, 1922), it rapidly adopted a strategy of foreign investment. It founded in 1903 the Società per la Forze Motrici dell'Anza (in addition to the acquisition by the BBC of the Tecnomasio Italiano Brown-Boveri in Milan in 1903–4). It acquired an interest in the Dynamo Società per la Imprese Elettriche in Milan in 1907, and took over the Forze Idrauliche della Maira concern in 1911. Through a merger, Motor's holding in Forze Idrauliche della Maira transformed to major shareholdings in the Società Riviera di Ponente, Ing R. Negri, Genoa. With the investments of other Swiss companies and private investors that also held shares in this company, the total Swiss investment in the Società Riviera di Ponente, Ing R. Negri amounted to 17 million lire (or a share of 15.5 per cent) (Himmel, 1922). In 1914 Motor also acquired an interest in the Italo-Swiss Finance Company along with three other Swiss companies – Electrobank, Société France-Suisse pour l'Energie Electrique and Société Financière Italo-Suisse. The four companies also held important shares in a number of other Italian electrical companies (Wavre, 1988).

13 As a consequence, the lowering of tariffs in the postwar period and the formation of the European Common Market affected BBC less than other companies, except for mass-produced items (Niehans, 1977).

14 The decreasing returns to scale with increasing plant size arose from the rising cost of domestic milk supplies which could not keep pace with the growing foreign demand.

15 In a strategic retreat, however, the Henri Nestlé Company sold all its American plants to a competitor firm, Borden, in 1902 (Wilkins, 1989).

16 The need for production decentralization was also balanced by efficiency considerations which dictate the centralization within a single corporate entity of the following activities: R&D, trademarks, production and shipping schedules and control of inventories, and the manufacture of machinery and equipment (previously). These activities were controlled from the company's headquarters in Vevey. The purchases of raw materials and production, however, were decentralized, with the subsidiaries playing an important role in the development of local agriculture, particularly in developing countries (Niehans, 1977).

17 The smallness of the Swiss domestic market was a powerful factor which stimulated the international expansion since the beginning of the twentieth century of Ciba-Geigy, Sandoz, and Hoffman-La Roche – the leading MNCs worldwide in pharmaceuticals and speciality chemicals. In more recent years, the Swiss market represented between 2 and 4 per cent of the world sales of these companies (Sally, 1993).

18 Although only one-third of the capital investments of these three Swiss chemical firms and an average of over a fifth of their employees are in Switzerland, their production and research are still largely concentrated in Switzerland. Sandoz had 95 per cent of its pharmaceutical production in Basel, and 40 per cent of Ciba-Geigy's chemicals production was in Switzerland. In addition, 55 per cent of Ciba-Geigy's and Sandoz's total R&D budgets were spent in Switzerland. This has contributed to the chemicals industry having the highest research intensity of all industries of the Swiss economy (Sally, 1993).

19 For example, Ciba-Geigy initiated its own production of an important intermediate product for dyestuffs which was previously bought from Bayer, the most important competitor in the product market (Niehans, 1977).

20 For example, a Zürich graphics company did have a subsidiary at Como, and Société pour la Fabricacion de la Pâte de Bois – a Swiss holding company – owned factories at Carmignano di Brenta and at Friola in Italy (Wavre, 1988).

21 These alternative strategies of internationalization are by no means new to Swiss MNCs that used them widely from the early 1890s onwards. For example, the

three chemical firms, Ciba, Sandoz and Geigy, jointly operated production facilities in the United States, the United Kingdom, Italy and elsewhere from the 1920s well into the 1960s. And in 1977, Sandoz established a joint venture with the French enterprise Rhône-Poulenc in the field of hospital supplies. As in the case of MNCs from other countries, Swiss MNCs tended to resort to such strategies in conditions of relative economic insecurity, rapid expansion and so on (Schröter, 1993).

22 This is shown in the fact that since the first half of the 1970s the Swiss firms in service industries such as banking, insurance, retailing, construction and energy have been growing more rapidly and performing better in the domestic stock markets compared to firms in metals, machinery, engineering, pharmaceuticals and chemicals whose performance were relatively lacklustre (Niehans, 1977).

23 The international dimension is crucial in reinsurance from both the underwriting and investment perspectives. From the underwriting perspective, efficient risk management requires diversification across different insurance lines and geographic areas. From the investment perspective, efficient portfolio management does not require that assets and liabilities be matched in every national market separately, but does require the ability of firms to shift assets freely to meet claims in various national markets (Niehans, 1977).

24 Public funding accounted for the remaining 20 per cent of total Swiss R&D, the lowest in the OECD, with the federal government's role limited largely to supporting basic research in universities, higher technical colleges and federal institutes of technology (Sally, 1993).

14 The emergence and evolution of multinational corporations from Hong Kong

Introduction

Although the origins of Hong Kong-based MNCs can be traced to the British colonial period and Hong Kong's position as an entrepôt for trade in South East Asia and China, indigenous Chinese firms based in Hong Kong emerged as MNCs in the early 1950s, thus reflecting a longer history by comparison to MNCs from Taiwan and South Korea whose emergence can be traced to the early 1960s. Hong Kong has grown to become a significant home country of FDI with an outward FDI stock of $154.9 billion in 1998, or some 3.8 per cent of the global stock of outward FDI. Indeed, it had become the world's tenth-largest source of FDI in that year based on the size of outward FDI stock after the United States (with a share of 24.1 per cent of the global stock of outward FDI), United Kingdom (12.1 per cent), Germany (9.5 per cent), Japan (7.2 per cent), the Netherlands (6.4 per cent), France (5.9 per cent), Switzerland (4.3 per cent), Italy (4.1 per cent) and Canada (3.8 per cent). In fact, Hong Kong is almost as important as Canada whose outward FDI stock was $156.6 billion in that year. Thus, Hong Kong had become a significant source of FDI in the world economy, particularly more so in relation to the stock of outward FDI from developing countries where Hong Kong is the single largest home country with a share of almost 40 per cent.[1]

Not only is Hong Kong comparable to Switzerland as a home country of FDI in terms of the size of outward FDI stock, the study of the emergence and evolution of Hong Kong-based MNCs is of interest as another case study of MNCs from a resource-scarce small country. The growth pattern of Hong Kong-based MNCs as it has been evolving over the last half century can thus be compared to those based in Switzerland that share a similar pattern of national economic development and whose longer MNC history has been analysed in the previous chapter in this part of the book.

Before the historical excursion into the history of Hong Kong-based MNCs, it is important to clarify two key points. Firstly, this Chapter relates to the history of MNCs based in Hong Kong which includes not only the outward FDI of indigenous Chinese firms in Hong Kong, but also outward FDI coursed through Hong Kong whose ultimate beneficial ownership can be traced to another country. This applies to the outward FDI by holding companies established in Hong Kong by Australia, the United Kingdom and other industrialized countries that have business interests in South East Asia as well as outward FDI by some of the trading and financial companies in Hong Kong founded and in many cases still managed and controlled by the British: John Swire and Sons (HK Ltd), Hutchison International Ltd, Jardine Matheson and Company and the Hong Kong and Shanghai Banking Corporation (Wells, 1978).[2] In addition, since the 1990s there have been significant amounts of indirect investments to China from Taiwan that were channelled through Hong Kong owing to the illegality of direct trade between Taiwan and mainland China (see Chapter 10). The inclusion in Hong Kong-based FDI of outward FDI whose ultimate beneficial ownership can be traced to a third country is important in the analysis of outward FDI emerging from Hong Kong and helps to partly explain the prominent position of Hong Kong as a home country of FDI particularly in recent years. It is for this reason that the term 'Hong Kong-based' firms or MNCs is used throughout this chapter and book.

Secondly, the government of Hong Kong does not collect data on inward and outward FDI in and from Hong Kong. Thus, the analysis of outward FDI from Hong Kong has had to rely on scattered reports and the scanty data and information provided by the host countries in which investment by Hong Kong-based MNCs had been significant (Chen, 1981; Wells, 1978). Given the large number of significant host countries of Hong Kong-based FDI and the different currencies and criteria used in the compilation of data on inward FDI in each host country which precludes any possibility of data aggregation across host countries, the analysis of Hong Kong-based MNCs is rendered somewhat difficult particularly with respect to the examination of the major types of outward FDI, the fundamental determinants of outward FDI and the major host countries, as well as the changes in these variables over time.

Bearing in mind these two caveats in the analysis of Hong Kong-based FDI, this Chapter aims to present a faithful account of the history of Hong Kong-based MNCs in three time frames: from the 1950s to the 1970s, the 1980s and the 1990s.

The emergence of Hong Kong-based MNCs from the 1950s to the 1970s

To understand the origins of Hong Kong-based MNCs is to understand the economic and political history of Hong Kong as a British colonial entrepôt. Indeed, from its inception as a British colony in 1841 until the mid-twentieth century, the integration of Hong Kong in the world economy had been dictated by its role as a trans-shipment point for British exports to China and, to a lesser extent, other parts of the region and as a hub for Chinese commodity and financial transactions with Europe and the United States (Henderson, 1989). With a share of two-thirds of the exports (or re-exports) of Hong Kong,

China remained the principal export market of Hong Kong until 1951 (Phelps Brown, 1971). The virtual elimination of the entrepôt trade with China as a result of the Chinese Revolution of 1949 and the Korean War of 1950–53 which led to the export embargo on all goods of Chinese origin to the United States and the prohibition imposed by the United Nations on the export of essential materials and strategic goods to China (Szczepanik, 1958; So, 1986) combined with social changes in Hong Kong consequent to the Chinese Revolution and the restructuring of the world economy after the Second World War (Henderson, 1989) had powerful influences on the industrial development of Hong Kong starting from the early 1950s and the eventual development of the earliest Hong Kong-based MNCs. The focus on the development of an export-oriented industrial economy in the 1950s in the face of the collapse of the entrepôt trade with China was facilitated by the transfer to Hong Kong from Shanghai of industrial capital and managerial expertise in textile production in response to the Chinese Revolution of 1949 with the military triumph of the Chinese Communist Party over the Kuomintang opposition.[3] Thus, with the installation of modern factories combining new and modern production machinery and cheap refugee labour, the textile 'barons' of Shanghai were instrumental in placing textiles production at the core of the industrial development of Hong Kong in the 1950s, an industry that had from its inception a high export propensity and formed the major basis of Hong Kong's export-led growth (see also Ho, 1992; Wong, 1988, 1991).[4] Such export expansion was facilitated by the existence in Hong Kong of some 1,000 to 1,500 trading houses involved previously in the entrepôt trade with well-entrenched export links with the British and other export markets (Szczepanik, 1958). In addition, it helped that, although the Shanghai entrepreneurs had predominantly been oriented in their production towards the large domestic market of mainland China, a few had established export markets in Asia through ties with overseas Chinese in the region (Wells, 1978). With the structural problems associated with the cotton textile industry in Lancashire in the mid-1950s and the consequent shortfall of domestic textile supply to the domestic demand for cheaper cotton textiles and clothing in Great Britain (Gregory, 1985), British department stores and clothing chains not only concluded supplier contracts with Hong Kong manufacturers encouraged by Commonwealth preferential import tariff arrangements but also assisted directly with the further development of Hong Kong's production capacity and with the improvement of the quality of their output (Phelps Brown, 1971). Since then, textiles and clothing had become a principal industry of Hong Kong in terms of employment and value of output and a major contributor to Hong Kong's manufactured exports. In combination with the emergence and growth in the late 1950s of domestic production and exports of electronics (Henderson, 1989), industrial productivity grew by an annual average rate of 20 per cent through the 1950s, and by the end of that decade the value of manufactured exports of Hong Kong exceeded that of entrepôt trade even with its revival in the 1950s (Cheng, 1985). By the

early 1960s Hong Kong had not only become the largest supplier of manufactured commodities in developing countries (Lin and Ho, 1980) but also a
significant exporter to developed countries such as Great Britain.

The rapid development of an export-oriented economy and the spectacular
export growth performance of Hong Kong firms led eventually to the imposition of trade barriers by threatened export markets. The defence of export
markets provided the impetus behind the initial international production of
Hong Kong firms in a manner and extent seen in few other developing
countries (Wells, 1978). Indeed, defensive FDI by Hong Kong firms reflected
export patterns. Thus, in the late 1950s exports by Hong Kong of simple
consumer goods such as kerosene lanterns, flashlights, umbrellas and simple
food products (biscuits, flour, etc.) faced tariffs and quotas designed to encourage their domestic production in low-income countries, including those
initiated by Hong Kong-based firms. As exports of textiles and electronics
suffered a similar fate in the 1960s and the mid-1970s respectively with the
imposition of trade quotas in the developed countries, defensive FDI by Hong
Kong-based firms in the form of international production of textiles and
eventually electronics was initiated in host countries where trade quotas
had not yet been imposed. Although international production by Hong
Kong-based firms had always been directed to lower-income developing
countries, the market orientation of their international production activities
in the 1950s geared towards the domestic markets of low-income countries
had shifted rapidly since towards exports to the major markets of the developed countries to overcome trade protectionism. Thus, in terms of the types of
international production financed by outward FDI, the pattern of Hong
Kong-based FDI changed rapidly from import substituting manufacturing
towards export platform manufacturing.

Based on the stock of Hong Kong-based inward FDI in host countries,
manufacturing was the most important sector of Hong Kong-based FDI at
the end of the 1970s, followed by the services sector and the primary sector.[5]
Manufacturing accounted for as much as 52 per cent of Hong Kong-based
FDI in Indonesia as of 1976, 98 per cent of Hong-Kong based FDI in
Malaysia at the end of 1979 and 55 per cent of Hong Kong-based FDI in
Taiwan as of 1981 (based on data in Chen, 1981, 1983b).[6] The predominance
of outward FDI by Hong Kong in manufacturing at the initial phase of its
expansion in international production provides further support to the close
link between trade and outward FDI in the early phases of international
production (Cantwell, 1989a).[7]

Although most of Hong Kong-based FDI at the end of the 1970s was
concentrated in these three countries and Singapore, some foreign subsidiaries
had also been established in other Asian countries such as the Philippines, Sri
Lanka, India, Pakistan, Thailand and China and in Africa, particularly in
Nigeria and Ghana. In addition, some Hong Kong-based FDI was made in
Canada, Switzerland, the United Kingdom and the United States (Chen,
1981, 1983b).[8]

Indeed, Hong Kong-based MNCs had already been exerting some significance in their principal host countries at the end of the 1970s as shown by the share of Hong Kong in the stock of inward FDI. In Indonesia, Hong Kong accounted for 12 per cent of inward FDI as of 1976, and was the second-largest source of FDI after Japan. In Malaysia, Hong Kong accounted for 11 per cent of the stock of inward FDI as of 1979, and was the fourth-largest source of FDI after Japan, Singapore and the United Kingdom. In Taiwan, Hong Kong accounted for 9 per cent of inward FDI as of 1981, and thus was the third-largest source of FDI after Japan and the United States (based on data in Chen, 1981, 1983b). In Singapore, Hong Kong accounted for 12 per cent of inward FDI as of 1980, and thus was the third-largest source of FDI after the United Kingdom and the United States (based on data in Department of Statistics, Singapore, 1992). It was only in the Philippines where Hong Kong accounted for a smaller share of a little more than 4 per cent of the stock of inward FDI as of 1980, owing to the dominant role of the United States and Japan (based on data in Alburo, 1988).

By 1980, the estimated stock of Hong Kong-based FDI remained low at $148 million which although far less significant than that of Singapore of $819 million was comparable to that of South Korea of $142 million and higher than that of Taiwan of $97 million.[9] Five types of outward FDI undertaken by Hong Kong-based MNCs was evident around the end of the 1970s. In order of declining importance, these were: first, export platform manufacturing FDI in developing countries; second, import substituting FDI in manufacturing in developing countries; third, services FDI in developing countries; fourth, outward FDI to secure supplies of essential raw materials; and fifth, FDI in developed countries. Each of these types of Hong Kong-based FDI in the 1970s is discussed below.

Export platform FDI in manufacturing in developing countries

As indicated above, although import substituting FDI in developing countries was the dominant type of Hong Kong-based FDI in the 1950s, export platform FDI in developing countries grew rapidly in the 1960s and became the most important type of Hong Kong-based FDI. Such export platform FDI was initiated by textile manufacturers in the 1960s and also by electrical and electronics manufacturers in the 1970s.

In the 1960s, the major determinant of international production of textiles by Hong Kong-based firms was to overcome the quotas imposed by developed countries on Hong Kong's textile exports. Indeed, such quotas predate the establishment of the Multi-Fibre Arrangement in the early 1970s (Chen 1981, 1983b). The principal host countries were Singapore (where some 15 textile plants were established by Hong Kong-based firms in 1963 and 1964), Taiwan, Indonesia, Macao and Thailand (Luey, 1969) where quotas on textile exports had not yet been imposed or were less severe and where ethnic ties to Chinese communities could be cultivated in the establishment and maintenance of foreign operations (Chen, 1981, 1983b; Wells, 1978).[10] Since the United Kingdom was the principal export market for Hong Kong's textiles, the attractiveness of Singapore as a host country was enhanced owing to its preferential trading arrangements with the United Kingdom as a member of the Commonwealth and also because of the presence of shipping and financial facilities (Chen, 1981, 1983b).

Although the major emergence of international production by Hong Kong-based firms in textiles in the 1960s was rooted fundamentally in the quotas imposed by developed countries on Hong Kong's textile exports, the growth of international production in textiles beginning in the late 1960s and 1970s became even more imperative owing to the additional motivation to minimize production costs, to relieve domestic supply shortages of semi-skilled and unskilled labour, and to overcome the increasing competition between domestic firms in Hong Kong as well as the growth of emerging competition from firms based in the Asian newly industrialized countries (NICs). Increasing domestic competitive forces have been brought forth by the lack of opportunities for investment expansion within the limited domestic market of Hong Kong and by the specialization of Hong Kong in narrow lines of industrial activities. Thus, as labour and land prices in Hong Kong began to rise sharply and competition among domestic firms intensified, Hong Kong was losing rapidly its comparative advantage as a location for labour intensive production.

As confirmed by both the 1979 and 1982 surveys conducted by Chen (1981, 1983b) of Hong Kong-based firms engaged in defensive export oriented FDI, the choice of location of international production activities became more driven by the presence of abundance of unskilled and semi-skilled labour, low labour and land costs, the presence of good infrastructure, and political stability. The attractiveness of a host country was, of course, enhanced if it extended export incentives (as in the case of Malaysia), has no quotas imposed on its exports or enjoys a preferential trade access to the major export markets of Hong Kong (as in the case of Singapore). But a persistent feature of the location of international production particular to the Asia-Pacific region was the powerful alliances established by Hong Kong-based firms with the Chinese business community in different countries of South East Asia and China which provided business information as well as financial and marketing assistance. Indeed, the operations of Hong Kong-based firms in ASEAN had always been socially and culturally embedded in networks of relationships or *guanxi* (Granovetter, 1985, 1991; Granovetter and Swedberg, 1992). These regional business networks based on contacts spanning almost a century and close familial relations have provided strength to Chinese business organizations and paved the way for their economic hegemonic role in the business and commerce in the region (see also Wong, 1991). The emergence and growth of Singapore-based MNCs and Taiwanese MNCs in South East Asia only served to reinforce that hegemonic role (see Chapters 10 and 15).

The requirement to maintain the competitiveness of their exports became an important driving force for the increasing relocation of labour intensive production from Hong Kong to Asia beginning in the late 1960s, particularly also since keen competition was emerging from other Asian NICs. Since the mid-1960s, South Korea and Taiwan had exhibited similar patterns of industrial development as Hong Kong and, in fact, had become significant producers and exporters of similar manufactured products (Lee, 1989). Their lower labour and land costs combined with government assistance provided these countries with greater comparative advantages in the production and export of labour intensive products. Thus, trade barriers, declining comparative advantage of the home country in labour intensive production, the intensifying competition between domestic firms in Hong Kong and the emergence of competition from other Asian NICs worked to make export-oriented manufacturing an increasingly important activity by Hong Kong-based MNCs in South East Asia since the 1970s, and in China around the end of the 1970s. This included the expansion of existing FDI in Singapore, Taiwan, Indonesia, Macao and Thailand and the establishment of new textiles plants around South East Asia including in Malaysia, China and in Mauritius.[11] The further expansion of international production in textiles in Taiwan had the additional advantage of challenging the emerging competition posed by the emergence of Taiwanese firms in labour intensive industries by producing directly in Taiwan where Hong Kong-based firms can similarly derive competitiveness from low wages and low land costs while capitalizing on their ownership advantages such as longer experience, higher productivity, and better management expertise (Chen, 1981). In Malaysia, Hong Kong-based MNCs were favoured by the special incentives provided by the Malaysian government to foreign firms engaged in export-oriented FDI as mentioned, while the declaration of a new and more open economic policy in China (Mun and Chan, 1986) fostered the establishment of 500 Hong Kong-based firms in that country in the early 1980s enticed by the need to overcome the severe shortage and consequent high prices of industrial land in Hong Kong and the presence of an abundant supply of labour in China which enables Hong Kong to use China as a base for the production of labour intensive products or more labour intensive production process (Chen, 1983b).[12] Indeed, Hong Kong-based firms counted as among the earliest and most significant investors in China – another manifestation of the close economic links between these two formerly separate countries attributable to their close historical ties, common ethnic and cultural heritage and geographical proximity. However, the smaller manufacturing concerns of Hong Kong-based firms in textiles, clothing, watches and electronics in the export processing zones in areas close to the Hong Kong–Macao border and in Fujian were often production arrangements in the form of subcontracting and compensation trade rather than FDI (Chen, 1981).[13] In addition, Hong Kong-based firms established export-oriented manufacturing in Mauritius during the late 1970s and early 1980s enticed by the preferential trade access of Mauritius to the European Community through the Lomé Convention (Wells, 1983; Currie, 1986).

Indeed, textiles predominated the international production of Hong Kong by the end of the 1970s. Textiles and clothing accounted for more than 50 per cent of Hong Kong-based FDI in manufacturing in Indonesia in 1980, 50 per cent of Hong Kong-based FDI in all industries in Malaysia (or 51 per cent of Hong Kong-based FDI in manufacturing) in 1979, 17 per cent of Hong Kong-based FDI in all industries in Taiwan (or 30 per cent of Hong Kong-based FDI in manufacturing) in 1981.[14] Undeniably, Hong Kong was a significant source of inward FDI in the textiles industry in some of these countries as seen, for example, in Malaysia where Hong Kong accounted for one-third of the stock of inward FDI in the textiles industry in 1979, and in Taiwan where Hong Kong accounted for 29 per cent and 44 per cent of the stock of inward FDI in the textiles industry and clothing and footwear industry respectively in 1981 (based on data in Chen, 1983b). In addition, the active participation of Hong Kong-based firms in small- and medium-sized projects in China had been crucial to the success of the export processing zones in that country. Hong Kong accounted for some 94 per cent of the realized FDI in the export processing zones in China at the end of 1981 (Chen, 1983b).

The emergence of Hong Kong-based export-oriented MNCs in the electrical and electronics industry in the 1970s had its origins in the development of the industry in Hong Kong starting from 1959 with the domestic assembly of transistor radios for Sony Corporation of Japan under subcontracting arrangements. This propelled the growth of Hong Kong's first electronics company, the Champagne Engineering Corporation, which began to assemble over 4,000 radios a month for Sony, and some 11 other locally based companies which became more competitive producers of radios relative to those of Japan by 1961. The rapid growth of domestic competition and the 15-fold growth of Hong Kong's exports of transistor radios to the United States between 1960 and 1961 led to the imposition of a ban by the Japanese government on the exports of transistors to Hong Kong which was replaced by imports from Britain and the United States (Chen, 1971). The continuing growth of production and exports through the 1960s and 1970s led to the imposition of trade barriers against the exports of electrical and electronics products of Hong Kong around the mid-1970s (Wells, 1978). This in turn triggered a strong 'centrifugal force' (Yeung, 1995) as an increasing number of Hong Kong-based firms emerged as MNCs by the late 1970s. Trade barriers and the minimization of production costs became the primary determinants of the emergence and growth of export-oriented FDI in the electrical and electronics products industry starting in the 1960s particularly in Malaysia, Taiwan and Singapore, and became the second most important industry of international production of Hong Kong by the end of that decade.

Electrical and electronics accounted for 8 per cent of Hong Kong-based FDI in all industries in Malaysia in 1979 and 6 per cent of Hong Kong-based

FDI in all industries in Taiwan (or 10 per cent of Hong Kong-based FDI in manufacturing) in 1981.[15] Although their significance as a source of inward FDI in the industry from the perspective of host countries was not as significant as that in textiles and clothing, Hong Kong had become a noticeable source of inward FDI particularly in Malaysia where it accounted for 11 per cent of the stock of inward FDI in the electrical and electronics industry in 1979. In addition, Hong Kong became the most dominant source of export-oriented FDI in China as well as accountable for the smaller projects organized in the form of subcontracting and compensation trade for the production of textiles, clothing, watches and electronics in the export processing zones in areas close to the Hong Kong–Macao border and in Fujian, as mentioned above (Chen, 1981). By contrast, Hong Kong accounted for less than 2 per cent of the stock of inward FDI in the industry in Taiwan in 1981 owing to the dominant role of American and Japanese MNCs (Chen, 1983b).

The ownership advantages of Hong Kong MNCs in export-oriented industries over local firms in host developing countries derive from their longer experience in production and operations, superior management skills, more advanced technologies and better connections with export markets (Chen, 1983b; Wells, 1978, 1984). Indeed, considering that the origins of the Hong Kong-based textile 'barons' could be traced as far back to the emergence of an indigenous textiles industry in Shanghai in the inter-war period (see Chapter 9), extensive production experience combined with organizational expertise in the management of large-scale textile and clothing projects are their principal competitive assets over indigenous firms in developing countries.[16] These competitive assets have proven sustainable because although some elements of their production know-how are embodied in the machinery, both production and organizational expertise are more tacit and embodied in the minds of the managers and engineers that have continually been engaged in the cumulative and interactive process of technology creation and their use in production. In the context of the evolutionary theory of innovation in MNCs of Cantwell (1995), these competitive assets have fashioned the distinctive path of technological development of Hong Kong-based firms which other firms cannot easily replicate. Even in cases when production know-how is largely embodied in machinery, Hong Kong-based firms have been able to sustain an ownership advantage since a potential new producer in a developing country would face difficulty in the use of new or second-hand machinery.[17] New machines purchased at arms-length from the international machinery market are designed for long production runs and inappropriate for production conditions in developing countries (Wells, 1978), while the know-how to use second-hand machines to the extent that these could be obtained from the poorly developed second-hand machinery market (Dilmus, 1974) also poses high risks. The advantages of Hong Kong-based firms have derived not only from their access to machinery and their familiarity with second-hand machinery but from the ability to design special pieces of equipment for low-volume production that have either been produced in Hong Kong or obtained through bulk purchase from foreign machinery manufacturers.[18]

Reinforcing the sustainable ownership advantages of Hong Kong-based firms based on longer experience in production and operations, superior management skills and more advanced technologies is the relationship with foreign buyers with which the firms have established a reputation for reliability. This has proven to be the most important competitive tool of Hong Kong-based export-oriented firms *vis-à-vis* indigenous firms in developing countries further down in the pecking order even as these firms have over time developed a greater comparative advantage in labour intensive production (Wells, 1978).[19] Indeed, Hong Kong-based firms employ a variety of marketing channels and are far are more sophisticated in their marketing skills than could be expected of a developing country. Besides selling to regular clients and buying groups abroad, Hong Kong-based firms advertise in trade journals, participate in trade fairs, have their own sales/marketing team, and a high proportion of firms have brand names of their own (Chen, 1983b).

Hong Kong-based MNCs in export-oriented industries have also been able to sustain competition posed by other foreign-based MNCs in host developing countries due to their better understanding of production conditions in developing countries, lower costs for managerial and technical staff, greater flexibility and adaptability, and closer language and cultural affinity (Chen, 1983b). Such differences in the proprietary or firm-specific advantages of developing country firms *vis-à-vis* developed country firms can be understood within the framework of the theory of localized technological change advanced by Lall (1981, 1982, 1983b, 1983c). In this framework, the advantages of firms based in developing countries derive from their ability to innovate on essentially different lines from those of the more advanced countries, i.e. innovations that are based on lower levels of research, technology, size and skills. Based on the evolutionary theory of technological change of Atkinson and Stiglitz (1969), Nelson and Winter (1977, 1982) and Arthur (1988, 1989), the theory of localized technological change argues that technical change is firm specific, path dependent and irreversible because older technologies cannot be efficiently reproduced or transferred once an entire industry has progressed to new technologies and become firmly established. This helps to explain why developed country firms whose competitive assets derive from 'frontier' technologies, large-scale production and sophisticated marketing cannot replicate without high costs and risks the competitive assets of developing country firms based on widely diffused technologies, small-scale production, special knowledge of marketing relatively undifferentiated products or special managerial or other skills. The lower production costs of developing country firms derive from the use of more appropriate production techniques that are more labour intensive and smaller scale responsive to the factor conditions and market sizes of developing countries, lower costs of expatriate managers and technical specialists or lower building costs (Wells, 1973, 1978; Lecraw, 1976). Owing to their lower costs of production, Hong Kong-based firms have succeeded in gaining world market shares through price competition (Wells, 1973). These assets have not always been sufficient to sustain international production as shown in the joint ventures concluded by Hong Kong-based firms with firms from the developed countries to fulfil their need for technical assistance, overcome the lack of an established trade name and limited consumer marketing skills. Such is the case, for example, with the 45 per cent equity interest of a Japanese synthetic fibre manufacturer in a large Hong Kong-based textiles firm which had the objective of enabling the Japanese firm to expand its operations as part of a vertically integrated network throughout South East Asia and enabling the Hong Kong-based firms to gain access to synthetic fibres produced by the Japanese firm for its spinning and weaving operations (Wells, 1978).

Import substituting FDI in manufacturing in developing countries

Although export platform manufacturing FDI in developing countries has been the most important type of outward FDI of Hong Kong-based firms in the period until the end of the 1970s in industries in which these firms capitalize on their extensive production experience particularly in textiles and clothing, its outward FDI in manufacturing in developing countries also encompass international production geared to serve host country markets in which foreign plants had been established. Indeed, there is evidence to suggest that the earliest FDI by Hong Kong had been of this type. In the late 1950s as Hong Kong-based firms found their exports of simple manufactured items such as kerosene lanterns, flashlights and umbrellas threatened by tariffs and quotas designed to encourage local production in low income developing countries, Hong Kong-based firms initiated outward FDI in those very countries previously served through exports in order to defend markets gained through exports (Wells, 1978). Such investments grew in the 1960s and 1970s as tariff barriers associated with the pursuit of import substitution industrialization was imposed by many countries of South East Asia (UN, ESCAP, 1988).

Since the 1960s and 1970s import substituting outward FDI by Hong Kong-based firms had also begun to be significant in industries that had been of limited or no significance to the economy of Hong Kong owing to its limited size, lack of natural resources or because of environmental considerations (as in the case of chemicals) and thus domestic based firms had little or no domestic production experience in these industries. Thus, unlike the defensive export platform FDI propelled by defensive considerations to overcome trade barriers and high domestic production costs by international production in developing countries that had not been affected by trade barriers or affected less severely, or that have lower land costs and wages, import substituting manufacturing FDI has been driven by profits to be derived from the exploitation of comparative advantages of host countries and to serve and develop local market demand in host countries. This second type of import substituting manufacturing FDI represents an aggressive strategy adopted by Hong Kong-based firms in joint investment ventures with local firms in host countries by which new markets and new lines of activities were developed for their parent firms in Hong Kong and an assured supply of raw materials for the plastics and furniture industries in Hong Kong was secured (Chen, 1981, 1983a, 1983b).

This type of FDI has concentrated in different kinds of industries. The production of metal products, chemicals, food, minerals and metals and basic metals and paper accounted collectively for some 50 per cent of Hong Kong's FDI in manufacturing in Indonesia in 1980, while the production of food, drink and tobacco, chemicals, wood, fabricated metals, rubber and a whole range of other industries apart from textiles and electrical and electronics which accounted collectively for some 40 per cent of Hong Kong's FDI in all industries in Malaysia (or more than 40 per cent of FDI in manufacturing) in 1979. In Taiwan, Hong Kong-based firms have focused on chemicals, plastics and rubber, pulp and paper, basic metals and metal products, non-metallic minerals, and machinery and equipment and a range of other industries apart from textiles and electrical and electronics that accounted collectively for some 31 per cent of Hong Kong's FDI in all industries (or 57 per cent of FDI in manufacturing) in 1981.[20]

In some host countries such as Indonesia, Hong Kong's FDI in the import substituting manufacture of chemicals, food, clothing and printing has gained increasing importance over time and became the more dominant pattern of Hong Kong's manufacturing FDI over the 1980s. This development owed largely to the objective of the Indonesian government to support the growth of infant domestic firms in export-oriented, labour intensive production for which Indonesia had been developing a comparative advantage. For that reason, the growth of export-oriented manufacturing by Hong Kong-based firms in Indonesia in such industries as textiles, electronics, watches and clocks, toys and plastics began to wane in the 1970s.

Services FDI in developing countries

As mentioned, services was the second most important economic sector of Hong Kong-based MNCs at the end of the 1970s. The sector accounted for as much as 29.9 per cent of the stock of Hong Kong-based FDI in Indonesia as of 1976, and 43.8 per cent of the stock of Hong Kong-based FDI in Taiwan as of 1981. It was only in Malaysia where services accounted for a meagre 2 per cent of the stock of Hong Kong-based FDI at the end of 1979 (based on data in Chen, 1981, 1983b). Although data is not available on inward FDI in Singapore, it is probable that the the share of Hong Kong's FDI in manufacturing declined by the end of the 1970s, with the growth in importance of services (Chen, 1983b).

This type of FDI has concentrated in different kinds of industries in different host countries. Hong Kong-based FDI in services was concentrated in trade/ hotels, construction and leisure services in Indonesia in 1976, transportation, services, banking and insurance, and trade in Taiwan in 1981, construction, tourism and trade in Singapore at the end of the 1970s, and in construction and tourism in China at the end of the 1970s organized in the form of large projects in joint investment ventures with local firms. In addition, what little services FDI there was from Hong Kong in Malaysia in 1979 was directed largely to hotels and tourism (based on data and information in Chen, 1983b).

The importance of services in Hong Kong-based FDI can be explained in the context of the process of deindustrialization and industrial restructuring in the Hong Kong economy since the early 1970s which has transformed the British colony into a leading trading and financial centre in the region (Ho, 1992). Indeed, although manufacturing still accounted for 27.6 per cent of gross domestic product at factor cost in 1979, the diversification of manufacturing companies into property and other non-manufacturing activities combined with the relocation of their manufacturing activities to cheaper production sites in South East Asia and China is serving to further enhance the dominant role of services in the industrial structure of Hong Kong.

Outward FDI to secure supplies of essential raw materials

Outward FDI of this type is the fourth most important type of Hong Kong-based FDI at the end of the 1970s. The lesser importance of outward FDI of this type was associated with the position of the primary sector as the least most important economic sector of Hong Kong-based MNCs at the end of the 1970s. The sector accounted for 18.2 per cent of the stock of Hong Kong-based FDI in Indonesia as of 1976. In the Philippines, some 58 per cent of Hong Kong based-FDI in the period between 1978 and 1982 was in agriculture, thus emphasizing the sourcing of raw materials as a major motive for Hong Kong-based FDI in that country (ESCAP/UNCTC, 1986). However, the resource-rich country of Indonesia was the most important host country for Hong Kong-based FDI of this type in all primary sector industries but with a particular concentration on forestry and agriculture. The investments to extract timber in Indonesian or Malaysian Borneo – the main sources of Hong Kong's hardwood – had several determinants. A few Hong Kong-based firms established foreign resource extractive operations to obtain timber as a raw material for the furniture manufacturing industry of Hong Kong whose origins could be traced back to the production by highly skilled artisan labour of traditional carved wooden furniture for both domestic and export markets since the early 1930s (Cooper, 1981).

While to some extent this type of FDI represented the backward vertical integration of manufacturing companies in resource extraction as an important means to stabilize supplies and costs associated with the instability of market prices for tropical timber (Wells, 1978), in some cases the timber extractive investments in Indonesia was accounted for by individuals or business groups which although based in Hong Kong either do not have parent companies in Hong Kong or have parent companies in Hong Kong in entirely different lines of business (Chen, 1981). These independent firms would be equivalent to the 'free-standing companies' coined by Wilkins (1986a, 1988a) in referring to the thousands of British companies established prior to 1914 which did not undertake any prior production in the United Kingdom before investing abroad but were registered in the United Kingdom and floated on the London capital market primarily for the purpose of undertaking business exclusively or mainly abroad. In the case of Hong Kong, the resource extractive FDI in timber were mostly accounted for by investment companies based in Hong Kong that treat FDI as form of portfolio investment (Chen, 1981; Wells, 1978) to gain monopoly rents or minimize business risks through the diversification or entry into new business activities.

The role of ethnic Chinese business contacts in the establishment and maintenance of Hong Kong-based FDI of this type was just as important, if not more important than that of other types of Hong Kong-based FDI in South East Asia and China. In Indonesia in particular, the investments in timber extraction consisted of informal joint ventures with Indonesian military officials or with Indonesians of Chinese extraction, both of which enabled the investments not to be registered as foreign and thus circumventing the official prohibition of foreign participation in most timbering operations in Indonesia (Wells, 1978).[21]

FDI in developed countries

Outward FDI of this type is the fifth and least most important type of Hong Kong-based FDI at the end of the 1970s. There are several determinants of Hong Kong-based FDI of this type. Perhaps the most important was for the purpose of acquiring advanced foreign technologies and to obtain more secure access to the parts, components and semi-manufactures necessary for the assembly production in Hong Kong of some electronics products and watches. The foreign subsidiaries established by Hong Kong-based firms in Switzerland for the manufacture of watches, in the United States for the production of watches and electronics (to include the acquisition by the Hong Kong-based firm, Stelux, of 29 per cent of the United States-based Bulova Watch Company in 1976) and in the United Kingdom for the production of textiles (Chen, 1981, 1983b) are excellent cases in point of outward FDI of Hong Kong-based FDI in developed countries driven by these determinants. Outward FDI in these cases had been largely in the form of acquisitions which enabled Hong Kong-based firms to gain rapid market access and benefit from the established reputation (or brand name), production facilities, marketing outlets and other assets of their acquired firms.

Another determinant of outward FDI of this type was the horizontal integration of manufacturing companies in their major export markets in the developed countries for the purpose of defending markets for goods made in

Hong Kong or other low-wage production sites by Hong Kong-based firms in the face of tariffs or to save on high shipping costs. An example of this was the outward FDI by Hong Kong-based furniture manufacturers that established final assembly plants for furniture in the United States with components shipped from Hong Kong or from foreign affiliates of Hong Kong-based firms. Apart from overcoming the higher shipping costs and tariffs on furniture, the establishment of the final assembly plant facilitated the determination of evolving consumer requirements for furniture in the United States. Another example was the acquisition of production facilities in England by a Hong Kong-based textile manufacturer for the purpose of obtaining the marketing channels of a failing British company. In both these examples, the assembly, production and marketing outlets established by Hong Kong-based firms in the developed countries enabled the firms to exploit more fully their advantages in low-cost production in South East Asia (Wells, 1978).

Other Hong Kong-based FDI in developed countries are determined by non-economic factors. For example, some Hong Kong-based investments in the textiles industry of Canada have claimed that the principal determinant of the investment was to acquire Canadian citizenship (Chen, 1983b).

The growth of Hong Kong-based MNCs in the 1980s

The 1980s was an era of rapid growth of Hong Kong-based outward FDI. In terms of actual FDI outflows based on balance-of-payments data, the annual average FDI outflows of Hong Kong was about $2.4 billion in the period between 1986 and 1991 compared to that of Taiwan of $3.2 billion and that of South Korea of $923 million. The estimated stock of Hong Kong-based FDI reached $13.2 billion in 1990 which when compared to its stock of $148 million in 1980 represents an annual average rate of growth of 57 per cent. Thus Hong Kong by 1990 had surpassed even Singapore as a home country of FDI whose stock of outward FDI reached $7.8 billion (Department of Statistics, 1996). Indeed, Hong Kong by 1990 had become a significant home country of FDI with a share of some 0.8 per cent of the global stock of outward FDI (based on data in UNCTAD, 1999). In fact, the 1980s may have marked the decade in which Hong Kong became a net outward investor in terms of FDI flows. Between 1984 and 1988, the annual average FDI outflows of Hong Kong at $2.5 billion was higher compared to that of FDI inflows at $1.8 billion (based on data in UNCTC, 1992). This trend was sustained in the later years between 1986 and 1991 when the annual average FDI outflows of Hong Kong at $2.4 billion was higher compared to that of its FDI inflows over the same period at $1.7 billion (based on data in UNCTAD, 1999).

Based on the stock of Hong Kong-based inward FDI in host countries, manufacturing remained the most important sector of Hong Kong-based FDI at the end of the 1980s, followed by the services sector and the primary sector.[22] Manufacturing accounted for 100 per cent of Hong Kong-based FDI in Malaysia as of 1987, 54 per cent of the stock of Hong Kong-based FDI in Taiwan as of 1989, 46.5 per cent of Hong-Kong based FDI in the Philippines in the period between 1981 and 1989, and 29.5 per cent of Hong-Kong based FDI in Thailand in 1987 (based on data in Wong, 1992). The principal destinations of Hong Kong-based FDI at the end of the 1980s was China and South East Asia particularly Indonesia, Thailand, Taiwan, Singapore, South Korea, the Philippines and Malaysia (see also World Bank, 1989; Wong, 1992 and Yeung, 1994).[23] The growth since the mid-1980s of Hong Kong FDI in the Asian region, including China combined with the parallel growth of FDI in the region by the other Asian newly industrialized countries – Singapore, South Korea and Taiwan – as well as Japan have helped to create an intra-regional dimension to trade and FDI that is increasingly becoming more emphasized over time. However, although most of Hong Kong-based FDI at the end of the 1980s was in the Asia-Pacific region, some Hong Kong-based firms have also invested in Mauritius, Mexico and in countries of the Organization of Economic Cooperation and Development (World Bank, 1989) to include Canada, Switzerland, the United Kingdom and the United States (see also Wong, 1992).

Indeed, Hong Kong-based MNCs had continued to exert some significance in their principal host countries as shown by their share in inward FDI at or around the end of the 1980s. In China, Hong Kong was the largest source of FDI (World Bank, 1989) including larger projects organized in the form of joint investment ventures with local firms, and was responsible for a wide range of non-equity production arrangements including subcontracting, compensation trade and joint production.[24] In Thailand, Hong Kong accounted for some 14 per cent of the flows of inward FDI between 1980 and 1990, and was the third-largest source of FDI after Japan and Singapore.[25] In Indonesia, Hong Kong accounted for 11 per cent of the stock of inward FDI as of 1979, and was the second-largest source of FDI after Japan – a position essentially unchanged by 1990.[26] In the Philippines, while Hong Kong accounted for a smaller share of a little more than 4 per cent of the stock of inward FDI as of 1980 owing to the far more dominant role of the United States and Japan, by 1991 Hong Kong accounted for 6.3 per cent of the stock of inward FDI despite the fact that the United States and Japan remained the dominant sources of inward FDI in that country.[27]

It was perhaps only in Malaysia and Singapore that the relative importance of Hong Kong as a source country of FDI declined between the end of the 1970s and the end of the 1980s. In Malaysia, although Hong Kong accounted for 11 per cent of the stock of inward FDI as of 1979, and was the fourth-largest source of FDI after Japan, Singapore and the United Kingdom, it accounted for less than 5 per cent of the flows of inward FDI between 1980 and 1990, and thus became the sixth-largest source of FDI after Taiwan, Japan, Singapore, the United Kingdom and the United States.[28] Similarly, although Hong Kong accounted for 12 per cent of the stock of inward FDI in Singapore as of 1980, and was the third-largest source of FDI after the United Kingdom and the United States, by 1989 Hong Kong accounted for less than 7 per cent of the stock of inward FDI and was only the fourth-largest source of FDI after Japan, the United States and the United Kingdom.[29]

As at the end of the 1970s, there were five types of outward FDI undertaken by Hong Kong-based MNCs that could be identified around the end of the 1980s. In order of declining importance, these were: first, export platform manufacturing FDI in developing countries; second, import substituting FDI in manufacturing in developing countries; third, services FDI in developing countries; fourth, FDI in developed countries and fifth, outward FDI to secure supplies of essential raw materials. Thus in the decade of the 1980s resource extractive FDI became the least most important type of Hong Kong-based FDI. The discussion below focuses on the first and fourth types of Hong Kong-based FDI where important changes have taken place in the 1980s.

Export platform FDI in manufacturing in developing countries

This FDI type remained the most important type of Hong Kong-based FDI at the end of the 1980s. The determinants of this type of international production in previous decades stemming from trade barriers, declining comparative advantage of the home country in labour intensive production, the intensifying competition between domestic firms in Hong Kong and the emergence of competition from other Asian NICs continued to be relevant in explaining the pre-eminent role of export-oriented manufacturing by Hong Kong-based MNCs in China and in South East Asia throughout the 1980s.

Export-oriented production by Hong Kong-based firms in China of various forms in the 1980s have been concentrated in highly labour intensive assembly production of mainly travel goods, handbags, toys and footwear of which some 80 per cent were shipped back to Hong Kong as re-exports. The much faster annual rate of growth of re-export trade from China to Hong Kong at around 28 per cent per year in relation to Hong Kong's own manufactured exports at around 2 to 3 per cent per year made re-export trade the most dynamic force in the export-led growth of Hong Kong in the 1980s (Census and Statistics Department, *Hong Kong External Trade*, March 1991). The large-scale relocation of production to China by Hong Kong's labour intensive manufacturers and the further expansion of existing investments and production arrangements in that country had been motivated largely by efforts to expand the volume of exports of labour intensive goods for which Hong Kong's firms have a competitive advantage in combination with the lower land costs as well as abundant labour in China.

By comparison, export platform manufacturing FDI by Hong Kong-based MNCs in South East Asia have tended to display a wider product range, including higher value added and technologically more sophisticated products such as electronics and electrical appliances in addition to more standardized products (World Bank, 1989). Despite the far more dominant role of China in Hong Kong's export platform FDI in manufacturing in the 1980s, the continuing trade privileges enjoyed by many countries of South East Asia under the Generalized System of Preferences (with the exception of Singapore) combined with the gradual shift in Hong Kong's emphasis from the production of cheap goods to higher end products requiring higher skilled workforce, the political upheavals in China as exemplified in the pro-democracy crackdown in 1989, the difficulties of doing business in China including the lack of basic infrastructure and the presence of bureaucratic red tape (Chia, 1989) and the continuing threat to China's most-favoured-nation trading status with the United States are factors that would ensure that South East Asia remains an important location for Hong Kong-based FDI of this type.[30]

Indeed, the magnitude of Hong Kong-based FDI of this type and the high degree of multinationality of Hong Kong-based MNCs involved was evident in the size of foreign employment in relation to domestic employment in Hong Kong. Around the end of the 1980s, Hong Kong-based firms employed around three and one-half times more people abroad than in Hong Kong which is probably the highest degree of internationalization of production seen in any home country.[31]

FDI in developed countries

Hong Kong-based FDI in the developed countries increased over the 1980s propelled by the same set of determinants as that in the period until the end of the 1970s. This type of Hong Kong-based FDI had increased considerably over the 1980s with the uncertainty surrounding the return of Hong Kong to mainland China in 1997. Outward FDI in the manufacturing sector in particular was motivated by a continuing interest in preserving market access, improving industry reputation and gaining access to established brand names, and upgrading technology (World Bank, 1989). Market diversification propelled the growth of Hong Kong-based FDI in the United States and in Europe in view of the growing trend toward regional economic integration in the North America Free Trade Agreement and the Single European Market, while the FDI by Hong Kong-based clothing manufacturers in Canada, for example, had been geared to overcome quotas restrictions imposed by the United States against exports of clothing from Hong Kong as well as to benefit from NAFTA (Wong, 1992). Indeed, the continuing trend towards trade protectionism would continue to influence the expansion of Hong Kong-based FDI directly in the United States and Europe or indirectly through Hong Kong-based FDI in third countries.

Perhaps the most notable Hong Kong-based FDI in manufacturing in the developed countries in the 1980s was the acquisition in 1989 by Semi-Tech Microelectronics (Far East) Ltd of Singer Sewing Machine Company, the owner of the Singer brand name and a vast network of sewing machine factories and consumer durable shops.[32] Indeed, the Singer Sewing

Machine Company highlighted in Chapter 2 of this book as a special case study of one of the earliest American manufacturing MNCs was no longer an American-owned company by 1989.

However, in the 1980s Hong Kong-based FDI in the services sector of developed countries also grew in a major way particularly in banking and financial services, hotels, trading and property. Indeed, owing to its welcoming attitude to Hong Kong-based investors and the enticement of investment-linked immigration, Canada was an important destination of some of the capital flight from Hong Kong prior to 1997 directed primarily into property.[33] More importantly, there had been major acquisitions by large and prominent Hong Kong-based firms in the developed countries and in the United States and the United Kingdom in particular in the period since 1986 and this involved mainly the services sector (banking and finance, hotels and distribution).[34]

The growth of Hong Kong-based FDI in services more generally and in developed countries in particular which have increased in importance since the mid-1980s is indicative of Hong Kong's historical position as a British colonial entrepôt facilitated by the local establishment of large British companies in the banking and finance, trading and services industries starting in 1841 with the inception of Hong Kong's colonial history. The eventual deindustrialization of Hong Kong's economy with the emphasis towards the development of the services sector have led to its evolution as an international financial centre which had positive knock-on effects on the development of some local firms (Chen, 1989). The growth of significant outward FDI in banking and finance thus resulted from the position of Hong Kong as an investment base for many foreign-controlled corporations such as the Hong Kong and Shanghai Banking Corporation and Jardine Matheson as well as the home base of many locally controlled and indigenously Chinese companies such as Y. K. Pao Group and Li Kashing Group that equally became significant investors abroad in the banking and finance industries (see UN, TCMD, 1993a).

The further expansion of Hong Kong-based MNCs in the 1990s

The 1990s represented a continuation of the rapid growth of Hong Kong-based outward FDI. In terms of actual FDI outflows based on balance-of-payments data, the annual average FDI outflows of Hong Kong was about $20.3 billion between 1992 and 1998, an amount more than eight times the annual average FDI outflows of $2.4 billion in the period between 1986 and 1991. The estimated stock of Hong Kong-based FDI reached $154.9 billion in 1998 which when compared to its stock of $13.2 billion in 1990 represents an annual average rate of growth of 36 per cent – a rapid rate of growth indeed but not as rapid as the 57 per cent annual average rate of growth registered between 1980 and 1990. Indeed, Hong Kong by 1998 had become an even more significant home country of FDI with a share of some 3.8 per cent of the global stock of outward FDI compared to its 0.8 per cent share in 1990 (based on data in UNCTAD, 1999).

The pattern of outward FDI of Hong Kong in the 1990s remained essentially similar to that pertaining in the 1980s. Thus, there had been relative stability since the emergence of Hong Kong-based MNCs in the role of manufacturing as the pre-eminent economic sector of Hong Kong-based FDI at the end of the 1990s, followed by the services sector and the primary sector. The principal destinations of Hong Kong-based FDI at the end of the 1980s was China and South East Asia, particularly Indonesia, Thailand and Singapore (see also Yeung, 1995). However, although most of Hong Kong-based FDI at the end of the 1990s as in decades past was in the Asia-Pacific region, some Hong Kong-based firms have also invested in Mauritius, Mexico and in the developed countries.

Indeed, Hong Kong-based MNCs had continued to exert some significance in their principal host countries as shown by their share in inward FDI in the 1990s. In China, Hong Kong accounted for 61 per cent of the stock of inward FDI in 1993, and China in turn accounted for some 75 per cent of total outward FDI stock from Hong Kong (Low *et al.*, 1995). In Thailand, Hong Kong accounted for some 11 per cent of the stock of inward FDI in 1994, and was the second-largest source of FDI after Japan (based on data in Yeung, 1995).

However, the relative importance of Hong Kong in the inward FDI of Malaysia, Singapore and Indonesia declined in the 1990s. In Malaysia, although Hong Kong accounted for 11 per cent of the stock of inward FDI as of 1979, it accounted for less than 4 per cent of the stock of inward FDI in 1992, and thus was only the sixth-largest source of FDI after Japan, Singapore, Taiwan, the United Kingdom and the United States.[35] Similarly, although Hong Kong accounted for 12 per cent of the stock of inward FDI in Singapore as of 1980, it accounted for less than 7 per cent of the stock of inward FDI in 1991 and thus was only the fifth-largest source of FDI after Japan, the United States and the United Kingdom and the Netherlands.[36] In Indonesia, although Hong Kong accounted for 11 per cent of the stock of inward FDI in non-oil sectors as of 1979, it accounted for 8 per cent of the stock of inward FDI as of 1994, and thus was the third-largest source of FDI after Japan and Taiwan.[37]

As in decades past, there were five types of outward FDI undertaken by Hong Kong-based MNCs that could be identified at the end of the 1990s. In order of declining importance, these were: first, export platform manufacturing FDI in developing countries; second, import substituting FDI in manufacturing in developing countries; third, services FDI in developing countries; fourth, FDI in developed countries and fifth, outward FDI to secure supplies of essential raw materials. Thus the types of Hong Kong-based FDI and their relative importance in relation to one another not only remained essentially stable in the 1980s and 1990s, but was perhaps reinforced in the 1990s. The brief discussion below focuses on some aspects of first and third types of Hong Kong-based FDI where important changes have taken place in the 1990s particularly with reference to the expansion of Hong Kong-based FDI in China.

Export platform manufacturing FDI in developing countries

The dominant role of this type of Hong Kong-based FDI continued in the 1990s, precipitated by the high rates of GDP growth in Hong Kong since the recession of 1985 which have in turn led to high inflation rates, rising wages and an increasingly tight labour market made worse by high rate of emigration owing to uncertainties surrounding the return of Hong Kong to China in 1997.[38] Ironically, the surge of export platform manufacturing FDI by Hong Kong-based firms was directed towards China with the result that the number of direct and indirect employment accounted for by Hong Kong-based manufacturers in Guangdong province reached 3 million by 1992 (Federation of Hong Kong Industries, 1992).

Services FDI in developing countries

With the further commitments made by China to economic reform in the early 1990s, including opening its doors wider to foreign investment, and the realization of the link between deeper capital commitments in China and the viability of their domestic business in Hong Kong, major Hong Kong-based companies have been enticed for the first time in the 1990s to invest in the development of property and public utilities in China including roads, railway networks, power plants, port development and management and telecommunications.[39] This is for the purpose of assisting China to accumulate the basic social capital that is a prerequisite to private productive investments, to forge stronger and closer geographical and economic links between Hong Kong and the heartland of China and to ensure Hong Kong's role in China's long-term development.

Conclusion

This chapter examined the history of Hong Kong-based MNCs over the last half century. The major types of FDI undertaken by these firms, the determinants of outward FDI and their industrial and geographical patterns have remained essentially stable since the period of their emergence until the end of the 1990s. Five types of outward FDI undertaken by Hong Kong-based MNCs were evident around the end of the 1970s. In order of declining importance, these were: first, export platform manufacturing FDI in developing countries; second, import substituting FDI in manufacturing in developing countries; third, services FDI in developing countries; fourth, outward FDI to secure supplies of essential raw materials; and fifth, FDI in developed countries. By the end of the 1980s, these five major types of Hong Kong-based FDI remained valid except that the relative importance of the fourth and fifth FDI types had been transposed owing to the faster growth of outward FDI in developed countries which led to this type of FDI becoming the fourth most significant type of outward FDI and resource extractive FDI becoming the fifth and least most significant type. Such pattern in Hong Kong-based FDI was sustained until the end of the 1990s.

The stock of outward FDI based in Hong Kong reached $154.9 billion in 1998 which represents some 3.8 per cent of the global stock of outward FDI. Indeed, Hong Kong is almost as important a home country for FDI as Canada whose outward FDI stock was $156.6 billion in that year. Hong Kong thus ranked as the tenth-largest source country of FDI in the world economy after the United States, United Kingdom, Germany, Japan, the Netherlands, France, Switzerland, Italy and Canada. The fact that its international production activities remained predominantly in the export-oriented manufacture of labour intensive products has not, in any way, made Hong Kong-based MNCs less successful controllers and coordinators of an international network of income-generating assets. Their mastery of technologies and long production experience in fairly narrow and defined areas of activity – the assembly and manufacture of light consumer goods for exports whose production are suited to lower-cost and labour-abundant developing countries – are strongly reinforced by their large reservoir of entrepreneurial and export marketing know-how and access to global distribution channels. Many of their firms have developed long-term, reliable and stable supplier relationships with their customers in their export markets (Wells, 1978, 1984). In such almost perfectly competitive industries where competitive advantage is based solely on price competition, Hong Kong-based firms have erected a powerful barrier to entry in export marketing, relationships with suppliers and access to global distribution channels. It is these managerial and marketing advantages rather than the possession of strong technological advantages that have given Hong Kong-based MNCs a unique and distinctive competitive edge to sustain a leading role in overseas labour intensive manufacturing over time and in spite of the growth of other developing countries that have developed greater comparative advantages in labour intensive manufacturing.

The pattern and growth of domestic industrial development by Hong Kong has been a cause and effect of the pattern and growth of their MNCs. The lack of explicit support by the government of Hong Kong for domestic industrial diversification and technological development has had three important implications. The first is deindustrialization associated with the declining contribution of the manufacturing sector to gross domestic product from 27.7 per cent in the period between 1970 and 1974 to 22.4 per cent in the period between 1980 and 1986 (based on data provided in Hong Kong Government, *Estimates of Gross Domestic Product*) associated partly with the relocation of manufacturing activities abroad through FDI. The second implication arises from the first and derives from the increasing dominance of the services sector in gross domestic product from 70.4 per cent in the period between 1970 and 1974 to 76.8 per cent in the period between 1980 and 1986 (based on data provided in Hong Kong Government, *Estimates of Gross Domestic Product*). Such growing dominance of services in domestic economic activity is also reflected in the increased importance of Hong Kong-based outward FDI in services. The third implication is the stable pattern of Hong Kong-based FDI with the persistently pre-eminent role of labour intensive production in a manner not observed in any other country including Switzerland and Singapore, both small-sized countries with similar natural resource scarcity where there had been upgrading and deepening of the domestic industrial structure. To the extent that some industrial upgrading in manufacturing did occur in Hong Kong it revolved around the traditional industries of textiles and clothing in which domestic firms shifted in specialization from low- to high-quality products, but not in the development of heavy and chemical industry nor more advanced consumer durable or producer goods industries. The constraint imposed by the lack of domestic industrial upgrading has perhaps placed greater pressure on the internationalization of production by Hong Kong than a similar economy of an equally small size but with greater technological depth and which has managed to constantly upgrade to higher value added manufacturing and services industries.

In the labour intensive industries in which Hong Kong has been losing comparative advantage throughout the course of its history owing to trade barriers, rising production costs and shortages of domestic labour supply, international production has enabled Hong Kong-based firms to continue to reap the benefits of their expertise in the management and organization of labour intensive production and to maintain access to their export markets. But just as the limited domestic industrial development has not led to the evolution of international production of Hong Kong in more complex manufacturing activities, outward FDI as such has not also affected domestic industrial development of the Hong Kong economy in a significant way. This is despite the high degree of internationalization of Hong Kong-based firms and the increasing trend towards the relocation of labour intensive industries with the loss of comparative advantage. The limited evidence in support of the role of outward FDI in domestic industrialization is found in the clothing companies from Hong Kong whose domestic production have become geared increasingly towards flexible, higher quality and higher value added production for established export markets in Europe and North America on the basis of the supply of fabrics from their spinning and weaving subsidiaries in Taiwan. In addition, the closely integrated networks of production that had been established between parent firms in Hong Kong and their subsidiaries/suppliers in China have enabled the more sophisticated and complex parts of the production processes to be conducted in Hong Kong and the more labour intensive stages of the production processes to be conducted in China. It is perhaps their less significant outward FDI in import substituting manufacturing and resource extraction that enabled Hong Kong-based firms to develop new areas of expertise that could not have otherwise occurred owing to the limited size and resource base of the Hong Kong economy and the narrow industrial development policies that it has pursued. Perhaps this may change with efforts made in the 1990s to increase government collaboration with industry to foster domestic industrial and technological development.[40]

Notes

1 The analysis in this paragraph is based on data contained in UNCTAD (1999).
2 Many of the British trading companies in Hong Kong such as Jardine, Swire, and Wheelock Warden have invested abroad in manufacturing (Chen, 1983b). In addition, the Hong Kong and Shanghai Banking Corporation has initiated major foreign acquisitions in the 1980s and 1990s through its holding company, HSBC Holding plc (UN, TCMD, 1993a). Estimates by Chen (1983a, 1983b) show that outward FDI by Hong Kong Chinese firms (including joint ventures with non-Chinese partners) in manufacturing would have probably accounted, by a very rough estimate, for 33 to 44 per cent of Hong Kong's total overseas FDI (including investment in both manufacturing and services) in 1981.
3 Indeed, by the 1930s Shanghai had already developed a 'modern' cotton textile industry which as noted in Chapter 9 of this book determined the expansion of international production of Japan in cotton spinning in China in the inter-war period. The manufacturing base of Hong Kong from its inception had almost entirely been geared towards production for world markets. This is an industrialization process unique to Hong Kong and observed in few other countries including Singapore (Henderson, 1989). Although also a country of small size, Singapore underwent a phase of industrialization based on import substitution, however brief (see Wong, 1979).
4 This is not to say that textiles was the earliest industry to develop in Hong Kong. Since the early 1930s, traditional carved wooden furniture had been produced for both domestic and export markets by highly skilled artisan labour, and basic cotton clothing was made for export to other parts of the British empire in Asia, particularly to Malaysia (Cooper, 1981; Mok, 1969). Such manufactures, however, accounted for a small share of Hong Kong's exports (Henderson, 1989). In addition, some plastics and metals firms were established in the late 1940s and 1950s. Although these industries were export oriented, the export propensity was relatively lower in comparison to textiles, clothing and electronics (Chen, 1983b). Thus, Hong Kong had a head start among the Asian newly industrialized countries in export-oriented industrial development. By comparison, as evident in Chapters 10 and 11 of this book, the process of export-oriented industrialization in Taiwan and South Korea was initiated in a major way in the early 1960s. See also Deyo (1987) and Gold (1986).
5 This statement had been deduced based on available host country data provided in Chen (1981, 1983b) of the stock of Hong Kong-based inward FDI in Indonesia, Malaysia and Taiwan at or around the end of the 1970s. The percentage shares of the primary, secondary and tertiary sectors in Hong Kong-based FDI in Indonesia as of 1976 was 18.2 per cent, 51.9 per cent and 29.9 per cent, respectively. The percentage shares of the secondary and tertiary sectors in Hong Kong-based FDI in Malaysia at the end of 1979 was 98 per cent and 2 per cent, respectively. The percentage shares of the primary, secondary and tertiary sectors in Hong Kong-based FDI in Taiwan as of the end of 1981 was 1.2 per cent, 55 per cent and 43.8 per cent, respectively.

6 Data based on Yoshihara (1976) show that at least 85 per cent of total Hong Kong-based FDI in Singapore in 1973 was in food and beverages, textiles and clothing, chemicals and electrical products and electronics. And according to Wells (1978), Hong Kong-based firms were responsible for more than 50 factories in one export area alone of Singapore in 1977.

7 In a study of the sectoral pattern of the growth of exports and outward FDI in the developed countries, Cantwell (1989a) found a greater correlation in the sectoral distribution of the growth of exports and outward FDI for the relatively newer investors such as Germany and Japan. By comparison, there is a much greater disparity in the sectoral distribution of growth of exports and outward FDI and, hence, greater scope for substitution between trade and outward FDI for the more mature and established international investors such as the United Kingdom and the United States.

8 Despite the seemingly wide geographical diversity of Hong Kong-based FDI, the bulk is concentrated in South East Asia and China where Hong Kong-based MNCs can profit from Chinese business networks and entrepreneurs. The role of the 'Chinese' factor in explaining the internalization of Chinese business firms and that of Chinese business networks on the location of international production was examined by Redding (1990, 1991) and Yeung (1994). The presence of such business networks had helped to reduce the pyschic distance of international production particularly during the period of emergence. For example, a Hong Kong-based toy manufacturer set up its first foreign operation in 1951 in Singapore encouraged by the chairman's personal friend who emigrated to Singapore during the communist takeover in mainland China. Similarly, a Hong Kong-based metal trading company participated in a minority share joint venture in Singapore on the basis of friendship and trust relations which was regarded to be far more important than any economic criteria in determining the decision to engage in international production and the location of international production. In the framework of Yeung (1994), these would be examples of inter-firm networks of business and network relationships or *guanxi* established by Hong Kong-based MNCs.

9 Data on Hong Kong, South Korea and Taiwan were based on UNCTAD (1999). The data on Singapore was obtained from Department of Statistics (1991) which indicated that the stock of outward FDI from Singapore reached S\$1,677.7 million in 1981. This was converted to US dollars using the end-of-period exchange rate in 1981 obtained from the International Monetary Fund, *International Financial Statistics Yearbook 1998*, Washington, DC: IMF.

10 During the early twentieth century, turmoil and foreign occupations in mainland China forced an outward exodus of the Chinese workforce, particularly young labourers from the southern provinces. Many of these became permanent residents in the Chinese diaspora in South East Asia (Wong, 1991). The contacts and the ethnic ties by the Chinese business community in South East Asia determined the success of Chinese-led international business in the region.

11 Singapore continued to be an important location for export-oriented international production by Hong Kong through the 1970s, despite the higher industrial wage rates in that country by comparison with that in other Asian countries such as Taiwan, Indonesia, Korea, Thailand and the Philippines because of its higher labour productivity. In addition, land prices were much lower in Singapore and the government of Singapore provided favourable considerations to the use of industrial land by foreign investors. However, since the end of the 1970s the growth of Hong Kong-based FDI in Singapore declined due in part to the increasing attractiveness of other Asian countries as host countries and in part to the new emphasis being accorded by the government of Singapore on capital and technology intensive industries in which Hong Kong-based firms were not yet ready to take part (Chen, 1981).

12 Lower wages are not an important determinant of Hong Kong's production in China given that the labour processing charges fixed by the Chinese authorities were only marginally below the wage cost for the same job in Hong Kong, while workers in China were generally less productive (Chen, 1981). Thus the unit costs of production (the ratio of wage rates to local productivity) could have been higher in China than in Hong Kong.

13 Some $200 million of the total $280 million of Hong Kong-based FDI in China at the end of 1981 was in the export processing zones of China (Chen, 1983b). In subcontracting, Chinese factories usually perform the processing in accordance with the specifications of the foreign firms which supply raw materials and/or intermediate products. Most of the foreign firms engaged in subcontracting activities in China are Hong Kong based, and almost all the factories are in the border districts nearest Hong Kong. Compensation trade in China first began in early 1978 and was intended for larger industrial projects. Under this cooperative arrangement between China and foreign firms, China usually provided land and labour and foreign firms supplied raw materials, parts, machinery and equipment, and technical and managerial personnel as well as training for the Chinese labour. While requiring substantial initial investment, it had the advantage that foreign firms were not constrained by existing levels of skills and technology, and China is able to acquire modern technologies and management skills – one of the most important motives behind China's liberalization of foreign trade and FDI. Many of the Hong Kong-based projects in China for the production of textiles, clothing, watches and electronics have been set up in the form of compensation trade (Chen, 1981). However, even though some of the production arrangements by Hong Kong-based firms were non-equity forms of investments, the Hong Kong-based firms, being the monopsonistic buyers, exercised control over the production arrangements without the need of internalization provided by FDI. Evidence for this is found in the fact that over 80 per cent of the output of its subcontracting projects in China were shipped back to Hong Kong as re-exports (World Bank, 1989). Thus, while the manufactured exports of Hong Kong was growing at just 2–3 per cent per year in the late 1980s, the outward-processing-related exports of Hong Kong's firms in China increased by around 28 per cent (Census and Statistics Department, *Hong Kong External Trade*, March 1991). A major explanation for the high share of the production output accounted for by Hong Kong-based firms in China that is re-exported to Hong Kong is the vertically integrated networks of production established between Hong Kong-based parent firms and their Chinese suppliers. Unlike the affiliates of Hong Kong-based firms in South East Asia in manufacturing that typically produce final products, the various production arrangements accounted for by Hong Kong-based firms in China are typically responsible for only one stage of the production process, usually the labour intensive stage, much like the outward FDI of many American and European MNCs in developing countries. Owing to the limited technological and skill capacities existing in China, the more sophisticated parts of the production processes are performed in Hong Kong (Chen, 1981). In no other host country has Hong Kong developed a closely integrated network of international production.

14 Owing to the lack of a consistent data series on inward FDI in Singapore, no data could be found of inward FDI in Singapore at the end of the 1970s. However, based on estimates contained in Yoshihara (1976), Hong Kong-based FDI in Singapore amounted to $75 million in 1973, of which textiles and clothing accounted for 61 per cent of Hong Kong-based FDI in manufacturing. The relative importance of textiles and clothing had thus increased considerably from its 39 per cent share in Hong Kong-based FDI in manufacturing in Singapore in 1966 (based on data in Luey, 1969).

15 Owing to the lack of a consistent data series on inward FDI in Singapore, no data could be found of inward FDI in Singapore at the end of the 1970s. However, based on estimates contained in Yoshihara (1976), Hong Kong-based FDI in Singapore amounted to $75 million in 1973, of which electrical and electronics accounted for 7.5 per cent of Hong Kong-based FDI in manufacturing. The relative importance of electrical and electronics had thus increased from its 5.5 per cent share in Hong Kong-based FDI in manufacturing in Singapore in 1966 (based on data in Luey, 1969).

16 Based on a 1982 survey conducted by Chen (1983b) of 32 Hong Kong-based MNCs which accounted for a high proportion of Hong Kong's total outward FDI, the efficiency and dynamism of management expertise of Hong Kong-based firms did not derive from having a high educational attainment (only one-third of the managerial staff of the 32 Hong Kong-based MNCs had received post secondary education), but rather from experience and the ability to learn and acquire the necessary skills rapidly. There was in-service training courses for managerial staff in half of the firms surveyed.

17 It was estimated that some 5–25 per cent of Hong Kong-based MNCs in manufacturing other than plastics production sourced their machinery and equipment from Hong Kong. By comparison, some 60 per cent of Hong Kong-based MNCs firms in plastics production utilized machinery and equipment from Hong Kong. Such high proportion was due to the capability of the machinery manufacturing industry of Hong Kong to produce very high quality and sophisticated plastics machinery for blow moulding, injection moulding and extrusion (Chen, 1983b).

18 Indeed, in some cases several Hong Kong-based firms were motivated to search for foreign business opportunities in developing countries in South East Asia and West Africa with abundant and low-cost labour to make use of labour intensive machinery and equipment no longer suitable for the domestic conditions of Hong Kong owing to rising wages and which could not be sold arms-length in the poorly developed international markets for second-hand machinery. As a result, much of the equipment was repainted, labelled as new, and in some cases imported illegally to developing countries where outward FDI was made (Wells, 1978).

19 It is the superior connections to export markets that have proven harder to replicate by domestic firms in developing countries as shown in the case of Indonesia. Despite the restrictions imposed by the Indonesian government to the entry of new Hong Kong-based firms in the textiles and clothing industry in the 1970s in light of the growing comparative advantage of Indonesia in labour intensive production, Indonesian firms met limited success (Wells, 1978).

20 Owing to the lack of a consistent data series on inward FDI in Singapore, no data could be found of inward FDI in Singapore at the end of the 1970s. However, based on estimates contained in Yoshihara (1976), Hong Kong-based FDI in Singapore amounted to $75 million in 1973, of which food and drink, chemicals and other industries apart from textiles and clothing and electrical products and electronics accounted collectively for 31.5 per cent of Hong Kong-based FDI in manufacturing. The relative importance of these industries had declined from their 55.7 per cent share in Hong Kong-based FDI in manufacturing in Singapore in 1966 (based on data in Luey, 1969).

21 This would constitute an example of extra-firm and inter-firm networks in the typology of Yeung (1997) in describing the three dimensions of networks of relationships or *guanxi* in which the operations of Hong Kong-based firms in South East Asia are socially and culturally embedded. The extra-firm networks are all kinds of political relationships or connections at the highest level of government to overcome regulatory barriers or institutional obstacles to the conduct of business. This is done by coopting influential politicians in local subsidiaries of Hong Kong-based MNCs in foreign countries or simply by calling on political connections available

to the Hong Kong parent firms. The inter-firm networks are the overseas Chinese networks which form the social and cultural basis of Chinese business whose FDI are helping to transform the Chinese-based inter-firm networks into a Chinese commonwealth (Kao, 1993).

22 This statement had been deduced based on available host country data provided in Wong (1992) on the stock of Hong Kong-based inward FDI in Thailand, Taiwan, Philippines and Malaysia at or around the end of the 1980s. The percentage shares of the primary, secondary and tertiary sectors in Hong Kong-based FDI in Thailand as of 1987 was 8.5 per cent, 29.5 per cent and 62 per cent, respectively. The percentage shares of the primary, secondary and tertiary sectors in Hong Kong-based FDI in Taiwan as of 1989 was 0.3 per cent, 54.1 per cent and 45.6 per cent, respectively. The percentage shares of the primary, secondary and tertiary sectors in Hong Kong-based FDI in the Philippines between 1981 and June 1989 was 6.5 per cent, 46.5 per cent and 46.1 per cent, respectively. The secondary sector accounted for 100 per cent of Hong Kong-based FDI in Malaysia as of 1987. In addition, qualitative data contained in Yeung (1994) indicate that in the period between 1980 and 1988 there were 42 projects of Hong Kong-based firms in the manufacturing sector of Indonesia and 15 in the services sector, compared with 14 projects in agriculture and none in mining. In Singapore, on the other hand, there seemed to be a marked concentration of Hong Kong-based FDI in services particularly in commerce, restaurants and hotel services and finance.

23 A survey conducted by World Bank (1989) of 2,000 Hong Kong-based manufacturing firms in November 1988 indicated that the firms accounted for $12 billion in cumulative FDI abroad as of mid-1988. Of this amount, outward FDI in China made up approximately $6.5 billion or more than half, while third countries accounted for the rest. A similar conclusion is reached by Yeung (1994) who stated that more than two-thirds of Hong Kong's FDI *outflows* in the 1980s have been directed to East, South and South East Asia, and another third to the United States. Within the Asia-Pacific region, a large proportion had been invested in China and particularly in the Guangdong Province. In June 1991, some $16 billion of FDI in that province (or four-fifths of its total FDI) were sourced from Hong Kong.

24 The announcement of China's Law on Joint Ventures on 8 July 1979 led to some larger projects by Hong Kong-based firms in China organized in the form of joint investment ventures. Of the 29 projects set under the joint venture scheme, 16 had Hong Kong participation in the early 1980s. However, the smaller projects continued to be arranged in the form of compensation trade, cooperation and subcontracting. In the early 1980s another form of of FDI in China had been introduced by the Chinese government, namely joint production by which the control over production and the share in profits are determined beforehand by contracts and agreements instead of by relative shares of partner companies in total capital investment. Most of these joint production projects with foreign firms were in Guangdong and accounted for by investors from Hong Kong (Chen, 1983b).

25 Calculated based on data on inward FDI flows in Thailand contained in ADB (1988) for data for the years between 1980 and 1985 and in Tan *et al.* (1992) for data for the years between 1986 and 1990.

26 Calculated based on data on inward FDI in Indonesia contained in Lee (1990) for data for 1979, and in ADB (1988) and Tan *et al.* (1992) for data for the years between 1980 and 1990.

27 Calculated based on data on inward FDI in the Philippines contained in Alburo (1988) for data for 1980, and in ADB (1988) and Yeung (1995) for data for 1991.

28 Based on data and information on inward FDI in Malaysia contained in Chen (1981, 1983b) for data for 1979, and in ADB (1988) and Tan *et al.* (1992) for data for the years between 1980 and 1990.

29 Based on data on inward FDI in Singapore contained in Department of Statistics, Singapore (1992).

30 See also Wong (1992); and 'Fresh pastures: Hongkong firms shift investment away from China', *Far Eastern Economic Review*, 20 September 1990.

31 Such figure was arrived on the basis that there were some 2 million people employed by 2,000 Hong Kong-based firms in China in 1988/1989 (Thoburn *et al.*, 1990) and 1 million people employed in other foreign countries (UN, TCMD, 1993a), and the domestic employment in Hong Kong was around 870,000 (Thoburn *et al.*, 1990).

32 See 'Singer in harmony: Hongkong firm aims to capitalise on brand name', *Far Eastern Economic Review*, 27 September 1990.

33 See 'Right this way: Hong Kong money in Canada', *The Economist*, 23 March 1991.

34 See UN, TCMD (1993a) for examples. The global expansion of Hong Kong and Shanghai Banking Corporation (HSBC), particularly in the developed countries, was in support of its role as one of the world's largest financial institutions. The bank controlled more assets than any other foreign bank in the United States. Its aggressive investment strategies in the United States had taken the form of acquisitions: it purchased Marine Midland Bank as well as Golden Pacific and Global Union – American financial institutions that cater to the American Chinese market. See 'Hong Kong bank wants a bigger share of American market', *Asian Finance* (Hong Kong), 15 March 1987. Through its holding company HSBC Holding plc, the HSBC completely acquired Midland Bank in the United Kingdom in 1992. See UN, TCMD (1993a).

35 Based on data and information on inward FDI in Malaysia contained in Chen (1981, 1983b) for data for 1979, and in Yeung (1995) for data for 1992.

36 Based on data contained in Department of Statistics, Singapore (1992) for data for 1980, and in Yeung (1995) for data for 1991.

37 Calculated based on data on inward FDI in Indonesia contained in Lee (1990) for data for 1979, and in Yeung (1995) for data in 1994.

38 See also 'Asia its oyster: Hong Kong', *The Economist*, 8 December 1990.

39 See 'A bit belatedly, large Hong Kong companies try to gain investment foothold in South China', *The Asian Wall Street Journal Weekly*, 11 May 1992; also 'Ties that bind: Hongkong firms seek closer China links', *Far Eastern Economic Review*, 30 July 1992.

40 Towards this end, the Hong Kong University of Science & Technology was established in 1991, and the government invested $57 million in an Industrial Technology Centre. By creating a firm foundation in basic science, it is expected that the university and research centre will help foster Hong Kong's and South China's industrial competitiveness in more advanced industries such as advanced electronics and telecommunications. See 'Is "the MIT of Asia" growing in Hong Kong?', *Business Week Special Report*, 7 December 1992.

15 The emergence and evolution of multinational corporations from Singapore

Introduction

Owing to their status as British colonial entrepôts, the origins of both Singapore-based MNCs and Hong Kong-based MNCs can be traced to the British colonial period with the outward FDI of British companies based in these countries and some Chinese immigrants in Singapore. Thus, MNCs based in the Asian newly industrialized city states have a far longer history by comparison to MNCs based in the Asian newly industrialized countries with larger domestic markets, Taiwan and South Korea. With an outward FDI stock of $47.6 billion in 1998 or some 1.2 per cent of the global stock of outward FDI, Singapore was not as significant a source country of FDI as Hong Kong with $154.9 billion but considerably more significant than Taiwan with $38 billion and South Korea with $21.5 billion. Indeed, Singapore constituted the second-largest home country based in developing countries in 1998 after Hong Kong with a share of some 12.2 per cent of the stock of outward FDI from developing countries in that year.[1]

The study of the emergence and evolution of Singapore-based MNCs is of interest as another case study of MNCs from a resource-scarce developing country with a small domestic market. The growth pattern of Singapore-based MNCs as it has been evolving since the British colonial period can be compared most directly to those of Hong Kong, another newly industrialized country in Asia, or more broadly to that of Swiss MNCs whose home countries share similar patterns of national economic development and whose histories have been analysed in the previous chapters in this part of the book.

Before the excursion into the history of Singapore-based MNCs, it is important to clarify two key points. This chapter refers to the history of MNCs based in Singapore which includes not only the outward FDI by locally owned or indigenous Chinese firms in Singapore, but also outward FDI by wholly or majority-owned foreign affiliates in Singapore whose ultimate beneficial ownership can be traced to another country. Owing to its history as a British colonial entrepôt and a Crown colony since 1867, Singapore was a base for the economic expansion of the British empire in Malaya, South East Asia and China until the Second World War as well as a base for Britain's military expansion in East Asia until the late 1960s. The dominant role of foreign based MNCs in the modern economic history of Singapore meant that the role of Singapore as a base for the growth of their outward FDI in South East Asia and China prevailed since the Second World War. Since the 1970s, the development of Singapore as an international financial centre, the presence of a strong domestic currency, domestic political and macro-economic stability, the relatively liberal financial regime as well as a safe haven for some excess savings of countries in South East Asia has also made the country an important financial base from which foreign companies launch outward FDI in the region (see also Lim, 1990). Indeed, the share of wholly and majority-owned foreign companies in the stock of outward FDI from Singapore reached 47.7 per cent in 1981 and although such share had been declining in the first half of the 1980s to reach a trough of 25.9 per cent in 1985, it climbed rapidly since to 33.6 per cent in 1989 and 56.4 per cent in 1993.[2] Conversely, the share of wholly and majority-owned local companies or indigenous Chinese MNCs based in Singapore has been declining in significance. Their investments are accounted for not only by indigenous Chinese firms, some of which are family owned, but perhaps more importantly by the state-owned corporations and government-linked companies, companies established abroad originally by Chinese families that transferred their operations to Singapore in the 1970s as well as companies formerly foreign owned but subsequently acquired to become Singapore-based MNCs.[3] The dominant role of foreign companies and state-owned companies in Singapore made the role of indigenous private enterprises in Singapore's industrialization and outward FDI far less significant. Both the significant role of foreign companies as well as the peculiar nature of local companies based in Singapore help to explain the large amounts of outward FDI emerging from Singapore, particularly in neighbouring countries. It is for this reason that the term 'Singapore-based' firms or MNCs is used throughout this chapter and book.

Secondly, the government of Singapore has not published data on outward FDI from Singapore until the 1990s.[4] Thus, as in the case of Hong Kong the analysis of outward FDI from Singapore has had to rely at times on scattered reports and the scanty data and information provided by the host countries in which investment by Singapore-based MNCs had been significant. Given the large number of significant host countries of Singapore-based FDI and the different currencies and criteria used in the compilation of data on inward FDI in each host country which precludes any possibility of data aggregation across host countries, the analysis of Singapore-based MNCs is rendered difficult particularly with respect to the examination of the major types of outward FDI, the fundamental determinants of outward FDI and the major host countries, as well as the changes in these variables over time.

Bearing in mind these two caveats in the analysis of Singapore-based FDI, this chapter aims to provide a faithful account of the history of Singapore-based MNCs as it relates to the unique pattern of the domestic industrial development of Singapore in three time frames: from the British colonial period to the 1970s, the 1980s and the 1990s.

The emergence of Singapore-based MNCs from the British colonial period to the 1970s

The history of Singapore-based MNCs is imbued inextricably in its British colonial heritage. The endowment of Singapore by Sultan Hussain of Johore to the British East India Company in 1819, the formation of the Straits Settlements uniting Singapore with Penang and Malacca (both British territories on the Malay peninsula) and the designation of this settlement as a Crown colony in 1867 had important implications on the emergence of the first Singapore-based MNCs. As its importance as a British colonial entrepôt in South East Asia exceeded that of Penang – Britain's main port in that region – particularly since the opening of the Suez Canal in 1869, Singapore rapidly became a base for the economic expansion of the British empire in Malaya, South East Asia and China until the Second World War as well as base for Britain's military expansion in East Asia until the late 1960s.[5] The prosperity of many British trading companies derived not only from mercantile activities in the free trade port but also from more productive endeavours in the Malay peninsula to include rubber plantations (the wild Brazilian rubber plant was cultivated in Singapore's botanical gardens and then transplanted to Malaya) and tin mining, with the latter activity dominated largely by Chinese immigrants in Singapore that arrived from China as a result of major socio-economic upheavals and lengthy insurrections in that country between 1850 and 1878 (Mirza, 1986). Although this may have been the earliest outward FDI from Singapore in history, it may not have been considered as FDI given the union of Singapore with Penang and Malacca, both territories on the Malay peninsula, under the Straits Settlement formed by the British East India Company in 1826. In any event, investments in Malaya, South East Asia and China by British companies and Chinese immigrants based in Singapore flourished throughout the British colonial period. In fact, many Singapore-based companies had been well established in peninsular Malaysia before the separation of Singapore from Malaysia in 1965 (Pang and Komaran, 1985). In South East Asia, among the earliest Singapore-based FDI was that of the Khong Guan Biscuit Factory in biscuits manufacturing in Indonesia during the 1950s, and by other Singapore-based manufacturers such as Ho Rih Hua that had investments in Thailand, and Lau Ing Woon's brothers that expanded in South East Asia (Chan and Chiang, 1994).

By 1959, it was recognized that entrepôt trade which had accounted for a considerable proportion of domestic employment and about half of Singapore's total trade could not continue in its role as an engine of economic growth. This was owing to two factors: first, the changing pattern of trade away from colonial trade with the withdrawal of British colonial rule in the period around the Second World War. Indeed, the British imperial epoch was over with the capitulation of the Singapore's military forces to the Japanese in 1942, despite the resumption of British control in 1945 (Mirza, 1986). Second, unemployment which increased considerably from 5 per cent in 1957 to 13.5 per cent by 1960 (Lecraw, 1985) became a major economic problem. The rate of unemployment was rising much faster than the annual population growth of 3.6 per cent between 1947 and 1957 or 4.3 per cent if migration is included. These two factors led *ipso facto* to industrialization becoming the most rational development strategy for the country (Wong, 1979).

The history of modern Singapore is intertwined closely with the development of the People's Action Party (PAP) of Lee Kuan Yew as an anti-colonial movement (Mirza, 1986). Based on the party's manifesto that the development of manufacturing industries is the long-term solution to unemployment, the liberal faction of the party proceeded to the reins of power in 1959. The party chose to rely on foreign capital as a tool to attain rapid industrialization and economic growth and thus legitimize its political domination (Yeung, 1998). This was owing not so much to the weak industrial bourgeoisie and the lack of any significant domestic manufacturing base as argued by Yeung (1998), Rodan (1989), Yoshihara (1976) and Hughes and Yon (1969) because Singapore had already developed a fairly modern though nascent manufacturing base providing direct employment to 26,697 workers towards the end of the 1950s (Wong, 1979). Thus, there had neither been a shortage of domestic entrepreneurship nor a lack of supply of hard-working labour which have posed bottlenecks in the industrial development of other countries in South East Asia. The constraint to the industrial development of Singapore rather lay in the inadequate domestic managerial and technical know-how and knowledge about export market development (Wong, 1979) as well as financial resources for industrialization (Yeung, 1998).[6] Moreover, the PAP-ruled state was suspicious of indigenous capitalists for fear of their pro-Communist and pro-China attitudes (McVey, 1992; Menkhoff, 1993) and thus the party led by Lee Kuan Yew neglected the role of local Chinese entrepreneurs in Singapore's industrialization for social, economic and political reasons, despite their arguably infant state in the early 1960s (Régnier, 1993).

As a result, institutional structures such as the Economic Development Board and the Jurong Town Corporation were created in 1961 and 1968, respectively, and other measures were implemented including the offering of generous incentives schemes to pave the way for the dominant role that foreign companies would play in domestic industrial development.[7] Among these measures was the state regulation of the labour market through the Trade Union (Amendment) Bill in 1966, the Employment Act in 1968 and the Industrial Relations (Amendment) Act (Rodan, 1989; Huff, 1995). These labour market regulations resulted in the creation of a highly disciplined and depoliticized labour force with limited rights to strike and bargain for wage increases or to engage in radical political dissent (Lecraw, 1985), thus facilitating the important role of Singapore in the emerging industrial division of labour spearheaded by global MNCs.

Nevertheless, the growth of indigenous firms was encouraged through government incentives such as the Small Industries Finance Scheme (SIFS), technical assistance, general assistance in establishing joint ventures with foreign companies as well as assistance provided by the state-owned trading company, Intraco, to indigenous firms in servicing foreign markets. In addition, state-owned companies or statutory boards invested in infrastructure development (the port, roads, and industrial and housing estates), in industries that were natural monopolies (utilities, water, transportation and port services) and in other industries in which private sector sector investment was not forthcoming to the desired extent (steel, petrochemicals and shipyards) (Lecraw, 1985) or as a means of overcoming the dearth of entrepreneurial, technological and capital resources in the domestic economy (Mirza, 1986). Furthermore, public investment in the manufacturing sector started as early as 1963 with the establishment of seven public enterprises: Sugar Industry of Singapore Ltd, National Grain Elevator Ltd, Singapore Textile Industries, Ltd, United Industrial Corporation Ltd, Singapore Polymer Corporation Pte Ltd, Jurong Shipyard Ltd and Ceramics (M) Pte Ltd (Rodan, 1989). Eventually, state-owned enterprises came to permeate such diverse manufacturing industries as food products, textiles, wood and wood products, printing and publishing, chemicals and petrochemicals, iron, steel and metal products, electrical and electronics products, engineering products and shipbuilding and repair. The services sector has similarly been directly or indirectly regulated and promoted by statutory boards and enterprises such as the Monetary Authority of Singapore, the Singapore Tourism Promotion Board and the Development Bank of Singapore. Other state-owned enterprises also participated in all the major service industries such as banking and finance, distribution, communication, business services, tourism and real estate (based on information in Mirza, 1986).[8] While such process of 'state entrepreneurship' (Mirza, 1986) has constituted an important driving force behind the restructuring and diversification of the Singapore economy in higher technology and higher value added industries which served to attract growing inflows of FDI by foreign MNCs in Singapore, some of the state-owned enterprises spun-off from statutory boards to include the Keppel Group, the Sembawang Group and Temasek Holdings became the government-linked companies that were the major driving force behind the regionalization drive of Singapore in the 1990s (Yeung, 1998). This had helped to create the image of 'Singapore Inc.' (Davies, 1983).

The annual average annual FDI outflows of Singapore based on balance-of-payments data reached $69.2 million between 1972 and 1979, a level seven times higher than the annual average FDI outflows of South Korea of $9.8 million between 1970 and 1979 and almost 29 times higher than that of Taiwan of $2.4 million in the same period.[9] With an outward FDI stock of some $819 million by 1981, Singapore had been by far the most significant home country of FDI among the four Asian newly industrialized countries.[10] Indeed, the size of outward FDI by Singapore was remarkable among developing countries particularly since the country had a population of only 2 million in the early 1980s (Lecraw, 1985).

To the extent that financial factors played a role in explaining the growth of outward FDI from Singapore in the 1970s this derives from the sharp appreciation of the Singapore dollar by more than 30 per cent against the United States dollar between the end of 1969 and the end of 1979. However, unlike in the case of Japan since the 1960s and Korea and Taiwan in the 1980s, the accumulation of large financial surpluses and foreign exchange reserves in Singapore had not been due to the growth of net exports in the 1970s but from the rapid growth in the domestic savings rate fostered by the government policy of compulsory contribution to the Central Provident Fund (the national savings scheme).[11] This led the share of savings to GDP to increase two and one-half times from 11.5 per cent in the period between 1960 and 1969 to 28.8 per cent in the period between 1970 and 1979.[12] To the extent that the high domestic savings rate may have provided a financial basis to the growth of Singapore-based outward FDI during the 1970s, the large pool of potential investment capital amassed by the national savings scheme may have favoured the state-owned enterprises rather more than private sector enterprises whose potential financial resources for investment was channelled instead to the Central Provident Fund, thus restraining their capacity to finance domestic and overseas expansion. The savings-investment process of Singapore has thus served to crowd out the growth of local Chinese entrepreneurship (Tan, 1991). Apart from funding the domestic and foreign expansion of state-owned companies, the amassed financial resources by the state-enforced savings scheme may have also been recycled through the massive growth of foreign portfolio investments for which the government of Singapore had been responsible. Indeed, the Government of Singapore Investment Corporation had over $15 billion in foreign portfolio investments mainly in the United States in early 1983 (*Fortune*, 21 March 1983).

Some 77 per cent of the stock of outward FDI by Singapore in the end of the 1970s and early 1980s had been concentrated in the developing countries of Asia, with some 64 per cent in South East Asia. Less than 5 per cent was directed to the developed countries mainly in the United Kingdom and the United States, but there had also been some investments in the Netherlands, other Europe and Japan (based on data in Department of Statistics, 1991).

In terms of industrial distribution, outward FDI by Singapore by the end of the 1970s had been present in all the three major economic sectors – primary, secondary and tertiary. Outward FDI in manufacturing was spread over a broad range of industries that were either dominated by local capital or foreign capital. This included low technology, low value added industrial and product niches and industries typically intensive in the use of labour and natural resources to include food and drink, clothing, printing and publishing, plastic products, leather and rubber products, cement and concrete products (such as bricks, tiles and clay products, earthenware, glass and other non-metallic mineral products), fabricated metal products and transport equipment that were local capital-dominated industries in Singapore in the late 1970s and early 1980s. In addition, outward FDI in manufacturing also emerged in textiles, wood products, paper products and in higher technology and higher value added industries such as petroleum products and electrical and non-electrical machinery that were foreign capital-dominated industries in Singapore in the late 1970s and early 1980s.[13] The different industries in which foreign or local capital dominated reflected the difficulties faced by local firms to penetrate the higher value added industries owing to their lack of technology, experience, entrepreneurship and size (Mirza, 1986).

The locally based firms that have engaged in outward FDI in manufacturing have been of several types. This included the state-owned companies such as Intraco Ltd (a trading company), Keppel Shipyard (Pte) Ltd, and Sembawang Shipyard Ltd (both shipbuilding and repair companies), National Iron & Steel Mills Ltd (an iron and steel products company), and Acma Electrical Industries Ltd (a manufacturer of refrigerators and home appliances), among other state-owned companies. Among the privately owned companies that made outward FDI in manufacturing were Yeo Hiap Seng (a formerly family-owned food and drinks company that eventually became a public company), The Soap Manufacturing Company and Khong Guan Flour (local companies owned largely by Chinese families), and Wah Chang International and Jack Chia MPH (both migrating MNCs that had been established abroad originally by Chinese families that transferred their operations to Singapore in the 1970s). Both Wah Chang International and Jack Chia MPH are conglomerate companies with the former having interests in construction, shipbuilding and engineering and the latter having interests in pharmaceuticals, perfumeries, publishing and property (Pang and Komaran, 1985).

Apart from outward FDI in manufacturing, outward FDI from Singapore was also significant in services in the period around the late 1970s and early 1980s. This was indicative of the dominant position of services in the GDP of Singapore at more than 71 per cent in 1979 (based on data in Huff, 1995) owing to the small size of the economy and its historical position as a trade entrepôt (Mirza, 1986). For example, Singapore-based firms have invested in civil engineering and construction projects in Sri Lanka, construction, engineering and support services related to shipping and oil production in India, hotels and construction projects in China, construction and engineering projects in the Middle East, and trading and property development in Australia and North America (Pang and Komaran, 1985), among other examples. The prominent position of engineering and construction in outward FDI reflected the leading role of state-owned companies in infrastructure development in Singapore particularly with the construction and buildings boom in the 1960s and 1970s.[14] This is complemented by skills in architecture and town planning which enabled some Singapore-based firms to extend their business throughout the region of South East Asia, South Pacific and China capitalizing on their experience in planning the efficient tropical metropolis of their home country and in the management of complex design work such as that involved in the construction of large hotels and shopping complexes (Hill and Pang, 1991).

Similarly, the prominent role of Singapore-based MNCs in the marine industry (ship repair services, shipbuilding, oil rig construction, as well as marine engineering and related industries) signifies the importance of these industries in the economy of Singapore accounting for some 11 per cent of value added and employment in manufacturing and some 5 per cent of merchandise exports in 1982. The marine industry is dominated by indigenous firms which either wholly or majority own all but 40 of the 260 establishments in the industry (Mirza, 1986). Although some of the major local shipbuilding and repair companies trace their origins back to the British colonial period, the majority have been established since the 1960s and their development owed much to the successful transfer of more advanced technologies and organisational skills from their joint ventures with foreign shipbuilding companies, particularly those of Japan (Yeow, 1984). This enhanced the many skills and facilities already possessed by local firms in the marine industries and built on the excellent infrastructure and communications of Singapore as an established trading post and the rapid expansion of related industries (shipping, petroleum processing, oil services, etc.).

In addition to outward FDI in services, there had also been Singapore-based FDI in the primary sector to include oil exploration in China and Australia by such firms as Wah Chang International and Chuan Hup (Pang and Komaran, 1985), agricultural-based and mining activities in the Philippines (UN, ESCAP, 1988) as well as the cultivation and processing of palm oil in China and the extraction and processing of marble in Malaysia by such Singapore-based firms as Intraco, Keck Seng, Guthrie and Sim Lim Group (Tolentino, 1993), among other examples. The significant presence of vertically integrated outward FDI in the petroleum industry despite its inherent scarcity in petroleum resources reflects the position of Singapore as the largest oil services centre in Asia ranging from oil rig construction to tanker bunkering and state-of-the-art refinery and petrochemical manufacturing. In such capacity, Singapore performs an important role in refining and providing oil-related services to crude oil sourced from the Middle East, Indonesia, Malaysia and Brunei for the Japanese, other Asia and world markets (Lim, 1990).

Notwithstanding the presence of Singapore-based FDI in resource extractive activities, to obtain raw materials and intermediate was not the major determinant of the internationalization of Singapore-based firms (Pang and Komaran, 1985). Similarly, despite the pressures imposed by an increasingly tight labour market and rising wages in Singapore since the early 1970s, this had not been a major factor in the relocation of labour intensive operations offshore owing to the ready access to foreign labour under the government's liberal worker policy (Chia, 1989). Instead, in a manner typical of MNCs based in small countries such as Sweden and Switzerland but not Hong Kong whose outward FDI was fundamentally of the export platform type, the major determinant of Singapore-based firms to engage in outward FDI in manufacturing and services stemmed from the need to find new markets and investment opportunities and to sell their technological expertise (see Chapters 4, 13 and 14).[15] The main constraint to the growth of Singapore-based firms had been the limited size of the domestic market combined with growing protectionist policies in major export markets in the region.

Indeed, outward FDI starting in the mid-1970s had been a response to the policies instituted by the governments of Indonesia, Malaysia, Thailand and the Philippines to encourage the upgrading of their natural resource and agricultural products prior to export. This served to undermine the location-specific advantages of Singapore in resource-based industries such as the refining of petroleum and the processing of rubber, timber, vegetable oil and food and thus coerced both foreign-based MNCs and indigenous firms to retrench existing investments in these industries in Singapore and to re-direct these and subsequent investments outside of Singapore. Thus, even indigenous firms have been able to utilize their firm-specific advantages in product and process technologies as well as sourcing and marketing skills to compete in foreign markets through outward FDI (Lecraw, 1985).

Besides limited growth in the domestic market and to sell technological expertise, another important determinant inducing international production particularly for state-owned firms in mature industries had been the need for diversification towards high technology industries such as computers, microelectronics, robotics, biotechnology and genetic engineering. However, given that developed countries accounted for less than 5 per cent of the stock of Singapore-based FDI in the late 1970s and early 1980s, this factor carried far less weight in the determination of the outward FDI of Singapore in this period. However, it had been an important factor behind the outward FDI of the state-owned National Iron & Steel Mills that invested some $3 million in a few small venture capital companies in California as a means to gain access to advanced foreign technologies and their applications. Similarly, a major motivation behind the outward FDI by locally owned companies in printing and publishing that were concentrated in the developed countries had been to keep abreast of rapidly changing technologies in the industry. This was the primary motive behind the acquisition in the early 1980s of a large British publishing group by a locally owned publishing conglomerate based in Singapore in exchange for 100 million Singapore dollars (see Hill and Pang, 1991). In the quest to support domestic industrial diversification, the original remit of the Economic Development Board to promote inward FDI in Singapore had been expanded to assist Singapore-based firms to engage in outward FDI (Pang and Komaran, 1985).

The expansion of Singapore-based FDI in the 1980s

The 1980s represented an era of even more rapid growth of Singapore-based FDI. Based on balance-of-payments data, the annual average FDI outflows of Singapore was about $215.3 million in the period between 1980 and 1989 compared to $69.2 million between 1972 and 1979, and the estimated stock of Singapore-based FDI reached $7.8 billion in 1990 compared to $819 million by 1981.[16] Thus, from being the most significant home country of FDI among the four Asian newly industrialized countries in 1980, Singapore had been surpassed by Hong Kong and Taiwan whose stock of outward FDI reached $13.2 billion and $12.9 billion in 1990, respectively. However, Singapore remained a far more important home country of FDI relative to South Korea whose stock of outward FDI was $2.3 billion in 1990 (based on data in UNCTAD, 1999).

To the extent that financial factors played a role in explaining the growth of outward FDI from Singapore in the 1980s this derived from the continuing appreciation of the Singapore dollar by more than 12 per cent against the United States dollar between the end of 1979 and the end of 1989. As in the 1970s, the accumulation of large financial surpluses and foreign exchange reserves had not been created as much from net exports that reached an annual average level of 321.5 million Singapore dollars between 1980 and 1989 by comparison to net imports of 1,268.9 million Singapore dollars between 1970 and 1979 and 408 million Singapore dollars between 1968 and 1969 but rather from the continuing growth in the domestic savings rate fostered by the government policy of increased compulsory contribution to the Central Provident Fund (the national savings scheme).[17] Thus, savings as a share of GDP increased further from 28.8 per cent in the period between 1970 and 1979 to 42.7 per cent in the period between 1980 and 1992 (based on data in Huff, 1995). As a result, the annual average foreign exchange reserves of Singapore increased almost fourfold from $3.0 billion between 1970 and 1979 to $11.9 billion between 1980 and 1989. However, the bulk of such reserves had been recycled as in previous decades in the form of portfolio investments abroad to include investments in developed country bond and government security markets, property, liquid debt instruments and in investments resulting in less than 5 per cent of the equity ownership of a number of large companies in the United States (Hill and Pang, 1991) or New Zealand.[18] To the extent that such reserves may have financed outward FDI, it may have favoured the state-owned corporations that gained rather more from the national savings scheme compared to the private sector that had been deprived of potential investment capital (see also Yeung, 1998).

The even more rapid expansion of outward FDI from Singapore in the 1980s relative to previous decades had been accompanied by significant changes in its industrial and geographical patterns. This owed much to the swift process of domestic industrial restructuring towards higher value added manufacturing starting around the late 1970s as well as in services associated closely with efforts of Singapore to promote itself as an international centre for offshore banking, finance and other services, especially trading, transport and communications from around the early 1970s (see Mirza, 1986). Indeed, more advanced manufacturing (to include the continued expansion of petroleum and petrochemicals), trade, tourism, transport and communications, and knowledge intensive service industries (computer, financial, medical and consultancy services) became the five pillars of growth (Lecraw, 1985).

Typical of the pattern of development of the manufacturing sector in most countries, Singapore had begun to alter its emphasis away from labour intensive industries in response to much lower unemployment rate of around 4 per cent in the early 1970s from 13.5 per cent in 1960 and the resultant labour shortages and rapid rise in real wages *despite* efforts by the government to limit wage increases by wage controls and encouraging immigration which provided the domestic economy with access to foreign labour (Lecraw, 1985). The attainment of full employment around the mid-1970s, the presence of already high levels of foreign workers combined with mounting trade protectionism in export markets against Singapore's labour intensive products in the late 1970s and early 1980s triggered the implementation of policies to encourage the development of domestic manufacturing industries that embody higher value added and human capital and technological intensity, as well as emphasize the further development of the service industries that accounted for more than 71 per cent of GDP in 1979. This pattern of domestic industrial upgrading in Singapore's manufacturing sector was not evident in Hong Kong where the more *laissez faire* stance of the state led to the limited industrial upgrading of its manufacturing sector (see Chapter 14). By 1979 the Economic Development Board of Singapore designated 11 manufacturing industries for promotion: automotive components, machine tools and machinery, medical and surgical apparatus and instruments, specialty chemicals and pharmaceuticals, computers, computer peripheral equipment and software development, electronic instrumentation, optical instruments and equipment (including photocopying machines), advanced electronic components, precision engineering products, and hydraulic and pneumatic control systems (Lecraw, 1985).

Such process of domestic industrial upgrading required an even greater reliance on foreign MNCs to provide the package of assets: technology, capital, management, and access to markets as well as increasing state entrepreneurship. Thus, the provision of more attractive investment incentives was complemented by extensive programmes for industrial training and skill upgrading to enhance the productivity of labour, the implementation of mandatory rapid increases in wages between 1979 and 1981 (thus reversing the policy of wage suppression in the 1960s and much of the 1970s) and mandatory increased payments by firms and labour in the state enforced savings and pension plan, the Central Provident Fund (see Rodan, 1989; Henderson, 1989; Wong, 1979). In addition, forward and backward integration by firms was encouraged to enhance value added per unit of output of goods produced in Singapore as well as the use of locally produced inputs and capital equipment. Not only did all these policy measures lead to the growth of domestic production and exports in more advanced industries but also enabled the diversification of export markets away from a concentration in developing countries towards high- and middle-income countries.

The central role of foreign MNCs in domestic industrial upgrading was evident in their increasingly dominant role in employment, value added, output and direct exports in the manufacturing sector between 1968 and 1989. In terms of total employment in the sector, these firms accounted for some 26 per cent in 1968, a share which increased further to 57.5 per cent in 1979 and 60 per cent in 1989. In terms of value added in the sector, these firms accounted for some 44 per cent in 1968, a share which increased further to 67.3 per cent in 1979 and 74 per cent in 1989. In terms of output in the sector, these firms accounted for some 46 per cent in 1968, a share which increased further to 73.8 per cent in 1979 and 76 per cent in 1989. In terms of direct exports in the sector, these firms accounted for some 54.4 per cent in 1963, a share which increased further to 85.2 per cent in 1979 and 86 per cent in 1989.[19]

Foreign MNCs also accounted for a considerable proportion of total paid-up capital in financial institutions and other business or producer services (such as consultancies and property), distributive services (including trade, transportation and communications) and social and personal services. For example, in financial institutions in 1980 foreign firms accounted for the majority of paid-up capital in merchant banks (87 per cent), commercial banks and Asian currency units (76 per cent) and insurance companies (67 per cent), while local firms dominated brokerage (99 per cent), finance companies (86 per cent), investment companies (82 per cent), discount houses (68 per cent) and other financial institutions (88 per cent). In producer services other than banking and finance in 1980, foreign firms accounted for the majority of paid-up capital in advertising and market research (77.9 per cent) and engineering, architectural and technical services (70.4 per cent), while local firms dominated in property and housing development (88.6 per cent), business and management consultancy (71.4 per cent) and legal, accounting and data processing services (55.1 per cent). In social and personal services excluding government social services in 1980/81, foreign firms accounted for the majority of paid-up capital in social services (education and medical services) (55.9 per cent) and personal and household services (52 per cent), while local firms dominated leisure and cultural services (91.6 per cent) and restaurants and hotels (86.7 per cent). It was only in distributive services in 1980/81 that local firms accounted for the majority of the paid-up capital in the sector as a whole and in the component industries to include air transport (99.7 per cent), land transport (97.4 per cent), telecommunications (88.5 per cent), sea transport (80.1 per cent), retail trade (74.7 per cent), other transport (74.5 per cent), and wholesale trade (65.4 per cent).[20] This symbolizes the strengths of local companies in a broad range of financial services other than merchant banking, commercial banks and Asian currency units and insurance, other producer services and social and personal services and especially distributive services – a legacy of its historical position as an entrepôt for trade in South East Asia with large internationally oriented service industries (Lecraw, 1985; Mirza, 1986).

Thus, by the end of the 1980s the economy of Singapore had been successfully transformed with a dual concentration in high value added manufacturing and as an international centre for offshore banking, finance and other services (Ho, 1993, 1994; Huff, 1995). Its economy had progressed beyond its traditional role during the British colonial period as an entrepôt for trade in South East Asia to fulfil a more complex and sophisticated role in modern times as the hub of trade, FDI and thus growth in South East Asia (see also Lim, 1990). However, the more dominant role of foreign MNCs in the economy of Singapore also meant that the share of wholly and majority-owned foreign companies in the stock of outward FDI from Singapore increased significantly from a trough of 25.9 per cent in 1985 to 33.6 per cent in 1989 (based on data in Department of Statistics, 1991).

In terms of geographical destination, some 51 per cent of the stock of outward FDI by Singapore in 1990 had been concentrated in the developing countries of Asia, with some 26 per cent in South East Asia. This was a considerable decline from the corresponding shares of 77 per cent and 64 per cent in 1981. By contrast, more than 23 per cent was directed to the developed countries by 1990 compared to less than 5 per cent in 1981, showing the increased importance of outward FDI in New Zealand, Europe (particularly the Netherlands and the United Kingdom) and the United States (based on data in Department of Statistics, 1996) and largely directed towards property and property development (Tolentino, 1993).[21] The importance of New Zealand as a host country for Singapore-based FDI proceeds from its geographical proximity and familiarity of Singapore with the business and legal practices of British Commonwealth countries. The latter reason also served to favour the United Kingdom, while the attraction of the United States was indicative of that country's status as Singapore's largest trading partner and as an important source of new technologies (Aggarwal, 1987).

Indeed, Singapore-based MNCs had continued to exert some significance in their principal host countries as shown by their share in inward FDI at or around the end of the 1980s. In Thailand, Singapore accounted for some 20 per cent of the flows of inward FDI between 1980 and 1990, and was the second-largest source of FDI after Japan.[22] In Malaysia, Singapore had been the largest source of FDI in Malaysia at the end of 1987 (Chia, 1989) attesting the locational advantages associated with geographical proximity and close historical and business ties. However, Singapore accounted for less than 8 per cent of the flows of inward FDI between 1980 and 1990, and thus was the third-largest source of FDI after Taiwan and Japan.[23] Singapore-based FDI in Malaysia in the 1980s have concentrated in a broad range of labour intensive and capital intensive industries to include food, drink and tobacco, non-metallic mineral products, textiles, electrical and electronics products, fabricated metals, basic metal products, chemical products, wood products, rubber products and transport equipment (Chia, 1989). There have also been substantial Singapore-based FDI in hotels and tourism, trade and financial services (UN, ESCAP, 1988). Given the wide spread of industries of Singapore-based FDI in Malaysia in the 1980s in industries dominated by both foreign and local capital in Singapore, the outward FDI in Malaysia have been accounted for both by locally owned firms as well as foreign MNCs based in Singapore. Singapore-based FDI in Malaysian manufacturing have been located predominantly in the southern Malaysian state of Johor, Singapore's immediate hinterland, spurred by the conscious attempt by the Johor authorities at 'economic twinning' between the two cities. Apart from the obvious attraction of space and lower land rentals, the major incentives for Singapore-based manufacturers in labour intensive industries had been the lower wages and abundant labour availability. The relocation of labour intensive manufacturing industries of Singapore to Johor as a lower cost production base parallels that of the relocation of labour intensive manufacturing industries of Hong Kong to Shenzhen and Guangdong Provinces (Hill and Pang, 1991).

It was perhaps in Indonesia and the Philippines where Singapore exerted far less significance as a source country of FDI. In Indonesia, Singapore accounted for some 1 per cent of the stock of inward FDI as of 1979, and some 3 per cent by 1990.[24] This was indicative partly of the lingering reservations of Singapore's business community about Indonesia and partly by Indonesia's continuing mistrust of Chinese-based investments. In the Philippines, Singapore accounted for 0.8 per cent of the stock of FDI in 1991.[25] This is owing largely to that country's farther geographic and cultural distance compared to the rest of countries in South East Asia as well as its lacklustre economic performance since the late 1970s (Hill and Pang, 1991).

Singapore investors have also been attracted to China in the 1980s despite the long geographical distance, the lack of diplomatic relations and bureaucratic difficulties involved in investing and doing business in China. In mid-1988, Singapore-based FDI in China stood at $830 million, making Singapore the fourth-largest foreign investor in China, after Hong Kong, Japan and the United States. Their investments in Hong Kong have also grown rapidly, amounting to $306.9 million in 1989 directed towards manufacturing, services and property in part to service China (see also Chia, 1989).

The continuing importance of South East Asia and China to Singapore-based FDI in the 1980s, or at least that part attributable to indigenous Chinese firms, was broadly parallel to the expansion of outward FDI from Taiwan and Hong Kong to avail of the socially and culturally embedded networks of relationships or *guanxi* with the Chinese business community in different countries of South East Asia and China which provided business information as well as financial and marketing assistance (Granovetter, 1985, 1991; Granovetter and Swedberg, 1992). These regional business networks based on contacts spanning almost a century and close familial relations have provided strength to Chinese business organizations and paved the way for their economic hegemonic role in the business and commerce in the region (see also Wong, 1991). The emergence and growth in South East Asia of indigenous Chinese MNCs based in Singapore only serves to reinforce that hegemonic role.

In addition, that part of Singapore-based FDI attributable to indigenous Chinese firms display a distinct preference towards outward FDI through joint ventures or contractual resource transfers which tended to reflect not as much on the shortage of funds or host country regulations but rather the risk averse approach of indigenous Chinese firms based in Singapore and their lack of strong ownership-specific advantages given their concentration in industries with mature technologies (food, drink, basic metals, electrical and electronics, plastics, textiles and clothing).[26] While some of the cash-rich, Singapore-owned conglomerate companies have displayed a higher propensity to engage in outward FDI through the establishment of wholly or majority-owned foreign affiliates, the mode of outward FDI of indigenous firms based in Singapore would remain eclectic reflecting the particular circumstances of the investment and the host country as well as the ownership-specific advantages of firms (see also Pang and Komaran, 1985).

In terms of industrial distribution, outward FDI by Singapore by the end of the 1980s had been as at the end of the 1970s present in all the three major economic sectors – primary, secondary and tertiary. Manufacturing and services continued to be the most important economic sectors of economic activity relative to the primary sector.[27] Owing to the success of domestic industrial upgrading of the Singapore economy, outward FDI in manufacturing at the end of the 1980s was spread over a far broader range of older and newer industries by comparison to the end of the 1970s with the range of industries continuing to encompass those that were dominated by either local capital or foreign capital in Singapore. The 1980s also witnessed the further growth of outward FDI in services by major state-owned enterprises such as Singapore Airlines (which has expanded into finance) and Keppel Shipyards (presently involved in retail banking, international finance and insurance) and changes in Singapore legislation in the 1980s meant that even the Monetary Authority of Singapore would assume an increasingly significant commercial as well as a regulatory role in the future (Mirza, 1986).[28]

Although the expansion of Singapore-based FDI in the 1980s continued to be driven by the need to overcome the limited size of the domestic market and to expand production and service capacity as well as by the need to acquire more advanced forms of technologies in the developed countries, a major factor propelling the growth of outward FDI in the 1980s and particularly since 1986 had been the accelerated pace of wage increases in Singapore and the tightening labour market problem which, owing to tighter foreign labour and immigration restrictions, could no longer be relieved by unskilled labour importation as in the 1970s (see also Aggarwal, 1986, 1987; Chia, 1989; Hill and Pang, 1991).[29] Although foreign capital had played a far more dominant role in the Singapore economy than in Japan which instead nurtured the growth of domestic infant industries through import substitution industrialization, the position of Singapore in the 1980s can nevertheless be compared to the position of Japan in the period immediately following the Second World War when rapid domestic industrial restructuring of the Japanese economy had been initiated and there was the expansion of Japanese FDI in manufacturing (see Chapter 9). In both countries, outward FDI was a crucial instrument or catalyst for the rapid process of domestic industrial upgrading in a manner broadly consistent with the views of Kojima and Ozawa (1985). In particular, the loss of comparative advantages of Singapore in labour intensive production precipitated the relocation of labour intensive industries to labour abundant developing countries, thus enabling Singapore-based manufacturers in these industries to continue to appropriate the returns to their accumulated managerial and technical expertise while also helping to encourage the diversification of the domestic economy in more modern manufacturing and service industries.

Outward FDI from Singapore in the 1980s had also been favoured by a shift in the policy stance of the government away from a focus on attracting inward FDI in the Singapore economy to the active encouragement of outward FDI, given its key role in the domestic industrial upgrading and economic growth of Singapore. Such encouragement to outward FDI included the expansion of the remit of the Economic Development Board to include the facilitation of outward FDI as well as the provision of various investment incentives.[30]

The growth of Singapore-based FDI in the 1990s

The 1990s had been an era of the most rapid growth of Singapore-based FDI. Based on balance-of-payments data, the annual average FDI outflows of Singapore was $3.4 billion between 1990 and 1998 compared to $215.3 million in the period between 1980 and 1989 and $69.2 million between 1972 and 1979, and the estimated stock of Singapore-based FDI reached $47.6 billion in 1998 compared to $7.8 billion in 1990 and $819 million by 1981.[31] Thus, among the four Asian newly industrialized countries Singapore continued to be surpassed in importance as a home country of FDI by Hong Kong whose stock of outward FDI reached $154.9 billion in 1998. However, Singapore remained far a more important home country of FDI relative to Taiwan and South Korea whose stock of outward FDI was $38 billion and $21.5 billion in 1998, respectively (based on data in UNCTAD, 1999).

To the extent that financial factors played a role in explaining the growth of outward FDI from Singapore in the 1990s this proceeds from the continuing appreciation of the Singapore dollar by almost 12 per cent against the United States dollar between the end of 1989 and the end of 1997. By contrast to the previous decades, the accumulation of large financial surpluses and foreign exchange reserves in the 1990s had been created more significantly from net exports which grew to annual average level of almost 13.3 billion Singapore dollars between 1990 and 1996, a level more than 40 times larger than the annual average level of net exports of 321.5 million Singapore dollars between 1980 and 1989 and net imports of 1,268.9 million Singapore dollars between 1970 and 1979 and 408 million Singapore dollars between 1968 and 1969.[32] Such significant growth in net exports reinforced the continuing growth in the domestic savings rate fostered by the government policy of increased compulsory contribution to the Central Provident Fund (the national savings scheme) which as mentioned in the previous section increased the share of savings to GDP from 28.8 per cent in the period between 1970 and 1979 to 42.7 per cent in the period between 1980 and 1992 (based on data in Huff, 1995). As a result, the annual average foreign exchange reserves of Singapore reached $52.9 billion between 1990 and 1997 by comparison to $11.9 billion between 1980 and 1989 and $3.0 billion between 1970 and 1979. However, the bulk of such reserves had continued to be recycled through the growth of foreign portfolio investments and the growth of FDI by state-owned corporations which increased rapidly in significance in the 1990s (see Yeung, 1998).

The rapid expansion of Singapore-based FDI in the 1990s was a result of the emphasis placed on outward FDI as a means to overcome the vulnerabilities of Singapore to external shocks as displayed by the global economic recession in the mid-1980s which affected Singapore severely and, more importantly, to the dominant role of foreign capital in the domestic economy. A significant stock of outward FDI by local firms was envisaged to contribute to the greater resilience of the economy during economic recessions, reduce the dependence of Singapore on foreign capital for long-term economic growth as well as on developed countries for markets and divert concentration away from the domestic market at the expense of a potential gain through participation in the regional market boom. However, the stunted growth of private entrepreneurship owing to past suspicions regarding the pro-Communist and pro-China attitudes of indigenous capitalists by the PAP-ruled state, the growth in dominance of foreign-based MNCs as well as state-owned companies in major domestic economic activities and the savings-investment process which worked against the financial interests of private firms led to the continuing less significant role of privately owned firms in the outward FDI of Singapore in the 1990s. This was despite the trend towards the privatization of state-owned companies since the mid-1980s to allow for greater private sector participation. Thus, apart from wholly- or majority-owned foreign affiliates based in Singapore which accounted for 56.4 per cent of the stock of outward FDI of Singapore in 1993, a significant proportion of outward FDI generated by locally owned firms based in Singapore were those of state-owned corporations or government-linked corporations (GLCs).[33] Strengthened by the institutional support provided by state agencies and key politicians, these companies led the process of regionalization of Singapore-based firms in the 1990s whose FDI projects were mostly related to infrastructural developments located in Asia (Yeung, 1998). Nevertheless, efforts had been implemented to promote the long-term goal of enabling the private sector to play a more prominent role behind regionalization of Singapore-based firms.[34]

A second major determinant of Singapore-based FDI which explain the growth of their FDI in the developed countries has been the need to diversify in high technology industries. Indeed, there had been a significant increase in technology-related ventures by Singapore's state-owned and government-linked companies in the Silicon Valley of the United States to support the dynamic comparative advantage of Singapore in semiconductors production. This certainly explains the outward FDI of some $300 million by Singapore Technologies Holding – one of the three largest state-owned holding companies in Singapore – in a range of fledging computer companies located in the Silicon Valley in the early 1990s.[35]

In terms of geographical destination, some 53 per cent of the stock of outward FDI by Singapore in 1993 had been directed to the developing countries of Asia (of which some 28 per cent in South East Asia), a significant decline from their 66 per cent share in 1989 and 77 per cent share in 1981. Thus, although Asia continued to be a significant host region for Singapore-based FDI, the importance of South East Asia in particular declined from its share of 64 per cent in 1981. This is owing to the considerable expansion of outward FDI in Hong Kong which reached 4 billion Singapore dollars in 1993 by comparison to South East Asia which attracted 5.9 billion Singapore dollars. By contrast, some 23 per cent was directed to the developed countries by 1993 representing a considerable increase from 16 per cent in 1989 and less than 5 per cent share in 1981. The most important developed host countries and regions were the United States, New Zealand and Europe (including the Netherlands and the United Kingdom) that had always been the most significant host countries of Singapore-based FDI (based on data in Department of Statistics, 1996).

The continued significance of South East Asia for the outward FDI of Singapore had been facilitated by the sub-regional 'growth triangle' formed by Singapore, the southern peninsular Malaysian state of Johor, and the Riau island of Indonesia situated just south of Singapore in early 1990. A programme of coordinated public and private sector development projects in the growth triangle would combine the industrial expertise, technology, skills, infrastructure and services of Singapore with cheaper land and labour in Johor and Riau to accelerate the economic development of all countries. Joint infrastructure development projects and industrial incentive programmes have served to attract local and foreign factories in Singapore confronted with labour shortages, rising wages and a strong domestic currency to relocate to Johor and Riau which also offer cheaper storage and abundant land to Singapore-based companies (Lim, 1990; Lee, 1991).

As for industrial distribution, financial and manufacturing firms accounted for between 72 and 74 per cent of foreign direct equity investment of Singapore in 1992 and 1993. The most important industries of outward FDI in declining order of significance were finance, manufacturing, commerce, property and business services (based on data in Department of Statistics, 1996).

Conclusion

This chapter analysed the emergence and evolution of Singapore-based MNCs since the British colonial period until the present time. It showed that the history of Singapore-based MNCs is imbued inextricably with both the outward FDI of locally owned or indigenous Chinese firms in Singapore as well as the outward FDI of wholly or majority-owned foreign affiliates in Singapore. Indeed, the share of wholly and majority-owned foreign companies in the stock of outward FDI from Singapore reached 47.7 per cent in 1981 and although this share had been declining in the first half of the 1980s to reach a trough of 25.9 per cent in 1985, it climbed rapidly since to 33.6 per cent in 1989 and 56.4 per cent in 1993. Such high shares of foreign affiliates in the outward FDI of Singapore was closely associated with the history of Singapore as a British colonial entrepôt and a Crown colony since 1867 as well as the dominant role of foreign-based MNCs in the modern economic history of Singapore.

In terms of industrial distribution, outward FDI by Singapore by the end of the 1970s had been present in all the three major economic sectors – primary, secondary and tertiary. Outward FDI in manufacturing was spread over a broad range of industries that were either dominated by local capital or foreign capital. This included low technology, low value added industrial and product niches and industries typically intensive in the use of labour and natural resources to include food and drink, clothing, printing and publishing, plastic products, leather and rubber products, cement and concrete products (such as bricks, tiles and clay products, earthenware, glass and other non-metallic mineral products), fabricated metal products and transport equipment that were local capital-dominated industries in Singapore in the late 1970s and early 1980s. In addition, outward FDI in manufacturing also emerged in textiles, wood products, paper products and in higher technology and higher value added industries such as petroleum products and electrical and non-electrical machinery that were foreign capital-dominated industries in Singapore in the late 1970s and early 1980s.

Apart from outward FDI in manufacturing, outward FDI from Singapore was also significant in services in the period around the late 1970s and early 1980s. This was indicative of the dominant position of services in the GDP of Singapore at more than 71 per cent in 1979 (based on data in Huff, 1995) owing to the small size of the economy and its historical position as a trade entrepôt (Mirza, 1986). The most important service sectors of Singapore-based FDI had been civil engineering and construction, the marine industry (ship repair services, shipbuilding, oil rig construction, as well as marine engineering and related industries), hotels and construction, trading and property development.

The even more rapid expansion of outward FDI from Singapore in the 1980s relative to previous decades had been accompanied by significant changes in its industrial and geographical patterns. This owed much to the swift process of domestic industrial restructuring towards higher value added manufacturing starting around the late 1970s and in services associated closely with efforts of Singapore to promote itself as an international centre for offshore banking, finance and other services especially trading, transport and communications from around the early 1970s (see Mirza, 1986). Manufacturing and services continued to be the most important economic sectors of Singapore-based MNCs relative to the primary sector. Owing to the success of domestic industrial upgrading of the Singapore economy, outward FDI in manufacturing at the end of the 1980s was spread over a far broader range of older and newer industries by comparison to the end of the 1970s with the range of industries continuing to encompass those that were dominated by either local capital or foreign capital in Singapore. The 1980s also witnessed the further growth of outward FDI in services by major state-owned enterprises such as Singapore Airlines (which has expanded into finance) and Keppel Shipyards (presently involved in retail banking, international finance and insurance), among other state-owned companies. By the 1990s, the most important industries of Singapore-based outward FDI in declining order of significance were finance, manufacturing, commerce, property and business services.

As for geographical destination, some 77 per cent of the stock of outward FDI by Singapore in the end of the 1970s and early 1980s had been concentrated in the developing countries of Asia, with some 64 per cent in South East Asia. Less than 5 per cent was directed to the developed countries mainly in the United Kingdom and the United States but there had also been some investments in the Netherlands, other Europe and Japan. By 1989, some 67 per cent of the stock of outward FDI by Singapore continued to be directed towards the developing countries of Asia with some 39 per cent in South East Asia, while 16 per cent was directed to the developed countries, showing the increased importance of outward FDI in New Zealand, Europe (particularly the Netherlands and the United Kingdom) and the United States and directed largely towards property and property development. By 1993, the share of the stock of outward FDI by Singapore directed towards the developing countries of Asia further declined to 54 per cent with some 28 per cent in South East Asia, while the share directed to developed countries increased further to 23 per cent.

In a manner typical of MNCs based in small countries such as Sweden and Switzerland but not Hong Kong whose outward FDI was fundamentally of the export platform type, the major determinant of outward FDI by Singapore-based firms in manufacturing and services stemmed from the need to find new markets and investment opportunities and to sell their technological expertise. The main constraint to the growth of Singapore-based firms had been the limited size of the domestic market combined with growing protectionist policies in major export markets in the region. Besides limited growth in the domestic market and to sell technological expertise, another important determinant inducing international production particularly for state-owned firms in mature industries had been the need for diversification towards high-technology industries such as computers, microelectronics, robotics, biotechnology and genetic engineering. However, given that developed countries accounted for less than 5 per cent of the stock of Singapore-based FDI in the late 1970s and early 1980s, this factor carried far less weight in the determination of the outward FDI of Singapore in this period.

Although these determinants continued to be valid in explaining the growth of outward FDI in the 1980s, the expanding levels of outward FDI in that decade and particularly since 1986 had been spurred by the accelerated pace of wage increases in Singapore and the tightening labour market problem which owing to tighter foreign labour and immigration restrictions could no longer be relieved by unskilled labour importation as in the 1970s. By the 1990s, the rapid growth of Singapore-based FDI particularly within its region had been associated by the role of outward FDI as a means to overcome the vulnerabilities of Singapore to external shocks as displayed by the global economic recession in the mid-1980s which affected Singapore severely and, more importantly, to the dominant role of foreign capital in the domestic economy. A second major determinant of Singapore-based FDI in the 1990s which explain the growth of their FDI in the developed countries had been the need to diversify in high-technology industries. Indeed, there had been a significant increase in technology-related ventures by Singapore's state-owned and government-linked companies in the Silicon Valley of the United States to support the dynamic comparative advantage of Singapore in semiconductors production.

Notes

1 Based on data contained in UNCTAD (1999).
2 Based on data contained in Department of Statistics (1991, 1996).
3 Thus, as in the history of British MNCs, there was evidence in the case of Singapore-based MNCs of the presence of a *migrating multinational* coined by Jones (1986a). In the case of Britain, this was a company whose headquarters was based originally in one foreign country, invested in Britain and then evolved to become a British-headquartered MNC over time. Such was the case of Borax Consolidated Ltd (Travis and Cocks, 1984) and British-American Tobacco (BAT) (Jones, 1986a). This was similarly the case of Jack Chia MPH, Wah Chang International and Prima Flour that were companies established abroad by Chinese families that transferred their operations to Singapore in the 1970s and eventually became Singapore-based MNCs. In addition, among the formerly foreign-owned companies that eventually become Singapore-based firms and MNCs were Wearne Brothers (originally a distributor of motor vehicles and equipment) and Fraser & Neave Ltd (a soft drinks company). Both companies were acquired

by the Overseas Chinese Banking Corporation, one of the largest banks in Singapore (Pang and Komaran, 1985).

4 These published data are contained in Department of Statistics (1991, 1996). These contain annual data on outward FDI from Singapore as from 1976. The data on outward FDI refer to the amount of paid-up shares of overseas subsidiaries and associated companies held by companies in Singapore. Direct equity investment refers to direct investment plus the reserves of the overseas subsidiaries and associates attributable to these companies. For overseas branches, the net amount due to the local parent companies is taken as an approximation of the magnitude of direct investment.

5 Prior to the opening of the Suez Canal in 1869, ships often took the shortest route from Europe to East Asia via Java and Sumatra bypassing Singapore. With the opening of the Suez Canal, the strategic location of the island port of Singapore on the new route enabled it to serve as a *conduit* for exports from East Asia to Europe and manufactured goods in the opposite direction. Singapore's entrepôt trade thus increased rapidly despite the challenge of Hong Kong after 1842 (Mirza, 1986).

6 The lack of resources to finance industrialization is, in fact, debatable. Although the ratio of savings to GDP was low at 6.7 per cent in the period between 1960 and 1966 thus limiting the amount of reserves to finance investment (based on data in Huff, 1995), most local firms were engaged in trading or banking at the cornerstone of Singapore's industrialization in the 1960s (Lecraw, 1985; Yoshihara, 1976; Hughes and Yon, 1969). In addition, Singapore at the end of the 1950s had inherited from its entrepôt tradition an invaluable access to financial resources and market information as well as various forms of contacts within the region (Wong, 1979). Thus, it is difficult to comprehend why the Chinese immigrants in Singapore did not establish powerful alliances or *guanxi* with the Chinese business community in different countries of Asia to avail of business information as well as financial and marketing assistance. As seen in Chapters 10 and 14, these ethnic alliances enabled the Chinese to play an economic hegemonic role in the business and commerce of South East Asia, and the growth of Taiwanese MNCs and Hong Kong-based MNCs served to reinforce that hegemonic role.

7 In 1961, the Economic Development Board (EDB) of Singapore was established as a one-stop investment promotion agency to assist foreign firms in Singapore. This had been the main preoccupation of EDB until the 1990s. It has played a key role in shaping the Singapore economy and developing the industrial sector through its efforts in investment promotion and training of manpower (Low *et al.*, 1993). The establishment of the Jurong Town Corporation (JTC) in 1968 represented another institutional structure to support the state's strategy to rely on foreign capital for industrial development. It was primarily responsible for the construction and management of industrial estates as low-cost production sites for foreign manufacturing firms, the first of which was located in the Jurong Area. Both EDB and JTC have been instrumental in attracting a large inflow of FDI into Singapore since the 1960s (Yeung, 1998).

8 By 1983, the state had invested directly in 58 diverse companies with a total paid-up capital of 2.9 billion Singapore dollars. These companies wholly or partially owned some 490 firms in Singapore (Huff, 1995). The state-owned firms and statutory boards amassed profits of 5 to 7 billion Singapore dollars in 1983 equivalent to a considerable third of total GDP or half of indigenous GDP (Mirza, 1986).

9 Based on data contained in the FDI database of UNCTAD.

10 Data obtained from Department of Statistics (1991) indicated that the stock of outward FDI from Singapore reached 1,677.7 million Singapore dollars in 1981. This was converted to United States dollars using the end-of-period exchange rate in 1981 obtained from the International Monetary Fund, *International Financial Statistics Yearbook 1998*, Washington, DC: IMF. The value of outward FDI stock

from Singapore in United States dollars – $819 million in 1981 – was then compared with the stock of outward FDI from Hong Kong ($148 million in 1980), South Korea ($142 million in 1980), and Taiwan ($97 million) based on data contained in UNCTAD (1999).

11 In fact, the annual average net exports of goods and services of Singapore worsened from −408 million Singapore dollars between 1968 and 1969 to −1,268.9 million Singapore dollars between 1970 and 1979 before increasing to 321.5 million Singapore dollars between 1980 and 1989 and 13,280 million Singapore dollars between 1990 and 1996. The annual average foreign exchange reserves nevertheless increased from $762 million between 1968 and 1969 to $3.0 billion between 1970 and 1979, $11.9 billion between 1980 and 1989 and $52.8 billion between 1990 and 1997. This was much higher compared to the case of South Korea where the annual average foreign exchange reserves increased from $468.6 million between 1968 and 1969 to $1.4 billion between 1970 and 1979, $5 billion between 1980 and 1989 and $21.7 billion between 1990 and 1997. Based on data contained in the International Monetary Fund, *International Financial Statistics Yearbook 1998*, Washington, DC: IMF.

12 Government policy since the 1960s to increase the share of savings to GDP went hand in hand with policies to increase the share of investment to GDP. The gamut of measures implemented to enforce this included low corporate and personal taxation, investment incentives for foreign and locally owned private firms, government investment, forced savings via the government-run pension plan to which employers and employees must contribute, the creation of an attractive investment climate through infrastructure and human resource development and tight control of labour costs and practices. The savings-investment gap was financed by borrowing from abroad and, most importantly, by inward FDI (Lecraw, 1985). Given this gap between domestic savings and domestic investment in the 1970s, it may in fact be debatable if the large amount of savings generated by the compulsory savings scheme could have financed outward FDI given that it had not even been enough to finance domestic investment.

13 The analysis was based on information on the significant manufacturing industries of outward FDI around the end of the 1970s and early 1980s contained in Lecraw (1985) and Pang and Komaran (1985) and information on the industries dominated by local and foreign capital in 1982 contained in Mirza (1986). The finding that outward FDI by Singapore in the manufacturing sector had been in industries that were either dominated by local capital or foreign capital is at odds with the finding of Lecraw (1985) who argued that outward FDI by Singapore in the manufacturing sector had been concentrated largely in industries in which the share of inward FDI in Singapore was relatively low.

14 Data and information contained in Mirza (1986) indicate that construction and property are important sectors of state-owned enterprises in Singapore. There were 27 such enterprises in these sectors in the mid-1980s, of which 17 were already incorporated by 1979.

15 The internationalization of the state-owned company Acma Electrical Industries through outward FDI and arms-length technology contracts is an excellent case in point. Protected by tariffs in the 1960s and 1970s, Acma developed ownership-specific advantages in the production of refrigerators. However, the lifting of tariffs in the late 1970s led to the loss of its price competitiveness *vis-à-vis* imports, thus forcing the firm to seek foreign markets and sell its acquired technology. The range of its international business activities varied from the conclusion of technology contracts to produce refrigerators in Pakistan and Sri Lanka to the provision of technological assistance in Nigeria and Kenya for the assembly of refrigerators and airconditioners. It was only in China that the firm engaged in FDI through the establishment of a refrigerators plant, and had

intentions to build another three factories in other provinces (Pang and Komaran, 1985).

16 Data obtained from Department of Statistics (1996) indicated that the stock of outward FDI from Singapore reached 13,621.7 million Singapore dollars in 1990. This was converted to United States dollars using the end-of-period exchange rate in 1990 obtained from the International Monetary Fund, *International Financial Statistics Yearbook 1998*, Washington, DC: IMF.

17 Based on data contained in International Monetary Fund, *International Financial Statistics Yearbook 1998*, Washington, DC: IMF.

18 However, in certain cases the investment resulting in around 5 per cent of the equity ownership in a foreign holding company in turn resulted in the acquisition of significant equity stakes in companies of the holding company. For example, in one of its biggest investments abroad Singapore invested $465 million in May 1991 to acquire 5 per cent of Brierley Investments, a holding company in New Zealand. Such investment led Singapore to acquire significant equity stakes in a wider range of companies held by Brierley Investments in the United States and United Kingdom including Cummins Engine, La Quinta Motor Inns and Playboy Enterprises as well as 30 per cent of Brierley's biggest holding, Mount Charlotté Investments which owned 104 hotels in Britain, including London's White Hotel. See 'Singapore goes global', *Fortune*, 15 July 1991.

19 Based on data contained in Department of Statistics, *Report on the Census of Industrial Production, 1976–1985*, Singapore: DOS and Economic Development Board, *Report on the Census of Industrial Production, 1986–1989*, Singapore: EDB. Foreign firms are defined to be firms that are wholly owned and majority owned by foreigners.

20 Based on data contained in Mirza (1986).

21 This included the acquisition of the Thanksgiving Tower in Dallas, Texas for $160 million in 1989. However, there were large investments in other industries such as food products as seen in the $52 million acquisition of the Chun King Oriental food line from RJR Nabisco in 1989. See 'Singapore goes global', *Fortune*, 15 July 1991.

22 Calculated based on data on inward FDI flows in Thailand contained in ADB (1988) for data for the years between 1980 and 1985 and in Tan *et al.* (1992) for data for the years between 1986 and 1990.

23 Based on data and information on inward FDI in Malaysia contained in ADB (1988) and Tan *et al.* (1992) for data for the years between 1980 and 1990.

24 Calculated based on data on inward FDI in Indonesia contained in Lee (1990) for data for 1979, and in ADB (1988) and Tan *et al.* (1992) for data for the years between 1980 and 1990.

25 Calculated based on data on inward FDI in the Philippines contained in Yeung (1995) for data for 1991.

26 Over half of the 117 Singapore-owned firms with foreign operations in 1985 were minority joint ventures, and only 35 had wholly or majority-owned foreign sub-sidiaries. In China, the government defined specific modes of participation by foreign firms: processing and assembly, compensation trade, co-production, joint ventures and technology contracts (Pang and Komaran, 1985).

27 Based on data on the sectoral distribution of Singapore-based FDI in Indonesia, Malaysia, Philippines, Thailand and China at or around the end of the 1980s contained in UN, ESCAP (1988) and other sources. However, outward FDI in the primary sector continues to be important. For example, ECI Mineral of Singapore concluded a joint venture with the Burmese state-owned No. 3 Mining Enterprise to produce and market drilling-mud-grade baryte powder. Two other Singapore-based firms, Natsteel Trade International and Inotech Industries, have also concluded joint ventures in Burma to produce construction materials. See 'Singapore firms set up joint ventures in Burma', *Far Eastern Economic Review*, 18 April 1991.

28 Private sector enterprises have also engaged in large-scale outward FDI in services. This was the case, for example, with Hotel Properties Ltd (HPL), Singapore's premier hotel and leisure concern. However, the direct investments abroad of the company were channelled through a series of holding companies registered mostly in Hong Kong and the British Virgin Islands. Through these foreign holding companies, HPL owns *inter alia* most of the Four Seasons hotels in the United States and Canada, the Sydney Hilton, a half stake in the Inn on the Park in London, and the Australian franchise for Hyundai cars. See 'Suite dreams: Singapore hotelier in expansion drive', *Far Eastern Economic Review*, 8 August 1991.

29 In a crude regression model relating outward FDI stock of Singapore (the dependent variable) and the monthly wage earnings in manufacturing in Singapore (the independent variable lagged by one year) in the period between 1980 and 1987, I had found that ß, the value of the coefficient of the independent variable, is positive as predicted confirming the positive association between the two variables. Moreover, 92 per cent of the variation in the values of outward FDI stock of Singapore in the period between 1981 and 1988 was accounted for, or explained by, a linear relationship with the values of the monthly wage earnings in manufacturing in Singapore and the regression equation was significant at the 99 per cent level. It would have been preferable to relate outward FDI stock of Singapore in manufacturing (the dependent variable) to the monthly wage earnings in manufacturing (the independent variable). However, such outward FDI data by sector was unavailable from the Department of Statistics of Singapore.

30 The Economic Development Board (EDB) initiated in 1988 the International Direct Investment Programme as a means of extending financial inventives to outward FDI that seek to gain access to advanced foreign technology, create high value added engineering and technical employment, provide access to overseas markets and expand the scope of domestic activities in Singapore. EDB offices abroad had also been given the task of assisting Singapore-based companies to find investment opportunities abroad. In addition, companies are also allowed tax write-offs for losses incurred in overseas ventures and in April 1989 it was announced that government equity partnership was available for Singapore-based firms wanting to engage in outward FDI (Chia, 1989).

31 Based on UNCTAD FDI database.

32 Based on data contained in International Monetary Fund, *International Financial Statistics Yearbook 1998*, Washington, DC: IMF.

33 GLCs are state-owned firms that had been privatized since the mid-1980s but in which the state still retains significant influence over their management control (Yeung, 1998).

34 In full recognition of the important role of small- and medium-sized enterprises (SMEs) in surviving the economic recession in the mid-1980s, policy initiatives to promote the growth of the private sector became more concrete than ever before. This included the Local Industry Upgrading Program initiated in 1986 in which participating foreign-based MNCs in Singapore provide focused assistance to their local suppliers to upgrade their operations and become more competitive (Economic Development Board, 1991), the SME Master Plan published in 1989 and the revised Local Enterprise Finance Scheme announced in 1992 which provides low-cost loans for the purchase of equipment and industrial facilities needed for overseas operations. The most important document, however, has been the Strategic Economic Plan published in 1991 in which the state regarded locally based MNCs as one of the strategic tools in enabling Singapore to attain a fully developed country status. By 1992, there were over 60 such schemes and programmes addressing a broad spectrum of business needs (Economic Development Board, 1993). There were also a number of tax incentives to encourage local enterprises to invest abroad, and training programmes for key operators,

supervisors and engineers to receive training in Singapore as well as study missions organized by industry associations or government agencies as well as the Economic Development Board and Trade Development Board to develop an understanding of operating environments in foreign countries and the range of business opportunities available (Yeung, 1998).

35 See 'Silicon implants: Singapore gambles on US technology ventures', *Far Eastern Economic Review*, 6 February 1992. However, it had been reported that Singapore Technology Holdings Corporation was making losses in its operations in the United States in the production of wafers for integrated circuits (Hu, 1995).

16 Conclusion

The emergence and evolution of multinational corporations from the resource-scarce small countries

This part of the book analysed the emergence and evolution of resource-scarce small countries. In particular, it compared the growth of the MNCs from Switzerland, Hong Kong and Singapore. As in the previous parts of the book, the main rationale was to determine whether there exists a common pattern in the emergence and evolution of MNCs from resource-scarce small countries. The general conclusion that can be drawn is that there are some comparable patterns in the emergence and evolution of MNCs from the three countries that share a common pattern of national economic development based on their small size and natural resource scarcity which led to the high degree of outward orientation of their economies both though exports and outward FDI. Multinational corporations based in these countries share a common origin in international production in simple manufacturing that were intensive in the use of labour or natural resources (see Table 16.1). Despite the close similarity in the type of FDI in which MNCs from the three resource-scarce small countries emerged, there were important differences with respect to the types of firms that undertook the initial FDI, the actual development paths of outward FDI as it related to local industrialization in each country (Table 16.2) and the form of technological accumulation of leading national firms in relation to the natural course of outward FDI in each country (Table 16.3). This chapter seeks to reflect on the similarities and differences in the emergence and evolution of MNCs from the resource-scarce small countries.

The emergence of multinational corporations

The history of MNCs based in the three resource-scarce small countries have a common basis in simple manufacturing and particularly in the textiles industry in the case of both Switzerland and Hong Kong. The emergence of MNCs was associated closely with the important role of the textiles industry in domestic production and exports of the economies of Switzerland in the nineteenth century, in Hong Kong since the 1950s, and later in Singapore in the 1970s and 1980s. In Switzerland and Hong Kong, the outward FDI was initiated by individual adventurers or families with established interests in

Table 16.1 Resource-scarce small countries: variations in the early stages of outward direct investment across different countries

Examples of countries	Dominant form of earliest outward FDI	Type of locally based MNC
Switzerland	Trade, commerce and banking	Prominent families
	Insurance	Small insurance companies
	Textiles (cotton, silk)	Individual adventurers or families with established interests in the industry; small- and medium-sized firms
Hong Kong	Manufacture of simple consumer goods (kerosene lanterns, flashlights, umbrellas and simple food products)	Manufacturing firms, small- and medium-sized
	Textiles	Manufacturing firms, small- and medium-sized
Singapore	Productive endeavours in the Malay peninsula to include rubber plantations, etc	British trading companies
	Productive endeavours in the Malay peninsula in tin mining	Chinese immigrants in Singapore
	Simple manufacturing in South East Asia to include food products	Indigenous manufacturing companies

Source: Author's compilation based on the analysis contained in the country chapters.

the industry, and/or small- and medium-sized manufacturing companies. Perhaps a major difference between Switzerland and Singapore on the one hand and Hong Kong on the other lies in the domestic management of labour shortages and rising wages in labour intensive industries generally, the textiles industry inclusive. While outward FDI was but one aspect of a range of solutions adopted by Switzerland and Singapore to deal with its domestic labour market problem to include the employment of migrant workers, automation and specialization in more differentiated segments of the market or domestic industrial upgrading in more advanced manufacturing industries, there was little evidence that solutions other than outward FDI had been pursued in Hong Kong.

The determinants of international production in the textiles industry in the three countries have a common basis in the rapid loss of comparative advantage of the home country as a location for labour intensive production due to high production costs (associated with rising land costs and wages with industrial development and growth of labour productivity and shortage of semi-skilled and unskilled labour) and the imposition of tariffs and quotas in major export markets. Some subtle differences can, however, be detected in the relative importance of these factors as well as other factors for MNCs based in the three countries. In the period between 1750 and 1850–60, the lower cost of the factors of production in Italy primarily was probably the most critical factor in the relocation of production by individual Swiss adventurers or families with established interests in cotton manufacture and silk spinning (Wavre, 1988). The growth of these investments since 1850–60 owed much to the new Italian tariffs of 1 March 1888 which led to the significant decline in the value of Swiss exports to Italy, including cotton fabric. The imposition of more rigorous labour legislation in Switzerland in 1898, with the prohibition of employment of school children also forced the relocation to Italy of Swiss companies in silk spinning in the period around the turn of the twentieth century. On the other hand, the establishment of new foreign-based factories in the cotton industry just across the Swiss border in Southern Germany was caused by the formation of the German customs union in 1834 which threatened the continued growth of Swiss firms in a highly export-oriented industry.

By contrast, the initial thrust to the emergence of Hong Kong MNCs in textiles and clothing in the 1960s was provided by trade barriers imposed by important export markets in both developed and developing countries. The imposition of trade barriers was a response by developed countries to the phenomenal growth of Hong Kong's manufactured exports and an instrument of import substitution industrialization in developing countries.[1] Although the emergence of Hong Kong-based MNCs is rooted fundamentally in the imposition of tariffs and quotas which impinged upon the growth of its exports of manufactured products (the foundation of the Hong Kong economy), the further expansion of their international production activities beginning in the late 1960s became even more imperative owing to the *additional* motivation to minimize production costs, to relieve domestic supply shortages of semi-skilled and unskilled labour as well as to overcome increasing competition between domestic firms based in a small home market as well as the competition posed by the emergence of firms based in other Asian NICs that became significant producers and exporters of similar manufactured products (Table 16.2). Unlike that of Swiss MNCs and Singapore-based MNCs whose international production have largely been geared to local markets in host countries, export-oriented international production had always been the main focus of Hong Kong-based MNCs since their emergence associated with their mastery of technologies and long production experience in the assembly and manufacture of light consumer goods for exports and a large reservoir of entrepreneurial and export-marketing know-how and access to global distribution channels. Moreover, the main type of firm that initiated the manufacturing FDI of Hong Kong was small-and medium-sized manufacturing companies, almost all of which were privately owned (Pang and Komaran, 1985).

Table 16.2 Resource-scarce small countries: actual development paths for outward direct investment, and their association with local industrialization across different countries

Examples of countries	Link between domestic development and the growth of outward FDI	Type of locally based MNC
Switzerland	Historical position of the country as a trading, commercial and financial centre	Prominent families in trading, commerce and banking
		Companies in insurance and reinsurance
	Growth of domestic production and exports in the textiles industry. International production fostered by the formation of the German customs union in 1843, foreign tariffs, high domestic wages and land costs, the imposition of more rigorous labour legislation in Switzerland in 1898 and the availability of raw materials	Individual adventurers or families with established interests in industry
		Small- and medium-sized manufacturing firms
	Domestic industrial upgrading in processed metals	Manufacturing firms, typically resource-based but fully internationally integrated
	Accumulation of expertise in mechanical engineering and electrical equipment allied to accumulated strengths of domestic firms in textiles and processed metal products. International production prompted by a strategy to penetrate foreign markets in durable goods, non-tariff trade barriers associated with buying practices of major customer groups abroad and differences in technical standards as well as high transportation costs	Manufacturing firms, horizontally integrating

	Accumulation of expertise in the processing of agricultural commodities into manufactured food products protected by trademarks and patents, and the marketing of these products to the retail trade. International production prompted by a strategy to penetrate foreign markets, overcome trade barriers and to guarantee a steady supply at the retail level of essential food commodities by having a local presence	Manufacturing firms, horizontally integrating
	Accumulation of expertise in the production of speciality chemicals and pharmaceutical products. International production prompted by a strategy to penetrate foreign markets, reduce risks and to overcome tariff and non-tariff barriers protection accorded to pharmaceuticals and chemicals in foreign markets	Manufacturing firms, horizontally integrating
	Expertise and long experience in the management of hotels	Hotels
	Established reputation in banking, finance and insurance	Banks, finance and insurance companies
Hong Kong	The transfer to Hong Kong from Shanghai of industrial capital and managerial expertise in textiles production led to the installation of modern factories combining new production machinery and cheap refugee labour. This made textiles become the core of the industrial development of Hong Kong in the 1950s, an industry that had from its inception a high export propensity and formed the basis of Hong Kong's export-led growth. Such export expansion was facilitated by the existence in Hong Kong of some 1,000 to 1,500 trading houses previously involved in the entrepôt trade with well entrenched export links with British and other export markets. Exports of textiles faced tariffs and quotas in the 1960s	Textiles and clothing companies

Table 16.2 (continued)

Examples of countries	Link between domestic development and the growth of outward FDI	Type of locally based MNC
	The emergence and growth in the late 1950s of domestic production and exports of electronics. International production initiated by the defence of export markets which imposed protectionism in the mid-1970s as well as the minimization of production costs	Electrical and electronics companies
	Since the late 1960s and 1970s international production as a means to minimize production costs, to relieve domestic supply shortages of semi-skilled and unskilled labour, and to overcome the increasing competition between domestic firms in Hong Kong as well as the growth of emerging competition from firms based in the Asian NICs	Manufacturing firms in textiles, clothing, electrical and electronics, plastics and toys and other labour intensive industries
	International production as a means to develop industries that have been of little or no significance to the economy of Hong Kong owing to its limited size, lack of natural resources or environmental considerations	Manufacturing firms in chemicals, metal products, food, paper, rubber, wood, etc.
	Historical position of the country as an entrepôt facilitated by the local establishment of large British companies in the banking and finance, trading and services industries starting in 1841. Deindustrialization and the country's small size led to the development of the services sector and the transformation of the former British colonial entrepôt into a leading trading and financial centre	Services firms (trading, banking, finance, insurance, construction, tourism, etc.)

	The requirement of timber as a raw material for the domestic furniture manufacturing industry	Manufacturing firms engaged in backward vertical integration; free-standing companies or investment companies that treat FDI as a form of portfolio investment
	The requirement of the domestic economy and firms for advanced foreign technologies and to obtain more secure access to the parts, components and semi-manufactures for the assembly production in Hong Kong of some electronics products and watches	Manufacturing firms
Singapore	Its history as a British colonial entrepôt and a British Crown colony since 1867 that was used as a base for the economic expansion of the British empire in Malaya, South East Asia and China until the Second World War	British trading companies
	The dominant role of foreign based MNCs in the modern economic history of Singapore led to the use of Singapore as a base for the growth of their outward FDI in South East Asia and China	Foreign affiliates undertaking outward FDI in the region
	The development of Singapore as an international financial centre, the presence of a strong domestic currency, domestic political and macroeconomic stability, the relatively liberal financial regime as well as a safe haven for some excess savings of countries in South East Asia has also made the country an important financial base from which foreign companies launch outward FDI in the region	Foreign affiliates and other foreign companies undertaking outward FDI in the region
	The dominant role of state-owned companies and government-linked companies in the domestic economy of Singapore owing ot the neglect of local Chinese entrepreneurs for social, economic and political reasons	State-owned companies Government-linked companies

Table 16.2 (continued)

Examples of countries	Link between domestic development and the growth of outward FDI	Type of locally based MNC
	Development of local firms in low technology, low value added industrial and product niches and industries intensive in the use of labour and natural resources to include food and drink, clothing, printing and publishing, plastic products, leather and rubber products, cement and concrete products, fabricated metal products and transport equipment which made these industries local capital-dominated industries in the late 1970s and early 1980s. International production was a means to overcome the main constraint to growth posed by the limited size of the domestic market and the growing protectionist policies in major export markets in the region	State-owned companies in manufacturing and services Privately owned manufacturing companies and conglomerate companies
	The development of textiles, wood products, paper products, petroleum products and electrical and non-electrical machinery as foreign capital-dominated industries in Singapore in the late 1970s and early 1980s	Foreign affiliates undertaking outward FDI in the region
	The small size of the domestic economy and its historical position as a trade entrepôt led to the increasing dominant role of services in the GDP of Singapore in the late 1970s and early 1980s	State-owned companies in services and privately owned conglomerate companies engaged in outward FDI in infrastructure development, construction and marine industry (ship repair services, oil rig construction, marine engineering, etc.)
	The prominent position of Singapore in oil refining and petrochemicals manufacturing and as the largest oil services centre in Asia. This fostered the growth of outward FDI in oil exploration in China and Australia	State-owned companies and privately owned conglomerate companies

The need for diversification towards high technology industries such as computers, microelectronics, robotics, biotechnology and genetic engineering	State-owned companies Privately owned companies
Swift process of domestic industrial restructuring towards higher value added manufacturing starting around the late 1970s. This led to the growth of domestic and international production in more advanced manufacturing industries. The accelerated pace of wage increases and tightening domestic labour market associated with more restrictive foreign labour and immigration regulations in the 1980s that affected adversely labour intensive manufacturing industries	Foreign affiliates Some privately owned companies
The development of Singapore as an international centre for offshore banking, finance and other services especially trading, transport and communications from around the early 1970s	Foreign affiliates, foreign companies, state-owned companies and privately owned companies engaged in outward FDI in banking, finance and insurance and other producer services (consultancies and property), distributive services (trade, transportation and communications) and social and personal services

Source: Author's compilation based on the analysis contained in the country chapters.

Table 16.3 Resource-scarce small countries: technological accumulation and the national course of outward direct investment

	Stages of national development		
	(1)	*(2)*	*(3)*
Form of technological competence of leading indigenous firms	Basic engineering skills, complementary organizational routines and structures	More sophisticated engineering practices, basic scientific knowledge, more complex organizational methods	More science-based advanced engineering, organizational structures reflect needs of coordination
Type of outward direct investment	Trading, commerce, banking, insurance, labour intensive and natural resource intensive manufacturing	Resource-oriented, market-targeted or export-oriented investments in more advanced manufacturing and services	Research-related investment and integration into international networks
Industrial course of outward direct investment	Simple manufacturing and services	Growth of fabricating industries targeted to local markets or more complex export-oriented manufacturing; growth of services	More sophisticated manufacturing and services systems, international integration of investment
Stage of development			
Switzerland	between 1750 and 1850–60	between 1860–70 and the Second World War	since the Second World War
Hong Kong	late 1950s and mid 1970s	since 1980s, but yet to fully develop	currently still unreached
Singapore	between 1960s and 1970s	since 1980s, but yet to fully develop	currently still unreached

Source: Author's compilation based on the analysis contained in the country chapters.

Unlike Switzerland in the nineteenth century and Hong Kong since the 1960s, textiles and clothing was never a prominent feature of the export pattern of Singapore. The dominance of food products in some of the earliest outward FDI from Singapore in South East Asia in the 1950s was superseded by the end of the 1970s in outward FDI in a broad range of manufacturing industries that were either dominated by local capital or foreign capital. This included low technology, low value added industrial and product niches and typically industries intensive in the use of labour and natural resources to include food and drink, clothing, printing and publishing, plastic products, leather and rubber products, cement and concrete products, fabricated metal products and transport equipment that were local capital-dominated industries in Singapore in the late 1970s and early 1980s. In addition, outward FDI in manufacturing also emerged in textiles, wood products, paper products and in higher technology and higher value added industries such as petroleum products and electrical and non-electrical machinery that were foreign capital-dominated industries in Singapore in the late 1970s and early 1980s. The different industries in which foreign or local capital dominated reflected the difficulties faced by local firms to penetrate the higher value added industries owing to their lack of technology, experience, entrepreneurship and size (Mirza, 1986). Thus, there were at least five types of firms that led Singapore-based manufacturing FDI at the end of the 1970s and early 1980s: foreign affiliates in Singapore undertaking outward FDI in the region, foreign companies using Singapore as a financial base to launch outward FDI, state-owned companies or government-linked companies in manufacturing and services, privately owned manufacturing companies, and privately owned conglomerate companies.

Despite the pressures imposed by an increasingly tight labour market and rising wages in Singapore since the early 1970s, this had not been a major factor in the relocation of labour intensive operations offshore in the 1970s owing to the ready access to foreign labour under the government's liberal worker policy (Chia, 1989). Instead, in a manner typical of MNCs based in small countries such as Sweden and Switzerland (later) but not Hong Kong whose outward FDI was fundamentally of the export platform type, the major determinants of international production by Singapore-based firms at the end of the 1970s had been to overcome the main constraints to growth posed by the limited size of the domestic market by the search for new markets and investment opportunities abroad as well as to resolve the growing protectionist policies in major export markets in the region.

There is, however, an analogous pattern of development in the geographical destination of the early outward FDI in manufacturing in the three countries. In all the three countries, outward FDI was focused on host countries that had close geographical proximity and cultural affinity to the home country. These were Germany, France and Italy in the case of Swiss FDI (Schröter, 1993). Their close geographical and cultural links to Switzerland has facilitated the consideration of these countries as an extension of the home country market both in terms of their importance as the first export markets of Switzerland and as a site for the location of production of the early Swiss MNCs. Similarly, Indonesia, Malaysia, Singapore, Taiwan and Thailand and other countries of South East Asia, South Asia as well as Africa have been important foreign sites for production by Hong Kong's textile manufacturers since the 1960s, and in China since 1978 (Mun and Chan, 1986). Some 77 per cent of the stock of outward FDI by Singapore at the end of the 1970s and early 1980s had also been concentrated in the developing countries of Asia, with some 64 per cent in South East Asia and in particular in Malaysia as the most prominent host country (based on data in Department of Statistics, 1991). The dominant role of South East Asia and China in the outward FDI of both Hong Kong and Singapore is explained not only in terms of close geographical proximity but also by the familiarity of both these countries with the economic, social and political conditions of Asian countries, gained in part through the historical roles of Hong Kong and Singapore as entrepôts for trade and commerce as well as financial centres in South East Asia. In addition, the outward FDI attributable to indigenous Chinese firms in Hong Kong and Singapore can be regarded as parallel to the expansion through outward FDI of indigenous Chinese firms in Taiwan to avail of the socially and culturally embedded networks of relationships or *guanxi* with the Chinese business community in different countries of South East Asia and China which provided business information as well as financial and marketing assistance (Granovetter, 1985, 1991; Granovetter and Swedberg, 1992). These regional business networks based on contacts spanning almost a century and close familial relations have provided strength to Chinese business organizations and paved the way for their economic hegemonic role in the business and commerce in the region (see also Wong, 1991). The emergence and growth in South East Asia of indigenous Chinese MNCs based in Taiwan, Hong Kong and Singapore only served to reinforce that hegemonic role.

The evolution of MNCs

Despite the common origins of the MNCs based in the three resource-scarce small countries in simple manufacturing, the evolution of their outward FDI while sharing some similarities particularly with respect to MNCs based in Switzerland and Singapore also display some notable differences. The closest similarity in the pattern of MNC evolution in the three countries derive from the rapid growth in dominance of the services sector in their respective home economies and in outward FDI. While in the case of Switzerland this phenomenon can be explained in the context of its post industrialization status, it is certainly also the case that Switzerland, Hong Kong and Singapore have large internationally oriented service industries because of their small size and historical position as entrepôts for trade in Europe and South East Asia both of which favoured a greater emphasis on the development of the services sector in their domestic industrialization strategies which led to the prominent role of the sector in gross domestic product and outward FDI.

The closest parallelism can be drawn in the growth and development of outward FDI in trading, banking, finance and insurance in all three countries. In the case of Switzerland, this arises from the country's well-entrenched historical position as a trading, commercial and financial power (Bergier, 1968). This was favoured by the country's central geographical location on major European trade routes and political neutrality that allowed its firms to benefit from the maintenance of commercial contacts with each of the major

European power centres (France, Germany and Great Britain) even during times of conflict. Its central location in Europe and well-entrenched position in trading and commerce in turn helped to foster its expertise in banking and insurance (Porter, 1990). The importance of general business services and personal services for Switzerland generally and as important industries of Swiss FDI also proceed from the highly advanced pattern of domestic demand for these services associated with the high per capita income of the Swiss economy and its position as a location of the regional headquarters of foreign firms and international organizations.

There had been two waves in the growth of the services sector in the history of Swiss MNCs: the first wave is identified with the pioneering roles of trading, commerce and banking in Swiss FDI by prominent Swiss families and in insurance by small companies. The second wave began in the decade of the 1960s with the multinationalization of firms in the banking, finance, insurance and other services industries. As a result, the share of the services sector in the stock of recent Swiss FDI has increased steadily from a share of more than 30 per cent in 1986, some 32 per cent in 1987 (Bürgenmeier, 1991), 38 per cent in 1988 and 56 per cent in 1997 (UNCTAD, 1999).

In an analogous fashion, the services sector has also been a prominent feature in the growth of Hong Kong MNCs since the 1980s and enjoys a steady growth. These were mainly in hotels, property, trading, leisure services, construction, and banking and financial services (Chen, 1983b). In the period since 1986, their FDI in services (mainly in banking and finance, hotels and distribution) in the developed countries (primarily in United States, United Kingdom and France) have grown in a significant way through large-scale acquisitions of domestic firms in those countries. The growth of significant outward FDI in banking and financial sectors by Hong Kong in particular is resonant of the growth and development of Swiss MNCs in these sectors and draws from a similar historical position of Hong Kong as an entrepôt for trade in its region facilitated by the local establishment of large British companies in the banking and finance, trading and services industries starting in 1841 with the inception of Hong Kong's colonial history. The process of deindustrialization combined with the country's small size also led to the growth of the domestic economy as a services-based economy and its transformation from a British colonial entrepôt into a leading trading and financial centre in the region (Ho, 1992). The growth of Hong Kong's outward FDI in the other services industries (hotels, property, leisure services, construction) is linked to the development of Hong Kong as a services-based economy (Table 16.2).

The services sector has likewise been fairly significant for the economy of Singapore and its MNCs. As in Hong Kong, the importance of FDI in services is closely associated with a similar historical position of Singapore as a British colonial entrepôt and a Crown colony since 1867 which made the country a base for the economic expansion of the British empire in Malaya, South East Asia and China until the Second World War as well as base for Britain's military expansion in East Asia until the late 1960s. Indeed, many British companies, including Boustead and Co., Simon and Paterson and Guthrie and Co., can trace their origins and early prosperity back to the beginnings of Singapore's role as an entrepôt (Mirza, 1986). In more modern times, Singapore has progressed beyond its traditional role as a British colonial entrepôt to fulfil a more complex and sophisticated role as a centre for trading, transportation, communications, industrial, commercial and financial services in South East Asia (Lim, 1990). Among the service industries in which Singapore-based MNCs have emerged and evolved were infrastructure development, construction, trading, marine industry (ship repair services, oil rig construction, marine engineering, etc.), property and property development, finance, and management and consultancy services. Unlike in the case of Switzerland and Hong Kong where almost all of the outward FDI in services have been generated by firms in the private sector, there were at least five types of firms that initiated and led Singapore-based services FDI in the 1980s and 1990s: foreign affiliates in Singapore undertaking outward FDI in the region, foreign companies using Singapore as a financial base to launch outward FDI, state-owned companies or government-linked companies in manufacturing and services, privately owned manufacturing companies, and privately owned conglomerate companies. Indeed, Singapore's conglomerate firms – Jack Chia-MPH, Wah Chang International, Haw Par Brothers, Intraco, Joo Seng Group and Keck Seng Group – have engaged in international operations as trading companies in the initial phases of their companies' history, long before their diversification towards the manufacturing sector (Lim, 1984).

Another distinctive feature in the growth of MNCs from the three resource-scarce small countries – that is unlike that of MNCs from the resource-scarce large countries – is the relatively lower significance of outward FDI in resource extraction particularly for MNCs from Hong Kong and Singapore. The outward FDI of resource-scarce *large* countries in resource-related activities was provoked by a necessity: the demand of domestic processing industries and consumers for minerals, energy, other raw materials and agricultural products that either did not exist in the home country or were available in inadequate amounts to support industrialization or meet domestic consumption. By contrast, a similar shortage of most natural resources and the need to import most raw materials as well as energy in Hong Kong and Singapore did not precipitate high levels of outward FDI to search for, and gain control over, sources of natural resources in resource-rich foreign countries apart from the resource-based investments to extract timber in Indonesia by Hong Kong-based firms to support the growth of the domestic furniture industry (described further below) and the earlier productive endeavours in the Malay peninsula in rubber plantations by the British trading companies based in Singapore and in tin mining by Chinese immigrants in Singapore during the British colonial period. This stems from the specialization of their economies in a narrower range of industries that did not involve the large-scale development of domestic processing industries. The marked exception is Switzerland that spawned highly successful firms such as Alusuisse-Lonza, von Roll, Georg Fischer and other firms in the processed metal products industries which enabled those industries to count as one of a broad range of industries in which Switzerland developed a highly competitive position. Unlike in the resource-abundant small country of Sweden, these strengths in metals processing cannot be attributed to the relatively few natural resources that Switzerland possesses, with the exception of its capability to generate relatively inexpensive hydroelectric power useful in the reduction of alumina to aluminium (see the case study of Alusuisse in Chapter 13).

By contrast, to the extent that there had been resource-based investments by Hong Kong-based firms which represented the backward vertical integration of manufacturing companies in resource extraction as an important means to stabilize supplies and costs associated with the instability of market prices for tropical timber (Wells, 1978), in most cases the timber extractive investments in Indonesia were accounted for by individuals or business groups which although based in Hong Kong either did not have parent companies in Hong Kong or had parent companies in Hong Kong in entirely different lines of business (Chen, 1981). Indeed, the resource extractive FDI in timber extraction that were mostly accounted for by investment companies based in Hong Kong had tended to regard FDI as a form of portfolio investment (Chen, 1981; Wells, 1978) to gain monopoly rents or minimize business risks through the diversification or entry into new business activities.

Beyond the analysis of the common grounds in the emergence and evolution of MNCs from the resource-scarce small countries lies the examination of major differences that exist between the three countries. One such difference arises from the wider breadth of industries that Swiss and Singapore-based MNCs have evolved into by comparison to MNCs from Hong Kong. Considering the manufacturing sector alone, Swiss MNCs have emerged first in textile industries of cotton, silk and straw in the eighteenth century as mentioned, and then evolved in the nineteenth and twentieth century in processed metal products, machinery, electro-technical and electro-chemical industries, processed food products (milk products, baby foods, chocolates, jams and preserves), speciality chemicals and pharmaceuticals, paper and graphics and so forth. Similarly, as mentioned there had been a wide breadth of manufacturing industries in which Singapore-based MNCs have emerged by the late 1970s and the swift process of domestic industrial restructuring towards higher value added manufacturing industries starting around the late 1970s meant that the outward FDI in manufacturing at the end of the 1980s and through the 1990s was spread over a far broader range of older and newer industries even if it continued to encompass those industries in Singapore that were dominated by either local capital or foreign capital.

Despite the wide scope of domestic industrial diversification and outward FDI in manufacturing in both Switzerland and Singapore, there were fundamental differences in the types of firms that initiated international production and the determinants of such international production in the two countries. The evolution of Swiss FDI in manufacturing has been led in the main by privately owned domestically based manufacturing firms engaging in horizontal integration in foreign markets in response to both a home-country specific factor (to overcome the limited size of the domestic market through the search of new markets and business opportunities abroad) as well as host country-specific factors (tariff and non-tariff trade barriers in major export markets, high transportation costs, the need for a local presence to guarantee effective market penetration and servicing, reduce risks, etc.). In Singapore, on the other hand, there were at least four types of firms that led Singapore-based FDI in the wide range of manufacturing industries at the end of the 1980s and through the 1990s: foreign affiliates in Singapore undertaking outward FDI in the region, state-owned companies or government-linked companies in manufacturing and services, privately owned manufacturing companies, and privately owned conglomerate companies. Although the expansion of Singapore-based MNCs in manufacturing in the 1980s continued to be driven as in the case of Swiss MNCs by the need to overcome the limited size of the domestic market and to expand production and service capacity, a major factor propelling the growth of outward FDI in the 1980s and particularly since 1986 had been the accelerated pace of wage increases in Singapore and the tightening labour market problem which owing to tighter foreign labour and immigration restrictions could no longer be relieved by unskilled labour importation as in the 1970s (see also Aggarwal, 1986, 1987; Chia, 1989; Hill and Pang, 1991). Thus, it seems that home country-specific factors played a more important role in explaining the growth of Singapore-based FDI in manufacturing than was the case for Switzerland.

By contrast, Hong Kong has a much narrower range of manufacturing industries that constitute its manufacturing sector. Clothing and textiles accounted for more than 52 per cent of Hong Kong's exports of manufactured goods in 1961, and such share had declined to only 44 per cent by 1987. To the extent that there had been some diversification of the manufacturing sector of Hong Kong, it is evident in the increased importance of two other labour intensive industries in the country's exports of manufactured goods: electronics whose share increased from 2.5 per cent in 1961 to almost 23 per cent in 1987 and watches, clocks and other precision instruments whose share increased from 0.7 per cent in 1961 to 8.7 per cent in 1987 (based on data in *Hong Kong Trade Statistics*, various issues). The case of Hong Kong is particularly distinctive in its concentration in the export-oriented manufacture of labour intensive products in both domestic and international production, with some limited evidence towards specialization in the domestic textiles and clothing industry in the more differentiated segments of the market.

Since the differences in the breadth of industries that MNCs from the three resource-scarce small countries does *not* cut along the divide between developed and developing countries, a more fundamental explanation has to be sought outside the framework of the stage of economic development attained by the home country. Indeed, other more relevant explanations for the difference may have to be gleaned from the demand conditions in the home country, government macro-organizational policies pursued in achieving industrial development generally (including the more *dirigiste* role of the government of Singapore versus the more *laissez faire* role of the government of Hong Kong in industrial development) and other country-specific factors (such as cultural diversity) which has tended to differentiate the three countries other than the different stages of their economic development.

The peculiar nature of Swiss demand conditions has supported a remarkably broad base of competitive industries for a small nation which has contributed both to the resilience of the Swiss economy, and the wide scope and continuity in the pattern of its MNC growth. To overcome the small size of the domestic market, Swiss firms deliberately chose to operate in highly specialized and highly productive industry segments. The multi-cultural environment of Switzerland along with the emphasis placed on regarding the neighbouring countries of Germany, France and Italy as extensions of the home country market has provided an invaluable window through which Switzerland discerned evolving product needs and widened the range of sophisticated buyers in close contact with Swiss industry, both of which enabled domestic firms based in Switzerland to develop extraordinary strengths in a much broader range of industries than would be expected for a small nation.[2] Cultural diversity may account for the wider breadth of advantage in Swiss domestic industry and MNC activity not only in comparison to the lesser developed, small countries of Hong Kong and Singapore but also in comparison to another developed, small country such as Sweden that similarly lacks such industrial diversity. The greater industrial diversity of Switzerland also helps to explain its greater need to overcome the limitations posed by the domestic market owing to its small size or the lack of natural resources through international production which facilitates access to larger foreign markets and primary raw materials.

Apart from the broader breadth of Swiss industrial development, another feature that distinguishes Switzerland from the lesser developed resource-scarce small countries such as Hong Kong and Singapore is the source of competitiveness of the country and its constituent firms. The products of Swiss industry – whether manufacturing or services – have generally been focused on quality or based on extensive R&D and technical expertise (Schröter, 1993). The broad breadth of Swiss industries that are technologically advanced helps to explain both the much larger scale of their outward FDI, and the greater propensity of the leading companies in those industries to engage in international production should exports prove incapable to fulfil, or continue to fulfil, foreign demand. By contrast, the products of indigenous firms based in Hong Kong and Singapore derive their competitiveness from lower cost labour or access to such lower cost labour. The indigenous firms in the Asian city states demonstrate relatively weaker capabilities in capital intensive industries and more technology intensive industries. This arises from the more non-interventionist policy of the government of Hong Kong in industrial development as mentioned but even the more *dirigiste* role of the government of Singapore has not made the benefits of rapid domestic industrial upgrading affect local firms. A major explanation for this can be found in the dominant role assigned to foreign capital to attain rapid industrialization and economic growth in Singapore particularly in more advanced manufacturing and services industries to the detriment of local entrepreneurs or indigenous capitalists whose potential contribution was neglected owing to social, economic and political reasons (Régnier, 1993).

In Hong Kong, not only has the non-interventionist policy of the government in industrial development tended to mitigate the importance of expenditures on R&D by both the public and the private sectors but also the peculiar pattern of domestic demand as well as export market demand for lower-quality, mass produced consumer goods. Notwithstanding the narrow scope of its domestic manufacturing sector, however, Hong Kong became the world's tenth-largest source country of FDI in the world economy in 1998 after the United States, United Kingdom, Germany, Japan, the Netherlands, France, Switzerland, Italy and Canada and the leading source of outward FDI from developing countries. The fact that its international production activities remained predominantly in the export-oriented manufacture of labour intensive products has not, in any way, made Hong Kong-based MNCs less successful controllers and coordinators of an international network of income-generating assets. Their mastery of technologies and long production experience in the assembly of labour intensive goods strongly reinforced by their ownership advantages in export marketing, stable relationships with major suppliers and access to global distribution channels have provided Hong Kong-based firms a unique, distinctive and sustainable competitive edge to maintain a leading role in labour intensive manufacturing over time, and in spite of the growth of other developing countries that have developed greater comparative advantages in labour intensive manufacturing.

Thus, in relating stages of national development to the form of technological competence of leading indigenous firms, the type of outward FDI and its industrial course over time, MNCs from Hong Kong and Singapore are at a much earlier stage of development compared to the more advanced MNCs from Switzerland, despite the similarity in some patterns of their MNC growth (see Table 16.3).

Notes

1 Exports of textiles and clothing from Hong Kong were subject to the quotas by developed countries prior to the establishment of the Multi-Fibre Arrangement (Chen 1981, 1983b). Similarly, in the late 1960s, tariff barriers were imposed by many South East Asian countries that pursued import substitution industrialization (UN, ESCAP, 1988).

2 This helps to explain the unique strengths of the French part of Switzerland in consumer goods, and that of the German part in precision machinery and chemically related industries. Similarly, owing to its extensive business dealings with Germany itself, Switzerland has been prone to adopt German technical and environmental standards, not having a national standards setting agency in its own country. Since German standards are tough, further beneficial pressures on the demand side are created in affected Swiss industries (Porter, 1990). Indeed, the home market of both the Swiss machine-building industry and the chemical industry includes Germany. It is for these reasons that Switzerland is considered a special case of Porter's home nation-based theory (Borner *et al.*, 1991).

Part V
Conclusion

17 The emergence and evolution of multinational corporations

Implications for theory

This book analysed the emergence and evolution of MNCs. In particular, it sought to test the hypothesis whether variations exist in the pattern of the early stages of outward FDI across different types of countries and in their developmental paths over time that are determined by distinctive patterns of national economic development. Such unique patterns of national economic development derive from different endowments of natural resources, different sizes of the domestic market and different types of development path pursued in achieving industrial development. Three country groups were described: resource-abundant countries, resource-scarce large countries (with resource intensive production) and resource-scarce small countries (with non-resource intensive production). The organization of countries in these groups facilitates the identification of the dominant form of earliest outward FDI and the developmental course or path of outward FDI over time with the changing forms of technological competence of leading indigenous firms.

The research hypothesis was tested on the basis of 11 case studies of MNCs based in 11 countries. These countries were Brazil, Germany, Hong Kong, Japan, Singapore, South Korea, Sweden, Switzerland, Taiwan, United Kingdom and the United States. These 11 home countries, comprising six developed countries and five developing countries, accounted collectively for two-thirds of the global stock of outward FDI in 1998. The six developed countries accounted for an equivalent share of the stock of outward FDI of the developed countries in 1998, while the five developing countries accounted for 69.5 per cent of the stock of outward FDI of the developing countries (based on data in UNCTAD, 1999). The selection of these 11 home countries of widely divergent national characteristics was necessary to demonstrate the relevance of the inter-relationships between the emergence and the evolutionary process of outward FDI and distinctive patterns of national economic development and to advance the development of a comprehensive theory of the emergence and evolution of MNCs on the basis of general principles.

This concluding chapter sets out those general principles. The reader is invited to refer back to the concluding chapters of the various parts of the book for more detailed and refined analysis of the emergence and evolution of

MNCs based on distinctive patterns of national economic development. (Chapters 6, 12 and 16).

Variations in the early stages of outward direct investment across different types of country

The findings of the research confirm the general validity of the hypothesis that patterns of national economic development determine the emergence and evolution of MNCs. However, the precise form of the relationship varies among the country groups.

Resource-abundant countries

The earliest outward FDI of MNCs based in the resource-abundant countries has tended to predominate in one of two principal activities. The first principal activity has been in resource extraction and sometimes resource processing in agriculture, forestry, petroleum or minerals in resource-rich host countries. The emergence and expansion of outward FDI in the extraction and processing of natural resources can be linked to the presence in the home country of rich and abundant natural resources which enabled indigenous firms to develop management and organizational skills and technologies in natural resource extraction and processing which were exploited profitably abroad. The second principal activity in their earliest outward FDI has been the establishment of sales and production subsidiaries in large foreign markets by firms in the engineering industries comprising metal manufactures, machinery and transport equipment in which firm-specific knowledge has been generated from an abundance of mineral resources in the home country (Table 17.1). The research findings on the early stages of outward FDI by MNCs in the resource-abundant countries thus support the validity of the hypothesis contained in Table 1.1 describing the dominant form of the earliest outward FDI in these countries.

Resource-scarce large countries

The dominant form of the earliest outward FDI of MNCs based in the resource-scarce large countries has also been in resource extraction and sometimes resource processing in agriculture, forestry, petroleum or minerals in resource-rich host countries. However, the driving factor behind this early outward FDI in MNCs based in the resource-scarce large countries differed fundamentally from that in MNCs based in the resource-abundant countries. Outward FDI in resource-based activities by MNCs in the resource-scarce large countries was provoked by the demand of domestic processing industries for minerals, energy and other raw materials and the demand of domestic consumers for agricultural produce. Owing to the resource scarcity of these countries, these vital needs of industries and consumers either did not exist in the home country or were available in inadequate amounts to support industrialization and domestic consumer demand. The shortage of natural resources in the large economies of the United Kingdom, Germany, Japan, South Korea and Taiwan became the driving force to search and gain control over new or additional sources of natural resources in resource-rich foreign countries. This explains much of the FDI by firms from the United Kingdom and Germany in the nineteenth century, as well as those of Japan in the 1950s and South Korea and Taiwan in the 1960s and 1970s.

So important was this objective for the United Kingdom that among the oldest British MNCs were 'free-standing' firms, i.e. firms which did not undertake any prior production in the home country before investing abroad but were created primarily for the purpose of undertaking resource-related business exclusively or mainly abroad. This was a prominent feature of the emergence of British MNCs (Wilkins, 1988a) which was not evident in the emergence of MNCs from Germany, Japan, South Korea and Taiwan where for the most part resource-related overseas activities were undertaken by manufacturing firms, trading companies and even banks (as seen in the historical oil-related activities of Deutsche Bank).

Apart from resource-related activities, manufacturing was also an important sector in the origins of MNCs from the five resource-scarce large countries although the manufacturing FDI that emerged across the five countries in the kinds of industries that spawned the earliest MNCs, the kinds of firms that initiated international production, the determinants of such international production and their principal host countries.

The international production of a significant number of British manufacturing MNCs emerged in branded consumer goods (Dickens, 1992; Chandler, 1986) which was a reflection to a large extent of the comparative advantages of the United Kingdom in labour intensive, capital neutral and human capital-scarce products (Crafts and Thomas, 1986) and the technological hegemony of the United Kingdom in the industries associated with the First Industrial Revolution. These were textiles or textiles-related industries and consumer goods industries intensive in the use of natural resources. By contrast, the industries that spawned the earliest German FDI in manufacturing were the new, skills intensive, technically advanced and fast growing industries of the Second Industrial Revolution in the late nineteenth century: chemicals, pharmaceuticals, machinery, electro-technical and motor vehicles. The emergence of Japanese MNCs, Taiwanese MNCs and South Korean MNCs in manufacturing bear a closer resemblance to the pioneering British MNCs in their concentration in industries reflecting a comparative advantage in labour intensive, capital neutral and human capital-scarce products rather than in the higher technological intensity of the early German MNCs. The emergence of the outward FDI of Japan in manufacturing perhaps bears the closest similarity to that of the United Kingdom in their common basis in textiles or textiles-related industries.

Table 17.1 Variations in the early stages of outward direct investment across different types of country

Categorization of national development	Country case studies	Dominant form of earliest outward FDI	Type of locally based MNC
Resource-abundant countries	United States Sweden Brazil	Resource extraction in agriculture, minerals or petroleum in resource-rich host countries	Resource-based firms Manufacturing firms, backwardly integrating
		Resource processing, distribution and marketing	Resource-based firms
		Installation of mercantile houses overseas as well as foreign branches of local merchants overseas	Colonial merchants
		The establishment of foreign sales branches to promote exports	Trading companies. Manufacturing firms, forwardly integrating
		Local market-oriented production in large host countries	Manufacturing firms, serving host country markets
		Railroads construction	Railroad companies, horizontally integrating Agricultural companies, forwardly integrating

(continued)

Table 17.1 (continued)

Categorization of national development	Country case studies	Dominant form of earliest outward FDI	Type of locally based MNC
Resource-scarce large countries	United Kingdom Germany Japan Taiwan South Korea	Resources extraction and processing and associated service investments to support domestic industrial expansion and to meet domestic consumer needs	Free-standing firms Firms in manufacturing, petroleum and trading, backwardly integrating
		Agriculturally based investments based on entrepreneurial perceptions of profitable investment opportunities	Free-standing firms Large integrated firms
		Trading investments in major export markets to promote the growth of exports	Trading firms Manufacturing firms, forwardly integrating
		Local market-oriented manufacturing production in response to trade barriers	Manufacturing firms, serving host country markets
		International production in small- to medium-scale labour intensive manufacturing in response to rise of domestic wages and other production costs in the home country	Manufacturing firms, serving local or export markets

(*continued*)

Table 17.1   (continued)

Categorization of national development	Country case studies	Dominant form of earliest outward FDI	Type of locally based MNC
		Banking investments to gain access to international capital markets and to promote the growth of exports	Banks
Resource-scarce small countries	Switzerland Hong Kong Singapore	International production in labour intensive manufacturing for local or export markets	Individual adventurers or families; small- and medium-sized firms in manufacturing
		Trade, commerce and banking	Prominent families
		Insurance	Small companies

Source: Author's compilation based on the analysis contained in the country chapters.

As to the kinds of firms that initiated international production in the resource-scarce large countries, although some of the historical British manufacturing FDI were initiated by free-standing companies and trading companies such as Jardine Matheson & Company that established the Ewo Cotton Spinning and Weaving Company in Shanghai before 1914 (Wilkins, 1986b), direct capital exports of the United Kingdom consisted more of the establishment of foreign subsidiaries and branches by manufacturing companies already operating in their home countries – essentially the kind of foreign activity of the modern classic MNCs which mainly predominates in modern times (Dunning and Archer, 1987). The German firms that initiated manufacturing FDI were also predominantly large manufacturing firms in industries producing the newest and most technologically advanced products whose emergence can also be attributed to the rapid development of administrative hierarchies in German business organizations (Wilkins, 1988b). As in the German case, manufacturing firms were also solely responsible for the earliest Taiwanese manufacturing FDI. These firms were often the largest firms in their respective industries (Schive and Hsueh, 1985) but were perhaps not as large as the pioneering German or British manufacturing companies. By contrast, two types of firms initiated Japanese and South Korean manufacturing FDI: manufacturing companies and trading companies. The emergence of Japanese FDI in manufacturing were initiated by the cotton spinning companies and trading companies thus bearing a closer similarity to the types of firms that generated the earliest British manufacturing FDI.

The determinants of international production by firms from the five resource-scarce large countries in both the traditional and modern industries have a common basis in the inability of important foreign markets to be supplied, or supplied as cheaply, through exports. The precise reason for the inadequacy of arms-length trade to service foreign markets differed across the five countries. In the case of the United Kingdom, import restrictions imposed by the host governments more often than not was the key element that rendered exports uncompetitive in foreign markets. Indeed, imported manufactured products faced higher tariffs in the United States, Canada and most European countries after 1880 (Jones, 1996). By contrast, the determinants of the earliest international production by German manufacturing MNCs was spurred by a more complex set of factors. In the chemicals industry, the major determinants were the imposition of tariffs and non-tariff barriers to trade in Russia and the United States as well as stringent patent legislation in foreign countries that required the initiation of local production after a period of time and in an extreme case the French patent law of 1844 which did not allow the patenting of any type of medicine. In the electro-technical industry, the establishment of foreign factories was often initiated as a consequence of the need to have a local presence to service the needs of foreign markets more adequately (as in the case of the establishment and maintenance of a telegraph network in Russia by the German firm, Siemens), the growing pressures from the Russian administration which insisted on domestic production for continuing state orders of public utilities [the case of Allgemeine Elektricitäts-Gesellschaft (AEG)], the subsidies extended by the host country states to domestic producers that use parts and components made solely in the host country, and the need to respond to the requirement for domestic production of components used for racing cars participating in certain racing events in Great Britain (the case of Bosch that initiated international production in France and the United Kingdom of magneto ignitions for cars and trucks). In addition, the high cost of freight and the 45 per cent tariff on the value of magnetos led to the construction of a magneto factory in Springfield, Massachusetts in 1910 where production started in 1912 (Hertner, 1986a).

The determinants of the switch from exports to international production in the case of the pioneering Japanese MNCs in manufacturing in the cotton textiles industry stemmed from the Shimoneseki Treaty of 1895 which allowed foreigners to manufacture in Chinese treaty ports for the first time. The consequent establishment of foreign-owned cotton spinning factories in China by British trading companies such as Jardine Matheson & Company and three foreign textile firms from Great Britain, the United States and Germany (Wilkins, 1986b) combined with the incipient growth of Chinese investments in spinning mills, and the sale in China of cheap Indian yarn posed a threat to the continued growth of the Chinese export market for both Japanese cotton spinners and trading companies. In addition, much of the Japanese FDI in China in the early twentieth century may have been made initially to ascertain local costs of production in China and to keep close watch over the Chinese textile market. This contrasts with the case of the pioneering British and German manufacturing MNCs in which there was little reason to suppose that lower production costs abroad influenced the initiation of outward FDI before 1914 since international production seldom occurred in very labour intensive industries, and hence low labour costs abroad were rarely an enticement.

The predominant motivations behind international production of the earliest Taiwanese MNCs and South Korean MNCs were on a class of their own. Given their concentration on processing industries and heavy and chemical industries, the earliest manufacturing projects initiated by Taiwanese and South Korean firms was determined by the search for markets, the security of export markets and access to abundant raw material supplies, cheaper energy and land costs.

In general, the host countries of the earliest FDI in manufacturing in the five countries were their major export markets, a finding that provides broader empirical support for the product cycle model of Vernon (1966) in explaining the initiation of international production by American MNCs in their major export markets. The strong ownership-specific advantages of British and German MNCs and the high income elasticity of demand for their export products formed the basis for the establishment of international production in developed host countries. By contrast, the geographical destination of the earliest international production activities by Japan, Taiwan and South Korea was directed towards developing countries of South East Asia and China predominantly, and can be explained by the early stage of national development at which Japan, Taiwan and South Korea engendered MNCs of their own and thus the more simple ownership-specific advantages possessed by the leading domestic firms.

Apart from outward FDI in resource-based activities and in manufacturing, the scarcity of natural resources of the five countries and the consequent large dependence on international trade as both an engine and handmaiden of economic growth led to a common emergence of a domestically based infrastructure in support of trade – banks, shipping companies, marine insurance companies and, above all, trading companies – which emerged rapidly as MNCs.

Although the research findings on the early stages of outward FDI by MNCs in the resource-scarce large countries validates the hypothesis contained in Table 1.1, which describes the dominant form of the earliest outward FDI in these countries to be in local market-oriented and trade-related activities, this would need to be broadened *inter alia* to include resource-based extraction and processing (see Table 17.1 for a complete picture).

Resource-scarce small countries

The dominant form of the earliest outward FDI of MNCs based in the resource-scarce small countries has been in labour intensive manufacturing (particularly textiles) for local or export markets and in services (in particular trade, commerce, banking and insurance) associated with the strategic geographical locations of their home countries in Europe and Asia and the development of their domestic economies as centres for trading, commerce and finance. The emergence of MNCs was closely associated with the important role of the textiles industry in domestic production and exports of the economies of Switzerland in the nineteenth century, in Hong Kong since the 1950s, and later in Singapore in the 1970s and 1980s. In Switzerland and Hong Kong, the outward FDI was initiated by individual adventurers or families with established interests in the industry, and/or small- and medium-sized manufacturing companies. The determinants of international production in the textiles industry in the three countries have a common basis in the rapid loss of comparative advantage of the home country as a location for labour intensive production due to high production costs (associated with rising land costs and wages with industrial development and growth of labour productivity and shortage of semi-skilled and unskilled labour) and the imposition of tariffs and quotas in major export markets. There is also an analogous pattern in the geographical destination of the outward FDI in the textiles industry in the three countries. In all three countries, the outward FDI in the textiles industry in particular, and labour intensive manufacturing industries in general, was focused on host countries that had close geographical proximity and cultural affinity to the home country. These were Germany, France and Italy in the case of Swiss MNCs, and countries of South East Asia and China in the case of MNCs based in Hong Kong and Singapore.

However, despite the similarity in the industrial and geographical patterns of early outward FDI in the textiles industry in particular and labour intensive industries in general, the market orientation of these investments has differed between MNCs based in Switzerland and Singapore on the one hand and MNCs based in Hong Kong on the other. Since the early pattern of outward FDI in the textiles industry by Switzerland has tended to be concentrated in the very export markets threatened by import duties (Italy) and the formation of a customs union (Germany), and a similar protectionist trend in South East Asia had tended to precipitate the initiation of Singapore-based FDI in manufacturing, the early forms of international production of these firms have been geared largely to local markets in host countries. On the other hand, international production of MNCs based in Hong Kong in neighbouring countries has been essentially of the export platform type as a means of overcoming high domestic production costs and to overcome the high trade barriers imposed by their major export markets in the developed countries. Export-oriented international production had always been the main focus of Hong Kong-based MNCs since their emergence associated with their mastery

of technologies and long production experience in the assembly and manufacture of light consumer goods for exports and a large reservoir of entrepreneurial and export-marketing know-how and access to global distribution
channels.

In sum, although the research findings on the early stages of outward FDI
by MNCs in the resource-scarce small countries confirm the validity of the
hypothesis contained in Table 1.1 which describes the dominant form of the
earliest outward FDI in these countries to be in export-oriented and service-
based activities, this would need to be expanded to include import substitution as an early form of international production (see Table 17.1).

Development paths for outward direct investment, and their association with local industrialization across different types of country

Resource-abundant countries

The emergence and evolution of manufacturing firms and MNCs based in the
three resource-abundant countries derive from cumulative strengths in the
engineering industries comprising metal products, machinery and transportation equipment (Table 17.2). This finding holds regardless of the size of the
home countries concerned or its stage of development.[1] The basis of their
affinity is the abundance of mineral resources in the three countries which
fostered a well-entrenched tradition of industrialization based on metals processing, the technical and metallurgical know-how of which spilled over into
related sectors of the engineering industry such as machinery and transport
equipment.

Apart from the common emergence of manufacturing MNEs in the engineering industry broadly defined, the process of international expansion of
firms in this industry has been fairly common, despite the seemingly disparate
attempts to model such process in the case of American metallurgical companies and Swedish engineering companies by Wilkins (1970) and Johanson
and Wiedersheim-Paul (1975), respectively (see Chapter 6). Four stages were
described by Wilkins (1970). In the *first stage*, the domestic concern sold
abroad through independent agents (through an export person or export or
commission houses in the home country) or on occasion filled orders directly
from abroad. In general, companies frequently started to export using the
facilities of international trading firms. In the *second stage*, the company
appointed a salaried export manager, an existing export agency and its contacts, or independent agencies in foreign countries to represent the company.
In the *third stage*, the company either installed one or more salaried representatives, a sales branch or a distribution subsidiary abroad, or it purchased a
formerly independent agent located in a foreign country. At this point, for the
first time, the company made a foreign investment. In the *fourth stage*, a finish-

ing, assembly or manufacturing plant might be established to meet the needs of a foreign market (Wilkins, 1970).

Despite the broad analogy in the process of international expansion of American and Swedish firms in the manufacturing sector, two main factors differentiate the American and Swedish models. Firstly, the first stage described in the American model of exports being handled by independent agents or international trading firms was often bypassed in the Swedish case. This was owing to the needs of high-quality Swedish steel mills and the new industrial Swedish companies based on mechanical engineering to establish direct contacts with foreign markets by employing company-appointed representatives or agents or travelling salesmen (Hörnell and Vahlne, 1986) owing to increased product differentiation (e.g. production of special steel versus ordinary steel), and the necessity to have intimate knowledge of market developments and to adapt products to customers' particular needs (Carlson, 1977). Secondly, by comparison to the United States that is another resource-rich country of a larger size, Sweden and other small countries in general have had earlier forays into international markets through exports (Swedenborg, 1979) associated with their offensive and aggressive strategy to overcome the limited size of their domestic markets.

Another common pattern in the growth of MNCs from the three resource abundant countries lies in the importance of the mining sector, once again a feature of the presence of rich and abundant mineral resources in the three countries which enabled American, Swedish and Brazilian firms to develop management and organizational skills and technologies in mineral resource extraction and processing which were exploited profitably abroad. Indeed, mineral extraction and processing featured at some stage in the history of the growth of MNCs based in the three resource-abundant countries; however, this type of investment assumed the highest prominence in the case of the United States and one that emerged at an early stage of development of American MNCs.

Despite the presence of a seemingly common pattern of emergence of MNCs based in the three resource-abundant countries, some fundamental differences arise in the pattern of evolution between MNCs based in the resource-abundant large countries of United States and Brazil and MNCs based in the resource-scarce small country of Sweden. Multinational corporations from the resource-abundant large countries have evolved in a wider breadth of industries by comparison to MNCs from the resource-abundant small countries. Considering the manufacturing sector alone, American MNCs have emerged first in metallurgical industries as mentioned, and then evolved in the 1920s in industries that competed on the basis of product differentiation (such as food and drink, textiles and clothing), as well as in industries with distinctive products. American industries with worldwide technological leadership gained from the transfer abroad of techniques in product design, engineering and organization of production (electrical industry, motor vehicle industry, certain metal products, petroleum), as have companies with

Table 17.2 Actual development paths for outward direct investment, and their association with local industrialization across different types of country

Categorization of national development	Link between domestic development and the growth of outward FDI	Type of locally based MNC
Resource-abundant countries	Related diversification from resource extraction towards downstream processing (i.e. metal processing, wood products, petroleum processing and petrochemicals, agribusiness, etc.)	Resource-based firm Manufacturing firm
	International production in metallurgical industries resulting from the rapid development of domestic technologies, sometimes linked with mass production	Manufacturing firms, typically exporting firms
	Growth of international production in other manufacturing industries as a result of expansion of domestic firms and industries with trade-marked or branded merchandise or firms with distinctive products and techniques in product design, engineering and organization of production or advanced marketing methods. International production was prompted by prospects of profitable business opportunities in the host country which could not be fulfilled by exports	Manufacturing firms, typically exporting firms
	Further industrial upgrading towards the services sector in the domestic economy and associated growth of exports and outward FDI in services	Services firms in a diverse range of industries

(*continued*)

Table 17.2 (continued)

Categorization of national development	Link between domestic development and the growth of outward FDI	Type of locally based MNC
Resource-scarce large countries	Changing comparative advantage of the home country for labour intensive activities as wages rise following productivity growth	Manufacturing firms. small- and medium-sized
	Domestic industrial upgrading and export growth in processing industries, in industries related to new consumer needs and in fabricating industries embodying greater capital- and technology intensity and sometimes linked with mass production	Free-standing firms Manufacturing firms
	Further industrial upgrading towards the services sector in the domestic economy and associated growth of exports and outward FDI in services	Services firms, including trading companies, banks and financial institutions
Resource-scarce small countries	Position of the home country as a trading and financial centre	Prominent families, trading companies, banks and financial institutions
	Growth of domestic production and exports in simple manufacturing typically labour intensive industries leading to international production due to trade barriers in major export markets, rising production costs, shortage of domestic labour or restrictive legislation in the home country	Manufacturing firms, small- and medium-sized

(*continued*)

Table 17.2   (continued)

Categorization of national development	Link between domestic development and the growth of outward FDI	Type of locally based MNC
Resource-scarce small countries (*continued*)	Domestic industrial upgrading in a broad range of industries both capital intensive and technology intensive, some of which are protected by trademarks and patents	Manufacturing firms, horizontally integrating
	Limited opportunities for growth due to the small size of the domestic market	Manufacturing firms and services firms, searching new markets
	Development of their home countries as service economies	International service firms, banks and financial institutions

Source: Author's compilation based on the analysis contained in the country chapters.

advanced marketing methods (motor vehicle industry, metal products, petroleum). The kinds of American manufacturing enterprises that invested abroad in later decades closely resembled the investors of earlier years: these were leading firms in their industries in the United States that had advantages in technology, unique products and a long history of international economic orientation. The fact that an industry was technologically advanced did not *ipso facto* guarantee large FDI, but it generally meant that leading companies in that industry would in time, after finding exports could not continue to fulfil foreign demand, show an interest in extending their investments abroad. The American companies whose entrepreneurs exhibited far-sighted leadership grew rapidly and made the most far-ranging investments in foreign countries (Wilkins, 1974). These features describe consistently the growth of American MNCs over time, particularly in the manufacturing sector.

By comparison, Swedish manufacturing MNCs have had a narrower industrial focus in the engineering industry generally and, to a lesser extent, in the pulp and paper industry. Despite their narrower industrial focus, Swedish firms in these industries are technologically intensive, have distinctive products and a longer history of international economic orientation compared to the United States owing to their small country status. Many Swedish firms have become world leaders in their product niches on the basis of product design, engineering and organization of production and advanced marketing methods.

Although Brazilian manufacturing MNCs – like those of the United States and other large countries generally – have also been involved in a wide breadth of industries to include food products, textiles, clothing and footwear, paper packaging, wood and furniture, bicycles, lifts, electrical products, steel products and capital goods, motor vehicle parts, and aircraft, these firms do not compete on the same basis as American or Swedish manufacturing MNCs. Although Brazilian manufacturing firms in the metalworking, mechanical engineering and electrical equipment industries have fairly advanced foundry skills and skills in the organization of production, these firms did not feature prominently in trademarked or branded merchandise widely advertised in Brazil. In fact, as the experiences of some Brazilian MNCs such as Copersucar and Gradiente Electronics showed, some of the outward FDI was geared to penetrate the markets of developed countries by acquiring an established trade name abroad. Neither has any Brazilian industry or firm attained significant worldwide technological leadership (the closest it had achieved was the approach to the world technological frontier where this was fairly stable in some metalworking and mechanical engineering industries) nor developed sophisticated technological advantages and advanced marketing methods.

The research findings on the pattern of evolution of MNCs based in the resource-abundant countries thus support the validity of the hypothesis contained in Table 1.2 describing the *potential* development path for outward direct investment, and their association with local industrialization for this group of countries. The more detailed *actual* development path is shown in Table 17.2.

Resource-scarce large countries

Despite the common origins of the MNCs based in the five resource-scarce large countries in certain types of outward FDI, the evolution of their outward FDI while sharing some similarities particularly with respect to MNCs based in Japan, Taiwan and South Korea have also some notable differences. At the broad sectoral level, the rapidly growing importance in recent decades of the services sector in the domestic economies of the United Kingdom, Germany and Japan has been reflected in the sectoral pattern of their outward FDI. By contrast, the manufacturing sector remains the dominant sector of economic activity in the outward FDI of both Taiwan and South Korea, a reflection once again of the earlier stage of their national economic development.

However, it is in the manufacturing sector where some of the most important changes have taken place in the evolution of MNCs based in the five resource-scarce large countries in the course of their history. International production by British manufacturers has essentially remained oriented towards the mature, relatively low technology sectors whose competitiveness emanate from the large British market with high income: large size, established technological strengths, product differentiation, quality, and marketing and managerial skills and experience. These characteristics favoured the continuing pre-eminent role of consumer goods firms as British MNCs during the inter-war years and for much of the period since the Second World War. As a result, British MNCs still have not featured prominently in the more modern and growth-oriented fabricating industries, but the few that emerged have been the most active in pursuing more rationalized or efficiency seeking FDI since the 1970s associated with an increasing amount of intra-firm trade and product and process specialization.

By contrast, the high technological intensity that has always consistently described German manufacturing MNCs throughout the course of their history stems from the position of Germany as the birthplace of modern science in the late nineteenth century (Porter, 1990). This helped the country to develop a deep scientific and technical knowledge base drawing on an abundance of skilled workers and professionals which proved instrumental in their efforts to upgrade domestic industry in the new, skills intensive, technically advanced and fast growing industries of the Second Industrial Revolution in the late nineteenth century.

The evolution of Japanese MNCs in manufacturing is in many respects *sui generis* in the growth of modern MNCs. Although the origins of Japanese MNCs can be traced to the late nineteenth century, the early pattern of Japanese FDI remained essentially the same from the period prior to 1914 through to the inter-war period. The early stage of domestic industrial development in which Japan generated MNCs of their own explains why Japanese MNCs from the period of their emergence until the Second World War did not derive their ownership-specific advantages from technological strengths and organizational competence, the possession of brand and trademarks and the ability to supply high quality, differentiated goods which described the manufacturing MNCs from the United Kingdom and Germany since their emergence in the nineteenth century. However, the period since the Second World War was associated wtih the rapid domestic industrial transformation of the Japanese economy accompanied by the rapid evolution of Japanese FDI in manufacturing. In this period more than ever before, Japan's outward FDI was a crucial instrument or catalyst for the rapid process of domestic industrial upgrading (Ozawa, 1985; Kojima and Ozawa, 1985). The rapid industrial transformation from a concentration on the primary sector towards the secondary and tertiary economic sectors, and within the secondary sector from labour intensive light manufacturing and heavy and chemical manufacturing to knowledge intensive, fabricating assembly-based industries and mechatronics-based, flexible manufacturing of highly differentiated goods accompanied by a continual rapid rise in labour productivity and wages at the end of each phase have led to shifting patterns of production and trade competitiveness for Japanese manufacturing companies as well as evolving patterns of Japanese MNC activities.

To the extent that Taiwan manufacturing FDI and South Korean manufacturing FDI could be compared with that of Japan since the Second

World War, the first and second phases of Japanese FDI in labour intensive manufacturing in textiles, sundries and other low wage goods (the first
phase) and in heavy and chemical industries during the Ricardo-Hicksian
trap stage of Japanese FDI (the second phase) had been transposed in the
case of the history of both Taiwanese and South Korean manufacturing
MNCs. This is associated with a prolonged dependency or sustained comparative advantage of both the Taiwanese and South Korean economy on
labour intensive production until the late 1980s. Although both Taiwan
and South Korea had reached the third phase of Japanese FDI in assembly-
based, subcontracting-dependent, mass production in foreign markets in the
same set of consumer durables industries – cars, consumer electronics and
semiconductors – that have been at the core of the global strategies of the
large Taiwanese manufacuturing firms and the South Korean large conglomerate companies (*chaebols*) in the 1980s and 1990s. The South Korean *chaebols*
much like their Japanese counterparts – the *keiretsus* – spearheaded the large-
scale import substituting FDI of their countries geared to overcome trade
restrictions in the developed countries. The emergence and growth of the
chaebols and the *keiretsus* had been fostered by their respective governments
in their effort to accelerate domestic industrial development in large-scale,
complex and technologically advanced industries through high industrial
concentration. Indeed, a common theme in the three resource-scarce large
countries of East Asia is the leading role of their governments in directing
shifts in their countries' dynamic comparative advantage (see also Aggarwal
and Agmon, 1990). Towards this end, the development of indigenous skills
and technological capacities was emphasized, and growth was based on
knowledge-based industries (Crawford, 1987). Where South Korea and
Taiwan differ is in the pace in which domestic industrial development
evolved towards more advanced industries. The smaller size of its domestic
market and less interventionist role of the government of Taiwan in indigenous technological development at an early stage rendered the process of
industrial upgrading to proceed at a slower pace in Taiwan compared to
that in South Korea. Indeed, the technological depth and competitive prowess in which South Korea's industry has developed is not observed in any
other developing country (UN, TCMD, 1993a).

Despite the arrival of two Asian newly industrialized countries in the third
evolutionary phase of Japanese FDI in manufacturing, there are important
differences between Japanese MNCs on the one hand and the South Korean
and Taiwanese MNCs on the other. The broader range of more technologically advanced consumer durable goods produced by Japanese *keiretsus* in
their international production networks worldwide from around the mid-
1970s onwards combined with their advanced marketing capabilities (high
brand name recognition, product differentiation, extensive distribution channels, high advertising and promotion expenditures, etc.) contrasted with the
much narrower product range, less technologically advanced and far less

advanced marketing capabilities of South Korean and Taiwanese MNCs in major foreign markets in the 1980s and 1990s.

These research findings on the pattern of evolution of MNCs based in the resource scarce large countries are broadly consistent with the hypothesis contained in Table 1.2 describing the *potential* development path for outward direct investment, and their association with local industrialization for this group of countries. The *actual* development path is shown in Table 17.2 which incorporates some modifications to consider *inter alia* further industrial upgrading towards the services sector in the domestic economy consistent with the post-industrial status of most developed countries leading to the growth of exports and outward FDI in services.

Resource-scarce small countries

A common pattern of growth of MNCs from the three resource-scarce small countries lies in the importance of the services sector in outward FDI – a feature of the development of their home countries towards service-based economies. The closest parallelism can be drawn in the growth and development of outward FDI in trading, banking, finance and insurance in all three countries. In the case of Switzerland, this arises from the country's well-entrenched historical position as a trading, commercial and financial power (Bergier, 1968). This was favoured by the country's central geographical location on major European trade routes and political neutrality that allowed its firms to benefit from the maintenance of commercial contacts with each of the major European power centres (France, Germany and Great Britain) even during times of conflict. Its central location in Europe and well-entrenched position in trading and commerce in turn helped to foster its expertise in banking and insurance (Porter, 1990). The importance of general business services and personal services for Switzerland generally and as an area of Swiss FDI stem from the highly advanced pattern of domestic demand for these services associated with the high per capita income of the Swiss economy and its position as a location of the regional headquarters of foreign firms and international organizations.

In an analogous fashion, the services sector has also been a prominent feature in the growth of Hong Kong MNCs since the 1980s and enjoys a steady growth. These were mainly in hotels, property, trading, leisure services, construction, and banking and financial services (Chen, 1983b). In the period since 1986, their FDI in services (mainly in banking and finance, hotels and distribution) in the developed countries (primarily in United States, United Kingdom and France) have grown in a significant way through large-scale acquisitions of domestic firms in those countries. The growth of significant outward FDI in banking and financial sectors by Hong Kong in particular is resonant of the growth and development of Swiss MNCs in these sectors and draws from a similar historical position of Hong Kong as an entrepôt for trade in its region facilitated by the local establishment of large

British companies in the banking and finance, trading and services industries starting in 1841 with the inception of Hong Kong's colonial history. The process of deindustrialization combined with the country's small size also led to the growth of the domestic economy as a services-based economy and its transformation from a British colonial entrepôt into a leading trading and financial centre in the region (Ho, 1992). The growth of Hong Kong's outward FDI in the other services industries (hotels, property, leisure services, construction) is linked to the development of Hong Kong as a services-based economy.

The services sector has likewise been fairly significant for the economy of Singapore and its MNCs. As in Hong Kong, the importance of FDI in services is closely associated with a similar historical position of Singapore as a British colonial entrepôt and a Crown colony since 1867 which made the country a base for the economic expansion of the British empire in Malaya, South East Asia and China until the Second World War as well as base for Britian's military expansion in East Asia until the late 1960s (Mirza, 1986). In more modern times, Singapore has progressed beyond its traditional role as a British colonial entrepôt to fulfil a more complex and sophisticated role as a centre for trading, transportation, communications, industrial, commercial and financial services in South East Asia (Lim, 1990). Among the service industries in which Singapore-based MNCs have emerged and evolved were infrastructure development, construction, trading, marine industry (ship repair services, oil rig construction, marine engineering, etc.), property and property development, finance, and management and consultancy services. Unlike in the case of Switzerland and Hong Kong where almost all of the outward FDI in services have been generated by firms in the private sector, there were at least five types of firms that initiated and led Singapore-based services FDI in the 1980s and 1990s: foreign affiliates in Singapore undertaking outward FDI in the region, foreign companies using Singapore as a financial base to launch outward FDI, state-owned companies or government-linked companies in manufacturing and services, privately owned manufacturing companies, and privately owned conglomerate companies. Indeed, Singapore's conglomerate firms – Jack Chia-MPH, Wah Chang International, Haw Par Brothers, Intraco, Joo Seng Group and Keck Seng Group – have engaged in international operations as trading companies in the initial phases of their companies' history, long before their diversification towards the manufacturing sector (Lim, 1984).

Another distinctive feature in the growth of MNCs from the three resource-scarce small countries – that is unlike that of MNCs from the resource-scarce large countries – is the relatively lower significance of outward FDI in resource extraction, processing and associated service investments particularly for MNCs from Hong Kong and Singapore. This stems from the specialization of their economies in a narrower range of industries that did not involve the large-scale development of domestic processing industries. The marked exception is Switzerland that spawned highly successful firms such as Alusuisse-

Lonza, von Roll, Georg Fischer and other firms in the processed metal products industries which enabled those industries to count as one of a broad range of industries in which Switzerland developed a highly competitive position. Unlike in the resource-abundant small country of Sweden, these strengths in metals processing cannot be attributed to the relatively few natural resources that Switzerland possesses, with the exception of its capability to generate relatively inexpensive hydroelectric power useful in the reduction of alumina to aluminium (see the case study of Alusuisse in Chapter 13). The seemingly important role of some domestic resource intensive production in Switzerland does not support the generalization in Table 1.1 that resource-scarce small countries pursue non-resource intensive production in all cases, and hence such distinction has been deleted.

Notwithstanding the shared pattern of emergence of MNCs based in the three resource-scarce small countries in labour intensive industries and the textiles industry in particular and their common evolution in outward FDI in services, the further evolution of Swiss and Singapore-based MNCs in a wider breadth of manufacturing industries is distinctive and particularly remarkable in relation to MNCs from Hong Kong. Swiss MNCs have emerged first in textile industries of cotton, silk and straw in the eighteenth century, and then evolved in the nineteenth and twentieth century in processed metal products, machinery, electro-technical and electro-chemical industries, processed food products (milk products, baby foods, chocolates, jams and preserves), speciality chemicals and pharmaceuticals, paper and graphics and so forth. Similarly, there had been a wide breadth of manufacturing industries in which Singapore-based MNCs have emerged by the late 1970s and the swift process of domestic industrial restructuring towards higher value added manufacturing industries starting around the late 1970s meant that the outward FDI in manufacturing at the end of the 1980s and through the 1990s was spread over a far wider range of older and newer industries even if it continued to encompass those industries in Singapore that were dominated by either local capital or foreign capital.

Despite the wide scope of domestic industrial diversification and outward FDI in manufacturing in both Switzerland and Singapore, there were fundamental differences in the types of firms that initiated international production and the determinants of such international production in the two countries. The evolution of Swiss FDI in manufacturing have been led in the main by privately owned domestic-based manufacturing firms engaging in horizontal integration in foreign markets in response to both a home-country specific factor (to overcome the limited size of the domestic market through the search of new markets and business opportunities abroad) as well as host country-specific factors (tariff and non-tariff trade barriers in major export markets, high transportation costs, the need for a local presence to guarantee effective market penetration and servicing, reduce risks, etc.). In Singapore, on the other hand, there were at least four types of firms that led Singapore-based FDI in the wide range of manufacturing industries at the end of the 1980s and through the 1990s: foreign affiliates in Singapore undertaking outward FDI in the region, state-owned companies or government-linked companies in manufacturing and services, privately owned manufacturing companies, and privately owned conglomerate companies. Although the expansion of Singapore-based MNCs in manufacturing in the 1980s continued to be driven as in the case of Swiss MNCs by the need to overcome the limited size of the domestic market and to expand production and service capacity, a major factor propelling the growth of outward FDI in the 1980s and particularly since 1986 had been the accelerated pace of wage increases in Singapore and the tightening labour market problem which owing to tighter foreign labour and immigration restrictions could no longer be relieved by unskilled labour importation as in the 1970s (see also Aggarwal, 1986, 1987; Chia, 1989; Hill and Pang, 1991). Thus, it seems that home country-specific factors played a more important role in explaining the growth of Singapore-based FDI in manufacturing than was the case for Switzerland.

By contrast, the much narrow range of manufacturing industries that constitute Hong Kong's manufacturing sector explain the narrower breadth of industries that Hong Kong MNCs emerged and evolved. The domestic economy of Hong Kong continues to be dominated by the labour intensive production of clothing and textiles and to the extent that there had been some industrial diversification of the manufacturing sector, it is evident in the increased importance of two other labour intensive industries in the country's domestic production and exports of manufactured goods: electronics and watches, clocks and other precision instruments whose share increased over the last 40 years.

Since the difference in the breadth of industries that MNCs from the three resource-scarce small countries does *not* cut along the divide between developed and developing countries, a more fundamental explanation has to be sought outside the framework of the stage of economic development attained by the home country. Indeed, other more relevant explanations for the difference may have to be gleaned from the demand conditions in the home country, government macro-organizational policies pursued in achieving industrial development generally (including the more *dirigiste* role of the government of Singapore versus the more *laissez faire* role of the government of Hong Kong in industrial development) and other country-specific factors (such as cultural diversity) which has tended to differentiate the three countries other than the different stages of their economic development (see Chapter 16 for further elaboration).

However, the source of competitiveness of the country and its constituent firms seems to cut along the divide between developed and developing countries. The products of Swiss industry – whether manufacturing or services – have generally been focused on quality or based on extensive R&D and technical expertise (Schröter, 1993). The broad breadth of Swiss industries that are technologically advanced helps to explain both the much larger scale of their outward FDI, and the greater propensity of the leading companies in those industries to engage in international production should exports prove incapable to fulfil, or continue to fulfil, foreign demand. By contrast, the products of indigenous firms based in Hong Kong and Singapore derive their competitiveness from lower cost labour or access to such lower cost labour. The indigenous firms in the Asian city states demonstrate relatively weaker capabilities in capital intensive industries and more technology intensive industries.

These research findings on the pattern of evolution of MNCs based in the resource-scarce small countries substantiate the hypothesis contained in Table 1.2 describing the *potential* development path for outward direct investment, and their association with local industrialization for this group of countries. The more detailed and complete description of the *actual* development path is, however, contained in Table 17.2.

Technological accumulation and the national course of outward direct investment

Table 17.3 charts the course of the form of technological competence of leading indigenous firms according to stages of national development in relation to the corresponding types of outward FDI by distinctive groups of countries, its industrial course and the current stage of development of MNCs based in the 11 countries. Since the table was constructed on the basis of the research findings, it represents the authenticated version of the hypothetical Table 1.3 formulated at the beginning of the research.

A particularly striking feature of the research results reflected in the table is that regardless of observed variations in the pattern of the early stages of outward FDI across different types of countries as well as their developmental paths over time that are determined by distinctive patterns of national economic development, there seems to be some convergence in the type of outward FDI and the industrial course of outward FDI in the most advanced stages. Regardless of country type, MNCs from the more industrialized countries are commonly involved in more research-related investments in manufacturing and services and in the integration of their international networks in the type and industrial course of their outward FDI at later stages of their evolution. Such convergence has resulted partly from the cross-penetration of national markets by MNCs and the growing importance of the economies of large-scale production, cross-investments and intra-industry trade and production. Nevertheless, MNCs based in different countries have differed in the extent of their progression along a continuum between being nationally embedded MNCs and becoming more globally oriented MNCs (see also Lane, 1998).

The convergence in the type and industrial course of outward FDI of MNCs from the industrialized countries should not be confused with the notion that MNCs from different national origins have specialized in the same set of industries at the later stage of their evolution. Indeed, although there are certain industries that have truly become international industries by attracting sizeable amounts of outward FDI from MNCs of different national origins, the historical research on the emergence and evolution of MNCs in this study has shown that because firm-specific knowledge has been generated and sustained in industries in which each country has a comparative advantage to a relatively large extent, MNCs in different countries that have different comparative advantages have tended to emerge and specialize in different industries. Furthermore, the industrial course of their outward FDI over time have tended to be correlated closely with the upgrading of their home country's domestic industrial structure consistent with the attainment of dynamic comparative advantage. Since a large part of firm-specific knowledge is learning by doing, there is no reason to justify any claim that the industrial distribution of FDI undertaken by MNCs of different nationality or its evolutionary course over time should necessarily be the same, despite the trend towards globalization of industries, cross-investments, intra-industry trade and production, and the common propensity of MNCs to engage in more research-related investments in manufacturing and services and to integrate their international networks in the later stages of their evolution. Indeed, there have been long waves in the industrial specializations of countries reflecting the stability in national industrial strengths which span many decades and sometimes even centuries.

If the above finding is true, then it would imply that the core competencies or ownership advantages of firms and MNCs during the period of their emergence and early evolution reflect the distinctive patterns of national economic development of their home countries, i.e. their natural resource endowments, the size of their domestic market and the type of development path pursued in achieving industrial development. However, during the later period of their evolution, the advantages of firms and MNCs become *also* determined to an increasing extent by firm-specific factors such as the extent of multinationality of firms and the nature of technological development which tend to favour the retention of advantages within each firm. In the framework of the eclectic paradigm of international production, more mature MNCs would derive their asset ownership advantages – their ownership or exclusive or privileged access to proprietary or intangible assets – not only from their home base but increasingly from their cross-border network of international production. This is the case particularly of knowledge- or learning-based firms with several home bases, as demonstrated in this research by some Swedish companies that benefit from cross-border product, process or technological specialization and learning from producing in different environments. Not only does the capacity of MNCs to generate asset ownership advantages expand, the firm also gains sequential ownership advantages of the transaction cost minimizing kind derived from the common governance of separate but inter-related activities located in different countries, and the way in which their assets are coordinated with assets of other firms and with the locational advantages of countries (Dunning, 1998). The specificities of innovation across locations and

Table 17.3 Technological accumulation and the national course of outward direct investment

	Stages of national development		
	(1)	*(2)*	*(3)*
Form of technological competence of leading indigenous firms	Basic engineering skills, complementary organizational routines and structures	More sophisticated engineering practices, basic scientific knowledge, more complex organizational methods	More science-based advanced engineering, organizational structures reflect needs of coordination
Type of outward direct investment			
Resource-abundant countries	Trading, railroads, early resource- and market-seeking, services	More advanced resource-oriented or market-targeted investment in manufacturing and services	Research-related investment and integration into international networks
Resource-scarce large countries	Early resource-seeking, market-seeking or export-oriented invest-ments in manufacturing and services	More advanced resource-oriented or market-targeted or export-oriented investments in manufacturing and services	Research-related investment and integration into international networks
Resource-scarce small countries	Trading, commerce, banking, insurance, labour intensive and natural resource intensive manufacturing	Resource oriented, market-targeted or export-oriented investments in more advanced manufacturing and services	Research-related investment and integration into international networks
Industrial course of outward direct investment			
Resource-abundant countries	Trading, railroads, resource based (extractive MNCs or backward vertical integration), manufacturing, construction and consulting engineering	More forward processing of resources, or growth of fabricating industries for local markets or exports; growth of services	More sophisticated manufacturing and services systems, international integration of investments

(continued)

Table 17.3 (continued)

	Stages of national development		
	(1)	*(2)*	*(3)*
Resource-scarce large countries	Resources based (extractive MNCs or backward vertical integration), simple manufacturing, trading, banking, insurance, transportation and construction	More forward processing of resources, or growth of fabricating industries for local markets or exports; growth of services	More sophisticated manufacturing and services systems, international integration of investments
Resource-scarce small countries	Services based and simple manufacturing	Growth of fabricating industries targeted to local markets or more complex export-oriented manufacturing; growth of services	More sophisticated manufacturing and services systems, international integration of investments

Stage of development in the dawn of the twenty-first centry and the third millennium

	Brazil	United States
	Taiwan	Sweden
	South Korea	United Kingdom
	Hong Kong	Germany
	Singapore	Japan
		Switzerland

Source: Author's compilation based on the analysis contained in the country chapters.

firms increases the complexity of the process of technological development of firms, while the intra-firm utilization of a distinctive type of technology generated within each firm explain the extension of the MNC network across national borders and the direct control of the MNC over such network as a whole (Cantwell, 1989a). Firm-specific factors thus account for the increase in the ownership advantages of MNCs over time, the growth in the complexity of the determinants of such ownership advantages and the convergence in the type and industrial course of outward FDI by MNCs of different national origins in the later stages of their evolution.

Indeed, firm-specific factors play an important role in enabling the exploitation of country-specific advantage as well as in enhancing it. By broadening its geographical scope, MNCs gain from the use of their unique line of technological development in new environments, while the exposure to new environments in turn extends the firm's unique path of technology generation in new growth directions. As a case in point, the exploitation abroad of ownership advantages of British MNCs deriving from capital intensity – a relative source of advantage that is correlated with their national origin – has become possible with the large size and increasing efficiency of British MNCs in capital raising and allocation. These are firm-specific factors that increase directly with increasing geographical scope and extent of multinationality. At the same time, such country-specific advantage has been enhanced through the access gained by firms to international capital markets (Clegg, 1987). The convergence in the type and industrial course of outward FDI by firms of different national origins over time can be regarded in the context of the increased capacity and incentive of national groups of MNCs based in the more industrialized countries to increasingly resemble one another's ownership advantages as the degree and extent of their multinationality increases.

Table 17.3 also sets out the current stage of development of MNCs from countries that constituted the case studies of the research. Multinational corporations based in different country groups have evolved at different paces regardless of distinctive patterns of national development. It is seen that the current stage of development achieved by MNCs from the 11 country case studies tend to be determined by the divide along developed and developing countries. Thus, in relating stages of national development to the form of technological competence of leading indigenous firms, the type of outward FDI and its industrial course over time, MNCs based in the developing countries remain at a much earlier stage of development compared to the more advanced MNCs from the developed countries, despite the more rapid pace in their emergence and in the evolutionary path of their outward FDI. At the dawn of the twenty-first century and the third millennium, there remains considerable scope for MNCs based in developing countries to catch up in the dynamic and cumulative process of international production.

Notes

1 This important finding serves to substantiate the hypothesis that there is no need to differentiate between resource-abundant large countries and resource-abundant small countries in analysing the emergence and evolution of their MNCs.

References

Aggarwal, R. (1986) 'Managing for Economic Growth and Global Competition: Strategic Implications of the Life Cycle of Economies', *Advances in International Comparative Management*, 2: 19–44.

Aggarwal, R. (1987) 'Foreign Operations of Singapore Industrial Firms: A Study of Emerging Multinationals from a Newly Industrializing Country', in Dutta, M. (ed.), *Asia-Pacific Economies: Promises and Challenges, Research in International Business and Finance*, vol. 6, part B. Greenwich, Connecticut: JAI Press.

Aggarwal, R. and Agmon, T. (1990) 'The International Success of Developing Country Firms: Role of Government-Directed Comparative Advantage', *Management International Review*, 30(2): 163–180.

Aitken, H. G. J. (1985) *The Continuous Wave: Technology and American Radio, 1900–1932*, Princeton: Princeton University Press.

Alburo, F. A. (1988) Transnational Corporations and Structural Change in the Philippines. University of the Philippines Discussion Paper No. 8811. Quezon City: University of the Philippines, School of Economics.

Aliber, R. Z. (1970) 'A Theory of Direct Foreign Investment', in Kindleberger, C. P. (ed.), *The International Corporation: A Symposium*, Cambridge, Massachusetts: MIT Press.

Archer, H. J. (1986) An Eclectic Approach to the Historical Study of UK Multinational Enterprises. PhD dissertation, University of Reading.

Archer, H. (1990) 'The Role of the Entrepreneur in the Emergence and Development of UK Multinational Enterprises', *The Journal of European Economic History*, 19(2): 293–310.

Arthur, W. B. (1988) 'Competing Technologies: An Overview', in Dosi, G., Freeman, C., Nelson, R. R., Silverberg, G. and Soete, L. L. G. (eds), *Technical Change and Economic Theory*, London: Frances Pinter.

Arthur, W. B. (1989) 'Competing Technologies, Increasing Returns, and Lock-In by Historical Events', *The Economic Journal*, 99.

Ashworth, W. (1960) *An Economic History of England, 1870–1939*, London: Methuen.

Asian Development Bank (ADB) (1988) Foreign Direct Investment in the Asia and Pacific Region. Paper presented at a symposium in Manila, the Philippines, January.

Atkinson, A. B. and Stiglitz, J. E. (1969) 'A New View of Technological Change', *The Economic Journal*, 79(3).

Baba, Y. (1987) Internationalisation and Technical Change in Japanese Electronics Firms or Why the Product Cycle Doesn't Work. Mimeo, Science Policy Research Unit, University of Sussex, January.

Baer, W. (1965) *Industrialization and Economic Development in Brazil*, Chicago: Homewood.

Bain, F. and Thornton Read, T. (1934) *Ores and Industry in South America*, New York.

Bairoch, P. (1973) 'European Foreign Trade in the XIXth Century: The Development of the Value and Volume of Exports', *The Journal of European Economic History*, 2(1): 5–36.

Baklanoff, E. N. (1971) 'Brazilian Development and the International Economy', in Saunders, J. (ed.), *Modern Brazil: New Patterns and Development*, Gainesville: University of Florida Press.

Bamberg, J. H. (1994) *The History of the British Petroleum Company*, vol. 2, Cambridge: Cambridge University Press.

Bamford, J. (1993) The Changing Dynamics of Third World Multinationals: The Case of Taiwan's Direct Investment into the World's Advanced Industrialized Countries. Master's degree dissertation, Harvard University.

Beck, F. (1922) *Die Handelsbeziehungen Zwischen Italien und der Schweiz*, Weinfelden.

Bergier, J.-F. (1968) *Problèmes de L'Histoire Economique de la Suisse*, Berne.

Bergsman, J. (1970) *Brazil: Industrialization and Trade Policies*, London: Oxford University Press.

Bloom, M. (1992) *Technological Change in the Korean Electronics Industry*, Paris: OECD.

Blumenthal, T. and Teubal, M. (1975) Factor Proportions and Future Oriented Technology: Theory and An Application to Japan. Mimeo, Tel Aviv University David Horowitz Institute.

Bonnant, G. (1976) 'Les Colonies Suisses d'Italie à la Fin du XIX Siècle', *Schweizerische Zeitschrift für Geschichte*, 26.

Borner, S. (1986) *Internationalization of Industry: An Assessment in the Light of A Small Open Economy (Switzerland)*, Berlin: Springer-Verlag.

Borner, S., Porter, M., Weder, R. and Enright, M. (1991) *Internationale Wettbewerbsvorteile ein Strategisches Konzept für die Schweiz*, Frankfurt and New York.

Borner, S. and Wehrle, F.(1984) *Die Sechste Schweiz: Überleben auf dem Weltmarkt*, Zurich and Schwäbisch Hall: Orell Füssli.

Borrmann, A. and Jungnickel, R. (1992) 'Foreign Investment as a Factor in Asian Pacific Integration', *Intereconomics*, 27(6): 282–288.

Bostock, F. and Jones, G. (1994) 'Foreign Multinationals in British Manufacturing, 1850–1962', *Business History*, 36(1): 89–126.

Brauchlin, E. A. (1986) 'Role and Structure of Swiss Multinationals', in Macharzina, K. and Staehle, W. H. (eds), *European Approaches to Industrial Management*, Berlin: Walter de Gruyter.

Brody, H. (1986) 'Taiwan: From Imitation to Innovation', *High Technology*, 6(11): 24–27.

Buckley, P. J. and Casson, M. C. (1976) *The Future of the Multinational Enterprise*, London and Basingstoke: Macmillan.

Buckley, P. J. and Roberts, B. R. (1982) *European Direct Investment in the U.S.A. Before World War I*, London: Macmillan.

Bürgenmeier, B. (1986) 'Determinants of Swiss Investment Abroad: An Empirical Study', in Keller, C., Matejka, H. and Sézénasi, K. (eds), *Technology, Policies and Economics*, Geneva: Graduate Institute of International Studies.

Bürgenmeier, B. (1991) 'Swiss Foreign Direct Investment', in Bürgenmeier, B. and Mucchielli, J. L. (eds), *Multinationals and Europe 1992: Strategies for the Future*, London: Routledge.

Cantwell, J. A. (1989a) *Technological Innovation and Multinational Corporations*, Oxford: Basil Blackwell.

Cantwell, J. A. (1989b) 'The Changing Form of Multinational Enterprise Expansion in the Twentieth Century', in Teichova, A., Lévy-Leboyer, M. and Nussbaum, H. (eds), *Historical Studies in International Corporate Business*, Cambridge: Cambridge University Press.

Cantwell, J. A. (1991) 'A Survey of Theories of International Production', in Pitelis, C. and Sugden, R. (eds), *The Nature of the Transnational Firm*, London and New York: Routledge.

Cantwell, J. A. (1995) 'Multinational Corporations and Innovatory Activities: Towards a New Evolutionary Approach', in Molero, J. (ed.), *Technological Innovation, Multinational Corporations and New International Competitiveness*, Chur: Harwood, 1995.

Cantwell, J. A. (1997) 'Globalization and Development in Africa', in Dunning, J. H. and Hamdani, K. A. (eds), *The New Globalism and Developing Countries*, Tokyo and New York: United Nations University Press.

Cantwell, J. A. and Iammarino, S. (1998a) 'MNCs, Technological Innovation and Regional Systems in the EU: Some Evidence in the Italian Case', *International Journal of the Economics of Business*, 5(3): 383–408.

Cantwell, J. A. and Iammarino, S. (1998b) Multinational Corporations, and the Location of Technological Innovation in the UK Regions. University of Reading Discussion Papers in International Investment and Management, No. 262. Reading, UK: University of Reading, Department of Economics.

Cantwell, J. A. and Iammarino, S. (1999) 'Regional Systems of Innovation in Europe and the Globalisation of Technology', in Bartzokas, A. (ed.), *Technology Policy and Regional Integration*, London: Routledge.

Cantwell, J. and Tolentino,. P. E. E. (1990) Technological Accumulation and Third World Multinationals. University of Reading Discussion Paper in International Investment and Business Studies, No. 139. Reading, UK: University of Reading, Department of Economics..

Cardoso, F. H. (1973) 'Associated-Dependent Development: Theoretical and Practical Implications', in Stepan, A. (ed.), *Authoritarian Brazil*, New Haven: Yale University Press.

Cardoso, F. H. and Falleto, E. (1979) *Dependency and Development*, Berkeley: University of California Press. Translated by Marjory Mattingly Urquidi.

Carlson, S. (1977) 'Company Policies for International Expansion: The Swedish Experience', in Agmon, T. and Kindleberger, C. P. (eds), *Multinationals from Small Countries*, Cambridge, Massachusetts: MIT Press.

Carstensen, F. V. (1984) *American Enterprise in Foreign Markets: Singer and International Harvester in Imperial Russia*, Chapel Hill: University of North Carolina Press.

Cassis, Y. (1990) 'Swiss International Banking, 1890–1950', in Jones, G. (ed.), *Banks As Multinationals*, London: Routledge.

Caves, R. E. (1971) 'International Corporations: The Industrial Economics of Foreign Investment', *Economica*, 149: 1–27.

Caves, R. E. (1982) *Multinational Enterprise and Economic Analysis*, Cambridge: Cambridge University Press.

Chalmin, P. (1990) *The Making of A Sugar Giant. Tate & Lyle 1859–1989*, Char: Harwood.

Chan, K. B. and Chiang, S.-N. C. (1994) *Stepping Out: The Making of Chinese Entrepreneurs*, Singapore: Simon & Schuster.

Chandler, A. D. Jr (1980) 'The Growth of the Transnational Industrial Firm in the United States and the United Kingdom: A Comparative Analysis', *The Economic History Review*, 33(3): 396–410.

Chandler, A. D. Jr (1986) 'The Evolution of Modern Global Competition', in Porter, M. E. (ed.), *Competition in Global Industries*, Boston: Harvard Business School Press.

Chandler, A. D. Jr (1990) *Scale and Scope*, Cambridge, Massachusetts: Harvard University Press.

Channon, D. F. (1973) *The Strategy and Structure of British Enterprise*, London: Macmillan.

Chao, K. (1977) *The Development of Cotton Textile Production in China*, Cambridge, Massachusetts: Harvard University Press.

Chapman, S. D. (1985) 'British-Based Investment Groups Before 1914', *The Economic History Review*, XXXVIII (2): 230–251.

Chen, C-H. (1986) 'Taiwan's Foreign Direct Investment', *Journal of World Trade Law*, 20(6): 639–664.

Chen, E. K. Y. (1971) The Electronics Industry of Hong Kong: An Analysis of its Growth. M.Soc.Sc. thesis, University of Hong Kong.

Chen, E. K. Y. (1981) 'Hong Kong Multinationals in Asia: Characteristics and Objectives', in Kumar, K. and McLeod, M. G. (eds), *Multinationals from Developing Countries*, Lexington, Massachusetts: D.C. Heath and Company.

Chen, E. K. Y. (1983a) *Multinational Corporations, Technology and Employment*, New York: St Martin's Press.

Chen, E. K. Y. (1983b) 'Multinationals from Hong Kong', in Lall, S. (ed.), *The New Multinationals*, Chichester: John Wiley & Sons.

Chen, E. K. Y. (1989) 'The Changing Role of Asian NICs in the Asia-Pacific Region: Towards the Year 2000', in Shinohara, M. and Lo, F-C. (eds), *Global Adjustment and the Future of the Asian-Pacific Economy*, Tokyo: Institute of Developing Economies.

Cheng, T. Y. (1985) *The Economy of Hong Kong*, Hong Kong: Far East Publications.

Chia, S. Y. (1989) Asian NIEs as Traders and Investors. Paper presented at the conference on The Future of the Asia-Pacific Economies (FAPE III) Bangkok, 8–10 November.

Choi, D. W. and Kenny, M. (1995) The Globalisation of Korean Industry: Korean Maquiladoras in Mexico. Mimeo, University of Southern California.

Clegg, J. (1987) *Multinational Enterprise and World Competition: A Comparative Study of the USA, Japan, the UK, Sweden and West Germany*, London: Macmillan.

Clough, S. (1964) *The Economic History of Modern Italy*, New York: Columbia University.

Coase, R. H. (1937) 'The Nature of the Firm', *Economica*, 4(4): 386–405.

Cochran, S. (1980) *Big Business in China*, Cambridge, Massachusetts: Harvard University Press.

Cohen, B. I. (1975) *Multinational Firms and Asian Exports*, New Haven: Yale University Press.

Cooper, E. (1981) *Woodcarvers of Hong Kong*, Cambridge: Cambridge University Press.

Coram, T. C. (1967) The Role of British Capital in the Development of the United States. MSc thesis, University of Southampton.

Corley, T. A. B. (1989) 'The Nature of Multinationals, 1870–1939', in Teichova, A., Lévy-Leboyer, M. and Nussbaum, H. (eds), *Historical Studies in International Corporate Business*, Cambridge: Cambridge University Press.

Cottrell, P. L. (1991) 'Great Britain', in Cameron, R. and Bovykin, V. I. (eds), *International Banking: 1870–1914*, Oxford: Oxford University Press.

Crafts, N. F. R. and Thomas, M. (1986) 'Comparative Advantage in UK Manufacturing Trade 1910–1935', *Economic Journal*, 96: 629–645.

Crawford, M. H. (1987) 'Technology Transfer and the Computerisation of South Korea and Taiwan – Part I: Developments in the Private Sector', *Information Age*, 9: 10–16.

Currie, J. (1986) 'Export-Oriented Investment in Senegal, Ghana and Mauritius', in Cable, V. (ed.), *Foreign Investment: Policies and Prospects*, London: Commonwealth Secretariat.

Dahlman, C. (1984) 'Foreign Technology and Indigenous Technological Capability in Brazil', in Fransman, M. and King, K. (eds), *Technological Capability in the Third World*, New York: St Martin's Press.

Davies, K. (1983) 'Singapore Adjusts to World Recession', *The Banker*, July.

Davies, P. N. (1990) *Fyffes and the Banana: Musa Sapientum*, London: Athlone.

Department of Statistics (1991) *Singapore's Investment Abroad, 1976–1989*, Singapore: Department of Statistics.

Department of Statistics (1992) *Foreign Equity Investment in Singapore, 1980–1989*, Singapore: Department of Statistics.

Department of Statistics (1996) *Singapore's Investment Abroad, 1990–1993*, Singapore: Department of Statistics.

Deyo, F. (ed.) (1987) *The Political Economy of the New Asian Industrialism*, Ithaca: Cornell University Press.

Diaz-Alejandro, C. F. (1977) 'Foreign Direct Investment by Latin Americans', in Agmon, T. and Kindleberger, C. P. (eds), *Multinationals from Small Countries*, Cambridge, Massachusetts: MIT Press.

Dicken, P. (1992) *Global Shift*, second edition, London: Paul Chapman.

Dilmus, J. D. (1974) *Used Machinery and Economic Development*, East Lansing: Michigan State University.

Dörrenbächer, C. and Wortmann, M. (1991) 'The Internationalization of Corporate Research and Development', *Intereconomics*, 26(3): 139.

Dunning, J. H. (1958) *American Investment in British Manufacturing Industry*, London: Allen & Unwin.

Dunning, J. H. (1981a) *International Production and the Multinational Enterprise*, London: Allen & Unwin.

Dunning, J. H. (1981b) 'Explaining Outward Direct Investment of Developing Countries: In Support of the Eclectic Theory of International Production' in Kumar, K. and McLeod, M. G. (eds) *Multinationals from Developing Countries*, Lexington, Massachusetts: D.C. Heath and Company.

Dunning, J. H. (1982) 'Explaining the International Direct Investment Position of Countries: Towards a Dynamic or Developmental Approach', in Black, J. and Dunning, J. H. (eds), *International Capital Movements*, London: Macmillan.

Dunning, J. H. (1983) 'Changes in the Level and Structure of International Production: The Last 100 years', in Casson, M. C. (ed.), *The Growth of International Business*, London: Allen & Unwin.

Dunning, J. H. (1985) 'The United Kingdom', in Dunning, J. H. (ed.), *Multinational Enterprises, Economic Structure and International Competitiveness*, Chichester: John Wiley & Sons.

Dunning, J. H. (1986a) 'The Investment Development Cycle and Third World Multinationals', in Khan, K. M. (ed.), *Multinationals of the South: New Actors in the International Economy*, London: Frances Pinter.

Dunning, J. H. (1986b) *Japanese Participation in British Industry*, London: Croom Helm.

Dunning, J. H. (1986c) 'The Investment Development Cycle Revisited', *Weltwirtschaftliches Archiv*, 4(122): 667–677.

Dunning, J. H. (1988) 'The Eclectic Paradigm of International Production: A Restatement and Some Possible Extensions', *Journal of International Business Studies*, 19(1): 1–25.

Dunning, J. H. (1992) *Multinational Enterprises and the Global Economy*, Wokingham: Addison-Wesley.

Dunning, J. H. (1998) 'The Changing Geography of Foreign Direct Investment: Explanations and Implications', in Kumar, N. in collaboration with Dunning, J. H., Lipsey, R. E., Agarwal, J. P., and Urata, S., *Globalization, Foreign Direct Investment and Technology Transfers*, London and New York: Routledge in association with the UNU Press.

Dunning, J. H. and Archer, H. (1987) 'The Eclectic Paradigm and the Growth of UK Multinational Enterprise 1870–1983', *Business and Economic History*, 16: 19–49.

Dunning, J. H. and Cantwell, J. A. (1982) Inward Direct Investment from the US and Europe's Technological Competitiveness. University of Reading Discussion Paper in International Investment and Business Studies, No. 65. Reading, UK: University of Reading.

Dunning, J. H. and Cantwell, J. A. (1990) 'The Changing Role of Multinational Enterprises in the International Creation, Transfer and Diffusion of Technology', in Arcangeli, F., David, P. A. and Dosi. G. (eds), *Technology Diffusion and Economic Growth: International and National Policy Perspectives*, Oxford: Oxford University Press.

Dunning, J. H. and Cantwell, J. A. (1991) 'Japanese Direct Investment in Europe', in Bürgenmeier, B. and Mucchielli, J. L. (eds), *Multinationals and Europe 1992: Strategies for the Future*, London: Routledge.

Dunning, J. H. and Norman, G. (1985) 'Intra-Industry Production as a Form of International Economic Involvement', in Erdilek, A. (ed.), *Multinationals as Mutual Invaders: Intra-Industry Direct Foreign Investment*, London: Croom Helm.

Duus, P. (1984) 'Economic Dimensions of Meiji Imperialism: The Case of Korea, 1895–1910', in Myers, R. H. and Peatti, M. R. (eds), *The Japanese Colonial Empire 1895–1945*, Princeton, New Jersey: Princeton University Press.

Eakin, M. C. (1989) British Enterprise in Brazil, Durham: Duke University Press.

ECLAC (Economic Commission for Latin America and the Caribbean) (1992) *Social Equity and Changing Production Patterns: An Integrated Approach*, Santiago, Chile: ECLAC.

Economic Development Board (1991) *Singapore Enterprise*, Singapore: EDB.

Economic Development Board (1993) *Growing With Enterprise: A National Report*, Singapore: EDB.

Encarnation, D. J. (1992) *Rivals Beyond Trade*, Ithaca and London: Cornell University Press.

Ernst, D. (1994) What are the Limits to the Korean Model? Paper prepared for The Berkeley Roundtable on the International Economy, University of California at Berkeley.

Esho, H. (1985) 'A Comparison of Foreign Direct Investment from India, S. Korea and Taiwan by Size, Region and Industry', *Journal of International Economic Studies*, 1: 1–37.

Euh, Y-D. and Min, S. H. (1986) 'Foreign Direct Investment From Developing Countries: The Case of Korean Firms', *The Developing Economies*, 24(2): 149–168.

Evans, P. (1976) 'Foreign Investment and Industrial Transformation: A Brazilian Case Study', *Journal of Economic Development*, IX(3).

Federation of Hong Kong Industries (1992) *Hong Kong's Industrial Investment in the Pearl River Delta: 1991 Survey Among Members of the Federation of Hong Kong Industries*, Hong Kong: Federation of Hong Kong Industries.

Feldenkirchen, W. (1987) 'The Export Organisation of the German Economy', in Yonekawa, S. and Yoshihara, H. (eds), *Business History of General Trading Companies*, Tokyo: University of Tokyo Press.

Fieldhouse, D. K. (1978) *Unilever Overseas: The Anatomy of a Multinational, 1895–1965*, London: Croom Helm.

Findlay, R. (1978) 'Relative Backwardness, Direct Foreign Investment, and The Transfer of Technology: A Simple Dynamic Model', *Quarterly Journal of Economics*, XCII(1): 1–16.

Flowers, E. B. (1976) 'Oligopolistic Reactions in European and Canadian Direct Investment in the United States', *Journal of International Business Studies*, 7: 43–55.

Franko, L. G. (1976) *The European Multinationals*, New York: Harper.

Franko, L. G. (1983) *The Threat of Japanese Multinationals*, Chichester: John Wiley & Sons.

Fransman, M. (1984) 'Promoting Technological Capability in the Capital Goods Sector: The Case of Singapore, *Research Policy*, 13(1): 33–54.

Gereffi, G. and Evans, P. (1981) 'Transnational Corporations, Dependent Development, and State Policy in the Semiperiphery: A Comparison of Brazil and Mexico, *Latin American Research Review*, XVI(3): 31–64.

Giddy, I. H. (1978) 'The Demise of the Product Cycle Model in International Business Theory', *Columbia Journal of World Business*, 13(1).

Gold, T. (1986) *State and Society in the Taiwan Miracle*, Armonk, N.Y.: M. E. Sharpe.

Graham, E. M. (1975) Oligopolistic Imitation and European Direct Investment. PhD dissertation, Harvard Graduate School of Business Administration.

Graham, E. M. (1978) 'Transatlantic Investment by Multinational Firms: A Rivalistic Phenomenon?', *Journal of Post-Keynesian Economics*, 1(1): 82–99.

Graham, E. M. (1985) 'Intra-Industry Direct Investment, Market Structure, Firm Rivalry and Technological Performance', in Erdilek, A. (ed.), *Multinationals as Mutual Invaders: Intra-Industry Direct Foreign Investment*, London: Croom Helm.

Granovetter, M. (1985) 'Economic Action and Social Structure: The Problem of Embeddedness', *American Journal of Sociology*, 91: 481–510.

Granovetter, M. (1991) 'The Social Construction of Economic Institutions', in Etzioni, A. and Lawrence, P. R. (eds), *Socio-Economics: Toward A New Synthesis*, Armonk, N.Y.: M. E. Sharpe.

Granovetter, M. and Swedberg, R. eds., (1992) *The Sociology of Economic Life*, Boulder: Westview Press.

Gregory, C. (1985) British Labor in Britain's Decline. PhD dissertation, Harvard University Press.

Guimaraes, E. A. (1986) 'The Activities of Brazilian Firms Abroad', in Oman, C. (ed.), *New Forms of Overseas Investment by Developing Countries: The Case of India, Korea and Brazil*, Paris: OECD.

Haber, L. F. (1971) *The Chemical Industry 1900–1930: International Growth and Technological Change*, Oxford: Clarendon Press.

Håkanson, L. (1990) 'International Decentralization of R&D—The Organizational Challenges', in Bartlett, C. A., Doz, Y. and Hedlund, G. (eds), *Managing the Global Firm*, London and New York: Routledge.

Håkanson, L. and Nobel, R. (1989) Overseas Research and Development in Swedish Multinationals. Research Paper 89/3. Stockholm: Institute of International Business.

Harvey, C. (1981) *The Rio Tinto Company: An Economic History of a Leading International Mining Concern. 1873–1954*, Penzance: Alison Hodge.

Harvey, C. and Press, J. (1990) 'The City and International Mining, 1870–1914', *Business History*, 32(3): 98–119.

Harvey, C. and Taylor, P. (1987) 'Mineral Wealth and Economic Development: Foreign Direct Investment in Spain, 1851–1913', *The Economic History Review*, XL(2): 185–208.

Hassbring, L. (1979) *The International Development of the Swedish Match Company, 1917–1924*, Stockholm: Liberförlag.

Hedlund, G. (1984) 'Organizing in Between: The Evolution of the Mother–Daughter Structure of Managing Foreign Subsidiaries in Swedish MNFs', *Journal of International Business Studies*, 15(3).

Heer, J. (1966) *World Events 1866–1966: The First Hundred Years of Nestlé*, Rivaz: Nestlé.

Heiduk, G. and Hodges, U. (1992) 'German Multinationals in Europe: Patterns and Perspectives', in Klein, M. W. and Welfens, P. J. J. (eds), *Multinationals in the New Europe and Global Trade*, Berlin: Springer Verlag.

Heitger, B. and Stehn, J. (1990) 'Japanese Direct Investment in the EC – Response to the Internal Market 1993?', *Journal of Common Market Studies*, 29(1): 1–15.

Henderson, J. (1989) *The Globalisation of High Technology Production*, London and New York: Routledge.

Hertner, P. (1986a) 'German Multinational Enterprise Before 1914: Some Case Studies', in Hertner, P. and Jones, G. (eds), *Multinationals: Theory and History*, Aldershot: Gower.

Hertner, P. (1986b) 'Financial Strategies and Adaptation to Foreign Markets: The German Electro-Technical Industry and its Multinational Activities,' in Teichova, A., Lévy-Leboyer, M., and Nussbaum, H. (eds), *Multinational Enterprise in Historical Perspective*, Cambridge: Cambridge University Press.

Hill, H. and Pang E. F. (1991) 'Technology Exports from a Small, Very Open NIC: The Case of Singapore', *World Development*, 19(5): 553–568.

Himmel, E. (1922) *Industrielle Kapitalanlagen der Schweiz im Ausland*, Zurich: Langensalza.

Hirschmeier, J. and Yui, T. (1975) *The Development of Japanese Business, 1600–1973*, London: Allen & Unwin.

Ho, K. C. (1993) 'Industrial Restructuring and the Dynamics of City-State Adjustments', *Environment and Planning A*, 25: 47–62.

Ho, K. C. (1994) 'Industrial Restructuring, the Singapore City-State, and the Regional Division of Labour', *Environment and Planning A*, 26: 33–51.

Ho, Y. P. (1992) *Trade, Industrial Restructuring and Development in Hong Kong*, London: Macmillan.

Hobday, M. (1995) 'East Asian Latecomer Firms: Learning the Technology of Electronics', *World Development*, 23(7).

Hobsbawn, E. (1968) *Industry and Empire*, Baltimore: Penguin Books.

Hörnell, E. and Vahlne, J-E. (1986) *Multinationals: The Swedish Case*, London: Croom Helm.

Hou, C-M. (1965) *Foreign Investment and Economic Development in China 1840–1937*, Cambridge, Massachusetts: Harvard University Press.

Houston, T. and Dunning, J. H. (1976) *UK Industry Abroad*, London: *Financial Times*.

Hu, Y. S. (1995) 'The International Transferability of the Firm's Advantages', *California Management Review*, 37(4): 73–87.

Hufbauer, G. C. (1965) *Synthetic Materials and the Theory of International Trade*, London: Duckworth.

Hufbauer, G. C. (1970) 'The Impact of National Characteristics and Technology on the Commodity Composition of Trade in Manufactured Goods', in Vernon, R. (ed.), *The Technology Factor in International Trade*, New York: Columbia University Press.

Huff, W. G. (1995) 'The Development State, Government, and Singapore's Economic Development since 1960', *World Development*, 23: 1421–1438.

Hughes, H. and Yon, P. S. (eds.) (1969) *Foreign Investment and Industrialisation in Singapore*, Madison: The University of Wisconsin Press.

Hymer, S. and Rowthorn, R. (1970) 'Multinational Corporations and Industrial Organisation: The Non-American Challenge', in Kindleberger, C. P. (ed.), *The International Corporation: A Symposium*, Cambridge, Massachusetts: MIT Press.

Jeon, Y-D. (1992) 'The Determinants of Korean Foreign Direct Investment in Manufacturing Industries', *Weltwirtschaftliches Archiv*, 128: 527–542.

Jo, S-H. (1981) 'Overseas Direct Investment by South Korean Firms: Direction and Pattern', in Kumar, K. and McLeod, M. G. (eds), *Multinationals From Developing Countries*, Lexington, Massachusetts: D.C. Heath and Company.

Joe, S. (1990) The Development of Taiwan's Overseas Manufacturing investment. Paper prepared for the Annual Meeting of the Academy of International Business, Toronto, Ontario.

Johanson, J. and Wiedersheim-Paul, F. (1975) 'The Internationalization of the Firm – Four Swedish Cases', *The Journal of Management Studies*, 12(3): 305–322.

Johnson, H.G. (1958) *International Trade and Economic Growth*, London: Allen & Unwin.

Jones, G. (1984a) 'The Growth and Performance of British Multinational Firms Before 1939: The Case of Dunlop', *The Economic History Review*, 36(1): 35–53.

Jones, G. (1984b) 'The Expansion of British Multinational Manufacturing 1890–1939', in Inoue, T. and Okochi, A. (eds), *Overseas Business Activities*, Tokyo: University of Tokyo Press.

Jones, G. (ed.), (1986a) *British Multinationals: Origins, Management and Performance*, Aldershot: Gower.

Jones, G. (1986b) 'The Performance of British Multinational Enterprise, 1890–1945', in Hertner, P. and Jones, G. (eds), *Multinationals: Theory and History*, Aldershot: Gower.

Jones, G. (1986c) *Banking and Empire in Iran*, Cambridge: Cambridge University Press.

Jones, G. (1988) 'Foreign Multinationals and British Industry before 1945', *The Economic History Review*, XLI(3): 429–453.

Jones, G. (1996) *The Evolution of International Business*, London: Routledge.

Jong S. K. (1990) Influence of Transaction and Production Costs on Internationalization of Korean Textile and Apparel firms. Paper prepared for the Annual Meeting of the Academy of International Business, Toronto, Ontario.

Juhl, P. (1985) 'The Federal Republic of Germany', in Dunning, J. H. (ed.), *Multinational Enterprises, Economic Structure and International Competitiveness*, Chichester: John Wiley & Sons.

Jun, Y. W. (1990) Korean Overseas Investment: Patterns, Characteristics and Strategic Behaviours. Paper prepared for the ESCAP/UNCTC Joint Unit on Transnational Corporations, Bangkok.

Kao, J. (1993) 'The Worldwide Web of Chinese Business', *Harvard Business Review*, March–April, pp. 24–26.

Kapellas, J. D. and Liu, H. F. (1990) 'The Growth of ROC Trade and Investment in Southeast Asia', *Journal of Southeast Asia Business*, 6(4): 87–95.

Katz, J. (1984) 'Technological Innovation, Industrial Organisation and Comparative Advantages of Latin American Metalworking Industries', in Fransman, M. and King, K. (eds), *Technological Capability in the Third World*, New York: St Martin's Press.

Kawabe, N. (1989) 'Japanese Business in the United States Before the Second World War: The Case of Mitsui and Mitsubishi', in Teichova, A., Lévy-Leboyer, M. and Nussbaum, H. (eds), *Historical Studies in International Corporate Business*, Cambridge: Cambridge University Press.

Kim, K-S. (1990) Korean Overseas Investment. Mimeo, United Nations Centre on Transnational Corporations, July.

Kim, Y. (1996) 'Technological Capabilities and Samsung Electronics' International Production Network in Asia', in Ernst, D., Borros, M. and Haggard, S. (eds), *Asian Production Networks in Electronics: Their Impact on Trade and Technology Diffusion*, Oxford: Oxford University Press.

Kim, Y. (1997) 'Global Competition and Latecomer Production Strategies: Samsung of Korea in China', *Asia-Pacific Business Review*, 4(2/3): 84–108.

Kinugasa, Y. (1984) 'Japanese Firms' Foreign Direct Investment in the United States – The Case of Matshushita and Others', in Okochi, A. and Inoue, T. (eds), *Overseas Business Activities*, Tokyo: University of Tokyo Press.

Kitson, M. and Michie, J. (1995) 'Trade and Growth: A Historical Perspective', in Michie, J. and Grieve Smith, J. (eds), *Managing the Global Economy*, Oxford: Oxford University Press.

Knickerbocker, F. T. (1973) *Oligopolistic Reaction and the Multinational Enterprise*, Boston: Harvard University Press.

Kojima, K. (1973) 'A Macroeconomic Approach to Foreign Direct Investment', *Hitotsubashi Journal of Economics*, 14(1): 1–21.

Kojima, K. (1975) 'International Trade and Foreign Investment: Substitutes or Complements', *Hitotsubashi Journal of Economics*, 16(1).

Kojima, K. (1978) *Direct Foreign Investment: A Japanese Model of Multinational Business Operations*, London: Croom Helm.

Kojima, K. (1982) 'Macroeconomic versus International Business Approach to Direct Foreign Investment', *Hitotsubashi Journal of Economics*, 23(1): 1–19.

Kojima, K. (1990) *Japanese Direct Investment Abroad*, Tokyo: International Christian University Social Science Research Institute.

Kojima, K. and Ozawa, T. (1984) 'Micro- and Macro-Economic Models of Direct Foreign Investment', *Hitotsubashi Journal of Economics*, 5: 1–20.

Kojima, K. and Ozawa, T. (1985) 'Toward a Theory of Industrial Restructuring and Dynamic Comparative Advantage', *Hitotsubashi Journal of Economics*, 26(2): 135–145.

Koo, B-Y. (1984) Outward Investment by Korean firms: Their Forms and Characteristics. Paper presented at the OECD meeting of collaborating researchers on New Forms of Investment in Developing Countries, Paris, June.

Korea Institute for Foreign Investment (1989) A Study on Korean Overseas Investment. Mimeo, May.

Kravis, I. B. and Lipsey, R. E. (1992) 'Sources of Competitiveness of the United States and of its Multinational Firms', *The Review of Economics and Statistics*, LXXIV(2): 193–201.

Kumar, K. and Kim, K. Y. (1984) 'The Korean Manufacturing Multinationals', *Journal of International Business Studies*, XV(1): 45–61.

Kuwahara, T. (1982) 'The Business Strategy of Japanese Cotton Spinners: Overseas Operations 1890–1931,' in Okochi, A. and Yonekawa, S. (eds), *The Textile Industry and its Business Climate: Proceedings of the Fuji Conference*, Tokyo: University of Tokyo Press.

Kuwahara, T. (1989) 'The Japanese Cotton Spinners' Direct Investments into China Before the Second World War', in Teichova, A., Lévy-Leboyer, M. and Nussbaum, H. (eds), *Historical Studies in International Corporate Business*, Cambridge: Cambridge University Press.

Kwag, D-H. (1987) 'Korea's Overseas Investment', *Korea Exchange Bank Monthly*, September: 3–13.

Lall, S. (1981) *Developing Countries in the International Economy: Selected Papers*, London and Basingstoke: Macmillan.

Lall, S. (1982) 'The Export of Capital from Developing Countries: India', in Black, J. and Dunning, J. H. (eds), *International Capital Movements*, London: Macmillan.

Lall, S. (1983a) Trade in Technology by a Slowly Industrialising Country: India. Paper prepared for the conference on International Technology Transfer: Concepts, Measures and Comparisons organized by the US Social Science Research Council, New York, 2–3 June.

Lall, S. (1983b) 'The Rise of Third World Multinationals from the Third World', *Third World Quarterly*, 5(3) July.

Lall, S. (1983c) 'The Theoretical Background', in S. Lall (ed.), *The New Multinationals*, Chichester: John Wiley & Sons.

Lamoriello, F. (1992) 'East is West: Japanese Investment in Europe', *Journal of European Business*, pp. 8–9.

Lane, C. (1998) 'European Companies Between Globalization and Localization: A Comparison of Internationalization Strategies of British and German MNCs,' *Economy and Society*, 27(4): 462–485.

Lecraw, D. (1976) Choice of Technology in Low-Wage Countries: The Case of Thailand. PhD dissertation, Harvard University.

Lecraw, D. (1985) 'Singapore', in Dunning, J. H. (ed.), *Multinational Enterprises, Economic Structure and International Competitiveness*, Chichester: John Wiley & Sons.

Lee, K-H. (1989) 'The Emerging Pacific Century: Myth or Reality?', in Kaynak, E. and Lee, K-H. (eds), *Global Business: Asia-Pacific Dimensions*, London: Routledge.

Lee, T. Y. (1990) 'NIC Investment in ASEAN: The Pattern in the Eighties', in Ying, S. L. (ed.), *Foreign Direct Investment in ASEAN*, Kuala Lumpur: Malaysian Economic Association.

Lee, T. Y. (1991) *Growth Triangle: The Johor-Singapore-Riau Experience*, Singapore: Institute of Southeast Asian Studies.

Leff, N. (1968) *The Brazilian Capital Goods Industry 1929–1964*, Cambridge, Massachusetts: Harvard University Press.

Leontief, W. W. (1954) 'Domestic Production and Foreign Trade: The American Capital Position Reexamined', *Economia Internazionale*, 7(1).

Lewis, C. (1938) *America's Stake in International Investment*, Washington, DC: Brookings Institution.

Lim, L. Y. C. (1990) 'Singapore in Southeast Asia', *Journal of Southeast Asia Business*, 6(4): 65–74.

Lim, L. Y. C. and Pang E. F. (1982) 'Vertical Linkages and Multinational Enterprises in Developing Countries', *World Development*, 10(7): 585–595.

Lim, M. H. (1984) Survey of Activities of Transnational Corporations from Asian Developing Countries. Paper submitted to the ESCAP/UNCTC Joint Unit on Transnational Corporations, Bangkok.

Lin, T. B. and Ho, Y. P. (1980) *Export-Oriented Growth and Industrial Diversification in Hong Kong*, Hong Kong: Economic Research Centre, Chinese University of Hong Kong.

Linder, S. B. (1961) *An Essay on Trade and Transformation*, New York: John Wiley & Sons.

Lindgren, H. (1979) *Corporate Growth. The Swedish Match Industry in its Global Setting*, Stockholm: Liberförlag.

Lindqvist, M. (1991) *The Internationalization Process of Young Technology-based Firms*, Stockholm: Institute of International Business.

Low, L., Ramstetter, E. D. and Yeung, W-C. H. (1995) Accounting for Outward Foreign Direct Investment from Hong Kong and Singapore: Who Controls What? Paper presented at the conference on Geography and Ownership as Bases for Economic Accounting, National Bureau of Economic Research, Washington, DC, 19–20 May 1995.

Low, L., Toh, M. H., Soon, T. W., Tan, K. Y. and Hughes, H. (1993) *Challenge and Response: Thirty Years of the Economic Development Board*, Singapore: Times Academic Press.

Luey, P. (1969) 'Hong Kong Investment', in Hughes, H. and Seng, Y. P. (eds), *Foreign Investment and Industrialization in Singapore*, Madison: University of Wisconsin.

Lundström, R. (1986) 'Swedish Multinational Growth Before 1930', in Hertner, P. and Jones, G. (eds), *Multinationals: Theory and History*. Aldershot: Gower.

Maddison, A. (1969) *Economic Growth in Japan and the U.S.S.R.*, London: Allen & Unwin.

Masnata, A. (1924) *L'Emigration des Industries Suisses*, Lausanne.

Mason, F. R. (1920) *American Silk Industry and the Tariff*, Cambridge, Massachusetts: American Economic Association.

Mason, M. (1987) 'Foreign Direct Investment and Japanese Economic Development, 1899–1931', *Business and Economic History*, 16: 93–107.

Mason, M. (1992) 'The Origins and Evolution of Japanese Direct Investment in Europe', *Business History Review*, 66(3): 435–474.

Mason, R. H. (1980) 'A Comment on Professor Kojima's Japanese Type versus American Type of Technology Transfer', *Hitotsubashi Journal of Economics*, 20(2): 42–52.

McKenzie, F. A. (1901) *The American Invaders*, New York.

McVey, R. (1992) 'The Materialization of the Southeast Asian Entrepreneur', in McVey, R. (ed.), *Southeast Asian Capitalists*, Ithaca: Cornell University Southeast Asia Program.

Menkhoff, T. (1993) *Trade Routes, Trust and Trading Networks – Chinese Small Enterprises in Singapore*, Saarbrucken: Verlag Breitenback.

MITI (1982) *Tsusho Hakusho* [White Paper on International Trade], Tokyo: Government Printing Office.

Mirza, H. (1986) *Multinationals and the Growth of the Singapore Economy*, London: Croom Helm.

Mitsui & Co. (1977) *The 100 Year History of Mitsui & Co., Ltd. 1876–1976*, Tokyo.

Mok, C. H. (1969) The Development of the Cotton Spinning and Weaving Industries in Hong Kong, 1946–1966. MA thesis, University of Hong Kong.

Morris, J. (ed.) (1991) *Japan and the Global Economy*, London: Routledge.

Mun, K. C. and Chan, T. S. (1986) 'The Role of Hong Kong in US-China Trade', *Columbia Journal of World Business*, 21: 67–73.

Murphy, B. (1973) *A History of the British Economy 1086–1970*, London: Longman.

Nabseth, L. (1974) 'Summary and Conclusions', in Nabseth, L. and Ray, G. F. (eds), *The Diffusion of New Industrial Processes: An International Study*, Cambridge: Cambridge University Press.

Nankobogo, F. (1989) Analyse des Conditions d'Efficacité de l'Enterprise Transnationale dans des Contextes Culturels Différenciés. Thesis, University of Geneva.

Nelson, R. R. and Winter, S. G. (1977) 'In Search of a Useful Theory of Innovation', *Research Policy*, 6(1): 36–76.

Nelson, R. R. and Winter, S. G. (1982) *An Evolutionary Theory of Economic Change*, Cambridge, Massachusetts: Harvard University Press.

Newfarmer, R. (1980) *Transnational Conglomerates and the Economics of Dependent Development: A Case Study of the International Electrical Oligopoly and Brazil's Electrical Industry*, Greenwich, Connecticut: JAI Press.

Nicholas, S. (1982) 'British Multinational Investment Before 1939', *The Journal of European Economic History*, 11(3): 605–630.

Nicolaides, P. and Thomsen, S. (1991) 'Can Protectionism Explain Direct Investment?', *Journal of Common Market Studies*, 29: 635–643.

Niehans, J. (1977) 'Benefits of Multinational Firms for a Small Parent Economy: The Case of Switzerland', in Agmon, T. and Kindleberger, C. P. (eds), *Multinationals from Small Countries*, Cambridge, Massachusetts: MIT Press.

Nishimizu, M. and Robinson, S. (1984) 'Trade Policies and Productivity Change in Semi-Industrialized Countries', *Journal of Development Economics*, 16(1/2): 177–206.

Nolan, P. and Harvie, D. (1995) 'Labour Markets: Diversity of Restructuring', in Coates, D. (ed.), *Economic and Industrial Performance in Europe*, Aldershot: Edward Elgar.

Nordström, K. A. (1991) *The Internationalization Process of the Firm: Searching for New Patterns and Explanations*, Stockholm: Institute of International Business.

O'Brien, T. F. (1989) 'Rich Beyond the Dreams of Avarice: The Guggenheims in Chile', *Business History Review*, 63(1): 122–159.

O'Farrell, P. N., Wood, P.A. and Zheng, J. (1996) 'Internationalization of Business Services: An Interregional Analysis', *Regional Studies*, 30(2): 101–118.

Olsson, U. (1977) *The Creation of a Modern Arms Industry. Sweden 1939–1974*, Gothenburg: Institute of Economic History of Gothenburg University.

Olsson, U. (1993) 'Securing the Markets. Swedish Multinationals in a Historical Perspective', in Jones, G. and Schröter, H. G. (eds), *The Rise of Multinationals in Continental Europe*, Aldershot: Edward Elgar.

Ozawa, T. (1971) 'Transfer of Technology from Japan to Developing Countries', United Nations Institute for Training and Research Report, No. 7, New York.

Ozawa, T. (1974) *Japan's Technological Challenge to the West, 1950–74: Motivation and Accomplishment*, Cambridge, Massachusetts: MIT Press.

Ozawa, T. (1977) 'Japan's Resource Dependency and Overseas Investment', *Journal of World Trade Law*, 11(1): 52–73.

Ozawa, T. (1979a) *Multinationalism, Japanese Style: The Political Economy of Outward Dependency*, Princeton, New Jersey: Princeton University Press.

Ozawa, T. (1979b) 'International Investment and Industrial Structure: New Theoretical Implications from the Japanese Experience', *Oxford Economic Papers*, 31: 72–92.

Ozawa, T. (1982) 'A Newer Type of Foreign Investment in Third World Resource Development', *Rivista Internazionale di Scienze Economiche e Commerciali*, 29(12): 1133–1151.

Ozawa, T. (1985) 'Japan', in Dunning, J. H. (ed.), *Multinational Enterprises, Economic Structure and International Competitiveness*, Chichester: John Wiley & Sons.

Ozawa, T. (1991) 'Japanese Multinationals and 1992', in Bürgenmeier, B. and Mucchielli, J. L. (eds), *Multinationals and Europe 1992: Strategies for the Future*, London: Routledge.

Ozawa, T. (1992) 'Foreign Direct Investment and Economic Development', *Transnational Corporations*, 1(1): 27–54.

Pang, E. F. and Komaran, R. V. (1984) Hong Kong and Singapore Multinationals: A Comparison. Preliminary draft on the research project on New Forms of Investment in Developing Countries, phase II, OECD, May.

Pang, E. F. and Komaran, R. V. (1985) 'Singapore Multinationals', *Columbia Journal of World Business*, 20(2): 35–43.

Pang, E. F. and Lim, L. Y. C. (1989) 'High-Tech and Labour in the Asian NICs', *Labour and Society*, 14: 1–15.

Pasinetti, L. L. (1981) 'International Economic Relations', in Pasinetti, L. L., *Structural Change and Economic Growth: A Theoretical Essay on the Dynamics of the Wealth of Nations*, Cambridge: Cambridge University Press.

Patrick, H. T. (1967) 'Japan 1868–1914', in Cameron, R. with the collaboration of Crisp, O., Patrick, H. T. and Tilly, R., *Banking in the Early Stages of Industrialisation: A Study in Comparative Economic History*, New York: Oxford University Press.

Peres Núñez, W. (1993) 'The Internationalization of Latin American Industrial Firms', *CEPAL Review*, 49: 55–74.

Phelps Brown, E. H. (1971) 'The Hong Kong Economy: Achievements and Prospects', in Hopkins, K. (ed.), *Hong Kong: The Industrial Colony*, Hong Kong: Oxford University Press.

Pohl, H. (1989) 'The Steaua Romana and the Deutsche Bank: 1903–1920', *Studies on Economic and Monetary Problems and on Banking History*, No. 24, Mainz: v. Hase & Koehler Verlag.

Porter, A. (1986) *Victorian Shipping, Business and Imperial Policy: Donald Currie, the Castle Line, and Southern Africa*, Woodbridge and New York.

Porter, M. E. (1990) *The Competitive Advantage of Nations*, New York: Free Press.

Posner, M.V. (1961) 'International Trade and Technical Change', *Oxford Economic Papers*, 13(3): 323–341.

Pugel, T. A. (1981) 'Determinants of Foreign Direct Investment: An Analysis of U.S. Manufacturing Industries', *Managerial and Decision Economics*, 2(4): 220–228.

Pugel, T. A. (1985) 'The United States', in Dunning, J. H. (ed.), *Multinational Enterprises, Economic Structure and International Competitiveness*, Chichester: John Wiley & Sons.

Randerson, C. and Dent, C. M. (1996) 'Korean and Japanese Foreign Direct Investment in Europe: An Examination of Comparable and Contrasting Patterns', *Asian Studies Review*, 20(2): 45–69.

Ranis, G. (1992) (ed.) *Taiwan: From Developing to Mature Economy*, Boulder: Westview Press.

Read, R. (1983) 'The Growth and Structure of Multinationals in the Banana Export Trade', in Casson, M. (ed.), *The Growth of International Business*, London: Allen & Unwin.

Read, R. (1986) 'The Banana Industry: Oligopoly and Barriers to Entry,' in Casson, M. (ed.), *Multinationals and World Trade*, London: Allen & Unwin.

Reddaway, W. B. in collaboration with Potter, S. J. and Taylor, C. T. (1968) *Effects of UK Direct Investment Overseas: Final Report*, Cambridge: Cambridge University Press.

Redding, G. (1990) *The Spirit of Chinese Capitalism*, Berlin: De Gruyter.

Redding, G. (1991) 'Weak Organizations and Strong Linkages: Managerial Ideology and Chinese Family Business Networks', in Hamilton, G. G. (ed.), *Business Networks and Economic Development in East and South East Asia*, Hong Kong: University of Hong Kong Centre of Asian Studies.

Redies, T. (1990) 'Japanese Foreign Direct Investment in the 1980s: An Exercise in Financial Power', in Nemetz, N. (ed.), *The Pacific Rim: Investment Development and Trade*, Vancouver: UBC Press.

Régnier, P. (1993) 'Spreading Singapore's Wings Worldwide: A Review of Traditional and New Investment Strategies', *The Pacific Review*, 6: 305–312.

Remer, C. F. (1933) *Foreign Investments in China*, New York: Macmillan.

Rodan, G. (1989) *The Political Economy of Singapore's Industrialization: National State and International Capital*, London: Macmillan.

Roemer, J. E. (1976) Japanese Direct Foreign Investment in Manufactures: Some Comparison with the US Pattern', *Quarterly Review of Economics and Business*, 16(2): 91–111.

Rosenberg, N. (1976) *Perspectives on Technology*, Cambridge: Cambridge University Press.

Sally, R. (1993) 'The Basel Chemical Multinationals: Corporate Action Within Structures of Corporatism in Switzerland', *West European Politics*, 16(4): 561–580.

Schell, W. (1990) 'American Investment in Tropical Mexico: Rubber Plantations, Fraud and Dollar Diplomacy, 1897–1913', *Business History Review*, 64(2): 217–254.

Schive, C. (1982) A Summary Report of Domestic Multinationals Investment Commission. Mimeo, Ministry of Economic Affairs, Taiwan.

Schive, C. and Hsueh, K-T. (1985) Taiwan's Investment in ASEAN Countries and its Competitiveness. Mimeo, April.

Schmitz, C. (1986) 'The Rise of Big Business in the World Copper Industry, 1870–1930', *The Economic History Review*, 39(3): 392–410.

Schröter, H. G. (1986) 'A Typical Factor of German International Market Strategy: Agreements between the U.S. and the German Electrotechnical Industries up to 1939', in Teichova, A., Lévy-Leboyer, M. and Nussbaum, H. (eds), *Multinational Enterprise in Historical Perspective*, Cambridge: Cambridge University Press.

Schröter, H. G. (1988) 'Risk and Control in Multinational Enterprise: German Business in Scandinavia, 1918–1939', *Business History*, 62(3): 420–443.

Schröter, H. G. (1993) 'Swiss Multinational Enterprise in Historical Perspective', in Jones, G. and Schröter, H. G. (eds), *The Rise of Multinationals in Continental Europe*, Aldershot: Edward Elgar.

Schwarzenbach, A. (1934) 'Die Schweizerische Seidenindustrie', *Zeitschrift für Schweizerische Statistik und Volswirtschaft*, 70.

Scott, J. D. (1958) *Siemens Brothers 1858–1958*, London.

Sekiguchi, S. (1979) *Japanese Foreign Direct Investment*, London: Macmillan.

Sekiguchi, S. and Krause, L. B. (1980) 'Direct Foreign Investment in ASEAN by Japan and the United States', in Garnaut, R. (ed.), *ASEAN in a Changing Pacific and World Economy*, Canberra: Australian National University Press.

Sercovich, F. C. (1983) The Sale of Brazilian Technology Abroad: Scope, Performance and Implications. Mimeo.

Sercovich, F. C. (1984) 'Brazil', *World Development*, 12(5/6): 575–599.

Servan-Schreiber, J. J. (1967) *The American Challenge*, London: Hamish Hamilton.

Sestáková, M. (1989) 'Financial Operations of US Transnational Corporations: Developments after the Second World War and Recent Tendencies', in Teichova, A., Lévy-Leboyer, M. and Nussbaum, H. (eds), *Historical Studies in International Corporate Business*, Cambridge: Cambridge University Press.

Shapiro, H. (1994) *Engines of Growth: The State and Transnational Auto Companies in Brazil*, Cambridge: Cambridge University Press.

Shin,Y. H. and Lee, Y-I. (1995) 'Korean Direct Investment in Southeast Asia', *Journal of Contemporary Asia*, 25(2): 179–196.

So, A. Y. (1986) 'The Economic Success of Hong Kong: Insights from a World-System Perspective', *Sociological Perspectives*, 29(2): 241–258.

Söderström, C. (1980) *Swedish Enterprises Operating Abroad. The Seventies – A Decade of Change*, Stockholm: National Central Bureau of Statistics.

Sölvell, Ö., Zander, I. and Porter. M. E. (1991) *Advantage Sweden*, Stockholm: Norstedts.

Stead, W. T. (1902) *The Americanization of the World*, London.

Stein, S. (1957) *The Brazilian Cotton Manufacture: Textile Enterprise in an Underdeveloped Area*, Cambridge, Massachusetts: Harvard University Press.

Stobaugh, R. B. Jr (1968) The Product Life Cycle, U.S. Exports and International Investment. DBA dissertation, Harvard Business School.

Stopford, J. (1974) 'The Origins of British-Based Multinational Manufacturing Companies', *Business History Review*, 48(3): 303–345.

Stopford, J. (1976) 'Changing Perspectives on Investment by British Manufacturing Multinationals', *Journal of International Business Studies*, 7(2): 15–27.

Strange. R. (1993) *Japanese Manufacturing Investment in Europe*, London: Routledge.

Streit, C. (1949) *Union Now: A Proposal For An Atlantic Federal Union of the Free*, New York: Harper.

Svedberg, P. (1981) 'Colonial Enforcement of Foreign Direct Investment', *Manchester School of Economic and Social Studies*, 50: 21–38.

Swedenborg, B. (1979) *The Multinational Operations of Swedish Firms: An Analysis of Determinants and Effects*, Stockholm: Industrial Institute for Economic and Social Research.

Szczepanik, E. (1958) *The Economic Growth of Hong Kong*, Oxford: Oxford University Press.

Tan, H. (1991) 'State Capitalism, Multi-national Corporations and Chinese Entrepreneurship in Singapore', in Hamilton, G. G. (ed.), *Business Networks and Economic Development in East and South East Asia*, Hong Kong: University of Hong Kong.

Tan, H. Y., Toh, M. H. and Low, L. (1992) 'ASEAN and Pacific Economic Co-operation', *ASEAN Economic Bulletin*, 8: 309–332.

Teichova, A. (1983) 'The Mannesmann Concern in East Central Europe in the Inter-War Period', in Teichova, A. and Cottrell, P. L. (eds), *International Business and Central Europe, 1918–1939*, Leicester: Leicester University Press.

Thoburn, J. T., Leung, H. M., Chau, E. and Tang, S. H. (1990) *Foreign Investment in China Under the Open Policy: The Experience of Hong Kong Companies*, Aldershot.

Thomsen, S. (1993) 'Japanese Direct Investment in the European Community: The Product Life Cycle Revisited', *The World Economy*, 16(3): 301–315.

Thwaite, B. H. (1902) *The American Invasion*, London.

Tilly, R. (1991) 'International Aspects of the Development of German Banking, 1870–1914', in Cameron, R. and Bovykin, V. I. (eds), *International Banking: 1870–1914*, Oxford: Oxford University Press.

Ting, W-L. (1980) 'A Comparative Analysis of the Management Technology and Performance of Firms in Newly Industrialising Countries', *Columbia Journal of World Business*.

Ting, W-L. and Schive, C. (1981) 'Direct Investment and Technology Transfer from Taiwan', in Kumar, K. and McLeod, M. G. (eds), *Multinationals from Developing Countries*, Lexington, Massachusetts: D.C. Heath and Company.

Tolentino, P. E. E. (1987) The Global Shift in International Production: The Growth of Multinational Enterprises from the Developing Countries. PhD dissertation, University of Reading.

Tolentino, P. E. E. (1993) *Technological Innovation and Third World Multinationals*, London: Routledge. Foreword by John Dunning.

Tolentino, P. E. E. (1995) Third World Multinationals: Emergence and Evolution. Paper prepared for the 1995 Annual Meeting of the Academy of International Business, Seoul, South Korea, 15–18 November.

Tolentino, P. E. E. (1997) Review of *Korean Enterprise: The Quest for Globalization* by Ungson, G. R., Steers, R. M. and Park, S-H, *Transnational Corporations*, 6(3): 157–162.

Travis, N. J. and Cocks, E. J. (1984) *The Tincal Trail: A History of Borax*, London: Harrap.

Tsurumi, Y. (1976) *The Japanese are Coming: The Multinational Spread of Japanese Firms*, Cambridge, Massachusetts: Ballinger.

Tweedale, G. (1987) *Sheffield Steel and America: A Century of Commercial and Technological Interdependence, 1830–1930*, Cambridge: Cambridge University Press.

Ungson, G. R., Steers, R. M. and Park, S-H. (1997) *Korean Enterprise: The Quest for Globalization*, Boston: Harvard Business School Press.

United Nations Centre on Transnational Corporations (UNCTC) (1986) *Transnational Corporations in the International Semiconductor Industry*, New York: United Nations Publications.

United Nations Centre on Transnational Corporations (UNCTC) (1988) *Transnational Corporations in World Development: Trends and Prospects*, New York: United Nations Publications.

United Nations Centre on Transnational Corporations (UNCTC) (1991) *World Investment Report, 1991: The Triad in Foreign Direct Investment*, New York: United Nations Publications.

United Nations Centre on Transnational Corporations (UNCTC) (1992) *World Investment Directory 1992: Foreign Direct Investment, Legal Framework and Corporate Data. Volume I: Asia and the Pacific*, New York; United Nations Publications.

United Nations Conference on Trade and Development (UNCTAD) (1999) *World Investment Report 1999: Foreign Direct Investment and the Challenge of Development*, New York and Geneva: United Nations.

United Nations, ESCAP (1986) 'Hong Kong Transnational Corporation Investments in Developing Asian Economies', *Asia-Pacific TNC Review*, January: 23–30.

United Nations, ESCAP (1988) *Transnational Corporations from Developing Asian Economies: Host Country Perspectives*, Bangkok: ESCAP/UNCTC Joint Unit on Transnational Corporations.

United Nations, Transnational Corporations and Management Division (TCMD) (1993a) *Transnational Corporations from Developing Countries*, New York: United Nations.

United Nations, Transnational Corporations and Management Division (TCMD) (1993b) *World Investment Directory. Volume III: Developed Countries*, New York: United Nations.

US Department of Commerce, Office of Business Economics (1950) *Foreign Investments of the United States, 1950*, Washington, DC: Government Printing Office.

US Department of Commerce, Office of Business Economics (1960) *U.S. Business Investments in Foreign Countries*, Washington, DC: Government Printing Office.

Urata, S. (1998) 'Japanese Foreign Direct Investment in Asia: Its Impact on Export Expansion and Technology Acquisition of the Host Economies', in Kumar, N. in collaboration with Dunning, J. H., Lipsey, R. E., Agarwal, J. P. and Urata, S., *Globalization, Foreign Direct Investment and Technology Transfers*, London and New York: Routledge in association with UNU Press.

Ursacki, T. and Vertinsky, I. (1994) 'Long Term Changes in Korea's International Trade and Investment', *Pacific Affairs*, 67(3): 385–400.

Van Hoesel, R. (1997) 'The Emergence of Korean and Taiwanese Multinationals in Europe: Prospects and Limitations', *Asia Pacific Business Review*, 4(2/3): 109–129.

Vaupel, J. W. and Curhan, J. P. (1969) *The Making of a Multinational Enterprise*, Cambridge, Massachusetts: Harvard University Press.

Vaupel, J. W. and Curhan, J. P. (1974) *The World's Largest Multinational Enterprises*, Cambridge, Massachusetts: Harvard University Press.

Vernon, R. (1966) 'International Investment and International Trade in the Product Cycle', *The Quarterly Journal of Economics*, 80(2): 190–207.

Vernon, R. (1971) *Sovereignty at Bay*, Harmondsworth: Penguin Books.

Vernon, R. (1974) 'The Location of Economic Activity', in Dunning, J. H. (ed.), *Economic Analysis and the Multinational Enterprise*, London: Allen & Unwin.

Vernon, R. (1979) 'The Product Cycle Hypothesis in a New International Environment', *Oxford Bulletin of Economics and Statistics*, 41(3): 255–267.

Vernon, R. (1983) 'Organisational and Institutional Responses to International Risk', in Herring, R. (ed.), *Managing International Risk*, Cambridge: Cambridge University Press.

Villela, A. (1983) 'Multinationals from Brazil', in Lall, S. (ed.), *The New Multinationals*, Chichester: John Wiley & Sons.

Von Niephans, T. (1976) 'Die Compagnie d'Assurance Suisse à Livorno', *Revue Suisse d'Economie Politique et de Statistique*, pp. 87–99.

Von Weiher, S. and Goetzeler, H. (1977) *The Siemens Company: Its Historical Role in the Progress of Electrical Engineering, 1847–1980*, Berlin: Siemens.

Wang, C. and Hsu, J. (1992) The Impacts of the European Integration on Taiwan and Taiwan's Response. Paper presented at the 1992 Sino-European conference on Economic Development: Impacts of the European Integration and the Responses of Non-EC Countries, Taipei: Chung Hua Institute for Economic Research, 6–7 May.

Wavre, P-A. (1988) 'Swiss Investments in Italy from the XVIIIth to the XXth century', *The Journal of European Economic History*, 17(1): 85–102.

Wells, C. (1988) 'Brazilian Multinationals', *Columbia Journal of World Business* (Winter): 13–23.

Wells, L. T. Jr (ed.), (1972) *The Product Life Cycle and International Trade*, Boston: Harvard University Press.

Wells, L. T. Jr (1973) 'Economic Man and Engineering Man', *Public Policy*, Summer: 319–342.

Wells, L. T. Jr (1978) 'Foreign Investment from the Third World: The Experience of Chinese Firms from Hong Kong', *Columbia Journal of World Business*, (Spring): 39–49.

Wells, L. T. Jr (1983) *Third World Multinationals: The Rise of Foreign Investment from Developing Countries*, Cambridge, Massachusetts: MIT Press.

Wells, L. T. Jr (1984) 'Multinationals from Asian Developing Countries', in Moxon, R. W., Toehl, T. W. and Truitt, J. F. (eds), *International Business Strategies in the Asian Pacific Region, Research in International Business and Finance*, vol. 4, part A, Greenwich, Connecticut: JAI Press.

Westphal, L. E., Rhee, Y. W., Kim, L. and Amsden, A. H. (1984) 'Republic of Korea', *World Development*, 12(5/6): 505–533.

White, E. (1981) 'The International Projection of Firms from Latin American Countries', in Kumar, K. and McLeod, M. G. (eds), *Multinationals from Developing Countries*, Lexington, Massachusetts: D.C. Heath and Company.

Wikander, U. (1980) *Kreuger's Match Monopolies, 1925–1930. Case Studies in Market Control through Public Monopolies*, Stockholm: Liberförlag.

Wilkins, M. (1970) *The Emergence of Multinational Enterprise: American Business Abroad from the Colonial Era to 1914*, Cambridge, Massachusetts: Harvard University Press.

Wilkins, M. (1974) *The Maturing of Multinational Enterprise: American Business Abroad from 1914 to 1970*, Cambridge, Massachusetts: Harvard University Press.

Wilkins, M. (1986a) 'Defining a Firm: History and Theory', in Hertner, P. and Jones, G. (eds), *Multinationals: Theory and History*, Aldershot: Gower.

Wilkins, M. (1986b) 'Japanese Multinational Enterprise Before 1914', *Business History Review*, 60(2): 199–231.

Wilkins, M. (1988a) 'The Free-Standing Company, 1870–1914: An Important Type of British Foreign Direct Investment', *The Economic History Review*, 41(2): 259–285.

Wilkins, M. (1988b) 'European and North American Multinationals, 1870–1914: Comparisons and Contrasts', *Business History*, XXX(1): 8–45.

Wilkins, M. (1989) *The History of Foreign Investment in the United States Before 1914*, Cambridge, Massachusetts: Harvard University Press.

Williams, I. A. (1931) *The Firm of Cadbury*, London.

Williamson, O. (1975) *Markets and Hierarchies: Analysis and Anti-Trust Implications*, New York: Free Press.

Williamson, O. (1986) *Economic Organization*, Brighton: Wheatsheaf.

Wilson, C. (1954) *The History of Unilever*, vols. 1–2, London: Cassell.

Wohlert, K. (1989) 'Multinational Enterprise – Financing, Trade, Diplomacy: The Swedish Case', in Teichova, A., Lévy-Leboyer, M. and Nussbaum, H. (eds), *Historical Studies in International Corporate Business*, Cambridge: Cambridge University Press.

Wong, J. (1979) *ASEAN Economies in Perspective: A Comparative Study of Indonesia, Malaysia, the Philippines, Singapore and Thailand*, London: Macmillan.

Wong, S. L. (1988) *Emigrant Entrepreneurs: Shanghai Industrialists in Hong Kong*, Hong Kong: Oxford University Press.

Wong, S. L. (1991) 'Chinese Entrepreneurs and Business Trust', in Hamilton, G. G. (ed.), *Business Networks and Economic Development in East and South East Asia*, Hong Kong: Centre of Asian Studies.

Wong, Wilfred (1992) Direct Foreign Investment: The Hong Kong Experience. Paper presented at the OECD Informal Dialogue with the Dynamic Asian Economies on Foreign Direct Investment Relations, Bangkok, 26–27 March.

World Bank (1989) Foreign Direct Investment from the Newly Industrialized Economies. Industry and Energy Department Working Paper, No. 22, December. World Bank: Industry and Energy Department.

World Bank (1998) *Global Development Finance 1998. Analysis and Summary Tables*, Washington, DC: The World Bank.

Wray, W. D. (1984) *Mitsubishi and the N. Y. K., 1870–1914: Business Strategy in the Japanese Shipping Industry*, Cambridge, Massachusetts: Harvard University Press.

Wright, H. K. (1971) *Foreign Enterprise in Mexico: Laws and Policies*, Chapel Hill: University of North Carolina Press.

Wurm, C. (1993) *Business, Politics and International Relations*, Cambridge: Cambridge University Press.

Yamazaki, H. (1989) 'Mitsui Bussan during the 1920s', in Teichova, A., Lévy-Leboyer, M. and Nussbaum, H. (eds), *Historical Studies in International Corporate Business*, Cambridge: Cambridge University Press.

Yasumuro, K. (1984) 'The Contribution of Sogo Shosha to the Multinationalization of Japanese Industrial Enterprises in Historical Perspective', in Okochi, A. and Inoue, T. (eds), *Overseas Business Activities*, Tokyo: University of Tokyo Press.

Yeow, T. H. (1984) *Shipbuilding and Shiprepairing Industry in Singapore*, Geneva: UNCTAD.

Yeung, H. W-C. (1994) 'Hong Kong Firms in the ASEAN Region: Transnational Corporations and Foreign Direct Investment', *Environment and Planning A*, 26(12): 1931–1956.

Yeung, H. W-C. (1995) 'The Geography of Hong Kong Transnational Corporations in the ASEAN Region', *Area*, 27(4): 318–334.

Yeung, H. W-C. (1997) 'Business Networks andTransnational Corporaitons: A Study of Hong Kong Firms in the ASEAN Region', *Economic Geography*, 73(1): 1–25.

Yeung, H. W-C. (1998) 'The Political Economy of Transnational Corporations: A Study of the Regionalization of Singaporean Firms', *Political Geography*, 17(4): 389–416.

Yoffie, D. B. (1993) 'Foreign Direct Investment in Semiconductors', in Froot, K. A. (ed.), *Foreign Direct Investment*, Chicago: University of Chicago Press.

Yonekawa, S. (1985) 'The Formation of General Trading Companies: A Comparative Study', *Japanese Yearbook on Business History*, pp. 1–31.

Yonekawa, S. and Yoshihara, H. (eds.) (1987) *Business History of General Trading Companies*, Tokyo: University of Tokyo Press.

Yoshihara, K. (1976) *Foreign Investment and Domestic Response: A Study of Singapore's Industrialization*, Singapore: Eastern University Press.

Young, S., McDermott, M. and Dunlop, S. (1991) 'The Challenge of the Single Market', in Bürgenmeier, B. and Mucchielli, J. L. (eds), *Multinationals and Europe 1992*, London: Routledge.

Index

Governance and Policy in Sport Organizations

Now in a fully updated and expanded fifth edition, this textbook introduces the power and politics of sport organizations to the readers. It explores the managerial activities essential to good governance and policy development and looks at the structure and functions of individual organizations within the larger context of the global sport industry.

Full of real-world examples, cases, and data, this book examines the dilemmas faced by sport managers, administrators, and policymakers in their everyday work, helping readers to understand the importance of good governance and sound policy frameworks in any successful sport organization. Introducing core managerial functions and surveying every sector of contemporary sport from school and community sport to professional leagues and international megaevents, this edition includes brand-new chapters focused on diversity, equity, and inclusion; on esports; and on governance in times of crisis, covering issues such as COVID-19, climate change, scandal, and security risks.

Helping readers to see a big picture across the contemporary sport industry, at all levels, and to find their place in it as future sport managers, this textbook is essential for all courses on sport governance, sport policy, or sport development.

This book is accompanied by a suite of useful ancillary materials, including an instructors' guide, test bank, and PowerPoint slides.

Mary A. Hums is a Professor of Sport Administration at the University of Louisville, USA. Her main research interests include policy development in sport organizations, especially in regard to the inclusion of people with disabilities and also sport and human rights. She is also a member of the North American Society for Sport Management (NASSM), the International Society of Olympic Historians (ISOH), and the International Olympic Academy Participants Association (IOAPA). She was named a Zeigler Lecturer, NASSM's most prestigious academic honor.

Yannick Kluch is an Assistant Professor and Director of Inclusive Excellence in the Center for Sport Leadership at Virginia Commonwealth University, USA. His research and consulting work are focused on utilizing sport as a vehicle for social change as well as on eliminating barriers to social justice. Specifically, his areas of expertise include sociocultural studies of sport, athlete activism, inclusive sport policy, human rights, and equity, diversity, and inclusion in sport. In 2020, he was one of only four experts appointed to the inaugural Team USA Council on Racial and Social Justice by the United States Olympic and Paralympic Committee. For more information, please visit his website at www.yannickkluch.com.

Samuel H. Schmidt is an Assistant Professor of Sport Management at the University of Wisconsin – La Crosse (UWL), USA. His research areas include athlete activism and sport and social movements, and his sport industry experience has primarily focused on facilities and game day operations and sports information. At UWL, Schmidt serves on the Committee for Academic Policies and Standards for the university, among various other service duties. He also serves as an Editorial Board Member for the *Journal of Amateur Sport*.

Joanne C. MacLean is a President and Vice Chancellor of the University of the Fraser Valley (UFV), Canada. She also serves as Chair of the USports Board of Directors, the governance body for university sport in Canada. She served as Dean of the Faculty of Health Sciences at UFV from 2012 to 2018, prior to being selected as the institution's first female President. From 2002 to 2012 she was a Professor of Sport Management at Brock University, where she held the positions of Department Chair and Interim Dean. She also has extensive experience in the administration, governance, and human resource management of university sport within Canada.

GOVERNANCE AND POLICY IN sport ORGANIZATIONS

FIFTH EDITION

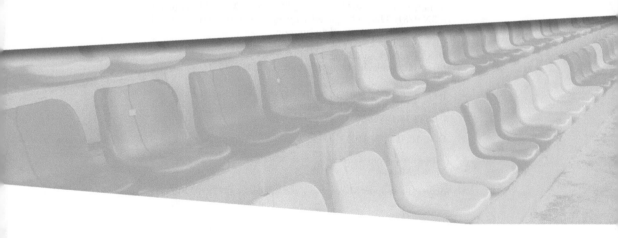

**Mary A. Hums, Yannick Kluch,
Samuel H. Schmidt, and
Joanne C. MacLean**

Routledge
Taylor & Francis Group

NEW YORK AND LONDON

Designed cover image: Michael H / Getty Images

Fifth edition published 2023
by Routledge
605 Third Avenue, New York, NY 10158

and by Routledge
4 Park Square, Milton Park, Abingdon, Oxon, OX14 4RN

Routledge is an imprint of the Taylor & Francis Group, an informa business

First edition published by Holcomb Hathaway 2003
Fourth edition published by Routledge 2018

British Library Cataloguing-in-Publication Data
A catalogue record for this book is available from the British Library

Library of Congress Cataloging-in-Publication Data
Names: Hums, Mary A., author. | Kluch, Yannick, author. |
Schmidt, Sam (Samuel H.), author. | MacLean, Joanne, 1959– author.
Title: Governance and policy in sport organizations /
Mary A. Hums, Yannick
Kluch, Sam Schmidt and Joanne C. MacLean.
Description: Fifth edition. | Abingdon, Oxon; New York, N.Y. : Routledge,
2023. | Includes bibliographical references and index.
Identifiers: LCCN 2022047516 | ISBN 9781032300474 (hardback) |
ISBN 9781032300429 (paperback) | ISBN 9781003303183 (ebook)
Subjects: LCSH: Sports administration.
Classification: LCC GV713 .H86 2023 |
DDC 796.06/9—dc23/eng/20221108
LC record available at https://lccn.loc.gov/2022047516

ISBN: 978-1-032-30047-4 (hbk)
ISBN: 978-1-032-30042-9 (pbk)
ISBN: 978-1-003-30318-3 (ebk)

DOI: 10.4324/9781003303183

Typeset in Sabon
by codeMantra

Access the Support Material: www.routledge.com/9781032300429

Contents

Foreword

Governance and Policy in Sport Organizations, Fifth Edition, is a brilliant resource for understanding the nuance of governing sport in a diverse, fast-paced, 21st-century environment.

Central to the brilliance of Governance and Policy in Sport Organizations is the group of authors assembled to create this text. The lead author, who is the originator and remains the driver of this text, Dr. Mary Hums has been at the forefront of international governance for decades. Most impressive is her ability to weave the structural and technical aspects of sport governance which are outlined in the introduction as being rooted in "power, authority, control, and high-level policy making" with the very core of any organization – its people. Over the years, this weaving of governance and people has been done with every chapter drawing on the ways people and their decision-making influence organizational outcomes. Specifically, Dr. Hums and colleagues have been able to successfully introduce the topics into sport texts through the chapters on Paralympic Sport and Ethics. These two chapters are routine for sport governance, independent of the edition.

The Fifth Edition of Governance and Policy in Sport Organizations boasts two new authors, Dr. Yannick Kluch and Dr. Samuel H. Schmidt, as well as the inclusion of the chapter titled, "Diversity, Equity, and Inclusion in Sport Governance". Drs. Kluch and Schmidt bring a wealth of research, teaching, and practitioner experience in diversity, equity, and inclusion (DEI). The advantage of having authors that have expertise in research, teaching, and practice is on display in their execution of this chapter being both informative and prescriptive. The authors' framing of the data provides a rich understanding of the state of diversity in various sport governance environments. This is followed by prescriptive examples and frameworks for building DEI strategic plans, complete with case examples from the field. Never has this level of intentionality been provided in a sport governance book. The authors are compelling practitioners, students, researchers, and others who read this text to consider the role DEI plays in their organizations and they provide the tools needed for readers to immediately engage in organizational DEI planning and improvement. I serve as Vice Chancellor for Equity and Inclusion and Chief Diversity Officer at the University of Massachusetts Amherst, a large research-intensive flagship institution. As a DEI practitioner and researcher, who has studied DEI for decades, I am most moved by the authors' ability to have a chapter of work dedicated to DEI while providing DEI context-specific text throughout the entire textbook. Diversity, equity, and inclusion are not adjacent to sport governance – it is at the very core of every sport governance structure. Through their efforts to amplify the importance of DEI as a subject worthy of its own chapter, but omnipresent in every other, Hums, Kluch, Schmidt, and MacLean have modeled the level of intentionality needed to transform sport governance in a 21st-century environment.

Nefertiti A. Walker, PhD
Professor, Vice Chancellor, and Chief Diversity Officer
University of Massachusetts Amherst, USA

Preface

As with previous editions, the fifth edition of *Governance and Policy in Sport Organizations* is designed for use in governance or policy development courses with upper-level undergraduate students. The text can also be used with some graduate-level courses that introduce students to the business and policy development aspects of the sport industry.

Sport management students entering the workforce will work in various sport industry segments. An important part of their training is learning the structure and function of the various sport organizations they will work within or interact with. Successful sport managers understand the big picture of how their sport organizations are structured. They also know what issues their organizations – and they as managers – will have to confront. This book challenges students to integrate management theory with governance and policy development practices. It discusses where the power lies in an organization or industry segment and how individual sport organizations fit into the greater industry. The book also interweaves ethical issues throughout, as sport managers need to make sure their decisions are just, fair, and inclusive. Diversity, equity, and inclusion are also integral parts of the chapters.

The fifth edition of this book offers three new chapters which are meant to address innovative and evolving areas of the sport industry. First, the authors added a chapter focused on diversity, equity, and inclusion. While this has always been interwoven throughout past editions, it was time to feature it prominently and present this important material in its own stand-alone chapter. Next, a chapter on crisis management was added. Since the last edition, our world has been impacted by a number of tumultuous situations. Topping the list, of course, is the COVID-19 pandemic, which even at the writing of this text, continues to wreak global havoc. The sport industry was impacted swiftly and directly and sport managers found themselves having to make lightning-fast decisions about the health and safety of fans, athletes, and employees. Other crises also must be dealt with ranging from terrorist attacks to climate change. Sport managers need guidance on how to best respond to, but also think proactively about, the challenges these issues represent. Finally, as the business of sport evolves, new forms of entertainment and competition become more popular and so the third new chapter in this edition is on esports. Now valued as a billion-dollar industry, esports present a whole new landscape to the sport industry and their governance is somewhat complex and evolving. The new chapter outlines the current status of esports governance and its relationship to traditional sport governing bodies.

One more new addition to the book is a short feature in each industry segment chapter introducing an organization that exhibits best practices in diversity, equity, and inclusion in sport governance. These examples will help readers see what can be done to make sport a more welcoming environment for everyone, especially groups that have historically been excluded, marginalized, or minoritized in the sport industry.

This fifth edition of the book reflects industry changes and offers real-world examples. Numerous sport organization websites are highlighted throughout the text, to prompt in-class discussions, online investigation, and further research. These websites are displayed in the margins of the text and thus are easily referred to. This edition also features a new set of Industry Portraits and Perspectives boxes, with contributions from sport managers in a wide variety of sport industry segments. For this feature, administrators answered our questions about their jobs and their organizations to give students a glimpse into the practical concerns of sport managers and the impact of governance and policy in the sport industry.

In this edition, we continue to explore current topics that impact policy development. Some of these are ongoing and others have emerged since the last edition. Issues such as these will continue to emerge and evolve, and this text provides a springboard for class discussions and projects.

Knowledge of sport beyond North America is essential for a well-informed manager and is required for success in the increasingly global sport industry. Although this book focuses primarily on North American sport organizations, it presents sufficient information on international sport organizations to provide students with working models and an understanding of these organizations. For example, we highlight how North American scholastic, intercollegiate, and professional sports differ from those in other regions of the world. In addition, our backgrounds as authors – two from the United States, one from Germany, and one from Canada – contribute to the international scope of this book. This combination brings a unique skill set and knowledge base to the text. For example, Mary Hums is an internationally recognized researcher and advocate in the area of management issues affecting disability sport, having worked in four Paralympic Games and also the Para-Pan American Games and contributing her expertise to the Paralympic Sport chapter. Yannick Kluch has worked with both the NCAA and the USOPC on matters related to diversity, equity, and inclusion. Samuel Schmidt brings experience from working in intercollegiate athletics on the campus level and also esports. Joanne MacLean is an experienced University athletic administrator and served as Canada's Chef de Mission for the FISU Games, lending her experience to the chapters on intercollegiate athletics and major games in amateur sport.

Governance and Policy in Sport Organizations is written and organized with the goal of being teacher- and student-friendly so that instructors will be comfortable with the topic and can present the material to students in a clear, organized fashion. To this end, we now include useful information on accreditation standards, on the book's organization and pedagogy, and about the instructor materials available to adopters. The fourth edition of this book has now been used by instructors teaching in the online format. The feedback has been very positive, as instructors have noted the organization of the chapters and the end-of-chapter questions and case studies work successfully with online students. With so many Sport Management programs adding online classes, this fifth edition will continue to prove useful for in-person, hybrid, or online-only instruction.

COSMA CURRICULAR APPLICATION

Today, Sport Management education programs can be fully accredited by COSMA – the Commission on Sport Management Accreditation. COSMA defines governance and policy as "Methods of oversight for and control over sports and recreation programs in schools and communities, both nationally and internationally, within amateur and professional sport" (COSMA, 2022, p. 16). Governance is included in the COSMA documents as one of the Common Professional Components, in Section 3.2 of the *Accreditation Principles Manual & Guidelines for Self Study*. Governance is included as a Foundation of Sport Management along with Management Concepts and International Sport.

ORGANIZATION OF THE CHAPTERS

The book is divided into two main sections. The purpose of this division is to first establish the theoretical knowledge bases related to governance and policy development that sport managers need to operate their sport organizations on a daily basis, and then to present ways the theoretical bases play out in practical sport governance environments.

The first section, Chapters 1 through 4, presents the basics of specific managerial activities necessary for governance and policy development in sport organizations. This section includes material that is more theoretical in nature, covering an introduction to sport governance; the management functions of planning, organizing, decision-making; an ethics chapter; and now a new chapter on diversity, equity, and inclusion. Because sport managers face ethical dilemmas on a regular basis, the book devotes a chapter to ethical decision-making and the importance of corporate social responsibility.

The second section of the book, Chapters 5 through 15, details the governance structures of various sport industry segments, including

- scholastic sport
- amateur sport in the community
- campus recreation
- intercollegiate athletics

- major games
- Olympic sport
- Paralympic sport
- North American professional sport leagues
- professional individual sport
- professional sport leagues outside of North America
- esports – a new chapter in this edition

The content of this section is much more applied in nature. These chapters on the specific industry segments include sections on history, governance structures, and current policy issues. Organizational policies often develop as a reaction to current issues faced by sport organizations and sport managers. Because these issues change and evolve over time, for each industry segment the text presents a selection of current policy issues and the strategies that sport managers are implementing to deal with these issues. Often, such organizational policy decisions have ethical underpinnings as well. Throughout the current policy sections, the book addresses the ethical questions sport managers confront when developing policies. Each chapter also contains a case study related to the chapter content, and ethical concerns are integrated into many of these case studies.

Finally, in Chapter 16, we introduce a new chapter on crisis management. This chapter is meant to help aspiring sport managers understand the process of making complicated decisions under difficult, and often highly publicly scrutinized, circumstances.

THE BOOK'S PEDAGOGY

Each chapter includes Chapter Questions for use either as an in-person class or online class homework assignments or for class discussion. In addition, chapters contain many updated case studies that have been used successfully in sport governance classes and have proven useful for students. A section titled For Additional Information contains updated links directing readers to sources that expand on what the chapter covers. New to this edition, industry segment chapters also contain a Diversity, Equity, and Inclusion box which highlights a best practice in that area. The industry segment chapters also include Industry Portraits and Perspectives boxes that provide insights into the job responsibilities of and issues faced by sport managers in that segment. Students really do learn the best from successful people who are making their living in the sport industry. These boxes can be used to stimulate further discussion as readers consider the way the professionals responded to an issue and how they might respond differently. As these industry professionals were agreeable to helping students by participating in the interviews, students might also wish to reach out to them for advice after reading their interviews. Please note, too, that for easy reference the book's front matter includes a table of the many acronyms and abbreviations that students will encounter here and in the field.

SUPPORT MATERIAL

Ancillary materials are available to instructors who adopt this text. These ancillary materials include an Instructor's Manual, PowerPoint presentations, and a test bank. The Instructor's Manual contains student learning objectives, exam questions (multiple choice, short answer, and true/false, also offered in test bank form), and suggestions for additional assignments. The PowerPoint presentations focus on key points from the chapters to help instructors deliver the material as effectively as possible. With the ebook version of the text, students can directly access the websites with active links.

We hope that instructors and students will find this book to be an interesting and useful tool for learning the fundamentals of sport governance and its relationship to current policy and ethical issues facing today's sport managers. It is a book designed to help readers understand the big picture of the sport industry and their place in it as future sport management professionals.

Acknowledgments

A book project is truly a labor of love, and it cannot be undertaken and successfully completed without a great deal of help from others. Thus, many thanks are in order.

We are delighted to once again be working on this new edition with Routledge and with Simon Whitmore in particular. It is an honor that an internationally well-respected and well-known publishing company would deem this book worthy to undertake. We would also like to thank our previous publishers, Colette Kelly, Gay Pauley, and all the good people at Holcomb Hathaway for their support and insightful comments throughout the years in getting us to the point that a fifth edition was in demand.

Our sincere thanks to the many reviewers who have offered us thoughtful, thorough, and relevant feedback for each edition, making a significant contribution to the book's usefulness. Every edition is better because of your willingness to put your time, energy, and thoughtfulness into our book.

Mary A. Hums
Yannick Kluch
Samuel H. Schmidt
Joanne C. MacLean

I would like to specifically acknowledge my Sport Administration colleagues along the way at the University of Louisville. Without their ongoing and daily support and encouragement, as academics but more importantly as friends, this project would never have reached completion. I am especially thankful for all they have done for me here at the University of Louisville. I would also like to thank the student research assistants who helped me with this book in one or all of its editions: Michael Clemons, Sarah Williams, Seonghun Lee, Yung Chou (Enzo) Chen, Morgan Fishman, Robert Sexton, Mark Perry, and Kathleen Sipe. They put in numerous hours searching for information and working on APA citations. I would like to give particular recognition to Michael Clemons for his assistance with manuscript preparation for this fifth edition – it was superb and he was amazing – thank you!

My personal thanks to Dr. Packianathan "Chella" Chelladurai for his help in inspiring me to write this textbook in the first place and even more for his valuable guidance and mentorship in my academic life. I would also like to thank all of our professional colleagues who have adopted the book and made it a success. I sincerely appreciate your support. And to all the students who read it – a big thank you! Without students, we wouldn't have had the opportunity to write this book.

Finally, I would like to thank my family and friends here and around the world for their ongoing support and encouragement. Once again I cannot say enough about the good work of my co-author over the years – Joanne MacLean, who is now President of the University of the Fraser Valley, and will be transitioning off the book as her Presidential duties are now her professional priorities. I hope UFV knows how lucky they are to have her steering their ship! Finding someone who shares a common work ethic and dedication to a task on a long-term project such as this is a blessing. One of the main reasons I asked Joanne if she wanted to team with me when we initiated this project was because I knew she was a "do-er," and through each edition, she has been a valuable and supportive colleague and friend. She motivated me and kept me going on more occasions than she will ever know! I enjoy working with people like her – people who make me better.

Now I have two new co-authors – Yannick Kluch and Samuel Schmidt – and their enthusiasm and fresh takes on ideas are inspiring to me. Both Yannick and Samuel share a dedication to promoting diversity, equity, and inclusion and have infused new materials in these areas and on new topics like esports and crisis management into the text. Their creativity and innovative thinking helped this fifth edition take on another new life and I look forward to working with them in the future!!

And, of course, thanks to Amy for her ongoing support as I work through yet another edition of the book. I know I can always count on you to be there!!!

Thanks to all of you!

Mary A. Hums

This project came to me at a time I was facing a professional crossroads. I had just left my first tenure-track position in communication studies and was getting ready for a new role in a sport management department, trying to figure out how I could combine in meaningful ways my interdisciplinary research agenda on social and racial justice, diversity/equity/inclusion consulting work in the sports industry, and commitment to teaching future change-makers both in sport and beyond. One such meaningful way manifested when Dr. Mary Hums invited me to join this project alongside Dr. Samuel Schmidt and Dr. Joanne MacLean. What better way to drive change than teaching future decision-makers how to govern sport in ethical, inclusive, and equity-minded ways? So first of all, I would like to thank my co-authors for inviting me to this project and for their collaboration, hard work, and enthusiasm for this book. Part of the academic journey is finding your long-term collaborators – your team – and I am lucky to call my co-authors part of mine. Mary and Samuel, I am so grateful for your mentorship, friendship, and collaboration – on this book and for the years to come. And, of course, thank you to Joanne for the outstanding work she has done on previous editions of this book!

I also want to thank my other regular collaborators, whom I admire, look up to, and learn from every day. Thank you to Aimee Burns, Raquel Wright-Mair, Tomika Ferguson, Elizabeth Taylor, Eric Martin, Andrew Mac Intosh, Amy Wilson, Jean Merrill, Eli Wolff, Nina Siegfried, Nicholas Swim, Robert Turick, Anthony Weems, Debbie Sharnak, Scott Brooks, Stacey Flores, Joseph Cooper, Shannon Jolly, Emma Calow, Travis Scheadler, Lilian Feder, Javonte Lipsey, Evan Frederick, Raymond Schuck, David Cassilo, Shaun Anderson, Lydia Bell, Kelsey Gruganis, James Bingaman, Travis Bell, Evan Brody, Julie Richmond, Lamont Williams, and Akilah Carter-Francique. You all are brilliant, kind-hearted scholars (and, more importantly, friends) whom I am lucky to have in my corner!

Special thanks also go to my colleagues in the Center for Sport Leadership at VCU: Carrie LeCrom, Brendan Dwyer, Greg Burton, Abby Bergakker, Beatriz Ferreira, Samanha Coles, and Mya Thompson. I appreciate your support for this project! I also would like to express my gratitude to my friends at my past professional and/or academic homes at Bowling Green State University (where I discovered my passion for sport management and social justice work!), Rowan University, and the NCAA national office. Thank you – you know who you are!

In addition, I want to thank my students – past, current, and future ones – for nurturing my passion for teaching and education, for engaging on topics related to sport governance, and for inspiring me to co-write this book.

Finally, none of my work would be possible without the support and encouragement of my family and friends, and those who have become both, in the United States, in Germany, and across the globe. Thank you to Felix, Dani, Rosalie, and Isabella back home in Germany for their unconditional love and support! Thank you to my parents for instilling in me the importance of pursuing knowledge and living my dream. Thank you to my US family (Steve, Melinda, Ashley, Flo, David, Nanci, Taylor, Anna, Jedidiah, Ezra, McKenzie, Ben, and Mason) for turning what was once a foreign country into what is now my home. I can't imagine life without you all!

Yannick Kluch

I would like to thank my co-authors Mary Hums and Yannick Kluch. They are inspirations to me and continue to serve as role models in advocacy and academia. Mary has been an incredible mentor of mine since 2013 and I still use many "Hums-isms" in my daily life. Yannick is someone I look up to as his work is cutting-edge and making the world a more equitable place. My former colleagues in the University of Louisville Sport Administration program deserve immense praise for the mentorship and guidance they've provided me after all these years.

In addition to the mentors in my life, I want to acknowledge all the students that I've had the honor to teach in the past, present, and future. I've found that not many things in this world lift my spirits up like a good class filled with discussion, learning, and critical thinking. Students provide me with purpose and clarity on days that can be difficult.

I want to also acknowledge my family and the love they provide me. That includes my wife, Linh, for all she does both seen and unseen. As someone who faces many battles, her strength and perseverance are traits few have in this world. She is dedicated to her craft and making the world a more inclusive and equitable place for all, traits I am inspired by each day. Finally, I would like to acknowledge my dogs Louis and Taku for reminding me that breaks from work are needed (for walks of course).

Samuel H. Schmidt

As I reflect on the several editions of this book, I am grateful for the experiences and interactions that fuelled my interest in good governance in the sport industry and my drive to express governance concepts in this book. I appreciate the many colleagues at three Canadian universities who supported me as a scholar, and most especially Drs. Corlett, Boucher (RIP), Weese, Thibault, Kerwin, and Cousens. I am also thankful to have experienced local, national, and international sport management through provincial and national sport organizations, and three FISU Games. I am eternally grateful to Dr. Mary Hums, for our friendship beginning years ago in Graduate School at The Ohio State University, and for the opportunity to collaborate and consider best practices for student learning in the higher education classroom. Dr. Hums is an inspiration to me, and it has been a pleasure and honor to collaborate on this book. Finally, everlasting thanks to my family and friends for their support and interest in my work, and to Maureen for making everything in life better. You are my world!

Joanne C. MacLean

Credits List

ermission to reprint the following third-party material has kindly been granted as follows:

Exhibit 3.5: Excerpt from Essay "Using the Power in Your Privilege" (reproduced with permission from Kim Miller)

Exhibit 5.5: Wisconsin Interscholastic Athletic Association Committee Organization (reproduced with permission from Wisconsin Interscholastic Athletic Association)

Exhibit 5.6: OSAA Gender Identity Participation Policy (reproduced with permission from Oregon Scholastic Athletic Association)

Exhibit 8.6: Organizational Structure of the University of Oklahoma Athletic Department (reproduced with permission from University of Oklahoma)

Exhibit 8.7: Organizational Structure of Bowie State University (reproduced with permission from Bowie State University)

Exhibit 11.3: IPC Governance Structure (reproduced with permission from International Paralympic Committee)

Exhibit 13.3: PGA of America Member Classifications (reproduced with permission from PGA of America)

Image Credits

Abbreviations

AACSB	American Association of Colleges and Schools of Business
AAC	Athletes' Advisory Council
AANAPISI	Asian American Native American Pacific Islander-Serving Institution
AAU	Amateur Athletic Union
ABA	American Basketball Association
ACC	Atlantic Coast Conference
ACT	American College Test
AD	Athletic Director
ADA	Americans with Disabilities Act
ADHD	attention deficit hyperactivity disorder
AFC	American Football Conference
AFL	American Football League
AGM	Annual General Meeting
AIAW	Association for Intercollegiate Athletics for Women
AIM	American Indian Movement
AIOWF	Association of International Olympic Winter Sports Federations
AOC	Affiliate Organization Council
APR	Academic Progress Rate
ARISF	Association of IOC Recognised International Sports Federations
ASCOD	African Sports Confederation for the Disabled
ASOIF	Association of Summer Olympic International Federations
BCCI	Board of Cricket for Cricket in India
BIPOC	Black, Indigenous, and People of Color
BUCS	British Universities and Colleges Sport
BWBL	Baltic Women's Basketball League
CAF	Confédératión Africaine de Football
CBA	collective bargaining agreement
CBLOL	Campeonato Brasileiro de League of Legends
CCES	Canadian Centre for Ethics in Sport
CDL	Call of Duty League
CEO	Chief Executive Officer
CEWL	Central European Women's League
CFAR	Council of Faculty Athletics Representatives
CGF	Commonwealth Games Federation
CIAA	Central Intercollegiate Athletic Association
CISS	International Committee of Silent Sports
COC	Canadian Olympic Committee
COI	Committee on Infractions
CONCACAF	Confederation of North, Central American and Caribbean Association Football
CONMEBOL	Confederación Sudamericana de Fútbol
COP	Council of Presidents
COSMA	Commission on Sports Management Accreditation
CPCDE	Committee to Promote Cultural Diversity and Equity
CPISRA	Cerebral Palsy International Sport and Recreation Association
CSA	Council for Student-Athletes
CS:GO	Counter Strike: Global Offensive
CSR	corporate social responsibility
CWA	Committee on Women's Athletics
DI	Division I
DI-FBS	Division I Football Bowl Series
DI-FCS	Division I Football Championship Series
DII	Division II
DIII	Division III
DEI	Diversity, Equity, and Inclusion
DIECE	Diversity, Inclusion, and Equity Council of Excellence
EEWBL	European Women's Basketball League
EPC	European Paralympic Committee
EPL	English Premier League

FA	Football Association of England		LCL	League of Legends Continental League
FAR	Faculty Athletic Representative		LCO	League of Legends Circuit Oceania
FBS	Football Bowl Subdivision		LEP	League of Legends European Championship
FCS	Football Championship Subdivision			
FEI	International Fencing Federation		LGBTQ/ LGBTQ+	lesbian, gay, bisexual, transgender, and queer
FIBA	Fédération Internationale de Basketball Association			
			LJL	League of Legends Japan League
FIFA	Fédération Internationale de Football Association		LLA	Liga Latinoamérica
			LOL	League of Legends
FINA	Fédération Internationale de Natation/ International Swimming Federation		LPGA	Ladies Professional Golf Association
			LPL	League of Legends Pro League
FISU	Fédération Internationale du Sport Universitaire		MEMOS	Executive Masters in Sports Organization Management
FISU	Federation Internationale du Sport Universitaire		MiLB	Minor League Baseball
			MLB	Major League Baseball
GEF	Global Esports Federation		MLBPA	Major League Baseball Players Association
GM	general manager		MLS	Major League Soccer
GPA	grade point average		MOBA	multi-player online battle arena
HBCU	Historically Black College and University		MOIC	Minority Opportunities and Interests Committee
HIS	Hispanic serving institution			
IAAUS	Intercollegiate Athletic Association of the United States		MSI	minority-serving institution
			MVP	most valuable player
IBSA	International Blind Sports Association		NAC	National Administrative Council
ICC	International Cricket Council		NACAC	National Association for College Admissions Counseling
ICSD	International Committee of Sports of the Deaf			
			NACDA	National Association of Collegiate Directors of Athletics
IESF	International Esports Federation			
IF	International Federation		NACE	National Association of Collegiate Esports
IGF	International Golf Federation		NADOHE	National Association of Diversity Officers in Higher Education
IJF	International Judo Federation			
INTERPOL	The International Criminal Police Organization		NAIA	National Association of Intercollegiate Athletics
IOC	International Olympic Committee		NAIB	National Association of Intercollegiate Basketball
IOSD	International Organization of Sport for the Disabled			
			NASCAR	National Association for Stock Car Auto Racing
IPC	International Paralympic Committee		NASSM	North American Society for Sport Management
IPL	Indian Premier League			
ISA	International Surfing Association		NBA	National Basketball Association
ISOD	International Sport Organization for the Disabled		NBCA	National Basketball Coaches Association
			NBL	National Basketball League
IWAS	International Wheelchair and Amputee Sports Federation		NBPA	National Basketball Players Association
			NCAA	National Collegiate Athletic Association
JCC	Jewish Community Centers		NCC	National Coordinating Committee
KHSAA	Kentucky High School Athletic Association		NCCAA	National Christian College Athletic Association
KeSPA	Korean e-Sports Association		NFC	National Football Conference
LCK	League of Legends Championship Korea		NFHS	National Federation of State High School Associations
LCS	League Championship Series			

NFL	National Football League
NFLPA	National Football League Players Association
NGB	national governing body
NGBC	National Governing Bodies Council
NHL	National Hockey League
NHLPA	National Hockey League Players Association
NIA	National Intramural Association
NIL	Name, Image, and Likeness
NIRSA	National Intramural-Recreational Sports Association
NJCAA	National Junior College Athletic Association
NOC	National Olympic Committee
NPC	National Paralympic Committee
NSGB	National Sport Governing Body
NSO	National Sport Organization
NTC	National Tennis Center
NWBA	National Wheelchair Basketball Association
OCOG	Organizing Committee for the Olympic Games
ODI	One-Day International (cricket)
OFC	Oceania Football Confederation
OGR	Olympic Golf Rankings
OL	Overwatch League
OSAA	Oregon School Activities Association
OSU	The Ohio State University
OVS	Olympic Virtual Series
PA	Players Association
PAC	Paralympic Advisory Council
PBA	Professional Bowlers Association
PBR	Professional Bull Riders
PCS	Pacific Championship Series
PGA	Professional Golfers' Association
PASO	Pan American Sports Organization
PASPA	Professional and Amateur Sports Protection Act
PLEDIS	Premier League Equality, Diversity and Inclusion Standard
PSA	public service announcement
R&A	Royal & Ancient (golf)
RBI	Reviving Baseball in the Inner Cities
ROA	Return on Athletics
SA	Surfing Australia
SAT	Scholastic Aptitude Test
SEC	Southeastern Conference
SOI	Special Olympics, Inc.
SPR	Seattle Parks & Recreation
SWA	Senior Woman Administrator
T&CP	Teaching and Club Professionals
TIDES	The Institute for Diversity and Ethics in Sport
T2	Twenty20
TCL	Turkish Championship League
TCU	Tribal College and University
TOP	The Olympic Partner
UC	University of California, Berkeley
UCLA	University of California, Los Angeles
UEFA	Union of European Football Associations
UM	University of Michigan
UN	United Nations
UNESCO	United Nations Educational, Scientific and Cultural Organization
USADA	United States Anti-Doping Agency
USC	University of Southern California
USCAA	United States Collegiate Athletic Association
USEF	United States eSports Federation
USFL	United States Football League
USGA	United States Golf Association
USLTA	United States Lawn Tennis Association
USNLTA	United States National Lawn Tennis Association
USOC	United States Olympic Committee
USOPC	United States Olympic and Paralympic Committee
USTA	United States Tennis Association
UT	University of Tennessee
UTSNZ	University and Tertiary Sport New Zealand
VCS	Vietnam Championship Series
VETS	Voices of Employees that Served
WA1T	We Are One Team
WABA	Women's Adriatic Basketball Association
WADA	World Anti-Doping Agency
WFL	World Football League
WHA	World Hockey Association
WIAA	Wisconsin Interscholastic Athletic Association
WIHSEA	Wisconsin High School Esports Association
WNBA	Women's National Basketball Association
WNBPA	Women's National Basketball Players Association
XFL	Xtreme Football League
YMCA	Young Men's Christian Association
YWCA	Young Women's Christian Association

Introduction to Sport Governance

The National Collegiate Athletic Association (NCAA) works to make decisions about appropriate athlete compensation for use of their name, image, and likeness (NIL). A Major League Baseball (MLB) player gets suspended for domestic abuse. A city's parks and recreation department creates new programs for people with disabilities and those who are overweight or obese. A campus recreation center creates new

DOI: 10.4324/9781003303183-1

1

health and safety protocols for exercise classes and equipment use in the wake of the COVID-19 pandemic. The International Olympic Committee (IOC) wrestles with the postponement of the Tokyo 2020 Olympic Games. The International Paralympic Committee (IPC) changes its athlete classification rules to work to ensure fair competition. These types of occurrences in the sport industry deal with issues of membership, regulation, programming, or organizational structure. All relate to governance and policy development in sport organizations. Many organizations in the industry act as governing bodies and make decisions about their everyday operations. This chapter sets the groundwork for you to see governance and policy development in action in the sport industry.

WHAT IS THE SPORT INDUSTRY?

Today's sport industry is continually expanding and evolving on a global scale. Sport is distinctive and remarkable in magnitude and influence, reaching billions of participants and followers. The mass media devote special coverage to the sporting world, recounting competitive, recreational, and leisure-time activities for a variety of age groups and participant levels. The latest stories and trends are the focus of many blogs, Twitter feeds, Instagram postings, and other forms of social media. Scholars study sport from each angle as well, including sport history, sport psychology, and sport management.

In your sport management academic career, by now you are well aware of the industry segments of this global industry, including professional sport, intercollegiate athletics, the Olympic and Paralympic Movements, recreational sport, facility management, event management, sport for people with disabilities, health and fitness, sport club management, interscholastic sport, sport marketing, and legal aspects of sport. Considering the numerous segments comprising the sport industry, what is the size and monetary value of this business we call sport?

SCOPE OF THE INDUSTRY

The sport industry and marketplace continue to show strong numbers, even though the COVID-19 pandemic had an impact. The North American sports market size is predicted to grow to $83.1 billion by 2023 (Gough, 2021), but this growth rate is not just a North American phenomenon. According to Global News Wire (2021, para, 2):

> The global sports market is expected to grow from $388.28 billion in 2020 to $440.77 billion in 2021 at a compound annual growth rate (CAGR) of 13.5%. The growth is mainly due to the companies rearranging their operations and recovering from the COVID-19 impact, which had earlier led to restrictive containment measures involving social distancing, remote working, and the closure of commercial activities that resulted in operational challenges. The market is expected to reach $599.9 billion in 2025 at a CAGR of 8%.

Despite its size, the sport industry is a people-oriented, service-oriented, and diverse industry. The importance of treating one's customers and employees in a positive and culturally respectful manner has likely been widely discussed throughout all of your coursework. Most Sport Management courses deal with a micro approach to the industry, focusing on various specific areas such as marketing, facility management, event management, or financial issues. Governance courses, by contrast, take a macro view of sport organizations and will help you understand the big picture of how the various sport industry segments work together. Successful sport managers need conceptual skills (Chelladurai, 2014) to see the big picture, an important element in governance. Governance is more closely related to courses dealing with management, organizational behavior, or most especially, legal aspects of sport, which deal specifically with governance-related and policy-oriented issues. What, then, is the definition of *governance*?

DEFINITION OF GOVERNANCE

A common dictionary definition of *governance* is "the exercise of authority." Sometimes people mistakenly think of "governance" as meaning the "government" or elected political officials. It is much more than that. Sport managers must, of course, be aware of legislation or court decisions that affect them. In the sport industry, governance includes regulatory and service organizations. Governance is associated with power, authority, control, and high-level policy-making in organizations. People involved in governance make decisions that set the tone for the entire organization.

Sport governance occurs mainly on three different levels – local, national, and international. An example of a local governance organization is the Kentucky High School Athletic Association at the state level. On the national level, examples include the National Basketball Association (NBA), the National Federation of State High School Associations, USA Soccer, Hockey Canada, and the Korean Baseball Organization (in South Korea). International organizations include the IOC, IPC, Fédération Internationale de Football Association (FIFA, the international governing body for soccer), Fédération Internationale de Basketball Association (FIBA, the international governing body for basketball), and World Athletics (the international governing body for track and field, known internationally as athletics).

Governance structures within such organizations, while often similar, are not universally the same. Governance structures in North American sport differ from governance structures in European sport. For example, professional leagues in Europe use the promotion and relegation system in which teams that finish at the bottom of a Division I league are relegated to Division II, while the top teams in Division II are promoted to Division I. This system is not used in professional sport leagues in North America. Even within North America, differences exist between governance structures in Canada and the United States. NCAA rules and regulations governing collegiate sport in the

United States are different from the U Sports rules and regulations governing collegiate sport in Canada. The role of a country's government in sport also differs. Jamaica, for example, has a Ministry of Culture, Gender, Entertainment and Sport (Ministry of Culture, Gender, Entertainment and Sport, 2017). Slovenia has a Ministry of Education, Science and Sport (Ministry of Education, Science and Sport, 2017). The Republic of South Africa has its Ministry of Sport, Arts, and Culture (Government of South Africa, 2021). Again, this differs from the United States, where no federal government office exists to oversee sport in the country. Wouldn't it be interesting if the US government had a cabinet position for a Department of Sport, similar to having a Department of State or a Department of Transportation? We would then have a person designated as the Secretary of Sport (and start thinking about this – who might a good person for that position be?).

In the professional sport industry segment, Moorman, Sharp, and Claussen (2020) describe *governance* as being "roughly divided into governance of team sports by professional sport leagues and governance of individual sports by players associations operating professional tours" (p. 274). King (2017) sees sport governance as coming from three perspectives. First, governance describes a system in which an organization is steered. Second, governance can be seen as managing and delivering sport through networks. Finally, governance can take a focus on management guided by ethical and legal standards. For international sport managers, according to Thoma and Chalip (1996), governance involves making effective choices among policy alternatives. They suggest three techniques sport managers can use in the international setting: ideology analysis, political risk analysis, and stakeholder analysis.

As you can see from the discussion above, sport governance is often easy to identify but difficult to define. Several authors have offered their thoughts on the subject. According to O'Boyle (2013, p. 1), governance, broadly speaking, is "The process of granting power, verifying performance, managing, leading, and/or administering within an organization." Ferkins, Shilbury, and McDonald (2009) state that sport governance is "the responsibility for the functioning and overall direction of the organization and is a necessary and institutionalized component of all sport codes from club level to national bodies, government agencies, sport service organizations and professional teams around the world" (p. 245). As noted above, King (2017) acknowledges that governance includes both political and administrative aspects. For the purposes of this textbook, the authors of this book define *sport governance* as follows:

> Sport governance is the exercise of power and authority in sport organizations, including policy making, to determine organizational mission, membership, eligibility, and regulatory power, within the organization's appropriate local, national, or international scope.

We will study each element of this definition in this book.

WHAT IS AN ORGANIZATION?

In order for sport governance to take shape, we must first have an organization that needs governing. Certainly, plenty of groups are involved with sport, but what truly identifies a group as an organization rather than just a group of people?

Chelladurai (2014) lists several attributes of an organization, including the following:

1. identity
2. program of activity
3. membership
4. clear boundaries
5. permanency
6. division of labor
7. hierarchy of authority
8. formal rules and procedures

To illustrate how an organization is different from just a group of people, let's attempt to apply these attributes to professional sport organizations and then to the group of people you met in college and regularly go out with on weekends.

1. **IDENTITY.** Teams establish a public identity by their name, whether that name is the Bucks, the Sparks, the Packers, the Lightning, or Bayern Munich. Teams also have an established corporate identity separate from the players or the fans. In contrast, your group of friends usually does not have a name and is not an established business entity.
2. **PROGRAM OF ACTIVITY.** A program of activity implies that an organization has a certain set of goals it wishes to achieve and that these goals are tied to its mission statement and its successful business operation. A basketball team wants to win a championship, and the people who work in its front office want to maximize revenues. A group of friends may have goals (for example, to go to an event), but it does not have a written set of goals it wishes to accomplish or a written mission statement.
3. **MEMBERSHIP.** Organizations have set rules about membership. In order to play professional basketball, one would need a certain amount of experience at a certain level of play to be considered for "membership" in the league. This would typically include having played on the collegiate or international club level. A group of friends really does not have membership rules. They could be people you know from meeting them face-to-face or might be your followers on Instagram or Twitter or people with whom you have regular Zoom gatherings.
4. **CLEAR BOUNDARIES.** We know who plays on the Bucks or the Sparks because they have published league-approved rosters. Any claim that "I play for the Bucks" can be easily verified. Groups of

friends change from year to year, often from week to week or day to day, depending on what is happening in your life at the current time.

5. **PERMANENCY.** True organizations have relative permanency, although the organizational members come and go. For example, Bayern Munich's current players are not the same people as when the franchise was new, but the Bayern Munich organization still exists. A group of friends also changes over time but is more likely than an organization to cease to exist as a group because people move away or, over time, simply drift away from each other's life circles.

6. **DIVISION OF LABOR.** Within organizations, labor is divided among members. Tasks are determined, and then people are assigned to the tasks. Organizations clearly illustrate division of labor with organizational charts. The front office of a basketball team has specialization areas such as marketing, media relations, community relations, and ticket sales. However, a group of friends has no organizational chart with assigned duties, except perhaps for the designated driver.

7. **HIERARCHY OF AUTHORITY.** An organizational chart also reflects an organization's hierarchy of authority. Who reports to whom is clear from the lines and levels within the chart. At the top of the chart, a basketball team has a General Manager in charge of the day-to-day operations of the club; all others answer to them. In a hierarchy, people higher up are responsible for the actions of the people below them. In a group of friends, one seldom has any personal responsibility for the actions of others in the group.

8. **FORMAL RULES AND PROCEDURES.** Organizations have formal rules and procedures, such as constitutions, bylaws, and operational manuals. Examples of some of these include office dress codes, social media policies, policies on accepting gifts from clients, sexual harassment policies, or policies on acceptable use of organizational property. Friends don't have policy manuals outlining how the group will operate, what brand of smartphone everyone will use, or who can come along on Saturday night.

As you see from these very simplified examples, organizations are formalized entities with rules about mission, membership, structure, operation, and authority. You will read about these fundamental elements of governance in the chapters focusing on the different segments of the sport industry.

REGULATORY POWER

Another significant aspect of governance is that organizations have regulatory power over members, an ability to enforce rules and impose punishments (sanctions) if necessary. Different governing bodies possess this sanctioning power to different degrees. For example, state high school athletic associations establish rules on age limits, academics, and transferring schools. Failing to follow these rules may prevent an athlete from participating in regular season games or post-season tournament play. The IOC has the power to ban athletes from competing in future Olympic Games if they test positive for performance-enhancing drugs, and MLB can impose a luxury tax on teams whose payrolls exceed a certain amount. The NCAA determines recruiting rules for coaches, which, if violated, carry sanctions such as loss of scholarships for an ensuing season.

EXTERNAL AND INTERNAL INFLUENCES ON SPORT ORGANIZATIONS

Sport organizations do not exist in a vacuum. As part of the greater society in which they exist, they must anticipate changes in both their external and internal environments, preferably before they *must* react. Chelladurai (2009) subdivides the organization's external environment into two categories: (1) the task or operating environment, sometimes referred to as the *proximal* (close) environment, and (2) the general environment, also called the *distal* (further removed) environment. Sport managers must be cognizant of what is happening in their external environments and adapt accordingly. Their internal environments are created through each organization's specific policies and procedures (Chelladurai, 2014).

As open systems with inputs, throughputs, and outputs, sport organizations are in a constant state of interaction with their various environments (Minter, 1998). Governance structures, therefore, must adapt to changes in an organization's internal and external environments. For example, as society increasingly disapproves of the use of performance-enhancing drugs by athletes, more and more sport organizations are toughening their policies and procedures regarding the use of banned substances; as a result, we see the formation of organizations such as the World Anti-Doping Agency (WADA). As another example, as more athletes at all levels are standing up for social causes, leagues and teams are responding by making sure the athletes can voice their opinions and not setting policies that would infringe on their right to free speech. These *internal* reactions to trends in the sport organization's *external* environment are made to adapt with the times.

WHAT DETERMINES GOOD GOVERNANCE?

As we begin to learn about what governance is and what it looks like in action, it is important to note some basic principles that make up good governance. Believe me when I write this, all too often we see examples of sport-governing bodies making policy and governance decisions which fall short of what might be seen as good governance. Three examples would be the ill-fated attempt by the Ladies Professional Golf Association (LPGA) to require players to speak only English (Dorman, 2008), the NCAA's previous rules which did not allow college student-athletes the opportunity to profit off the use of their own image and likeness (Blinder, 2021), or FINA's (Fédération Internationale de Natation, the international federation for swimming) attempt to reject allowing swim caps called Soul Caps which protected hairstyles such as afros, dreadlocks, weaves, and extensions in the pool (Evans, 2021). We can readily find examples of missteps in governance, but on the other hand, what does good governance look like?

Good governance involves adherence to a set of principles that any sport manager can put into practice in their organization (Halleux, 2017; IOC, 2008). A most frequently used set of good governance principles in sport organizations include transparency, solidarity, democracy, and accountability.

An example of utilizing good governance can be illustrated by how sport-governing bodies support their athletes who speak out on social issues. Exhibit 1.1 illustrates how four of these pillars – transparency, solidarity, democracy, and accountability – can serve as a base for sport organizations to support these athletes (Hums et al., 2020, para 21).

> Democracy implies a freedom of expression. Solidarity reinforces support and unity. Accountability means people are responsible for the ramifications of their actions. Finally, transparency includes not only letting the light shine into an organization from the outside but also letting the light shine out from the inside. When these pillars of good governance are present, athletes can feel their voices are not just allowed to be heard, but encouraged to be heard.

Now think about soccer star Megan Rapinoe who knelt during the US National anthem before a US Women's National Team match, US Olympian Gwen Berry who stood with a raised fist on the medal podium at the Pan American Games, or Ethiopian marathoner Feyisa Lilesa who crossed the finish line with his arms crossed as a sign of solidarity with people who were protesting government actions in his home country and how their sport-governing bodies responded after each of them made their statements. Good governance or not?

For a governing body to operate successfully, it must be structured in such a way that important information can flow throughout

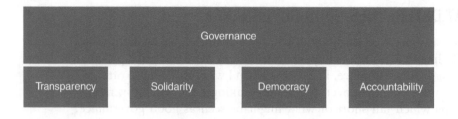

exhibit **1.1**   Pillars of Good Governance in Action

Source: Hums, et al. (2020).

the organization and be disseminated externally to others needing this information. But what types of information are most important? According to Grevemberg (1999, p. 10), sport organizations need to be mindful of what he calls the "Five Rs":

1. *Regulations* – systems that report organizational governance structures, constitutions, legal control mechanisms, event selection criteria, and codes of eligibility, conduct, and ethics;
2. *Rules* – systems that report technical rules for the officiating and management of the respective sport's competitive events;
3. *Rankings* – systems that report and place athlete/team performances based on results and competitive criteria into numerical order from first to last place;
4. *Records* – systems that report the best performances ever accomplished by athletes/teams within competitions, time periods, or overall;
5. *Results* – systems that report the final standings and performance statistics from competitions.

If sport organizations can consistently apply the Five Rs across all their operations, they will find that governance can be consistent and efficient.

ORGANIZATIONAL UNITS AND DOCUMENTS IN SPORT GOVERNANCE

How many of you have ever taken a language class? Maybe you took Spanish or Mandarin or Arabic. When you started the class, you had to take small steps. First you learned the alphabet and then small words like numbers or food. Finally, you put together sentences and paragraphs and were able to know how to ask directions, meet new people, and find the washroom. So it is with learning governance. Governance has a certain language, with basic terms and concepts we must know in order to understand governance and the inner workings of sport organizations. In terms of sport governance, organizations are made up of distinct units with varying degrees of authority and responsibility. In the past, these groups met in-person but the COVID-19 pandemic forced sport-governing bodies to work virtually. Sport managers will now have to decide how to strike a new balance between in-person meetings and virtual meetings. On the one hand, nothing can truly replace being in the same room with others during discussions as one can do more than just hear a debate. Body language and side-bar conversations are part of meetings as well. On the other hand, virtual meetings cut down on travel time and expense and may allow for participation by more people in the meeting. How business is done will be evolving even as this textbook is being written!

In addition to actual meetings, sport organizations usually maintain a set of documents dealing with governance structures. This section briefly introduces some of these organizational units and documents.

GENERAL ASSEMBLIES

Many sport organizations (e.g., the IOC, IPC, and NCAA) are voluntary in nature. Nevertheless, these organizations employ paid staff members. The size of the headquarters staff can vary widely. The NCAA headquarters and the IOC headquarters both employ over 600 people, while the IPC employs around 100, for example. The paid staff members handle the day-to-day operations of organizations, although they are not the ones who actually govern the organizations.

The primary governing body for many sport organizations is usually called a General Assembly; it may also be called a Congress or a General Business Meeting. In many sport organizations, the members of this governing body are volunteers. According to Chelladurai and Madella (2006), "Voluntary organizations are truly political systems; power is continuously exerted by professionals and volunteers to influence decisions and actions in order to satisfy personal or group needs" (p. 84).

A General Assembly for a sport organization usually convenes on a regular basis (often yearly). The members of the General Assembly, selected in accordance with an organization's constitution and bylaws, vote as a group on legislation, rules, policies, and procedures. This type of governing body generally elects officers such as the President, Vice Presidents, and Secretary. It also utilizes several standing committees assigned specific tasks, as will be discussed later. Its meetings, or sessions, are generally conducted using a common set of rules of operation. One of the most commonly used sets of guidelines for running a meeting is called *Robert's Rules of Order* (Robert et al., 2020). The National Hockey League (NHL) constitution, for example, directs that Robert's Rules of Order be used at their organizational meetings (NHL. n.d.). Those of you who are in a fraternity or sorority may recognize these from house meetings you attended. These rules are often referred to as Parliamentary Procedure, and most organizations designate a person who is familiar with Robert's Rules as the organizational Parliamentarian. It is the Parliamentarian's job to make sure the group members properly follow Robert's Rules of Order so that discussions can proceed in an orderly manner. Meetings organized using Robert's Rules usually follow this order of events:

Robert's Rules of Order
https://robertsrules.com/

1. call to order
2. approval of the minutes of the previous meeting
3. committee reports
4. old business
5. new business
6. announcements
7. a call to adjourn the meeting

During the meeting, when someone wants to make a suggestion for action by the group, they make what is called a "motion to take the action." Another person from the group must then second the motion. Then, the meeting chair allows debate on the action to begin. During this time, the motion can be amended and if so, the members must then vote on the amended motion. When the debate is complete, the chair calls for a vote. For in-person meetings, votes are taken by a count of hands, by voice, or sometimes by a paper ballot, depending on the topic. At virtual meetings, votes can be tallied directly in the chat area or by use of online polls. If no decision can be reached because, for example, the members need more information, a motion can be tabled – that is, set aside for action at the next meeting. When all the business of the group is complete, someone makes a motion to adjourn the meeting, and after another member seconds that motion, the meeting ends. Using these rules for the standard operation of a business meeting helps ensure fairness and enables all members to voice their opinions in an orderly manner (Introduction to Robert's Rules of Order, n.d.). Typically, the agendas for General Assembly meetings and the issues on which they vote come from a body known as an Executive Committee or Management Council.

GOVERNING BOARDS, EXECUTIVE COMMITTEES, AND MANAGEMENT COUNCILS

Many sport organizations are run by use of a board structure. These boards have different names such as a Board of Directors, Board of Overseers, or Governing Board, but all share common basic purposes. What are the main responsibilities of any board? According to Board Effect (2019), Board members have ten basic responsibilities:

1. Establish the organization's vision, mission, and purpose.
2. Hire, monitor, and evaluate the chief executive.
3. Provide proper financial oversight.
4. Ensure the organization has adequate resources.
5. Create a strategic plan and ensure that it is followed.
6. Ensure legal compliance and ethical integrity.
7. Manage resources responsibly.
8. Recruit and orient new board members and assess board performance.
9. Enhance the organization's public standing.
10. Strengthen the organization's programs and services.

While these are all actions a board can take, boards cannot act completely unfettered. For example, boards are limited to taking actions only over duties which are under their jurisdiction as defined by the organization's bylaws and constitution. In addition, and although it should go without saying, boards are not free to take any actions which would violate the law. Boards typically have officers. Standard officers on a board would include a President, perhaps several Vice Presidents, a Secretary, and a

Treasurer. There may be other titles as well but these are the most common. It is also imperative that the make-up of a board's membership be as diverse as possible while avoiding tokenism. According to Creary et al. (2019, para. 4), "social diversity (e.g., gender, race/ethnicity, and age diversity) and professional diversity are both important for increasing the diversity of perspectives represented on the board." Sport England and UK Sport recently took a step to ensure this in their organization by making changes to their Code of Sport Governance to increase diversity on their board and senior leadership teams. According to Tim Hollingsworth, Sport England chief executive, "It is a further step towards greater diversity of background, experience and understanding of sport and activity environments having a seat at the table at the very top of sporting organisations" (Pavitt, 2021, para. 15).

Who typically are the people who serve on boards? Usually, they are people who have power and influence by virtue of reputation, financial capacity, or expertise/experience in the area in which the organization operates. For example, sport-governing bodies will often have retired athletes on their boards such as Olympians Sergey Bubka (athletics – pole vaulting) and Sebastian Coe (athletics – 1500 meters) who are on the World Athletics Executive Board. Jeanie Buss, successful businesswoman and executive with the Los Angeles Lakers, is on the Board for LA2028. People like these often populate boards of sport-governing bodies.

Descriptions of the duties and responsibilities of board members are typically found in either an organization's bylaws or constitution. These documents will describe how someone becomes a member of the Board, the length of their term, and how many terms they can serve. In the sport industry, people who serve on boards are typically volunteers. As boards are generally where policies are developed and approved, it is interesting to note that in the sport industry, sport managers are not always the people who make big policy decisions. Rather, sport managers must put into practice the policies the board approves even though the sport managers are not always the ones creating the highest levels of organizational policy.

Boards are in place across all different levels of the sport industry from major international governing bodies to local-level organizations. The IPC has the IPC Governing Board. The United States Olympic and Paralympic Committee (USOPC) has its Board of Directors. As an example on the local level, the lead author of your textbook is a member of the Board of Directors of her city's Miracle League, a nonprofit organization that offers baseball for young people with disabilities.

Boards form the highest level of governance in many organizations, yet they may not always be where the power lies in an organization. Oftentimes that rests in a subset of the board often called the Executive Committee. Executive Committees are small subsets of an organization's board or General Assembly. Members of the General Assembly select a group, usually from five to 20 members, to serve on the organization's Executive Committee. Many believe that the Executive Committee is where the "real power" in a sport organization lies. This group usually generates the agenda action items on which the General Assembly votes. If the Executive Committee does not endorse an idea, it will almost never be brought to the General Assembly for a vote. In addition, this group meets formally more frequently, often two or three times a year, in order to deal with issues that may come up between General Assembly meetings. Given the changes in how workplaces operate in the post-COVID world, a great deal of that work is now accomplished virtually via Zoom, Teams, or other web conferencing platforms, conference calls, or through e-mail.

STANDING COMMITTEES

Sport organizations also designate standing committees with specific responsibilities within their governance structures. The type and number of standing committees vary by organization. The USOPC, for example, has standing committees for the following areas (USOPC, 2021):

Athlete and NGB Services
Compensation
Ethics
Finance
Audit and Risk
NGB Oversight & Compliance
Nominating and Governance

As a comparison, here are the standing committees for FIFA, the International Federation for football [soccer] (FIFA, 2022).

Development Committee
Finance Committee
Football Stakeholders Committee
Medical Committee
Member Associations Committee
Organising Committee for FIFA Competitions
Referees Committee

AD HOC COMMITTEES

At times, sport organizations face issues that need to be dealt with on a short-term basis. For example, perhaps the organization is planning to host a special fundraising event or play a home game at a site other than its usual home arena. Because the event may just occur one time, the organization will assemble an *ad hoc*, or temporary, committee that is in charge of the event. Unlike standing committees that deal with ongoing concerns, once the event is over, the ad hoc committee usually ceases to exist. An ad hoc committee usually only operates for a short period of time, generally less than one year. Occasionally, a topic initially addressed by an ad hoc committee will become an ongoing concern and the organization will then establish a standing committee to address it.

EXECUTIVE STAFF

The people who serve on a General Assembly, Governing Board, or Executive Council are almost always volunteers. Their business expenses may be paid, but they are not employees of the sport organization. The people who are employed by the sport organization to run the daily operations are called Executive or Professional Staff. People in these positions have titles such as Executive Director, General Manager, Marketing Director, Sport Administrator, Social Media Coordinator, or Event Coordinator. These individuals are paid sport management professionals, employed by the governing body. They work in the organization's headquarters, as opposed to volunteers who may be located anywhere in the world. For example, the Executive Staff of the NCAA works daily in offices in Indianapolis, Indiana. The Athletic Directors on NCAA committees, however, are located at their home institutions across the nation and come together only at designated times during the year. As another example, the staff of the World Curling Federation work at the headquarters of the organization in Perth, Scotland, but the organization is managed by a Board with members in different countries (World Curling Federation, n.d). Similarly, the Executive Staff for the IPC work in the organization's headquarters in Bonn, Germany. They run the organization on a daily basis, planning events, handling financial matters, and marketing upcoming events. The volunteers, however, who work with a specific sport such as Ice Sledge Hockey, may be located in Canada, Norway, or the United States. They will meet together at a Sport Technical Committee meeting, for example, at a designated time and place each year or may hold a business session during the Paralympic Games.

CONSTITUTIONS AND BYLAWS

Almost all sport organizations have documents outlining the basic functions of the organization, usually called the constitution and bylaws. An organization's constitution acts as a governing document that includes statements about the organization's core principles and values. Bylaws, also governing documents, are more operational in nature, outlining how an organization should conduct its business in terms of elections, meetings, and so on. For examples of what these types of documents include, see Exhibit 1.2, a Table of Contents from the Bylaws of the USOPC.

Examples of these different organizational units and documents will be discussed throughout each of the industry segment chapters in this text.

Bylaws of the USOPC (Effective March 11, 2021)

exhibit **1.2**

SECTION 1 NAME, OFFICES, AND DEFINITIONS
SECTION 2 THE MISSION
SECTION 3 THE CORPORATION BOARD
SECTION 4 OFFICERS
SECTION 5 COMMITTEES AND TASK FORCES
SECTION 6 THE CEO
SECTION 7 THE OLYMPIC AND PARALYMPIC ASSEMBLY
SECTION 8 MEMBERS
SECTION 9 ATHLETES' RIGHTS
SECTION 10 COMPLAINTS OF NON-COMPLIANCE AGAINST AN NGB
SECTION 11 APPLICATION TO REPLACE AN NGB
SECTION 12 CODE OF CONDUCT FOR VOLUNTEERS, STAFF, AND MEMBER ORGANIZATIONS
SECTION 13 THE OFFICE OF ATHLETE OMBUDS
SECTION 14 ATHLETES' ADVISORY COUNCIL AND U.S. OLYMPIANS AND PARALYMPIANS ASSOCIATION
SECTION 15 NATIONAL GOVERNING BODIES COUNCIL
SECTION 16 AFFILIATE ORGANIZATIONS
SECTION 17 INDEMNIFICATION
SECTION 18 ORGANIZATION OF DELEGATION EVENTS IN THE UNITED STATES
SECTION 19 DELEGATION EVENT-RELATED MATTERS
SECTION 20 FINANCIAL MATTERS
SECTION 21 IRREVOCABLE DEDICATION AND DISSOLUTION
SECTION 22 MISCELLANEOUS
SECTION 23 AMENDMENTS OF THE BYLAWS

Source: USOPC (2021).

WHY STUDY GOVERNANCE?

Given all the areas to study within the academic discipline of Sport Management, why study governance? Three main reasons come to mind: (a) you need to understand the "big picture," (b) you need to understand how governance fits within the Sport Management curriculum, and (c) you definitely will use your knowledge of sport governance in whichever industry segment you work.

UNDERSTANDING THE BIG PICTURE

In studying governance, you will truly be challenged to put together all the pieces of the sport industry. As mentioned above, studying sport governance requires the ability to see the big picture, to understand how individual sport organizations fit into the greater industry, and to see the similarities and differences among the various industry segments. Sport governance also prepares you for the global sport industry you will be entering. Sport managers who lack the ability to see how their organizations fit in to the global picture guarantee the ultimate failure of their organizations. With an understanding of sport governance, you will see how the governing structures of seemingly dissimilar industry segments such as intercollegiate athletics and the Olympic Movement have much more in common than you would think.

GOVERNANCE IN THE SPORT MANAGEMENT CURRICULUM

As the number of Sport Management academic programs has increased greatly in the past few decades, so have issues of program quality. From the very beginning, governance has been an integral part of Sport Management education.

Today, Sport Management education programs can be fully accredited by an organization known as COSMA – the Commission on Sport Management Accreditation (COSMA). According to NASSM (n.d., para. 1), COSMA "is a specialized accrediting body that promotes and recognizes excellence in sport management education in colleges and universities at the baccalaureate and graduate levels." Governance is now included in the COSMA documents as one of the common professional components. In addition, many Sport Management programs are now housed in Business Schools, which have their specific accreditation agency – the American Association of Schools and Colleges of Business (AASCB).

It is important to note that sport management education is not limited to schools in North America. Programs are common around the world now and thrive across all regions of the world – Africa, Asia, Europe, Oceania, and the Americas. This reinforces why learning about governing structures on all levels will help develop your understanding of how this worldwide industry operates.

COSMA
https://www.cosmaweb.org/

American Association of Schools and Colleges of Business (AASCB)
https://www.aacsb.edu/

USING KNOWLEDGE OF GOVERNANCE IN YOUR CAREER

Understanding governance structures is important for any career in sport management. If you work at a bank, you need to know the rules for your workplace and probably some basic federal and state laws. But for the most part, especially at an entry-level position, you will not be interacting with the people who write the policies for your bank or for the broader banking industry. In sport, however, especially because of the ramifications of enforcement, you will need to be keenly aware of governance structures and issues. You will also need to understand different contexts of governance. There will be governance issues dealing with a given sport and its rules as well as governance issues dealing with the business side of sport. You will need to know where the power lies in your organization, and studying governance can help you understand this. You will need to know which governing bodies you will deal with in your industry segment. If you work in a college athletic department, even as you start perhaps by doing an internship, you must understand how you relate to governance structures of your university, your conference, and the NCAA. (Even interns can commit NCAA rules violations!) In professional sport, if you deal with players associations or players unions, you will need to know how they relate to the decisions you make. In high school sport, the power rests at the state level in organizations such as the Indiana State High School Athletic Association or the Tennessee Secondary School Athletic Association. A recreation director in a City Parks and Recreation Department may answer to the mayor or the City Council. If you work for a sport federation such as FIBA, you will have to be knowledgeable about the federation's rules and regulations dealing with athlete eligibility and interactions with other sport organizations such as the WADA.

Studying sport governance gives you a perspective on where you fit into your sport organization and where your sport organization fits into its industry segment. For example, if you work in the front office of an NBA team, you will need to understand various levels of governance in your job, from the club's front office to the League Office to the Commissioner's Office. In Olympic and Paralympic Sport, you may be involved in interactions among IFs, National Governing Bodies, National Olympic Committees, and Organizing Committees of the Olympic Games. In intercollegiate athletics, you will need to know basic NCAA compliance rules to avoid placing your school in danger of NCAA sanctions. If you work for the Commonwealth Games Federation, you will need to understand eligibility rules for athletes, so that any athletes you are responsible for do not jeopardize their eligibility. In sport, you are likely to have more direct interactions with governing bodies and policy makers than in almost any other industries. You need to understand who has the power and where the power lies in any sport organization you work or interact with.

Next, sport-governing bodies need to build their policies and purposes on a strong foundation of diversity, equity, and inclusion. What area of our lives is really more diverse than sport after all? Teammates, coaches, fans, the media, all are stakeholders in the sport industry and all come from a broad spectrum of society as a whole. Sport managers working in governing bodies setting policies must be sure that these policies are as inclusive as possible. As Levine (2020, para. 1) points out, "Smart leaders know that their companies benefit from attracting, developing and retaining a diverse workforce. In addition, they understand that creating and maintaining a welcoming and inclusive culture makes talented people want to join and stay." Beyond the measurable business benefits of promoting diversity, equity, and inclusion – it is just the right thing to do. For this reason, this textbook devotes a designated chapter to learning about diversity, equity, and inclusion in sport governance.

Finally, the importance of sport managers acting in an ethical manner in any sport governance situation cannot be overstated. Sport managers face ethical dilemmas on a daily basis. How they deal with them is a measure of their own ethical nature and that of their organization. For this reason, this textbook – and similar to having a chapter on diversity, equity, and inclusion – designates a chapter to ethical decision-making and the importance of corporate social responsibility and discusses various ethical issues sport managers may face in different industry segments.

SUMMARY

T he sport industry continues to grow and develop on a global scale. Studying sport governance allows you to take a big-picture approach to this global industry. Learning about the governing structures and documents for sport organizations illustrates where power and authority exist within the industry. This area of study is sufficiently important to be discussed in the COSMA accreditation documentation, reemphasizing the importance of understanding this complex, fascinating aspect of the sport industry.

In your previous Sport Management classes, you learned about basic managerial activities and functions. For purposes of this textbook, we will focus on four of these important areas in Chapters 2 through 4 – planning; decision-making; diversity, equity, and inclusion; strategic management and policy development; and ethical decision-making. These activities are the heart and soul of the governance process and have separate chapters devoted to them, further explaining their roles in the governance of sport organizations.

The remainder of the text guides you through selected industry segments and explains how sport governance is implemented in those segments, using numerous examples to illustrate governance in action. It is our hope that you will enjoy these challenging and interesting areas of study within sport management.

caseSTUDY

INTRODUCTION TO GOVERNANCE

Recently, the NBA has experienced an increase in inappropriate fan interactions with players on the court. Players are just an arm's length from courtside fans at times and have endured abusive language and objects being thrown at them. The NBA Commissioner's Office is considering putting together an ad hoc committee to develop policies concerning player safety and fan control.

1. Who should the Commissioner have on their committee? Choose any three of the following and explain your rationale for having them on the ad hoc committee: player representative, team General Manager, arena security specialist, member of the media, referee, arena design specialist, representative from the Commissioner's Office, season ticket holder, someone else not listed here.

2. What type of information should your group gather to help in making this decision?

3. Once you make the decision and draft a new policy, who (people, organizations) should get to review/approve it before it is implemented?

4. In your opinion, what would be a sound policy for the NBA to implement and enforce in order to address this situation? How would it deal with both fans and players?

CHAPTER questions

1. Choose a sport organization and then use Chelladurai's model of organizational attributes from this chapter to define the different elements of that organization.

2. Find two sample sport organization constitutions or bylaws. Compare the two for content. Explain why you think they are different or similar.

3. Using the definition of sport governance from this chapter, choose a sport organization and identify the different parts of the definition in that organization.

FOR ADDITIONAL INFORMATION

1. The organization known as Play the Game often includes insightful articles on major governance issues in international sport: www.playthegame.org

2. This website includes useful information about the governance structure of the Olympic and Paralympic Movement: https://www.teamusa.org/About-the-USOPC/Structure:/

3. This source from Sport England sets forth a Code for Sport Governance: https://www.sportengland.org/campaigns-and-our-work/code-sports-governance

REFERENCES

Blinder, D. (2021, June 20). College athletes may earn money from their fame, NCAA rules. *New York Times*. https://www.nytimes.com/2021/06/30/sports/ncaabasketball/ncaa-nil-rules.html

Board Effect. (2019). *10 basic responsibilities of board members.* https://www.boardeffect.com/blog/10-basic-responsibilities-board-members/

Chelladurai, P. (2009). *Managing organizations for sport and physical activity: A systems perspective* (3rd ed.). Holcomb Hathaway.

Chelladurai, P. (2014). *Managing organizations for sport and physical activity: A systems perspective* (4th ed.). Holcomb Hathaway.

Chelladurai, P., & Madella, A. (2006). *Human resource management in Olympic sport organizations.* Human Kinetics.

Creary, S.J., McDonnell, M.H., Ghai, S., & Scruggs, J. (2019, March 27). When and why diversity improves your board's performance. *Harvard Business Review*. https://hbr.org/2019/03/when-and-why-diversity-improves-your-boards-performance

Dorman, L. (2008). Golf tour's rule: Speak English to stay in play. *New York Times.* https://www.nytimes.com/2008/08/27/sports/golf/27golf.html

Evans, A. (2021, July 2). *Soul Cap: Afro swim cap Olympic rejection 'heartbreaking' for Black swimmers.* BBC. https://www.bbc.com/news/newsbeat-57688380

Ferkins, L., Shilbury, D., & McDonald, G. (2009). Board involvement in strategy: Advancing the governance of sport organizations. *Journal of Sport Management, 23,* 245–277.

FIFA. (2022). *Committees.* https://www.fifa.com/about-fifa/organisation/committees

Global News Wire. (2021). *Global sports market report (2021–2030) – COVID19 impact and recovery.* https://www.globenewswire.com/fr/news-release/2021/03/18/2195540/28124/en/Global-Sports-Market-Report-2021-to-2030-COVID-19-Impact-and-Recovery.html

Gough, C. (2021). *North American sports market size 2009–2023.* Statista. https://www.statista.com/statistics/214960/revenue-of-the-north-american-sports-market/

Government of South Africa. (2021). *Sport, Arts, and Culture [Ministry of].* https://www.gov.za/about-government/contact-directory/ministers/ministers/arts-and-culture-ministry

Grevemberg, D. (1999, May). *Information technology: A solution for effective Paralympic Sport administration.* Paper presented at the VISTA 1999 Paralympic Sport Conference, Cologne, Germany.

Halleux, V. (2017). *Good governance in sport. European Parliament Think Tank.* https://www.europarl.europa.eu/thinktank/en/document.html?reference=EPRS_BRI(2017)595904

Hums, M.A., Wolff, E.A., & Siegfried, N. (2020, June 23). *Making a case for athlete advocacy: Humanitarians, Olympism, and good governance.* Play the Game. https://playthegame.org/news/comments/2020/1006_making-a-case-for-athlete-advocacy-humanitarians-olympism-and-good-governance/?fbclid=IwAR0zjeMLmJD0YREyUqR8pTAEVProxK6hmgnkrcr3sPKW7mufsp0PVtA5lNk

Introduction to Robert's Rules of Order. (n.d.). www.robertsrules.org/rulesintro.htm

IOC. (2008). *Basic universal principles of good governance of the Olympic and sports movement.* https://stillmed.olympic.org/media/Document%20Library/OlympicOrg/IOC/Who-We-Are/Commissions/Ethics/Good-Governance/EN-Basic-Universal-Principles-of-Good-Governance-2011.

pdf#_ga=2.236488201.924206999.1591117105-65331301.1591117105

King, N. (2017). *Sport governance: An introduction.* Routledge.

Levine, S.R. (2020, January 16). Diversity confirmed to boost innovation and financial results. *Forbes.* https://www.forbes.com/sites/forbesinsights/2020/01/15/diversity-confirmed-to-boost-innovation-and-financial-results/?sh=481daae1c4a6

Ministry of Culture, Gender, Entertainment and Sport. (2017). *Home page.* http://mcges.gov.jm/

Ministry of Education, Science and Sport. (2017). *In focus.* http://www.mizs.gov.si/en/

Minter, M.K. (1998). Organizational behavior. In J.B. Parks, B.R.K. Zanger, & J. Quarterman (Eds.), *Contemporary sport management* (pp. 79–89). Human Kinetics.

Moorman, A.M., Sharp, L.A., & Claussen, C.L. (2020). *Sport law: A managerial approach* (4th ed.). Routledge.

NASSM. (n.d.). *Program accreditation.* http://www.nassm.com/InfoAbout/NASSM/ProgramAccreditation

NHL. (n.d.). *Constitution of the National Hockey League.* https://www.lakelawgroup.com/wp-content/uploads/2017/02/constitution-NHL-.pdf

O'Boyle, I. (2013). Managing organizational performance in sport. In D. Hassan & J. Lusted (Eds.). *Managing sport: Social and cultural perspectives.* (pp. 1–16). Routledge.

Pavitt, M. (2021, July 4). *Sport England makes governance code changes to boost diversity and athlete welfare.* Inside the Games. https://www.insidethegames.biz/articles/1109771/code-of-sports-governance-changes

Robert, H.M., Honemann, D.H., Balch, T.J., & Seabold, D.E. (2020). *Robert's Rules of Order newly revised* (12th ed.). https://robertsrules.com/books/newly-revised-12th-edition/

Thoma, J.E., & Chalip, L. (1996). *Sport governance in the global community.* Fitness Information Technology.

United States Olympic and Paralympic Committee (USOPC). (2021). *Bylaws of the United States Olympic and Paralympic Committee.* https://www.teamusa.org/-/media/About-the-USOC/Board-Docs/062920-USOPC-Bylaws-Effective-June-18-2020-FINALua.pdf

World Curling Federation. (n.d.). *About.* https://worldcurling.org/about/

Managerial Functions Related to Governance

Sport managers carry out a wide assortment of managerial activities and functions on a daily basis. The four functions of management have been defined as planning, organizing, leading, and evaluating (Chelladurai, 2014). Sport managers dealing with governance issues must be able to carry out all these functions, but two functions, planning and organizing, are more critical to governance than others and will be

DOI: 10.4324/9781003303183-2

discussed in this chapter. Decision-making, a subset of leading, is also essential to sport managers dealing with governance issues, and it will be discussed here as well. Sport managers perform these functions daily. This chapter provides a brief overview of these important managerial day-to-day activities and their relationship to sport governance.

PLANNING

The Importance of Planning for Sport Organizations

Sport organizations need to plan because the sport industry presents a complex environment. Whether it is the Olympic Games bringing together nations or a high school softball tournament featuring local teams, the sport industry requires interaction and cooperation in order for teams, leagues, tours, and events to be successful. What are the specific purposes of planning? According to the Department of Communities Tasmania (2019, p. 1), planning serves to help a sport organization:

define its vision/mission, values, and main objectives,
have greater control over its direction,
be proactive rather than reactive,
build teamwork,
improve financial performance,
learn from and avoid past mistakes,
evaluate performance.

Given these benefits of planning, why then do some people still resist efforts to plan (Bryce, 2008)?

Resistance to Planning

People tend to develop comfort zones. They do things a certain way because "We've always done it that way." To these individuals, trying to implement a plan to do something different or new is a challenge they do not wish to undertake. Sometimes people who have worked with an organization for many years respond to planning initiatives with "Why should we do this? We've done this a million times before, and we know no one really looks at these things. Then five years later they ask us to do it again, and they ignore us again." Sometimes, longtime employees may resist changes to operations they personally developed. Finally, there are those who simply lack the ability to plan and are intimidated by the process. Good sport managers make planning a priority and learn how to deal effectively with those employees who resist the importance of planning.

Types of Plans

Sport managers must be able to develop both short-term and long-term plans. They must develop timetables so projects and events will take place as smoothly as possible. *Short-term planning* refers to planning projects and events that will occur within the next one to three years. For example, with the Olympic and Paralympic Games, test events in all venues are run within a year or two of the start of the Games. At a new stadium, contracts with concessionaires and security are finalized in the year before the stadium opens.

In contrast, *long-term planning* involves planning that extends three or more years. For example, the International Olympic Committee (IOC) awards the Olympic and Paralympic Games to a host city multiple years before the flame is lit to open the Games in that city. The upcoming Summer Games scheduled for Paris in 2024 and Los Angeles in 2028 were announced in 2017 and Brisbane was just announced as the host for the 2032 Summer Games (Diaz, 2021). Sport teams deciding to build new stadiums must begin working with architects and contractors at least five years before the first beverage is poured at a game. For example, groundbreaking for construction of the new SoFi Stadium in Los Angeles began in 2016 and the venue opened in 2020. This does not include the years of planning that preceded the first shovel being turned. Within these long-term plans, short-term plans must be implemented. Returning to our stadium example, the people who work in presentation (coordinating the music, announcements, etc., during an event) need to be able to work with the sound system (short-term planning). However, before a sound system in a venue can be checked, all the proper infrastructure for the power must be in place (long-term planning). All these plans must be carefully sequenced for the sport organization to be successful.

Long-term and short-term plans are not the only types of plans sport managers need to develop. For example, there are standing plans and single-use plans (Chelladurai, 2014). *Standing plans* refer to plans that are put in place and then referred to continuously as certain events repeat. For example, a facility manager working at a Major League Baseball team stadium needs to have parking and traffic plans to use at every home game for the season. A standing plan for a sport-governing body could be illustrated by having elections for their officers on a set schedule as dictated in their organizational bylaws such as the IOC electing a new President for an eight-year term followed by a re-election vote for a four-year term. *Single-use plans* refer to plans developed for events that may occur just once. For example, ESPN2's College Game Day may come to your city for the first time and plans need to be made for this one-time event. Another interesting example here could be the Fédération Internationale de Football Association's (FIFA) decision to hold the World Cup in Qatar during the winter months as opposed to its traditional summer schedule.

All of these plans, however, can be altered unexpectedly as the sport industry experienced during the COVID-19 pandemic. The National Basketball Association (NBA) set up a bubble in Orlando to play out the season. The National Collegiate Athletic Association (NCAA) canceled championships. Local parks and recreation departments closed parks, pools, and playgrounds bringing summer programming for children to a halt. Postponement of the Olympic and Paralympic Games disrupted all the short-term and long-term planning preceding the event. Even as the final days wound down before the Tokyo Games opened, adjustments were still being made on numbers of spectators allowed in venues due to surges in COVID cases in Japan. Talk about crisis management for sport-governing bodies! How should sport managers navigate environments such as these? This edition of your textbook has now incorporated a new chapter on Sport Governance in Times of Crisis – Chapter 16. The authors think you will find this chapter to be useful and challenging. Sport managers must learn to be flexible while keeping in mind the integrity of the events they are planning to carry out and this is no easy task.

The Planning Process

Sport organizations need to follow a set process to establish effective short-term and long-term plans. For some sport organizations, this process may begin with a vision statement. Other organizations will start with a mission statement and then take the following steps (adapted from VanderZwaag, 1998):

1. vision/mission statement
2. goals
3. objectives
4. tactics
5. roles
6. evaluation

The next sections examine each step in the planning process and use examples to illustrate the different steps. The focus is on the sport organization's front-office planning, rather than on-the-field plans such as the Milwaukee Bucks' objective to win the NBA Championship or the University of Arizona women's basketball team's objective to win the NCAA tournament. As a sport manager, you most likely will be working in off-the-field careers, so the focus of this chapter is on goals and objectives dealing with business-related matters such as increasing ticket sales or securing sponsorship packages. For sport organizations to effectively move into the future, they must establish both long- and short-term plans. Because sport governance issues are generally broad in nature and affect the entire sport organization, any course of action dealing with governance issues must be carefully planned.

Vision Statements

A number of sport organizations begin their planning process by developing what is known as a vision statement. These can be as simple as one sentence. According to Peek (2020, para. 8), vision statements are "future-based and are meant to inspire and give direction to the employees of the company, rather than to customers."

Vision statements are different from mission statements in that vision statements focus on the organization's future aspirations and values while mission statements focus on an organization's purpose.

The vision for the International Paralympic Committee (IPC) states "make an inclusive world through Para sport" (International Paralympic Committee, n.d.a, para. 1). Commonwealth Sport Canada (2020, para. 1) lists the following as its vision "Commonwealth sport is an integral part of Canadian high performance sport and significantly contributes to the social development of commonwealth countries." As you can see, these are quite brief and aspirational in nature. Mission statements, as you will see next, are more concrete.

Mission Statements

Sport organizations are a lot like sailboats. Without a rudder to steer, it does not matter how much wind there is – the boat will not go in its intended direction. It will still float, but it will not get where the crew wants it to go. What gives a sport organization its direction?

Direction is established early in the planning process with the organization's mission statement. As previously stated, a *mission statement* focuses on an organization's purpose. More specifically, it

1. describes who we are
2. describes what we do
3. uses concise terms
4. uses language that is understandable to people inside and outside the sport organization
5. communicates the organization's purpose, philosophy, and values.

A well-written mission statement does not need to be a lengthy document. It may be only 30–40 words, or two or three sentences long, although some organizational mission statements will incorporate a few short paragraphs. In a perfect world, a sport organization's mission statement should fit on the back of a business card. All organizational planning documents should flow from the mission statement. As you read the mission statements in this book, keep in mind they are living documents that are subject to change.

In this section, when learning about planning from a sport-governing body perspective, we will use the IPC as an example. We have already listed their vision statement in the preceding section so now let's look at the mission statement which can be found in their most current strategic plan. Simple and straightforward, the mission reads, "To lead the Paralympic Movement, oversee the delivery of the Paralympic Games and support members to enable Para athletes to achieve sporting excellence" (IPC, 2019, p. 7). (And it really could fit on the back of a business card!)

www

International Paralympic Committee
https://www.paralympic.org/

Goals

Let's now build on that mission statement starting with the next step in the planning process – setting goals. Different textbooks use differing definitions for *goals* and *objectives*. Sometimes the terms are even used interchangeably. In this textbook, however, goals are defined as broad, qualitative statements that provide general direction for a sport organization. "Setting goals helps define the direction that a business will take. Goals should align with your business' mission and vision statements" (Norman, 2019, para. 3). We will work with two sample goals and use some actual language from the IPC's strategic plan documents.

> **Goal #1:** Broaden the number of countries seeking to compete at the Games to ensure athletes from all regions in the world are well represented (IPC, 2019, p. 13)
>
> **Goal #2:** Promote gender balance in leadership positions across the Paralympic Movement (IPC, 2019, p. 11)
>
> These goals apply to very different parts of the IPC and provide good examples to look at some specific objectives.

Objectives

As opposed to goals, which are qualitative in nature, *objectives* are defined as quantitative statements that help a sport organization determine if it is fulfilling its goals. They are measurable, realistic, and clear (Norman, 2019). Another way to think of objectives is that they are SMART – Specific, Measurable, Achievable, Realistic, and Timely (Siddiqui, 2015). Because objectives can be measured, they are useful tools in evaluating both employee and organizational performance. To be measurable, objectives always contain quantifiable measures such as numbers, percentages, or monetary values. Objectives are tied directly to achieving specific goals. For example:

> **Goal #1** – Broaden the number of countries seeking to compete at the Games to ensure athletes from all regions in the world are well represented (IPC, 2019, p. 13)
>
> **Objective** – Increase the number of countries competing in the next Summer Paralympic Games by 6.
>
> **Goal #2** – Promote gender balance in leadership positions across the Paralympic Movement (IPC, 2019, p. 11)
>
> **Objective** – Reach a 50/50 gender balance at the IPC senior level and Board membership

How did we choose the number six? In each Summer Paralympic Games since 2000, on average nine new countries have been added. The number dipped at the last Games making six perhaps a more realistic number going forward. While a 50/50 gender balance may be a challenging objective, it is one the IPC is committed to achieve. Objectives must be challenging but not unattainable. Measurable objectives are important in two ways: first, they can be used in employee and organizational evaluations and, second, they are necessary because it is difficult to manage something you cannot measure! How should the IPC set out to meet these objectives? They need to take specific actions known as tactics.

Tactics

Once sport managers establish their goals and objectives, they must determine specifically how to achieve them. These specific how-to steps are called tactics. (Some textbooks use the term *strategies*, but to avoid confusion with strategic planning principles, *tactics* is the term used here.) *Tactics* are the specific "how to" actions sport managers take to achieve organizational objectives. For example:

>**Goal #1** – Broaden the number of countries seeking to compete at the Games to ensure athletes from all regions in the world are well represented (IPC, 2019, p. 13)
>
>**Objective** – Increase the number of countries competing in the next Summer Paralympic Games by 6
>
>**Tactic 1** – Provide financial assistance for sport development through grant opportunities
>
>**Tactic 2** – Train more coaches, classifiers, and officials
>
>**Goal #2** – Promote gender balance in leadership positions across the Paralympic Movement (IPC, 2019, p. 11)
>
>**Objective** – Reach a 50/50 gender balance at the IPC senior level and Board membership
>
>**Tactic 1** – Review and analyze relevant policies that may affect rates of women elected or appointed to leadership positions within the Paralympic Movement (e.g., nomination, appointment, and election procedures)
>
>**Tactic 2** – Produce and share material on gender equality for all IPC members (e.g., data sheets, event advertisements)

These tactics for participation make sense when you think about what is necessary to grow sport participation – you need the financial assets to do so and the people to carry out the program. Also, data about representation of women in leadership positions help to support the case for more women in leadership positions. Next, departments and people must take responsibility for carrying out the tactics. Remember learning in Chapter 1 that the most important asset in any sport organization is people?

Roles

After the tactics have been determined, the responsibilities for carrying out those tactics must be assigned. Roles refer to the organizational units and people specifically responsible for carrying out the sport organization's tactics and the behaviors needed to achieve success (VanderZwaag, 1998). For example:

> **Goal #1** – Broaden the number of countries seeking to compete at the Games to ensure athletes from all regions in the world are well represented (IPC, 2019, p. 13)
>
> **Objective** – Increase the number of countries competing in the next Summer Paralympic Games by 6
>
> **Tactic 1** – Provide direct Financial Support Grant money and support Games Capacity Programme
>
> **Roles** – Agitos Foundation, Finances Department
>
> **Tactic 2** – Train more coaches, classifiers, and officials
>
> **Roles** – Sports Volunteer Workforce Professional Development Coordinator, Classification Senior Manager
>
> **Goal #2** – Promote gender balance in leadership positions across the Paralympic Movement (IPC, 2019, p. 11)
>
> **Objective** – Reach a 50/50 gender balance at the IPC senior level and Board membership
>
> **Tactic 1** – Review and analyze "relevant policies that may affect rates of women elected or appointed to leadership positions within the Paralympic Movement (e.g., nomination, appointment, and election procedures)" (IPC, n.d.b, p. 4).
>
> **Roles** – IPC Women in Sport Committee
>
> **Tactic 2** – "Produce and share material on gender equality for all IPC members (e.g., data sheets, event advertisements)" (IPC, n.d.b, p. 3)
>
> **Roles** – Membership Engagement Department

Evaluation

In the final step in the planning process, sport managers must evaluate the planning process to see if they are fulfilling the organization's mission statement by successfully completing the stated goals, objectives, tactics, and roles. For example:

> **Goal** – Broaden the number of countries seeking to compete at the Games to ensure athletes from all regions in the world are well represented (IPC, 2019, p. 13)
>
> **Objective** – Increase the number of countries competing in the next Summer Paralympic Games by 6
>
> **Tactic 1** – Provide financial assistance for sport development through grant opportunities
>
> **Roles** – Agitos Foundation, Finances Department
>
> **Tactic 2** – Train more coaches, classifiers, and officials
>
> **Roles** – Sports Volunteer Workforce Professional Development Coordinator, Classification Senior Manager

Evaluation – Compare the actual number of new countries to the objective of six more

Goal #2 – Promote gender balance in leadership positions across the Paralympic Movement (IPC, 2019, p. 11)

Objective – Reach a 50/50 gender balance at the IPC senior level and Board membership

Tactic 1 – Review and analyze "relevant policies that may affect rates of women elected or appointed to leadership positions within the Paralympic Movement (e.g., nomination, appointment, and election procedures)" (IPC, n.d.b, p. 4)

Roles – IPC Women in Sport Committee

Tactic 2 – "Produce and share material on gender equality for all IPC members (e.g., data sheets, event advertisements)" (IPC, n.d.b, p. 3)

Roles – Membership Engagement Department

Evaluation – Compare actual percentage of women in leadership positions to the objective of 50/50 representation

Evaluating here is pretty straightforward. Determine the actual number of new countries and numbers of women in leadership positions and compare to the targets in the objectives. If the numbers are met, reward responsible employees or committee members appropriately. If the objective is not met, the organization needs to take corrective action with the employees or committee members who did not meet expectations.

This brief example from the IPC shows how the planning process flows from one step to the next in a sport-governing body. It also shows that the planning process does not consist of a number of separate, fragmented steps but rather is a seamless garment. The process is part of the big picture of the entire organization. Note, too, that all the steps in the process can be traced directly back to, and should be consistent with, the organization's mission statement.

The Role of Planning in Governance

Sport governance is complex and ever-changing. Sport organizations can be as small as a city soccer league or as massive as the Olympic and Paralympic Movements. Whatever the size of the sport organization, those in charge of the governance structures must plan accordingly. Examples of organizations that planned both well and poorly come from the Olympic Games. Local organizing committees for the Athens 2004 and the Rio 2016 Summer Games ended up running millions of dollar and euro deficits from construction cost overruns. Meanwhile, the London 2012 Olympic and Paralympic Games came in under budget. An organization without a well-thought-out and organized plan complete with a mission statement, goals, objectives, tactics, roles, and an evaluation system is destined to fail. Remember: if you fail to plan, you plan to fail.

ORGANIZING

The traditional view of organizing revolves around staffing. We usually think of establishing tasks, determining who will be responsible for those tasks, and then placing those people into a hierarchy, commonly illustrated by an organizational chart. It is "the process of coordinating and allocating a firm's resources in order to carry out its plans. Organizing includes developing a structure for the people, positions, departments, and activities within the firm" (BC Campus, n.d., para. 1).

An organizational chart is a diagram illustrating all positions and reporting relationships within an organization. Sport organizations vary in size. Some organizations such as an indoor baseball or softball facility employ a small number of people. Large, complex organizations like the NCAA or the IOC employ hundreds of workers. Many organizations are departmentalized into subunits according to the division of labor within the organization and the responsibility of members within each subunit. Exhibit 2.1 is an example of an organizational chart subdivided by business function. Other charts can be organized by strategic business unit structure (Exhibit 2.2) or geographic region (Exhibit 2.3).

In all cases, well-established structures are important for sport organizations. According to Daft (2015), "Organizations are (1) social entities that (2) are goal directed, (3) are designed as deliberately structured and coordinated activity systems, and (4) are linked to the external environment" (p. 13).

Structural Features of Sport Governance Organizations

Governing organizations generally have several hierarchical levels of work units and subunits. Paid staff members usually maintain the organization's headquarters and take care of the day-to-day operations of the organization. As stated previously, for high school sport, these would be the paid sport managers working for a state or provincial athletic association such as the Colorado High School Activities Association or the New Hampshire Interscholastic Athletic Association. Major professional sport leagues operate league offices such as the NBA League Office and the National Hockey League Office, both in New York City. The headquarters for the IOC resides in Lausanne, Switzerland, while the IPC is housed in Bonn, Germany. The sport managers employed in these offices have titles such as Executive Director, League Vice President, and Marketing Director. These employees keep the organization moving along, and their responsibilities include budgeting, staffing, scheduling tournaments and events, marketing, and social media. However, this level of the organization does not typically set policies dealing with governance issues. Rather, these sport managers implement the policies determined by another level of the structure.

Most major sport organizations are nonprofit organizations with voluntary membership – not individual people, but institutions or nations. The membership of these organizations – not the paid staff – determines their policies, rules, and regulations. The vehicle for setting these policies usually takes the form of regularly scheduled,

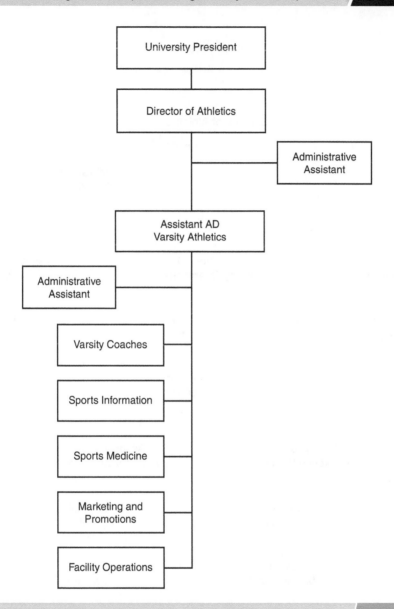

often annual, meetings of the membership. Sometimes called General Assemblies or Annual Business Meetings, members meet in person or virtually and vote to establish new policies and rules or to modify existing ones. You learned about these back in Chapter 1. The International Gymnastics Federation (FIG) provides an example with its regularly scheduled Congress. According to the FIG Statutes, "the Congress is the highest regulatory and supreme authority of the FIG. It is the biennial general assembly of the delegates of the affiliated member Federations and takes place in even years" (FIG, n.d., para. 1). While professional sports leagues are for profit, they still have meetings at which they make

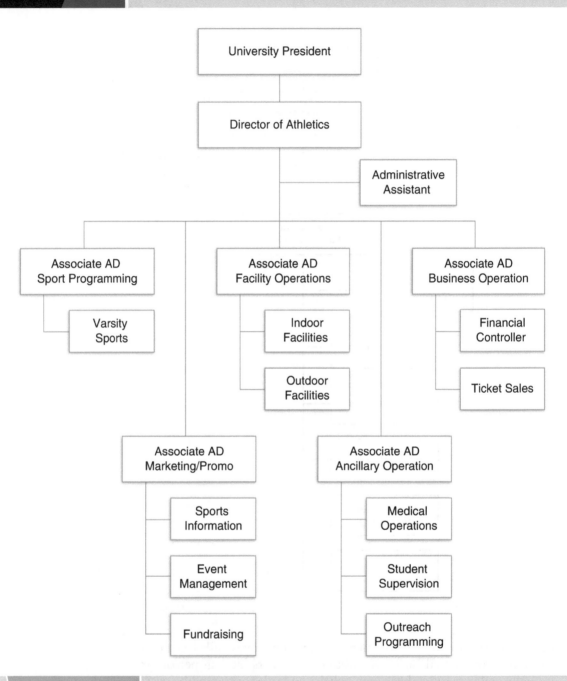

governance decisions. Every year, for example, at the Baseball Winter Meetings, representatives of minor league baseball teams vote on conducting the business of minor league baseball.

These General Assemblies and Annual Business Meetings often appoint a President for the organization as well as an Executive Committee or Executive Council. Sometimes the organization needs to

Organizational Chart for Sporting Goods Company Organized by Strategic Business Unit *exhibit* 2.3

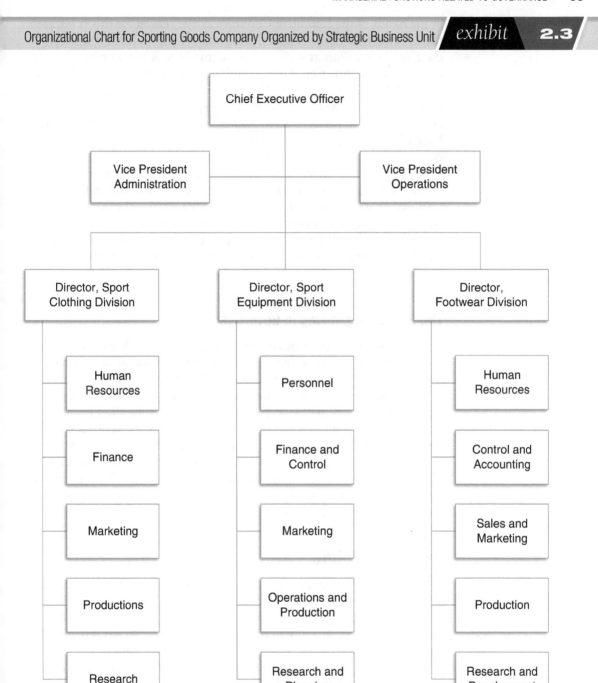

make adjustments between General Assembly meetings, and so it vests the responsibility and authority to do so in the hands of the President and the Executive Committee. Executive Committees also generate ideas for the Annual General Assemblies to consider. The Executive Committee for the IOC has ten members selected from the Annual General Assembly. For most sport organizations, the true power rests in the hands of the Executive Committee. The reason for this is because the Executive Committee sets the agenda for the General Assembly. If the Executive Committee decides an action item is not worthy of going forward to the full General Assembly, the item never goes before the General Assembly for a vote. The opposite is true as well and Executive Board members may have a specific item they feel is critical to the organization and needs to go to a vote. You can clearly see here how the Executive Committee can control the direction of an entire organization because of the agenda items they favor or disfavor.

Naturally, the governance structures within each industry segment differ. This section merely gives an overview of some important organizational characteristics and shows how organizational charts for sport governance organizations contain elements different from a traditional sport front-office organizational chart. In each industry segment chapter, the organizational structures for that industry segment are explained in more detail.

The Role of Organizing in Governance

The organizational structures in any sport-governing body dictate the flow of information and the setting of policies and rules and act as philosophical statements about the organization. Power is distributed differently within these structures. Sometimes power rests with the membership, and other times, it rests with the Executive Committee or the President. It is important to note that the organizational charts for sport-governing structures parallel traditional organizational charts with one difference: in traditional organizational charts, we see people or titles in certain places; in governance structures, we see governance units. This difference illustrates how governing bodies transcend individual responsibility and also shows us the big picture involved in governance.

DECISION-MAKING

The decision-making process is essential in sport governance. We all make decisions every day. Some decisions are very simple. You chose what to wear to class on Monday. What did you consider when deciding what clothes to wear? You probably considered the weather, what was clean, what matched, whom you might sit next to in class, if you had to go to work right after class, or if you had to give a presentation. Even this simplistic example shows the two basic parts of decision-making – gathering and analyzing information. Sport managers must also make decisions on an everyday basis (after they decide what to wear to work).

Routine and Complex Decisions

Sport managers face a variety of decisions in the workplace, some of which are routine and some of which are not routine. *Routine decisions*, sometimes referred to as *programmed decisions*, are straightforward, repetitive, and mundane (Business Case Studies, 2021). During a Houston Rockets basketball game, if one of the light banks goes dark in the arena, calling an electrician is an easy decision. When the Athletic Director's (AD's) suite at a Washington State University football game is running low on soft drinks (a recurring situation), the decision to order more drinks is routine.

All decisions are not so easy. Many problems are unique. These types of problems, also called nonprogrammed decisions, emerge from complex situations or situations where managers find themselves facing an issue for the first time. We need to look no further than the chaos that was the sport industry during the ongoing COVID-19 pandemic. Sport managers working with spectator sports had to first decide whether a game or event should take place, be postponed, or be canceled. Depending on that decision, some sample follow-up decisions come into play. If canceled, then game-related contracts and ticket holders need to be dealt with. If postponed, then new timelines need to be established. Since the actual decision was for the Games to take place, a decision was needed as to if spectators will be allowed and if so, how many and how to keep them as safe as possible. Sport managers working in the recreational sport industry also faced a set of decisions. For example, should a YMCA even be open? If so, what activities can be permitted – basketball and volleyball leagues or exercise classes? And what about the pool? These examples illustrate the complex decision-making situations sport managers at different levels faced during the COVID-19 pandemic. As we move out of this time, sport managers will now have to decide what a new "normal" will look like for their athletes, workers, fans, and participants.

The Rational Model

One item is of great importance to sport managers when they make decisions: They must follow an organized, thoughtful process. Decisions of great magnitude cannot be left to chance. Rather, the sport manager's thought process must be detailed and organized. One such process, the Rational Model based on Robbins (1990), is outlined here:

1. Identify the REAL problem.
2. Identify the decision objective.
3. Gather all pertinent information.
4. Identify any hurdles.
5. Brainstorm for alternatives.
6. Narrow down the options.
7. Examine the pros and cons of each option.
8. Make the decision.
9. Evaluate the decision.

To illustrate this decision-making process, assume that you are the Associate AD for Academic Services at Big State University, home of the Fightin' Saugers. Recently, your Coordinator of Academic Performance informed you that your baseball team has not met the minimum Academic Progress Rate (APR) requirements. As a reminder,

> [APR] holds institutions accountable for the academic progress of their student-athletes through a team-based metric that accounts for the eligibility and retention of each student-athlete for each academic term ... The APR system includes rewards for superior academic performance and penalties for teams that do not achieve certain academic benchmarks.
>
> (NCAA, n.d.)

Your baseball team has a strong record of on-the-field success, having recently appeared in several NCAA Regionals and even two Super Regionals so keeping up on the academic side is important as it is for all teams. So now you need to decide what steps to take to get the team back into compliance to avoid any penalties.

Let's apply the Rational Model:

1. *Identify the REAL problem.* The real problem is that the baseball team is out of compliance with NCAA's APR guidelines.
2. *Identify the decision objective.* The decision objective is to bring the team back into compliance.
3. *Gather all pertinent information.* This is where you need to find out as much information (both tangible and intangible) about the situation as possible so you can help the team do better academically. While this list is not exhaustive, you would want to know about the following: team and player grade point averages (GPAs), team majors, coaches' attitudes toward academics, academic performance of players the coaches are recruiting, resources available in the Academic Service Center such as number of academic advisors and tutors available for the team, the budget for Academic Services, number of study rooms and computers in the Academic Center, and so on.
4. *Identify any hurdles.* Be aware not only of tangible hurdles, such as budgetary constraints, but also of intangible hurdles, such as people's attitudes toward academics. Hurdles often become apparent when you start to generate that list of information. These could include low-budget allocations which might result in too few academic advisors, limited numbers of tutors, or the Academic Center needing more computers or study rooms. The coaches' and the players' attitudes toward balancing academics and athletics can come into play here.
5. *Brainstorm for alternatives.* Simply put, "brainstorming is a method for inspiring creative problem solving by encouraging group members to toss out ideas while withholding criticism or judgment. Brainstorming, in its many forms, has become a standard tool for ideation (development of new ideas)" (Rudy, 2020, para. 2). In brainstorming, people list all ideas now and sort through them later. Remember – in brainstorming all ideas are welcome at first until they are discussed, even if some may seem a bit unrealistic. In this case, the options could include asking

the university for additional funding, increasing sponsorship dollars for the Academic Center, finding new donors who want to specifically support academics, hiring more tutors, hiring more academic advisors, requiring a certain amount of hours of study hall for the team, talking to the players' professors, dropping baseball, or firing the coach. Not all the ideas may be reasonable, but when brainstorming, remember to put all ideas on the table.

6. *Narrow down the options.* Then narrow down your list of options to three or four. Let's say that the best options appear to be finding additional funding sources such as increasing sponsorship dollars for the Academic Center, finding new donors who want to specifically support academics, hiring more tutors, hiring more academic advisers, and requiring a certain amount of hours of study hall for the players.

7. *Examine the pros and cons of each option.* Carefully weigh the pros and cons of these options. Finding additional funding sources such as increasing sponsorship dollars for the Academic Center might be difficult since sponsors typically make those decisions early in the year and you might have missed the window for them being able to enter into an agreement. Finding new donors who want to specifically support academics is a good possibility as people like being given the opportunity to donate for a very specific cause. Hiring more tutors and hiring more academic advisers could be done within budgetary limitations. Requiring a certain amount of hours of study hall for the players costs nothing but you might want to make decisions on whether this would apply to everyone or just players at certain levels (e.g., freshmen only or students with certain GPAs).

8. *Make the decision.* You decide on a combination approach, using the tactics of finding new donors who want to specifically support academics, hiring five more tutors, and hiring one more full-time academic adviser. These options go together as acquiring new donor money could be used to offset the cost of the new personnel.

9. *Evaluate the decision.* These tactics will have to be evaluated over the coming years. Once your plans are implemented, you can see how the team's GPA has hopefully now reached an acceptable level.

This simplified version of a very complex decision-making process illustrates how an organized approach to decision-making can help a sport manager decide on a course of action.

The SLEEPE Principle

Another decision-making method, which takes a more global view of the organization and the implications of sport managers' decisions, is called the SLEEPE Principle. The decisions sport managers make are often publicly scrutinized by the media, fans, and the general public. Therefore, when sport managers make decisions, they must be able to use their conceptual skills to see the big picture. By using the SLEEPE method, decision makers can analyze decisions, especially decisions affecting policy development or interpretation. Originally set up as the SLEEP model by W. Moore at Ohio State University in 1990, this decision-making model has since been modified (Hums, 2006; Hums & Wolff, 2017) and applied in the sport industry to help sport managers see the big picture by analyzing the many different ramifications of their decisions. The components of the SLEEPE Principle are as follows:

S – Social
L – Legal
E – Economic
E – Ethical
P – Political
E – Educational

Using this model helps sport managers understand how their decisions will be viewed in different ways by various constituencies in society. The following example dealing with the Olympic Games will illustrate this:

The COVID-19 pandemic caused the postponement of the 2020 Tokyo Olympic Games. Once it was decided that the Games would take place, a decision needed to be made as to whether or not spectators should be allowed in the venues. As we know, the decision was ultimately not to allow the fans in, but what did sport managers need to take into consideration when making this decision? Let's use the SLEEPE Principle to analyze this situation:

S – SOCIAL. Look at the social ramifications of the decision: Allowing fans would have meant hearing all the cheering and chanting by fans from all over the world rooting for their home country athletes. Fans would have planned vacations to visit Japan and attend the Games. Yet public opinion was not in favor of fans in the stands (or even hosting the Games at all). Not allowing fans changes the atmosphere in the stadiums and arenas for the athletes. While they are still competing for their country at the highest level, they would definitely miss the support of their home fans as they try to bring home the gold. And as for those fans who could not attend, will they now even choose to watch the Games from home, having counted on being there in person, or will viewership and social media activity actually increase?

L – LEGAL. From the legal standpoint, allowing fans may have opened the organizers up to liability if they did not properly ensure the health and safety of those who did attend, including athletes, team personnel, spectators, and members of the media. Not allowing fans likely meant alterations in contracts with, for example, concessions companies who had planned on selling their products to fans in the stands. Fans may also experience legal difficulties in attempting to get their money refunded from secondary market ticket sellers.

E – ECONOMIC. Revenues from the Olympic Games are essential for the operation of the IOC. Allowing fans means all the income from ticket sales would be received as planned from what are almost always sell-outs for high capacity events. Fans who attend events would also purchase concessions and merchandise, generating more revenue streams. Obviously, not having fans in the stands has the direct opposite effect, resulting in loss of revenue. In addition, local businesses and restaurants which counted on the spectators visiting the city would not earn that revenue and may have to lay off staff or even shut down completely.

E – ETHICAL. Next are the ethical considerations, the ones which produce the most discussion. Allowing fans would keep the spirit of the Games intact but in the midst of a pandemic and a city ultimately in emergency lockdown, is allowing fans the right thing to do and is it worth the risk? On the other hand, as the world slowly moves out of the pandemic, the Games can be seen as a positive step, a rallying point that can move not just sport, but the world at large, forward. The Games could be seen as a beacon of hope. Seeing people in what has always been seen as a "normal" event (going to a game) can show that we can and will move forward and sport can be a vehicle for that positive movement.

P – POLITICAL. Politically there are a number of constituencies to consider. In this context, the term *political* is not limited to elected officials only; it is broader, including any groups or stakeholders who may exert some type of political power or influence in a given situation. Allowing fans means that local elected officials will shoulder some of the responsibility for their safety. Protests will abound as a public, already unhappy that the Games are even taking place, now see the potential for competitions with fans to become "super-spreader" events. The athletes would be a group here that would also have political influence. They would want as many fans as possible to attend so that they can have the full Olympic experience. Without the athletes, there are no Games. Not allowing fans would take some of the political pressure off of these elected officials who made the decision for the Games to even take place.

E – EDUCATIONAL. Finally, there is the educational component to consider. Here, sport managers need to reflect on what they have learned by going through the decision-making process in this situation. No one in our lifetimes has experienced anything like the COVID-19 pandemic. It challenged every level of society including the sport industry, forcing managers to make decisions in a very complicated and fluid environment. In order to make good decisions, it is important for managers to have a structured decision-making process. Using such a structure in a very complex decision-making situation will inform future good practice when tough decisions arise again.

This example illustrates the complexity and public nature of many decisions sport managers face on a daily basis and how these decisions take on even more difficulty in the face of an international public health crisis. During regular times, sport managers must also make decisions on other issues. Athlete misconduct, substance abuse, equity, diversity, inclusion, and violence, for example, are all present in the sport industry just as they are in general society. Because of the far-reaching ramifications of their decisions, forward-thinking sport managers must learn to examine all potential results of and reactions to their decisions before they make them.

This section outlined various decision-making models sport managers can incorporate when solving problems. Often, sport managers' decisions will be ethical in nature. Do the same decision-making models apply when the problem has ethical implications? To address the importance of dealing with ethical issues, Chapter 4 focuses on ethical issues and ethical decision-making.

The Role of Decision-Making in Sport Governance

Sport managers dealing with governance issues are faced with decisions that have far-reaching implications. Their decisions, from simple to complex, shape the direction of the organization. The decisions sport managers make are open to public scrutiny and media discussion. As such, sport managers must make sure they have a concrete method for analyzing any decisions they need to make.

SUMMARY

S port managers need to be able to perform the major management functions of planning, organizing, leading, and evaluating. This chapter focused on planning, organizing, and one subset of leading – decision-making. It is important to have a solid foundation in these areas before further examining specific industry segments.

Planning is the basis for everything a sport manager does. Sport managers must make both short- and long-term plans. The planning process is sequential: organizational goals, objectives, tactics, roles, and evaluation all flow from the mission statement. Sport organizations are organized with different levels of responsibility. Determining the tasks an organization needs to accomplish and the people needed to accomplish those tasks is essential for organizational success. Sport managers must make decisions every day. Some of their decisions are routine; others are unique. Two structured methods to help sport managers make solid decisions are the Rational Model of decision-making and the SLEEPE Principle. By mastering these important skills, sport managers can successfully conduct the business of governance and policy development in their sport organizations.

caseSTUDY

MANAGERIAL FUNCTIONS

You have been hired to work in the Marketing Department at Middle States University in a Group of Five conference school. You have been tasked with working on increasing attendance at women's sports. Your softball team has been fairly successful over the years and you read about the recent record number on viewers for the Women's College World Series so have decided that softball is the sport you want to promote first. Currently, attendance at the games is quite low, even though the team does well with a few conference championships in its record. You recently hired Head Coach Sue Banker, who came over from a successful Power 5 school where she had been an Assistant Coach for the past three years.

Use the Rational Model of decision-making to determine your course of action for how to best drive attendance for the softball team.

CHAPTER questions

1. Locate organizational charts for three different sport organizations. Compare and contrast the titles and structures of each. Why are some aspects similar and others different?

2. Choose one of the following:
   - minor league baseball team
   - college or university athletic department
   - high school athletic department
   - charity 5K run/walk
   - campus recreation department
   - sporting goods store

   After you choose one of these sport organizations, develop the following:
   a. mission statement
   b. one goal
   c. two objectives for that goal
   d. two tactics for each objective
   e. the roles for each tactic

3. For a sport organization of your choosing, identify two situations that would involve routine decision-making and two situations that would involve complex decision-making.

FOR ADDITIONAL INFORMATION

For more information, check out the following resources:

1. This website takes you to the Third Version of the Tokyo 2020 Playbooks. These Playbooks outline the game plan to ensure all Olympic and Paralympic Games participants and the people of Japan stay safe and healthy. They were jointly developed over time during the COVID-19 pandemic by Tokyo 2020, the IOC, and the IPC in close collaboration with the Government of Japan and the Tokyo Metropolitan Government https://olympics.com/tokyo-2020/en/news/third-and-final-version-of-the-tokyo-2020-playbooks-published

2. This link takes you to the IPC's full Strategic Plan 2019–2022 https://www.paralympic.org/sites/default/files/document/190704145051100_2019_07+IPC+Strategic+Plan_web.pdf

3. This link contains the IOC's Annual Reports so you can see how a governing body reports what they have done on an annual basis: https://olympics.com/ioc/documents/international-olympic-committee/ioc-annual-report

REFERENCES

BC Campus. (n.d.). *Organizing.* https://opentextbc.ca/businessopenstax/chapter/organizing/

Bryce, T. (2008). *Why we resist planning.* http://www.articles.scopulus.co.uk/Why%20We%20Resist%20Planning.htm

Business Case Studies. (2021). *Decision making.* https://businesscasestudies.co.uk/decision-making-2/

Chelladurai, P. (2014). *Managing organizations for sport and physical activity: A systems perspective* (4th ed.). Routledge. https://doi.org/10.4324/9781315213286

Commonwealth Sport Canada. (2020). *Mission, vision, values.* https://commonwealthsport.ca/about-cgc/mission-vision-values.html

Daft, R.I. (2015). *Organizational theory and design* (12th ed.). Cengage Learning.

Department of Communities Tasmania. (2019). *Strategic and operational planning for sport organizations.* https://www.communities.tas.gov.au/__data/assets/pdf_file/0022/18265/Strategic-and-Operational-Planning-for-Sporting-Organisations.pdf

Diaz, J. (2021, July 21). *Australia to host Olympics for the third time in 2032 after Brisbane wins its bid.* NPR. https://www.npr.org/sections/tokyo-olympics-live-updates/2021/07/21/1018699263/australia-to-host-the-olympics-for-the-third-time-in-2032-after-brisbane-wins-it

FIG. (n.d.). *Congress.* https://www.gymnastics.sport/site/pages/about-congress.php

Hums, M.A. (2006, May). *Analyzing the impact of changes in classification systems: A sport management analysis model.* Paper presented at the VISTA 2006 International Paralympic Committee Congress, Bonn, Germany.

Hums, M.A., & Wolff, E.A. (2017). Managing Paralympic sport organizations: The STEEPLE framework. In S. Darcy, S. Frawley, & D. Adair (Eds.), *Managing the Paralympics* (pp. 155–174). Palgrave Macmillan.

International Paralympic Committee. (n.d.a). *About the International Paralympic Committee.* https://www.paralympic.org/ipc/who-we-are

International Paralympic Committee (n.d.b). *Strategy 2019–2022: Women in Sport Committee.* https://www.paralympic.org/sites/default/files/2019-08/190729115757953_WiSC%2BStrategy.pdf

International Paralympic Committee. (2019). *International Paralympic Committee Strategic Plan 2019–2022.* Author.

NCAA. (n.d.). *Division I Academic Progress Rate (APR).* https://www.ncaa.org/about/resources/research/division-i-academic-progress-rate-apr

Norman, L. (2019). *What is the business difference between objectives and goals?* http://smallbusiness.chron.com/business-difference-between-objectives-goals-21972.html

Peek, S. (2020, May 7). *What is a vision statement?* Business News Daily. https://www.businessnewsdaily.com/3882-vision-statement.html

Robbins, S.P. (1990). *Organizational theory: Structure, design and applications* (3rd ed.). Prentice Hall.

Rudy, L.J. (2020). *What is the definition of brainstorming? (For groups and individuals).* https://business.tutsplus.com/tutorials/what-is-the-definition-of-brainstorming--cms-27997

Siddiqui, F. (2015). *Defining the terms: Vision, mission, goals and objectives.* https://www.linkedin.com/pulse/defining-terms-vision-mission-goals-objectives-fareed

VanderZwaag, H.J. (1998). *Policy development in sport management.* Praeger.

Diversity, Equity, and Inclusion in Sport Governance

If you are a sports fan and have an active Twitter or TikTok account, chances are you can pick up your phone right now and find a story related to diversity, equity, or inclusion on ESPN or any other major sports media site: The United States Women's Soccer National Team fighting for decades to receive equal pay. Athletes across countries and professional leagues including those competing in the Men's National Basketball Association (NBA), Women's National Basketball Association (WNBA), and National

DOI: 10.4324/9781003303183-3

Football League (NFL) in the United States or Europe's Premier League protesting racial injustice in the wake of the Black Lives Matter movement. The hiring of trailblazing "firsts" and the shattering of race and gender glass ceilings in the process, as was the case when Duke University hired Nina King as its Director of Athletics and the Miami Marlins hired Kim Ng as General Manager, both of whom were the first Women of Color to hold their respective positions. Indeed, advances in diversity, equity, and inclusion (or, sometimes, the lack thereof) have become increasingly common in our news cycles. Beyond the news stories, however, matters related to diversity, equity, and inclusion often deeply affect the ways in which leaders govern, or fail to govern, sport organizations effectively. Who makes the hiring decisions for the most powerful leaders of a sport organization? Who is being hired, and who is left out of the conversation? How do common practices in sport governance either advance or serve as roadblocks to a diverse workforce, equitable opportunities, and an inclusive work environment? What role can policy play in eliminating barriers to diversity, equity, and inclusion in sport? These are just a few of the questions you should be thinking about if you consider a career in (or related to) the management of sport. Because diversity, equity, and inclusion have become increasingly important to successful sport governance, this chapter will provide an introduction to how they affect decision-making, governing, and policy-making in the context of sport.

WHY CENTER DIVERSITY, EQUITY, AND INCLUSION IN SPORT GOVERNANCE?

Take a look around your physical or virtual classroom. Do you see a pattern of who is represented and, perhaps more importantly, who does not yet have a seat at the table? Would you consider your class to be a diverse group? Once you reflect on the (visible) diversity in your class, you will likely spot a pattern. This is not surprising because studies of undergraduate Sport Management education have consistently pointed to a lack of (visible) diversity in academic programs, especially when it comes to race and gender. For example, in 2011 researchers found that among US undergraduate Sport Management students, only 20–30% of students were women and even fewer (only 14%!) were People of Color (Hancock & Hums, 2011). A separate study conducted in 2018 found similar results: Compared to other majors across campus, Sport Management programs tended to be less diverse, with more than two-thirds of the sample identifying as white (66.4%) and men (63.4%) (Barnhill et al., 2018).

A similar pattern can be seen in the sport industry at large. In the United States, The Institute for Diversity and Ethics in Sport (TIDES) has published an annual Racial and Gender Report Card tracking the race and gender demographics for major US sport governing organizations since 2002. These annual report cards are a "definitive assessment of hiring practices of women and people of color in most of the leading professional and amateur sports and sporting organizations in the United

www

Institute for Diversity and
Ethics in Sport (TIDES)
https://www.tidesport.org

www

2021 TIDES Gender and
Racial Report Card
https://www.tidesport.org/
complete-sport

States" (The Institute for Diversity and Ethics in Sport, 2022, para. 1). In order to receive an A in the respective area, the gender and racial make-up of a sport organization's key stakeholder groups (e.g., employees, athletes, leadership) must match that of larger US society: At least 30% must be People of Color to receive an A+ on the Race section, and a minimum of 45% must be women for an A+ in the Gender category. Out of the grades reported on the 2021 Gender and Racial Report Card, only one organization received an A or higher for its hiring practices: the Women's National Basketball Association (WNBA). College Sport scored the lowest with a C, with other organizations falling somewhere in between: the National Basketball Association (NBA) scored a B+, the NFL and Major League Soccer (MLS) scored a B, and Major League Baseball earned a C+ (The Institute for Diversity and Ethics in Sport, 2021a). Exhibit 3.1 provides a detailed overview of the Gender and Race statistics for the 2021 report card. How do you think your major or academic program compares to these trends in the sports industry?

aphic Comparison of Race, Ethnicity, and Gender Across US Professional Sport Leadership | *exhibit* | **3.1**

	2020 US POPULATION	PRESIDENTS/CEOS					GMS				
		NBA	MLB	MLS	NFL	WNBA	NBA	MLB	MLS	NFL	WNBA
White	57.8%	93.5%	93.3%	86.2%	90.6%	75%	60%	86.7%	78.1%	84.4%	75%
Black or African American	12.1%	6.5%	3.3%	3.4%	3.1%	25%	26.7%	3.3%	6.3%	15.6%	25%
Hispanic/ Latinx	18.7%	0%	0%	6.9%	3.1%	0%	3.3%	3.3%	12.5%	0%	0%
Asian	5.9%	0%	0%	3.4%	3.1%	0%	0%	6.7%	3.4%	0%	0%
Hawaiian/ Pacific Islander	0.2%	0%	0%	0%	0%	0%	0%	0%	0%	0%	0%
Am. Indian or Alaska Native	0.7%	0%	0%	0%	0%	0%	0%	0%	0%	0%	0%
Two or More Races	4.1%	0%	0%	0%	0%	0%	10%	0%	0%	0%	0%
Total People of Color	42.2%	6.5%	3.3%	13.8%	9.4%	25%	40%	13.3%	18.8%	15.6%	25%
Women	50.46%	8.7%	0.0%	13.8%	3.1%	66.7%	0%	3.3%	0%	0%	33.3%

(Continued)

	US POPULATION	TEAM MANAGEMENT/SENIOR ADMIN.					HEAD COACHES				
		NBA	MLB	MLS	NFL	WNBA	NBA	MLB	MLS	NFL	WNBA
White	57.8%	68%	79%	78.8%	79.6%	58.2%	70%	80%	53.6%	84.4%	58.3%
Black or African American	12.1%	15%	5.2%	3.9%	10.7%	23.5%	23.3%	3.3%	7.1%	9.4%	41.7%
Hispanic/ Latinx	18.7%	7.6%	9.9%	8.9%	4.3%	5.9%	3.3%	13.3%	32.1%	3.1%	0%
Asian	5.9%	5.1%	2.7%	2.5%	3.6%	5.3%	3.3%	6.7%	0%	0%	0%
Hawaiian/ Pacific Islander	0.2%	categories combined on Report Card	0.1%	0.4%	0.1%	0%	0%	0%	0%	0%	0%
Am. Indian or Alaska Native	0.7%	0.2%	0.2%	0%	0%	0.6%	0%	0%	0%	0%	0%
Two or More Races	4.1%	3.4%	1.6%	1.2%	1.5%	4.7%	0%	3.3%	3.6%	3.1%	0%
Total People of Color	42.2%	31.3%	19.8%	17%	20.1%	40%	30%	20%	42.8%	15.6%	41.7%
Women	50.46%	37.9%	28.5%	24.1%	25.3%	49.4%	0%	0%	0%	0%	41.7%

Source: United States Census Bureau (2022); The Institute for Diversity and Ethics in Sport (2021a).

Of course, these numbers focus solely on race and gender demographics and only provide a partial look at the current diversity, equity, and inclusion trends in sport governance. However, they do illustrate one of the larger issues: a persistent lack of racial and gender diversity at the highest level of US sport governance (i.e., the leadership ranks). That is why TIDES issues the following call to those governing sport:

> The Institute for Diversity and Ethics in Sport (TIDES) firmly believes that diversity, equity and inclusion both on the playing field and off are vital to the sustainable growth of sport not just in America, but around the globe. … This is true for both the business operations and sports operations side of the front office/athletic department. Generally, aside from head coaches, general managers, team presidents and college athletic directors, the key decision makers within this space are less visible to the public eye yet they help influence trends within

the industry. It is, therefore, critical that professional leagues and the NCAA increase diverse and inclusive hiring practices when hiring league employees, front office and team professionals, and university administrators. Moreover, it is the responsibility of leagues, teams, colleges and universities to have meaningful diversity initiatives and sustainable programs put in place to help promote and create this growth – and ultimately shatter the barriers to upward mobility.

(The Institute for Diversity and Ethics in Sport, 2021a, pp. 1–2)

As you can see, making sure diversity, equity, and inclusion are anchored in sport governance is not just the responsibility of those in the leadership ranks of sport. Every member of the industry has a part to play in making sure sport reflects the diversity of the larger society it operates in by creating equitable opportunities and inclusive environments.

When discussing diversity, equity, and inclusion in sport, you will likely hear some professionals argue that there is a business case to be made for diversity, equity, and inclusion; that is, increased diversity can lead to better business outcomes (Özbilgin et al., 2014). It is important to stress that making a business case for diversity, equity, and inclusion implies that if these areas were not good for business, we should not be promoting them. However, because every person should have a chance at making it to the top levels of sport governance, and feeling free to be their authentic selves while doing so, advancing diversity, equity, and inclusion is everyone's responsibility. As a future professional in sport, you have a role to play in making sure the individuals you work with feel like they are seen, heard, and valued. In fact, a commitment to diversity, equity, and inclusion is a crucial component of good governance. As Dame Katherine Grainger, Chair of UK Sport, reminds us:

Away from specific measures of diversity, we must all continue to embed good governance practices at the heart of our operations. Governance cannot simply be a tick box exercise, it needs to be part of the culture and fabric of every sporting organisation and our decision making.

(Sport England & UK Sport, 2019, p. 2)

In Chapter 2, we used the metaphor of a sailboat to describe the management of sport organizations. Advancing diversity, equity, inclusion is more like being in a paddle boat going upstream: Unless you are actively paddling upstream, the boat will go backward. Similarly, unless we actively work toward diversity, equity, and inclusion, we will continue to reinforce barriers to progress. As such, we have an ethical commitment to actively advance diversity, equity, and inclusion in and through sport governance.

So far, we have loosely used the terms *diversity, equity*, and *inclusion* as a group. However, to be most effective in advancing these areas we must understand how they differ from one another. That is why we will cover definitions for each of these concepts, and related ones, in the next section.

DEFINING AND APPLYING KEY TERMS

How would you define diversity, equity, and inclusion? Take a few moments to think about what each of these terms means to you. If you struggle to come up with a definition (or if you Googled a definition), you will see that you are not alone. You will see there is no one widely accepted definition for each of these terms, but there are certainly commonalities between the many definitions. Check out Exhibit 3.2 for the diversity and inclusion statement of Women's National Basketball Players Association (WNBPA), which is included on the same page as their mission statement. The WNBPA is the Players Association (players' union) for the WNBA and you will learn more about these types of organizations in the chapter on Professional Sport in North America.

Throughout this book, we draw from the following definitions for diversity, equity, and inclusion as we look at how they impact the governance of sport:

> **DIVERSITY.** Diversity refers to the demographic mix of a group. It accounts for "the various ways in which difference is constructed in society, often based on social identity categories such as race, ethnicity, gender, sexuality, socioeconomic class, nationality, or religion" (Kluch et al., 2023, p. 301). Diversity includes more than these identity categories and can also take into account differences in (dis)ability and neurodiversity, upbringing, education levels, personality traits, cultural background, ideological worldviews, and belief systems (Harrison & Klein, 2007). Two points need to be highlighted in regard to the concept of diversity. First, diversity is a descriptive concept that captures matters of representation. It provides us with facts about the composition of a group such as a sport organization's Board of Directors, a specific Committee, or

---

exhibit **3.2**   Mission and Diversity Statement of the WNBPA

Mission Statement

To unite, in one labor organization, all WNBA players eligible for membership in a manner that not only promotes a high sense of loyalty among all members but is also diverse, inclusive and a direct reflection of our core principles.

Our Statement on Diversity & Inclusion

Our diversity extends to include all eligible players regardless of race, ethnicity, gender, sexual orientation, gender identity or expression, age, religion or spirituality, political or ideological viewpoints.

Honoring diversity ensures that multiple perspectives are engaged across the membership.

Source: WNBPA (2022).

overall staff. Second, diversity is a group concept, meaning it looks at the make-up of a group rather than any individual within this group. No one individual is more diverse than another, but some groups are more diverse than others.

EQUITY. While diversity looks at numbers that capture the representation of different identities and backgrounds within a group, *equity* looks at why individuals from certain groups are represented or not, focusing on matters of fairness and access. The University of California at Berkeley defines equity as the "guarantee of fair treatment, access, opportunity, and advancement for all ..., while at the same time striving to identify barriers that have prevented the full participation of marginalized groups" (UC Berkeley Strategic Plan for Equity, Inclusion, and Diversity, 2009, p. 5). The terms *equity* and *equality* are often used interchangeably, but they mean two different things. Equality means giving everyone the same, while equity means taking into account the unique needs of each individual and meeting those needs to reach the same outcome. Imagine your family is going on a bike trip to a friend's house across town. To get there, you will likely not all use the exact same bike, but instead, each uses a bike fitting your specific needs. For example, you may have a younger sibling in need of a smaller bike with training wheels or a family member in a wheelchair in need of adaptive equipment. You all have the same destination, but may need different tools to get there – that is what equity is about. See Exhibit 3.3 for a popular meme explaining the difference between equality and equity.

INCLUSION. Whereas diversity is concerned with *who* is in the room and equity looks at *why* individuals are represented or not (and how they can get there!), inclusion is primarily focused on *how* individuals are treated. Inclusion is the intentional action of "creating environments in which any individual or group can feel welcomed, respected, supported, and valued" (UC Berkeley Strategic Plan for Equity, Inclusion, and Diversity, 2009, p. 5). An inclusive organization is one that values differences as its strength, makes sure everyone feels seen and all voices are heard, and creates a sense of belonging for its members, with a particular focus on those who have historically been marginalized or excluded from fully participating in said organization. What does it mean to belong? That is a key question guiding inclusion efforts. For example, for college students having a sense of belonging often means "perceived social support on campus, a feeling or sensation of connectedness, the experience of mattering or feeling cared about, accepted, respected, valued by, and important to the group (e.g., campus community) or others on campus (e.g., faculty, peers)" (Strayhorn, 2012, p. 3). Harvard University (2022) posits, "belonging means that everyone is treated and feels like a full member of the larger community, and can thrive" (pp. 1–2). Belonging also means being able to live as one's authentic self and not having to hide specific parts of one's identity.

exhibit **3.3** Equality versus Equity: Meme

EQUALITY **EQUITY**

Credit: Interaction Institute for Social Change | Artist: Angus Maguire. *Source:* Interactioninstitute.org and madewithangus.com. Available under Creative Commons licence: Attribution-ShareAlike 4.0 International (CC BY-SA 4.0), see https://creativecommons.org/licenses/by-sa/4.0/ for details.

www
NFL Commitment to Diversity
https://www.nfl.com/careers/diversity

Let's put these definitions in a sport governance context by looking at one of the major US sport organizations: the NFL. For this example, we will focus on head coaching and senior leadership positions within the organization, as those have the most decision-making power:

Diversity: The 2021 TIDES Gender and Racial Report Card (2021) presents alarming numbers in terms of racial and gender diversity of those at the top of the organization's hierarchy. The statistics for team owners (3.1% People of Color, 21.9% Women; Letter Grade for both areas: F); team CEOs/presidents (9.4% People of Color, 3.1% women; Letter Grade for both areas: F); general managers (15.6% People of Color; Letter Grade: C+; no woman currently holds this position); and C-suite executives (17.3% People of Color, Letter Grade: B; 28.6% women, Letter Grade: C-) all reveal that both People of Color and women are underrepresented in (or, to use more accurate wording, have historically been excluded from) those positions. A similar demographic gender and racial make-up is found among head coaching ranks, where the organization received a C+ for racial hiring in 2021, with 15.6% of NFL head coaches being People of Color. It did not receive a grade in the gender category, as currently no head coaching positions are held by women. Kenneth Shropshire, former CEO of the Global Sport Institute at Arizona State University, captured the current state of racial hiring best when he said:

> The abysmal hiring record of Black head coaches indicates that most franchises continue to struggle with their identity and role in social progress. To be blunt, this isn't an issue of just color; this is an issue of Black leadership and White decision-making within the sport.
> (National Football League, 2021, p. 22)

Equity: Given its record in diversity hiring, the NFL has attempted to create more equitable hiring policies through the adoption of the Rooney Rule in 2003, which was further refined over the years. The rule is an affirmative action policy that "aims to increase the number of minorities hired in head coach, general manager, and executive positions" (NFL, 2022, para. 3). It acknowledges that both People of Color and women have historically been excluded from these decision-making positions and thus requires hiring committees to interview at least one member of a minoritized group for vacant positions. In 2021, the league further refined the rule requiring every team to interview at least two minority candidates, both of whom must be external to the organization, for head coaching positions as well as one external candidate from a minority group for coordinator positions (National Football League, 2022). Questions still remain about the effectiveness of the rule, however, as the representation of historically excluded and marginalized groups in these positions remains low.

Inclusion: Remember that inclusion is about making individuals and groups feel like they belong. Whereas organizations often prioritize diversity (for example, by asking: "How can we get more historically excluded groups to the table?"), they should be thinking about inclusion first (for instance, by asking: "How can we make sure members of historically excluded groups can thrive as their authentic selves and feel like they belong once they get here?"). To create a more inclusive environment, the NFL has a number of employee resource groups (e.g., Women's Interactive Network, Black Engagement Network, NFL Pride, Asian Professional Exchange, and the Parent Initiative Network), hosts cultural celebrations (e.g., Asian American Pacific Islander Heritage Month, Black History Month, Latinx Heritage Month, Pride Month, and Women's History Month), and has multiple partnerships with organizations such as RISE, a nonprofit aimed at improving race relations in sport, and GLAAD, a leading lesbian, gay, bisexual, transgender, and queer (LGBTQ+) advocacy organization, to provide training and resources focused on diversity, equity, and inclusion (The Institute for Diversity and Ethics in Sport, 2021e).

WWW

The NFL's Rooney Rule
https://operations.nfl.com/inside-football-ops/inclusion/the-rooney-rule/

Of course, there are other terms related to diversity, equity, and inclusion that play a role in sport governance. Some of the ones we use throughout the chapter and book are captured in Exhibit 3.4. For example, diversity, equity, and inclusion efforts are often aimed at removing systemic barriers to social justice. These deeply ingrained institutional barriers shape our everyday life – through our educational institutions, religious institutions, the media, and of course the institution of sport – and create the proverbial uneven playing field in the first place. We will cover some of the most prominent barriers later in the chapter, but it is worth noting here that these barriers often emerge from structural inequities, such as those created by systemic racism, sexism, heterosexism, ableism, and other forms of discrimination. These structural barriers have systemically privileged some groups and disadvantaged, marginalized, or excluded others in sport governance, some of which we turn our attention to next.

exhibit 3.4 Key Terms for Discussing Diversity, Equity, and Inclusion

TERM	DEFINITION
Ableism	Discrimination or prejudice, whether intentional or unintentional, against persons with disabilities.
Affirmative Action	An action or policy that considers attributes of historically marginalized individuals such as race, color, religion, sexual orientation, or national origin, especially in relation to employment and education.
Bias	A preference for or against something or someone whether conscious or unconscious.
Cisgender	Someone whose sex assigned at birth aligns with their gender identity.
Gender Identity	Gender is the internal sense of being a woman, man, neither, both, or another gender.
Heterosexism	A form of bias and discrimination that favors people who are exclusively romantically and/or sexually attracted to people of the opposite sex/gender.
Minoritized/ Marginalized	When underrepresented groups are made to feel "less than."
Intersectionality	A concept describing the interconnection of oppressive institutions and identities.
Oppression	Use of power to privilege one group over another.
Prejudice versus Discrimination	An unfair feeling or dislike for another group is prejudice. Prejudice leads to discrimination, the unfair treatment of someone.

(Continued)

TERM	DEFINITION
Privilege	Refers to certain social advantages, benefits, or degrees of prestige and respect that an individual has by virtue of belonging to certain social identity groups. Within American and other Western societies, these privileged social identities – of people who have historically occupied positions of dominance over others – includes whites, males, heterosexuals, Christians, and the wealthy, among others.
Racism	Prejudice, discrimination, or antagonism directed toward someone of a different race based on the belief that one's own race is superior. Racism involves one group having the power to carry out systemic discrimination through the institutional policies and practices of society and by shaping the cultural beliefs and values that support those racist policies and practices.
Institutional Racism (Ableism, Heterosexism, Sexism, etc.)	The ways in which the structures, systems, policies, and procedures of institutions are founded upon and then promote, reproduce, and perpetuate advantages for the dominant group and the oppression of disadvantaged and underrepresented groups.
Social Justice	Promoting a just society by valuing diversity and equal access for all social groups.
Tokenism	Making symbolic and minimal gestures in offering opportunities to underrepresented groups.
Transgender	Someone who does not identify as the gender that aligns with the sex they were assigned at birth
Historically Excluded/ Underrepresented	Refers to groups of people who traditionally and currently are represented in lower proportional numbers than those groups of higher proportional numbers

Source: Rowan University Division of Diversity, Equity and Inclusion (2020).

PRIVILEGED AND HISTORICALLY EXCLUDED/ MARGINALIZED GROUPS IN SPORT GOVERNANCE

The statistics on NFL leadership shared in the previous section are unfortunately not outliers when looking at people who govern sport across the globe. Did you know that all Presidents of the IOC have been cisgender white men (i.e., white men whose gender identity aligns with their sex assigned at birth)? Did you know that only one woman each has served as President of the NCAA (Judith Sweet) and the Commonwealth Games (Dame Louise Livingstone Martin), and no woman has ever served as President of FIFA? Did you know that in recent years there has only been one openly gay man (Carl Nassib) competing in any of the five major North American men's sport leagues, with some of the leagues never having had an openly gay or trans man compete in their ranks (the NHL)? Did you know that whereas over 80% of players in the NBA are People of Color, only 6.5% of NBA Team Presidents and CEOs are People of Color (The Institute for Diversity and Ethics in Sport, 2021d)? Did you know that Andrew Parsons, the current president of the International Paralympic Committee (IPC), does not have a disability?

You may be surprised by some of these facts, but they do not happen by accident. Rather, there are systems in place that make it easier for some identity groups to attain leadership roles in sport than others. Throughout this chapter, we distinguish between two types of groups when it comes to diversity, equity, and inclusion in sport governance: (a) members of privileged groups and (b) members of groups who have historically been excluded and marginalized in sport. The term *privilege* is a loaded term, but its meaning is important to understand. Privilege refers to "certain social advantages, benefits, or degrees of prestige and respect that an individual has by virtue of belonging to certain social identity groups" (Rowan University Division of Diversity, Equity and Inclusion, 2020, p. 2). Within the United States and other Western countries, "these privileged social identities – of people who have historically occupied positions of dominance over others – includes whites, males, heterosexuals, Christians, and the wealthy, among others" (Rowan University Division of Diversity, Equity and Inclusion, 2020, p. 2). Privilege does not mean that a person holding a privileged identity has always had it easy and faced no challenges in life. Instead, it means that holding that specific social identity (e.g., being white) has not made life harder for that person. Check out the powerful excerpt from an essay titled "Using the Power in Your Privilege" written by Kim Miller, former Vice President for Programs at RISE, a nonprofit focused on improving race relations in sport, in Exhibit 3.5.

exhibit   **3.5**   Excerpt from Essay "Using the Power in Your Privilege"

Written by Kim Miller, Vice President for Human Resources at the Miami Dolphins and former Vice President for Programs at RISE, a nonprofit focused on improving race relations in sport

Many would prefer to avoid the topic of privilege. Like conversations on politics and religion, talking about privilege can sometimes make us uncomfortable, frustrated and defensive.

The notion of privilege forces us to reconsider the role that hard work and other choices we've made have truly contributed to our success. That can be a sobering thought, and perhaps why it makes us uncomfortable.

What's true, though, is that to varying degrees, we all have privilege. However, the disadvantages people face as a result of not having certain privileges can vary greatly and have severe consequences. … Privilege can be the difference between life and death during an encounter with police. It can also determine who gets the job interview, promotion or raise.

Fortunately, the more we understand privilege, its consequences and its power, the better positioned we are to use that privilege for good and to address critical issues in society.

Privilege can be defined as the rights, advantages and protections available to a particular individual or groups of people. It is often not something that can be earned, but instead is granted based on an innate aspect of your identity that affords you membership in a specific group. …

Many of the privileges available to certain groups are rooted in historical structures of race, gender or sexual orientation. However, privilege isn't only about demographics. If it were, then someone like me, a Black woman, could argue that I don't have privilege. But when I reflect and dig a bit deeper, I know there are ways I'm also privileged.

I was fortunate to grow up in a two-parent household, which gave me advantages over some of my peers. … Benefiting from your privilege doesn't necessarily mean you didn't work hard or are undeserving. Thinking about privilege this way is what tends to make people defensive or uncomfortable about the subject. However, we should be careful not to overlook privilege as simply a fact of life we have no control over. Privilege gives us the power and influence to assist those who do not have similar advantages.

Look at NBA All-Star Kyle Korver. His piece in The Players' Tribune, Privileged, gave readers a personal look at his experiences as a white male. Korver shares his perspective on racism and privilege and how he feels a responsibility to act because of the privilege he has. The story received national attention and sparked a dialogue on privilege and racism in America, and serves as a great example of how one person can use their privilege in a positive way to impact others.

Addressing privilege is not a responsibility that can be left to only one type of person – one race or one gender. There are times when we all have privilege.

Athletes, young and old, have unique privileges. They have a voice and a platform to influence their community. When they harness their privilege effectively, they can be tremendously impactful. Throughout history, athletes have used their privilege to spur social change. They have stepped up to help pave the way for desegregation, criminal justice reform, policies to address pay inequities, and more.

I hope you'll take time to explore your own privilege, but know that acknowledging it is just the start. Go beyond reflection by considering the ways you can use the power associated with your privilege to give access and provide opportunities to others. The action you take can make a difference that not only changes someone's individual circumstance, but may also lead to a more equitable society.

Source: Miller (2020).

In her essay, Miller raises an important point: Privilege is not absolute, but rather a person can have privilege in one area (e.g., socioeconomic background) while lacking privilege in others (e.g., race and gender identity). In sport governance, privilege manifests in who has (easier) access to governing positions (e.g., sport committee composition), who is being governed (e.g., demographic make-up of sport teams), how they are governed (e.g., bias in distribution of resources), and how the process of governance and policy-making affects specific populations (e.g., outcomes of affirmative action policies). For example, how could a woman ever get selected to be on an International Olympic Committee working group if she is from a home country which restricts her ability to work with her own National Olympic Committee?

As you can see, some groups have been historically excluded from and marginalized in sport governance. These groups are sometimes also referred to as *minorities*, *minoritized*, or *historically underrepresented*, the latter of which describes

> groups who have been denied access and/or suffered past institutional discrimination … revealed by an imbalance in the representation of different groups in common pursuits such as education, jobs, and housing, resulting in marginalization for some groups and individuals and not for others, relative to the number of individuals who are members of the population involved.
>
> (Emory University, 2022, para. 38–39)

We primarily use the terms *historically marginalized* and *excluded* to account for the various ways in which these groups are systemically disadvantaged in sport governance. Exhibit 3.6 provides an overview of some of these groups. It is important to note here that while some global forces impact which groups gain privilege and which do not (such as settler-colonialism and racism), what groups can be considered histori-cally marginalized and excluded can differ across cultural contexts.

exhibit	**3.6**	Overview of Privileged and Underrepresented, Historically Excluded, and Marginalized Groups

GROUP	PRIVILEGED	UNDERREPRESENTED, HISTORICALLY EXCLUDED, AND/OR MARGINALIZED
Gender	Cisgender men	Women, transgender, and gender nonconform-ing persons
Race	white	Black, Indigenous, and People of Color
Socioeconomic Class	Middle and upper class	Poor, working class
Nationality	US/Developed nations	Developing nations
Ethnicity	European	All other ethnicities
Sexual Orientation	Heterosexual/Straight	Lesbian, gay, bisexual, asexual, nonbinary people
Religion	Christian	All other religions, non-religious people
Physical Ability	Able-bodied	Persons with disabilities
Age	Youth	Elderly persons
Language	English	All other languages

Source: We Are One Team (WA1T) Team Player Program/Center for Leadership, Bowling Green State University.

In North American sport, for example, the following groups have historically been marginalized in or excluded from governing sport:

- **WOMEN:** Although women make up almost half of the world's population, when it comes to their representation in sport governance, especially at the leadership levels, women continue to be drastically excluded. In one of the most comprehensive studies on gender diversity in the governance of sport associations across the globe, Adriaanse (2016) found that women continue to be underrepresented in the most influential positions of sport governance: on boards of directors (with women representing only 19.7% of board directors), as board chairs (10.8%), and as chief executives (16.3%). No wonder, then, that there is the perception that "the leadership in international sport is an exclusive club of men" (Lapchick, 2016, p. 1). On the US side, women constitute 8.7% of presidents/CEOs in the NBA, zero percent of presidents/CEOs in MLB, 13.8% of presidents/CEOs in MLS, 3.1% of presidents/CEOs in the NFL, and 14% of Athletic Directors in NCAA Division I athletic departments (The Institute for Diversity and Ethics in Sport, 2021a). (In the WNBA, on the other hand, 66.7% of presidents and CEOs are women – a bright spot!) We can also see lack of representation and appreciation for women's sport in the media, where only roughly five percent of coverage is dedicated to women's sport, a number that has not changed in 30 years (Cooky et al., 2021). This is at least partly surprising because interest in women's sport among general sport fans has grown drastically in recent years, with now as many as eight in ten general sport fans expressing interest in women's sport (Nielsen Sports, 2018). In terms of the environments women find themselves in once they break into sport, research has shown over and over again that women in sport lack opportunities for advancement (Hancock et al., 2018), receive lower entry salaries while working longer hours (Harris et al., 2015), are less satisfied due to repeated instances of discriminatory behavior (Cunningham et al., 2005), and are subject to sexist comments and sexual harassment by their colleagues (Taylor et al., 2018).

- **BLACK, INDIGENOUS, AND PEOPLE OF COLOR:** A second group that has historically been marginalized and excluded in sport governance are racially and ethnically minoritized groups, such as those whose members identify as Black, Indigenous, and People of Color (BIPOC). (The term "BIPOC" has been critiqued by some as creating a hierarchy between racially minoritized groups; see Deo, 2021). In the United States, this group includes Black or African Americans, Asian or Asian Americans, Native Americans, Pacific Islander Americans, multiracial Americans, or Hispanic and Latino/a/x Americans. Other examples of indigenous communities beyond North America are the Aborigines in Australia or the Maori in New Zealand. According to 2020 US Census data (United States Census Bureau, 2022), 38.3% of the US population identified as members of racially minoritized groups, which would mean that in a world with no barriers to diversity, equity, and inclusion sport organizations should have a similar percentage of People of Color represented in governance. However, as you may recall from earlier in the chapter, the numbers of People of Color in sport leadership are nowhere close to mirroring the overall population. What does this tell us? Systemic barriers continue to make it harder for People of Color to attain governance positions in sport. This is particularly problematic when sports like American football or basketball are governed by white people but rely heavily on the labor of Black athletes on the field or court. Research shows that structural racism creates hostile work environments for People of Color as it tokenizes them and positions them as outsiders, pays them less in salary, gives fewer promotions, and reinforces occupational stereotyping (Simpkins et al., 2022). For example, McDowell and Carter-Francique (2017) used a theory called *intersectionality* as a framework useful in determining how multiple forms of oppression intersect when trying to understand the workplace experiences of African American women in Athletic Director positions. They found that these women faced both gender and racial stereotypes. They frequently had their authority called into question, were accused of having been hired to fill a quota, and sometimes had to conceal their authentic selves to meet dominant cultural expectations. Clearly, these findings show the unique challenges that African American sport professionals in general, and African American women in particular, face when entering a career in sport governance.

■ **PEOPLE WITH DISABILITIES:** Approximately 15% of the world population have a disability, but people with disabilities remain starkly underrepresented in sport governance (Misener & Darcy, 2014). While some outlets showcase the athletic prowess of athletes with disabilities, such as the Paralympic Games (see Chapter 9), people with disabilities face unique challenges when it comes to their full inclusion in society:

> Persons with disabilities often face societal barriers and disability evokes negative perceptions and discrimination in many societies. As a result of the stigma associated with disability, persons with disabilities are generally excluded from education, employment and community life which deprives them of opportunities essential to their social development, health and well-being. In some societies persons with disabilities are considered dependent and seen as incapable, thus fostering inactivity which often causes individuals with physical disabilities to experience restricted mobility beyond the cause of their disability.
> (United Nations, 2022, para. 2)

At the grassroots sport level, people with disabilities are less likely to participate in sport activities compared to their non-disabled peers (United States Government Accountability Office, 2010).

www
United States Census Bureau
https://www.census.gov/

www
United Nations Women: Intersectionality Resource Guide & Toolkit
https://www.unwomen.org/en/digital-library/publications/2022/01/intersectionality-resource-guide-and-toolkit

Playing experience is often an important gateway into a career in sports. In fact, you are likely reading this book because you are considering a career in sport inspired by your own involvement in athletics throughout your life. People with disabilities are less likely to get hired because of ableist stigma, which can prompt employers to mistakenly think hiring people with disabilities will result in additional costs due to workplace accessibility requirements (Waterhouse et al., 2010).

- **LGBTQ+ PEOPLE:** The smallest out of the four historically excluded and marginalized groups covered here is the LGBTQ+ community. According to a Gallup survey conducted in 2020, 5.6% of the US population identify as members of this community. Among Generation Z, one in six adults considers themselves to be members of the LGBTQ+ community, indicating that identification as LGBTQ+ is increasingly common among younger generations (Jones, 2021). Compare that to some of the statistics about the participation of LGBTQ+ youth in sport:
  - 84% of US Americans have either witnessed or experienced anti-LGBTQ+ attitudes in the context of sport;
  - Only 24% of a LGBTQ+ youth play on a sports team, compared to 68% of a national sample of all youth;
  - Four in five LGB athletes and 82% of transgender youth are not open about their sexuality or gender identity to their coaches;
  - 11% of LGBTQ+ youth do not feel safe in the locker room. (Human Rights Campaign, 2017)

On the sport business side, LGBTQ+ people continue to face discrimination due to persistent homophobia, biphobia, and transphobia in the institutions governing sport. LGBTQ+ employees in sport often conceal their sexuality in order to avoid the stigma attached to LGBTQ+ identities in sport (Krane & Barber, 2005). While some LGBTQ+ people report positive experiences in sport, there is also evidence that they are negatively stereotyped, denied leadership positions, receive less support, and get poorer performance evaluations compared to their straight peers (Melton & Cunningham, 2014). In addition to discrimination experienced at work, LGBTQ+ people are also targeted by legislation at the state or federal level. For example, by mid-2022 more than 30 US states had either passed or introduced legislation preventing transgender girls and women from competing in sports – part of a larger, broad-scale attack on the rights of transgender people in the country.

Of course, these are just some of the issues facing members of the historically excluded and marginalized groups described above. Many more groups deserve to be highlighted, some of which are included in Exhibit 3.6. For example, think about how a Muslim athlete must feel on an all-Christian team, especially when a coach leads a team in a Christian prayer. They may be facing particular challenges during the month of Ramadan, unbeknownst to their Christian peers. Similarly, increasing attention is being paid to the role of socioeconomic background in sport management, as the practice of requiring unpaid internships institutionalizes the privilege of higher socioeconomic classes and makes it harder, and sometimes impossible, for others to break into the industry (Walker et al., 2021).

www
NCAA Office of Inclusion
https://www.ncaa.org/sports/
inclusion

That is why diversity, equity, and inclusion efforts focus on supporting historically marginalized and excluded groups, and this work often starts with sport organizations identifying the groups that have been marginalized in their respective organizational context. For example, the NCAA Office of Inclusion has identified five core areas they concentrate their efforts on: international student-athletes, LGBTQ+ people, student-athletes with disabilities, race and ethnicity, and women (NCAA Office of Inclusion, 2022). Via strategic planning, programming, resources, and resource allocation, sport organizations can map out ways to dismantle the barriers these groups face. To illustrate how each segment of the industry engages in diversity, equity, and inclusion initiatives, we have added DEI spotlights to each of the industry segment chapters in this book (Chapters 5–15). Each spotlight illustrates one or more ways in which members from historically marginalized and excluded groups are supported across the different segments of the sports industry. Now that we know what groups must benefit from diversity, equity, and inclusion efforts, let's turn our attention to some of the most persistent barriers they face when it comes to fully participating in the governance of sport.

COMMON BARRIERS TO DIVERSITY, EQUITY, AND INCLUSION

If you recall the graphic shown in Exhibit 3.3, you may remember that the two people pictured were standing on boxes to be able to look over a fence and watch the baseball game taking place on the other side of the fence. The fence was what prevented them from watching the game – it served as a physical barrier to their access to the game and broader inclusion to the spectacle that is game day. That is why there are now versions of the graphic that have a third picture among the other two: one in which the fence – the systemic barrier – is removed entirely (Froehle, 2016). Much like the literal fence in the graphic, there are barriers in sport that prevent historically excluded and marginalized groups from fully participating in the governance of sport. For those committed to advancing diversity, equity, and inclusion, the ultimate goal is to identify, and eventually remove, the barriers to diverse representation, equitable treatment, and inclusive environments. In this section, we start with identifying some of these barriers, before turning to strategies to dismantle them in the final section.

You have probably picked up on some of the most persistent barriers by now: a lack of representation at the leadership level, discriminatory structures and behaviors by those governing sport, and limited access to sport governing spaces in the first place. To understand what barriers are most persistent in sport, Kluch et al. (2022) studied the experiences of those often tasked with dismantling them: diversity, equity, and inclusion professionals (yes, that's a whole emerging job field in itself!). Diversity, equity, and inclusion staff have first-hand experience with the roadblocks that negatively impact their work. Focusing on the context of college athletics, Kluch et al. (2022) identified five primary barriers for diversity, equity, and inclusion in sport – (a) structural barriers, (b) cultural barriers, (c) conceptual barriers, (d) emotional, and (e) social/relational barriers (even though we will combine the latter two into one category for this chapter). The barriers are illustrated in Exhibit 3.7.

Let's take a closer look at each of these barriers in detail.

- **STRUCTURAL BARRIERS:** The first type of barriers are structural barriers, or "systemic barriers woven deeply into the institutional fabric" of the sport governance organization (Kluch et al., 2022, p. 7). These barriers are among the hardest to overcome because they often seem natural to us and are hard to spot in the first place. Here is a list of common structural and systemic barriers:

  - **Lack of resources.** For diversity, equity, and inclusion work to be successful, resources need to be allocated to it. These resources can include staff, funding/financial support, physical or virtual space to conduct the work, and so on. However, often sport organizations do not provide sufficient resources to meaningfully support diversity, equity, and inclusion initiatives.

  - **Lack of infrastructure.** In Chapter 2, we outlined some of the structural elements of a sport governance organization. Oftentimes, these organizations lack sufficient infrastructure to advance diversity, equity, and inclusion work. Infrastructure includes committees and task forces dedicated to diversity, equity, and inclusion, staff leading and/or supervising DEI efforts, accountability measures for those who might be blocking this work, and continuous strategic planning for diversity, equity, and inclusion action. If an organization does not have any of these in place, diversity, equity, and inclusion work is destined to fail.

  - **Lack of diverse representation.** A common saying in diversity, equity, and inclusion work is "you can't be what you can't see." Indeed, a lack of diverse role models at the top of

Barriers to Diversity, Equity, and Inclusion in Sport   *exhibit* **3.7**

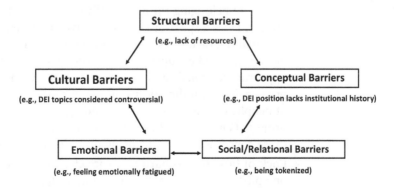

Source: Kluch et al. (2022).

the governance hierarchy (e.g., the executive suite) trickles down to the entry levels of the sport industry. If members of historically excluded and marginalized groups cannot see anybody like them in sport leadership positions, they are less likely to choose a career in sport.

- **Lack of access.** At the grassroots level, powerful cultural stereotypes, one-dimensional representations in media, socioeconomic barriers, and socio-environmental factors often determine *who* is drawn to *what* sports. For example, research shows that Black men are often socialized into sports like football and basketball (Brooks & McKail, 2008). These factors can make it harder for members of historically excluded and marginalized groups to enter the sport industry.
- **Legislation and policy.** We often think of policy and legislation as neutral tools guiding the governance process. However, policies are rarely neutral and instead either work against or toward advancing diversity, equity, and inclusion (Kendi, 2019). Policies need to be reviewed continuously to make sure they do not hinder members of historically excluded and marginalized groups from participating in sport governance – and ideally should make it easier for them to get a seat at the table (think about the impact Title IX has had on gender equity!). That is why both the NCAA and the NAIA have recently passed legislation that requires a minimum number of women and racial minorities on their main governance committees.

- **CULTURAL BARRIERS:** Structural barriers are very powerful because they create a culture that can hinder diversity, equity, and inclusion efforts. For example, if there is little diversity within a governance organization, that organization will be less likely to see the value diversifying its own ranks can bring. Cultural barriers therefore capture "the culture created in the spaces ... [and] to what extent that culture hindered ... endeavors to drive [diversity, equity, and inclusion] action" (Kluch et al., 2022, p. 7). Here are some common ones:

  - **Discriminatory practices and procedures.** Accepted organizational practices can be major barriers to diversity, equity, inclusion efforts, especially when it comes to nurturing a sense of belonging for historically excluded and marginalized groups. Does a sport organization have any formal or informal opportunities for mentoring historically excluded or marginalized groups? Are professional development opportunities consistently awarded to those with the most privilege? Are funding decisions made by the same group of people, over and over again, leading to the same types of projects being funded? All of these practices raise red flags to those concerned with advancing diversity, equity, and inclusion.
  - **Lack of buy-in.** It is hard to create change if you are the only one pushing for it, which is why buy-in for diversity, equity, and inclusion efforts from key stakeholders is crucial. Is the CEO of the organization on board with diversity, equity,

and inclusion work? If those with the most power within the governance process do not value diversity, equity, and inclusion, it is going to be almost impossible to move the needle. Regular inclusive leadership training for sport leaders can help generate the buy-in that is so crucial to driving diversity, equity, and inclusion.

- **Non-engagement on social issues.** Sport is often perceived to be an apolitical or neutral domain. However, if sport were neutral then none of the social ills we see in society would find their way into sport. But because sport is a microcosm of society, they do. If an organization rejects any engagement on social topics, whether it is publicly (e.g., putting out statements when injustices occur) or internally (e.g., team meetings), they run the risk of alienating members of groups disproportionately affected by those topics.

- **Performative work.** In the aftermath of the murder of George Floyd by Minnesota police, protests emerged across the country to call out police brutality and racial injustice, but "from toothless statements to silence, pro-sports teams have not acted like the community leaders they claim to be" (LeBlanc, 2021, para. 1). Rather, they engaged in performative work; that is, they put out statements but often didn't actually commit to doing the work needed to remove the systemic barriers that facilitate police brutality and racial injustice in the first place. For diversity, equity, and inclusion work to be successful, it needs to be substantive, strategic, and planned out rather than performative.

- **Prioritizing diversity over inclusion, with no eye toward equity.** Diversity, equity, and inclusion work can also be weakened if an organization prioritizes diversity over inclusion, meaning that it is more interested in bringing members of historically excluded and marginalized groups into the organization than creating an environment in which these individuals can actually thrive. If the culture of the organization is hostile toward marginalized groups, those with historically excluded and marginalized identities will leave sooner rather than later, and the cycle begins anew. Sometimes diversity is a top priority without taking into account the unique needs of marginalized groups (e.g., additional funding for professional development, mentoring), so they are set up to fail from the start.

- **CONCEPTUAL BARRIERS:** When Kluch et al. (2022) studied the experience of diversity, equity, and inclusion professionals, they found that one challenge for these staff members was that their job field was fairly new. They faced conceptual barriers, or "barriers that are rooted in a lack of consistent DEI industry standards as well as institutional and industry-wide history of DEI positions" (Kluch et al., 2022, p. 8). Beyond their study site (college athletics), sport organizations can lack consistent language, parameters, or evaluation processes for diversity, equity, and inclusion efforts.

  - **Not situating efforts in organizational context.** Note how we defined key terms early on in the chapter to show what diversity, equity, and inclusion mean in the context of this book on sport governance. Sometimes organizations use these terms, but they fail to clearly articulate what they mean within their sphere of influence. For example, what constitutes an underrepresented group will differ at a predominantly white college campus from a minority-serving institution. Effective diversity, equity, and inclusion work must take into account the unique institution's engagement (or lack of engagement) on these issues and clearly articulate what success means in their respective organizational contexts.

- **"Tick-box" approaches.** Have you noticed how conversations on diversity often focus on race and gender? That is because these two metrics have historically been tracked in the United States and are often the main characteristics used to determine whether or not a group is diverse. Some organizations use these identity categories to "check the boxes" when it comes to diversity but fall short of doing the transformative work needed to make institutions more equitable and inclusive.

- **No benchmarking and strategic planning.** Oftentimes, sport leaders with no inclusive leadership expertise reduce diversity, equity, and inclusion efforts solely to matters of representation and trainings. In reality, such efforts must encompass transforming policies and procedures, building resources, creating supplier programs to enhance diversity, retaining talent, strategic planning, and addressing biases.

- **Non-inclusive language.** There is power in language, and inclusive language is crucial for creating environments of belonging. Sometimes, the language used is not conducive to creating environments where everyone can feel as if they are seen and valued. For example, sport has a history of using gendered language (e.g., sports*man*ship, chair*man*). Instead, organizations should pay close attention to the use of language. That is why the NCAA's Division III recently voted to make all policies and formal correspondence gender neutral (e.g., by changing pronouns from *he/she* to *they*; NCAA, 2019).

- **EMOTIONAL & SOCIAL/RELATIONAL BARRIERS:** Picture the following scenario: You have an important exam tomorrow and a to-do list that seems to be getting longer by the minute. You feel overwhelmed, stressed, and the more time passes, the more frustrated you get. These feelings are quite common among those working toward diversity, equity, and inclusion, especially because they are more likely to be members of historically excluded and marginalized groups themselves and thus have to personally deal with discrimination frequently. Emotional barriers are those that negatively impact one's mental health and well-being, making it harder to pursue inclusive excellence (Kluch et al., 2022). They are often closely related to a fifth type of barrier, social and relational barriers, which refer to "feelings of social connectedness and … [the] ability to form relationships that can benefit work related to DEI" (Kluch et al., 2022, p. 9). Both emotional and social/

relational barriers focus on interpersonal interactions and the sense of belonging among those driving diversity, equity, and inclusion efforts. Inclusion, after all, is a team sport – so a strong network of collaborators is crucial for anchoring diversity, equity, and inclusion in the governance of sport. Some of these barriers include:

- **Workload, burnout, and emotional fatigue.** Oftentimes, diversity, equity, and inclusion tasks are put on the shoulders of those who have historically been excluded and marginalized within the organization (there's even a term for this: cultural taxation). However, those efforts are everyone's responsibility, and overburdening marginalized groups can lead to intense workload, burnout, and emotional fatigue. Research suggests that members of marginalized groups are more likely to experience mental health issues (Lipson et al., 2019), so relying on them to dismantle the barriers that oppress them can further lead to psychological harm for these groups.

- **Microaggressions.** "You throw like a girl." "No, but where are you *really* from?" "Class at 8 am? That's so gay!" You surely have heard these sayings before. They are called microaggressions, or "brief and commonplace daily verbal, behavioral, and environmental indignities, whether intentional or unintentional, that communicate hostile, derogatory, or negative … slights and insults to the target person or group" (Sue, 2010, p. 5). What may seem like a joke or inappropriate comment here and there can cause serious psychological harm to those on the receiving end of the microaggression. Hearing these indignities over and over again can negatively affect one's mental health and well-being, and thus significantly impact an individual's ability to be their authentic self and drive diversity, equity, and inclusion action within an organization.

- **Bias.** Biases are preferences we have for or against something or someone, whether that be conscious or unconscious (Rowan University Division of Diversity, Equity and Inclusion, 2020). For example, while white men only make up a little over 30% of the US population, they represent more than 95% of athletic team ownership positions and 92% of Forbes 500 executive CEOs (The Institute for Diversity and Ethics in Sport, 2021a; Zweigenhaft, 2021). The hiring decisions for these positions are often made by predominantly white groups, revealing racial bias in hiring for leadership positions. Our conscious and unconscious bias affects who we hire, how we interact with them at work, what policies we create, and whom we support in our organizations. If bias remains unchecked, barriers to diversity, equity, and inclusion all but certainly remain intact. You can review how racial bias may affect the hiring process in Exhibit 3.8.

- **Tokenism.** Bias can also lead to tokenism, which is the practice of giving the impression of being inclusive of historically excluded and marginalized groups while not actually doing the work needed to create truly inclusive environments. For example, Vaccaro and Newman (2016) studied underrepresented students in the classroom and found that they were often treated differently, tokenized, and expected to speak for their entire group as the only representative from that group such as asking a student from Japan, "What did Japanese people think about the Olympic and Paralympic Games in Tokyo being held without fans?" When sport organizations engage in tokenism, they are not really invested in diversity, equity, and inclusion but rather engage in symbolic and performative efforts that cause harm to the tokenized individuals and groups.

As you can see in Exhibit 3.7, these barriers are not mutually exclusive but rather work together to make advancing diversity, equity, and inclusion in sport governance a massive undertaking. As Kluch et al. (2022) explain, those driving diversity, equity, and inclusion efforts

> may have felt emotionally fatigued (emotional barrier) due to being tokenized (social/relational barrier), which led them to stay away from DEI topics considered controversial (cultural barrier) and made it hard for them to advocate for resources (structural barrier) given their position was fairly new in the department and lacked institutional history (conceptual barrier).
>
> (p. 7)

Perhaps now you can start to see why it is so hard to change the system that has created the barriers in the first place. But there is good news: We can all do our part to help dismantle these barriers, both in individual and institutional ways. That's what we will turn our attention to in the final section of this chapter. Get ready to be a part of the change!

---

exhibit   **3.8**   How Can Racism Affect the Hiring Process?

RACISM IN THE HIRING PROCESS

Screening Job Applicants	Interviewing	Making the Hiring Decision
• Screening out job applicants with indigenous, Black, or "foreign" sounding names • Screening out job applicants deemed to be "over qualified"	• Avoidance behavior in the interview, including frowning, making less eye contact, leaning away from the candidate, and cutting the interview short • Asking interview questions that are culturally based or which allow for the subjective assessment of the answer • Assessing candidates on the interviewers "overall impression" of them	• Wanting to hire the candidate that is most like oneself • Conducting police record checks when it is not a bona fide job requirement • Not hiring racialized people for positions they are fully qualified for

Source: Turner Consulting Group (2022).

STRATEGIES FOR ADVANCING DIVERSITY, EQUITY, AND INCLUSION IN GOVERNANCE

Dismantling the barriers to diversity, equity, and inclusion outlined in the previous section requires careful, strategic action across all areas of sport governance. As someone at the beginning of their career in sport, you may be wondering what you personally can do to help right some of the wrongs mapped out in this chapter. In their analysis of allyship in global sport, Jolly and colleagues (2021) asked that exact question. They argued that there are two ways of driving social justice efforts in the global arena of sport, of which considerations of diversity, equity, and inclusion are a part: through individual and institutional actions. Individual strategies are those you can engage in with your everyday actions as a member of the sport community, like educating yourself (which you already started doing by reading this chapter!), regularly reflecting on your bias (e.g., via the free Harvard Implicit Association Tests), and creating environments where all members feel seen, heard, and valued. In Exhibit 3.9, we present the Human Rights Campaign's eight strategies for being a champion for LGBTQ+ inclusion, many of which refer to individual strategies for advancing diversity, equity, and inclusion.

Because we focus on the governance of sport, though, the remainder of this section will outline institutional strategies for centering diversity, equity, and inclusion in decision-making, policy, and the governance process. Institutional strategies happen at the structural level, where "individuals in an organisation have to create a methodical and strategic action plan to use their institutional power to challenge the status quo and deconstruct

www

Implicit Association Tests (Harvard University)
https://implicit.harvard.edu/implicit/aboutus.html

Individual Strategies for Diversity, Equity, and Inclusion: The Human Rights Campaign's CHAMPION for LGBTQ Inclusion Model *exhibit* **3.9**

CREATE safe spaces for all athletes, coaches, and fans at all times – on the field, in the locker room, in the stands, at home, and in the community.
HOLD players, coaches, teammates, and fans accountable for non-inclusive language or actions. Creating an inclusive team environment is everyone's responsibility.
ARM yourself with information and resources about the laws and policies that impact your LGBTQ players, staff, and their families.
MODEL inclusive behaviors by weaving respect, diversity, and inclusion into your team's culture on Day 1.
PROMOTE allyship and respect on and off the field of play. Your team culture isn't just formed in practice.
IMPLEMENT LGBTQ inclusion policies. Make sure you have an inclusive non-discrimination statement, trans-inclusive participation policies, competent data collection policies, LGBTQ-inclusive travel and uniform policies, and a fan code of conduct.
ORGANIZE inclusion trainings for players, coaches, and parents. Like in sports, it's hard to improve without practice and learning.
NEVER ASSUME someone's sexual orientation, gender identity, pronouns, or experiences.

Source: Human Rights Campaign (2017).

systemic barriers that undermine equality and equity" (Jolly et al., 2021, p. 236). Let's take a look at how those who govern sport can advance diversity, equity, and inclusion within their organizations (and beyond!).

Conduct Regular Diversity, Equity, and Inclusion Assessments

First and foremost, diversity, equity, and inclusion efforts should be data-driven. That is why those efforts should start or continue with comprehensive assessment of where the organization is at when it comes to diversity, equity, and inclusion, so that benchmarks for success can be developed. This data can be collected in multiple ways: employee experience surveys, formal and informal interviews (e.g., exit interviews with those leaving the organization), focus groups, town halls, and so on. When assessing their organization's strengths and weaknesses in terms of diversity, equity, and inclusion, those committed to advancing diversity, equity, and inclusion should ask the following questions to measure the success of their efforts:

- What resources are allocated to diversity, equity, and inclusion (e.g., funding, staff)?
- Do the demographics of the organization match the demographics of the larger society? What groups are overrepresented and underrepresented at the various levels of the organization (e.g., entry-level positions, executive team)?
- At the leadership level, how many members are from historically excluded or marginalized groups?
- How long do members of historically excluded or marginalized groups stay with the organization? Why do they leave?
- What mentorship programs are in place? How much time is spent on mentorship, especially for those from historically excluded and marginalized groups?
- What is the mechanism in place to report diversity, equity, and inclusion incidents (e.g., repeated microaggressions)? How many incident reports have been filed?
- What diversity, equity, and inclusion trainings are available? What is the impact of the training?
- What accountability mechanisms do we have in place? How do we hold senior leadership, in particular, accountable for driving diversity, equity, and inclusion efforts? (adjusted from Hall, 2021)

www
USOPC Diversity, Equity, and Inclusion Initiatives
https://www.teamusa.org/about-the-usopc/diversity-equity-inclusion

www
USOPC Diversity, Equity, and Inclusion Scorecards
https://www.teamusa.org/About-the-USOPC/Diversity-Equity-Inclusion/D-and-I-Scorecards

Regular diversity, equity, and inclusion assessments are common practices in sport governing bodies. To measure progress for diversity specifically, the USOPC requires each NGB to submit a diversity, equity, and inclusion scorecard tracking demographics for race, gender identity, disability, and military status every year. This information maps out benchmarks for diversity and is an important assessment tool for the USOPC and NGBs to "easily identify opportunities to become more diverse and inclusive as it relates to athletes, coaches, staff, board of directors and membership" and "assist in creating diversity plans and identifying the best use of resources for DE&I success" (USOPC, 2022, para. 4).

Develop Diversity, Equity, and Inclusion Strategic Plans to Guide Your Efforts

Assessment is only the first step in driving diversity, equity, and inclusion action. Once you have assessed a sport organization's climate, successes, and areas for improvement in diversity, equity, and inclusion, that information should be used to inform a comprehensive strategic plan focused on anchoring diversity, equity, and inclusion in organizational practices. You may recall from Chapter 2 how important planning is to organizational success. Sometimes, such planning takes place at the league level. For instance, the English Premier League established its Equality, Diversity and Inclusion Standard (PLEDIS) to aid clubs in their diversity, equity, and inclusion efforts. At both the league and individual organizational levels, a solid diversity, equity, and inclusion strategic plan should, at minimum, contain the following components (National Wheelchair Basketball Association, 2021):

- **DEI Objectives:** Overarching goals the organization hopes to achieve when it comes to diversity, equity, and inclusion. These should be realistic and measurable, both in time needed to complete them and in outcome.
- **Tactics:** Projects, initiatives, and action items that will help achieve the larger objective.
- **Target groups:** Groups that will benefit from the accomplishment of the objective. These are usually historically excluded or marginalized groups such as the ones outlined earlier in the chapter.
- **Time frame:** The specific timeline for achieving the objective and/or initiatives/action strategies. Most strategic plans are a minimum of one year in length, with most mapping out multi-year plans (e.g., three to five years).
- **Success/evaluation measures:** The data points that will be used to evaluate whether or not the objective was achieved.

www

Premier League established its Equality, Diversity and Inclusion Standard (PLEDIS) https://www.premierleague. com/equality-diversity-and- inclusion-standard

Some more in-depth diversity, equity, and inclusion strategic plans also include roles (see Chapter 2), point person(s) in charge of overseeing the respective objective of the plan, information on what resources are needed to complete the plan, the mission or vision of the overall organization, a specific diversity, equity, and inclusion mission statement, and the organization's definitions of key terms.

To give an example, check out one of the objectives and corresponding other components in the 2021–24 diversity, equity, and inclusion strategic plan of the National Wheelchair Basketball Association:

DEI Objective: Promote the retention and recruitment of women to participate in the NWBA and sport of wheelchair basketball.

Tactics: Create the NWBA Women's and Diversity Committee to ensure alignment across all levels of the organization on women's initiatives; Host opportunities to develop, classify and train women in NWBA High Performance Program; Promote participation of women across divisions of play.

Target group(s): Women, Women of Color

Time frame: 2021–24

Success/evaluation measures: Increase of the number of new women athletes by 8–11% by 2024; Increase player retention by 25% by 2024.

In a slightly different format, the NBA's Portland Trail Blazers have developed a comprehensive strategic plan across seven areas of impact: overall vision, measurement and accountability, inclusive culture, talent management, brand, service and sales, social responsibility, and supplier diversity. For each of these, the organization's diversity, equity, and inclusion staff developed objectives and action strategies. For example, for the *measurement and accountability* impact area, they identified creating an environment of accountability for DEI as a goal/objective and implementing diversity, equity, and inclusion check-ins at department and leadership meetings as an action tactic (Portland Trail Blazers, 2021).

www

Portland Trail Blazers DEI
Strategic Plan
https://www.ripcityresource.
com/pdf/21_DEI_Strategic-
Report_v4.pdf

Create the Infrastructure for Diversity, Equity, and Inclusion Efforts

The Trail Blazers example provides an apt transition to the next strategy to advance diversity, equity, and inclusion in sport governance – creating the infrastructure for diversity, equity, and inclusion across the organization. A strong infrastructure for diversity, equity, and inclusion efforts ensures that all three can inform the governance process. It ensures the voices of historically excluded and marginalized groups have input on the governance of sport and that those who counteract diversity, equity, and inclusion are held accountable for doing so. This infrastructure can be created in a variety of ways, many of which you will read about in the chapters to come:

- **Committees.** One of the most effective ways to establish institutional infrastructure for diversity, equity, and inclusion efforts is by implementing it into the governance process. This can be done by creating committees and/or task forces responsible for diversity, equity, and inclusion efforts. For example, the NCAA has the Minority Opportunities and Interests Committee (MOIC), the Committee on Women's Athletics (CWA), the Committee to Promote Cultural Diversity and Equity, and the Gender Equity Task Force. All of these have their own charter, requirements for composition, and ability to provide recommendations to the NCAA to enhance diversity, equity, and inclusion. For example, the adoption of Athletics Diversity and Inclusion Designee legislation (see Chapter 8 on

Intercollegiate Athletics), which requires each NCAA school to appoint a designee as a point of contact for diversity, equity, and inclusion, was an initiative that came out of the MOIC. At the professional level, MLS is one of the first leagues to create a Board of Governors Diversity, Equity, and Inclusion Committee to provide strategic support for the league's initiatives (The Institute for Diversity and Ethics in Sport, 2021c).

- **Advisory Councils, Alliances, Coalitions, and Task Forces.** Sometimes, urgent diversity, equity, and inclusion action needs to be taken, and creating a committee or task force will take too long to get off the ground because there are usually rules about how they need to be integrated into the governance structure. That's why some sport governing bodies have utilized coalitions or advisory councils as tools to drive diversity, equity, and inclusion. For example, in response to the murder of George Floyd and renewed calls for racial justice in 2020, the USOPC launched the Team USA Council on Racial and Social Justice to identify barriers to progress within the US Olympic and Paralympic movements (see Chapter 10). Similarly, the NHL established the Hockey Diversity Alliance and the WNBA launched a Social Justice Council. The NBA convened a cross-departmental Social Impact Task Force to "harness the full power of the organization to create sustained change by addressing racial inequality, creating greater economic opportunity, advocating for needed reform and promoting civic engagement" (The Institute for Diversity and Ethics in Sport, 2021d, p. 36). The organization also established a National Basketball Social Justice Coalition in collaboration with the National Basketball Players Association.

- **DEI Staff or Designations**. One of the most powerful ways to institutionalize diversity, equity, and inclusion efforts is by committing resources to them, including staff solely focused on these efforts. Most major sports leagues currently have staff focused on diversity, equity, and inclusion. Sometimes, organizations use designations to advance diversity, equity, and inclusion. For example, the NCAA created the Primary Woman Administrator designation in 1981 to increase the representation of women in departmental leadership (Hoffman, 2010). This designation was later renamed to Senior Woman Administrator (SWA) and today every member school in the NCAA must designate an SWA on their staff. The SWA is not a specific position but rather a designation that can be given to any individual who is the most senior-ranking woman in the department. For example, if the most senior-ranking woman in the department is the Senior Associate Athletic Director for Compliance, that person would serve as the SWA. Often, the SWA is on the senior leadership team of the athletic department. Internationally, the UK's Premier League requires a senior member of staff in each club to assume the role of Equality Lead Officer.

WWW
NHL Hockey Diversity Alliance
https://hockeydiversityalliance.org

WWW
WNBA Social Justice Council
https://www.wnba.com/social-justice-council-overview/

WWW
National Basketball Social Justice Coalition
https://coalition.nba.com

WWW
MLB Players Alliance
https://www.playersalliance.org

- **Funding.** Have you heard the saying "put your money where your mouth is"? It can be used to refer to organizations that say they are committed to diversity, equity, and inclusion but do not provide crucial funding to actually facilitate change. Funding, and other forms of financial support such as grants, is one of the most important tools to create an infrastructure for diversity, equity, and inclusion. This funding can be internal (i.e., to benefit the organization's own diversity, equity, and inclusion initiatives) or external (i.e., to support the work of nonprofit organizations). For example, in 2021 more than 50% of the MLB's charitable support went to nonprofits focused on racial justice (The Institute for Diversity & Ethics in Sport, 2021b). The league also provides funding through its Jackie Robinson Foundation and committed up to $150 million over a ten-year period to the MLB Players Alliance, a nonprofit organization for Black MLB players (The Institute for Diversity & Ethics in Sport, 2021b).

Notice how a number of the initiatives described in this section focus on providing either resources in support of or an outlet for those on the margins to be heard (e.g., Black athletes). It is crucial to center the voices of historically excluded and marginalized groups when creating the infrastructure for diversity, equity, and inclusion because they have often experienced first-hand the barriers described in the previous section. Only when their voices are institutionalized can diversity, equity, and inclusion efforts be most effective.

Anchor Diversity, Equity, and Inclusion in Practices, Policies, and Procedures

As part of building the infrastructure for a diverse, equitable, and inclusive sport governance organization, it is crucial to review how current practices, policies, and procedures may support or counteract structural diversity, equity, and inclusion efforts. One of the most public examples of how policy can create inequities is what has been going on in the world of US soccer. For decades, the men's national team was paid more money than the women's national team, despite the latter being at the top of their game. This issue was finally resolved in 2022 when the US women's national soccer team agreed to a new collective bargaining agreement with the US Soccer Federation after six years of fighting for equal pay (Hernandez, 2022). Similarly, you may remember the backlash the NCAA faced when videos documenting the great inequities between the men's and women's Final Four tournaments went viral on social media. Let's review how organizations like the US Soccer Federation and the NCAA can anchor their commitment to diversity, equity, and inclusion in their organizational practices, policies, and procedures:

- **Review (and Rework) Organizational Fabric.** Diversity, equity, and inclusion should take center stage in the organizational fabric. For example, mission statements, vision, and core values should communicate a clear commitment to diversity, equity, and inclusion – so that these organizations can be held accountable when they fall short on demonstrating their commitment to these areas. Similarly, policies need to be reviewed on a regular basis to make sure they do not constitute barriers to diversity, equity, and inclusion.

- **Internal and External Communication.** Do an organization's external communication materials show diversity in pictures in brochures, on its website, social media sites, and so on? Do internal documents use inclusive language? Do organizations acknowledge how both their internal (e.g., staff) and external stakeholders (e.g., fans) may be impacted by issues happening in society? Do marketing materials provide visibility to people and groups advancing diversity, equity, and inclusion? Are those in charge of communicating the organization's values internally and externally a diverse group? These are important questions to ask when it comes to communicating organizational commitment to diversity, equity, and inclusion. For the 2022 Beijing Winter Paralympic Games, the UK's TV station Channel 4 took an unprecedented step to demonstrate such a commitment: The entire team in front of the camera consisted of people with physical disabilities – a global first!

- **Accountability Mechanisms.** Policy rarely changes behaviors, but it creates a mechanism for accountability. Accountability, especially for those leading sport governance organizations, is crucial in advancing diversity, equity, and inclusion. For example, in one recommendation for transforming the US Olympic and Paralympic Movements, the Team USA Council on Racial and Social Justice called on the USOPC and NGBs to "establish a culture of accountability handling issues of access, diversity, equity, and inclusion (e.g., via a restorative justice philosophy), which holds members of the organization accountable for use of language and practices that exclude and harm (e.g., hate speech)" (USOPC, 2021, para. 12). Tying diversity, equity, and inclusion to performance management is one such strategy to create accountability mechanisms (e.g., by having a diversity/equity/ inclusion section in annual performance reviews).

- **Strategic Partnerships and Coalition Building.** No organization can likely do diversity, equity, and inclusion work alone. Strategic partnerships are an important aspect of coalition building for maximum impact. Partnerships with community groups or nonprofit organizations committed to diversity, equity, and inclusion are a good way to support these groups while advancing the organization's diversity, equity, and inclusion initiatives. For example, MLS has partnerships with 100 Black Men of America, the Anti-Defamation League, Athlete Ally, Autism Speaks, the National Coalition of 100 Black Women, Special Olympics Unified Sports, Street Soccer USA, the Women's Sport Foundation, and You Can Play (The Institute for Diversity and Ethics in Sport, 2021c). Strategic partnerships can also include contracting with suppliers for a sport organization, such as companies created for or led by members of historically excluded and marginalized groups (e.g., Black-Owned Businesses). Some major US sport organizations also work with player associations (e.g., National Women's Basketball Players Association) or player collectives (e.g., MLS's Black Players for Change) to provide a collective voice to athletes.

Next, we will turn our attention to strategies for increasing diversity across the organization, an undertaking that must take place at all levels of sport governance but is particularly important at the leadership level.

Increase Diversity across the Organization, with Particular Focus on Leadership

Leaders have the most power when it comes to sport governance, and diversity at the highest level of governance is crucial for having diverse representation, creating equitable policies, and fostering inclusive environments. (Remember: If you can see one – you can be one!) That is why the NFL introduced the Rooney Rule in the early 2000s. The league has refined it significantly since then to extend its reach beyond the leadership ranks of the organization and to provide incentives for hiring more members of historically excluded and marginalized groups. The success of the rule has been called into question by critics, however, because the track record of hiring in the NFL does not show much progress. In 2022, for example, NFL coach Brian Flores filed a class-action lawsuit against the League alleging racial discrimination (Archy, 2022). Flores's lawsuit shows that the NFL has a long way to go for its leadership to truly reflect the demographics of US society. So how can sport governance organizations increase diversity across the board, with a particular focus on the leadership level? Let's look at some strategies below:

- **Hiring Incentives.** The Rooney Rule is not the only example of a policy providing incentives for diversifying staff. Led by Gloria Nevarez, the first Latinx commissioner at the Division I level, the NCAA's West Coast Conference established the "Russell Rule" in 2020. Named after lifelong social justice activist and NBA champion Bill Russell, the rule "requires each member institution to include a member of a traditionally underrepresented community in the pool of final candidates for every athletic director, senior administrator, head coach and full-time assistant coach position in the athletic department" (West Coast Conference, 2020, para. 2). A year after the introduction of the rule, the conference shared some initial successes: 52.4% of hires throughout the year were from traditionally underrepresented groups, and 96.4% of the total hires met the Russell Rule hiring commitment (West Coast Conference, 2022).
- **Pipeline Programs.** Leaders in sport governance know that the best way to diversify future leadership is to start diversifying at the

entry level. That's why quite a few sport organizations established pipeline programs aimed at helping members of historically excluded and marginalized groups enter the organization (and industry at large). Among these programs are the MLB Diversity Fellowship and Diversity in Ticket Sales Training programs, NASCAR's Diversity Internship and Drive for Diversity programs, or the NBA's HBCU Fellowship Program, aimed at increasing career opportunities for HBCU graduates.

■ **Hiring Guides and Retention Initiatives.** Beyond incentives and pipeline programs, organizations should provide resources focused on inclusive hiring and retention. For example, providing search committees with bias training and an inclusive hiring guide can help mitigate any potential harmful biases members may bring to the table. It may also help diversify applicant pools by developing a comprehensive outreach strategy to reach members of historically excluded and marginalized groups. While a great deal of effort tends to focus on recruiting for diversity, equal attention (if not more) should be paid to retention. Members of historically excluded and marginalized groups will only stay with an organization if they feel like they belong and can be their authentic selves, so intentional initiatives to both recruit and retain diverse talent are crucial.

■ **Professional Development and Mentoring Programs.** One way to retain talent is by providing professional development and mentoring programs. These are particularly important for members of historically excluded and marginalized groups who are often entering potentially harmful spaces and run the danger of being tokenized. These programs can connect staff with other members of the organization holding similar identities or utilize external offerings for professional development. The NCAA's Leadership Development department, for example, offers multiple opportunities for historically excluded and marginalized groups to advance in their careers. One such program is the Dr. Charles Whitcomb Leadership Institute, which "provides tailored programming to assist ethnic minorities in strategically mapping and planning their careers in athletics administration by providing professional development programming over the course of a calendar year" (NCAA, 2022a, para. 1).

Diversity can only thrive if the environment it is situated in is inclusive. People governing sport must pay attention to the environments they create, which should always aim at fostering communities of belonging.

Facilitate Communities of Belonging

We close this chapter with a keyword that we have woven consistently throughout the past pages: belonging. Inclusive cultures are ones where everyone can belong, be their authentic selves, and build community. Sport governance organizations need to be intentional about creating communities of belonging for members of historically excluded and marginalized groups. Some strategies to accomplish this are listed below:

- **Employee Engagement Groups:** To create community among historically excluded and marginalized groups and provide resources, sport organizations often create Employee Engagement Groups (sometimes called Employee Resource Teams or Affinity Groups) focused on a shared identity or experience. For example, the NBA offers seven Employee Resource Groups (with group focus in parentheses): the NBA Women's Network (women), Conexión éne-bé-a (Latinx/Hispanic), NBA Pride (LGBTQ+ community), Dream in Color (Black employees), Asian Professional Exchange (Asian and Asian American culture awareness), NBA Young Professionals (early career staff), and NBA Voices of Employees that Served (VETS; veterans). These groups are usually open to the target population and its allies, and they can provide a crucial space for community building.

- **Cultural Celebrations:** Celebrating diversity is an important aspect to creating inclusive environments. That is why sport organizations should celebrate the various identities and cultures. Often, these cultural celebrations are a sign of appreciation of the value diversity brings to our lives. They often take place during larger month-long celebratory initiatives (e.g., Black History Month, Women's History Month, Pride Month). For example, 65 teams of Minor League Baseball (MiLB) staged Pride Nights during Pride Month (June) in 2019, which became "the largest documented Pride celebration in professional sports" (Hill, 2019, para. 4). MiLB has additional initiatives for celebrating cultural backgrounds and social identities. For example, the league annually holds the Copa de la Diversión initiative, where teams adopt culturally-relevant personas for a series of games each season, in an effort to connect the sport and its teams to Hispanic and Latinx communities. However, such celebration should not only be tied to one event or month but rather organizations should engage in ongoing efforts of celebrating cultural differences.

www

Minor League Baseball's
Copa de la Diversión
https://www.milb.com/
fans/copa

- **Trainings, Awareness Campaigns, and Education:** An organization committed to fostering belonging needs to provide sufficient trainings and educational resources for its members to learn how they can each contribute to creating an inclusive environment. Such trainings can include anti-bias trainings (check out the free Harvard Implicit Association tests!), guest speakers, broader diversity/equity/inclusion trainings, and trainings focused on supporting specific groups within an organization (e.g., Black employees). For example, the NCAA Office of Inclusion recently released a set of strategies for addressing racial injustice (NCAA, 2022b) and holds an annual diversity and inclusion social media campaign. Awareness campaigns can be powerful tools for education, but they are only effective if they are followed by action. For instance, the Premier League's No Room for Racism initiative uses awareness via Public Service Announcements, instructions on how to report acts of racism, educational resources, and action plans to tackle racism within the league.

■ **Industry and/or Community Resources.** No sport governance organization exists in a vacuum, and the broader community the organization is housed in can provide valuable resources for community building. Many of the biggest sport governing organizations are headquartered in major cities with rich diversity around the globe (IOC in Lausanne, Switzerland; FIFA in Zurich, Switzerland; NFL, NBA, NHL, MLS, and WNBA in New York City, United States). Organizations need to be intentional about connecting their staff, especially those from historically excluded and marginalized groups, with community resources or groups within their immediate surroundings (e.g., by hosting resource fairs). The COVID-19 pandemic has also forced organizations to find ways to build community virtually. For example, a group of diversity, equity, and inclusion professionals recently launched the Diversity, Inclusion, and Equity Council of Excellence (DIECE) as a platform to connect sport professionals with a commitment to inclusive excellence.

WWW
No Room for Racism
(Premier League)
https://www.premierleague.com/NoRoomForRacism

We can likely all think of a time when we felt like we did not belong. These feelings are more common among those of you with historically excluded and marginalized identities, who have to navigate institutions permeated by systemic racism, sexism, heterosexism, ableism, or other forms of systemic injustice. The strategies above are a starting point to nurture a stronger sense of belonging for everyone.

WWW
Diversity, Inclusion
and Equity Council of
Excellence (DIECE)
https://www.dieceathletics.com

Throughout this section, we used a variety of examples from sport governance organizations to illustrate effective strategies to promote diversity, equity, and inclusion. The fact we chose different sport organizations throughout rather than just one is no accident. In fact, no one sport organization gets diversity, equity, inclusion work one hundred percent right (some are close, like the WNBA!), and there are no cookie-cutter approaches to creating diverse, equitable, and inclusive environments. Instead, working toward diversity, equity, and inclusion requires regular organizational assessments, continuous strategic planning, and intentional action. And, of course, it requires that all of us do our part in making sport governance a more inclusive, equitable, and diverse space.

SUMMARY

Matters of diversity, equity, and inclusion have become increasingly visible in the governance of sport. These matters deal with the demographic make-up of groups (diversity), the fairness (or lack thereof) in treatment, access, and outcomes (equity), and the creation of environments where everyone can belong and be their authentic selves (inclusion). Global calls for racial justice in 2020 and beyond, in addition to the inequities revealed by the COVID-19 pandemic, have shown that promoting diversity, equity, and inclusion is everyone's responsibility. An ethical case can be made for it – it's the right thing to do. In contrast to those holding privileged identities, persistent systemic barriers have made it harder for members of historically excluded and marginalized groups to enter, advance in, and find spaces of belonging and equitable treatment in sport governance. These groups include women, People of Color, members of the LGBTQ+ community, and people with disabilities (among others, of course). These groups face barriers on a variety of levels, from structural barriers (a lack of representation or resources) and cultural barriers (a lack of appreciation for diversity, equity, and inclusion) to conceptual barriers (a lack of institutional history of doing effective diversity, equity, and inclusion work within the organization) and emotional, social, and relational barriers (burnout and emotional fatigue).

Intentional, systematic, and strategic action is required at various levels of the sport governance process to remove these barriers. Those in leadership positions are particularly important in driving diversity, equity, and inclusion action at both individual and institutional levels. Diversity, equity, and inclusion can be advanced in sport governance by many strategies: conducting regular diversity, equity, and inclusion audits; developing strategic plans; creating the infrastructure for diversity, equity, and inclusion within the governance structures (e.g., committees, resources); anchoring diversity, equity, and inclusion in the organization's practices, policies, and procedures; increasing diversity across the organization, and especially the senior leadership level; and facilitating communities of belonging. Diversity, equity, and inclusion is a team effort – it requires all of us to work toward a more just sport industry.

caseSTUDY

THE WNBA – LEADING THE WAY

A significant portion of this chapter focuses on what is *wrong* with sport organizations when it comes to governing with a focus on diversity, equity, and inclusion. This case study will require you to look at an organization that has long led the way in modeling inclusive excellence, social justice leadership, and diverse representation across the organization. You guessed it: the WNBA. For the annual ESPY awards, you are asked to write a documentary feature illustrating the league's leadership in promoting diversity, equity, and inclusion (and, more broadly, social justice) in the United States and beyond. What aspects of the league's efforts will you showcase to a global audience? You may want to look at any of the following aspects, but this list is by no means exhaustive – feel free to be creative!

- Diversity and hiring practices within the league, league leadership, and key stakeholder groups (check out the WNBA TIDES report card for the most recent numbers!);

- Policies, practices, and/or procedures for increasing diversity, equity, and inclusion;
- Proactive diversity, equity, and inclusion initiatives within the league;
- Political organizing, social movement engagement, and athlete activism;
- Trailblazers within the league, across identity categories;
- Infrastructure for anchoring diversity, equity, and inclusion in governance; and
- Strategic partnerships.
- Who are some athletes you might want to feature and why?
- Who are some league executives you might want to feature and why?

CHAPTER questions

1. Explain in your own words why diversity, equity, and inclusion are everyone's responsibility. What can different stakeholders within sport governance (CEOs, coaches, staff, interns, etc.) do to remove some of the barriers to diversity, equity, and inclusion we identified in this chapter? Who has the most power in driving change and why?

2. Imagine you were to be hired as a diversity, equity, and inclusion consultant for one of the sport governance organizations below.
   - Angel City Football Club
   - National Federation of State High School Associations
   - National Hockey League
   - USA Fencing
   - University of Washington Athletic Department
   - Fédération Internationale de Natation/International Swimming Federation (FINA)

   Map out a diversity, equity, and inclusion evaluation for the organization. Answer the following questions:

   a. When it comes to diversity, equity, and inclusion, what are two of the organization's strengths and two areas for improvement?

   b. Across the organization and its key stakeholders, what groups can be considered privileged (overrepresented) and excluded or marginalized (underrepresented)? Are there imbalances in representation at the different levels? How diverse is the organizational leadership team?

   c. What governance structures are in place to advance diversity, equity, and inclusion?

    d. What diversity, equity, and inclusion initiatives does the organization engage in, both internally (e.g., employee resource groups) and externally (e.g., cultural celebrations), to foster a more diverse workforce, equitable experience, and inclusive environment?

    e. How can the organization measure success when it comes to diversity, equity, and inclusion?

3. Throughout the chapter, we have focused on four historically excluded groups in sport governance: women; Black, Indigenous, and People of Color; LGBTQ+ individuals; and people with disabilities. What challenges may members of other historically excluded and marginalized groups face (see Exhibit 3.6), including those holding multiple of these identities (e.g., Black women with disabilities)? How can sport managers meet their needs when it comes to diversity, equity, and inclusion?

4. Look up how your favorite sports team or league has responded to the growing calls for social justice rocking the sports industry in summer of 2020. You can look at a high school team, recreational league, college team, pro-sports team or league, national sports governing body, or international sport organization.

    a. How has the organization changed since then, especially when it comes to diversity, equity, and inclusion?

    b. Was the organization's response substantial or more "just for show" in nature? How so?

    c. What specific individual and institutional strategies for advancing diversity, equity, and inclusion has the organization engaged in?

    d. How can the organization anchor diversity, equity, and inclusion more in its governance structures, policies, and/or organizational praxis?

FOR ADDITIONAL INFORMATION

1. Check out these two case studies focusing on diversity, equity, and inclusion initiatives within the NCAA:

    Kluch, Y., & Rentner, T. (2022). "As Falcons, We Are One Team!" Launching a grassroots institutional change initiative to promote diversity and inclusion through sport at a NCAA Division I institution. *Sport Management Education Journal, 16*(1), 95–104. https://doi.org/10.1123/smej.2020-0050

    Kluch, Y., & Wilson, A. (2020). #NCAAInclusion: Using social media to engage student-athletes in strategic efforts to promote diversity and inclusion. *Case Studies in Sport Management*, 33–43. https://doi.org/10.1123/cssm.2019-0027

2. Want to learn how to be an antiracist in sport? Check out these helpful resources: https://antiracisminsport.ca/resources/

3. Kyle Korver's powerful essay "Privileged," where he reflects on the privilege attached to whiteness in the United States: https://www.theplayerstribune.com/articles/kyle-korver-utah-jazz-nba

4. Learn how to support transgender individuals in (and beyond) sport at: www.transathlete.com

5. LGBTQ+ individuals continue to struggle for visibility in sports, but this website is dedicated to highlighting their stories: www.outsports.com

6. The National Association of Diversity Officers in Higher Education (NADOHE) recently released a framework for advancing anti-racism on campuses: https://nadohe.memberclicks.net/assets/2021/Framework/National%20Association%20of%20Diversity%20Officers%20in%20Higher%20Education%20-%20Framework%20for%20Advancing%20Ant-Racism%20on%20Campus%20-%20first%20edition.pdf

7. For social justice-focused work in sport governance, visit the websites of the following institutes and centers:

   ■ Centre for Social Justice in Sport and Society, Leeds Beckett University (UK): https://www.leedsbeckett.ac.uk/research/centre-of-social-justice-in-sport-and-society/

   ■ Centre for Sport and Human Rights (Switzerland): https://www.sporthumanrights.org

   ■ Institute for Sport and Social Justice (US): https://sportandsocialjustice.org

   ■ Institute for the Study of Sport, Society, and Social Change (ISSSC), San Jose State University (US): https://www.sjsu.edu/wordstoaction/

   ■ Tucker Center for Research on Girls & Women in Sport, University of Minnesota (US): https://www.cehd.umn.edu/tuckercenter/

8. Interested in learning more about the WNBA's efforts for social justice? Check out this The Undefeated feature on the evolution of the league's efforts: https://www.youtube.com/watch?v=JgjAxKHEDZc&t=3s

REFERENCES

Adriaanse, J. (2016). Gender diversity in the governance of sport associations: The Sydney scoreboard global index of participation. *Journal of Business Ethics, 137*, 149–160. https://doi.org/10.1007/s10551-015-2550-3

Archy, A. (2022, April 8). *2 former coaches join Flores in his discrimination suit against the NFL.* NPR. https://www.npr.org/2022/04/08/1091559405/coaches-join-nfl-discrimination-lawsuit

Barnhill, C.R., Czekanski, W.A., & Pfleegor, A.G. (2018). Getting to know our Students: A snapshot of Sport Management students' demographics and career expectations in the United States. *Sport Management Education Journal, 12*(1), 1–14. https://doi.org/10.1123/smej.2015-0030

Brooks, S.N., & McKail, M.A. (2008). A theory of the referred Worker: A structural explanation for Black male dominance in basketball.

Critical Sociology, 34(3), 369–387. https://doi.org/10.1177/0896920507088164

Cooky, C., Council, L.D., Mears, M.A., & Messner, M.A. (2021). One and done: The long eclipse of women's televised sports, 1989–2019. *Communication & Sport, 9*(3), 347–371. https://doi.org/10.1177/21674795211003524

Cunningham, G.B., Sagas, M., Dixon, M., Kent, A., & Turner, B.A. (2005). Anticipated career satisfaction, affective occupational commitment, and intentions to enter the sport management profession. *Journal of Sport Management, 19,* 43–57. https://doi.org/10.1123/jsm.19.1.43

Deo, M.E. (2021). Why BIPOC fails. *Virginia Law Review Online, 107,* 115–142. https://www.virginialawreview.org/wp-content/uploads/2021/06/Deo_Book_107.pdf

Emory University. (2022). *Common terms.* https://equityandcompliance.emory.edu/resources/self-guided-learning/common-terms.html

Froehle, C. (2016, April 14). *The evolution of an accidental meme.* Medium.com. https://medium.com/@CRA1G/the-evolution-of-an-accidental-meme-ddc4e139e0e4#.pqiclk8pl

Hall, S.H. (2021, November 30). *9 ways to measure the success of your DEI strategy in 2022.* https://seniorexecutive.com/ways-to-measure-the-success-of-your-dei-strategy-in-2022/

Hancock, M.G., Darvin, L., & Walker, N.A. (2018). Beyond the glass ceiling: Sport management students' perceptions of the leadership labyrinth. *Sport Management Education Journal, 12*(2), 100–109. https://doi.org/10.1123/smej.2017-0039

Hancock, M.G., & Hums, M.A. (2011). *If you build it, will they come?* Paper presented at the Proceedings of the North American Society for Sport Management Twenty-Sixth Annual Conference.

Harris, K.F., Grappendorf, H., Aicher, T.J., & Veraldo, C.M. (2015). "Discrimination? Low pay? Long hours? I am still excited:" Female sport management students' perceptions of barriers toward a future career in sport. *Advancing Women in Leadership, 35,* 12–21.

Harrison, D.A., & Klein, J.K. (2007). What's the difference? Diversity constructs as separation, variety, or disparity in organizations. *Academy of Management Review, 32*(4), 1199–1228. https://doi.org/10.5465/amr.2007.26586096

Harvard University. (2022). *Glossary of diversity, inclusion and belonging (DIB) terms.* https://edib.harvard.edu/files/dib/files/dib_glossary.pdf

Hernandez, J. (2022, Mary 18). *The U.S. men's and women's soccer teams will be paid equally under a new deal.* NPR. https://www.npr.org/2022/05/18/1099697799/us-soccer-equal-pay-agreement-women?t=1658667385089

Hill, B. (2019, June 6). *Out at home: Pride Nights across the minors.* MiLB.com. https://www.milb.com/news/minor-league-baseball-teams-stage-lgbtq-pride-nights-307738616

Hoffman, J. (2010). The dilemma of the Senior Woman Administrator role in intercollegiate athletics. *Journal of Issues in Intercollegiate Athletics, 3,* 53–75.

Human Rights Campaign. (2017). *Play to win: Improving the lives of LGBTQ youth in sports.* https://hrc-prod-requests.s3-us-west-2.amazonaws.com/files/assets/resources/PlayToWin-FINAL.pdf

Jolly, S., Cooper, J.N., & Kluch, Y. (2021). Allyship as activism: Advancing social change in global sport through transformational allyship. *European Journal for Sport and Society, 18*(3), 229–245. https://doi.org/10.1080/16138171.2021.1941615

Jones, J.M. (2021, February 24). *LGBT identification rises to 5.6% in latest U.S. estimate.* Gallup. https://news.gallup.com/poll/329708/lgbt-identification-rises-latest-estimate.aspx

Kendi, I.X. (2019). *How to be an antiracist.* One World.

Kluch, Y., Anderson, S., & Ferguson, T. (2023). How can organizations better support athletes? A case study of the impact of COVID-19 on minoritized communities in intercollegiate sport. In T. Rentner & D. Burns (Eds.), *Social issues in sport communication: You make the call.* (pp. 300–301). Routledge.

Kluch, Y., Wright-Mair, R., Swim, N., & Turick, R. (2022). "It's like being on an island by yourself": Diversity, equity, and inclusion administrators' perceptions of barriers to diversity, equity, and inclusion work in intercollegiate athletics. *Journal of Sport Management* (ahead of print). https://doi.org/10.1123/jsm.2021-0250

Krane, V., & Barber, H. (2005). Identity tensions in lesbian intercollegiate coaches. *Research Quarterly for Exercise and Sport, 76*(1), 67–81.

Lapchick, R. (2016). *Gender Report Card: 2016 International Sports Report Card on Women Leadership Roles.* The Institute for Diversity and Ethics in Sport. https://www.tidesport.org/_files/ugd/7d86e5_47dc0dc55b294fe185d1392e676b6a51.pdf

LeBlanc, C. (2021, December 14). *When it comes to supporting protests, most teams are bad sports.*

https://www.fatherly.com/news/professional-sports-teams-george-floyd-racism-police-brutality-response

Lipson, S.K., Raifman, J., Abelson, S., & Reisner, S.L. (2019). Gender minority mental health in the U.S.: Results of a national survey on college campuses. *American Journal of Preventive Medicine, 57*(3), 293–301. https://doi.org/10.1016/j.amepre.2019.04.025

McDowell, J., & Carter-Francique, A. (2017). An intersectional analysis of the workplace experiences of African American female athletic directors. *Sex Roles, 77*, 393–408. https://doi.org/10.1007/s11199-016-0730-y

Melton, E.N., & Cunningham, G.B. (2014). Examining the workplace Experiences of sport employees who are LGBT: A social categorization theory perspective. *Journal of Sport Management, 28*(1), 21–33. https://doi.org/10.1123/jsm.2011-0157

Miller, K. (2020, April 26). *Using the power in your privilege.* Rise to Win. https://risetowin.org/perspectives/2020/4-26/using-the-power-in-your-privilege/index.html

Misener, L., & Darcy, S. (2014). Managing disability sport: From athletes with disabilities to inclusive organisational perspectives. *Sport Management Review, 17*, 1–7. http://dx.doi.org/10.1016/j.smr.2013.12.003

National Football League. (2021). *2021 NFL diversity and inclusion report.* https://operations.nfl.com/media/4989/nfl-occupational-mobility-report-volume-x-february-2021.pdf

National Football League. (2022). *The Rooney Rule.* https://operations.nfl.com/inside-football-ops/diversity-inclusion/the-rooney-rule/

National Wheelchair Basketball Association. (2021). *NWBA DE&I action plan, 2021–2024.* https://cdn1.sportngin.com/attachments/document/5c4c-2718342/2021-2024_NWBA_DEI_Action_Plan.pdf#_ga=2.54744035.993197177.1658581724-1191124138.1658581703

NCAA. (2019). *Division III moves to gender-neutral language.* https://www.ncaa.org/news/2019/1/23/division-iii-moves-to-gender-neutral-language.aspx

NCAA. (2022a). *Dr. Charles Whitcomb Leadership Institute.* https://www.ncaa.org/sports/2013/11/21/dr-charles-whitcomb-leadership-institute.aspx

NCAA. (2022b). *Strategies: Addressing racial injustice.* https://ncaaorg.s3.amazonaws.com/inclusion/ethnic/INC_AddressingRacialInjustice.pdf

NCAA Office of Inclusion. (2022). *NCAA inclusion statement.* https://www.ncaa.org/sports/2016/3/2/ncaa-inclusion-statement.aspx

Nielsen Sports. (2018). *The rise of women's sports: Identifying and maximizing the opportunity.* https://nielsensports.com/wp-content/uploads/2021/01/Rise-of-Womens-Sports-1.pdf

Özbilgin, M., Gulce Ipek, A.T., & Sameer, M. (2014). *The business case for diversity management.* Association of Chartered Certified Accountants & the Economic and Social Research Council. https://www.hrpsor.hr/wp-content/uploads/2020/02/pol-tp-tbcfdm-diversity-management_20161.pdf

Portland Trail Blazers. (2021). *DEI strategic plan: 2021–2024.* https://www.ripcityresource.com/pdf/21_DEI_Strategic-Report_v4.pdf

Rowan University Division of Diversity, Equity and Inclusion. (2020). *The language of identity.* https://sites.rowan.edu/diversity-equity-inclusion/_docs/the-language-of-identity-080720.pdf

Simpkins, E., Velija, P., & Piggott, L. (2022). The sport intersectional model of power as a tool for understanding intersectionality in sport governance and leadership. In P. Velija, & L. Piggott (Eds.), *Gender equity in UK sport leadership and governance* (pp. 37–50). Emerald Publishing Limited.

Sport England & UK Sport. (2019). *Diversity in sport governance: Annual survey 2018/19.* https://www.uksport.gov.uk/-/media/files/resources/executive-summary---diversity-in-sport-governance-report-final.ashx

Strayhorn, T.L. (2012). *College students' sense of belonging: A key to educational success for all students.* Routledge.

Sue, D.W. (2010). *Microaggressions in everyday life: Race, gender, and sexual orientation.* John Wiley & Sons.

Taylor, E. A., Smith, A. B., Welch, N. M., & Hardin, R. (2018). "You should be flattered!": Female sport management faculty experiences of sexual harassment and sexism. *Women in Sport and Physical Activity Journal, 26*(1), 43–53. https://doi.org/10.1123/wspaj.2017-0038

The Institute for Diversity and Ethics in Sport. (2021a). *2021 Racial and Gender Report Card.* https://www.tidesport.org/_files/ugd/403016_ede01db0e78446e7960974504587709f.pdf

The Institute for Diversity and Ethics in Sport. (2021b). *The 2021 Racial and Gender Report Card: Major League Baseball.* https://www.tidesport.org/_files/ugd/403016_5c311ff6920442b780924552fd8fdc88.pdf

The Institute for Diversity and Ethics in Sport. (2021c). *The 2021 Racial and Gender Report Card: Major League Soccer.* https://www.tidesport.org/_files/ugd/138a69_eb4dd72c61624316aab09e4779ddcbcc.pdf

The Institute for Diversity and Ethics in Sport. (2021d). *The 2021 Racial and Gender Report Card: National Basketball Association.* https://www.tidesport.org/_files/ugd/138a69_4b2910360b754662b5f3cb52675d0faf.pdf

The Institute for Diversity and Ethics in Sport. (2021e). *The 2021 Racial and Gender Report Card: National Football League.* https://www.tidesport.org/_files/ugd/326b62_5afc0093dedf4b53bdba964fa0c1eb0c.pdf

The Institute for Diversity and Ethics in Sport. (2022). *The Racial and Gender Report Card.* https://www.tidesport.org/racial-gender-report-card

Turner Consulting Group. (2022). *Racism in the hiring process.* https://twitter.com/DiversityMusing/status/1529928132576395265

UC Berkeley Strategic Plan for Equity, Inclusion, and Diversity. (2009). *Executive summary.* https://diversity.berkeley.edu/sites/default/files/executivesummary_webversion.pdf

United Nations. (2022). *Disability and sports.* https://www.un.org/development/desa/disabilities/issues/disability-and-sports.html

United States Census Bureau. (2022). *Race and ethnicity in the United States: 2010 census and 2020 census.* https://www.census.gov/library/visualizations/interactive/race-and-ethnicity-in-the-united-state-2010-and-2020-census.html

United States Government Accountability Office. (2010). *Students with disabilities: More information and guidance could improve opportunities in physical education and athletics. Report to Congressional Requesters.* http://www.gao.gov/assets/310/305770.pdf

USOPC. (2021). *Team USA Council on Racial and Social Justice releases third recommendation on institutional awareness and cultural change.* https://www.teamusa.org/Media/News/USOPC/100721-Team-USA-Council-on-Racial-and-Social-Justice-Releases-Third-Recommendation

USOPC. (2022). *Diversity, equity & inclusion scorecards.* https://www.teamusa.org/About-the-USOPC/Diversity-Equity-Inclusion/D-and-I-Scorecards

Vaccaro, A., & Newman, B.M. (2016). Development of a sense of belonging for privileged and minoritized students: An emergent model. *Journal of College Student Development, 57*(8), 925–942. https://doi.org/10.1353/csd.2016.0091

Walker, N.A., Agyemang, K.J., Washington, M., Hindman, L.C., & MacCharles, J. (2021). Getting an internship in the sport industry: The institutionalization of privilege. *Sport Management Education Journal, 15*(1), 20–33. https://doi.org/10.1123/smej.2019-0061

Waterhouse, P., Kimberley, H., Jonas, P., & Glover, J. (2010). *What would it take? Employer perspectives on employing people with a disability.* National Centre for Vocational Education Research. https://www.ncver.edu.au/research-and-statistics/publications/all-publications/what-would-it-take-employer-perspectives-on-employing-people-with-adisability

West Coast Conference. (2020, August 3). *Russell Rule diversity hiring commitment.* https://wccsports.com/news/2020/8/2/general-russell-rule-diversity-hiring-commitment.aspx

West Coast Conference. (2022, March 16). *TIDES WCC Racial and Gender Report Card tracks diversity hiring commitment.* https://wccsports.com/news/2022/3/16/general-tides-wcc-racial-and-gender-report-card-tracks-diversity-hiring-commitment.aspx

WNBPA. (2022). *Mission statement.* https://wnbpa.com/about/mission-statement/

Zweigenhaft, R.L. (2021). *Diversity among Fortune 500 CEOs from 2000 to 2020: White women, hi-tech South Asians, and economically privileged multilingual immigrants from around the world.* https://whorulesamerica.ucsc.edu/power/diversity_update_2020.html

Ethics in Sport Organizations

Sales representatives alter receipts to get more money than they are entitled to. Managers lie to their bosses about using company-owned cars. Bosses lie to their employees about company policies. Politicians make false claims in order to rile up their base. Accountants alter the books to cover up questionable spending practices. Major corporations are forced to close down because of income mismanagement.

DOI: 10.4324/9781003303183-4

Large manufacturers violate the human rights of their workers. Wealthy government officials ignore the needs of their lower-income–level constituents to line their pockets with financial support from political action committees. Organizations use production processes and build facilities that are not environmentally friendly. Such negative news from the corporate and political worlds call into question the ethics we see practiced in business and industry on a daily basis.

It is of the utmost importance that leaders in sport organizations behave in an ethical manner. Leaders set the tone for their sport organizations. We need leaders with formal authority in sport organizations, but more so, according to Dov Seidman, author of the book *How* and CEO of LRN (quoted in Friedman, 2017), what really makes any organizational system work, is

> "when leaders occupying those formal positions—from business to politics to schools to sports—have moral authority. Leaders with moral authority understand what they can demand of others and what they must inspire in them. They also understand that formal authority can be won or seized, but moral authority has to be earned every day by how they lead. And we don't have enough of these leaders." In fact, we have so few we've forgotten what they look like. Leaders with moral authority have several things in common, said Seidman: "They trust people with the truth—however bright or dark. They're animated by values—especially humility—and principles of probity, so they do the right things, especially when they're difficult or unpopular."
>
> (paras. 13–14)

The world of sport is a place where we want to believe in fair play and good conduct. But is this world somehow immune to the ethical issues confronting managers and business people in general society? Unfortunately, the answer is "No," as illustrated by the following examples:

- National Football League (NFL) teams fail to hire qualified minority head coaches.
- National Collegiate Athletic Association (NCAA) football players participate in sexual violence and their coaching staff and university disregard the victims.
- Major League Baseball (MLB) players use banned substances to improve performance.
- A state high school athletic association policy on transgender athletes forces those athletes into competing under their incorrect gender.
- Fans shout racial slurs at opposing players.

- International officials accept bribes to swing the votes for the selection of a host city for the World Cup.

- Investigations reveal state-sponsored systematic doping programs which protect "dirty" athletes.

- A well-known international gymnastics coach is accused of sexual misconduct involving athletes.

Sadly, these types of incidents appear with regularity in the sports pages and are often the lead stories on the evening sports broadcasts. All the above scenarios involve behavior that is considered unethical. As sport managers, you will confront situations that will present you with ethical dilemmas. How, then, should you respond to them?

SPORT AS A MIRROR OF SOCIETY

Every society faces its own unique issues, including violence, substance abuse, domestic abuse, racism, sexism, homophobia, ableism, economic downturns, differential treatment based on religion, bullying, and corporate cheating. These issues also appear in all levels and facets of industry, including the sport industry. It has been said that sport is a mirror, a microcosm, or reflection of society, not just in the United States, but other nations as well (Eitzen & Sage, 2009; Gargan, 2015; Gibbs, 2016; Maguire & Nakayama, 2006). It should come as no surprise, therefore, that sport managers face the same issues.

According to DeSensi and Rosenberg (2010), the following are ethical considerations for sport managers:

- professionalism
- equity
- legal and financial management
- personnel concerns
- governance and policies
- league and franchise issues
- matters of social justice (p. 2)

Notice that one of these points mentions governance. Many of the ethical considerations listed above have obvious parallels as societal issues. So how do these societal issues manifest themselves in sport? The simple diagram in Exhibit 4.1 illustrates how societal issues are reflected in sport.

The first example depicted in the model is social activism. People as a whole are becoming more aware of social issues present in society and are making public statements about them. Look no further than the Black Lives Matter movement or the 2017 Women's March in Washington, DC. Sport figures are now standing up for just causes as well. While the most notable athlete to take this brave step was the great Muhammad Ali, others have since followed suit. NFL quarterback Colin Kaepernick led the way with his decision to kneel during the US national anthem as a statement against the treatment of African Americans. Megan Rapinoe and the entire US Women's National Soccer Team challenged society on gender equity in and out of sport and recently won a multi-million dollar equal pay settlement. Hudson Taylor, a former Maryland University wrestler, founded Athlete Ally as a group to stand up for the rights of the lesbian, gay, bisexual, trans, and queer community (Athlete Ally, 2020). Athletes are finding their voice and realizing the powerful platform they have to exert influence in society.

The second example shown in Exhibit 4.1 – racism – still exists in society, as evidenced by the need for the Black Lives Matter movement. In sport, we see the world's most popular sport still subject to intense racism against players. In 2019, international soccer stars Romelu Lukaku, Mario Balotelli, and Paul Pogba all publicly cited acts of racism against them by fans. Balotelli even threated to leave his club game after being targeted by racial abuse by spectators (Lapchick, 2020). In fact, Kick It Out, an English anti-racism and pro-inclusion group, noted that demonstrations of discrimination (i.e., gender, sexual orientation, religion, and race) increased 32% from the 2018 to 2019 season. Of the 422 acts of discrimination, 65% were race-based (Davies, 2020).

www

Kick It Out
https://www.kickitout.org/

The third example illustrated in Exhibit 4.1 is sexual assault, a daily occurrence in society, often depicted by the media on television and in films. Sexual assault, the societal issue, appears on the left-hand side of the model. As the issue is filtered through sport, we see the disaster that was Baylor University football and the systematic cover-up by coaches and administrators to protect 19 players at the expense of the 17 survivors (Schnell, 2017). The most notorious example is the case of Larry Nassar who sexually assaulted over 150 girls and women over two decades as the long-time doctor for USA Gymnastics and Michigan State University Gymnastics. Many of the survivors accused USA Gymnastics, the US Olympic Committee, and Michigan State University of turning a blind eye toward Nassar's abuse after the victims spoke out (Levenson, 2018).

This diagram could include other examples, but its purpose is not to provide a comprehensive list. Rather, its purpose is to ask, "As sport managers, when we are confronted by these issues, how will we respond?" Responding to these incidents is never easy, because sport managers feel pressure from many constituencies to do the right thing. But what is "the right thing," and more important, how should a sport manager decide on carrying out "the right thing?" Remember, too, as mentioned in Chapter 2, the decisions sport managers make will be publicly scrutinized by the media, fans, and casual consumers of the news. As a sport manager, you will be faced with ethical dilemmas on a regular basis. But will you know how to recognize an ethical dilemma?

| Societal Issues Reflected in Sport | *exhibit* | **4.1** |

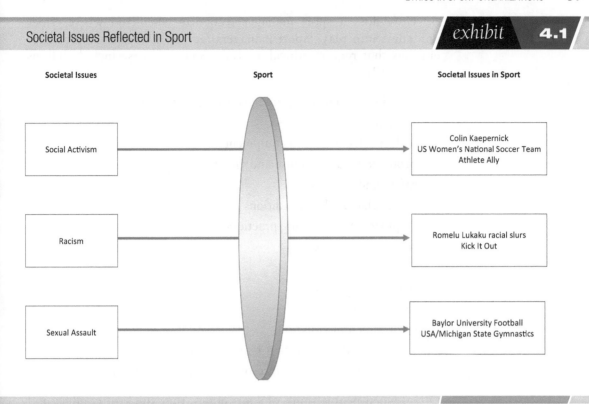

Societal Issues   Sport   Societal Issues in Sport

Social Activism → Colin Kaepernick / US Women's National Soccer Team / Athlete Ally

Racism → Romelu Lukaku racial slurs / Kick It Out

Sexual Assault → Baylor University Football / USA/Michigan State Gymnastics

ETHICS DEFINED

Where do ethical codes in sport management emanate from, and who is responsible for ethical behavior? The United Nations Educational, Scientific and Cultural Organization (UNESCO) (n.d.) has a Code of Sports Ethics and states that responsibility for ethical behavior lies with those who are involved with sports for children and young adults, including (a) governments at all levels and the agencies that work with them; (b) sports and sports-related organizations including sports federations and governing bodies; and (c) individuals including parents, teachers, coaches, referees, officials, sport leaders, athletes who serve as role models, administrators, journalists, doctors, and pharmacists. Clearly, sport managers in general and those working specifically in sport governing bodies are responsible for ethical behavior. While this may sound far away from you as you read this book, ultimately it means as a sport manager you will be responsible for ethics in your sport organization. All sport managers are responsible for ethical behavior, and the higher one moves up in the organization the greater that responsibility becomes. As Lussier and Kimball (2009) point out, "An organization's ethics are the collective behavior of its employees. If each employee acts ethically, the actions of the organization will be ethical, too. The starting place for ethics, therefore, is you" (p. 35).

How does a sport manager recognize when and where ethics will come into play? Sport managers face certain issues and considerations that require ethical analysis. Some of these include (Hums et al., 2012):

- non-discrimination/equity in sporting participation
- fair play
- inclusive facilities and equipment
- protection against abuse and violence
- safety and security
- sport-related labor conditions
- employment and hiring practices
- right to due process
- access to information
- freedom of speech
- right to privacy
- environmental violations
- displacement of persons for sporting events
- access and availability of resources/financial spending

Often, when people think of sport ethics, they think of conduct and fair play *on* the field. As stated by the Council of Europe Committee of Ministers (2010), "the ethical considerations that underpin fair play are not an optional element but an essential component to all sporting activities" (p. 1). Fair play extends to the management of sport as well. From the information above, it is obvious that ethics and ethical concerns spill over into the business aspects of sport. The business venue is where sport managers will encounter myriad ethical issues and dilemmas.

ETHICAL DILEMMAS

 hat is an ethical dilemma? According to Mullane (2009), an *ethical dilemma* occurs:

> when important values come into conflict, and the decision maker (the leader, in many cases) must make a choice between these values ... To further complicate things, ethical dilemmas usually involve multiple stakeholders (those affected by the ultimate decision), and the outcome is marred by uncertainty.
>
> (pp. 2–3)

According to Wallace (2012), a manager faces a significant ethical conflict when the following exist: "(1) significant value conflicts among differing interests, (2) real alternatives that are equally justifiable, and (3) significant consequences to stakeholders in the situation" (para. 2). For example, a star athlete is arrested for driving while intoxicated one week before the final game for the league championship. Should the player be allowed to play?

- Are value conflicts present? Some people would say the athlete should not play because of the arrest; others would say he has been arrested but not convicted, so let him play.

- Do real alternatives exist that are equally justifiable? Real alternatives may include playing the athlete, not playing the athlete at all, or limiting playing time.

- Are there significant consequences to stakeholders, including owners, sponsors, or investors? If the athlete does not play and the team loses the championship, significant revenue could be lost; if the athlete does play, negative publicity will be significant.

When faced with an ethical dilemma, what is a sport manager to do? What should guide the sport manager when making a decision about an ethical dilemma like this?

ETHICAL DECISION-MAKING MODELS

As discussed in Chapter 2, a sport manager must have an organized and sequential method for making decisions. It is no different when the problem is ethical in nature. The literature on ethical decision-making is loaded with different models for managers to use. Some very practical models and guidelines have been suggested (McDonald, 2001; Thornton et al., 2012; Zinn, 1993). These models examine a variety of factors and involve multiple steps. For our purposes, we will identify a straightforward model sport managers can apply in the workplace. The model presented below by Hums et al. (1999, p. 64) and Hancock and Hums (2011) is an adaptation of Zinn's model:

1. **Identify the correct problem to solve.** When making any type of decision, the decision maker must first identify the *real* problem. Identifying a symptom of the problem and acting on that will not resolve the problem itself.

2. **Gather all pertinent information.** Good decision makers try to be as informed as possible. Is it realistic to think you can gather every piece of information needed? Probably not, but sport managers need to make a good faith effort to find all the information possible to guide them in their decision-making.

3. **Explore codes of conduct relevant to one's profession or to this particular dilemma.** More and more sport organizations are developing codes of conduct. Take a look at codes from yours and other sport organizations to see if they provide guidance for your decision.

4. **Examine one's own personal values and beliefs.** We all come to the workplace with our own unique sets of values. Be sure you understand your values and how they could impact your decisions. This does not mean every decision has to be in line with your own

values. It means you must be keenly aware of your own values and how they may influence your decisions.

5. **Consult with peers or other individuals in the industry who may have experience in similar situations.** Sport managers throughout the industry are facing increasing numbers of ethical issues. Perhaps some trusted colleagues have faced a similar dilemma. Talk with them to discuss how they went about solving the issue.

6. **List decision options.** Good decision makers learn to look at as many options as possible so they can make the best choice.

7. **Look for a win-win situation if at all possible.** This is a difficult but critical step. Ethical dilemmas arise when there are questions about the right thing to do. Try to make a decision that maximizes the outcome for the parties involved.

8. **Ask the question, "How would my family feel if my decision and how and why I arrived at my decision appeared on social media tomorrow?"** Remember, as a sport manager, your decisions will be publicly analyzed and criticized. Be sure you have done all the right things in making your decision and that there is nothing about your decision you could not be up front about.

9. **Sleep on it.** Do not rush to a decision. In other words, think hard about the situation and the options and consequences you face. You need to make a well-thought-out decision.

10. **Make the best decision possible, knowing it may not be perfect.** At some point, you will have to make your decision. Knowing you have followed the steps listed above will help you reach the best decision possible. In ethical decision-making, reasonable people will often reasonably disagree over decisions.

11. **Evaluate the decision over time.** Often overlooked by managers, it is important to reflect on the decision later to see how it is working, or how changes could be made to improve upon it. This step is especially important if the issue or a similar one arises again.

Another useful decision-making technique for examining ethical decisions is the SLEEPE (Social, Legal, Economic, Ethical, Political, and Educational) Principle (presented in Chapter 2). This model helps the sport manager look at the big picture before making a decision. Ethical decisions, by their very nature, are bound to have far-reaching and complex ramifications. The SLEEPE Principle helps a sport manager think in broad terms about the ramifications of ethical decisions. In addition, this model has ethical considerations already built in as the second *E* in SLEEPE. Regardless of which model a sport manager chooses to use, each provides a structure to help make decisions (Hums, 2006; 2007).

Up to this point, we have concentrated mainly on ethical situations and how individual sport managers will respond to them. Now we must expand that view and look at sport organizations as a whole and their corporate stance on ethical issues. One way to assess the ethical nature of a sport organization is to examine what kind of "citizen" the sport organization represents. This idea of a sport organization as a "citizen" can be looked at through the concept of corporate social responsibility (CSR).

CSR

CSR is a term that often appears in business ethics literature. What does it mean to be a responsible corporate citizen? In general, CSR is defined by the United Nations Industrial Development Organization as:

> a management concept whereby companies integrate social and environmental concerns in their business operations and interactions with their stakeholders. CSR is generally understood as being the way through which a company achieves a balance of economic, environmental and social imperatives ("Triple-Bottom-Line Approach"), while at the same time addressing the expectations of shareholders and stakeholders.
>
> (UNIDO, n.d., para 1)

wwww

United Nations Industrial
Development Organization
https://www.unido.org/

Another definition comes from Archie Carroll, one of the pioneers of CSR research, who defined CSR as: "The economic, legal, ethical, and philanthropic expectations that society has of organizations at a given point in time" (Carroll & Brown, 2018, p. 45). CSR remains hard to define, however. In a review of CSR definitions between 1980 and 2003, a researcher identified 37 different definitions of CSR (Dahlsrud, 2008). Yet, the benefits for organizations that embrace CSR are staggering. CSR has been found to (a) build public trust, (b) enhance positive relationships with the community and consumers, (c) help organizations become more sustainable, (d) increase profits, and (e) encourage professional and personal growth in employees (Lin, 2019).

Sport management researchers have begun to study the topic of CSR in sport (Babiak & Trendafilova, 2011; Babiak & Wolfe, 2009; Bradish & Cronin, 2009) and have called for sport managers to be more aware of the power of sport to do good (Darnell, 2012; Hums, 2010; Hums & Hancock, 2012; Hums & Wolff, 2014; Thibault, 2009). Social responsibility should become part of who you are and how you do your job on a daily basis once you become a sport manager. It helps to have a few ideas on how to make sure your sport organization acts in a socially responsible manner.

How can CSR be measured? There is no singular measure of good citizenship because, by its very nature, it can be ascertained only from the perspectives of multiple stakeholders. One of the earliest and still often cited measurements of CSR comes from Carroll (1991), who explained how corporations can operate at four different CSR levels:

1. economic – focus only on economic concerns
2. legal – follow the letter of the law
3. ethical – follow the spirit of the law
4. philanthropic – act as a leader in promoting CSR

He examined corporations and the levels where they existed relative to various organizational stakeholders, including owners, customers, employees, the community, and the public at large (Carroll, 1991). Carroll created a pyramid where the economic level represents the lowest level of CSR and its base. At its peak is the philanthropic level, which represents the area organizations should strive to achieve, but only after satisfying the below components. The concept of CSR is usually applied in business settings; rarely has Carroll's model been applied in a sport industry setting.

Applying CSR in a Sport Setting

www

Title IX and Sex Discrimination

https://www2.ed.gov/about/ offices/list/ocr/docs/tix_ dis.html

For our purposes, let's examine a college athletic department and its compliance with Title IX. By law, a college athletic department can comply with Title IX by meeting any one prong of the so-called three-prong approach:

1. proportionality – the number of female and male athletes is substantially proportionate to their respective enrollments
2. a continuing history of expanding athletic opportunities for the underrepresented sex
3. demonstrating success in meeting the interests and abilities of underrepresented sex (U.S. Department of Education, n.d.)

Given this information about Title IX compliance, let's apply the four levels of the CSR model to an intercollegiate athletic department.

1. **ECONOMIC LEVEL.** If the department is operating at the economic level, this means it is most interested in achieving purely financial goals. Here the athletic department would basically ignore Title IX, doing nothing to comply until forced to do so by outside influences. Using the logic that complying with Title IX is too costly, the athletic department would not take any steps to comply with the law unless it became costlier not to comply.
2. **LEGAL LEVEL.** If the athletic department operates at the legal level, it will attempt to meet the minimum legal criterion for compliance. At this stage, organizations strive to meet legal minimums and basically follow the letter of the law. Most certainly, it would only attempt to fulfill one of the three prongs currently used to determine compliance. Since Title IX is not a quota system, this athletic department may rely on the so-called proportionality rule. It may make the case for continuing progress in developing opportunities for female athletes, or it could show it is meeting the needs and interests of the female student population. Athletic departments at this level are likely to drop men's sports to be in compliance, a quick and efficient way to come into minimum compliance with the proportionality prong. These same departments may state the reason for dropping the sports was because they needed more resources for women's sport, when in fact the reason was the football "arms race," a football program's increased need for expenditures to keep up with competing teams.
3. **ETHICAL LEVEL.** The athletic department operating at the ethical level would follow not just the letter of the law, but the

spirit of the law as well. The department would not just meet but may exceed the legal minimums because it values providing equal opportunities for women. The department would add emerging sports that attract female participants, such as rowing or lacrosse. Rather than dropping men's sports, the department would find alternative sources of funding or ways to redirect the budget so that female opportunities are increased without adversely affecting opportunities for males.

4. **PHILANTHROPIC LEVEL.** Athletic departments operating at the philanthropic level become active advocates for Title IX. They develop model programs for compliance and strive to fulfill all three prongs of the law, instead of the minimal one prong for compliance. These departments actively offer help to athletic administrators at other universities as their departments work to comply with Title IX. By presenting their programs as models and perhaps acting in a consultative mode to help other institutions comply with the law, athletic departments can operate at the philanthropic level.

Sport Organizations and CSR in Practice

More and more sport organizations, governing bodies, and clubs are beginning to integrate elements of CSR into their everyday business operations (Rankin, 2017). For example, Nike (much maligned for its labor practices, including unsafe working conditions and low wages) recently issued its *2020 Impact Report*. The company self-reviews goals set out in 2015 over the following categories: community impact, manufacturing, product, materials, carbon and energy, waste, water, and chemistry (Nike, 2020) and aimed at new goals for 2025. West Ham United won the Best Corporate Social Responsibility Scheme at the 2020 Football Business Awards and runner up at the 2021 Awards. The English football club created the "Players' Project" in 2018, with their men's, women's, and academy players donating more than 600 hours of their time to key areas in community work each year. The club assisted with over 30 projects ranging from health, community sport, football development, education, and employability. Additionally, the club has donated an estimated 28 million pounds by the end of 2021 (West Ham United, 2020).

In 2014, The International Olympic Committee (IOC) turned its sights on CSR being an important part of the Olympic and Paralympic idea. During a meeting in December 2014, the IOC approved the "Olympic Agenda 2020," a set of recommendations for ensuring the Olympic values and strengthening the role of sport in society. Two of those recommendations were to ensure all Olympic and Paralympic Games Organizers incorporate sustainability into all aspects of the Olympic Games and ensure the IOC itself is being sustainably conscious in the Olympic Movement's daily operations (IOC, 2020). The Tokyo 2020 organizers have claimed that their Olympic and Paralympic Games were the most environmentally friendly to date. Some of Tokyo's CSR

www

FY20, NIKE, Inc. Impact Report
https://purpose.nike.com/ fy20-nike-impact-report

efforts included making the Olympic medals from recycled materials; podiums were made from recycled plastic waste; athlete beds were made from recyclable cardboard, recycling 65% of waste from the Games and ensuring gender equality, diversity, and inclusion with the Games being the first gender-balanced Games ever (IOC, 2021). All eyes are set to the future, however, as 2024 (France) and 2028 (Los Angeles) were the first bids to heavily feature sustainability as part of the bid process (Avison, 2016).

Major professional sport organizations also act in a socially responsible manner. For example, the NFL has its annual Walter Payton Man of the Year Award.

"Each team nominates one player who has had a significant positive impact on his community. Representing the best of the NFL's commitment to philanthropy and community impact, 32 players are selected as their team's Man of the Year and become eligible to win the national award." (NFL, n.d., para. 1–2) The Women's National Basketball Association (WNBA) created the program "Take a Seat, Take a Stand" where $5 of every ticket was donated to one of six national nonprofit partners. MLB's Reviving Baseball in the Inner Cities (RBI) program has grown over the last few years. Many universities support local community activities through programs such as the University of Notre Dame's partnership with its campus Center for Social Concerns, where athletes get involved with projects on campus, in the local community, and in other areas around the country.

www

Take a Seat, Take a Stand
https://www.wnba.com/
takeastand/

Codes of ethics vary in content, length, and complexity. Exhibit 4.2 offers a sample code of coach's ethics from the USA Basketball's youth development program (USA Basketball, n.d.). Exhibit 4.3 shows the table of contents of the IOC Code of Ethics (IOC, 2020). Whether the code is long or short, concise or complex, what is most important is that sport organizations are beginning to develop codes.

---

exhibit **4.2** USA Basketball Coaches Code of Ethics

To maintain a USA Basketball Coach License or otherwise risk forfeiture, I agree to the USA Basketball Coaches Code of Conduct, which states as follows.

I will:

- Conduct myself in a dignified manner relating to emotions, language, attitude and actions
- Act at all times to protect the principles of fun, safety and development of all athletes
- Demonstrate respect for the ability of opponents as well as for the judgment of referees, officials and opposing coaches
- Display control and professionalism at all times under any circumstance
- Respect the rights, dignity and worth of every person, including opponents, other coaches, officials, administrators, parents, athletes, and spectators
- Refrain from physical contact with athletes except where necessary for the development of the athletes' skill(s) or athletic ability

- Be aware and understand the role and influence of a coach as an educator, imparting knowledge of skill as well as proper personal, academic, and social behavior
- Be reasonable in my demands on athletes' time, energy and enthusiasm
- Provide an opportunity for all athletes to play the sport
- Agree to uphold the Progressive Coaching principles found within the USA Basketball Player Development Model
- Ensure that equipment and facilities meet safety standards and are appropriate to the level of the athletes
- Seek to learn the latest coaching practices that take into account the principles of growth and development of athletes
- Agree to abide by all applicable USA Basketball rules and regulations, including USA Basketball's SafeSport policy

---

IOC Code of Ethics   *exhibit*   **4.3**

International Olympic Committee Code of Ethics

Preamble

A Fundamental Principles

B Integrity of Conduct

C Integrity of Competitions

D Good Governance and Resources

E Candidatures

F Confidentiality

G Reporting Obligation

H Implementation

Source: IOC (2022).

ETHICS AND SPORT GOVERNANCE ORGANIZATIONS

Why is it important to include a chapter in this book about ethics? Individual sport managers and sport organizations look to their governing bodies for guidance on a wide range of topics, including legal issues, safety issues, and personnel issues – and also ethical issues such as good business practices.

Sport governing bodies can set the tone from the top down regarding ethical issues. The stance taken by a governing body will influence decisions you make as a sport manager in any organization under that governing body's umbrella. An example of this relationship is the effect on International Federations and National Sports Organizations when the IOC makes a ruling on banned substances or participation by transgender athletes. State and provincial high school athletic associations are taking stronger stances against hazing of minors, so individual schools are also instituting such programs. Governing bodies' rulings on ethical issues will hopefully result in behavioral changes and choices by their constituencies as well.

Sport managers need to understand the effects of their decisions and the number of people affected by their decisions (Crosset & Hums, 2015). According to DeSensi (2012):

> Is it too much to expect of sport managers that they become ethically, morally and socially responsible professionals? My response is no; I believe they should be at the forefront of this issue … The moral climate of sport has always been in need of improvement, and who better than sport managers to assume this responsibility?
>
> (p. 130)

Knowing you will face ethical dilemmas, you will need to employ an ethical decision-making model to deal with them. You will have the opportunity to consider ethical questions in some case studies in this book. You will shape the culture of your organization, impacting its ethical climate and its stance on being a good corporate citizen. In other words, the driving force for a sport organization's level of CSR rests with its employees and, most certainly, with its managers. That means you!

SUMMARY

We see numerous examples of ethical dilemmas in society today. The sport industry is part of that greater society; therefore, we encounter ethical issues and dilemmas in the sport industry as well. When sport managers are faced with ethical dilemmas, they have to make decisions about them. To make sound decisions, sport managers must follow a systematic ethical decision-making process.

When examining ethics, remember to take into account ethical activity on the organizational as well as on the individual level. One measure of ethics in a sport organization is CSR, whether it is reflected in a sport organization's stance on a particular issue, its involvement in charitable community events, or its adoption of a code of ethics.

Finally, it is important to examine ethics in the context of governance because of the influence governing bodies have over individual sport organizations. By exhibiting ethical behaviors at the top levels, governing bodies set the ethical tone for their membership.

case STUDY

ETHICS

1. Use the ethical decision-making model we just discussed in class to deal with the following situations:

    a. You are the Director of the Tiger Athletic Fund, the fund-raising arm of your NCAA Football Bowl Subdivision school, Central City University. One of your major long-time donors, Ray Mosswell, whose name adorns the football practice facility, recently made a significant investment in a construction company owned by an international business with an abysmal human rights abuse record, having recently been tied to human trafficking. Needless to say, the story made national headlines and has lit up social media. It is time to renew Ray's annual donation to Tiger Athletics and funding is tight after COVID-19. What is your decision regarding the future of your relationship between Ray and Central City University?

b. You are the Athletic Director (AD) at Marion Heights College. A big baseball fan, you often go to the games and the players know and respect you. One of the players, Matt, just received national attention. Matt is a highly skilled player who is projected to be taken in the early rounds of the draft. Matt, who is White, recently had old social media posts from four years ago (when he was 14) of him using racial slurs and making degrading comments toward people of color. Matt comes to you saying he was "just a dumb kid who didn't know what those words meant" when he was younger and never would think or say those things today. You have to decide what should be done about Matt (e.g., suspension, expulsion, nothing) and Matt's next steps.

c. You are the AD at Little Creek High School, a public high school in an urban setting. One day, the parents of a student who just transferred to Little Creek scheduled a meeting with you to discuss athlete eligibility. Their sophomore transgender student is in the process of transitioning from female to male and would like to participate in the boys' track and field team.

   i. What do you need to do to approach the situation and people with respect before the meeting?

   ii. What information do you need to gather in order to make a decision about the student's participation? What other sport governing bodies will you contact for assistance?

   iii. Would your decision on a transgender student participating be influenced by whether the student was transitioning from male to female as opposed to female to male?

d. You are the Assistant Director of Scholastic Marketing for Adidas. It has come to your attention that numerous high school accounts which you manage still use Native American mascot names which many people feel are offensive. Recently, you received a letter from the American Indian Movement (AIM) indicating that if Adidas continues to supply schools using these mascots with their products, they will call for a national boycott of the brand. You know that Native Americans' annual buying power exceeds $90 billion and many of those dollars could go to sporting equipment and apparel, but also that those accounts you manage supply significant revenue for Adidas. How will you respond to AIM?

CHAPTER questions

1. Using the model presented in Exhibit 4.1, choose three additional societal issues and illustrate how they are seen in sport.

2. Place the sport organizations/persons listed below on Carroll's CSR Pyramid (Economic to Philanthropic). Provide examples as to why you placed them where you did. Remember, an organization/person can be in more than one place on the model. For example, if Lance Armstrong were on this list, one could say at times he belongs on the economic level due to his wanting to win at all costs by doping and yet he could be on the Philanthropic level due to his work in raising funds to fight cancer.

NCAA	Megan Rapinoe
IOC	Tom Brady
NFL	LeBron James
WNBA	Adam Silver

3. Find codes of conduct or codes of ethics from three different sport organizations, each from a different industry segment. Compare the three for similarities and differences. Create and prepare to defend the content of your own code of conduct statement in a class discussion.

4. The Tokyo 2020 Olympic and Paralympic Games claimed to be the "Green Games," as they were the most sustainable and ethical to date. Yet, many nonprofits and journalists questioned if the Games really were green. Research what practices Tokyo 2020 engaged with to ensure sustainability and critically analyze how effective their efforts were.

FOR ADDITIONAL INFORMATION

For more information check out the following resources:

1. This article discusses athlete activism and sport governance: Hums, M.A., Wolff, E.A., & Siegfried, N. (2020, June 23). Making a case for athlete advocacy: Humanitarians, Olympism, and good governance. *Play the Game*. https://playthegame.org/news/comments/2020/1006_making-a-case-for-athlete-advocacy-humanitarians-olympism-and-good-governance/

2. This is an excellent podcast that tackles head on all the big ethical issues in sport! Podcast: Edge of Sports with Dave Zirin: https://www.edgeofsports.com/

3. This website explains the WNBA's Social Justice Council Overview & Mission: https://www.wnba.com/social-justice-council-overview/

4. This discusses green initiative at the Tokyo summer Olympic and Paralympic Games: The 'Green' Olympic Games in Tokyo 2021: https://actionrenewables.co.uk/news-events/post.php?s=the-green-olympic-games-in-tokyo-2021

5. Here you will find information on North American pro sport's most green arena: Climate Pledge Arena: Sustainability: https://climatepledgearena.com/sustainability/

REFERENCES

Athlete Ally (2022). *About Athlete Ally.* https://www.athleteally.org/about/

Avison, B. (2016, February 17). *IOC welcomes "most sustainable ever" bids for 2024 Games.* Host City. https://www.hostcity.com/news/event-bidding/ioc-welcomes-%E2%80%9Cmost-sustainable-ever%E2%80%9D-bids-2024-games

Babiak, K., & Trendafilova, S. (2011). CSR and environmental responsibility: Motives and pressures to adopt green management practices. *Corporate Social Responsibility and Environmental Management, 18*(1), 11–24.

Babiak, K., & Wolfe, R. (2009). Determinants of corporate social responsibility in professional sport: Internal and external factors. *Journal of Sport Management, 2*(6), 717–742.

Bradish, C., & Cronin, J.J. (2009). Corporate social responsibility in sport. *Journal of Sport Management, 23,* 691–697.

Carroll, A.B. (1991). The pyramid of corporate social responsibility: Toward the moral management of organizational stakeholders. *Business Horizons,* 39–48.

Carroll, A.B., & Brown, J.A. (2018). Corporate social responsibility: A review of current concepts, research and issues. In J. Weber, & D. Wasieleski (Eds.), *Corporate Social Responsibility* (Chapter 2, pp. 39–69). Emerald Publishing Co.

Council of Europe Committee of Ministers. (2010). *Code of sport ethics.* https://rm.coe.int/16805cecaa

Crosset, T., & Hums, M.A. (2015). Ethical principles applied to sport management. In L.P. Masteralexis, C.A. Barr, & M.A. Hums (Eds.), *Principles and practice of sport management* (5th ed., pp. 131–148). Jones & Bartlett.

Darnell, S.C. (2012). *Sport for development and peace: A critical sociology.* Bloomsbury Academic.

Dahlsrud, A. (2008). How corporate social responsibility is defined: An analysis of 37 definitions. *Corporate Social Responsibility and Environmental Management, 15,* 1–13.

Davies, G. (2020, February 1). *Racism in soccer an 'epidemic' that mirrors disturbing trends in Europe: Advocates.* ABC News. https://abcnews.go.com/Sports/racism-soccer-epidemic-mirrors-disturbing-trends-europe-advocates/story?id=67850877

DeSensi, J. (2012). The power of one for the good of many. In A. Gillentine, R.E. Baker, & J. Cuneen (Eds.), *Critical essays in sport management* (pp. 125–132). Holcomb Hathaway.

DeSensi, J., & Rosenberg, D. (2010). *Ethics and morality in sport management* (3rd ed.). Fitness Information Technology.

Eitzen, D.S., & Sage, G.H. (2009). *Sociology of North American sport* (8th ed.). Paradigm Publishers.

Friedman, T. (2017, June 21). Where did "We the People" go? *The New York Times.* https://www.nytimes.com/2017/06/21/opinion/where-did-we-the-people-go.html

Gargan, M. (2015, November 19). *Sport holds up a mirror to society.* Irish Catholic. https://www.irishcatholic.com/sport-holds-up-a-mirror-to-society/

Gibbs, L. (2016, June 4). *Muhammad Ali: The original activist-athlete.* Think Progress. https://archive.thinkprogress.org/muhammad-ali-the-original-activist-athlete-a13ac939f310/

Hancock, M., & Hums, M.A. (2011). Participation by transsexual and transgender athletes: Ethical dilemmas needing ethical decision making skills. *ICSSPE Bulletin*, 68. http://connection.ebscohost.com/c/articles/95769419/participation-by-transsexual-transgender-athletes-ethical-dilemmas-needing-ethical-decision-making-skills

Hums, M.A. (2006, May). *Analyzing the impact of changes in classification systems: A sport management analysis model.* Paper presented at the VISTA 2006 International Paralympic Committee Congress, Bonn, Germany.

Hums, M.A. (2007, June). *The business of the Paralympic Games: Economics and ethics.* Paper presented at 7th International Conference on Sports: Economic, Management and Marketing Aspects, Athens, Greece.

Hums, M.A. (2010). The conscience and commerce of sport: One teacher's perspective. *Journal of Sport Management, 24,* 1–9.

Hums, M.A., & Hancock, M. (2012). Sport management: Bottom lines and higher callings. In A. Gillentine, B. Baker, & J. Cuneen (Eds.), *Critical essays in sport management: Exploring and achieving a paradigm shift* (pp. 133–148). Holcomb Hathaway.

Hums, M.A., & Wolff, E.A. (2014, April 3). *Power of sport to inform, empower, and transform.* Huffington Post. www.huffingtonpost.com/dr-mary-hums/power-of-sport-to-inform-_b_5075282.html

Hums, M.A., Barr, C.A., & Guillion, L. (1999). The ethical issues confronting managers in the sport industry. *Journal of Business Ethics, 20,* 51–66.

Hums, M.A., Wolff, E.A., & Morris, A. (2012). Human rights in sport checklist. In K. Gilbert (Ed.), *Sport, peace, and development* (pp. 243–254). Common Ground.

IOC. (2020). *Olympic Agenda 2020: Closing report.* https://stillmed.olympics.com/media/Document%20Library/OlympicOrg/IOC/What-We-Do/Olympic-agenda/Olympic-Agenda-2020-Closing-report.pdf?_ga=2.236010177.1951034275.1643578764–1752807646.1643578764

IOC. (2021, July 22). *All you need to know about Tokyo 2020 sustainability.* https://olympics.com/ioc/news/all-you-need-to-know-about-tokyo-2020-sustainability

IOC. (2022, January). *Ethics.* https://stillmed.olympics.com/media/Document%20Library/OlympicOrg/Documents/Code-of-Ethics/Code-of-Ethics-ENG.pdf?_ga=2.201381585.1951034275.1643578764–1752807646.1643578764

Lapchick, R. (2020, February 19). *Racism reported in sports decreasing, but still prevalent.* ESPN. https://www.espn.com/espn/story/_/id/28738336/racism-reported-sports-decreasing-prevalent

Levenson, E. (2018, January 24). *Larry Nassar sentenced to up to 175 years in prison for decades of sexual abuse.* CNN. https://www.cnn.com/2018/01/24/us/larry-nassar-sentencing/index.html

Lin, Y. (2019, April 26). *5 benefits of corporate social responsibility.* Energy Link. https://goenergylink.com/blog/5-benefits-of-corporate-social-responsibility/

Lussier, R.N., & Kimball, D.C. (2009). *Applied sport management skills.* Human Kinetics.

Maguire, J.A., & Nakayama, M. (2006). *Japan, sport and society: Tradition and change in a globalizing world.* Routledge.

McDonald, M. (2001). *A framework for ethical decision-making: Version 6.0 Ethics shareware.* www.ethics.ubc.ca/upload/A%20Framework%20for%20Ethical%20Decision-Making.pdf

Mullane, S.P. (2009). *Ethics and leadership—White Paper.* University of Miami Johnson Edosomwan Leadership Institute.

NFL. (n.d.). *Walter Payton Man of the Year.* https://www.nfl.com/honors/man-of-the-year/

Nike. (2020). *FY20 NIKE, Inc. Impact Report.* https://purpose-cms-preprod01.s3.amazonaws.com/wp-content/uploads/2021/04/26225049/FY20_NIKE_Inc_Impact_Report2.pdf

Rankin, N. (2017). *Sport and CSR: Lessons learnt.* Sportanddev. https://www.sportanddev.org/en/article/news/sport-and-csr-lessons-learnt

Schnell, L. (2017, February 3). How many more sickening Baylor details does college football need before it changes? *Sports Illustrated.* https://www.si.com/college-football/2017/02/03/baylor-sexaul-assault-scandal-art-briles-assistants

Thibault, L. (2009). Globalization of sport: An inconvenient truth. *Journal of Sport Management, 23,* 1–20.

Thornton, P.K., Champion, Jr., W.T., & Ruddell, L.S. (2012). *Sports ethics for sport management professionals.* Jones & Bartlett.

UNESCO. (n.d.). *Unesco – fair play – the winning way.* Oro, Plata y Bronce. https://www.oroplataybronce.com/unesco-fair-play-the-winning-way/

UNIDO. (n.d.). *What is CSR?* https://www.unido.org/our-focus/advancing-economic-competitiveness/competitive-trade-capacities-and-corporate-responsibility/corporate-social-responsibility-market-integration/what-csr

USA Basketball. (n.d.). *Coaches code of ethics.* https://www.usab.com/youth/development/coach/coaches-code-of-conduct.aspx

U.S. Department of Education. (n.d.). *Intercollegiate athletics policy: Three-part test – Part three.* https://www2.ed.gov/about/offices/list/ocr/docs/title9-qa-20100420.html

Wallace, D. (2012). *Definition of an ethical dilemma.* https://www.scribd.com/doc/56047818/Definition-of-an-Ethical-Dilemma

West Ham United. (2020, December 18). *Hammers win Sports Business Award for community work.* https://www.whufc.com/news/articles/2020/december/18-december/hammers-win-sports-business-award-community-work

Zinn, L.M. (1993). Do the right thing: Ethical decision making in professional and business practice. *Adult Learning, 5,* 7–8, 27.

5

Scholastic Sport

On any given Friday night in the fall, the air is filled with the sounds of fans screaming, helmets pounding, and bands playing. On Saturdays in winter, the ball bounces off the court, and skates glide over the ice. In the spring, the starter's gun marks the beginning of the 100-meter dash, and we hear the familiar sound of softballs and baseballs being hit and caught. On fields, courts, and rinks everywhere, it is time for the weekend

DOI: 10.4324/9781003303183-5

ritual – high school sports – although the COVID-19 pandemic certainly disrupted this for a while. All across the land, young athletes compete for their schools, their communities, and themselves. Many of you remember those days, no doubt. This chapter will focus on high school sport, a sizable section of the sport industry involving thousands of schools and participants.

Although it may surprise you, this is primarily a North American model. In many nations of the world, young people compete for their local municipality's club teams, not for their high school teams. For example, a young athlete living in Wetzlar, Germany, will compete in gymnastics for the local sports club, not for Wetzlar High School. The high school may not even field any teams, since the schools do not compete against each other in sports like schools in North America. In Munich, a talented young football (soccer) player might compete on a developmental team of the Bayern Munich professional football club. The best young athletes are generally selected early for development by professional clubs. In other countries, high school students participate in sport on the local level – and sometimes even on the national level – as with the Koshien national high school baseball championship in Japan. While most of this chapter focuses on the North American model, remember that this is not the only competitive model for youth sport. "The opportunity provided by member schools to girls and boys to represent their school and community through participation in interscholastic athletics is a privilege to young people in American education" (PIAA, 2017, para. 1). How, then, did it happen that North American high school sport developed into the product we see today? To understand this evolution, let's take a look at the history and evolution of sport in high schools.

HISTORY OF HIGH SCHOOL SPORT

T he history of high school sport is long and storied. It began simply to develop healthy habits in youngsters and has gained widespread popularity among spectators, with important steps along the way. For purposes of this textbook, we will examine mainly how the governance of high school sport evolved.

Early Development

In the late 1800s, sport was seen as a vehicle to help solve societal ills, such as delinquency and poor health, so schools began to promote sport (Seymour, 1990). In its early days, high school sport was initiated, organized, and operated by students, similar to the way intercollegiate sport started. Also similar to intercollegiate sport, the need for adult supervision and direction soon became apparent, because the adults working with high school sport at the time did not care for the direction it was heading. In the 1890s, the popularity of high school football soared, and its abuses mirrored those of college sport at the time – overemphasis on winning, using ineligible players, and misman-agement of finances (Rader, 1999). To uphold a certain moral image, administrators felt it necessary to extend their authority over interscholastic sport.

In the early 1900s, rules were put in place for high school athletes that defined minimum course loads and satisfactory progress in school, as well as participation eligibility certification. These rules were a progression from those outlined by the Michigan State Teachers' Association's Committee on High School Athletics in 1896 (Forsythe, 1950). As athletics became more integral in a student's academic experience, government-funded educational institutions assumed increased control over the governance of high school athletics (Vincent, 1994).

Development of the National Federation of State High School Associations (NFHS)

In 1920, representatives of five Midwestern states – Illinois, Indiana, Iowa, Michigan, and Wisconsin – met in Chicago to discuss concerns about collegiate and non-school sponsorship of high school events. The result was a plan to ensure the well-being of high school student-athletes in competitive situations. These five state associations banded together to form the Midwest Federation of State High School Athletic Associations. Eventually, more state associations joined this group, and in 1923, they changed their name to the National Federation of State High School Athletic Associations (NFHS, 2021b). By the 1930s, the group assumed responsibility as the rules-writing and rules-publishing body for high school sports. The organization grew throughout the 20th century, adding members until 1969 when all 50 states and the District of Columbia belonged. In the 1970s, the fine arts were added under the organization's umbrella, and the term *Athletic* was removed from the organization's name. The official name became the "National Federation of State High School Associations," as it remains today, and since 1997, the organization has gone by the abbreviation "NFHS" (NFHS, 2021b). With its headquarters located in Indianapolis, Indiana, organizational development continued from the 1980s to the present, with increased educational programming, incorporation of debate and spirit programs, and ongoing rules interpretations and publications.

National Federation of State High School Associations
https://www.nfhs.org/

Value of High School Sport Today

The values and benefits of high school sport have been well defined by its advocates. For example, the mission statement of the Colorado High School Activities Association states, "In pursuit of educational excellence, the Colorado High School Activities Association strives to create a positive and equitable environment in which all qualified student participants are challenged and inspired to meet their greatest potential" (CHSAA, 2020, p. 3).

Colorado High School Activities Association
https://chsaanow.com/

The organization goes on to explain how its members will implement this mission: To fulfill this mission, the Colorado High School Activities Association will:

- Act as an integral component of the educational process.
- Administrate, interpret, and seek compliance with the CHSAA Bylaws as needed to promote competitive equity within Colorado activities and athletics.
- Provide diverse and equitable opportunities for participation that encourage all qualified students to take part in the activity/athletic experience.
- Provide an environment that enhances personal development through sporting behavior, character education, teamwork, leadership, and citizenship while increasing values that partner with the educational standards of the State of Colorado.
- Recognize the outstanding accomplishments of Colorado athletes, teams, coaches, and administrators through our academic and activity awards programs (CHSAA, 2020, p. 3).

As another example, the Illinois High School Association's mission statement reads, "The IHSA governs the equitable participation in interscholastic athletics and activities that enrich the educational experience" (IHSA, 2013, para. 1). As a final example, here is the mission statement of the Minnesota State High School League, "The Minnesota State High School League provides educational opportunities for students through interscholastic athletic and fine arts programs and provides leadership and support for member schools" (MSHSL, 2020, p. 2).

For all these reasons, high school sport has become an important component of many students' total educational experience. To make sure worthwhile activities happen in a well-planned and organized environment, governance structures must be in place.

Illinois High School Association
https://www.ihsa.org/default.aspx

Minnesota State High School League
https://www.mshsl.org/

GOVERNANCE

Scholastic sport governance occurs on a number of different levels. As mentioned previously, at the national level, there is the NFHS, a service organization that includes members from the United States and Canada. But unlike some national-level sport governing bodies, this is not where the real power lies as the NFHS is a service organization. In high school sport, the real power and authority rest at the state level, where the regulatory power lies. According to Wong (1994), "the power and authority in high school athletics are in the individual state organizations, which determine the rules and regulations for the sport programs and schools within that state" (p. 22). There is also governance on the local level, meaning the individual school or school district. Let's look at the organizational structures at these different levels and the scope of their authority.

National Federation of State High School Associations

The NFHS (2021a) is a member-governed, nonprofit national service and administrative organization of high school athletics, fine arts, and performing programs. From its offices in Indianapolis, the NFHS serves its 50-member state high school athletic and activity associations, plus those of the District of Columbia. The organization also has a number of affiliate members in Canada and other US territories. The organization publishes playing rules for numerous sports for boys and girls and provides programs and services that its member state associations can use in working with the 19,500 member high schools and approximately 12 million young people involved in high school activity programs (NFHS, 2021a).

From this information, it is important to note two main points. First, the NFHS is considered a service organization. In contrast to the National Collegiate Athletic Association (NCAA), which has strong sanctioning power over members, the purpose of the NFHS is to provide services to its members. Also, the NFHS is not involved solely in athletic competition. Subgroups within the NFHS also include Music Directors and Adjudicators; Speech, Debate, and Theater Directors and Judges; and Spirit. Thus, the organization has a broad base across many high school extracurricular activities. This textbook, however, focuses on only those aspects of the NFHS dealing directly with interscholastic athletics.

Mission

The mission statement of the NFHS (presented in Exhibit 5.1) states that its purpose is to promote activities that contribute positively to a student's educational experience. Also apparent is that the organization seeks to develop students into people who will be contributing members of society due to the good lessons they learned from their sport experience. From this mission statement, it is clear that high school sport is meant to help interscholastic athletic administrators make sure students become positive citizens through their athletic experiences.

Mission Statement of the NFHS · *exhibit* · 5.1

The NFHS serves its members by providing leadership for the administration of education-based high school athletics and activities through the writing of playing rules that emphasize health and safety, educational programs that develop leaders, and administrative support to increase opportunities and promote sportsmanship.

We Believe:

Student participation in education-based high school athletics and activities:

- Is a privilege.
- Enriches the educational experience.
- Encourages academic achievement.
- Promotes respect, integrity, and sportsmanship.
- Prepares for the future in a global community.

- Develops leadership and life skills.
- Fosters the inclusion of diverse populations.
- Promotes healthy lifestyles and safe competition.
- Encourages positive school/community culture.
- Should be fun.

The National Federation does the following:

- Serves as the national authority that promotes and protects the defining values of education-based high school athletics and activities in collaboration with its member state associations.
- Serves as the national authority on competition rules while promoting fair play and seeking to minimize the risk of injury for student participants in education-based high school athletics and activities.
- Promotes lifelong health and safety values through participation.
- Develops and delivers impactful, innovative, and engaging educational programs to serve the changing needs of state associations, administrators, coaches, directors, officials, students, and parents.
- Provides professional development opportunities for member state association staff.
- Promotes cooperation, collaboration, and communication with and among state associations.
- Collects and provides data analysis in order to allow its membership to make informed decisions.

Source: NFHS (2022).

Membership

Who belongs to the NFHS? NFHS membership is made up of state associations, not individual people. (This is similar to the NCAA where universities and conferences are the members and not individual athletic directors). As mentioned earlier, the active members of the NFHS are the 50 state high school athletic/activity associations, plus the District of Columbia. There are also affiliate members, including associations in the US territories, and Canada (NFHS, 2021e). The affiliated members from outside the United States include Canadian School Sport Federation from Alberta, British Columbia, Manitoba, New Brunswick, Nova Scotia, Ontario, Quebec, and Saskatchewan. Other affiliates include Guam and the Bahamas, as well as several state independent schools athletic associations (e.g., Georgia, North Carolina, Virginia) (NFHS, 2021e).

Financials

How does the NFHS finance itself? Approximately one-third of the organization's income comes from sales revenue and educational programming (NFHS, 2020a). In addition, the organization earns funds from membership dues; contributions, royalties, and sponsors; meetings and conferences; and a few lesser sources. How is the money spent? The major expense categories for the NFHS are salaries and benefits followed by education and professional development, professional organizations, and rules-making products/publications (NFHS, 2020a).

Organizational Structure

The organizational structure of the NFHS indicates that the membership drives the governance of the organization. As illustrated in Exhibit 5.2, the member state associations form the base of support for the chart. The

National Council is the legislative body of the NFHS and is responsible for enacting amendments to the constitution and bylaws in addition to other duties. The National Council consists of one representative from each high school association which holds membership. The National Council meets two times a year and each member association has one vote. The Board of Directors of the NFHS is made up of 12 members, one from each geographic section and four additional at-large members. The board elects the President, the President-Elect, and hires the Executive Director who administers the organization's affairs (NFHS, 2020b). The next group involved in the governance of the NFHS is the 41-member Executive Staff, the organization's paid employees, and includes positions such as Executive Director and Directors of Sports; Sports and Sport Medicine; Sports, Sanctioning, and Student Services; Publications and Communications; Information Services; Educational Services; Sports; Sports, Events, and Development; Performing Arts and Sports; and Sports and Officials Education. Several Assistant Directors and a Chief Operating Officer are also employed, as well as a General Counsel to handle legal questions and issues (NFHS, 2021d). Finally, as with most organizations, a series of committees work in designated areas, such as the National Athletic Directors Conference Advisory Committee, Education Committee, Sports Medicine Advisory Committee, Technology Committee, general committees, designated Sports and Activities Committees, and special committees (NFHS, 2020b).

State High School Athletic Associations

Each state has its own high school athletic association. As you will see, they have different names. For example, we see the Georgia High School Association, the Idaho High School Activities Association, the Indiana High School Athletic Association, the Maine Principals' Association, the Texas University Interscholastic League, and the Wisconsin Interscholastic Athletic Association. Despite their different names, these organizations share common missions and authorities.

State high school associations serve several important functions. First, they are the regulatory bodies for high school sport in a particular state. As noted earlier, the power in high school athletics resides on this level. According to Sharp et al. (2021),

> Authority to govern interscholastic athletics within a state is granted to the state association by the state legislature or by judicial decision. Each state's high school athletic association is responsible for implementing and enforcing regulations governing interscholastic athletics participation of the member high schools.
>
> (p. 297)

This statement clearly establishes that the regulatory power in high school sport governance lies at the state level. Second, they are responsible for organizing state championships, always the highlight of the year for any sport. Maybe some of you had the opportunity to compete in some of these when you were in high school. This author did and won a

exhibit   5.2   NFHS Organizational Chart

ORGANIZATIONAL CHART

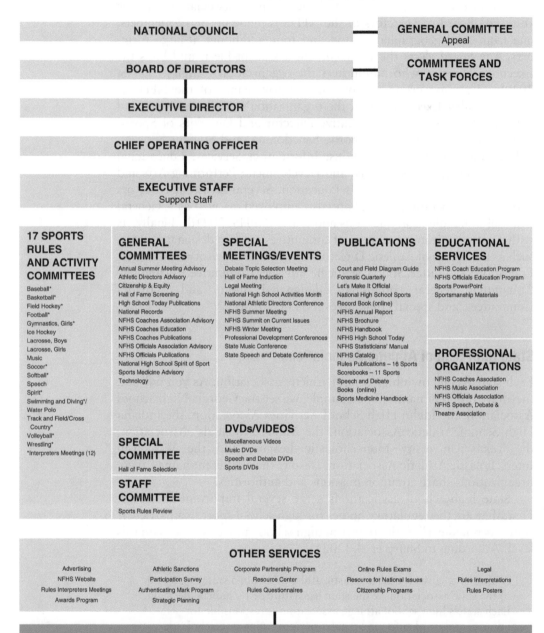

NATIONAL COUNCIL — **GENERAL COMMITTEE**
Appeal

BOARD OF DIRECTORS — **COMMITTEES AND TASK FORCES**

EXECUTIVE DIRECTOR

CHIEF OPERATING OFFICER

EXECUTIVE STAFF
Support Staff

17 SPORTS RULES AND ACTIVITY COMMITTEES
Baseball*
Basketball*
Field Hockey*
Football*
Gymnastics, Girls*
Ice Hockey
Lacrosse, Boys
Lacrosse, Girls
Music
Soccer*
Softball*
Speech
Spirit*
Swimming and Diving*/
Water Polo
Track and Field/Cross
  Country*
Volleyball*
Wrestling*
*Interpreters Meetings (12)

GENERAL COMMITTEES
Annual Summer Meeting Advisory
Athletic Directors Advisory
Citizenship & Equity
Hall of Fame Screening
High School Today Publications
National Records
NFHS Coaches Association Advisory
NFHS Coaches Education
NFHS Coaches Publications
NFHS Officials Association Advisory
NFHS Officials Publications
National High School Spirit of Sport
Sports Medicine Advisory
Technology

SPECIAL COMMITTEE
Hall of Fame Selection

STAFF COMMITTEE
Sports Rules Review

SPECIAL MEETINGS/EVENTS
Debate Topic Selection Meeting
Hall of Fame Induction
Legal Meeting
National High School Activities Month
National Athletic Directors Conference
NFHS Summer Meeting
NFHS Summit on Current Issues
NFHS Winter Meeting
Professional Development Conferences
State Music Conference
State Speech and Debate Conference

DVDs/VIDEOS
Miscellaneous Videos
Music DVDs
Speech and Debate DVDs
Sports DVDs

PUBLICATIONS
Court and Field Diagram Guide
Forensic Quarterly
Let's Make It Official
National High School Sports
Record Book (online)
NFHS Annual Report
NFHS Brochure
NFHS Handbook
NFHS High School Today
NFHS Statisticians' Manual
NFHS Catalog
Rules Publications – 16 Sports
Scorebooks – 11 Sports
Speech and Debate
Books (online)
Sports Medicine Handbook

EDUCATIONAL SERVICES
NFHS Coach Education Program
NFHS Officials Education Program
Sports PowerPoint
Sportsmanship Materials

PROFESSIONAL ORGANIZATIONS
NFHS Coaches Association
NFHS Music Association
NFHS Officials Association
NFHS Speech, Debate &
Theatre Association

OTHER SERVICES

Advertising	Athletic Sanctions	Corporate Partnership Program	Online Rules Exams	Legal
NFHS Website	Participation Survey	Resource Center	Resource for National Issues	Rules Interpretations
Rules Interpreters Meetings	Authenticating Mark Program	Rules Questionnaires	Citizenship Programs	Rules Posters
Awards Program	Strategic Planning			

MEMBER STATE ASSOCIATIONS

Source: NFHS (2020a).

state championship in volleyball. Finally, they maintain the educational philosophy for high school athletics in their respective states.

State-level governance is often vested with power from the state legislature. State associations have the authority to revoke eligibility for individual students and to disqualify schools from participating in events if the schools break state association rules. In disputes about eligibility and other questions about the interpretation of rules, the US state high school associations are named in any resulting lawsuits. The reason for this is that in most cases the state association has been found to be a state actor (Brentwood Academy v. Tennessee Secondary Sch. Athletic Ass'n, 2001); that is, an organization working as if it was empowered by the government to act. You may remember from your Legal Aspects of Sport class that this makes state associations subject to the requirements of the US Constitution. Whenever a high school athlete feels their constitutional rights have been violated because of an association's rule, that athlete names the high school association in the suit. Thus, when state associations craft policies, they must be mindful not to enact policies or procedures that could be construed as infringing on a student's fundamental rights, such as the right to due process if a student is denied eligibility for some reason.

Mission

As mentioned earlier, while the associations' names differ from state to state, common ideals are reflected in each association's mission statement. Sample mission statements from the New Jersey State Interscholastic Athletic Association (NJSIAA) and the Oregon State Athletic Association are presented in Exhibits 5.3 and 5.4, respectively. While these mission statements are somewhat different, one can see similarities between them. Shared themes include the place of athletics in an educational setting, the values and benefits students derive from high school sport, and the provision of service to their members.

www
New Jersey Interscholastic Athletic Association
https://www.njsiaa.org/

www
Oregon State Athletic Association
https://www.osaa.org/

Membership

High school associations generally are voluntary, nonprofit organizations whose members are the public and private secondary schools in that particular state. In some cases, junior high schools and middle schools may also belong to the association. The size of each association varies, depending on the number of high schools in the state. The membership of high school associations is similar to that of the NCAA, where institutions (high schools), not individual people, are the members of the organization.

The sources of funding for athletic associations vary by state. The Georgia High School Association's primary source of revenue is state tournaments/playoffs, followed by corporate/vendor partnerships, and then community coach registrations. Its main expenses are state tournaments/playoffs and salaries and wages (GHSA, 2020). The Washington Interscholastic Athletic Association indicates its largest revenue source

www
Georgia High School Association
https://www.ghsa.net/

www
Washington Interscholastic Athletic Association
https://www.wiaa.com/

exhibit **5.3**   Mission Statement of the NJSIAA

NJSIAA MISSION STATEMENT

The NJSIAA, a private, voluntary association, is committed to serving all types of student-athletes, its member schools, and related professional organizations by the administration of education-based interscholastic athletics, which support academic achievement, good citizenship, and fair and equitable opportunities. We believe our member schools, along with their leagues and conferences, share the following convictions:

- A safe and healthy playing environment is essential to our mission.
- Participation in interscholastic athletics enhances the educational experience of all students.
- Interscholastic athletics is a privilege.
- Excellence in both academics and athletics is pursued by all.
- Interscholastic participation develops good citizenship, promotes healthy lifestyles, fosters involvement of a diverse population, and promotes positive school/community relations.
- Rules promote fair play and minimize risk.
- Cooperation among members advances their individual and collective well-being.
- Training of administrators and coaches promotes the educational mission of the interscholastic experiences.
- Properly trained officials/judges enhance interscholastic competition.
- The NJSIAA is the recognized state authority on interscholastic athletic programs.

Source: NJSIAA. (n.d.).

exhibit **5.4**   OSAA Mission Statement

Our Mission

The mission of the OSAA is to serve member schools by providing leadership and state coordination for the conduct of interscholastic activities, which will enrich the educational experiences of high school students. The OSAA will work to promote interschool activities that provide equitable participation opportunities, positive recognition, and learning experiences to students while enhancing the achievement of educational goals.

Source: OSAA (2022).

is state tournament income and then fees from member schools. The largest expense items are operations-related and state-wide activities costs (WIAA, 2021) and these are common expenses across the state associations. As budgets become tighter, associations must come up with more creative means of financing their programs.

Organizational Structure

STATE ASSOCIATIONS. During the year, an Executive Committee or Board of Directors meets to deal with any ongoing issues. This Executive Committee is made up of Superintendents, Principals, and

Athletic Directors (ADs) from various high schools around the state. In addition, paid Executive Staff members work in the association headquarters year-round. These headquarters are usually located in the capital of the state. This provides a central location for members and also access to state elected officials and education policy makers if/when issues related to high school sport in the state are being discussed at legislative sessions. While the titles may vary, often the highest ranking paid staff member is typically called the Commissioner or the Executive Director. There are also several Associate or Assistant Commissioners or Associate or Assistant Executive Directors, each of whom has distinct responsibilities for certain sports and other areas, such as eligibility, rules interpretation, officials, coaches' clinics, sportsmanship and ethics programs, and trophies and awards. The organizational chart for the Wisconsin Interscholastic Athletic Association is presented in Exhibit 5.5.

Wisconsin Interscholastic
Athletic Association
https://www.wiaawi.org/

SCHOOL DISTRICTS. School districts have various responsibilities, including dealing with high school athletic programs in a limited fashion. For example, local school boards can set the guidelines for hiring a coach. Here is an example from the Jefferson County Public Schools (JCPS, 2018, para. 1), the school district that includes 22 high schools including the schools in Louisville, Kentucky.

Jefferson Country Public
Schools
https://www.jefferson.
kyschools.us/

All coaches who have been recommended to be hired by a school must have the following:

- A completed online coaching application
- A contract signed by the coach, AD, and principal
- Proof of cardiopulmonary resuscitation (CPR) and automated external defibrillator (AED) certification
- A state and federal criminal records check with fingerprints
- A Child Abuse and Neglect (CAN) check
- A passing score on the NFHS coaches' test course
- Completion of the KHSAA Sports Safety Course
- I-9 Form
- Character First training
- Two forms of identification submitted to the JCPS Activities and Athletics Office
- A Direct Deposit Form
- Submission of official college transcript with 64+ college hours or completion of the GE 40 Form and all required NFHS courses
- Employment Information Release Authorization

It is important again to note that school district responsibilities differ from state to state and even from county to county within the same state. School districts can also pass rules about athletics, but no

exhibit **5.5**

WISCONSIN INTERSCHOLASTIC ATHLETIC ASSOCIATION Committee Organization

A charter member of the NFHS in 1923

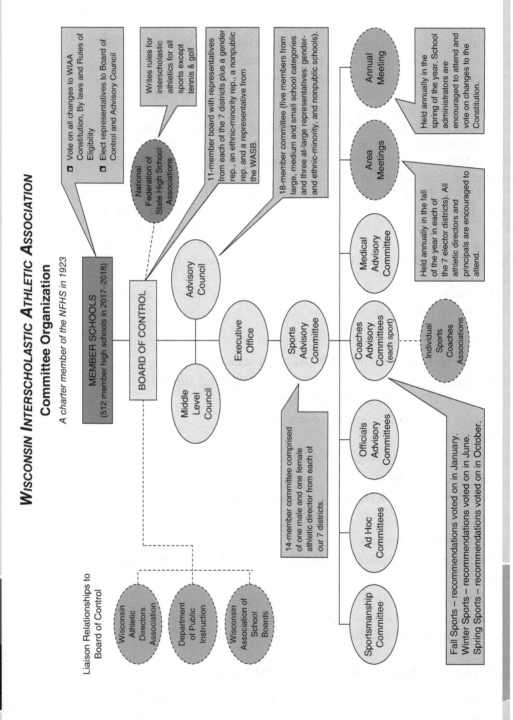

Source: WIAA (n.d.).

rule can be less stringent than the rules of their state association. For example, if a state association says athletes need to have a 2.0 grade point average (GPA) to compete, a school district cannot set their GPA at 1.8.

INDIVIDUAL SCHOOLS. Each high school has someone who serves as the AD. Sometimes this person also works as a coach, although that is not ideal. According to Angst (2019, para. 1), ADs' duties vary and may include the following:

- Providing guidance and direction for a school's sports program
- Preparing budgets and allocating spending on items such as coaches' salaries, team travel, equipment purchases, and facility upkeep
- Coordinating with coaches about the scheduling of games and practices
- Collaborating with conferences and leagues about scheduling issues
- Speaking with league officials about subjects such as postseason play
- Determining the time allocated for a field, court, or weight room
- Working with coaches and perhaps a travel coordinator to plan trips
- Coordinating officials and umpires at games and budgeting for their pay
- Filing reports on the status of each team and its successes and shortcomings
- Mediating any disputes between athletes and coaches or between coaches

The AD is the person who puts forward coaching candidates for the school board's approval. In some states, the AD must also have current teaching credentials. For all of these tasks, the average high school AD's salary varies from state to state. ADs in New Jersey average $67,260, in Wisconsin $64,363, and in California $70,240. Other states are lower, such as Missouri at $39,535, Wyoming at $41,167, and Utah at $35,269 (Zippia Careers, 2021).

In addition, at the individual school level, Principals also have influence in running high school athletic departments. Typically, the AD reports directly to the Principal on any matters concerning budget, scheduling, or personnel. If there are conflicts within the department that the AD cannot resolve, then responsibility may fall to the Principal. The Principal and AD need to work as a team to make sure the department runs smoothly and efficiently. This was never as evident as when the COVID-19 pandemic caused havoc on high school sport.

Spotlight: Diversity, Equity, & Inclusion in Action

National Federation of State High School Associations

As one of its basic beliefs, the NFHS states that student participation in high school sport "fosters the inclusion of diverse populations" (NFHS, 2022a, para. 2). In keeping with this belief, the NFHS announced that in Fall 2022 it will undertake its first-ever book study to support its commitment to diversity, equity, and inclusion (DEI). "With the goal of gaining a better understanding of issues related to diversity, equity and inclusion, participants in this national initiative will be studying a book entitled, From Athletics to Engineering: 8 Ways to Support Diversity, Equity and Inclusion for All" (NFHS, 2022b, para. 3). This action is consistent with the organization's adoption of Diversity, Equity, and Inclusion as one of the NFHS's five strategic priorities in its 2021–25 Strategic Plan. The organization will intentionally involve all 51 member state associations along with stakeholders such as students, coaches, administrators, parents, and others in order to hear from its various constituencies with the goal of making high school sport as accessible to as many people as possible. A number of state high school athletic associations have already incorporated diversity, equity, and inclusion in their governing documents and this action by the NFHS will help support that effort.

CURRENT POLICY AREAS

A number of policy areas are prominent in the governance of high school athletics, including eligibility, risk minimization, gender identity, gender equity, participation by athletes with disabilities, and student-athlete mental health. Sport managers working in this industry segment must be aware of the constantly changing tides of public opinion about issues surrounding school-age children and young adults as these will impact their policy-making decisions. Several of these issues result from legal challenges to existing or proposed policies. As pointed out earlier in the chapter, because state associations are considered state actors, they cannot violate any students' fundamental rights. The same is true for public school employees, including coaches and ADs. Students who believe their rights have been violated can initiate a lawsuit against a coach, an AD, a local school board, and the state high school athletic association. The outcomes of these cases can directly affect policies in high school sport, as sometimes the courts mandate that an association change a rule that violated a student's rights.

Eligibility

The policy area that is always in the spotlight is eligibility. Some of you may have attended a high school where an athlete was denied the opportunity to play for a particular reason – maybe their grades were not high enough or maybe they moved in from a different school district. Perhaps you knew a student who took an eligibility case to the legal system. First, you must remember that playing high school sport is *not* a right guaranteed by law. Attending school until you are a certain age is your right. Participating in interscholastic sport is *a privilege, not a right* (Czekanski et al., 2019; Menke v. Ohio High School Athletic Association, 1981). However, although athletic participation is a privilege, it cannot be taken away arbitrarily. Sport governing bodies must still ensure due process is followed in any decisions they make about eligibility otherwise they may end up in litigation (Czekanski et al., 2019). Remember, state associations are state actors, and while their actions are not supposed to violate anyone's constitutional rights, sometimes they do, resulting in litigation. Eligibility discussions most often relate to questions of ethics, academic eligibility, public versus private schools, transfer rules, and age limits for participation.

Eligibility Rules

Eligibility rules in general generate some interesting ethical questions. For example, a high school athletic association may have certain rules about "No pass, no play." While this seems acceptable on the surface, what is the effect on academically struggling students for whom sport is the primary motivation to stay in school? If that chance is taken away, are they more likely to drop out? Although transfer rules are based on the educational premise that a student learns best by staying in the same school for an academic career, should students be "punished" for family problems that may result in relocation to another school district? It is important to make sure the outcome of such rules is, in fact, as fair as possible to the student.

Academic Eligibility

All high school associations have policies governing academic eligibility. The reason for this is clear. High school sports are meant to be an extension of a student's educational experience.

According to the Kentucky High School Athletic Association (KHSAA, 2019, p. 6), a number of reasons exist for academic eligibility standards:

1. Interscholastic athletic activity programs are an extension of the classroom, and academic standards help ensure the balance between participation in the activity and appropriate academic performance;

2. Interscholastic athletic and activity programs assist in the educational development of all participants;

3. Academic standards promote the objective of graduation from the institution and that student participants are truly representing the academic mission of the institution;

4. Overall, academic standards promote educational standards, underscore the educational values of participating in activities, encourage appropriate academic performance, and allow the use of interscholastic participation as a motivator for improved classroom performance;

5. Participants in the interscholastic athletic program are expected to be student-athletes;

www

Kentucky High School
Athletic Association
https://khsaa.org/

6. High school sports are not intended to be a "farm team" for college and professional sports, but a complementary activity to the total learning experience;

7. Standards shall be in place to ensure that in addition to sports participation, a student shall be on schedule to graduate with his/her class; and

8. As class systems change (block and other alternative schedules), these requirements shall be continually reviewed to make certain that all students are meeting the necessary requirements to graduate from high school and be positive contributors to society.

A focus of discussion in this area is the so-called "No pass, no play" rule (Hayward, 2014). While the specifics differ from state to state, basically this type of rule makes players ineligible to participate for a set number of weeks if they do not meet certain academic standards. For example, students who fail a class in a given term may be ineligible to play sports the following term. Proponents indicate that such rules keep students focused on academics rather than athletics. For athletes seeking college scholarships, these rules also help them stay on a course to meet the NCAA Eligibility Center standards. Opponents of such rules mention that for some students, being able to play sports is what motivates them to stay in school, and without the opportunity to participate, those students may drop out. Opponents also point out that this rule may lead students to choose a less challenging curriculum so they do not put themselves in academic jeopardy. Some states and districts are now revisiting and modifying these rules.

This matter is determined in a simple way in most European countries.

> Being enrolled in and regularly attending classes in a school of secondary education is a primary criterion for eligibility. The second category is age, which cannot of course be waived. When these two conditions are met, the student is eligible to participate in school sport competitions.
>
> (Hums et al., 2011, p. 105)

Transfer Rules

High school associations have transfer rules for various reasons. Transferring schools is generally not easy on high school-age students, either educationally or socially. If you have ever transferred schools, you may remember the difficulties of making new friends, having new teachers, and learning new rules. Transferring raises questions such as will my GPA transfer with me or how will transferring high schools impact my records when I apply for college? Transfer rules work to maintain the ideal of students starting and ending their academic careers in the same school.

Transfer rules have the goal to keep students from moving from school to school for nonfamily reasons. In other words, these rules are in place to keep students from simply enrolling in whichever school has a successful sports program or a well-known coach. School district boundaries are drawn to ensure fair distribution of students across districts. These distributions are tied to the amount of funding the districts receive from the state. Students "jumping" out of district to play sports interrupt this balance. Another reason for transfer rules is to deter coaches from recruiting students away from other schools. High school athletes, especially the talented ones, have enough pressure on them already. People are wondering where they will go to college or maybe go pro directly out of high school, their daily lives are under constant social media scrutiny, they are likely traveling regionally and nationally to competitions with elite teams in their sport, sport apparel companies may already be approaching them, and they still have to maintain their grades for eligibility. (And now throw in the trickle-down effect of the new collegiate Name, Image, and Likeness situation!) Attempting to recruit these students to rival high schools only adds another layer of pressure to their lives.

Risk Minimization

Participating in any sport or physical activity carries with it some element of risk of injury. That is just the nature of such activities. Responsible sport managers know, however, that it is their job to keep participants as safe as possible, particularly when those participants are children or minors. High school sport administrators have been given a series of directives from the National Federation to help them minimize risk in certain sports. While the concept of risk minimization usually makes us immediately think of football or hockey, one area which has garnered considerable interest of late relates to spirit activities (e.g., cheer and dance). The NFHS recently outlined guidelines emphasizing safety in certain aspects of spirit. For example, there are now rules about inversions, release stunts, and tosses as well as allowing athletes to wear religious head coverings without having to get prior state association approval (NFHS, 2021c). In other sports, baseball has pitch limits and required rest periods. Water polo adopted new rules allowing athletes to wear approved soft padded caps.

The list of risk minimization steps related to COVID protocols is still evolving. Mask wearing, social distancing, and hand hygiene are now standard procedures for high school athletics settings. One could easily make the case that many of these protocols should remain in place once the pandemic ends just to ensure the health and safety of athletes, coaches, and fans in protecting themselves and others from other forms of communicable illnesses such as new COVID variants or the seasonal flu. While trying to minimize these risks, what are some important lessons we have learned? According to Dr. Karissa Niehoff, Executive Director of the NFHS, these are some lessons to take away from these challenging times (Niehoff, 2021):

The importance of participation
The mental and emotional health of students is tied to participation

We must be thankful and appreciative
There is more than one way to accomplish goals and dreams
We must show respect for opponents, officials, and others
We can do anything
Fans and the community at large are essential for high school
sports and the performing arts.

These are important lessons for high school athletic administrators to
remember as they move their programs safely forward.

Transgender Athletes

We are all born the way we are and should be able to live our lives
in our true identities. However, in recent years, state legislation in the
United States has targeted a group of high school students that is systemically discriminated against - transgender individuals. For example,
many state legislatures are now working to pass transphobic legislation which would prevent trans-athletes from participating on high
school sport teams that match their gender identity. On the other hand,
"Fifteen states and Washington, D.C., currently have Trans-inclusive
state athletic association guidance, and years of open participation by
transgender students in those places have produced no evidence of pur-
ported harms to cisgender people" (Goldberg & Santos, 2021, para. 2).
These authors continue by discussing the current status (at the time this
textbook was being written) of policies related to transgender people in
sport:

- The Biden administration recently interpreted the US Supreme
  Court's decision on *Bostock v. Clayton County* as prohibiting
  gender identity discrimination under Title IX. This interpretation
  will protect transgender students from discrimination, including in
  the context of school sports participation.
- 15 states and Washington, D.C. – together home to more
  than 6.8 million high school students and approximately
  42% of transgender high school-age youth – have policies
  allowing transgender students to participate in school sports
  without requirements of medical or legal transition.
- Other states either offer no guidance or have policies requiring
  transgender students to experience administrative scrutiny or
  undergo medical procedures in order to participate. Six states
  limit participation solely based on athletic institutions' individual
  definitions of "biological sex" rather than scientific evidence or
  individual identity.
- The NCAA, the International Olympic Committee (IOC), and
  various professional and amateur leagues have long allowed
  transgender athletes to participate in accordance with their
  gender identity, even though both the NCAA and the IOC

recently changed their frameworks for the participation of transgender athletes. The NCAA's change in policy has been heavily critiqued for its exclusionary nature while the IOC's emphasis on inclusion has been applauded by experts (Kliegman, 2022).

■ In 2020, 20 states introduced high school transgender sports bans, one of which passed in Idaho and has since been enjoined, with the court noting an "absence of any empirical evidence" justifying the law. In 2021 so far, dozens of states have introduced similar bills, even where lawmakers cannot cite a single case of a transgender girl participating in sports (Goldberg & Santos, 2021, paras. 6–10).

How should high school athletic administrators act in such a fluid environment dealing with a topic which elicits strong emotion on all sides? Here are suggestions for tactics:

Thoroughly educate yourself and all members of your department on the scientific facts about gender identity

Learn the appropriate language (pronouns, names) to use when working with people of all gender identities

Together with any transgender or nonbinary athlete, develop a communication plan for opposing teams and also officials for upcoming competitions

Monitor any legislative activity in your state so that you know the current status of any potential new laws which might affect your state high school athletic association

Know and understand the language contained in your state athletic association's handbook regarding gender identity. A useful example of an inclusive policy is provided in Exhibit 5.6 – the handbook language of the Oregon State Activities Association (OSAA)

Be supportive and open and let your athletes know you are someone they can talk with to discuss this topic in a safe space

Educate yourself on the basic human rights of people of all gender identities so that you can make informed decisions for all of your athletes and their families

Make no mistake – this is a topic where sport and politics intersect on a very human level. Organizations such as Athlete Ally (2018) provide solid information to help athletic administrators and coaches if they have questions as to how to proceed. An athletic administrator who is properly educated in the facts about gender identity will be the one to be a true leader in making sure all athletes have the opportunity to participate in high school sport in a safe and respectful manner.

www

Athlete Ally
https://www.athleteally.org/

exhibit   **5.6**   OSAA Gender Identity Participation Policy

39. GENDER IDENTITY PARTICIPATION (Winter 2019)

The OSAA endeavors to allow students to participate in the athletic or activity program of their consistently asserted gender identity while providing a fair and safe environment for all students. As with Rule 8.2 regarding Duration of Eligibility/Graduation, rules such as this one promotes harmony and fair competition among member schools by maintaining equality of eligibility and increasing the number of students who will have an opportunity to participate in interscholastic activities.

This policy was developed in consultation with the Oregon Department of Education (ODE). The OSAA recognizes that this policy will need to be reviewed on a regular basis based on improved understanding of gender identity and expression, evolving law, and societal norms....

B. Participation.

For both historical reasons, as well as reasons related to compliance with Title IX, interscholastic athletics and activities have typically been divided by gender, with a few exceptions. Formulating new processes to address concerns about participation regardless of a student's gender identity requires a new approach to eligibility, an approach reflected in these policies. In interpreting these policies, the OSAA recognizes the value of activities and sports for all students and the potential for inclusion to reduce harassment, bullying, and barriers faced by certain students.

1. As is true with all eligibility determinations, the student's member school will be the first point of contact for determining the student's eligibility. When a student registers for athletics or activities the student shall indicate the student's gender during that registration process, consistent with other school enrollment procedures. Athletics and activities personnel should refer to member school processes for registration/ enrollment information. Disputes regarding these gender identity determinations will be resolved solely at the member school level; because of the diversity of private and public school rules that may bear on such determinations, and gender identity issues being particularly sensitive, the OSAA will not hear any appeal of a member school's determination made under this section.
2. Subject to section B(1), once a transgender student has notified the student's school of their gender identity, the student shall be consistently treated as that gender for purposes of eligibility for athletics and activities, provided that if the student has tried out or participated in an activity, the student may not participate during that same season on a team of the other gender.
3. Subject to section B(1), once a nonbinary or intersex student has notified the student's school of their gender identity, the student shall be treated as either gender for purposes of eligibility for athletics and activities that are gender-segregated or gender-specific, provided that If the student has tried out or participated in athletics or an activity that is gender-specific or gender-segregated, the student may not participate during that same season on a team of the other gender.
    1. Q. If a transgender student is transitioning from one gender to another, what is the procedure for that student to access athletics and activities?
        A. When a student or the student's parent or guardian, as appropriate, notifies the school administration that the student will assert a gender identity that differs from previous representations or records,

the OSAA will recognize a school's decision to modify the student's eligibility, consistent with the student's gender identity, subject to section B(2).

2. Q. What communication or support plans need to be put in place when a transgender or nonbinary student is participating in athletics or activities?

    A. Privately ask the student what is needed for support. All students may ask for privacy in locker rooms and restrooms as well as possible accommodations when traveling with a team. Schools should refer to their district policies when developing support plans.

3. Q. What if a nonbinary or intersex student experiences gender fluidity during a season that is documented at school by pronoun change or is of a transitional nature?

    A. As a student transition, communication should be documented within school registration information consistent with other school procedures. If the activity in which the student is trying out for is gender-segregated or gender-specific, then the student shall commit to the team with which they register for the entirety of that season, subject to section B (3).

4. Q. Can a nonbinary student access more than one sport or activity during the same season?

    A. Yes, provided that a student may not participate in gender-segregated and/or gender-specific sports/ activities at the same time but is otherwise eligible to participate in all sports/activities that are not gender-segregated or gender-specific.

5. Q. What is the procedure for athletics or activities that are not gender-segregated or gender-specific?

    A. If a sport or activity is not gender-segregated or gender-specific (for example, speech, football, etc.) students would not need to elect any specific gender in order to participate.

Source: OSAA (2019).

GENDER EQUITY

Whenever we hear the term *gender equity*, one phrase should come to mind immediately – Title IX. Although most of the publicity generated around Title IX during its now 50+ year existence has involved college athletics, this piece of legislation also applies to high school sport. The full title of this historic legislation in the United States is Title IX of the Educational Amendments of 1972. It reads as follows: "No person in the United States shall, on the basis of sex, be excluded from participation in, be denied the benefits of, or be subjected to discrimination under any education program or activity receiving Federal financial assistance" (US Department of Justice, n.d., Overview of Title IX section).

While great progress has been made for girls in high school sport, there is still work to be done. In 1972, only 295,000 girls competed in high school sports, a mere 7.4% of all high school athletes, compared to 3.67 million boys. By the 2018–19 school year, the number of girls had swelled to 3.4 million, while the number of boys was 4.5 million. Yet, 40% of teen girls are not actively participating in sports and more than one million more boys are participating than girls (Women's Sports Foundation, 2021).

www

Women's Sports Foundation
https://www.
womenssportsfoundation.org/

Discussions about gender equity and fairness in participation opportunities will continue, even as the courts and legislative agencies interpret and reinterpret the laws. The importance of sport participation for girls cannot be overstated. Benefits from participation include reducing the risk of chronic illness, lowering obesity levels, lowering instances of smoking and drug use, improving academic performance, raising levels of self-esteem and self-confidence, teaching how to work as a team and set goals, and increasing the likelihood of getting involved in their community (Burlington Basketball, 2021). There are, of course, many more reasons but even this short list illustrates the importance of sport participation for girls. Athletic administrators should consider how they respond to Title IX and whether they are providing opportunities not just because the law mandates it but also because it is the right thing to do.

Participation by Athletes with Disabilities

Athletes with disabilities have successfully challenged age limit rules that restrict their full access to participation in high school sport. Beyond that, athletes with disabilities are now gaining opportunities to participate more equally alongside their classmates.

Most notable was a young track athlete from Maryland, Tatyana McFadden. McFadden, born in Russia with spina bifida, attended Atholton High School and wanted to compete on the school's track team. A wheelchair racer who had won silver and bronze medals in the 2004 Paralympic Games in Athens, Greece, McFadden was initially blocked from participation. She won the right to compete with able-bodied athletes in track meets after she successfully sued the Howard County Public School System in federal court in 2006. Ultimately, Maryland passed the landmark Fitness and Athletics Equity for Students with Disabilities Act in 2008 (Graham & Seidel, 2010):

> The Maryland Public Secondary Schools Athletic Association changed its laws to accommodate those athletes. New language was added ...
> to the MPSSAA [Maryland Public Secondary Schools Athletic Association] bylaws, allowing students with disabilities to participate in school sports programs as long as they meet preexisting eligibility requirements, are not ruled to present a risk to themselves or others, and do not change the nature of the game or event.
> (Graham & Seidel, 2010, para. 1&2)

McFadden has since gone on to be one of the most medaled athletes in the Paralympic Games, having won multiple gold, silver, and bronze medals, including four golds at the Rio 2016 Summer Paralympic Games.

Other states are now following Maryland's lead in providing opportunities for high school athletes with disabilities. According to Ohio University (2020, para. 2), "The National Federation of State High

School Associations (NFHS) said equality, equity, and fairness are the ultimate goals for the 30 or so state associations currently offering state championships in inclusive sports."

Opportunities for high school students with disabilities took an enormous step forward with the issuance of a Dear Colleague Letter from the Department of Education (DOE) early in 2013. Similar to how Title IX expanded opportunities for girls and women in sport (Crenshaw, 2013), the DOE's guidelines stress that students with disabilities must be treated equitably with regard to interscholastic sport participation opportunities. In summary:

> On January 24, 2013 the Office for Civil Rights issued a Dear Colleague Letter clarifying a school's obligations under the Rehabilitation Act of 1973 to provide extracurricular athletic opportunities for students with disabilities. It creates a clear road for how schools can integrate students with disabilities into mainstream athletic programs and create adapted programs for students with disabilities.
>
> (Active Policy Solutions, 2013, p. 1)

Participation opportunities made available by schools to allow students with disabilities to participate in athletics to the greatest extent possible can include:

- *Mainstream programs* – school-based activities that are developed and offered to all students.
- *Adapted physical education and athletic programs* – programs that are specifically developed for students with disabilities.
- *Allied or unified sports* – programs that are specifically designed to combine groups of students with and without disabilities together in physical activities.

This Dear Colleague Letter represents a watershed moment for young athletes with disabilities.

Student-Athlete Mental Health

It is important to realize that high school student-athletes' health involves more than just their physical well-being. Statistics show that teens are at risk for anxiety, depression, and behavioral problems (CDC, 2021). High school athletes fall squarely into the age group which deals with such issues. While sports participation has many positive outcomes for high school students, sport participation can also contribute to anxiety and depression (Flanagan, 2019). Overtraining; injuries; social media scrutiny; and pressure from parents, peers, family, and local community people all can make sport participation tough for high school athletes. We have seen a number of high-profile athletes speak out about their mental health – Olympic swimmer Michael Phelps, tennis star Naomi Osaka, the NBA's Kevin

Love, and world-renowned gymnast Simone Biles to name just a few. These athletes serve as role models to let others, especially younger athletes, realize the importance of mental health. What can athletic administrators do to help ensure their student-athletes' mental health is a priority? Morin (2020) suggests a list of actions that includes encouraging a young athlete to live a balanced life, teaching relaxation techniques, avoiding actions that make pressure situations worse, recognizing and helping an athlete work through injuries appropriately, not allowing sports to become the young athlete's identity, and watching for substance abuse. High school athletic administrators can also look to college sports for some guidance.

> College-sports programs [have] started to address the problem by adding therapists to athletic offices, screening players for anxiety and depression, and educating staff about how to identify mental-health issues. These steps could ratchet up the pressure on high schools to follow suit.
>
> (Flanagan, 2019, paras. 11–12)

No matter their age we must always remember, "Athletes, and their physical and mental health, are not commodities" (Crouse, 2021, para. 16).

SUMMARY

Sport governance takes place at a variety of levels, from international to national to state to local. Remember that the real power in scholastic sport lies with the state associations, although at times the courts will mandate their activities if any policies violate students' fundamental rights. The rules-setting and regulatory powers truly reside in these associations. The NFHS acts as a service organization for its state members. School districts, principals, and ADs can set rules on their levels, but they must be in accord with the state rules. In the United States, high school associations have repeatedly been identified as state actors and therefore subject to the US Constitution for their actions.

The types of governance and policy issues scholastic sport administrators must deal with are vast and complex. The governance decisions they make when setting policy will interact with state as well as federal legislative bodies and laws such as Title IX. They will often come in contact with the judicial system as well, and so they must be prepared for their decisions to be questioned in courts of law. Despite these considerations, scholastic sport administrators have the opportunity to provide programming that has a positive impact on the lives of thousands of young athletes. It is an exciting and personally fulfilling segment of the sport industry.

HENRY POPE *Director of Athletics*

Birmingham Alabama City Schools

As the district-level athletic administrator, I serve as the compliance officer for the district member schools regarding the rules and laws of our state and national association. I am also responsible for ensuring each school, and we as a district, effectively execute the proverbial "Fourteen Duties of the Athletic Administrator."

One of the most important issues confronting sport managers and athletic administrators today is obvious in our position: the shortage of athletic game officials. Athletic participation positively impacts the lives of students and adults, but the decline in teaching sportsmanship negatively impacts participation by umpires, referees, and judges. Alabama is experiencing a great shortage of baseball umpires, so much so that we had to reschedule middle school and junior varsity contests because there were not enough officials to provide even one umpire (much less two) for a varsity contest scheduled that week. Physical and verbal attacks on officials drive them out of sports and leave our programs scrambling to ensure proper coverage.

The next is a bit more general for consumers: supply shortages. The shortage/delay in mineral items (metal, plastic, rubber, etc.) creates a scarcity of items required for safety and protection (helmets, shoulder pads) and even balls for competition. Student safety is paramount in our profession, and supply shortages delay field replacements (artificial turf) and gym floor surfaces upkeep. As the AD, providing safe competition sites for the scholar-athletes during practices and contests is one of our duties and scarcity is a challenge.

We are all taught that athletic participation is privilege, not a right. That pertains to both athletes and coaches. For years, coaches have taken advantage of opportunities to coach multiple sports, but when do they have the opportunity to grow and develop as a coach?

Our district has implemented a new policy regarding coaching overlapping sports to ensure no scholar-athlete experiences a disservice or participates in unsupervised activities due to no coaches being available at the start of a new sport season because they are finishing a current sport season. "No one may serve as coach of overlapping sports." This allows the coach to decompress as well as participate in clinics and learning opportunities to develop professionally and remain abreast of changes within their sport. It also ensures our scholars are ready on Day One to begin official start dates of practice and have off-season development time with their coaches.

As member schools of the Alabama High School Athletic Association (AHSAA), we are under the umbrella of the National Federation of High School Associations (NFHS). Association members ratify the AHSAA bylaws through annual proposals and votes. Just like in politics, we all have a voice, but we also make the rules which govern us. No one wants the "Wild West of Sports" in their association.

This can also be a negative just as in politics because the majority is the voice most heard. That places an independent school at a great disadvantage because of travel restrictions due to distance or finances which may be adversely affected by the decisions of the more financially affluent majority or those with access to unlimited resources.

One of the greatest positives of our structure is having a state governance association to ensure rules are followed, including a process to discipline those who violate those rules. However, having an association that understands how times change and the need to ratify bylaws to meet the needs of all member schools is as important as having a Central Board of Control for grievances and appeals made up of members from each district.

As with all things athletic, high school sports benefit from decisions made in collegiate and professional sports, but we also suffer due to them as well. The "Name, Image, and Likeness" (NIL) rulings are one of the greatest future obstacles facing high school sports today. We will have to monitor and mitigate the locker room distractions, the impact on recruiting due to the financial status of one district versus a rural or inner city school, and must work to ensure student-athletes retain their amateur status.

Before you choose a career in high school sports, understand that the financial compensation will never match the work required to lead or serve your students. You must know that this career will be most gratifying when you are a part of the overall development of students becoming great citizens who display sportsmanship.

When challenges arise and you believe that your work is not meeting your expectations, always realize there is a student coming to school every day and working diligently to succeed because of the impact you are making in their life. Keep working and keep pushing because you are accomplishing the ultimate goal of making a positive difference in the lives of our scholars!

case STUDY

SCHOLASTIC SPORT

You work for your state's high school athletic association as the organization's social media coordinator. The Commissioner has assembled an ad hoc committee to choose one new sport to add to the list of the state's approved sports. The ad hoc committee has come up with a list of the following sports to be added:

a. eSports
b. trap and skeet shooting
c. skateboarding
d. taekwondo

1. Go online and watch two videos about each sport.
2. List two positive attributes which would make adding each sport a good choice.
3. List two potential drawbacks to adding each sport.
4. Outside of your committee, who would you ask for input on the choice of sport?
5. The Committee has decided to allow the public in the state to vote for their choice. How would you get information out to them? What information would you make available? Who would be eligible to vote and how would the voting take place?

CHAPTER questions

1. If you played high school sport, choose two benefits you think you gained from that experience and provide a specific example of each. (E.g., For one of our book authors, I played 4 sports in high school, so I learned time management. I also played on a state championship volleyball team and we won despite being a school with a small enrollment of only 550 students. I learned the value of working together as a team to overcome the odds!)

2. Where does the main power lie in high school sport? Why?
3. High school sport has a number of goals, including education and participation. Should one of these goals be the preparation of college athletes?
4. Think back on your days in high school. Did your school have a good AD? If yes, what were two reasons for your answer? On the other hand, did you have an AD who was not so good? What were two reasons for your answer?

FOR ADDITIONAL INFORMATION

1. This link takes you to an article about the member associations of the NFHS: https://www.nfhs.org/articles/state-high-school-associations-come-in-all-shapes-and-sizes/
2. This link takes you to an article explaining some of the complexities of high school athletes transferring schools: https://www.nfhs.org/articles/helping-students-parents-to-understand-transfer-rules/
3. Here is the NFHS' current stance on name, image, and likeness for high school athletes: https://www.nfhs.org/articles/nil-rulings-do-not-change-for-high-school-student-athletes/

REFERENCES

Active Policy Solutions. (2013). *Q and A: Disability in Sport Dear Colleague Letter*. http://www.activepolicysolutions.com/wp-content/uploads/2013/01/OCR-Dear-Colleague_Q-and-A_2-15-13-revised-for-website.pdf

Angst, F. (2019, September 23). *What does an athletic director do?* https://www.thebalancecareers.com/athletic-director-job-profile-3113301

Athlete Ally. (2018). *Resources*. https://www.athleteally.org/resources/

Brentwood Academy v. Tennessee Secondary Sch. Athletic Ass'n, 531 U.S. 288 (2001)

Burlingtom Basketball. (2021, April 19). *The many benefits for girls in sport*. https://burlingtonbasketball.ca/the-many-benefits-for-girls-in-sports/

CDC. (2021). *Data and statistics on children's mental health*. https://www.cdc.gov/childrensmentalhealth/data.html

CHSAA. (2020). *Constitution and bylaws of the Colorado High School Activities Association*. https://chsaanow.com/wp-content/uploads/bylaws/2020-21-chsaa-bylaws.pdf

Crenshaw, S. (2013). *Sports are a civil right for the disabled, according to US Education Department directive*. https://www.al.com/sports/2013/01/sports_are_a_civil_right_for_d.html

Crouse, L. (2021, July 27). Simone Biles just demonstrated a true champion mindset. *New York Times*. https://www.nytimes.com/2021/07/27/opinion/culture/simone-biles-just-demonstrated-a-true-champion-mind-set.html

Czekanski, W.A., Siegrist, A., & Aicher, T. (2019). Getting to the heart of it all: An analysis of due process in interscholastic athletics. *Journal of Legal Aspects of Sport, 29*, 152–170. https://doi.org/10.18060/22311

Flanagan, L. (2019, April 17). Why are so many teen athletes struggling with depression? *The Atlantic*. https://www.theatlantic.com/education/archive/2019/04/teen-athletes-mental-illness/586720/

Forsythe, L.L. (1950). *Athletics in Michigan schools: The first hundred years*. Prentice-Hall.

GHSA. (2020). *GHSA 2020–2021 budget*. https://www.ghsa.net/sites/default/files/documents/financials/GHSA-Budget-2020-21.pdf

Goldberg, S., & Santos, T. (2021, March 18). *Fact sheet: The importance of sport participation for trans youth*. Center for American Progress. https://www.americanprogress.org/issues/lgbtq-rights/reports/2021/03/18/497336/fact-sheet-importance-sports-participation-transgender-youth/

Graham, G., & Seidel, J. (2010, March 25). Public schools open sports to athletes with disabilities. *Baltimore Sun*. https://www.baltimoresun.com/sports/bs-xpm-2010-03-25-bal-va-rule25mar25-story.html

Hayward, L. (2014, December 24). No pass no play heads into fourth decade. *Midland Reporter-Telegram*. www.mrt.com/sports/article/No-Pass-No-Play-heads-into-fourth-decade-7419312.php

Hums, M.A., MacLean, J.C., & Zintz, T. (2011). *La gouvernance au coeur des politiques des organisations sportives*. Traduction et adaptation de la 2e édition américaine. Groupe De Boeck.

IHSA. (2013). *About the IHSA*. https://www.ihsa.org/About-the-IHSA/ConstitutionBylawsPolicies

KHSAA. (2019). *By-laws of the Kentucky High School Athletic Association governing high school participation (grades 9–12)*. https://khsaa.org/common_documents/handbook/bylaws.pdf

Kliegman, J. (2022, July 6). Understanding the different rules and policies for transgender athletes. *Sports Illustrated*. https://www.si.com/more-sports/2022/07/06/transgender-athletes-bans-policies-ioc-ncaa

Menke v. Ohio High School Athletic Association, 2 Ohio App.3d 244, 245 (1981).

Morin, A. (2020, July 5). *How to help a teen athlete deal with sports pressure*. Very Well Family. https://www.verywellfamily.com/how-to-help-a-teen-athlete-deal-with-sports-pressure-4057989

MSHSL. (2020). *Official handbook*. https://www.mshsl.org/sites/default/files/2020-07/handbook_2020-21_generalinfo_web_2.pdf

NFHS. (2020a). *Annual report 2019–2020*. https://www.nfhs.org/uploadedfiles/3dissue/AnnualReport/2019-20annualreport/index.html

NFHS. (2020b). *NFHS handbook 2019–2020*. https://www.nfhs.org/media/1020439/2019-20-nfhs-handbook.pdf.

NFHS. (2021a). *About us*. https://www.nfhs.org/who-we-are/aboutus

NFHS. (2021b). *Centennial celebration*. https://www.nfhs.org/100years.

NFHS. (2021c). *Reducing injury risk during inversions, release stunts and tosses focus of 2021–22 high school spirit rules changes*. https://www.nfhs.org/articles/reducing-injury-risk-during-inversions-release-stunts-and-tosses-focus-of-2021-22-high-school-spirit-rules-changes/

NFHS. (2021d). *Staff*. https://www.nfhs.org/who-we-are/Staff

NFHS. (2021e). *State association listing*. https://www.nfhs.org/resources/state-association-listing

NFHS. (2022). *Mission statement*. https://www.nfhs.org/who-we-are/missionstatement.

Niehoff, K. (2021, April 14). *What have we learned during the pandemic in high school sports, performing arts?* NFHS. https://www.nfhs.org/articles/what-have-we-learned-during-the-pandemic-in-high-school-sports-performing-arts/

NJSIAA. (n.d). *Inside NJSIAA*. https://www.njsiaa.org/inside-njsiaa

Ohio University. (2020). *Implementing and maintaining inclusive sports programs*. https://onlinemasters.ohio.edu/blog/implementing-and-maintaining-inclusive-sports-programs/

OSAA. (2019). *Gender identity participation*. https://www.osaa.org/docs/handbooks/GenderIdentityParticipationBP.pdf

OSAA. (2022). *About the OSAA*. https://www.osaa.org/about

PIAA. (2017). *Introduction*. www.piaa.org/about/introduction.aspx

Rader, B.C. (1999). *American sports: From the age of folk games to the age of televised sports* (4th ed.). Prentice Hall.

Seymour, H. (1990). *Baseball: The people's game*. Oxford University Press.

Sharp, L.A., Moorman, A.M., & Claussen, C.L. (2021). *Sport law: A managerial approach* (4th ed.). Routledge.

US Department of Justice. (n.d.). *Title IX legal manual*. https://www.justice.gov/crt/title-ix#II.%C2%A0%C2%A0%20Synopsis%20of%20Purpose%20of%20Title%20IX,%20Legislative%20History,%20and%20Regulations

Vincent, T. (1994). *The rise of American sport: Mudville's revenge*. University of Nebraska Press.

WIAA. (2021). *Washington Interscholastic Athletic Association annual budget: Fiscal year 2021–2022*. https://www.wiaa.com/ConDocs/Con132/WIAA%202021-22%20Budget.pdf

WIAA. (n.d.). *WIAA committee organization*. https://www.wiaawi.org/Portals/0/PDF/organization_grid.pdf

Women's Sports Foundation. (2021). *Our research*. https://www.womenssportsfoundation.org/what-we-do/wsf-research/

Wong, G.M. (1994). *Essentials of sport law*. Praeger.

Zippia Careers. (2021). *Athletic Director: Best states*. https://www.zippia.com/athletic-director-jobs/best-states/

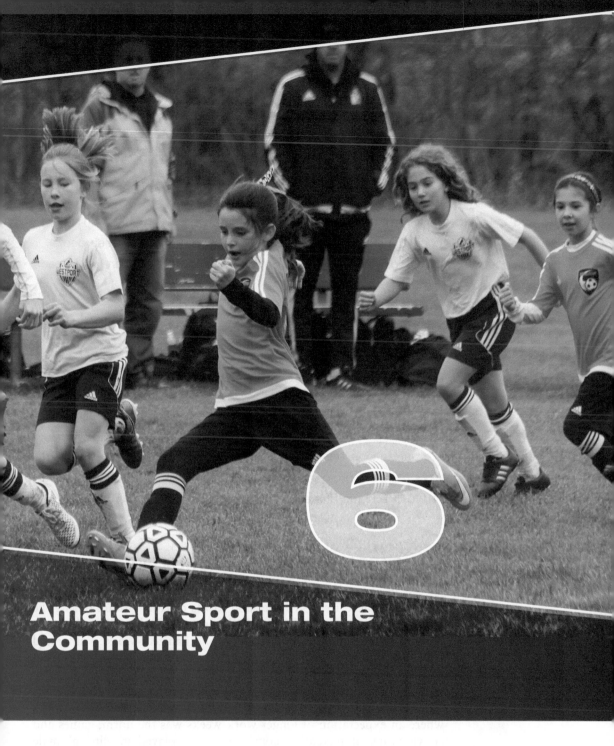

6

Amateur Sport in the Community

The term *amateur sport* describes a diverse set of individual and group sporting activities engaged in by millions of people worldwide. Different people play for different reasons including enjoyment, group affiliation, fitness, healthy living, and the joy of competition. Amateur athletes do not get paid for their efforts. Rather, a great many amateur sporting activities involve participants who simply choose to play. Participants

DOI: 10.4324/9781003303183-6

range from young children to senior citizens to people with disabilities, and their involvement is usually in addition to their primary responsibilities with jobs or schools. Amateur sports include highly competitive events like the National Collegiate Athletic Association (NCAA) championships and the Little League Baseball World Series. They also include less-competitive activities such as organized Quidditch tournaments and father–daughter golf events. At times, extensive media coverage presents the glitz and glitter of highly competitive amateur sporting activities, such as high school and college championship games, but a local weekend beach volleyball tournament with thousands of participants may go unnoticed. As sport management students you need an understanding of how community amateur sport entities are organized and governed because many of you will, at some point in your career, either be in a leadership position in such organizations or work with these events in some capacity as a volunteer or paid staff member. You may be responsible for setting policy and ensuring the effective pursuit of organizational goals in this extensive segment of the sport industry.

Organizations delivering amateur sport in the community for youth and adults are numerous and offer a variety of activities. In fact, amateur sport in the community has a rich history and an abundance of community structures delivering opportunities for participation. The organizations that govern and establish policy for amateur sport are normally categorized as public or nonprofit. To begin this chapter, we will discuss how amateur sport for members of the community first developed and became organized.

HISTORY OF COMMUNITY AND YOUTH SPORT

The roots of modern-day amateur sport in North America might be traced to the villages and towns of rural Britain during the industrialization of the 18th and 19th centuries (Kidd, 1999). As farming and other industries became more mechanized, workers migrated from the countryside to cities, and later to North America, bringing their enjoyment of physical activity with them. Traditionally, farm workers competed in folk games and other precursors to today's athletic events. Although such activities were scarce in overcrowded cities where an expectation of longer work weeks was the norm, games and active forms of recreation continued to be played in elite, all-male schools (Morrow & Wamsley, 2009). As sport for the elite became more popular and better organized, an interest in participating quickly spread to upper-class girls and women and to working-class boys and men. In a similar way, participating in amateur sport emerged and gained momentum in many parts of the world.

In the United States and Canada, adaptation of British sports and development of new games began in the early 1800s. Indigenous people were accomplished runners, climbers, swimmers, and canoeists, and they participated in many tests of skill and strength. The indigenous Algonquian tribe invented lacrosse (then called stick ball) for entertainment, religious reasons, or to prepare for combat (Kinney, 2021). Settlers from England and other parts of Europe brought their own games and co-opted the indigenous games and sports they found on the already occupied land. While working-class men and women had little leisure time to devote to athletic contests, they still participated and enjoyed watching during holidays and other special events. By the mid-19th century, North Americans were engaging in a wide variety of athletic contests. An increase in population, the changing nature of work in an increasingly mechanized industrial world, and decreased working hours provided increased leisure time and paved the way for the development of public and nonprofit sport organizations.

PUBLIC SPORT ORGANIZATIONS

As the urban population increased during the mid- to late 19th century, housing density amplified in urban centers and local governments began to scrutinize municipal sporting activities and private sporting clubs. Originally, their interest was in regulating leisure practices. Governments declared public holidays, dedicated land for parks and sporting activities, and enacted laws prohibiting activities they considered immoral or improper, such as racing and gambling on Sundays. Municipalities subsidized sport competitions such as rifle shooting and banned the rowdiness and immorality thought to be associated with highly publicized prize fights. As the numbers and types of activities and publicity increased, generating a greater number of eager participants, public sport organizations developed. Fledgling sport leagues were formed. Such leagues were managed by groups of individuals who set schedules, adapted and enforced rules and regulations, and promoted and publicized events such as baseball, football, rugby, and track and field. The leagues frequently led to the development of municipal groups, where teams were assembled from the top local talent to represent the entire community. Such a team would then enter into competition by challenging another team from a nearby town. While travel was difficult and kept to a minimum in the early days, teams often endured substantial travel distances in an effort to reap the glory of victory for their hometowns. Competitions between teams from different communities led to the development of sport festivals and jamborees, the precursors to today's state games in the United States and provincial games in Canada. As sport gained interest and became organized by community groups, the need for more opportunities and diversity in sport offerings soon resulted in the development of nonprofit sport organizations.

NONPROFIT SPORT ORGANIZATIONS

A *nonprofit sport organization* delivers programs and services for a particular sport or group with no intent to gain profit. These types of organizations range from very small (Portland's Rose City

Hockey all girls club) to large professionally run associations (the North Texas Youth Football Association). Nonprofit sport organizations emerged as an alternative to programs such as recreational sport leagues run by city recreation departments and in addition to other programs developed with the express intent of making money. In the beginning, nonprofit organizations filled the gap in programming between the two and provided opportunities for participation in sporting events regardless of class or financial background. For-profit organizations offered programming based on business strategies, inevitably providing only the most popular activities. It was impossible for public recreation departments to offer all possible types of sports. Therefore, interested individuals formed their own organizations according to their own interests. Nonprofit sport organizations emerged all over North America for sporting interests as diverse as waterskiing, bicycling, walking, and badminton.

Public and nonprofit sport organizations began developing organizational structures, constitutions, positions of leadership with duties and responsibilities, and programs. What are the governance structures for these types of organizations, and how do they develop policy?

Rose City Hockey Club
http://www.
rosecityhockeyclub.com/

North Texas Youth Football Association
https://www.leaguelineup.
com/welcome.asp?url=ntyfa

GOVERNANCE

As illustrated by the history of community sport, amateur athletic organizations are structured in a manner consistent with their purpose and mission. While private for-profit ventures exist (such as climbing gyms, gymnastic centers, and figure-skating clubs), most amateur sport in the community is publicly run with funding from some level of government or delivered by nonprofit service organizations. What types of groups fall within these categories, and how are they organized to deliver amateur sport within the community? The following sections will identify the missions, funding, membership, and organizational structures of community amateur sport organizations.

THE GOVERNANCE OF PUBLIC SPORT ORGANIZATIONS

The three main types of public sport organizations delivering amateur sport in our communities are:

1. City Parks and Recreation Departments
2. recreational sport leagues
3. state games and provincial games

Urban Ascent Rock Climbing Gym, Boise Idaho
https://www.urbanascent.
com/

Gold Medal Gymnastics Center, Long Island NY
https://gmgc.com/

Ann Arbor Figure Skating Club, Michigan
https://www.annarborfsc.org/

City Parks and Recreation Departments

City Parks and Recreation Departments have traditionally housed community sport, recreation, and physical activity programs. Cities provide a wide array of services to their citizens, including utilities such as sewers, water, gas, and electric. Cities also provide public transportation and care for infrastructures such as roads and bridges. In addition to these basic services, many cities also take it upon themselves to offer a wide array of sport and recreation facilities and programming.

MISSION. City Parks and Recreation Department mission statements are as varied as the activities they offer. They usually include themes such as opportunities for leisure-time activities, learning and playing in a safe environment, provision of a wide variety of facilities, and statements of inclusivity and support for diversity. Two sample mission statements, one of the Parks and Recreation Department of the City of Houston (Texas) and one of the City of Onalaska (Wisconsin), are presented in Exhibits 6.1 and 6.2.

MEMBERSHIP. Generally, these programs are open to any and all residents of a particular city. Established policies deal with participation by non-residents and guests. The people who take part in physical activities offered by city parks and recreation programs have a wide variety of activities to choose from. Depending on location and facilities, offerings could include swimming, soccer, softball, fitness programs, martial arts, tennis, hiking and biking trails, and many others. Activities are not limited to the traditional team sport offerings, as departments try to keep up with trends by offering popular activities. For example, the City of Seattle Parks and Recreation (SPR) Department manages the following:

> a 6,414-acre park system of over 489 parks and extensive natural areas. SPR provides athletic fields, tennis courts, play areas, specialty gardens, and more than 25 miles of boulevards and 120 miles of trails. The system comprises about 12% of the city's land area. SPR also manages many facilities, including 27 community centers, eight indoor swimming pools, two outdoor (summer) swimming pools, four environmental education centers, two small craft centers, four golf courses, an outdoor stadium, and much more.
>
> (City of Seattle, 2022)

www
City of Houston Parks & Recreation
https://www.houstontx.gov/parks/

www
City of Onalaska Parks & Recreation
https://cityofonalaska.com/parkrec

www
Seattle Parks and Recreation Department
http://www.seattle.gov/parks/

Mission Statement of the City of Houston Parks and Recreation Department	*exhibit*	**6.1**

To enhance the quality of urban life by providing safe, well-maintained parks, and offering affordable programming for our community.

Source: City of Houston (2022).

Mission Statement of the City of Onalaska, Parks & Recreation Department	*exhibit*	**6.2**

The City of Onalaska, Park & Recreation Department is committed to enhancing the quality of life for the citizens of Onalaska through promotional development, maintenance of public recreation through enrichment opportunities, parklands, related facilities, and the preservation of natural areas.

Source: City of Onalaska, Wisconsin (2022).

City of Kissimmee Parks & Recreation
https://www.kissimmee.gov/departments/parks-recreation

City of Snellville Parks & Recreation
https://www.snellville.org/parks-recreation

City of San Diego Parks & Recreation
https://www.sandiego.gov/park-and-recreation/

Boise Timbers | Thorns Soccer Club
https://www.boisetimbersthorns.org/

Glencoe Baseball Association
https://www.glencoebaseball.org/

FINANCIALS. Because these facilities and staff are provided by the city, city residents' tax money underwrites a good portion of the costs. As a result, some facility use and programming may be offered free of charge, such as swimming at the neighborhood public pool. Other services may require a fee. For example, a city-sponsored softball league may require teams to pay an entrance fee to cover the costs of umpires, field maintenance, and equipment for each game.

The size of a city parks and recreation budget will vary from city to city. For example, the City of Kissimmee, Florida, had a 2020 budget of $6,502,074 (of the total $107 million for the entire City of Kissimmee general fund), of which personal services (employees' wages, salaries, and benefits; $4.93 million), operating expenses ($1.16 million), and capital outlay ($414 million) made up the expenditures. The City of Kissimmee received money for their parks and recreation department from ad valorem taxes (which often include property taxes), utility taxes, charter school fees, impact fees, sales taxes, grants, and a utility surcharge (City of Kissimmee, 2021). A much smaller city, Snellville, Georgia, budgeted for $927,765 in expenses in 2021 (City of Snellville, 2022).

ORGANIZATIONAL STRUCTURE. A City Parks and Recreation Department is one of a number of departments within the organizational structure of a city. Exhibit 6.3 shows how parks and recreation fits into the overall organizational chart for the large city of San Diego, California. Exhibit 6.4 illustrates how the organizational chart for the City Parks and Recreation Department can become very complex in larger cities like San Diego. By contrast, the parks and recreation department for a smaller city (population 23,500) such as Kerrville, Texas, is presented in Exhibit 6.5.

Recreational Sport Leagues

Recreational sport leagues provide opportunities for regular participation in sport for both children and adults. Leagues might be established by an interested group of individuals who wish to play basketball on a regular basis, or they may be run by community recreation staff in city parks and recreation facilities. Leagues are commonly available in a wide variety of sports, like football, baseball, hockey, curling, volleyball, soccer, and bowling. Most recreational sport leagues are considered a public service. They are organized to provide individuals the opportunity to participate in their sport of choice.

MISSION. A recreational sport league's mission statement will likely include some language about what sport the league is delivering and at which level of competition. The league may be highly competitive or designed with more of a recreational focus on fun. Fair play, respect, and ethical conduct are common components of recreational sport league mission statements. See Exhibits 6.6 and 6.7 for the mission statements of the Boise Timbers | Thorns Soccer Club in Boise, Idaho, and of the Glencoe Baseball Association in Glencoe Park District, Illinois.

MEMBERSHIP. Recreational sport leagues are organized for a vast array of participants. Some activities target children and youth groups,

Organizational Chart for the City of San Diego

exhibit **6.3**

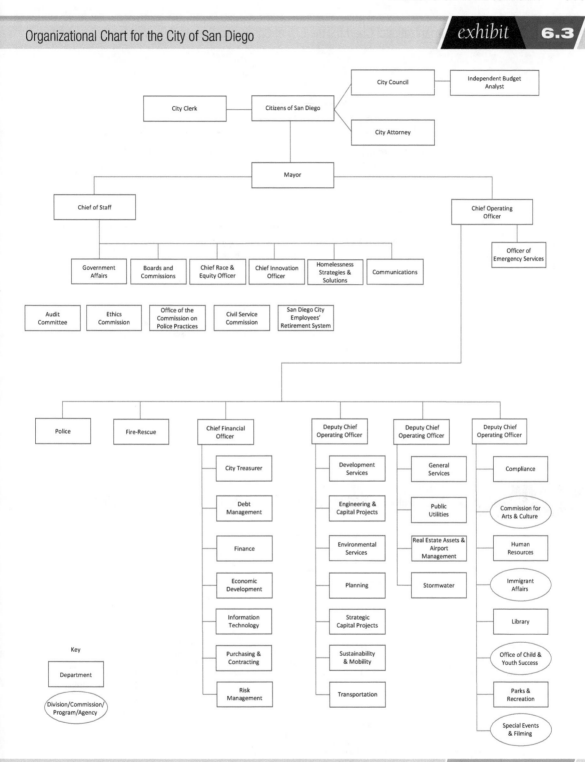

Source: City of San Diego (2022a).

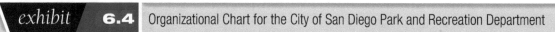

exhibit 6.4 Organizational Chart for the City of San Diego Park and Recreation Department

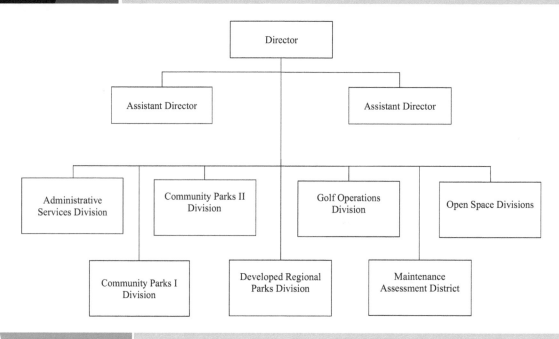

Source: City of San Diego (2022b).

exhibit 6.5 Organizational Chart for the City of Kerrville, Texas, Parks & Recreation Department

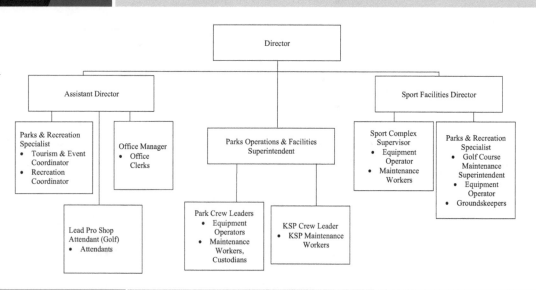

Source: City of Kerrville, Texas (2021).

other leagues are run specifically for teens and young adults, and still others include adults, both young and old. Leagues might be gender-specific or co-ed. The type of membership is mainly determined based on the mission of the organization.

FINANCIALS. Funding for recreational sport leagues is often provided through municipal sources and supplemented by participation fees. The salaries of administrators designing and delivering the league may be paid through city government departments. The costs of facilities are also borne by the community. Sponsors have become a major part of funding for recreational leagues, too, and range from local "mom and pop" stores, to individuals, or corporations. Participants will often pay a small fee to generate some revenue toward paying for administrative costs, officials, and equipment. Of course, the "pay for play" fee could be more substantial if the particular sport has expensive requirements, such as ice rink rental for hockey leagues. The major expense categories for these local sport leagues include facility rental, officials, purchase of equipment, and promotion and publicity.

ORGANIZATIONAL STRUCTURE. A group of officers is usually elected to help organize and govern the activities of the league. These voluntary positions might include a President, Vice President, and Chairs of a few committees specific to the particular sport. For example, one might expect a beach volleyball league to be governed by a League Executive Committee that comprises the President, Vice

Mission Statement of the Boise Timbers | Thorns Soccer Club Boise Timbers | Thorns Mission Statement	*exhibit* 6.6

The Boise Timbers | Thorns, building on a proven history, is a family-focused destination club for youth soccer in the region. With an emphasis on character, education, and innovation, we provide high-quality opportunities and pathways for players and coaches to reach their highest potential both on and off the field as mission.

Source: Boise Timbers | Thorns (n.d.).

Mission Statement of the Glencoe Baseball Association, Illinois	*exhibit* 6.7

The Glencoe Baseball Association's (GBA's) primary mission is to provide an environment for Glencoe youth to have an enjoyable, safe, team-oriented, and community-based experience while learning and playing the game of baseball. At the same time, the GBA is committed to teaching our youth the life lessons of good sportsmanship and teamwork. Through the game of baseball, the GBA intends to help youth to set and work toward common goals and to build character, leadership ability, and confidence.

Source: Glencoe Baseball Association (2021).

President, Chair of Scheduling, and Chair of Facilities. The President runs meetings and provides overall direction and leadership, while the Vice President might be responsible for handling league promotion, complaints, and discipline. The Scheduling Chair develops and communicates all league scheduling, and the Facilities Chair schedules event locations and coordinates with facility staff for equipment and any necessary personnel. The elected members of the Executive Committee develop policies and discuss issues and ideas regarding league operation at an Annual Meeting.

State Games and Provincial Games

State games (United States) and provincial games (Canada) are amateur sport festivals held every year or two. Individuals and teams may have to qualify to attend the games by successfully advancing through regional competitions, by gaining entry through a lottery, or on a first-come, first-served basis. These games are usually multisport events, held in both summer and winter. For example, the Iowa Games Annual Sport Festival is held in winter and summer locations each year. Over 60 summer and 24 winter sports are organized for both adult and youth athletes, reaching 14,000 summer and 4,000 winter participants. The Iowa Games are a multisport festival of Olympic-style competition for Iowa's amateur athletes, and in this case, age, ability, and gender are not considered criteria for participation (Iowa Games, 2017).

WWW
Iowa Games
https://www.iowagames.org/

MISSION. The mission of state and provincial games focuses on delivering well-organized amateur sport competition for athletes of a variety of ages. The idea of the games is to offer citizens the opportunity to compete, gain experience, and come together in a festival atmosphere. The games are usually multisport and designed for participants, coaches, officials, spectators, volunteers, and sport managers—an experience for everyone. See Exhibit 6.8 for the mission statement of the Iowa Games.

MEMBERSHIP. The members of summer or winter games hosted by a US state or Canadian province include mostly participants and volunteers. The organization does not have a group of individual members outside the event but is composed of a paid professional staff and a volunteer group that functions as a Board of Directors to organize the events.

FINANCIALS. State and provincial games are funded through both public and private sources, entry fees, and money raised through sponsors and marketing initiatives. The Iowa Games, for example, is

exhibit **6.8** Mission Statement of the Iowa Games

The mission of the Iowa Games is to provide sports and recreation opportunities for all Iowans through Olympic-style festivals, events, and programs.

Source: Iowa Games (2017).

a project of the Iowa Sports Foundation made possible because of the financial support from "corporate sponsorships, donations, grants and entry fees" (Iowa Sports Foundation, 2022, para. 3). Entry fees typically depend on the sport. Adult volleyball, for instance, is $140 for a team, whereas fencing is $25.

ORGANIZATIONAL STRUCTURE. The paid staff usually comprises an Executive Director, one or more Directors of Sports, Event Operations, and Finance, and a number of assistant positions who help to organize specific components of the games. The Board of Directors, all volunteers, includes a Chair or President, Vice President, Treasurer, Secretary, and a number of board members who may be responsible for specific aspects of the events.

www

Iowa Sports Foundation
https://www.
iowasportsfoundation.org/

THE GOVERNANCE OF NONPROFIT ORGANIZATIONS INVOLVED IN COMMUNITY SPORTS

The term *nonprofit* aptly describes many community organizations involved in sport. These organizations deliver activities and services with no intent of making a profit. They may be large or small and may have a simple or an intricate organizational structure. Prominent examples of nonprofit amateur sport organizations include the Y [formerly the Young Men's Christian Association (YMCA)], the Young Women's Christian Association (YWCA), the Boys & Girls Clubs, and the Jewish Community Centers (JCC). Other local nonprofit community groups also provide opportunities for amateur sport.

THE Y [FORMERLY YMCA]

The Ys are service organizations and collectively represent the leading nonprofit committed to strengthening community through youth empowerment and personal and community wellness. The Ys operate in over 10,000 communities (cities and surrounding areas) in the United States alone (The Y, 2022b). The organization recently rebranded by shortening its name to the "Y" as a way of emphasizing its programs' impact on youth and healthy living. The name change resulted from more than two years of research orchestrated by the national YMCA of America, which revealed that most people do not understand the organization's activities and mission. The Ys provide programming for children and adults, for men and women of all races, religions, abilities, ages, and incomes. They have a significant history in basketball, volleyball, and racquetball and were the original leaders in camping and fitness, as well as in providing children with swimming lessons (The Y, 2022b). The Ys serve more than 64 million people in 120 countries around the world. In the United States, there are just under 2,700 Ys reaching 11 million people.

MISSION. The Y's mission revolves around putting Christian values into practice to build a strong spirit, mind, and body. But, what are

www

The Y
https://www.ymca.org/

those values? The Ys values are four pillars that are taught throughout their educational classes: caring, honesty, respect, and responsibility. The Y believes that each member has a responsibility to treat each individual, regardless of age, backgrounds, or walks of life with care, honesty, and respect. A large component of the Y's mandate is delivered via amateur sports programs, including leagues, instructional classes, family nights, youth sports programming, mentoring, and exchange programs. Each Y strives to nurture the healthy development of children and teens and strengthen families in order to make the community a better place. The mission involves the development of the "whole body" through programs that often incorporate physical activity. See Exhibit 6.9 for the mission statement of the Y (The Y, 2022a).

MEMBERSHIP. The Ys are a part of community life in neighborhoods and towns across North America. In the United States, over four million youth and seven million adults enjoy their services each year. Several types of memberships are available, from family, to couple, to individual. In keeping with their values, the Y will offer membership based on flexible pricing. For instance, if a family's income is $0–$26,000, the family can receive 90% off the price of a membership. The scale slides down (e.g., 80%, 70%, 60%, etc.) and the more a family makes, the less the membership is discounted. After a family income reaches $86,000, there is no discount for a membership (The La Crosse YMCA, n.d.). Individuals can also have the ability to join only to participate in specific programs or groups.

FINANCIALS. Each local Y is an independent, charitable, and nonprofit organization required to pay dues to the National Council. Of the total revenue the National Council collects per year, most of the revenue comes from contributions and donations from individuals, foundations, corporations, and trusts (~33% of total revenue), member YMCAs that pay a due to the National Council (~29%), in-kind donations (oftentimes media, public service announcements; ~19.5%), and government grants (~7%). Conversely, awards and grants to member associations (~32%), personnel costs (~24%), and advertising (~20%) make-up the largest expenses. In 2020, the National Council of the Y brought in $169 million in revenue and spent $153 million in expenses in 2020 (The Y, 2021). As a typical local-level example, the YMCA of Greater Rochester, New York, derives most of their revenue from membership fees (25.7%), program fees (23.8%), and their annual campaign (20.6%). The majority of their expenses are employee expenses (49%), facilities and occupancy (16.9%), and program expenses (7.7%; YMCA of Greater Rochester, 2021).

ORGANIZATIONAL STRUCTURE. The Ys rely on volunteers and full-time paid professional staff who help set policy that is then implemented by both employees and volunteers. Most operate with a volunteer Board of Directors, steered by an Executive Committee elected from board members. Other committees work on specific types of programs or initiatives, like youth sports, clubs and camps, and family nights. The local board has jurisdiction over the development of policy for the independent Y, as long as the independent Y meets

| Mission Statement of the Y | *exhibit* | 6.9 |

To put Christian principles into practice through programs that build healthy spirit, mind, and body for all.

Source: The Y (2022a).

the following requirements as outlined in the national constitution (The Y, 2022):

1. Annual dues are paid by the local Y to the national office, the Y of the United States.
2. The Y refrains from any practices that discriminate against any individual or group.
3. The national mission is supported.

Other decisions, including program offerings, staffing, and style of operation, are the purview of the local Y. The organizational structure of a typical Y is shown in Exhibit 6.10.

YWCA

The YWCA seeks to respond to the unique needs of local communities through advocating for justice, health, human dignity, freedom, and care of the environment. It has been working to raise the status of women since its inception in the 1850s (YWCA, 2022a). The YWCA aims to provide safe places for girls and women no matter what their situation. The YWCA USA is focused on three high-impact areas: (a) racial justice and civil rights, (b) empowerment and economic advancement for women and girls, and (c) health and safety of women and girls (YWCA, 2022a).

www

YWCA
https://www.ywca.org/

MISSION. Nationally, the YWCA is committed to a strategic framework of consultation and collaboration with local YWCAs around the United States. The YWCA embraces its legacy as a pioneering organization that squarely confronts social justice issues to make lasting, meaningful change. The organization's mission is to eliminate racism, empower women, stand up for social justice, help families, and strengthen communities (YWCA, 2022a).

MEMBERSHIP. The YWCA has more than 25 million members in 120 countries, including 2.3 million members and participants in 225 local associations in the United States. The YWCA is one of the oldest and largest women's organizations in the nation, serving women, girls, and their families.

FINANCIALS. The YWCA's revenues and expenses are similar to the Y's based on the missions and setup of the organizations. The YWCA USA's revenues include mostly contributions and donations (72%), support fees from local YWCAs (16%), and investments (11%).

In terms of expenses, local initiatives (support for local YWCA's to further their missions; 50%), advocacy (14%), communications (13%), and management and administrative fees (12%) make-up most of the YWCA USA's costs. In total, the YWCA brought in $13,021,483 in revenues and $9,725,598 in expenses for the year 2020 (YWCA USA, INC, 2020).

ORGANIZATIONAL STRUCTURE. The national office is charged primarily with conducting advocacy at the national level and for marketing and branding. At the YWCA Annual Meeting in May 2012, a transition from the prior regional structure to a national federated

---

exhibit **6.10** MetroWest YMCA Organization Chart

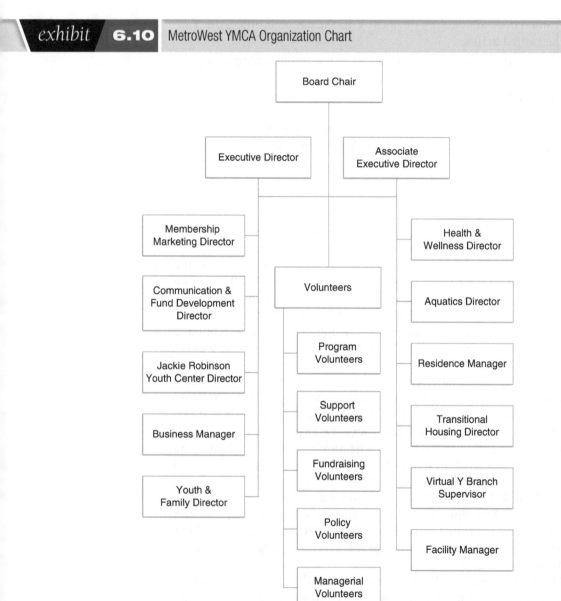

Source: MetroWest YMCA (n.d.).

structure was approved, followed by the adoption of new bylaws in November 2012:

> The goal of the new organizational structure is to leverage the collective strength of local YWCAs and the national organization as a national movement. Under the new structure, YWCA USA provides a range of technical assistance and capacity-building services to local YWCAs, safeguard the integrity of the YWCA brand, and maintain a strong national presence that will position the iconic, multi-tiered organization for increased stability and sustainability for the future.
>
> (YWCA, 2022b, para. 7)

All members of the YWCA USA are eligible to attend an Annual Meeting and any Special Meetings called to elect board members and conduct business. Voting is conducted through two votes afforded by all the members of local associations. The Board of Directors manages the affairs of the national YWCA, focusing on mission, programs, national advocacy, member communication, branding, and strategic collaboration. Local YWCAs manage their own programming and personnel, and areas critical to their specific mission.

BOYS & GIRLS CLUBS

The Boys & Girls Clubs of America is a national association of community boys and girls clubs. Such clubs offer programs and services to promote and enhance the development of boys and girls in a safe, healthy environment. Sports programming is one component of the services provided by Boys & Girls Clubs.

www

Boys & Girls Clubs of America
https://www.bgca.org/

MISSION. The Boys & Girls Clubs of America acts to provide boys and girls with the following:

- Safer Childhoods: Providing safe and fun places for kids to grow and thrive.
- Life-Enhancing Programs: Delivering engaging programs focused on academics, health, and leadership.
- Caring Mentors: Offering trained staff who guide, coach, and motivate kids to be successful (Boys & Girls Club of America, 2022a).

The Boys & Girls Clubs of America strives to provide opportunities for young people, especially those from disadvantaged backgrounds. See Exhibit 6.11 for its mission statement.

MEMBERSHIP. The Boys & Girls Clubs has more than 4,700 club facilities in the United States that serve 4.3 million young people, two million through membership, and 2.3 million through community outreach. Of the club members, approximately 55% are male and 45% are female with 36% of the members being ages six–nine, 30% being 10–12, and 19% ages 13–15. A race and ethnicity breakdown consists of 30% White, 27% Black, 23% Latino, 6% two or more races, 3% American Indian or Alaska Native, 3% Asian, and 1% Native Hawaiian or Pacific

To enable all young people, especially those who need us most, to reach their full potential as productive, caring, and responsible citizens.

Source: Boys & Girls Clubs of America (2022 b).

Islander (Boys & Girls Clubs of America, 2021). Membership rates are kept to a minimum ($10–$50 per year), and no one is turned away from programming because of an inability to pay. On average, about $500 is spent per youth in a given year. The clubs provide a variety of unique sport programming, tournaments, and the JrNBA and JrWNBA programs. Clubs are located in schools, youth centers, and public housing and are found in cities, rural communities, and on native lands. Other specialized learning programs, such as the Athletic Director University program, teach youngsters leadership skills to organize and administer their own leagues, clubs, and teams.

FINANCIALS. Boys & Girls Clubs are nonprofit organizations that rely heavily on gifted support. In fact, 49% of their revenues are generated through contributions from individuals and corporations. Government grants another 34%, investments 11%, and membership dues provide 3% of the total revenues of $362 million in 2021. The Boys & Girls Clubs spend 48% of their expenses on assistance for clubs and establishment of new clubs, 42.5% on leadership training, support, and development of youth programs, 4.5% on fundraising, and 4% on management and general costs. The Boys & Girls Club had a total of $302 million in expenses for the year 2021 (Boys & Girls Clubs of America & Subsidiaries, 2022).

ORGANIZATIONAL STRUCTURE. There may be as many as 25 clubs in a region, each operating with a small professional staff and many volunteers. The Boys & Girls Clubs of Metro Atlanta, for example, are organized with a large Board of Directors who are volunteers from the community elected to oversee the strategic operations of the organization (Boys & Girls Clubs of Metro Atlanta, 2022). The board is elected and is composed of 10 unit board presidents from regions within the geographical area and 41 board members (Boys & Girls Clubs of Metro Atlanta, 2022). The Board of Directors and the association members are responsible for developing policies such as the division of revenues between each of the metro clubs. Policies may determine that the use of revenues collected by individual clubs be defined without restriction by the club, but that revenues distributed to individual clubs by the overarching Atlanta parent group be used in specific areas such as programming, professional staff, and facilities.

The organizational structure of a typical Boys & Girls Club is shown in Exhibit 6.12.

www

Boys & Girls Clubs of Metro Atlanta
https://bgcma.org/

Typical Board of Directors for a Boys & Girls Club of America

exhibit **6.12**

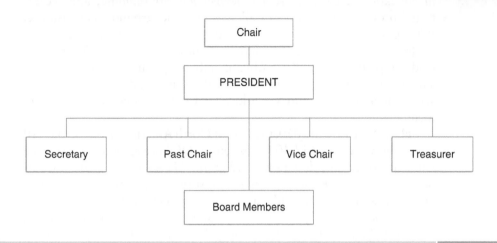

Source: Boys & Girls Club of Oshkosh (2022).

JEWISH COMMUNITY CENTERS

JCCs are nonprofit organizations under the authority of the Jewish Community Center Association of North America, an umbrella organization including more than 172 JCCs servicing 1.5 million people each year (JCCA, 2022d). The Jewish Community Center Association offers a wide range of services and resources to help its affiliates provide educational, cultural, social, Jewish identity-building, and recreational programs for people of all ages and backgrounds.

Each year, the JCC Association puts together a sporting festival for Jewish teens known as the JCC Maccabi Games. These games are an Olympic-style sporting competition held each summer on multiple sites in North America and are the largest organized sports program for Jewish teenagers in the world (JCCA, 2022b). In addition, the JCC Association helps sponsor the Maccabiah Games, which is essentially the quadrennial Jewish Olympic Games held in Israel the year following the Summer Olympic Games.

MISSION. JCC programs, services, and connections are intended to support a diverse and thriving Jewish people—and as appropriate, the broader community—in making fulfilling and healthy life choices. Basic tenets of JCC programs include valuing Jewish wisdom, respecting difference, building communities, encouraging creativity, and realizing aspirations in exciting and powerful new ways (JCCA, 2022c). The JCC of Greater Baltimore, the nation's first JCC, defines its mission as to promote and strengthen Jewish life and values through communal programs and activities for individuals and families (JCC of Greater Baltimore, 2022).

MEMBERSHIP. The JCC Association serves more than 172 JCCs in North America, reaching an estimated 1.5 million people annually. Each

www

Jewish Community Center Association of North America
https://jcca.org/

www

JCC Maccabiah Games
https://www.jccmaccabigames.org/

of these organizational members pays annual dues to the association. The association offers its members a wide range of services and resources in educational, cultural, social, Jewish identity-building, and recreational programs for people of all ages and backgrounds. For instance, JCCs put on artistic performances; day and overnight camps; fitness, wellness, sports, and aquatic activities; and training and education on leadership. Individual JCCs offer memberships in various categories. The choices include individual memberships, family memberships, senior memberships, and some local JCCs even offer full community memberships.

FINANCIALS. In 2020, the JCC of North America saw $13,731,910 in revenues and $10,613,404 in expenses. Of those revenues, most came from contributions, donations, and grants (51%) and dues from member JCCs across the country (16%). Most of the expenses included program enrichment (37%) and management and general fees (21%; JCC of North America, 2021). Focusing on a specific JCC member institution, the JCC of San Francisco revenues were $27.5 million and expenses were $31.4 million. Program revenue (38%), fitness center (35%), and contributions and grants (19%) made up most of the revenues. Conversely, the fitness center (25%), youth and family services (19%), management and general fees (16%), and adult services (11%) made up most of the expenses at the local JCC level (JCC of San Francisco, 2021).

WWW
JCC of San Francisco
https://www.jccsf.org/

ORGANIZATIONAL STRUCTURE. The JCC Association of North America is organized with an Executive Committee and a Board of Directors. The Executive Committee, formed by elected officers, consists of a President, a Chair, seven Vice-Chairs, a Secretary, two Associate Secretaries, and seven Honorary Chairs. The Board of Directors is very large, consisting of 50 members (JCCA, 2022a). The JCC Association has three North American offices, as well as one in Jerusalem, Israel. The North American offices are located in New York City; Austin, Texas; and St. Paul, Minnesota.

The JCC of San Francisco is led by an Executive Director and an executive staff, a group of elected officers and a volunteer Board of Directors. Along with the Executive Director, the elected officers include a President, a Vice President, a Treasurer, and a Secretary. The volunteer Board of Directors is also large, consisting of 16 members (JCC of San Francisco, 2022).

COMMUNITY GROUPS

Many individuals and groups run nonprofit organizations for sport and physical activity. These may be community based but not run by community government agencies such as City Parks and Recreation Departments, and they usually focus on a single sport or physical activity. Community groups such as the Pacers and Racers Running and Walking Group located in New Albany, Indiana, and the Mountain View Masters Swim and Social Club located in Mountain View, California, have organized themselves to participate in activities related to their sport. The San Juan Sledders Snowmobile Club of

Durango, Colorado, is another example of a small nonprofit amateur sport organization (San Juan Sledders, 2017). This group marks and grooms winter trails in the area, encouraging their use for multiple outdoor winter sports such as dog sledding, cross-country skiing, snowshoeing, and snowmobiling. The Traverse City Curling Club in Michigan was opened in 2014 as a nonprofit community sport organization when more than 500 people showed up for curling demonstrations and practice (Traverse City Club Curling, 2022).

MISSION. The mission of a nonprofit community-based amateur sport organization involves delivering a particular activity by providing facilities and related services, such as arranging for individual participation or competition. Many community groups involve leagues or clubs through which regular activities are scheduled. Encouraging safe, ethical conduct and a good learning environment for all participants is a common theme. The mission statement of the Dallas Nationals Baseball Club is presented in Exhibit 6.13.

MEMBERSHIP. Volunteers run community nonprofit amateur sport organizations. Members of the organization are solicited to help organize the activities. In turn, they gain the benefit of services such as events, tournaments, and championships. Membership dues are usually collected on a yearly basis.

FINANCIALS. Nonprofit community groups are typically funded through participant membership fees, sponsorships, and other methods of fundraising. They may also be eligible for grants extended by

WWW
Pacers and Racers Running and Walking Group
https://pacersandracers.com/running-group/

WWW
Mountain View Masters Swim and Social Club
https://mvm.org/

WWW
San Juan Sledders Snowmobile Club
https://www.sanjuansledders.com/

WWW
The Traverse City Curling Club
https://tccurling.org/

Mission Statement of the Dallas Nationals Baseball Club *exhibit* **6.13**

The DALLAS NATIONALS BASEBALL CLUB mission is to prepare young men for the next level of competition both on and off the diamond. Our coaches and staff are selected specifically to improve baseball skills and knowledge; however, since there is more to life than baseball, we will also work to build character at every opportunity.

While we definitely aren't the oldest program in the area, our staff is made up of the best coaches and instructors from all of the other organizations that got tired of seeing a tainted product on the diamond and decided to get together and build a program that was respected and produced. The Dallas Nationals coaching staff is committed to assisting each young man to fulfill their baseball dreams and aspirations at each and every level of play from recreational to select, JV to varsity, or varsity to college. We are extremely proud to have many success stories at each level; even sending Dallas National athletes to college on scholarships to play baseball at the highest levels.

Since most tournaments we enter will be played on weekends and we do not like our families to choose between church and baseball, the Dallas Nationals will have optional devotionals on Sunday led by parents and staff. The devotionals are by no means a replacement to your current church home, but rather a viable option for praise and worship in order to create community within the organization and study the Bible together. Bible study will be non-denominational, but Christ-centered.

Thank you for taking the time to learn more about the program available through the DALLAS NATIONALS BASEBALL CLUB. We look forward to working with you toward your son's athletic goals and providing him the opportunity to succeed in our national pastime!

Source: Dallas Nationals Baseball Club (2022).

municipal government agencies or private foundations. For example, a "fun run" organized by a city running club might involve municipal support (facilities and race-day logistics), sponsorships (printing costs, post-race refreshments, and awards), and a small entry fee to cover the costs of race T-shirts and timing.

ORGANIZATIONAL STRUCTURE. Depending on its size, a community sport organization might be loosely configured or highly structured. A smaller organization such as the San Juan Sledders Snowmobile Club might have a small Executive Committee consisting of a few members in positions such as the President, Vice President, and Treasurer. Other members of the group are then given tasks, but often they hold no specific titles. More often, the President will provide overall leadership, and the members of the group will complete tasks for specific events. An expectation exists for a fair division of labor among all group members. Once initial policies are established for managing events, the group focuses on service as opposed to being a rules-making or sanctioning body. Conversely, in a larger organization such as the Dallas Nationals Baseball Club, a structured managerial group may provide leadership with a President, several Vice Presidents, and a number of Committee Chairs.

Spotlight: Diversity, Equity, & Inclusion in Action

Sport for Social Development in Indigenous Communities (SSDIC) Program

Sport for Social Development in Indigenous Communities (SSDIC) Program

Canada's history is similar to the United States in that European settlers came to North America, colonized the land, and severely threatened or extinguished indigenous culture, language, and social systems. In order to reconcile the colonialism of Canada, Sport Canada created the Sport for SSDIC Program. The SSDIC Program is designed to reduce barriers to sport for indigenous communities for roughly 150 communities and 33,000 indigenous youth (Beyond Sport, 2021). The targeted outcomes for the SSDIC are to improve health, education, and employability, while reducing at-risk behavior (Government of Canada, 2021).

The SSDIC Program offers three forms of funding:

- Stream One: $5.3 million is made available annually to the 13 Provincial/Territorial Aboriginal Sport Bodies and Aboriginal Sport Circles for development of sport.

- Stream Two: $3.6 million is available for indigenous governments, communities, and organizations to promote sport in indigenous communities.

- Stream Three: $2.5 million is available annually to ensure indigenous women and girls have meaningful sport activities.

Eligible organizations can apply for one of the streams given their mission and objectives. The review committee is made up of federal regional office members and indigenous stakeholders to ensure regional and indigenous knowledge in the project evaluation. The SSDIC Program is an initiative designed from the Truth and Reconciliation Commission of Canada's report that urged all levels of government to work together to change policies and programs to create a more inclusive and harmonious country (Beyond Sport, 2021).

CURRENT POLICY AREAS

Several policy areas are prominent for managers of amateur sport organizations in the community. It is important that policies be defined for each of the areas described in this section because the issues are critical to the effective delivery of amateur sport, especially as related to children and children's programming. Effective policy enables effective decision-making.

Fundraising

Policy enabling decision-making for fundraising is a prominent issue for amateur sport organizations in the community. Community-based amateur sport organizations rely on raising capital for both programming and infrastructure. Public sport organizations compete with every other level of social programming for funding and government grants often fall short of their needs. As important as fundraising is to public sport organizations, policy that enables effective financial management is even more important with nonprofit sport organizations. Managers in the nonprofit sector are continually concerned with maintaining strict measures for raising money, along with ensuring that budgets are balanced and spending is effective and shared among as many programs as possible. For instance, the Boys & Girls Clubs of America rely heavily on donations from both the private and the public sectors. While it takes money to run a Boys & Girls Club (about $500 per youth per year), the alternative costs of keeping a young adult in jail for a year are many times higher.

The fundraising policy areas receiving most of the attention are the following:

1. **Identifying Fundraising Sources.** Most organizations set specific policies that identify sources of funds worth pursuing. Most funding sources include private donations, government grants, and corporate sponsorships. Specific policies might list donors in order of priority based on which are most likely to enter a partnership and in specific categories of new contacts to pursue. In addition, such policies will certainly outline categories of unacceptable sources of funding. For instance, no Boys & Girls Club would solicit funds from a tobacco company. On the contrary, the club will work hard to disassociate unhealthy practices such as smoking from the club's programming for children and teenagers. The policy might also define special event types of fundraising (charity dinners, silent auctions, golf tournaments, etc.) and set some parameters on exactly what events will be hosted and what goals will define success.
2. **Soliciting Donations and Sponsorships.** Other policies will define exactly how donations and sponsorships are solicited and by whom. Will potential donors be called, contacted by mail, e-mail, phone, and personal communications? Will the

club website be used to initiate fundraising? How will money be received, and how will records be kept? It is important now more than ever that donating should be easy for the individual or corporation. Developing effective policies will answer questions such as these.

3. **Servicing and Maintaining Donor Relationships.** Ensuring donors and sponsors are "serviced" is very important as well. This means giving back to the donor or sponsor. Once programming has support, every effort must be made to inform, involve, and thank the sponsors for their involvement. Courtesy and reciprocity are critical to building relationships and are the best ways to maintain donor or sponsor involvement from year to year. Newsletters, invitations to see programs in action, thank you letters from participants, and summarized information outlining the positive outcomes resulting from the donation are all effective means of servicing. The Boys & Girls Clubs of America use an extremely prominent National Board of Governors and Officers as well as alumni in their fundraising efforts and most certainly want to maintain these high-level benefactors who include high-profile individuals such as Condoleezza Rice, Ken Griffey Jr., Jennifer Lopez, and Denzel Washington (Boys & Girls Club of America, 2022c).

Inactivity of Youth

Currently, the inactivity of youth generally and the declining numbers of participants in amateur sport have been targeted as trends that community-level sport programs can help reverse. Policy is often set to deal with recurring issues and to enable decision-making to effect change regarding such issues. All too often, children and youth are less physically active than is optimal for health and wellness because of the prevalence of passive activities such as videos and computer games. Research shows that only 24% of children 6–17 years of age participate in 60 minutes of physical activity per day pre-COVID (Centers for Disease Control and Prevention, 2020). Post-COVID, however, children ages 5–13 became more likely to decrease their physical activity and increase their sedentary behavior as the pandemic continued. These changes could create permanently entrenched behaviors, leading to a myriad of health issues (Dunton et al., 2020).

Some community-based amateur sport organizations are setting policy to ensure that programming exists for all youth age groups. The goal is to provide a variety of fun-filled, supportive sport and physical activity experiences for youth. Such policy promotes skill development, healthy lifestyles, and improved self-esteem. For example, some municipalities have enacted policies to ensure opportunities exist for age-group participants in sports that require the use of already overbooked areas and fields. At times, the policies specify participation for particular groups, such as a girls- or boys-only activity. In many community amateur athletic organizations, such as the City of Columbus, Ohio Recreation and Parks Department, T-ball programs and Little League Baseball are offered for girls as well as boys. Beyond gender, issues of race, ethnicity, disability, and urban versus rural further complicate the issue of youth inactivity. The reasons impacting this issue are complicated and involve a mix of societal issues such as culture, income, family responsibilities, and the impact of technology.

Parental Involvement

Parental involvement has been a major topic of youth sport over the past several years (Positive Coaching Alliance, 2022). Stories of parents being "muzzled" or banned from events are too frequently presented in the media. Adult misbehaviors have become more commonplace: splashing hot coffee in the face of an official; verbally abusing officials, coaches, and kids; and fighting, threatening, and other forms of confrontations. According to Dan Bylsma, former National Hockey League (NHL) player and former coach of the Pittsburgh Penguins and Buffalo Sabres, two questions must be answered: Why have parents become so invested in the progress of their children in sports to the exclusion of other arguably more important endeavors, such as academics? And why does this parental involvement contradict what is best for the child? Amateur sport in the community is about fun, teamwork, dedication, and respect for authority. Excessive parental involvement in amateur sport, especially involvement that overshadows other important aspects of growing up, such as doing homework and chores and gaining experience in a number of different activities, teaches children the wrong lessons. Consider the values a child learns when *thousands* of dollars are spent on hockey travel by a parent who would not consider spending tens of dollars on a math tutor (Atkinson, 2014).

Positive Coaching Alliance
https://positivecoach.org/

   If the purpose of youth sports is to have fun, increase athleticism and physical fitness, and learn the value of teamwork and discipline, then some adults are teaching the wrong lessons. Sport managers and program administrators are working to reverse such involvement by setting policy that curtails "parental over-involvement." Examples include spectator codes of conduct, parental contracts agreeing to acceptable conduct and involvement, and conferences and seminars for parents. i9 Sports, the largest multisport provider focused on high-quality, community-based kids sports leagues in the United States, has multiple resources to help parents know how to be involved in an appropriate manner. The organization created a "10 ways to be a successful sports parent" resource specifically for parents. Tips include modeling behaviors you want your child to develop, celebrating competition above winning, fostering independence, making the coach your friend and not your foe, and among other tips (i9sports, 2021).

YouthFirst
https://www.youthfirst.us/

Amateur sport groups in the community are well advised to have policies governing parental involvement. Research suggests that parents can have a profound impact on the experience of youth sport participants and that top-down policy to address issues in parental behavior is warranted by sport organizations (Elliott & Drummond, 2015). Dan Doyle (2013, p. 319) suggests, "parents must help young athletes understand the meaning of gamesmanship as it applies to their sport(s), and that maintaining one's integrity begins with adhering not only to the rules of the sports, but to the spirit of the rules." How amateur sport organizations can be proactive (or have to be reactive) to parental involvement certainly is an area to consider in this industry.

Violence in Sport

Violent behaviors associated with amateur sport are not restricted to parents. Overly aggressive and violent acts by participants and spectators are regularly reported in the media. This includes both physical and verbal acts of aggression. Reducing sport violence involves curtailing both athlete and spectator aggression. Policies dealing with reducing athlete violence strive to achieve the following:

- provide proper, nonaggressive role models for young athletes
- develop rules that allow low tolerance for acts of violence
- apply severe and swift penalties for violence involving the actions of athletes, referees, and coaches
- apply severe and swift penalties for coaches who support and promote violent or aggressive play
- remove stimuli that provoke aggression
- organize referee, coach, parent, and athlete workshops
- provide ample positive reinforcement for appropriate displays of behavior in sport
- teach and practice emotional control (Kids First Soccer, 2016)

Amateur athletic organizations can curtail spectator violence through policies that deal with the following items:

- banning alcoholic beverages
- making it a family affair
- ensuring that the media are not contributing to the buildup of tension
- focusing on achieving excellence rather than fighting the enemy
- fining unruly spectators (Kids First Soccer, 2016)

US Center for SafeSport
https://uscenterforsafesport.
org/

For example, SafeSport is a program developed by the US Center for SafeSport that aims to foster a national sport culture of respect and safety. SafeSport attempts to equip sport organizations with tools to address problems before abuse occurs (US Center for SafeSport, 2022). For example, a recreation facility may institute an Anti-Violence Policy to raise awareness among spectators and parents of their role in creating a positive environment and to give volunteers and staff the mandate and power to deal with violent and antisocial behavior. Such a policy may define *violent behavior* as the following:

- loud verbal assaults
- intimidation and threats
- aggressive actions such as approaching another individual or throwing articles
- striking another individual
- attempting to incite violence

Policy will dictate that individuals engaging in any of the above activities are immediately ejected from the facility by designated leaders and banned from the facility for a period of time defined by the recreation facility staff.

Selecting Youth Sport Coaches

Good programming for kids depends on having suitable supervision and instruction, typical roles of the coach. Tens of millions of youth participating in amateur athletics means that millions of youth sport coaches are needed, and the majority of whom are volunteers. Unfortunately, the large number of volunteer coaches required sometimes results in the hiring of untrained, unprepared coaches. Experts have determined that between 90 and 95% of youth sports coaches have little to no training (Hall, 2019). Far more dangerous is the potential for placing a pedophile or some other criminal in contact with children and youth. To ensure this does not happen, all sport organizations need to use specific criteria for hiring youth coaches and should utilize reference and criminal background checks before hiring them. All sport organizations must have a personnel policy that contains procedures and requirements for youth sport coaches and volunteers that includes items such as the following:

- required coach training and background
- background information disclosure
- police record check
- coaching expectations and code of conduct
- coaching your own child
- understanding the goals and objectives of the association
- feedback on coaching performance
- dismissing a coach
- an individual's right to appeal

Responding to COVID-19

COVID-19 certainly impacted every form of sport, and amateur sport is no exception. The YWCA, Y, JCC of North America, and other amateur sport organizations create in-person programming specifically for women, children, and their members. When COVID-19 forced the world to shut down, many of these organizations struggled with serving their constituents when participants were not allowed in the facilities. As such, leaders within amateur sport organizations needed to develop innovative policies and practices to serve their members. The YWCA Belarus provided psychological support online and by telephone to women facing domestic violence as violence toward women skyrocketed during the pandemic (Harden, 2020). Many Ys were involved in the US.-sponsored Emergency Child Care, food distribution, virtual learning, and free child care programs during vaccine appointments (The Y, 2022c). JCCs across the world created hundreds of hours of virtual content, like religious celebrations and online physical activity classes, to stay involved with their members during the pandemic (JCCA, 2022e). Determining what best policies, programs, and practices for each amateur sport organization in response to COVID-19 (and move forward) is an important step for many managers as the pandemic continues to evolve.

SUMMARY

Thousands of amateur sport organizations are community-based. Their mandate is to provide a broad spectrum of sports and physical activities for a broad range of demographics. Such organizations can be categorized as public or nonprofit, depending on their purpose and sources of funding. Amateur sport at the community level has a rich history and is considered one of the foundations of a society in which happiness, health, and well-being are core values. Such organizations include leagues, clubs, special-interest groups, and organizations such as the Y.

The managers of community-based amateur sport organizations deal with a wide variety of governance and policy issues. Funding is a key area because fundraising is at the core of the operation of the organization. Programming is dependent on funding and interest and the inactivity of youth are areas of concern in policy development. Inappropriate conduct or interference by parents and violence are other current policy issues confronting sport managers in amateur sport. Despite these issues, community-based sport managers provide programming that positively impacts the lives of millions of participants. It is an exceptionally important component of the sport industry.

portraits + perspectives

HOWARD Z. WEINSTEIN *CPRP, CPO, AFO Recreation Administrator*

Town of Norwood, Massachusetts

I am the Recreation Administrator for the Town of Norwood in Massachusetts. In this capacity, I oversee a town-funded budget of $500,000 and a General Revolving fund of $1,000,000 which is program revenue driven. I also process our bills and payrolls. Beyond my financial responsibilities, I staff and schedule the Community Center and provide guidance on community center maintenance issues. In addition to the Community Center, I am responsible for our community outdoor pools with 50 staff and schedule our 27 athletic fields for youth and adult groups.

Issues in community recreation are always changing and you must be able to pivot and address issues as they arrive. Right now, from my viewpoint, there are two major issues. The first is inclusion – this is a positive issue to address. As a local community becomes a melting pot of diversity, community recreation must be willing to learn and be able to offer programs/services that will be of interest to a diversified community. The second issue is social media – before Facebook, Twitter, etc. you could address "issues" face-to-face or over the phone. Now people will voice issues on social media. As management, you want to be engaging with the public but, at the same time, you cannot have a "knee-jerk" reaction to every issue posted on social media.

Over my 23-year career, I have been involved in so many policy discussions! Plus, keep in mind, a policy is not set in stone. Even in the last two years with COVID, your normal policies had to be adapted. We had to establish a "resident only" rule for our programs and community center. Policies for using athletic fields, pools, etc. had to be adjusted for COVID. It was a true test on how to adapt policies so you could still provide services.

It is funny to think how governance structures differ from town to town, even from state to state. In Norwood, we have been very fortunate that our town officials have been very supportive of recreation. Unlike other towns that have a "recreation board" that inserts themselves into daily operations, we are not "handcuffed" to introduce changes to programming or policies. Myself, I feel that I have been spoiled compared to other towns' community recreation directors.

Every town recreation service will have its own different challenges. However, I believe it goes back to something I mentioned earlier – INCLUSION. As your community changes, you must be able to identify new needs, new wants, and new resources. A diverse population makes a better community. As a manager, if you do not grow along with your community, they will not support you.

If you want to "work" in community recreation, know that people who are in community recreation do not think of it as a "job" because they love what they do. Community recreation is not about awards or bright lights. It is about knowing that a little child in a program of yours has a safe and fun place to go. In addition, you must be willing to adapt, pivot, and try new things. Community recreation is not a 9–5 job. Do not be scared to think big – you can always scale back!

case STUDY

UNBECOMING CONDUCT IN YOUTH SPORT

As the Director of Children's Sport Programming for the town of Clarington, you are organizing a Soccer League for children from six to ten years old. Experience tells you that the parents and kids of Clarington are a competitive group. At the winter hockey leagues, several groups of parents were banned, and suspensions for violent behavior were common among the participants. In an effort to be proactive and eliminate such behaviors in the Soccer League, you have developed a Code of Conduct for both participants and spectators.

1. Describe your Code of Conduct for directing the behavior of (a) participants, (b) parents, and (c) general spectators.
2. How do you plan to communicate the Code of Conduct?
3. How do you plan to enforce the Code of Conduct?
4. How might you go about getting both participants and spectators to buy in to the Code?
5. What ethical dilemmas might you face implementing the policy, and how will you solve them?

CHAPTER questions

1. What is the difference between a public and a nonprofit sport organization? How do the governance structures of the two categories differ? Why do different types of recreational sport organizations exist? With a partner, use your device to look up local examples of each a public and nonprofit sport organization. Determine with your partner which structure better fits the needs of the organization, and why.

2. Using the Internet, locate a community sport organization, for example, your community's Little League Baseball organization. With which category of those mentioned in this chapter does it most closely align? Summarize its governance structure by describing the following: mission, financials, membership, and organizational structure. As a class, compare some of the different organizations searched by students, and identify the components of a well-governed community sport organization from the examples you researched.

3. The local Greater Wyoming Valley Area YMCA (located in Wilkes-Barre, Pennsylvania) is looking to provide new programming that is going to attract younger participants (ages 9–17). They already have an aquatic program that includes both competitive and open play for young swimmers. They also have a "Camp Kresge" that includes a day & overnight summer camp, children's camping weekends, and various outdoor education field trips. They have standard youth leagues for basketball, soccer, and t-ball, also. (1) What other programs would you consider offering to engage with the 9–17 year-old demographic? (2) Esports is an option that the YMCA is considering offering. Research esports (check out the chapter on eSports later in this book), decide if you would include esports programming as part of the YMCA program offering, and suggest what games would you include if you did offer esports?

FOR ADDITIONAL INFORMATION

For more information check out the following resources:

1. Here you can learn about some real-world responses to the COVID -19 pandemic: How YWCAs are responding to COVID-19. https://shespeaksworldywca.org/how-ywcas-are-responding-to-covid-19/

2. This site discusses important aspects of being a successful youth sport coach: Become a Certified Coach: The American Coaching Academy: https://americancoachingacademy.com/certification/

3. This is a well-written code of conduct for youth sport settings: National Alliance for Youth Sport: Code of Ethics: https://www.nays.org/coaches/training/code-of-ethics/

4. Amateur sport in the community is not just a North American phenomenon. Here is a European example: Older people, sport and physical activity (SportScotland): https://sportscotland.org.uk/documents/research_reports/older_people_report_final.pdf

5. Here you can learn about what i9 Sports has to offer: YouTube: i9 Sports. https://www.youtube.com/user/i9SportsFranchise

REFERENCES

Atkinson, J. (2014). How parents are ruining youth sports. *Boston Globe*. https://www.bostonglobe.com/magazine/2014/05/03/how-parents-are-ruining-youth-sports/vbRln8qYXkrrNFJcsuvNyM/story.html

Beyond Sport. (2021, July 9). *The Canadian Government commits to more sport opportunities for Indigenous youth*. https://www.beyondsport.org/articles/the-canadian-government-commits-to-more-sport-opportunities-for-indigenous-youth/

Boise Timbers | Thorns. (n.d.). *About us*. https://www.boisetimbersthorns.org/about-us

Boys & Girls Club of Oshkosh. (2022). *Board of directors*. https://bgcosh.org/about-us/board-of-directors/

Boys & Girls Clubs of America. (2021). *Club impact*. https://ywca.bgca.org/about-us/club-impact

Boys & Girls Clubs of America. (2022a). *About us*. https://www.bgca.org/about-us

Boys & Girls Clubs of America. (2022b). *Our mission & story*. https://www.bgca.org/about-us/our-mission-story

Boys & Girls Clubs of America. (2022c). *Alumni Hall of Fame*. https://www.bgca.org/about-us/alumni-hall-of-fame

Boys & Girls Clubs of America & Subsidiaries. (2022). *Consolidated financial statements*. https://www.bgca.org/about-us/annual-report/annual-report-financials

Boys & Girls Clubs of Metro Atlanta. (2022). *Boys & Girls Clubs of Metro Atlanta—A positive place for kids*. https://www.bgcma.org

Centers for Disease Control and Prevention. (2020, April 21). *Physical activity facts*. https://www.cdc.gov/healthyschools/physicalactivity/facts.htm

City of Houston. (2022). *About us*. https://www.houstontx.gov/parks/aboutus.html

City of Kissimmee, Florida (2021). *Annual operating budget*. https://www.kissimmee.gov/home/showpublisheddocument/7506/637709342086800000

City of Kerrville, Texas. (2021). *Parks and Recreation Department organizational chart*. https://www.kerrvilletx.gov/DocumentCenter/View/37218/PARD-Org-Chart-Effective-2021

City of Onalaska, Wisconsin. (2022). *Parks & Recreation*. https://cityofonalaska.com/parkrec

City of San Diego. (2022a). *City of San Diego organizational structure*. https://www.sandiego.gov/sites/default/files/city-org-chart.pdf

City of San Diego. (2022b). *Department organizational structure*. https://www.sandiego.gov/park-and-recreation/general-info/orgchart

City of Seattle. (2022). *Seattle Parks and Recreation*. https://www.seattle.gov/parks/about-us

City of Snellville. (2022). *Budget & finance.* https://www.snellville.org/administration/budget-finance

Dallas Nationals Baseball Club. (2022). *About us: Our mission.* https://www.dallasnationals.com/page/show/1293582-about-us

Doyle, D. (2013). *The encyclopedia of sports parenting.* Skyhorse Publishing.

Dunton, G.F., Do, B., & Wang, S.D. (2020). Early effects of the COVID-19 pandemic on physical activity and sedentary behavior in children living in the U.S. *BMC Public Health, 22*(1), 1–13. https://doi.org/10.1186/s12889-020-09429-3

Elliott, S., & Drummond, M. (2015). The (limited) impact of sport policy on parental behaviour in youth sport. A qualitative inquiry into junior Australian football. *International Journal of Sport Policy and Politics, 7*(4), 519–530. https://doi.org/10.1080/19406940.2014.971850

Glencoe Baseball Association. (2021). *GBA mission statement.* https://www.glencoebaseball.org/aboutgba

Government of Canada. (2021, November 25). *Sport for Social Development in Indigenous Communities – Sport Support Program.* https://www.canada.ca/en/canadian-heritage/services/funding/sport-support/social-development-indigenous-communities.html

Harden, C. (2020, April 6). *How YWCAs are responding to COVID-19.* She Speaks. https://shespeaksworldywca.org/how-ywcas-are-responding-to-covid-19/

Hall, B. (2019, April 10). *A shocking number of youth sports coaches are unqualified for the gig.* Stack. https://www.stack.com/a/a-shocking-number-of-youth-sports-coaches-are-unqualified-for-the-gig/

i9 Sports. (2021, December 10). *10 ways to be a successful sports parent.* https://www.i9sports.com/blog/10-ways-to-be-a-successful-sports-parent

Iowa Games. (2017). *About.* https://www.iowagames.org/about/

Iowa Sports Foundation. (2022). *What is the Iowa Sports Foundation?* https://www.iowasportsfoundation.org/About

JCCA. (2022a). *Board of directors.* https://jcca.org/about-us/board-of-directors/

JCCA. (2022b). *JCC Maccabi.* https://jcca.org/what-we-do/jcc-maccabi/

JCCA. (2022c). *JCC movement statement of principles for the 21st century.* https://jcca.org/about-us/statement-of-principles/

JCCA. (2022d). *Mission statement.* https://jcca.org/about-us/mission-statement/

JCCA. (2022e). *What does the Coronavirus crisis mean for the JCC movement?* https://jcca.org/what-does-the-coronavirus-crisis-mean-for-the-jcc-movement/

JCC of Greater Baltimore. (2022). *About us.* https://www.jcc.org/about-us

JCC of North America. (2021). *Financial statements.* https://jcca.org/app/uploads/2021/06/JCCA_Financial_Statements_2020.pdf

JCC of San Francisco. (2021). *Impact report.* https://www.jccsf.org/about/financial-statements/annual-report/

JCC of San Francisco. (2022). *Who we are.* https://www.jccsf.org/the-center/who-we-are/

Kidd, B. (1999). *The struggle for Canadian sport.* University of Toronto Press.

Kids First Soccer. (2016). *Aggression and violence in sport.* https://www.kidsfirstsoccer.com/violence.htm

Kinney, E. (2021, November 2). Popular sports with Indigenous American roots. *The Los Angeles Loyolan.* http://www.laloyolan.com/sports/popular-sports-with-indigenous-american-roots/article_83025b20-93c2-5e07-bf6a-8453a969b4d6.html

Positive Coaching Alliance. (2022, June 9). *Coaching parents for a positive sports culture.* https://positivecoach.org/the-pca-blog/coaching-parents-for-a-positive-sports-culture/

MetroWest YMCA. (n.d.). *Association organization chart.* https://www.metrowestymca.org/sites/default/files/METROWEST%20YMCA%20ORGANIZATION%20CHART%20012618.pdf

Morrow, D., & Wamsley, K. (2009). *Sport in Canada: A history.* Oxford University Press.

San Juan Sledders Snowmobile Club. (2017). *San Juan Sledders Snowmobile Club.* https://www.sanjuansledders.com

The La Crosse YMCA. (n.d.). *Flexible pricing.* https://www.laxymca.org/flexible-pricing/

The Y. (2021). *Committed to community.* https://www.ymca.org/sites/national/files/2021-06/2020%20National%20Council%20of%20YMCA%20of%20the%20USA%20FS%5B98%5D.pdf

The Y. (2022a). *Our mission statement.* https://www.ymca.org/who-we-are/our-mission

The Y. (2022b). *Who we are.* https://www.ymca.org/who-we-are

The Y. (2022c). *YMCA pandemic response.* https://www.ymca.org/who-we-are/our-impact/pandemic-response

Traverse City Club Curling. (2022). *Club history.* https://tccurling.org/index.php/about-the-club/club-history

US Center for SafeSport. (2022). *Training and reporting to prevent abuse.* https://uscenterforsafesport.org/

YMCA of Greater Rochester. (2021). *2020/2021 annual report.* https://rochesterymca.org/wp-content/uploads/2021/12/Annual-Report-20-21.pdf

YWCA. (2022a). *About YWCA USA.* https://www.ywca.org/about/

YWCA. (2022b). *FAQs.* https://www.ywca.org/about/faq/

YWCA USA, INC. (2020). *Financial statements and independent auditor's report.* https://www.ywca.org/wp-content/uploads/YWCA-USA-Final-6-30-20-Financial-Statements.pdf

Campus
Recreation

Campus recreation is the umbrella term used to describe a variety of recreation and leisure activity programming on university and college campuses throughout the United States and Canada. Recreation Departments exist on virtually every college and university campus where students attend classes in person. This segment of the industry has extensive numbers of facilities and personnel, and many of you may want

DOI: 10.4324/9781003303183-7

to include Campus Recreation Departments in your career plans. Understanding the organization, governance, and policy issues pertinent to this extensive segment of the sport industry will help you prepare for management positions located on college campuses or with umbrella organizations helping to lead the campus recreation industry.

Historically, college and university administrators have accepted responsibility not only for education but also for the general welfare of their students. When promoting health and well-being was identified as critical for student welfare, Campus Recreation Departments became essential components of institutions of higher learning. In general, the mandate of Campus Recreation is to offer opportunities to participate in both organized and open-facility recreational activities. This mandate gained momentum because college campuses are often community-oriented and can serve large populations of students, faculty, and staff members. Today, Campus Recreation Departments aim to enrich student life and are often considered tools for recruiting and retaining college students through their facilities and activities.

The size of Campus Recreation Departments varies depending on the campus setting and the size of the institution; however, their purpose is often strikingly similar. Campus recreation provides opportunities to engage in sport and leisure activities. The prime target audience of such programming is the student body. However, programming is usually also accessible to faculty and staff members and sometimes their families. In addition, many activities are made available to the community at large (Wallace Carr et al., 2013). The basic premises underlying campus recreation programming are enjoyment, fun, and promoting a healthy lifestyle through physical activity. The missions and visions of such programs (as referred to in Chapter 2) are represented by slogans such as "Life in Motion," "Engage, Educate, and Empower," "Live Play! Learn," "Recreation, Fitness, and Fun … Steps to Life Long Activity," and "Active Living for Health and Happiness."

Many Campus Recreation Department mission statements link the importance of campus facilities to the operation of recreational programming. In addition, the vision of the Campus Recreation Department involves inclusivity, since it operates as a vital part of the university community at large. The core values of NIRSA – the national campus recreation organization we will learn about in this chapter – are presented in Exhibit 7.1.

With this basic understanding of campus recreation, let's look at the roots of Campus Recreation Departments, why they were developed, and how they are organized and governed.

NIRSA's Mission Statement

NIRSA is a leader in higher education and the advocate for the advancement of recreation, sport, and well-being by providing educational and developmental opportunities, generating and sharing knowledge, and promoting networking and growth for our members.

NIRSA's Vision Statement

NIRSA is the premier association of leaders in higher education who transform lives and inspire the development of healthy communities worldwide.

NIRSA's Strategic Values

Equity, Diversity & Inclusion
Global Perspective
Health & Well-Being
Leadership
Service
Sustainable Communities

Sources: NIRSA (2022c, 2022d).

THE HISTORY OF CAMPUS RECREATION

Recreation and leisure activities undoubtedly contributed to the early growth of competitive athletics (Langley & Hawkins, 2004). The interest in playing, learning to engage in new activities, or simply getting active is well documented in the history of sport and the pursuit of good health. Early on, goals were likely pure enjoyment, opportunities to socialize with friends, and relief from the boredom of work, study, or everyday life. If physical activity is viewed on a continuum from informal play to highly competitive sport (LaVoi & Kane, 2014), it is easy to understand that the early history of recreation on campus is interwoven with campus sport as we know it today.

The First Campus Recreation Programs

Early sport and leisure activities originated with English Sport Clubs and German turnverein (gymnastics clubs). Around the midpoint of the 19th century, North Americans were looking for opportunities to be physically active other than in highly competitive sports or in the rigid routine of gymnastics. This interest in pursuing physical and recreational activities and sport was naturally present on college campuses. The campus was an ideal setting for spontaneous games, with divisions already defined by academic class, major, or residence housing. The term *intramural*, used to describe the first programs of campus recreation, comes from the Latin words for "within" (*intra*) and "wall" (*murus*), that is, within the walls of an organization (Franklin, 2013).

Competition between different classes soon became commonplace. Colleges and universities embraced the notion that programming for leisure and recreational pursuits, along with competitive athletics, was an important component in the overall education and well-being of their students. In 1904, Cornell University (CU) developed a system of what is known today as instructional sport for students not participating at the varsity level. During the next decade, the surge of student interest in recreational sports resulted in the development of a department to manage such student programming. The Ohio State University (OSU) and the University of Michigan (UM) each established organized intramural departments in 1913 (Mueller & Reznik, 1979). CU quickly followed, along with other colleges and universities in the United States and Canada.

A Rationale for Campus Recreation

When considering the history of higher education, intramural sport is perhaps one of the oldest organized campus activities (Wallace Carr et al., 2013), having marked 100 years in 2013. A significant factor in legitimizing campus recreation occurred in the early part of the 20th century. In 1918, the National Education Association (NEA) in the United States coined the phrase "worthy use of leisure time" as one of the Seven Principles of Education (Colgate, 1978). In essence, the positive use of leisure time became a pillar of a well-rounded education. The greater meaning was attached to educating the whole person and to the importance of out-of-classroom educational experiences (Smith, 1991). Over time, the notion that healthy individuals are active, involved, and accomplished in activities of both mind and body became another cornerstone supporting the need for open recreation and intramural programming on campus. As the world became more technologically sophisticated, the amount of leisure time increased and the demand for recreational activities on campus continued to grow. Recreational activities were social in nature and offered opportunities both for affiliation with other students living on or off campus and for friendly competition. As we all know, all of us are tied more closely these days to our various devices and social media platforms. Campus recreation programs are now having to creatively pivot to capture the attention of students as the notion of free time evolves for people on college campuses.

Each of the reasons for the establishment and growth of the Campus Recreation Department remains today. In addition, the extensive facilities for recreation and the breadth of such programming in today's universities are recruitment tools for potential students. Students and their parents are naturally drawn to institutions with excellent facilities and programs, providing a natural link between the Campus Recreation Department and the overall mission of the university (Loomer, 2019; Popke, 2019). Research indicates that student retention is favorably influenced by getting students involved in extracurricular activities such as those housed within campus recreation. Finally, and perhaps most importantly, campus recreation programs began to flourish because of student interest. The student body's interest has grown and changed over the years, leading to a proliferation of facilities and new program offerings. In 1928, the UM was the first institution to devote a building primarily to intramurals (Mueller & Reznik, 1979). Today, most campuses have facilities solely dedicated to recreational use.

The Formation and Evolution of National Intramural Recreation Associations

Following World War II, while athletic and physical education groups were holding annual meetings and looking to associate with one another for a variety of purposes, recreation programmers had not yet organized themselves in a similar way. To fill this void in the United States, Dr William N. Wasson of Dillard University in New Orleans engaged a research study to compare intramural programs at Black Colleges, resulting in a number of Historically Black Colleges and Universities (HBCUs) forming the National Intramural Association (NIA) in 1950 (NIRSA, 2022b). The mandate of the NIA was to provide an association for professionals working in college and university intramural sports programs in the United States to share ideas, develop policy, and encourage professional development. In 1975, the NIA membership voted to change its name to the National Intramural Recreational Sports Association (NIRSA). The membership felt NIRSA more aptly described the expanded and diversified role of Recreation Departments on college campuses. Such units organize and deliver programming far beyond the boundaries of intramural sports, and the scope and mission of NIRSA have expanded phenomenally to what is currently an extensive national association. Given that NIRSA was founded by African American leaders, it is fitting that inclusion and diversity are fundamental values of NIRSA that endure today. In 2012, NIRSA officially expanded its name to *NIRSA: Leaders In Collegiate Recreation* in order to promote and communicate the organization's diverse and holistic approach to collegiate recreation (NIRSA, 2022b). The governance structure of NIRSA is described later in this chapter.

The next section describes the governance structure of NIRSA in further detail and defines how Campus Recreation Departments are organized on college and university campuses.

www
Dillard University
https://www.dillard.edu/

www
NIRSA
https://nirsa.net/nirsa/about/

GOVERNANCE

This section examines the national association to which most campus recreation professionals in North America belong. We answer the following questions about NIRSA: What is the mission of the organization? How is it funded? Who are its members? How is the organization structured?

National Intramural Recreational Sports Association

MISSION. The purpose of NIRSA has expanded greatly since it was founded in 1950. Beyond intramurals, areas of interest include aquatics, integrated wellness and fitness, informal recreation, instructional programs, outdoor recreation, programs for people with disabilities, special events, Sport Clubs, extramurals (regional and national sports tournaments), and student leadership and development. NIRSA strives to provide its members with research, teaching, presenting, and publishing opportunities as well. NIRSA is a professional association dedicated to promoting quality recreational sports programs. The association is equally committed to providing continuing education and career development for recreational sport professionals and students. NIRSA's mission statement, vision statement, and strategic values are presented in Exhibit 7.1.

MEMBERSHIP. Today, NIRSA has an extensive reach and membership across North America, but this was not always the case. From the first organizational meeting in 1950 that included 20 individuals from 11 HBCUs, the association now boasts a membership of over 4,500 professionals and students from colleges, universities, correctional facilities, military installations, and Parks and Recreation Departments. NIRSA initiatives and programming reach millions of recreational sport enthusiasts, including an estimated 8.1 million college students. NIRSA also reaches professionals and students in Canadian schools, colleges, and universities (NIRSA, 2022a).

Membership is offered under several designations: institutional, professional, professional life, student, student leadership, emeritus, retired, honorary, and state association categories, as well as the associate level for commercial organizations that provide products or services to the organization and state association level. Within each category, NIRSA offers many opportunities to get involved. With its sole focus on the advancement of Intramural-Recreational Sport programs and their professionals and students, NIRSA provides access to program standards, a Code of Ethics, an extensive resource library, career opportunity services, and the Sports Officials' Development Center. In addition, NIRSA's message is delivered to professionals through state and regional conferences, symposia and workshops, and the Annual National Conference and Exhibit Show. NIRSA institutional members represent large and small and public and private two- and four-year colleges and universities. For those with an institutional-level membership, access is provided to nationally sponsored programs and events ranging from individual and team sport events to fitness and wellness exhibitions and special publications.

FINANCIALS. NIRSA is a nonprofit organization. However, material and financial growth led the organization to reorganize into three independent legal entities, each of which has a significant role in managing NIRSA finances. In addition to the parent NIRSA organization, the NIRSA Foundation and the NIRSA Services Corporation were formed (NIRSA, 2021a).

NIRSA has approximately 4,500 individual and organization members. Membership fees constitute a major source of funding. The NIRSA Foundation is also a nonprofit organization mandated to support the NIRSA mission; it receives donations to NIRSA. The NIRSA Services Corporation is a for-profit entity supporting NIRSA's mission and vision. It develops assets through partnerships and manages NIRSA endorsed and sponsored programs (NIRSA, 2021a).

ORGANIZATIONAL STRUCTURE. In 2010, NIRSA began a governance transition process that streamlined the number of positions on the Board of Directors and identified the Board's role in visionary leadership. The Board is composed of seven members, including the following elected positions: President, President Designee, President-Elect, Annual Director, and three At-Large Directors. The individuals elected to these positions hold one- to three-year terms, staggered to ensure continuity on the Board of Directors from year to year. The Board is assisted by a nonvoting position of Secretary. The Board is responsive to the members of the organization and works in concert with the Member Network and the Assembly.

The Member Network is the primary vehicle for member communication, representation, networking, and professional development. It comprises one regional representative from each NIRSA region, one student from each NIRSA region, the designated NIRSA Student Leader, the Past-President's Member Network representative, one member of the Board of Directors, and the Chair of the Member Network. The NIRSA Assembly facilitates the germination of ideas and national discussion in order to ensure contemporary relevance and knowledge sharing. Its membership is at the discretion of the Board of Directors and is meant to be a broad representation of NIRSA's constituencies including individuals working in the profession, students, a Past-President's representative, and a member of the Board of Directors.

The Executive Director (ED) of NIRSA, the only paid officer of the association, answers to the Board of Directors and is responsible for the daily activities of the organization. The NIRSA National Center, located in Corvallis, Oregon, has a professional staff grouped into areas including Executive, Professional Development, Finance, Communications, Human Resources, Leadership Programs, Operations, Membership, Philanthropy, Corporate Relations, and Sport Programs. Approximately 30 individuals are employed in managerial positions at NIRSA Headquarters.

NIRSA is primarily a service organization, dedicated to continuing education for its members and the promotion of quality recreational sport programs. NIRSA provides its members with knowledge, ideas, and a community for solving problems. Its prime policy role is the development of program standards for events and activities within recreation programs. Ensuring the safety of participants and the quality of programming have been the focuses of such policy development. Another focus has been the development of codes of ethical practices for professionals and participants within recreational sport settings. The NIRSA Membership Code of Ethics encompasses respect and fairness, integrity and responsibility, and development. The Preamble, in part, reads:

> Ethical conduct is a key pillar of how we approach our work. In making the choice to affiliate with this professional association, individuals assume the responsibility to conduct themselves in accordance with the ideals and standards espoused by NIRSA. When members' actions are aligned with these ethical principles, the profession and our values are strengthened — communities of higher learning are enriched.
>
> (NIRSA, 2021c, para. 3)

www

Portland State University Campus Recreation
https://www.pdx.edu/recreation/about-us#mission

www

University of Nebraska Student Affairs – Campus Recreation
https://crec.unl.edu/mission-vision-and-guiding-principles

Campus Recreation Department Structure

**University of Arizona
Campus Recreation**
https://rec.arizona.edu/about

**East Carolina University
Campus Recreation and
Wellness**
https://crw.ecu.edu/
about-us/our-story/
mission-vision-and-goals/

Virtually every college and university in North America has a Campus Recreation Department. The campus is viewed as a community, and the pursuit of fitness and play through sport and physical activity is an important part of any community in promoting health and happiness.

MISSION. The structure and function of the Campus Recreation Department is directly linked to the unit's mission and goal statements, which are based upon participants' and clients' goals. A typical mission statement for a Campus Recreation Department might read: "We are committed to providing the finest programs and services in order to enrich the university learning experience and to foster a lifetime appreciation of and involvement in recreational sport and wellness activities for our students, faculty, and staff." To view a few sample mission statements, see Exhibit 7.2.

---

exhibit **7.2** Sample Campus Recreation Department Mission Statements

Portland State University

Campus Rec provides an inclusive environment where recreation and wellness opportunities inspire, empower, and educate people to be positive contributors to the global community.

Source: Portland State University –Campus Recreation (2022).

University of Nebraska-Lincoln

Campus Recreation is committed to enriching the educational experience by promoting lifelong healthy living. We are dedicated to celebrating differences, providing active and innovative recreational opportunities, cultivating a community of supportive relationships, and developing global leaders.

Source: University of Nebraska Student Affairs –Campus Recreation (2022).

University of Arizona

We provide diverse opportunities for balanced and healthy lifestyles to the University of Arizona community through inclusive and quality programs, collaborations, and facilities.

Source: University of Arizona –Campus Recreation (2022).

East Carolina University

To engage the Pirate Community in diverse recreational and wellness experiences by providing exceptional facilities, programs, and services.

Source: East Carolina University –Campus Recreation and Wellness (2022).

MEMBERSHIP. Campus Recreation Departments exist primarily to serve the members of their campus communities. As mentioned previously, the constituents of a modern campus recreation program include the students, faculty, and staff. However, this group may well be broadened to include alumni, families of faculty and staff, and community members interested in recreation opportunities (Wallace Carr et al., 2013).

FINANCIALS. The financial operations of the small versus the large Campus Recreation Department will vary. Some institutions support programming and facilities through central budgets housed in Student Life or Student Affairs. Budgetary support ranges from full to partial funding by the institution; more commonly, university budgets provide one of several sources of the overall recreation budget. Another common source of revenue for recreation is a required recreation fee charged to every university student to help support student programs and facilities. Such fees are collected in addition to tuition and other required academic and non-academic fees at the beginning of each semester or quarter. Another revenue source includes "pay for play" fees collected from the participants in a league, class, or special event. Budgets also often include revenue components including rentals, facility membership or daily use fees, advertising, and other marketing initiatives. In this case, the operation is run on a break-even basis, equating operational and program spending to the revenues generated through some combination of the sources defined above.

Other campus recreation programs are run as *profit centers*. Many institutions, regardless of size, have built multi-million-dollar facilities to service the needs of their current constituents. This proactive stance recognizes the role campus facilities can play in attracting and retaining future students, faculty, and staff. Profit centers generate revenue to offset the costs beyond those related to operations, for example, to pay a facility mortgage. Often, using the facility involves a membership fee. Students may pay through the recreation fee charged within the tuition package, and faculty and staff may be required to pay monthly membership fees. Opportunity for memberships may also be extended to alumni, family members, and community users on separate fee schedules. The proposal for building such a facility sometimes involves a student referendum for an additional building fee that might extend from as few as five years to as many as 20 years. In this case, an additional facility or building fee is charged to all students for the duration of the agreed-upon timeframe.

Some institutions look beyond their students to additional sources of revenue to build recreation facilities. An alternative model for financing the construction or renovation of facilities involves developing partnerships. In such cases, the university partners with the community, with local governments, or with the private sector to raise capital. Agreements for use and profit allocation are developed in return for building capital. The facility is run as a business with market rates charged for use. The Director of Recreation must ensure certain profit levels are maintained through memberships and sources of program revenue. Significant sources of revenues in the millions of dollars can accrue from rental payments, instructional programs, and sport camps, to name a few.

Many institutions also seek corporate sponsors to fund sport leagues and specific programs.

For example, Marshall University offers the following benefits for becoming a corporate partner – event partnerships; tabling, products sampling and promotional sponsorships; program sponsorships; recreation digital sponsorship; Rec Radio; promotional items; signage opportunities; and program guide advertisements (Marshall University Recreation Center, 2020). Cleveland State University enlisted corporate partners Medical Mutual, Sedgwick, VSP, Skyway, and Nosotros Rock Climbing Gym for their annual WellFest event (Cleveland State University, 2021).

ORGANIZATIONAL STRUCTURE. The structure of the Recreation Department is also partially determined by its size. A small college may have a fairly simple organizational structure due to fewer constituents, limited facilities, smaller levels of programming, and less need for full-time staff. However, a large college will have a complex organizational structure that provides extensive levels of programming for multiple constituents through state-of-the-art facilities. Let's have a look at the administration and operation of two examples.

www
Marshall University Campus Recreation
https://www.marshall.edu/campusrec/

www
Cleveland State University Recreation Center
https://www.csuohio.edu/recreationcenter/recreationcenter

Small Recreation Programs

At a small school, the Campus Recreation Director may be the only full-time employee with direct responsibilities for recreational programming. They may have an assistant but often manage the area alone, with support from the Director and from employees who manage the facilities. In small programs, it is easy to comprehend the large role student employees play in organizing and delivering the campus recreation program. Of necessity, in the beginning, such programming was student-run, and recreational programming today is still largely student-run (Brown, 1998). Full-time university employees direct the overall program, set policy, and manage finances, but the actual development and delivery of programming is often led by students. These programs allow students to gain valuable management and leadership skills. Clearly, these students have their fingers on the pulse of their classmates' interests when it comes to assessing programming. Student leaders help govern the program as well, often by way of management teams and advisory councils that feed information to the full-time university employees. For example, student-led committees dealing with areas such as intramurals, participant conducts, special events, clubs, and officials may report through a Student Supervisors' Council to the Campus Recreation Director. Student input and leadership is the foundation of the program. Student employees and supervisors, along with the Director, are likely to be heavily involved in the development and implementation of policy.

Large Recreation Programs

The scope of a large campus recreation program can differ significantly from that of the small college presented above. Many colleges and universities have large student, faculty, staff, and alumni populations. Some are housed in large urban centers. In such cases, the administration and operation of the Campus Recreation Department is generally large and complex, with many full-time professional staff and several programming divisions.

At a larger university, the structure of the Campus Recreation Department is extensive and is compartmentalized into several operational and management areas based on the defined programming. Students help run the respective areas through positions in each of the program's areas of delivery, and a large professional staff manages the department. Associate Directors responsible for different types of programming are common, each reporting to the Director. The different programming areas depend on the campus constituents and their needs and environment. The following areas are most common: intramural sport, extramural sport, outdoor adventure, Sport Clubs, fitness and wellness activities, special events, community programming including sport camps, instructional programming, aquatics, dance, martial arts, family recreation, informal recreation, adapted recreational sports, equipment rentals, facility operations, marketing, technology, business operations, and student personnel.

Advisory committees composed of students, faculty, staff, alumni, retirees, and designated groups' (intramural sports, residence halls, fraternities, and sororities) representatives provide input to a wide spectrum of programming and management issues. These committees contribute to policy development that might ultimately funnel to the Director; for example, a policy regarding penalties imposed on teams that are late or fail to show up for an intramural event. Another level of advisory committee, perhaps called the Advisory Committee on Recreational Sports, takes on the responsibility for overall issues of program and facility equity and direction. This Advisory Committee may be led by a Director of Recreational Sport. It meets regularly and may be composed of several faculty members, an equal number of students appointed by Student Association(s), and a representative from central administration, perhaps appointed by the Office of the Vice President for Student Affairs. The Advisory Committee plays a role in policy development, usually considering issues related to campus recreation as a whole such as equity, finance, future directions, and public relations.

PROGRAMMING IN CAMPUS RECREATION

Regardless of the size of the Campus Recreation Department or its system of funding, an important issue on every campus involves the scope of programming. When the question of scope arises, almost without exception, programming needs to be *extensive and relevant.* Campus recreation programs certainly live up to the "something for everyone" theme, with offerings ranging from intramural team sport leagues to instructional activities to wellness programs to events for people with disabilities. Even small programs run by a single professional staff member with limited access to facilities can deliver extensive and relevant campus recreation programs.

Campus recreation began as intramural sport, and this area still draws strong interest today. Intramurals include not only team sports leagues but also individual sports. Intramurals encompass men-only, women-only, and co-ed leagues in virtually every sport of student interest. The traditional team sports like basketball, softball, and flag football remain popular but other options such as ultimate frisbee, spikeball, or wheelchair basketball can generate interest. The same can be said for typical individual sports like tennis or swimming but also newer additions like mini golf.

Sport Clubs are also popular. Clubs are student-run organizations that provide in-depth opportunities to learn and participate in a particular activity. The extensive array of possibilities ranges from archery to bowling to fencing to racquetball to ultimate frisbee. Each club is formed, developed, governed, and administered by students. Student leadership and continuity is a key to success. The Campus Recreation Department typically sets rules and regulations and provides administrative assistance. Furthermore, most Sport Clubs belong to governing bodies that define rules and regulations specific to their sport. These clubs may be recreational or competitive in nature. The Competitive Sport Clubs generally compete on a national level against other universities and some have governing bodies. The American Collegiate Hockey Association (ACHA), for example, has been the national governing body for non-varsity collegiate hockey since 1991 (ACHA, 2021).

Instructional activities include classes in fitness, dance, yoga, "learn to" activities, certification courses, and facility time for self-directed activities. Self-directed activities involve sport for fitness and fun, such as individual weight training or open swimming, pickup basketball, tennis, or health and safety courses.

Another popular area of campus recreation programming is the pool. Aquatic activities range from self-directed open fitness swimming to learn-to-swim to aqua-fitness programming. Aqua-fitness classes take place in the water and usually follow the same format as regular fitness activities. Intramural inner tube water polo is also popular with participants and is now being offered on some campuses.

When the fitness industry exploded in North America in the 1970s, demand for aerobic classes, yoga, spin, and fitness classes soon peaked on campus, and the trend continues today. The public has become better informed of the health benefits and personal satisfaction associated with recreation and leisure activities, so campus recreation has expanded its wellness offerings with such classes as nutrition, aromatherapy, reflexology, tai chi, Pilates, and power yoga. Events for individuals with disabilities (for example, wheelchair basketball) or to populations of international students (for example, a cricket festival) are also commonly offered.

Most of the activities and events described above take place indoors. The final area of campus recreation programming involves outdoor

www

American Collegiate Hockey Association
https://www.achahockey.org/

Spotlight: Diversity, Equity, & Inclusion in Action

NIRSA

The national organization associated with campus recreation – NIRSA – has its very roots in the notions of diversity, equity, and inclusion. The organization held its foundation Intramural Summit at Dillard University in 1950. The Summit included intramural directors representing 11 HBCUs and the group worked together to form the NIA (NIRSA, 2022b). The organization, which later changed its name to what we know today as NIRSA, includes equity, diversity, and inclusion as one of its strategic goals. Given the changing face of college campuses,

> Students and employees are becoming more diverse on a broad range of dimensions including gender, sex, sexual orientation, language, age, ability status, national origin, religion, socio-economic status, as well as race, ethnicity, and heritage. Those who manage programs and services, as well as those who help to develop the talents of students and the workforce, need to be prepared to address the environmental factors that influence performance and affect overall wellbeing.
>
> (NIRSA, 2022a, para. 2)

The organization publishes a resource guide for campus recreation personnel that includes information on the various aspects of diversity and also case studies which can help campus recreation programs develop best practices in promoting equity, diversity, and inclusion (NIRSA, 2019).

recreation and adventure activities. This includes activities in the great outdoors to take advantage of the lakes, mountains, and forest terrain of the surrounding areas. Cycling, kayaking, canoeing, hiking, mountain biking, and scuba diving are popular pursuits, with classes for beginners to advanced participants. Some outdoor adventure activities include an element of risk, such as outdoor rock climbing or open-water kayaking. Such outdoor programming led by trained students and professionals is generally organized into three categories – trips, outdoor education, and equipment rental.

It is easy to understand why campus recreation programmers have coined the slogans "Something for everyone" and "We do more than just play games." They clearly live up to these mottoes, providing an extensive and diverse array of recreation and leisure time activities.

CURRENT POLICY AREAS

Like any other department in an institution of higher education, the recreation unit must develop and update policies. The Campus Recreation Director leads the debate on issues of policy, gathering input from both full- and part-time recreation employees and participants. Some issues are impacted by policy enacted by the institution as a whole. Some of the current policy areas that impact campus recreation leaders today are discussed in this section.

Resource Constraints

In the sport industry, managers are always dealing with balancing limited resources, and this is true for campus recreation as well. The three main resources sport managers work with are money, personnel, and facilities. Let's start with money and budgets. As mentioned earlier in the chapter, the main source for funding campus recreation programs comes from student activity fees. While this makes good sense since students are the primary people who use the facilities and programs (although some never do), care must be taken to not overtax them. In addition, as we move forward, there are going to be fewer typical college-aged people. The eight-year time period ending in 2019, for example, saw an 11% decline in college student enrollment (Nadworny & Larkin, 2019). While several factors contribute to this, two major ones are (a) the fact that there are fewer high school students and (b) the cost of college remains on the rise. These factors mean there will be fewer students to collect activity fees from, resulting in some tough budget decisions.

The next resource is personnel. In order to offer safe and quality programs, you need properly trained personnel. Lifeguards, fitness instructors, and personal trainers, for example, all need to be properly certified by organizations such as the Red Cross. Finding people, primarily students, who meet that criteria can be challenging and then training them can also be costly. Yet, this is necessary to ensure appropriate programming.

Finally, there are facilities. Campus recreation administrators may want to offer an exciting and broad array of activities, but will their campus facilities be able to meet those needs? Aquatics programs are very popular, but the cost of maintaining a pool can be astronomical. Oftentimes, a campus only has one pool and it typically falls under the responsibility of the Athletic Department and the competitive swim teams. While this may shift some financial burden away from Campus Rec, it also will limit availability for use by the general campus population since the swim teams will control schedules those teams' practices and meets will be given priority in scheduling. There may also be demand for new activities (for example, climbing walls) which require building new or updating existing facilities, so decisions must be made about the feasibility of making these changes.

Activity Trends

Deciding on which activities to offer, mentioned above as an issue in funding campus recreation programs, is an important area in policy development. Fads and trends are normal in recreation, and they shift quickly. Campus recreation leaders need to have their fingers on the pulse of shifting interests, and they can gather the information they need through three important mechanisms. First, they can administer participant and nonparticipant questionnaires to collect interests, likes, and dislikes. Second, Recreation Department student employees can gather informal data about interests. Finally, recreation administrators can observe what events are popular, what events fail to draw participants, and what other colleges and universities are offering in the way of programming. Through program evaluation, managers can gather information needed to make appropriate changes. This is critically important to meet the interests and needs of the people who would be using the campus recreation facilities, and how these interests and needs change. Ensuring that facility development includes adaptable space for multiple uses is a key planning issue given that participation fads change. After all, the campus recreation program is nothing without participants.

In the past years, participation in esports has increased at a powerful rate. Reports indicate that 21% of the Internet population participates in esports and the industry revenue was projected to reach $159 billion in 2020 (Fazio, 2021). This trend has not been lost on NIRSA or campus recreation programs when it comes to program offerings (NIRSA Communications, 2021). However, offering esports programming has its challenges. For example, where should esports programs really be housed – campus rec, athletics, or Student Affairs? There is also a question of facilities when hosting an esports event. Finally, esports culture can have downsides such as the sometimes toxic environment in online interactions. All is not negative, though, as esports provides a number of benefits as well. When COVID-19 forced us to limit direct personal interactions, esports were able to maintain a stronghold because being in-person was not as essential. The popularity of esports also offers opportunities for sponsors because of the large number of views available (Howard, 2021). There is no denying esports are here to stay and offer savvy campus recreation professionals the opportunity to create participation opportunities on their campuses.

Risk Management

Campus Recreation Directors must be concerned with risk management (Zabonick-Chonko, 2016). Of course, some risk is associated with crossing the street or playing a game of baseball. One may be run over by a speeding car or hit on the head by the baseball. Society attempts to minimize the possibility of being run over by a car by posting speed limits, building sidewalks, and setting up traffic lights and stop signs. Similarly, recreation administrators attempt to minimize risk in physical activity by requiring participants to wear protective equipment, ensuring that participants are taught proper techniques, and strictly enforcing safety rules. If the baseball player is hit in the head with a ball, hopefully their helmet will protect them. It is impossible to eliminate risk; the challenge is to manage the risk. In other words, the recreation professional is responsible for setting up and delivering programs that reduce overt risks and communicate other levels of risks to participants. According to Zabonick-Chonko

(2016), to achieve this, policy is required to create a culture of risk management. Managers need to:

- develop a written risk-management plan to show evidence of proactive prevention
- keep accurate and detailed records on participants and injuries that occur
- ensure that the rules of play are properly enforced and communicated
- train and certify leaders in the areas they teach and supervise
- ensure proper supervision of areas needing supervision and restrict access to some unsupervised areas
- check equipment regularly to ensure it is in good working order
- require paperwork that provides lists of participants, identification information, and possible health information
- develop and communicate emergency response procedures
- train event leaders in first aid and cardiopulmonary resuscitation (CPR)
- develop and implement informed-consent forms and health-related questionnaires for all but the very minimal risk activities (Miller, 1998).

For example, an aerobics class should be led by a trained, certified instructor. They should have a class list, and participants should be required to complete a medical screening questionnaire such as the Physical Activity Readiness Questionnaire (PAR-Q) to identify any potential risk factors. In another activity, such as an outdoor adventure trip with white-water kayaking, medical clearance may be required, and program leaders will use waiver forms to have participants acknowledge the risks involved and to take responsibility for such risks. Waivers and informed-consent forms are commonly employed and should be a required component of the policy surrounding risk management for the Campus Recreation Department, especially where off-site travel or high-risk activities are involved.

Medical Issues

Currently, COVID-19 still presents challenges for campus recreation programs. NIRSA provides a webpage with updates and resources for campus recreation (NIRSA, 2021b). Here you can find guidelines for reopening, information from the American Red Cross on COVID-19 and aquatics facilities, and survey results on the financial impact of COVID-19 and the state of campus recreation related to COVID-19. This illustrates how NIRSA is a service organization as it provides up-to-date information for its membership on dealing with the COVID-19 pandemic on campus.

Beyond COVID-19, however, there are the more typical and ongoing medical concerns that campus recreation administrators face. Pre-activity screening is important when dealing with medical conditions. However, it will not prepare employees in a Recreation Department for the acute medical emergencies that may result during sporting events and other physical activities. Many programs are considering or creating concussion protocols for intramural and sport club participants. Of course, it is inevitable that one player will step on another's foot while playing intramural basketball and sprain or break an ankle. The Campus Recreation Department must have set policy on procedures to help the injured participant. Having emergency supplies and equipment on-site is important, and many college recreation facilities have automated external defibrillators (AEDs) in case of cardiac incidents. Having an emergency action plan and trained supervisors who know what to do is also critically important. Some activities require collecting and keeping information on-site in the event of an emergency. The group leaders on a daylong cycling tour should have information on a participant who is allergic to insect stings, including medical insurance information and contact numbers and names. Leaders must also be trained in managing information and protecting the privacy of participants. Generally, all student employees are mandated to have FERPA (Family Educational Rights and Privacy Act) training given their handling of sensitive information. Medical information needs to be collected and collated into a manageable, perhaps laminated form that can be safely and securely taken care of by a campus recreation professional. Having on-site information and knowing what to do in the event of an emergency are critical areas for policy development and a clear responsibility of the Recreation Department in terms of employee training.

Recreational Opportunities for Individuals with Disabilities

The most critical issue with respect to recreational opportunities for students with disabilities is gaining an accurate demographic picture of who has needs and defining those needs (Martinez, 2017; Spencer, 2019). To achieve this, campus recreation professionals must set policy enabling the collection of important information regarding maximizing opportunity for students with disabilities. This is readily achieved through a Disabled Student Advisory Council, which sets a schedule for defining appropriate activities, defines a mechanism for publicizing activities and facility schedules, creates opportunities for training with respect to facilities and equipment, and provides ongoing leadership in the assessment of the effectiveness of the overall effort. Colleges and universities across the United States must work harder to provide recreational programming designed specifically for individuals with disabilities. Martinez (2017) found four themes relative to successfully including disabled students in recreational programming in a study examining 14 Big Ten Universities:

(1) website information and ease of navigation for disabled programming, (2) appropriate use of language that is inclusive to disabled students, (3) access to facilities and accessibility information, and (4) support for accessibility and inclusion. After examining campus recreation programs which are successful at including people with disabilities, Spencer (2019) found that alignment to the university mission, listening to a diversity and inclusion committee, staff training in cultural awareness, Americans with Disabilities Act (ADA) compliance/accessibility, and the help of community partners make campus recreation programs more welcoming to people with disabilities. Intramural sports like sitting volleyball, goal ball, and unified sports where people with and without disabilities participate together are gaining in popularity. Fitness spaces should now be geared toward being accessible with machines that are adaptable or cater directly to patrons with disabilities.

Transgender Participants

University of North Carolina – Chapel Hill Campus Recreation
https://campusrec.unc.edu/

University of Houston Campus Recreation
https://uh.edu/recreation/

Just as in high school sport, transgender individuals should have the opportunity to participate in sport and physical activity. NIRSA has always prided itself in its inclusivity and wanting all participants to be welcome. A number of campus recreation programs have developed policies regarding participation by transgender individuals. For example, the University of North Carolina – Chapel Hill has this language on their website regarding participation in intramural sports (UNC Campus Recreation, n.d.):

> All eligible Intramural Sports participants may participate in accordance with their personal gender identity. A participant's gender identity is applied when there are gender-specific rules or player ratio requirements in co-rec leagues. Once an individual identifies with a particular gender, they will remain in that classification until the end of that sport's season. An individual's gender identity must be consistent across all Intramural Sports teams and leagues.
>
> (para. 1–2)

The University of Houston transgender policy goes into a bit more detail, stating (University of Houston, 2021):

> University of Houston students, faculty/staff and spouse/partners of students, faculty/staff shall be permitted to participate in Departmental programs, activities, and services (to include Intramural Sports, Sport Clubs, Aquatics, Outdoor Adventure, and Fitness) in accordance with the person's gender self-identification or expression. Sport Club participants will be reviewed on a case-by-case basis based on the applicable National Governing Body's policies and procedures. The UH Campus Recreation and Wellness Center will ensure that every student, faculty, staff, and guest

have access to a locker room and bathroom facilities in a safe, comfortable, and convenient environment. Transgender individuals shall not be forced to use the locker room corresponding to their gender assigned at birth. Such accommodations could include but are not limited to a private area in the locker room or use of the family changing room.

(Transgender Policy section)

In summation and in keeping with its commitment to inclusion, NIRSA makes the following statement, "As an Association dedicated to reaffirming the rights of students who are often at the margins of society, it's imperative that we continue to look inward and evaluate how we can create equitable opportunities for recreation" (Guzman, 2017, para. 6).

SUMMARY

Recreational programming is thriving on college campuses. Universities have embraced the notion that higher education involves much more than lectures and exams and that the quality of student life is an important concern. The Campus Recreation Department helps to promote the overall goals of the institution by offering student activities that promote health, happiness, and affiliation. The diversity of programming is often extensive and usually student-led. As a result of the considerable recreational scheduling on the college campus, several organizations have been established at the community, state or provincial, and national levels to promote recreation and offer recreation leaders sources of both professional development and practical resources. The largest national association is NIRSA in the United States. It includes elected recreation officials from state associations, many of whom are campus recreation professionals and students.

The Campus Recreation Department can be organized as a unit with a small number of professional staff, housed within the university Athletics Department, or it can function as a stand-alone unit within the Student Affairs operation of the university, with many departments led by recreation professionals employed on a full-time basis. In either case, the organizational structure will rely on an extensive group of student employees and student volunteers who deliver a vast array of programming. Recreation professionals manage the affairs of the department by defining mission, vision, and goals; by managing facility operations and finances; and by setting and enforcing policy. Many policy areas draw the attention of the Campus Recreation Department leaders, from defining how the unit is funded and maximizing funding sources to program offerings and participant eligibility to access for people with disabilities. The Campus Recreation Department plays an extensive role in the delivery of recreational opportunities for the constituents on the modern university campus.

portraits + perspectives

KAT HALBLEIB *Associate Director of Facilities*

University of Louisville Campus Recreation Department

The main job responsibility for my management position is to ensure that the administration of the intramural sports component within the Campus Recreation Department is performed at a high level. This includes recruiting, hiring, training, scheduling, and supervising undergraduate supervisors, officials, and scorekeepers; adjusting the annual intramural calendar to keep things current; ensuring that playing surfaces are safe for play; scheduling all games for meets, tournaments, and leagues; overseeing all maintenance for intramural fields; and enforcing good sporting conduct policies.

Two current issues that I see within the campus recreation field are student participation and lack of professionals in the field. It is becoming clear that students are not participating in the variety of opportunities we offer as a department. Students are participating in other activities across campus and there is still some hesitancy to come back to the recreation center due to COVID-19. Along with student participation, the lack of professionals within the field is problematic. Throughout the hiring process for positions, many people do not have experience in campus recreation or in sports in general. We are seeing a lot of people exit jobs in campus recreation to take higher paying positions and then there is no one to replace them.

A policy/governance piece that I have been involved with is ensuring there is transparency within the department. I oversee all the daily deposits for the department and fill out all the necessary forms (online/paper), so if an audit is initiated everything is available for the auditor. Transparency is a priority for our department.

Since we are under the direction of the University, our governance structure can be somewhat gray. But our overarching association is NIRSA. Our state association is KIRSA (Kentucky Intramural and Recreational Sports Association). Both of these associations are leaders in higher education and are advocates for advancing recreation, sport, and well-being. One positive is that these organizations provide communication channels for people working in campus recreation at different universities to interact with one another to create better opportunities and events for students. A negative is that typically they do not initiate contact with universities but are available when a specific department/university reaches out.

One of the major future challenges for sport managers working in campus recreation today is that the incoming workforce views their positions as stepping stones. Gone are the days when employees would stay in a role for 30+ years. There are pros and cons to this challenge, but this is something that sport managers are going to have to accept moving forward.

The first piece of career advice I would give to students wanting to work in campus recreation is to be intentional about attending networking opportunities. It is not about who you know, but who knows you. My second piece of advice would be to get involved with the Campus Recreation Department at your school. There are plenty of positions that students can hold and you have the opportunity to gain vast experience and responsibilities which will aid you in your sport industry career path.

case STUDY

FACILITY DEVELOPMENT

As the newly hired Assistant Director of Campus Recreation at Big State University, your first assignment is to upgrade program offerings in your new facility. You have the following facilities:

1 gym – big enough for eight basketball courts
1 pool measuring 25 meters
1 cardiovascular room with treadmills and stationary bikes
1 weight room with free weights and a weight-machine system
2 activity rooms with 10-foot ceilings
4 multipurpose grass fields

Right now your program offerings include the following:

Basketball	Volleyball	Softball	Flag football
Badminton	Dodgeball	Baseball	Soccer
Yoga	Pilates	Spin classes	Dance fitness
Open Swim	Karate	Judo	Walking class

Your student body enrollment of 20,000 includes many nontraditional students, and residence halls coexist with fraternities and sororities on campus.

1. What are two new activities you would add to your program? Please indicate if they would be for men, women, or as co-ed. What are two reasons for adding each of the sports you chose?
2. What would be one tactic you would use to recruit more facility and program staff to work in your campus recreation program?
3. What would be one way you could make sure your facility and program offerings are inclusive of students with disabilities?

CHAPTER questions

1. Using the Internet, search for the committee structure of NIRSA. Build an organizational chart of the committees showing how they link together and where they report. How are the ideas generated and the problems solved at the committee level turned into policy? Trace and describe one example of such policy development by reading the committee meeting minutes as posted on their website.
2. Investigate the campus recreation program at your institution. How is it structured, and how is policy developed? Who has the authority to make decisions? How is the program financed? How would you go about creating a new program activity?

3. Varsity athletics and campus recreation often compete for facilities and resources on campus. Develop an organizational structure with the best chance of downplaying this internal rivalry.

FOR ADDITIONAL INFORMATION

For more information check out the following resources:

1. This website goes into detail about NIRSA's core values: https://nirsa.net/nirsa/strategic-values/
2. Scroll down on this website from NIRSA for a video about the first 100 years of campus recreation: https://nirsa.net/nirsa/about/history/
3. This site contains links to a ranking of the top 35 most luxurious student recreation centers. https://www.collegerank.net/features/best-student-recreation-centers/
4. If you want to see the types of jobs available in campus recreation, have a look at this website: https://www.bluefishjobs.com/

REFERENCES

ACHA. (2021). *ACHA history*. https://www.achahockey.org/acha-history.

Brown, S.C. (1998). Campus recreation. In J.B. Parks, B. Zanger, & J. Quarterman (Eds.), *Contemporary sport management* (pp. 139–154). Human Kinetics.

Cleveland State University. (2021). *WellFest*. https://www.csuohio.edu/recreationcenter/wellfest

Colgate, J.A. (1978). *Administration of intramural and recreational activities: Everyone can participate*. John Wiley.

East Carolina University Campus Recreation and Wellness. (2022). *Mission, vision, and goals*. https://crw.ecu.edu/about-us/our-story/mission-vision-and-goals/

Fazio, J.V. (2021). *Esports: What we should expect in 2021*. National Law Review. https://www.natlawreview.com/article/esports-what-we-should-expect-2021

Franklin, D. (2013). Evolution of campus recreational sports: Adapting to the age of accountability. In NIRSA Publication, *Campus recreational sports: Managing employees, programs, facilities, and services*. Human Kinetics.

Guzman, R. (2017, July 10). *What's new in RSJ?* NIRSA. https://nirsa.net/nirsa/2017/07/10/whats-new-in-the-rsj/

Howard, B. (2021, May 5). Overcoming esports challenges 2021. *Campus Rec Magazine*. https://campusrecmag.com/overcoming-esports-challenges-2021/

Langley, T.D., & Hawkins, J.D. (2004). *Administration for exercise-related professions* (2nd Ed.). Thomson/Wadsworth.

LaVoi, N.M., & Kane, M.J. (2014). Sociological aspects of sport. In P.M. Petersen & L. Thibault (Eds.), *Contemporary sport management* (5th ed., pp. 426–449). Human Kinetics.

Loomer, M. (2019). How to use the powerful recruiting tool. *Campus Rec Magazine*. https://campusrecmag.com/how-to-use-the-powerful-recruiting-tool/

Marshall University Recreation Center. (2020). *Corporate sponsorship program*. https://www.marshall.edu/campusrec/files/2020/09/corporate-sponsorship-fall-2020.pdf

Martinez, A.X. (2017). *How inclusive are campus recreation programs?* National Center on Health, Physical Activity and Disability. https://www.nchpad.org/1504/6460/How~Inclusive~are~Campus~Recreation~Programs~

Miller, R.D. (1998). Campus recreation risk management. *NIRSA Journal, 22*(3), 23–25.

Mueller, P., & Reznik, W. (1979). *Intramural-recreational sports programming and administration* (5th ed.). John Wiley.

Nadworny, E., & Larkin, M. (2019). *Fewer students are going to college. Here's why that matters*. NPR. https://www.npr.org/2019/12/16/787909495/fewer-students-are-going-to-college-heres-why-that-matters

NIRSA. (2019). *Equity, diversity, and inclusion: A resource guide for leaders in collegiate recreation*. https://nirsa.net/nirsa/portfolio-items/equity-diversity-inclusion-resource-guide/

NIRSA. (2021a). *About*. https://nirsa.net/nirsa/about/

NIRSA. (2021b). *Corona virus updates and resources for campus recreation*. https://nirsa.net/nirsa/covid19/

NIRSA. (2021c). *NIRSA member code of ethics*. https://nirsa.net/nirsa/wp-content/uploads/nirsa-member-code-of-ethics.pdf

NIRSA. (2022a). *Equity, diversity, inclusion*. https://nirsa.net/nirsa/strategic-values/#equity-diversity-inclusion

NIRSA. (2022b). *History*. https://nirsa.net/nirsa/about/history/

NIRSA. (2022c). *NIRSA's mission and vision*. https://nirsa.net/nirsa/about/mission-vision/#:~:text=Our%20Vision,development%20of%20healthy%20communities%20worldwide

NIRSA. (2022d). *NIRSA's strategic values*. https://nirsa.net/nirsa/strategic-values/

NIRSA Communications. (2021, June 18). *The rise of esports in campus rec*. NIRSA. https://nirsa.net/nirsa/2021/06/18/the-rise-of-esports-in-campus-recreation/

Popke, M. (2019, October 2). New trends in the campus recreation center's design. *Athletic Business*. https://www.athleticbusiness.com/recreation/new-trends-in-the-campus-recreation-centers-design.html

Portland State University Campus Recreation. (2022). *About us*. https://www.pdx.edu/recreation/about-us#mission

Smith, P. (1991). Positioning recreational sport in higher education. In R.L. Boucher & W.J. Weese (Eds.), *Management of recreational sports in higher education* (pp. 5–12). WCB Brown & Benchmark.

Spencer, T.C. (2019). *Campus recreation inclusion for people with disabilities: A qualitative investigation of current inclusive practices* (Publication No. 3313) [Doctoral dissertation, University of Louisville]. ThinkIR: The University of Louisville's Institutional Repository.

UNC Campus Recreation. (n.d.). *Commitment to inclusion: Transgender participation policy*. https://campusrec.unc.edu/program/transgender-guidelines/

University of Houston. (2021). *Campus recreation departmental & facility policies*. https://uh.edu/recreation/facilities/policies/#transgender-policy

University of Arizona Campus Recreation. (2022). *About*. https://rec.arizona.edu/about

University of Nebraska Student Affairs -Campus Recreation. (2022). *Mission, vision, and guiding principles*. https://crec.unl.edu/mission-vision-and-guiding-principles

Wallace Carr, J., Robertson, B., Lesnik, R., Byl, J., Potter, C.J., & Ogilvie, L. (2013). Unique groups. In Human Kinetics, *Introduction to recreation and leisure* (2nd ed.). Human Kinetics.

Zabonick-Chonko, R. (2016, March 1). A culture of risk management. *Campus Rec Magazine*. www.campusrecmag.com/a-culture-of-risk-management/

Intercollegiate Athletics

Click on a college website, scroll through your Twitter feed, or turn on the television and you will immediately see the interest in North American college athletics. National championship events such as "The Final Four" in men's and women's college basketball are known worldwide. The spectacle of college sport will likely continue to grow and endure the test of time, due in part to the mass media's role in strengthening its appeal

DOI: 10.4324/9781003303183-8

by bringing events and personalities directly into our homes and onto our mobile devices. The appeal is strongest in the United States, but colleges and universities in many countries around the world also sponsor competitive athletic opportunities for their students. In the United States and Canada, colleges and universities support extensive competitive athletic programs. This type of competition is commonly known as *intercollegiate athletics*. In the United States specifically, intercollegiate athletics has undergone drastic changes over the past few years – prompting those governing intercollegiate athletics to adapt to an ever-evolving landscape brought on by state legislation (e.g., name, image, and likeness), global health crises (e.g., the COVID-19 pandemic), and calls for student-athlete empowerment.

The appeal of intercollegiate athletics is unquestionable and at the same time paradoxical. From one perspective, the loyalty of cheering college students with faces painted in school colors, fully caught up in the excitement of events and intense rivalries, sometimes with national distinction at stake, is completely understandable. If you ask any sport fan about some of their favorite sport memories, they will likely share moments tied to intercollegiate athletics in some form. But viewed from another perspective, intercollegiate athletics is woven with problems. From the consumer viewpoint, the quality of play may not compare with professional leagues. In addition, a long history of abuses, excesses, and cheating has plagued intercollegiate athletics, challenging the very core concepts of sport in general and amateurism specifically. Ideals such as fairness, honesty, character development, and competitive balance have been questioned. Some critics of intercollegiate athletics in the United States, in addition, have questioned the dual role of the college athlete (or, to use terms the National Collegiate Athletic Association (NCAA) uses, the student-athlete). In some cases, the very existence of a high-profile sport enterprise in connection with an educational institution has been called into question.

Some of these issues, along with the increasing costs of programming, have resulted in some schools dropping programs. To manage in times of fiscal restraint resulting from decreased resources, increased costs, and the global health crisis brought about by the COVID-19 pandemic, some colleges have resorted to dropping teams. For example, by 2020 more than 300 teams across the NCAA, NAIA, and community and junior colleges had been dropped, with schools blaming the negative financial impact – more than $120 billion – facing US higher education as a result of the pandemic (Anderson, 2021). Among the teams cut were those competing for some of the most prominent athletic powerhouses, such as teams at the University of Iowa (men's gymnastics, men's tennis, and men's swimming and diving) and Michigan State University (men's and women's swimming and diving). Others, like Stanford University and Clemson University, announced that they would cut teams but later reversed course due to public backlash (Shapiro, 2021).

Historically, schools have also used Title IX, the federal law that prohibits sex discrimination in education, as a reason for dropping teams. In these instances, schools argued that to comply (or at least to project the image of compliance) with Title IX legislation, universities chose to drop teams. For example, in 2015, the University of Tennessee (UT)-Chattanooga dropped men's indoor and outdoor track and field teams, a surprising move given the long history of the program (Shahen & Heron, 2015). UT-Chattanooga is just one example of dozens of Division I schools that have eliminated men's teams like wrestling, gymnastics, and swimming to comply with Title IX (Thomas, 2011). However, the reality of discontinuing sports to comply with Title IX is perhaps masking the real issue of poor fiscal decision-making on behalf of Athletic Department personnel. Football teams may lose millions of dollars and yet no one suggests dropping football. This only notably happened when the University of Alabama – Birmingham briefly dropped its football program but then brought it back a few years later (Dickey, 2018). Unfortunately, the financial decisions intentionally supported by colleges are sometimes labeled Title IX issues, when in reality Title IX is about the fairness of opportunity for both men and women. Some institutions have also been padding women's rosters with practice players who never compete or count women athletes competing on teams such as cross country and track and field twice, to increase the proportion of women reportedly involved in intercollegiate athletics (Thomas, 2011).

Yet despite these and many other issues, consumer enthusiasm for intercollegiate athletics continues to grow as evidenced by ticket sales and television revenues (Meyer & Zimbalist, 2017). While *growth* may have served as the perfect one-word descriptor of 20th-century and early 21st-century college athletics in the United States, *change* is a more suitable word to describe intercollegiate sport since the turn of the century. Interest in supporting the local team gathered momentum to a point where now more than a thousand colleges and universities offer intercollegiate sport in the United States alone. Sport in the NCAA specifically has turned into a billion-dollar industry, attracting millions of fans to state-of-the-art athletic facilities, mobile devices, and television screens. There is no doubt the sheer magnitude of intercollegiate athletics may be one reason for its enduring and expanding appeal.

Different people bring differing perspectives on the phenomenal growth of intercollegiate athletics over the past century. A historian might suggest that the leadership of President Theodore Roosevelt and a group of college presidents provided the original momentum for the growth of college athletics when they intervened in college football to promote more extensive rules and safety requirements in 1905. A sociologist might identify the place of sport in US society and the feelings of personal success and hometown pride when the local college team wins (MacGaffey, 2013). A psychologist might point to improved psychological health with individuals identifying strongly with a local sport team (Wann, 2006). The economist might suggest that colleges and universities need the revenue generated by athletics, whether from television, recruiting students, or developing and managing the image of the institution. A critical theorist might critique the exploitation of athletes in revenue-producing sports, the majority of whom are Black, to generate wealth for the institutions they represent. Other viewpoints exist, but one thing is sure: intercollegiate athletics is a major component of the sport industry of North America.

This chapter focuses on the many differences between the governance of organizations delivering collegiate sport in the United States (the NCAA and the National Association of Intercollegiate Athletics [NAIA], and Junior, Small and Christian Colleges) – such as size, financial capacity, committee structures, scope of operations, sports supported, and rules and underlying philosophies. Distinctions also exist in the governance and policy development between collegiate sport organizations in the United States and other countries around the world. Although collegiate sport exists in a number of countries worldwide, and athletes from these organizations represent their countries at the World University Games (Federation Internationale du Sport Universitaire [FISU] Games; see the chapter on Major Games), their college sport programs may differ substantially from those in the United States. For instance, the British Universities and Colleges Sport (BUCS) has over 165 member universities, while University and Tertiary Sport New Zealand (UTSNZ) has eight member universities and two affiliated organizations, compared to the larger NCAA which has over 1,200 members. Some countries within Africa, Asia, and South America may have very limited sport offerings, with emphasis on sports unknown or less popular in North America, such as footvolley (mix of football and beach volleyball), jujitsu, biribol (volleyball in a swimming pool), kabaddi (a contact team sport combining elements of wrestling and rugby), and running events. The big business nature of United States collegiate sport fueled by considerable fan interest and television reach (and revenue) is not necessarily mirrored in other countries around the world. U Sports, the organization governing university sport in Canada, has a far smaller membership of only 56 institutions and concomitant economic impact. The organizations that manage sport and create the governance structures and policies for operations are unique to the political, social, and cultural environments, as well as historical events, of each location. Therefore, the umbrella organizations delivering collegiate sport in countries around the world vary significantly in their size, capacity, programs, rules and regulations, and structures.

Let's turn our attention now to the governance of collegiate athletics in the United States. Exactly how is it organized? How are rules made, and who decides the issues of the day? A brief look at the history and evolution of intercollegiate sport will answer those questions.

WWW
British Universities and Colleges Sport (BUCS)
https://www.bucs.org.uk

WWW
University & Tertiary Sport New Zealand (UTSNZ)
https://www.utsnz.co.nz

WWW
U Sports
https://usports.ca/en

HISTORY OF US AMERICAN INTERCOLLEGIATE ATHLETICS

Often the largest and most popular events are born of the humblest beginnings, and this is exactly the case with intercollegiate athletics. The idea for athletic competition did not come from educators, nor was it a part of the curriculum. Rather, it originated with the student body.

The Beginning

College athletics began as recreational activities organized by students to meet their desire for both physical and social activities (Davenport,

1985). Although faculty members were not involved, they accepted the idea that students needed some diversions from classroom activity. It is easy to understand how college athletics developed. Two groups of students got together to play a game in the late afternoon sun; later, over dinner, the victors boasted of their success. Perhaps their classmates listened in and decided to show up for the next game to cheer on their friends. Next, for even more bragging rights, the victorious group then challenged the college in the next town. This, basically, is the story of the first intercollegiate competition, when a crew (rowing) race was organized between Harvard and Yale in 1852 (Scott, 1951).

Original Events

The next organized intercollegiate activity was baseball. The first baseball game was between Amherst College and Williams College in 1859 (Davenport, 1985). Such student-led activities gained significant interest among spectators and some notice from college faculty and administration. The administration noticed that winning athletic contests helped recruit students to campus and provide some positive attention for the college. Only 10 years later, on November 6, 1869, the first intercollegiate football game was played between Rutgers and Princeton (Davenport, 1985). Challenges for competition became more and more common. This growth was not always viewed positively, however. Administrators were concerned about the unproductive nature of athletic contests, and significant resistance was voiced against the emerging popularity of intercollegiate football.

Despite the attitude of some university administrators, tremendous interest in collegiate football was evident by the 1890s. A win-at-any-cost mentality developed, and to please the spectators in the overflowing college grandstands, players and coaches without affiliation to the college were inserted in the line-up. Street brawls became common after games:

> In 1893 New York was thrown into a virtual frenzy by the annual
> Thanksgiving game between Yale and Princeton. Hotels were jammed
> ... Clergymen cut short their Thanksgiving Day services in order to get
> off to the game in time. Clearly, football had arrived.
> (Rudolph, 1990, p. 375)

Sports were becoming so popular on college campuses that they were likened to small business enterprises (Davenport, 1985).

Birth of College Sport Organizations

Up to this point, college athletic activities were organized and operated by students, and merely tolerated by the university administration. But it was becoming evident that athletic teams served as a unifying function of the college. Heroes emerged; public interest grew, as did the public relations opportunities. All these factors, along with the potential for revenue generation, resulted in university administrators changing their position. College presidents and their inner circles realized successful sports teams could generate additional resources for their cash-strapped institutions, as well as draw both political favor and alumni support. College administrators, especially college faculty members, moved to take over management and control of athletic programs. Athletic department personnel as we know them today did not exist yet.

On January 11, 1895, a historic meeting of faculty representatives was held in Chicago to develop eligibility and participation rules for football. This was the inaugural meeting of the Intercollegiate Conference of Faculty Representatives, the forerunner to the Big Ten (Davenport, 1985). Soon thereafter, personnel in other regions of the United States also met and copied many of the rules developed at this initial meeting. Faculty exercised control over schedule development and equipment purchase. Playing rules and regulations were enforced, and some eligibility and financial restrictions were put in place.

At about the same time, an alarming number of football players were seriously injured as a result of common strategies such as gang tackling and mass formations. In 1905, 18 athletes were killed and 159 seriously injured while playing collegiate football (Gerdy, 1997), prompting US President Theodore Roosevelt to intervene. He called representatives from Harvard, Yale, and Princeton to two White House conferences to discuss the problems. At the request of Chancellor Henry M. MacCracken of New York University, representatives of 13 institutions met in New York City in December 1905 (Smith, 2011). The original intent of this meeting and a follow-up meeting later that month was to resolve issues related to football. However, the result was much more significant. More university administrators shared concerns, and as a result, 62 members founded the Intercollegiate Athletic Association of the United States (IAAUS) to oversee and regulate all college sports. The association was officially constituted on March 31, 1906 and represented the foundation for the future NCAA.

Evolution of College Sport Organizations

The development of the IAAUS represented a pivotal moment in the history of US intercollegiate athletics and marked the beginning of an era in which collegiate sport instituted rules, regulations, supervision, and philosophical direction. Faculty members in physical education departments were hired to coach teams and administer programs.

In 1910, the IAAUS renamed itself the NCAA. During its initial years, the NCAA was composed only of faculty members from its affiliate institutions who acted as a discussion group and rules-making body. Collegiate sports continued to grow, and more rules committees followed. The evolution of the NCAA continued, and in 1921, the first national championship was held in track and field. Other sports and more championships were gradually added over the years. By 1973, the membership was divided into three legislative and competitive divisions (Divisions I, II, and III) based on institutional size. Subsequently, Division I members voted to subdivide football into Divisions I-A and I-AA. More recently, Division I football now uses the categories Football Bowl Subdivision (FBS) and Football Championship Subdivision (FCS). As the NCAA has grown in the number of events and size of membership, a shift in power away from the collection of colleges to the centralized authority of the NCAA has taken place. Today, the NCAA is a large, powerful organization operated by a National Office in Indianapolis, which is home to more than 500 full-time employees and delivers 90 championships in 24 sports across three divisions (NCAA, 2022l).

Growth of Women's Sport

In the United States, women's participation in intercollegiate sport was conspicuously missing in the beginning. Little in the way of formal competition existed for women until 1971, when women physical educators established the Association for Intercollegiate Athletics for Women (AIAW). Several national championships were sponsored, and women's intercollegiate athletics gained momentum, quickly becoming an important component of college athletics. This interest in women's athletic opportunities prompted the NCAA to expand its structure to include programming for women just 10 years after the AIAW was established. The first NCAA programming for women occurred in 1980 when Divisions II and III took a leadership role by adding 10 national championships for women. This historic action prompted an extensive governance plan to be passed in 1981–1982, including establishing 19 additional women's championships, along with services and representation in decision-making for administrators of women's athletics.

These changes in women's sport opportunities were also elevated by the passing of crucial landmark legislation for gender equity at the federal level: Title IX, which was a component of the Education Amendments that came into law in 1972. As you may recall, Title IX states, "no person in the United States shall, on the basis of sex, be excluded from participation in, be denied the benefits of, or be subjected to discrimination under any education program or activity receiving federal financial assistance" (Title IX, 2017, para. 2). While we will cover a more in-depth look into gender equity as a current policy area later in the chapter, it is worth noting here that the passage of Title IX, which celebrated its 50th anniversary in 2022, profoundly changed sport opportunities for girls and women in the United States by barring discrimination based on sex in the country's schools, colleges, and universities (Staurowsky et al., 2022). The numbers speak for themselves: whereas only 29,977 women athletes competed at the college level from 1971 to 1972, that number has skyrocketed to 215,486 women competing on teams sponsored by institutions in the NCAA during the 2020–21 academic year (Staurowsky et al., 2022). That means that the percentage of women athletes competing on intercollegiate sport teams has increased from 15% in 1971–1972 to 44% in 2020–2021 (Staurowsky et al., 2022).

Historically Black Colleges and Universities & Other Minority-Serving Institutions

Historically Black Colleges and Universities (HBCUs) are typically liberal arts institutions that were established before 1964 with the intention of serving the African American community in the United States.

Institutions with large African American student populations founded after the *Brown v. Board of Education* ruling that outlawed racial segregation are known as "predominantly Black," but not "historically Black."

Today, 105 HBCUs exist as public and private institutions, including community colleges, four-year institutions, and medical and law schools, enrolling more than 200,000 students. Over the years, HBCU graduates have gone on to make names for themselves in all spectrums of society including athletics, where some notable graduates include NFL Most Valuable Player (MVP) Steve McNair (Alcorn State University) and Eddie Robinson (Grambling State University), one of the winningest coaches in college football history (408 wins at one institution). HBCUs are affiliated with both the NCAA Division I and II, and the NAIA. Examples of HBCU institutions with traditions of excellence in college athletics include Morgan State University located in Maryland, Norfolk State University and Hampton University in Virginia, Howard University in Washington, DC, Grambling State University in Louisiana, and Florida A&M in Florida. At the conference level, Division II's Central Intercollegiate Athletic Association (CIAA) is the oldest HBCU conference in the country, and its men's basketball tournament draws over 100,000 fans annually.

Although the name might lead one to believe that only Black student-athletes attend HBCUs, today their student populations, while predominantly Black, do contain a more representative picture of society. As such, most HBCU athletic mission and vision statements emphasize this diverse aspect of their population, such as the one from Florida A&M University in Exhibit 8.1.

However, in comparison to non-HBCU colleges and universities, HBCU athletic budgets are much lower. In 2019, only six HBCUs ranked among the top 200 (out of over 300) athletic budgets in Division I (USA Today, 2020). Some small schools, such as Division I FCS Southern

---

exhibit **8.1** Mission and Vision Statements of Athletics at Florida A&M University

Mission Statement

Through excellence with caring, FAMU Athletics creates champions in competition and in life through academic success, integrity, and diverse and inclusive experiences.

Vision Statement

FAMU Athletics is committed to being a nationally recognized intercollegiate athletics program that fosters a diverse, inclusive, and caring environment of excellence in academics, athletics, and community engagement.

Source: Florida A&M University Athletics. (2020).

University, rely heavily on one or two games to generate the majority of the revenue for their athletic departments. These so-called "guarantee games" take place when small schools travel to face national powerhouses. While these games usually guarantee a substantial payday for the small school, it is hard to compete against more resource-rich institutions. For example, in

> one recent season, South Carolina State lost to Clemson, 59-0 but received a check for $300,000; Delaware State lost to Missouri, 79-0 ($550,000); Prairie View lost to Texas A&M, 67-0 ($450,000), and for similar amounts, Morgan State lost to Marshall 62-0, North Carolina Central to Duke 49-0, Florida A&M to Miami 70–3, Savannah State to Southern Mississippi 56-0, and on and on, year after year.
>
> (Gilmore, 2022, para. 18)

Since the economic downturn in 2008, many HBCU institutions have cut programs, left vacant positions unfilled or eliminated them altogether, and slashed team travel budgets (Trahan, 2016).

HBCUs are considered Minority-Serving Institutions (MSIs). MSIs are institutions that provide minority groups with access to higher education. In addition to HBCUs, there are also Tribal Colleges and Universities (TCUs), Hispanic Serving Institutions (HSIs), and Asian American Native American Pacific Islander-Serving Institutions (AANAPISI). What all of these MSIs have in common is that they serve minority populations; however, there are unique criteria each type of MSI must fulfill to qualify as an MSI, such as the percent of students enrolled who are from a historically excluded group. Some of the schools with the biggest and most well-known athletic programs are MSIs today, including the University of Minnesota, the University of Texas at Austin (both serving Asian American Native American Pacific Islander students), and the University of Arizona (serving Hispanic students).

GOVERNANCE

The growth, popularity, and subsequent reform of college sport dictated a more formal approach to managing and governing intercollegiate athletics. The NCAA is the largest and oldest organization formed for this purpose, and its history is closely intertwined with the growth of intercollegiate athletics. However, other organizations also govern intercollegiate athletics. In the United States, the NAIA is another umbrella organization of like-minded institutions, often compared in philosophical orientation to NCAA Division II schools. The National Junior College Athletic Association (NJCAA) and the United States Collegiate Athletic Association (USCAA) each exist to oversee the athletic programs of junior and small colleges, respectively. Finally, the National Christian College Athletic Association (NCCAA) administers intercollegiate competition between Christian schools.

National Collegiate Athletic Association

National Collegiate Athletic
Association
www.ncaa.org

The NCAA has global recognition, thanks to television and marketing efforts. It is technically a voluntary association of colleges and universities, run by a President and staffed by about 500 employees at the National Office in Indianapolis. Members of the NCAA consider issues and policies affecting more than one region, thus making them national issues. In college athletics, the power lies on the national level.

Mission

The NCAA is devoted to the administration of intercollegiate athletics for its membership. The core purpose of the NCAA is to govern competition in a fair, safe, and equitable manner. According to the NCAA (2022g), the goals of the organization are specific to supporting student-athletes with academic services, opportunities and experiences, financial assistance, wellness and insurance, and personal and professional development. The association supplies a governance structure to provide rules and establish consistent policy through which all NCAA member institutions operate. NCAA literature states three priorities that guide its purpose and operations – academics, fairness, and well-being (NCAA, 2022g). These are further explained in Exhibit 8.2.

Membership

The NCAA comprises member institutions whose representatives retain voting privileges on setting policy and directing the future of intercollegiate athletics. Currently 1,200 institutions, conferences, and affiliated organizations are members of the NCAA (NCAA, 2022o).

It is important to note that NCAA members are institutions, not individuals. Institutions are afforded membership by virtue of their mission in higher education, along with other membership criteria. All sizes and types of institutions are eligible for membership, as long as they are accredited by the recognized agency within their academic region, offer at least one sport for both men and women in each of the three traditional sport seasons, abide by the rules and regulations set forth by the NCAA (as certified by the CEO), and agree to cooperate fully with NCAA enforcement programs (NCAA, 2022o).

NCAA member institutions belong to one of three divisions – Division I, II, or III. The main criteria used for establishing an institution's divisional classification are its size, number of sports offered, financial base and sport-sponsorship minimums, focus of programming, football and basketball scheduling requirements, and availability of athletic grants-in-aid (NCAA, 2022o). Division I (DI) football institutions are further subdivided into I–FBS and I–FCS. DI–FBS programs must meet minimum paid-football attendance criteria. Institutions competing in Division I but which do not offer football are categorized simply as Division I. Divided fairly evenly among each division are the 1,098 active institutional members: Division I has 350, Division II has 310, and Division III has 438 (NCAA, 2022o).

The NCAA is a member-led organization focused on cultivating an environment that emphasizes academics, fairness and well-being across college sports.

Academics

To get the most out of college, student-athletes have to succeed on the court and in the classroom. The NCAA provides opportunities to learn, compete and grow on and off the field. The ultimate goal of the college experience is graduation, and college athletes are graduating at rates that are higher than ever.

Fairness

With so much changing in college sports, rule changes are focused on improving the student-athlete experience. The NCAA is committed to providing a fair, inclusive and fulfilling environment for student-athletes and giving them a voice in the decision-making process.

Well-Being

In 1906, the NCAA was founded to keep college athletes safe. The Association is still working hard to protect them physically and mentally. Through its Sport Science Institute, the NCAA provides recommendations and guidelines to ensure college athletes are getting the best care possible.

Source: NCAA. (2022g).

Finally, athletic conferences and affiliate organizations are also members of the NCAA. There are 102 voting athletic conferences such as the Big Ten, Southeastern Conference, the Ivy League, the East Coast Conference, and the New England Collegiate Conference. A number of organizations are considered affiliated organizations including the National Association of Collegiate Directors of Athletics (NACDA) and the National Association for College Admissions Counseling (NACAC). Affiliate members must be nonprofit organizations whose function relates to NCAA sport and that involve coaches or university administrators.

Financials

The NCAA is a nonprofit organization, yet it is also a billion-dollar enterprise. Given the breadth of focus described above, it requires substantial revenue to fund an incredibly wide-ranging agenda. That revenue comes from two primary sources: (a) the NCAA Division I Men's Basketball Championship television and marketing rights and (b) championships ticket sales (NCAA, 2022p). Despite suffering heavy losses in revenue at the height of the COVID-19 pandemic, mostly due to the cancelation of its men's March Madness tournament (its main revenue source) and other championships, the NCAA today is in a healthy financial situation. It generated a record $1.15 billion in revenue for the 2021 fiscal year (Berkowitz, 2022)! These revenues are generated from television rights, championships, royalties, investments, sales and services, and philanthropic contributions. The association's expenses include sport sponsorship and scholarship funds, the Division I Basketball Performance Fund, championships, and the Student-Athlete Assistance Fund, among numerous other categories.

Association-Wide and Division-Specific Structure

As with most other self-governing organizations, the NCAA began and existed for many years with a governance structure allotting one vote to each member institution. At an annual national convention, members debated issues and voted on matters of policy. This organizational structure was altered on August 1, 1997. In general, the reform provided each division greater autonomy for managing division-specific matters and gave university presidents more involvement in and control of developing legislation. For instance, in Division I the one-vote-per-institution principle was replaced with a system based on conference representation. Rather than every member voting on each issue at an annual convention, the Division I Board of Directors is charged with managing all legislation, strategy, and policy for the division. The Board of Directors has 24 members: 20 university presidents (one from each of the 10 FBS conference and 10 representing the remaining 22 conferences on a rotating basis), one Athletic Director (AD), one Faculty Athletics Representative (FAR), one Senior Woman Administrator (SWA), and one student-athlete. Of course, the Board of Directors is not able to complete all of the Division's business.

DIVISIONAL GOVERNANCE STRUCTURE. Prior to 2008, the Division I Management Council reported to the Board of Directors. However, the Management Council was then replaced with two 31-member councils: the Leadership Council and the Legislative Council. These changes were designed to increase efficiency and provide more support to the Board of Directors. Today one Division Council exists with a variety of committees that assist with issues of strategy and policy development. The Council is comprised of 41 athletic administrators and faculty athletic representatives (FARs), academics from the college faculty appointed to provide their perspective on athletic policy making, and student-athletes.

Committees report to the Council in the following areas with variable membership numbers presented in brackets: student-athlete experience (11), strategic vision and planning (10), legislative (19), nominating (12), competition oversight (19), men's basketball oversight (12), football oversight (15), women's basketball oversight (12), and student-athlete advisory (32).

One unique change in Division I governance occurred in 2015, when the NCAA granted autonomy to five "power" conferences including the Atlantic Coast Conference (ACC), Big Ten, Big 12, Pac-12, and Southeastern Conference (SEC), allowing them to pass legislation through their own voting process (Trahan, 2015). Legislation is then adopted as rule changes by all Division I NCAA member institutions. By granting autonomy to the power conferences, member schools have more power to control operational decisions. While the power is passed on, certain rules are expected to remain unchanged (Trahan, 2015).

Divisions II and III are structured similarly to Division I but have several important distinctions for conducting division-specific business. Each has a Board of Directors made up of institutional CEOs, but the body is called the Division's Presidents Council, not a Board. The Division II Presidents Council includes 16 presidents, while the Division III Presidents Council is comprised of 18 presidents of institutions belonging to Division III. Both Divisions II and III also have Management Councils (with 28 and 21 members, respectively), but Division III has broadened the representation of this group by adding university presidents (2) to the athletic administrators, FARs, and student-athletes comprising the Division I and II councils. Although Division II and III are structured in a similar way to Division I, one very important distinction remains: legislation in both divisions is considered by the traditional one-school, one-vote method, as opposed to the conference representation used by Division I. Both Divisions II and III have a committee structure to deal with issues specific to their business and sports. Exhibit 8.3 illustrates the NCAA Division I structure.

www

Complete Graphic of NCAA Division I Governance Structure

https://www.ncaa.org/sports/2015/10/28/how-the-ncaa-works.aspx

NCAA Division I Governance Structure

exhibit **8.3**

Division I Board of Directors

The Division I Board of Directors is the highest governing body for the Division, charged with developing strategy and policy and overseeing management and legislation for the division.

monitors legislation

Committee on Academics
= primary academic authority of Division I. It is subject to review by the Board of Directors and can recommend legislation to the Division I Council.

Committees with Direct Reporting Line to Board of Directors

| Committee on Infractions | Infractions Appeals Committee | Finance Committee |

monitors legislation

Division I Council– PRIMARY AUTHORITY FOR LEGISLATION

Division I Council has primary legislative authority in the division. However, its work is reviewed by the Board of Directors.

Division I committees debate ideas and recommend whether or not the Division I Council should introduce them as legislation.

Division I Committees

| **Legislative Committee** Committee on Student Athlete Reinstatement | Nominating Committee | Student-Athlete Advisory Committee | Competition Oversight Committee | Student-Athlete Experience Committee |
| Committee on Legislative Relief *& other committees report to Legislative Committee* | Strategic Vision and Planning Committee | Women's Basketball Oversight Committee | Men's Basketball Oversight Committee | Football Oversight Committee |

Autonomy for Power 5 Conferences
The ACC, Big Ten, Big 12, Pac-12 and SEC have autonomy to make rules in specific categories (e.g., meals and nutrition).

Source: NCAA. (n.d.).

ASSOCIATION-WIDE GOVERNANCE STRUCTURE. As you can see, the NCAA is a large enterprise. Each division has many members and layers of committee structures for managing the business of intercollegiate sport. As with any large conglomerate, however, there is always the need to oversee association-wide issues that go beyond the scope of the divisional governing entities. Whereas each of the three divisions manages its day-to-day needs, there are some areas that affect NCAA athletics in its entirety. That is why there are a variety of association-wide committees comprised of members from all three divisions. These committees work collaboratively with the divisions to shape overarching NCAA legislation and policy. The highest such committee within the NCAA is the Board of Governors.

The Board of Governors used to be comprised of 16 presidents from NCAA member institutions as well as five independent members, for a total of 21 voting members on the board. However, in 2022 the NCAA membership approved a new constitution – the first major revision of the NCAA constitution since 1997 – that reduced the Board of Governors to nine members (four members from Division I; 1 from Division II; 1 from Division III; 2 independent members; 1 graduated student-athlete). There are also a number of ex-officio (nonvoting) members on the Board, including the NCAA President, the Chairs of the Division I Council and the Division II and III Management Councils, the president of an HBCU, as well as one student-athlete each for the divisions not represented by the voting student-athlete member. The Board of Governors is commissioned to ensure each division operates consistently with the overall principles, policies, and values of the NCAA. More specifically, the Board is charged with strategic planning for the Association, which includes overseeing the overall budget, resolving issues that affect all members of the association, and initiating legislation. The Board of Governors has the power to implement legislation that all divisions must adhere to. For example, in 2020 the Board of Governors created a Confederate flag policy that prevented any NCAA championship to be held in states where the state government displayed the symbol (NCAA, 2020).

In addition to the Board of Governors, 14 association-wide committees exist to ensure that the principles, policies, and values of the NCAA on common issues like medical and safety concerns are articulated and communicated. They include the Board of Governors Ad Hoc Committee on Sports Wagering, Board of Governors Committee to Promote Cultural Diversity and Equity, Board of Governors Student-Athlete Engagement Committee, Committee on Competitive Safeguards and Medical Aspects of Sports, Committee on Sportsmanship and Ethical Conduct, Committee on Women's Athletics, Constitution Committee, Honors Committee, Minority Opportunities and Interests Committee, Olympic Sports Liaison Committee, Playing Rules Oversight Panel, Postgraduate Scholarship Committee, Research Committee, and Walter Byers Scholarship Committee. See Exhibit 8.4 for a summary of the governance structure of the NCAA.

Major initiatives in enforcement and governance in the NCAA have occurred since 1981. An agenda of reform and strategy of college presidential involvement in the affairs of the organization, along with a philosophical push to put academics first in the athlete-education dyad was emphasized through changes to rules, governance structures, and institutional change. The NCAA's focus on Academic Progress Rate (APR), for example, is indicative of the action taken to improve graduation rates. Although many argue that the changes have not worked or do not go far enough, that significant changes have been made is evident given the size and complexity of the NCAA Division Manuals and sport-specific Rule Books. New rules have involved governing athlete eligibility and financial aid, cost containment, recruitment, coach salaries, drug testing, championships, women's issues, and student-athlete welfare. Television has contributed to the growth and has publicized issues such as academics and amateurism, expectations put on student-athletes, growth and finance, diversity and inclusion, and external interventions by the government and courts of law. As the organization and its members' goals and values evolve, one area has not faded: problems within the NCAA continue to be visible and critics are more vocal and active than ever before.

Over the past few years specifically, intercollegiate athletics has undergone some of the most transformative changes in decades. Among the most significant changes are the Association's changing (and, arguably, not voluntary) stance on student-athletes' ability to make money off of their name, image, and likeness (NIL; see the policy section of this chapter). The advances of NIL policies across intercollegiate athletics are indicative of broader calls for change to the institution that is US collegiate sport. For example, in late 2020 Senators Cory Booker (Democrat of New Jersey) and Richard Blumenthal (Democrat of Connecticut) introduced the College Athletes Bill of Rights, a congressional proposal aimed at reshaping collegiate sports by, among other items, making schools share revenue with its athletes.

www

Complete Graphic of NCAA Association-Wide Structure
https://ncaaorg.
s3.amazonaws.com/
champion-magazine/
HowNCAAWorks/AW_
HowNCAAWorks.pdf

NCAA Association-Wide Structure

exhibit **8.4**

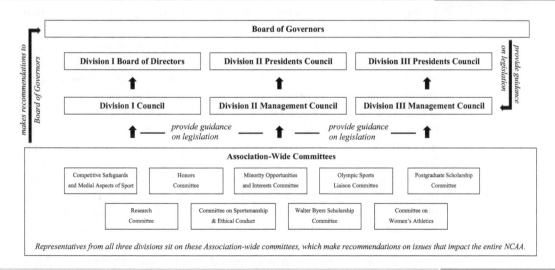

Another legislative blow was dealt to the NCAA by the Supreme Court in 2021, when the Court ruled against the NCAA in the *NCAA vs. Alston* case. The Court found that NCAA rules restricting education-related benefits for athletes were in violation of federal anti-trust laws. And in 2022, California Governor Gavin Newsom introduced Senate Bill 1401, titled the *College Athlete Race and Gender Equity Act*, which would require schools in the state to share part of their annual revenue with student-athletes in revenue-producing sports such as football. While it is too early to see how each of these developments will affect intercollegiate athletics in the long term, one thing remains certain: potentially transformative *change* is about to come in the billion-dollar industry that is US intercollegiate athletics.

National Association of Intercollegiate Athletics

National Association of
Intercollegiate Athletics
www.naia.org

The NAIA is another national association governing intercollegiate athletics in the United States. More than 250 mostly small-size institutions comprise the NAIA, many of which emphasize the link between education and athletics more strongly than revenue generation. Initially formed to regulate intercollegiate basketball, the association was called the National Association of Intercollegiate Basketball (NAIB) until 1952, when the organization changed its name to the NAIA.

Mission

NAIA institutions view athletics as "co-curricular," that is, as part of the overall educational process, something that goes hand in hand with the pursuit of academic goals. They believe involvement in athletics will enrich the student-athlete's college experience, balancing success in both the classroom and on the field of play. The NAIA National Office in Kansas City is the hub of organization and planning for national championships, and it defines the rules, regulations, and structures that govern member institutions.

Fundamental to the NAIA mission is their Champions of Character initiative. Champions of Character is a program that attempts to instill and align the core values of integrity, respect, responsibility, sportsmanship, and servant leadership into the behaviors of athletes, coaches, and administrators. Every year, each school in the NAIA has the opportunity to demonstrate its commitment to the Champions of Character philosophy by completing a Champions of Character Scorecard. As part of the Scorecard, schools can earn points for initiatives related to character training, academic performance, and conduct in competition. For the 2020–21 academic year, more than 150 out of the 252 member institutions were named Champions of Character Five-Star Award winners – the highest honor in the Champions of Character initiative.

In 2018, the NAIA also launched Return on Athletics (ROA), a resource aimed to "maximize the business performance of small college athletics departments" (NAIA, 2022a, p. 2). The goal of ROA is to provide NAIA member institutions with robust financial data that can aid in institutional (financial) decision-making and "optimize the financial return of athletics programs" (NAIA, 2022a, p. 3).

Membership

Like in the NCAA, NAIA members are institutions. NAIA membership is rather diverse, with an assorted group of institutions around the United States and a few members from Canada. The NAIA has over 250 member colleges and universities in 21 conferences across the United States (NAIA, 2022a). It has two categories of membership: active and associate members (NAIA, 2022a). Active members consist of four-year colleges and universities and upper-level two-year institutions in the United States or Canada that award undergraduate degrees. These institutions must be fully accredited by one of six institutional accrediting bodies from across the United States (such as the Southern Association of Colleges and Schools Commission on Colleges or the New England Association of Schools and Colleges Commission on Institutions of Higher Education), must abide by the constitution and bylaws of the NAIA, must be accepted for membership by the Council of Presidents, and must pay membership fees. Associate members are required to meet the same standards as active members except for full accreditation by one of the six accrediting bodies. These institutions must, however, be committed to the development of a fully accredited baccalaureate (undergraduate) program. Associate members are not eligible for postseason competition, nor do their institutional representatives vote on issues, serve on committees, or participate in national awards programs.

Individual institutions become members of one of the NAIA's organized regions and conferences. Issues that impact the overall association or that deal with national championships are deliberated at national meetings or in committee forums.

Financials

The NAIA is a nonprofit association funded through membership fees, sponsorship, championship revenues, and merchandise sales. It collects fees for running national championships and other special programs and shares net revenues with its membership.

Organizational Structure

While the organization is led by a President & CEO who manages the large NAIA National Office in Kansas City, Missouri, the NAIA is organized to govern its business through a series of councils and committees. The association deliberately places the membership at the top of the organizational chart to emphasize the importance of each institution and its student-athletes for whom programming is organized. The Council of Presidents (COP) responds to the membership.

Similar to the NCAA, the NAIA relegates control and responsibility of a school's intercollegiate athletic program to the President or CEO of the institution. Each of the 21 conferences within the NAIA elects one institutional CEO to represent the conference membership on the COP. Several other independent, at-large, or ex officio (nonvoting) Presidents are appointed members based on expertise. In addition, two at-large positions are reserved for female and/or minority Presidents. An elected Executive Committee, made up of a chair, three other members, and the NAIA administrative head, manages the business of the Council between meetings. The COP employs and supervises the President & CEO of the organization, and has final authority for all fiscal matters, membership applications, and council recommendations. Each member of the Council holds one vote. Three governing councils report to the COP and develop policy for the NAIA (NAIA, 2022f): The National Administrative Council, the Council of FARs, and the Council for Student-Athletes. Each of these Councils is tasked with collaborating with a variety of associations and other stakeholder groups. For example, the NAC works with the Athletics Directors Association, the Sports Information Directors Association, the Association of Independent Institutions, and NAIA Coaches Associations. The CFAR collaborates with the FAR Association and the Registrars Association, while the CSA is charged with working with the Association of Student-Athletes and the Athletic Trainers Association. Each of the three policy-making Councils also has a variety of standing committees to oversee the various issues affecting NAIA institutions.

A National Coordinating Committee (NCC) serves as the communications link between each of these three groups and reviews the operational policies developed for these groups, among other duties. The NCC is comprised of the chair and chair-elect of each of the three major administrative groups outlined below as well as five at-large members elected by NAIA members. Similar to the COP, the NAIA in 2021 revised its constitution to make it mandatory for the NCC and three major Councils to reserve multiple positions for women and/or minority representatives (two out of five at-large members on NCC; a minimum of 2 of the at-large positions on NAC and CSA, respectively; 2 on CFAR).

1. **NATIONAL ADMINISTRATIVE COUNCIL (NAC).**
    The NAC is responsible for all sport-related business. Its members are comprised of those individuals at each institution who hold the position of AD, Sports Information Director, and Coach. An elected President and Vice President, chosen from among the Association Chairs, and a Vice Chair govern it. The purpose of the NAC is to develop rules for national championships and postseason play and to oversee each of the associations.

2. **COUNCIL OF FACULTY ATHLETICS REPRESENTATIVES (CFAR).** This Council oversees, evaluates, and implements NAIA academic standards. It performs this role via

interactions with college registrars and FARs. This committee is comprised of faculty members, staff liaison members, and one student-athlete (a nonvoting member). The CFAR has a mandate for setting operational policies for academic standards within the organization and liaising with the Registrars and FARs.

3. **COUNCIL FOR STUDENT-ATHLETES (CSA)**. The CSA promotes the student and academic priorities clearly endorsed by the NAIA. It is made up of administrators and student-athletes who develop policies, services and programming to support the athlete experience. Topics of interest include the health and safety of student-athletes and leadership and professional development opportunities. The CSA is also charged with initiating policies for the Champions of Character program.

Committees managing specific elements of NAIA business, such as the National Conduct and Ethics Committee, the Constitution and Bylaws Committee, the Champions of Character Advisory Committee, and the National Eligibility Committee, serve each of the preceding three Councils. Each group, along with the COP, is committed to educational athletics and the true spirit of competition as described by five basic principles: (1) respect, (2) integrity, (3) responsibility, (4) servant leadership, and (5) sportsmanship (NAIA, 2022c).

Currently, the NAIA offers 28 national championships. Teams qualify for their championships through regional conferences. The NAIA sponsors a slate of 17 different sports (baseball, basketball, bowling, competitive cheer, competitive dance, cross country, football, golf, indoor track and field, lacrosse, outdoor track and field, soccer, softball, swimming and diving, tennis, volleyball, and wrestling), with competition scheduled for men in 13 and women in 12 sports. Two sports are co-ed. The organization has created two divisions for basketball for both men and women, allowing individual schools to select their level of competitive entry.

Governance Structure of the NJCAA *exhibit* **8.5**

NJCAA Governance Structure:
  NJCAA Board of Regents
  NJCAA Presidents' Commission
  NJCAA Region Leadership
  NJCAA Sport Committees
NJCAA Councils
  Equity, Diversity, & Inclusion Council
  Student-Athlete Welfare & Safety Council
  Divisional Oversight Committees
NJCAA Coaches Association

Source: NJCAA. (n.d.).

Other College Athletic Associations in the United States

While the NCAA and NAIA are the largest and best-known associations governing collegiate athletics in the United States, they are not alone. The NJCAA (see Exhibit 8.5) was established in 1938 and governs intercollegiate competition for two-year colleges, the USCAA hosts championships for small colleges in nine sports, and the NCCAA was founded in 1968 to promote intercollegiate sport participation with

exhibit **8.6** Organizational Structure of the University of Oklahoma Athletic Department

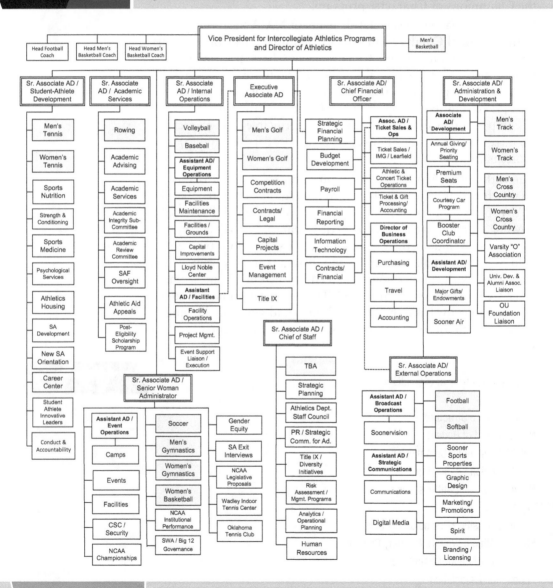

Source: University of Oklahoma (n.d.)

Organizational Structure of Bowie State University *exhibit* **8.7**

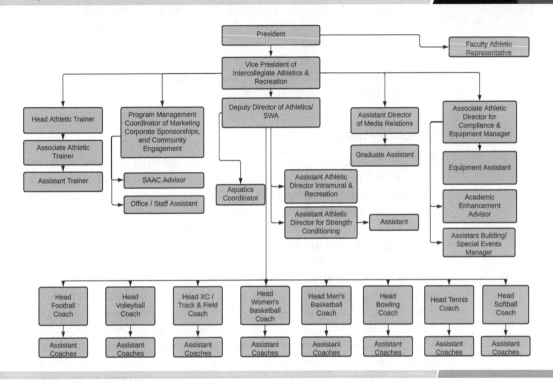

Source: Bowie State University (n.d.).

a Christian perspective. These organizations are typically smaller in reach and each has specific missions, memberships, and financials that you can check out on their websites. You can see their governance structures on the websites provided in the boxes in the text above.

Individual Campus Athletics Management

The organizations discussed above operate on the national level and are comprised of institutional members providing competitive opportunities for their athletes and coaches. A series of rules and regulations, policies, procedures, and bylaws help regulate the competition, focusing on everything from the underlying purpose and philosophy of the competition to how events are operated. Although a great deal of regulatory power exists with the leaders of the national associations and coalitions of larger institutions, the day-to-day responsibility for the intercollegiate athletic program resides on campus in the college Athletic Department.

College athletic departments can be very large or very small, as illustrated in Exhibits 8.6 and 8.7. Larger colleges and universities employ a wide variety of sport professionals to manage, deliver, and supervise intercollegiate athletics: administrators, coaches, trainers, facility and

WWW

National Junior College
Athletic Association
www.njcaa.org

WWW

United States Collegiate
Athletic Association
www.theuscaa.com

WWW

National Christian College
Athletic Association
www.thenccaa.org

event managers, and faculty representatives all play important roles in the intercollegiate program. Of course, all campus initiatives fall under the auspices of the President's Office. Over the years, the college or university President has had varying degrees of involvement in the athletic program. Today, direct involvement by the college President is more a rule than an exception. University Presidents are directly involved in the NCAA and NAIA, responsible as the Board of Governors for the strategic direction and financial oversight.

University President

The immense popularity of intercollegiate athletics became apparent early in its history as large crowds gathered to watch games and their outcomes became front-page news. The traditional attitude of college administrators toward student-life initiatives has been supportive but with a hands-off approach. Historically, Presidents have not meddled in the Athletic Department business; instead, they hoped for positive results and no scandals. However, the proliferation of highly publicized violations and reported abuses garnered increased presidential interest in intercollegiate athletics. After nearly a decade of highly visible scandals, the Knight Foundation created a Commission on Intercollegiate Athletics (Knight Commission on Intercollegiate Athletics, 2017). The purpose of the commission was to propose an agenda of reform, outlined in reports originally issued in 1991 (*Keeping Faith with the Student-Athlete*), in 2001 (*A Call to Action: Reconnecting College Sports and Higher Education*), 2010 (*Restoring the Balance: Dollars, Values and the Future of College Sport*), 2020 (*Transforming the NCAA D-I-Model*), and 2021 (*Advancing Racial Equity in Sports* and *Connecting Athletics Revenues with the Educational Model of College Sports*) (Knight Commission on Intercollegiate Athletics, 2022).

One of the proposals of the Knight Commission was to increase presidential control over college athletics. The fear that intercollegiate athletics had become too large, was riddled with unethical behaviors, and was separate from the academic mission of the institution led to the proposal for greater presidential and senior academic management involvement. As a result, the NCAA, for example, holds Presidents accountable for member conduct and requires that they be involved in developing and managing policy through the Board of Directors and the COP. In essence, presidential involvement is the result of negativity, corruption, and deceit. However, increased presidential involvement has also resulted from the potential institutional benefits of athletics participation, not the least of which is money. Developing an institutional image of excellence, increasing alumni involvement, and recruiting new students have always been positive goals for university Presidents, but the potential for revenue generation has led to a more serious focus.

Gerdy (1997) summarizes the enhanced role of the university President in college athletics:

> This explosion of media attention, coupled with the growing need to market higher education to an increasingly skeptical public, suggests

www

Knight Commission
www.knightcommission.org

that presidents, and the academic community they represent, take a more active role in managing that exposure. Thus, the heretofore hands-off, keep-me-off-the-front-page approach to managing this powerful university resource no longer serves greater institutional purposes. In short, athletics' visibility and public influence must be looked upon as something not simply to be tolerated but rather harnessed and exploited for larger university gain.

(p. 9)

The level of presidential involvement varies from institution to institution. Many Presidents have supported the formation of an athletic board and the appointment of a FAR to help scrutinize policy, direction, and operations of the Athletic Department and advise on athletic activities and issues.

In January 2006 the Summit on the Collegiate Athlete Experience was held, bringing together coaches, journalists, and athletes to discuss the collegiate athlete experience, substance abuse, and recruiting ethics. The results of the Summit, along with NCAA statistics regarding rules violations, prompted the Knight Commission to strongly urge college Presidents to support academic reform in college athletics (Knight Commission on Intercollegiate Athletics, 2017). Since then, the Commission has regularly called for change within the intercollegiate athletics community. For example, in 2010 it argued for improved transparency through the reporting of assessment measures and accountability metrics and rewarding practices that make academic achievement a priority (Knight Commission on Intercollegiate Athletics, 2017). In a 2020 report targeting NCAA Division I specifically, the Knight Commission urged the NCAA to allow athletes to make money off of their NIL, change its revenue distribution system, and reform the Division I Governance model (Knight Commission on Intercollegiate Athletics, 2020). However, according to Robinson (2014), in the past "many of the recommended reforms are vague, difficult to implement, or impossible to enforce. It's unlikely that these reforms would make any difference to 'business as usual' in athletics departments" (para. 6) where incentives exist to cheat.

The Athletic Board

The Athletic Board is a committee normally placed at the top of the Athletic Department hierarchy between the AD and the college President. Its purpose is to advise the AD and possibly the President and to oversee Athletic Department operations. The Athletic Board is generally composed of a large number of constituents, including faculty, students, alumni, community businesspersons, and university administrators. The group meets regularly each semester to oversee policy development and implementation and to scrutinize budgets. Since the AD is *responsible* to the university President, the Athletic Board in effect serves as part of the checks and balances on Athletic Department operations, a sounding board for the development of new initiatives, and a mechanism for control of operations. The AD is *responsive* to the Athletic

Board and works in consultation with it but usually reports directly to the university President (or indirectly via a Dean or Vice President). The Athletic Board at the University of Kentucky is a committee of the Board of Trustees comprised of a mixed group of board members who may be community members, students, or faculty and staff. At Arizona State University, the Athletic Board is comprised of students and faculty members. In theory, the Athletic Board wields considerable power (particularly in personnel matters), but in reality, it is typically relatively powerless, receiving general information and rubber-stamping reports that are peripheral to the major issues facing the unit and its personnel. An individual who is typically on the Athletic Board and who is important in the on-campus Athletic Department operation is the FAR.

Faculty Athletics Representative

NCAA Bylaws stipulate that each institution appoints a FAR. The FAR appointee is one of only five institutional representatives authorized to request an NCAA legislative interpretation on behalf of the institution (along with the President, director of athletics, SWA, and compliance coordinator) (NCAA Division I Interpretations Committee, 2021). The FAR is appointed by the President and is responsible for ensuring that the operation of the Athletic Department remains within the educational context of the institution. The individual appointed to this position is normally a faculty member, designated to liaise between the Athletic Department and the institution's administration. The FAR also represents the institution in affairs related to academic matters discussed at the NCAA and within the university's member conference. The FAR addresses a variety of topics but typically oversees policy regarding the educational priorities of athletes, scheduling and time commitments, graduation rates, academic support programs, policy enforcement, and the appropriate balance between athletics and academics for an institution's student-athletes. The FAR often reports to the President and works in concert with the Athletic Board and the AD.

In some cases, this reporting structure can marginalize the FAR, because members of the Athletic Department may construe their presence and mandate as intrusive. When the FAR reports directly to the President, struggles over jurisdiction and perspective between the value of athletics versus the importance of education may occasionally result. Most often, however, the FAR works diligently to reduce such conflict and plays an important contributing role in delivering athletic programming in the higher education setting.

Athletic Director

The AD is the head of the Athletic Department. As a leader, the AD is responsible for the unit and must understand the policies, activities, and actions of those working within each department such as Compliance, Marketing, and Sports Information. Recently, in many large athletic departments, the AD's activities have become more externally driven, with fundraising responsibilities, community and alumni relations, and capital-building projects now essential parts of their portfolio. Some would even say that the Division I AD is really the CEO of a multi-million-dollar corporation. Ultimately, this individual is responsible for planning, organizing, leading, and evaluating all departmental programs and personnel. Depending on the scope of these external responsibilities, an Associate AD may manage the day-to-day operations of the unit. In any event, the specific responsibilities of the AD include finance, facilities, medical services, travel, event management, compliance, media, scheduling, marketing, ticket sales, public relations, personnel, communication, risk management, television, and student-athlete services. Effective ADs possess a great deal of business sense, critical-thinking and problem-solving ability, and strong communication skills. They must ensure the Athletic Department's mission coexists effectively with the mandate of higher education. Striving for excellence in two domains means conflicts will certainly arise between athletics and academics. For example, how should the AD respond when a coach develops a schedule requiring a student-athlete to be away from campus for ten days in the middle of the semester? The games may be highly beneficial from a competitive athletics standpoint but taking a student out of class for this length of time might prove disastrous for their academic performance. The AD must balance academic and athletic goals, in this case by decreasing the length of the trip. Educational concerns must supersede athletic goals, because colleges and universities are first and foremost institutions of higher learning. Without a doubt, there are ADs who lose sight of this founding principle in their quest for athletic excellence. Having the management skills to plan and organize is important, but it is perhaps more important that the leader of the Athletic Department effectively promotes the educational priorities of the institution and the personal achievement of the student-athletes. The most important role for the AD is the role of educator. This is also true for coaches, the other main constituent of the Athletic Department.

Coaches

Coaches truly are managers in that they plan, organize, lead, and evaluate their teams like small, independent organizations. In managing and building the athletic program, they establish liaisons with constituent groups such as student-athletes, students in general, high school recruits, the media, parents, alumni, Athletic Department administrators and colleagues, sponsors, members of the college community, and professional colleagues. Today's college coaches are skilled in a particular sport and knowledgeable generally about teaching, developing tactics and strategies, communicating, motivating athletes, and building support for their program. They represent an integral component of the structure of the college Athletic Department because they drive the main business of the organization. Coaches hold a critical position of influence and authority. In their highly visible positions, they greatly affect the welfare of their student-athletes and the images of their institutions.

Spotlight: Diversity, Equity, and Inclusion in Action

Athletics Diversity and Inclusion Designee (ADID)

In 2020, all three NCAA Divisions passed legislation requiring each member school to appoint an Athletics Diversity and Inclusion Designee (ADID). The ADID serves as the "conduit for information related to national-, local- and campus-level issues of diversity and inclusion and supports diverse and inclusive practices related to athletics" (NCAA, 2022m, para. 1). The ADID must be a staff member appointed by the school's President or Chancellor. Although the ADID does not have to be housed within athletics, it must have access to student-athletes, coaches, and athletic administrators.

These designations are an important step toward diversity and inclusion in college sport; however, they do not come without criticism. This is perhaps best seen in some of the criticisms of the SWA designation in Athletic Departments. Some of the most commonly pointed out flaws of designations such as the SWA and the ADID are that they do not require the hiring of specific staff (which can mean these responsibilities are just added to existing administrators' responsibilities), their roles and responsibilities are often vague and vary depending on the institution, and they have limited power in facilitating meaningful change on campus (Hoffman, 2010).

In some Athletic Departments, especially at the Division II or Division III levels, coaches sometimes wear many hats. For example, in addition to coaching, they may also hold administrative duties related to compliance, student-athlete development, or other areas fundamental to the health of the organization. At the same time, coaches hold an important voice in the critical policy areas currently under review within intercollegiate athletics.

CURRENT POLICY AREAS

Many areas within intercollegiate athletics require the development of policy to ensure fairness. Countless policy areas are debated in college athletics regardless of division or affiliation. The sheer numbers of individuals involved with this industry segment and the issues evolving from the unique sport environment require a problem-solving, action-oriented culture. The focus on developing new policies and amending the old policies helps administrators effectively manage the evolving, dynamic environment of intercollegiate athletics. Let's investigate some recurring and new governance policy areas facing collegiate athletics today.

Eligibility

Eligibility defines who is allowed to play. In the early days of college sport, concerns arose when an individual not affiliated in any way with the college emerged as the star of the team. If collegiate athletics is about representing one's particular institution, it follows that the members of the team must also be students at the college. *Eligibility*, then, is the global term used to define the rules for entering, in-progress, and transferring student-athletes.

Initial Eligibility

To be eligible to compete in intercollegiate athletics, a high school graduate must possess the required grades in a set of core courses acceptable to the college or university for entrance. The NCAA uses an external body called the NCAA Eligibility Center to process student-athlete eligibility to enforce common standards related to academics and amateurism (NCAA, 2022b). Students register with the Eligibility Center (normally during their junior year of high school) by submitting high school transcripts and standardized test scores (such as the Scholastic Aptitude Test [SAT] or American College Test [ACT]). The Eligibility Center then approves an application when an institution requests information about a particular student. Currently, the NCAA Divisions I and II require 16 core courses to be completed satisfactorily along with a minimum standardized test score defined in conjunction with grade point average (GPA). Division I uses a sliding scale of GPA, SAT, and ACT scores to identify minimum standards of initial eligibility. The minimum GPA for student-athletes to compete at the Division I level is 2.3 (NCAA, 2022b) and 2.2 at Division II (NCAA, 2022e). In Division II, according to the Full Qualifier Sliding Scale, for example, a minimum SAT score of 920 or a sum score of 70 on the ACT is required to compete on an athletic team with a 2.2 GPA (NCAA, 2022e). Division III schools set their own eligibility criteria (NCAA, 2022b).

The NAIA (2021) requires two of the following three items for eligibility: (1) a minimum 2.0 GPA (on a 4.0 scale), (2) graduation in the upper half of the high school graduating class, and (3) a minimum score of 18 on the ACT or 970 on the SAT. Students who gain admission to a college but have scores below these NCAA or NAIA standards are ineligible to compete for their first full year of attendance (two semesters, three quarters, or the equivalent).

Both the NAIA and the NCAA temporarily changed their requirements for initial eligibility due to the COVID-19 pandemic, which brought with it restrictions for taking tests such as the ACT and SAT. For example, the NAIA allowed first-year students to meet the initial eligibility requirements through the 2.0 GPA requirement alone (NAIA, 2022b). Similarly, the NCAA dropped the standardized testing requirement through at least the 2023–24 academic year (NCAA, 2022j).

Academic Progress

To maintain eligibility, student-athletes must take and achieve passing grades for a specific number of courses or maintain a specific GPA.

In the NCAA, student-athletes must demonstrate steady progress toward graduation. They must be (a) pursuing full-time study, (b) in good standing, and (c) making satisfactory progress toward a degree. In addition, NCAA Division I member institutions must be accountable for the academic success of their student-athletes, and each Division I sports team receives an APR score. Essentially, the APR is conceived as a team-based metric developed to track, manage, and enforce accountability for academic achievement by NCAA athletes. One retention point is allotted to each scholarship athlete for staying in school and one eligibility point is given to an athlete for being academically eligible. The points are added together for each team, divided by the number of points possible, and multiplied by 1,000, yielding the team's APR score. For instance, an APR of 950 means that the student-athletes on the respective team have earned 95% of the points possible (NCAA, 2018a). In addition to an annual APR, the NCAA also tracks schools' four-year APRs. According to NCAA data released in 2022, the average four-year APR (tracking the 2017–18 through 2020–21 academic years) was 984 – a one-point increase from the previous report (NCAA, 2022h).

NCAA Eligibility Center
https://web3.ncaa.org/
ecwr3/

While teams with high APRs receive public recognition from the NCAA, teams that score below 930 can be sanctioned, for example, by receiving reductions in scholarships or losing postseason access (NCAA, 2022d). For the 2020–21 academic year, 21 teams at 13 schools were penalized for not meeting the NCAA's minimum APR threshold (with additional schools facing penalties for not meeting APR minimum scores set by their institution or conference) (Heath, 2022). Due to the COVID-19 pandemic, the NCAA suspended APR penalties and public recognition awards through the Spring 2023 competition season. The APR is a data-driven initiative indicative of the NCAA trend toward utilizing available data in implementing legislation reforms.

Transfer Students

Rules regarding transfer students are an important aspect of intercollegiate athletics' eligibility policy. Transfers to institutions for the sole purpose of playing intercollegiate athletics used to be discouraged because such moves contradict the philosophy of student first, athlete second. The purpose of attending a college or university is to get an education, earn a degree, and become a contributing member of society. For that reason, intercollegiate athletics traditionally required that student-athletes, especially those from revenue-producing sports such as basketball and football, sit out for one year before being eligible to compete for their new school. In an attempt to modernize its rules, however, the NCAA Division I Council adopted a one-time transfer blanket policy in 2021 allowing athletes across all sports to transfer once without being required to sit out their first year. To make it easier for student-athletes to find a new school should they choose to transfer, the NCAA launched the Transfer Portal in 2018. Student-athletes can post notices of their intent to transfer here, and administrators and coaches at NCAA member institutions can search for players to recruit. Of course, a series of related policies determine when and how student-athletes can communicate with other schools (and vice versa), what academic information is needed to transfer, and so on. In addition to the NCAA rules, some conferences may have additional policies in place that regulate transferring. For example, in some cases student-athletes could be prohibited from transferring to rival schools for a certain period.

The NAIA requires a 16-week residency period before competition after transfer from a four-year institution unless the institution from which the student-athlete transfers provides a no-objection release, the athlete has a minimum 2.0 GPA, and they meet all academic requirements, in which case immediate play is allowed. Transfers from four-year institutions must also have completed a certain amount of academic course hours (24 semester hours or 36 quarter hours) to transfer. The NAIA allows a transfer from a two-year institution with no residency requirement after filing the appropriate disclosure forms. The small colleges competing in the NJCAA, USCAA, and NCCAA each enforce transfer rules to declare student-athletes eligible to compete, some utilizing academic progress and probationary provisions.

Eligibility is a common policy issue in intercollegiate athletics because competing institutions have always valued fairness and a balanced starting point in the competition. These are achieved when the respective eligibility committees define benchmark standards to guide the agreed-upon concept that collegiate athletics is for full-time college students progressing toward a legitimate diploma or degree. Such eligibility requirements differentiate collegiate athletics from other avenues of amateur and professional sport.

Amateurism

Over the years, intercollegiate athletic administrators have worked hard to ensure college competition is identified as an amateur, as opposed to professional, sport. This premise is increasingly open for debate, however, given the requirements placed on student-athletes and the big business nature of college athletics, especially at the NCAA Division I level. In its 2021–22 Division I Manual, the NCAA outlines its *Principle of Amateurism*:

> Student-athletes shall be amateurs in an intercollegiate sport, and their participation should be motivated primarily by education and by the physical, mental and social benefits to be derived. Student participation in intercollegiate athletics is an avocation, and student-athletes should be protected from exploitation by professional and commercial enterprises.
>
> (NCAA, 2022a, p. 3)

From the NCAA's perspective (and therefore the perspective of its member institutions), amateurism is "a bedrock principle ... maintaining amateurism is crucial to preserving an academic environment in which acquiring a quality education is the first priority" (Louisiana State University Athletics Compliance, 2022, para. 1). In any event, student-athletes are not directly paid salaries by their universities for their athletic services and it is the responsibility of member institutions to validate the amateur status of incoming athletes, which can be challenging. In general, amateurism requirements do not allow for salary or prize money for play or contracts with professional teams (NCAA, 2021b). Their involvement is voluntary and non-contractual, and their primary purpose for attending the institution is educational. Does this ring true? Perhaps not always, and this is the specific reason amateurism is hotly debated by administrators at the institutional, regional, and national levels. College athletics is said to be amateur in the context of education, but the demands and the stakes suggest a more professional sport environment to some critics who maintain that colleges should pay athletes for their performance, just as professionals are compensated for the entertainment they provide.

Regardless of the debate, collegiate athletic administrators continue to embrace amateurism and have invoked rules to prohibit professionals from the competition. Student-athletes are not paid directly by their universities to compete, but many receive athletic scholarships or grants-in-aid to help offset the costs of a college education. In 2015, the 65 NCAA schools in the ACC, Big Ten, Big 12, Pac-12 and SEC approved a rule change to allow student-athletes to receive funding apportioned to the cost of attending the institution. The Cost of Attendance rule, since adopted by other Division I schools and conferences, allows schools to provide more dollars for college athletes for elements of attending college that is formally defined by federal guidelines, as established by school financial aid officers. A student's Cost of Attendance can be adjusted based on individual circumstances such as transportation, childcare needs, and unusual medical expenses (NCAA, 2017). However, an "athlete turned pro" is prohibited by league policy from returning to college competition in their sport. The amateur policy of the NCAA describes many activities that may result in an athlete losing amateur status (NCAA, 2021b):

- playing for pay
- accepting a promise to pay
- signing a contract to play professionally
- prize money above actual and necessary expenses
- competing for a professional team (tryouts, practice, or competition)
- entering into an agreement with an agent
- benefiting from an agent or prospective agent
- delaying initial full-time enrollment to compete in organized sport

Without a doubt, college athletic administrators still believe it is important for college athletes to be students first. However, recent advances in legislation surrounding a student-athletes' ability to profit off of their name, image, and likeness call into question the amateurism model.

Name, Image, and Likeness and Financial Treatment of Student-Athletes

Fair financial treatment of athletes is an ongoing focus of policy debate. Providing athletes with financial aid, commonly known as an athletic scholarship, has been a source of debate within college athletics for decades. NCAA Divisions I and II have set policies for awarding a certain number of athletic scholarships, whereas Division III and the NAIA have been more restrictive in their rules. For example, NCAA Division I and II schools sponsor a certain number of athletic scholarships per sport, whereas Division III schools do not offer athletic-based awards. In the United States, the source of funding for these awards has not been the most important issue. Rather, intercollegiate governing bodies such as the NCAA have developed policies to ensure fair practices among institutions regarding how much financial aid can be given to a particular athlete (to ensure a distinction from paid professional athletes), to balance awards given to men and women, and to regulate when such an award can be withdrawn. Some questionable practices, such as overpaying or unjustly withdrawing an athletic scholarship, have prompted a focus on these issues. Implementation of the Cost of Attendance policy mentioned earlier in the chapter was a move taken to ensure that athletes are not treated unfairly compared to the regular student body.

One of the key topics at the core of discussions about the financial situation student-athletes find themselves in is whether or not they should be allowed to make money off of their name, image, and likeness (NIL). The NCAA has long argued against student-athletes profiting from their NIL, but recent developments have forced the Association's hand in changing its stance. In late 2019, the state of California passed the Fair Pay to Play Act which was set to take effect by 2023. This legislation would allow student-athletes to make money by selling their name, image, and likeness. Despite the NCAA calling for such legislation to be reconsidered given its threat to the amateurism model the association relied on since its inception, more States followed suit. By mid-2021 NIL laws took effect in over 15 states, including Alabama, Connecticut, Florida, Georgia, Kentucky, Ohio, and Texas – with many more under consideration across the country. The pressure from some state legislation scheduled to take effect July 1, 2021, led the NCAA to suspend its previous NIL rules in June of 2021 and establish an "interim" policy allowing student-athletes to be paid for their name, image, and likeness. This interim policy stated that student-athletes "can engage in NIL activities that are consistent with the law of the state where the school is located" (Hosick, 2021, para. 4) and was set to be in place until federal legislation was passed or new NCAA rules were to be adopted.

Interestingly, the NAIA was more proactive in allowing student-athletes to profit from their NIL. In October 2020, the NAIA was the first college sport association to officially allow NIL payments, but even before then, it had policies in place allowing student-athletes to receive such compensation as long as they did not reference their status as athletes (NAIA, 2020). Because the NAIA was years ahead of the NCAA in adopting NIL legislation, it is not surprising that an NAIA student became college sport's first athlete to make money off of their NIL. Aquinas College's Chloe Mitchell, a volleyball player with more than two million followers on TikTok at the time the NAIA adopted its NIL legislation in 2020, is considered the first collegiate athlete to enter endorsement deals in this new NIL era (Hruby, 2021). Since then, student-athletes have taken advantage of their ability to enter NIL deals. For example, University of Alabama quarterback Bryce Young is said to have made almost a million dollars in NIL deals before his first season as a starter even began (Holland, 2021). However, such astronomic numbers are the exception. Opendorse, a company tracking NIL deals, found that by summer 2022 the average compensation for NIL deals was $3,711 for NCAA Division I student-athletes, $204 for Division II student-athletes, and $309 for Division III student-athletes (Opendorse, 2022). For many of these athletes, social media is an important component in NIL activities when it comes to promoting themselves and their personal branding. Next, we will look at social media as an important policy area in college sport.

NAIA Name, Image, and
Likeness
https://www.naia.
org/membership/
name-image-likeness

NCAA Name, Image, and
Likeness Resource Page:
NIL Taking Action
https://www.ncaa.org/
sports/2021/2/8/about-
taking-action.aspx

Social Media

Intercollegiate Athletic Departments and athletes actively use Facebook, Twitter, Instagram, Snapchat, TikTok, and other social media platforms to publicize their activities and their brands. Many Athletic Departments are using social media to increase fan engagement, interaction, and feedback, and to build brand loyalty. Building the fan base by retaining current fans and recruiting new ones is core to using social media technologies today. Athletic Departments are hiring whole units of media directors and graphic designers to effectively communicate their messages (maybe types of jobs you might want to pursue – start looking for those internships soon!).

However, a double standard has developed in many Athletic Departments related to social media use by athletes. This became most evident in the summer of 2020 when an increasing number of athletes took to social media to join global calls for justice after the continued murders of Black Americans by law enforcement (Black et al., 2022). University Athletic Departments have enacted policies to control and limit an athlete's use of social media, characterizing it as risky to the image and brand of their team, Athletic Department, or university. While athletes' social media use will likely be monitored by Athletic Department personnel for content, preventing athletes from using social media is undoubtedly a concern for their personal freedom. Governing bodies such as the NCAA have hesitated to create policy on this issue, likely because of the difficulty inherent to monitoring it and the legal issues associated with doing so. Coaches have voiced concerns within departments because of the potential competitive impact if an athlete announces an injury on Twitter, or a recruit changes their mind about attending a school because of some off-hand, heated comments. Reasons such as these prompted many Athletic Departments and teams to create social media usage policies. Importantly, athletes need to be trained about the publicity "fishbowl" that is college athletics, where what they do and say is magnified and of interest to millions. Depending on their Athletic Department, athletes are being coached to use social media minimally and appropriately and to understand the negative impact it can have on future jobs. Policies that balance the positives and negatives of this issue are essential. You may remember that for many students, social media became an important tool to connect with peers during the COVID-19 pandemic, a global health crisis that brought an unprecedented set of challenges for those governing intercollegiate athletics.

COVID-19, Health, and Well-Being

As you may recall from earlier in the chapter, the NCAA was founded in response to the deaths of college football players and the resulting need to keep college athletes safe. Today, well-being remains one of the top priorities of the Association. Beyond the NCAA, other governing bodies frequently discuss policies with the goal of improving the health and well-being of student-athletes. Policy areas related to health and well-being focus on the regulation of conditions affecting student-athletes' physical health (such as injuries, performance, and nutrition), their mental well-being (such as transition out of sport or mental health conditions generally), as well as issues related to both (such as sexual assault and interpersonal violence). For the NCAA, the organization's Sport Science Institute serves as the primary unit charged with supporting holistic well-being. Its mission is "to promote and develop safety, excellence, and wellness in college student-athletes, and to foster life-long physical and mental development" (NCAA Sport Science Institute, 2022, para. 2).

No recent crisis has centered on the topic of health more than the COVID-19 pandemic. Aside from the emotional toll the pandemic likely took on members of the intercollegiate athletics community, it also brought with it unprecedented challenges for those governing college sport. To attempt to navigate the emerging crisis, the NCAA established a COVID-19 advisory panel of experts in March 2020. Shortly after, the men's and women's basketball tournaments were canceled – for the first time in NCAA history (NCAA, 2022n). All other NCAA sports were canceled as well. A year later, men's and women's basketball returned to the national stage, with all rounds of the respective Division I tournaments held in a tournament "bubble" in one geographic region. Aside from regulations determining if, where, and how championships were held, the NCAA also needed to develop policies related to previously unexplored topics to ensure the safety of the student-athletes: infection control, testing protocols, health services, and resocialization post-pandemic, to name a few. And, as you may remember from earlier in the chapter, the COVID-19 pandemic also led to changes in policy in other domains (such as initial eligibility). For example, to account for the potential loss of a season during the pandemic, the NCAA granted every student-athlete an extra year of eligibility. The NAIA and NJCAA followed suit.

The COVID-19 pandemic also brought into the spotlight another policy area receiving more attention in college sport in recent years: student-athletes' mental health. In its Student-Athlete Well-Being Study, the NCAA found increased mental health concerns among student-athletes, with such concerns being 1.5 to 2 times higher than prior to the onset of the pandemic (NCAA, 2022k). Even before the pandemic, being a student-athlete often meant juggling many competing responsibilities, making involvement in athletics a stressor that can jeopardize a student-athletes' well-being. Add in the alarming number of student-athletes dying by suicide in the first half of 2022 alone (Hensley-Clancy, 2022), and it becomes apparent why policy supporting athletes who struggle with their mental health is so crucial. Providing access to mental health resources is now anchored in the NCAA constitution (NCAA, 2021a).

NCAA Sport Science
Institute
https://www.ncaa.org/
sports/2021/5/24/sport-
science-institute.aspx

Gender Equity, Diversity, and Inclusion

Virtually every administrative body governing intercollegiate athletics has a policy pertaining to gender equity. As mentioned earlier, in the United States Title IX provides the impetus for committees and task forces to discuss and implement change to achieve equitable intercollegiate athletic programs for both men and women. The NCAA uses the following definition of gender equity, as delivered by the Gender Equity Task Force in 1992:

> An athletics program can be considered gender equitable when the participants in both the men's and women's sports programs would accept as fair and equitable the overall program of the other gender. No individual should be discriminated against on the basis of gender, institutionally or nationally, in intercollegiate athletics.
> (NCAA Committee on Women's Athletics and
> Gender Equity Task Force, 2022, para. 7)

University Presidents and ADs determine specific compliance with Title IX on individual campuses. However, athletes and parents have initiated numerous lawsuits that challenge actual Title IX compliance. Title IX is also being used by men and those involved in smaller, non-revenue-producing sports to challenge practices such as alleged sex discrimination in circumstances where men's sport programs are underfunded or have been dropped. The specific duty to comply with equitable practices rests on individual campuses and must be the joint responsibility of the President, AD, and Athletic Board. Gender equity has been a long-standing contentious issue in college athletics, but the topic was brought into the national spotlight during the 2021 NCAA Division I Men's and Women's Basketball Tournaments when a video posted to TikTok by University of Oregon women's basketball player Sedona Prince went viral. In the video, Prince called out the inequities between both tournaments. For example, whereas the men had a fully equipped workout room to use during the tournament, the women were only provided with a single rack of weights. The video triggered an extensive gender equity review by an independent law firm hired by the NCAA. Not surprisingly, the review revealed persistent inequities between the men's and women's tournaments. It revealed that the NCAA's organizational structure, media arrangements, and revenue distribution framework prioritize men's basketball (Kaplan Heckler & Fink, 2021).

The 2021 March Madness example shows that inequities play out in many different, often systemic, ways. Participation opportunities for men have historically exceeded those for women. Many sports, such as football and ice hockey, existed for men only. The budgets allocated to men's teams and salaries paid to coaches of men's teams are significantly higher than for women's teams. The decision-makers in athletic departments, university central administration, and coaching have predominantly been men. Many more men than women coach women's teams. These and other imbalances in gender that elevate men's over women's sport continue to create issues of inequity that must remain a focus of the reform required in collegiate athletics.

Gender equity is just one of the components that make up matters of diversity and inclusion as an important policy area in intercollegiate athletics. Both the NCAA and the NAIA have engaged in strategic efforts to increase diversity and inclusion over the past 10 years. For example, in 2016 the NCAA established the Board of Governors Committee to Promote Cultural Diversity and Equity, tasked with making recommendations to advance diversity and inclusion in the Association (NCAA, 2022c). All three NCAA divisions also require active member institutions to submit regular diversity, equity, and inclusion reviews. And in 2021, the NAIA amended its constitution to make sure its major governance units (such as the COP and National Administrative Council) reserve positions for women and representatives from minority groups (NAIA, 2021).

Gambling

Sports gambling is a multi-billion-dollar business. Recent growth, coupled with the expansion of and media hype surrounding college sports, has resulted in a noticeable increase in gambling associated with collegiate sport. This was likely elevated by the Supreme Court's 2018 overturning of the Professional and Amateur Sports Protection Act (PASPA), the legislation that effectively outlawed sports betting in the United States. Within only a year after overturning PASPA, more than 18 states introduced laws legalizing sports betting, hence creating a new area of sports gambling in the country.

Despite these changes in the federal and state legislative landscapes, the NCAA has held onto its long-standing policy prohibiting Athletic Department staff and student-athletes, as well as those with athletics-related responsibilities (such as the FAR or Athletic Board members), from engaging in any form of wagering activities. The policy prohibits gambling on amateur, intercollegiate, and professional sports and stipulates that an individual involved in collegiate sport must not knowingly provide information to individuals involved in organized gambling, solicit a bet, accept a bet, or participate in any gambling activity through any method (NCAA, 2022a). The only exception to the gambling ban is traditional wagers between institutions – think of those usually associated with bowl games or high-profile rivalries.

Why is the NCAA so strict when it comes to sports gambling? Administrators and members of the NCAA are concerned that it threatens the very existence of college sports (NCAA, 2018b). For example, Ellenbogen et al. (2008) studied over 20,000 NCAA student-athletes and reported that 62% of men and 43% of women reported gambling activities, with 13% of men engaged in weekly gambling. These authors suggest that:

> Gambling among student athletes represents a multifaceted problem, particularly when examining sport wagering. If students incur significant losses or develop associations with other gamblers, they may be pressured to use or share information conceding collegiate games, or possibly alter their performance to influence the outcomes of games.
>
> (p. 249)

The harshness of the policy results from the NCAA's belief that illegal sports wagering is big business, and big business attracts organized crime. The involvement of impressionable college-age students is a concern. Not only is the welfare of student-athletes jeopardized, the very integrity of sport contests can be undermined. This concern is not unique to the NCAA. The NAIA also believes gambling undermines the values of NAIA athletics and the organization has a zero-tolerance policy for any form of sports wagering. Administrators and governing bodies have moved quickly to set policies prohibiting any association between gambling and college athletics. Enforcing the rules is the next hurdle.

Enforcement

Because intercollegiate athletic organizations are collectives of member institutions, legislation is created *by* the members of the organization *for* the members of the organization. Enforcement Services, also called Compliance, is the department that ensures institutions are abiding by the rules, thus maintaining the integrity of the rules and fair play among all participants. The intent of enforcement programs is to reduce violations by education, discovery, and the disbursement of appropriate penalties.

College Athletic Departments in the United States are expected to monitor their rules compliance and self-report any violation. In addition, the NCAA has an Enforcement and Infractions staff of approximately 20 specialists who work to ensure a level playing field through rules enforcement. The importance of enforcement policies is underscored by the fact that each NCAA division has a Committee on Infractions (COI), an independent group of lawyers, law school professors, and members of the public who assess penalties for those who break the rules. Unfortunately, this arrangement is frequently perceived as inadequate by members of the NCAA because of the huge workload assumed by only approximately 20 people. The inquiry process involves field investigations, formal correspondence of inquiry, the development of a case summary, hearings before the COI, and, if necessary, a ruling regarding the violation and penalty (NCAA, 2022i). An appeal process is also provided, and attempts are made to ensure due process is followed in any investigation.

Enforcement is not accorded as much focus in the NAIA, where self-reporting is virtually the sole means of policing infractions. The organization also has a committee to deal with allegations of impropriety. Two staff members in a unit called Legislative Services assist the NAIA membership in interpreting NAIA legislation and overall compliance with NAIA rules. This lack of additional enforcement staff is primarily a financial issue, given the lack of resources available for full-time enforcement officers. For many sport organizations, funding is problematic, and intercollegiate athletics is no different.

NCAA Conference Realignment

While all had been quiet for a while on conference realignment, recent announcements have brought this issue to the forefront again. As of this writing, The University of Texas and the University of Oklahoma have shifted their conference membership from the Big 12 to the SEC. This move was followed approximately a year later by the news that UCLA and USC will leave the Pac-12 to become members of the Big Ten. While these are the most notable moves to date, other conferences are in transition as well. Will the so-called Power Five still be that?

These moves have primarily been driven by football and men's basketball and the potential for revenue generation by new marquee match-ups. The trickle-down effect is now in process. Accepting new members into a conference means that those conferences are relinquishing their memberships in a previous conference, oftentimes with long and storied histories. Questions still remain – which other schools will be on the move? Conferences like the American Conference and Conference USA have particularly been affected (Moss, 2022). The word is still out on the impact on the ACC, and what action the University of Notre Dame (football conference independent) will take remains to be seen.

In the past, geography often came into play for conference membership. The bottom line now, however, remains monetary. According to Marshall (2022, paras. 11–12):

> The SEC has a $3 billion deal with ESPN that's set to kick in in 2024 and the Big Ten is currently negotiating a massive media rights deal. The Pac-12 has floundered when it comes to TV as the conference's network has struggled to gain footing while many of its games are played late at night.
> With costs to run college athletic programs have climbed in recent years, exacerbated by the pandemic, moving to an even bigger conference provides more financial stability. For the Big Ten, adding UCLA and USC gives the conference a foothold in the nation's second-largest media market.

While the primary focus of these moves has been on football and men's basketball, the ramifications run deep for the Olympic sports. Think about a softball or golf team from UCLA having to travel to Rutgers in New Jersey for games. Or Maryland's women lacrosse team flying to play against USC. The travel time for athletes in sports such as these will greatly increase, impacting the academic side of these universities as well (Coaston, 2022; Georgia State University, 2022). ADs, university Presidents, and Conference Commissioners will be facing numerous complicated decisions. This would be a good time for them to use the SLEEPE principle (discussed in Chapter 2 of this textbook) as they make their governance decisions!

SUMMARY

For many people, the beginning of the college sports season is synonymous with the new school year. This association indicates the appeal of college athletics, not just for the participants and student body but also for the wider public who attend games and tune in to the mass media. Widespread interest in intercollegiate athletics has resulted in colleges offering a vast array of teams and developing excellent state-of-the-art facilities for participants and spectators. It has also resulted in the need for college administrators to actively supervise the intercollegiate sport enterprise, its governance, operations, and policy development. From the humble beginning of a crew race between Harvard and Yale in 1852, thousands of competitions among institutions take place today.

To compete with other colleges and universities with similar philosophies and values, institutions become members of governing associations such as the NCAA, the NAIA, and the NJCAA in the United States. Members of these organizations meet to set policy, procedures, rules and regulations, and legislation regulating competitions, and to deal with current issues.

The policies regulating competition are debated from year to year and hinge on the current political and financial environments of the institutions involved. Eligibility seems to be a perennial concern. The preservation of amateurism is also a timeless issue. The issues of name, image, and likeness (and the financial treatment of student-athletes in general), social media, and health and well-being in the context of a pandemic have surfaced in more recent decades. The implementation of practices working toward gender equity, diversity, and inclusion is also debated frequently. Setting policy regarding gambling and enforcement are even newer issues in intercollegiate athletics as well.

Intercollegiate athletics and its governance will continue to be hot topics on university campuses, and the governing structures will play an ever-increasing and important role in ensuring a safe and fair environment for competitions between colleges for participants and spectators alike.

NICK OJEA *Executive Senior Associate Athletic Director*

Grand Canyon University

I have now closed out year seven at Grand Canyon University (GCU). My responsibilities have evolved over time from my previous primary functions of Compliance, Sports Medicine, Sports Performance oversight, and sport administration duties to now my current role serving as our Executive Senior Associate Athletic Director. I now oversee the majority of our internal support operations including Human Resources, Compliance, Student-Athlete Development, Sports Medicine, Sports Performance and serve as sport administrator for Women's Basketball and Women's Volleyball.

Looking ahead on intercollegiate athletics governance, I believe the impending NCAA transformation rule changes and Name, Image, and Likeness will significantly impact how we conduct our day-to-day business. The industry has seen a great deal of transition in year one of Name, Image, and Likeness. As the dust settles, we are learning the trends, patterns, and opportunities that student-athletes have engaged in. Like anything new, it will continue to evolve and grow, impacting recruiting, the transfer portal, donor relations, and academics. The transformational changes within our governance structure will more than likely impact institutions financially ... tough decisions and reallocation of resources are imminent.

When I arrived at GCU, the institution was entering year three of the Division 1 transition process. There were many areas where we worked to establish consistency with Division 1 requirements. The Compliance department needed restructuring starting with a policies and procedures manual. The implementation and training required a lot of assistance from across campus with our campus partners (e.g., financial aid, admissions, and academic records). We had to implement an admissions process for athletics, scholarship distribution protocol, and other institutional control safeguards. We were very fortunate to have great support and collaboration from our campus partners.

Overall, I believe the NCAA governance structure as a framework makes sense. Safeguards and reporting structures are in place to support academic and athletic excellence. There is a need for some changes to reflect the current trends we are seeing and the evolving roles/responsibilities of the NCAA staff. Over time questions have arisen over whether certain decision-making processes should be more localized or conference based rather than at the NCAA level. These adjustments could aid in more efficient decision-making and implementation when rule changes occur or trends shift ... or in some cases of enforcement procedures. As a result, these adjustments could allow the NCAA Staff to focus on critical areas throughout the membership to enhance the student-athlete experience. I go back to the work being done by the NCAA Transformation Committee. Many critical issues are on the table and will be addressed one by one. The future of college athletics will evolve into a new model. Administrators will need to be flexible and nimble in their decision-making.

I would say if you want to be an intercollegiate athletic administrator, remember you should make the choice to work in college athletics because it is a career – not a job. The demands of the industry will mean that at times you will need to make sacrifices for the greater good. We are here to serve and develop our student-athletes. Be flexible and welcome change. As the industry changes, local decision-making occurs. Never be complacent and always focus on continuous improvement.

caseSTUDY

COLLEGE ATHLETES AND COMMUNITY SERVICE

College Athletic Departments are constantly looking for ways to create a positive public image. One way they can do so is by involving members of the Athletic Department, particularly student-athletes, in community service projects.

1. If you were the AD, what might be some approaches your department could use to get your student-athletes involved in community service?

   a. Who should choose the activity? Student-athletes? Coaches? The AD? A team of people?

   b. What types of community organizations would make good partners?

   c. How can you get coaches on board to help?

   d. When you or a friend of yours was a student-athlete in high school or college, did you (or they) ever do community service? How was the experience?

2. Sometimes coaches use community service projects as punishments for athletes who have broken team rules. The coaching staff for a sport will tell the student-athlete they have to perform a certain number of community service hours to make up for their poor behavior. The student-athlete may have no interest whatsoever in really helping the community, just in getting in their hours as mandated. Is requiring someone to do community service an appropriate punishment for breaking team rules? Why or why not?

3. In what ways is requiring someone to do community service supporting or contradicting the organizational philosophy of major intercollegiate athletics governing bodies (e.g., NCAA or NAIA)?

CHAPTER questions

1. Compare and contrast the organizational structures of the NCAA, the NAIA, and the NJCAA. What is different in these three organizations, and why?

2. How might a policy help an athlete deal with the struggle of balancing requirements and expectations of academics versus athletics? Write a policy encouraging the balance between both components of the term *student-athlete*.

3. Suppose you are the AD of a large Division I university with teams competing in the NCAA. It has come to your attention that the men's basketball coach has broken a series of recruiting rules to attract a 7-foot center to the team. In the end, the coach was unsuccessful in recruiting the athlete, but self-disclosure rules still exist in the NCAA. Using the SLEEPE Principle presented in Chapter 2, analyze the situation to help understand each of the ramifications of your decision. In the end, what will you do?

4. Throughout this chapter, we have identified *change* as an apt word to describe the current state of intercollegiate athletics, especially in North America. What do you see as the most significant changes affecting college sport over the past decade? What changes do you anticipate in the future?

FOR ADDITIONAL INFORMATION

For more information check out the following resources:

1. Here is information on the NAIA Champions of Character Program: https://www.naia.org/champions-of-character/index
2. This is a solid resource from the NAIA – its ROA Resource: https://www.naia.org/return-on-athletics/prospects/index
3. NCAA Leadership Development: https://www.ncaa.org/sports/2021/2/8/ncaa-leadership-development.aspx
4. HBO documentary on the Student-Athlete Experience: Documentary *Student Athlete*, produced by HBO Sports in collaboration with LeBron James, Maverick Carter and Steve Stoute. https://www.hbo.com/movies/student-athlete
5. The Institute for Diversity and Ethics in Sport (TIDES) produces an annual Racial and Gender Report Card on College Sport. Here is the 2021 version: https://www.tidesport.org/_files/ugd/403016_14f7be7c35154a668addb71b75b7e14f.pdf
6. On the 50th anniversary of Title IX, this provides some good food for thought: Staurowsky, E. J., Flowers, C. L., Busuvis, E., Darvin, L., & Welch, N. (2022). *50 Years of Title IX: We're Not Done Yet*. Women's Sports Foundation. https://www.womenssportsfoundation.org/wp-content/uploads/2022/05/Title-IX-at-50-Report-FINALC-v2-.pdf

REFERENCES

Anderson, P. (2021, May 6). Cutting sports in the context of Title IX, COVID-19. *Athletic Business*. https://www.athleticbusiness.com/operations/legal/article/15161296/cutting-sports-in-the-context-of-title-ix-covid-19

Berkowitz, S. (2022, February 2). NCAA revenue returned to $1.15 billion in 2021, but prospect of pandemic impacts looms. *USA Today*. https://www.usatoday.com/story/sports/2022/02/02/ncaa-revenue-up-but-pandemic-impacts-loom-basketball-tournament/9313735002/

Black, W.L., Ofoegbu, E., & Foster, S.L. (2022). #TheyareUnited and #TheyWantToPlay: A critical discourse analysis of college football player social media activism. *Sociology of Sport Journal* (published online ahead of print). https://doi.org/10.1123/ssj.2021-0045

Bowie State University. (n.d.). *Department of Athletics and Recreation organizational chart*. https://bowiestate.edu/about/administration-and-governance/org-chart-athletics-recreation.pdf

Coaston, J. (2022). U.C.LA. and U.S.C., welcome to the Big Ten. I already hate you. *New York Times*. https://www.nytimes.com/2022/07/09/opinion/big-ten-ucla-usc.html

Davenport, J. (1985). From crew to commercialism: The paradox of sport in higher education. In D. Chu, J.O. Segrave, & B.J. Becker (Eds.), *Sport and higher education* (pp. 5–16). Human Kinetics.

Dickey, J. (2018, December 18). A brush with death saved UAB football. Should other schools try it out? *Sports Illustrated*. https://www.si.com/college/2018/12/18/uab-blazers-return-boca-raton-bowl

Ellenbogen, S., Jacobs, D., Derevensky, J., Gupta, R., & Paskus, T.S. (2008). Gambling behavior among college student-athletes. *Journal of Applied Sport Psychology, 20*, 349–362. https://doi.org/10.1080/10413200802056685

Florida A&M University Athletics (2020). *Strike: Strategic plan to transform Florida A&M Athletics*. https://famuathletics.com/

documents/2020/12/6//FAMUAthletics_
StrategicPlan.pdf?id=803

Georgia State University. (2021, November 11). *Research finds flawed premises underpinning years of college conference realignments.* https://news.gsu.edu/2021/11/11/college-conference-realignments/

Gerdy, J.R. (1997). *The successful college athletic program: The new standard.* Oryx Press.

Gilmore, A. (2022, February 3). HBCUs: At the financial and competitive crossroads of college sports. *The Journal of Blacks in Higher Education.* https://www.jbhe.com/2022/02/hbcus-at-the-financial-competitive-crossroads-of-college-sports/

Hensley-Clancy, M. (2022, May 19). Reeling from suicides, college athletes press NCAA: 'This is a crisis'. *The Washington Post.* https://www.washingtonpost.com/sports/2022/05/19/college-athletes-suicide-mental-health/

Holland, H. (2021, July 20). Nick Saban of Bryce Young's NIL deals: "It's almost seven figures." *Sports Illustrated.* https://www.si.com/college/alabama/bamacentral/bama-central-nick-saban-bryce-young-seven-figures-nil-deal-july-20-2021

Hosick, M.B. (2021, June 30). *NCAA adopts interim name, image and likeness policy.* NCAA. https://www.ncaa.org/news/2021/6/30/ncaa-adopts-interim-name-image-and-likeness-policy.aspx

Hruby, P. (2021, April 8). *"I just filmed it, put it up, got paid": How Chloe Mitchell became the first college athlete to profit from NIL.* Hreal Sports. https://hrealsports.substack.com/p/i-just-filmed-it-put-it-up-got-paid

Heath, S. (2022, June 14). *DI student-athletes demonstrate academic success, resilience amid challenges.* NCAA. https://www.ncaa.org/news/2022/6/14/media-center-di-student-athletes-demonstrate-academic-success-resilience-amid-challenges.aspx

Hoffman, J. (2010). The dilemma of the Senior Woman Administrator role in intercollegiate athletics. *Journal of Issues in Intercollegiate Athletics, 3,* 53–75.

Kaplan Hecker & Fink. (2021). *NCAA external gender equity review – Phase I: Basketball championships.* https://kaplanhecker.app.box.com/s/6fpd51gxk9ki78f8vbhqcqh0b0o95oxq

Knight Commission on Intercollegiate Athletics. (2017). *Commission reports.* www.knightcommission.org/presidential-control-a-leadership/commission-reports

Knight Commission on Intercollegiate Athletics. (2020). *Transforming the NCAA D-I model.*

https://www.knightcommission.org/wp-content/uploads/2021/02/transforming-the-ncaa-d-i-model-recommendations-for-change-1220-022221-update-01.pdf

Knight Commission on Intercollegiate Athletics. (2022). *Commission reports and official recommendations.* https://www.knightcommission.org/commission-reports-official-recommendations/

Louisiana State University Athletics Compliance. (2022). *Amateurism.* http://www.compliance.lsu.edu/amateurism

MacGaffey, J. (2013). *Coal dust on your feet: The rise, decline, and restoration of an anthracite mining town.* Bucknell University Press.

Marshall, J. (2022, July 3). College sports: Conference realignment makes another seismic shift. *The Atlanta Voice.* https://theatlantavoice.com/college-sports-conference-realignment-makes-another-seismic-shift/

Meyer, J., & Zimbalist, A. (2017). Reforming college sports. The case for a limited and conditional antitrust exemption. *The Antitrust Bulletin, 62*(1), 31–61. https://doi.org/10.1177/0003603X16688829

Moss, T. (2022, June 30). *College basketball realignment tracker: Keeping track of NCAA Division I conference changes.* ESPN. https://www.espn.com/mens-college-basketball/story/_/id/32855347/college-basketball-realignment-tracker-keeping-track-ncaa-division-conference-changes

NAIA. (2020, October 6). *NAIA passes landmark Name, Image and Likeness legislation.* https://www.naia.org/general/2020-21/releases/NIL_Announcement

NAIA. (2021). *2021–2022 official & policy handbook.* https://www.naia.org/legislative/2021-22/files/NAIA-2021-Official-Handbook.pdf

NAIA. (2022a). *2022–23 membership basics.* https://www.naia.org/why-naia/pdf/NAIA_Member_Basics_2022-2023.pdf

NAIA. (2022b). *COVID-related decision regarding eligibility.* https://www.naia.org/covid19/files/COVID-Related_Decisions_Regarding_Eligibility.pdf

NAIA. (2022c). *Five core values.* https://www.naia.org/champions-of-character/five-core-values

NAIA. (2022f). *NAIA governance structure.* https://www.naia.org/legislative/governance-structure

NCAA. (2017). *Cost of attendance q & a.* www.ncaa.com/news/ncaa/article/2015-09-03/cost-attendance-qa

NCAA. (2018a). *NCAA Division I Academic Progress Rate (APR): Public use dataset codebook.* https://ncaaorg.s3.amazonaws.com/research/gradrates/data/2018RES_2018APRDataSharing Codebook.pdf

NCAA. (2018b). *NCAA supports federal sports wagering legislation.* https://www.ncaa.org/news/2018/5/17/ncaa-supports-federal-sports-wagering-regulation.aspx

NCAA. (2020). *NCAA Board of Governors expands Confederate flag policy to all championships.* https://www.ncaa.org/news/2020/6/19/ncaa-board-of-governors-expands-confederate-flag-policy-to-all-championships.aspx

NCAA. (2021a). *NCAA constitution.* https://ncaaorg.s3.amazonaws.com/governance/ncaa/constitution/NCAAGov_Constitution121421.pdf

NCAA. (2021b). *NCAA guide for the college-bound student-athlete: 2021–2022.* http://fs.ncaa.org/Docs/eligibility_center/Student_Resources/CBSA.pdf

NCAA. (2022a). *2021–22 Division I manual.* https://web3.ncaa.org/lsdbi/reports/getReport/90008

NCAA. (2022b). *Academic standards for initial-eligibility.* https://www.ncaa.org/sports/2013/11/25/academic-standards-for-initial-eligibility.aspx

NCAA. (2022c). *Board of Governors Committee to promote cultural diversity and equity.* https://www.ncaa.org/sports/2020/2/24/board-of-governors-committee-to-promote-cultural-diversity-and-equity.aspx

NCAA. (2022d). *Division I Academic Progress Rate (APR).* https://www.ncaa.org/sports/2013/11/20/division-i-academic-progress-rate-apr.aspx

NCAA. (2022e). *Division II academic requirements.* http://fs.ncaa.org/Docs/eligibility_center/Student_Resources/DII_ReqsFactSheet.pdf

NCAA. (2022f). *How the NCAA works: Association-wide.* https://ncaaorg.s3.amazonaws.com/champion-magazine/HowNCAAWorks/AW_HowNCAAWorks.pdf

NCAA. (2022g). *Mission and priorities.* https://www.ncaa.org/sports/2021/6/28/mission-and-priorities.aspx

NCAA. (2022h). *National and sport-group APR averages and trends.* https://ncaaorg.s3.amazonaws.com/research/academics/2022RES_APRAveragesTrends.pdf

NCAA. (2022i). *NCAA Division I infractions 2020–21 annual report.* https://ncaaorg.s3.amazonaws.com/infractions/d1/2021D1Inf_AnnualReport.pdf

NCAA. (2022j). *NCAA Eligibility Center COVID-19 response FAQ.* http://fs.ncaa.org/Docs/eligibility_center/COVID-19_Spring2023_Public.pdf

NCAA. (2022k). *NCAA student-athlete well-being study.* https://ncaaorg.s3.amazonaws.com/research/other/2020/2022RES_NCAA-SA-Well-BeingSurvey.pdf

NCAA. (2022l). *Overview.* https://www.ncaa.org/sports/2021/2/16/overview.aspx

NCAA. (2022m). *The Athletics Diversity and Inclusion Designee (ADID).* https://ncaaorg.s3.amazonaws.com/committees/ncaa/moic/MOIC_AthleticsDiversityInclusionDesignation Resource.pdf

NCAA. (2022n). *Timeline – 2020s.* https://www.ncaa.org/sports/2021/6/14/timeline-2020s.aspx

NCAA. (2022o). *What is the NCAA?* https://www.ncaa.org/sports/2021/2/10/about-resources-media-center-ncaa-101-what-ncaa.aspx

NCAA. (2022p). *Where does the money go?* https://www.ncaa.org/sports/2016/5/13/where-does-the-money-go.aspx

NCAA. (n.d.). *How the NCAA works – Division I.* https://ncaaorg.s3.amazonaws.com/champion-magazine/HowNCAAWorks/D1_HowNCAAWorks.pdf

NCAA Committee on Women's Athletics and Gender Equity Task Force. (2022). *Report of the NCAA Committee on Women's Athletics and Gender Equity Task Force, July 20, 2021, Joint Video Conference.* https://ncaaorg.s3.amazonaws.com/committees/ncaa/comwa/July2021CWA_GETFReport.pdf

NCAA Division I Interpretations Committee. (2021). *NCAA Division I Interpretations Committee: Policies and procedures.* https://ncaaorg.s3.amazonaws.com/committees/d1/lri_interp/D1IC_PoliciesAndProcedures.pdf

NCAA Sport Science Institute. (2022). *About the SSI.* https://www.ncaa.org/sports/2016/8/23/about-the-ssi.aspx?id=116

NJCAA. (n.d.). *NJCAA governance structure.* https://www.njcaa.org/about/governance_structure

Opendorse. (2022). *Compensation per athlete by division, through May 21, 2022.* https://opendorse.com/nil-insights/

Robinson, J. (2014, July 23). *I take a look at three reform-minded athletics reports and find a few (very few) good ideas.* The James G. Martin Center for Academic Renewal. https://www.jamesgmartin.center/2014/07/i-take-a-look-at-three-reform-minded-athletics-reports-and-find-a-few-very-few-good-ideas/

Rudolph, F. (1990). *The American college and university: A history*. University of Georgia Press.

Scott, H.A. (1951). *Competitive sports in schools and colleges*. Harper and Brothers.

Shahen, P., & Heron, M. (2015, January 27). *Battle continues to save UT men's track & field program*. Local 3 News. https://www.local3news.com/local-news/whats-trending/battle-continues-to-save-utc-mens-track-field-program/article_34865b3b-8ab1-5492-9976-e0c6f273f71b.html

Shapiro, M. (2021, May 18). Report: Stanford reverses decision to eliminate 11 varsity sports. *Sports Illustrated*. https://www.si.com/college/2021/05/18/stanford-reverses-decision-eliminate-11-varsity-sports

Smith, R.A. (2011). *A history of big-time college athletic reform*. University of Illinois Press.

Staurowsky, E.J., Flowers, C.L., Busuvis, E., Darvin, L., & Welch, N. (2022). *50 years of Title IX: We're not done yet*. Women's Sports Foundation. https://www.womenssportsfoundation.org/wp-content/uploads/2022/05/Title-IX-at-50-Report-FINALC-v2-.pdf

Thomas, K. (2011, April 26). Gender games. Colleges teams, relying on deception, undermine Title IX. *New York Times*. www.nytimes.com/2011/04/26/sports/26titleix.html

Title IX. (2017). *Title IX and sex discrimination*. U.S. Department of Education. www2.ed.gov/about/offices/list/ocr/docs/tix_dis.html

Trahan, K. (2015, January 17). *The 4 things to know about the new NCAA's autonomy structure*. SB Nation. www.sbnation.com/college-football/2014/8/7/5966849/ncaa-autonomy-power-conferences-voting-rules

Trahan, K. (2016, May 12). *Should Grambling State, Southern, and other HBCUs drop out of Division I football?* Vice. https://www.vice.com/en/article/yp8vg7/should-grambling-state-southern-hbcus-drop-division-i-football

University of Oklahoma. (n.d.) *University of Oklahoma Athletic Department Organizational Chart*. https://ou.edu/content/dam/irr/docs/Fact%20Book/Fact%20Book%202019/19_1_31_chart_athletics.pdf

USA Today. (2020). *NCAA finances*. https://sports.usatoday.com/ncaa/finances

Wann, D.L. (2006). Examining the potential causal relationship between sport team identification and psychological well-being. *Journal of Sport Behavior, 29*(1), 79–95.

The Major Games

Think of amateur sport as a highway. The highway is a stretch of road spanning informal, recreational opportunities (such as pickup basketball and Sunday afternoon touch football) to elite, multi-event competitions (such as World Championships and the Olympic Games). Lanes are open for participants, coaches, officials, and spectators. Participants can easily enter and exit the highway as their interests and

DOI: 10.4324/9781003303183-9

abilities dictate. Events are organized all along the highway, filling specific needs for competition for all age groups and ability levels of the athletes. Along the road, amateur sport evolves from recreation into the elite competition. Such events exist for a variety of age groups at the local, national, and international levels. And not all athletes are amateur in status, as professional athletes compete in major games in some sports as well. For example, teams compete for National Championships. Athletes are selected to represent their country on national teams competing in the World Championships, World University Games (also called the FISU [Fédération Internationale du Sport Universitaire] Games), and Pan American/Para-Pan American Games. Such competitions lead to the pinnacle events staged every four years – the Olympic and the Paralympic Games (discussed in their own chapters). The athletes are from high schools and colleges, and from amateur and professional leagues. The purpose of this chapter is to investigate the governance structures of the organizations delivering the major games in elite athletics. Given the scope and importance of the Olympic and Paralympic Games, the organization and governance of those two major games will be discussed in separate chapters.

How did other major games come to exist, and how are they organized and governed? This chapter approaches these questions in two ways: first, we discuss the governance of different organizations that provide athletes for major games. Second, we present several major games to illustrate the governance structures of the actual events. Let's look briefly at the evolution and history of the major games.

HISTORY OF THE MAJOR GAMES

Major advances in technology and urbanization led historians to describe the 19th century as "the age of progress" (Riess, 1995). Sport progressed at a phenomenal rate as well: "[T]he international foundation was truly laid for the gigantic proportions of sport today" (Glassford & Redmond, 1988, p. 140). Inventions such as the camera (1826), the railroad (1830), the electric lamp (1881), the motorcar (1885), and the radio (1901) were among the profound technological changes contributing to the evolution of sport (Glassford & Redmond, 1988).

Urbanization was also a major factor in the growth of the sport industry. The city became the site of massive stadiums and other facilities and the focal point for crowds of participants and spectators. Tournaments, festivals, and special events became more commonplace as both leisure time and general affluence increased (Kidd, 1999). Technology, especially related to easing long-distance transportation, provided the opportunity for both national and international competition. Before long, governments focused on the idea of sport as an alternative to war, whereby political ideologies and national strength could be displayed by winning international sporting competitions (Riordan & Kruger, 1999).

The advancement of political ideology through sport likely occurred around the time Baron Pierre de Coubertin revived the ancient Olympic Games in the late 19th century. Baron de Coubertin's dream was that sport could be used to increase goodwill among nations of the world. He reinstated Olympic competition when, in 1894, officials from 12 countries endorsed a modern cycle of Olympic Games (Glassford & Redmond, 1988).

Baron de Coubertin's Olympic Games were not entirely original. Games in England's Cotswolds and the Highland Games of Scotland were staged in the 19th century. The concept of major games and festivals spread quickly. The Far Eastern Championship Games were organized in 1913 as regional games after the rebirth of the modern Olympic Games. Teams from China, Japan, the Philippines, Thailand, and Malaysia participated (Glassford & Redmond, 1988). Similar games were established in Central America, where teams from Puerto Rico, Cuba, Mexico, and other Latin American countries participated. The first British Empire Games (later renamed the Commonwealth Games) took place in Hamilton, Canada, in 1930. Other countries organized regional games such as State Games and National Championships. In addition, international competitions such as the Asian Games, the Pan American Games, the Goodwill Games, and other special group events developed. For instance, the International Student Games were first held in 1924 (renamed the Universiade in 1959), and in 1960 the first Paralympic Games were held for individuals with specific physical disabilities. Today, games exist for every age group in virtually every sport. The World Little League Baseball Championships, America's Cup yachting competition, World Championships for speed skating, and the Deaflympics are examples of major amateur sporting events that dominate today's sporting calendar. Who organizes these events, and how are they operated? Next, we investigate the organizations and governance structures of several major games in amateur sport.

GOVERNANCE

The governance of amateur sport differs in countries around the world. Government focus on policy involving amateur sport via non-profit and voluntary organizations became more prominent in the latter half of the 20th century (Langlois & Menard, 2013). Even so, the degree to which a government is involved in sport policy differs depending on a nation's social, cultural, and political perspectives. In the United States, sport is intensely popular and a cause for national unity. However, US public policy has historically claimed (some say rhetorically) that sport is independent of government (Chalip & Johnson, 1996). In countries around the world, promoting national unity and identity are central themes in government involvement in sport-policy development. Some level of government involvement helps shape the policies governing the athletes representing their nations at the major games of amateur athletics.

Three branches of government exist in the United States: the legislative branch is responsible for policy-making, the executive branch implements laws and public policies, and the judicial branch interprets the law. Each branch plays an important role in policy development, along with state and local governments. In fact, many state and local governments are influenced by national policies. The policy developed at each level of government has implications for amateur sport. At the national level, laws specific to sport have been enacted. For example, the Amateur Sports Act of 1978 promotes, coordinates, and sets national goals for amateur sport in the United States through the development of national governing bodies (United States Amateur Sports Act, 1978). Another example is the Stevens Amendment of 1998, which changed the Amateur Sports Act so that it became known as the Ted Stevens Olympic and Amateur Sports Act. The new law strengthened athletes' rights, provided procedures for dispute resolution, and incorporated the Paralympic Movement into the Act by updating provisions for athletes with disabilities (US Senate S.2119, 1998). Policies affecting sport might also result from the application of laws not written specifically for sport, such as the Americans with Disabilities Act of 1990, which was established to prevent discrimination on the basis of disability, or through federal government agencies such as the President's Council on Physical Fitness and Sport, which sets policy on issues related to physical fitness and sport (Americans with Disabilities Act, 1990; Chalip & Johnson, 1996).

www

Ted Stevens Olympic and Amateur Sports Act
https://www.govtrack.us/congress/bills/105/s2119

State and national organizations exist to provide rules, regulations, promotion, and competition for specific sports. In the next sections, three examples of state and national organizations are presented. We start with the Amateur Athletic Union (AAU). It is one of the largest multisport organizations in the United States, incorporating both state and national offices with the mandate to promote and develop amateur opportunities in a variety of sports. Second, Sport Canada is discussed. Through Sport Canada, the Canadian government focuses on encouraging all Canadians (from amateur to elite) to get involved with sport. Finally, USA Basketball is presented. This is an example of a national sport organization that organizes basketball in the United States by operating in conjunction with 20 affiliate associations. How are these groups organized and how is policy developed?

THE AAU

The AAU is a multisport organization dedicated to promoting and developing amateur sport and physical fitness programs. It was founded in 1888 to establish standards for amateur sport participation (AAU, n.d.). In the early days, the AAU represented all amateur sports at International Federation meetings and was responsible for organizing national teams to represent the United States at international competitions, including the Olympic Games. As mentioned earlier, in 1978 the US Senate and the US House of Representatives enacted the Amateur Sports Act, the purpose of which was to coordinate amateur sport throughout the United States. This was done, in part, by establishing individual organizations for the purpose of developing specific sports. The Amateur Sports Act had a profound effect on the mandate of the AAU and caused the organization to refocus its purpose away from representing US teams internationally and toward the development and provision of sports programs for a wider spectrum of participants (AAU, n.d.). At this point, the AAU introduced the "sports for all, forever" philosophy. Today, the AAU offers a broad spectrum of activities, from baseball to wrestling, with 39 sports in between.

MISSION. The AAU promotes and delivers amateur sport widely within the United States. It is a network of local chapters that provides programs for children, men, and women in a large number of activities. The breadth of its mandate is illustrated by the inclusiveness of its programming. The mission statement of the AAU is presented in Exhibit 9.1.

MEMBERSHIP. Athletes, coaches, volunteers, and officials make up the membership of the AAU. The organization has thousands of members (700,000 participants and 150,000 volunteers) and offers programming for both youth and adult participants. Membership costs (discussed below) are kept purposefully low to allow as much inclusion as possible for potential participants.

FINANCIALS. The AAU is a non-profit organization funded through membership dues and donations. Yearly member dues are modest. As of this writing, any youth can belong for only $14, and the Extended Coverage Membership option (which allows participation in non-AAU licensed events) is only an additional $16. Dues for coaches and adults are only slightly higher. Sponsorships and partnerships are solicited, such as the partnership established between the AAU and Walt Disney World in 1996, which precipitated the relocation of the AAU National Office to Lake Buena Vista, Florida. Each year more than 60 AAU national events are held at the ESPN Wide World of Sports Complex near the Disney Resorts in Florida (AAU, n.d.).

ORGANIZATIONAL STRUCTURE. Fifty-five district offices for associations make up the AAU, each representing either a state (for example, Oklahoma) or a region (for example, New England) of the United States. The AAU is managed by a small Executive Committee, which is comprised of a group of officers elected by their Congress: the President, First and Second Vice President, Secretary, and Treasurer. Each officer is elected for a four-year term. The Congress is the primary actor for the business of the AAU and is composed of district representatives elected at the local, National Sport Committee, or National Officers levels. The Congress consists of approximately 600 members. The Congress constitutes a 36-member Board of Directors, consisting of designated members and those elected to act on AAU business between meetings of the Congress. National Sport Committees responsible for a particular sport define and direct policy related to that sport. The entire operation is managed by full-time staff members led by the Executive Director. In addition, a host of committees deal with AAU activities such as Finance, Insurance, Legislation, Nominations and Elections, Redistricting, Registration, and Administrative. Policy is developed through committees, analyzed and voted on by the Executive Committee, and then voted upon by the Board of Directors at annual national meetings. Much of the policy discussion involves the development of rules, regulations, and hosting guidelines for events. Exhibit 9.2 depicts the organizational structure of the AAU (AAU, 2022).

Amateur Athletic Union
https://aausports.org/

exhibit **9.1** Mission of the AAU

The mission of the AAU is "To offer amateur sports programs through a volunteer base for all people to have the physical, mental, and moral development of amateur athletes and to promote good sportsmanship and good citizenship."

Source: AAU (n.d.).

exhibit **9.2** Organizational Structure of the AAU

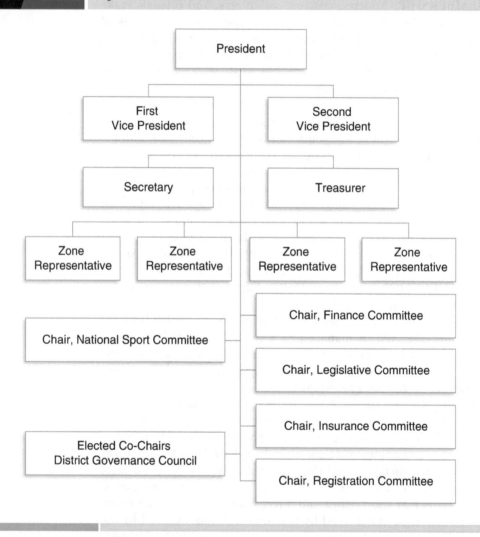

Source: AAU (2022).

SPORT CANADA

Sport Canada
https://www.canada.ca/en/
services/culture/sport.html

Sport Canada is a branch within the Department of Canadian Heritage of the Canadian federal government. It is an excellent example of how the government is involved in amateur sport in many countries of the world, such as Australia and Great Britain, although the United States has no such government department. Sport Canada is responsible for elite sport programming and sport policy development and is dedicated to valuing and strengthening the Canadian sport experience. "Sport Canada provides leadership and funding to help ensure a strong Canadian sport system which enables Canadians to progress from early sport experiences to high performance excellence" (Sport Canada, 2020, para. 2). Sport Canada has three programs of support: (1) the Athlete Assistance Program, that provides funding to athletes to support them during training and competition; (2) the Hosting Program, that provides support for the hosting of Canada Games and international sports events; and (3) the Sport Support Program, which involves funding for developing coaches' and athletes' technical abilities at the highest international levels, and promoting and increasing the number of Canadians involved in sport at all levels.

MISSION. The mission of Sport Canada is to promote sport opportunities for all Canadians. The organization coordinates and encourages excellence among Canadian athletes on the world stage and works with a variety of partners, including national sport organizations, coaches, and other levels of government to provide the necessary environment for high-performance athletes to achieve. Sport Canada's mission also promotes sport as a source of pleasure, personal satisfaction, and a means of achieving good health. Each of these pursuits is captured in the Canadian Sport Policy, a document that originally defined the goals for sport to achieve in Canada by 2012 (Canadian Sport Policy, 2012; set to be renewed in 2023) and which was subsequently updated and reviewed (Sutcliff Group, 2016). These goals include introducing sport to everyone in Canada, providing opportunities for recreational sport, improving competitive sport, improving high-performance (i.e., international competition) sport, and using sport as a tool for social and economic development. The mission statement of Sport Canada is presented in Exhibit 9.3.

MEMBERSHIP. Sport Canada is an umbrella organization that supports the mandate of high-performance sport. As such, it does not have a membership like national or state sport organizations have but instead is comprised of civil servants (employees) of the Canadian federal government. It is important to note that Sport Canada is separate and distinct from Canadian National Sport Organizations (NSOs) [such as Canada Basketball or Canada Skateboard] and the Canadian Olympic Committee (COC). The NSOs are linked to Sport Canada because they provide some degree of funding in return for compliance with policy and directives as set by Sport Canada. The COC, however, is completely separate from Sport Canada. While consultation and an open chain of communication are encouraged by both organizations, no formal relationship or reporting structure exists.

FINANCIALS. Sport Canada receives its funding from the Canadian federal government. It then establishes funding priorities and guidelines for the Canadian sport system. Sport Canada finances three programs: the Athlete Assistance Program, the Hosting Program, and the Sport Support Program. The Athlete Assistance Program, which provides living

The mission of Sport Canada is to enhance opportunities for all Canadians to participate and excel in sport.

Source: Sport Canada (2020).

and training allowances for high-performance athletes participating in international sport, spent $32.1 million on athletes in 2018–19. The Hosting Program, which is designed to provide financial assistance for the hosting of international events, provided only $25.7 million in payments in 2018–19. However, the Hosting Program received $142.8 million in 2013–14 and $281.9 million in 2015–16. This large amount was spent specifically for hosting the Pan American and Parapan American Games in Toronto in 2015. Finally, the Sport Support Program was transferred $156.9 million (the most ever) in 2018–19 (Ménard, 2020).

ORGANIZATIONAL STRUCTURE. Sport Canada is led by a Director General who reports through a Deputy Minister to the Minister of Sport and Persons With Disabilities. The organization is subdivided into program areas including Policy and Planning, Sport Support, Athlete Assistance, Hosting, Canadian Sport Centres, and Business Operations. The areas of sport policy and sport programs are further subdivided to deal with specific areas of focus, for instance, the national sport policy and policies covering Sport For Women and Girls, Doping, Tobacco Sponsorship, Persons With Disabilities, and Aboriginal Peoples Participation in Sport. Within Sport Canada, the policy is set regarding eligible forms of funding for athletes, NSO requirements regarding gender and language equity, and intergovernmental strategy and communication. A variety of program managers and sport consultants handle the duties within each subunit of Sport Canada.

NATIONAL SPORT GOVERNING BODIES

www
USA Basketball
https://www.usab.com/

www
Canada Snowboard
https://www.
canadasnowboard.ca/en/

www
Hockey Australia
https://www.hockey.org.au/
home/

Within each country, one national-level sport organization generally is recognized as the regulatory body for a particular sport. Sometimes these are governmental units, and other times they are free-standing sport organizations. These organizations have names like USA Baseball, Canada Snowboard, or Hockey Australia. Depending on the nation, these organizations are called national governing bodies or NSOs. To illustrate the governance structure, USA Basketball provides an excellent example.

USA Basketball is the national governing body (NGB) for basketball in the United States. As such, its employees oversee the development of the game of basketball from the grassroots through the elite levels in the United States.

MISSION. USA Basketball is the international US representative to the United States Olympic and Paralympic Committee (USOPC), and acts as the Fédération Internationale de Basketball Association (FIBA) member in the United States (USA Basketball, 2021a). FIBA is the international governing body for basketball. USA Basketball is responsible for selecting, training, and fielding national teams to compete in international FIBA competitions and in the Olympic Games. The association is also responsible for the development, promotion, and coordination of basketball in the United States, and identifies 11 components of its stated organizational purpose to do the following (USA Basketball, 2021b):

1. Develop interest and participation and take responsibility for basketball in the United States.
2. Coordinate between organizations to minimize conflicts in scheduling basketball events and practices.
3. Inform athletes about policy.
4. Sanction competitions.
5. Provide participation for amateur athletes.
6. Provide equitable support for both men and women.
7. Encourage and support individuals with disabilities.
8. Provide and coordinate technical, coaching, training, and performance analysis.
9. Encourage, support, and disseminate research in basketball safety and sports medicine.
10. Commit to equal opportunity and fair treatment to job applicants and employees without regard to race, color, religion, sex, national origin, age, physical handicap, sexual orientation or marital status, and actively involve qualified minorities and women to occupy positions at all levels of the Association.
11. Disseminate and distribute rules and/or any change of rules from USA Basketball to the International Olympic and Paralympic Committees.

The mission statement of USA Basketball is presented in Exhibit 9.4.

MEMBERSHIP. There are five membership types at USA Basketball: professional, collegiate, scholastic, youth, and associate. Professional membership includes those organizations delivering a national, professional competitive basketball program, such as the National Basketball Association (NBA) and Women's National Basketball Association (WNBA). Collegiate membership includes NSOs delivering basketball in university, college, and collegiate-level programs, like the National

---

Mission Statement of USA Basketball   *exhibit* **9.4**

The purpose of this Association is to act as the NGB for the sport of basketball in the United States, and in such connection, to be recognized as such by the USOPC and to act as the FIBA member in the United States.

Source: USA Basketball (2021b).

Collegiate Athletic Association (NCAA) and the National Association of Intercollegiate Athletics (NAIA). The scholastic membership category involves national organizations in school sport, like the National Federation of State High School Associations. Youth membership currently includes only the AAU, a community-based non-scholastic and non-collegiate sport organization that delivers national basketball programs for youth. Finally, associate memberships include other organizations that conduct significant basketball programs in the United States. Examples of associate members include Athletes in Action, National Basketball Players Association, USA Deaf Sports Federation, and United States Armed Forces. Members are non-voting, except that they have the right to elect or select certain members of the Board of Directors.

FINANCIALS. USA Basketball is a non-profit organization. Members within the organization pay annual dues of $250 (Youth), $500 (Associate), or $2,000 (Professional, Collegiate, Scholastic; USA Basketball, 2021b). In a non-COVID impacted year (2019), the majority of funding is derived from revenues associated with licensing and marketing for the national teams ($6.9 million), licensing and funding from the youth division ($3.4 million), and support from the USOPC ($1.2 million; USA Basketball, 2020). Meanwhile, the majority of expenses in a non-COVID-impacted year include the Men's Senior National Team ($4.9 million), the youth division ($3.3 million), and the Junior Men's National Teams ($1.3 million). Those numbers can change, however, based on which teams play their international competitions (for instance, the Senior Women's National Team was the largest expense in 2020).

ORGANIZATIONAL STRUCTURE. Members are allowed to go to the "USA Basketball Assembly," an annual meeting to discuss issues of common interest to the sport of basketball. The USA Basketball Assembly receives reports on the "State of USA Basketball" from the Board of Directors on past and future activities. The Assembly has no rule-making, budgetary, legislative, or other authority but rather acts in an advisory capacity to the Board of Directors.

USA Basketball is governed by a Board of Directors. The Board is led by the Chairperson, and the immediate Past Chair holds a position in the year immediately following the Olympic Games to ensure continuity. The Board consists of 15 voting members: 3 from the NBA; 3 from the NCAA; 5 Athletic Directors; 1 Director to represent a non-NBA, WNBA, and NCAA basketball league; and 3 At-Large Directors. The Board meets at least four times per year and has primary responsibility for developing policy and approving actions regarding the competitive basketball programs of the association (USA Basketball, 2021b). The Board of Directors provides leadership for the organization's Executive Director/CEO and professional staff. The policy is defined by the Board and the Executive Committee and via committee work. Standing committees of the association include the Nominating and Governance Committee, Finance and Audit Committee, and the Ethics Committee. In addition, there are an Athlete Advisory Council and USOPC Athlete Advisory Council that act on behalf of players. The organizational structure of USA Basketball is presented in Exhibit 9.5.

Annual Assembly

Board of Directors

USA Basketball Committees

USA CEO & Staff Members

Source: USA Basketball (2021b).

The governing bodies discussed above are all involved in organizing athletic competitions for elite-level athletes or sending elite athletes to major competitions. State and provincial organizations feed into NGBs and NSOs, respectively. NSOs are aided by other organizations such as the AAU, Sport Canada, national coaches' associations, and the NCAA via their roles in training elite athletes, coaches, and officials. One component of their collective missions is to enhance the ability of athletes to perform on the world stage at international competitions and major games. Next, we investigate the governance structures of some of the major games of amateur athletics.

ORGANIZATIONS THAT MANAGE MAJOR GAMES IN AMATEUR SPORT

Major games are national or international events run as single-sport or multisport championships. International world championships are common to many sports, for example, the FIBA World Basketball Championships and the FIFA Soccer World Cup. Also common are major international multisport games for which participation is restricted by eligibility criteria (such as country of origin, age group, or disability). This includes the Pan American Games, Commonwealth Games, World University Games, Deaflympics, and Special Olympics World Games. How are these major games organized? The next sections will address these multisport competitions.

WWW

Fédération Internationale de Basketball Amateur
https://www.fiba.basketball/

WWW

Fédération Internationale de Football Association
https://www.fifa.com/

PAN AMERICAN GAMES

The Pan American Games are a celebration of sport, competition, and international friendship for nations of the Americas in the Western Hemisphere. The Games have taken place on a strict

WWW

Pan American Games
https://www.panamsports.org/

quadrennial cycle since the first competition in 1951 in Buenos Aires, Argentina. They are typically scheduled for the summer in the year preceding the Summer Olympic Games.

MISSION. The Pan American Games are first and foremost an international multisport competition. However, since the event's inception, the organizers have sought a broader purpose. The Pan American Games "fuels the development of sport and supports the work of our 41 member nations to inspire the next generation of athletes in our continent" (Pan Am Sports, 2021a, para. 1). They are also a celebration of the Americas' community and of each country's dedication to their fellow nations making up the Americas. The motto of the Pan American Sports Organization (PASO) incorporates Spanish, Portuguese, English, and French: "América, Espírito, Sport, Fraternité," which translates loosely as "The American spirit of friendship through sports." The mission statement of the Pan Am Games is presented in Exhibit 9.6.

MEMBERSHIP. Athletes from countries in the Americas are eligible to compete in the Pan American Games. This includes North, Central, and South America, as well as Caribbean nations. Currently, 41 nations belong to PASO and are divided as follows: 3 North American members, 19 Caribbean members, seven Central American members, and 12 South American members.

FINANCIALS. The Pan American Games are an enormous undertaking, third in scope after only the Olympic Games and Asian Games. As one can imagine, the revenues and expenses vary from year to year, depending on when the Pan Am Games are hosted (the summer preceding Summer Olympics). For instance, in 2019 (when the XVIII Pan Am Games were hosted in Lima, Peru), the organization received $17.6 million in revenue from those XVIII Games, $8 million from the IOC's Olympic Solidarity Program, and $4 million from the XIX Pan Am Games (the 2023 Pan Am games set to be hosted in Santiago, Chile). Expenses in 2019 include $15.2 million to the Olympic Solidarity Continental Programs (which help the NOCs meet their demands), $11 million for the organizing committee for the 2019 Pan American Games

---

exhibit 9.6 Mission Statement of the Pan American Games

Panam Sports fuels the development of sport and supports the work of our 41 member nations to inspire the next generation of athletes in our continent. Panam Sports fuels the development of sport and supports the work of our 41 member nations to inspire the next generation of athletes in our continent. We work closely with all members of the sport community including

1. National Olympic Committees (NOCs)
2. International Sport Federations (IFs)
3. Pan American Sport Confederations
4. The athletes and Organizing Committees for all regional Games

Ensuring the success and celebration of our flagship event, the Pan American Games.

Source: PASO (2021a).

(which was only $169,286 in 2018), and $2.6 million in Administration fees (Pan Am Sports, 2021c). While Pan Am Sports only reported $11 million to the organizing of the 2019 Pan Am Games, overall costs included $1.2 billion; $430 million for organization, $470 million in infrastructure, $180 million for construction of the Pan Am Village, and $106 million for other expenses (Perelman, 2019).

ORGANIZATIONAL STRUCTURE. PASO governs the games, awards the hosting rights, and sets the policies and direction for the competition. PASO headquarters is located in Mexico City and is presided over by a President, Executive Board, and Voting members. Voting members, which are all the 41 nations that participate in the Pan Am Games, have voting rights at Pan Am Sports General Assemblies. These members typically vote on who gets the right to host the Pan Am Games, any amendments to the Constitution, and any elections or appointments needed (Pan Am Sports, 2021b). The Executive Board consists of the President, three Vice Presidents, the Secretary-General, the Treasurer, nine representatives of the Voting members, the President of the Athletes Commission, the President of the Association of Pan American Sport Confederations, and the Chair of the Pan Am Sports Legal Commission. The Executive Board essentially runs the PASO as they are responsible for all affairs of Pan Am Sports, guarantee and protect the celebration and continuity of the Games, and manage finances, among many other tasks. Finally, the President chairs Executive Board meetings and General Assemblies for PASO, as well as implements strategic plans and policies (Pan Am Sports, 2021b). As with most major games, an organizing committee is created prior to the event and is dissolved after. The Organization Committee is responsible for the organization and completion of the event in the host city.

COMMONWEALTH GAMES

The Commonwealth Games are a multisport competition bringing together countries from around the world that are united by history, as opposed to geography (as in the Pan American Games). The Commonwealth Games involve competition among countries that once belonged to the British Empire (see Exhibit 9.7) and for nations and territories that subsequently joined the British Commonwealth after the empire (Dheensaw, 1994). The countries comprising the Commonwealth share a history and acceptance of a common past. The Commonwealth Games continue to bring athletes and spectators of these nations together every four years for a significant festival of sport.

MISSION. The vision of the Commonwealth Games Federation (CGF) is "building peaceful, sustainable and prosperous communities globally by inspiring Commonwealth Athletes to drive the impact and ambition of all Commonwealth Citizens through Sport" (Commonwealth Games Federation, 2020a, p. 32). The organization seeks to achieve this vision by hosting a world-class multisport event for people of Commonwealth nations. The idea of a friendly festival of competition, held on a four-year cycle, is part of their mission statement, presented in Exhibit 9.8.

www

Commonwealth Games
https://thecgf.com/

exhibit **9.7** Countries Expected to Compete in the 2022 Commonwealth Games

AFRICA				
Botswana	Kenya	Mozambique	Seychelles	The Gambia
Cameroon	Lesotho	Namibia	Sierra Leone	Uganda
Eswatini	Malawi	Nigeria	South Africa	Zambia
Ghana	Mauritius	Rwanda	Tanzania	
AMERICAS				
Bahamas	Bermuda	Falkland Islands	St. Helena	
Belize	Canada	Guyana		
OCEANIA				
Australia	Kiribati	Niue	Samoa	Tuvalu
Cook Islands	Nauru	Norfolk Islands	Solomon Islands	Vanuatu
Fiji	New Zealand	Papua New Guinea	Tonga	
CARIBBEAN				
Anguilla	British Virgin Islands	Jamaica	St. Vincent & The Grenadines	
Antigua & Barbuda	Cayman Islands	Montserrat	Trinidad & Tobago	
Bahamas	Dominica	St. Kitts & Nevis	Turks & Caicos Isles	
Barbados	Grenada	St. Lucia		
EUROPE				
Cyprus	Gibraltar	Isle of Man	Malta	Scotland
England	Guernsey	Jersey	Northern Ireland	Wales
ASIA				
Bangladesh	India	Maldives	Singapore	
Brunei Darussalam	Malaysia	Pakistan	Sri Lanka	

Source: Commonwealth Games Federation (2021).

MEMBERSHIP. Six thousand and six hundred athletes from 72 nations worldwide are expected to compete in the 2022 Commonwealth Games in Birmingham, England. These nations and territories are located in parts of Africa, the Americas, Asia, the Caribbean, Europe, and Oceania. Some countries are eligible to compete in both the Pan American and Commonwealth Games, including Canada and many Caribbean countries.

FINANCIALS. The economics of the Commonwealth Games is very similar to the Pan American Games. Revenues are generated through television, sponsorship, and advertising, and ticket and merchandise

| Mission Statement of the CGF | *exhibit* | **9.8** |

To be an athlete-centered, sport-focused Commonwealth Sports Movement, with integrity, global impact and embraced by communities that accomplish the following:

- We deliver inspirational and innovative Commonwealth Games and Commonwealth Youth Games built on friendships and proud heritage, supported by a dynamic Commonwealth Sports Cities Network;
- We nurture and develop one of the best governed and well-managed sports movements in the world;
- We attract and build on public, private, and social partnerships that widely benefit Commonwealth athletes, sports and communities;
- We champion, through our brand, Commonwealth athlete, citizen, and community engagement in everything we do.

Source: Commonwealth Games Federation (2020a).

sales. Perhaps more importantly, the Games play an economic role and create a legacy for the hosting community. One of the first host committees to embrace this approach was Manchester, England in 2002. Part of Manchester's platform for hosting was based on the economic impact to be gained for the region. The organizing committee identified the crucial role the Games would play in the continued physical, economic, and social regeneration of Manchester, bringing regional economic benefits. The 2006 Commonwealth Games were also successful in creating a financial legacy for the City of Melbourne, Australia (Insight Economics, 2006). The 2014 Commonwealth Games hosted by Glasgow, Scotland, also focused on legacy, sustainability, accessibility, and the environment. Forty-six major sponsors enabled the Games to exceed the overall commercial target of roughly $167.62 million (Glasgow2014, 2014). Gold Coast City, Australia hosted the 2018 Commonwealth Games, with the goal to showcase the city and lifestyle of this place of natural beauty. The Gold Coast Games cost an estimated A$1.86 billion, but many still believe the economic boost from the Games outweighed the costs (Cunningham, 2019).

ORGANIZATIONAL STRUCTURE. The CGF is the umbrella organization responsible for regulating the competition. It is led by a President and three Vice Presidents, along with an Executive Board. The Executive Board is composed of the President and three Vice Presidents, as well as six elected Regional Vice Presidents (representing Africa, the Americas, Asia, the Caribbean, Europe, Oceania, and Australia, but not the host region). In addition, an Athletes Representative is appointed by the Board (Commonwealth Games Federation, 2020a). The Executive Board and Officers (e.g., Ethics Officer) help to set policy enacted by a CGF professional staff led by the CEO.

As with the Pan American Games, the local hosting community develops an organizing committee to deliver the competition. This group forms to bid for the event and, if successful, operates for several years prior to staging the Commonwealth Games. The committee then dissolves in the year after the games, once final financial and operational reports are completed.

WORLD UNIVERSITY GAMES

The World University Games (also called the Universiade or FISU Games) is a sporting and cultural festival held every two years for university-level athletes and governed by FISU, the International University Sports Federation.

MISSION. Founded in 1949, FISU is responsible for supervising the summer and the winter Universiades, as well as World University Championships in select sports. FISU's mission is "To provide opportunity for all students to participate in physical activity, while acting for their health and wellbeing and thereby helping them become tomorrow's leaders" (FISU, 2021a, p. 4). Five values guide FISU: excellence, teamwork, innovation, the joy of sport, and integrity (see Exhibit 9.9). As evident by the name of the event, Universiades is designed specifically for student-athletes. As such, the focus of the event is on self and cultural learning, the joy of sport, and education. The event is the largest in the globe for student-athletes as it brings together thousands of athletes from over 150 countries to compete in 25 compulsory sports (FISU, 2022a).

MEMBERSHIP. FISU comprises 164 National University Sports Federations (FISU, 2022b). One of these is the United States Collegiate Sports Council, which is composed of representatives of the various administrators and support staff of the NCAA, NAIA, and their member institutions. Membership in FISU is divided by countries within the five world continents – Africa (41), Americas (28), Asia (39), Europe (47), and Oceania (9). National university sports federations gain membership by paying fees and providing proof of eligibility. The Universiades are open to all student-athletes between the ages of 17 and 25 who are eligible to compete in university sport at home and who have not been out of school for more than a year.

FINANCIALS. FISU is funded through marketing activities, television revenue, organizing and entry fees, and subscriptions. The Universiade is run as a multisport festival, bid for by a host country, and run as a business. Revenues are generated in a manner similar to the Pan American and the Commonwealth Games, with government funding, corporate sponsorship, television-rights fees, entry fees, and sales making up the largest components of the budget. For instance, FISU receives EURO 20 per athlete and official from member associations (EURO 40 if not a member) and countries must make a deposit of EURO 5,000 per team for basketball, volleyball, and water polo (FISU, 2021b). Summer World University Games are generally one of the largest sporting festivals in the world, in the top three in size with the Olympic Games, involving as many as 174 nations (in Daegu, Korea, in 2003) and 11,785 participants (in Kazan, Russia, in 2013). With the vast number of participants, one could expect there to be a boon to the

FISU sets the values that shape and underpin all the work of Federations and FISU members.

- *Excellence*: In mind, body and lifestyle. A passion for excellence in sport and education
- *Teamwork*: Utilize individual capabilities in a coordinated effort to develop and promote the university sports movement.
- *Innovation*: Embrace new event formats and cutting-edge technology to enhance entertainment value.
- *Joy of Sport*: Create excitement across student-athlete sport events and a global fan base.
- *Integrity*: All that FISU does is honest, transparent, and promotes fair play. Athletes deserve the highest levels of integrity and ethics from those working in sport.

Source: FISU (2022a).

local economy of the hosting nation. FISU claims that the Universiade generated a EUR 180 million impact for the Honam region when it hosted the Gwangju 2015 Games (FISU, 2022c).

ORGANIZATIONAL STRUCTURE. FISU is composed of a General Assembly in which each of the 164 member nations are represented. The General Assembly elects an Executive Committee to act on its behalf between meetings of the Assembly. The Executive Committee of FISU is composed of 28 positions and most members are elected for four-year terms. It is led by the President and comprised of a First Vice President, four Vice Presidents, a Treasurer, the Senior Executive Committee Member, 15 Committee Members, and one delegate for each of the five continental associations. This committee meets twice per year and periodically at the call of the President, and is the main policy-making group within FISU. In addition, 13 working committees (e.g., Media and Communication Committee, and Finance Committee). The host city for the Universiade will also create an organizing committee to plan and manage the staging of the World University Games.

WORLD GAMES FOR THE DEAF: THE DEAFLYMPICS

The first International Games for the Deaf (renamed World Games for the Deaf in 1969) were held in 1924 in Paris, France (Carbin, 1996). Just prior to the inaugural Games, a group of European men with hearing impairments organized the International Committee of Silent Sports, abbreviated CISS. Today, CISS refers to the International Committee of Sports for the Deaf (CISS, 2017). This organization

www

Deaflympics
https://www.deaflympics.com/

oversees the World Games for the Deaf (known as the *Deaflympics*) and the Deaf World Championships.

MISSION. The motto of the International Committee of Sports of the Deaf (ICSD) is "Equality Through Sports." The organization brings athletes with hearing impairments together to compete in a range of athletic events, offering them the opportunity to celebrate their achievements and uniqueness as athletes with hearing impairments. The ICSD mission statement is presented in Exhibit 9.10. It stresses the value of competition, equality through sports, and adhering to the ideas of the Olympic Games (ICSD, 2020).

MEMBERSHIP. The CISS has five membership categories: full, associate, provisional, regional confederations, and honorary life. Full members are national associations who control Deaf Sports in their respective countries. Full members have voting rights and make up the Congress (roles of the Congress later). Associate members consist of international deaf sports federations that are responsible for only the technical aspect of one sport. Provisional members are either a national deaf sports federation or international deaf sports federation that is aiming to be a full or association member, but are awaiting approval to full member by the Congress. Regional confederations are a group of 10 or more full members that are geographically similar. There are currently four regional confederations – Asia/Pacific, Africa, Europe, and Pan America – who set the dates for the championships, competition programs, doping-control organization, and any modifications to ICSD rules and requirements. Finally, honorary life memberships are awarded to individuals who have done incredible work for the ICSD in the past (ICSD, 2021).

FINANCIALS. The ICSD is mostly funded through participation in events, memberships, and the media. In 2019, the Deaflymipcs were held in Italy and brought in $121,310 in participation fees (up from $21,210 the year preceding). Membership fees were roughly $48,000 and the media accounted for $19,000 in revenue. Total income for 2019 was $256,910, which more than doubled the year prior when there was no Deaflympics. For expenses, personnel charges (payments to administrators) were $94,456, accounting fees accounted for $47,176, travel cost was $42,245, and lodging accounted for $24,248 for a total of $271,199 in operational costs. As you can tell, most of the costs for the ICSD are administrative in nature (ICSD, 2019).

---

exhibit **9.10** Mission of the Deaflympics

To cherish the value of the spirit of Deaflympics where Deaf athletes strive to reach the pinnacle of competition by embracing the motto of **PER LUDOS AEQUALITAS (Equality through sports)** and adhering to the ideals of the Olympics.

Source: ICSD (2021).

ORGANIZATIONAL STRUCTURE. Each full member has a vote at the Congress. The Congress is responsible for determining the direction of Deaf Sports, electing the Executive Board, approving financial statements, approving rules, and many more charges. In addition to the Congress, the Executive Board and Executive Team oversee the ICSD. The Executive Board consists of one President, two Vice Presidents, four members at large, and the President's appointee from each regional confederation in a non-voting capacity. The Executive Board implements policies agreed upon by the Congress, approves the rules of procedure and bylaws, approves the calendar of events, and many more tasks. All members of the Executive Board must be a deaf person fluent in international signs. The executive team consists of just the President and Vice President of World Sports and has limited charges (ICSD, 2021).

SPECIAL OLYMPICS WORLD GAMES

Most people are familiar with Special Olympics. Perhaps you have volunteered at local- or state-level events. But did you know that Special Olympics also holds major international summer and winter events? The Special Olympics World Games is a multisport festival held every four years in both summer and winter for individuals with all levels of cognitive and developmental disabilities (Special Olympics, 2022b). It is important to note that the Special Olympics and the Paralympic Games are two separate organizations that are both recognized by the IOC. The Special Olympics provides sport opportunities for individuals with cognitive and developmental disabilities, while the Paralympic Games provide sports opportunities for elite athletes with specific physical disabilities. The Special Olympics World Games take place the year before the Olympic Games, while the Paralympic Games are conducted immediately following the Olympic Games.

MISSION. The mission of the Special Olympics World Games includes providing an exceptional sporting experience for participants with disabilities from around the world. The motto of the 16th Special Olympics World Games in Berlin, Germany in 2023 is "Nothing about us without us having a say" which organizers chose because the event will be a global demonstration of inclusion. Germany will welcome 7,000 Special Olympics athletes from roughly 170 countries to compete in 24 sports (Special Olympics, n.d.). The mission statement of the Special Olympics is presented in Exhibit 9.11.

MEMBERSHIP. The Special Olympics World Games have participants rather than members. The 2019 Games in Abu Dhabi, United Arab Emirates showcased the athletic skills, courage, and dignity of thousands of athletes with cognitive and developmental disabilities from around the world. Approximately 7,500 athletes, 3,000 coaches, 20,000 volunteers, 500,000 spectators, and thousands of families, journalists, and spectators attended the Games (Special Olympics World Games Abu Dhabi, 2019).

www

Special Olympics World Games
https://www.specialolympics.org/our-work/games-and-competition/world-games

exhibit   **9.11**   Mission of the Special Olympics

The mission of Special Olympics is to provide year-round sports training and athletic competition in a variety of Olympic-type sports for children and adults with intellectual disabilities, giving them continuing opportunities to develop physical fitness, demonstrate courage, experience joy and participate in a sharing of gifts, skills, and friendship with their families, other Special Olympics athletes, and the community.

Source: Special Olympics (2022a).

Per the Special Olympics official rules, every person with an intellectual disability of at least eight years of age is eligible to participate in the Special Olympics events (Special Olympics, 2012).

FINANCIALS. The Special Olympics, just like every other sport organization, experienced an impact due to COVID-19. Revenues for the non-profit organization totaled $128 million in 2020, compared to $142 million in 2019. While direct mail contributions (a special form of active marketing for donations and volunteers) increased by $2 million from 2019 ($47 million) to 2020 ($49 million), individual and corporate contributions and sponsorships fell from $55 million in 2019 to $29 million in 2020. The other largest revenue source for the organization is federal grants, which totaled $24 million in 2020. Expenses for the Special Olympics in 2020 totaled $111 million, compared to $133 million in 2019. While the organization was making less, it was also spending less. Program assistance dropped from $73 million in 2019 to $57 million in 2020. Public education and communications stayed roughly the same at $32 million, but fundraising decreased by $3 million between the non-pandemic to pandemic year to a total of $13 million in 2020 (Special Olympics, 2020).

ORGANIZATIONAL STRUCTURE. The Special Olympics Movement is run by Special Olympics, Inc. (SOI). SOI is in charge of establishing and enforcing all official policies and requirements of Special Olympics, overseeing the conduct of the Special Olympics, and providing training and support for other accredited programs (SOI-licensed organization that focus on training and competition programs for Special Olympic participants) throughout the world. SOI is governed by the Board of Directors (Special Olympics, 2012). The Board of Directors consists of between nine to 38 elected directors elected to the Board by the Nominating Committee. In addition, SOI has officers elected by the Board, including the Chair, three Vice Chairs, a President, a CEO, a Secretary, and a Treasurer. Finally, the Board of Directors take part in standing committees: Executive, Finance, Audit and Risk, Compensation, Nominating, and International Advisory committees (Special Olympics, 2019).

Spotlight: Diversity, Equity, & Inclusion in Action

Commonwealth Games Pride House

Half of the 70 countries worldwide that criminalize homosexuality are Commonwealth sovereign states (Staples, 2020). This deeply disturbing fact has sparked those at the Commonwealth Games to enact change. For the first time in Commonwealth Games history, the Birmingham 2022 Commonwealth Games hosted a Pride House (Birmingham 2022, 2022). Situated in the heart of that UK city's Gay Village, the Pride House aims to explore diversity across the Commonwealth, showcase talent, promote freedom of expression, and deliver a legacy that supports LGBTQ+ inclusion in sports. The Pride House Birmingham was delivered by Pride Sports, a UK-based sports development organization (Pride House Birmingham 2022, n.d.). While Pride Sports aims to make the United Kingdom more inclusive of LGBTQ+ individuals, Pride House Birmingham attempted the same with all the participants and spectators of the Birmingham 2022 Commonwealth Games. Prior to the Games, the Pride House hosted webinars on including LGBTQ+ people in sport, LGBTQ+ inclusive coaching, including trans-athletes, and communicating LGBTQ+ inclusion. The House has three aims: (1) Celebrate – queer sport, culture, and the Commonwealth; (2) Participate – encouraging LGBTQ+ participation in sport and physical activity; and (3) Educate – offer educational programming under the theme "everyone is welcome in Birmingham" (Pride House Birmingham 2022, n.d.).

CURRENT POLICY AREAS

Hosting any of the major games described above is a significant undertaking. Each event requires considerable organizational efforts, large financial support, and thousands of paid and volunteer workers. The organizing committee may plan for three or four years to ensure a safe and effective competition. Some issues organizers will surely deal with are internal to the particular event, such as fundraising and security. Many others are externally imposed, often by the governing international federation, and include issues such as doping control and the influences of the media. Other concerns stem from our global society and safety, and the very definition of *amateurism*. Each of these issues involves current policy areas.

Sport and International Politics

The association between sport and politics, and the subsequent political maneuvering that might occur through major games, is a policy area of interest to event participants and organizers. Recent issues of the international politics of sport include the advance of political ideologies such as democracy versus communism; capitalism and international relations; religion, gender, and disability sport; sport and terrorism; international travel bans; and most recently the outbreak of the war between Russia and Ukraine (Hassan, 2012; Levermore & Budd, 2004; Longman, 2017). Sport has traditionally been used for nation-building. It symbolizes the values of success, "of our ways compared to your ways," "of our people over your people." Sport illustrates power, wealth, business might, and general superiority. It has even been used as a show of moral authority and political legitimization (Allison, 2005; Houlihan & Lindsey, 2013).

Political factors have influenced the location of game sites. For instance, international federations have chosen host sites on the basis of generating economic support and facilitating legacy to an underdeveloped part of the world. Boycotting tactics – a nation refusing to send athletes and teams or dignitaries (called a diplomatic boycott) to an event in protest over another country's domestic or foreign policy – have been used as a form of political maneuvering. In this case, national policy might directly influence sport policy, and a nation may decide the extent to which sport will be used on the world stage to further other national objectives. The degree to which sport is used to enhance a national political agenda directly affects the political maneuvering associated with major games. Another factor might include the involvement of business and the commercialization of sporting events. Today, the long-term involvement of businesses in sport might actually weaken the ability of a government to manipulate and exploit sport as a means for promoting a diplomatic agenda, as the international sport agenda becomes more and more dominated by big business and sponsors as opposed to state politics (Allison, 2005).

Financing the Games and Economic Impact

Fundraising and marketing have become increasingly important to hosting major games in amateur athletics (Coates & Wicker, 2017). While several levels of a government might commit to contributing some financial support, such an offering is seen as a component of a larger financial landscape. Therefore, since hosting requires significant resources, particularly when new facilities are required, developing the financial backing to deliver the event is critical for success.

The importance of marketing for a major amateur athletic event is a given (Smith, 2017). While significant positive results can accrue from attracting donor funds and marketing the event (even to the extent that one without the other may well be impossible), other issues arise from marketing and fundraising practices. Members of organizing committees are debating methods of increasing the value of television and sponsorship packages, not to mention the impact of social media. Increased commercialization of amateur sport results in a shift of power and control toward sponsors. Struggles develop as a result of exclusive sponsorship categories. For example, water and isotonic drinks are obviously an important sponsorship category associated with major games. When exclusive sponsorship rights are awarded to one company, the extent of the sponsorship agreement might become a source of problems. Does the sponsorship agreement extend to all other products associated with the company? This becomes an issue because today's multinational conglomerates produce a vast array of products. A balance between the sponsor's needs and the best interests of the games is required. Without fundraising, sponsorship, and marketing, the very existence of major games could be jeopardized, given the significant requirement for operating revenues.

Today's global economy drives both cost and value. On the one hand, the hosting of most major games runs in millions of dollars or euros. The value of selling certain properties associated with major games, such as title sponsorship or television rights, slides on a scale depending on the location and economic factors. Organizers are constantly concerned with keeping costs down and value up. While costs can remain fairly neutral once established, the values of the properties of major games are more difficult to pinpoint. These values depend on many factors, and they change as a result of economic and political factors. For instance, how does the value of television rights for hosting the Pan American Games differ between a host site in North America versus South America or the Asian Games being held in China versus South Korea? Many factors, such as the number of potential television carriers, the size of target markets, the time of year, competition with other established events, and the ability of the host to attract other corporate partners affect the value of the television package. Of course, the organizing committee enacts policies to drive the value of the contract as high as possible. These policies might suggest the importance of publicity to increase the television audience, scheduling games and events at the best time of day to ensure the highest possible television numbers, and so on.

Broadcasting and sponsorship revenues come with an associated cost. When outside groups buy services or properties, conflict may arise over how the event is run. Policies to define rights and privileges associated with each partnership are critical to successfully hosting major games. For instance, the organizing committee must define a specific television policy that establishes explicit guidelines for how the event will be scheduled, with game-day timelines determined in advance. The policy will also suggest how changes to the timelines can be made, naming the groups or individuals who must be consulted.

Exclusivity is a term used for selling sponsorships that involve dividing the event into sponsorship categories and allowing for each category to be sold only one time, thereby providing one sponsor with "exclusive rights" without competition for its product. For instance, selling soft drink sponsorship exclusively to Pepsi would preclude any sponsorship with Coke or any other soft drink company. It may be difficult to decide which categories of sponsorship should be sold exclusively. Often, the sponsorship policy defines exactly which sales categories will be sold with exclusivity. Preferably, the marketing and sales personnel of the organizing committee will carefully define and communicate this practice to sponsors in advance, thereby lessening the potential for conflict between sponsors, controlling for ambush marketing attempts, and enhancing the sponsors' interest in being associated with the event.

Amateur athletic organizers rely heavily on marketing and sponsorship to deliver an event of the magnitude of FISU or the Commonwealth Games. The comments made by Slack (1998) still apply: "in no previous period have we seen the type of growth in the commercialization of sport that we have seen in the last two decades. Today, sport is big business and big businesses are heavily involved in sport" (p. 1). Such reliance, however, is of concern to event organizers. In a best-case scenario, major games could run as an entity by itself. Given this is not the case, international federations and major games organizing committees set policy to encourage revenue generation beyond corporate sponsorship and advertising. For example, the Finance Committee of major games would define sources of funding to manage the event. Those sources will be as diverse as possible to decrease the threat of reliance on any one funding category. Government funding, television rights, categories of exclusive and non-exclusive sponsors and corporate partners, pure advertising, ticket sales, merchandising, entry fees – all are sought by organizing committees to diversify revenue sources, prevent running an overall deficit, and deflect undue influence and/or control of the games by outside groups.

Finding the resources to deliver the games is imperative. A highly touted outcome of hosting the games is now commonly referred to as an economic impact. The economic impact is defined as the complex measurement of factors associated with hosting the games that leave a legacy for the area in terms of infrastructure and urban renewal, tourism and construction jobs, factors of the economy, and so on. The 2018 Commonwealth Games was reported to have a 1.2 billion-euro impact on Australia's gross domestic product. This was the largest to date, surpassing Manchester's 2002 Games (1.1 billion euros), Melbourne's 2006 Games (1 billion euros), and Glasgow's 2014 Games (0.8 billion euros; Commonwealth Games Federation, 2020b).

Safety

Imagine being the Director of Security for a large international event in this environment. Your task is to ensure the safety of 5,000 competitors from 145 countries and the 400,000 people who will gather to enjoy the competition and related cultural events. This is a monumental task and the focus of extensive debate and policy development. As the Director of Security, you have to navigate safety from physical attacks like a terrorist attack. Simultaneously, you have to make decisions about the best practices when it comes to protecting athletes, volunteers, coaches, and anyone else associated with the games as it relates to viral diseases like COVID-19. Beyond these is also the increased possibility of cyberattacks during the event. Sport and major games are not immune to the greater issues that affect millions of people around the globe. For major games organizers, safety has to be the number one most important factor when putting on a major event.

The changes to daily life associated with a world on alert are manifested in many ways. Citizenship, travel, security, privacy, and global politics each take on heightened meaning. The impact on event management is particularly important. The tragic hostage crisis in the Olympic Village during the Munich Olympic Games of 1972 that resulted in the deaths of 11 Israeli athletes and coaches, 5 Palestinian terrorists, and 1 German police officer opened the world to the idea that terrorism can happen in sport. The potential for terrorist action at major games is of real and continued concern for game organizers. One cannot forget that global pandemics make life difficult for event organizers, too. Many major games had to make a difficult decision about postponing or canceling their multimillion dollar events due to health safety risks associated with COVID-19. For instance, the 2022 Gay Games were postponed until 2023 due to COVID-19 restrictions (Dowdeswell, 2021) in hopes the organization can still put on the event. However, FISU had to cancel the 2021 Winter Universiade Games in Switzerland due to travel restrictions (Morgan, 2021). The decision to postpone and/or cancel has a multi-million dollar impact on the organizers and the cities, not to mention the impact it has on the athletes hoping to participate in what, for some of them, may be once in a lifetime event. Changes may also take place due to climate change, where start times for events may be shifted a cooler time for athlete safety and the threat of increasingly severe weather (ex – hurricanes) may require postponing or canceling events.

Enacting policy regarding safety at major games involves the collaboration of several levels of administration. Security, law enforcement, and now public health officials from local and governmental offices provide the foundation. Sport federations might provide expertise on past experience that proves valuable for future actions. The organizing committee ensures the coordination of all agencies and the implementation of the related policies. Other levels within the government of the host country help with coordination. The governments of competing countries may offer assistance and will undoubtedly require assurances of readiness. In the end, the policies they establish will define parameters for safe and secure travel, admittance, contact, and conduct of participants, spectators, volunteers, and affiliates of the major games.

The successful bid by Gold Coast City, Australia, to host the 2018 Commonwealth Games identified security as the main planning theme. Historical evidence of managing security, having the capacity and systems in place to coordinate massive numbers of visitors and venues, and experience in command and control are consistent requirements of major games hosting. Expect future bids to now have a statement regarding the transmission of infectious diseases. At the Commonwealth Games, as with Pan American Games, World University Games, and other events and championships, safety is of paramount concern, and significant resources, planning, and collaboration with local and national law enforcement are required.

Performance Enhancement

The use of drugs to enhance performance and influence the outcomes of athletic contests is termed *doping*. Worldwide, sport agencies and federations view doping as cheating and prohibit the use of performance-enhancing drugs (PEDs). Athletic competition is about pitting the natural athleticism and skills of an individual or group against another. Fairness requires each individual or group to compete within a common set of parameters. PEDs are considered shortcuts around the rigors of training and preparation. As such, doping is considered artificial and is thus banned as a means of achieving a competitive edge. In addition, many doping practices are dangerous and in direct opposition to the concept of "healthy mind, healthy body" which is the benefit of sport and physical activity. To combat the issue of doping in sports, national associations such as the United States Anti-Doping Association (USADA), UK Anti-Doping, and the Canadian Centre for Ethics in Sport (CCES) have been organized to work in conjunction with Olympic Committees and the World Anti-Doping Agency (WADA). Policy on doping in sport has been defined to

- protect those who play fair
- educate about the dangers and consequences of doping
- research and publish prohibited substances and methods lists
- coordinate anti-doping activities globally
- deter those who might cheat

www

Gold Coast City Bid for the 2018 Commonwealth Games
https://thecgf.com/sites/default/files/2018-03/Gold_Coast_VOL_1.pdf

- apply common sanctions for doping infractions and provide detailed procedures for establishing a breach in the rules (WADA, n.d.)

WWW
United States Anti-Doping Agency
www.usantidoping.org

WWW
UK Anti-Doping
http://ukad.org.uk/

WWW
Canadian Centre for Ethics in Sport
www.cces.ca

WWW
World Anti-Doping Agency
www.wada-ama.org/

All major games, international federations, and the Olympic Movement have provided a unified approach to setting policy that outlines banned substances and practices and outlawing anyone contributing to doping in sport. WADA provides for this required unified approach to developing a doping-control policy, referred to as the WADA (n.d.). Testing procedures, penalties, laboratory analyses, results management, protests and appeals, and reinstatement procedures are basic elements of doping-control policy. The ultimate goal of the policy is to create anti-doping rules, set mandatory international standards for banned substances and testing procedures, and model best practices. The issue is defined as a current policy area because it remains a dynamic issue. The use of banned substances and subsequent reports of positive tests remain a common occurrence at major games such as the Commonwealth Games and Pan American Games. For example, a 16-year-old Nigerian weightlifter was stripped of her gold medal at the Glasgow 2014 Commonwealth Games after testing positive for two banned substances (Butler, 2014). Fifteen athletes tested positive at the 2019 Pan American Games in sports ranging from track cycling, judo, and bowling (Pavitt, 2019). Testing and strict anti-doping procedures are enacted at all major games by organizing committees. However, the will to win and the stakes for winning on the world stage help promulgate a win-at-any-cost attitude, which results in the development of new performance-enhancing techniques and substances. Thus, policymakers at all national and international levels continue to focus on this issue to curb such behaviors.

SUMMARY

The major games of amateur sport have a rich and diverse history. The Olympic Games are still the world's largest and most prestigious sport festival, but in the years between Olympic Games, many other events are organized and attended by nations worldwide. Major games are organized mostly for amateur competitors, and NGBs and NSOs help initiate and manage the competitors selected to represent their nations. In the United States, the AAU plays a major role in developing and organizing competitive athletics. State and national governing bodies oversee the national-level competition and send representatives to international games, set policies, and provide funding for teams to compete at world championships.

Major international games include the Pan American Games, the Commonwealth Games, the FISU Games, the Deaflympics, and the Special Olympic World Games. Such events require extensive planning and organizing and are major financial undertakings. Organizing committees spend years preparing and managing many policy areas in an effort to ensure a safe, effective sporting competition. Current policy issues include political maneuvering, funding, safety, and doping control. The stakes are high for participants and organizers, given the enormity of the overall profile, size, and financial commitment involved in major games.

JEFF MANSFIELD *Member at Large*

Board of the USA Deaf Sports Federation

I serve as a Member at Large on the Board of the USA Deaf Sports Federation. The United States' National Deaf Sports Federation (USADSF) is recognized by the ICSD. USADSF is the Deaflympic equivalent of the USOPC. I chair USADSF's development committee, strategic planning committee, and serve as the USADSF liaison to the USOPC. USADSF is also a member of USOPC's Affiliated Organizations Council.

In my view, the two most important issues confronting sports leaders working in major games management today are the sustainability of the Games and inclusion and equality for all, as defined by the framework of the United Nations Convention on the Rights of Persons with Disabilities (UN CRPD) Article 30.5. Hosting any major Games, whether Olympic, Paralympic, or Deaflympic, requires enormous capital, political, and social investment. The privilege to host major games is often a source of national and regional pride as well as a statement of international cooperation and the betterment of society. However, such grand statements often compete with the practical day-to-day needs of a city, region, or country, pitting the local constituency against the aspirations and ambitions of political figureheads. Vast sums of capital are required to construct or prepare venues for competition, significant disruptions to municipal as well as environmental agendas occur, and legacy planning often gives citizens, cities, and regional governments pause in hosting major games despite the promise of an economic and tourism boost. With respect to UN CRPD Article 30.5, which outlines the right for deaf and people with disabilities to participate in and access sports "on an equal basis with others," systemic gaps and inequities in sports environments remain at every level, from community-based programs to major games, including the Deaflympics, Paralympic Games, and Olympic Games.

I drafted Anti-Nepotism and Role Conflict articles for inclusion in the ICSD Constitution, which were approved by the 47th ICSD Congress in Verceia, Italy, and have been modified and adopted into the current Constitution. These were proposed to prevent the potential for and the appearance of corruption, collusion, and favoritism in deaf sports, which can be mired at times by personal allegiances and "small-world" politics in which individuals, families, and associates unfairly accumulate power and influence.

For the last decade-plus, the management of the Deaflympics has been reactive and opportunistic, which has been key to remaining agile and ensuring that the Games take place despite uncertainty and numerous obstacles. However, to secure the long-term stewardship of the Deaflympics, its managers would be well-served to cultivate stronger relationships between the ICSD, IOC, and IPC and to be more strategic in its efforts to augment the Deaflympic movement.

As the second oldest Olympic-style competition (after the Olympic Games), the Deaflympics, founded in 1924 by the deaf Frenchman Eugène Rubens-Alcais and the deaf Belgian Antoine Dresse as the International Silent Games, is a major deaf- and disability-led international event conceived by the very community it serves. As such, it endures as a seminal example of *self-determination and social agency*. This unique distinction sets the Deaflympics apart from the Paralympic Games, first conceived by an able-bodied spinal-cord doctor as the Stoke Mandeville Games. The ICSD Congress has resisted several overtures from the Olympic and Paralympic community over the years to merge the Deaflympic program with the Paralympic Games (which would have resulted in fewer sports and the marginalization of sign language). Although such resistance by the deaf sports community was seen by some at the time as short-sighted, in fact, it was rooted in upholding the rights of autonomy and self-determination for deaf people and people with disabilities deaf and disabled people's, which the United Nations has since acknowledged as core principles of the UN CRPD (see UN CRPD 30.5(b)).

One issue at the heart of this matter is the incongruity between the IOC, IPC, and ICSD governance at the national level, which compromises the IOC's ability to deliver on the promise of inclusion as outlined in the Olympic Charter, Principle 4. This tension often plays out on the national level. In addition, the lack of a robust structural relationship between an NGB and NSO in the United States as well as elsewhere often results in Deaf and Hard of Hearing athletes experiencing diminished access to competitive and developmental pathways to elite performance.

For those looking to enter a career in major games management, I cannot emphasize enough the importance of building genuine relationships and of showing up with humility, curiosity, and empathy. I also encourage emerging leaders in this field to foreground equity and disability justice frameworks in their work on a daily basis. I do believe in sport's remarkable power to evoke the universal and create bridges between people in the service of a better, more just, and verdant world. Realizing this vision is a collaborative and collective effort that will require each of us to honor our intersectionalities and care for one another through sports as well as life.

case STUDY

MAJOR GAMES IN AMATEUR SPORT

You work for your local area sports commission. You are putting together your strategic plans for the next ten years and have decided to put in a bid for the next Pan American Games. You are located in a major metropolitan area with a population of approximately two million residents. Your city has a large university with excellent sport facilities and a college with good outdoor facilities. Your community also has one AAA minor league baseball team; considerable other sports facilities, both private and public; and extensive park areas that could serve as potential venues.

Using the bids created by the communities of Ciudad Bolívar, Venezuela, La Punta San Luis, Argentina, Lima, Peru, and Santiago de Chile, Chile, for the 2019 Pan Am Games that are presented on the Bids for the 2019 Games website for assistance, answer the following questions:

1. Make a list identifying each area of information that will be required, forming an outline of the sections of the bid document.
2. With which governing bodies (local, national, international) will you need to communicate?
3. Exactly which sports will be on the Games program, and what is your plan for selecting the venues you would like to use for each sport?
4. Whom will you work with to ensure the safety of athletes, coaches, volunteers, and fans as it relates to terrorist threats and COVID-19 protocols?
5. What local community groups will you actively pursue to assist with your bid, and what will their specific roles be?

CHAPTER questions

1. Choose any two of the organizations that host major games presented in this chapter (e.g., FISU or SOI). What are their missions? Discuss how those missions are similar and different. Then, find an example of an initiative put on by the organization that supports the mission.

2. How do major games market their product and entice sponsors? Using the websites of any three major games, review the fundraising practices of the organizing committees.

3. What is WADA and why does it exist? Explain WADA's goal. What role do major game organizers play in helping WADA achieve its mission? Within a group of classmates, debate whether WADA is working to ensure that athletes "play true."

FOR ADDITIONAL INFORMATION

For more information check out the following resources:

1. This website discusses some financial issues associated with a major games: Inside the Games: Report finds Toronto 2015 Pan American Games came in over budget: https://www. insidethegames.biz/articles/1038327/report-finds-toronto-2015-pan-american-games-came-in-over-budget

2. This article addresses the relationship between sporting events globalization: The Sport of Globalization: Globalization. It's a home run: https://writingmerrimack.wordpress.com/2013/10/14/the-sport-of-globalization-its-a-home-run/comment-page-1/

3. This video provides a look at what the AAU does: You Tube Video: 2020 AAU Summer Recap: https://www.youtube.com/watch?v=XvIOnOTysQM

4. Here you can read about how the Deaflympcs responded to the COVID pandemic: Deaflympics: ICSD Update on COVID-19 https://www.deaflympics.com/latest-covid-19-update

5. This is an in-depth overview of the FISU Games: FISU: The winter Universiade: https://www.fisu.net/sport-events/fisu-world-university-games/winter-fisu-world-university-games

REFERENCES

AAU. (n.d.). *About AAU*. https://aausports.org/page.php?page_id=99844

AAU. (2022). *2022 AAU codebook*. https://image.aausports.org/codebook/codebook.pdf

Allison, L. (2005). The curious role of the USA in world sport. In L. Allison (Ed.), *The global politics of sport: The role of global institutions in sport* (pp. 101–117). Routledge.

Americans With Disabilities Act. (1990). *ADA of 1990*. www.ada.gov/

Birmingham 2022. (2022, March 31). *Pride House Birmingham will explore diversity across the Commonwealth*. https://www.birmingham2022.com/news/2556327/you-dont-always-realise-the-level-of-impact-youre-making

Butler, N. (2014, August 1). *Nigerian weightlifter stripped of gold medal after positive doping test confirmed*. Inside the Games. https://www.insidethegames.biz/articles/1021638/nigerian-weightlifter-stripped-of-gold-medal-after-positive-doping-test-confirmed

Canadian Sport Policy. (2012). *The Canadian sport policy*. https://sirc.ca/wp-content/uploads/files/content/docs/Document/csp2012_en.pdf

Carbin, C.F. (1996). *Deaf heritage in Canada: A distinctive, diverse and enduring culture*. McGraw-Hill Ryerson.

Chalip, L., & Johnson, A. (1996). Sport policy in the United States. In L. Chalip, A. Johnson, & L. Stachura (Eds.), *National sport policies: An international handbook* (pp. 404–430). Greenwood Press.

CISS. (2017). *International Committee of Sports for the Deaf*. www.ciss.org/

Coates, D., & Wicker, P. (2017). Financial management. In R. Hoye & M.M. Parent (Eds.), *The Sage handbook of sport management* (pp. 117–137). Sage.

Commonwealth Games Federation. (2020a). *Constitutional documents of the Commonwealth Games Federation*. https://thecgf.com/sites/default/files/2020-12/Constitutional%20Documents%20of%20the%20Commonwealth%20Games%20Federation%202020.pdf

Commonwealth Games Federation. (2020b). *New report reveals Commonwealth Games consistently provides over £1 billion boost for host cities*. https://thecgf.com/news/new-report-reveals-commonwealth-games-consistently-provides-over-ps1-billion-boost-host-cities

Commonwealth Games Federation. (2021). *Teams and countries*. https://thecgf.com/countries

Cunningham, E. (2019, April 11). *Benefits from Gold Coast 2018 to outweigh costs*. https://www.sportcal.com/Insight/Features/125370

Dheensaw, C. (1994). *The Commonwealth Games*. Orca.

Dowdeswell, A. (2021, September 15). *Gay Games postponed to 2023 due to Hong Kong COVID-19 restrictions*. Inside the Games. https://www.insidethegames.biz/articles/1113027/gay-games-hong-kong-covid-postponed

FISU. (2021a). *FISU global strategy 2027*. https://www.fisu.net/fisu/about-fisu/fisu-global-strategy-2027

FISU. (2021b). *Regulations for the FISU World University Games*. https://www.fisu.net/medias/fichiers/wug2021_regulations_20200817_releasing_version.pdf

FISU. (2022a). *About us*. https://www.fisu.net/fisu/about-fisu

FISU. (2022b). *Member associations (NUSF)*. https://www.fisu.net/fisu/member-associations-nusf

FISU. (2022c). *Hosting*. https://www.fisu.net/sport-events/hosting

Glasgow2014. (2014). *Glasgow Commonwealth Games*. www.glasgow2014.com/media-centre/press-releases/glasgow-2014-says-big-thank-you-xx-commonwealth-games-sponsor-family

Glassford, R.G., & Redmond, G. (1988). Physical education and sport in modern times. In E.F. Zeigler (Ed.), *History of physical education and sport* (pp. 103–171). Stipes.

Hassan, D. (2012). Sport and terrorism: Two of modern life's most prevalent themes. *International Review for the Sociology of Sport, 47*(3), 263–267.

Houlihan, B., & Lindsey, I. (2013). *Sport policy in Britain*. Routledge.

ICSD. (2019). *Report of the statutory auditor on the financial statements*. http://deaflympics.com/pdf/2019-icsd-audit.pdf

ICSD. (2020). *Mission statement*. https://www.deaflympics.com/icsd/mission-statement

ICSD. (2021). *Constitution*. https://www.deaflympics.com/icsd/constitution

Insight Economics. (2006). *Triple bottom line assessment of the XCIII Commonwealth Games*. https://sport.vic.gov.au/__data/assets/pdf_file/0028/55594/download.pdf

Kidd, B. (1999). *The struggle for Canadian sport*. University of Toronto Press.

Langlois, M.-C., & Menard, M. (2013). *Sport Canada and the public policy framework for participation and excellence in sport*. https://publications.gc.ca/collections/collection_2016/bdp-lop/bp/YM32-2-2013-75-eng.pdf

Levermore, R., & Budd, A. (2004). *Sport & international relations: An emerging relationship*. Routledge.

Longman, J. (2017, January 28). Trump's immigration order could have big impact on sports. *New York Times*. https://www.nytimes.com/2017/01/28/sports/trump-refugee-ban.html?_r=0

Morgan, L. (2021, November 29). *Emergence of new COVID-19 variant forces cancellation of Winter Universiade in Lucerne*. Inside the Games. https://www.insidethegames.biz/articles/1116119/lucerne-2021-cancelled-covid19-variant

Ménard, M. (2020, January 23). *Sport Canada and the public policy framework for participation and excellence in sport*. https://lop.parl.ca/staticfiles/PublicWebsite/Home/ResearchPublications/BackgroundPapers/PDF/2020-12-e.pdf

Pan Am Sports. (2021a). *About PANAM sports*. https://www.panamsports.org/about-panamsports/

Pan Am Sports. (2021b). *Constitution of the Pan American sports organization*. http://www.panamsports.org/constitution/

Pan Am Sports. (2021c). *Statement of financial position*. https://www.panamsports.org/wp-content/uploads/2020/09/Financial-Report-ODEPA-2019-EN.pdf

Pavitt, M. (2019, September 29). *Silva stripped of Lima 2019 gold as Pan Am Sports confirm 15 doping cases*. Inside the Games. https://www.insidethegames.biz/articles/1085176/silva-stripped-gold-lima-2019-15-doping

Perelman, R. (2019, August 8). *LANE ONE: Lima has done well, but what is the future of the Pan American Games?* The Sport Examiner. http://www.thesportsexaminer.com/lane-one-lima-has-done-well-but-what-is-the-future-of-the-pan-american-games/

Pride House Birmingham. 2022. (n.d.). *Who we are*. https://pridehousebham.org.uk/who-we-are/

Riess, S.A. (1995). *Sport in industrial America 1850–1920*. Harlan Davidson.

Riordan, J., & Kruger, A. (1999). *The international politics of sport in the 20th century*. Routledge.

Slack, T. (1998). Studying the commercialisation of sport: The need for critical analysis. *Sociology of Sport Online, 1*(1), 1–16.

Smith, A.C.T. (2017). Sport marketing. In R. Hoye & M.M. Parent (Eds.), *The Sage handbook of sport management* (pp. 138–159). Sage.

Special Olympics. (n.d.). *Special Olympics World Games Berlin 2023*. https://www.specialolympics.org/stories/news/special-olympics-world-games-berlin-2023

Special Olympics. (2012). *Special Olympics official general rules*. https://dotorg.brightspotcdn.com/ef/76/6da131bc4d8ba82cb5ff40de975f/amended-general-rules-v2.pdf

Special Olympics. (2019). *Special Olympics, Inc. bylaws*. https://media.specialolympics.org/soi/legal/Bylaws-Amended-and-Restated-2019-2-26.pdf?_ga=2.149055862.134799168.1641478800-39084598.1641478800

Special Olympics. (2020). *Special Olympics, Inc. and affiliates*. https://dotorg.brightspotcdn.com/f0/24/fdd2fdc64054bc5ea4529ec4e8cd/audit-and-financial-statement-2020.pdf

Special Olympics. (2022a). *Our mission*. https://www.specialolympics.org/about/our-mission?locale=en

Special Olympics. (2022b). *World Games*. https://www.specialolympics.org/our-work/games-and-competition/world-games

Special Olympics World Games Abu Dhabi 2019. (2019). https://www.abudhabi2019.org/

Sport Canada. (2020). *The role of Sport Canada*. https://www.canada.ca/en/canadian-heritage/services/role-sport-canada.html

Staples, L. (2020, January 8). Inside the fight for LGBT+ rights across the Commonwealth. *Independent*. https://www.independent.co.uk/news/long_reads/lgbt-gay-rights-commonwealth-mauritius-homosexuality-british-empire-a8912641.html

Sutcliff Group. (2016). *Canadian sport policy formative evaluation and thematic review of physical literacy and LTAD: Final report*. https://sirc.ca/wp-content/uploads/2019/12/formative_evaluation_thematic_review_physlit_ltad_2016.pdf

United States Amateur Sports Act. (1978). *United States Amateur Sports Act*. https://www.govtrack.us/congress/bills/95/s2727

US Senate S.2119 (105th): Olympic and Amateur Sports Act. (1998). *Committee clears legislation, nominations*. https://www.govtrack.us/congress/bills/105/s2119/summary

USA Basketball. (2020). *Consolidated financial statements*. https://www.usab.com/about/about-usa-basketball/governance.aspx

USA Basketball. (2021a). *Inside USA Basketball. Who we are*. https://www.usab.com/about/about-usa-basketball.aspx

USA Basketball. (2021b). *USA Basketball constitution*. https://www.usab.com/-/media/fb7219252d54457aae19f765aff3f1cf.pdf

WADA. (n.d.). *What we do*. https://www.wada-ama.org/en/what-we-do

Olympic Sport

Imagine what it must be like to strive to be the best in the world in your chosen sport: the years of preparation, the excitement of the competitions, the media coverage, the social media attention, the applause of fans, the travel, the agony of defeat, and the thrill of victory. Now imagine the feelings of competing at the Olympic Games, often described by athletes as the adventure of a lifetime. Without doubt, the Olympic

DOI: 10.4324/9781003303183-10

Games are one of the most significant sporting competitions in the world, scheduled every four years for both summer and winter events. Athletes at virtually every level dream of one day competing for their nation on the world stage in the Olympic Games. Winning an Olympic Gold Medal holds tremendous meaning worldwide. Not only does it signify the accomplishment of being the best in the world for the athlete(s) involved, Olympic Gold means instant recognition, fame, nation building, and sometimes financial success. For world leaders, Olympic Gold Medals are symbolic of success throughout the society the winner represents and can be used to promote the legitimization of political ideologies. No wonder the Olympic Games are held in such high regard and taken so seriously by nations around the world.

Citius, Altius, Fortius – Communiter ("Faster, Higher, Stronger – Together") is the motto of the modern Olympic Games. The word "Communiter" ("Together") is a fairly new addition to the motto. It was only added in July 2021 – at the height of the COVID-19 pandemic – to recognize "the unifying power of sport and the importance of solidarity" (IOC, 2022v, para. 2). The Summer and Winter Olympic Games alternate every two years so that four years (a *quadrennial*) constitutes a full cycle. Both Games are regularly viewed by billions of people. For example, the PyeongChang 2018 Games enjoyed the most extensive coverage ever produced for the Winter Olympic Games, reaching a record Winter Games audience of 1.92 billion people (that's 28% of the world's population!) (IOC, 2018b). The PyeongChang 2018 Games also doubled digital viewership from the previous Winter Games in Sochi, with 3.20 billion video views and more than 16 billion minutes viewed (IOC, 2018b). A global audience of almost half the world's population, 3.05 billion people, watched some portion of the 2020 Tokyo Summer Games (held, due to the COVID-19 pandemic, in 2021). Digital consumption of the 2020 Games reached new levels with 6.1 engagements across the nine Olympic social media handles, and the Games were the most watched Olympic Games ever on digital platforms, with 28 billion digital video views – a 139% increase from the 2016 Rio Summer Games (IOC, 2021f). For the most recent Games, the Beijing 2022 Winter Games, people from more than 220 countries and territories across the world had access to the over 6,000 hours of coverage produced for the event (IOC, 2022i).

The incredible global reach of the Olympic Games, especially given the popularity of digital consumption, makes them more than just a sporting event. They are a media extravaganza, a cultural festival, an international political stage, an economic colossus, and a location for developing friendships. Everyone strives for excellence, from the competing athletes to the host city. The Olympic Games are a showcase, and "Faster, Higher, Stronger – Together" reflects the essence of the event. This chapter looks at the history of the Games, their organization and governance, and the policy issues currently confronting organizers.

HISTORY OF THE OLYMPIC GAMES

T he history of the Olympic Games can be divided into two distinct timeframes. The Games originated in Ancient Greece, were discontinued for at least 1,500 years, and then were reinstituted in the late 19th century. In the early Olympic Games, the Ancient Greeks competed for the glory of their gods. Much later in history, in the so-called modern era of the Games, the ancient festival was reintroduced and evolved into the event we know today.

The Ancient Olympic Games

Early Greek civilizations loved athletics and assimilated strength and vigor with rhythm, beauty, and music in their style of games and pursuits (Nelson, 2007). The Greeks participated in contests and athletic events like chariot racing, boxing, wrestling, footraces, discus throwing, and archery. The first Olympic Games were held in 776 BC in Olympia, Greece and were celebrated again every four years until their abolition by the Roman Emperor Theodosius in 393 AD (Young, 2004). The four years between Games were called an *Olympiad*, a system upon which time was calculated in ancient Greek history (The Olympic Museum Educational and Cultural Services, 2013). Specific events changed over the centuries, but footraces, the pentathlon, boxing, and various types of chariot races were common. The ancient Olympic Games were restricted to free Greek men. It was not until the modern Olympic era that women were included (either as competitors or as spectators) and that people of different nations were allowed to compete.

The Modern Era of the Olympic Games

From 1859 until the actual revival of the Olympic Games in 1896, the idea of reinstituting the festival of the Olympiad was discussed by both Greek nationalist Evagelis Zappas and English citizen William Penny Brookes (Toohey & Veal, 2007). Baron Pierre de Coubertin of Paris visited with Brookes and is the individual now credited with successfully launching the modern Olympic Games. Baron de Coubertin believed strongly in the healthy mind–healthy body connection (de Coubertin & Müller, 2000). He envisioned amateur athletes from all around the world competing in a festival of sports similar to those of ancient Greece. In 1894, the Baron presided over a congress held at the Sorbonne in Paris. Representatives from 13 countries attended the meeting, and another 21 wrote to support the concept of reviving the Olympic Games (Toohey & Veal, 2007). The assembled nations unanimously supported the revival of the Greek Olympic festival, to be held every four years, and to which every nation would be invited to send representatives. The modern Olympic Games were reborn in 1896 and were held in Athens, Greece. Two hundred and forty-one athletes (all men) from 14 nations participated in nine sports (43 events) in the first modern Olympiad (IOC, 2022b).

Many traditions taken for granted in the Olympic Games today were born during the early modern Olympic Games, including the opening ceremony and the parade of nations into a stadium; the medal ceremonies and the flag raising of the Gold Medal-winning athlete; housing the participants in an Olympic Village at the site of the Games; and beginning and ending the Games with the lighting and the extinguishing of the Olympic flame, brought to the site from the ancient site in Olympia, Greece. Quickly, the Olympic Games became a world focus, and today athletes from all over the world compete in various sports for the glory of representing their nation. Since the revitalization of the modern Games, the Olympic Games have grown in size and complexity, requiring an increasingly sophisticated international governance structure. Exactly how are the Olympic Games governed, and how is an Olympic Games planned, organized, and managed?

GOVERNANCE

Mention "the Olympic Games" and competition, ceremony, and colors of the world come to mind, along with memories of spectacle and stories of unimaginable achievements. But from a sport management perspective, what makes these Games happen? In reality, an enormous amount of planning and coordination is required, in addition to volumes of policy that set standards for what and how the Games take place. Three main levels of organizational influence direct the Olympic Games as we know them. First, the Olympic Games are organized through the jurisdiction of the International Olympic Committee (IOC), led by its President. Second is the Session (an annual General Assembly or annual meeting). Finally, there is the Executive Board (similar to an Executive Committee). Bids to host the Olympic Games are made by National Olympic Committees (NOCs) from interested countries. Once a bid has been awarded to a particular country, the responsibility for organizing an Olympic Games falls upon the Organizing Committee for the Olympic Games (OCOG). Each, in turn, contributes significantly to the staging of the Olympic Games. A final important group of constituents in the Olympic Movement are the International Federations (IFs), which oversee each of the sports recognized by the IOC.

International Olympic Committee (IOC)

International Olympic
Committee
https://olympics.com/ioc/
overview

The IOC, founded on June 23, 1894, is a group of officials governing Olympic organization and policy. The members are elected at the Session from the worldwide sport community. The headquarters of the IOC is located in Lausanne, Switzerland. It is a nonprofit organization independent of any government or nation (just like the NCAA). Having a NOC does not guarantee that a country will be eligible to have an IOC representative. However, an effort is made to ensure that IOC membership represents geographical regions of the world. In addition, while countries that have hosted Olympic competitions are eligible to have two IOC members, some other countries choose not to fill their IOC seat, and they are not required to do so. Some countries have more than two IOC members because those additional members head an International Sports Federation for one of the sports on the program of the Olympic Games. Some IOC members have no nationality or NOC affiliation at all. This change was made possible through an amendment to the Olympic Charter during the 2021 Session in Tokyo, which led to refugee Yiech Pur Biel being elected as an IOC member in 2022 – a first in Olympic history (IOC, 2022k). Keep in mind that all IOC members are elected to their positions and serve as representatives to the IOC to promote Olympism. They are not required to reside in the country for which they are a delegate to the IOC.

MISSION. The roles of IOC members are specific: first, they are expected to serve the Olympic Movement (that is, to promote the tenets of Olympism as outlined in the Olympic Charter) by helping to organize and govern policy relative to the staging of the Olympic Games. They are also expected to further the cause and understanding of all things associated with the Olympic Movement in their respective countries. IOC members are not representatives of their *nation* to the IOC. Rather, they are representatives of the *IOC* to their nations. They are expected to care first and foremost about what is best for the Olympic Games and work only on promoting Olympism and furthering the Olympic Movement as a whole. As one of the largest sport governing bodies in the world, the IOC is driven by its vision of "building a better world through sport" (IOC, 2022j, para. 3). The fundamental mission of the IOC includes ensuring the regular celebration of the Olympic Games, putting athletes at the center of the Olympic Movement, and promoting Olympic values (and sport in general) in society (IOC, 2022j). The IOC is also committed to making sure that the integrity of sport is strengthened and clean athletes, sport organizations, and stakeholders are supported (IOC, 2022j).

To achieve these goals, the IOC has developed a Strategic Roadmap, *Olympic Agenda 2020+5*, which presents strategic recommendations to safeguard the Olympic Movement and strengthen sport in society. *Olympic Agenda 2020+5* is the successor of the IOC's previous strategic plan *Olympic Agenda 2020*, which achieved key goals such as changing the Games hosting application procedures, reducing the costs of bidding for the Games, launching an Olympic channel, fostering gender equality, and strengthening IOC governance principles (IOC, 2022n). Building on this previous strategic plan, *Olympic Agenda 2020+5* identifies five key trends as "likely to be decisive in the post-coronavirus world" and "areas where sport and the values of Olympism can play a key role in turning challenges into opportunities" (IOC, 2022o, para. 3). The five trends are (1) need for more solidarity among societies, (2) growing digitalization, (3) the urgency of sustainable development, (4) need for more credibility of institutions and organizations within the Olympic Movement, and (5) economic and financial resilience (IOC, 2022p). To address these trends, *Agenda 2020+5* maps out 15 recommendations for the IOC and the Olympic Movements at large, which are presented in Exhibit 10.1.

MEMBERSHIP. Historically, IOC members were elected by the other members of the committee, a practice that labeled the committee

Olympic Charter
https://olympics.com/ioc/olympic-charter

Olympic Agenda 2020 Closing Report
https://olympics.com/ioc/documents/international-olympic-committee/olympic-agenda-2020

Olympic Agenda 2020+5
https://olympics.com/ioc/olympic-agenda-2020-plus-5

exhibit   **10.1**   15 Recommendations of the Olympic Agenda 2020+5

1. Strengthen the uniqueness and the universality of the Olympic Games
2. Foster sustainable Olympic Games
3. Reinforce athletes' rights and responsibilities
4. Continue to attract the best athletes
5. Further strengthen safe sport and the protection of clean athletes
6. Enhance and promote the Road to the Olympic Games
7. Coordinate the harmonization of the sports calendar
8. Grow digital engagement with people
9. Encourage the development of virtual sports and further engage with video gaming communities
10. Strengthen the role of sport as an important enabler for the UN Sustainable Development Goals
11. Strengthen the support to refugees and populations affected by displacement
12. Reach out beyond the Olympic community
13. Continue to lead by example in corporate citizenship
14. Strengthen the Olympic Movement through good governance
15. Innovate revenue generation models.

Source: IOC (2022o).

as elitist, incestuous, and existing for the gratification of its members. In the beginning, the committee was an extended group of friends and business associates of the original members, mostly from the upper class of society. Following the corruption allegations associated with the 2002 Salt Lake City Winter Olympic Games bid, the IOC changed some of its procedures. For instance, the IOC is now composed of 103 members (IOC, 2022l). Members of the IOC are allowed to serve until age 70 (except for members who started terms between 1966 and 1999 who may serve to age 80), although some choose to retire earlier. Positions are still elected by the members of the General Assembly, individual members, active Olympic athletes elected by their peers at the Olympic Games, members from IFs, and NOCs. Restrictions limit the numbers of a particular group being from the same country or Federation.

FINANCIALS. The IOC generates extensive revenues through its ownership of the rights to the Olympic Games and associated marks and terminology. These include the Olympic symbol, consisting of the five interlocking Olympic rings, the Olympic motto, anthem, flag, and the Olympic flame and torch. Permission to use these symbols is granted to the host organizing committee, and NOCs are permitted to use the rings in developing their own national Olympic symbol. In the United States, for example, a special statute requires the United States Olympic and Paralympic Committee's (USOPC) consent to all commercial uses of Olympic-related marks and terminology. Countries hoping to host the Olympic Games guarantee a percentage of the money they raise be turned over to the IOC in return for the rights to host. The size of the IOC's share became an issue after the 1984 Summer Olympic Games in Los Angeles. These games generated a surplus of $225 million. The IOC was unsuccessful in getting a share of the revenue, but it intensified its resolve to get a fair share of Olympic revenues from future hosting rights and acted to establish its own sources of income through marketing the Olympic symbols, with amazing success (Senn, 1999). According to the IOC (2022a), the organization's revenue is mostly generated through Broadcasting Rights (61%) and The Olympic Partner (TOP) Marketing Rights (30%), but other licensing rights (5%) and revenue streams (4%) such as ticket sales also contribute (IOC, 2022a). The IOC manages broadcast and sponsorship programs and the OCOGs manage domestic sponsorships, ticketing, and licensing within the host country, under the direction of the IOC. To give you an idea of the economic scope of the IOC, its total revenue for the 2017–2020/21 Olympiad was US$7.6 billion, which was an increase from the US$5.7 billion the IOC made in revenue during the previous Olympiad spanning from 2013–2016 (IOC, 2022c). As you can see, the organization is in a healthy financial situation despite the financial hardships resulting from the COVID-19 pandemic (IOC, 2022a). As a nonprofit association, the IOC distributes 90% of its revenue back to the Olympic Movement in some form (e.g., staging the Olympic Games and Youth Olympic Games, financially supporting the World Anti-Doping Agency, the NOCs, and IFs, and athlete development initiatives), while 10% is kept for operating the IOC (IOC, 2022a).

ORGANIZATIONAL STRUCTURE. Three components are central to IOC governance and the development of policy: the Session, the Executive Board, and the Office of the President. The IOC also occasionally creates IOC Commissions for the purpose of advising these three governance units.

1. *The Session.* The Session, comparable to a General Assembly or parliament, is a regularly scheduled meeting of all IOC members. As the "supreme organ of the IOC" (IOC, 2021a), the most important decisions affecting the Olympic Movement are made by the IOC Session. For example, the Session is in charge of adopting, modifying, and interpreting policy relating to the Olympic Charter. The Olympic Charter, the official governing document of the Olympic Movement, includes the purpose and description of the ideals of Olympic participation, along with the rules and regulations for Olympic events, membership in and recognition by the IOC (IOC, 2021d). The Olympic Charter provides the framework which governs the organization and operation of the Olympic Movement and stipulates conditions for hosting the Olympic Games. Elections for accepting new IOC members as well as for the IOC President, Vice-Presidents, and members of the IPC Executive Board are also held during the Session. Meetings are held annually unless unusual circumstances dictate the calling of a special meeting.

   The Session is also responsible for two other vital tasks – approving the choice of the host cities for upcoming Games and approving the sports that will be competed in as part of the Olympic Program for upcoming Games.

2. *The Executive Board.* The Executive Board is a smaller subset of the Session and is responsible for the management and overall direction of the IOC between meetings of the Session.

WWW

IOC Olympic Marketing Fact File (2021 Edition)
https://stillmed.olympics.com/media/Documents/International-Olympic-Committee/IOC-Marketing-And-Broadcasting/IOC-Marketing-Fact-File-2021.pdf?_ga=2.266644045.256442538.1656929888-547494262.1656929888

It was first conceived by Baron de Coubertin in order to share the responsibility for directing the IOC and to prepare for an orderly succession of leadership (Senn, 1999). Executive Board membership (Exhibit 10.2) includes 15 positions: the President, four Vice-Presidents, and 10 additional members elected by the Session. Each Executive Board member's term of office is four years, with the exception of the President, who is elected to an eight-year term. Members of the IOC Executive Board can serve no more than two consecutive terms, and members can be elected again after a two-year waiting period. The board meets regularly, at the call of the President or at the request of a majority of its members. The Executive Board of the IOC has the following specific responsibilities (IOC, 2022h):

- holds overall responsibility for administration of IOC
- ensures that the Olympic Charter is observed, implemented, and promoted
- administers the IOC and appoints the Director General, who oversees the daily business affairs of the IOC
- manages IOC finances and financial reporting
- formulates bylaw or rule changes for implementation by the IOC Session
- approves the organizational chart and internal operations of the IOC
- makes recommendations for elections
- establishes the agenda for all IOC meetings
- enacts all regulations for the proper organization of the Olympic Games
- organizes meetings with IFs and NOCs
- maintains the records of the IOC
- creates and allocates IOC honorary distinctions

3. *The Office of the President*. The President of the IOC is elected from members of the Session. The term of office is initially eight years, and the incumbent President may be re-elected for one subsequent four-year term. In the modern era, the Olympic Games have had only eight Presidents (see Exhibit 10.3). It is a critical position with power and responsibility for directing the general course of the IOC. The President is the official spokesperson of the IOC and presides over the Executive Board. It is the President's role to convene the Executive Board and lead the business of the IOC. In addition, the President has the ability to establish IOC Commissions and working groups as necessary. The President is an ex officio member of all commissions.

4. *IOC Commissions*. IOC Commissions are important tools when it comes to policy-making and governance in the IOC. They serve as units providing expertise around a particular topic of interest to the IOC. As such, IOC Commissions advise the Session, the

www

IOC Commissions
https://olympics.com/ioc/commissions

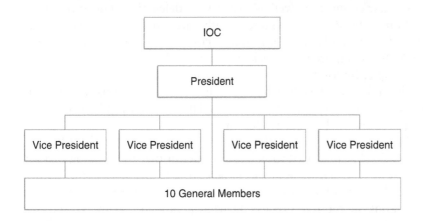

Source: IOC (2022h).

IOC Presidents — exhibit 10.3

President	Country of Origin	Years of Service
Dimitrius Vikelas	Greece	1894–1896
Pierre de Coubertin	France	1896–1825
Henri Baillet-Latour	Belgium	1825–1842
J. Sigfrid Edstrom	Sweden	1842–1846 (acting) 1946–1852
Avery Brundage	United States	1952–1972
Lord Killanin	Ireland	1972–1980
Junan Antonio Samaranch	Spain	1980–2001
Jacques Rogge	Belgium	2001–2013
Thomas Bach	Germany	2013–

Source: Olympic Museum (2022).

Executive Board, and the President. They are made up of both IOC members and external experts. As of 2021, there were 31 IOC Commissions, including the Athletes' Commission, the Ethics Commission, the Finance Commission, the Future Host Commissions, Olympic Channel Commission, Olympic Education Commission, the Sustainability and Legacy Commission, and the Women in Sport Commission (IOC, 2022f).

You may be surprised to hear that the officers of the IOC do not actually organize the Olympic Games. Rather, the IOC works with the groups responsible for Olympic sport and hosting activities within individual nations, NOCs, and OCOGs.

National Olympic Committees (NOCs)

NOCs control operations and policy relative to the Olympic Games for a particular country, as well as oversee the delegation sent to represent a nation at the Olympic Games. With rare exceptions, only athletes certified by an NOC are permitted to compete at the Olympic Games. The NOC is required to check participant eligibility rules as defined by Olympic, International Sport Federation, and NOC policies. NOCs have been described as "the basic building blocks in the structure of the Olympic Games" (Senn, 1999, p. 11). There are currently 206 NOCs recognized by the IOC (IOC, 2022m).

As part of the substantial changes mapped out in the *Olympic Agenda 2020*, the IOC responded to criticism that the bidding process for hosting the Games was too expensive and narrow. For example, until 2019 NOCs had to select one host city to focus their attention on throughout the bidding process. In what it calls the "New Norm," the IOC has since mapped out a new process to reduce costs for submitting host proposals and provide more flexibility for NOCs interested in hosting the Olympic Games (including the ability to look into multiple cities potentially co-hosting). Through a multi-step, shared dialog approach, the IOC works with NOCs in determining if a nation is well-suited to host the Games (see Policy section of this chapter).

Changes in the bidding process have been long in the making. For example, in 2017, the IOC made history by selecting two Summer Games hosts at the same Session in Lima, Peru. Both the 2024 and 2028 Summer Olympic Games hosts were elected, with Paris, France, unanimously elected to host the 2024 Games and Los Angeles, United States, unanimously elected to host the 2028 Summer Games. The IOC described this historic approach as a tripartite agreement, bringing stability to the Olympic Movement by announcing Olympic host cities well in advance to the athletes of the world. In 2021, Brisbane was announced as the host of the 2032 Summer Games, making the city – and the broader Queensland region – "the first future host to have been elected under, and to have fully benefited from, the new flexible approach to electing Olympic hosts" (IOC, 2021b, para. 6).

United States Olympic & Paralympic Committee (USOPC)

The USOPC was known as the United States Olympic Committee (USOC) until 2019 when it changed its name to the USOPC to acknowledge its governance of the Paralympic Movement in the United States (see Chapter 11). The USOPC is a federally chartered nonprofit organization which governs, manages, promotes, and liaises within and outside the United States for all activities of the Olympic, Paralympic, Youth Olympic, Parapan American, and Pan American Games. In 1978, the US Congress passed the Amateur Sports Act, which was amended in 1998 and is now called the Ted Stevens Olympic and Amateur Sports Act. The amended law includes activities associated with the Paralympic Games and addresses athletes' rights and other matters. The law specifically mandates the USOPC to govern all US activities for the Olympic and Paralympic Games. Pursuant to this federal act, the USOPC has the exclusive right to use the Olympic symbols, marks, images, and terminology in the United States. Further, it has the power to authorize the use of the Olympic marks with sport governing bodies sending athletes to represent the United States at competitions and to act as a coordinating body for Olympic sport within the country. The USOPC is composed of a group of individuals and organizations whose common goals are athletic excellence and achievement on the world stage and promoting nation building through the achievement of athletes.

Although more than a century old at the time, the year 2003 represented a historic time for the USOPC. In 2002, the USOPC's President was forced to resign because of misstatements on her resume. In 2003, the organization was confronted with allegations of violations of its Code of Ethics by its CEO, which led to infighting between the CEO and the President of the organization. The good name of the USOPC was tarnished, and the image of the association was at an all-time low. The US Congress even voiced concern: three Senators requested an independent commission be appointed to investigate the practices of the organization and recommend change. In addition, the USOPC appointed a Governance and Ethics Task Force to recommend a course of action for changing the practices, mandate, and expectations of the USOPC. The sizes of the Board of Directors and Executive Committee were particularly criticized, along with the breadth of the USOPC's all-encompassing mandate, which extends very broadly beyond training athletes, building facilities, and designing equipment.

The independent commission and the Governance and Ethics Task Force focused their recommendations on these three major issues (Sandomir, 2003):

1. narrowing the USOPC mandate to focus on training athletes for national and international competition related to the Olympic and Paralympic Games
2. ensuring ethical, responsible, and transparent business and financial practices
3. creating a workable governance structure that better defines the responsibility of volunteers and professional staff and that reduces and changes the numbers and constituents involved in decision-making

They recommended that (1) the mission, goals, and objectives of the organization be focused to ensure that the ideals of the Games be preserved and reflected in practice and conduct; (2) the governance structure of the USOPC be clearly redefined concerning responsibilities, authority, and accountability; (3) the overall governance structure of the USOPC be streamlined and downsized; and (4) that ethical policy and compliance with ethical policy be instituted (US Senate Report, 2003).

In the fall of 2020, even more substantive changes came to the USOPC in the wake of organization's complicity in the USA Gymnastics sexual abuse scandal involving national team doctor Larry Nassar, who had sexually abused US gymnasts for more than three decades. The scandal brought into the public eye the role of institutions such as the USOPC in protecting athletes (or, to be more accurate, their failure to do so). To protect future athletes from suffering such systemic abuse, the US Congress passed "The Empowering Olympic, Paralympic and Amateur Athletes Act of 2020" as an amendment to the Ted Stevens Act. The new law granted Congress the power to decertify Olympic and Paralympic governing bodies, mandated increased representation of athletes in USOPC governance, and provided more funding for the protection of athletes through the US Center for SafeSport (Conrad, 2021). In addition, it established the Commission on the State of US Olympics and Paralympics, tasked with conducting an extensive review of the governance of the US Olympic and Paralympic Movements. In response to the new law, Congressional investigations, and a report from the Borders Commission, an independent panel charged with reviewing the USOPC's governance strategy, policies, and procedures, the USOPC announced another set of sweeping governance reforms. The results from each of these initiatives facilitated drastic changes to the governance of the organization, including refining its mission, size and composition of the Board of Directors, election procedures, and role of the Chairperson, Board, CEO, and Assembly.

MISSION. The recent name change from USOC to USOPC reflects a larger shift in culture of the USOPC in the past years. Responding to calls to serve the holistic needs of its athletes, the USOPC updated its mission statement to the following in 2019: "Empower Team USA athletes to achieve sustained competitive excellence and well-being" (USOPC, 2022a, para. 3). The mission statement is also accompanied by a vision statement ("inspire and unite us through Olympic and Paralympic sport"; USOPC, 2022a, para. 2) and global purpose statement ("join our global peers in building a better, more inclusive world through sport"; USOPC, 2022a, para. 1). It is no accident that the USOPC's mission, vision, and global purpose statements connect to the themes and the meanings of the Olympic Games as outlined by the IOC. They speak to the ideals of Olympism, the promotion of ethical conduct, and peace among nations achieved through sport competitions. The USOPC hopes to drive national unity and pride within the United States through the accomplishments of US athletes in competition with their peers from other countries, while making sure athletes compete in a safe and healthy environment. Six core principles guide the USOPC in achieving its mission, which are presented in Exhibit 10.4.

MEMBERSHIP. Members of the USOPC are properly qualified organizations who have the authority for managing sports represented in the Olympic, Paralympic, Youth Olympic, Parapan American, and Pan American Games, or those that widely promote the participation and engagement in amateur sport in the United States. A number of requirements of membership are set out in the USOPC bylaws and membership remains fairly static year after year. There are 50 US National Sport Governing Body' (NSGB) members of the USOPC: 37 Olympic summer sport NSGBs, eight Olympic winter sport NSGBs, and five Pan American sport NSGBs. In addition, the USOPC's Affiliate

| USOPC Core Principles | *exhibit* **10.4** |

- We promote and protect athletes' rights, safety, and wellness.
- We champion the integrity of sport.
- We respect the important role of our member organizations and support their success.
- We set clear standards of organizational excellence and hold ourselves and all member organizations accountable.
- We engage as a trusted and influential leader to advance the global Olympic and Paralympic Movements.
- We honor and celebrate the legacy of Olympic and Paralympic athletes.

Source: Team USA (2022).

Organization Council (AOC) brings together members in the following categories: community-based multisport organizations, education-based multisport organizations, armed forces organizations, recognized sport organizations, and adaptive sport organizations/other sport organizations. Organizations that are strictly commercial or political are not eligible for membership. The power to elect members of the USOPC resides with its Board of Directors (USOPC, 2021).

FINANCIALS. The USOPC has the use and authority to authorize the use of the Olympic marks within the United States. It licenses that right to sponsors, thereby generating significant revenues in support of its mission. According to its 2021 financial statements outlined in its annual Impact Report, the organization's main sources of revenue are sponsorship and licensing (45%), television broadcast revenue (42%), contributions (11%), investment income (1%), and other revenue (1%) (USOPC Impact Report, 2021). Further, significant revenue comes to the USOPC from the IOC. These revenues are largely the result of lucrative television and sponsorship deals from broadcasting the Olympic Games. The organization has extensive reach in the corporate world, with a corporate partnership and advertising program contributing high yearly revenues. The USOPC also established the US Olympic & Paralympic Foundation (USOPF) as its primary fundraising arm in 2013. Because the USOPC does not receive any financial support from the US government (a key difference to other NOCs, which receive financial support from their federal governments), the organization relies on philanthropic support from the US public. In 2021, the consolidated financial statements for the USOPC and USOPF show revenues in excess of $459 million (USOPC Impact Report, 2021).

The USOPC's spending is primarily focused on three main areas: (1) supporting athlete excellence (which includes giving money to the National Governing Bodies (NGBs) that govern them), (2) sport advancement, and (3) community growth. Accounting for almost 60% of all USOPC expenses at over $200 million in 2021, the *athlete excellence* category includes athlete stipends and funding for NGBs, athlete services, and competition preparation. For example, the USOPC gave

WWW
USOPC 2021 Impact Report
https://2021impactreport.
teamusa.org/index.html#gsc.
tab=0

WWW
**USOPC 2021 Impact Report:
Financial Summary**
https://2021impactreport.
teamusa.org/financials-
reports-and-disclosures/
financials.html#gsc.tab=0

$110 million in high-performance grants to NGBs, directly impacting athletes (USOPC Impact Report, 2021). Nearly $43 million was spent on *sport advancement*, which includes initiatives to support athletes' safety and access to sport (USOPC Impact Report, 2021). Finally, the USOPC spent $26.5 million on *community growth* initiatives (e.g., digital support for NGBs). After facing less revenue in 2019 and 2020 due to the COVID-19 pandemic and falloffs in TV revenue during the two non-Olympic years (with $194 million in 2019 and $179 million in 2020, respectively), the USOPC made a surplus of $113 million in 2021 (USOPC Impact Report, 2021). As you can see, the organization is now in a robust financial situation heading into the next Olympiad.

ORGANIZATIONAL STRUCTURE. The USOPC employs more than 350 staff members. The organization is structured such that an appointed group of officers forms the Board of Directors and various other committees provide direction to the staff who implements policy. In addition, volunteers involved with the USOPC include some of the most influential leaders in both sport and business from around the United States.

Officers. Two of the most important leaders of the USOPC are the Chair of the Board of Directors (an elected volunteer) and the CEO (a paid employee). Both fulfill leadership roles in establishing USOPC policy and are the principal spokespersons for the organization. The CEO is also responsible for day-to-day operations, strategic policy initiatives and directions, and management of the professional staff (USOPC, 2022b). Day-to-day operations are led by the USOPC Executive/Senior Level Officials and Managers, a group of 34 professional administrators in addition to the CEO, overseeing staff in areas such as sport performance, marketing, diversity/equity/inclusion, finance, and human resources. Per the USOPC bylaws, in addition to the CEO, the organization's general counsel and chief financial officer serve in official officer capacities, with the former serving as Secretary and the latter serving as Treasurer (USOPC, 2021).

Board of Directors. The task force investigating USOPC reform began working on a new leadership structure for the organization in February 2003. The organization was viewed as being too large and overly bureaucratic. To overcome these issues, the task force recommended that the Board of Directors be reduced in size (from 125 members to 11, which has since been amended to 16 voting members), the numbers of standing committees be reduced, the Executive Committee be eliminated from the governance structure, delineation of roles and responsibilities be enacted, and a US Olympic and Paralympic Assembly be created. Like many other sport organizations, a two-tiered governance structure is used to manage USOPC affairs. The Board of Directors has the ultimate authority and responsibility for the finances, policy development, election of officers, and activities of the USOPC. The Board has the authority to amend the constitution and bylaws of the USOPC, admit and terminate members, and receive and review reports from committees and members. The Board of Directors meets a minimum of four times each year, one meeting of which must be held in association with the Olympic and Paralympic Assembly. Exhibit 10.5 provides a list of the 2022 Board of Directors, included to show the Board's composition and the Board Members' professional backgrounds.

Membership of the 2022 USOPC Board of Directors, Reflecting Diverse Backgrounds / *exhibit* 10.5

NAME	POSITION	PROFESSIONAL BACKGROUND
Susanne Lyons	Chair & Independent Director	Business Executive
Anita DeFrantz	IOC Member, *ex officio*	Olympian, President of Kids in Sport
David Haggerty	IOC Member, *ex officio*	President, International Tennis Federation
Rich Bender	NBC Council-Elected Director	Executive Director, USA Wrestling
Cheri Blauwet	Independent Director	Paralympian, Sports Medicine Director
Beth Brooke	Independent Director	Co-Chair, International Council on Women's Business Leadership
Gordon Crawford	USOPF Chair, *ex officio* (*non-voting member*)	Chair, USOPF
Muffy Davis	IPC Member, *ex officio*	Paralympian, Member of IPC Governing Board
Donna de Varona	At-Large Athlete-Elected Director	Olympian, Member of IOC Communications Commission, Founder of Women Athletes Business Network
James Higa	Independent Director	Executive Director, Philanthropic Ventures Foundation
Steve Mesler	AAC-Elected Director	Olympian, CEO of Classroom Champions
John Naber	At-Large Athlete-Elected Director	Olympian, Broadcasting Professional
Dexter Paine	NGBC-Elected Director	Chair, Paine Schwartz Partners
Daria Schneider	AAC-Elected Director	Head Coach, Harvard Fencing
Brad Snyder	AAC-Elected Director	Paralympian, Retired US Naval Officer
Kevin White	NGBC-Elected Director	Former Athletic Director, Duke University
Robert L. Wood	Independent Director	Board Member, Praxair Corporation, MRC Global, and Univar
Sarah Hirshland	USOPC CEO, *ex officio* (*non-voting member*)	CEO, USOPC

Source: USOPC (2022c).

Olympic and Paralympic Assembly. The Olympic and Paralympic Assembly is held once per year and is an event where all constituent groups of the USOPC gather to discuss the achievements of the organization and communicate to the Board of Directors. According to the USOPC Bylaws (USOPC, 2021), the purpose of the Olympic and Paralympic Assembly is to facilitate communication among the Board and USOPC members and constituent groups, and the Advisory Committees of the USOPC. Information on USOPC organizational and financial performance, preparations for the Olympic, Pan American, Parapan America, and Paralympic Games, and actions taken are presented and discussed. Although input is sought, the Olympic Assembly does not conduct or perform any governance functions (USOPC, 2021).

Constituent councils and other committees. Three constituent councils serve "as sources of opinion and advice" (USOPC, 2022c, para. 1) and deal with specific areas of importance, interest, and concern to the USOPC. These are the (1) Athletes' Advisory Council (AAC), (2) NGB Council (NGBC), and (3) AOC. The AAC brings together athlete representatives from Olympic, Paralympic, and Pan American sport to communicate active athletes' insights to the USOPC. Similarly, the NGBC represents the views of the NSGBs to the organization. The AOC, previously called the Multi-Sport Organization Council, comprises representatives from the 35 affiliate organizations of the USOPC. In addition to the three constituent councils, the USOPC created the Paralympic Advisory Council in 2011 to focus on enhancing programming and resources related specifically to the Paralympic Movement. Additionally, a variety of standing committees and task forces help drive the work of the USOPC. Standing committees include those focused on athlete and NGB services, ethics, finance, and NGB oversight and compliance, among others (USOPC, 2021). Task forces can be focused on particular topics of interest to the organization. For example, in 2020, the USOPC launched an inaugural Team USA Council on Racial and Social Justice to identify ways in which the organization may work to overcome barriers to racial and social justice in the Olympic and Paralympic Movements and beyond. You can find a more in-depth overview of the Council's work in our diversity, equity, and inclusion (DEI) spotlight in this chapter.

Organizing Committee for the Olympic Games (OCOG)

OCOGs are another vital component of the Olympic structure. An OCOG is formed within a community after it has successfully won the bid to host the Olympic and Paralympic Games. (As part of the bid process, the host city is responsible for staging both the Olympic and Paralympic Games.) The work for the OCOG begins many years (perhaps 10–15) in advance of the actual event. The predecessor to this committee prepares the bid and plans for all aspects of hosting. With the support of the host country's NOC, the bid is submitted to the IOC and judged on many criteria. If the bid is unsuccessful, the bid committee finalizes its affairs and dissolves. The bid committee that is awarded the hosting rights becomes an OCOG and continues and intensifies planning. Some examples of successful OCOG bids are the Paris 2024 (France) Organizing Committee which is organizing the 2024 Summer Games, the Milano Cortina 2026 (Italy) Organizing Committee for the 2026 Winter Games, and the Los Angeles 2028

www
Paris 2024 OCOG
https://www.paris2024.org/en/

www
Milano Cortina 2026 OCOG
https://www.milanocortina2026.org

www
Los Angeles 2028 OCOG
https://la28.org

(United States) Organizing Committee for the Summer Olympic Games in 2028.

Given that the Summer and Winter Olympic Games occur in a different location every four years, OCOGs are developed to manage one event at a time. As mentioned in the discussion on NOCs, bids for hosting the Games are made well in advance of the event to allow for facility development and proper event planning. Developing a bid is a process that takes years from idea to concept to plan to operationalize. Although the bid is presented to the IOC by a city or region interested in hosting, the OCOG must demonstrate the support of various levels of government and the NOC. Once the application to bid is supported, the OCOG will move into action to prepare for the Games and will stay active for about a year after the event to finalize all financial accounting, resolve ongoing legal issues, and file final reports. In total, members of the OCOG are likely to be involved in some stage of bidding, planning, executing, or reporting for 10–15 years. At any given time, four OCOGs (two for the Winter Games and two for the Summer Games) are in some stage of this process. Many OCOG employees come from the host city, while some sport and event managers travel the world with the Olympic Games, moving from position to position as specialists in some capacity with the organizing committees.

MISSION. The mission of an OCOG clearly reflects the ideals of the Olympic Movement: to be the best, to host the best, to show the world the best Olympic Games ever. When the President of the IOC speaks at the Olympic Games closing ceremonies, the local organizers listen carefully. They want to hear something to the effect of "This was the best Olympic Games ever," which is the goal of every OCOG. The mission of each OCOG is often best captured by its vision or Olympic motto. An example of a vision statement for the 2020 Tokyo, Japan Summer Games is presented in Exhibit 10.6. In addition to the vision, an OCOG often chooses a Games Motto to capture the essence of what the Games represent to the host country and what sentiment the host nation wants to share with the rest of the world. Often, these mottos speak to the unique cultural moment of the times the Games are held. For instance, the Games Motto for Tokyo 2020 was "United by Emotion." At a time of growing political tensions and a brutal, raging pandemic, the motto served as a reminder that the Olympic and Paralympic community "will come together and understand there is more that unites than divides us" (Tokyo 2020, 2022, para. 9). (Of course, these mottos are often very positive, and some host nations have been called out for their hypocrisy given their horrific track records on human rights, as you will see later in the Policy section of this Chapter.)

Putting the mission and motto of such a high-scale sport mega event into action requires an enormous number of administrative tasks. Some of the main administrative tasks of each OCOG include ensuring that competitions adhere to the rules set by international sports federations, choosing (or creating) competition sites, providing required equipment, arranging accommodation for athletes, organizing medical services, engaging with mass media, celebrating cultural events reflecting

| exhibit | 10.6 | Example vision statement of an Olympic Games Organizing Committee: Tokyo 2020 Summer Olympic Games (held in 2021 due to COVID-19 pandemic) |

Sport has the power to change the world and our future.
The Tokyo 1964 Games completely transformed Japan. The Tokyo 2020 Games, as the most innovative in history, will bring positive reform to the world by building on three core concepts:

"Striving for your personal best (Achieving Personal Best)"
"Accepting one another (Unity in Diversity)"
"Passing on Legacy for the future (Connecting to Tomorrow)"

Source: Tokyo 2020 (2019).

the spirit of the Games, and writing regular reports, including a Final Report due no later than two years after conclusion of the Games (IOC, 2022r).

MEMBERSHIP AND ORGANIZATIONAL STRUCTURE. OCOG members include both paid professional staff and volunteers. If some day you want to work at the Games, an OCOG is where you would want to find yourself working in some capacity. The OCOG is usually led by a volunteer President and Board of Directors, and a staff CEO. The OCOG is then subdivided into areas of responsibility such as finance, facility development, sports, technical liaison, marketing and sponsorship, volunteers, security, ticket sales, merchandising, television, doping control, and so forth.

Each of the governing structures listed above plays a role in setting or enforcing policy for the Olympic Games. The IOC is charged with this mandate, and the NOCs and OCOGs help enact it. Policy is often under public scrutiny, depending upon the prevailing issues of the day. For example, the 1999 scandal associated with IOC members accepting bribes from the Salt Lake City Organizing Committee resulted in a focus on procedures for choosing the host city. The presence of doping, as illustrated by positive drug tests at virtually every recent Olympic Games, continues to focus attention on testing procedures. The threat of terrorism has resulted in heightened security measures and the scrutiny of security procedures. The participation of the Refugee Olympic Team at the Rio 2016 Summer Games served to highlight the magnitude of the worldwide refugee crisis. Issues related to sustainability and athletes' rights are also commonly debated. In 2020, with the world grappling with the emerging COVID-19 pandemic, policy related to athlete health and safety took center stage. These and other current policy issues are discussed later in this chapter.

FINANCIALS. The Olympic Games are run like a business. The goal is to have surplus funds available after the Games in order to leave a legacy for the next host city and country. An inordinate amount of money goes into hosting, and it is raised through various sources, including government grants and funding, television rights, corporate sponsorship and advertising, licensing and sale of merchandise, ticket sales, and other marketing and special event functions. For example, the total costs associated with hosting the postponed 2020 Summer Games in Tokyo (delayed to 2021 due to the COVID-19 pandemic) was about $13 billion, while the cost for hosting the 2022 Beijing Winter Games was estimated at $38.5 billion (Teh & Stonington, 2022). In contrast, the 2014 Olympic Winter Games in Sochi, Russia cost about $51 billion, the costliest Olympic Games in history (Ahmed & Leahy, 2016).

It is also worth noting that hosting an Olympic Games has the potential for positive financial impact for the host city generally referred to as economic impact. Economic impact is defined as additional expenditures in a geographical area generated by actions such as spectators coming to an event who stay in a hotel, eat in restaurants, buy tickets to events, travel, and make purchases. It includes expenditures made by organizers of the event as well. Consider the jobs involved in building new facilities and community infrastructure. The British government issued a report, although with some skepticism from economists, indicating that the UK economy saw a £9.9 billion boost in trade and investment from hosting the 2012 London Olympic and Paralympic Games (BBC News, 2013). The question of short-term financial gains from tourism and job creation is often offset, however, by enormous levels of long-term debt incurred from cost of facility construction (Zimbalist, 2016) and diversion of funds from other needed local construction projects (e.g., schools).

International Sport Federations

A final group of constituents in the Olympic Movement is IFs. These federations are international nongovernmental organizations that oversee the sports recognized by the IOC. Once recognized by the IOC, they must conform with the Olympic Charter. IFs are tasked with the day-to-day administration of their respective sports at all levels, which includes development of athletes competing in the sport, promoting the sport globally, organizing regular competitions, and guaranteeing a fair playing field in the sport (IOC, 2022e). Examples of IFs include the International Basketball Federation, International Fencing Federation, International Judo Federation, International Swimming Federation, and World Athletics.

A process of continuous communication between the IOC and IFs is crucial to make sure the sports governed at the world level adhere to consistent standards when it comes to topics like safety, fairness, and inclusion. In order to provide a vehicle to discuss common issues, determine event calendars, and communicate with the IOC, IFs have organized themselves through three associations: (1) the Association of Summer Olympic International Federations for sports competing at the Summer Olympic Games, (2) the Association of International Olympic Winter Sports Federations for sports competing at the Winter Olympic Games, and (3) the Association of IOC Recognised International Sports Federations (ARISF). IFs often provide valuable insights on policy areas in Olympic sport, which we will turn our attention to in the next section.

www

International Sports Federations in Olympic Sport
https://olympics.com/ioc/international-federations

Spotlight: Diversity, Equity, and Inclusion in Action

Team USA Council on Racial and Social Justice

In the United States, continued calls for racial justice in the wake of brutal murders of Black Americans at the hand of law enforcement led to increased scrutiny on sport organizations' stances on such systemic injustices. These calls impacted the USOPC, which launched its inaugural Team USA Council on Racial and Social Justice in 2020. The athlete-led Council brought together more than 40 current and former Team USA athletes, NGB representatives, USOPC representatives, and outside experts to "address the rules and systems in the U.S. Olympic and Paralympic Movements that create barriers to progress ... with the aim of eradicating social injustice and cultivating change through strengthened athlete voices" (USOPC, 2020, para. 1).

The Council's work led to some historic changes in Olympic sport, most notably a change in the USOPC's stance on athlete protests and demonstrations. However, advocating for athletes' right to protest was only one facet of the Council's work. Over the span of two years, the Council published four sets of recommendations focused on (1) athlete protests and demonstrations, (2) athlete expression and advocacy, (3) institutional awareness and cultural change, and (4) antiracism and acts of discrimination. Via 32 recommendations across these four focus areas, the Council established a roadmap for the USOPC, NGBs, and other stakeholders (e.g., sponsors, fans) to anchor social justice in their organizational praxis.

CURRENT POLICY AREAS

Baron Pierre de Coubertin envisioned the modern Olympic Games as the focus of the world spotlight, and he purposely kept the four-year time span between Games so the spotlight would continue to burn bright long into the future. Given the continued interest and prestige of the Games today, de Coubertin and his collaborators would be well pleased. Every so often during the interlude between Olympic Games, the media focuses on an issue related to international sport and ultimately the Olympic Games. These areas of intense interest and speculation are a good starting point for discussing current Olympic policy issues.

Choosing a Host City or Region

www

IOC Future Host Election
https://olympics.com/ioc/
future-host-election

Imagine this: the President of the IOC moves toward the microphone, paper in hand, to announce the successful bid for hosting the next Summer Olympic Games. You and other members of the Organizing Committee sit in the audience of the capacity-crowd press conference, thinking back over the long years of work spent on the bid: hundreds of meetings; millions of dollars or euros raised and a good sum spent; countless hours of attempting to convince citizens, city workers, and government officials of the value of bringing the Olympic Games home; visits by Olympic and IOC members; a massive enterprise coordinated. Yet, you still wait to hear if you won the bid to host the Olympic Games. The President of the IOC announces, "And the winner is...."

At the press conference to award the bid to host the Olympic Games, the stakes are immense because of the time and money invested as well as the potential for gain in achieving the bid. It is a nation's chance to gain the world spotlight. Given this situation, it is easy to understand why the IOC procedures for awarding the bid to host are considered an important area for policy development. Policies must stipulate exactly how the decision will be made, what criteria will be used, and what the timeline will be. In this case, policies provide a framework for the bidding committee. What is important to the IOC in terms of staging the Olympic Games? Who will make the decision, and when will the decision be made? How can we position our bid to be held in the best possible regard by members of the IOC?

In recent years, the answer to these questions – and the overall process of selecting hosts for the Olympic Games – has not come without controversy. Rumors of IOC members accepting bribes of money, trips, gifts, and promises had been debated for years, but the December 1998 scandal that erupted as a result of alleged bribes associated with the 2002 Salt Lake City Olympic bid resulted in a thorough scrutiny and revision of the policies mandating the conduct of IOC members. In the past, the IOC did not condone IOC Bid Selection Committee members accepting gifts, but it also did little to monitor the policy. In effect, it produced a scenario where it appeared that the votes of IOC members had to be secured through bribes in order to win the bid. Organizing committees spent considerable time, money, and thought on planning how to best influence members of the IOC. However, the public outcry resulting from the Salt Lake City corruption allegations resulted in the IOC sanctioning those involved. Investigators discovered that the Salt Lake City Bid Committee paid hundreds of thousands of dollars in cash, gifts, travel, and medical aid to IOC members. In the end, four IOC violators resigned, six were expelled, and ten received official warnings by the IOC President.

The corruption scandal related to the Games host selection was one of the developments that added pressure to an IOC already facing increased criticism for its host selection process. The process was perceived as too costly, too complex, and left too many losing bidders who would then be discouraged to bid again. In response to this criticism, the IOC mapped out a comprehensive reform of the host selection policy in its *Olympic Agenda 2020*. The goal of the new approach to host selection was to invite interested parties to develop projects that best fit their own needs. Rather than adapting the host city/region to the Games, the new approach would require the Games to adapt to the city/region. The host-centric approach, so the IOC argued, would make it more appealing for interested parties to apply to host the Games given they can "create Olympic projects that are less expensive and that maximise operational efficiencies, while also unlocking greater value for future hosts, with a strong emphasis on legacy and sustainability" (IOC, 2022d, para. 5).

With a new approach, titled "The New Norm," the IOC outlined a set of 118 reforms that re-envision the delivery of the Games. These reforms aim to reduce cost for bidding, provide more flexibility

wwww

IOC's "The New Norm"
Reforms for Host Selection
https://olympics.com/ioc/
news/the-new-norm-it-s-a-
games-changer

for planning (e.g., getting rid of rigid bidding deadlines, considering multiple cities as potential host cities within a country), and guarantee opportunities for ongoing dialogue between the organization and parties interested in hosting the Games, such as cities, regions, countries, or NOCs. Among the most significant changes were the creations of two Future Host Commissions, one for the Summer Games and one for the Winter Games, each tasked with overseeing interest in the Games and advising the IOC Executive Board. The Commissions also guide interested parties in developing host proposals that aligned with their long-term development goals. The Commissions also plays a crucial role in the new streamlined approach for the bidding process:

1. the *Informal Exchange* phase, during which interested cities/regions can learn more about the selection process. The IOC will provide contextual advice on how to develop a strong bid (the Future Host Commission is not involved at this stage).

2. the *Continuous Dialogue* phase, a non-committal phase where Future Host Commissions work with interested cities/regions to explore how feasible hosting the Games would be in the specific cities/regions, aligning ideas for the bid with local long-term goals for development of the cities/regions (e.g., highlighting opportunities and challenges). This phase includes developing an initial strategy for venues, funding, games delivery, and alignment with Olympic Agenda 2020+5 recommendations, but no submission of formal documents to the IOC is required. However, at the end of this phase, the IOC Executive Board can invite interested hosts to open a targeted dialogue (see next phase) based on recommendations made by the respective Future Host Commission. Up until this point, interested hosts do not have to specify the specific edition they intend to apply for (e.g., 2034 Games).

3. the *Targeted Dialogue* phase, where interested parties go through a defined process to explore hosting a specific edition of the Games. This is where the Future Host Commission works with the potential hosts to conduct detailed assessments of their plans for hosting the Games (this may include a visit to the bidding city/region) and submits a report to the IOC Executive Board that addresses all essential elements for hosting the mega event. Based on this report, the IOC Executive Board then decides what potential host(s) to put up for election by the IOC Session. If an election is called by the IOC Session, the preferred host(s) participate in a consultation meeting to discuss their plans and make a final presentation to the Session (immediately following a separate final presentation of the Future Host Commission).

You may be wondering what criteria interested hosts will be judged on. During the Targeted Dialogue phase, all preferred hosts must submit the future host questionnaire to the IOC, which covers the following (IOC, 2021e):

- Vision, games concept, and legacy (includes venue master plan, funding, dates of the Games, and alignment with broader developmental goals of city/region);
- Games experience for athletes and fans;
- Plans for the Paralympic Games;
- Sustainability goals;
- Support for the Games (includes support from the public, political figures, etc.) and Games governance (includes considerations in regard to human rights, safety, security, transport, and legacy);
- Economic analysis of the Games.

As you can see, candidature processes continue to evolve as the IOC attempts to reduce the costs of hosting the Games and provide more flexibility in the bidding process. Whether more cities and regions become interested in hosting remains to be seen, but the IOC has argued that their new approach is already working. For the Olympic Winter Games 2026, for example, the introduction of the non-committal dialogue stage reduced the cost for candidature budgets by 80% compared to the average for 2018 and 2022 (IOC, 2022d). The focus of the IOC on reducing cost for simply bidding on the Games shows that rising cost continues to be an important policy area in Olympic sport.

Costs of Hosting the Olympic Games, Corporate Sponsorship, and Media Rights

As you can see, one of the motivations for changing the Games' bidding process was to reduce cost. If simply bidding for the Games was considered too expensive, you can imagine how hosting the actual Games easily becomes a multi-billion endeavor. For example, the OCOG for the 2020 Tokyo Summer Games, delayed by a year by the COVID-19 pandemic causing construction delays and expiration of TV contracts, announced it spent $12.7 billion dollars on the Games – almost twice as much from what they calculated when submitting the bid proposal (Lewis, 2021). At almost $8 billion, most of the expenses were venue-related, such as renovating permanent venues or creating temporary infrastructure, but the OCOG also spent a significant amount on services related to competition operations, transport, and security ($4.8 billion) and COVID-19 countermeasures ($0.3 billion) (IOC, 2022u).

While the officially reported number is close to the average cost of hosting ($12 billion; Flyvbjerg et al., 2021), it does not include other Games-related costs for Japan, such as the construction of infrastructure to host the Games. These costs are rarely included in the OCOG's financial reporting yet add significant expenses to hosting the Games. For the

2020 Tokyo Summer Games, the estimated overall cost was upwards of $25 billion, including $2.7 billion caused by postponing the event by a year (Vyshnavi, 2021). The cost for the 2022 Beijing Winter Games was estimated at $38.5 billion, which is more than 10 times the reported amount (Teh & Stonington, 2022). Neither the Tokyo nor the Beijing Games are outliers. In a 2020 study, researchers at the University of Oxford looked at the cost overruns of the Summer and Winter Games since 1960. They found that on average the proposed host budget was surpassed by 172% (Flyvbjerg et al., 2021). The Rio 2016 Summer Olympic Games had an overrun of 352% (Lewis, 2021). These numbers speak a clear language: The costs of delivering the Games are often higher than expected, continue to rise exponentially, and, of course, depend upon building the required new infrastructure. For instance, here are the approximate final operating budgets and total costs in millions of US dollars for hosting some past Summer Olympic Games prior to Tokyo 2020 and Paris 2024: Tokyo 1964 – $72/$72; Moscow 1980 – $231/$1,350; Atlanta 1996 – $1,800/$2,000; Beijing 2008 – $44,000/$44,000 (Campbell, 2016; Hodgkinson, 2007; Pravda Report, 2008).

Given the enormous bill that comes with hosting, financing an Olympic Games may be the biggest issue facing an OCOG (Ahmed & Leahy, 2016). Without sponsors and revenue from television and other media rights, the Olympic Games would not exist in their current form given the size and complexity of the sport spectacular. Some countries rely on their government for funding, others get funding from public sources such as lottery returns, and many will raise money privately through corporate sponsorship. The United Kingdom, Brazil, and Japan relied heavily on revenues raised from partnerships for hosting the 2012 London, 2016 Rio Summer, and 2020 Tokyo Games. For example, the primary sources of revenue for Tokyo 2020 were local sponsorships ($3.4 billion), contributions from the IOC ($0.8 billion), and sponsorships through TOP program's global sponsors ($0.5 billion), the IOC's official business partners (IOC, 2022u). The global sponsors are a group of 14 well-known multinational companies: Airbnb, Alibaba Group, Allianz, Atos, Bridgestone, Coca Cola, Deloitte, Intel, Omega, Panasonic, P&G, Samsung, Toyota, and Visa.

Despite support from corporate sponsors, the IOC committed to reducing the rising costs in its strategic plan *Olympic Agenda 2020*, a strategic priority it considered achieved in the agenda's closing report in 2022 (IOC, 2022n). These initiatives are the result of the IOC receiving fewer bids to host (there were only two candidates to host the 2022 Winter Games) and of potential host cities (Rome and Budapest for Summer 2024) withdrawing due to the costs of competitive bids (Goldblatt, 2016). To better manage costs and maximize sponsorship and broadcast revenues, the IOC has developed extensive sponsorship policies. However, sponsorship is a complicated area. The stakes are high because of the potential for revenue generation. The issue is complicated because of the need to define exactly who owns the rights to the various Olympic symbols. Three layers of organizations have an interest or a right to sponsorship associated with the Olympic Games: the IOC, the NOC, and the OCOG.

The IOC has a written policy to define who has the right to market and sell which sponsorships. Beyond outlining who has the right to which properties, Olympic sponsorship policy is intended to set guidelines for the practices of the different levels of governance. For example, are title sponsorships allowed? Could the Olympic Games be called the Coca Cola Olympic Games? Obviously, that practice is not allowed by the IOC. However, defining the limits of acceptance regarding sponsors is a hot topic for debate.

Of all the sponsorship, advertising, and marketing opportunities available at the Olympic Games, no property for potential revenue generation is larger than television and media rights. In the aftermath of the 1984 Los Angeles Olympic Games, when significant profits were realized from hosting, the IOC began intense scrutiny of its own funding portfolio. It focused first on hosting guarantees and revenue-sharing methods and second on the revenues to be generated from selling television broadcast rights. The IOC decided to retain the right to negotiate television contracts and to share the revenues among the IOC, NOCs, OGOCs, and IFs. For example, in 2014 the IOC sold the broadcasting rights to the Olympic Games through 2023 to NBC for $7.75 billion (Armour, 2014). For the 2020 Tokyo Games alone, the IOC made up to $4 billion in television rights (Wade & Komiya, 2021). It is easy, given this tremendous potential revenue stream, to understand the importance of the policy issue regarding who owns the television rights. The IOC has acted to retain the rights of negotiation and dispersal of television revenues. Its purpose included control, consistency, and ensuring value. By setting policies ensuring central control of this important negotiation, the IOC has ensured the potential for developing consistent, long-term contracts of the highest possible value. Of course, OCOGs would prefer to hold the rights themselves, and they can be expected to push for further debate on this issue. In addition, in the United States, the USOPC takes the position (and the IOC recognizes) that the Ted Stevens Olympic and Amateur Sports Act grants it certain rights to participate in (and to share the proceeds of) the US television-rights broadcast negotiations.

New Olympic Sports

The size of the Olympic festival has been a topic of ongoing debate. The masses of visitors and spectators at the Sydney 2000 Olympic Games resulted in further scrutiny and policy development by the IOC, and the concern is revisited with each Olympic hosting. The issue of the number of competitors and sports has been added to the debate. Of course, many IFs lobby intensely to have their sports included in the Olympic Games. How many sports should the Olympic Games include? How many competitors are too many? How can the IOC balance the interest in Olympic competition and its mandate to provide opportunities for nations all around the world (206 NOCs), with the management issues that arise from competitions that are simply too large? Should every nation be permitted to send Olympic participants in all sports, even though the individuals may not meet world standards?

As part of *Olympic Agenda 2020* and *Olympic Agenda 2020+5*, the IOC has made it a priority to "strengthen the uniqueness and the universality of the Olympic Games" (IOC, 2022p, p. 4). Because the program is at the heart of the Olympic Games, this includes regularly reviewing the sports offered to make sure the Games stay relevant and reflect the zeitgeist of the time. As a result, the IOC has developed policies to define exactly which sports will be competed in during the Olympic Games and how to add new sports. Currently, there are 45 summer and 16 winter sports with over 400 events represented in the Summer Games, Winter Games, and Youth Olympic Games (IOC, 2022t). During the 2020 Tokyo Summer Games, 28 of those sports were included in the program, compared to seven included in the Olympic Winter Games in Beijing in 2022 (IOC, 2022q). Within the Olympic Charter, policy is written to define the following:

- Olympic sports
- disciplines (different events within sports)
- events (competitions resulting in medals; summer – 310, winter – 100)
- criteria for admitting each sport
- approximate numbers of athletes (summer – 10,500, winter – 2,900), coaches/support personnel (summer – 5,000, winter – 2,000)

Often, the IOC will name an addition to the sport program at a particular Olympic Games in accordance with the wishes of the OCOG. In fact, as a result of *Agenda 2020*, OCOGs are now able to make proposals for what sports and events should be included for the specific Games they are hosting. As you might imagine, a significant amount of lobbying occurs in an attempt to have a sport recognized for Olympic competition. Policy is required to define the criteria, procedure, and timing of decisions relative to the sports program of an Olympic Games. Analyzing the program for each Summer and Winter Games is the responsibility of the Olympic Programme Commission, which will then form proposals to be considered by the IOC Executive Board and voted on at the IOC Session. The Olympic Programme Commission is charged with reviewing and analyzing the slate of sports and defining the permissible number of athletes in each Olympic sport, with a focus on events delivered. It is also responsible for developing recommendations on the principles and structure of the Olympic program.

www

Olympic Programme
Commission
https://olympics.com/
ioc/olympic-programme-
commission

The ultimate goal of the Committee's work is to modernize the Olympic Games and make sure they remain popular and relevant. Proposals to add sports are evaluated on 35 criteria across the following five categories: Olympic administrative aspects (e.g., competition format, venues), the value the sport may bring to the Olympic Movement, institutional components, popularity of the sport, and overall business model (IOC, 2022q). For example, the 2020 Tokyo Summer Games featured five new sports: baseball/softball (returning after not having been included in the 2012 and 2016 Games), karate, skateboarding, sport climbing, and surfing. Baseball/softball and karate will not be featured at the 2024 Paris Summer Games, but the Games will see athletes compete in breakdancing for the first time – an attempt to broaden the Games' appeal to younger audiences. On the Winter Games side, snowboarding was recently added to the program to attract young viewers and thereby boost television ratings and sponsorship revenues.

Athlete Safety

Athlete safety has been an important policy area in Olympic sport, but it has come into particular focus more in recent years due to multiple crises permeating the institutions governing global sport in unprecedented ways. The onset of the COVID-19 pandemic, a once-in-a-century global health crisis, forced the 2020 Tokyo Summer Games OCOG to postpone the Games by one year. A first in Olympic history! The postponement of the Games created a plethora of logistical challenges: TV contracts expired, construction of crucial sites was delayed, and the Japanese public became increasingly critical of the country's hosting of the Games due to rising costs burdening taxpayers. In addition, policy needed to be created to make sure athletes could compete in a safe environment and the Japanese public was protected once the Games resumed a year later. For example, the decision was made not to allow any spectators at the Games. Decisions like this one were the result of the work of multiple task forces and daily coordination meetings between the IOC and the Tokyo 2020 OCOG, which were created as a means to continuously monitor the situation and identify countermeasures to COVID-19 (IOC, 2022s). These countermeasures regulated safety in the face of the pandemic across six key pillars: (1) travel restrictions and entry into Japan, (2) physical distancing to stop the spread of the disease, (3) protective equipment and cleaning measures, (4) a process to test for and track COVID-19 as well as providing spaces for isolation as needed, (5) information provision, and (6) administering vaccines (IOC, 2022s).

While the world was grappling with the global health crisis, a separate development rocked the foundation of Olympic sport: Multiple revelations of systemic physical, psychological, and sexual abuse of athletes across multiple cultural contexts. In the United States, more than 300 survivors revealed ongoing, decades-long sexual abuse of gymnasts by Larry Nassar, a gymnastics team doctor whose predatory behavior was enabled by organizations such as USA Gymnastics, the USOPC, and the FBI, all of which were aware of his behavior yet did not take sufficient action to protect the athletes. In a highly publicized trial, Nassar was sentenced to a minimum 40 to a maximum of 125 years in prison for his crimes. The testimony of the brave survivors painted a shocking picture of the institutions sworn to protect them: They stood by idly, caring more about medals than the well-being of its athletes. The scope of the abuse, and the systemic nature of its attempted cover-up, even led US Congress to intervene – leading to the transformative changes in US Olympic sport governance mentioned earlier in this chapter (e.g., the passing of the Empowering Olympic, Paralympic, and Amateur Athletes Act of 2020).

Beyond the United States, Olympic sport organizations in other countries have faced similar scrutiny related to their role in overlooking abuse. For example, in 2022, over 80 Canadian bobsleigh and skeleton athletes penned a letter calling for the resignations of high-ranking officials in their NGB, Bobsleigh Canada Skeleton, because the organization had created a culture of physical and racial abuse (Murray, 2022). In the same year, Canadian gymnasts called on Sport Canada to launch an independent investigation into their NGB, which they accused of creating a toxic culture that covered up repeated abuse. Across the Atlantic, German magazine *Der Spiegel* revealed that national team swim coach Stefan Lurz had sexually assaulted multiple women on the women's national team and that the German Swim Association had knowledge of the predator's crimes yet did nothing to protect the athletes. Institutions and lawmakers alike have responded to these revelations: In the United States, the US Center for SafeSport was created as part of the Protecting Young Victims from Sexual Abuse and Safe Sport Authorization Act of 2017. The Center's mission is "ending sexual, physical, and emotional abuse on behalf of athletes everywhere" (US Center for SafeSport, 2022, para. 2). Athleten Deutschland, an advocacy group representing elite sport athletes in Germany, is currently calling for the establishment of a similar Center in their country. Instances like these show the importance of creating policy that protects athletes, so that future generations no longer have to endure harmful and abusive behavior by people in the very institutions supposed to protect them.

Politics at the Olympic Games

The originators of the Olympic Games sought to avert governmental interference by forming the IOC as a group independent of the funding requirements, politics, and power of a particular nation's government. In reality, however, it has been impossible to keep politics out of the Olympic Games. The IOC is a very political organization in which alliances are regularly formed to accomplish some vision or goal. Even more so, political involvement in the Olympic Games occurs as a result of the gains possible for a national government and its ruling ideology. The Olympic Games have a diverse following of people around the world, and governments naturally try to exploit the Games for their own purpose. This is sometimes referred to as "sportswashing." In this way, the Olympic Games provide an avenue to unite people, to develop a national consciousness, and to provide ammunition to suggest that "our ways" are better than "your ways." Governments have used sport to send political messages to another country by sending a team to compete prior to an Olympic Games. For example, when the US table tennis team went to China in the early 1970s (the so-called "Ping-Pong Diplomacy"), it signaled a renewed interest in discussing foreign policy between the two nations. There are many other examples of political motives driving decisions associated with athletic competitions. Take the country of Taiwan, for example. When Taiwan competes at the Olympic and Paralympic Games, the athletes must compete under the country name "Chinese Taipei" and are not allowed to display their country's flag or play their country's anthem. This is due to the influence China exerts on the international political and sport stage.

In 2016, the IOC established the Refugee Olympic team for the Rio Summer Games. For the first time, athletes with refugee status were chosen to compete under the Olympic flag in order to raise awareness of the magnitude of the worldwide refugee crisis. The UN Refugee Agency estimates that more than 65 million people are refugees or displaced people, forced from their homes by war, famine, and natural or civilization-created disasters (Jones, 2016). The IOC, NOCs, IFs, and the UN Refugee Agency collaborate to identify Olympic-level athletes with refugee status. For Rio 2016, 10 athletes were selected from a larger pool of almost 1,000 and were supported by funding, coaches, and medical staff provided by the IOC. Just four years later, 29 athletes from 11 countries competed in 12 sports as part of IOC Refugee Olympic Team for the Tokyo 2020 Summer Games. These athletes come from countries such as Afghanistan, Congo, Iran, Syria, or Venezuela. The IOC's efforts to include refugee athletes make a political statement, sending a signal of hope to all refugees of the world of the human capacity to overcome tragedy and contribute to society (Jones, 2016).

Sometimes, the pressure on the IOC to act on political matters is so intense that the organization cannot refuse to enter political terrain. This was the case in 2022, for example, when the Russian and Belarusian governments launched an invasion to Ukraine only a few days after the conclusion of the Olympic Games and, by doing so, violated the Olympic Truce. In response to the Russian and Belarusian invasion, the IOC recommended that all international sports federations and event organizers ban Russian and Belarusian athletes and officials from competing at international competitions (IOC, 2022g). The IOC also recommended that no sport events should be held in Russia or Belarus and stripped organizational power away from high-ranking officials affiliated with the countries, including from Russian President Vladimir Putin and Dmitry Chernyshenko, Deputy Prime Minister of the Russian Federation. (It's worth noting here that even prior to the invasion, calls for excluding Russia from the Olympic Movement had increased given the country's state-sponsored doping program, which to this day constitutes one of the biggest scandals in Olympic sport history. A more in-depth overview of the scandal is provided in Chapter 16.

As you can see, the political ramifications of the Olympic Games are most certainly an issue for policy definition, whether the political issues are related to the internal workings of the IOC or external to the governments of the participating nations. Each group will set policies to manage their own interests and perspectives. Perhaps organizing committees and their sponsors engage in political maneuvering. Or perhaps all parties are worried about the dangers of terrorist violence, the embarrassment of positive tests for performance-enhancing drugs, legal injunctions over team membership, or the authority of IFs. One thing is for sure: policies to deal with issues of power and politics in the Olympic Games will be necessary into the foreseeable future. As Senn (1999) put it almost two decades ago, "Those who refuse to recognize the politics of the Games put themselves at the mercy of the people and organizations who actively participate in the political competition" (p. 296).

Protest, Freedom of Expression, and Athlete Rights and Representation

What images come to mind if you were asked to name some of the most iconic moments in Olympic sport history? Chances are high that the following image is among them: US Olympians Tommie Smith and John Carlos, having just won first and third place in the 200-meter sprint event at the 1968 Mexico City Olympic Games, step onto the podium and raise their black-gloved fists in the air during the playing of the US national anthem to protest racial injustice on the world's biggest stage. What is now considered one of the most important statements for human rights in the history of the Olympic Movement was met with harsh consequences by the IOC in 1968. The IOC President at the time, USA's Avery Brundage, deemed the protest a violation of the supposedly apolitical nature of the Games and ordered Smith and Carlos to be suspended from the Olympic team and sent home. The US Olympic Committee also criticized the athletes' actions and left Carlos and Smith on the side lines of the US Olympic Movement for decades. It was not until 2019 that the USOPC inducted both athletes into its Hall of Fame and acknowledged the important role they played in sport and society (Armour, n.d.).

The IOC had rules in place to limit athlete protests during the Games since the 1950s, and the IOC's swift response to the US athletes' protest was partly because they saw their vision of the Games as an apolitical arena publicly jeopardized. In its Rule 50, the Olympic Charter states: "No kind of demonstration or political, religious or racial propaganda is permitted in any Olympic sites, venues, or other areas" (IOC, 2021d, p. 95). The wording "racial propaganda" was added to the charter in 1975, a policy change that historians and scholars argue was a direct response to what happened in Mexico City in 1968 (Sharnak & Kluch, 2021). Similar to the IOC, NOCs have tried to limit athletes' ability to protest during the Games. For example, in as late as 2019 the USOPC sanctioned two athletes, hammer thrower Gwen Berry and fencer Race Imboden, for protesting in support of racial justice during the Pan American Games in 2019. Both athletes were put on probation, because they protested on the podium – a space that is often considered sacred when it comes to activist messages.

Leading up to the 2020 Tokyo Games, the IOC's controversial Rule 50, and the sanctioning of Berry and Imboden on the USOPC side in the same year Smith and Carlos were inducted into its Hall of Fame, came under fire in the midst of the (re)emergence of the Black Lives Matter movement and global calls for racial justice. The IOC responded by pledging to review the rule, and launched a consultation process with athletes and NOCs in 2020. While the IOC published new guidelines on Rule 50 ahead of the Tokyo Games, providing some outlets for athletes to share activist messages, experts critiqued the guidelines as not going far enough and continuing to ban protests in the most public spaces such as the podium (Muhammad Ali Center, 2021). On the US side, however, historic change was happening. In December of 2020, the newly formed Team USA Council on Racial and Social Justice (see DEI spotlight) released its first recommendation, calling on the USOPC to no longer sanction athletes protesting in support of racial and social justice. In a surprising move, the USOPC followed the recommendation, lifted the sanctions on Berry and Imboden, and committed to supporting US athletes heading into the Games should they choose to protest. Given the USOPC's and IOC's conflicting approaches to athlete protests,

> a showdown between the IOC and advocates for a more athlete-centered approach to global sport governance [was] all but certain on Rule 50 in what has become one of the most prominent battlegrounds for racial justice and human rights in modern sport.
> (Sharnak & Kluch, 2021, para. 13)

Multiple US athletes protested on the podium during the Tokyo Games (track and field athlete Raven Saunders and fencer Race Imboden), neither of which faced major sanctions from the IOC – a stark contrast to the organization's response in 1968.

The increased scrutiny of Rule 50, and limitations to athletes' freedom of expression that come with it, is also reflective of a broader policy area related to athlete rights and representation that has shaped much of the most recent developments in global sport governance. Noting the crucial voice of athletes, organizations in the Olympic Movement have started to give more power to athletes in the governance of Olympic sport. For example, the IOC adopted the *Athletes' Rights and Responsibilities Declaration* at the 133rd IOC Session in 2018, a document outlining "a common set of aspirational rights and responsibilities for athletes within the Olympic Movement and within the jurisdiction of its members" (IOC, 2018a, para. 1). The IOC also added an amendment to the Olympic Charter that would boost athlete representation across Olympic sport, most notably by requiring NOCs to allow for athlete representation in decision-making (IOC, 2021c). Some NOCs have already undergone separate initiatives to increase athlete representation. For example, as part of the most sweeping governance reform in decades, the USOPC amended its bylaws in 2019 to require a threshold of 33% athlete representation on USOPC and US NSGB boards, which led to the addition of two athlete representatives (an increase from three to five athlete members) on the USOPC Board of Directors. Representation of athletes in decision-making is crucial for governance in Olympic sport and their insights can provide valuable input on many policy areas, including the important area of human rights that we will now turn our attention to.

WWW

Athletes' Rights and Responsibilities Declaration
https://olympics.com/athlete365/who-we-are/athletes-declaration/

Human Rights

One of the reasons why the backlash against IOC Rule 50 has been so intense is because critics argue that it violates athletes' freedom of expression, a fundamental human right anchored in the United Nation's Universal Declaration of Human Rights, the milestone document outlining the rights of every person. But even beyond freedom of expression, Olympic sport has become the site for the struggle over human rights. In the wake of the Rio 2016 Summer Games, human rights activists around the world called on the IOC to include more comprehensive human rights policy in candidate city bids and advocated for the importance of enforcing internationally recognized human rights standards (Gibson, 2016). Rio 2016 illustrated the large numbers of families who were relocated against their wills for Games facility construction and the ensuing major clashes with police that followed. These calls got even more intense leading up to the Tokyo 2020 Summer Games (see above) and, especially, the 2022 Beijing Winter Games as the Chinese government was credibly accused of committing crimes against humanity targeting the country's Uyghur Muslim population and other ethnic minoritized populations. In the months leading up to the Games, more than 240 nongovernmental organizations and human rights groups released statements documenting some of the human rights abuses happening in China as the country was getting ready to serve as the world stage for the Games (Human Rights Watch, 2022):

- Arbitrary detention and torture of the Uyghur population (and other Turkic groups);
- Arbitrary detention and torture of human rights defenders;
- Targeting of democratic forces (e.g., media) in Hong Kong;
- High-tech surveillance enabling Chinese authorities to monitor all communication;
- Suppressing people exercising their rights to peaceful assembly and freedom of expression

While China denied any wrongdoing, the well-documented violations led to some countries warning their athletes against speaking up on human rights at the Beijing Games, as they feared for their safety and well-being. Of course, the Beijing Games were not the only Games in recent history scrutinized for shortfalls in human rights. Both Japan (Tokyo 2020) and Russia (Sochi 2014), for example, faced backlash for their treatment of LGBTQ+ people (Reid, 2014).

Despite having a commitment to human rights referenced in its Charter, the IOC has historically been hesitant to call out host country's human rights abuses. This is mostly because the IOC is rooted in the idea that the Olympic Games operate outside of the political sphere (which, as you can see from our previous section, is not entirely true). It gets even more complicated when one adds in the fact that some of the host countries accused of human rights abuses also provide major financial contributions to the IOC.

After decades of advocacy work from civil society organizations, the IOC in 2019 announced it would review its human rights policies and launched an independent expert investigation into how the organization can better protect human rights globally. The investigation culminated in a report proposing a comprehensive IOC Human Rights Approach, mapping over 30 recommendations on how to anchor human rights within and beyond the organization, based on the following five pillars (Zeid & Davis, 2020, p. 24):

1. Articulating the IOC's human rights responsibilities;
2. Embedding respect for human rights across the organization;
3. Identifying and addressing human rights risks;
4. Tracking and communicating on progress;
5. Strengthening the remedy ecosystem in sports.

Following the release of the report, the IOC made important human rights commitments in its *Agenda 2020+5*, including the creation of a Human Rights Unit within the organization. This was just one of the initiatives to "continue to embed in a more systematic and comprehensive way human rights due diligence in its operations so as to reduce and mitigate risks of negative impacts on people, as well as ensure remediation in a proactive way" (IOC, 2022p, p. 32). This commitment was further refined in 2022, when the organization released an introductory document outlining the first stages of a new IOC Human Rights Strategic Framework, aligned with the UN Guiding Principles on Business and Human Rights.

Aside from the IOC, world leaders sometimes decide to take matters into their own hands and make important statements in support of human rights. For example, ahead of the 2022 Beijing Winter Olympic Games multiple countries (including the United States, UK, Canada, and Australia) announced they would engage in a diplomatic boycott of the Games (Reuters, 2022). The diplomatic boycott meant that these countries would not send high-level officials to Beijing. Whether or not such boycotts actually create change is debatable, but they surely show the immense stage the Olympic Games provide to send a message that potentially reaches billions of people.

www

Recommendations for an IOC Human Rights Strategy
https://stillmedab.
olympic.org/media/
Document%20Library/
OlympicOrg/News/2020/12/
Independent_Expert_Report_
IOC_HumanRights.pdf

www

IOC Human Rights Framework
https://stillmed.olympics.
com/media/Documents/
Beyond-the-Games/
Human-Rights/Introduction-
IOC-Human-Rights-
Strategic-Framework.
pdf?_ga=2.67085482.5484
29037.1657449042-
547494262.1656929888

www

United Nations Guiding Principles on Business and Human Rights
https://www.ohchr.
org/sites/default/files/
documents/publications/
guidingprinciplesbusinesshr_
en.pdf

SUMMARY

T he Olympic Games are the pinnacle of world sporting events. Even professional athletes dream of winning an Olympic Gold Medal. Participating at an Olympic Games is described as an experience of a lifetime, and athletic careers are routinely described in terms of Olympic Medals won and number of Games attended.

All things Olympic are governed by the IOC, an elected group of Olympic enthusiasts charged with overseeing Olympic events and promoting the Olympic Charter. Policies of the Olympic Charter are developed by IOC committees and approved by the Session of the IOC, which meets at least once per year. NOCs control operations and policies for Olympic participation within the country and are responsible for choosing between possible cities or regions interested in bidding to host the Olympic Games. Host countries create Organizing Committees for the Olympic Games well in advance of hosting to build facilities and plan for hosting athletes and visitors. IFs govern Olympic sports on a day-to-day basis, and they work closely with the IOC to develop talent competing at all levels of sport, including the world level. Hosting the Olympic Games is a complex undertaking, involving bidding for the Games, rising costs of hosting, the number of Olympic sports, protecting athletes, and considering broader concerns for safety, political maneuvering, athlete expression and representation, and strengthening human rights commitments.

caseSTUDY

OLYMPIC SPORTS

You are lucky enough to be involved with putting together the Olympic bid for the Summer Olympic Games in 2036. You have a choice of the following cities and regions: Central Florida, Chicago, Denver, New York City/New Jersey, Portland, Dallas/Fort Worth, Phoenix, Washington, D.C. For your choice of cities, you have been asked to help complete the Olympic Games Future Host Questionnaire provided by the IOC.

1. Choose a city.
2. Research the criteria that are required components of the Olympic Games Future Host Questionnaire. Here is the link you will need: https://olympics.com/ioc/documents/olympic-games/future-olympic-hosts
3. For your choice of city, explain as thoroughly as possible what information will go into the sections you are writing. You may need to do some research to learn more about the specific city. Explain why your bid information is worthy of IOC support.
4. As we explained in this chapter, the IOC has recently shifted its focus in the bidding process toward a more shared dialog approach. Imagine you have completed a first draft of your host questionnaire. What questions would you want to ask the IOC to further strengthen your proposal?
5. One of the reasons for changing their host selection process was because the IOC wanted to underline that "The Games adapt to the city/region, the city/region does not adapt to the Games" (IOC, 2021e, p. 6). How will you make sure your proposed Games will adapt to the city/region of your choice? How will the Games fit into the broader strategic planning and unique cultural aspects of the city/region?
6. What procedures might you implement to ensure that all procedures and actions detailed in the bid document meet the highest legal and ethical standards?

CHAPTER questions

1. Compile a chart that depicts the levels of all organizations involved in delivering Olympic sport. Because the IOC has ultimate authority, put it at the top of the chart and work down to national sport organizations.

2. Does digital media (e.g., social media, streaming services) have a positive or negative effect on the Olympic Games? Make a list of both positive and negative effects before making your final decision.

3. How do sports become Olympic events? Review the list of current Olympic sports. Are they popular and interesting? Are there any sports that might be replaced? If so, which ones, and what would replace them? What rationale would you provide the IOC for adding or dropping a sport? Discuss your ideas as a whole class activity.

4. Consider the costs of hosting an Olympic Games. Using an Internet search engine, compare the predicted costs of hosting identified in Tokyo's 2020 or Beijing's 2022 bid to actual costs involved in hosting. If there was a change in predicted versus actual costs, where did it come from?

5. The IOC has always used a commitment to neutrality as a reason not to get involved in political conflicts, yet critics have called the IOC's inaction when it comes to crimes against humanity and violations of human rights as an "easy out." What is the role of the various organizations involved in making the Games a reality (IOC, NOCs, IFs, etc.) when it comes to human rights, athlete rights, and/or social justice more broadly?

6. What might the next IOC document following *Olympic Agenda 2020+5* look like? What will be strategic focus areas of the IOC in the future?

FOR ADDITIONAL INFORMATION

For more information, check out the following resources:

1. Olympic Games Bids website: www.gamesbids.com

2. The tradition and meaning of the Olympic rings: https://olympics.com/ioc/Olympic-rings

3. Olympic Games Costs and Finances: https://www.factretriever.com/OlympicGames-costs-facts

4. The Youth Olympic Games, YouTube: https://www.youtube.com/user/YouthOlympics

5. US Olympic and Paralympic Brand Guidelines https://www.teamusa.org/brand-usage-guidelines

6. Olympic and Paralympic Analysis 2020: Mega events, media, and the politics of sport. Comprehensive collection of essays on the Tokyo Games, including topics related to Tokyo & Mega-Events, Media Coverage and Representation, Performance and Identity, Fandom and National Identity, and Politics of Sport. https://olympicanalysis.org

REFERENCES

Ahmed, M., & Leahy, J. (2016). Rio 2016: The high price of Olympic glory. *Financial Times*. https://www.ft.com/content/594d2320-5326-11e6-9664-e0bdc13c3bef

Armour, N. (n.d.). Opinion: 'We were wrong,' as USOPC finally do right by Tommie Smith, John Carlos. *USA Today*. https://eu.usatoday.com/story/sports/columnist/nancy-armour/2019/09/23/olympics-tommie-smith-john-carlos-get-recognition-they-deserve/2423576001/

Armour, N. (2014, May 7). NBC Universal pays $7.75 billion for Olympics through 2032. *USA Today*. https://eu.usatoday.com/story/sports/olympics/2014/05/07/nbc-olympics-broadcast-rights-2032/8805989/

BBC News. (2013). *London 2012 Olympic Games 'have boosted UK economy by £9.9bn.'* www.bbc.com/news/uk-23370270

Campbell, J. (2016, September 25). *Costs to host Olympic Games skyrockets*. http://abcnews.go.com/US/story?id=95650&page=1

Conrad, M. (2021). The COVID-19 pandemic, the empowering Olympic, Paralympic, and Amateur Athletes Act, and the dawn of a new age of U.S. OlympicrReform. *Journal of Legal Aspects of Sport, 13*(1), 1–59. https://doi.org/10.18060/24919

de Coubertin, P., & Müller, N. (2000). *Olympism: Selected writings*. International Olympic Committee.

Flyvbjerg, B., Budzier, A., & Lunn, D. (2021). Regression to the tail: Why the Olympics blow up. *Environment and Planning A: Economy and Space, 53*(2), 233–260.

Gibson, O. (2016). Olympic Games 2016: How Rio missed the gold medal for human rights. *The Guardian*. https://www.theguardian.com/sport/2016/aug/02/olympic-games-2016-rio-human-rights

Goldblatt, D. (2016). *The Games: A global history fo the Olympics*. Macmillan Publishers International.

Hodgkinson, M. (2007, February 8). London 2012 must learn the £1bn Sydney hangover. *The Telegraph*. https://www.telegraph.co.uk/sport/olympics/2307426/London-2012-must-learn-from-the-1bn-Sydney-hangover.html

Human Rights Watch. (2022, January 27). *Beijing Olympics begin amid atrocity crimes: 243 global groups call for action on rights concerns*. https://www.hrw.org/news/2022/01/27/beijing-olympics-begin-amid-atrocity-crimes

IOC. (2018a). *Athletes' Rights and Responsibilities Declaration*. https://olympics.com/athlete365/who-we-are/athletes-declaration/

IOC. (2018b). *Global broadcast and audience report: June 2018*. https://stillmed.olympics.com/media/Document%20Library/OlympicOrg/Games/Winter-Games/Games-PyeongChang-2018-Winter-Olympic-Games/IOC-Marketing/Olympic-Winter-Games-PyeongChang-2018-Broadcast-Report.pdf?_ga=2.90677593.256442538.1656929888-547494262.1656929888

IOC. (2021a). *Factsheet: IOC sessions*. https://stillmed.olympics.com/media/Document%20Library/OlympicOrg/Factsheets-Reference-Documents/Olympic-Movement/IOC-Sessions/2021-20-IOC_Sessions.pdf?_ga=2.99928669.256442538.1656929888-547494262.1656929888

IOC. (2021b). *IOC elects Brisbane 2032 as Olympic and Paralympic host*. https://olympics.com/ioc/news/ioc-elects-brisbane-2032-as-olympic-and-paralympic-host

IOC. (2021c). *IOC Executive Board proposes change to boost athlete representation*. https://olympics.com/ioc/news/ioc-executive-board-proposes-change-to-boost-athlete-representation

IOC. (2021d). *Olympic charter*. https://stillmed.olympics.com/media/Document%20Library/OlympicOrg/General/EN-Olympic-Charter.pdf?_ga=2.22449343.548429037.1657449042-547494262.1656929888

IOC. (2021e). *The approach to Olympic host elections*. https://stillmed.olympics.com/media/Documents/Olympic-Games/Future-Host/Approach-to-Olympic-host-elections.pdf?_ga=2.63794219.548429037.1657449042-547494262.1656929888

IOC. (2021f). *Tokyo 2020 audience & insights report*. https://stillmed.olympics.com/media/Documents/International-Olympic-Committee/IOC-Marketing-And-Broadcasting/Tokyo-2020-External-Communications.pdf?_ga=2.90677593.256442538.1656929888-547494262.1656929888

IOC. (2022a). *Annual report 2021*. https://stillmed.olympics.com/media/Documents/International-Olympic-Committee/Annual-report/IOC-Annual-Report-2021.

pdf?_ga=2.893426.256442538.1656929888-547494262.1656929888

IOC. (2022b). *Athens 1896.* https://olympics.com/en/olympic-games/athens-1896

IOC. (2022c). *Funding.* https://olympics.com/ioc/funding

IOC. (2022d). *Future host selection.* https://olympics.com/ioc/future-host-election

IOC. (2022e). *International Sports Federations.* https://olympics.com/ioc/international-federations

IOC. (2022f). *IOC Commissions.* https://olympics.com/ioc/commissions

IOC. (2022g). *IOC EB recommends no participation of Russian and Belarusian athletes and officials.* https://olympics.com/ioc/news/ioc-eb-recommends-no-participation-of-russian-and-belarusian-athletes-and-officials

IOC. (2022h). *IOC Executive Board.* https://olympics.com/ioc/executive-board

IOC. (2022i). *IOC marketing: Media guide Olympic Winter Games Beijing 2022.* https://stillmed.olympics.com/media/Documents/Olympic-Games/Beijing-2022/Media-Guide/IOC-Beijing-2022-Media-guide.pdf?_ga=2.90677593.256442538.1656929888-547494262.1656929888

IOC. (2022j). *IOC principles.* https://olympics.com/ioc/principles

IOC. (2022k). *IOC Session elects five new members.* https://olympics.com/ioc/news/ioc-session-elects-five-new-members

IOC. (2022l). *Members.* https://olympics.com/ioc/members

IOC. (2022m). *National Olympic Committees.* https://olympics.com/ioc/national-olympic-committees

IOC. (2022n). *Olympic Agenda 2020 closing report.* https://stillmed.olympics.com/media/Document%20Library/OlympicOrg/IOC/What-We-Do/Olympic-agenda/Olympic-Agenda-2020-Closing-report.pdf?_ga=2.66429549.256442538.1656929888-547494262.1656929888

IOC. (2022o). *Olympic Agenda 2020+5.* https://olympics.com/ioc/olympic-agenda-2020-plus-5

IOC. (2022p). *Olympic Agenda 2020+5: 15 recommendations.* https://stillmed.olympics.com/media/Document%20Library/OlympicOrg/IOC/What-We-Do/Olympic-agenda/Olympic-Agenda-2020-5-15-recommendations.pdf?_ga=2.99985885.256442538.1656929888-547494262.1656929888

IOC. (2022q). *Olympic Programme Commission.* https://olympics.com/ioc/olympic-programme-commission

IOC. (2022r). *Organising Committees for the Olympic Games.* https://olympics.com/ioc/olympic-games-organising-committees

IOC. (2022s). *Report of the Coordination Commission for the Games of the XXXII Olympiad Tokyo 2020.* https://stillmed.olympics.com/media/Documents/International-Olympic-Committee/Sessions/139th-Session/Tokyo-CoCom-Report.pdf?_ga=2.136744884.548429037.1657449042-547494262.1656929888

IOC. (2022t). *Sports.* https://olympics.com/en/sports/

IOC. (2022u). *Tokyo 2020 Organising Committee publishes final balanced budget.* https://olympics.com/ioc/news/tokyo-2020-organising-committee-publishes-final-balanced-budget

IOC. (2022v). *What is the Olympic motto?* https://olympics.com/ioc/faq/olympic-symbol-and-identity/what-is-the-olympic-motto

Jones, I. (2016). *Refugee Olympic athletes deliver message of hope for displaced people.* IOC. https://www.olympic.org/news/refugee-olympic-athletes-deliver-message-of-hope-for-displaced-people

Lewis, L. (2021, December 22). Tokyo Olympics cost almost twice as much as predicted. *Financial Times.* https://www.ft.com/content/1783b893-431a-4054-b677-f7667563aae2

Muhammad Ali Center. (2021). *Open Letter to IOC & IPC Leadership.* https://alicenter.org/programs-athletes-social-change-rule-50-expert-letter/

Murray, N. (2022, March 15). *From concussions to coercion: Why Canada's Olympic sliders say their safety is at risk.* CBC. https://www.cbc.ca/news/canada/bobsleigh-canada-skeleton-athlete-safety-racial-abuse-1.6383602

Nelson, M. (2007). The First Olympic Games. In G. P. Schaus & S. R. Wenn (Eds.), *Onward to the Olympics: Historical perspectives on the Olympic Games* (pp. 47–58). Wilfrid Laurier University Press.

Olympic Museum. (2022). *Presidents of the IOC since 1894.* http://olympic-museum.de/president/pres_ioc.html

Pravda Report. (2008, June 8). *Beijing Olympic Games to cost China 44 billion dollars.* https://english.pravda.ru/sports/106003-beijing_olympics/

Reid, G. (2014). *The Olympics have left Russia but don't forget the LGBT Russians.* Human Rights Watch. https://www.hrw.org/news/2018/02/08/Olympics-have-left-sochi-don't-forget-lgbt-russians

Reuters. (2022). *Most countries won't join diplomatic boycott Beijing Games - IOC.* https://www.reuters.com/lifestyle/sports/most-countries-wont-join-diplomatic-boycott-beijing-games-ioc-2021-12-15/

Sandomir, R. (2003, April 13). Olympics; Drastic U.S.O.C. revision proposed. *New York Times.* https://www.nytimes.com/2003/04/13/sports/olympics-drastic-usoc-revision-proposed.html

Senn, A.E. (1999). *Power, politics, and the Olympic Games.* Human Kinetics.

Sharnak, D., & Kluch, Y. (2021, August 22). *Rule 50 and racial justice: The long history of the IOC war on athletes' free expression.* History News Network. https://historynewsnetwork.org/article/181025

Team USA. (2022). *About the United States Olympic and Paralympic Committee.* https://www.teamusa.org/about-the-usopc

Teh, C., & Stonington, J. (2022, January 30). *Beijing says the cost of hosting the 2022 Winter Games is among the cheapest ever at $3.9 billion. But the real cost might be more than $38.5 billion, 10 times the reported amount.* Business Insider. https://www.insider.com/real-cost-of-beijing-games-10-times-chinas-reported-figure-2022-1

The Olympic Museum Educational and Cultural Services. (2013). *The Olympic Games in antiquity.* https://stillmed.olympics.com/media/Document%20Library/OlympicOrg/Documents/Document-Set-Teachers-The-Main-Olympic-Topics/The-Olympic-Games-in-Antiquity.pdf?_ga=2.94858591.256442538.1656929888-547494262.1656929888

Tokyo 2020. (2019). *Tokyo 2020 guidebook.* https://web.archive.org/web/20200421045205/https://gtimg.tokyo2020.org/image/upload/production/ecriqx8baaxuo1xcnax2.pdf

Tokyo 2020. (2022). *Games motto.* https://web.archive.org/web/20201001121956/https://tokyo2020.org/en/games/vision-motto/

Toohey, K., & Veal, A.J. (2007). *The Olympic Games: A social science perspective* (2nd ed.). Cabi.

US Center for SafeSport. (2022). *Our story.* https://uscenterforsafesport.org/about/our-story/

US Senate Report. (2003). *US Olympic Committee Reform Act of 2003.* https://books.google.ca/books?id=oStJ-OnsJ7AC&pg=RA4-PA4&lpg=RA4-PA4&dq=usoc+governance+and+ethics+task+force&source=bl&ots=EywVRK9_kE&sig=hRQ9uQrD5ECyFQh43gqIUQ8mR6M&hl=en&sa=X&ved=0ahUKEwjs6r7f4OXTAhVny1QKHXHvD6QQ6AEIPTAF#v=onepage&q=usoc%20governance%20and%20ethics%20task%20force&f=false

USOPC. (2020, August 28). *U.S. Olympic & Paralympic Committee and Athletes' Advisory Council convene Team USA Council on Racial and Social Justice.* https://www.teamusa.org/News/2020/August/28/USOPC-And-AAC-Convene-Team-USA-Council-On-Racial-And-Social-Justice

USOPC. (2021). *USOPC bylaws effective March 11, 2021.* https://www.teamusa.org/footer/legal/governance-documents

USOPC. (2022a). *About the U.S. Olympic & Paralympic Committee.* https://www.teamusa.org/about-the-usopc

USOPC. (2022b). *Executive team: Sarah Hirshland.* https://www.teamusa.org/About-the-USOPC/Leadership/Executive-Team/Sarah-Hirshland

USOPC. (2022c). *Leadership.* https://www.teamusa.org/about-the-usopc/leadership

USOPC Impact Report. (2021). *Financial summary.* https://2021impactreport.teamusa.org/financials-reports-and-disclosures/financials.html#gsc.tab=0

Vyshnavi, Pv. (2021, July 27). *List of brands sponsoring the Tokyo 2020 Olympics.* https://startuptalky.com/tokyo-2020-olympic-sponsors/

Wade, S., & Komiya, K. (2021, July 6). *Tokyo Olympics shaping up as TV-only event with few fans.* AP News. https://apnews.com/article/tokyo-business-health-coronavirus-pandemic-2020-tokyo-olympics-1ea9a7973b39c84b219934149cca160b

Young, D.C. (2004). *A brief history of the Olympic Games.* Wiley-Blackwell.

Zeid, R.A.H., & Davis, R. (2020). *Recommendations for an IOC Human Rights strategy.* https://stillmedab.olympic.org/media/Document%20Library/OlympicOrg/News/2020/12/Independent_Expert_Report_IOC_HumanRights.pdf

Zimbalist, A. (2016). *Circus maximus: The economic gamble behind hosting the Olympic Games and the World Cup.* Brookings Institution Press.

Paralympic Sport

When we hear the word *athlete*, certain images come to mind. We envision people who are strong and fast, can throw or run great distances, shoot three-pointers, or ski downhill at incredible speed. When we read about athletes running the 100 meters in just under 11 seconds, high jumping more than 6 feet (1.97 meters), or lifting 600

DOI: 10.4324/9781003303183-11

pounds (272 kilograms), we know they most certainly are elite athletes. All these *are* accomplishments of elite athletes—elite athletes with physical disabilities who competed in the Paralympic Games, an event that draws thousands of elite athletes with disabilities from more than 140 nations every four years, enjoys millions of corporate sponsorship dollars, lights up social media, and includes worldwide media coverage that reaches more than four billion fans. Who are these athletes, and what are the Paralympic Games?

DEFINING DISABILITY SPORT

The term *disability sport* may initially bring to mind the image of a Special Olympian. However, that is not the only form of sport for individuals with disabilities. DePauw and Gavron (2005) define *disability sport* as "sport designed for or specifically practiced by athletes with disabilities" (p. 8). People who participate in the Special Olympics are typically living with cognitive or developmental disabilities. This chapter does not deal with that form of participation. For the purposes of this chapter, the focus is on highly competitive, international, elite-level disability sport for people with specific physical disabilities, specifically, the Paralympic Games.

PARALYMPIC ATHLETES

The Paralympic Games showcase elite-level athletes with disabilities. Incorporating the same ideology as the Olympic Games in celebrating the accomplishments of elite international athletes, the Paralympic Games take place two weeks after the Olympic Games conclude. Competitions occur in the same cities and venues and are staged by the same organizing committee. When a host city is awarded the Olympic Games, the Paralympic Games are an obligatory part of the host city bid process.

www
International Paralympic
Committee
https://www.paralympic.org/

The vision statement of the International Paralympic Committee (IPC) is to "make for an inclusive world through Para sport" (IPC, n.d.a, para. 1). The sports on the official Paralympic Games Programme are presented in Exhibit 11.1. The athletes who compete in the Paralympic Games have a range of disabilities, including visual impairments, cerebral palsy, amputations, spinal cord injuries, and, on a very limited basis, athletes with intellectual disabilities. Not all disability types are eligible to compete in the Paralympic Games, however. Note that athletes with hearing impairments do not compete in the Paralympic Games. They compete in separate World Games and other competitions for the deaf, including the Deaflympics. Athletes with cognitive disabilities such as Down syndrome also do not compete in the Paralympic Games. The Special Olympics was established to provide opportunities for people with cognitive and developmental disabilities. The mission of the Special Olympics includes skill development and social interaction, not the development of international elite-level athletes.

(AS OF TOKYO 2020 AND BEIJING 2022)

SUMMER SPORTS – 22

Archery	Goalball	Taekwondo
Athletics	Judo	Triathlon
Badminton	Powerlifting	Wheelchair basketball
Boccia	Rowing	Wheelchair fencing
Canoe	Shooting Para sport	Wheelchair rugby
Cycling	Sitting volleyball	Wheelchair tennis
Equestrian	Swimming	
Football 5-a-side	Table tennis	

WINTER SPORTS – 6

Alpine skiing	Cross-country skiing	Snowboard
Biathlon	Para ice hockey	Wheelchair curling

Source: IPC (n.d.g).

HISTORY OF THE PARALYMPIC GAMES

Sport for people with disabilities existed for many years before the founding of the Paralympic Games and began to grow after World War II, when sport was used to rehabilitate the many injured military and civilian persons. In 1944, Sir Ludwig Guttmann opened the Spinal Injuries Center at Stoke Mandeville Hospital in England and incorporated sport as an integral part of the rehabilitation process (IPC, n.d.f). On July 28, 1948, the first Stoke Mandeville Wheelchair Games were held and athletes competed in archery competitions. The date was significant because it corresponded to the Opening Ceremonies for the Summer Olympic Games in London that same day. Disability sport soon expanded beyond archery and beyond only athletes who used wheelchairs. The International Sport Organization for the Disabled (ISOD), formed in 1964, brought together athletes who had a variety of disabilities such as visual impairments, cerebral palsy, and amputations. Additional sport organizations for people with disabilities were formed as well, such as the Cerebral Palsy International Sport and Recreation Association (CPISRA) and the International Blind Sports Association (IBSA), organized in 1978 and 1980, respectively. To help these and other sport organizations for athletes with disabilities coordinate their activities, the IPC was founded in 1989 in Düsseldorf, Germany. The IPC represents a multi-disability international sport organization recognized by the IOC (IPC, n.d.a).

The Summer Paralympic Games began with the 1960 Games in Rome, Italy. The first Winter Paralympic Games were held in 1976 in Sweden. The first Paralympic Games under the direct management of the IPC were the Winter Paralympic Games in Lillehammer in 1994. As the statistics show, the Paralympic Games remain strong in the number of both athletes and nation members participating. The 2008 Paralympic Summer Games in Beijing, China, had 3,951 athletes from 146 nations; 4,237 athletes from 164 countries competed at the London 2012 Paralympic Games; and 4,403 athletes from 162 nations plus a Team of Refugee Paralympic athletes competed in the 2021 Summer Games in Tokyo, Japan.

The IPC was faced with an enormous decision on the eve of the 2022 Winter Games in Beijing. Literally days before the Games were set to open, Russia invaded Ukraine. The IPC leadership had to decide what to do and in a very short period of time – postpone the Games? Sanction Russia? The situation represented a case study in crisis management (and this textbook now has a chapter devoted specifically to that topic). The initial decision was to allow athletes from Russia (and its partner in the invasion, Belarus) to still compete in the Games. After an uproar of opposition from numerous countries' delegations, the IPC changed course and banned those countries from the Games (Martinez & Mann, 2022). This would prove to be one of the first dominoes to fall in the subsequent banning of both countries from major sporting events including the FIFA World Cup and numerous world championships. Russia also subsequently lost hosting rights to major events including the ECL final match scheduled for St. Petersburg and the Sochi Formula 1 Grand Prix event as the international sporting community acted to condemn the Russian invasion and war in Ukraine.

GOVERNANCE

Just as with the Olympic Movement, the Paralympic Movement and the Paralympic Games fit into a complex set of governance structures. A number of governing bodies are involved with Paralympic Sport, including the IPC, National Paralympic Committees (NPCs), and International Federations (IFs).

International Paralympic Committee

The IPC, the supreme authority of the Paralympic Movement, is the major international sport governing body for elite sports for athletes with disabilities. The IPC organizes, supervises, and coordinates the Paralympic Games and other multi-disability competitions at the elite sports level, including regional and world championships. Sometimes people think the IPC is governed by the International Olympic Committee. This is not the case. The IOC governs Olympic sport while the IPC governs Paralympic sport. The two organizations are separate sport governing bodies, although they do cooperate with each other when need be.

VISION/MISSION. The IPC has a vision statement and a mission statement. (See Exhibit 11.2) The vision statement for the IPC contains an overall picture of IPC philosophy (IPC, n.d.a). Its mission statement is a bit more detailed and provides a picture of what the IPC does and strives for.

International Paralympic Committee Vision and Mission Statements *exhibit* **11.2**

VISION STATEMENT:

Make for an inclusive world through Para sport.

MISSION STATEMENT:

To lead the Paralympic Movement, oversee the delivery of the Paralympic Games and support members to enable Para athletes to achieve sporting excellence.

Source: IPC (n.d.e).

MEMBERSHIP. The members of the IPC include the International Organizations of Sport for the Disabled (IOSDs), the NPCs, the International Paralympic Sport Federations (IPSFs), and regional/continental Paralympic organizations (Hums & Pate, 2018).

The IOSDs are as follows:

CPISRA	Cerebral Palsy International Sport and Recreation Association
IBSA	International Blind Sports Federation
VIRTUS	World Intellectual Impairment Sport
IWAS	International Wheelchair and Amputee Sports Federation

In addition to these international sport organizations, NPCs from different nations are also full members of the IPC. Currently, approximately 182 NPCs are members of the IPC. Five regional organizations are members as well (the African Paralympic Committee, the Americas Paralympic Committee, the Asian Paralympic Committee, the European Paralympic Committee, and the Oceania Paralympic Committee). These full members have voting rights at the IPC's General Assembly. In addition to the above-mentioned full members, the IPSFs have voting and speaking rights at the IPC's General Assembly. Besides being the supreme authority for the supervision and organization of the Paralympic Games, the IPC also fulfills an important role as the IF for several sports, although it is currently moving away from that model and shifting responsibility for these sports to existing IFs.

FINANCIALS. For the 2020 financial period, the IPC reported,

> The biggest source of revenue was in marketing and broadcasting fees from Paralympic Games organising committees, which amounted to €14.2 million (US$16.5 million). Sponsorship and fundraising efforts also contributed to that figure, which represents an increase of 14.1% on 2019. The remainder of the revenue came from membership fees, grants, broadcasting projects, special project funding and other minor revenue streams.
>
> (SportsPro, 2021, para. 3)

WWW
CPISRA   Cerebral Palsy International Sport and Recreation Association
https://cpisra.org/

WWW
IBSA   International Blind Sports Federation
https://ibsasport.org/

WWW
VIRTUS World Intellectual Impairment Sport
https://www.virtus.sport/

WWW
IWAS   International Wheelchair and Amputee Sports Federation
https://iwasf.com/

WWW
African Paralympic Committee
http://www.africanparalympics.org/

WWW
Americas Paralympic Committee
https://www.paralympic.org/americas-paralympic-committee

WWW
Asian Paralympic Committee
https://asianparalympic.org/

www
**European Paralympic
Committee**
https://www.
europaralympic.org/

www
**Oceania Paralympic
Committee**
https://www.
oceaniaparalympic.org/

www
Agitos Foundation
https://www.paralympic.org/
agitos-foundation

Income used by organizing committees to operate the Games themselves is derived from a combination of government support, Olympic support, and sponsorship deals. Currently, the IPC has 14 Worldwide Paralympic Partners: Airbnb, Alibaba Group, Allianz, Atos, Bridgestone, CocaCola, Intel, Omega, Ottobock, Panasonic, P&G, Samsung, Toyota, and VISA. In addition, BP and Citi are International Partners (IPC, n.d.c). Other companies sign on to be sponsors of a specific Games, for example, British Airways and Sainsbury's during the 2012 Summer Games in London, or Bradesco or Correios for the 2016 Rio Games. Tokyo 2020 saw a number of domestic sponsors as well, including Cannon and Nippon Telegraph and Telephone Corporation (Dixon, 2021). Finally, the Agitos Foundation was established to help support specific sport and athlete development projects.

ORGANIZATIONAL STRUCTURE. The structure of the IPC consists of the General Assembly, the Governing Board, the Management Team, a number of Standing Committees, and an Athlete Council. Exhibit 11.3 illustrates the general structure of the IPC.

The General Assembly is the governance body for the IPC and its highest authority, defining the organization's vision and general direction (IPC, n.d.e). Members of the General Assembly meet at least once every two years to discuss and vote on policy matters of concern to the Paralympic Movement. At the General Assembly, each full member has one vote. The members include International Sports Federations, National Paralympic Committees, IOSD, and several Regional Organizations. The General Assembly is responsible for the following (IPC, n.d.e, General Assembly section):

- Elect the President, the Vice President, and 10 Members at Large
- Consider and approve the IPC's budget and the IPC membership fee policy
- Consider and approve the policy and procedures for the nomination and election of Governing Board members
- Consider motions from members and through the Governing Board from Standing Committees
- Hear and receive the reports of the Governing Board and CEO
- Approve and admit full members of the IPC
- Consider and approve the financial reports and audited accounts and thereby discharge the Bodies of the Organization
- Consider and approve the bylaws outlining the Members' rights and obligations
- Approve amendments to the IPC constitution
- Receive and approve the minutes of the previous General Assembly

The Governing Board is primarily responsible for the implementation of policies and directions set by the General Assembly. Additionally, the Governing Board provides recommendations on membership to the General Assembly, including motions received from members. The Board meets three times a year (either virtually or in person) and has

IPC Governance Structure

exhibit **11.3**

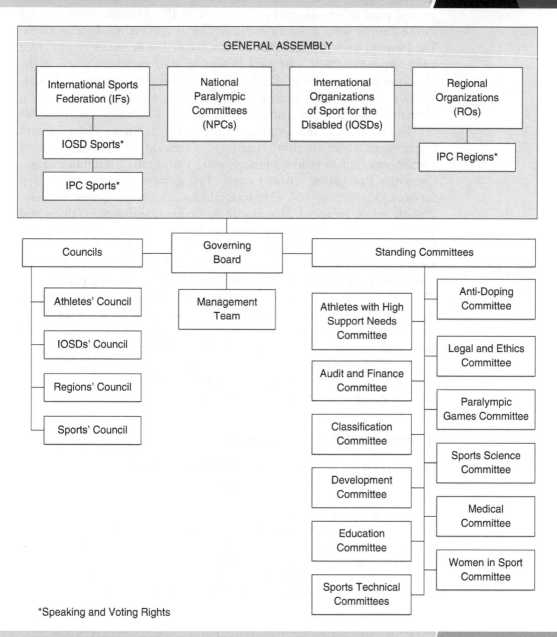

*Speaking and Voting Rights

Source: IPC (n.d.e).

14 members. Ten of these are at-large members elected every four years from the General Assembly and the other two are the President and Vice President. (IPC, n.d.d). The IPC also maintains a number of Standing Committees. These are Audit and Finance, Classification, Education, Legal and Ethics, Medical, Paralympic Games, and Women in Sport (IPC, n.d.b).

The IPC Management Team "consists of the professional staff working under the direction of the CEO" (IPC, n.d.h, para. 3). These are the paid sport managers who work at IPC Headquarters, similar to the people who work in Indianapolis at the NCAA headquarters. The Headquarters for the IPC has been located in Bonn, Germany since 1999. Approximately 110 paid Executive Staff members work at the Headquarters, including the CEO, who oversees numerous departments including Athletes Programs, Finances, Government Relations, Human Resources, Paralympic Sport and IF Relations, Legal Affairs & Governance, Anti-Doping, Corporate Services, Membership Engagement, Membership Programs, Corporate Development and Technology, Information Management, Corporate Communications, Content, Paralympic Brand and Engagement, NPC Marketing Services, Marketing and Commercial, Broadcasting, Client Services, Classification, Protocol, Hospitality & Events, Paralympic Games, and Sport Information Technology (IPC, n.d.h). In addition, specific sports fall under here including World Para Athletics, World Para Dance Sports, World Para Ice Hockey, World Para Powerlifting, World Para Snow Sports, World Para Swimming, and World Shooting Para Sports (WSPSs). Managing the Paralympic Games is as complex as managing any large multisport event and sport managers need to be aware of the social, technological, economic, ethical, political, legal, and educational aspects of running this growing global event (Hums & Wolff, 2017).

The IPC also has an Athletes' Council. This group of athletes are elected by their peers and their role is to bring the voice of Para athletes into the governance structure of the IPC.

WWW
IPC Athletes' Council
https://www.paralympic.org/
athletes-council

National Paralympic Committees

Every nation participating in the Paralympic Games must have an NPC. In terms of sport governance, an NPC is

> a national organisation recognised by the IPC as the sole representative of the Paralympic Movement in that country or territory to the IPC, and recognised as such by the respective national sports council or similar highest sports authority within a country. NPCs undertake the co-ordination within the respective country or territory and are responsible for relations and communications with the IPC.
>
> (IPC, 2019, p. 2)

WWW
United States Olympic and Paralympic Committee
https://www.teamusa.org/
about-the-usopc

WWW
Canadian Paralympic Committee)
https://paralympic.ca/
node/146

In the United States, the United States Olympic and Paralympic Committee (USOPC) officially changed its name to be more inclusive in 2019 (USOPC, 2019). Most nations have a separate NPC, and the US is now one of only five countries [Norway, Saudi Arabia, South Africa, The Netherlands, United States] that house their NOC and NPC in the same sport governing body (Allentuck, 2019). For example, in Canada the Canadian Paralympic Committee is responsible for the Paralympic Movement, Australia has Paralympics Australia, and Greece has the Hellenic Paralympic Committee. In 2001, the then-named USOC established US Paralympics, a division of the then-USOC, to manage Paralympic sport. "U.S. Paralympics, a division of the nonprofit USOPC, is dedicated to becoming the world leader in the Paralympic Movement and promoting excellence in the lives of people with disabilities, including physical disabilities and visual impairments" (USOPC, 2021a). Their vision, posted on the Team USA website, reads "to inspire and unite us through Olympic and Paralympic sport" (USOPC, 2021b, para. 2).

US Paralympics and the appropriate National Governing Bodies (NGBs) help prepare and select the athletes who will represent the United States at the Paralympic Games. It works to meet the urgent need to provide opportunities for people with disabilities, including an increasing number of military veterans, to participate in sport. Offering programs from the grassroots (Paralympic sports clubs) to the elite level, US Paralympics works to be a world leader in promoting excellence in the lives of people with disabilities through sport (USOPC, 2021a).

International Federations

IFs serve as the representatives of sports that compete under the IPC's governance. The IPC recognizes the 15 IFs that hold jurisdiction over Paralympic sports (IPC, n.d.c) In the past, the IPC has also operated as the IF for several Paralympic sports. That model is currently being re-evaluated and the IPC will be relinquishing its role as an IF in a phase-out period with the goal of those sports to be independently governed by the target year of 2026 (Lloyd, 2021). In addition, the IPC also recognizes 14 IFs which are not eligible to be IPC members. These federations represent sports that are not currently on the Paralympic Games program but still work with athlete development. Examples of these include sports of climbing, handball, and surfing.

WWW

Paralympics Australia
https://www.paralympic.
org.au/

WWW

**Hellenic Paralympic
Committee**
https://www.paralympic.org/
greece

WWW

US Paralympics
https://www.teamusa.
org/team-usa-athlete-
services/paralympic-sport-
development

Spotlight: Diversity, Equity, & Inclusion in Action

United States Olympic and Paralympic Committee

In 2019, the United States Olympic Committee (USOC) took the forward-thinking step of rebranding and changed its name to the USOPC. This makes the USOPC one of only a handful of National Olympic Committees to act as the joint national organization for Olympic and Paralympic sport, joining the likes of Norway, Saudi Arabia, South Africa, and The Netherlands in this select group. According to USOPC CEO Sarah Hirshland,

> The decision to change the organization's name represents a continuation of our long-standing commitment to create an inclusive environment for Team USA athletes … Paralympic athletes are integral to the makeup of Team USA, and our mission to inspire current and future generations of Americans. The new name represents a renewed commitment to that mission and the ideals that we seek to advance, both here at home and throughout the worldwide Olympic and Paralympic movements.
>
> (USOPC, 2019, para. 5)

The name change closely followed the decision to pay Paralympic medalists the same amount they pay Olympic medalists – $37,500 for a gold medal, $22,500 for silver and $15,000 for bronze (Allentuck, 2019). The rebranding of the USOPC marked a significant international statement on inclusion by one of the world's largest and most influential National Olympic Committees.

CURRENT POLICY AREAS

Like any major international sport entity, the Paralympic Movement faces a wide variety of issues for which it must formulate policies. The new strategic plan developed by the IPC Governing Board and management team focuses on five strategic priorities. These are to

> Strengthen the effectiveness of the Paralympic Movement at all levels, Enhance the Paralympic Games experience and further its reach as a celebration of human diversity, Drive a cultural shift through Para sport for a truly inclusive society, Continuous pursuit of excellence in what we do and how we do it, and Develop and deliver a new brand statement that globally positions our vision and mission.
> (Around the Rings, 2021, para. 12)

This strategic plan illustrates how a sport governing body identifies and addresses important issues. The plan provides a backdrop for this section on current policy areas the IPC will be addressing while moving forward.

Classification

Classification is an ever-present policy issue that the IPC reassesses periodically. Classification determines eligibility for para sport and ensures athletes compete against others who are similarly challenged, but this is not as simple as it sounds (Palmer, 2021). Over 4000 athletes needed to be classified to compete at the Tokyo Summer Games. Athletes with a number of physical disabilities are allowed to compete at the Paralympic Games, but there can be quite a bit of variability within those disability categories. While classification is a bit more straightforward when it comes to athletes who are, for example, amputees, it is more difficult when it comes to classifying athletes with coordination issues such as those with cerebral palsy (Wade, 2021). Classifying athletes contains a human element on the part of the classifiers and at times their decisions may be seen as arbitrary or unscientific. Athletes may be forced into new classification categories if the classification codes change and occasionally allegations surface of athletes taking measures to perform differently during classification to compete in a category that would provide them a competitive advantage (France24, 2021). For a brief time before the Tokyo Games, it appeared the International Wheelchair Basketball Federation (IWBF) was deemed out of compliance with the IPC Athlete Classification Code and had to create an action plan to ensure its continuation on the Paralympic Games program (Lloyd, 2021). The IPC is currently undertaking its periodic review of the classification system, and it remains to be seen what changes may be on the horizon for the 2024 Summer Games in Paris. Given the nature of disability sport, this is a policy issue the IPC will continue to wrestle with.

Athlete Expression and Section 2.2

The Olympic Movement has seen its share of athletes speaking out on issues of social justice. In so doing, these athletes have run aground of the IOC's Rule 50 which sets guidelines for acceptable political statements. There is a parallel universe in Paralympic sport with the IPC's Section 2.2. This policy reads (IPC, 2020, p. 10):

2.2 Discrimination and demonstrations
2.2.1 No unlawful discrimination is allowed on the grounds of disability, race, skin colour, national, ethnic or social origin, age, sex, gender, sexual orientation, language, political or other opinion, religion or other beliefs, circumstances of birth, or other improper ground against any person, group of persons, or country or territory.
2.2.2 No kind of demonstration, protest, or political statement is permitted in any Paralympic venues or other areas related to the Paralympic Games, except to the extent permitted in any supplementary regulations issued by the IPC in relation to this Article 2.2.2.
2.2.3 National Paralympic Committees, International Federations, and other relevant bodies and authorities involved in the organization of the Paralympic Games should adopt and implement policies and regulations (including disciplinary procedures) that prevent unlawful discrimination and protect the principle of neutrality in sport.

The Summer Games in Tokyo saw US rower Charley Nordin make a statement by wearing a t-shirt demanding justice for Oscar Grant, a young man killed by a BART (Bay Area Rapid Transit) police officer in California. While the USOPC took no action against Nordin, the IPC sanctioned him to issue a public apology (IPC Legal and Ethics Committee, 2022) – which is surely a less strict punishment compared to, for example, the IOC's sanctions on Tommie Smith and John Carlos in 1968. Similar to Rule 50, there are those who advocate for the abolition of Section 2.2 as it violates the basic human rights for athletes' freedom of expression. "As the Paralympic platform becomes bigger, we need to more rigorously scrutinize the practices, procedures, and policies that hinder athletes from turning their visibility into action for systemic change – starting with the IPC's outdated Section 2.2" (Siegfried et al., 2021a, para. 9). Given the comparably light punishment of Nordin by the IPC, it will be interesting to see how the IPC responds to not only similar acts of athlete expression in the future, but also to the greater issue of the rights of Paralympians to freely express their opinions while at the Paralympic Games going forward.

Integration of Athletes with Disabilities into Able-Bodied Sport

This complex policy area is one in which the solutions are evolving differently in different parts of the world. While not exactly an issue directly related to the Paralympic Movement, it is an important ongoing discussion in sport for people with disabilities and so is brought up in this chapter. The extent to which athletes with disabilities have been integrated into able-bodied sport can be measured by the following organizational components: (1) governance, (2) media and information distribution, (3) management, (4) funding and sponsorship, (5) awareness and education, (6) events and programs, (7) awards and recognition,

(8) philosophy, and (9) advocacy (Hums et al., 2003; Hums et al., 2009; Wolff, 2000). Each of these components can be examined as presented in Exhibit 11.4.

How can these components relate to policy development within a sport organization? For media and information distribution, sport organizations can make sure athletes with disabilities are presented on a consistent basis in social media and in press releases. Under management, sport organizations could establish hiring procedures creating opportunities for people with disabilities to be represented in the pool of candidates for open management positions. Under events and programs, sport organizations could establish participation categories for athletes with disabilities, just as there are categories of participation for men and women. For funding, sport organizations could establish funding opportunities specifically for athletes with disabilities so when organizational representatives meet with potential donors, disability sport is part of the conversation. These examples show how the criteria for inclusion can influence organizational policy-making, and they also provide a solid framework for assessing the status of athletes and people with disabilities within sport organizations.

---

exhibit 11.4 Nine Organizational Component Model for Analyzing the Integration of Athletes with Disabilities

1. Governance – examine how organizational policies and procedures deal with athletes with disabilities
2. Media and information distribution – look at the representation of athletes with disabilities in organizational publications or media guides
3. Management – examine the number of persons with disabilities working in management positions or sitting on governance boards
4. Funding and sponsorship – determine from the budget how much money raised by the organization is going to support athletes with disabilities
5. Awareness and education – consider how informed and knowledgeable people within the organization are about disability sport
6. Events and programs – determine the number of competitive opportunities the organization provides for athletes with disabilities
7. Awards and recognition – evaluate how the organization publicly recognizes the accomplishments of its athletes with disabilities
8. Philosophy – review the organization's mission statement and determine how athletes with disabilities are reflected in it
9. Advocacy – determine whether a sport organization is actively promoting sport for people with disabilities via special programming

Sources: Hums et al. (2003; 2009); Wolff (2000).

Organizations that embrace diversity are acting as socially responsible partners in society. Including athletes with disabilities in the management structure of sport organizations increases the size of the networks involved and allows for more voices at the table.

Paralympic Military Program

Given the reality of war and conflict around the world, a growing number of wounded soldiers and civilians are seeking ways to reclaim their lives as best they can. One way people feel more fully alive after having been injured is through sport and physical activity. A program that specifically supports wounded servicemen and servicewomen is the US Olympic Committee Paralympic Military Program (Ward & Hums, in press).

Launched by the then-named USOC in 2004, the Paralympic Military Program uses Paralympic sport opportunities to support wounded, ill, and injured American service members and veterans, including those with amputations, traumatic brain injuries, and visual impairments. Through camps and clinics held year-round across the country, service members and veterans are introduced to Paralympic sport techniques and opportunities, including local and regional competitions, and are also connected to ongoing Paralympic sport programs in their communities (TeamUSA, 2021). At the 2012 London Paralympic Games, Team USA featured 20 military athletes as US Paralympic Team members. Four years later at the Rio 2016 Paralympic Games, Team USA included 33 military athletes. In 2021, 18 military veterans and 3 active-duty athletes competed for Team USA at the Tokyo 2020 Paralympic Games (TeamUSA, 2021). The Veterans Administration (VA) supports many of these athletes through the Veterans Monthly Assistance Allowance, which provides a monthly stipend allowing them to train and compete in Paralympic sports (Molina, 2021). One area the USOPC needs to explore, however, is how to include military veterans living with post-traumatic stress disorder (PTSD). A recent study examined the prevalence of PTSD among the military veteran population, with estimates ranging from a modest 1.09% to a high of 34.84% (Xue et al., 2015). More recently, a 2019 study estimated that 354 million military veterans globally live with PTSD and/or major depression (Hoppen & Morina, 2019). Overall, the USOPC has been instrumental in helping disabled military veterans find purpose in their lives. It is interesting to see how the Paralympic Movement has come full circle. The first Stoke Mandeville games were held to help wounded World War II veterans, and today we see the benefits of sport for wounded veterans from the wars and occupations in Iraq and Afghanistan. The spirit of those first games lives on today.

www

Paralympic Military Program
https://www.teamusa.org/Team-USA-Athlete-Services/Paralympic-Sport-Development/Programs-and-Events/Military

Creating Opportunities for Underrepresented Athletes

The IPC has been making an effort of late to increase competition opportunities for two groups that have been underrepresented at the Paralympic Games – athletes with high support needs and women. Who are athletes with high support needs? These athletes

can have a variety of physical, vision and intellectual impairments
that require additional support at competitions. Support may involve

directly assisting athletes during competition, or with everyday living needs. For example, an athlete with limited hand function may need assistance to load a pistol in a shooting event or get dressed for the day. They may also have a greater degree of physical impairment (such as cerebral palsy or muscular dystrophy) and use powerchairs.

(The Conversation, 2021, para. 1)

The IPC released its plan to increase the number of medal events for these athletes at Paris 2024 by more than half over Tokyo 2020. This will mean more opportunities for athletes specifically competing in boccia, judo, and para-rowing (IPC, 2021).

Over the years, the IPC has also acknowledged the need to open up more opportunities for women with disabilities to compete and has established a standing committee named the Women in Sport Committee to work in this area. While progress has been made, it has been slow. Looking ahead to Paris 2024, the Games

www

Women in Sport Committee
Strategy 2019–2022
https://www.paralympic.
org/sites/default/files/2019-
08/190729115757953_
WiSC%2BStrategy.pdf

will include a record 235 medal events for women, eight more than at Tokyo 2020. This number also represents a 28% increase on the 183 medal events for female athletes at the Athens 2004 Paralympic Games. In addition to more medal events, there will be at least 1,859 slots for female athletes, 77 more than Tokyo 2020.

(IPC, 2021b, paras. 4–5)

Paris 2024 will serve to showcase increased opportunities for both of these groups and broaden the scope of participation in the Games. It is hoped that this increased visibility will encourage more athletes with high support needs and also more women with disabilities to find their athletic potential.

SUMMARY

The Paralympic Games vision is to "Make for an inclusive world through Para sport." The Games are an ever-growing, international, multi-disability, multisport competition. Attracting thousands of athletes from numerous nations, the Games showcase the best elite athletes with disabilities. As a sport property, the Games are gaining the interest of corporations eager to connect with millions of consumers around the world. The governing structure of the Paralympic Games parallels that of the Olympic Games in many ways, as both are large multisport international events. Just as women and racial and ethnic minorities contribute to diversifying the sport industry, people with disabilities do as well. This large part of the sport industry is often overlooked despite its size and importance, and sport managers are advised to keep an eye on the growth and development in this area.

The Paralympic Games, as a growing competition, must deal with evolving policy issues and refocused strategic goals. As the Games grow, pressure will increase to maximize corporate sponsorship opportunities. The social media presence of the Games is expanding with each new iteration of the event. It is an exciting time for the Paralympic Games, given their future growth potential.

portraits + perspectives

TYLER ANDERSON *World Shooting Para Sport Senior Manager*

International Paralympic Committee

My role revolves around managing all aspects of WSPS. I create the competition calendar from host agreement to event execution. This includes different levels of events and a global spread ensuring access to all regions. I oversee the rulebook including rulebook reviews and ensure that our rules are followed. I facilitate the qualification to major events such as the Parapan American Games and the Paralympic Games. In addition, I organize education for our competition officials, coaches, and classifiers. I oversee the classification system (applies to all Para sport) in WSPS. Finally, I have additional duties dealing with media, anti-doping, marketing (sponsorships), etc.

The first current issue sport managers in Paralympic sport must deal with is finance. This is a consistent issue across Para sport specifically. It takes money to organize a sport and without the proper financial and support systems in place, there can be negative consequences for the Para sport in question. A second major issue is Paralympic Games inclusion. This is the most important factor for any sport that is a part of the Olympic/Paralympic Movement. It is critically important to remain a part of the Games programme, so all sports must continue to adapt and evolve to remain relevant in today's ever-changing sport environment.

Right now I am involved in an exciting chapter in Para sport and that is the transfer of governance of the ten sports currently governed by the IPC. These sports will either transfer to their able body IF or another entity or perhaps try to succeed independently. This is exciting because of how unique it is. I do not think a transition of this magnitude has happened before or will happen again.

When it comes to the governance of Paralympic sport, based on the membership of the IOC, IPC, or the IFs, there is very good representation specifically for smaller countries. They have a seat at the table and voting rights based on their membership which means they can be heard and their voice matters in governance decisions about the sport. In addition, we see synergy – the integration of Para into the able body IF means synergies can be shared with the able body sport and Para sport which can lead to greater resources being available than the Para sport could have on their own. Para sport also has a positive impact on the able body sport which is a win/win scenario and how any successful partnership should operate. There are always downsides in the governance of international sport. For example, there is corruption – in Olympic sport this is a trend we see across Olympic IFs and it is not clear how to combat this phenomenon.

In Para sports integration can be a massive challenge. Usually when there is integration at the IF level it means that by default the National Federations (NFs) are also integrated. This can be a very good thing if the NF is supportive of Para but if the NF is not then it can ruin the Para sport in that country due to neglect.

Para sport by its nature is known for the ability to be inclusive for athletes who can otherwise be excluded. As the Paralympic Movement continues to grow, the sports are becoming more professional and the quality is increasing. This means that we are transitioning more and more to high performance which also means more money is involved. This needs to be considered and planned for carefully to avoid unintended negative consequences such as cheating to win as we have seen in Olympic sport.

If you would want to work in Paralympic sport I would say you need to be adaptable. The Movement and those who work in it are constantly changing so you must be ready to change, too. You should constantly be questioning the way you do things and if they can be improved. Organizations in the Paralympic Movement can drastically change quickly and you must be adaptable to your new situation. You also need to be willing to learn. This seems obvious but you cannot be afraid to admit what you do not know and ask others for help. Being "coachable" is extremely important no matter what level of your career you are at. I heard once that everyone in the world is better than you at something which means never be afraid to learn from others who are above or below you in your position. You must always treat other people the way you want to be treated. You never know what people you will cross in your career and how they will impact your career in the short or long term.

case STUDY

SPORT FOR INDIVIDUALS WITH DISABILITIES

You are working with USA Hockey, the NGB for ice hockey in the United States. Recently, the USOPC has asked you to consider the inclusion of the Paralympic men's ice sledge hockey team into USA Hockey. At the 2018 Paralympic Games in Pyeongchang, South Korea, and again at the 2022 Paralympic Games in Beijing, China, the US men's team brought home the gold medal.

The ultimate goal after making this assessment is to integrate the activities of the Paralympic sledge hockey team within USA Hockey. What types of events or activities for the Paralympic sledge hockey team could USA Hockey sponsor? What types of publicity could it provide? In other words, what are some concrete strategies USA Hockey could use to help sledge hockey grow and prosper? How should you go about starting your task?

1. The best place to start is to use the nine-component model designed by Hums et al. (2003); Hums et al. (2009); and Wolff (2000) and presented in Exhibit 11.4. Using each component, show how USA Hockey could incorporate the ice sledge hockey team into its organization.
2. Think of other sport organizations that could assist you in your tasks, such as the NHL, Hockey Canada, or any other sport organization that could potentially be working with hockey players with disabilities.
3. Some people would argue that athletes with disabilities should retain a separate identity from able-bodied athletes and compete only with other athletes with disabilities. Do you agree or disagree, and why? What benefits do disability sport and athletes with disabilities offer to able-bodied sport organizations?

CHAPTER questions

1. The IPC is considering expanding the Summer and Winter Paralympic Games by one sport each. Which sports would you choose to add, and why?
2. The US Paralympic men's soccer team is playing a series of exhibition matches against the Brazilian Paralympic men's soccer team. Your college or university has been awarded one of the matches. Develop a series of marketing strategies to promote the event. What community groups and sponsors will you want to involve?

3. Go to https://www.paralympic.org/sponsors

    Here you will see the IPC's Worldwide Paralympic Partners.

    a. Name two companies that are Worldwide Paralympic Partners of the IPC.

    b. If you were to suggest two new partners the IPC could try and sign on, what company would it be, and what would be one reason for each company being a good fit *specifically* for the IPC/Paralympic Games?

    [Note: Example to help guide you – Choosing Nike and then saying it is because Nike makes great equipment is too generic and therefore incorrect. Nike makes great equipment for any sport or athlete. You need to tie your reasons for choosing your companies *specifically* to the *IPC/Paralympic Games*.]

4. A number of Paralympic athletes are talented enough to compete in the Olympic Games. What are your thoughts on Paralympic athletes competing in the Olympic Games? What effect could competing in the Olympic Games possibly have on the Paralympic Games, where they also compete? Where do you weigh in on the role technology plays in Paralympians' performances?

5. Go online and choose one Paralympic athlete and watch a video of their story:

    i. Which athlete did you choose and why?

    ii. Include the link and title for their video.

    iii. Paralympians always need to find more sponsorship money. You are now the agent for the athlete you chose. Choose one company you would approach to sponsor your athlete and explain why you chose this company *specifically* for your *Paralympic* athlete.

    [Note: Example to help guide you – Choosing Nike and then saying it is because Nike makes sponsors great athlete is too generic and therefore incorrect. Nike sponsors great athletes in many sports. You need to tie your reason for choosing your company *specifically* to your Paralympic athlete]

FOR ADDITIONAL INFORMATION

1. Youtube: We're the Superhumans - Awesome video combining Paralympians and music: https://www.youtube.com/watch?v=IocLkk3aYlk&t=40s

2. This is the Athlete Hub for the Paralympic Games where you can read Paralympic Athlete Biographies, Interviews, and Medalist Information. There is also information on the Athlete Council: https://www.paralympic.org/athletes

3. This link takes you to the full IPC Strategic Plan 2019–2022: https://www.paralympic.org/sites/default/files/document/190704145051100_2019_07+IPC+Strategic+Plan_web.pdf

4. This is a great video on Team USA's Melissa Stockwell, a wounded veteran and her Paralympic journey. She eventually won gold in Tokyo in Para-Triathlon: https://www.cbsnews.com/news/melissa-stockwell-tokyo-paralympics/

5. At this site, you can read the history of Paralympic Movement https://www.paralympic.org/ipc/history

6. Check out the WeThe15 Campaign to end discrimination of people with disabilities in society: https://www.wethe15.org

REFERENCES

Allentuck, D. (2019, June 29). Paralympians see a big welcome in a small title change. *New York Times*. https://www.nytimes.com/2019/06/29/sports/olympics/usoc-paralympians-.html

Around the Rings. (2021, July 12). *The IPC reveals new strategic direction*. https://www.infobae.com/aroundtherings/anoc-noc/2021/07/12/the-ipc-reveals-new-strategic-direction/

DePauw, K., & Gavron, S. (2005). *Disability sport* (2nd ed.). Human Kinetics.

Dixon, E. (2021, January 4). *Tokyo domestic sponsors agree to extensions for 2021*. https://www.sportspromedia.com/news/tokyo-2020-domestic-sponsors-yoshihide-suga-thomas-bach-ioc-covid/

France24. (2021, August 30). *Paralympic disability categories under fire over fairness*. https://www.france24.com/en/live-news/20210830-paralympic-disability-categories-under-fire-over-fairness

Gibson, O. (2016). Olympic Games 2016: How Rio missed the gold medal for human rights. *The Guardian*. https://www.theguardian.com/sport/2016/aug/02/olympic-games-2016-rio-human-rights

Goldblatt, D. (2016). *The Games: A global history of the Olympics*. Macmillan Publishers International.

Hoppen, T.H., & Morina, N. (2019). The prevalence of PTSD and major depression in the global population of adult war survivors: A meta-analytically informed estimate in absolute numbers. *European Journal of Psychotraumatology, 10*(1), 1–11. https://doi.org/10.1080/20008198.2019.1578637

Hums, M.A., & Pate, J.R. (2018). The International Paralympic Committee as a governing body. In *The Palgrave Handbook of Paralympic Studies* (pp. 173–196). Palgrave Macmillan. https://doi.org/10.1057/978-1-137-47901-3_9.

Hums, M.A., & Wolff, E.A. (2017). Managing Paralympic sport organisations: The STEEPLE framework. In S. Darcy, S. Frawley, & D. Adair (Eds.), *Managing the Paralympics* (pp. 155–174). Palgrave Macmillan.

Hums, M.A., Legg, D., & Wolff, E.A. (2003, June). *Examining opportunities for athletes with disabilities within the International Olympic Committee*. Paper presented at the Annual Conference of the North American Society for Sport Management, Ithaca, NY.

Hums, M.A., Moorman, A.M., & Wolff, E.A. (2009). Emerging disability rights in sport: Sport as a human right for persons with disabilities and the 2006 Convention on the Rights of Persons with Disabilities. *Cambrian Law Review, 40*, 36–48.

Lloyd, O. (2021, August 3). *International Wheelchair Basketball Federation implements key rule changes in bid to stay in the Paralympic Movement*. Inside the Games. https://www.insidethegames.biz/articles/1111167/iwbf-implement-key-rule-changes

IPC. (n.d.a). *About us*. https://www.paralympic.org/ipc/who-we-are

IPC. (n.d.b). *Committees*. https://www.paralympic.org/ipc-committees

IPC. (n.d.c). *Federations*. https://www.paralympic.org/ipc/federations

IPC (n.d.d). *IPC governing board*. https://www.paralympic.org/ipc-governing-board

IPC. (n.d.e). *IPC operational structure*. https://www.paralympic.org/ipc-structure

IPC. (n.d.f). *Paralympics history*. https://www.paralympic.org/ipc/history

IPC. (n.d.g). *Paralympic sports*. https://www.paralympic.org/sports

IPC. (n.d.h). *Team IPC*. https://www.paralympic.org/the-ipc/management-team

IPC. (2019, October). *IPC handbook – Rights and obligations of IPC members*. https://www.paralympic.org/sites/default/files/2019-11/Rights%20and%20Obligations%20IPC%20Members_2019.pdf

IPC. (2020). *IPC handbook – Paralympic sport chapter*. https://www.paralympic.org/sites/default/files/2021-08/Sec%20i%20chapter%203%20Paralympic%20Games%20Principles_2020.pdf

IPC. (2021, November 19). *Paris 2024 Paralympic medal events programme and athlete quotas announced*. Press release. https://www.paralympic.org/news/paris-2024-paralympic-medal-events-programme-and-athletes-quotas-announced

IPC Legal and Ethics Committee. (2022). *Decision of the IPC Legal and Ethics Committee: Complaint concerning alleged breaches of the IPC Code of Ethics*. IPC. https://www.paralympic.org/sites/default/files/2022-01/2022_01_07_LEC%20Decision_%20SoC%20IPC%20v%20Mr%20Charley%20Nordin_.pdf

Lloyd, O. (2021, December 11). *IPC to cease acting as IF for 10 sports by 2026*. Inside the Games. https://www.insidethegames.biz/articles/1116688/paralympic-order-ipc-general-assembly

Martinez, A., & Mann, B. (2022, March 3). *Paralympics reverses decision, denies Russia and Belarus access to Games*. NPR. https://www.npr.org/2022/03/03/1084154170/paralympics-reverses-decision-denies-russia-and-belarus-access-to-games

Molina, M. (2021, August 22). *Meet the 21 military veterans representing Team USA at the Paralympic Games*. https://blogs.va.gov/VAntage/93655/meet-the-21-military-veterans-representing-team-usa-at-the-paralympic-games/

Palmer. (2021, April 1). *IPC to allow classification at Tokyo after placing rule on hold*. Inside the Games. https://www.insidethegames.biz/articles/1106113/ipc-classification-policy- suspended

Siegfried, N., Kluch, Y., Hums, M.A., & Wolff, E.A. (2021a). *Activism starts with representation: IPC Section 2.2 and the Paralympics as a platform for social justice. Olympic and Paralympic Analysis 2020: Mega events, media, and the politics of sport*. https://olympicanalysis.org/section-5/activism-starts-with-representation-ipc-section-2-2-and-the-paralympics-as-a-platform-for-social-justice/

Siegfried, N., Green, E.R., Swim, N., Montanaro, A., & Greenwell, T.C. (2021b). An examination of college adaptive sport sponsorship and the role of cause-related marketing. *Journal of Issues in Intercollegiate Athletics, 14*, 483–500.

SportsPro. (2021, November 2). *IPC revenue falls 26.7% due to Tokyo 2020 delay*. https://www.sportspromedia.com/news/ipc-paralympics-tokyo-2020-financial-statement-profit-revenue-covid/

TeamUSA. (2021). *Military veterans*. https://www.teamusa.org/Team-USA-Athlete-Services/Paralympic-Sport-Development/Programs-and-Events/Military/Military-Paralympians

The Conversation. (2021, September 6). *After the Paralympics: New initiative to get more Canadians involved in power chair sports*. https://theconversation.com/after-the-paralympics-new-initiative-to-get-more-canadians-involved-in-power-wheelchair-sports-167013

USOPC. (2019). *US Olympic Committee changes name to US Olympic and Paralympic Committee*. https://www.teamusa.org/news/2019/june/20/us-olympic-committee-changes-name-to-us-olympic-paralympic-committee

USOPC. (2021a). *Paralympic sport development*. https://www.teamusa.org/Team-USA-Athlete-Services/Paralympic-Sport-Development

USOPC. (2021b). *Vision*. https://www.teamusa.org/about-the-usopc

Wade, S. (2021, August 27). *Paralympic classes: A confusing maze in search of fairness*. AP News. https://apnews.com/article/swimming-track-and-field-sports-b0728a158445bff870f14505d675bd0e

Ward, J.M., & Hums, M.A. (in-press). Major para-sport games. In R. Hardin & J. Pate, *Introduction to parasport and recreation*. Human Kinetics.

Wolff, E.A. (2000). *Inclusion and integration of soccer opportunities for players with disabilities within the United States Soccer Federation: Strategies and recommendations*. Senior Honors Thesis, Brown University.

Wong, G. (2002). *Essentials of sport law*. (3d ed.). Praeger.

Xue, C., Ge, Y., Tang, B., Liu, Y., Kang, P., Wang, M., & Zhang, L. (2015). A meta-analysis of risk factors for combat-related PTSD among military personnel and veterans. *PloS One, 10*(3). https://doi.org/10.1371/journal.pone.0120270

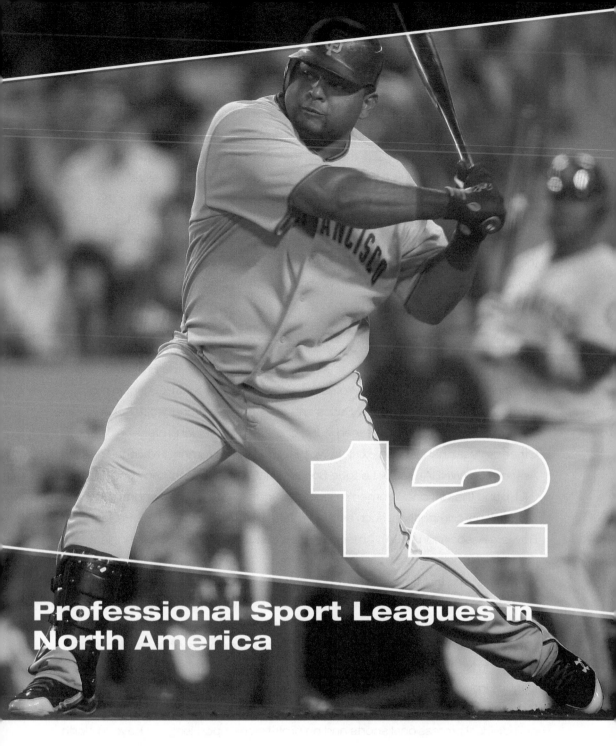

12

Professional Sport Leagues in North America

Many people dream of hitting a home run in the bottom of the ninth inning to win the World Series, making the game-winning shot at the buzzer in the National Basketball Association (NBA) Finals, throwing a touchdown pass to a wide-open receiver in the Super Bowl, or scoring the game-winning goal in the final game of the Stanley Cup. In North America, we have grown up watching these exciting moments in-person at the stadium, streamed online, on TV, or through highlights on ESPN or TSN. All these

DOI: 10.4324/9781003303183-12

spectacular plays belong to the most visible of all our sport industry segments – professional sport. Yet most of us have warning track power at best, shoot the occasional air ball, are unable to throw a true spiral, and cannot even skate backward. The odds against making it as a professional athlete in the Big Four – Major League Baseball (MLB), the NBA, the National Football League (NFL), and the National Hockey League (NHL) – are astronomical, but these athletes' games would not happen and their exploits would remain relatively unknown if not for the sport managers who work behind the scenes (Maryville University, 2022; Masteralexis et al., 2019). Without these men and women working tirelessly behind the scenes, there would be no season tickets for sale, no sponsorships available, no games broadcast on TV, no merchandise emblazoned with logos, no social media to follow, and no new high-tech stadiums to visit.

Professional sport in North America takes many forms. We think first of the before-mentioned Big Four. Professional sport also takes place in other forms: tours or organized series, such as we see in professional golf with the Professional Golfers' Association (PGA) and Ladies Professional Golf Association (LPGA); or in motor sports with the NASCAR Xfinity Series and the Camping World Truck Series. These tours and series will be addressed in the chapter in individual professional sport. Other professional leagues exist as well, including the Women's National Basketball Association (WNBA), Major League Soccer (MLS), the National Lacrosse League (NLL), and the National Women's Soccer League (NWSL). Other sports also have professional competition, such as the Professional Bull Riders (PBR), Professional Bowlers Association (PBA), American Track League (ATL), and Bass Anglers Sportsman Society (BASS). We cannot overlook the growing popularity of esports as well and for the first time, this textbook features a chapter on esports.

How are these professional sports organized? Who are the people delivering the product to the public? Who are the power players and groups governing these sports? This chapter focuses on the governance structures of professional sport leagues in North America. The authors recognize there are a wide variety of these leagues in North America, but the main focus of the chapter is on the traditional Big Four – MLB, the NBA, the NFL, and the NHL. Other chapters in this book focus on individual professional sports and on professional sport leagues beyond North America.

HISTORY AND DEVELOPMENT OF PROFESSIONAL SPORT

While a comprehensive history of the development of professional sport in North America is beyond the scope of this book, it is important to note key dates for certain events in the histories of each of the Big Four; for example, when the leagues took the form they have today. It is also important to note governance issues, such as when rival leagues formed to compete with existing leagues.

Major League Baseball

MLB has the longest history of any professional sport in North America. The first professional club, the Cincinnati Red Stockings, was founded in 1869. The first professional sports league was baseball's National League, established in 1876, followed by the American League in 1901. The signing of the National Agreement between the National League and the American League in 1903 established much of the basic governance structure for MLB (Scully, 1989). Early on, players formed a number of rival leagues such as the Players' League and the Federal League. Players objected to management's strict rules, such as the reserve system that "reserved" a player to his club and prohibited other clubs from negotiating with him, thus controlling players' salaries. While these leagues presented brief challenges to professional baseball's structure, all eventually failed (Abrams, 1998). As you will discover, this long-standing unchallenged status is quite different from the history of the other Big Four leagues.

Currently, MLB has 30 teams. The American League has 15 (five in the East, five in the Central, and five in the West Division). The National League has 15 (five in the East, five in the Central, and five in the West Division).

Major League Baseball
mlb.com

National Football League

The first professional football league, the American Professional Football Association, was established in 1920. It changed its name to the National Football League in 1922 (NFL, 2013). The NFL has experienced a long history of rival leagues, dating all the way back to its first inter-league battle in 1926 with the American Football League (AFL) (Quirk & Fort, 1992). Different forms of the AFL emerged and challenged the NFL, and finally in 1966 the leagues signed an agreement establishing an inter-league championship, and full amalgamation began in 1970. This event marked the first time that TV income played a critical role in the survival of a rival league because the AFL had an existing TV contract that by 1969 gave each team approximately $900,000 (Quirk & Fort, 1992). Since then, the NFL has been challenged by the World Football League (WFL) in 1974–75, the United States Football League (USFL) in 1983–85, and the Xtreme Football League (XFL) in 2001 with a brief return in 2020 and another return slated for 2023. Additionally, the USFL began operations again in 2022 with a new eight team line-up.

Currently, the NFL has 32 teams with two conferences, the American Football Conference (AFC) and the National Football Conference (NFC). The AFC North, AFC South, AFC East, and AFC West each have four teams, as do the NFC North, NFC South, NFC East, and NFC West.

National Football League
nfl.com

www

National Basketball
Association
nba.com

National Basketball Association

The first professional basketball league was established in 1924 as the American Basketball League. It disbanded in 1947, and another league emerged, the Basketball Association of America (BAA; Quirk & Fort, 1992). At the same time, a league called the National Basketball League (NBL) was also in existence. By 1949, the BAA absorbed the remaining NBL teams, and the organization renamed itself the National Basketball Association (NBA Hoops Online, n.d.). In 1967, the rival American Basketball Association (ABA) formed, which had sufficient financial backing and talent to pose a threat to the NBA. In 1976, however, an agreement was reached between the two leagues, resulting in four franchises from the ABA—Denver, Indiana, New York, and San Antonio—moving to the NBA (Quirk & Fort, 1992). No rival leagues have attempted to compete with the NBA since then.

Currently, the NBA has 30 teams in two conferences. The Eastern Conference has five teams each in the Atlantic, Central, and Southeast Divisions. The Western Conference has five teams each in the Northwest, Pacific, and Southwest Divisions.

www

National Hockey League
nhl.com

National Hockey League

The NHL started in 1917 with four teams located in Canada—Toronto, Ottawa, and two teams in Montreal. The league expanded to the United States in 1924 (Quirk & Fort, 1992). The NHL faced its biggest challenge from rival leagues from the World Hockey Association (WHA), a league that began in 1972. The WHA attempted to establish itself in medium-size Canadian markets without NHL franchises as well as in major United States cities in direct competition with existing NHL teams. After a seven-year war with the NHL, four remaining franchises—Winnipeg, Quebec, Edmonton, and Hartford—moved to the NHL in 1979 (Quirk & Fort, 1992).

Currently, the NHL has 32 teams in two conferences. The Eastern Conference has eight teams each in the Atlantic and Metropolitan Divisions. The Western Conference has eight teams each in the Central and Pacific Divisions. The newest NHL team, the Seattle Kraken, opened play in the 2021–2022 season as a member of the Pacific Division in the Western Conference.

GOVERNANCE

Each professional sport league has various levels of governance structures. The governance structures exist at the league level and the front-office level. The league level includes the league offices and the Office of the Commissioner, which constitute the structures on the management side of professional sport. The players' side is governed by Players Associations. Although the governance structures of the various leagues are not identical, they share common components. According to Gladden and Sutton (2011), these components are:

1. a league Commissioner
2. a Board of Governors or a committee composed of team owners

3. a central administrative unit that negotiates contracts and agreements on behalf of the league and assumes responsibility for scheduling, licensing, record keeping, financial management, discipline and fines, revenue sharing payments, marketing and promotional activities, developing and managing special events, and other functions such as coordinating publicity and advertising on behalf of the teams as a whole. (p. 125)

In addition to these components, governance issues are also dealt with at the team level. Next, we will briefly examine each of these components.

Commissioner's Office

Major professional sports leagues are led by a Commissioner. The first Commissioner in professional sport was Kenesaw Mountain Landis, who became Major League Baseball Commissioner in 1921 (Abrams, 1998). As Commissioner, he ruled with an iron fist, basing his decisions on his interpretation of what was in the game's best interest. The Commissioner is in some ways an employee of the owners, because the owners have the power to hire and terminate them. However, in other ways the Commissioner is the owners' boss, having disciplinary power over them (Wong, 2002).

The Office of the Commissioner is typically created and defined within a league's collective bargaining agreement (CBA) and its constitution and bylaws (Pacifici, 2014). The role of the Commissioner in professional sport has evolved over time, but some of the basic powers of the office have remained throughout the years in different sports. In general, the discretionary powers of the Commissioner include (Yasser et al., 2003):

- approval of player contracts
- resolution of disputes between players and clubs
- resolution of disputes between clubs
- resolution of disputes between players or clubs and the league
- disciplinary matters involving owners, clubs, players, and other personnel
- rule-making authority. (p. 381)

The power to act in the best interest of a sport has carried over from Landis' days, and according to Renicker (2016),

> Since that initial language was implemented, commissioners in other sports have utilized the clause during their tenure. As former MLB Commissioner Bud Selig stated, "[T]he intent of the best interest clause was to protect the integrity of and ensure public confidence in the game."
>
> (p. 1066)

As Wharton (2011) notes, however, "the scope of the commissioner's power—as it has developed over the years—varies from sport to sport. The NHL does not have a 'best interests' clause as baseball does, a league spokesman said. The NFL and NBA do" (para. 12). The Professional Baseball Agreement contains the following language:

Sec. 2. The functions of the Commissioner shall include:

a. To serve as Chief Executive Officer of Major League Baseball. The Commissioner shall also have executive responsibility for labor relations and shall serve as Chairman, or shall designate a Chairman, of such committees as the Commissioner shall name or the Major League Clubs shall from time to time determine by resolution.

b. To investigate, either upon complaint or upon the Commissioner's own initiative, any act, transaction or practice charged, alleged or suspected to be not in the best interests of the national game of Baseball, with authority to summon persons and to order the production of documents, and, in case of refusal to appear or produce, to impose such penalties as are hereinafter provided.

c. To determine, after investigation, what preventive, remedial or punitive action is appropriate in the premises, and to take such action either against Major League Clubs or individuals, as the case may be.

d. From time to time, to formulate and to announce the rules of procedure to be observed by the Commissioner and all other parties in connection with the discharge of the Commissioner's duties. Such rules shall always recognize the right of any party in interest to appear before the Commissioner and to be heard.

e. To appoint a President of each League to perform such functions as the Commissioner may direct.

f. To make decisions, or to designate an officer of the Commissioner's Office to make decisions, regarding on-field discipline, playing rule interpretations, game protests and any other matter within the responsibility of the League Presidents prior to 2000. (MLB, 2020, pp. 1–2)

While these basic duties remain relatively unchanged, the office itself has changed. "The modern commissioner must have expertise in TV contracts, labor relations, player health, handling Congress, handling lawsuits, criminal law, public relations, finding new revenues in new media and, as ever, owner relations. That last one might be the trickiest" (Allen & Brady, 2014, para. 6). Gary Bettman, NHL Commissioner as of this writing, had the following to say about how the office duties have evolved:

- "For better or worse, the world seems more litigious," he says. "That is something you have to manage. And being a lawyer certainly helps."
- "You need to be facile and adaptable with respect to the changes in media that are happening at breathtaking speed," he says. "You have to be on top of the industry."
- "And finally you have to be prepared to look at A.) new business opportunities, such as international and the shrinking of the world

and B.) changing old business into new businesses, such as how you distribute your licensed products," he says. "Retail in a brick-and-mortar building used to be the only way for fans to go, and now you can get anything on the Internet in 24 hours." (as quoted in Allen & Brady, 2014, para. 16)

Commissioners' actions often go unnoticed and are also frequently misunderstood by the public, as they are typically handled internally with the league. However, some notable exceptions include MLB Commissioner Rob Manfred's lack of disciplinary action against players in the Houston Astros' World Series cheating scandal, NFL Commissioner Roger Goodell's actions related to the New England Patriots' "Deflate-gate" case and on the Ray Rice domestic abuse incident, and NBA Commissioner Adam Silver's multi-million-dollar fine and lifetime ban imposed on Donald Sterling for racist remarks. As evidenced in this section, the Commissioner's Office exercises its regulatory power in professional sports leagues through decisions concerning fines, suspensions, and disciplinary actions. Some regulatory power also rests at the league level and with the Players Associations.

Board of Governors or Owners Committee

Despite a wide range of powers, the Commissioner is not necessarily the final decision maker in issues involving governance of professional sport leagues. Although the Commissioner is very influential, the owners still have the ultimate say in policy development (Robinson et al., 2001). We have all read about the annual league owners' meetings, where policies, rules, and business decisions concerning league operations are addressed. Each league has a committee structure made up of owners who ultimately make decisions on league-wide matters. In Major League Baseball, this group is called the Executive Committee and has the following powers:

Sec. 2. The Executive Council shall have jurisdiction in the following matters:

a. To cooperate, advise and confer with the Commissioner and other offices, agencies and individuals in an effort to promote and protect the interests of the clubs and to perpetuate Baseball as the national game of America, and to surround Baseball with such safeguards as may warrant absolute public confidence in its integrity, operations and methods.

b. To survey, investigate and submit recommendations for change in, elimination of, addition to or amendments to any rules, regulations, agreements, proposals or other matters in which the Major League Clubs have an interest and particularly in respect to:

1. Rules and regulations determining relationships between players and clubs and between clubs, and any and all matters concerning players' contracts or regulations; and

2. Rules and regulations to govern the playing of World Series games, All-Star Games and any other contests or games in

which Major League Clubs participate and/or games that may be played for charitable purposes.

c. In the interim between Major League Meetings, to exercise full power and authority over all other matters pertaining to the Major League Clubs, not within the jurisdiction granted to the Commissioner under this Constitution, including the adoption, amendment or suspension of Major League Rules, for said interim; provided that all actions of the Executive Council pursuant to this paragraph (c) shall be noticed for action at the next regular or special Major League Meeting for approval or other disposition.

d. To submit to the Major League Clubs recommendations as to persons to be considered for election as Commissioner whenever a vacancy may exist in that office.

e. To review and to either approve or disapprove, in whole or in part, the proposed budget submitted annually by the Commissioner for the financing of the Commissioner's Office and requests by the Commissioner for authority to incur expenses in excess thereof. (MLB, 2020, pp. 5–6)

At this level, policy making often takes place within the frameworks of each league's constitution and bylaws. As we will see shortly, the players' side has a governance structure as well. Although policy making occurs at this level, daily league operations occur at another league-wide governance level. Each league has a league office employing paid sport managers to handle these tasks.

Central Administrative Unit: League Office

As mentioned earlier, this governance level deals with league-wide operations. A unique aspect of professional sport, as opposed to other businesses, is that the teams must simultaneously compete and cooperate (Mullin et al., 2014). League offices schedule games, hire and train officials, discipline players, market and license logoed merchandise, and negotiate broadcast contracts (Sharp et al., 2010). League offices are usually organized by function, with a range of different departments. For example, the NBA League Office includes the following departments: Commissioner's Office, Basketball Operations, Basketball Strategy and Analytics, Communications, Content, Digital Media, Facilities and Administration, Finance, Global Innovation Group, Global Media Distribution, Global Partnerships, Human Resources, Information Technology, International Regional Offices, Legal, Marketing, Media Operations and Technology, NBA 2K League, NBA G League, Referee Operations, Security, Social Responsibility and Player Programs, Team Marketing and Business Operations, and WNBA (NBA, n.d.).

Individual Team Level

The day-to-day operations of a professional sports franchise take place on the individual team level. The two major groups that are responsible for daily operations—the owners and the front-office staff—are discussed in the following sections.

Owners

What drives the people who own major professional sport franchises? Today's owners are multibillionaires, many of whom are even members of the Forbes 400 richest individuals in the nation. Some of the more recognizable team owners over the years are Microsoft's Paul Allen (Seattle Seahawks), Jerry Jones (Dallas Cowboys), Magic Johnson (Los Angeles Dodgers), music mogul Jay-Z (Brooklyn Nets), and entrepreneur Mark Cuban (Dallas Mavericks). What motivates a person with money to purchase a professional sports franchise? Reasons include the excitement of being involved in professional sport and the publicity and spotlight that accompany owning a team, especially a winning team. But do not be fooled—these individuals did not accumulate wealth without keeping a sharp eye on the bottom line. According to Quirk and Fort (1999), "As important as winning is to them, it might well be a matter of ego and personal pride that they manage to do this while pocketing a good profit at the same time" (p. 97). While teams may lose money on a daily operational basis, franchises appreciate in value over time. For example, in 1979 the Lakers were sold as a part of a larger deal that included the Kings and the Forum. The price then was $16 million. The team is now worth $5.5 billion (Linn, 2021). Owners know that in the long run, few franchises have ever been sold for less than their purchase price, basically ensuring future long-term capital gains.

While some owners, such as Mark Cuban, want to be a bit more closely involved with the daily operations of their franchises, for the most part owners leave those daily chores to the people who work in the front offices. The owners' place in the policy-making process lies mainly at the league level, as discussed earlier. Some owners may impose policies on their front-office staff and players, but that is not necessarily the norm. For example, Steven Bisciotti, owner of the NFL's Baltimore Ravens, refused to silence Brendon Ayanbadejo in his support for same-sex marriage in Maryland, despite being asked to do so by a state legislator.

Front-Office Staff

The front office is the place where the day-to-day operations and business decisions are made for the individual professional sports franchise. This is where, as a future sport manager interested in working in pro sport, you may find yourself employed. Similar to league offices, the front-office staff are usually departmentalized by function. The Tennessee Titans provide an example of a typical NFL front office structure. Departments include Broadcasting; Business Intelligence and Analytics; Coaching; Community Impact; Corporate Partnerships; Creative, Content, & Event Production; Equipment; Finance; Football Administration; Football Communications; Grounds;

Human Resources; Information Technology; Legal; Marketing and Communications; Medical Staff; Player Engagement; Player Personnel; Security; Staditm Events and Experience; Stadium Operations; Team Operations; Tickets; and Video (Tennessee Titans, 2021).

Let's compare this to an NHL front office by looking at the Tampa Bay Lightning. Their departments include Arena Management, Broadcast, Business Strategy and Analytics, Executive Suite Sales, Finance, Group Sales Lightning Foundation and Community Relations, Lightning Vision, Marketing, Membership Sales and Service, Partnership Development and Activation, People Operations and Legal, Public Relations, Tampa Bay Entertainment Properties, Technology and Innovation, and Ticket Office (Tampa Bay Lightning, 2020).

As students who may want to potentially work in pro sport, you can see the wide variety of opportunities for internships and careers. The entities and governance levels described so far in this chapter all dealt with the management side of professional sport. However, governance structures also exist on the players' side of professional sport. These organizations, the players' unions, are commonly referred to as *Players Associations*.

Players Associations

Each of the Big Four professional leagues has what is known as a Players Association. These Players Associations, or PAs as they are sometimes called, are the players' unions. Professional baseball has the longest history of labor organization of all the professional sports. A player named John Montgomery Ward led the earliest unionization efforts. He founded what was considered the first players' union in 1885 – The Brotherhood of Professional Baseball Players - and later established the Players League in 1890 (Abrams, 1998), although both efforts ultimately failed. Baseball players saw themselves as skilled tradespeople, similar to those workers who filled the factories of that era. Professional baseball witnessed these and other unsuccessful attempts at unionization until the Major League Baseball Players Association (MLBPA) was established in 1953. In 1966, former United Steelworkers employee Marvin Miller became its first Executive Director (Pessah, 2016). Miller negotiated the players' first CBA in 1968. The National Hockey League Players Association (NHLPA) began in 1967 when Player Representatives from the original six clubs met, adopted a constitution, and elected a president (NHLPA, 2022b). In 1954, NBA All Star Bob Cousy began organizing the players, ultimately forming the National Basketball Players Association (NBPA; NBPA, 2022a). The NFL players' efforts were first organized in 1956, when a group of NFL players authorized a man named Creighton Miller and the newly formed National Football League Players Association (NFLPA) to represent them (NFLPA, 2022a).

MISSION STATEMENTS. Each PA has its own mission statement. The mission statement of the NBPA is presented in Exhibit 12.1. This mission statement makes it clear that the number-one priority of any PA is its members—the players—and protecting their rights.

MLB Players Association https://www.mlbplayers.com/

NHL Players Association nhlpa.com

National Basketball Players Association nbpa.com

NFL Players Association nflpa.com

The NBPA mission is to ensure that the rights of NBA players are protected and that every conceivable measure is taken to assist players in maximizing their opportunities and achieving their goals, both on and off the court.

Source: NBPA. (2022a).

PAs share the common goals of representing players in matters related to what are known as terms and conditions of employment such as wages, hours, working conditions, and players' rights. They help players with any type of dispute or problem they may encounter with management. They also deal with insurance benefits, retirement, and charitable opportunities, just as non-sport labor unions do. The NFLPA, for example, is a member of the American Federation of Labor–Congress of Industrial Organizations (AFL-CIO), a major non-sport union representing workers from a variety of industries. Here is what the NFLPA does:

- Represents all players in matters concerning wages, hours and working conditions andprotects their rights as professional football players
- Makes sure that the terms of the CBA are met
- Negotiates and monitors retirement and insurance benefits
- Provides other member services and activities
- Provides assistance to charitable and community organizations
- Enhances and defends the image of players and their profession on and off the field. (NFLPA, 2022b, para. 3)

MEMBERSHIP. The membership of any PA may include more than just active players. With the MLBPA, all players, managers, coaches, and trainers holding a signed contract with MLB are eligible for membership (MLBPA, n.d.). There is also an umpires' union named the MLB Umpires Association. For basketball, the NBPA represents players while the National Basketball Coaches Association (NBCA) is the labor union that represents NBA coaches (NBCA, 2022).

FINANCIALS. PAs rely on two primary sources of revenues. The first is individual membership dues. For example, in 2021 the MLBPA players' dues were set at $85 per day during the season (MLBPA, 2022a), while NHLPA members pay $30 per day during the regular season (NHLPA, 2022a). The second revenue source is each association's licensing division. For example, National Football League Players Incorporated, which is known as NFL PLAYERS, is the licensing and marketing arm of the NFLPA (NFLPA, 2022c). Created in 1994, its mission is "taking the helmets off" the players and marketing them as personalities as well as professional athletes.

ORGANIZATIONAL STRUCTURE. PAs share relatively common governance structures, with Player Representatives, an Executive Board, and an Executive Committee. However, ultimate power rests with the players themselves. Every year each team elects, by secret ballot, a Player Representative (called a "Player Rep") and an Alternate Player Representative to serve on a Board of Player Representatives (for the NFL) or on an Executive Board (for MLB and the NHL). Player Reps generally serve one-year terms and act as liaisons between the union and the team members. According to the NBPA (2022b), the Player Reps have the following responsibilities:

- Serving as the team's delegate during all Player Representative meetings
- Each Representative is entrusted to speak on behalf of his teammates and report back any relevant and necessary information addressed during meetings
- Nominating and electing Executive Committee Members
- Selecting the Executive Director of the union. (para. 5)

The Boards of Player Reps select the members of an Executive Committee. For the NBPA, the Executive Committee consists of nine players, and for the NHLPA the Executive Committee consists of 32 players, one representing each NHL team.

PAs also employ full-time sport managers to staff their offices. The departments of the MLBPA include Legal, Finance, Operations, International and Domestic Player Operations, Player Services, Information Technology, and the Office of the Executive Director (MLBPA, 2022b). The NHLPA main office includes people in the following positions: Executive Director, General Counsel, Divisional Player Representatives, European Affairs Representative, Chief Financial Officer, Senior Counsel, Salary Cap & Marketplace/Senior Counsel, Business Operations/Associate Counsel, Senior Director of Marketing & Community Relations, Senior Director of Player Insurance & Pensions, Director of Strategic Initiatives, Senior Manager of Communications & Media Relations, and Manager of Corporate Communications (NHL, 2022a).

CURRENT POLICY AREAS

As with any segment of the sport industry, professional sport has a myriad of policy areas to discuss. Among these are mental health concerns; players running afoul of the law; player, coach, and front office personnel misconduct; player safety; and responding to athlete activism.

Professional Athletes and Mental Health

For too many years, athletes toiled in silence when it came to their mental health. Living in a hyper-competitive and very public world, discussing any sign of perceived weakness was discouraged. Now, of

Spotlight: Diversity, Equity, & Inclusion in Action

NHL – Hockey is For Everyone

The National Hockey League (NHL) operates a program called Hockey is for Everyone. According to the NHL (2022b, paras. 1–2),

> Hockey is for Everyone™ uses the game of hockey - and the League's global influence - to drive positive social change and foster more inclusive communities. We support any teammate, coach or fan who brings heart, energy and passion to the rink. We believe all hockey programs - from professionals to youth organizations - should provide a safe, positive and inclusive environment for players and families regardless of race, color, religion, national origin, gender identity or expression, disability, sexual orientation and socio-economic status.

A quick look at the website for the program shows how the NHL wants to make hockey accessible to as many people as possible. There are links to Celebrating Black History, Celebrating Gender Equality, Celebrating Asian Pacific Islander Heritage, Celebrating Hispanic Heritage, and Pride. Indigenous Hockey and Disabled Hockey also feature prominently. The month of February marks Hockey is for Everyone month and the NHL and the NHLPA use that opportunity to promote diversity and inclusion across all levels of the sport. Since the program began in 1998, over 120,000 boys and girls have participated in Hockey for All's grassroots youth programming (NHL Public Relations, 2018).

course, it is common knowledge that mental health needs to be talked about openly – there is no shame for those living with mental health challenges. Recently, more professional athletes have spoken out about mental health issues and more sport organizations are taking actions to work with their athletes. Tennis star Naomi Osaka was among the first to speak openly in this area. Others include the NBA's Kevin Love and soccer star Andy Robertson. One estimate is that 35% of professional athletes experience issues with their mental health ranging from depression and anxiety to eating disorders and burnout (Marius, 2021). So how have major sport organizations responded?

The NBPA established a Mental Health and Wellness Department. The Department believes that "mental health and wellness is at the core of maintaining balance between a player's professional, business, personal, family, and spiritual lives" (NBPA, 2022c, para. 1). They provide a directory of vetted and licensed mental health providers and a safe space for questions and answers about getting help. The NFL has partnered with the NFLPA and

> In collaboration with the NFLPA, the NFL is committed to helping build a positive culture around mental health by providing players and the NFL family with comprehensive mental health and wellness resources and equipping them with the tools to succeed, on and off the field, over the course of their lives.
>
> (NFL Enterprises, 2022, para. 1)

The league offers a number of player resources including NFL Total Wellness, NFL Player Care Foundation, and NFL Life Line (NFL

WWW
Mental Health and Wellness Department
https://nbpa.com/mentalwellness

WWW
NFL Total Wellness
https://operations.nfl.com/inside-football-ops/players-legends/nfl-total-wellness/

Enterprises, 2022). These actions are long overdue and hopefully leagues and teams will continue to formulate policies and programs which will encourage professional athletes to be more mindful of their mental health.

Ongoing Labor Management Conflict

As pointed out earlier in this chapter, there is a long-standing history of conflict between players and management in professional sport. The roots of these issues can be traced back to the initial player reservation system which was set up with William Hulbert's formation of the National League. We witnessed how acrimonious player-management relations can be with the recent MLB lockout. This labor action was not a strike by the players but a lockout by the owners. In a lockout, the owners literally lock the players out from any official team facilities. After 99 days, the lockout ended and the 2022 MLB season proceeded with a full slate of 162 games. In the end, in addition to the on-field adoption of the universal designated hitter, some of the negotiated business outcomes included the following:

> The minimum salary will jump to $700,000 in 2022 (and increase $20,000 each year), and the luxury tax thresholds, which effectively penalize owners for overspending, will go from starting at $210 million in 2021 to $230 million in 2022 (and rise to $244 million in 2026).
> A $50-million bonus pool will be created for top young players not yet eligible for salary arbitration; a lottery system will be added for the top six spots of the amateur draft as a way to stop teams from losing to gain the No. 1 pick; the postseason will be expanded to 12 teams (the owners wanted 14, which the union argued watered it down); and there will be new ways to prevent clubs from manipulating young players' service time (those who finish first or second in the annual Rookie of the Year voting will be credited with a full year).
>
> (Wagner, 2022, paras. 28–29)

Now that MLB has settled its labor disputes for the time being, which leagues will be next? Here is a list of major North American leagues and when their current CBAs expire:

NBA	Through 2023–24 season
NHL	September 2026
MLB	Through 2026
WNBA	Through 2027
MLS	January 31, 2028
NFL	After 2030 season

So, you can see the next league up will be the NBA. It will be interesting to see what issues surface as most important to players and owners in their negotiations.

Players and the Legal System

Unfortunately, professional athletes' names are linked to run-ins with the law almost daily. The stories of current and former NFL players DeShaun Watson, Richard Sherman, Alvin Kamara, and Johnny Manziel; MLB's Marcel Ozuna and Trevor Bauer; and the list of NBA players involved in a healthcare scam are all too familiar to sports fans in North America. Since 2000, the NFL has seen 982 player arrests (Boxscore, 2022).

Dealing with off-the-field incidents is difficult from the league perspective. First, the incidents are widely covered in the media. Second, taking punitive action may give the appearance of guilt even if the legal system has not yet passed judgment on an accused athlete. Third, if a team or league controls information about an incident, it is often accused of covering up information. Finally, the issue of how the PA will react to punishing players for off-the-field incidents remains. The Personal Conduct Policy for NFL Players includes this statement:

> It is a privilege to be part of the National Football League.
> Everyone who is part of the league must refrain from "conduct detrimental to the integrity of and public confidence in" the NFL. This includes owners, coaches, players, other team employees, game officials, and employees of the league office, NFL Films, NFL Network, or any other NFL business.
> Conduct by anyone in the league that is illegal, violent, dangerous, or irresponsible puts innocent victims at risk, damages the reputation of others in the game, and undercuts public respect and support for the NFL. We must endeavor at all times to be people of high character; we must show respect for others inside and outside our workplace; and we must strive to conduct ourselves in ways that favorably reflect on ourselves, our teams, the communities we represent, and the NFL.
> (NFL, 2018, p. 1)

The policy further defines *prohibited conduct* as including, but not limited to:

Actual or threatened physical violence against another person, including dating violence, domestic violence, child abuse, and other forms of family violence;

Assault and/or battery including sexual assault or other sex offenses;

Violent or threatening behavior toward another employee or a third party in any workplace setting;

Stalking, harassment, or similar forms of intimidation;

Illegal possession of a gun or other weapon (such as explosives, toxic substances, and the like), or possession of a gun or other weapon in any workplace setting;

Illegal possession, use, or distribution of alcohol or drugs;

Possession, use, or distribution of steroids or other performance-enhancing substances;

Crimes involving cruelty to animals as defined by state or federal law;

Crimes of dishonesty such as blackmail, extortion, fraud, money laundering, or racketeering;

Theft-related crimes such as burglary, robbery, or larceny;

Disorderly conduct;

Crimes against law enforcement, such as obstruction, resisting arrest, or harming a police officer or other law enforcement officer;

Conduct that poses a genuine danger to the safety and well-being of another person; and

Conduct that undermines or puts at risk the integrity of the NFL, NFL clubs, or NFL personnel. (NFL, 2018, p. 2)

Players convicted of criminal activity are subject to discipline as determined by the Commissioner, including possible fines and suspensions. Players always have the right to appeal any sanctions.

Responding to allegations of player misconduct is a delicate area for sport managers, and one where the SLEEP Principle is very important in deciding on a course of action because the decisions have social, legal, ethical, economic, and political ramifications. Athletes are role models, and criminal activities like issues involving drugs and violence, reflect negatively on the players and the leagues. Thus, codes of conduct are necessary.

Player, Coach, and Front Office Employee Misconduct

Not all misconduct involves run-ins with the legal system. Sometimes people violate codes of conduct relative to their league. Major League Baseball faced a major scandal when the depth of cheating by the Houston Astros, who won the World Series in 2017, became public knowledge. In the wake of it all, the Astros fired manager AJ Hinch and GM Jeff Luhnow and MLB levied a $5 million fine against the team along with loss of draft picks. No players were punished, however, and the Astros' title was not revoked (Vigdor, 2021). As of this writing the team, and even players who have since been traded to new teams, still hear booing and fans yelling "cheaters!"

The NFL faced strong public scrutiny in the handling of the DeShaun Watson case. Watson was accused by 20+ women of sexual misconduct during massage sessions. Eventually, the NFL and the NFLPA reached an agreement whereby Watson would serve an 11 game suspension without pay, was fined $5 million, and was required to undergo mandatory evaluations by behavioral experts (Trotter, 2022).

The NFL had faced a prior significant on-the-field scandal with the "Bounty Gate" allegations and punishment of the New Orleans Saints. Everyone recognizes football is a violent contact sport, but in the eyes of many, players on the New Orleans Saints crossed the line. According to ESPN (2013), from 2009–11 between 22 and 27 Saints players were part of a so-called "bounty system" where they were given bonuses for hard hits and intentionally injuring opposing players. Commissioner Goodell stepped in and handed out punishments and suspensions not only to players, but to the front office as well.

> [Head Coach Sean] Payton (one season), general manager Mickey Loomis (eight games) and assistant head coach Joe Vitt (six games) were punished for not doing enough to stop the bounty system after repeated NFL warnings. The Saints also were fined $500,000 and stripped of 2012 and 2013 second-round draft choices.
>
> (Marvez, 2012, para. 25)

It is important to note how the Commissioner chose to sanction not just players, but also the coaching staff and the front office. This should serve as a signal to sport managers that they are ethically responsible for the actions of their employees.

Actions by coaches and front office staff that affect the integrity of the game remain in the news. Chris Correa, a front office employee of the St. Louis Cardinals, was sentenced to 46 months in federal prison for five counts of unauthorized access to a protected computer when he hacked into the Houston Astros internal database (Reiter, 2018). Media members also have been subject to discipline for inappropriate on-air comments. Thom Brennaman, long time Cincinnati Reds announcer, was sanctioned for a homophobic remark and Mike Milbury, an NBC hockey announcer, was let go by the network after sexist remarks. These types of activities negatively impact the integrity of each sport. Of course, no one is naïve enough to believe players and other team employees will not do everything possible to help their teams win. However, it is up to the sport managers in each organization to set the tone for ethical behavior from the top down.

Player Safety

Player safety is an issue that came to the forefront, particularly with the concerns over concussions in the NFL. Over the years, the NFL has instituted numerous safety-related rules. Within the past five years, rules have been put in place dealing with chop blocks, horse collar tackles, defenseless players, and peel back blocks. Perhaps the most hotly debated rules, however, were the ones dealing with targeting.

The NBA instituted safety rules that established a larger area that must be clear underneath the baskets and also set limits on the number of photographers along the baseline (Freeman, 2014). MLB has established new safety rules as well, starting with banning home plate collisions (Buster Posey Rule) and then later the league banned players sliding into second from kicking, shoving, or throwing their bodies into an oncoming infielder (Chase Utley Rule). As you can imagine, players are not always supportive of some of these changes, as they change the nature of the game they have played all their lives, and of course the Players Associations have weighed in on their appropriate implementation. Professional leagues recognize the importance of keeping their stars healthy and on the field.

In addition to managing player safety on the field of play, the NBA in particular has seen a number of instances where fans have encroached onto the court or behaved in such a way as to endanger the players. Racial slurs and strings of profanities have been directed at players and their families in the stands and objects have been thrown at players. Protesters have rushed onto the court in attempts to publicize their cause or voice their opposition to the actions of team owners. The NBA has a Fan Code of Conduct and fans who violate it can be removed and ultimately banned from arenas. Exhibit 12.2 includes the text of this code.

exhibit 12.2   NBA Fan Code of Conduct

The National Basketball Association seeks to foster a safe comfortable and enjoyable sports and entertainment experience in which:

Players and fans appreciate each other

Guests will be treated in a professional and courteous manner by all team personnel

Fans will enjoy the basketball experience free from disruptive behavior including foul or abusive language and obscene gestures

Guests will consume alcoholic beverages in a responsible manner. Intervention with an impaired intoxicated or underage guest will be handled in a prompt and safe manner

Guests will sit only in their ticketed seats and show their tickets when requested

Guests who engage in fighting, throwing objects, or attempting to enter the court will be immediately ejected from the arena

Guests will smoke in designated smoking areas only

Obscene or indecent messages on signs or clothing will not be permitted

Guests will comply with requests from arena staff regarding arena operations and emergency response procedures

Guests will comply with all COVID-19 health and safety protocols

A list of prohibited items at NBA arenas provided in PDF

The arena staff has been trained to intervene when necessary to help ensure that the above expectations are met and guests are encouraged to report any inappropriate behavior to the nearest usher, security guard, or guest services staff member. Guests who choose not to adhere to this code of conduct will be subject to penalty including, but not limited to, ejection without refund, revocation of their season tickets and or prevention from attending future games. They may also be in violation of local ordinances resulting in possible arrest and prosecution

The NBA thanks you for adhering to the provisions of the NBA code of conduct

Source: NBA Media Ventures. (2022b).

This is not just an NBA concern, of course. In a game between the New York Yankees and the Cleveland Guardians, Yankees' fans rained debris down on Cleveland outfielders. The NHL's Nazem Kadri received threats so vicious yet realistic that authorities had to investigate them (Shaw, 2022). On the one hand, leagues and teams want fans to be as close to the action and the athletes in person or via social media as possible to enhance their experience at a game or match. But doing so presents challenges as fans have seemingly become more unruly and bold in their actions at games. More than ever, sport managers in governing bodies such as the NBA or MLB need to establish policies on how to deal with fans who pose a threat to athletes or to other spectators.

Responding to Athlete Activism

From the days of the great Muhammad Ali and Wyomia Tyus, through the iconic stance of Tommie Smith and John Carlos with raised fists on the Olympic medal stand, to Colin Kaepernick's choosing not to stand for the national anthem to protest police brutality targeting Black Americans, and the WNBA's Atlanta Dream's political activism, brave athletes have spoken up in support of just causes in the face of public backlash. These activist athletes have provided new ways to engage in social issues, forcing responses from their professional organizations. Going back to 2012, LeBron James and the rest of the Miami Heat players posted a team photo on James' Twitter page with the hashtags #WeAreTrayvonMartin and #WeWantJustice. The photo was in response to the death of an unarmed Black teenager, Trayvon Martin, just a few miles away from the Miami organization. The Heat front office issued a statement in support of the athletes stating, "We support our players and join them in hoping that their images and our logo can be part of the national dialogue and can help in our nation's healing" (Lawrence, 2012, para. 12). In a similar instance, the St Louis Rams responded to the St Louis County Police Department after activism by their players. In 2014, five Rams' players came out for introductions in a "Hands up, Don't shoot" gesture protesting the police shooting of Black teen Michael Brown. The St Louis County Police Department, in response to the demonstration, demanded an apology from the players and the St Louis Rams organization. The Rams front office stated they respected the concerns the police department raised, but would neither apologize, nor force their players to apologize (Yan, 2014).

So much more has happened in the time since these events took pace, most notably the murders of George Floyd and Breonna Taylor, both of which sparked national and international outrage. After Floyd's death in particular, sports persons including LeBron James, Naomi Osaka, Steve Kerr, JJ Watt, Serena Williams, Dwayne Wade, Malcom Jenkins and numerous others spoke out on the injustices they saw. The WNBA dedicated its 2020 season to Breonna Taylor.

Professional athletes have also spoken out in support of the LBGQT+ community, including players like soccer star Megan Rapinoe and the NFL's Carl Nassib. Not only players, but also front office staff

MAYA BRANCH *Senior Manager, Corporate and Community Partnerships*

Atlanta Dream

I execute contractual obligations for the team's corporate partners. Each partner spends a specific amount to sponsor the team, with team elements to deliver in return. I collaborate with each department to deliver satisfactory ROI to our partners. I build relationships with corporate partners to grow revenue for the team and community partners to grow our presence in the community.

The first issue sport managers in professional sport need to focus on is fan engagement. Fans are so important! They fill the arena to cheer on their favorite team/player. If the fans are unhappy, they will stop attending/supporting the team and the team's revenue will be impacted. The team's performance will also be impacted because who wants to play in an empty arena, am I right?

The next issue is investing in women sports. By working in multiple leagues, I see the difference of investing in women sports. The male leagues aren't facing the same issues the women are with finances, lack of resources, support, etc. That makes a difference from the players/staff you sign and employ to how you market and engage with fans. We were #1 in the league this year for fan experience during our games. We had eight sellouts and a Rookie of the Year. Each year we've progressed in the way we do things. Our 2019 and 2020 seasons, we were last in the league in just about everything. The change and growth started when our previous team owners sold the team in March of last year. The new owners have given us the resources, staff, and tools to be successful. Their investment has caused major companies like Microsoft, Crown Royal, and many others to take interest in us.

The WNBA is the most progressive league when it comes to its players and social justice. We saw that with the Atlanta Dream and its active role in the elections in the state of Georgia. The league itself has established a Social Justice Council which supports education and activism on social issues. The WNBA is behind, however, on sports betting and cryptocurrency. The league has been slow to allow these companies to sponsor teams but these companies are pouring millions into other leagues. The WNBA is also behind on the resources and tools needed to progress at a faster rate. Take, for example, metrics and impressions. The league values our assets lower than the teams expect. More and more companies are wanting to invest in women sports. With that said, the values must increase in order to generate more revenue. More revenue means more resources, better players, etc.

Looking at future issues, sport managers working in professional sport need to be concentrating more on the Crypto and Meta world space. The future of both of these spaces is unlimited due to a number of variables and unknowns.

If you want to work in sports, do it for your love of the game and not for money.

Be flexible. There are only so many teams/agencies in your city. Be open to relocating.

Network and be patient. Due to the large number of people wanting to work in professional sport, often times it's who you know that can land you a job. That has been the case with most of my jobs.

have chosen to come out – most notably Billy Bean, MLB's VP of Social Responsibility and Inclusion and Ryan Resch, the Phoenix Suns vice president of basketball strategy and evaluation (Chiari, 2022; Webb & Belmonte, 2019). A high number of MLB, WNBA, and MLS teams have also hosted Pride Nights (ESPN, 2022).

We have also seen professional sports teams get actively involved in the democratic process. The WNBAs' Atlanta Dream took an active role in supporting candidate Rev. Raphael Warnock in his successful bid for election to US Senate (Gregory, 2021). In addition, over 20 NBA teams allowed their arenas to be used as polling places and election centers for the 2020 US Presidential election (NBA Media Ventures, 2022a). Using these facilities in this manner makes good sense. They are located in large cities, are spacious, and are typically located in areas readily accessible by public transport. These factors add up to making the democratic process easier for citizens to engage in. While many more examples could be listed here, it is important to note the role that teams and leagues played in supporting their athletes in standing up for their beliefs in social justice movements.

SUMMARY

Professional sport in North America takes many forms. Most prominent are the Big Four—MLB, the NBA, the NFL, and the NHL. These major sport organizations have different governance levels, including Commissioners, league offices, and individual franchise levels on the management side, and PAs on the players' side. The policy issues facing managers in the professional sport industry segment are numerous and become more complex when athletes belong to a players' union, so both sides must be cognizant of their leagues' respective collective bargaining agreements when developing policy and deciding on governance issues.

caseSTUDY

NORTH AMERICAN PROFESSIONAL SPORT

Sport organizations have increasingly engaged in cultural celebrations, such as Women's History Month, Black History Month, or Pride Month. Major League Baseball celebrates summer holidays like Memorial Day or July 4th with themed days at the ballpark.

Pick your favorite professional sports team and do some research on what audiences they could engage with in more meaningful ways. Then, plan a cultural celebration that highlights a historically excluded or marginalized group.

1. Choose one specific team
2. Choose a cultural celebration relevant to the team and the larger community where the team is located
3. For the celebration, plan the following.
    a. What on-site giveaway would you suggest?
    b. Would you plan any specific ceremonial activity (e.g., throwing out the first pitch at an MLB game) and who would you invite to take part?
    c. Which sponsors would you get involved?
    d. What types of special hats/uniforms/uniform patches could players wear that day?
    e. How would you promote the day on social media?

CHAPTER questions

1. Which of the Big Four do you consider to be the model example of a professional sport league? Why?
2. You have been hired to work in the Social Media Department of an NFL franchise. In your position, how will you interact with the various governance levels in the NFL, both directly and indirectly?
3. Who are the Commissioners of each of the Big Four, and what are their employment backgrounds? Which do you consider to be the most powerful, and why?
4. Choose one of the Big Four leagues. If you were Commissioner of that league, what are two changes you would implement to improve the league?
5. Choose a North American professional sport league other than the NBA, NFL, NHL, or MLB and answer the following about that league.
   a. Name of league
   b. When established
   c. Location of headquarters
   d. Name of person in charge (e.g., Commissioner, President, Executive Director)
   e. Explain one major current issue facing the league
   f. Explain one future issue the league needs to consider going forward
   g. Explain one ethical issue sport managers in this organization must deal with

FOR ADDITIONAL INFORMATION

1. If you wondered what a Collective Bargaining Agreement looks like, here is the link to the NFL's current CBA: https://nflpaweb. blob.core.windows.net/media/Default/NFLPA/CBA2020/NFL-NFLPA_CBA_March_5_2020.pdf
2. Here is the Major League Baseball CBA which was agreed upon to end the 2022 lockout: https://www.mlbplayers.com/_files/ugd/b0a 4c2_95883690627349e0a5203f61b93715b5.pdf
3. This link takes you to the home page of the NBPA: https://nbpa. com/
4. This article compares different Commissioners in North American sport leagues: https://www.theguardian.com/sport/2022/mar/08/ ranking-north-americas-best-sports-commissioners-and-rob-manfred
5. Here is some career advice if you want to be a General Manager: https://www.jobsinsports.com/blog/2021/11/03/ how-to-become-an-nfl-gm/

REFERENCES

Abrams, R.I. (1998). *Legal bases: Baseball and the law*. Temple University Press.

Allen, K., & Brady, E. (2014). A commissioner's job description changes over time. *USA Today*. https://www.usatoday.com/story/sports/mlb/2014/08/13/commissioner-bud-selig-adam-silver-roger-goodell-gary-bettman/14016343/

Boxscore. (2022, March 14). *A new report highlights the NFL's worst off the field defenders*. https://boxscorenews.com/a-new-report-highlights-the-nfls-worst-offfield-offenders-p164843-330.htm

Chiari, M. (2022, June 25). *Suns' Ryan Resch becomes NBA's 1st publicly out Basketball Operations Exec*. Bleacher Report. https://bleacherreport.com/articles/10039887-suns-ryan-resch-becomes-nbas-1st-publicly-out-lgbt-basketball-operations-exec?utm_source=cnn.com&utm_medium=referral&utm_campaign=editorial

ESPN. (2013, February26). *Saints bounty scandal*. http://www.espn.com/nfl/topics/_/page/new-orleans-saints-bounty-scandal

ESPN. (2022). *ESPN's guide to Pride Nights in the major professional sports leagues*. https://www.espn.com/espn/story/_/id/34030356/espn-guide-pride-nights-major-professional-sports-leagues

Freeman, E. (2014, August 26). *NBA will institute new baseline safety rules, which have been a long time coming*. Yahoo Sports. https://sports.yahoo.com/blogs/nba-ball-dont-lie/nba-will-institute-new-baseline-safety-rules--which-have-been-a-long-time-coming-004102107.html

Gladden, J., & Sutton, W.A. (2011). Professional sport. In P.M. Pedersen, J.B. Parks, J. Quarterman, & L. Thibault (Eds.), *Contemporary sport management* (4th ed., pp. 122–141). Human Kinetics.

Gregory, S. (2021, January 7). 'We did that': Inside the WNBA's strategy to support Raphael Warnock—and help Democrats in the Senate. *Time*. https://time.com/5927075/atlanta-dream-warnock-loeffler/

Lawrence, M. (2012, March 24). NY Knicks' Amar'e Stoudemire, Heat's LeBron James, Dwyane Wade show support for Trayvon Martin after tragic shooting. *New York Daily News*. www.nydailynews.com/sports/basketball/knicks/ny-knicks-amar-stoudemire-heat-lebron-james-dwyane-wade-show-support-trayvon-martin-tragic-shooting-article-1.1050095

Linn, J. (2021). Clippers and Lakers team worth revealed. *Sports Illustrated*. https://www.si.com/nba/clippers/news/clippers-and-lakers-team-worth-revealed

Marius, M. (2021, July 27). How have leading athletes addressed their struggles with mental health? *Vogue*. https://www.vogue.com/article/how-have-other-leading-athletes-addressed-their-struggles-with-mental-health

Marvez, M. (2012, May 3). *NFL suspends four players in New Orleans Saints bounty scandal*. FOX Sports. http://msn.foxsports.com/nfl/story/jonathan-vilma-scott-fujita-anthony-hargrove-will-smith-suspended-in-new-orleans-saints-bounty-scandal-050212

Maryville University. (2022). *What does a sport manager do?* https://online.maryville.edu/online-bachelors-degrees/sport-business-management/careers/what-does-sports-manager-do/

Masteralexis, L.P., Barr, C.A., & Hums, M.A. (2015). *Principles and practice of sport management* (5th ed.). Jones & Bartlett.

MLB. (2020). *Major League constitution*. https://s3.documentcloud.org/documents/6784510/MLB-Constitution.pdf

MLBPA. (2022a). *FAQ*. https://www.mlbplayers.com/faq

MLBPA. (2022b). *MLBPA staff*. https://www.mlbplayers.com/departments

Mullin, B.J., Hardy, S., & Sutton, W.A. (2014). *Sport marketing* (4th ed.). Human Kinetics.

NBA. (n.d.). *Departments*. https://careers.nba.com/departments/

NBA Hoops Online. (n.d.). *History*. https://nbahoopsonline.com/History/

NBA Media Ventures. (2022a). *NBA arenas and facilities being used for 2020 election*. https://www.nba.com/nba-arenas-polling-place-voting-center-2020-election

NBA Media Ventures. (2022b). *NBA Fan Code of Conduct*. http://nba.com/nba-fan-code-of-conduct

NBCA. (2022). *Who we are*. https://nbacoaches.com/about/

NBPA. (2022a). *About & history*. https://nbpa.com/about/

NBPA. (2022b). *Mental Health and Wellness Department*. https://nbpa.com/mentalwellness

NBPA. (2022c). *NBPA leadership*. https://nbpa.com/leadership

NFL. (2013). *NFL chronology of professional football*. http://static.nfl.com/static/content/public/image/history/pdfs/History/2013/353-372-Chronology.pdf

NFL. (2018). *National Football League personal conduct policy*. https://nflcommunications.com/Documents/2018%20Policies/2018%20Personal%20Conduct%20Policy.pdf

NFL Enterprises. (2022). *Player health and wellness – mental health*. https://www.nfl.com/playerhealthandsafety/health-and-wellness/mental-health

NFLPA. (2022a). *History*. https://www.nflpa.com/about/history

NFLPA. (2022b). *How the NFLPA works*. https://nflpa.com/about

NFLPA. (2022c). *NFL Players Inc*. https://nflpa.com/players-inc

NHL. (2022a). *NHLPA office*. https://records.nhl.com/organization/nhlpa-offices

NHL. (2022b). *Hockey is for Everyone*. https://www.nhl.com/community/hockey-is-for-everyone

NHL Public Relations. (2018). *Hockey is for Everyone Month begins*. https://www.nhl.com/news/hockey-is-for-everyone-month-begins/c-295488554

NHLPA. (2022a). *FAQs*. https://www.nhlpa.com/the-pa/what-we-do/faq

NHLPA. (2022b). *What we do*. https://www.nhlpa.com/the-pa/what-we-do

Pacifici, A. (2014). Scope and authority of sports leagues disciplinary power: Bounty and beyond. *Berkeley Journal of Entertainment and Sports Law, 3*(1), 93–115.

Pessah, J. (2016). *The game: Inside the secret world of Major League Baseball's power brokers*. Little, Brown & Company.

Quirk, J., & Fort, R. (1992). *Pay dirt: The business of professional team sports*. Princeton University Press.

Quirk, J., & Fort, R. (1999). *Hard ball: The abuse of power in pro team sports*. Princeton University Press.

Reiter, B. (2018, October 4). What happened to the Houston Astros hacker? *Sports Illustrated*. https://www.si.com/mlb/2018/10/04/chris-correa-houston-astros-hacker-former-cardinals-scouting-director-exclusive-interview

Renicker, C. (2016). A comparative analysis of the NFL's disciplinary structure: The Commissioner's power and players' rights. *Fordham Intellectual Property, Media and Entertainment Law Journal, 26*(4), 1051–1113.

Robinson, M.J., Lizandra, M., & Vail, S. (2001). Sport governance. In B.L. Parkhouse (Ed.), *The management of sport: Its foundation and application* (3rd ed., pp. 237–269). McGraw-Hill.

Scully, G.W. (1989). *The business of Major League Baseball*. University of Chicago Press.

Sharp, L.A., Moorman, A.M., & Claussen, C.L. (2010). *Sport law: A managerial approach* (2nd ed.). Holcomb Hathaway.

Shaw, R.A. (2022, May 24). "Criminal in any other setting": Do fans pose a threat to NBA and MLB stars? *The Guardian*. https://www.theguardian.com/sport/2022/may/24/nba-mlb-fans-players-confrontations-sports

Tampa Bay Lightning. (2020). *Lightning ownership and front office*. https://www.nhl.com/lightning/team/front-office

Tennessee Titans. (2021). *Front office*. https://www.tennesseetitans.com/team/front-office-roster/

Trotter, J. (2022, August 18). *Cleveland Browns QB Deshaun Watson suspended 11 games, fined $5 million after settlement between NFL, NFLPA*. ESPN. https://www.espn.com/nfl/story/_/id/34412381/source-cleveland-browns-qb-deshaun-watson-suspended-11-games-fined-5m-settlement-nfl-nflpa

Vigdor, N. (2021, November 3). The Houston Astros' cheating scandal: Sign stealing, buzzer intrigue and tainted pennants. *New York Times*. https://www.nytimes.com/article/astros-cheating.html

Wagner, J. (2022, March 10). Play ball! Lockout ends as MLB and union strike a deal. *New York Times*. https://www.nytimes.com/2022/03/10/sports/baseball/mlb-lockout-ends.html#:~:text=It%20took%2099%20days%20of,day%20scheduled%20for%20April%207

Webb, B., & Belmonte, A. (2019, November 11). *MLB player-turned-LGBTQ advocate discusses 'living a secret life'*. Yahoo Sports. https://www.yahoo.com/video/mlb-is-leading-the-way-for-lgbtq-inclusion-billy-bean-183252357.html

Wharton, D. (2011). Commissioners walk a fine line. *Los Angeles Times*. http://articles.latimes.com/2011/may/15/sports/la-sp-0515-commissioner-power-20110515

Wong, G.M. (2002). *Essentials of sport law* (3rd ed.). Praeger.

Yan, H. (2014, December 2). *St. Louis County police, Rams spar over reported apology*. CNN. www.cnn.com/2014/12/02/us/ferguson-nfl-st-louis-rams/index.html

Yasser, R., McCurdy, J., Goplerud, P., & Weston, M.A. (2003). *Sports law: Cases and materials* (5th ed.). Anderson Publishing.

Professional Individual Sports

The previous chapter focused on the governance of North American professional team sports. As mentioned in that chapter, not all professional sports are team sports. Individual sports also have a professional component. The governance of these sports' competitions differs from the governance of professional team sports and leagues. For example, individual sports such as professional golf require athletes to complete

DOI: 10.4324/9781003303183-13

extensive qualifying requirements before they can compete in select events. Other individual sports such as professional tennis place a heightened focus on grassroots development, similar to national governing bodies (NGBs) in the Olympic Movement. Overall, individual sport organizations develop and focus their rules and structures based on individual athletes' needs rather than on a team.

Each year in the United States alone, an estimated 37 million people play golf (National Golf Foundation, n.d.) and over 21.6 million people play tennis (USTA, 2021a). These numbers reflect participation on the recreational level, but as we all know, professionals compete internationally in organized professional tours in both sports. NGBs such as the US Golf Association and the US Tennis Association play integral roles in the governance and growth of these sports. Automobile racing such as NASCAR (National Association for Stock Car Auto Racing) represents another individual professional sport. Speedways dot the countryside, ranging from small-town tracks like the Salem Speedway in Salem, Indiana, or Dixie Speedway in Woodstock, Georgia, to superspeedways like Talladega in Alabama or Daytona in Florida, which accommodate over 100,000 spectators.

Internationally, individual sports are also on the rise, and their governing bodies operate differently depending on the nature of sport. For instance, Surf Australia is supported by the Australian Sports Commission and Australian Olympic Committee. Similar to the United States, the purpose of Australia's involvement with such non-mainstream sports is to facilitate grassroots development and to promote individual lifestyle sports rather than focusing solely on the popularity of the sport.

Although there are many differences between these professional sports and professional leagues and teams, one aspect remains the same – the need for governing structures to establish rules, regulations, and policies. This chapter examines golf, tennis, automobile racing, and surfing and each sport's history, governance structure, and current policy issues.

PROFESSIONAL GOLF

In the US, four major organizations are involved in the governance of golf: the United States Golf Association (USGA), the Professional Golfers' Association of America (PGA of America), the PGA TOUR, and the Ladies Professional Golf Association (LPGA). Although all four work closely together, they have different purposes (see Exhibit 13.1). The USGA oversees the regulations and rules of golf and equipment standards. The PGA and LPGA serve professional golfers, and the PGA TOUR organizes national tour events.

Professional Golf Organization Responsibilities in the United States *exhibit* **13.1**

USGA

Establishes Rules of Golf

Determines equipment standards

Sets handicaps and course rating systems

Sponsors turf management research

Operates 14 national championships

PGA OF AMERICA

Focuses on professional instruction and golf management

Operates 5 major championships:

Ryder Cup

PGA Championship

PGA Grand Slam of Golf

Senior PGA Championship

Women's PGA Championship

Sponsors a juniors golf program

PGA TOUR

Focuses on professional play

Hosts 48 events for 3 tours:

PGA Tour

Champions Tour

Nationwide Tour

LPGA

LPGA Professionals

Focuses on female golf instruction

LPGA Tour

Focuses on female professional play

Operates 34 tournaments

Operates the Epson Tour with 20 events

Sponsors a juniors golf program

Source: LPGA. (n.d.a).

USGA

History

United States Golf
Association
https://www.usga.org/

The USGA, formally known as the Amateur Golf Association of the United States, is a non-profit organization established in 1894 as the central body of golf in the United States.

Governance

MISSION. The USGA's mission is "to champion and advance the game of golf" (USGA, n.d.a, para 1). The USGA has six main tasks: (a) host championships; (b) write, interpret, and educate on the golf rules; (c) promote outstanding playing experiences through turf management; (d) deepen engagement with golfers through programs, education, and services; (e) expand participation and advocacy; and (f) celebrate the game's history. Currently, the organization also hosts 14 national championships, including four open championships (US Open, US Women's Open, US Senior Open, and US Senior Women's Open) and 10 amateur championships each year. Starting in 2022, the USGA will also begin offering the US Adaptive Open for golfers with disabilities (O'Neill, 2021).

At the international level, the Royal and Ancient Golf Club of St Andrews was founded in Scotland in 1754 to serve the United Kingdom and other countries. Until recently, this organization oversaw all the golfers and tours around the world except those in the United States. In 2004, a major reorganization took place that formed a separate entity called the R&A, which is independent from the Royal and Ancient Golf Club of St Andrews, to take over joint administration of the rules of golf with the USGA. The R&A is responsible for administering the rules of golf for over 30 million golfers in 128 countries in Europe, Africa, Asia-Pacific, and the Americas, while the R&A and USGA jointly develop and issue the rules of golf for the United States and Mexico. In addition, the R&A organizes the Open Championship (British Open) as well as other amateur and junior events sanctioned by golf governing bodies around the world.

R&A
https://www.randa.org/

MEMBERSHIP. As of 2017, the USGA listed membership as over 700,000, with 8,000 member clubs. USGA membership is open to both amateur and professional golfers – in essence, to anyone interested in playing golf. Golf clubs, public and private, can also be members of the USGA. In fact, USGA member club representatives control over 10,600 golf courses nationwide, and more than 680 golf clubs hold qualifying rounds for USGA or state golf championships. Finally, USGA membership includes approximately 130 men's and women's state and regional golf associations that provide services to millions of golfers across the United States (USGA, 2017).

FINANCIALS. The USGA's main sources of revenue are generated from broadcast rights of their championships and championship revenues. In 2020, a year impacted heavily by COVID-19, the USGA made $90 million in media rights (down from $113,594 million in

2019). The USGA only made $3 million from their championship reve-
nues in 2020, a stark decrease from the $45 million made in champion-
ship revenues in 2019. Why the grand change? COVID-19 caused the
USGA to eliminate all but four of their championships and to have no
fans present at the events. Other revenues for the USGA include USGA
memberships, corporate sponsorships, and service/other revenues.
Championships are an important part of the revenue for the USGA.
In a typical year, the USGA makes 75% of its revenue from the US
Open alone (Nichols, 2020). To compensate for the lost revenues, the
USGA decreased their expenses in golfer engagement (–30%), governance (–23%), golf course maintenance and sustainability (–34%), and
administration and other costs (–26%; USGA, 2021). The USGA pro-
vided $18.5 million in grants and championship prize money, including
$12.5 million for the 2021 US Open (Boone, 2021). From the revenues
generated, the USGA spends a large portion of its earnings supporting
grassroots-level development programs such as The First Tee; LPGA–
USGA Girls Golf; Drive, Chip & Putt Championship; and PGA Junior
League Golf. The USGA also provides financial support for the Special
Olympics, which hosts the Special Olympics Golf National Invitational
Tournament annually (USGA, n.d.b).

ORGANIZATIONAL STRUCTURE. The leadership of the USGA
is organized as shown in Exhibit 13.2.

PGA OF AMERICA AND PGA TOUR

History

The PGA of America (henceforth PGA) was founded in 1916 in New
York City with 35 charter members. These golf professionals and ama-
teurs believed the formation of a golf association would improve golf
equipment sales. The inaugural PGA Championship was held from
October 10 through October 14 that same year at Siwanoy Country
Club in Bronxville, New York, and the organizers awarded the winning
trophy and $2,580 in prize earnings. In 1917, the USGA extended priv-
ileges to the PGA and allowed the organization to choose Whitemarsh
Valley Country Club in Pennsylvania to host the US Open (PGA, n.d.b).
With the continuous support of professional golfers, the PGA published
the first issue of the *Professional Golfer of America* in 1920. The close
relationship between the PGA and the USGA flourished when the USGA
adopted the PGA's suggestion to host the US Open annually in June.
They also worked together in adopting the new steel iron club technology, and the USGA legalized the PGA line of irons in 1926. The two
organizations still maintain this relationship today as PGA and PGA
TOUR golfers compete under the rules and guidelines established by the
USGA (PGA, n.d.b).

As an individual sport, professional golf is divided into two distinct
organizations: the PGA and the PGA TOUR. The PGA serves male and
female professional instructors, players, and local clubs, while the PGA
TOUR is the tournament division organizing men's professional golf
tours in North America. In 1968 the PGA TOUR separated from the

WWW

Professional Golfers'
Association of America
(PGA)
https://www.pga.com/

WWW

PGA TOUR
https://www.pgatour.com/

exhibit **13.2**   Organization of the USGA Leadership

EXECUTIVE COMMITTEE

The USGA's 15-member volunteer group that serves as the Association's executive policy-making board.

EXECUTIVE LEADERSHIP TEAM

The USGA's management team that directs and oversees the Association's day-to-day operations.

WOMEN'S COMMITTEE

The 14-member committee that helps conduct the USGA's women's championships.

Source: USGA. (n.d.c).

PGA to operate the Tournament Players Division. Although the separation was considered risky at the time, it seemed inevitable due to the two distinct groups of professional golfers involved in the sport (Gorant, 2018). One group of golf professionals includes players who compete regularly on national tours (e.g., Dustin Johnson, USA; Rory McIlroy, Ireland; Hideki Matsuyama, Japan; John Rahm, Spain), while the second group represents professional golfers operating or teaching golf at local country clubs and golf facilities. The separation became necessary to better serve golf professionals with different needs. For example, players on national tours are trained toward perfecting the game of golf and mastering the skills necessary to win tour events. In comparison, teaching professionals are trained to help amateur golfers learn and better understand the fundamentals and mechanics of the golf swing and other elements of the game. Some golf professionals are also trained to operate, maintain, and design golf courses.

Since making this distinction, the PGA and the PGA TOUR organize and operate separate major tournaments. The PGA operates five major golf championships: the Ryder Cup, the PGA Championship, the PGA Grand Slam of Golf, the Women's PGA Championship, and the Senior PGA Championship (PGA, n.d.c) while the PGA TOUR hosts 47 annual events for three tours: the PGA Tour for qualified professionals, the Champions Tour for players 50 and over, and the Korn Ferry Tour for professionals who have not qualified for the Tour card or are not on the PGA TOUR (PGA TOUR, n.d.a). An athlete who demonstrates top performance in the Korn Ferry Tour may compete in the PGA TOUR the following year. In fact, Phil Mickelson and Tiger Woods competed in the (then) Nationwide Tour prior to joining the PGA TOUR. When players turn 50, they are eligible to compete in either the PGA TOUR or the Champions Tour. Details of the tour qualifications are discussed in the membership section.

Governance

MISSION. The mission of the PGA of America is "to help you navigate your golf journey so that you can take it as far as you want to go" (PGA, n.d.c, para. 1), while the PGA TOUR's mission is "By showcasing golf's greatest players, we engage, inspire and positively impact our fans, partners and communities worldwide" (PGA TOUR, 2021a, para. 1). As stated earlier, the PGA and PGA TOUR both serve professional golfers with distinctively different purposes. The PGA provides services and support for teaching professionals, while the PGA TOUR only serves professional athletes who play for national tours.

MEMBERSHIP. The PGA is currently comprised of 29,000 qualified men and women professionals teaching and managing the game of golf in its 41 PGA sections (PGA, n.d.c). The PGA membership license features 24 different categories depending on the type of qualification each professional has earned as shown in Exhibit 13.3 (PGA, 2022).

Conversely, PGA TOUR membership is exclusive to professional golfers who have earned a PGA TOUR card by finishing in the top 25 on the Korn Ferry Tour in regular season points, finishing in the top 25 after the three-event Korn Ferry Tour Finals, winning three Korn Ferry Tour tournaments in a season, or gaining a PGA TOUR Special Temporary Membership through sponsor exemptions and Monday qualifiers. Players receiving exemptions to the above qualifications include former major champions, former multiple tournament winners, and those listed in the top 50 in lifetime career earnings or listed numbers 126–150 on the money list the previous year among 39 total exemptions (PGA TOUR Media Guide, 2021).

FINANCIALS. While COVID-19 certainly impacted every sport in 2020, the PGA and the entire golf industry were met with less of an impact than other sports or leagues. In 2020, the PGA saw revenues (in millions) of $123,722 from championships, $24,417 from partnership development, $18,826 in golf course operations, $9,227 from PGA REACH, $9,701 from membership benefit programs, $5,402 from member championships and $2,348 from member dues, to name the highest revenue generating activities (Professional Golfers' Association of America, 2020). Total revenues in 2020 equated $204,900, an increase from 2018s revenue of $195,682 (Professional Golfers' Association of America, 2018). Championships represented the PGA of America's highest expense items in 2020 ($97,624), followed by corporate services ($44,078), golf course operations ($19,147), partnership development ($14,041), section affairs ($11,916), marketing and communication ($11,533), and PGA REACH ($10,545; Professional Golfers' Association of America, 2020). The total expenses for the PGA of America in 2020 ($253,063) experienced a 9% increase from 2018 expenses ($231,838; Professional Golfers' Association of America, 2018).

Similar to the PGA, the PGA TOUR generates revenues from tournament operations, sponsorships, licenses, merchandise sales, membership dues, and network media rights deals. The PGA TOUR signed a media rights deal with CBS, NBC, and ESPN, an agreement that will run from *exhibit* **13.3**   PGA of America Member Classifications

CLASSIFICATION	DESCRIPTION
Member: A-1 Associate: B-1	**Head Professional at a PGA Recognized Golf Course** The term "Head Golf Professional" shall refer to an individual whose primary employment is: the ownership and operation of a golf shop at a PGA Recognized Golf Facility; or the supervision and direction of the golf shop and supervision of teaching at a "PGA Recognized Golf Facility."
Member: A-2 Associate: B-2	**Head Professional at a PGA Recognized Golf Range** The term "Head Golf Professional" shall refer to an individual whose primary employment is: the ownership and operation of a golf shop at a PGA Recognized Golf Facility; or the supervision and direction of the golf shop and supervision of teaching at a "PGA Recognized Golf Facility."
Member: A-3 Associate: Not Applicable	**Exempt PGA Tour, Champions Tour, Nationwide Tour, LPGA Tour, and Futures Tour players**
Member: A-4 Associate: B-4	**Director of Golf at PGA Recognized Golf Facilities** The term "Director of Golf" shall refer to an individual who directs the total golf operation of a PGA Recognized Golf Facility, including the golf shop, golf range, golf car operations (if applicable) and supervision of the Head Golf Professional.
Member: A-5 Associate: Not Applicable	**Past Presidents of the Association**
Member: A-6 Associate: B-6	**Golf Instructor at a PGA Recognized Facility** PGA Members employed as golf instructors, golf teachers, or golf coaches, including both in-person and online.
Member: A-7 Associate: B-7	**Head Professional at a PGA Recognized Facility Under Construction** Individuals employed as Directors of Golf or Head Golf Professionals at PGA Recognized Golf Facilities under construction.
Member: A-8 Associate: B-8	**Assistant Golf Professional at a PGA Recognized Facility** The term "Assistant Golf Professional" shall refer to an individual who is primarily employed at a PGA Recognized Golf Facility and spends at least 50% of the time working on club repair, merchandising, handicapping records, inventory control, bookkeeping and tournament.
Member: A-9 Associate: B-9	**Employed in Professional Positions in Management, Development, Ownership Operation and/or Financing of Facilities** Individuals who are employed in professional positions in management, development, ownership, operation and/or financing of facilities. (Employment at more than two facilities: Individuals who are involved in the management of more than two facilities, regardless of positions, titles or responsibilities shall be classified A-9 or B-9.)

(Continued)

PGA of America Member Classifications (*continued*)   *exhibit* **13.3**

CLASSIFICATION	DESCRIPTION
Member: A-10 Associate: B-10	**Golf Clinician** The term "Golf Clinician" shall refer to an individual whose main source of income is golf shows or clinics.
Member: A-11 Associate: B-11	**Golf Administrator** Individuals who are employed by the Association, a Section or the PGA Tour in an administrative capacity and individuals who are employed full-time as employees of golf associations recognized by the Board of Directors.
Member: A-12 Associate: B-12	**College or University Golf Coach** Individuals who are employed as golf coaches at accredited colleges, universities and junior colleges.
Member: A-13 Associate: B-13	**General Manager** Individuals who are employed as General Managers/Directors of Club Operations who have successfully completed the requirements set forth by the PGA Board of Directors. (General Managers/Directors of Club Operations shall manage the entire golf facility including golf operations, golf course maintenance, clubhouse administration, food and beverage operation and other recreational activities at the facility.)
Member: A-14 Associate: B-14	**Director of Instruction at a PGA Recognized Facility** PGA Members who are employed as Director of Instruction who is managing, supervising and directing the total teaching program, whether in-person or online, and individuals who instruct PGA Professionals how to teach/coach.
Member: A-15 Associate: B-15	**Ownership or Management of a Retail Golf Facility** Individuals whose primary employment is ownership or management of golf products or services at a "PGA Recognized Retail Facility" provided such employment specifically excludes primary employment as a clerk.
Member: A-16 Associate: B-16	**Golf Course Architect** Individuals who are primarily employed in the design of golf courses as architects or individuals who are primarily employed in an ownership or management capacity as golf course builders.
Member: A-17 Associate: B-17	**Golf Course Superintendent** Individuals primarily employed in the management of all activities in relation to the maintenance, operation, and management of a golf course. Individuals in this classification are required to satisfy the criteria of either a Golf Course Superintendent or Assistant Golf Course Superintendent as defined by the Golf Course Superintendent's Association of America.
Member: A-18 Associate: B-18	**Golf Media** Individuals primarily employed in the reporting, editing, writing or publishing of golf-related publications in any form of media (inclusive of, but not necessarily limited to, newspapers, magazines, and the Internet) or in the broadcasting or commentating about golf events on network television, cable networks, the Internet or any other form of related media.
Member: A-19 Associate: B-19	**Golf Manufacturer Management** Individuals primarily employed in an executive, administrative or supervisory position with a golf industry manufacturer or golf industry distributor.

(Continued)

exhibit **13.3**   PGA of America Member Classifications (*continued*)

CLASSIFICATION	DESCRIPTION
Member: A-20 Associate: B-20	**Golf Manufacturer Sales Representative** Individuals primarily employed by one or more golf manufacturing or distributing companies involved in the wholesale sales and distribution of golf merchandise or golf-related supplies to golf facilities, retail stores, or any other golf outlets.
Member: A-21 Associate: B-21	**Tournament Coordinator/Director for Organizations, Businesses or Associations** Individuals primarily employed in the coordination, planning and implementation of golf events for organizations, businesses, or associations.
Member: A-22 Associate: B-22	**Rules Official** Individuals primarily employed in the provision of services as a rules official for recognized golf associations, recognized golf tours, or recognized golf events.
Member: A-23 Associate: B-23	**Club Fitting/Club Repair** Individuals primarily employed in the business of club fitting must use a recognized fitting system or a comparable system, must have all the necessary equipment normally associated with club fitting and must have access to a PGA Recognized Golf Range or a range at a PGA Recognized Golf Course to monitor ball flight. Individuals primarily employed in club repair must have an established place of business with all necessary equipment normally associated with club repair or must service one or more golf tours or series of golf events.
Member: A-24 Associate: B-24	**Employed within the golf industry and not eligible for another Active classification** To be eligible to transfer to the A-24 / B-24 classification, the following criteria must be met: The PGA Member must be primarily employed in the golf industry, or have employment duties requiring the expertise of a PGA Member that prove vital to providing a golf-related service. For purposes of the regulation, the term "Golf Industry" is defined as a business that provides primarily golf-related products or golf-related services to consumers, wholesalers, distributors, retailers, PGA-affiliated facilities, or others.

Source: PGA (2022).

2022 to 2030, and an exclusive cable television agreement with the Golf Channel. While financials were not revealed, SportsBusiness Journal did report that the deal could be as much as $700 million per year, equating to close to $2.7 billion at the end of the contract (Beall, 2020). Sponsorship revenues come from major corporate partners, and the organization has negotiated merchandise license contracts with many major golf brands. PGA TOUR expenses take the form of prize money, salary and benefits, and tournament operations. As a major part of its mission, PGA TOUR events have donated more than a total of $3.05 billion ($204.3 million in 2019 alone) to help over 2,000 charities and countless individuals (PGA TOUR, 2020).

ORGANIZATIONAL STRUCTURE. The PGA's national office Board of Directors is elected by the organization's Board Members. The members serve a minimum of one year as an officer and become eligible for re-election and re-appointment after their first term. The national office also has a President, Vice President, Secretary, Chief Executive Officer, Chief Championships Officer, Chief Financial Officer, Chief Commercial Officer, Chief Innovation Officer, Chief People Officer, Chief Membership Officer, Chief Revenue Officer, General Counsel, and 19 Directors who establish association policies. The Directors include representatives from each of the PGA's 14 districts, three independent directors, a member of the PGA TOUR, and an at-large director (PGA, n.d.d).

The PGA TOUR is a tax-exempt membership organization with multiple Executive Officers including Commissioner, Chief Operating, Chief Media, Chief Administrative, Chief Legal, and also Chief Financial Officers. There are also Executive Vice Presidents for Player Relations, Licensing, Media Content and Communications, Marketing and Corporate Partnerships, Social Responsibility and Inclusion, and International (PGA Tour, n.d.b). In addition, the leadership for the PGA TOUR includes five board directors, four-player directors, and a 16-member Player Advisory Council.

LPGA

History

To meet the needs of female golfers, the LPGA was founded in 1950 by 14 pioneering women seeking to create a full professional tour. Over the years, the LPGA has grown to provide 34 official events in 2022 for a total of $85.7 million in official purses (LPGA, 2021). Similar to the PGA and the PGA TOUR, in 1959 the LPGA established the LPGA Teaching division, called the LPGA Professionals, and the LPGA Tour to serve two types of golf professionals. In 1980, the LPGA also created Duramed Futures Tour (currently known as the Epson Tour) to assist players at the developmental level. The creation of the event proved successful: more than 530 players have moved on to the LPGA Tour over the years (LPGA, n.d.a). To strengthen the grassroots development of women's golf, in 1991 the LPGA Foundation started to support junior golf programs. The LPGA and USGA also jointly created LPGA–USGA Girls Golf to increase their grassroots developmental programming. The LPGA has earned the distinction of being "one of the longest-running women's professional sports associations in the world" (LPGA, n.d.a, para. 1).

www

Ladies Professional Golf Association (LPGA)
https://www.lpga.com/

Governance

MISSION. The LPGA's mission is to "be a recognized worldwide leader in the world of sport by providing women the opportunity to pursue their dreams through the game of golf" (LPGA, n.d.c, para. 1). The organization's values include For Women of Golf, Play it Forward,

An Open Book, Act like a Founder, and Role Reversal. Separate from the PGA and PGA TOUR, the LPGA specifically focuses on serving all professional female golfers around the world, including teaching professionals and professional athletes on tour. The LPGA, PGA, and PGA TOUR have different missions and visions, yet the three organizations strive to improve the game of golf and increase the number of people watching and playing the sport.

MEMBERSHIP. The LPGA represents the ultimate governing body for female golf professionals. For female tour professionals, the organization administers an annual qualifying school (Q-School) and operates the Epson Tour, providing privileges for top finishers to join the LPGA Tour the following year.

For the LPGA Professionals members, the qualification and certification processes are similar to PGA members. One major difference between the LPGA and PGA is the type of licenses available to their respective members. Although the PGA provides 31 different membership categories, the LPGA only provides three membership types (Level I, II, and III) for teaching and operation. Female professionals who wish to obtain specialty licenses, for example as a college or university golf coach or golf course superintendent, must achieve certification through the PGA. In 2021, the LPGA served approximately 530 tour professional members and 1,800 LPGA Teaching and Club Professionals (T&CP) members (LPGA, n.d.a).

FINANCIALS. Similar to the PGA and PGA TOUR, the LPGA generates revenues from sponsorships, golf facility management, licenses, merchandise sales, membership dues, tournament operations, and network media rights deals. While it is unclear how much is made from TV money, in 2019, the PGA and LGPA Tours let the PGA negotiate the television contract jointly. Given the PGA TOUR wanted to negotiate a 70% increase in rights fees to $700 million, the LPGA TOUR was hoping to be part of that large revenue stream. It has been reported that the PGA has kept 90%–95% of the revenues from that deal, with the LPGA only receiving the remaining amount (Kaplan, 2021). The LPGA also receives sponsorship revenues from numerous major marketing partners (e.g., Callaway Golf, Getty Images, Rolex Watch USA, Inc., etc.) and has various official licensees (e.g., Adidas, Cutter & Buck, and Greg Norman; LPGA, n.d.d). For major expenses, the LPGA spends a great deal on prize money, including a record-setting $84 million in purses for all 34 events in 2022 (LPGA, 2021).

ORGANIZATIONAL STRUCTURE. As a non-profit organization, the LPGA is under the guidance of a Commissioner. The LPGA executive team also includes a Chief Communications Officer, Chief Tour Operations Officer, Chief Teaching Officer, Chief Business Officer, Chief Legal Officer, and Chief Financial Officer (LPGA, n.d.b). The Board of Directors is composed of six independent directors, including the LPGA Player Directors (Player Executive Committee), the National President of the LPGA Professionals, and the Commissioner of the LPGA (LPGA, n.d.a). Similar to the PGA TOUR, LPGA officials make decisions on player eligibility, suspensions, and disqualifications while adhering to USGA golf rules.

Spotlight: Diversity, Equity, & Inclusion in Action

Black Girls Golf

Black Girls Golf (BGG) started as an idea in 2011. BGG founder, Tiffany Mack Fitzgerald, noted how many men in corporate America would go golfing which provided ample professional and personal opportunities. Tiffany learned the game with a steep learning curve to help her professional career. Tiffany noted that not many women, especially women of color, were seen on the golf course. In 2013, she sent out a call to Black women to join her on the greens. When 26 women showed up, Tiffany knew she had to create BGG to help grow the game for Black women and girls.

The aim of the organization is to help Black women and girls learn, practice, and play golf (Black Girls Golf, n.d.a). The league is not competitive, but rather is a tool to teach golf and provide networking opportunities for its members. Memberships have various levels. The Free BGG Membership allows members to get digital learning resources, invitations to meet-ups and events, access to VIP registration for BGG-hosted events, and direct access to certified LPGA and PGA teaching professionals. The paid Premium BGG Membership also allows access to the BGG member directory to connect with other BGG members, special rates for BGG events and merchandise (Black Girls Golf, n.d.b). Some of BGG's partners include Dewar's, BMW, and the PGA.

CURRENT POLICY AREAS

The professional golf industry faces potential challenges and growth opportunities in several key areas: Olympic qualifiers, taking advantage of opportunities provided by COVID-19, and attracting younger audiences.

Olympic Qualification System

Golf returned to the Olympic Games at the 2016 Rio Olympic Games for the first time since 1904 and remained on the Olympic Games Programme for the 2020 Tokyo Games. The qualification system decided on by the International Olympic Committee (IOC) and the International Golf Federation (IGF) limited the field to 60 players over 72 holes of stroke play for the men's and women's competitions. Among the 60 players, 58 athletes must qualify through the Olympic Golf Rankings (OGR) and two spots are reserved for host country athletes. Regardless of player ranking, no more than four players are allowed to compete from one country (Murphy, 2021). The men's OGR recognizes most of the golf tours such as the PGA TOUR, European TOUR, PGA Tour of Australasia, Japan Golf Tour, Asian Tour, PGA TOUR Canada, and other major PGA-sanctioned tournaments. The women's OGR also recognizes official tournaments such as LPGA, Ladies European Tour, China LPGA, and most other LPGA-sanctioned tournaments. In each tournament, players accumulate points to determine their world ranking.

Due to the limitation of four athletes per country, however, some of the biggest names in golf are not able to compete. A variety of countries were represented, but not all of the best players around the world were included under the current format. Unlike other sports that allow players to compete through qualifying stages, golf adopted qualifying criteria similar to tennis. Compared to the US Open that usually accommodates 156 players, the 60-player limitation poses quite a challenge for athletes who wish to compete in the Olympic Games. Criticisms have been raised over the single format event (stroke play), which only offers six medals in total, as it also limits players' ability to compete in various golf events (Greenstein, 2016). The criticism is valid as some countries determine funding based on medal potentials.

Capitalizing on COVID-19

As mentioned, COVID-19 hit many sports hard. Yet, the sport of golf was able to thrive in the COVID-19 conditions. Due to the outdoor nature of the sport and the lack of transmission of COVID-19 outdoors, golf saw 2020 as "a year of resurgence" (Stachura, 2021, para. 1). The National Golf Foundation estimated 24.8 million golfers in the United States in 2020, an increase of 500,000 from 2019. This was the biggest jump in participants since Tiger Woods' 1997 Masters win. The surge was also seen specifically in women, where participation saw an 8% increase to six million (roughly 25% of all golfers). Equipment sales also were encouraging as the total sales of clubs and balls reached $2.9 billion (matching a non-COVID-19 impacted year in 2019). Even viewership of such events as the 2021 PGA TOUR increased by 30% in 2020 (Dixon, 2021). With the increase in participation, sales, and viewership, how can the PGA of America, PGA TOUR, and LPGA capitalize on the momentum and ensure golf is a trend and not a fad? One way is by working to attract younger audiences, which is addressed in the next section.

Response to LIV Golf

In spring 2021, reports surfaced of a new golf league that was meant to rival traditional golf leagues like the PGA Tour and PGA of America. The new league was LIV (meaning 54) Golf. The new league changed traditional golf in hopes of making the sport more appealing to viewers. Instead of individual play, the LIV tour created 12 teams to compete all season long. Major players like Phil Mickelson, Sergio Garcia, Bryson DeChambeau, Dustin Johnson, and Brooks Koepka are captains of the four-person teams that will compete individually but as a team during the season. That means that instead of hundreds of players competing for the series win, only 48 players are on the field for each tournament. Instead of the traditional four rounds, there are also only three rounds for a total of 54 holes of completion. This means there are no cuts, unlike major PGA and PGA of America events. One of the biggest changes was the shotgun start. Instead of every player starting at hole

1 at a specific time, all players start at different holes at the same time. This ensures a faster pace of play. There is a team championship on the 8th and final event of the season, which is a four-day, four-round seeded match play knock out (LIV Golf, 2022a). These changes were made to LIV Golf to make it more exciting for viewers. With the shotgun start, team events, and no cuts, the events will be faster paced and more appealing for a televised and social media audience. It is up to the PGA Tour and PGA of America to decide how to respond.

While the PGA Tour and PGA of America may feel compelled to change their rules to adapt to a tour that is more viewer-friendly, other policy issues also present themselves. LIV Golf was funded primarily by Saudi Arabia's Public Investment Fund, led by Crown Prince Mohammed bin Salman (Jerram, 2022). The Public Investment Fund is a sovereign wealth fund of the Saudi kingdom with roughly $600 billion in assets. LIV Golf was given $2 billion in initial funding to get the league started (Killingstad, 2022). While the Public Investment Fund has incredible amounts of money, the Saudi government has a reputation for having atrocious track record on human rights. This includes the association with the death of journalist Jamal Khashoggi (Johnston, 2022). LIV Golf was meant as a way to "sport wash", or clean up the image of the Saudi government through sport. Seventeen major golfers (Phil Mickelson, Sergio Garcia, Bryson DeChambeau, Dustin Johnson, and Brooks Koepka to name a few) joined the LIV, many unwilling to acknowledge the human rights violations occurred by the Saudi government. On June 9, PGA Tour commissioner Jay Monahan announced that the 17 golfers who competed in the first LIV Golf invitational had their memberships suspended from the Tour. This decision was not only met by applause by some but also concerns of anti-trust violations from others. However, the USGA did not suspend their players for the US Open. As of right now, the R&A's British Open seems willing to accept LIV Golf players, while the PGA of America's PGA Championship and Augusta's Masters are still considering options (Gardner, 2022). The LIV Golf series presents a point in the history of professional golf that will undoubtedly see extreme change.

UNITED STATES TENNIS ASSOCIATION (USTA)

History

The United States Tennis Association began in 1881 as the United States National Lawn Tennis Association (USNLTA), shortened its name to the United States Lawn Tennis Association (USLTA) in 1920, and finally became the USTA in 1975 (USTA, n.d.e). The governing body's original goals were to provide standardized playing rules while growing the sport. The organization and its sanctioned events evolved quickly as the doors opened to international players in 1886 and to women in 1889. Other changes over time included the Mixed Doubles Championships in 1892, the National Clay Court Championships in 1910, and the US Open in 1968 (USTA, n.d.e). Women received greater recognition in the 1970s when the USTA sanctioned the Virginia Slims Women's

www
United States Tennis Association
https://www.usta.com/en/home.html

tour and offered equal prize money to female and male competitors at the US Open (USTA, n.d.e). The governing body also sought to attract new adult and junior players by offering more activities at local parks and recreational facilities and introducing new programs such as the National Junior Tennis League in 1969 and Senior League Tennis in 1991. USTA membership grew to 250,000 in 1984 and doubled to 500,000 by 1993 (USTA, n.d.e). Capitalizing on the sport's increasing popularity, the governing body opened the $285 million Billie Jean King National Tennis Center (NTC) in New York City's Flushing Meadows Corona Park in 1995 (USTA, n.d.e).

Today, the USTA serves as the governing body for tennis in the United States and promotes tennis from the grassroots to the professional levels with three divisions: Community Tennis, Player Development, and Professional Tennis (USTA, n.d.c). Holding the title of the largest tennis organization in the world, the USTA includes over 680,000 individual members and over 7,000 organizational members (USTA, 2021). The Community Tennis division emphasizes the USTA's national grassroots efforts. Programs in this division include the USTA League, which offers tennis opportunities for 300,000 adult league members (USTA, n.d.a), and USTA Jr Team Tennis, which serves children and young adults participating in tournaments and other activities (USTA, n.d.b). The Player Development division provides coaching services and facilities for the nation's best junior players (ages 18 and younger) to fill the pipeline of top tennis performers from the United States (USTA, n.d.d). Finally, the Professional Tennis division arguably represents the most visible part of the USTA, as it hosts the US Open and other tennis tournaments leading up to the Grand Slam tournaments (the Australian Open, the French Open, Wimbledon, and the US Open). This arm of the USTA also assists in forming teams for the Olympic and Paralympic Games as well as the Davis Cup and Fed Cup, the premier international tennis team events for men and women, respectively. USTA leaders believe this division helps attract new players to the sport, as professional tennis increases fan exposure through television viewing and event attendance (USTA, n.d.c).

Governance

MISSION. According to its Constitution and Bylaws, the purposes of the USTA (2020, p. 92) are

> to promote the development of tennis as a means of healthful recreation and physical fitness
>
> to establish and maintain rules of play and high standards of amateurism and good sportsmanship
>
> to foster national and international amateur tennis tournaments and competitions
>
> to encourage, sanction, and conduct tennis tournaments and competitions open to athletes without regard to race, creed, color, or national origin and under the best conditions possible so as to effectively promote the game of tennis with the general public

to generally encourage through tennis the development of health, character, and responsible citizenship and

to carry on other similar activities permitted to be carried on by such a not-for-profit corporation

The governing body strives to increase the number and diversity of people watching and participating in the sport from the grassroots to the professional levels and uses numerous financial resources to help achieve its mission.

MEMBERSHIP. As stated earlier, membership fees represent an important revenue source for the USTA, as the Association received $17 million in related revenues in 2020 (USTA, 2021b). The governing body offers a variety of memberships for individuals and organizations. Individuals can take advantage of adult, junior, and family memberships, while organizations can obtain Community Tennis association, club, school, park and recreation department, or other USTA memberships (USTA, n.d.b). The governing body takes pride in its memberships and programs, noting that it uses revenues from membership dues and other sources to invest in community outreach activities such as improving public tennis courts and providing scholarships and athletic equipment to those in need. Individual members receive access to tournaments and leagues, and organization members receive benefits such as resources to conduct Community Tennis development workshops and host USTA-sanctioned tournaments (USTA, n.d.b).

FINANCIALS. In 2020, the USTA generated $225 million in revenue, a 55% decrease from 2019 (USTA, 2021b). While the impact of COVID-19 was devastating for the USTA in 2020, there were signs of growth as the USTA witnessed its revenues increase from $360 million in 2015 (USTA, 2016b) to $484 million in 2019. In 2020, however, the USTA was forced to cancel events or hold events without fans. This was the case with the US Open, which is by far the USTA's biggest revenue generator, accounting for 82% of revenue in 2019. Held in Queens, New York, the annual Grand Slam event takes place in late summer and attracts top players from around the world. Spectator numbers are an encouraging sign for the USTA, despite the pandemic. There were 631,134 fans in attendance at the 2021 US Open, 85% of the 2019 US Open's record-setting attendance of 737,919. In addition, 8.9 million unique devices visited USOpen.org and the US Open App over the tournament's two-week span (Oddo, 2021). ESPN secured exclusive rights to the US Open television broadcasts and online coverage from 2015 to 2025, and the cable network will pay the USTA $70 million annually for this exclusivity. The partnership marks a shift away from the organization's longstanding television broadcast relationship of 47 years with CBS (Ourand, 2013). Most of the governing body's remaining revenues come from Tour events, membership fees, and tennis facility programs (USTA, 2021b) and these revenues help offset organizational expenses. In 2019 (the last non-COVID-impacted year), the USTA spent $165 million on the US Open. Other large cost categories included $110 million for the Community Tennis division, $30 million for USA

Team and Tour events, and $26 million for administrative and support services (USTA, 2021). To see the most recent income statements for the USTA, check out Exhibit 13.4.

ORGANIZATIONAL STRUCTURE. This non-profit organization boasts a base membership of 700,000 individual members located in 17 regions around the United States and is run by a mixture of volunteer Executive Board Members, paid full-time staff, and other volunteers (USTA, n.d.c). The executive office includes the Executive Director and Chief Executive Officer, Chief Administrative Officer and General Counsel, Senior Executive Assistant, Chief Diversity and Inclusion Officer, Administrative Director to the Office of the President, and Director of Strategic Initiatives. In addition, there is the Chief Executive of Community Tennis, Chief Technology Officer, Chief Executive for Professional Tennis & US Open Tournament Director, Chief Operative Officer for the NTC, Chief Marketing Officer, Chief Revenue Officer, and Chief Financial Officer (USTA, 2021). The USTA has three divisions – Community Tennis, Player Development, and Professional Tennis, which also have leadership teams to manage their various initiatives and programs. In addition, each regional section has its own association, and 50 state associations operate alongside the regional sections and the national headquarters. Whether regional, district, or state, the associations are non-profit organizations run separately with their own boards of directors and staff members. The associations receive support from the Community Tennis Associations, which help the associations provide programs and initiatives to their respective members (USTA, n.d.c).

CURRENT POLICY AREAS

Despite its successes, professional tennis faces ever-changing and sometimes challenging policy issues. Attracting younger audiences, accommodating players' mental health, and the standard practices of major tournaments (Arnold, 2021) are issues the tennis industry faces.

Attracting Younger Audiences

As is the case with the golf associations, the USTA faces the challenge of getting young people involved in the sport at an early age in order to ensure the organization's and sport's continued popularity and growth. The USTA introduced the 10 and Under Tennis program at the grassroots level to spur interest in the sport among a younger audience. With this program, children use smaller racquets with foam or low-compression balls and play on smaller courts with lower nets. The changes were introduced after tennis analysts observed the challenges children face when playing on regulation courts with adult-sized equipment. Oversized nets and racquets might cause younger players to quickly become discouraged and lose interest in the sport. Conversely, smaller-sized equipment and courts allow them to experience more

United States Tennis Association Income Statement 2020–2021 *exhibit* **13.4**

UNITED STATES TENNIS ASSOCIATION INCORPORATED AND AFFILIATES		
CONSOLIDATED STATEMENTS OF FINANCIAL POSITION (US DOLLARS IN THOUSANDS)		
DECEMBER 31ST	2021	2020
Operating Revenues		
US Open	406,172	181,210
USA team events	1,972	1,495
Tour events	35,563	10,273
Membership	13,593	17,182
NTC tennis facility programs, other than US Open	5,223	2,305
Community Tennis leagues and tournaments	10,306	5,065
Investment Return	2,840	7,400
Other	1,370	945
Total Operating Revenues	**477,039**	**225,875**
Operating Expenses		
US Open	226,385	193,345
USA team events	2,858	3,248
Tour events	28,999	25,748
Membership	3,380	6,306
NTC Tennis facility programs	10,911	10,931
Community Tennis	96,434	96,772
Player development	13,761	15,093
Competitive Pathway and officials	6,871	5,819
Marketing, digital, and other program services	28,607	25,371
Total Program Services	**418,206**	**382,642**
Administrative and supporting services	24,181	24,753
Total Operating Expenses	**442,387**	**407,395**
Excess (Deficit) of Operating Revenues over Operating Expenses	**34,652**	**(181,520)**
Nonoperating other income and deductions		
Investment return	18,624	22
Equity in gain (loss) of unconsolidated investees	1,520	(288)
Total nonoperating other income and deductions	**20,144**	**(266)**
Excess (deficit) of revenues over expenses	**54,796**	**(181,786)**

Source: USTA. (2022, March 23).

success sooner and potentially sustain their participation through childhood and beyond. The initiative started in 2012 and attracted younger players to the sport. Yet the initiative also generated controversy among tennis traditionalists, who questioned the research behind the program. The USTA continues to support and promote the movement despite some opposition (Pilon & Lehren, 2014).

The USTA also wants to attract more millennial players and is using social media and on-campus college-based campaigns to increase exposure to the sport. These efforts include using Facebook and Twitter to highlight the collegiate national championship. Beyond this annual event, the USTA is promoting tennis as a sport for club and recreational players on campuses across the nation. Nearly 670 colleges and universities and over 40,000 students take part in this initiative (Vach, 2016). These efforts may help address tennis industry concerns about the need to attract more Generation Y and Z children and young adults, respectively, to the sport (Vach, 2016). The USTA believes increasing the levels of participation among these demographic groups will prove important in sustaining the sport and the USTA's membership base, which are factors critical to the organization's financial success as older players age and discontinue their participation.

WWW
USTA Facebook page
https://www.facebook.com/
USTA/

WWW
USTA Twitter feed
https://www.instagram.com/
usta

WWW
USTA Instagram feed
https://twitter.com/usta

WWW
USTA YouTube channel
https://www.youtube.com/
user/tennis

Using Technology Effectively

Coupled with the need to attract younger players, the USTA recognizes the importance of leveraging technology to its fullest (Vach, 2016). Part of those efforts includes using social media as discussed above. The USTA has Facebook, Twitter, Instagram, TikTok, and YouTube pages. Keeping in mind that these numbers are constantly in flux, the Facebook page has over 285,000 likes, the Twitter feed has over 435,600 followers, the Instagram account 185,000 followers, the TikTok page 18,800 followers, and the YouTube channel over 45,500 subscribers. These social media platforms allow the USTA to connect with users beyond its website, television broadcasts, and other outlets. In addition, the organization expects to benefit from its relationship with ESPN, which will promote coverage of the US Open on ESPN and ESPN2 as well as the WatchESPN app, which users can access through their smartphones, televisions, computers, tablets, and other devices. ESPN noted that online viewership of the US Open has increased, particularly among males aged 18–34, a key age demographic for the USTA (Nagle, 2015). Lastly the industry has experienced growth in online product sales. Tennis industry leaders have emphasized the importance of continuing to pursue opportunities in this area in order to increase tennis product sales and general sport consumption within the industry (Vach, 2016).

Athletes' Mental Health

In May 2021, four-time Grand Slam winner, highest-paid female athlete in the world, and #2 ranked female player in the world Naomi Osaka decided to not participate in the French Open. Osaka had struggled with depression since the US Open in 2018 and experienced waves of anxiety before speaking to the media (Blinder, 2021). The French Open has a policy requiring athletes to meet with media after matches. Osaka, noting that the mandatory media interactions and conferences were a "disregard for athletes' mental health" (Arnold, 2021, para. 3), opted not to participate in the media events after her first-round match. For her actions, tennis officials from the Australian Open, French Open, Wimbledon, and US Open wrote to her asking her to reconsider her position and then ultimately fined Osaka $15,000 (Roland Garros, 2021). Shortly after, Osaka withdrew from the French Open and the tournament was without one of its brightest stars. Osaka's situation highlights a difficult situation for event organizers who want athletes to speak to the media but want to keep the best interest of their athletes' mental health in mind.

NASCAR

History

The NASCAR is the sanctioning body for North American stock-car automobile racing and is the largest such organization in the United States. NASCAR began in 1948 when Bill France, Sr., along with race car drivers, racetrack owners, and racing enthusiasts, met in Daytona Beach, Florida, to form a new racing series (History Channel, 2017). The sport's popularity increased rapidly through the 1930s and 1940s, and France and others wanted to develop an organized structure to capitalize on that surging interest. During their 1948 meeting, the group settled on an organizational structure and declared France the organization's first Chief Executive Officer (Clarke, 2008). The first race sanctioned by NASCAR took place on February 15, 1948, and a Cup Series was introduced for the 1949 season. Changes occurred quickly in the 1950s and 1960s as more drivers gravitated to the sport. New racetracks emerged to accommodate the rising demand. Tracks were built not just in the South, where a large portion of NASCAR fans resided, but farther north in Michigan, Delaware, and Pennsylvania as the fan base expanded (NASCAR, 2017).

The 1970s ushered in additional changes, as Bill France, Sr., relinquished the helm to his son, Bill France, Jr., who led the organization until 2003. In 1971, tobacco company R.J. Reynolds became the title sponsor of NASCAR's premier racing series, and the name changed from the Grand National Series to the Winston Cup Series (Clarke, 2008). A strong corporate presence continued into the 1980s and 1990s, and NASCAR witnessed a significant increase in corporate sponsorships and advertising as other companies followed suit and initiated sponsorships with racetracks and drivers. In 2003, NASCAR moved from long-time title sponsor R.J. Reynolds to a sponsorship with the telecommunications industry's Nextel Corporation. When Nextel and Sprint merged, the Nextel Cup became the Sprint Cup (Clarke, 2008). In a massive shift for the industry, in 2019 NASCAR announced it would not have a title sponsor for their main series, now the NASCAR

WWW

NASCAR
https://www.nascar.com/

Cup Series, but moved toward a Premier Partner program (akin to the Olympic Movement) where the series would feature four brands: Busch Beer, Coca-Cola, GEICO, and XFINITY (NASCAR, 2019).

Beyond corporate sponsorships, NASCAR also experienced an expanded television presence. The organization entered a television partnership with FOX, NBC, and TNT in the 1990s. The contract proved a boon for all parties, as television viewership continued to grow, particularly for the Daytona 500, whose viewership increased 48% from 1993 to 2002. Racetrack owners reported a corresponding growth in attendance, as their facilities often hold from 100,000 to 200,000 fans – rivaling attendance at the National Football League's annual Super Bowl (Amato et al., 2005). Fans flocked to races, and attendance grew by 80% from 1993 to 1998. A large portion of this growth was attributed to the Sprint Cup Series, which witnessed a 57% increase during the same time period. The organization also created an online presence, introducing NASCAR.com in 1995, to reach fans before, during, and after the events (NASCAR, 2017).

Governance

MISSION. While NASCAR's mission statement was not publicly available, the privately owned organization lists its vision and values. NASCAR's vision is "As stewards of the sport, deliver the best motorsports racing in the world, provide outstanding entertainment experiences, and continue to build globally diverse community of loyal fans" (NASCAR, n.d.b, para. 1). NASCAR's values include being authentic, courageous, driven, inclusive, innovative, and stewarding. Headquartered in Daytona Beach, Florida, NASCAR serves as the sanctioning body for the NASCAR Cup Series as well as the XFINITY Series, Camping World Truck Series, eNASCAR, and other smaller series such as ARCA Menards Series, Whelen Modified Tour, PEAK Mexico Series, Whelen Euro Series, Pinty's Series, and Advanced Auto Parts Weekly series (NASCAR, n.d.a). The NASCAR Cup Series is undoubtedly NASCAR's most popular, but the organization also promotes other regional and local events. For example, the Whelen All-American Series represents a training ground for local drivers aspiring to one day compete at the sport's highest levels, and drivers can win local track, state, and national titles for their performances. As drivers improve on the local and regional circuits, they may seek greater opportunities with the Camping World Truck Series, XFINITY Series, and eventually the NASCAR Cup Series.

MEMBERSHIP. As a privately owned, family-run organization, NASCAR tries to consistently provide a lifestyle sports product, a goal that helped the company grow from a regional diversion into an international sports giant. Part of NASCAR's original success derived from its consistency through strict management controls. The governing body establishes guidelines for its owners, drivers, and support personnel both on and off the racetrack. In 2016, NASCAR announced a historic overhaul to the governance and membership structure for its premier racing series. First, a charter system was established, whereby 36 racing teams would receive automatic entry into the racing season's NASCAR Cup events. Next, the weekly race field was reduced from 43 to 40 spots. The charter members would assume 36 spots and four spots are available for non-charter members each week. Finally, each race charter has a nine-year life, and the 36 charter owners have the opportunity to sell their charter once every five years (NASCAR, 2020b). These developments marked a notable change from the previous structure, which gave race teams one-year contracts and required them to qualify for the weekly races based on times.

In addition to these changes, NASCAR created a Team Owner Council, affording team owners more involvement in the sanctioning body's decision-making processes and greater opportunities for revenue sharing. These benefits include allowing the chartered racing teams to sell their charters as noted above. Proponents of the new system believe benefits will include more stable revenue streams for race teams, as they will now have guaranteed entry into races throughout the season. The race teams can also capitalize on the potential long-term worth of their respective charters, assuming they increase in value over time. The new charter system has been utilized to attract sports stars from other industries. Michael Jordan joined NASCAR ownership with Denny Hamlin in 2021 as they signed Bubba Wallace to drive the number 23 car. The reports are that Jordan and Hamlin paid $13.5 million for a single charter (Walters, 2021).

FINANCIALS. As a privately held company, NASCAR is not required to disclose its complete financial statements to the public. NASCAR earns sizeable revenues through its television contracts. In 2013, NASCAR signed ten-year agreements with NBC and FOX, and will receive a combined $8.2 billion in revenues from the two broadcast networks during this time. The negotiated amount represents a 46% increase over the contracts NASCAR previously signed with ABC, ESPN, FOX, SPEED, and TNT (Smith, 2015).

Coupled with media rights, sponsorships represent an important revenue source for the governing body. NASCAR is the self-proclaimed leader in brand loyalty among all North American sports and is one of the most internationally recognized sport brands (NASCAR, 2020a). Corporate sponsors have often recognized the value of associating with NASCAR, and the organization has reaped the benefits of this interest. NASCAR signed a naming rights contract with Comcast's XFINITY, the title sponsor of its second-biggest race series. The rights were previously held by Nationwide, and NASCAR receives $20 million per year for a total of ten years starting in 2015 (Smith, 2015). Sprint, the telecommunications company, held the title sponsorship for the premier racing series from 2004 to 2016, paying NASCAR an estimated $50–$75 million each year for those rights. Monster Energy became the title sponsor for the 2017 NASCAR Cup Series (Heitner, 2016). While estimates for the title sponsorship was rumored to be $1 billion over ten years, Monster reportedly paid $20 million per year (Bromberg, 2019). In 2020, however, NASCAR ended its relationship with Monster and signed Busch Beer, Coca-Cola, GEICO, and XFINITY to be their four "Premier Partners" as mentioned earlier.

Race car owners and drivers also benefit from sponsorship agreements. A team's primary sponsor determines the race car's logos and color schemes, and drivers in the premier racing series receive up to 75% of their revenues from these sponsorships (Boudway, 2016). Yet securing sponsors has become increasingly harder for race teams since the 2008–09 recession. Many view the charter system as a way to lure sponsors back to NASCAR with guaranteed race entries for the charter teams. The previous system left sponsors and race teams unsure of whether their race cars would gain entry from week to week, much less year to year. Race teams now expect that sponsors will have a greater level of comfort with signing longer-term contracts under the new system, as they will have guaranteed entries for longer periods of time (Boudway, 2016). In addition to race car sponsorships, drivers receive purse earnings, depending on where they finish in each race. The 2020 Daytona 500 paid out $23.6 million to the participants of the race (~$2 million for the winner), up from $18 million in 2015 (Crabtree-Hannigan, 2020). These earnings, sponsorships, and other endorsements are used to defray team costs – as much as $25 million per year – for pit crews, race cars, travel, and related expenses (Boudway, 2016).

ORGANIZATIONAL STRUCTURE. In 2003, Bill France Jr.'s son, Brian France, became just the third NASCAR Chairman of the Board and Chief Executive Officer (Clarke, 2008). Following Brian France Jr.'s arrest for driving while intoxicated in 2018, Jim France (Bill's brother) took over as NASCAR chairman and CEO (Kelly, 2018).

Reporting to him are two Vice Chairmen, an Executive Vice President and Chief Racing Officer, a General Counsel, Senior Vice President and Chief Marketing Officer, Vice President of Analytics and Insights, and finally a Vice President for Partnership Marketing (The Official Board, n.d). The governing body has placed a heightened emphasis on safety and innovation, working to ensure the safety of drivers, pit crews, and spectators at sanctioned events. The organization issues specific guidelines and safety measures regarding modifications race car owners and mechanics are allowed to make on their vehicles. NASCAR officials monitor alterations closely and quickly issue citations to drivers and owners running afoul of the rules (Clarke, 2008).

CURRENT POLICY AREAS

NASCAR leaders are constantly focused on innovation and ways to improve their products and organization. Currently, they face challenges in three key areas: addressing declining attendance, reaching different demographics, and increasing diversity on and off the racetrack.

Declining Attendance

The number of people attending NASCAR events is declining – in stark contrast to the increases in revenues generated through its television broadcast rights and title sponsorships. For example, the 2016 attendance at the Indianapolis Motor Speedway race was approximately 50,000 spectators, a record low for this racetrack, which can hold up to 250,000 fans (USA Today, 2016). Industry analysts attribute these attendance figures to a number of factors, including continued challenges associated with the 2008–09 recession and its effects on employment rates, gas prices, and travel costs. Other factors include the aging and in some cases retirements of popular NASCAR drivers; more NASCAR fans using social media, apps, and the organization's website to track the sport in lieu of attending events; and shifting interests of younger sports fans in general (USA Today, 2016). Television viewership has also decreased. NASCAR experienced its greatest number of viewers in 2005 with 9.2 million fans tuning in each week. The viewership numbers have since declined to around 5.3 million per race, with the sharpest declines occurring from 2010 to 2014 (Wolfe, 2015). While viewership numbers and attendance have bounced back in the late 2010s and remained relatively steady through the pandemic, there are concerns that Formula One racing may pose a severe threat to NASCAR for American motorsports viewership (Caldwell, 2021).

Reaching Different Demographics

Beyond racetrack attendance and TV viewership, NASCAR fanship has also decreased from approximately 115 million fans in 2004 to 98 million in 2015, a decline of 15%. No other professional sport in recent history has experienced a similar decline (Mickle, 2015). Market data shows that things may be improving for NASCAR post COVID-19 pandemic. Research shows that 30% of race attendees are new and new attendees are twice as likely as returning fans to be under the age of 35 (Pockrass, 2021). Attendees under 35 are also twice as likely as other attendees to be Black or Hispanic, representing a much-needed shift in diverse demographics for the sport. In another effort to reach younger fans, NASCAR created eNASCAR in 2018. eNASCAR is NASCAR's version of "iRacing." iRacing is considered a simulation that has been called "extremely realistic" by NASCAR driver Clint Bowyer (Haislop, 2020). In March 2020, eNASCAR held a virtual event at Homestead-Miami Speedway that saw almost one million viewers and was the highest rated esports TV program in history. The eNASCAR iRacing Pro Invitational Series had 35 entries, including top NASCAR names like Dale Earnhardt Jr, Bubba Wallace, Jimmie Johnson, and Bobby Labonte (Haislop, 2020). This initiative has shown that NASCAR is attempting to dive into the world of esports and attract younger audiences who are more likely to engage with simulated races.

eNASCAR
https://www.enascar.com/

Increasing Racial Diversity On and Off the Racetrack

Part of NASCAR's attempt to reach different audiences includes focusing on increased racial diversity among its race car drivers, employees, and fans. NASCAR has traditionally been viewed as a sport catering to white fans from rural areas. Some have attributed these numbers to the Confederate flag, which traditionally has been flown by fans at NASCAR events and the lack of diverse drivers. Bubba Wallace, one of only three racially minoritized drivers, noted the way he was treated by his competitors and fans of the sport. The discrimination he experienced culminated in 2020 when, during a large movement against racial injustice in the United States in response to George Floyd's murder, he walked into his garage and found a pull rope fashioned like a noose (Paybarah, 2020). An FBI investigation revealed no hate crime occurred, despite claims by the public and the president of NASCAR Steve Phelps after the photo was released. This event occurred right after NASCAR banned the Confederate flag after years allowing the flag to fly in NASCAR venues across the country. NASCAR has focused on other actions to increase the racial diversity of its fans, with particular focus on Black and Hispanic attendees and drivers. These include the Drive for Diversity to spur greater fan attendance and participation by minoritized race car drivers and pit crews and partnering with NFL star Alvin Kamara to be a brand ambassador and help promote the sport to those from diverse backgrounds (Pockrass, 2021).

International Sport: Surfing Australia

History

Surfing Australia
https://surfingaustralia.com/

Surfing Australia (SA) was formed in 1963 and represents one of the 86 member countries of the International Surfing Association (ISA). SA was established to support the development of surfing in Australia. The structure of SA supports the growth of surfing in the Australian community, delivers high-quality events, and provides support and a performance center for the top surfers in Australia. SA is recognized by the Australian Sport Commission and the Australian Olympic Committee, and the organization is also a member of the Water Safety Council of Australia (Surfing Australia, 2021a).

MISSION. Unlike the individual sports in the United States, international individual sports are governed differently depending on each country's governance structure and sport governing bodies. The mission of SA is to "maximise its outcomes for the sport and develop our surfers at all levels" (Surfing Australia, 2021b, p. 1). The core values of SA are: (a) Real – we live the surfing lifestyle and we share the stoke; (b) Respectful – we are appreciative of our community and environment and celebrate our surfing history and culture; and (c) Progressive – we embrace change and innovation (Surfing Australia, 2021b).

MEMBERSHIP. The SA surfboard riders belong to a club that teaches and encourages skills needed to pursue a competitive career. Currently, 220 clubs are available in Australia for all levels of surfers to join, and these clubs include a combined 20,000 members. Depending on their respective levels, riders join clubs that are dedicated to all age groups at all levels or clubs that are more exclusively designated for professional surfers (Surfing Australia, 2021a).

FINANCIALS. Unlike the individual governing bodies in the United States, SA utilizes the digital media platform mySURF.tv as a primary source of communication rather than securing media rights with major broadcast outlets. Thus, media rights deals are not the biggest contributor to SA's financials. In the last non-COVID-impacted year (2019), SA saw AU7.65 million ($5.51 million) in total revenue. Their largest revenue drives were sponsorships and events (33%), the High Performance Program and Centre (32%), sport development programs (16%), and communication and digital media (8%). Their direct expenses equated to AU7.21 million ($5.23 million), most of which were direct expenses like event costs and equipment (66%; Surfing Australia, 2019). While COVID hit the organization hard in 2020, the most recent financial report indicates that revenues increased to AU7.49 million ($5.43 million) with a net profit of AU117,598 ($85,368) for the 2021 year (Surfing Australia, 2021).

CURRENT POLICY AREAS

At the 2020 Summer Games in Tokyo, Japan, surfing made its Olympic debut. This decision resulted from efforts made by the ISA and President Fernando Aguerre to include five new sports in the 2020 Olympic Games. A total of 40 athletes (20 men and 20 women) competed in the Games. SA embraced their inaugural national team, naming them "The Irukandjis" (named after the Irukandji jellyfish). The name was gifted to them by the local Yirrganydji people of North Queensland (Surfing Australia, 2021c). The Irukandjis were composed of four athletes (two men and two women) for the Games, with men's surfer Owen Wright taking home the bronze medal. However, adding surfing to this mega-event poses challenges for the organizers. These include venue locations and weather in order to provide an ideal environment for board riders. The IOC and ISA are still discussing the issue of the venue and the possibility of artificially made waves as well as the financials associated with the event. In order to prevent host nations from having to invest in new technologies to create waves, the IOC is carefully reviewing its options regarding the logistics of holding surfing events during the Games (Weisberg, 2016). As of this writing, the 2024 Paris Olympic and Paralympic Games include surfing as an official sport and the 2028 Los Angeles Olympic and Paralympic Games are leaning toward including it as well (Wharton, 2021).

SUMMARY

Many individual sports share some governance elements with sports that operate as leagues. For example, both have commissioners, boards of directors, and owners. The governing bodies for individual and league sports are responsible for setting rules and regulations, developing policies, and responding to current issues. Membership in the organizations is well-defined. One aspect that is different, particularly with golf and tennis, is the emphasis on the grassroots development of the sport. For these sports, the professional aspect is just one part of what governing bodies attend to. Cultivating grassroots participation is also of the utmost importance in order to identify

SCOTT GEARY *Executive Director*

Georgia Section of the PGA of America and Georgia PGA Foundation

I am the current Executive Director of both the Georgia Section of the PGA of America [501 c(6)] and the Georgia PGA Foundation [501 c(3) charitable arm of the Georgia PGA Section]. My major responsibilities for the Georgia PGA Section are to assist members with navigating their employment in the golf industry, develop and execute the strategic business plan, execute competitive tournaments, fundraise, and oversee the financial health of the entire Section. My major responsibilities for the Georgia PGA Foundation are to carry forth our mission to introduce and engage individuals of all ages and socioeconomic backgrounds to the game of golf by providing instruction, education, and playing opportunities through our three pillars of service – junior golf, diversity, equity, and inclusion, and military veterans. This is done through developing and executing the strategic business plan for the Foundation Board of Directors, fundraising, administering aligned programming, and ensuring the financial stability of the Foundation.

In the golf industry, I believe the two most important issues are a lack of work-life balance and demand outweighing QUALITY supply. These two issues feed off one another. For example, golf was the first sport to return following the pandemic as it was outdoors, allowing for social distancing. This led to a drastic increase in the number of new and current golfers playing more frequently. People consumed more golf through playing, lessons, apparel, equipment, etc. However, many golf course operators and golf business owners could not, or did not, adapt their business models to invest in their human capital (staff). Many golf industry employees were asked to do more for the same pre-pandemic compensation, leading to burnout and people leaving for other jobs. While the demand from the golf consumer remained, the supply needed to provide quality goods and services took a hit, with potential long-term effects. Golf is driven by individual consumers (the demand) and individual suppliers (the work force). As in all economics, there must be a balance when shifts take place.

Recently, our Governance Committee (a committee assigned by the President of our Board of Directors) was tasked with completing an entire overhaul of our Georgia PGA Section Constitution and ByLaws. As Executive Director, I am the staff liaison for the Governance Committee whose purpose is to consistently review these documents as our business strategy evolves. The committee submits recommended changes to the Board of Directors to review. Any changes to our Constitution require a full membership vote while any changes to our ByLaws can be done by the Board of Directors. This was a "heavy-lift," and took nearly a year to complete. We had to get with the entire membership, Board of Directors, and stakeholders to ensure that the organization of these documents made sense, allowed for future change, and aligned with our overall business strategy.

Currently, each of the 41 Sections of the PGA of America operates as its own business under the umbrella of the Association (PGA of America). Each Section has a slightly different board structure, but with a few constants:

1. Board members are always PGA Member professionals in good standing who are elected to their positions by their respective memberships.
2. Board members volunteer their time and efforts and are there to serve their members and help govern the Section business.
3. Each Section has three officers – a Secretary, Vice President, and President – who serve two-year terms and typically roll through these chairs to become President.

This structure provides a platform for individual members to learn about governance, grow as professionals, expand their networks, and ideally use their governance experience to further enhance their careers. In addition, it allows for people in my position to have the backing of a Board comprised of individuals who inherently understand the golf business.

One challenge I see with this structure is that it is a voluntary role. People in my position need board support and understanding to make certain decisions on behalf of the Section. Individuals on the Board often have previous engagements with their employer, owner, or operator, so Section business decisions and communication(s) must come secondary. While I believe that is how it should be, it creates difficulties to make change happen quickly.

I believe a potentially shrinking work force in the golf industry is certainly a future challenge. Not just due to the work-life balance issue discussed before but also because of potentially shrinking exposure to working in the golf industry to a younger and more diverse workforce. Golf has been labeled as a sport steeped with traditions (some good and some bad) for years. As such, golf has relied upon older generations to expose younger generations to opportunities of working in the industry. The younger workforce consumes information differently, is developing different wants/needs for their careers, and is vastly more diverse. The golf industry is an $84 billion+ industry and offers different career paths for everyone. And remember – You do not have to be a great golfer to work in this industry.

If you are considering working in golf, build your own personal Board of Directors! Golf offers the unique opportunity to get to know consumers, players, co-workers, and those around your more intimately than in team sports. Create genuine relationships and build a diverse group of individuals around you who can help get you to where you want to be, offer feedback/advice, and expose you to opportunities you may not expect.

Keep a Customer Service mindset! You will interact directly with individuals more often in golf than in team sports. You never know who you are going to meet and make an impression on who could be meaningful to your career. Going out of your way to help someone (even if a small inconvenience to yourself) can be the difference in where you end up.

and train the next generation of elite athletes. In this way, these sport organizations resemble NGBs in the Olympic Movement, which stresses both elite athlete development and grassroots participation.

A major difference between individual sports and league sports is the absence of unions and collective bargaining agreements. Also, the qualifying process differs markedly from the drafting process for league sports. From a spectator's perspective, team sports usually create a sense of community by encouraging spectators to be fans of a team. In individual sports, spectators often focus on a specific event (e.g., the Masters, Wimbledon) rather than a single player. Furthermore, individual sports might include senior tours or senior events promoting longevity in the careers of their individual athletes. It is important to recognize organizations that work with individual sports and realize sport governance in professional sport is not just about the Big Four.

case STUDY

FUTURE CHALLENGES FOR NASCAR

Industry analysts believe NASCAR faces a number of challenges. These include declining attendance and viewership numbers. The organization has traditionally benefited from a loyal fan base that has long supported its drivers and sponsors. But the demographic makeup of these supporters has skewed toward an older and less diverse audience. NASCAR has attempted to improve its diversity in different ways. The Drive for Diversity initiative was created to attract more women and

People of Color to the sport. NASCAR also has placed greater emphasis on its website and social media platforms such as Twitter, TikTok, and Instagram to reach younger audiences. However, challenges related to diversity and perceptions of the sport as one targeted toward older, white fans persist.

Given the high level of competition among sports to capture and retain an audience, NASCAR must work to address these challenges and grow its fan base. One study addressed the limited interest among younger sports fans in regard to NASCAR (Goldsmith & Walker, 2015). The researchers conducted an experiment whereby they asked sports fans, in this case, college students, to participate in a NASCAR fantasy league in order to determine whether playing this fantasy sport would increase students' interest in NASCAR over time. The results revealed fan interest did increase, as the students became more engaged with the race car drivers and events. The researchers concluded that NASCAR should identify ways to connect with younger audiences and that fantasy sports could serve as a potential conduit. NASCAR has also introduced eNASCAR as a way to attract younger generations who spend more time watching esports than traditional sports.

NASCAR has signed a sizeable television contract with NBC and FOX and wants to ensure it delivers viewers week in and week out to its broadcast partners. NASCAR also wants to provide the same assurances to their premier partners, as they have gone away from one main title sponsor to four primer sponsors for overall less money. The organization faces a delicate balance in attracting new fans without alienating its older and loyal fan base in the pursuit of younger and more diverse audiences. The next few years for NASCAR are important, especially as Formula 1 racing is starting to become popular in the United States.

1. What can NASCAR do to improve its attendance and viewership numbers? In addition to increases in numbers, what can the organization do to increase the diversity of its fan base? As you discuss strategies to enhance diversity among the fan base, keep in mind that different identity groups may require unique approaches to see how they can belong in the broader NASCAR community.

2. How can the organization leverage its new television contracts, sponsorships, and/or charter system to attract and retain fans?

3. How can NASCAR utilize their initiatives that reach younger audiences (e.g., fantasy sports, apps, eNASCAR) to ensure increase interest in the sport by Generation Y and Z?

4. What detailed plan would you recommend NASCAR develop and execute in order to address these challenges?

CHAPTER questions

1. The USTA has three important divisions but limited funds to support them. As the USTA leader, which division would you

emphasize and why? What are the advantages and disadvantages of highlighting one division versus another?

2. NASCAR has transitioned from a single title sponsor for their Series Cup to having four primer partners with less visibility. What are the advantages and disadvantages of this sponsorship system for the teams and NASCAR?

3. Professional tennis, golf, and stock car racing rely heavily upon sponsorship revenues. In challenging economic times, how can the three sports ensure ongoing support from sponsors? What alternatives can they pursue in the event of sponsorship declines?

FOR ADDITIONAL INFORMATION

For more information check out the following resources:

1. This site discusses increasing inclusion in NASCAR NASCAR: Martinelli, M.R. (2021, November 5). NASCAR vowed to be more inclusive, but is its progress moving fast enough? https://ftw.usatoday.com/2021/11/ nascar-diversity-inclusion-racism-kyle-larson-slur-change

2. Here you can find information about the growth of the tennis industry USTA (2021, August 23). Tennis industry data shows further growth of the sport in the U.S. https://www.usta.com/en/ home/stay-current/national/tennis-industry-data-shows-further-growth-of-the-sport-in-the-u-.html

3. LIV Golf controversies and PGA Tour response Chappell, B. (2022, June 9). *The Saudi-backed LIV Golf tees off, and the PGA Tour quickly suspends 17 players.* NPR. https://www.npr.org/2022/06/09/1103904851/ liv-golf-saudi-pga-sanctions-mickelson-jones

4. This site presents information about how the PGA of America is addressing diversity, equity, and inclusion Human Resource Executive: Cross, S. (2021, August 31). How the PGA of America expanded its diversity and inclusion. https://hrexecutive.com/ how-the-pga-of-america-expanded-its-diversity-and-inclusion/

5. Information on surfing's debut at the 2020 Tokyo Olympic Games Goodsir, C. (2021, November). Preparations underway for surfing at 2032 Brisbane Olympics. https://www.sen.com.au/news/2021/11/23/ preparations-underway-for-surfing-at-2032-brisbane-olympics/

REFERENCES

Amato, C.H., Peters, C.L.O., & Shao, A.T. (2005). An exploratory investigation into NASCAR fan culture. *Sport Marketing Quarterly, 14*, 71–83.

Arnold, C. (2021, May 31). *Naomi Osaka drops out of French Open after dispute over media appearances*. NPR. https://www.npr.org/2021/05/31/1001917952/naomi-osaka-drops-out-of-french-open-after-dispute-over-media-appearances

Beal, J. (2020, March 9). *PGA Tour announces nine-year media deals with CBS, NBC and ESPN*. Golf Digest. https://www.golfdigest.com/story/pga-tour-announces-nine-year-media-deals-with-cbs-nbc-and-espn

Black Girls Golf. (n.d.a). *About Black Girls Golf*. https://blackgirlsgolf.net/about-us/

Black Girls Golf. (n.d.b). *Membership info*. https://blackgirlsgolf.net/join-black-girls-golf/

Blinder, A. (2021, June 1). With her candor, Osaka adds to conversation about mental health. *New York Times*. https://www.nytimes.com/2021/06/01/sports/tennis/mental-health-osaka.html

Boone, K. (2021, June 20). 2021 *U.S. Open prize money, purse: Payouts, winnings for each golfer from huge $12.5 million pool*. CBS Sports Digital. https://www.cbssports.com/golf/news/2021-u-s-open-prize-money-purse-payouts-winnings-for-each-golfer-from-huge-12–5-million-pool/

Boudway, I. (2016, February 24). *What NASCAR learned from the NFL*. Bloomberg. www.bloomberg.com/news/articles/2016-02-24/what-nascar-learned-from-the-nfl

Bromberg, N. (2019, December 5). *NASCAR leverages existing relationships with companies for new Cup Series sponsor model*. Yahoo! Sports. https://www.yahoo.com/now/nascar-leverages-existing-relationships-with-companies-for-new-cup-series-sponsor-model-202953169.html

Caldwell, C. (2021, October 28). *2-Headed monster: Has NASCAR lost its stranglehold on American motorsports?* Frontstretch. https://www.frontstretch.com/2021/10/28/2-headed-monster-has-nascar-lost-its-stranglehold-on-american-motorsports/

Clarke, L. (2008). *One helluva ride: How NASCAR swept the nation*. Villard Books.

Crabtree-Hannigan, J. (2020, February 18). *Daytona 500 purse, payout breakdown: How much prize money will the winner make in 2020?* Sporting News. https://www.sportingnews.com/us/nascar/news/aytona-500-purse-payout-2020-prize-money-breakdown/17woqokonpjll1v5hhyrvzr5eh

Dixon, E. (2021, March 11). *PGA TOUR viewership on NBC up 30% in 2021*. SportsPro. https://www.sportspromedia.com/news/pga-tour-nbc-golf-channel-tv-viewership-audience-2021/

Gardner, S. (2022, June 10). LIV Golf players suspended by the PGA Tour, but what's their status for the majors? What we know. *USA Today*. https://www.usatoday.com/story/sports/golf/2022/06/10/liv-golf-membership-impact-majors-british-pga-us-open-masters/7570446001/

Goldsmith, A., & Walker, M. (2015). The NASCAR experience: Examining the influence of fantasy sport participation on 'non-fans'. *Sport Management Review, 18*(2), 231–243. https://doi.org/10.1016/j.smr.2014.06.001

Gorant, J. (2018, August 8). *War for the Tour: The day the PGA Championship nearly died*. GOLF Magazine. https://golf.com/news/tournaments/pga-championship-nearly-died/

Greenstein, T. (2016, August 15). How to make Olympic golf even better in 2020. *Chicago Tribune*. www.chicagotribune.com/sports/columnists/ct-golf-future-olympics-spt-0816-20160815-column.html

Haislop, T. (2020, April 13). *What is NASCAR iRacing? How the virtual races work, drivers, full schedule & more*. Sporting News. https://www.sportingnews.com/us/nascar/news/nascar-iracing-virtual-races-drivers-schedule/a53c8e5widhd14na1jinsz7rq

Heitner, D. (2016, July 18). How much money will NASCAR get from Sprint's replacement? *Forbes*. www.forbes.com/sites/darrenheitner/2016/07/18/will-nascar-get-100-million-per-year-from-sprints-replacement/#66668ac7210e

History Channel. (2017). *This day in history*. www.history.com/this-day-in-history/nascar-founded

Jerram, R. (2022, June 29). *LIV Golf: Everything you need to know about the rebel series*. https://www.todaysgolfer.co.uk/news-and-events/tour-news/everything-you-need-to-know-about-the-greg-norman-liv-golf-series-saudi-golf-league/

Johnston, M. (2022, June 8). *What to know about controversial Saudi-funded LIV Golf Invitational Series*. Sports Net. https://www.sportsnet.ca/golf/article/what-to-know-about-controversial-saudi-funded-liv-golf-invitational-series/

Kaplan, D. (2021, May 6). *PGA Tour acknowledges that it keeps more than 90% of revenue in joint TV deals with LPGA*. The Athletic. https://theathletic.com/2570326/2021/05/06/pga-tour-acknowledges-that-in-joint-tv-deals-with-lpga-it-keeps-more-than-90-of-revenue

Kelly, G. (2018, August 10). NASCAR rallies around Jim France; future of CEO position murky. *The Columbus Dispatch*. https://www.dispatch.com/story/news/regional/motor-racing/2018/08/10/nascar-rallies-around-jim-france-future-of-ceo-position-murky/6497044007/

Killingstad, L. (2022, June 12). *LIV Golf has money to blow*. https://frontofficesports.com/liv-golf-has-money-to-blow/

LIV Golf. (2022a). *Our format*. https://www.livgolf.com/format

LPGA. (n.d.a). *About LPGA*. https://www.lpga.com/about-lpga

LPGA. (n.d.b). *Culture – Our executive leadership team*. https://www.lpga.com/careers-about/culture/our-executive-leadership-team

LPGA. (n.d.c). *Culture – Our mission and values*. https://www.lpga.com/careers-about/culture/our-mission-values

LPGA. (n.d.d). *Official licensees of the LPGA*. https://www.lpga.com/licensees

LPGA. (2021). *LPGA Tour announces record-breaking 2022 schedule*. https://www.lpga.com/news/2021/the-lpga-tour-announces-record-breaking-2022-schedule

Mickle, T. (2015, June 26). Confederate flag at NASCAR earns a caution. *The Wall Street Journal*. www.wsj.com/articles/nascar-stance-earns-a-yellow-flag-1435359227

Murphy, B. (2021, July 28). *Olympics golf: Tokyo qualification, what to know for 2021*. NBCDFW. https://www.nbcdfw.com/news/sports/tokyo-summer-olympics/olympics-golf-tokyo-qualification-what-to-know-for-2021/2696428/

Nagle, D. (2015, September 15). *@ESPNTennis & US Open: Most-viewed in four years; WatchESPN viewership up 4X+*. ESPN Press Room. http://espnmediazone.com/us/press-releases/2015/09/espntennis-us-open-most-viewed-in-four-years-watchespn-viewership-up-4x/

NASCAR. (2017, February 28). *About NASCAR*. www.nascar.com/en_us/news-media/articles/about-nascar.html

NASCAR. (2019, December 5). *NASCAR introduces Premier Partners of NASCAR Cup Series: Busch Beer, Coca-Cola, GEICO, Xfinity*. https://www.nascar.com/news-media/2019/12/05/nascar-introduces-premier-partners-of-nascar-cup-series/

NASCAR. (2020a, January). *NASCAR is...* https://www.nascar.com/wp-content/uploads/sites/7/2020/01/2020-NASCAR-Sponsorship_Footer-Doc.pdf

NASCAR (2020b, September 22). *How the NASCAR charter system works*. https://www.nascar.com/news-media/2020/09/22/how-the-nascar-charter-system-works/

NASCAR (n.d.a). *NASCAR roots*. https://www.nascar.com/roots

NASCAR. (n.d.b). *Our vision, our values*. https://careers.nascar.com/vision-values/

National Golf Foundation. (n.d.). *Golf industry facts*. https://www.ngf.org/golf-industry-research/#golfers

Nichols, B.A. (2020, February 29). *USGA says U.S. Open generates $165 million annually. Here's where all the money goes*. Golfweek. https://golfweek.usatoday.com/2020/02/29/usga-us-open-generates-165-million-annually/

Oddo, C. (2021, September 12). *2021 US Open Finals, by the numbers*. US Open. https://www.usopen.org/en_US/news/articles/2021-09-12/us_open_final_by_the_numbers.html

O'Neill, D. (2021, December 3). USGA makes good on U.S. Adaptive Open. *Sports Illustrated*. https://www.si.com/golf/news/usga-makes-good-on-u-s-adaptive-open

Ourand, J. (2013, May 16). ESPN, USTA sign 11-Year Deal Worth More Than $770M To Put U.S. Open Solely On Cable. *Sports Business Journal*. https://www.sportsbusinessjournal.com/Daily/Issues/2013/05/16/Media/US-Open.aspx

Paybarah, A. (2020, June 26). NASCAR releases image of noose found in Bubba Wallace's garage. *New York Times*. https://www.nytimes.com/2020/06/26/sports/autoracing/nascar-noose-bubba-wallace.html

PGA. (2018). *Combined financial statements*. https://resources.pga.org/Document-Library/pga-2018-combined-financial-statements.pdf

PGA. (2020). *Combined financial statements*. https://resources.pga.org/Document-Library/2020%20PGA%20of%20America%20FS.pdf

PGA. (2022). *Associate registration eligibility*. https://www.pga.org/membership/associate-program/associate-registration-eligibility

PGA. (n.d.b). *History of the PGA*. www.pga.com/pga-america/history

PGA. (n.d.c). *The PGA of America.* https://www.pga. com/pga-of-america/about

PGA. (n.d.d). *PGA of America executive leadership and board of directors.* https://www.pga.org/ leadership/

PGA TOUR. (2020, January 28). *PGA TOUR, its tournaments surpass $3 billion in all-time charitable giving.* https://www.pgatour. com/tournaments/tour-championship/ news/2020/01/28/pga-tour-surpasses--3-billion-in-all-time-charitable-giving.html

PGA TOUR. (n.d.a). *About us.* https://www.pgatour. com/company/aboutus.html

PGA TOUR. (n.d.b). *Executive leadership.* https:// www.pgatour.com/company/executive-leadership.html

PGA TOUR Media Guide (2021). *Priority rankings.* https://www.pgatourmediaguide.com/player/ priority-rankings

Pilon, M., & Lehren, A.W. (2014, September 10). Modified training for children stirs new debates. *New York Times.* www.nytimes. com/2014/09/11/sports/tennis/modified-training-for-children-stirs-new-debates.html

Pockrass, B. (2021, August 9). *NASCAR hoping that uptick in fan attendance, improved demographics become trend.* FOX Sports. https://www. foxsports.com/stories/nascar/cup-series-fan-demographics-attendance-kyle-larson-dover-watkins-glen

Roland Garros. (2021, May 30). *Statement from Grand Slam tournaments regarding Naomi Osaka.* https://www.rolandgarros.com/en-us/ article/statement-from-grand-slam-tournaments-regarding-naomi-osaka

Smith, C. (2015, February 18). How NASCAR plans to turn its survival story into a decade of success. *Forbes.* www.forbes.com/sites/ chrissmith/2015/02/18/how-nascar-plans-to-turn-its-survival-story-into-a-decade-of-success/#33a5425d2ceb

Stachura, M. (2021, April 7). *The numbers are official: Golf's surge in popularity in 2020 was even better than predicted.* Golf Digest. https://www.golfdigest.com/story/national-golf-foundation-reports-numbers-for-2020-were-record-se

Surfing Australia. (2019). *Annual report, 2019.* https://surfingaustralia.com/governance#annual

Surfing Australia. (2021a). *Annual report, 2021.* https://surfingaustralia.com/governance#annual

Surfing Australia. (2021b). *Strategic plan.* https://surf-dev.com/wp-content/uploads/sites/6/2020/10/ Surfing-Australia-Strategic-Plan-Summary.pdf

Surfing Australia. (2021c). *The Irukandjis: Deadly in the water.* https://surfingaustralia.com/irukandjis/

The Official Board. (n.d.). *NASCAR.* https://www. theofficialboard.com/org-chart/nascar

USA Today. (2016, July 25). Five reasons for empty seats. *Indy Star.* https://www.indystar.com/story/ sports/motor/2016/07/25/5-reasons-nascars-empty-seats/87534784/

USGA. (n.d.a). *About the USGA.* https://www.usga. org/content/usga/home-page/the-usga--about-us. html

USGA. (n.d.b). *Community.* www.usga.org/serving-the-game/health-of-the-game/community.html

USGA. (n.d.c). *Our leadership.* https://www.usga.org/ about/our-leadership.html

USGA. (2017). *Club membership overview.* www. usga.org/content/usga/home-page/member-clubs/ club-membership-overview.html

USGA. (2021). *Consolidated financial statements and report of independent certified public accountants.* https://www.usga.org/content/dam/ usga/pdf/2021/2020%20United%20States%20 Golf%20Association%20FS.pdf

USTA. (n.d.a). *About recreational tennis: About USTA League.* https://www.usta.com/en/home/ play/adult-tennis/programs/national/about-usta-league.html

USTA. (n.d.b). *How membership benefits you.* https:// www.usta.com/en/home/membership/join-benefits.html

USTA. (n.d.c). *How the USTA works for you.* www.usta.com/About-USTA/Organization/ Organization/

USTA. (n.d.d). *Player & coach development.* www. playerdevelopment.usta.com/

USTA. (n.d.e). *USTA key dates.* https://www.usta. com/en/home/about-usta/usta-history/national/ usta-history.html

USTA. (2016). *Consolidated financial statements years ended December 31, 2015 and 2014.* https://www.usta.com/content/dam/usta/ pdfs/2015%2001_121514-association_final_ signed.pdf

USTA. (2020, January 1). *USTA corporate governance.* https://www.usta.com/content/dam/usta/2020-pdfs/2020%20USTA%20Constitution%20 Bylaws%20and%20Diversity%20and%20 Inclusion%20Statement.pdf

USTA. (2021a, February 11). *US tennis participation surges in 2020 Physical Activity Council (PAC) finds.* https://www.usta.com/en/home/stay-current/national/u-s--tennis-participation-surges-in-2020--pac-report-finds.html

USTA. (2021b, March 30). *United States Tennis Association incorporated and affiliates.* https://www.usta.com/content/dam/usta/pdfs/01_USTA%20Association%202020%20fst%201231%20EV%20Final%20SIGNED.pdf

USTA. (2022, March 23). *Consolidated financial statements years ended December 31, 2021 and 2020.* https://www.usta.com/content/dam/usta/2022-pdfs/USTA%20Association%202021%20fst%201231%20EV_Final%20draft_3.23.22_(secured)_FINAL%20SIGNED.pdf

Vach, R. (2016, March 23). *Blog: Tennis "State of the Industry" forum takeaways.* https://www.ustaflorida.com/blog-tennis-state-industry-forum-takeaways/

Walters, S. (2021, December 3). *Denny Hamlin comments on the cost of the NASCAR charter for 23XI Racing.* Racing News. https://racingnews.co/2021/12/03/denny-hamlin-comments-on-the-cost-of-the-nascar-charter-for-23xi-racing/

Weisberg, Z. (2016, June 2). *Surfing in the 2020 Olympics: Absolutely everything we know.* www.theinertia.com/surf/surfing-in-the-2020-olympics-absolutely-everything-we-know/

Wharton, D. (2021, December 9). Olympic officials look to continue surfing and skateboarding at 2028 L.A. Games. *Los Angeles Times.* https://www.latimes.com/sports/olympics/story/2021-12-09/olympics-look-to-continue-surfing-skateboarding-at-2028-la-games

Wolfe, A. (2015, June 26). Brian France tries to broaden NASCAR's appeal. *Wall Street Journal.* www.wsj.com/articles/brian-france-tries-to-broaden-nascars-appeal-1435340325

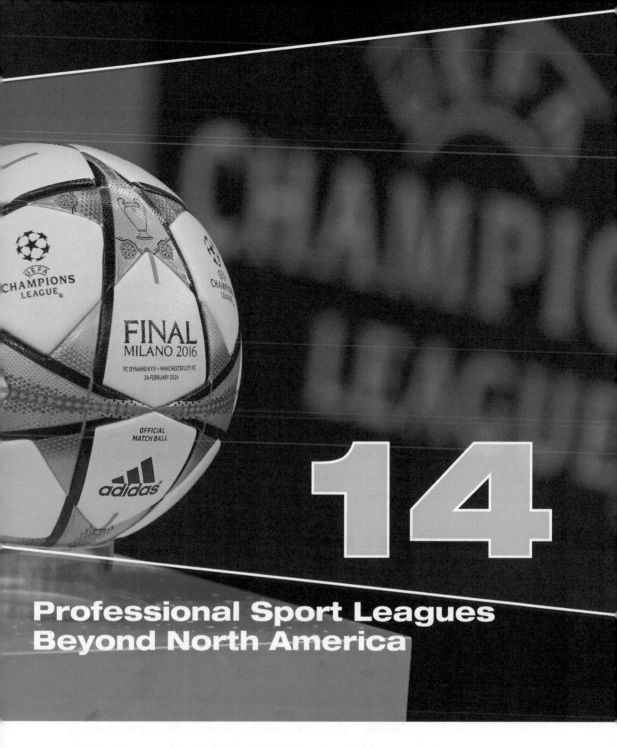

Professional Sport Leagues
Beyond North America

When you think of professional sport, the first leagues that come to mind are often the National Basketball Association (NBA), the National Football League (NFL), the National Hockey League (NHL), Major League Baseball (MLB), or the Women's National Basketball Association (WNBA). However, in many parts of the world, the topics of discussion are very different. Perhaps a debate rages over upcoming football matches

DOI: 10.4324/9781003303183-14

involving FC Barcelona or Real Madrid, the ongoing rivalry between Liverpool and Manchester United, or whether anyone can catch Max Verstappen in the next Formula 1 Grand Prix. If you are wondering what these examples refer to, then welcome to the exciting world of professional sport beyond North America, where *football* means *soccer*, basketball has a few different rules, motorsports does not mean NASCAR, and baseball is often completely unknown!

The goal of this chapter is to introduce you to some of the basics of the structures of professional sport outside of North America. As such, we will touch on some of the differences that exist, as well as some similarities. The chapter will primarily use football (soccer) as an example, but remember that many other types of professional leagues and sports exist internationally. Basketball has its EuroLeague and many countries have their own national-level leagues. Sports such as cricket, rugby, hockey, and team handball dominate in other countries, and there are motorsports such as Formula One. Women's sports are also alive and well internationally and we will introduce some of those leagues in this chapter, too. Hopefully, this chapter will spark an interest in you to enroll in an international sport class or to take advantage of the growing opportunities to do a Study Abroad experience during your time as a student. Even better, you might actually figure out a way to work or live in another country and learn another language at some point in your life. That would be the best outcome of all since you would get a chance to personally experience our global sport industry.

THE GLOBAL NATURE OF SPORT

As just mentioned, the sport industry is truly global in nature. Similar to other global products, such as Coca-Cola or McDonald's, sport transcends borders. The North American way of organizing and managing sport is not the only – and certainly not always the best – way to organize sport. Sport managers working in an international environment must learn to be respectful of local cultures, norms, and expertise. According to Pitts and Zhang (2016):

> Differences in such areas as culture, religion, tradition, politics, law, policy and regulation, communication, language, technology and environment in global, national, regional, and local communities make this task a very challenging one … Sport management professionals responsible for making the strategic, cultural, political, and economic decisions for sport organizations must be prepared for the challenges of the new sport landscape.
>
> (p. 8)

The only way to truly be successful in today's international sport marketplace is to actively engage in the global sport industry by reaching out to build bridges with international partners. Sport managers who fail to do so are setting their organizations up to fall behind the competition. According to Germany's former Chancellor Angela Merkel, "Whoever believes the problems of this world can be solved by isolationism and protectionism is making a tremendous error" (Collinson, 2017, para. 9). Working together across international differences builds a stronger sport industry for everyone.

This chapter introduces the basics of selected international sport leagues and events. For the most part, people in North America have a relatively narrow view of professional sport – the Big Four, the WNBA, NASCAR, golf, and tennis. Hopefully, this chapter will expand your horizons to professional sport around the world.

SELECTED DIFFERENCES IN PROFESSIONAL SPORT

The governance of professional sport internationally is very different from that in North America. We will highlight three of these governance-related themes here: (1) relationships to other sport governing bodies, (2) promotion and relegation, and (3) player movement systems.

Relationships to Other Governing Bodies

Relationships among several levels of sport governing bodies are involved in international sport. Specifically, leagues are tied to the National Governing Body (NGB) for a particular sport in each nation, the International Federation (IF) for that particular sport, and sometimes to a regional governing body. For example, the Premier League in England has ties to the Football Association (FA) of England, the Union of European Football Associations (UEFA), and FIFA (Fédération Internationale de Football Association) (Premier League, n.d.).

Promotion and Relegation

Another striking difference between professional leagues in North America and Europe is the system of promotion and relegation used in European leagues (Hums & Svensson, 2019; Szymanski & Valetti, 2003). Briefly, this system operates as follows:

> While there are several ways in which this scheme can operate, the most commonly used format is one where the worst performing teams in a league during the season, measured by the number of points won (a measure that is close to win percentage) are demoted to an immediately junior league to be replaced by the best performing teams in that league.
>
> (Szymanski, 2006, p. 685)

Here is an example from a hypothetical league setup. Let's say at the end of the 2018–19 season, the final standings looked like this:

A League	B League
Magdeburg	FC Hamburg
FC Berlin	Leipzig
Aufi	Wetzlar
Minden	Kroatische Geresdorf
Rhein-Neckar	Bergischer
Erlangen	Hanover

Using promotion and relegation, at the start of the 2019–20 season, the league membership would look like this:

A League	B League
Magdeburg	Rhein-Neckar
FC Berlin	Erlangen
Aufi	Wetzlar
Minden	Kroatische Geresdorf
FC Hamburg	Bergischer
Leipzig	Hanover

As you can see, the bottom teams from the A League were relegated to the B League, while the top teams in the B League were promoted to the A League. League membership can fluctuate from season to season under this system. Also, as opposed to North American leagues where the bottom teams might want to have that lower position to ensure a high draft pick, there is an incentive to avoid being relegated to a lower league.

As a useful comparison, what might this system look like if applied to MLB? Let's say that in a given year the team with the worst record in the National League was the Pittsburgh Pirates, and the team with the worst record in the American League was the Kansas City Royals. Next, let's say the team with the best record in the AAA-level International League was the Louisville Bats, and the team with the best record in the AAA-level Pacific Coast League was the Reno Aces. Applying the European scheme, the Bats would be promoted to the National League, while the Aces would be promoted to the American League. Consequently, the Royals would be relegated to the Pacific Coast League, and the Pirates would be relegated to the International League. "That sounds crazy," you say. "What about stadiums and season tickets and corporate sponsorships?" Although you may wonder how it could possibly work, promotion and relegation is a common practice in professional sport leagues outside of North America and has worked successfully for many years. In North American sport, teams with the worst records actually get rewarded with high draft picks the next season leading to the occasional allegations of "tanking" in order to draft the next big prospect. Promotion and relegation serve to deter that from happening. Of course, as sport is structured currently, this could not apply to the NFL or even the NBA as they often primarily use college sport as the "minor leagues" to develop talent.

Player Movement Systems

In North American leagues, players are routinely traded, reassigned, released, or perhaps put on waivers. The player transaction system is collectively bargained between players' associations and management, and federal and state labor laws are also involved. Deals are made involving players, players to be named later, future draft choices, and cash. Player movement systems are somewhat different when we look at professional sport outside of North America. In particular, we need to briefly touch on what is known as the transfer system. The system itself is quite complex (just like the complicated waiver systems in MLB), so the purpose here is just to make you aware of some of the basics.

Here in North America, children begin their sporting careers at rec leagues. The talented ones then move up through high school or Amateur Athletic Union (AAU) sport, and then to college, and a small fraction of them will make it to the professional level. In Europe, children will start their sporting careers playing for a local club. These clubs will have teams for various age levels and abilities and will develop talented players as they mature. Once these players become recognized for their talent, other higher-level clubs will be interested in signing them, and the players will transfer their rights to a new club. In simple terms:

> Here's how a basic transfer works. Club 1 has a player. Club 2 wants a player. Club 2 and Club 1 negotiate an agreeable price, the player and Club negotiate agreeable wages, and then the transfer happens – the player's registration is transferred from Club 1 to Club 2, and he signs a new contract – and all parties are happy.
>
> (Thomas, 2014, para. 13)

Of course, the process is not always this simple and sometimes will take weeks and months to hammer out.

What about the transfer fee – how is that set? There is no exact formula for this, but here are the basic elements that factor in (Poli et al., 2016):

> A first group of indicators concerns the characteristics of players such as age, position, length of contract remaining, and the residual book value. The latter variable is calculated from the transfer fee amount paid by the employer club, divided according to the percentage of years of contract since the signature.
> A second group of indicators takes into account the players' performances, notably in terms of the amount of time played in the different club competitions (domestic leagues, cups) or, eventually, in national teams. Recent performances are given more weight than previous ones.
> The last family of indicators refers to the level of the leagues where the footballers played their matches, as well as to the results obtained by the employer clubs. The level of the national team represented is also taken into account (para. 2).

Once the transfer fee is set, it will include payment to the new club and payment to the player who then gives a cut to his agent. Just as a point of reference, transfer fees have been skyrocketing of late with players such as Cristiano Ronaldo, Mbappé, and Jack Grealish all seeing their transfer fees top $100 million. The highest transfer fee to date was Brazilian soccer star Neymar's transfer in 2017. Neymar was transferred from Barcelona to Paris Saint-Germain (PSG) for $263 million (€222 million) … and then the PSG still had to pay Neymar's contract of $53 million (€45 million) per year! The total cost of acquiring Neymar via transfer fees and then his contract was $528 million (€450 million; Gaines, 2017).

Another player transaction that takes place in leagues outside of North America is a loan deal. This more frequently takes place with a younger player where the team who owns them would like to see them get the playing time they can't with their current club. The player could be loaned to a lower division team, a team in another country, and even occasionally another team in the same league. This gives the owning club the opportunity to observe the player and for the player to gain experience (Explaining Soccer, 2009). When a player is loaned, the team who actually owns the player typically pays their salary, although sometimes the salary is shared by both clubs. This is all determined by the negotiated loan deal. This is certainly something we are not accustomed to in North America, but it happens with American players. Linsey Horan, a midfielder for the USWNT, was loaned from the Portland Thorns to Olympique Lyonnais from January 2022 to June 2023 (Hruby, 2022). Could you imagine your favorite new young Golden State Warrior being loaned to the Clippers for a while to get more playing time?

Now that we have illustrated a few significant differences in policies in pro sport outside North America, let's have a look at a successful European football league – the English Premier League (EPL) – and see how its governance structure operates.

FOOTBALL: PREMIER LEAGUE

www
Premier League
www.premierleague.com

As an organized sport, football has been around for an extremely long time. Football played as a professional sport has a shorter past, however. In this section, we briefly trace the history of the Premier League; then we focus more directly on league operations and governance.

www
Fédération Internationale de
Football Association
www.fifa.com

CONFEDERATION OF NORTH, CENTRAL AMERICAN AND CARIBBEAN ASSOCIATION FOOTBALL

www
Asian Football
Confederation
http://www.the-afc.com

www
Confederation of African
Football
www.cafonline.com

Before discussing specific leagues, let's look at the overall governance of international football. The IF for football, FIFA, is the international governing body for the sport. FIFA consists of six regional organizations, including the Asian Football Confederation (AFC); the Confédératión Africaine de Football (CAF); the Confederation of North, Central American and Caribbean Association Football (CONCACAF); the Confederación Sudamericana de Fútbol (CONMEBOL); the Oceania Football Confederation (OFC); and UEFA. Within these regions, every nation has its own FIFA-recognized national football governing body, such as the US Soccer Federation, Canada Soccer, and the Deutscher

Fussball-Bund in Germany. Professional football leagues within specific nations must belong to their FIFA-recognized national football governing body to be eligible to advance in international play. For example, in the United States, Major League Soccer (MLS) belongs to the US Soccer Federation. The US Soccer Federation, in turn, is a member of CONCACAF, which is the regional division of FIFA that includes North America. Because of this arrangement, MLS, or any FIFA-recognized professional football league in any nation, must follow the rules and regulations set forth by FIFA. Although the leagues have their own individual league governance systems for daily operation, they must follow many basic policies handed down from FIFA on issues such as drug use, match scheduling, and player transfers. Let's turn now to an example of what is widely considered the world's most successful international league – the EPL.

MAJOR LEAGUE SOCCER

History

In the late 1980s, English football was in need of restructuring. A 1985 fire at a match at Bradford City Valley's Parade Grounds cost 56 lives when flames engulfed the wooden stadium (Conn, 2010). Around this same time, a number of violent incidents involving fans occurred. The sport's image was in tatters, and it also suffered from a lack of financial investment. Establishing itself as a business entity separate from the nation's FA allowed the FA Premier League to negotiate its own television and sponsorship contracts, the sources of income that have helped the league transition to its current success. The FA Premier League opening season took place in 1992–93 with 22 clubs (Premier League, 2018). The league attracts players and fans from all over the world and is a successful business entity. In 2007, it changed its name to simply Premier League.

Governance

How is the EPL, with its independently operating clubs, governed? According to the Premier League (n.d.), the following levels of governance exist:

> Each individual club is independent, working within the rules of football, as defined by the Premier League, The FA, UEFA, and FIFA, as well as being subject to English and European law.
>
> Each of the 20 clubs is a Shareholder in the Premier League. Consultation is at the heart of the Premier League and Shareholder meetings are the ultimate decision-making forum for Premier League policy and are held at regular intervals during the course of the season.
>
> The Premier League AGM (Annual General Meeting) takes place at the close of each season, at which time the relegated clubs transfer their shares to the clubs promoted into the Premier League from the Football League Championship.

Clubs have the opportunity to propose new rules or amendments at the Shareholder meeting. Each Member Club is entitled to one vote and all rule changes and major commercial contracts require the support of at least a two-thirds vote, or 14 clubs, to be agreed upon.

The Premier League Rule Book serves as a contract between the League, the Member Clubs, and one another, defining the structure and running of the competition (paras. 1–7).

From this description, you can see that the EPL has an annual general assembly which is referred to as the AGM. This is the same as the General Assembly the International Olympic Committee (IOC) holds or the National Collegiate Athletic Association's (NCAA) annual convention. There is also a governing document called the Premier League Rule Book. This would be analogous to the IOC's Olympic Charter or the NCAA's division manuals.

MISSION. The EPL does not have a published mission statement but does have a principal objective which is

> to stage the most competitive and compelling league with world-class players and, through the equitable distribution of broadcast and commercial revenues, to enable clubs to develop so that European competition is a realistic aim and, once there, they are playing at a level where they can compete effectively.
>
> (Premier League, 2022, para. 2)

The league also adheres to a set of core principles for professional football, including "an open pyramid, progression though sporting merit and the highest standards of sporting integrity" (Goal, 2021, para. 15).

MEMBERSHIP. The members of the privately-owned EPL are the teams themselves. The Premier League is owned by 20 shareholders – the member clubs (e.g., Arsenal, Leicester, Man United, Newcastle, and 16 others). Nations large and small around the world have professional football leagues with clubs as members. Some of the most notable, in addition to the Premier League in England, include the J. League in Japan, the Bundesliga in Germany, La Liga in Spain, Brasileirão in Brazil, Serie A in Italy, and MLS in the United States. This is in addition to multination leagues, such as the UEFA Champions League or Copa Libertadores. These are all the top-level leagues in their respective countries, so the UEFA Champions League consists of the top European teams and the Copa Libertadores features teams from all over South America in its annual tournament. Lower-level leagues exist as well. Different from the way MLB is organized, however, with designated minor league affiliates, these lower-level league teams are not directly tied to teams at the highest-level league. They are separate clubs. For example, the German Budesliga has 18 different clubs. The 2. Bundesliga also has 18 clubs operating independently from the Bundesliga teams. In Japan, there is the J2 League and Brazil has the Brasileirão Série B.

FINANCIALS. The sources of revenues for the EPL are similar to those in North America; they include merchandising programs, broadcast revenues, and corporate sponsorships, in addition to ticket sales. The following major corporations have official status with the Premier League:

Lead Partner	EA Sports
Official Bank	Barclays
Official Beer	Budweiser
Official Ball	Nike
Official Timekeeper	Hublot
Official Cloud Partner	Oracle
Official Engine Oil	Castrol

The Premier League distributes broadcast and central income to its individual clubs to help support clubs to develop and acquire talented players, build and improve stadiums, and support communities (Premier League, 2021). After the 2020–21 season, the Premier League paid the 20 individual clubs a total of £2.5 billion (approximately US$3 billion). Each club received the same amounts in the categories of UK television fees (£31 million), International television fees (£47 million), and central commercial (£5 million). Facility fees and merit payments differed by club. Manchester City (£152 million) received the highest amounts, while Sheffield United (£97 million) was at the bottom of the payment plan.

INDIAN PREMIER LEAGUE

The world's second most popular sport (behind soccer), is not American football, baseball, or basketball, but cricket. Cricket has an estimated 2.5 billion followers across the globe. There are three ways to play cricket: Test, One-Day International (ODI), and Twenty20 (T2). While the sport is can place its roots in England during the 16th century (if not earlier), the sport has taken on a global appeal today. Test cricket is the oldest form of cricket (circa 1850s), where teams can play for up to five days. It is called Test as teams would "test" one another's strength/skill by playing each other. However, spectators couldn't watch for five days nor was a tournament or league really viable under those rules. Thus, ODI cricket was born in the 1970s. The game could last up to nine hours, making spectatorship and tournaments much more viable. In 2003, T20 was invented to further help attract spectators. T20 takes about three hours to play and is much less concerned with tradition than Test and ODI forms (Beyer, 2022).

The Indian Premier League (IPL) was started in 2008, a year after India won the inaugural T20 World Cup. T20 cricket became a phenomenon in India after India defeated Pakistan in South Africa (TNN, 2013). Created by the Board of Control for Cricket in India (BCCI), the league started with eight teams in a franchise model akin to the NBA or NFL (Kohli, 2009). Teams would not be owned by the BCCI, like in other sport leagues in India, but by individual owners. The BCCI would be in charge of sharing revenue, tournament scheduling, providing regulations and match officials, maximizing media coverage, and guaranteeing territorial stability, while the owners would provide a franchise fee to gain entry into the IPL (Alter, 2007). The IPL's growth in less than two decades has been almost unprecedented. In fact, the IPL ($13.4 million) is now second behind the NFL ($17 million) in terms of per-match value, and just ahead of the EPL ($11 million; Rao, 2022). The league is estimated to be worth $7 billion and acquired 600 million viewers in 2021. The IPL's growth has led to the creation of a women's league with six teams, slated to begin in 2023 (Gentrup, 2022).

Just as the player transfer system is different from European soccer to the United States, the IPL's player movement structure is an auction. Debuting in 2008, the IPL auction is hosted annually by the BCCI. Each team is assigned a salary cap or budget for the auction. Hundreds of players go up for auction where the (now) 10 teams auction somewhere between 18 and 25 players. The names of the players are announced and teams begin to bid on the players one by one. When a player comes up whom a team likes and successfully outbid the other teams, they have the rights of the player. A player's salary is determined by the amount bid on them. Each team can retain or keep five players, a maximum of three Indian players and two international players (Hussain, 2022). In 2022, teams had a budget of 90 Crore ($ 11.55 million) to spend on players. If a player is not bid on in the early rounds, they could be recalled later for another bid. Players who are not bid on at all will try again next year (Harris, 2022). Imagine how many permutations and combinations general managers, coaches, strategists, and owners must prepare to successfully bid on the players they want!

MISSION. While there is no public mission statement for the IPL, the league's creator Lalit Modi (former Vice President for the BCCI), had a clear vision for the sport. Modi wanted to create a professional league in India on the basis of what he saw during his time in the United States. He focused his path on entertainment, marketing, licensing, and television. Modi would build the IPL on T20 cricket as that form of cricket would "entice an entire new generation of sports fans into the grounds throughout the country. The dynamic Twenty20 format has been designed to attract a young fan base, which also includes women and children" (Kohli, 2009, p. 10). He successfully created a franchised domestic cricket structure for India that is ever-growing.

MEMBERSHIP. As mentioned, the IPL is based on a franchise model. The franchise model was new to India when Modi initially brought the concept to the BCCI. The franchise model is different from the promotion/relegation model (like European soccer leagues are based on) as it offers stability for players, teams, and sponsors. Instead of needing to be concerned with "will we be in the top league next year?", players can have confidence that they will always compete in the IPL. In the UK, teams in the EPL that are relegated from the EPL to the Champions League lose an estimated £50 million in television rights alone (Smith, 2018). In the IPL, sponsors know that their logos and brands will be visible every time their team plays for the roughly 600 million viewers. The certainty that results from using a franchise model has attracted some powerful owners for IPL teams from Bollywood superstar Shah Rukh Khan to India's biggest conglomerate, Reliance Industries (Khandelwal, 2022).

FINANCIALS. In 2007, a year before the IPL began play, Modi was handed a $25 million check for recruitment and start-up costs. He would spend $20 million on the top 100 players in the world to come to the IPL. From there, Modi had to determine how much to charge for broadcasting fees and franchise fees. Modi decided on $1 million per game for each of the 59 games in the inaugural season for broadcasting rights. In terms of franchise fees, he hoped to raise a minimum $50 million for each of the eight franchises sold for operating costs (Kohli, 2009). He ended up getting an average of $90 million per team (Ozanian, 2022). A year into the league, franchises were valued at an average of $67 million. In 2022, that number rose to $1.04 billion (a growth rate of 24%). The current revenue model for IPL teams almost guarantees a profit. In 2021, the average revenue for an IPL team was $35 million, with an operating income of $9 million. Almost 80% of the IPL's revenue comes from central revenues negotiated by the BCCI, which includes media rights (~$6.02 billion) and central sponsorships (~$100 million). In all, 45% of the central revenue was shared equally among teams, 5% was given to the four playoff teams, and the BCCI keeps the other 50% of central revenues. About 20% of revenue comes from individual team sponsorships, tickets, and merchandise. The IPL is able to keep costs low as player payrolls are no more than 35% of revenue (compared to the NFL at 48% and NBA at 51%). The most valuable franchise is the Mumbai Indians with a value of $1.3 billion (Ozanian, 2022).

ORGANIZATIONAL STRUCTURE. While the BCCI operates the IPL, the Governing Council is in charge of many of the IPL's operations. The Governing Council for the IPL consists of six individuals: a Chair, Honorary Secretary, and Honorary Treasurer (both BCCI members), and three members (IPL, n.d.) The Governing Council meets annually to determine the league schedule, set policies for the league, approve expansion teams, and oversee the day-to-day operations of the leagues (Shah, 2022).

WOMEN'S PROFESSIONAL SPORTS

Women's professional sport has been represented on the international stage for years. Basketball, football (soccer), tennis, golf, volleyball, and more are played on the professional level across the world, with athletes competing not only in their home countries but also internationally, finding the most competitive events and tournaments in which to participate. With sponsorships, lucrative contracts, and international acclaim available, women's sport at the international level continues to increase in popularity and opportunity.

The highest level of professional basketball for women in Europe, the EuroLeague, began its inaugural season in 1958 (FIBA, 2017). EuroLeague is followed in succession by the next most competitive professional basketball league, EuroCup. Other leagues like the WABA League (Women's Adriatic Basketball Association), BWBL (Baltic Women's Basketball League), EEWBL (European Women's Basketball League), and CEWL (Central European Women's League) also take place around Europe. For women's basketball, FIBA oversees both the EuroLeague and the EuroCup tiers, bringing the annual champions together for the SuperCup championship (FIBA, 2017).

The draw for female athletes from the US to enhance or subsidize their incomes by playing a season internationally has increased significantly in the past few decades, as the best and highest paid level of basketball or football is often played in Europe, Russia, or China. The UEFA Women's Champions League made headlines in 2017 when it drew US national team members Carli Lloyd and Alex Morgan to Manchester City's Academy Stadium in a head-to-head match as they represented their respective professional clubs (Evans, 2017). In 2020, USWNT star Sam Mewis and 2019 World Cup breakout player Rose Lavelle were signed by Manchester City from their National Women's Soccer League teams (Oatway, 2021). It was common pre-2020 for WNBA players to play overseas to make more money. Brittany Griner of the Phoenix Mercury made $61,800 base salary in the fourth year of her initial contract with the team but made almost $1 million while playing for her EuroLeague team in Russia, UMMC Ekaterinburg (Metcalfe, 2017). All-time WNBA leading scorer, Diana Taurasi, opted to sit out the 2015 WNBA season and solely play for her club in Russia, as they were paying her approximately 15 times more than her WNBA salary, approximately $1.5 million (Mumcu, 2015). In 2020, the WNBA attempted to remedy this situation in their new collective bargaining agreement (CBA) with the WNBA Players Association. The WNBA would increase the minimum salary from $41,965 to $60,471 and the maximum salary from $117,500 to $228,094, with the assurance that WNBA players would limit their amount of time playing overseas. Still, an estimated 70 players of the ~140 in the WNBA play overseas to increase their earning power (Lowe, 2022).

The organizational structure of women's professional sport in Europe operates in many different ways, with some corporations representing a number of sport entities, most often including soccer and basketball for both men and women. For instance, the Olympique Lyonnais (Lyons) have a professional soccer team in the Division 1 Féminine (French Division 1 for Women), the OL Reign (of the NWSL), and a French basketball team in the French Pro A Women's League. The Lyons was also the first to create a mixed-gender academy as well (Olympique Lyonnais, n.d.).

CURRENT POLICY AREAS

Just as with professional sport in North America, international professional sport faces a broad range of policy issues. These issues include fighting racism in football, safety and security issues, and the rise and fall of the Super League.

Racism in Football

Unforgivable as it is as we begin to approach the mid-21st century, systemic racism is as rampant in many parts of the world as it is in the

Spotlight: Diversity, Equity, & Inclusion in Action

European Club Association

On February 24th, 2022, Russia launched a military offensive against Ukraine. Most of the world denounced Russia's invasion. In fact, the United Nations passed a resolution to condemn Russia by a 141–5 vote (Russia, Belarus, Syria, North Korea, and Eritrea voted against it), with a total of 35 abstentions (Granitz & Hernandez, 2022). The sporting world also responded. UEFA and FIFA both banned Russian national teams and clubs from competitions (including the 2022 World Cup). While banning and punishing Russia was a common tactic, many opted to assist Ukraine as at the time of this writing, 4,500 citizens had been killed in the war and 7.7 million were forced to flee (Berkeley, 2022).

The European Club Association (ECA) is an independent governing body that represents football clubs in Europe. As the "voice of the clubs," the ECA aims to protect and promote European club football (ECA, 2019). In response to Russia's attack on Ukraine and the subsequent war between the two countries, many around the world came to Ukraine's assistance. UEFA invited member ECA clubs to donate between €25,000 ($26,000) and €50,000 ($52,000) to Ukrainian relief funds, for a total of €1 million ($1.052 million). The funds would be specifically used for those impacted by Russia's military assault on Ukraine (Berkeley, 2022).

United States. The past years have seen a continuing number of incidents involving racist behavior on the part of football fans. Unfortunately, racism is still quite alive in some sectors of European football. Three Black players from the English national team – Marcus Rashford, Jadon Sanco, and Bukayo Saka – faced a bombardment of racial attacks following a loss to Italy in the UEFA Euro 2020 final (Sullivan, 2021). Former Inter Milan striker Romelu Lukaku was subject to racist chanting by the audience when he was about to take a penalty kick. The message he was given by fans of Inter Milan told him to not take it personally and that it was just part of the game (Panja, 2019). The examples are all too easy to list. Sport managers need to step up to fight against this onslaught of ignorant behavior and continue to educate the public with messages until the racist behaviors at sporting events end.

So what have sport organizations done to deal with this issue? At the 2010 World Cup in South Africa, the "Say No to Racism" message was expressed often and visibly. FIFA dedicated all four of the quarterfinal matches to spreading the word against racism. Players and officials held banners on the field reading "Say No to Racism" before these games, and team captains read pre-match pledges against racism. In addition, all the stadiums where matches were played were equipped with a racist incident monitoring system using trained security personnel (FIFA, 2011).

UEFA (2014, para. 5) has chosen to take a stand through its Unite Against Racism program:

> The No to Racism message aims to increase public awareness of intolerance and discrimination in football, as well as developing ideas and strategies on how to fight them. On the club competition

matchday dedicated to the campaign, team captains wear No to Racism armbands, anti-racism messages are played over clubs' public address systems and a video containing player testimonials backing the campaign is shown in stadiums. A No to Racism pennant is also prominently on show, held by players. At the start of every match, "No to Racism" banners are prominently displayed on the pitch.

What specific steps can sport managers take? In 2006, UEFA (2006) offered the following ten-point program for football associations and clubs which is still useful today:

1. Issue a statement saying that racism or any other kind of discrimination will not be tolerated, spelling out the action that will be taken against those who engage in racist chanting. The statement should be printed in all match programs and displayed permanently and prominently around the ground.
2. Make public address announcements condemning racist chanting at matches.
3. Make it a condition for season ticket holders that they do not take part in racist abuse.
4. Take action to prevent the sale of racist literature inside and around the ground.
5. Take disciplinary action against players who engage in racial abuse.
6. Contact other associations or clubs to make sure they understand the association's or club's policy on racism.
7. Encourage a common strategy between stewards and police for dealing with racist abuse.
8. Remove all racist graffiti from the ground as a matter of urgency.
9. Adopt an equal opportunities' policy in relation to employment and service provision.
10. Work with all other groups and agencies, such as the players' union, supporters, schools, voluntary organizations, youth clubs, sponsors, local authorities, local businesses, and the police, to develop proactive programs and make progress to raise awareness of campaigning to eliminate racial abuse and discrimination. (p. 18)

FIFA has just recently given referees the authority to stop a match if they deem racist incidents are taking place. At the beginning of the Confederations Cup qualifiers for FIFA's 2018 World Cup in Russia, referees were given the option to abandon a match in the face of abusive racist behavior by fans. In addition, FIFA released the following statement (Collins, 2017):

> FIFA is reinforcing its fight against discrimination in football with the introduction of a new anti-discrimination monitoring system for the 2018 FIFA World Cup™ qualifiers. The system includes the deployment of Anti-Discrimination Match Observers to monitor and report issues of discrimination at the games. It will be coordinated by

FIFA and implemented in collaboration with the Fare network, an organisation with long experience in the fight against discrimination in football and the deployment of match observers.

(para. 3)

The Premier League also has tried a number of efforts to combat racism through the creation or collaboration with anti-racism campaigns and organizations like "Show Racism the Red Card," "Kick it Out," and "No Room for Racism" (Hitchings-Hales, 2021). When the 2020 season opened, players wore kits featuring Black Lives Matter and teams and referees took a knee on the pitch before the matches kicked off (Bumbaca, 2020). Still, the systemic nature of racism becomes evident in these leagues.

Although this section focuses on racism, international football leagues and events also grapple with other forms of systemic social ills. Homophobia and heterosexism are evident as fans are sometimes heard using homophobic charts, slurs, and signs. Sexism is also still prevalent as seen with the low number of female referees officiating matches.

Ethical issues facing sport managers have been woven throughout this text. On this particular topic, sport managers have the opportunity to step up and do the right in taking steps to combat racism as well as heterosexism and sexism. Sport has the power to inform, the power to empower, and the power to transform (Hums & Wolff, 2014). Sport managers have the opportunity to use their platforms in sport to help fight the international scourge of intolerance.

Safety and Security Issues

Similar to what we see in North America, safety and security at major sporting events are a priority for sport managers in international settings as well. For example, Bayern Munich established tighter security measures at the beginning of its 2016–17 season. Now banned from stadiums are items including bottles, thermos flasks, and liquids. Handbags and backpacks – larger than A4 size – are also not allowed in their home stadium, Allianz Arena (Lovell, 2016). Bundesliga officials routinely review security measures. Since the 2016 attack in Paris, where explosions took place outside the Stade de France, coupled with the Berlin Christmas market attack that same year, officials have been even more mindful, and Germany's stadiums are considered among the safest in the world (Penfold, 2016).

Other sport organizations have taken note of the need for increased security measures as well. After the 2017 attack at the Manchester Arena after an Ariana Grande concert, the International Cricket Council (ICC) reviewed security policies for its ICC Championship Trophy (AS. com, 2017). Top sporting officials and security officials decided to ramp up security for the "El Classico" match between Real Madrid and Barcelona after the beforementioned Paris attacks (Reuters, 2017).

A very recent incident occurred prior to the 2022 UEFA Champions League final match held in Paris. Originally slated to be played Gazprom Arena in Saint Petersburg, in Russia, the match location was changed after the Russian invasion of Ukraine. As fans became unruly and tried to forcibly enter the Stade de France prior to the match, the start time was delayed by 30 minutes as riot police attempted to quell the crowd, even using tear gas. The incident itself was troubling enough and consequently prompted carry-over concerns about security for the upcoming Olympic and Paralympic Games set to be staged in Paris in 2024 (Breeden & Panja, 2022).

In addition to security, COVID-19 has impacted sports leagues across the globe. In March 2020, in the wake of a COVID-19 outbreak, the EPL decided to suspend its 2019–2020 season until May 19. On that date, the EPL initiated "Project Restart" that included multiple stages. Stage 1 was titled "Return to Training Protocol" where small-group training could begin. On May 27, 2020, Stage 2 allowed teams to begin contact training. The league would not start until June 17, and even then fans were not allowed into stadiums until May 2021 (Premier League, 2020). Other leagues in Belgium, France, and the Netherlands canceled their leagues completely. When fans were set to return during the UEFA Euro 2020 tournament (played June 11-July 11, 2021), The World Health Organization said that the mixing of crowds at the tournament, travel, and easing social restrictions had driven the number of new COVID-19 cases by 10% after a ten-week decline in cases (Skydsgaard & Gronholt-pedersen, 2021).

Whether it is fan violence, external threats, or COVID-19, sport managers must pay attention to the major risks associated with hosting events. The International Criminal Police Organization (INTERPOL) established a ten-year plan called Project Stadia that was designed to help sport managers of major sporting events manage the security challenges associated with those events (INTERPOL, 2022). Relying on the best practices of locations that have successfully hosted events, INTERPOL makes this information available to sport managers hosting events. Project Stadia includes expert groups, security observations, debriefing programs, training, conferences, and a stadia knowledge management system. Utilizing systems like Project Stadia can help sport managers navigate a new world with new security and safety risks.

The Rise and Fall of the European Super League

On April 18, 2021, it was announced that five English soccer clubs signed up to join a 20-team European Super League (Manchester United, Liverpool, Arsenal, Chelsea, and Tottenham Hotspur). Shortly after, another EPL member (Manchester City), three La Liga teams (Barcelona, Real Madrid, and Atlético Madrid), and three Serie A teams (Juventus, Inter, and Milan) joined in (Christenson, 2021). The European Super League was meant to bring Europe's biggest soccer clubs together as they renounced their membership in the EPL, La Liga, and Serie A and move to the new Super League. Fan backlash to the announcement was immediate and immense as supporters saw the move as motivated by financial greed. If the biggest clubs banded together to create this Super League, every other club would be left behind. It would be like the top three teams from each of the ACC, Big 10, Big 12, Pac-12, and SEC left their conferences to start a new one.

UEFA president, Aleksander Ceferin, condemned the move and made the statement that players who played in the Super League could neither play in the UEFA Euro Cup nor the World Cup. Even FIFA's President strongly disproved of the new league. Just two days later, Manchester City and Chelsea start to waver on their commitment to the new league. Manchester City became the first team to back out, followed by Liverpool, Arsenal, Manchester United, and Tottenham (Christenson, 2021). As of Summer 2022, three teams remain in the Super League: Real Madrid, Juventus, and Barcelona.

At this time, the nine teams that opted out of the league are in a difficult position. All of those teams have to repair the damaged relationship with their fan bases. Arsenal (owned by Los Angeles Rams own Stan Kroenke) wrote their fans saying, "We made a mistake, and we apologise for it. We know it will take time to restore your faith in what we are trying to achieve here at Arsenal…" (Arsenal, 2021, paras. 2–3). An executive for Manchester United even quit over the backlash he received for the move. Real Madrid president, Florentino Perez, still holds onto the dream of the Super League. He brought up the fact that the nine teams that "left" the Super League signed a binding contract with the other teams. He stated, "But the fact is, the clubs can't leave [the Super League]" (Buckingham, 2022, para. 7). Real Madrid, Juventus, and Barcelona have taken a breach of contract action to the European Court of Justice. While the Super League rose and collapsed within merely 48 hours, it seems that the Super League still has life left. What comes of it next is something to monitor.

SUMMARY

The world of international professional sport is complex. The interrelationships among professional leagues, IFs, regional federations, and NGBs are unlike any in North America. The promotion and relegation system does not operate in major North American leagues, although it would be interesting to see how that could play out. The player transfer system in European soccer or the Indian Cricket Play Auction is completely different mechanisms than those used in North America. Despite these differences, similarities exist, particularly when it comes to player movement between teams and corporate sponsorship concerns. New leagues are beginning and following blueprints from the success of other leagues to quickly rise in global prominence, as was seen with the IPL. Women's sports have solid footing in the international sportscape as well, with some leagues being lucrative enough to lure the top female athletes away from North American leagues such as the WNBA. Finally, sport managers working in this international realm need to be aware of and make proactive decisions in terms of fighting racism in football, security and safety issues, and the potential fallout from the European Super League. What is important to realize after reading this chapter is how global a product sport really is and how many opportunities exist beyond North American borders.

CRISTINA VANDERBECK *Senior Director*

PGA TOUR Latinoamérica

As a Senior Director for of one of the International Tours with the PGA TOUR my responsibilities include, but are not limited to, overseeing the developmental tour in Latin America. PGA TOUR Latinoamérica is one of two international developmental tours which provide professional golfers a pathway to the PGA TOUR. I am responsible for developing our calendar which consists of 12–14 events in roughly seven countries. While each event celebrates the culture of its host country, it is also my responsibility to ensure that it falls in line with all of TOUR's commercial, operational, and financial needs.

Similar to most industries, the sports industry is ever-changing and must constantly adapt to the world's trends. Whether it is a global pandemic, economic downturns, or political instability, the industry has to quickly adapt in order to fulfill its commitment to the sport, governing bodies, members, and partners. Then there is the competition aspect of the industry. Outside of what occurs on the field of play within each sport (what happens inside the course, field, and athletic venue), there is a high level of competition among different sports to continue to increase their fanbases, media coverage, and sponsorship dollars. I feel both of these issues go hand in hand, and I chose them because your success as a sports manager has a lot to do with how you navigate through both. In other words, how do you manage this ever-changing global environment and continue to increase overall engagement to keep your sport, league, team, or brand relevant.

Most recently we have had to make some adjustments to the level of access given to developmental players as they progress through the ranks of the PGA TOUR. As I previously explained, PGA TOUR Latinoamérica is one of the pathways for a professional golfer to become part of the PGA TOUR. Upon the conclusion of our season, our top 10 season-long performers receive some level of access to the Korn Ferry Tour (the professional Tour that precedes the PGA TOUR). Recent years have seen some discussion of how much access and how many spots should be awarded based on season-long performance. The most recent and improved access provided to the players was a result of serious and careful considerations based on data collected from all previous seasons.

The current governance at the PGA TOUR is that we have a Commissioner but there are differences from other leagues because we are a membership organization. You have to think of our players as franchise(s)/independent contractors, so the players can participate in the events of their choice as opposed to being obligated to participate in certain events. Again, different than most leagues, there is also no union, but the players have a powerful voice as they represent each other and are nominated as part of a player pack committee. PGA TOUR Latinoamérica, which is a department under PGA TOUR Inc., also works as a membership organization, where it can grow and expand our membership year after year, increasing the diversity of our players and continuing to expand the growth of the professional golf game. On the other hand, one challenge we face is that we would love to be able to count on the presence of certain well-established players at a given Tournament, but we are unable to do so. With that said, it is up to us, as executives of the Tour to make our product as attractive as possible to maintain membership engagement and position our events as the place where the best players would like to play.

A current and future challenge is staying relevant and growing our fanbase. We need to be adaptive to new technologies, media outlets, ever-growing competition, while staying true to the sport.

Looking back from where I started, I would tell anyone wanting to seek a career in sports to take the chance and take the opportunity. If you are able, be willing to move and discover a new area or a new sport. You never know where that would take you! Make connections, stay in touch. The world of sports may seem big, but it is actually very small because the people who work in it are truly connected.

Don't ever stop learning! You may think it is cheesy to say this in a textbook but staying current with your sport and being open to learning and further developing your own knowledge, management abilities, or other new skills will always set you apart and help bridge your success. Learning also keeps you humble and enhances your perspective. Second, I would say, never think a job is too little for you. You are working in sports so you are (very likely) a team player, and there is truly no job too little for anyone.

Lastly, if given the opportunity to work in this amazing industry work hard but have fun! Sports have this incredible ability to connect people from all over the world and celebrate incredible talent in so many different ways!

case STUDY

PROFESSIONAL SPORT BEYOND NORTH AMERICA

The time for you to start planning your internship is quickly approaching. After reading this book (especially this chapter), you realize opportunities exist for you to pursue internships in nations outside of North America. The opportunities include professional sports, national sport organizations, IFs, and major games and events.

1. List a number of countries you would like to live in or visit.
2. List a number of international organizations, sports, or events you would consider for an internship.
3. How can you find out more detailed information about these international sport organizations?
4. Ask if any of your classmates have either visited or lived in other parts of the world. What questions would you like to ask them about their experiences? Did they attend any sporting events? If so, what was the fan experience like compared to games here in North America?
5. Have you ever attended a sporting event in another country? If so, what was the fan experience like compared to games here in North America?
6. What are three specific industry experiences an international internship could offer you that a domestic internship could not?
7. How do you see the global face of the sport industry changing in the next ten years?

CHAPTER questions

1. Choose another professional sport league outside of North America and research its history, mission, membership, financials, and governance structure.
2. Choose a country outside of North America. Research the different professional sport offerings available in that nation.
3. In addition to the governance structures described in this chapter, some countries also have a governmental agency that oversees all sports, including professional sport. Sometimes this agency is named the Ministry of Sport. Think of it like our Cabinet positions like the Department of Defense or Department of Education. We have a Secretary of Defense and a Secretary of Education. What if we had a Department of Sport and a Secretary of Sport. What qualifications should that person have? Who would you suggest as the Secretary of Sport here in the United States?
4. What are your thoughts on promotion and relegation being used in North American professional leagues? MLB? MLS?

FOR ADDITIONAL INFORMATION

1. This website will give you more details in FIFA and how it operates: About FIFA. (2022). About FIFA: https://www.fifa.com/about-fifa

2. The importance of sport governance in international sport: YouTube: International Sports Convention's Sports Integrity and Governance: https://www.youtube.com/watch?v=ciV1UZGHtHw

3. Throughout class we learned about important governing documents. Here is an example of by-laws: EuroLeague 2021–22 Bylaws: Turkish Airlines Euroleague. (2021). 2021–2022 Turkish Airlines EuroLeague https://www.euroleague.net/rs/arly7cmx6hsh55up/84bd1f8d-134d-42a0-a8ee-cd688d29aaa2/282/filename/2021-22-euroleague-bylaws-linked.pdf 2016.pdf

4. This video provides some good basic information on cricket: Netflix: Explained (Season 1, Episode 1): Cricket.

5. Information on stadium safety in an international setting: FIFA Stadium Safety and Security Regulations: FIFA. (2022). FIFA stadium safety and security regulations: https://digitalhub.fifa.com/m/682f5864d03a756b/original/xycg4m3h1r1zudk7rnkb-pdf.pdf

REFERENCES

Arsenal. (2021, April 20). *An open letter to our fans*. https://www.arsenal.com/news/open-letter-our-fans

AS.com. (2017). *ICC to review security measures for Champions Trophy after Manchester terror strike*. https://en.as.com/en/2017/05/25/soccer/1495741401_007459.html

Alter, J. (2007, September 13). *Franchises for board's new Twenty20 league*. ESPN Cric Info. https://www.espncricinfo.com/story/franchises-for-board-s-new-twenty20-league-310819

Berkeley, G. (2022, June 18). *European Club Association members invited to apply to aid Ukrainian relief fund*. Inside the Games. https://www.insidethegames.biz/articles/1124603/eca-members-to-aid-ukrainian-relief-fund

Beyer, G. (2022, April 16). *A history of cricket: The world's second-most popular sport*. The Collector. https://www.thecollector.com/history-of-cricket-worlds-second-most-popular-sport/

Breeden, A., & Panja, T. (2022, May 30). Under fire for chaos at soccer final, France rejects blame for failures. *New York Times*. https://www.nytimes.com/2022/05/30/world/europe/ucl-game-delay-france.html

Buckingham, P. (2022, April 17). *Explained: The binding contract that means all six English clubs are still part of the Super League*. The Athletic. https://theathletic.com/3253386/2022/04/18/explained-the-binding-contract-that-means-all-six-english-clubs-are-still-part-of-the-super-league/

Bumbaca, C. (2020, June 17). *Soccer players wear Black Lives Matter kit, kneel at start of games as EPL returns*. USA Today. https://www.usatoday.com/story/sports/soccer/europe/2020/06/17/english-premier-league-soccer-return-black-lives-matter-protest-kneel/3208109001/

Christenson, M. (2021, April 20). *Timeline: European Super League's rise and fall – two and a half days*. The Guardian. https://www.theguardian.com/football/2021/apr/20/timeline-the-rise-and-fall-of-the-european-super-league-in-two-days

Collins, P. (2017, June 16). *FIFA will allow referees to end matches threatened by racial tension*. Good. https://sports.good.is/articles/fifa-referees-racism

Collinson, S. (2017, July 4). *The world looks past Donald Trump*. CNN. www.cnn.com/2017/07/04/politics/world-looks-past-donald-trump/index.html

Conn, D. (2010, May 11). Bradford remembered: The unheeded warnings that led to tragedy. *The Guardian*. www.guardian.co.uk/football/david-conn-inside-sport-blog/2010/may/12/bradford-fire-david-conn

ECA (2019). *About ECA*. https://www.ecaeurope.com/about-eca/

Evans, S. (2017, April 23). *Europe becomes a new destination for American women soccer stars.* Reuters. www.reuters.com/article/us-soccer-women-europe-americans-idUSKBN17Q043

Explaining soccer. (2009). *In soccer, what is a transfer and how does it work?* http://explainingsoccer.typepad.com/explaining-soccer/2009/07/in-soccer-what-is-a-transfer-and-how-does-it-work.html

FIBA. (2017). *SuperCup women*. www.fiba.com/supercup-women

Gaines, C. (2017, August 6). *Neymar's move to PSG will cost the French soccer giant more than $500 million.* Business Insider. https://www.businessinsider.com/how-much-neymar-salary-transfer-fee-psg-barca-2017-8

Gentrup, A. (2022, May 9). *IPL launching women's league as viewership rises.* Front Office Sports. https://frontofficesports.com/ipl-launching-womens-league-as-viewership-rises/

Goal. (2021). *What is the Premier League Owners Charter and why is it being introduced?* https://www.goal.com/en-us/news/what-is-the-premier-league-owners-charter-why-is-it-being/1nnq7d9bbi7ha1owbxj6cjmixa

Granitz, P. & Hernandez, J. (2022, March 2). *The U.N. approves a resolution demanding that Russia end the invasion of Ukraine.* Georgia Public Broadcasting. https://www.gpb.org/news/2022/03/02/the-un-approves-resolution-demanding-russia-end-the-invasion-of-ukraine

Hitchings-Hales, J. (2021, August 3). *Why aren't anti-racism efforts in football making any difference?* Global Citizen. https://www.globalcitizen.org/en/content/anti-racism-history-english-football-kick-it-out/

Harris, M. (2022). *How does the IPL auction work? – IPL 2022 auction rules all explained.* It's Only Cricket. https://www.itsonlycricket.com/how-does-ipl-auction-work

Hruby, E. (2022, January 27). *Thorns' Lindsey Horan moving to Lyons on loan through 2023.* Just Women's Sports. https://justwomenssports.com/nwsl-uswnt-soccer-horan-lyon-move-loan-thorns

Hums, M.A., & Svensson, P. (2019). International sport. In L.P. Masteralexis, C.A. Barr, & M.A. Hums, *Principles and practice of sport management* (6th ed.). Jones & Bartlett.

Hums, M.A., & Wolff, E.A. (2014, April 3). *Power of sport to inform, empower, and transform.* Huffington Post. www.huffingtonpost.com/dr-mary-hums/power-of-sport-to-inform-_b_5075282.html

Hussain, Z. (2022, February 13). Explained: How does the Indian Premier League's auction process work. *India Times*. https://www.indiatimes.com/explainers/news/how-indian-premier-league-auction-works-561905.html

INTERPOL. (2022). *Project Stadia*. https://www.interpol.int/en/How-we-work/Project-Stadia

IPL. (n.d.). *Governing Council*. https://www.iplt20.com/about/governing-council

Khandelwal, J. (2022, March 16). *Meet the top 5 richest team owners in the IPL.* The Cricket Lounge. https://thecricketlounge.com/2022/03/meet-the-top-5-richest-team-owners-in-the-ipl/

Kohli, R. (2009). The launch of the Indian Premier League. *Columbia Case Works-Columbia Business School*, 1–22.

Lovell, M. (2016, August 24). *Bayern Munich tighten security ahead of new Bundesliga season.* ESPN FC. www.espnfc.us/bayern-munich/story/2936379/bayern-munich-tighten-security-ahead-of-new-bundesliga-season

Lowe, S. (2022, March 5). Why Brittney Griner and other W.N.B.A. Stars play overseas. *New York Times*. https://www.nytimes.com/2022/03/05/sports/basketball/wnba-russia-brittney-griner.html

Metcalfe, J. (2017, March 12). Phoenix Mercury, Brittney Griner reach multi-year contract agreement. *Arizona Republic*. www.azcentral.com/story/sports/wnba/mercury/2017/03/12/phoenix-mercury-brittney-griner-reach-multi-year-contract-agreement/99052300/

Mumcu, C. (2015, August 31). Overseas opportunities could be a boon for WNBA, players. *Sports Business Journal*. https://www.sportsbusinessjournal.com/Journal/Issues/2015/08/31/Opinion/Ceyda-Mumcu.aspx

Oatway, C. (2021, May 17). *Manchester City can confirm that Sam Mewis and Rose Lavelle have returned to their respective NWSL clubs in North Carolina Courage and OL Reign following the conclusion of the 2020/2021 campaign.* Manchester City. https://www.mancity.com/news/womens/sam-mewis-rose-lavelle-departure-63756794

Olympique Lyonnais. (n.d.). *Our activities*. https://www.ol.fr/en/ol-group/company/our-activities

Ozanian, M. (2022, April 26). Indian Premier League valuations cricket now has a place among world's most valuable sports teams. *Forbes*. https://www.forbes.com/sites/mikeozanian/2022/04/26/ndian-premier-league-valuations-cricket-now-has-a-place-among-worlds-most-valuable-sports-teams/

Panja, T. (2019, September 4). In Italy, racist abuse of Romelu Lukaku is dismissed as part of the game. *New York Times*. https://www.nytimes.com/2019/09/04/sports/romelu-lukaku-inter-milan-racist-chants.html

Penfold, C. (2016, December 20). *Berlin attack forces reassessment of security at Bundesliga games*. Deutsche Welle. www.dw.com/en/berlin-attack-forces-reassessment-of-security-at-bundesliga-games/a-36843910

Pitts, B.G., & Zhang, J.J. (2016). Introduction: The WASM foundation stone. In B.G. Pitts & J.J. Zhang (Eds.), *Global sport management: Contemporary issues and inquiries* (pp. 3–18). Routledge.

Poli, R., Ravenel, L., & Besson, R. (2016). *Transfer values and probabilities: The CIES Football Observatory approach*. Football Observatory. www.football-observatory.com/IMG/sites/mr/mr16/en/

Premier League. (2018, January 1). *Premier League history*. https://www.premierleague.com/news/59001

Premier League. (2020, June 15). *How has the COVID-19 pandemic affected Premier League football?* https://www.premierleague.com/news/1682374

Premier League. (n.d.). *About the Premier League.* https://www.premierleague.com/about

Premier League. (2021, June 12). *Premier League value of central payments to clubs 2020/2021.* https://www.premierleague.com/news/2222377

Premier League. (2022). *Solidarity- what the Premier League does*. https://www.premierleague.com/about/solidarity

Rao, K.S. (2022, June 13). IPL media rights: At Rs 104 crore, IPL overtakes EPL in per match value. *Times of India*. https://timesofindia.indiatimes.com/sports/cricket/ipl/top-stories/ipl-media-rights-bcci-richer-by-rs-46000-crore-and-counting-digital-rights-soars-to-rs-50-crore-per-game/articleshow/92186238.cms

Reuters. (2017, November 19). *Stiff security measures in place for El Clasico*. http://gulfnews.com/sport/football/la-liga/stiff-security-measures-in-place-for-el-clasico-1.1622828

Shah, J. (2022, February 25). *Key decisions taken in IPL Governing Council meeting regarding TAT IPL 2022 season*. IPL. https://www.iplt20.com/news/3722/key-decisions-taken-in-ipl-governing-council-meeting-regarding-tata-ipl-2022-season/

Skydsgaard, N., & Gronholt-pedersen, J. (2021, July 1). *Euro soccer tournament under fire for helping spread COVID-19*. Reuters. https://www.reuters.com/business/healthcare-pharmaceuticals/euro-soccer-tournament-under-fire-for-helping-spread-covid-19-2021-07-01/

Smith, P. (2018, May 8). *The cost of relegation: what is the financial impact of dropping out of the Premier League?* Sky Sports. https://www.skysports.com/football/news/11661/11358620/the-cost-of-relegation-what-is-the-financial-impact-of-dropping-out-of-the-premier-league

Sullivan, B. (2021, July 12). *Three Black soccer players are facing racist abuse after England's Euro 2020 defeat*. NPR. https://www.npr.org/2021/07/12/1015239599/prince-william-and-boris-johnson-denounce-the-racist-abuse-of-englands-soccer-te

Szymanski, S. (2006). The promotion and relegation system. In W. Andreff & S. Szymanski (Eds.), *Handbook on the economics of sport* (pp. 685–688). Edward Elgar.

Szymanski, S., & Valetti, T. M. (2003). Promotion and relegation in sporting contests. In S. Szymanski (Ed.), *The comparative economics of sport* (pp. 198–228). Palgrave Macmillan.

Thomas, A. (2014, July 28). *The European soccer transfer market explained*. SB Nation. https://www.sbnation.com/soccer/2014/7/28/5923187/transfer-window-soccer-europe-explained

TNN. (2013, April 2). Indian Premier League: How it all started. *Times of India*. https://timesofindia.indiatimes.com/ipl-history/indian-premier-league-how-it-all-started/articleshow/19337875.cms

UEFA. (2006). *Tackling racism in club football: A guide for clubs*. www.uefa.com/MultimediaFiles/Download/uefa/KeyTopics/448328_DOWNLOAD.pdf

UEFA. (2014, January 2). *No to racism*. https://www.uefa.com/insideuefa/news/0211-0e75c25ed9d8-4ada33b00b6b-1000--no-to-racism/

Esports

It is Saturday, November 4th, 2017 and the *League of Legends* (LoL) World
Championship Final is about to be played. Teams from Europe, China, North America,
Korea, and Hong Kong/Taiwan and Macao joined other teams from Brazil, Eastern
Europe, Japan, Oceania, Turkey, Central America, Turkey, and Southeast Asia battled
for LoL supremacy in Beijing, China. The tournament, which is almost a month long,

DOI: 10.4324/9781003303183-15

sees 24 teams compete in a FIFA World Cup format where teams play a round-robin group stage and then the advancing teams compete in a single elimination bracket.

The final match included two teams from South Korea: SK Telecom T1 and Samsung Galaxy. SK Telecom T1, who has won three World Championships in the past four years, was led by Faker in the Mid-Lane, whom many consider to be the greatest of all-time. Samsung Galaxy, last year's runner-up, was a much more balanced team consisting of an incredible duo bottom lane with Ruler (Marksman) and CoreJJ (Support). Samsung Galaxy would avenge their defeat last year and sweep SK Telecom T1 by 3-0 in one of the biggest upsets in competitive LoL history.

What is more impressive than Samsung Galaxy's sweep of SK Telecom T1 are the facts and figures around the tournament. The final match was held in Beijing's Olympic "Bird's Nest" stadium, with a sold-out audience of 40,000 people in attendance (Pei, 2019). An additional 60 million viewers would also be watching the final match (and 80 million unique viewers for the tournament; Goslin, 2017). The event saw choreographed pre-game ceremonies, which included a performance by Taiwanese pop star Jay Chou and American musician Chrissy Costanza, CGI Dragon descending on the stadium, and almost 30 dancers. A broadcast with over 22 different casters, hosts, and personalities from across the globe helped first-time viewers and veterans of LoL make sense of all the matches they watched. The total prize pool for the tournament was $4.6 million USD, with $2.25 million coming from Riot Games (creators of LoL) and $2.35 million from fans of the video game that wanted to contribute to the overall monetary earnings.

The LoL World Championship numbers have increased steadily since 2017, just as the industry of esport has as well. The esport industry revenue was roughly 1.137 billion USD in 2021 and could potentially grow to between $1.866 billion to $2.285 billion (Newzoo, 2022). All this is helped by the fact that esport viewership grew from 435.7 million in 2020 to 532.0 million in 2022 (projected 640.8 million in 2025). This growth has continued and many are rushing to invest. Celebrities like David Beckham, Mark Cuban, Steph Curry, Chris Evert, Keisuke Honda, Zlatan Ibrahimovic, Michael Jordan, Shaquille O'Neal, Aerial Powers, and Bianca Smith are starting their own esport companies or investing in esports (Duran, 2021).

The following chapter dives into esports and its governance. First, we will go over the definition and history of esports. Then, we will look at how esport is governed at the high school, college, and professional/international levels. We examine current policy areas that affect and will continue to impact the industry. It is important to remember in this chapter that the esport industry is still very young. As this nascent industry is met with a tremendous boom in viewership and engagement it will continue to experience successes and failures along the way. Further, what may be relevant today will certainly change over time. As the industry grows and matures, without a doubt more corporatization and monetization of the product will occur. What that looks like? No one quite knows at this stage.

HISTORY OF ESPORT

Before diving into the history of esports, it is perhaps best to define it. "Esports, short for electronic sports, is a catch-all term that refers to competitive video gaming" (Darvin et al., 2020, p. 36). Instead of playing with bats, skates, and sticks as one would with "traditional sports," esport utilizes a computer, console, or some other sort of electronic device. Sometimes esport video games are based on traditional sports – *NBA 2K*, *Madden*, or *FIFA*. But, they don't have to. *League of Legends*, the world's most popular esport, is entirely based on fantasy characters. The characters fight to destroy one another's bases. *Super Smash Brothers*, another popular esport, is a fighting game based on Nintendo characters. Another popular esport, *Call of Duty*, is a first-person shooter where teams try to eliminate the other team or capture various objectives. While some may view these as just video games, they become esports when there is a level of organization attached to them.

Debate ranges as to when esports actually began. Some have traced esports origin back to pinball in the 1940s (Holden et al., 2020) and informal competitions surrounding those games. But the rise of electronic video games established a foothold in 1960s that would eventually lead to formal competitions. The first recorded video game competition was on October 19th, 1972 at Stanford University. Twenty-four competitors played a 1960s game called *Spacewar* with the top prize winner earning a one-year subscription to Rolling Stone magazine (Larch, 2022). Arcades exploded into US culture in the 1980s, allowing participants to compete with each other to determine who had the highest score. Games like *Asteroids*, *Starfire*, and *Pac-Man* dominated the scene. In 1980, a Space Invaders Championship involved over 10,000 participants. Different regions from across the United States served as preliminary qualifiers for the main event in New York City, with all expenses paid by Atari. The winner in New York City was 16-year-old Rebecca Heineman (Bell, 2020). Around this time, Walter Day from Ottumwa, Iowa, founded the first referee service for video games as national scoreboards for arcade games became more popular in 1982. Day also founded the US National Video Team and the North American Video Game Challenge in 1983 (Larch, 2022).

The next evolution of esports would be its biggest: online video games. In 1988, *Netrek* was released as the first multiplayer computer game. Based on the Star Trek universe, the game was played all over the world. The game was the first Multiplayer Online Battle Arena (MOBA; similar to *League of Legends* and *Dota2*). This game, along with the ability to play online or on a local area network (LAN), helped explode modern-day video games onto the scene. Games like *Doom*, *Quake*, *Counter-Strike*, *StarCraft*, and *Warcraft* became popular mainstays in competitive video gaming. These games allowed individuals or teams to vie in highly competitive matches. The result of this was the first esports leagues including the Professional Gamers League, Cyberathlete Professional League, and then the German Clan League (which eventually became today's Electronic Sports League; Larch, 2022).

www

Professional Gamers League
https://www.pglesports.com/

www

Electronic Sports League
https://www.eslgaming.com/

www

Korean e-Sports Association
http://www.e-sports.
or.kr/?ckattempt=1

www

World Cyber Games
https://www.wcg.com/

www

Twitch.tv
https://www.twitch.tv/

The 2000s brought about more growth and professionalization for esports. In South Korea, the Korean e-Sports Association (KeSPA) started in 2000 as an organization internal to the Ministry of Culture, Sports, and Tourism (K, 2020). The goal of the organization was to focus on marketing esports on television (something common today in South Korea). Due to the development of KeSPA, South Korea hosted the first "World Cyber Games" in 2000. Other major international tournaments/leagues would follow as the Electronic Sports World Cup was held in 2003, the Cyberathlete Professional League World Tour began in 2005, the Worldwide Webgames Championship in 2006, and then the Championship Gaming Series in 2007 (Larch, 2020). At this time, prize pools started to eclipse one million dollars. In 2010, Nintendo had its own Wii Games: Summer 2010 tournament. The tournament, held in California, kicked off with a grand ceremony reminiscent of the Olympics opening ceremony. Teams signed up and played five different games (*Wii Bowling, Wii Basketball, Wii Fit, Mario Kart Wii, and New Super Mario Brothers Wii*) to see who would gain the most points (Newton, 2010).

The 2010s followed the success of the 2000s through the help of mass viewership through Twitch.tv. Twitch.tv (originally Justin.tv) started in 2011 before rebranding to Twitch in 2014. The CEO at the time, Emmet Shear, was a fan of *StarCraft* and wanted to branch out from broadcasting general videos to competitive gaming. In 2012, the monthly user base for Twitch was 20 million, growing to 45 million in August 2013. Twitch quickly became the fourth largest source of Internet traffic and was dubbed the "ESPN of esports" by journalists at the time (Hoppe, 2018) and by 2014 focused specifically on esports. Live streaming professional leagues like *League of Legends*, tournaments like "The International" from *Dota2*, and events like "Twitch Plays Pokémon" helped the site continue to grow. Around 2014 Google and Amazon investigated acquiring Twitch, which ultimately Amazon did for $970 million (Hoppe, 2018). Although the number of competitors has grown (e.g., YouTube Live, Azubu), Twitch has a firm grasp on esport streaming dominance even today. In January 2022, Twitch reported 140 million active month users and 30 million active daily users who were watching over 2 billion hours of content (Dean, 2022).

A few esport competitions, leagues, and tournaments capitalized on Twitch's growth in the 2010s to become today's most popular esports. *League of Legends* created by Riot Games was one of those that quickly became the world's most popular esport in 2015 and continues on into the 2020s. The 2012 *League of Legends* World Championship saw 10,000 fans in the Galen Center in Los Angeles (and millions watching via Twitch), the 2014 championship was played in Seoul, South Korea, and witnessed 40,000 live fans, 2016's championship brought in 43 million viewers, and 2021 had 73 million viewers. It is not just the number of viewers that continues to increase. *Dota 2*'s The International was one of the first events to have a prize pool of over one million dollars ($1.6 million in 2011) and provided $40 million in 2021 as the total prize pool in their yearly tournament. The 2019 *Fortnite* World Cup Finals also had a total prize pool of $30.4 million US dollars (Heath, 2021).

Esports shows no signs of slowing down as it heads into the 2020s. Leagues, teams, players, coaches, sponsors, and events are all becoming more like traditional sports with each passing day (examine Exhibit 15.1 to view the different types of esports). High schools and colleges are starting to get serious about including esports as a way to boost enrollment or provide a community for students. As such, governing bodies like the High School Esports League and the National Association of Collegiate Esports were formed. In professional sport, the NBA created an esport league, the *NBA 2K League*, which operates much like the NBA with a draft, teams, jerseys (sponsored by Champion), and tournaments (Booton, 2019). Despite former ESPN President John Skipper said that he had no interest in esports, the network broadcasted esports on live television numerous times since 2015. The most recent of these took place in 2020 because esports continued to operate during the COVID-19 pandemic while traditional sports were shut down (Smith, 2021). The *Overwatch League* was one of the first to create franchises based in cities across the world (e.g., London Spitfire), despite the regular season being played in California (Howard, 2020). The International Olympic Committee (IOC) launched the Olympic Virtual Series ahead of the 2021 Tokyo Olympic Games (Bieler, 2021). While esports is not quite at the Olympic level, this move signifies with almost certainty that it will be by the 2030s. As esports expands from high school all the way to the international stage, there's no doubt that governance structures are needed for this growing industry.

Types of Esports	*exhibit* 15.1

TYPE OF ESPORT	EXAMPLE OF GAMES
Fighting Games	Super Smash Bros. Melee, Mortal Kombat
Racing Games	Mario Kart, iRacing
Sports Games	FIFA, Madden
Digital Card Games	Heathstone, Legends of Runeterra
Real-Time Strategy	StarCraft II, Age of Empires
First-Person Shooter	Call of Duty, Halo
Third-Person Shooter	Fortnite, PlayerUnknown Battlegrounds
Massive Online Battle Arena	League of Legends, Dota 2

Source: Maryville University. (2022).

GOVERNANCE STRUCTURES

t should be obvious from reading the history section of this chapter that a number of different organizations have jockeyed for position as the main governing body of esports. Various leagues, tournaments, and events have occurred over the industry's young history, with few taking hold for longer than a few years. The following section aims to examine esport governance at the high school, college, professional, and international levels. Keep in mind that there are often multiple governing bodies with the various esport games.

High School

High school esport governance is seemingly a bit of a wild west when it comes to who is in charge and what organizations make decisions. One major organization, PlayVS, jockeyed to provide high school esports with an infrastructure. PlayVS was a 2018 startup created by Delane Parnell focused on providing opportunities for high schoolers in esports. Parnell wanted to create a platform that would arrange teams into leagues, schedule matches, host matches online, compile statistics and records, organize and stream postseason competitions, and crown state champions (Ryan, 2022). In getting their seed funding (which ultimately ended up being $100 million USD), PlayVS signed with the National Federation of State High School Association (NFHS). NFHS was interested in getting esports into high school athletics and partnering with PlayVS was just the way to accomplish that task. The contract with NFHS gave PlayVS access to 16 million high school students. While PlayVS neither sets rules nor works at the individual state level, they partner with state associations to provide opportunities for competitions. Indeed, their mission states: "Our mission is to provide gamers with the most competition and the best competitive experiences" (PlayVS, n.d.).

One such state association that works with PlayVS is the Kentucky High School State Athletic Association (KHSAA). The KHSAA sanctioned esports in the Fall of 2018 with the inclusion of *League of Legends, Rocket League, and SMITE* (Kim, 2021). The Association created a "season 0" in hopes of understanding what esports might look like at the high school competitive level. The state was divided into four regional conferences with no more than 128 teams making playoffs. The cost was $65/season (fees coming from PlayVS) for each student and it was up to the school to field as many teams as they wanted in *League of Legends, Rocket League*, and *SMITE*. Jerseys were optional, but not computers, mouse, keyboard, and headsets (KHSAA, 2019). The first year of esports in Kentucky was a success and since then, the Association has added *Super Smash Brothers*, *FIFA 21*, and *Madden 21* (Kim, 2021). At that time, however, the KHSAA actually banned *Fortnite* from competition due to the shooting nature being inappropriate for the high school setting (Good & Hall, 2020). The KHSAA esports leagues continue to go strong and gain sponsors. The Spring 2022 KHSAA Esports State Championships were recently sponsored by

PlayVS
https://www.playvs.com/

the University of Kentucky Health Care at the Kentucky Federal Credit Union Esports Lounge (KHSAA, 2022).

However, not all state athletic associations have welcomed esports alongside their traditional sports. The Wisconsin High School Esports Association (WIHSEA) was created to do just that, seeing as how the Wisconsin Interscholastic Athletic Association (WIAA) has elected not to sponsor esports. The mission of the WIHSEA is: "The Wisconsin High School Esports Association governs, supports and promotes the growth of high school esports through community development, advocacy, equitable participation, and interscholastic competition to enrich the educational experience" (WIHSEA, n.d., para. 1). The organization partners with high schools across Wisconsin to get esports started at their schools. WIHSEA is a nonprofit overseen by a President, four board members, and four interns. The small organization started around 2014, evolving from a casual club to four divisions of 16 teams each in the playoffs. More than 100 schools from around Wisconsin participate in *Super Smash Brothers, Overwatch, Smite, Rocket League, Valorant, NBA 2k, Apex Legends*, and *Fortnite*. An important distinction from the KHSAA is the fact that all these students get to participate for free, instead of the $65/student cost. Unlike with esports in Kentucky, WIHSEA had to develop all its own rules, policies, and procedures without help from the state's athletic association, the WIAA. As such, students participating in both esports and other sports have to know and understand two sets of eligibility rules to compete in both WIHSEA and WIAA sports.

Wisconsin High School Esports Association
https://www.wihsea.org/

College

Back in 2014, Robert Morris University in Illinois announced it would offer athletic scholarships for *League of Legends* players. The scholarships could cover 50% of tuition and 50% of room and board (Murray, 2014). The new esport program, which would be a varsity athletic program at RMU, would be one of the first in the nation. This revolutionary step by Robert Morris would prove to be the beginning of a trend in the late 2010s up until today. Universities and colleges National Collegiate Athletic Association (NCAA) Division I to Division III (although esports are not NCAA sanctioned), the National Association of Intercollegiate Athletics (NAIA), National Junior College Athletic Association (NJCAA), public and private alike, are adding collegiate esports almost daily. Currently, at least 175 colleges and universities offer esports as a varsity sport, many of which run their esports through a governing body called the National Association of Collegiate Esports (NACE).

The National Association of Collegiate Esports
https://nacesports.org/

NACE is a nonprofit organization that serves over 5,000 students, offers $16 million in esport scholarships and aid, and hosts championship events for at least nine esport titles. NACE was officially formed in 2016 and was the result of research conducted by the NAIA. To this day, both entities have a close relationship and NACE operates out of NAIA offices in Kansas City, Missouri (Morrison, 2019). Only seven schools had varsity esports programs when NACE was formed. Today, NACE has over 170 member institutions accounting for over 94% of varsity esports programing (NACEsports, n.d.). It is currently the largest and oldest nonprofit membership association in college esport athletics.

MISSION.

> The purpose of NACE is to promote the education and development of students through intercollegiate esports participation. Member institutions, although varied and diverse, share a common commitment to high standards and to the principle that participation in organized esports competition serves as an integral part of the total educational process.
>
> (NACEsports, 2021, p. 1)

NACE has seven main charges: assist its membership in developing esports at their institution, establishing rules and standards for intercollegiate esports, respecting institutional diversity, providing national recognition for members, providing leadership opportunities at all levels of the Association, ensuring fiscal accountability, and conducting marketing and membership advancement.

MEMBERSHIP. Eligibility for member institutions starts first with being fully accredited by an authorized higher educational accrediting agency in the institution's region. From there, institutions can apply to be a member. Member institutions get to compete in NACE-sponsored events. In addition, each member gets one vote at the annual conference where voting on rules occurs. Member institutions shall, in response, ensure esports is a part of their total educational offering, encourage the broadest possible student involvement in esports, maintain high ethical standards through self-reporting, subscribe to the democratic principles of the Association, evaluate the esports program in terms of its educational purpose of the institution, engage in competition with other institutions, and promote gender equity and minority inclusion (NACEsports, 2021).

FINANCIALS. Not much information is available from NACE. As one can imagine, membership dues account for a large portion of revenues. Revenue from hosting events will also be a source of revenue as are sponsorships with brands like MSI, iBuyPower, Prosphere, and Hussey Seating. Expenses are mostly used for scholarships, as NACE provides over $16 million in scholarship aid. Hosting events, marketing, and administrative fees would be other organizational expenses.

ORGANIZATIONAL STRUCTURE. The organization starts at the top with the Executive Director and Board of Directors. The Executive Director is the head of NACE and oftentimes does the work that is not expressly stated by the Board of Directors. The Board of Directors has full authority on operational policies, fiscal matters, employment, membership eligibility, and supervision over Association sub-committees. Board members must be employed by a NACE member institution and hold one of the following positions at the institution: Chief Executive Officer, Vice President, Dean, Esport Director, Athletics Director, or any other senior administrator. The President/Chief Executive Officer of the NAIA is one of the 10 voting members of the Board. Each Board member serves three years for a term, not to exceed two consecutive terms. On the board, there are two officers, the chair and the chair-elect. Finally, the Executive Committee of the Board consists of the chair, the chair-elect, the NACE President/Chief Executive Officer, and two at-large members of the Board (NACEsports, 2021).

From there, NACE has several other councils populated by NACE members: Competition Council (sets policies, statistics, and awards for national competitions), Eligibility Enforcement Committee (interprets and enforces bylaws), Appellate Review Committee (acts as the final arbiter of all appeals), and the Student Representative Program (enhances total student experience for NACE students). The Board and other Councils and Committees meet once a year at the NACE National Convention. The event is held in Kansas City, Missouri (NACE National Office), in late July. Updates on the Association, Committee/Council appointments, financial statement distribution, and constitution and bylaw changes happen at that National Convention (NACEsports, 2021).

Professional

The evolution of professional esports from humble beginnings to NBA-sponsored teams has been quite fascinating. One of the first modern esport organizations, Team Liquid, started in 2000 as a media organization posting content about *StarCraft* before forming teams and eventually raising $35 million at a valuation of $415 million (Seck, 2022, May 5). Faze Clan, a popular esport organization that focuses mostly on shooters (i.e., *Call of Duty, Counter-Strike: Global Offensive*), was a social media content-producing organization on YouTube before fielding competitive teams. Today, they have celebrity members like Lil Yachty and Offset, partnered with soccer super-power Manchester City, and could be the first esport team to be valued at $1 billion (Lorenz, 2019). Many of these teams were established by former professional players who created a media site or wanted to start a team with very little business sense. Yet, today professional organizations have grown into multi-million-dollar entities and are even larger.

Like many of the esport organizations, the esport leagues themselves have had similar grass roots upbringings that evolved over time. *Call of Duty*'s league started in 2003 with Major League Gaming Corp, an organization that specifically runs and broadcasts esport events. It was not until the mid-2010s that a formalized Call of Duty World League would be established (Lopez, 2020). Riot Games, creator of *League of Legends*, was a gaming company that dove head first into esports in 2011. *League of Legends* would shortly thereafter become the world's most popular and watched esport. *Overwatch*, created by Blizzard, was designed heavily for esports (and monetization) in 2016. Even before the formation of the league, the cost of a franchise would be $20 million, sponsorship revenues would be over $200 million, and a broadcasting rights deal with Twitch and Disney was valued at $90

www

League of Legends Esports
https://lolesports.com/

million (Wolf, 2018b). In 2018, the *NBA 2K* League was formed by video game company Take-Two Interactive Software and the NBA. The inaugural season saw 17 (now 22) of the NBA's 30 franchises vying for a $1 million prize pool (Kennedy, 2018). Different leagues have grown in different ways, but all deserve close examination into how esports leagues have evolved and what their future holds.

League of Legends

Just as many of the organizations started off as professional gamers or those involved in esports, many of the esport leagues had a similar grassroots beginning. Take the biggest esport in the world from the mid-2010s until now, *League of Legends*. *League of Legends* was created by Riot Games. Riot was started by Brandon Beck and Mark Merrill in their dorm room at USC as they were discussing the issues with current video games like *Warcraft III*. They felt they could do better and started Riot Games in 2006. In 2007, they created *League of Legends*. The free-to-play game (still to this day) quickly made gains as in 2009 the game revenue was $1.29 million, up to $17.25 million in 2010, and $85 million in 2011 (Crecente, 2019). How does a free-to-play game lie *League of Legends* make money? The majority of their revenue comes from microtransactions, basically users pay a small amount of money for cosmetic changes (i.e., different "skins" or "outfits" for their characters). The game earns about $31 a second on microtransactions, or $2.64 million a day in 2021 (Cuofano, 2021). The creators of *League of Legends* decided to jump into professional esport play to help gain visibility for the game and to grow their fanbase.

In 2011, Riot Games hosted the first season championship at the DreamHack celebration in Sweden in June 2011. The event attracted more than 1.6 million viewers. Given the success, Riot Games launched the "League Championship Series" (LCS) which included eight teams in a North American league and eight teams in a European League to begin in 2013. The leagues featured guaranteed salaries for players, prize pools, full broadcasts with play-by-play casters and color commentators, a regular season, regional playoffs, and a World Championship featuring the top teams from North America, Europe, South Korea, China, and other smaller countries (Riot Games, 2013). The league has grown since then in infrastructure and popularity. The production established policies and practices to better serve their athletes, created a franchise system to mirror the NFL or NBA, attracted top-name sponsors like BW3 and State Farm, saw a players' union form, and developed an entire minor league system for future *League of Legends* professionals.

As mentioned, there is a World Championship featuring teams from around the globe. Yet, all those teams are not in the same leagues. Four major regions each have their own league: North America, Europe, South Korea, and China. North America's league is the LCS, as discussed above. The LCS is overseen by Riot Games, who appoints a commissioner in charge of oversight of the league's operations. Europe's league is the "League of Legends European Championship" (LEC).

Formerly the sister league of the LCS (used to be NA LCS and EU LCS), the LEC rebranded in 2018. The LEC is still run by Riot Games and has a commissioner, just both are situated in Europe. Both the LCS and LEC operate similarly but have creative control over their own leagues' content.

The South Korean league, the "League of Legends Championship Korea" (LCK) has been a powerhouse league since its inception. The main difference between the LCK and the LCS/LEC is the entity that oversees the leagues. While the LCS and LEC both are overseen by Riot Games, the LCK is administered by cooperation between Riot Games and KeSPA. Typically, however, the KeSPA takes main control of the operations for the leagues. For instance, the LCS and LEC have best-of-one matches for every team only two days per week. The LCK, however, has two best-of-three matches Wednesday-Sunday. This difference stems from the spectator patterns of the respective countries. Similar to the LCK set-up, China's "League of Legends Pro League" (LPL) is not entirely managed by Riot Games. It is governed by a joint venture between Riot Games and Tencent called Tengjing Sports. Tencent is a Chinese Tech giant that acquired Riot Games in 2015 and owns 40% shares in Epic Games (creator of *Fortnite*). Tengjing Sports is in charge of tournament organizing talent management, esports venues, and other *League of Legends* related content (Chen, 2019). LPL also typically hosts two games per night, but they run matches every day of the week.

In addition to these four major leagues, eight other leagues around the world also get a chance to compete in international tournaments. While the four major leagues get automatic bids for three or four teams to the World Championship, the other leagues typically have a pre-World Championship tournament just to be able to send their league winner to the World Championship. The Pacific Championship Series (PCS; comprising Taiwan, Hong Kong, Macao, and South East Asian teams) and the Vietnam Championship Series (VCS; comprising Vietnamese teams) are guaranteed one team at the World Championship due to their size and historical performance. The Campeonato Brasileiro de League of Legends (CBLOL; comprising Brazilian teams), League of Legends Circuit Oceania (LCO; comprising nations surrounding Australia), League of Legends Continental League (LCL; comprising Russian teams), League of Legends Japan League (LJL; comprising Japanese teams), Liga Latinoamérica (LLA; comprising Mexican teams), and Turkish Championship League (TCL; comprising Turkish teams) all are considered the wild card teams that have to compete for the chance to enter the World Championship. Each of the aforenoted leagues has at least eight teams, with a spring and summer split. See Exhibit 15.2 for the countries designated to each league.

Due to familiarity and information available, we will now focus on North America's LCS to examine how the league is set up.

MISSION. Most of the esports reviewed won't have mission statements, so one would look toward the values of the game maker. Riot Games does not have an explicit mission statement, but instead communicates its values as players experience first, dare to dream, thrive exhibit **15.2** *League of Legends* Leagues

NAME	ABBREVIATION	COUNTRIES WITHIN LEAGUE
League of Legends Champions Korea	LCK	South Korea
League Championship Series	LCS	North America
League of Legends European Championships	LEC	Western Europe
League of Legends Pro League	LPL	China
Pacific Championship Series	PCS	Taiwan, Hong Kong, Macau, & Southeast Asia
Vietnam Championship Series	VCS	Vietnam
Campeonato Brasileiro de League of Legends	CBLOL	Brazil
Liga Latinoamérica	LLA	Central & South America (not including Brazil)
League of Legends Circuit Oceania	LCO	Oceania
League of Legends Continental League	LCL	Russia
League of Legends Japan League	LJL	Japan
Turkish Championship League	TCL	Turkey

Source: Lolesports Staff (2021, March 30).

together, execute with excellence, and stay hungry; stay humble (*We Are Rioters*, n.d.).

MEMBERSHIP. Membership in the LCS is outlined in the LCS official rules. To play for an LCS team, one must be at least 17 years old (16 years old to play in the academy or minor league LCS). The LCS also has strict rules on who can join depending on the regions. Teams can hire both "Residents" (i.e., someone who is a lawful permanent resident of the United States or Canada) and "Non-residents" (i.e., someone from outside of the United State or Canada). "Non-residents" are further broken down into those from the LEC/LPL/LCK/VCS/PCS/LJL and those from "Emerging Regions" (i.e., players from countries in the LCO, LCL, LLA, CBLOL, TCL). Teams can have as many members who are "Residents" and from "Emerging Regions," but can only have two players from the LEC/LPL/LCK/VCS/PCS/LJL. The LCS does this to attempt to ensure that teams are representative of their regions. For instance, in 2013, team "LMQ" was established in the LCS. The team comprised entirely of players from China. Later that year, the LCS officially put a cap of two imports per team on every franchise in the LCS (Kay, 2021). Most teams in the LCS have utilized the import rule by having two players from the LEC/LPL/LCK and the rest from North America (LCS, 2020).

The league itself also has membership rules for organizations wanting to join. Prior to 2017, the LCS was based on a system similar to European soccer where there was promotion and relegation. [See the chapter on Professional Sport Beyond North America for an explanation of promotion and relegation.] The top seven teams each season would be safe while the bottom three of the LCS had to face off against the top three of the amateur scene. Whichever team won was moved into the LCS. In 2017, however, it was announced that the LCS was moving toward a franchising model (similar to the NFL). To gain entry into the LCS, a flat fee of $10 million was owed for teams that had already existed in the LCS and a $13 million fee for all newly formed organizations. In addition to the fee, organizations had to work through an application and interview process that included credit checks, deep dives into finances, and background checks for all owners (Khan, 2017). The move was meant to provide stability for the league. Sponsors could be confident teams were going to remain in the LCS, amateur and academy teams could be created with knowledge of long-term organizations remaining in the LCS, and the fans could develop relationships with organizations without fear of losing their favorite teams to relegation. With the new franchising model, teams are given 32.5% of all revenue; 32.5% of the pool went to Riot Games for broadcast production, live events, and operations of the leagues; and the final 35% of the pool went to players' salaries (Wills & Maturi, 2019). Player salaries have skyrocketed over the years, from an average of $105,000 in 2017 (prior to franchising) to $320,000 in 2019 (Wills & Maturi, 2019) and now hover around $400,000 (Smith, 2020).

FINANCIALS. Riot Games and the LCS are unfortunately private companies, so access to their financial statements is difficult. Still, there are some financial facts about the league that we can determine. One fascinating fact remains about the LCS: the LCS operates at a net loss in revenue with negative cash flow (Hasan, 2021). Indeed, since its inception, the LCS has not made a profit. That is okay for Riot Games though. While they are attempting to make it profitable, the LCS serves as a marketing tool for the game *League of Legends*. The LCS hopes to become profitable soon but is still investing millions without breaking even. As one can imagine, revenues most likely stem from those initial multi-million-dollar franchise fees, as well as revenue from events. Sponsors like Buffalo Wild Wings, Mastercard, State Farm, and Bud Light pay millions in exchange for their logos on the broadcast (Seck, 2022, January 12).

On the expense side, the broadcast, event organization, and production of events account for a large amount of money spent.

ORGANIZATIONAL STRUCTURE. To ensure stability, competitive integrity, and longevity of their league, the LCS requires an organizational structure for each team. Each team must have at least one (but not more than two) LCS Coaches with one coach being at the venue on game day. Teams must have a LACS Coach for the amateur team that they field. Teams must have a strategic coach, also. Sometimes the strategic coach maintains strategies for roster management, game play, and picking/banning specific champions. General Managers, similar to the NFL or WNBA, are also required for LCS teams. Roster sizes are between 10 players (assuming 5 for LCS and 5 for LACS) and 15 players. Each team is required to have a contract for their players and coaches (LCS, 2020).

The LCS has a commissioner, just like most professional leagues. Jackie Felling, former Director of Product for the Call of Duty League (CDL), was appointed LCS Commissioner in 2022 (Kelly, 2022). Felling's goals when she took over the LCS were to provide support for developing players, expanding broadcast production quality, and driving diversity and innovation. The Commissioner undoubtably has a large role in overseeing the development of the LCS as a whole. In addition to the Commissioner, there are jobs in analytics, such as broadcasting engineers, competition managers, graphics producers, marketing leads, production managers, game-day operations, and just about every other job associated with running a league.

NBA 2K League

www

NBA 2K League
https://2kleague.nba.com/

There are many other esports and their accompanying leagues around the world and in the United States. While the *League of Legends* esports leagues is the most popular around the world, we wanted to highlight an emerging esport that is not based on fantasy elements, but rather on a traditional sport. In 2017, the NBA and Take-Two Interactive Software, Inc. announced the launch of the "NBA 2K eLeague" (NBA Communications, 2017). This first official esports league operated by a US professional sports league consisted of teams operated by the NBA franchises. For instance, the Miami Heat's esport affiliate was the Heat Check Gaming. The league would offer a regular season with 5-on-5 matches, a bracketed playoff system, a championship, jerseys, roster spots, coaches, and a draft system that mirrors an NBA or WNBA draft. The initial season featured 17 teams, with over 100 players signed to the teams. Since 2018, the league has expanded to 24 teams (22 affiliated with NBA teams) and continues to grow.

MISSION. Like many esport ventures which are private entities, the NBA 2K League does not have a mission statement nor do they have any values that are publicly available. With the league being a joint venture between the NBA and Take-Two Interactive, either's mission would not be fair to examine without context. Per the NBA 2K League's website, the league is a "professional esports league featuring the best NBA 2K players in the world and is the first official esports league operated by a U.S. professional sports league" (NBA 2K League, 2022, para. 1).

MEMBERSHIP. In the inaugural season, 17 of 30 teams competed in the League. The price tag for a franchise was $750,000 and a three-year commitment (Wolf, 2018a). The decision was up to the individual franchises to determine if they wanted to put financial resources behind the teams or not. Seeing the success of the inaugural season, four teams added franchises into the League for the 2019 season (Hawks, Nets, Lakers, and Timberwolves). The following year, 2020, saw the Hornets Venom GT (sister team to the Charlotte Hornets) create another team. For the first time, however, a team outside of the NBA structure was added that same year. The Gen.G Tigers of Shanghai were added in 2020 (in a partnership with esport organization Gen.G) at a price tag of $25 million to be the 23rd team in the league (Mazique, 2019). The team resides in California, despite having the city moniker of Shanghai. The following year another international team was added as DUX Gaming entered into another agreement with the NBA 2K League for $25 million. DUX Gaming is based out of Madrid, but their team resides in Mexico (Gomez, 2021). At the start of the 2022 season, there are 24 teams in the NBA 2K League. Twenty-two of the NBA-affiliated teams and two international teams occupy this sport-based esport.

FINANCIALS. While it is difficult to know the revenues and expenses for the NBA 2K League, there is some information we can parse out. For instance, the initial fee for franchises was $750,000 per team (for 22 teams affiliated with the NBA). In addition, the partnership with Gen.G and DUX gaming netted another $50 million for the league. Those fees are primarily used to host the draft/in-game events, market the product, other operations, league administration salaries, and to create the infrastructure needed to develop the league. Given the most recent information, the 2021 NBA 2K league players earned around $35,000 for their six-month contracts. Those contracts are paid for by the NBA 2K League. First-round draft picks had a base salary of $35,000 and second-round draft picks had a salary of $33,000. Players who competed in the previous year and were retained by their teams earned $38,000. Players also received paid housing during the season, medical insurance, a retirement plan, and relocation expenses (NBA 2K League, 2021). For the 2022 season, a 65% increase in prize money totaling $2.5 million was announced. This money is divided among newly created 5v5 playoff and 3v3 tournaments, as well as the Finals valued at $1 million (Hitt, 2022). For the individual teams, sponsorships equate to the majority of a 2K team's revenue (upwards of 95%), with the remaining from camps or social media. In terms of expenses, the majority includes operations. While the league pays the players, the teams are in charge of housing, salaries of staff, and dues to the NBA 2K League (P. Glogovsky, personal communication, June 14, 2022).

ORGANIZATIONAL STRUCTURE. Like the LCS, the NBA 2K League has a commissioner. The commissioner for the 2022 season is Brendan Donohue, who has occupied the role since the league's inception. Donohue worked as the NBA Senior Vice President of Team Marketing and Business Operations prior to his role as commissioner.

He oversees the league's grand strategy and operations (Burns, 2017). Donohue has various managers working under him to build the infrastructure, marketing, and communications for the League. Each team in the League is meant to have a General Manager and Head Coach, just like in the traditional NBA. Some teams, like Bucks Gaming, have one person on payroll for just the esports side of the organization. He is the general manager who scouts players and coaches, manages the roster, is involved with communications, operations, budget management, and putting on community events. He is overseen by the Vice President of "New Business Ventures." Most NBA 2K teams have between seven or eight dedicated esport staff, while some have upwards of 10–20. It is not uncommon for staff to be shared between the 2K and G-League affiliates of the main NBA team (P. Glogovsky, personal communication, June 14, 2022).

International

International Esports Federation

www

International Esports
Federation
https://iesf.org/

Many esports (e.g., *League of Legends, Super Smash Brothers, Starcraft*) are international in nature, but are operated and organized by either the esports' creator or a separate organization. They focus specifically on one esport. The International Esports Federation (IESF) attempts to focus on the legitimization of esports across the globe. The IESF was founded in 2008 and featured nine member countries (Austria, Belgium, Denmark, Germany, South Korea, Switzerland, Taiwan, The Netherlands, and Vietnam) and is based in South Korea (Mackay, 2018). Think of the IESF as the IOC. The IOC hosts the Olympic Games, while the IESF hosts the Esports WE Championships. Both events bring in athletes who represent their home nation and both events feature a multitude of sports/games. Instead of track, soccer, and hockey, the Esports WE Championships hosts games like *Counter Strike: Global Offensive, League of Legends, Dota2*, and *Tekken 7*. Similarly, just as the United States Olympic and Paralympic Committee (USOPC) supports the US teams in the Olympic, Paralympic, Youth Olympic, Pan American, and Parapan American Games, the United States eSports Federation (USEF) is the official governing body for esports as a member of the IESF. In fact, of the 123 members of the IESF, most have their own governing body (e.g., Esports Chile, Electronic Sports Federation of India, and Esport-Bund Deutschland E.V.; IESF, n.d.).

MISSION. Per the IESF's statutes, the mission of the IESF is "to serve as the critical global organization representing, coordinating, harmonizing, and administrating esports while preserving the rights and providing a voice to all stakeholders of esports." (IESF, 2021, p. 2). It is clear from reviewing IESF's information that their aim is to legitimize and grow esports as a sport across the globe. In addition, the IESF wants to empower female athlete participation and end discrimination in the esport field. The IESF has an Anti-doping policy (discussed later) similar to the IOC's and uses the Court of Arbitration for Sports for arbitration when needed (IESF, 2021).

MEMBERSHIP. As previously mentioned, the IESF currently has 123 members. Africa has 19 members, the Americas comprise 19 members, Asia has 38 members, Europe is the biggest with 45 members, and Oceania has just two members (IESF, n.d.). For a country to be a member, it has to have an official governing body. Membership is broken down into Full Members and Associate Members. Both membership tyles are allowed to attend and speak at the General Meeting of IESF, participate in IESF activities, and place players in the IESF World Championship, but only Full Members have voting rights (IESF, 2021). Associate Members have only been approved by the Membership Committee, not the full General Meeting yet.

FINANCIALS. Revenues for the IESF come from membership dues, donations, the IESF's fundraising, events, and sponsors. Those revenues are used to advance the mission of the IESF. The expenses will go toward events and operations, funding for athletes, administrative costs, and marketing.

ORGANIZATIONAL STRUCTURE. The General Meeting is made up of all Full Members and acts as the supreme legislative body of the IESF. The General Meeting can propose rules and then will vote on them. The General Meeting elects the President. The President represents the IESF and oversees the Board of the IESF. The Board is the central executive body of the IESF and implements its organizational policies. The Board consists of the President, five elected members from the General Meeting, the General Secretary, the Chair of the Athletes Committee, and representatives of Confederations recognized by IESF (i.e., Pan-American Esports Confirmation). Two committees are part of IESF – the Membership Committee and the Athletes Committee. The Membership Committee sets rules based around membership, declares members as "Member not in Good Standing" if issues arise, and accepts new members. The Athletes Committee handles complaints raised by athletes, reviews rules as they pertain to athletes, and suggests policy changes to be approved by the General Meeting (IESF, 2021).

Global Esports Federation

The other major player in international esports it the Global Esports Federation (GEF). In a similar vein to the IESF, the GSF wants to promote the credibility, legitimacy, and prestige of esports. The GEF is much younger than the IESF, as it was founded in 2019. The governing body is located in Singapore, where the Singapore National Olympic Council Secretary General, Chris Chan, is president (Aziz, 2019). The Global Esports Games is the main event for the GEF, where teams compete for the Global Esports Cup. The GEF also puts on the Global Esports Tour (series of esport tournaments), a #WorldConnected series (esports events for the community), the Commonwealth Esports Championships (a partnership between the GEF and Commonwealth Games), and a GEFcon (convention for esport thought leaders; Global Esports Federation, n.d.).

Global Esports Federation
https://www.globalesports.org/

MISSION. The mission of the GEF is to "cultivate competition along with developing communication and the connection between sport, esports, and technology" (The Global Esports Federation, n.d., para. 2). The core objectives for the GEF are to (1) encourage and support the establishment of esport federations with appropriate regulations; (2) establish an athlete/player commission with a focus on athlete wellness, career support, and fair play; (3) develop governance structures and guidelines for the GEF; (4) establish and stage esport competitions and educational development programs; and (5) create, develop, and run the Global Esport Games (Global Esports Federation, n.d.).

MEMBERSHIP. In late 2021, the GEF accepted its 100th member with the United Arab Emirates (Lloyd, 2021). Per the GEF constitution, membership is comprising national governing bodies for esports, regional governing bodies for esports, international federations, independent entities and individuals, publishers and commercial partners, affiliates, honorary members, and any other entities seeking non-voting member status. The Board has sole discretion as to the confirmation of new members. Each member has the right to attend, speak, and vote at the General Assembly and nominate an individual for election to the Board. Non-voting members have the right to attend and speak at General Assembly (Global Esports Federation, 2020).

FINANCIALS. Activities of the GEF are funded by annual membership fees, donations, fees for services provided by GEF, fees for participation in GEF events, payment for the right to organize esports events held under the supervision of GEF, fines, and capital gains. In terms of how financial resources are utilized, they are meant for the organization of GEF events, participation of athletes and teams in GEF events, administration of GEF office, and worldwide promotion of esports (General Esports Federation, 2020).

ORGANIZATIONAL STRUCTURE. Like most organizations, the General Assembly (comprising all voting members) makes up the highest authority of the GEF. The General Assembly can amend the constitution, take action against members and non-voting members, and have discretion over new competitions the GEF might host. From the General Assembly, there is a Board, which is responsible for governing the GEF, overseeing and supervising activities of the GEF, and reporting to the General Assembly. The Board is made up of the President, up to six Vice-Presidents, the Secretary-General, and up to 18 ordinary Board Members. The Board has the power to admit members; sanction members; prescribe, adopt, amend, or repeal rules and regulations; interpret rules, and approve the annual budget. The President is meant to be the lead representative of the GEF and chairs meetings of the General Assembly, leads the Board, oversees commissions, authorizes transactions, and is involved in just about every other duty related to overseeing the GEF. The Secretary-General is the person focused on day-to-day administrative tasks. The Secretary-General manages office space, employs staff, carries out the annual plan, supports the President, and ensures compliance with all rules and regulations. There are also committees and commissions. The Executive Committee is made up of the President, Vice-Presidents, and Secretary General. This group

Spotlight: Diversity, Equity, & Inclusion in Action

Queer Women of Esports

An organization that highlights the advances in the idea of Diversity, Equity, and Inclusion in esports is Queer Women of Esports. Queer Women of Esports is a 501(c)(3) nonprofit organization whose mission is

> to provide opportunities and education to marginalized individuals who have been overlooked and under supported. Every program has the interests of our community at the core of its creation and aims to support anyone and everyone who wants to participate.
>
> (Queer Women of Esports, n.d., para. 1)

The organization notes that almost half of gaming participants are women and 10% identify as LGBTQIA+ yet esports does not share that same representation. Queer Women of Esports recognizes that the lack of access to resources, mentorship, gate-keeping, toxic environments, and overt harassment are barriers that have forced LGBTQIA+ and Black, Indigenous, and people of color (BIPOC) out of esports. Queer Women of Esports wants to push back on those barriers and make esports more inclusive through advocacy and education.

Queer Women of Esports's flagship program is their mentorship program. The mentorship program pairs queer individuals and people of marginalized genders together through a six-month program. The mentors are typically esport industry leaders (LGBTAIA+ and allies) who are available for advice, networking, and anything the mentee needs to be successful. The protegès work with their mentors on a capstone project focused on improving inclusivity in esports. Each project will receive $5,000 in a scholarship, with top projects receiving $15,000. All protegès and mentors are provided access to educational panels, workshops, and special networking opportunities through the online communication platform Discord.

deals with urgent matters. Commissions are also in place, like those focused on athletes, players, and community, education, culture, youth, financial, legal, administration, and event planning (Global Esports Federation, 2020).

CURRENT POLICY AREAS

Including Esports in Traditional Sports Governance

With the rise of esports in the 2010s, many eyes turned toward the NCAA and IOC to determine whether those governing bodies were going to include esports into their structures. In 2017, the NCAA announced it was going to research esports and determine if it should govern and create championships for esports. After 18 months of study, the NCAA decided to pass on getting involved in the governance of esports in 2019 (Smith, 2019). The Board of Governors unanimously voted to table the possibility of governing esports, citing concerns with Title IX, violence, and scholarships (Schad, 2019). A particular conundrum was how the NCAA would handle eligibility when many tournaments offer prize money for winning and participation and sponsors provide the top players and streamers money (Zavian, 2020). Adopting esports would force the NCAA to reconcile its stance on amateurism.

The IOC has had similar hesitancy to adopt esports onto the Olympic Programme, which is the slate of sports competed in at the Olympic Games. In 2017, esports was finally deemed a "sporting activity" by the IOC. That same year, however, IOC President Thomas Bach stated "We want to promote non-discrimination, non-violence and peace among people. This doesn't match with video games, which are about violence, explosions and killing" (Faber, 2021, para. 7). A year later, the IOC hosted a summit with the Global Association of International Sports Federations to discuss future engagement between esports, the gaming industry, and the Olympic Movement (IOC, 2018). While the consideration of esports in the Olympic Programme was not part of the forum, discussions over what a future among esports, gaming, and the Olympic Movement was heavily discussed.

In 2021, the IOC decided to dip a toe into the esport pool by launching the Olympic Virtual Series (OVS; Palmer, 2021). The OVS is designed to mobilize virtual sport, esports, and gaming enthusiasts to reach new Olympic audiences. The OVS featured virtual forms of baseball, cycling, rowing, sailing, and racing. The OVS was held just days before the Tokyo 2020 Olympic Games (IOC, 2021). In the future, virtual sport organizations in soccer, basketball, tennis, and taekwondo have expressed interest in joining the OVS (Palmer, 2021). French President Emmanuel Macron was hopeful that the Paris 2024 Olympic and Paralympic Games might host traditional esport games (i.e., *League of Legends* and *Dota 2*; Burke, 2022). Despite Macron's optimism, it appears there won't be traditional esports at the 2024 Games, but potentially the 2028 or 2032 Games.

As sport managers at the NCAA and IOC have yet to fully embrace esports, other sport managers and organizations have fully embraced esports. The Asian Games (non-Olympic) are set to include eight medaled events for esports, including *Hearthstone*, *League of Legends*, *Dota 2*, and *FIFA* in 2022 (Venkat, 2021). FIFA started offering esports (known as FIFAe) and hosted the first FIFAe esport championship in Copenhagen in July 2022 (Houston, 2022). While the NCAA has passed on esports, the NAIA (Morrison, 2019) and NJCAA (NACEsports, 2018) have welcomed esports with open arms. The case of esports provides an interesting look into the future of sport and sport governance.

Performance-Enhancing Drugs

A topic that traditional sports has struggled with for a long time is the use of illegal performance-enhancing drugs. Esports is no exception to this rule. While players may not be taking steroids or doping their blood, but they are taking Adderall. Adderall is often prescribed by doctors to treat or help manage attention deficit hyperactivity disorder (ADHD), along with Vyvanse and Ritalin. The drug has been used by esport athletes to sharper response time and reflexes during game play. Adderall use in esports has been a troubling reality. Professional *Counter-Strike* player Kore "Semphis" Friesen admitted that his entire team had taken Adderall while playing in a major tournament with a $250,000 purse. As a response, the *Counter-Strike* League started conducting drug-tests and banning players for using stimulants like Adderall. They are the only known league to drug-test (Hamstead, 2020). Other esports, like Epic Games' *Fortnite*, bans prescription drugs unless they are prescribed by a doctor during tournaments.

The difficulty esport leagues and tournaments have is where to draw the line. Drug-testing every player can be extremely expensive and difficult in online play. Outright banning ADHD could severely hinder those who live with ADHD and need the prescription for legitimate reasons. In that same vein, Adderall prescriptions are easy to acquire, so simply banning them unless it is prescribed to the individual might not prevent the abuse from happening (Ashley, 2021). Finally, getting rid of Adderall completely could be a slippery slope. Let's say Adderall is banned, would players turn to caffeine pills? Energy drinks? Would those be viewed as performance-enhancing in the same way as Adderall and Ritalin? So, what should be done about performance-enhancing drugs in esports? In an industry as young as esports, finding a solution to this "open secret" is crucial for the legitimacy of the industry.

Combating Misogyny (and Other Forms of Discrimination)

Undoubtably, misogyny and sexism are major issues in esports. This topic has been well-studied by academics (Darvin et al., 2020; Darvin et al., 2021). While there is evidence of prior misogyny, 2014 marked a big moment with "GamerGate." The #GamerGate movement was broadly related to esport journalist integrity. However, a closer examination into what was going on revealed members of the gaming community harassed, heckled, and threatened female esports journalists. The movement revealed how much misogyny was engrained into the gaming community (Romano, 2021). In an example, video game critic, Anita Sarkeesian, posted videos about the over-sexualization of women in video games and how video games reinforce sexist ideas toward women. As a response, male members of the gaming community threatened her life to the point she had to cancel a scheduled speech at Utah State University due to one individual threatening a mass shooting on campus. In addition, she had to leave her personal home due to safety concerns (which the FBI investigated; Dockterman, 2014). The comments and resulting the fear Sakeesian felt unnerved many women and People of Color across the gaming industry.

Fast-forward to 2017. To emphasis how prevalent the issue is, during one weekend in June 2020, more than 70 people in the gaming industry came forward with allegations of discrimination, harassment, and sexual assault (Lorenz & Browning, 2020). These survivors of sexual harassment and assault used their platforms to open up about abuse at the hands of top esport players, Twitch/YouTube personalities, and a CEO of a gaming organization. The response to these allegations of sexual misconduct was met with significantly more positive reactions

In my job, I wear two hats. First, I am the esports manager for the Bucks. For this position, my major responsibilities are business development, content ideation, event innovation and organization, and ultimately figuring out how to grow our business and extend the reach of Bucks' esports. Next, I am the general manager of Bucks Gaming, our professional NBA 2K team where I am responsible for managing the players and staff of Bucks Gaming. This includes roster management, community activations, partnership management, and more.

I think the two most important current issues facing sport managers working in esports are handling and managing a team that may be composed of players who have faced a structureless competitive lifestyle throughout their career. Thankfully, this is starting to change as esports is becoming a much more established sport and the infrastructure is starting to settle and become mainstream. Thus, players are experiencing more balanced competitive lifestyles and are being guided and led by educated and mature leaders to become not only good but also coachable players, good teammates, and good communicators.

Players and teams need to be managed in a way that will successfully alleviate any mental and emotional pressure/stress, thus resulting in less competitive burnout. What we often see in esports is an athlete's shelf life averaging out to be around four years. Figuring out how to provide a quality experience for players to increase career longevity will be vital for our industry moving forward.

I have been in discussion with the NBA about changing the format and competitive structure of the 2K League. As it currently stands, the season is six months long, during which players move to their team's market, get paid a six-month salary (by the NBA), and have their housing paid for. In the off-season, they are essentially unemployed and do not have any contractual obligations to their team or the NBA. Thus, this makes the lifestyle of a professional 2K player quite difficult and stressful. Ideally, the goal is to adopt a year-round structure with breaks between "splits" or circuits. It would be more expensive but is a much more ideal structure for every party involved.

Governance in esports is a very unique concept. Because of the ever-growing nature of the industry and how new titles and games are popping up everywhere, it becomes difficult to clearly define "esports" and where the industry will succeed the most. Every game has its own governance and competitive structure, some much more successful than others. At the collegiate level, there is no centralized governing body (think NCAA) for esports. Because of these two factors, esports is still very much wild west-like in terms of governance and structure. Some bodies and entities seem to have figured it out and set a great standard for the industry, but that doesn't come without spending millions of dollars that other titles and games simply cannot expend.

Sport managers working in esports will be challenged to become and stay profitable as an organization or team. The great majority of esports organizations operate at a loss. Figuring out how to maximize revenue and eventually become profitable is a key challenge for esports professionals today. The industry will take off at a global level once this issue is tackled.

If you want to work in esports, I offer the following advice. Network – I know it's cliché, but esports is a tight-knit community and knowing how to communicate with people will help with one's career trajectory and is massively important. Furthermore, getting to know other figures within the industry is a great way to get your name out there and meet people who can help shape your future and include you in potential opportunities

Get involved – The youthful nature of esports and the extended period of discovery we are in right now can be considered a weakness for the industry, but it also makes it a very rewarding industry to those who get involved. Whether it's a local esports organization you volunteer with for five hours a week, a college club you become a staff member of, esports events, seminars, or panels you attend, or even a team or event you start and manage yourself – esports tends to reward those who get involved, stay involved, and utilize their grassroots passion for the industry.

than #Gamergate and there was a change made. For instance, an esport organization Astro Gaming dropped three streamers who faced accusations of sexual assault and Facebook Gaming temporarily suspended a streamer after public allegations of domestic abuse surfaced (Lorenz & Browning, 2020).

Still, misogyny remains present in the gaming and esport community. Women and other minoritized groups face inappropriate comments about their appearance, performance, and qualifications. Katsumi, a player for esport organization Cloud9s all-female Valorant team (also known as Cloud9 White), noted how there are fewer women in the industry because misogyny has pushed them out. All of the players spent their middle school and high school years keeping their gender identity a secret for fear of backlash from other gamers (Giangreco, 2021). Cloud9 White, the all-women Valorant team, brings up an interesting debate within the esport community: should there be women-only teams, tournaments, and leagues? Proponents of the idea believe that women-only leagues create an inclusive environment for more women to compete in competitive gaming (Rorke, 2020). Still, the opposition will state that there is nothing keeping women from participating in any of the aforementioned leagues.

case STUDY

One area that different esport leagues have grappled with is to have their esport organizations be related to a location or not. For instance, *League of Legends* LCS (and other respective leagues) do not have their teams location-based. A popular team, Cloud9, is just named Cloud9. Most of the LCS teams are based out of California, where the LCS hosts its weekly matches during the regular season. For the playoffs, teams will compete in another location (e.g., Chicago, Houston, New York) so fans outside of California have the opportunity to witness *League of Legends* in person. *Counter Strike: Global Offensive* (CS:GO), another popular esport operates the same way. Teams are not city-based, but are just known by their esports organization names like Fnatic or Australia. Teams fly to different locations (Montreal, Rotterdam, Atlanta, etc.) and compete at major tournaments.

Other esports and leagues, like The Overwatch League (OL) and the Call of Duty League (CDL), have a system set up to closely resemble North American sports when it comes to being location-based. The OL was one of the first esports to create location-based teams (e.g., London Spitfire; Howard, 2019). The league started with 12 teams from Los Angeles to Seoul, Paris, Shanghai, and Vancouver all of which would have to be location-based. It was up to the teams and owners to decide where their location would be established (Barrett, 2017). While the first few seasons saw only matches in California, a plan was drawn up to do home-and-away matches for the OL in 2020. Unfortunately, COVID-19 derailed those plans, but there are hopes that the home-and-away model will come back and teams will be traveling around the

WWW
The Overwatch League
https://overwatchleague.com/en-us/

WWW
Call of Duty League
https://www.callofdutyleague.com/en-us/

globe to compete in the league. Another esport, Call of Duty, has established a similar set-up. In 2020, the CDL changed from an open circuit model to a franchise structure. With that franchise league format, teams would purchase city-based spots in the league. While popular teams like FaZe and Optic would remain in the CDL, they would be renamed to Atlanta FaZe and OpTic Texas (Hale, 2022). The price tag for those franchises was $25 million but secured territorial rights to those cities. The decision to use city-based teams was made after the success of the *Overwatch* league. The CDL currently has one team host all the teams during one of the four major tournaments and then the host rotates to a different team.

1. List several benefits and drawbacks for location-based esport teams.
2. List several benefits and drawbacks to non-location-based esports teams.
3. You are the new commissioner for the *League of Legends* LCS. You decide each team must have a location-based home base for the next season. In a league of ten teams, what ten cities would you select? Why would you select each city?

CHAPTER questions

1. How has the history of esport evolved with technological changes?
2. Should esports be sanctioned and organized by high school state athletic associations or should they remain independent of high school state athletic associations? Why or why not?
3. What are the benefits for a university of partnering with NACE, the most prominent collegiate sport governing body? Should the NCAA or the NAIA consider sanctioning esports as official sports?
4. How is *League of Legends* organized on a regional and international level?
5. Watch Netflix's "7 Days Out: World's Biggest Events" Episode 7 (focused on LCS). What are the similarities between traditional sports and esports when watching the show?

FOR ADDITIONAL INFORMATION

1. As professional gamers and esport owners gear up for a high-stakes tournament, one player's personal tragedy leaves the community reeling. Watch on Netflix: Netflix. 7 Days Out: World's Biggest Events (episode 7)
2. This article discusses the pros and cons of franchising: Call of Duty. https://archive.esportsobserver.com/opinion-call-of-duty-franchising/

3. This video lets you see a virtual cycling event: YouTube: Olympic Virtual Series Cycling Event – The Ultimate Chase Race. https://www.youtube.com/watch?v=pwzMjVuvz_k

4. This website takes you to a story of the NCAA and how it whiffed on esports and is paying the price. https://www.washingtonpost.com/video-games/esports/2020/08/06/ncaa-whiffed-esports-its-paying-price-can-still-learn-lesson/

5. Ever wonder how esport athletes train? Watch this video. YouTube: Esports: Inside the relentless training of professional gaming stars. https://www.youtube.com/watch?v=box4SFtGvA0

REFERENCES

Ashley, J. (2021, June 15). *Adderall in esports – How big is the problem and can it be fixed?* Esports.net. https://www.esports.net/news/adderall-esports-use-how-big-is-the-problem/

Aziz, S.A. (2019). E-sports: First global body to be headquartered in Singapore, SNOC sec-gen Chris Chan to be president. *The Straits Times*. https://www.straitstimes.com/sport/e-sports-first-global-body-to-be-headquartered-in-singapore-snoc-sec-gen-chris-chan-to-be

Barett, B. (2017, August 10). *Cloud9 will be London's Overwatch League team but no local stadium until after 2018.* PCGamesN. https://www.pcgamesn.com/overwatch/overwatch-league-london-cloud9-stadium

Bell, B.C. (2020, November 12). *40 years on, Videogames Icon Rebecca Heineman found herself beyond her 'escape'.* Outsports. https://www.outsports.com/2020/11/12/21561349/rebecca-heineman-atari-space-invaders-transgender-1980-champion-interplay-olde-skuul-bards-tale

Bieler, D. (2021, April 22). IOC announces inaugural slate of Olympic-licensed esports events. *Washington Post*. https://www.washingtonpost.com/video-games/esports/2021/04/22/ioc-olympics-esports/

Booton, J. (2019, February 13). *NBA 2K League outfits athletes and avatars with Champion gear.* Sport Techie. https://www.sporttechie.com/nba-2k-league-outfits-athletes-and-avatars-with-champion-gear/

Burke, P. (2022, May 2). *Macron believes Paris 2024 "historic opportunity" for esports in France.* Inside the Games. https://www.insidethegames.biz/articles/1122656/macron-esports-paris-2024

Burns, M.J. (2017, April 25). Senior NBA executive Brendan Donohue named managing director for NBA 2K Esports League. *Sports Business Journal*. https://www.sporttechie.com/senior-nba-executive-brendan-donohue-named-managing-director-for-nba-2k-esports-league/

Chen, H. (2019, January 11). *Tencent and Riot Games create joint Chinese esports venture: TJ Sports.* The Esports Observer. https://archive.esportsobserver.com/tencent-riot-games-tj-sports/

Crecente, B. (2019, October 27). League of Legends is now 10 years old. This is the story of its birth. *Washington Post*. https://www.washingtonpost.com/video-games/2019/10/27/league-legends-is-now-years-old-this-is-story-its-birth/

Cuofano, G. (2021, November 23). *How does Riot Games make money? The Riot Games Business model in a nutshell.* Four Week MBA. https://fourweekmba.com/riot-games-business-model/#:~:text=non%2Dmobile%20games.-,Riot%20Games%20revenue%20generation,tournament%20partnerships%2C%20and%20merchandise%20sales.

Darvin, L., Holden, J., Wells, J., & Baker, T. (2021). Breaking the glass monitor: Examining the underrepresentation of women in esports environments. *Sport Management Review, 24*(3), 475–499.

Darvin, L., Vooris, R., & Mahoney, T. (2020). The playing experiences of esports participants: An analysis of treatment discrimination and hostility in esports environments. *Journal of Athlete Development and Experience, 2*(1), 36–50.

Dean, B. (2022, January 5). *Twitch usage and growth statistics: How many people use Twitch in 2022.* Backlinko. https://backlinko.com/twitch-users

Dockterman, E. (2014, October 16). What is #Gamergate and why are women being threatened about video games? *Time*. https://time.com/3510381/gamergate-faq/

Duran, H.B. (2021, July 8). *A guide to: Celebrities in esports – football, basketball and sports*. Esports Insider. https://esportsinsider.com/2021/07/celebrities-esports-footballers-sports/

Faber, T. (2021, August 3). The Olympics need esports more than esports need the Olympics. *Financial Times*. https://www.ft.com/content/dbabdf17-2835-499e-890d-aa19a6b464e2

Giangreco, L. (2021, July 31). *'No matter what I'm doing, it's always controversial': Cloud9's all women esports team talks sexism in gaming*. Dot. LA. https://dot.la/female-esports-2654305078.html

Global Esports Federation. (n.d.). *Events*. https://www.globalsports.org/events

Global Esports Federation. (2020, September). *Constitution*. https://www.globalsports.org/resources

Gomez, E. (2021, September 29). *NBA 2K partners with DUX Gaming to add team from Mexico for 2022 season*. ESPN. https://www.espn.com/nba/story/_/id/32306190/nba-2k-partners-dux-gaming-expand-latin-america-mexican-team-begin-play-spring-2022

Good, O.S., & Hall, C. (2020, January 29). *Kentucky athletic association bans Fortnite from high school esports competition*. Polygon. https://www.polygon.com/fortnite/2020/1/29/21114359/fortnite-banned-kentucky-high-school-esports-playvs

Goslin, A. (2017, December 19). *More than 80 million people watched the Worlds 2017 semifinals*. Rift Herald. https://www.riftherald.com/lol-worlds/2017/12/19/16797364/league-of-legends-worlds-viewers-statistics

Hamstead, C. (2020, February 13). 'Nobody talks about it because everyone is on it': Adderall presents esports with an enigma. *Washington Post*. https://www.washingtonpost.com/video-games/esports/2020/02/13/esports-adderall-drugs/

Hasan, U. (2021, January 27). *How does the LCS make money?* EsportsHow.com. https://esporthow.com/how-does-the-lcs-make-money/

Hale, J. (2022, February 9). *All Call of Duty League franchise teams and owners*. Dexerto.com. https://www.dexerto.com/call-of-duty/all-confirmed-call-of-duty-league-franchise-teams-and-owners-969321/

Heath, J. (2021, December 29). *The top 10 highest prize pools in esports*. Dot Esports. https://dotesports.com/general/news/biggest-prize-pools-esports-14605

Hitt, K. (2022, March 16). NBA 2K League's 2022 season goes big with $2.5 million prize pool and player bonuses. *Sports Business Journal*. https://www.sportsbusinessjournal.com/Esports/Sections/Leagues/2022/03/NBA-2K-League-bigger-prizes.aspx

Holden, J.T., Edelman, M., & Baker, T. (2020). A short treatise on esports and the law: How America regulates its next national pastime. *University of Illinois Law Review, 2020*(2), 509–582.

Houston, M. (2022, May 22). *Copenhagen to host first-ever FIFAe Finals in July*. Inside the Games. https://www.insidethegames.biz/articles/1123465/fifae-finals-2022-copenhagen

Howard, B. (2019, September 14). *Why Overwatch League's local team model is the best*. The Gamer. https://www.thegamer.com/overwatch-leagues-local-team-model-best/

Howard, B. (2020, February 20). *The history of the Overwatch League*. Hot Spawn. https://www.hotspawn.com/overwatch/guides/the-history-of-the-overwatch-league

Hoppe, D. (2018, June 9). *The rise and importance of Twitch in Esports*. Gamma Law. https://gammalaw.com/the_rise_and_importance_of_twitch_in_esports/

IESF. (2021, November 17). *Statutes*. https://iesf.org/wp-content/uploads/2022/04/IESF-Statutes-2021.pdf

IESF. (n.d.). *Members*. https://iesf.org/about/members

IOC. (2018, July 21). *Olympic Movement, esports and gaming communities meet at the Esports Forum*. https://olympics.com/ioc/news/olympic-movement-esports-and-gaming-communities-meet-at-the-esports-forum

IOC. (2021, April 22). *IOC makes landmark move into virtual sprots by announcing first-ever Olympic Virtual Series*. https://olympics.com/ioc/news/international-olympic-committee-makes-landmark-move-into-virtual-sports-by-announcing-first-ever-olympic-virtual-series

Kay, M. (2021, March 6). *What are the import restrictions in competitive League of Legends*. Dot Esports. https://dotesports.com/league-of-legends/news/what-are-the-import-restrictions-in-competitive-league-of-legends

Khan, I. (2017, June 1). *Riot releases details on NA LCS franchising with $10M flat-fee buy-in*. ESPN. https://www.espn.com/esports/story/_/id/19511222/riot-releases-details-na-lcs-franchising-10m-flat-fee-buy-in

Kelly, M. (2022, February 15). *Jackie Felling joins LCS as new commissioner, promises expansion of player development systems*. Dot Esports. https://dotesports.com/league-of-legends/news/jackie-felling-joins-lcs-as-new-commissioner-promises-expansion-of-player-development-systems

Kennedy, A. (2018, April 4). *FAQ: Everything you need to know about the new NBA 2K League.* Hoops Hype. https://hoopshype.com/2018/04/04/faq-everything-you-need-to-know-about-the-new-nba-2k-league/

KHSAA. (2019, January 11). *Esports season zero playoff info, spring season update* [Press release]. https://khsaa.org/011119-esports-season-zero-playoff-info-spring-season-update/

KHSAA. (2022, April 28). *Spring 2022 KHSAA esports state championships presented by UK Healthcare slated for Saturday* [Press release]. https://blog.apastyle.org/apastyle/2010/09/how-to-cite-a-press-release-in-apa-style.html

Kim, D.J. (2021, January 7). KHSAA adds FIFA 21, Madden NFL 21 to sanctioned esports games. *Courier Journal.* https://www.courier-journal.com/story/sports/preps/kentucky/2021/01/07/khsaa-adds-fifa-nfl-madden-esports-games-kentucky-schools/6585283002/

K., J. (2020, March 26). *Brief history of esports in Korea & why it is so famous there.* KoreaGameDesk. https://www.koreagamedesk.com/brief-history-of-esports-in-korea-why-is-it-so-famous-there/

Larch, F. (2020, January 18). *History of esports: How it all began.* The Indoor and Outdoor Inspirer. https://www.ispo.com/en/markets/history-origin-esports

LCS. (2020). *Official rules, v. 20.1.* https://nexus.leagueoflegends.com/wp-content/uploads/2020/01/2020-LCS-Rule-Set-v20.1_6bkbwz26cgp7fngeat6a.pdf

Lloyd, O. (2021, August 9). *Global Esports Federation celebrates Emirates esports as its 100th member.* Inside the Games. https://www.insidethegames.biz/articles/1111464/gef-announce-100th-member-federation

Lolesports Staff. (2021, March 30). *MSI: 2021: Group draw and prize updates.* https://lolesports.com/article/msi-2021-group-draw-and-prize-updates/bltf3aff06c2db95be5

Lopez, J. (2020, February 25). *History of professional Call of Duty.* Hot Spawn. https://www.hotspawn.com/call-of-duty/guides/history-of-professional-call-of-duty

Lorenz, T. (2019, November 21). Can FaZe Clan build a billion-dollar business? *New York Times.* https://www.nytimes.com/2019/11/15/style/faze-clan-house.html

Lorenz, T., & Browning, K. (2020, June 23). Dozens of women in gaming speak out about sexism and harassment. *New York Times.* https://www.nytimes.com/2020/06/23/style/women-gaming-streaming-harassment-sexism-twitch.html

Mackay, D. (2018, November 14). *International eSports federation accepts United States as latest member.* The Sport Digest. http://thesportdigest.com/2018/11/international-esports-federation-accepts-united-states-as-latest-member/

Maryville University. (2022). *Different types of esports.* https://online.maryville.edu/blog/different-types-of-esports/

Mazique, B. (2019, September 26). NBA 2K League expands with reported $25 million deal with Geng.G. *Forbes.* https://www.forbes.com/sites/brianmazique/2019/09/26/nba-2k-league-expands-with-reported-25-million-deal-with-geng/?sh=70c1fac04985

Morrison, S. (2019, July 20). *NACE approves new rules governing college esports programs.* ESPN. https://www.espn.com/esports/story/_/id/27229071/nace-approves-new-rules-governing-college-esports-programs

Murray, R. (2014, November 7). *What it's like to be a video game athlete on college scholarship.* ABC News. https://abcnews.go.com/Technology/video-game-athlete-college-scholarship/story?id=26757455

NACEsports. (2018, March 13). *NACE and NJCAA partner to advance college esports.* https://nacesports.org/nace-njcaa-partner-advance-college-esports/

NACEsports. (2021, August). *Constitution and bylaws.* https://nacesports.org/wp-content/uploads/2022/02/NACE-Constitution-and-Bylaws-2-22.pdf

NACEsports. (n.d.). *About.* https://nacesports.org/about/

NBA 2K League. (2021, April 27). *NBA 2K League increase prize Pool to $1.5 million for 2021 season.* https://2kleague.nba.com/news/nba-2k-league-increases-prize-pool-to-1-5-million-for-2021-season/

NBA 2K League. (2022). *League info.* https://2kleague.nba.com/league-info/

NBA Communications. (2017, February 9). *NBA and Take-Two to launch 'NBA 2K eLeague'.* https://pr.nba.com/nba-2k-eleague-launch/

Newton, J. (2010). Wii *Games: Summer 2010 winners announced.* Nintendo Life. https://www.nintendolife.com/news/2010/09/wii_games_summer_2010_winners_announced

Newzoo. (2022). *Global esports & live streaming marketing report.* https://newzoo.com/insights/trend-reports/newzoo-global-esports-live-streaming-market-report-2022-free-version/

Palmer, D. (2021, April 21). *IOC makes biggest esports statement yet with launch of Olympic Virtual Series*. Inside the Games. https://www.insidethegames.biz/articles/1106962/ioc-olympic-virtual-series

Pei, A. (2019, April 14). *This esports giant draws in more viewers than the Super Bowl, and its expected to get even bigger*. CNBC. https://www.cnbc.com/2019/04/14/league-of-legends-gets-more-viewers-than-super-bowlwhats-coming-next.html

PlayVS. (n.d.). *Work at PlayVS*. https://www.playvs.com/careers

Queer Women of Esports. (n.d.-a). *Our programs*. https://www.queeresports.org/ourprograms

Riot Games. (2013, December 18). *Riot Games shares it vision for the future of eports, reveals initial details of Legends Championship Series*. https://web.archive.org/web/20150112133326/http://www.riotgames.com/sites/default/files/uploads/120806_NEWS_lol_champseriesannounce.pdf

Romano, A. (2021, January 7). *What we still haven't learned from Gamergate*. Vox. https://www.vox.com/culture/2020/1/20/20808875/gamergate-lessons-cultural-impact-changes-harassment-laws

Rorke, A. (2020, April 16). *Are female-only leagues good for esports?* Stevivor. https://stevivor.com/features/opinion/female-only-esports-good/

Ryan, K.J. (2022, January 25). *Why a titan of venture capital invested in a 3-person esports company*. Fast Company. https://www.fastcompany.com/90714948/why-nea-invested-in-delane-parnells-three-person-esports-company

Schad, T. (2019, May 21). *NCAA tables possibility of overseeing esports*. USA Today. https://www.usatoday.com/story/sports/college/2019/05/21/ncaa-and-esports-not-just-yet-organization-tables-possibility/3751122002/

Seck, T. (2022, January 12). SBJ Esports: Sponsors line up for new League of Legends season. *Sports Business Journal*. https://www.sportsbusinessjournal.com/SB-Blogs/Newsletter-Esports/2022/01/12.aspx

Seck, T. (2022, May 5). Team Liquid raises $35 million at $415 million valuation. *Sport Business Journal*. https://www.sportsbusinessjournal.com/Esports/Sections/Finance/2022/05/Team-Liquid-raises-$35-million-$415-million-valuation-Ares-Management.aspx

Smith, M. (2019, May 16). SBJ College: Esports not for NCAA. *Sports Business Journal*. https://www.sportsbusinessjournal.com/SB-Blogs/Newsletter-College/2019/05/16.aspx

Smith, N. (2020, June 19). LCS projects a profitable 2021 season, despite navigating covid-19 crisis. *Washington Post*. https://www.washingtonpost.com/video-games/esports/2020/06/19/lcs-projects-profitable-2021-season-despite-navigating-covid-19-crisis/

Smith, N. (2021, February 16). The rise, fall and resonance of ESPN Esports. *Washington Post*. https://www.washingtonpost.com/video-games/esports/2021/02/16/espn-esports/

The Global Esports Federation. (n.d.). *Global esports*. https://www.globalsports.org/about

Venkat, R. (2021, September 9). *Asian Games 2022: Esports to make debut; FIFA, PUBG, Dota 2 among eight medal events*. Olympics.com. https://olympics.com/en/news/fifa-pubg-dota-2-esports-medal-events-asian-games-2022

We Are Rioters. (n.d.). *Riot Games*. https://www.riotgames.com/en/who-we-are/values

WIHSEA. (n.d.). *Mission & history*. https://www.wihsea.org/home/mission-history

Wills, A., & Maturi, T. (2019). *League of Legends – Poster boy for esports franchising*. Norton Rose Fulbright. https://www.nortonrosefulbright.com/fr-ca/inside-sports-law/blog/2019/10/league-of-legends-poster-boy-for-esports-franchising

Wolf, J. (2018a, August 15). *NBA welcomes Hawks, Nets, Lakers, Wolves franchises to NBA 2K League*. ESPN. https://www.espn.com/nba/story/_/id/24381282/nba-welcomes-4-new-franchises-nba-2k-league

Wolf, J. (2018b, July 26). *Sources: Paris and Guangzhou teams expected to join Overwatch League*. ESPN. https://www.espn.com/esports/story/_/id/24200145/sources-paris-guangzhou-teams-expected-join-overwatch-league

Zavian, E. (2020, August 6). The NCAA whiffed on esports. It's paying a price but can still learn a lesson. *Washington Post*. https://www.washingtonpost.com/video-games/esports/2020/08/06/ncaa-whiffed-esports-its-paying-price-can-still-learn-lesson/

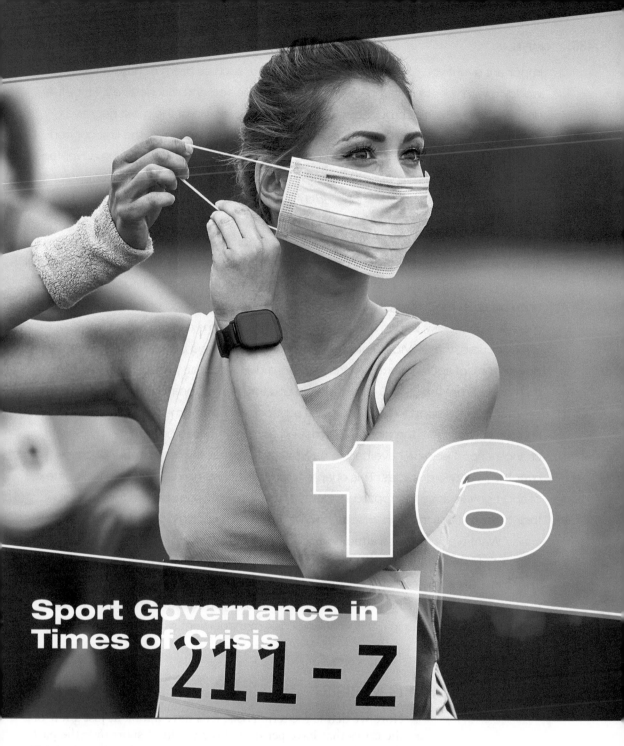

Sport Governance in
Times of Crisis

Think about the first time you realized that the COVID-19 pandemic would alter life as we knew it – what comes to mind? Perhaps you are thinking of countless class sessions on Zoom, where students and instructors alike tried to navigate a new type of learning while sitting in front of a screen, isolated from the rest of the world. Or a loved one, whose life was impacted by the pandemic. Or maybe you are thinking of an

DOI: 10.4324/9781003303183-16

important event you had been looking forward to, such as prom or graduation or a wedding with your best friends, that had to be canceled and rescheduled. Hopefully, despite the challenges brought on by the pandemic, some of the most negative memories are starting to become replaced by more positive ones: the first concert with friends after lockdown, the first holiday gathering with family, or taking advantage of all that college has to offer in an ongoing-pandemic world. Just like in our personal lives, the pandemic brought unprecedented challenges to the sports industry. Mega events such as the UEFA EURO Championship, the Olympic and Paralympic Games or the women's and men's Final Four had to be postponed or canceled, professional sport leagues ceased to play, sport organizations had to lay off staff or reduce salaries, comprehensive health protocols had to be developed and put in place, and once play resumed, stadiums, adhering to strict no-attendance policies, seemed more like ghost towns than epicenters of vibrant fandoms across the world.

Of course, the COVID-19 pandemic is just one example of how crises affect the governance of sport. In fact, perhaps no word describes the past few years better than *crisis*. From the COVID-19 pandemic and transformational changes in many segments of the sport industry (such as advances in name, image, and likeness rights in US college sport alone!) to pressing ecological crises brought on by climate change – sport leaders have had to navigate an unprecedented number of crises over the past two decades. This chapter discusses crisis management in the context of sport governance to prepare students and future decision-makers for leading their organizations through crises in an ever-changing sports industry. What is a crisis? What types of crises affect the governance of sport most? What are the different phases of a crisis? How can those governing sport lead through crises? These are just some of the questions we will cover in the pages to come. We chose to close our book with this important topic, because you will likely encounter crises throughout your career in sport – making crisis management skills a crucial trait of today's sport leaders.

WHAT IS A CRISIS?

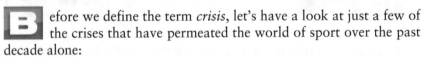efore we define the term *crisis*, let's have a look at just a few of the crises that have permeated the world of sport over the past decade alone:

- Just days before the opening of the 2022 Beijing Paralympic Games, Russia's dictator Vladimir Putin violated the Olympic truce as Russia and Belarus launched a military invasion of Ukraine, their neighboring country. The war was condemned by world leaders across the globe, and the International Paralympic Committee moved to ban Russian and Belarusian athletes from competing at the Games after multiple Paralympic Committees threatened not to compete. According to the IPC (2022), the move to ban the athletes from the countries that initiated the war was made "in order to preserve the integrity of these Games and the safety of all participants" (para. 11).

- In 2021, the (sporting) world watched in horror as the terrorist Taliban group seized power in Afghanistan after the United States withdrew its military forces from the country, severely restricting women's rights – including the rights of women athletes – in the process. While the global sport world watched from afar, the resilience of the Afghan women was a reminder that "contrary to many narratives in the mainstream media, the story of Afghan female athletes has been one of power and possibility" (Ahmed, 2021, para. 10). The humanitarian crisis prompted some sport governing bodies, such as FIFA and the World Cycling Union, to aid in the evacuation of women athletes competing in their sports from Afghanistan (Wong Sak Hoi & Hunt, 2022).

- In March 2020, sport came to a standstill due to the emerging COVID-19 pandemic, a once-in-a-century global health crisis. In the United States, the NBA was the first league to suspend its season, with others quickly following suit. By March 12, 2020, the NHL, MLB, the NCAA, and other major sport organizations had canceled or postponed their seasons and tournaments in what Zaru (2021) calls "the day the sports world stopped" (para. 14).

- In 2019, the World Anti-Doping Agency (WADA) moved to ban Russia from all sporting events for four years (a ruling later reduced to two years by the Court of Arbitration for Sport) after investigative journalists revealed systematic state-sponsored doping of Russian athletes. The Russian state was found to have supplied steroids and other performance-enhancing drugs to its athletes. The doping scandal, unprecedented in scope, resulted in more than 46 Olympic medals being stripped from the Russian Federation and brought into the public eye broader concerns about Russia's disregard for the integrity of sport (Keh & Panja, 2019).

WWW

World Anti-Doping Agency (WADA)
https://www.wada-ama.org/en

- During the 2018 and 2017 seasons, Major League Baseball's Houston Astros engaged in a series of rule violations by using a video camera located in the center field seats to engage in sign stealing. The illegal use of the camera system for the purpose of cheating was later confirmed through an independent MLB investigation, leading the league office to issue its most severe sanctions against a member club in baseball history (Verducci, 2020), including suspending members of the team leadership, stripping the team of first- and second-round draft picks for the 2020 and 2021 drafts, and fining the team its maximum amount allowed: $5 million (Verducci, 2020).

- In 2017, former USA Gymnastics national team doctor Larry Nassar was sentenced to life in prison after more than 365 survivors, most of whom were minors at the time, revealed that they were sexually abused by Nassar starting as early as the 1990s. Many of the organizations Nassar worked for enabled his decades-long abuse, including Michigan State University, USA Gymnastics, and the USOPC. The Nassar case is today considered one of the largest sexual abuse scandals in sports history (Graham, 2017), and it led to comprehensive structural changes in US Olympic and Paralympic sport governance (see Chapter 10).

- During its 2015–16 season, 25 Football League games in the United Kingdom were canceled because of extreme weather due to the effects of climate change (The Climate Coalition, 2018). The cancelation of sport events is a stark reminder that climate change, and other ecological crises, will likely alter the sports landscape and force sport governing bodies to adapt: "As heatwaves hospitalise players in sports from tennis to cricket, competitions are canceled due to extreme weather, and winter sports try to cope with less snow and ice, sporting bodies have begun eyeing ways to adapt to the changing climate" (Taylor, 2019, para. 7).

The Climate Coalition
https://www.
theclimatecoalition.org

- In 2015, FIFA came under scrutiny when the US and Swiss governments revealed investigations into widespread allegations of corruption across the organization, including accusations of bribery of FIFA officials by representatives working for Russia and Qatar, the eventual host nations of the 2018 and 2022 FIFA Men's World Cup, respectively. Ahead of the 65th FIFA Congress held mid-year, more than a dozen FIFA officials were arrested for corruption charges, forcing long-time President Sepp Blatter to resign and suspending the 2026 FIFA World Cup bidding process (US Department of Justice, 2015).

By no means is this an exhaustive list of all crises that have impacted the sports industry over the past decade, and you can probably think of other crises that have interrupted business as usual across the industry. However, this selection provides a good overview of the scope and impact crises can have in and beyond the sporting arena. So, what exactly is a crisis? Bundy et al. (2017) define a crisis as "an event perceived by managers and stakeholders as highly salient, unexpected, and potentially disruptive [which] can threaten an organization's goals and have profound implications for its relationship with stakeholders" (p. 1662). On a similar note, Coombs and Holladay (1996) define a crisis as "a threat or challenge to an organization's legitimacy – stakeholders question if an organization is meeting normative expectations" (p. 281). Let's use one of the examples shared above to illustrate the different components of these definitions: the 2015 FIFA Corruption Scandal. We chose this past example to show how an organizational crisis can impact an organization for decades to come – as was the case with this crisis, which still influences FIFA's strategic planning today.

Example: FIFA Corruption Scandal

- *Component 1: An event perceived by managers and stakeholders as highly salient, unexpected, and potentially disruptive.*
  Allegations of corruption in FIFA's organizational praxis were

nothing new for FIFA per se (the organization had long been accused of such unethical behavior). Allegations dated as far back as 1974 when then-President João Havelange took millions of dollars in bribes (Yglesias & Strober, 2015). It makes sense, then, that "the biggest surprise in all this isn't that FIFA officials allegedly engaged in this sort of corruption – it's that after years of widespread accusations, the US and Swiss governments are finally cracking down" (Yglesias & Strober, 2015, para. 15). Clearly, the investigations and arrests resulting from them interrupted FIFA's business as usual, and it became evident that change was needed.

- *Component 2: An event threatening an organization's goals.* On its website, FIFA states that "FIFA exists to govern football and to develop the game around the world" (FIFA, 2022, para. 1). Contradicting multiple principles of good governance, such as transparency and democracy, the acceptance of bribes by the organization's top decision-makers is a clear violation of its mission and goals.

- *Component 3: Profound implications for relationship with stakeholders, threatening the organization's legitimacy.* The fallout from the corruption scandal was swift: By October 2015, four of FIFA's major sponsors had called for then-President Blatter to resign from his position. Other key stakeholders, including governments (e.g., Australia, Brazil, and Germany), continental soccer organizations, and national soccer federations called for investigations and restructuring of the organization. These actions led to reforms such as the introduction of term limits for leadership, the separation of key managerial functions (e.g., executive and strategic functions), and the creation of new mechanisms to document the flow of money within the organization (FIFA, 2017). It also led FIFA to make public its strategic framework for governing the sport of soccer moving forward, including via its *The Vision 2020–23* strategic framework.

www

FIFA The Vision 2020–2023
https://publications.fifa.com/
en/vision-report-2021/

The FIFA example illustrates that "crises that impact sports organizations and athletes have the ability to cause harm by tarnishing a team or athlete's reputation or impairing their in-game performance … the negative fallout from recent sports-related crises shows their impact has progressed beyond the field, including the potential to damage … [an] entire organizational brand" (Brown-Devlin & Brown, 2020, p. 50). That is why crisis management is an important skill for those governing sport. *Crisis management* refers to the "coordination of complex technical and relational systems and the design of organizational structures to prevent the occurrence, reduce the impact, and learn from a crisis" (Bundy et al., 2017, p. 1664). Crisis management includes dealing with unexpected events that can potentially harm an organization, which requires working with the various organizational stakeholders to prevent, navigate, resolve, and grow from crises (Bundy et al., 2017). No organization is immune to crises, and most organizations will face multiple crises throughout their existence. Let's have a look at the different types of crises that can affect the governance of sport.

CAUSES AND TYPES OF CRISES

The COVID-19 pandemic. The USA Gymnastics abuse scandal. The academic integrity scandal at the University of North Carolina, Chapel Hill, where hundreds of collegiate athletes were enrolled in fake classes to make sure they remained eligible to play their sport. The New England Patriots' *Deflategate*. Hurricane Katrina's impact on sport in New Orleans, Louisiana. While representing vastly different crises, in each of these instances the crisis kickstarted a "process of transformation where the old system can no longer be maintained" (Venette, 2003, p. 43). When looking at the different types of crises that permeate sport, it is important to distinguish between *what* the cause of the crisis is, *how* the crisis emerges, and *what type* of crisis is taking shape. The causes of crises can involve a variety of factors, but often crises emerge due to one or more of the following components (Kovoor-Misra, 2019):

- Internal shortcomings: There are failures/dysfunctions within the organization.
- External factors: There are threats from outside the organization that jeopardize its performance, mission, goals, or overall efficiency.
- Handling of crisis: Those tasked with resolving the crisis mismanage the crisis.

Crises emerge in two primary ways: sudden crises and smoldering crises. *Sudden crises* are unexpected events that are often beyond the control of those leading organizations (Hayes James & Wooten, 2005). Even though decision-makers are not perceived to be at fault or responsible for this kind of crisis, "firm leadership is still expected to resolve the crisis, and any displays of empathy become short-lived if stakeholders perceive firm leadership as mishandling the execution of the crisis response" (Hayes James & Wooten, 2005, p. 143). Examples of sudden crises include the COVID-19 pandemic, the 9/11 terrorist attack, or natural disasters – all of which have impacted sport. *Smoldering crises*, on the other hand, are crises that are perceived to be the fault of organizational leadership (or, to be more accurate, the lack thereof). This kind of crisis refers to events that "as small, internal problems within a firm, become public to stakeholders, and, over time, escalate to crisis status as a result of inattention by management" (Hayes James & Wooten, 2005, p. 143). The vast majority of crises affecting the sport industry belong in that latter category, and prominent examples that come to mind are the FIFA corruption scandal mentioned earlier, the Jerry Sandusky child sexual abuse scandal at Penn State University, the Astros cheating scandal, and the gender equity investigation resulting from the inequities revealed during the 2021 NCAA women's and men's basketball tournaments (see Chapter 8). In each of these instances, the larger organization – whether FIFA, Penn State, the Astros, or the NCAA – should have identified early warning signs yet failed to do so, and sometimes the organization actively contributed to the crisis. The outcome was similar across each of these institutional contexts: The organizations faced intense public backlash, change in leadership, independent investigations, or other repercussions. Examples of sudden and smoldering crises are included in Exhibit 16.1.

TYPES OF CRISES

Aside from how quickly crises emerge, it is also important to understand the major types of crises that can affect organizations governing sport. Let's have a look at some of the major ones (Fontanella, 2022):

- **FINANCIAL CRISIS:** Financial crises are those that are linked to the assets of an organization. For example, an organization may have accrued too much debt and is unable to pay it off. This type of crisis usually is caused by a change in demand for a product or service. For example, the at-home fitness brand Peloton became known for its tech-connected exercise bikes during the COVID-19 pandemic, with its value estimated at $50 billion in late 2020. However, just a few years later when people returned to their gyms and old fitness habits, the company had to lay off 20% of its staff, saw its stock plummet, and lost $757 million in the first quarter of 2022 alone (Hartmans, 2022; Valinsky, 2022).

- **PERSONNEL CRISIS:** Personnel crises occur when people associated with the organization, such as a staff member or

---

Examples of Sudden and Smoldering Crises   *exhibit* **16.1**

SUDDEN CRISES	SMOLDERING CRISES
Natural Disasters	Product Defects
Terrorist Attacks	Rumors/Scandals
Workplace Violence	Workplace Safety
Product Tampering	Bribery
Sabotage	Sexual Harassment
Hostile Takeover	Consumer Activism
Technology Disruption	Mismanagement
	Whistle Blowing
	Class Action Lawsuits
	Labor Disputes

Source: James and Wooten (2005).

member of the Board of Directors, are involved in illegal or unethical misconduct. Such misconduct does not have to be tied to the professional world, as transgressions in a person's personal life can also reflect negatively on the organization the person is associated with. For example, in 2014 former NBA team owner Donald Sterling was banned from the NBA for life after private recordings were made public in which he made racist comments, prompting national media coverage and sending the league, consisting of a majority of Black players, into turmoil.

- **CRISIS OF ORGANIZATIONAL MISDEEDS:** This type of crisis refers to situations where organizations have wronged their constituents, customers, or employees. In organizational crises, "rather than creating mutually beneficial relationships, these businesses use their customers as a means of benefiting the company or abuse their employees to 'save face'" (Fontanella, 2022, para. 14). This type of crisis can range from deception (e.g., knowingly lying to the public) to management misconduct (e.g., leaders engaging in illegal activities) to misplaced organizational values (e.g., placing profit over people). One of the most prominent examples of an organizational crisis in sport is the USA Gymnastics sexual abuse scandal, where the organizations involved knew about Nassar's abuse for decades before being forced to take action and protect their athletes.

- **TECHNOLOGICAL CRISIS:** Have you ever worked on a class project for hours only to find your computer crash and all the work being lost? We have likely all been there: a technological crisis. At the organizational level, technological crises refer to crises caused by failures in technology. For example, shortly after the second-half kick-off of Super Bowl XLVII at the Superdome in New Orleans in 2013, half the stadium experienced a power outage so the building went dark. Doug Thornton, the Executive Vice President of Stadiums and Arenas at the time, later explained that "even though it wasn't our fault, it became our problem. That was the headline: BLACK EYE FOR THE CITY, BLACK CLOUD FOR THE SUPER BOWL. The whole thing bothered me for months" (Bishop, 2015, para. 7).

- **HEALTH CRISIS:** Health crises (sometimes also referred to as public health crises or, more broadly, humanitarian crises) are situations that impact a community's health and well-being. Health crises often result from diseases, poor policy-making, or industrial processes (Kohrt et al., 2019). The first example that comes to your mind here is probably the COVID-19 pandemic, and rightfully so. What an impact the pandemic had on sport, as we have laid out throughout the chapters of this book. From cancelations of mega events and participation in ghost games with no audience to the creation of complex health protocols and re-envisioning how athletic events are held (think of the NBA, WNBA, and NCAA bubble concept or the Beijing Winter Olympic

Games closed loop system as examples) – the pandemic was a health crisis unprecedented in scope in modern (sport) history.

- **NATURAL OR ENVIRONMENTAL CRISIS:** These crises result from ecological factors including extreme weather, natural disasters such as earthquakes or tsunamis, or other serious damage to the environment. The scientific community agrees that due to human activity the climate is changing faster than ever before, and sport contributes to climate change as sport organizations, participants, and fans leave carbon footprints that add to the climate crisis (Sport Ecology Group, 2019). Whether it's the 2020 California wildfires, deadly tornadoes in the Midwest, floods in the South, or hurricanes such as Hurricane Katrina or Hurricane Sandy, sport organizations will continue to be affected by the forces of nature and the devastation that can come with them.

- **CONFRONTATION CRISIS:** Remember when the Milwaukee bucks boycotted Game 5 of the First Round series versus the Orlando Magic in the COVID-19 bubble in 2020 to protest racial injustice? Or when the University of Missouri football team threatened not to engage in any football-related activities until university leadership stepped down due to their mishandling of racist incidents on campus? Both of those examples illustrate what is called a confrontation crisis. These occur when individuals or groups are discontent with the management or leadership of an organization, or an organization's handling of certain issues. Often, these groups present a list of demands or expectations to the organization, in hopes they are met and change will happen. Confrontation crises include boycotts, sit-ins, protests, resignations, or forms of public outcry.

- **VIOLENCE CRISIS:** This type of crisis occurs when violent acts are committed within an organization, against an organization, or on the grounds of an organization. It often includes a member of the organization, or a former member, committing acts of violence against the organization (or on organizational ground). A tragic example of such a crisis: In 2010, a player of the University of Virginia men's lacrosse team murdered his on-and-off girlfriend, who was a member of the women's lacrosse team at the university. The victim's mother later sued the lacrosse team's coaches, the university, and the university's athletic director (Ng, 2012).

- **CRISIS OF MALEVOLENCE:** Crises of malevolence happen when someone internal or external to an organization uses "criminal or illegal means to destabilize a firm, harm its reputation, extort it, or even destroy it … with the objective of harming an organization, its stakeholders, and its public image" (Fontanella, 2022, para. 50). Examples of this type of crisis are tampering with an organization's products to create harm, cybersecurity threats, terrorism, or spreading false rumors. For example, the unethical recruitment strategy known as negative recruiting based on sexual orientation has increasingly been used in US intercollegiate women's sport. Some coaches may use this strategy to steer recruits away from competing programs by "implying to a recruit that a rival college or university's coach is gay, or that an opposing team is 'full of lesbians' … to prey on unsubstantiated fears" (Women' Sport Foundation, 2011, p. 1). This homophobic practice is aimed at spreading rumors about the competing team – which can negatively impact that team's public image – and results in recruits choosing not to play for that program.

- **REPUTATIONAL CRISIS:** A reputational crisis refers to any "major event that has the potential to threaten collective perceptions and estimations held by all relevant stakeholders of an organization and its relevant attributes" (Sohn & Lariscy, 2014, p. 23). This type of crisis poses a potential for serious damage to an organization's reputation. For example, the newly-rebranded Washington Commanders had their fair share of reputational crises in the past five years alone. It only changed its racist former team name after intense public pressure in the wake of the murder of George Floyd and renewed calls for racial justice. Credible allegations of widespread sexual harassment within the organization as well as allegations of financial irregularities left the team's brand severely damaged, even prompting a Congressional investigation into the organization (Segal, 2022).

The Washington Commanders example brings us to an important realization: Sometimes multiple crises strike at once. While each crisis is unique and requires strong leadership by those governing sport, some commonalities exist between different types of crises. Knowing common characteristics of crises is crucial in identifying early signs of an impending crisis, which is why we turn our attention to some of the most common features of crises next.

COMMON CHARACTERISTICS OF CRISES

When reviewing the examples of crises shared throughout this chapter so far, can you identify some common threads among them? They appear vastly different in nature, yet there are some commonalities between them. For example, each of the crises significantly disrupted business operations of the organization(s) affected and led to a need for change, which is a common feature of crises. There is also usually an element of surprise involved, and crises quickly represent a threat to the organization, requiring it to make decisions in a relatively short amount of time (Seeger et al., 1998). A sense of urgency is common when a crisis situation emerges, coupled with a certain amount of uncertainty and lack of clarity on what to do next (Kovoor-Misra, 2019). That is why crises are often perceived as highly stressful for those involved. Crises emanating from within an organization possess another common feature: They often violate one or more of the principles of good governance identified in Chapter 1. Among those principles are transparency (e.g., violated by FIFA corruption scandal), solidarity (e.g., violated by Russian invasion of Ukraine), democracy (e.g., violated by Taliban ban of women's sport), and accountability (e.g., violated by USA Gymnastics enabling decades-long systemic abuse). Other common characteristics of crises include (adapted from Business Queensland, 2022):

- The potential of physical or psychological danger, as is the case with crises emerging from terrorist threats, natural disasters, or abuse, among others;
- The impacted organizations' staff may experience intensified pressures, tension, or stress;
- Key decision-makers and/or staff may not be available;
- It may become increasingly difficult, and sometimes impossible, to carry our business as usual, including common daily activities;
- Clear information on the evolving situation may not be available;
- Decisions on what to do may need to be made with limited time available;
- External pressures on the organization(s) may be intense, as there may be increased attention from key stakeholders, the media, or the staff itself;
- Information about the emerging situation may travel fast, which can shape the public perception of how the crisis is handled (or mishandled);
- A need for help from external entities may arise, such as emergency services, independent consulting firms, and so on.

As you can see, crises can severely impact sport organizations and disrupt the very foundation on which these organizations are built. That is why they are often perceived as something negative, as they can negatively affect an organization's reputation, the overall reputation of the sport the organization governs, or the health and satisfaction of the organization's members and stakeholders (Tennis New Brunswick, 2019). Another commonality among crises is that they generally follow a specific pattern, as each crisis usually consists of the same phases.

THE FIVE PHASES OF A CRISIS

Can a crisis be detected and, potentially, prevented? What is the lifecycle of a crisis? What happens once the crisis becomes inevitable? And what can we learn from it once the dust settles? Researchers of crisis management have grappled with questions such as these for decades. In one of the most widely-used studies analyzing crises across industries, Pearson and Mitroff (1993) sought to understand how crises, especially those induced by humans (rather than natural crises), should be managed. They conducted research with over 500 crisis management staff members at more than 200 companies and found striking similarities in how crises emerged and developed across both the public and private sectors. The researchers identified a minimum of five phases that make up the lifecycle of a crisis, which are visualized in Exhibit 16.2.

exhibit 16.2 The Five Different Phases of a Crisis

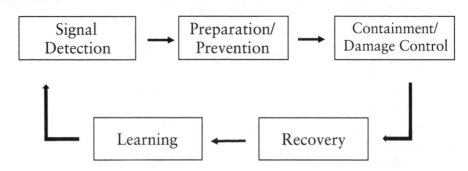

Source: Pearson and Mitroff (1993).

We will cover each of them briefly below, before applying them to an example from sport governance.

Phase 1: Signal Detection

Most crises, especially smoldering ones, do not come out of nowhere – they "leave a trail of early warning signals … in many cases, organizations not only ignore such signals, but may actually exert considerable efforts to block them" (Pearson & Mitroff, 1993, p. 53). Successful crisis management must proactively create the infrastructure and skill set among staff to (a) detect early warning signals and (b) communicate those warning signals to decision-makers if needed. As Pearson and Mitroff (1993) have shown, "organizations which prepare effectively for crises make a deliberate point to constantly probe and scrutinize their operations and management structures for potential errors or problems before they are too big to correct" (p. 53). During this phase, decision-makers should be asking the following questions to set the organization up for successful crisis management (adapted from James & Wooten, 2005):

- What are the most vulnerable areas of an organization?
- How can those vulnerable areas lead to a crisis?
- What practices, procedures, policies, and/or situations does the organization ignore that may result in a crisis?
- Has the organization created a culture where things that may be unpleasant or uncomfortable to address can be confronted?
- How do the systems in place contribute to potential crisis situations?

Phase 2: Preparation and Prevention

Organizations that respond successfully to crisis management situations prepare both systematically and strategically for potential crises. Such preparation requires an ongoing process of identifying potential shortcomings in organizational practice. The goal here is not to avoid all crises – that is impossible! Instead, the aim should be to do what is in one's power to prevent crises from happening and, should a crisis inevitably occur, to manage it effectively (Pearson & Mitroff, 1993). Such preparation can include institutionalizing crisis management procedures (e.g., by creating crisis teams), training staff and leadership on crisis management, and simulating crisis scenarios (Pearson & Mitroff, 1993). More specifically, the following questions should be asked during this phase (adapted from James & Wooten, 2005):

- Have decision-makers created plans for how to react to a crisis?
- Have sufficient resources been allocated to preventing crises?
- Is a strong infrastructure in place to facilitate crisis resolution, or is the organization's structure likely to hinder the resolution of a crisis?
- Is the organization physically prepared for crisis (e.g., space available to support staff in need)?
- Is the organization mentally ready to respond to crisis (e.g., by developing a readiness mentality)?

Phase 3: Containment and Damage Control

Of course, some crises happen due to circumstances outside of an organization's control and are inevitable (Pearson & Mitroff, 1993). Sometimes it happens: A crisis has hit an organization full force! What now? The ultimate goal of this phase is to limit the effects of the crisis and reduce the (potential) harm that may occur (Pearson & Mitroff, 1993). In fact, this is the phase most people think of when hearing the term *crisis management* (James & Wooten, 2005). Keep in mind that harm is usually caused to the organization itself and the people within that organization. Protecting the people, however, should always be the top priority, even if it may potentially cause (temporary) damage to the organization's reputation. Once people are protected (to the extent possible), think about how the reputation of the organization can be preserved and operations can be sustainably rebuilt (Marker, 2022). Organizations successfully managing crises will be prepared for crises by devoting "time and resources to assure that damage containment mechanisms and procedures are in place and effective" (Pearson & Mitroff, 1993, p. 53). Check out some of the questions that should guide organizational decision-making during this phase (adapted from James & Wooten, 2005):

- How can the organization be put in a position to strategically limit damage brought by a crisis?
- How (and by whom) is crisis-related information handled?
- Who are the key stakeholders affected by the crisis? What needs to be done to meet their needs during the crisis?
- What message needs to be communicated to stakeholders? What tools are utilized to communicate that message?

Phase 4: Recovery

During this phase, an organization must have a strategy in place for both short-term and long-term recovery (Pearson & Mitroff, 1993). Both short-term and long-term strategic planning for recovery must map out the ways in which the organization can return to business as usual (or a new normal). They must map out key activities, assign tasks and responsibilities, and develop measurements for success. The following questions should be in focus during the recovery phase (adapted from James & Wooten, 2005):

- What are the short-term goals for the organization post-crisis?
- What are the long-term goals for the organization post-crisis?
- What initiatives, interventions, or other activities must those leading the organization engage in to recover from the crisis?
- What metrics are used to determine whether or not the recovery strategy was a success?
- What communication channels are utilized to inform stakeholders about the end results of the recovery strategy? At what frequency does the organization communicate with stakeholders after the crisis?

Phase 5: Learning and Reflecting

A crisis should not end with the recovery; instead, it must conclude with an in-depth critical reflection of what happened, how the crisis was handled, and what lessons can be taken forward to navigate future crises. Organizations are often hesitant to engage in reflection because it may bring back negative memories associated with the crisis in the first place (Pearson & Mitroff, 1993). However, identifying both the positive and negative aspects of the crisis is a crucial component for successfully managing future crises. The following questions can aid in getting to a state of meaningful reflection (adapted from James & Wooten, 2005):

- What are the main takeaways from the crisis? What did the organization learn?
- How does leadership reflect on any mistakes or counterproductive behavior it engaged in?
- Is the organization committed to changing its practices, policies, procedures, and infrastructure to prevent future crises?
- Are experiences related to the crisis documented, so that an institutional memory can be created for future situations of this kind?
- Was each of the units within the organization utilized strategically? Is there room for improvement?

Exhibit 16.2 shows these phases as a cycle, illustrating how each phase will likely inform future crises. Let's apply these phases to a recent crisis in the governance of sport: the gender inequities revealed during the NCAA Division I men's and women's basketball tournaments in 2021. Remember: University of Oregon women's basketball player Sedona

Prince went viral with a video recording of the workout equipment provided to the women athletes at the tournament site, which consisted of a small rack of weights and yoga mats, compared to a fully equipped gym provided to the men. The example of the workout equipment, as soon became evident, was reflective of larger disparities between men and women in the tournament, including the in-game atmosphere and signage, availability of COVID testing, goodie bags provided to the athletes, and meal options available at the tournament sites. Here's how the crises progressed through the five phases identified in this chapter.

Phase 1: Signal Detection. If you follow women's college sport, you did not have to look far to spot the first signs of gender inequity. Up until last year, for example, the women's tournament was not allowed to use "March Madness" branding at all. Similarly, up until 2021 the NCAA only utilized one social media account, titled @FinalFour, as the official event feed for the tournament, yet only referred to the men's tournament. The staff for the women's tournament was half the size of that of the men's tournament. When it became clear that the 2021 tournaments had to be played in a bubble, the NCAA prioritized men's basketball as "its chief moneymaker" and "continued to view the less-developed women's game as one of many so-called 'non-revenue sports' as a cause rather than an asset, as an expense rather than the income stream that independent experts believe it can be" (Bushnell, 2022, para. 18). Despite growing interest in women's sport, the budgets available to the men's and women's tournament also differed starkly with the NCAA spending $53.2 million on the men's tournament and $17.9 million on the women's tournament in 2019 (Kaplan Hecker & Fink, 2021). Thus, it was barely surprising that an independent investigation revealed that "the NCAA's organizational structure and culture prioritizes men's basketball, contributing to gender inequity" (Kaplan Hecker & Fink, 2021, p. 7).

Phase 2: Preparation and Prevention. Not much is known about how the NCAA prepares for crises, but some existing institutional resources and structures likely assisted the national office in preparation of the increased spotlight it faced on matters of gender equity. Institutional infrastructure was in place to provide expertise on matters of gender equity. For example, the NCAA Gender Equity Task Force has existed since 1993 and the organization has three additional committees in its governance structure committed to matters of diversity, equity, and inclusion (see Chapter 8). The Gender Equity Task Force serves as a standing advisory group tasked with "engaging the membership, student-athletes, the governance structure, the media and affiliate organizations in identifying gender equity strategies to improve the professional and competitive environment for women in intercollegiate athletics at all levels" (NCAA, 2022, para. 2). Within the NCAA and its membership structure, trainings have also been made available focused on crisis management. For example, during the 2020 NCAA national convention, Division III offered an education session titled "Crisis Management: Being Your Best During Your Institution's Worst Days," which covered topics such as crisis communication, developing a crisis mindset, and crisis leadership (NCAA, 2020a).

www

NCAA Resource: Crisis Management – Being Your Best Self During Your Institution's Worst Days https://www.ncaa.org/sports/2019/9/23/2020-division-iii-convention-resources.aspx

Phase 3: Containment and Damage Control. In the immediate aftermath of Prince's recording and the public backlash, NCAA vice president of women's basketball, Lynn Holzman, issued a statement that the situation was being addressed. Less than a day after Prince's video went viral, the NCAA provided the athletes with a new weight room. The NCAA's Committee on Women's Athletics (CWA) also requested NCAA president Mark Emmert launch an independent investigation into the inequities of the signature tournament. The NCAA later followed up by hiring the firm Kaplan Hecker & Fink to conduct a comprehensive review of the organization's handling of the men's and women's tournament. In addition, multiple NCAA leaders, including Emmert and Dan Gavitt (vice president of men's basketball), took responsibility, issued public apologies, and committed to working toward greater gender equity in the future.

Phase 4: Recovery. In the year following the incident, the NCAA made some changes in regards to both tournaments. In addition to investing millions of dollars into the women's tournament, the NCAA undertook the following initial steps (Murphy, 2022):

- Increasing the number of teams competing in the tournament (from 64 to 68)
- Utilizing the March Madness brand and logo for both tournaments
- Creating identical gift packages for men's and women's basketball players
- Creating identical on-site lounge areas for both tournaments
- Adding additional promotional items (e.g., signage) for a better in-game atmosphere
- Expanding cross-promotion between both tournaments

In addition, Kaplan Hecker & Fink released the results of an initial comprehensive, 118-page report in 2022, which included 25 recommendations on how to anchor a commitment to gender equity within the NCAA's handling of the tournaments moving forward. These recommendations focused on providing structural support for gender equity (e.g., by developing an equal number of staff for both tournaments), improving transparency and accountability for gender equity (e.g., via regular audits), enhancing the value of the women's tournament through equitable marketing and sponsorships, recognizing gender equity in the distribution of revenue, and ensuring progress toward gender equity via regular audits (Kaplan Hecker & Fink, 2021).

Phase 5: Learning and Reflecting. Has the NCAA learned a lesson in how to treat women's and men's sport? Of course, the process of learning and reflecting is an internal one so it is hard to say for sure. However, some of the structural changes made to recover from the crisis certainly indicate a different approach to the Association's flagship tournament. Holzman, vice president of women's basketball, also stated that the crisis ensured "the proverbial curtain was pulled back" (Murphy, 2022, para. 6), indicating some reflection and learning is happening. The independent report can serve as another tool for learning and reflecting, as it documents the crisis and creates an institutional memory that can inform future decision-making. In addition, committee reports from the meetings of the Board of Governors Committee to Promote Cultural Diversity and Equity, the Gender Equity Task Force, the CWA, and the Minority Opportunities and Interest Committee following the crisis show that space was dedicated to discussing advances in gender equity as well as to providing input on the ongoing independent investigation (NCAA, 2021a; 2021b), all of which hint at a process of critical reflection.

We now have a better understanding of what a crisis is, what different types of crises exist, what some common features of crises are, and what phases the crises usually go through. In our final section, we turn our attention to how crises can be managed effectively in the governance of sport.

STRATEGIES FOR MANAGING CRISES IN SPORT GOVERNANCE

Y ou have probably picked up by now on the fact that a degree of uncertainty exists when it comes to crises. While it is impossible to prevent some crises from happening, there are strategies you can pursue to make sure you can manage crises effectively and efficiently once they hit your organization. Most organizations tend to be reactive when it comes to crises, but good crisis management requires a *proactive* approach. Would the 2021 NCAA March Madness tournament have become a site for discussions on gender equity had the NCAA spotted – and acted on – discrepancies in their treatment of women and men beforehand? Probably not, because then the issues would not have existed in the first place. In the following section, we share strategies for proactive and responsive crisis management, so that you are equipped to handle future crises as much as possible. Crisis management consultant Deb Hileman (2021) perhaps said it best: "Failing to plan *for* a crisis is planning to fail *in* a crisis. In other words, you can prepare and prevent, or repair and repent. The latter is considerably more expensive – both in dollars and reputation – than the former" (para. 2). This echoes what we learned in Chapter 2 generally – when you fail to plan, you plan to fail.

Throughout this section, we further narrow down the five stages of a crisis into three broader categories: pre-crisis strategies (accounting for the signal detection and preparation/prevention stages), mid-crisis strategies (representing the containments and damage control stage), and post-crisis strategies (focused on the learning phase). Exhibit 16.3 provides a general crisis management strategic checklist, four elements of which we will cover in more detail below. Let's dive right into it!

Identify Potential Risks (Pre-Crisis)

Identifying potential threats and risks to the organization is a crucial component of crisis management at the pre-crisis stage. Successful risk management analyses identify the most likely crises an organization *exhibit*   **16.3**   Crisis Management Strategic Checklist

STRATEGIC ACTIONS

1. Integrate crisis management into strategic planning processes.
2. Integrate crisis management into statements of corporate excellence.
3. Include outsiders on the Board and on crisis management teams.
4. Provide training and workshops in crisis management.
5. Expose organizational members to crisis simulations.
6. Create a diversity or portfolio of crisis management strategies.

TECHNICAL AND STRUCTURAL ACTIONS

1. Create a crisis management team.
2. Dedicate budget expenditures for crisis management.
3. Establish accountabilities for updating emergency policies/manuals.
4. Computerize inventories of crisis management resources (e.g., employee skills).
5 Designate an emergency command control room.
6. Assure technological redundancy in vital areas (e.g., computer systems).
7. Establish working relationship with outside experts in crisis management.

EVALUATION AND DIAGNOSTIC ACTIONS

1. Conduct legal and financial audit of threats and liabilities.
2. Modify insurance coverage to match crisis management contingencies.
3. Conduct environmental impact audits.
4. Prioritize activities necessary for daily operations.
5. Establish tracking system for early warning signals.
6. Establish tracking system to follow up past crises or near crises.

COMMUNICATION ACTIONS

1. Provide training for dealing with the media regarding crisis management.
2. Improve communication lines with local communities.
3. Improve communication with intervening stakeholders (e.g., police).

PSYCHOLOGICAL AND CULTURAL ACTIONS

1. Increase visibility of strong top management commitment to crisis management.
2. Improve relationships with activist groups.
3. Improve upward communication (including "whistleblowers").
4. Improve downward communication regarding crisis management programs/accountabilities.
5. Provide training regarding human and emotional impacts of crises.
6. Provide psychological support services (e.g., stress/anxiety/management).
7. Reinforce symbolic recall/corporate memory of past crises/dangers.

Source: Pearson and Mitroff (1993).

may face. For example, the IOC has a risk and assurance governance model in place to "reduce potential risks and to take advantage of opportunities, while also ensuring the fulfillment of its missions and objectives" (IOC, 2022, para. 1). Exhibit 16.4 provides an overview of the IOC Risk and Assurance Governance Model, which distinguishes between three primary groups charged with managing risks (IOC, 2022):

> **1st Line of Defense: Operational Functions.** These operational functions are embedded into the IOC's daily business operations. Each of the departments that make up the IOC are charged with identifying, reporting, evaluating, and responding to risks as they arise.
> **2nd Line of Defense: Managerial Functions.** This line is tasked with building and overseeing the first line of defense, making sure that sufficient frameworks, practices, policies, and procedures are put in place to identify and mitigate risks.
> **3rd Line of Defense: Independent Functions.** Reporting directly to IOC leadership, this level provides "assurance to the organisation's governing bodies and to the Director General on how effectively the organisation assesses and manages its risks, including the way the first and second lines of defence operate" (IOC, 2022, para. 6). The functions need to be independent so that objectivity can be guaranteed.

In addition to the three lines of defense, the IOC also utilizes external audits as part of its risk governance model. For example, when reporting financial information, the IOC relies on external audits in accordance with Swiss Auditing Standards to ensure the financial reports are free of errors (IOC, 2022).

The IOC Risk and Assurance Governance Model | *exhibit* **16.4**

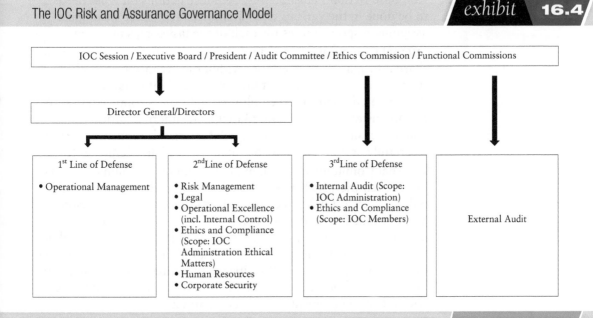

Source: IOC (2022).

Create a Comprehensive Crisis Management Plan (Pre-Crisis)

At the heart of proactive crisis management is a strong, comprehensive crisis management plan. Your crisis management plan should include, at minimum, the following components (adapted from Marker, 2022):

1. **Risk Analysis:** Information on the crises your organization is most likely to face (see above). This can include a description of specific scenarios an organization could find itself in. This section should cover a broad range, including the major types of crises identified earlier in this chapter as they apply to the specific organization.

2. **Activation Protocol:** Information on triggers that will activate a crisis response in your organization. Some organizations distinguish between different levels of the urgency of crises, each of which triggers a different crisis response. This protocol should also include directions for staff to respond to crises as well as establish some milestones that signal the end of a crisis.

3. **Chain of Command:** A clear statement on lines of authority for managing each respective crisis. This can, for instance, be in the form of a flowchart that visualizes who will take on what roles during the crisis.

4. **Command Center Plan:** Information on the base of operations during the crisis. This should include what units will need what supplies to operate during a crisis response.

5. **Response Action Plans:** Step-by-step instructions on what needs to be done in the various crisis scenarios. This must encompass assigning responsibilities for each of the tasks/steps included.

6. **Internal Communication Plan:** Infrastructure for the crisis team to share information with entities within the organization, including staff. Here, you must determine how crucial information is disseminated regularly to employees and other groups internal to the organization. Keep in mind that some internal information may be sensitive in nature (e.g., will the organization survive?), so a carefully crafted internal communication plan is important.

7. **External Communication Plan:** Plan outlining communication with external entities, such as key stakeholders, media, or the broader public. Marker (2022) recommends: "Appoint a spokesperson. Write detailed instructions, including whom you will notify (e.g., media outlets in a particular geographic area). Also, draft holding statements, the details of which you can fill in later, once you have the relevant information. Prioritize your strategic communication objectives and outline talking points. Make sure your plans align with other communication efforts" (para. 27).

8. **Resources:** List of items, supplies, expertise (internal or external), and information that must be available for the crisis to be resolved. This can include agreements with key stakeholder groups (e.g., contracts), physical resources (e.g., when natural disasters

strike), psychological resources (e.g., mental health support), and external expertise (e.g., independent consultants).

9. **Training:** Regular training(s) to educate organizational members on how to handle crises and practice for real-life crises. This may include actual drills (e.g., evacuations), practice scenarios where clear roles are assigned according to the organizational crisis flowchart (see above), or media training.

10. **Review:** Structured reflection process to learn from occurring crises. Keep in mind that a crisis management plan maps out what *should* happen during a crisis. When an organization actually goes through a crisis, it is important to review the initial plan and adjust it as needed. What worked and what didn't? Were there flaws in the plan? Those are questions that should be asked in a structured review process.

Let's look at an example from the world of sport: In March 2020, the Pennsylvania chapter of the Special Olympics, the world's largest sport organization dedicated to people with disabilities, released its updated crisis management plan. You can access the whole plan via the link on the side, but it's worth highlighting a few of the key items included in the plan:

www

Special Olympics
Pennsylvania Crisis
Management Plan (2020)
https://specialolympicspa.
org/images/Monthly_Update/
Resources/SOPA_Crisis_
Management_Plan_
March_2020.pdf

- The plan has five sections, covering crisis management team members, roles, and basic definitions of different levels of emergencies (Section 1), general crisis management guidelines, including information on emergency procedures (Section 2), specific crisis management guidelines with examples (Section 3), crisis communication procedures, including a crisis response flow chart, appointment of spokespersons, and sample public statements (Section 4), as well as a Crisis Fact Sheet for the Crisis Communications Administrator to complete and a Crisis Media Log (Appendices).

- The plan documents the composition of the crisis team and the roles of each team member (e.g., Crisis Communications Manager, Crisis Communications Administrator, Crisis Communications Team Member).

- In a section describing specific crises, the organization distinguishes between three levels of crises requiring action:

  ◦ Level 1: "a situation requiring internal emergency action which will not impact other venues or the function and image of Special Olympics Pennsylvania" (Special Olympics Pennsylvania, 2020, p. 11)

    Example(s): Delay of an event

  ◦ Level 2: "a situation requiring internal emergency action which may involve other venues or the function and image of Special Olympics Pennsylvania" (Special Olympics Pennsylvania, 2020, p. 11)

    Example(s): Lost athlete; behavioral crisis

> ⊐ Level 3: "a situation requiring both internal and external emergency action which may impact other venues or the function and image of Special Olympics Pennsylvania" (Special Olympics Pennsylvania, 2020, p. 12)
>
> Example(s): Criminal activity, health threat, bomb threat, and sexual abuse

- Each of the examples is also accompanied by specific step-by-step instructions on what needs to be done, by whom, and in what order.
- The plan includes a crisis response flow (see Exhibit 16.5), guidelines for the appointment of a spokesperson, and instructions

---

exhibit **16.5** Crisis Response Flow in Crisis Management Plan of Special Olympics Pennsylvania

Step 1: Identify the Crisis

> Using the examples in the "Levels of Emergencies" section, determine the nature and level of the crisis.
> Contact the member of the Senior Leadership Team most appropriate based on the nature of the crisis (examples: contact SVP of Programming for event incidents, contact Controller for financial incidents).
> As appropriate, request assistance from the Crisis Communications Team via the Crisis Communications Administrator.

Step 2: Assess and Review Crisis

> Once the Crisis Communications Administrator has been contacted, they will:
> > Gather and record all pertinent facts on a crisis fact sheet.
> > Assess the scope and nature of the crisis, including damage or potential damage.
> > Contact the Crisis Communications Team for a meeting (via phone or in person).
> > Contact additional parties as appropriate (staff, volunteers and/or external individuals) to secure their involvement.
> > Disseminate crisis fact sheets and other relevant information to relevant parties.

Step 3: Meeting of Crisis Communications Team

> The Crisis Communications Team meeting will involve:
> > A crisis debriefing, including new developments/updates, by the Crisis Management Administrator or those involved in the situation.
> > If necessary, contacting the Special Olympics North America Regional Office or Special Olympics headquarters.

Step 4: Planning Crisis Communications Response

> Once assembled, the Crisis Communications Team, under the leadership of the Crisis Manager, is immediately responsible for:
> > Planning a crisis communications response and ensuring execution.
> > Establishing communications strategies to address the crisis situation (e.g., selecting an appropriate spokesperson).
> > Creating key messages and public statements.
> > Identifying crisis situation response tactics to be implemented by others (employees/ or external individuals).
> > Monitoring the media via the public relations team

Source: Special Olympics Pennsylvania (2020).

on how to communicate with Special Olympics' key stakeholders, including staff and volunteers, constituent parties, and media.

- The plan makes available sample public statements and key messages, so that statements can be released quickly and in a consistent format throughout crises.
- The plan provides forms to document the crisis, allowing for institutional memory to be created – which can later inform learning and reflecting.

As you can see, having a crisis management plan in place is crucial to prepare for the crisis. Once a crisis emerges, a quick response is often needed. To facilitate immediate and efficient action, sport organizations rely on crisis management teams and strategic crisis communication, which we will cover next.

Form a Crisis Management Team and Communicate Transparently (Pre-Crisis/Mid-Crisis)

Sport organizations successfully managing crises usually have a crisis management team in place to make sure a crisis can be quickly resolved once it occurs. For example, when it was inevitable that the COVID-19 pandemic would alter life as we knew it, the IOC and 2020 Tokyo Summer Games Organizing Committee created the Coordination Commission for the Games of the XXXII Olympiad to continuously monitor the situation and identify countermeasures to COVID-19 (see Chapter 10). Roles on crisis management teams vary depending on the organization, but they usually include a crisis team chair, an activity recorder (documenting all steps taken), a spokesperson or media liaison, and officers in charge of aspects such as welfare, finances, and operations (Marker, 2022).

Once the crisis team is formed, the actual work begins. Let's have a look at a regional recreational organization: Tennis New Brunswick (the governing body for the sport of tennis in the Canadian province). The organization assigned the following responsibilities to the team formed for a specific crisis (Tennis New Brunswick, 2019):

- Making sure that facts of the situation are gathered as fast as possible and that the information collected is carefully verified and informs decision-making;
- Deciding what must be done in a responsible and quick manner;
- Ensuring that resources, both internal and external, are mobilized and focused on crisis communication;
- Creating infrastructure for fast distribution of accurate information, both internally and externally;
- Working collaboratively to determine crisis strategy and action items;
- Maintaining and reaffirming the values of the organization;

www

Tennis New Brunswick Crisis Management/ Communications Plan (2019)
https://tennisnb. ca/wp-content/ uploads/2021/04/Safe-Sport-Crisis-Management-Communications-Plan.pdf

- Allocating all available resources, as appropriate, to ensure effective crisis response;
- Documenting all crisis response activities;
- Engaging external relations/counsel as required (e.g., legal experts).

Notice how one item keeps coming up: communication. In fact, communication is key during crises – both internally and externally. Inside the organization, crisis managers want to make sure to keep employees as informed as possible to reduce the uncertainty or anxiety attached to the crisis. Externally, key stakeholder relationships may be on the line, so regular updates on how the crisis is addressed (and eventually resolved) are important to build trust, show transparency, and nurture strong relationships with key stakeholders. Here, it is important to appoint a spokesperson who can credibly speak on behalf of the organization and communicate what the organization is doing to address the situation.

Create an Infrastructure for Continuous Reflection (Post-Crisis)

Crises are often viewed in a negative light given the stress, uncertainty, and frequent anxiety associated with them. However, as James and Wooten (2005) remind us, "it is possible to use a crisis as an opportunity for creating a better organization ... [but it] requires that leaders adopt a learning mentality" (p. 149). Unfortunately, most organizations fail to engage in meaningful reflection, even though it is an essential step in successful crisis management (Pearson & Mitroff, 1993). So what does a learning mentality entail? James and Wooten (2005) explain:

> Learning entails examining the organization – its culture, policies and procedures – in such a way that the root causes of crises can be exposed. Learning entails facing information that might suggests that fault lies with the leadership of the firm. Learning entails encouraging and rewarding people who communicate truthful information about problems in the firm. Learning entails sharing information. Learning entails making changes to the organization that fundamentally revamp systems or remove people who are toxic to the organization.
>
> (p. 149)

You can provide an infrastructure for continued learning from the crisis by doing the following:

- Having the crisis team write up a post-crisis report with the main takeaways from the crisis, which should include items that worked well and ones that did not work so well;
- Dedicating regular space in team meetings to reflect on the crisis;
- Debriefing with and seeking input from key stakeholders about the crisis, including internal groups (e.g., employees, media relations department) and external ones (e.g., sponsors), as well as on any changes coming from the crisis (e.g., policy changes);

- Tying crisis management broadly, and debriefing on occurring crises, to performance management, especially for the leaders within the organization;

- Committing resources to reflection processes, especially when the crisis was associated with intense emotions and/or significantly tarnished the public image of the organization (e.g., external consultants).

Remember: Only through a continuous process of reflection and learning can a crisis turn into an opportunity!

We end this book with a chapter on crisis management because those governing sport had to navigate crises unprecedented in size and scope over the past few years. Whether it is the COVID-19 pandemic, institutions navigating (justified) calls for racial and social justice, the ever-so-present climate crisis, or transgressions in the behaviors and demeanor of those governing sport – crisis management has become a crucial skill set for decision-makers in the sport industry. We hope that you will use the knowledge gained in this chapter, and throughout the book, to advance your career in sport with a commitment to ethical, crisis-oriented, and inclusive governance.

SUMMARY

Crises are common occurrences in the governance of sport, but they have perhaps taken on a more prominent role given the various high-impact crises people governing sport have had to navigate at both societal (e.g., the COVID-19 pandemic) and institutional (e.g., sexual abuse scandals) levels. Because crisis management is a crucial skill set for sport managers and decision-makers, this chapter explained how crises and the management of crises affect the governance of sport. A crisis is an unexpected event (or a series of events) representing a threat to the legitimacy, goals, and/or efficiency of an organization and its relationship with stakeholders. Crises can be both sudden and smoldering, and there are ten types of crises that can affect the governance of sport: financial crises, personnel crises, crises of organizational misdeeds, technological crises, health crises, natural or environmental crises, confrontation crises, violence crises, crises of malevolence, and reputational crises.

While these seem like vastly different crises, there are some common characteristics among them, including a pattern of five different phases constituting a crisis: signal detection, preparation/prevention, containment/damage control, recovery, and learning/reflecting. Decision-makers in sport can engage in a variety of strategies to manage crises successfully and in proactive ways. First, identifying potential risks is of utmost importance in proactive crisis management, as is the creation of a comprehensive crisis management plan. Once a crisis arises, a crisis management team should be tasked with navigating the crisis, and internal and external communication should be frequent and transparent. Finally, a crisis turns into an opportunity for organizational change when those governing sport use it as a learning moment and engage in continued critical reflection.

caseSTUDY

ATHLETE ACTIVISM – CRISIS OR OPPORTUNITY?

Research from both the NCAA and scholars alike has shown a growing interest among athletes at the collegiate level in utilizing their platforms for social justice causes (NCAA, 2020b; Kluch, 2021). However, for the institutions governing college sport, such as the athletic departments and colleges and universities, activism can present a potential crisis. You are the Athletic Director at an NCAA institution of your choice and have heard from some of the coaches with whom athletes are discussing the option of staging a protest during the playing of the national anthem at one of the team's home events. You want to be proactive in turning athlete activism from a potential institutional crisis into a moment of opportunity to support the athletes' activist efforts. Utilizing the strategies mapped throughout the chapter, map out a crisis management plan for your department focused on college athlete activism. The following questions may help guide your creation of the draft crisis management plan:

1. **Scenarios:** What are the most likely outcomes should an athlete (or group of athletes) decide to protest? How can you best protect the athlete in each of the scenarios?
2. **Activation:** At what point will you determine if the protest constitutes a crisis?
3. **Backlash:** If there is backlash for the protest, what is the chain of command? Who needs to be involved, in what ways, at what point of the process?
4. **Stakeholder Engagement:** Who are the most important stakeholders (internally and externally) to engage before, during, and after the crisis? How so?
5. **Internal Communication:** What will you communicate to internal groups (athletes, coaches, administrators, university leadership, and/or campus community)?
6. **External Communication:** What will you communicate externally should there be backlash for the protest? Whom will you communicate what messaging to? How will you engage with media?
7. **Resources:** What resources will you need to protect the athlete(s)? What resources will you need to address the crisis from your institution's perspective?
8. **Learning:** How will you document the crisis? How will you utilize what happened to inform future crisis management? How will you make sure the "lessons learned" can inform future crisis management?

CHAPTER questions

1. Are there examples of crises (sudden and/or smoldering) emerging in sport at this moment in time? Using the common features of crises described in this chapter, explain how the example of your choice constitutes a crisis (or an emerging crisis).

2. In groups, analyze how each of the crises listed below reflects the lifecycle (i.e., different phases) of a crisis: Signal Detection – Preparation and Prevention – Containment and Damage Control – Recovery – Learning and Reflecting.

   USA Gymnastics Abuse Scandal   Deflategate

   COVID-19 Pandemic   Astros Signstealing Scandal

   Impact of Climate on 2026 Winter Olympic/Paralympic Games

3. Look up the crisis management plan of a sport governance organization of your choice. After reviewing the plan, discuss the following questions:

   a. According to the plan, what types of crises is the organization most likely to face?

   b. Does the crisis management plan have all the elements outlined in the strategies section above? Which ones are included, and which ones are missing?

   c. Would you consider the document a comprehensive plan? Why or why not?

   d. Does the plan include a crisis management team? What roles are assigned, and what is the organizational chain of command in times of crisis?

4. Compare how different sport organizations have reacted to similar types of crises (e.g., sexual abuse allegations, health crises, severe weather, etc.). How have they handled the crisis? What organizations have handled the crises well, and which ones not so much? How so?

FOR ADDITIONAL INFORMATION

1. Check out the following two case studies focused on crisis management in the NBA and its handling of the racist actions of Donald Sterling as well as the cancelation of the New York Marathon in the aftermath of Hurricane Sandy, respectively:

   Shreffler, M.B., Presley, G., & Schmidt, S. (2015). Getting Clipped: An evaluation of crisis management and the NBA's response to the actions of Donald Sterling. *Case Studies in Sport Management,* 4(1), 28–37. https://doi.org/10.1123/cssm.2014-0027

   Marks, W.W., Martin, T.R., & Warner, S. (2015). To run or not to run? A community in crisis. *Case Studies in Sport Management,* 4(1), 21–27. https://doi.org/10.1123/cssm.2014-0022

2. Sports documentaries are a common tool to shine light on some of the industry's biggest scandals and crisis. Here's a selection for you to watch:

   Athlete A (2020), documentary about the USA Gymnastics scandal. Access here: https://www.netflix.com/title/81034185

   Icarus (2017), documentary about the Russian state-sponsored doping scandal. Access here: https://www.netflix.com/title/80168079

   The Lost Winter: VICE Sports Climate Change Special (2015). Short Documentary on impact of climate change on winter sports. Access here: https://video.vice.com/en_us/video/the-lost-winter-vice-sports-climate-change-special-vice-sports/56676034353dec6d3747ab32?playlist=588b88cd98ae5a310dbb1aaf

3. The Sport Ecology Group, a community bringing together academics committed to sharing research on sport ecology-related topics, has a wide variety of resources focused on topics such as climate change, sustainability, and sport's impact on the environment (and vice versa): https://www.sportecology.org

4. Some scholars have started to identify sport-specific crises and crisis management. Check out these articles below for more information:

   Brown -Devlin, N., & Brown, K.A. (2020). When crises change the game: Establishing a typology of sports-related crises. *Journal of International Crisis and Risk Communication Research, 3*(1), 49–70. https://doi.org/10.30658/jicrcr.3.1.3

   Manoli, A.E. (2016) Crisis-communications management in football clubs. *International Journal of Sport Communication, 9*, 340–363. http://dx.doi.org/10.1123/IJSC.2016-0062

5. The Aspen Institute has created a series of webinars focused on navigating crises. Check out this webinar titled "Coronavirus and youth sports: How to manage the crisis": https://www.aspenprojectplay.org/webinars/how-to-manage-the-crisis

6. The Centre for Sport and Human Rights has released a number of articles looking at the COVID-19 crisis and its implications for human rights and vulnerable communities:

   Harvey, M. (2020, March 23). *Sport solidarity: How sport responds to crisis – Lessons for COVID-19.* https://sporthumanrights.org/library/sport-solidarity-how-sport-responds-to-crisis-lessons-for-covid-19/

   Rutherford, D. (2020, March 23). *Sport solidarity: Why local sport is critical to the COVID-19 response.* https://sporthumanrights.org/library/sport-solidarity-why-local-sport-is-critical-to-the-covid-19-response/

REFERENCES

Ahmed, S. (2021, September 21). *Opinion: Don't give up on Afghanistan's women athletes.* https://news.trust.org/item/20210921183123-yh7ti/?fbclid=IwAR3o___tSXs-ZMQng0LZXHNLbunOaL2dQZCqM9nj5Ty8EHYMXNkNyve_fl4

Bishop, G. (2015, December 22). Behind the Superdome blackout. *Sports Illustrated.* https://www.si.com/nfl/2015/12/22/super-bowl-xlvii-blackout-superdome

Brown-Devlin, N., & Brown, K.A. (2020). When crises change the game: Establishing a typology of sports-related crises. *Journal of International Crisis and Risk Communication Research, 3*(1), 49–70. https://doi.org/10.30658/jicrcr.3.1.3

Bundy, J., Pfarrer, M.D., Short, C.E., & Coombs, W.T. (2017). Crises and crisis management: Integration, interpretation, and research development. *Journal of Management, 43*(6), 1661–1692. https://doi.org/10.1177/0149206316680030

Bushnell, H. (2022, March 14). *Inside NCAA basketball's gender inequities and how they were exposed in 2021.* Yahoo! Sports. https://sports.yahoo.com/ncaa-basketball-gender-inequities-2021-exposed-march-madness-162712173.html

Business Queensland. (2022). *Characteristics of a crisis.* https://www.business.qld.gov.au/running-business/protecting-business/risk-management/incident-response/crisis

Coombs, W., & Holladay, S. (1996) 'Communication and attributions in a crisis: An experimental study in crisis communication'. *Journal of Public Relations Research, 8*(4), 279–295. https://doi.org/10.1207/s1532754xjprr0804_04

FIFA. (2017). *The journey: Milestones 2016.* https://publications.fifa.com/en/vision-report-2021/the-journey/the-journey-2016/

FIFA. (2022). *About FIFA.* https://www.fifa.com/about-fifa

Fontanella, C. (2022, February 24). *8 types of crisis your company could face (and protect against).* https://blog.hubspot.com/service/types-of-crisis

Graham, B.A. (2017, December 16). Why don't we care about the biggest sex abuse scandal in sports history? *The Guardian.* https://www.theguardian.com/sport/2017/dec/16/gymnastics-larry-nassar-sexual-abuse

Hartmans, A. (2022, February 12). *Peloton went from a pandemic-era success story worth $50 billion to laying off 20% of its workforce. Here's how the company's meteoric rise turned into an equally swift fall.* Business Insider. https://www.businessinsider.com/peloton-company-history-rise-fall-2022-2

Hileman, D. (2021, May 19). 15 *timeless principles of crisis management.* https://crisisconsultant.com/15-timeless-principles-of-crisis-management/

Hayes James, E., & Wooten, L. P. (2005). Leadership as (un)usual: How to display competence in times of crisis. *Organizational Dynamics, 34*(2), 141–152. doi: 10.1016/j.orgdyn.2005.03.005

IOC. (2022). *The IOC risk and assurance governance model.* https://olympics.com/ioc/integrity/ioc-governance-model-to-ensure-organisational-integrity

IPC. (2022). *IPC to decline athlete entries from RPC and NPC Belarus for Beijing 2022.* https://www.paralympic.org/news/ipc-decline-athlete-entries-rpc-and-npc-belarus-beijing-2022

Kaplan Hecker & Fink. (2021). *NCAA external gender equity review – Phase I: Basketball championships.* https://kaplanhecker.app.box.com/s/6fpd51gxk9ki78f8vbhqcqh0b0o95oxq

Keh, A., & Panja, T. (2019, December 8). Will Russia be thrown out of the Olympics on Monday? A primer. *New York Times.* https://www.nytimes.com/2019/12/08/sports/olympics/Wada-Russing-doping.html

Kluch, Y. (2021). "It's our duty to utilize the platform that we have": Motivations for activism among U.S. collegiate athletes. In R. Magrath (Eds.), *Athlete activism: Contemporary perspectives* (pp. 32–43). Routledge.

Kohrt, B. A., Mistry, A. S., Anand, N., Breecroft, B., & Nuwayhid, I. (2019). Health research in humanitarian crises: an urgent global imperative. *BMJ Global Health, 4*(6), 1–8. doi:10.1136/bmjgh-2019-001870

Kovoor-Misra, S. (2019). *Crisis management: Resilience and change.* Sage.

Marker, A. (2022, June 13). *Step-by-step guide to writing a crisis management plan.* https://www.smartsheet.com/content/crisis-management-plan

Murphy, D. (2022, May 15). *Sedona Prince, March Madness and the ongoing quest for gender equity at NCAA basketball tournaments.* ESPN. https://www.espn.com/womens-college-basketball/story/_/id/33482596/sedona-prince-march-madness-ongoing-quest-gender-equity-ncaa-basketball-tournaments

NCAA. (2020a). *Crisis management: Being your best during your institution's worst days.* https://ncaaorg.s3.amazonaws.com/governance/d3/convention/2020/2020D3Conv_CrisisCommunication.pdf

NCAA. (2020b). *NCAA student-athlete activism and racial justice engagement study.* https://ncaaorg.s3.amazonaws.com/research/demographics/2021RES_NCAASAActivismAndRJ_ES.pdf

NCAA. (2021a). *Report of the NCAA Board of Governors Committee to Promote Cultural Diversity and Equity, Committee on Women's Athletics, Gender Equity Task Force and Minority Opportunities and Interests Committee October 13, 2021, Joint videoconference.* https://ncaaorg.s3.amazonaws.com/committees/ncaa/culdiv/Oct2021CPCDECWAMOICGETF_JointReport.pdf

NCAA. (2021b). *Report of the NCAA Gender Equity Task Force October 7, 2021, Videoconference.* https://ncaaorg.s3.amazonaws.com/committees/ncaa/culdiv/Oct2021GETF_Report.pdf

NCAA. (2022). *Gender Equity Task Force.* https://www.ncaa.org/sports/2016/3/2/gender-equity-task-force.aspx

Ng, C. (2012, May 4). *Yeardley Love's mother sues lacrosse coaches over daughter's death.* ABC News. https://abcnews.go.com/US/yeardley-loves-mother-sue-lacross-coaches-daughters-death/story?id=16279118

Pearson, C.M., & Mitroff, I.I. (1993). From crisis prone to crisis prepared: A framework for crisis management. *The Executive, 7*(1), 48–59. http://www.jstor.org/stable/4165107

Seeger, M.W., Sellnow, T.L., & Ulmer, R.R. (1998). Communication organization and crisis. *Communication Yearbook, 21*, 231–275. https://doi.org/10.1080/23808985.1998.11678952

Segal, E. (2022, April 1). Congressional investigation expands into NFL's Washington Commanders. *Forbes.* https://www.forbes.com/sites/edwardsegal/2022/04/01/congressional-investigation-expands-into-nfls-washington-commanders/

Sohn, Y.J., & Lariscy, R.W. (2014). Understanding reputational crisis: Definition, properties, and consequences. *Journal of Public Relations Research, 26*(1), 23–43. https://doi.org/10.1080/1062726X.2013.795865

Special Olympics Pennsylvania. (2020). *Crisis management plan: Emergency procedures and guidelines for staff.* https://specialolympicspa.org/images/Monthly_Update/Resources/SOPA_Crisis_Management_Plan_March_2020.pdf

Sport Ecology Group. (2019). *10 things you should know about sport & climate.* https://www.sportecology.org/_files/ugd/a700be_44ffef2f99ee48af99fb983b420b1eef.pdf

Taylor, L. (2019, August 12). *Cancelled races, fainting players: How climate change affects sports.* https://www.weforum.org/agenda/2019/08/climate-change-effects-turns-up-heat-on-sports/

Tennis New Brunswick. (2019). *Tennis New Brunswick crisis management/communication plan: Safe sport.* https://tennisnb.ca/wp-content/uploads/2021/04/Safe-Sport-Crisis-Management-Communications-Plan.pdf

The Climate Coalition. (2018). *Game changer: How climate change is impacting sports in the UK.* https://static1.squarespace.com/static/58b40fe1be65940cc4889d33/t/5a85c91e9140b71180ba91e0/1518717218061/The+Climate+Coalition_Game+Changer.pdf

US Department of Justice. (2015, May 27). *Nine FIFA officials and five corporate executives indicted for racketeering conspiracy and corruption.* https://www.justice.gov/opa/pr/nine-fifa-officials-and-five-corporate-executives-indicted-racketeering-conspiracy-and

Valinsky, J. (2022, May 10). *Peloton is burning through cash and borrowing from Wall Street to stay afloat.* CNN. https://edition.cnn.com/2022/05/10/investing/peloton-third-quarter-earnings/index.html

Venette, S.J. (2003). *Risk communication in a high-reliability organization: APHIS PPQ's inclusion of risk in decision making.* North Dakota State University.

Verducci, T. (2020, January 13). Why MLB issues historic punishment to Astros for sign stealing. *Sports Illustrated.* https://www.si.com/mlb/2020/01/13/houston-astros-cheating-punishment

Women's Sport Foundation. (2011). *Negative recruiting/slander based on sexuality.* https://www.womenssportsfoundation.org/wp-content/uploads/2016/08/

recruiting-womens-sports-foundation-response-to-negative-recruiting_slander-based-on-sexuality-the-foundation-position.pdf

Wong Sak Hoi, G., & Hunt, J. (2022, May 9). *Afghan women footballers in exile fight to keep their sports dreams alive*. SWI. https://www.swissinfo.ch/eng/afghan-women-footballers-in-exile-fight-to-keep-their-sports-dreams-alive/47548822

Yglesias, M., & Stromberg, J. (2015, June 3*). FIFA's huge corruption and bribery scandal, explained*. Vox. https://www.vox.com/2015/5/27/8665577/fifa-arrests-indictment

Zaru, D. (2021, January 1). *8 sports moments that shook the US in 2020*. ABC News. https://abcnews.go.com/Sports/sports-moments-shook-us-2020/story?id=74934501

Index